T0180910

Lecture Notes in Computer Science 9626

More information about this series at http://www.springer.com/series/7409

Nicola Ferro · Fabio Crestani
Marie-Francine Moens · Josiane Mothe
Fabrizio Silvestri · Giorgio Maria Di Nunzio
Claudia Hauff · Gianmaria Silvello (Eds.)

Advances in Information Retrieval

38th European Conference on IR Research, ECIR 2016
Padua, Italy, March 20–23, 2016
Proceedings

Editors
Nicola Ferro
Department of Information Engineering
University of Padua
Padova
Italy

Fabio Crestani
Faculty of Informatics
University of Lugano (USI)
Lugano
Switzerland

Marie-Francine Moens
Department of Computer Science
Katholieke Universiteit Leuven
Heverlee
Belgium

Josiane Mothe
Systèmes d'informations, Big Data
et Recherche d'Information
Institut de Recherche en Informatique
de Toulouse IRIT/équipe SIG
Toulouse Cedex 04
France

Fabrizio Silvestri
Yahoo! Labs London
London
UK

Giorgio Maria Di Nunzio
Department of Information Engineering
University of Padua
Padova
Italy

Claudia Hauff
TU Delft - EWI/ST/WIS
Delft
The Netherlands

Gianmaria Silvello
Department of Information Engineering
University of Padua
Padova
Italy

ISSN 0302-9743 ISSN 1611-3349 (electronic)
Lecture Notes in Computer Science
ISBN 978-3-319-30670-4 ISBN 978-3-319-30671-1 (eBook)
DOI 10.1007/978-3-319-30671-1

Library of Congress Control Number: 2016932329

LNCS Sublibrary: SL3 – Information Systems and Applications, incl. Internet/Web, and HCI

Printed on acid-free paper

This Springer imprint is published by SpringerNature
The registered company is Springer International Publishing AG Switzerland

Preface

These proceedings contain the full papers, short papers, and demonstrations selected for presentation at the 38th European Conference on Information Retrieval (ECIR 2016). The event was organized by the Information Management Systems (IMS) research group[1] of the Department of Information Engineering[2] of the University of Padua[3], Italy. The conference was held during March 20–23 2016, in Padua, Italy.

ECIR 2016 received a total of 284 submissions in three categories: 201 full papers out of which seven papers in the reproducibility track, 66 short papers, and 17 demonstrations.

The geographical distribution of the submissions was as follows: 51 % were from Europe, 21 % from Asia, 19 % from North and South America, 7 % from North Africa and the Middle East, and 2 % from Australasia.

All submissions were reviewed by at least three members of an international two-tier Program Committee. Of the full papers submitted to the conference, 42 were accepted for oral presentation (22 % of the submitted ones) and eight as posters (4 % of the submitted ones). Of the short papers submitted to the conference, 20 were accepted for poster presentation (30 % of the submitted ones). In addition, six demonstrations (35 % of the submitted ones) were accepted. The accepted contributions represent the state of the art in information retrieval, cover a diverse range of topics, propose novel applications, and indicate promising directions for future research.

We thank all Program Committee members for their time and effort in ensuring the high quality of the ECIR 2016 program.

ECIR 2016 continued the reproducibility track introduced at ECIR 2015, which specifically invited the submission of papers reproducing a single paper or a group of papers from a third party, where the authors were not directly involved in the original paper. Authors were requested to emphasize the motivation for selecting the papers to be reproduced, the process of how results were attempted to be reproduced (successfully or not), the communication that was necessary to gather all information, the potential difficulties encountered, and the result of the process. Of the seven papers submitted to this track, four were accepted (57 % of the submitted ones).

A panel on "Data-Driven Information Retrieval" was organized at ECIR by Maristella Agosti. The panel stems from the fact that information retrieval has always been concerned with finding the "needle in a haystack" to retrieve the most relevant information from huge amounts of data, able to best address user information needs. Nevertheless, nowadays we are facing a radical paradigm shift, common also to many other research fields, and information retrieval is becoming an increasingly data-driven science due, for example, to recent developments in machine learning, crowdsourcing,

---

[1] http://ims.dei.unipd.it/

[2] http://www.dei.unipd.it/

[3] http://www.unipd.it/

user interaction analysis, and so on. The goal of the panel is to discuss the emergent trends in this area, their advantages, their pitfalls, and their implications for the future of the field.

Additionally, ECIR 2016 hosted four tutorials and four workshops covering a range of information retrieval topics. These were selected by workshop and tutorial committees.

The workshops were:

- Third International Workshop on Bibliometric-Enhanced Information Retrieval (BIR2016)
- First International Workshop on Modeling, Learning and Mining for Cross/Multilinguality (MultiLingMine 2016)
- ProActive Information Retrieval: Anticipating Users' Information Needs (ProAct IR)
- First International Workshop on Recent Trends in News Information Retrieval (NewsIR 2016)

The following ECIR 2016 tutorials were selected:

- Collaborative Information Retrieval: Concepts, Models and Evaluation
- Group Recommender Systems: State of the Art, Emerging Aspects and Techniques, and Research Challenges
- Living Labs for Online Evaluation: From Theory to Practice (LiLa2016)
- Real-Time Bidding Based Display Advertising: Mechanisms and Algorithms (RTBMA 2016)

Short descriptions of these workshops and tutorials are included in the proceedings.

We would like to thank our invited speakers for their contributions to the program: Jordan Boyd-Graber (University of Colorado, USA), Emine Yilmaz (University College London, UK), and Domonkos Tikk (Gravity R&D, Hungary). Short descriptions of these talks are included in the proceedings.

We are grateful to the panel led by Stefan Rüger for selecting the recipients of the 2015 Microsoft BCS/BCS IRSG Karen Spärck Jones Award, and we congratulate Jordan Boyd-Graber and Emine Yilmaz on receiving this award (unique to 2015, the panel decided to make two full awards).

Considering the long history of ECIR, which is now at it 38th edition, ECIR 2016 introduced a new award, the Test of Time (ToT) Award, to recognize research that has had long-lasting influence, including impact on a subarea of information retrieval research, across subareas of information retrieval research, and outside of the information retrieval research community (e.g., non-information retrieval research or industry).

On the final day of the conference, the Industry Day ran in parallel with the conference session with the goal of bringing an exciting program containing a mix of invited talks by industry leaders with presentations of novel and innovative ideas from the search industry. A short description of the Industry Day is included in these proceedings.

ECIR 2016 was held under the patronage of: Regione del Veneto (Veneto Region), Comune di Padova (Municipality of Padua), University of Padua, Department of Informantion Engineering, and Department of Mathematics.

Finally, ECIR 2016 would have not been possible without the generous financial support from our sponsors: Google (gold level); Elsevier, Spotify, and Yahoo! Labs (palladium level); Springer (silver level); and Yandex (bronze level). The conference was supported by the ELIAS Research Network Program of the European Science Foundation, University of Padua, Department of Informantion Engineering and Department of Mathematics.

March 2016

Nicola Ferro
Fabio Crestani
Marie-Francine Moens
Josiane Mothe
Fabrizio Silvestri
Giorgio Maria Di Nunzio
Claudia Hauff
Gianmaria Silvello

Organization

General Chair

Nicola Ferro	University of Padua, Italy

Program Chairs

Fabio Crestani	University of Lugano (USI), Switzerland
Marie-Francine Moens	KU Leuven, Belgium

Short Paper Chairs

Josiane Mothe	ESPE, IRIT, Université de Toulouse, France
Fabrizio Silvestri	Yahoo! Labs, London

Student Mentorship Chairs

Jaana Kekäläinen	University of Tampere, Finland
Paolo Rosso	Universitat Politècnica de València, Spain

Workshop Chairs

Paul Clough	University of Sheffield, UK
Gabriella Pasi	University of Milano Bicocca, Italy

Tutorial Chairs

Christina Lioma	University of Copenhagen, Denmark
Stefano Mizzaro	University of Udine, Italy

Demo Chairs

Giorgio Maria Di Nunzio	University of Padua, Italy
Claudia Hauff	TU Delft, The Netherlands

Industry Day Chairs

Omar Alonso	Microsoft Bing, USA
Pavel Serdyukov	Yandex, Russia

Test of Time (ToT) Award Chair

Norbert Fuhr — University of Duisburg-Essen, Germany

Best Paper Award Chair

Jaap Kamps — University of Amsterdam, The Netherlands

Student Grant Chair

John Tait — johntait.net Ltd., UK

Local Organization Chair

Gianmaria Silvello — University of Padua, Italy

Sponsorship Chair

Emanuele Di Buccio — University of Padua, Italy

Local Organizing Team

Antonio Camporese	University of Padua, Italy
Linda Cappellato	University of Padua, Italy
Marco Ferrante	University of Padua, Italy
Debora Leoncini	University of Padua, Italy
Maria Maistro	University of Padua, Italy

Website and Communication Material Chair

Ivano Masiero — University of Padua, Italy

Program Committee

Full-Paper Meta-Reviewers

Giambattista Amati	Fondazione Ugo Bordoni, Italy
Leif Azzopardi	University of Glasgow, UK
Roberto Basili	University of Rome Tor Vergata, Italy
Mohand Boughanem	IRIT, Paul Sabatier University, France
Paul Clough	University of Sheffield, UK
Bruce Croft	University of Massachusetts Amherst, USA
Arjen de Vries	Radboud University, The Netherlands
Norbert Fuhr	University of Duisburg-Essen, Germany
Eric Gaussier	Université Joseph Fourier, France
Cathal Gurrin	Dublin City University, Ireland
Gareth Jones	Dublin City University, Ireland

Joemon Jose	University of Glasgow, UK
Gabriella Kazai	Lumi.do, UK
Udo Kruschwitz	University of Essex, UK
Oren Kurland	Technion - Israel Institute of Technology, Israel
Henning Müller	University of Applied Sciences Western Switzerland, Switzerland
Wolfgang Nejdl	L3S and University of Hannover, Germany
Iadh Ounis	University of Glasgow, UK
Gabriella Pasi	University of Milano Bicocca, Italy
Paolo Rosso	Universitat Politècnica de València, Spain
Stefan Rüger	Knowledge Media Institute, UK

Full-Paper, Short Paper, and Demonstration Reviewers

Mikhail Ageev	Moscow State University, Russia
Dirk Ahlers	Norwegian University of Science and Technology, Norway
Ahmet Aker	University of Sheffield, UK
Elif Aktolga	University of Massachusetts Amherst, USA
M-Dyaa Albakour	Signal Media, UK
Omar Alonso	Microsoft, USA
Ismail Sengor Altingovde	Middle East Technical University, Turkey
Robin Aly	University of Twente, The Netherlands
Giambattista Amati	Fondazione Ugo Bordoni, Italy
Linda Andersson	TU Wien, Austria
Avi Arampatzis	Democritus University of Thrace, Greece
Jaime Arguello	University of North Carolina at Chapel Hill, USA
Seyed Ali Bahrainian	University of Lugano (USI), Switzerland
Krisztian Balog	University of Stavanger, Norway
Alvaro Barreiro	University of A Coruña, Spain
Roberto Basili	University of Rome Tor Vergata, Italy
Srikanta Bedathur Jagannath	IBM Research, India
Alejandro Bellogin	Universidad Autonoma de Madrid, Spain
Patrice Bellot	Université Aix-Marseille, France
Catherine Berrut	LIG, Université Joseph Fourier Grenoble I, France
Ralf Bierig	TU Wien, Austria
Toine Bogers	Aalborg University Copenhagen, Denmark
Alessandro Bozzon	Delft University of Technology, The Netherlands
Mateusz Budnik	University of Grenoble, France
Paul Buitelaar	Insight - National University of Ireland, Galway, Ireland
Fidel Cacheda	Universidad de A Coruña, Spain
Pavel Calado	INESC-ID, Instituto Superior Técnico, Universidade de Lisboa, Portugal
Fazli Can	Bilkent University, Turkey
Mark Carman	Monash University, Australia

Claudio Carpineto	Fondazione Ugo Bordoni, Italy
Marc Cartright	Google Inc., USA
Jean-Pierre Chevallet	Grenoble Alpes University, France
Luisa Coheur	Luisa Coheur, INESC-ID, Instituto Superior Técnico, Universidade de Lisboa, Portugal
Alfredo Cuzzocrea	ICAR-CNR and University of Calabria, Italy
Eva D'hondt	Laboratoire d'Informatique pour la Mécanique et les Sciences de l'Ingénieur (LIMSI), France
Walter Daelemans	University of Antwerp, Belgium
Martine De Cock	University of Washington Tacoma, USA
Pablo de la Fuente	Universidad de Valladolid, Spain
Thomas Demeester	Ghent University, Belgium
Emanuele Di Buccio	University of Padua, Italy
Giorgio Maria Di Nunzio	University of Padua, Italy
Vladimir Dobrynin	Saint-Petersburg State University, Russia
Huizhong Duan	WalmartLabs, USA
Carsten Eickhoff	ETH Zurich, Switzerland
David Elsweiler	University of Regensburg, Germany
Liana Ermakova	Institut de Recherche en Informatique de Toulouse (IRIT), Perm State National Research University, France
Hui Fang	University of Delaware, USA
Yi Fang	Santa Clara University, USA
Juan M. Fernández-Luna	University of Granada, Spain
Luanne Freund	University of British Columbia, Canada
Karin Friberg Heppin	University of Gothenburg, Sweden
Ingo Frommholz	University of Bedfordshire, UK
Patrick Gallinari	LIP6 - University of Paris 6, France
Kavita Ganesan	University of Illinois at Urbana Champaign, USA
Anastasia Giachanou	University of Lugano (USI), Switzerland
Giorgos Giannopoulos	Imis Institute, Athena R.C., Greece
Lorraine Goeuriot	Laboratoire d'informatique de Grenoble, France
Ayse Goker	Robert Gordon University, UK
Michael Granitzer	Universität Passau, Germany
Guillaume Gravier	IRISA and Inria Rennes, France
David Grossman	Georgetown University, USA
Antonino Gulli	Elsevier, The Netherlands
Matthias Hagen	Bauhaus-Universität Weimar, Germany
Allan Hanbury	TU Wien, Austria
Preben Hansen	Stockholm University, Sweden
Donna Harman	NIST, USA
Morgan Harvey	Northumbria University, UK
Claudia Hauff	Delft University of Technology, The Netherlands
Jer Hayes	IBM, Ireland
Ben He	University of Chinese Academy of Sciences, China
Daqing He	University of Pittsburgh, USA

Jiyin He	CWI, The Netherlands
Nathalie Hernandez	IRIT, France
Katja Hofmann	Microsoft, UK
Frank Hopfgartner	University of Glasgow, UK
Andreas Hotho	University of Würzburg, Germany
Gilles Hubert	IRIT, University of Toulouse, France
Dmitry Ignatov	National Research University Higher School of Economics, Russia
Shen Jialie	Singapore Management University, Singapore
Jiepu Jiang	University of Massachusetts Amherst, USA
Hideo Joho	University of Tsukuba, Japan
Jaap Kamps	University of Amsterdam, The Netherlands
Nattiya Kanhabua	Aalborg University, Denmark
Diane Kelly	University of North Carolina at Chapel Hill, USA
Liadh Kelly	Trinity College Dublin, Ireland
Yiannis Kompatsiaris	Information Technologies Institute, CERTH, Greece
Alexander Kotov	Wayne State University, USA
Udo Kruschwitz	University of Essex, UK
Monica Landoni	University of Lugano (USI), Switzerland
Martha Larson	Delft University of Technology, The Netherlands
Kyumin Lee	Utah State University, USA
Wang-Chien Lee	Pennsylvania State University, USA
Johannes Leveling	Elsevier, The Netherlands
Liz Liddy	Center for Natural Language Processing, Syracuse University, USA
Christina Lioma	University of Copenhagen, Denmark
Xiaozhong Liu	Indiana University Bloomington, USA
Elena Lloret	University of Alicante, Spain
Fernando Loizides	University of Wolverhampton, UK
David Losada	University of Santiago de Compostela, Spain
Bernd Ludwig	University of Regensburg, Germany
Mihai Lupu	TU Wien, Austria
Yuanhua Lv	Microsoft Research, USA
Craig Macdonald	University of Glasgow, UK
Andrew Macfarlane	City University London, UK
Walid Magdy	Qatar Computing Research Institute, Qatar
Marco Maggini	University of Siena, Italy
Thomas Mandl	University of Hildesheim, Germany
Stephane Marchand-Maillet	University of Geneva, Switzerland
Miguel Martinez-Alvarez	Signal Media, UK
Bruno Martins	NESC-ID, Instituto Superior Técnico, Universidade de Lisboa, Portugal
Yosi Mass	IBM Haifa Research Lab, Israel
Max Chevalier	IRIT, France
Edgar Meij	Yahoo Labs, UK
Marcelo Mendoza	Universidad Técnica Federico Santa María, Chile

Alessandro Micarelli	Roma Tre University, Italy
Dunja Mladenic	Jozef Stefan Institute, Slovenia
Josiane Mothe	Institut de Recherche en Informatique de Toulouse, France
Hannes Mühleisen	Centrum Wiskunde & Informatica (CWI), The Netherlands
Philippe Mulhem	LIG-CNRS, France
Dong Nguyen	University of Twente, The Netherlands
Boris Novikov	St. Petersburg University, Russia
Andreas Nürnberger	Otto von Guericke University of Magdeburg, Germany
Neil O'Hare	Yahoo Labs, USA
Michael O'Mahony	University College Dublin, Ireland
Michael Oakes	University of Wolverhampton, UK
Iadh Ounis	University of Glasgow, UK
Katerina Pastra	Cognitive Systems Research Institute, Greece
Virgil Pavlu	Northeastern University, USA
Pavel Pecina	Charles University in Prague, Czech Republic
Vivien Petras	HU Berlin, Germany
Karen Pinel-Sauvagnat	Institut de Recherche en Informatique de Toulouse (IRIT), France
Florina Piroi	TU Wien, Austria
Vassilis Plachouras	Thomson Reuters, UK
Barbara Poblete	University of Chile, Chile
Georges Quénot	Laboratoire d'Informatique de Grenoble, CNRS, France
Dmitri Roussinov	University of Strathclyde, UK
Alan Said	Recorded Future, Sweden
Michail Salampasis	Technological Educational Institute of Thessaloniki, Greece
Rodrygo Santos	Universidade Federal de Minas Gerais, Brazil
Markus Schedl	Johannes Kepler University Linz, Austria
Ralf Schenkel	Universität Passau, Germany
Pascale Sébillot	IRISA/INSA Rennes, France
Florence Sedes	IRIT, Paul Sabatier University, France
Giovanni Semeraro	University of Bari, Italy
Azadeh Shakery	University of Tehran, Iran
Jan Snajder	University of Zagreb, Croatia
Parikshit Sondhi	WalmartLabs, USA
Yang Song	Microsoft Research, USA
Simone Stumpf	City University London, UK
L. Venkata Subramaniam	IBM Research, India
Lynda Tamine	IRIT University Paul Sabatier, France
Bart Thomee	Yahoo Labs, USA
Ilya Tikhomirov	Institute for Systems Analysis, FRC CSC RAS, Russia

Marko Tkalcic	Johannes Kepler University, Austria
Dolf Trieschnigg	Mydatafactory, The Netherlands
Christos Tryfonopoulos	University of Peloponnese, Greece
Ming-Feng Tsai	National Chengchi University, Taiwan
Theodora Tsikrika	Information Technologies Institute, CERTH, Greece
Denis Turdakov	Institute for System Programming RAS, Russia
Ata Turk	Boston University, USA
Yannis Tzitzikas	University of Crete and FORTH-ICS, Greece
Marieke van Erp	Vrije Universiteit Amsterdam, The Netherlands
Jacco van Ossenbruggen	CWI & VU University Amsterdam, The Netherlands
Natalia Vassilieva	Hewlett Packard Labs, USA
Sumithra Velupillai	Stockholm University, Sweden
Suzan Verberne	Radboud University, The Netherlands
Stefanos Vrochidis	Information Technologies Institute, CERTH, Greece
Ivan Vulic	Cambridge University, UK
Jeroen Vuurens	Delft University of Technology, The Netherlands
V.G. Vinod Vydiswaran	University of Michigan, USA
Xiaojun Wan	Peking University, China
Hongning Wang	University of Virginia, USA
Jun Wang	University College London, UK
Lidan Wang	University of Illinois, Urbana-Champaign, USA
Wouter Weerkamp	904Labs, The Netherlands
Christa Womser-Hacker	University of Hildesheim, Germany
Tao Yang	Ask.com and UCSB, USA
David Zellhoefer	Berlin State Library, Germany
Dan Zhang	Facebook, USA
Lanbo Zhang	University of California, Santa Cruz, USA
Duo Zhang	University of Illinois, Urbana-Champaign, USA
Ke Zhou	Yahoo Labs, UK
Guido Zuccon	Queensland University of Technology, Australia

Reproducible IR Track Reviewers

Ahmet Aker	University of Sheffield, UK
Catherine Berrut	LIG, Université Joseph Fourier Grenoble, France
Fidel Cacheda	Universidad de A Coruña, Spain
Fazli Can	Bilkent University, Turkey
Luisa Coheur	INESC-ID, Instituto Superior Técnico, Universidade de Lisboa, Portugal
Pablo de la Fuente	Universidad de Valladolid, Spain
Thomas Demeester	Ghent University, Belgium
Norbert Fuhr	University of Duisburg-Essen, Germany
Guillaume Gravier	IRISA and Inria Rennes, France
Donna Harman	NIST, USA
Katja Hofmann	Microsoft, UK
Andreas Hotho	University of Würzburg, Germany

David Losada	University of Santiago de Compostela, Spain
Craig Macdonald	University of Glasgow, UK
Edgar Meij	Yahoo Labs, UK
Philippe Mulhem	LIG-CNRS, France
Karen Pinel-Sauvagnat	IRIT, France
Markus Schedl	Johannes Kepler University Linz, Austria
Ralf Schenkel	Universitaet Passau, Germany
Florence Sedes	IRIT, Paul Sabatier University, France
Suzan Verberne	Radboud University, The Netherlands
Wouter Weerkamp	904Labs, The Netherlands

Tutorial Selection Committee

Leif Azzopardi	University of Glasgow, UK
Alejandro Bellogin	Universidad Autonoma de Madrid, Spain
Ronan Cummins	University of Cambridge, UK
Julio Gonzalo	UNED, Spain
Djoerd Hiemstra	University of Twente, The Netherlands
Evangelos Kanoulas	University of Amsterdam, The Netherlands
Diane Kelly	University of North Carolina at Chapel Hill, USA
Jian-Yun Nie	Université de Montréal, Canada
Thomas Roelleke	Queen Mary University of London, UK
Falk Scholer	RMIT University, Australia
Fabrizio Sebastiani	Qatar Computing Research Institute, Qatar
Theodora Tsikrika	Information Technologies Institute, CERTH, Greece

Additional Reviewers

Aggarwal, Nitish
Agun, Daniel
Balaneshin-Kordan, Saeid
Basile, Pierpaolo
Biancalana, Claudio
Boididou, Christina
Bordea, Georgeta
Caputo, Annalina
Chen, Yi-Ling
de Gemmis, Marco
Fafalios, Pavlos
Farnadi, Golnoosh
Freund, Luanne
Fu, Tao-Yang
Gialampoukidis, Ilias
Gossen, Tatiana
Grachev, Artem
Grossman, David
Hasibi, Faegheh
Herrera, Jose
Hung, Hui-Ju
Jin, Xin
Kaliciak, Leszek
Kamateri, Eleni
Kotzyba, Michael
Lin, Yu-San
Lipani, Aldo
Loni, Babak
Low, Thomas
Ludwig, Philipp
Luo, Rui
Mota, Pedro
Narducci, Fedelucio
Nikolaev, Fedor
Onal, K. Dilek
Palomino, Marco

Papadakos, Panagiotis
Parapar, Javier
Petkos, Georgios
Raftopoulou, Paraskevi
Ramanath, Maya
Rasmussen, Edie
Rekabsaz, Navid
Rodrigues, Hugo
Sarwar, Sheikh Muhammad
Schinas, Manos
Schlötterer, Jörg
Şimon, Anca-Roxana
Sushmita, Shanu
Symeonidis, Symeon
Thiel, Marcus
Toraman, Cagri
Valcarce, Daniel
Vergoulis, Thanasis
Wang, Zhenrui
Wood, Ian
Xu, Tan
Yu, Hang
Zheng, Lu

Student Mentors

Paavo Arvola	University of Tampere, Finland
Rafael E. Banchs	I2R Singapore
Rafael Berlanga Llavori	Universitat Jaume I, Spain
Pia Borlund	University of Copenhagen, Denmark
Davide Buscaldi	Université Paris XIII, France
Fidel Cacheda	University of A Coruña, Spain
Marta Costa-Jussà	Instituto Politécnico Nacional México, Mexico
Walter Daelemans	University of Antwerp, The Netherlands
Kareem M. Darwish	Qatar Computing Research Institute, Qatar
Maarten de Rijke	University of Amsterdam, The Netherlands
Marcelo Luis Errecalde	Universidad Nacional de San Luís, Argentina
Julio Gonzalo	UNED, Spain
Hugo Jair Escalante	INAOE Puebla, Mexico
Jaap Kamps	University of Amsterdam, The Netherlands
Heikki Keskustalo	University of Tampere, Finaland
Greg Kondrak	University of Alberta, Canada
Zornitsa Kozareva	Yahoo! Labs, USA
Mandar Mitra	Indian Statistical Institute, India
Manuel Montes y Gómez	INAOE Puebla, Mexico
Alessandro Moschitti	Qatar Computing Research Institute, Qatar
Preslav Nakov	Qatar Computing Research Institute, Qatar
Doug Oard	University of Maryland, USA
Iadh Ounis	University of Glasgow, UK
Karen Pinel-Sauvagnat	IRIT, Université de Toulouse, France
Ian Ruthven	University of Strathclyde, UK
Grigori Sidorov	Instituto Politécnico Nacional México, Mexico
Thamar Solorio	University of Houston, USA
Elaine Toms	University of Sheffield, UK
Christa Womser-Hacker	University of Hildesheim, Germany

Test of Time (ToT) Award Committee

Maristella Agosti	University of Padua, Italy
Pia Borlund	University of Copenhagen, Denmark
Djoerd Hiemstra	University of Twente, The Netherlands
Kalervo Järvelin	University of Tampere, Finland
Gabriella Kazai	Lumi.do, UK
Iadh Ounis	University of Glasgow, UK
Jacques Savoy	University of Neuchatel, Switzerland

Patronage

Platinum Sponsors

Gold Sponsors

Google

UNIVERSITÀ
DEGLI STUDI
DI PADOVA

Palladium Sponsors

Silver Sponsors

Bronze Sponsors

Yandex

Keynote Talks

Machine Learning Shouldn't be a Black Box

Jordan Boyd-Graber

University of Colorado, Boulder CO 80309, USA

Machine learning is ubiquitous: detecting spam e-mails, flagging fraudulent purchases, and providing the next movie in a Netflix binge. But few users at the mercy of machine learning *outputs* know what's happening behind the curtain. My research goal is to demystify the black box for non-experts by creating *algorithms that can inform, collaborate with, compete with, and understand users* in real-world settings.

This is at odds with mainstream machine learning—take topic models. Topic models are sold as a tool for understanding large data collections: lawyers scouring Enron e-mails for a smoking gun, journalists making sense of Wikileaks, or humanists characterizing the oeuvre of Lope de Vega. But topic models' proponents never asked what those lawyers, journalists, or humanists needed. Instead, they optimized *held-out likelihood*. When my colleagues and I developed the *interpretability* measure to assess whether topic models' users understood their outputs, we found that interpretability and held-out likelihood were negatively correlated [2]! The topic modeling community (including me) had fetishized complexity at the expense of usability.

Since this humbling discovery, I've built topic models that are a collaboration between humans and computers. The computer starts by proposing an organization of the data. The user responds by separating confusing clusters, joining similar clusters together, or comparing notes with another user [5]. The model updates and then directs the user to problematic areas that it knows are wrong. This is a huge improvement over the "take it or leave it" philosophy of most machine learning algorithms.

This is not only a technical improvement but also an improvement to the social process of machine learning adoption. A program manager who used topic models to characterize NIH investments uncovered interesting synergies and trends, but the results were unpresentable because of a fatal flaw: one of the 700 clusters lumped urology together with the nervous system, anathema to NIH insiders [14]. Our tools allow non-experts to fix such obvious problems (obvious to a human, that is), allowing machine learning algorithms to overcome the *social* barriers that often hamper adoption.

Our realization that humans have a lot to teach machines led us to *simultaneous machine interpretation* [3]. Because verbs end phrases in many languages, such as German and Japanese, existing algorithms must wait until the end of a sentence to begin translating (since English sentences have verbs near the start). We learned tricks from professional human interpreters—passivizing sentences and guessing the verb—to translate sentences sooner [4], letting speakers and algorithms cooperate together and enabling more natural cross-cultural communication.

The reverse of cooperation is competition; it also has much to teach computers. I've increasingly looked at language-based games whose clear goals and intrinsic fun speed research progress. For example, in *Diplomacy*, users chat with each other while marshaling armies for world conquest. Alliances are fluid: friends are betrayed and enemies embraced as the game develops. However, users' conversations let us predict when friendships break: betrayers writing ostensibly friendly messages before a betrayal become more polite, stop talking about the future, and change how much they write [13]. Diplomacy may be a nerdy game, but it is a fruitful testbed to teach computers to understand messy, emotional human interactions.

A game with higher stakes is politics. However, just like Diplomacy, the words that people use reveal their underlying goals; computational methods can help expose the "moves" political players can use. With collaborators in political science, we've built models that: show when politicians in debates strategically change the topic to influence others [9, 11]; frame topics to reflect political leanings [10]; use subtle linguistic phrasing to express their political leaning [7]; or create political subgroups with larger political movements [12].

Conversely, games also teach humans *how computers think*. Our trivia-playing robot [1, 6, 8] faced off against four former Jeopardy champions in front of 600 high school students.[1] The computer claimed an early lead, but we foolishly projected the computer's thought process for all to see. The humans learned to read the algorithm's ranked dot products and schemed to answer just before the computer. In five years of teaching machine learning, I've never had students catch on so quickly to how linear classifiers work. The probing questions from high school students in the audience showed they caught on too. (Later, when we played again against Ken Jennings,[2] he sat in front of the dot products and our system did much better.)

Advancing machine learning requires closer, more natural interactions. However, we still require much of the user—reading distributions or dot products—rather than natural language interactions. Document exploration tools should describe in words what a cluster is, not just provide inscrutable word clouds; deception detection systems should say *why* a betrayal is imminent; and question answers should explain *how* it knows Aaron Burr shot Alexander Hamilton. My work will complement machine learning's ubiquity with transparent, empathetic, and useful interactions with users.

Bibliography

1. Boyd-Graber, J., Satinoff, B., He, H., Daumé III, H.: Besting the quiz master: crowdsourcing incremental classification games. In: Empirical Methods in Natural Language Processing (2012). http://www.cs.colorado.edu/~jbg/docs/qb_emnlp_2012.pdf
2. Chang, J., Boyd-Graber, J., Wang, C., Gerrish, S., Blei, D.M.: Reading tea leaves: how humans interpret topic models. In: Proceedings of Advances in Neural Information Processing Systems (2009). http://www.cs.colorado.edu/~jbg/docs/nips2009-rtl.pdf

---

[1] https://www.youtube.com/watch?v=LqsUaprYMOw
[2] https://www.youtube.com/watch?v=kTXJCEvCDYk

3. Grissom II, A., He, H., Boyd-Graber, J., Morgan, J., Daumé III, H.: Don't until the final verb wait: reinforcement learning for simultaneous machine translation. In: Proceedings of Empirical Methods in Natural Language Processing (2014). http://www.cs.colorado.edu/~jbg/docs/2014_emnlp_simtrans.pdf
4. He, H., Grissom II, A., Boyd-Graber, J., Daumé III, H.: Syntax-based rewriting for simultaneous machine translation. In: Empirical Methods in Natural Language Processing (2015). http://www.cs.colorado.edu/~jbg/docs/2015_emnlp_rewrite.pdf
5. Hu, Y., Boyd-Graber, J., Satinoff, B., Smith, A.: Interactive topic modeling. Mach. Learn. **95**(3), 423–469 (2014). http://dx.doi.org/10.1007/s10994-013-5413-0
6. Iyyer, M., Boyd-Graber, J., Claudino, L., Socher, R., Daumé III, H.: A neural network for factoid question answering over paragraphs. In: Proceedings of Empirical Methods in Natural Language Processing (2014). http://www.cs.colorado.edu/~jbg/docs/2014_emnlp_qb_rnn.pdf
7. Iyyer, M., Enns, P., Boyd-Graber, J., Resnik, P.: Political ideology detection using recursive neural networks. In: Proceedings of the Association for Computational Linguistics (2014). http://www.cs.colorado.edu/~jbg/docs/ 2014_acl_rnn_ideology.pdf
8. Iyyer, M., Manjunatha, V., Boyd-Graber, J., Daumé III, H.: Deep unordered composition rivals syntactic methods for text classification. In: Association for Computational Linguistics (2015). http://www.cs.colorado.edu/~jbg/docs/2015_acl_dan.pdf
9. Nguyen, V.A., Boyd-Graber, J., Resnik, P.: SITS: A hierarchical non-parametric model using speaker identity for topic segmentation in multiparty conversations. In: Proceedings of the Association for Computational Linguistics (2012). http://www.cs.colorado.edu/~jbg/docs/acl_2012_sitspdf
10. Nguyen, V.A., Boyd-Graber, J., Resnik, P.: Lexical and hierarchical topic regression. In: Proceedings of Advances in Neural Information Processing Systems (2013). http://www.cs.colorado.edu/~jbg/docs/2013_shlda.pdf
11. Nguyen, V.A., Boyd-Graber, J., Resnik, P., Cai, D., Midberry, J., Wang, Y.: Modeling topic control to detect influence in conversations using nonparametric topic models. Mach. Learn. **95**, 381–421 (2014). http://www.cs.colorado.edu/~jbg/docs/mlj_2013_influencer.pdf
12. Nguyen, V.A., Boyd-Graber, J., Resnik, P., Miler, K.: Tea party in the house: a hierarchical ideal point topic model and its application to Republican legislators in the 112th Congress. In: Association for Computational Linguistics (2015). http://www.cs.colorado.edu/~jbg/docs/2015_acl_teaparty.pdf
13. Niculae, V., Kumar, S., Boyd-Graber, J., Danescu-Niculescu-Mizil, C.: Linguistic harbingers of betrayal: a case study on an online strategy game. In: Association for Computational Linguistics (2015). http://www.cs.colorado.edu/~jbg/docs/2015_acl_diplomacy.pdf
14. Talley, E.M., Newman, D., Mimno, D., Herr, B.W., Wallach, H.M., Burns, G.A.P.C., Leenders, A.G.M., McCallum, A.: Database of NIH grants using machine-learned categories and graphical clustering. Nat. Methods **8**(6), 443–444 (2011)

A Task-Based Perspective to Information Retrieval

Emine Yilmaz

Deptartment of Computer Science, University College London
emine.yilmaz@ucl.ac.uk

The need for search often arises from a persons need to achieve a goal, or a task such as booking travels, organizing a wedding, buying a house, etc. [1]. Contemporary search engines focus on retrieving documents relevant to the query submitted as opposed to understanding and supporting the underlying information needs (or tasks) that have led a person to submit the query. Therefore, search engine users often have to submit multiple queries to the current search engines to achieve a single information need [2]. For example, booking travels to a location such as London would require the user to submit various different queries such as flights to London, hotels in London, points of interest around London as all of these queries are related to possible subtasks the user might have to perform in order to arrange their travels.

Ideally, an information retrieval (IR) system should be able to understand the reason that caused the user to submit a query and it should help the user achieve the actual task by guiding her through the steps (or subtasks) that need to be completed. Even though designing such systems that can characterize/identify tasks, and can respond to them efficiently is listed as one of the grant challenges in IR [1], very little progress has been made in this direction [3].

Having identified that users often have to reformulate their queries in order to achieve their final goal, most current search engines attempt to assist users towards a better expression of their needs by suggesting queries to them, other than the currently issued query. However, query suggestions mainly focus on helping the user refine their current query, as opposed to helping them identify and explore aspects related to their current complex tasks. For example, when a user issues the query "flights to Barcelona", it is clear that the user is planning to travel to Barcelona and it is very likely that the user will also need to search for hotels in Barcelona or for shuttles from Barcelona airport. Since query suggestions mainly focuses on refining the current query, suggestions provided commonly used search engines are mostly of the form "flights to Barcelona from <LOCATION>", or "<FLIGHT CARRIER NAME> flights to Barcelona" and the result pages provided by these systems do not contain any information that could help users book hotels or shuttles from the airport.

For very common tasks such as arranging travels, it may be possible to manually identify and guide the user through a list of (sub)tasks that needs to be achieved to achieve a certain task (booking a flight, finding a hotel, looking for points of interests, etc. when the user trying to arrange her travels). However, given the variety of tasks search engines are used for, this would only be possible for a very small subset of them. Furthermore, quite often search engines are used to achieve such complex tasks that often the searcher herself lacks the task knowledge necessary to decide which step to tackle next [2]. For example, a searcher looking for information about how to maintain a car with no prior knowledge would first need to use the search engine to identify the parts of the car that need maintenance and issue separate queries to learn about maintaining each part. Hence, retrieval systems that can automatically detect the task the user trying to achieve and guide her through the process are needed, where a search task has been previously defined as an atomic information need that consists of a set of related (sub)tasks [2].

With the introduction of new types of devices in our everyday lives, search systems are now being used via very different types of devices. The types of devices search systems are used over are becoming increasingly smaller (e.g. mobile phones, smart watches, smart glasses etc.), which limit the types of interactions users may have with the systems. Searching over devices with such small interfaces is not easy as it requires more effort to type and interact with the system. Hence, building IR systems that can reduce the interactions needed with the device is highly critical for such devices. Therefore, task based information retrieval systems will be even more valuable for such small interfaces, which are increasingly being introduced/used.

Devising task based information retrieval systems have several challenges that have to be tackled. In this talk, I will start with describing the problems that need to be solved when designing such systems, comparing and contrasting them we the traditional way in building IR systems. In particular, devising such task based systems would involve tackling several challenges, such as (1) devising methodologies for accurately extracting and representing tasks, (2) building and designing new interfaces for task based IR systems, (3) devising methodologies for evaluating the quality of task based IR systems, and (4) task based personalization of IR systems. I will talk about the initial attempts made in tackling in these challenges, as well as the initial methodologies we have built in order to tackle each of these challenges.

References

1. Belkin, N.: Some(what) grand challenges for IR. ACM SIGIR Forum **42**(1), 47–54 (2008)
2. Jones, R., Klinkner, K.L.: Beyond the session timeout: automatic hierarchical segmentation of search topics in query logs. In: Proceedings of ACM CIKM 2008 Conference on Information and Knowledge Management, pp. 699–708 (2008)
3. Kelly,D., Arguello, J., Capra, R.: NSF workshop on task-based information searchsystems. In: ACM SIGIR Forum, vol. 47, no. 2, December 2013

Lessons Learnt at Building Recommendation Services in Industry Scale

Domonkos Tikk

Gravity R&D Zrt, Budapest, Hungary
domonkos.tikk@gravityrd.com
http://gravityrd.com

Abstract. Gravity R&D has been providing recommendation services as SaaS solutions since 2009. Founded by top contenders in the Netflix Prize, the company can be considered as an offspring of the competition. In this talk it is shown how Gravity's recommendation technology was created from the big pile of task specific program codes to scalable services that serve billions of recommendation requests monthly. Having academic origin with strong research focus, the recommendation quality has always been the primary differentiating factor at Gravity. But we also learnt that machine learning competitions are different from scalable and robust services. We discuss some lessons learnt on this road to create a solution that can equally encompass complex algorithms, yet fast and scalable.

Keywords: Recommender systems • Scalability • Real-time • Matrix factorization • Context-aware recommenders • Neighbor based models

Gravity R&D experienced many challenges while scaling up their services. The sheer quantity of data handled on a daily basis increased exponentially. This presentation will cover how overcoming these challenges permanently shaped our algorithms and system architecture used to generate these recommendations. Serving personalized recommendations requires real-time computation and data access for every single request. To generate responses in real-time, current user inputs have to be compared against their history in order to deliver accurate recommendations.

We then combine this user information with specific details about available items as the next step in the recommendation process. It becomes more difficult to provide accurate recommendations as the number of transactions and items increase. It also becomes difficult because this type of analysis requires the combination of multiple complex algorithms that all may require heterogeneous inputs.

Initially, the architecture was designed for matrix factorization based models [4] and serving huge numbers of requests but with a limited number of items. Now, Gravity is using MF, neighborhood based models [5], context-aware recommenders [2, 3] and metadata based models to generate recommendations for millions of items within their databases, and now Gravity is experimenting with applying deep learning technology for recommendations [1]. This required a shift from a monolithic architecture with in-process caching to a more service oriented architecture with multi-layer caching. As a result of an increase in the number of components and number of clients, managing the infrastructure can be quite difficult.

Even with these challenges, we do not believe that it is worthwhile to use a fully distributed system. It adds unneeded complexity, resources, and overhead to the system. We prefer an approach of firstly optimizing current algorithms and architecture and only moving to a distributed system when no other options are left.

References

1. Hidasi, B., Karatzoglou, A., Baltrunas, L., Tikk, D.: Session-based recommendations with recurrent neural networks. CoRR (Arxiv) abs/1511.06939 (2015). http://arxiv.org/abs/1511.06939
2. Hidasi, B., Tikk, D.: Fast ALS-based tensor factorization for context-aware recommendation from implicit feedback. In: Flach, P., et al. (eds.) ECML PKDD 2012. LNCS vol. 7524, pp. 67–82. Springer, Berlin (2012)
3. Hidasi, B., Tikk, D.: General factorization framework for context-aware recommendations. Data Mining and Knowledge Discovery, pp. 1–30 (2015). http://dx.doi.org/10.1007/s10618-015-0417-y
4. Takács, G., Pilászy, I., Németh, B., Tikk, D.: Scalable collaborative filtering approaches for large recommender systems. J. Mach. Learn. Res. **10**, 623–656 (2009)
5. Takács, G., Pilászy, I., Németh, B., Tikk, D.: Matrix factorization and neighbor based algorithms for the Netflix Prize problem. In: 2nd ACM Conference on Recommendation Systems, pp. 267–274. Lausanne, Switzerland, 21–24 October 2008

Contents

Social Context and News

SoRTESum: A Social Context Framework for Single-Document Summarization 3
Minh-Tien Nguyen and Minh-Le Nguyen

A Graph-Based Approach to Topic Clustering for Online Comments to News. 15
Ahmet Aker, Emina Kurtic, A.R. Balamurali, Monica Paramita, Emma Barker, Mark Hepple, and Rob Gaizauskas

Leveraging Semantic Annotations to Link Wikipedia and News Archives . . . 30
Arunav Mishra and Klaus Berberich

Machine Learning

Deep Learning over Multi-field Categorical Data – A Case Study on User Response Prediction 45
Weinan Zhang, Tianming Du, and Jun Wang

Supervised Local Contexts Aggregation for Effective Session Search. 58
Zhiwei Zhang, Jingang Wang, Tao Wu, Pengjie Ren, Zhumin Chen, and Luo Si

An Empirical Study of Skip-Gram Features and Regularization for Learning on Sentiment Analysis. 72
Cheng Li, Bingyu Wang, Virgil Pavlu, and Javed A. Aslam

Multi-task Representation Learning for Demographic Prediction 88
Pengfei Wang, Jiafeng Guo, Yanyan Lan, Jun Xu, and Xueqi Cheng

Large-Scale Kernel-Based Language Learning Through the Ensemble Nyström Methods 100
Danilo Croce and Roberto Basili

Question Answering

Beyond Factoid QA: Effective Methods for Non-factoid Answer Sentence Retrieval 115
Liu Yang, Qingyao Ai, Damiano Spina, Ruey-Cheng Chen, Liang Pang, W. Bruce Croft, Jiafeng Guo, and Falk Scholer

Supporting Human Answers for Advice-Seeking Questions in CQA Sites . . . 129
Liora Braunstain, Oren Kurland, David Carmel, Idan Szpektor, and Anna Shtok

Ranking

Does Selective Search Benefit from WAND Optimization? 145
Yubin Kim, Jamie Callan, J. Shane Culpepper, and Alistair Moffat

Efficient AUC Optimization for Information Ranking Applications 159
Sean J. Welleck

Modeling User Interests for Zero-Query Ranking 171
Liu Yang, Qi Guo, Yang Song, Sha Meng, Milad Shokouhi, Kieran McDonald, and W. Bruce Croft

Evaluation Methodology

Adaptive Effort for Search Evaluation Metrics 187
Jiepu Jiang and James Allan

Evaluating Memory Efficiency and Robustness of Word Embeddings 200
Johannes Jurgovsky, Michael Granitzer, and Christin Seifert

Characterizing Relevance on Mobile and Desktop 212
Manisha Verma and Emine Yilmaz

Probabilistic Modelling

Probabilistic Local Expert Retrieval. 227
Wen Li, Carsten Eickhoff, and Arjen P. de Vries

Probabilistic Topic Modelling with Semantic Graph 240
Long Chen, Joemon M. Jose, Haitao Yu, Fajie Yuan, and Huaizhi Zhang

Estimating Probability Density of Content Types for Promoting Medical Records Search 252
Yun He, Qinmin Hu, Yang Song, and Liang He

Evaluation Issues

The Curious Incidence of Bias Corrections in the Pool 267
Aldo Lipani, Mihai Lupu, and Allan Hanbury

Understandability Biased Evaluation for Information Retrieval 280
Guido Zuccon

The Relationship Between User Perception and User Behaviour in Interactive Information Retrieval Evaluation 293
Mengdie Zhuang, Elaine G. Toms, and Gianluca Demartini

Multimedia

Using Query Performance Predictors to Improve Spoken Queries 309
Jaime Arguello, Sandeep Avula, and Fernando Diaz

Fusing Web and Audio Predictors to Localize the Origin of Music Pieces for Geospatial Retrieval 322
Markus Schedl and Fang Zhou

Key Estimation in Electronic Dance Music. 335
Ángel Faraldo, Emilia Gómez, Sergi Jordà, and Perfecto Herrera

Summarization

Evaluating Text Summarization Systems with a Fair Baseline from Multiple Reference Summaries 351
Fahmida Hamid, David Haraburda, and Paul Tarau

Multi-document Summarization Based on Atomic Semantic Events and Their Temporal Relationships 366
Yllias Chali and Mohsin Uddin

Tweet Stream Summarization for Online Reputation Management. 378
Jorge Carrillo-de-Albornoz, Enrique Amigó, Laura Plaza, and Julio Gonzalo

Reproducibility

Who Wrote the Web? Revisiting Influential Author Identification Research Applicable to Information Retrieval. 393
Martin Potthast, Sarah Braun, Tolga Buz, Fabian Duffhauss, Florian Friedrich, Jörg Marvin Gülzow, Jakob Köhler, Winfried Lötzsch, Fabian Müller, Maike Elisa Müller, Robert Paßmann, Bernhard Reinke, Lucas Rettenmeier, Thomas Rometsch, Timo Sommer, Michael Träger, Sebastian Wilhelm, Benno Stein, Efstathios Stamatatos, and Matthias Hagen

Toward Reproducible Baselines: The Open-Source IR Reproducibility Challenge. 408
Jimmy Lin, Matt Crane, Andrew Trotman, Jamie Callan, Ishan Chattopadhyaya, John Foley, Grant Ingersoll, Craig Macdonald, and Sebastiano Vigna

Experiments in Newswire Summarisation 421
Stuart Mackie, Richard McCreadie, Craig Macdonald, and Iadh Ounis

On the Reproducibility of the TAGME Entity Linking System 436
Faegheh Hasibi, Krisztian Balog, and Svein Erik Bratsberg

Twitter

Correlation Analysis of Reader's Demographics and Tweet Credibility Perception 453
Shafiza Mohd Shariff, Mark Sanderson, and Xiuzhen Zhang

Topic-Specific Stylistic Variations for Opinion Retrieval on Twitter 466
Anastasia Giachanou, Morgan Harvey, and Fabio Crestani

Inferring Implicit Topical Interests on Twitter. 479
Fattane Zarrinkalam, Hossein Fani, Ebrahim Bagheri, and Mohsen Kahani

Topics in Tweets: A User Study of Topic Coherence Metrics for Twitter Data 492
Anjie Fang, Craig Macdonald, Iadh Ounis, and Philip Habel

Retrieval Models

Supporting Scholarly Search with Keyqueries. 507
Matthias Hagen, Anna Beyer, Tim Gollub, Kristof Komlossy, and Benno Stein

Pseudo-Query Reformulation 521
Fernando Diaz

VODUM: A Topic Model Unifying Viewpoint, Topic and Opinion Discovery. 533
Thibaut Thonet, Guillaume Cabanac, Mohand Boughanem, and Karen Pinel-Sauvagnat

Applications

Harvesting Training Images for Fine-Grained Object Categories Using Visual Descriptions 549
Josiah Wang, Katja Markert, and Mark Everingham

Do Your Social Profiles Reveal What Languages You Speak? Language Inference from Social Media Profiles. 561
Yu Xu, M. Rami Ghorab, Zhongqing Wang, Dong Zhou, and Séamus Lawless

Retrieving Hierarchical Syllabus Items for Exam Question Analysis 575
John Foley and James Allan

Collaborative Filtering

Implicit Look-Alike Modelling in Display Ads – Transfer Collaborative Filtering to CTR Estimation 589
Weinan Zhang, Lingxi Chen, and Jun Wang

Efficient Pseudo-Relevance Feedback Methods for Collaborative Filtering Recommendation 602
Daniel Valcarce, Javier Parapar, and Álvaro Barreiro

Language Models for Collaborative Filtering Neighbourhoods 614
Daniel Valcarce, Javier Parapar, and Álvaro Barreiro

Adaptive Collaborative Filtering with Extended Kalman Filters and Multi-armed Bandits 626
Jean-Michel Renders

Short Papers

A Business Zone Recommender System Based on Facebook and Urban Planning Data. 641
Jovian Lin, Richard J. Oentaryo, Ee-Peng Lim, Casey Vu, Adrian Vu, Agus T. Kwee, and Philips K. Prasetyo

On the Evaluation of Tweet Timeline Generation Task 648
Walid Magdy, Tamer Elsayed, and Maram Hasanain

Finding Relevant Relations in Relevant Documents. 654
Michael Schuhmacher, Benjamin Roth, Simone Paolo Ponzetto, and Laura Dietz

Probabilistic Multileave Gradient Descent 661
Harrie Oosterhuis, Anne Schuth, and Maarten de Rijke

Real-World Expertise Retrieval: The Information Seeking Behaviour of Recruitment Professionals. 669
Tony Russell-Rose and Jon Chamberlain

Compressing and Decoding Term Statistics Time Series. 675
Jinfeng Rao, Xing Niu, and Jimmy Lin

Feedback or Research: Separating Pre-purchase from Post-purchase Consumer Reviews 682
Mehedi Hasan, Alexander Kotov, Aravind Mohan, Shiyong Lu, and Paul M. Stieg

Inferring the Socioeconomic Status of Social Media Users Based on Behaviour and Language 689
Vasileios Lampos, Nikolaos Aletras, Jens K. Geyti, Bin Zou, and Ingemar J. Cox

Two Scrolls or One Click: A Cost Model for Browsing Search Results 696
Leif Azzopardi and Guido Zuccon

Determining the Optimal Session Interval for Transaction Log Analysis of an Online Library Catalogue 703
Simon Wakeling and Paul Clough

A Comparison of Deep Learning Based Query Expansion with Pseudo-Relevance Feedback and Mutual Information 709
Mohannad ALMasri, Catherine Berrut, and Jean-Pierre Chevallet

A Full-Text Learning to Rank Dataset for Medical Information Retrieval 716
Vera Boteva, Demian Gholipour, Artem Sokolov, and Stefan Riezler

Multi-label, Multi-class Classification Using Polylingual Embeddings 723
Georgios Balikas and Massih-Reza Amini

Learning Word Embeddings from Wikipedia for Content-Based Recommender Systems 729
Cataldo Musto, Giovanni Semeraro, Marco de Gemmis, and Pasquale Lops

Tracking Interactions Across Business News, Social Media, and Stock Fluctuations 735
Ossi Karkulahti, Lidia Pivovarova, Mian Du, Jussi Kangasharju, and Roman Yangarber

Subtopic Mining Based on Three-Level Hierarchical Search Intentions 741
Se-Jong Kim, Jaehun Shin, and Jong-Hyeok Lee

Cold Start Cumulative Citation Recommendation for Knowledge Base Acceleration 748
Jingang Wang, Jingtian Jiang, Lejian Liao, Dandan Song, Zhiwei Zhang, and Chin-Yew Lin

Cross Domain User Engagement Evaluation 754
Ali Montazeralghaem, Hamed Zamani, and Azadeh Shakery

An Empirical Comparison of Term Association and Knowledge Graphs for Query Expansion 761
Saeid Balaneshinkordan and Alexander Kotov

Deep Learning to Predict Patient Future Diseases from the Electronic Health Records 768
Riccardo Miotto, Li Li, and Joel T. Dudley

Improving Document Ranking for Long Queries with Nested Query Segmentation 775
Rishiraj Saha Roy, Anusha Suresh, Niloy Ganguly, and Monojit Choudhury

Sketching Techniques for Very Large Matrix Factorization. 782
Raghavendran Balu, Teddy Furon, and Laurent Amsaleg

Diversifying Search Results Using Time: An Information Retrieval Method for Historians 789
Dhruv Gupta and Klaus Berberich

On Cross-Script Information Retrieval 796
Nada Naji and James Allan

LExL: A Learning Approach for Local Expert Discovery on Twitter 803
Wei Niu, Zhijiao Liu, and James Caverlee

Clickbait Detection 810
Martin Potthast, Sebastian Köpsel, Benno Stein, and Matthias Hagen

Informativeness for Adhoc IR Evaluation: A Measure that Prevents Assessing Individual Documents 818
Romain Deveaud, Véronique Moriceau, Josiane Mothe, and Eric SanJuan

What Multimedia Sentiment Analysis Says About City Liveability 824
Joost Boonzajer Flaes, Stevan Rudinac, and Marcel Worring

Demos

Scenemash: Multimodal Route Summarization for City Exploration. 833
Jorrit van den Berg, Stevan Rudinac, and Marcel Worring

Exactus Like: Plagiarism Detection in Scientific Texts. 837
Ilya Sochenkov, Denis Zubarev, Ilya Tikhomirov, Ivan Smirnov, Artem Shelmanov, Roman Suvorov, and Gennady Osipov

Jitter Search: A News-Based Real-Time Twitter Search Interface 841
Flávio Martins, João Magalhães, and Jamie Callan

TimeMachine: Entity-Centric Search and Visualization of News Archives . . . 845
Pedro Saleiro, Jorge Teixeira, Carlos Soares, and Eugénio Oliveira

OPMES: A Similarity Search Engine for Mathematical Content 849
Wei Zhong and Hui Fang

SHAMUS: UFAL Search and Hyperlinking Multimedia System 853
Petra Galuščáková, Shadi Saleh, and Pavel Pecina

Industry Day

Industry Day Overview 859
Omar Alonso and Pavel Serdyukov

Workshops

Bibliometric-Enhanced Information Retrieval: 3rd International BIR Workshop 865
Philipp Mayr, Ingo Frommholz, and Guillaume Cabanac

MultiLingMine 2016: Modeling, Learning and Mining for Cross/Multilinguality 869
Dino Ienco, Mathieu Roche, Salvatore Romeo, Paolo Rosso, and Andrea Tagarelli

Proactive Information Retrieval: Anticipating Users' Information Need 874
Sumit Bhatia, Debapriyo Majumdar, and Nitish Aggarwal

First International Workshop on Recent Trends in News Information Retrieval (NewsIR'16) 878
Miguel Martinez-Alvarez, Udo Kruschwitz, Gabriella Kazai, Frank Hopfgartner, David Corney, Ricardo Campos, and Dyaa Albakour

Tutorials

Collaborative Information Retrieval: Concepts, Models and Evaluation 885
Lynda Tamine and Laure Soulier

Group Recommender Systems: State of the Art, Emerging Aspects and Techniques, and Research Challenges 889
Ludovico Boratto

Living Labs for Online Evaluation: From Theory to Practice 893
Anne Schuth and Krisztian Balog

Real-Time Bidding Based Display Advertising: Mechanisms and Algorithms 897
Jun Wang, Shuai Yuan, and Weinan Zhang

Author Index 903

Social Context and News

SoRTESum: A Social Context Framework for Single-Document Summarization

Minh-Tien Nguyen[1,2](✉) and Minh-Le Nguyen[1]

[1] Japan Advanced Institute of Science and Technology (JAIST), 1-1 Asahidai, Nomi, Ishikawa 923-1292, Japan
{tiennm,nguyenml}@jaist.ac.jp

[2] Hung Yen University of Technology and Education (UTEHY), Hung Yen, Vietnam

Abstract. The combination of web document contents, sentences and users' comments from social networks provides a viewpoint of a web document towards a special event. This paper proposes a framework named *SoRTESum* to take advantage of information from Twitter viz. Diversity and reflection of document content to generate high-quality summaries by a novel sentence similarity measurement. The framework first formulates sentences and tweets by recognizing textual entailment (RTE) relation to incorporate social information. Next, they are modeled in a Dual Wing Entailment Graph, which captures the entailment relation to calculate the sentence similarity based on mutual reinforcement information. Finally, important sentences and representative tweets are selected by a ranking algorithm. By incorporating social information, *SoRTESum* obtained improvements over state-of-the-art unsupervised baselines e.g., Random, SentenceLead, LexRank of 0.51 %–8.8 % of ROUGE-1 and comparable results with strong supervised methods e.g., L2R and CrossL2R trained by RankBoost for single-document summarization.

Keywords: Data mining · Document summarization · Social context summarization · RTE · Ranking · Unsupervised learning

1 Introduction

Thanks to the growth of social networks e.g., Twitter[1], users can freely express their opinions on many topics in the form of tweets - short messages, maximum 140 letters. For example, after reading a web document which mentions a special event, e.g., Boston bombing, readers can write their tweets about the event on their timeline. These tweets, called social information [18] not only reveal reader's opinions but reflect the content of the document and describe facts about the event. From this observation, an interesting idea is that social information can be utilized as mutual reinforcement for web document summarization.

[1] http://twitter.com - a microblogging system.

N. Ferro et al. (Eds.): ECIR 2016, LNCS 9626, pp. 3–14, 2016.
DOI: 10.1007/978-3-319-30671-1_1

Given a web document, the summarization has to extract important sentences [10] by using statistical or linguistic information. Existing methods, however, only consider inherent document information as sentence or word/phrase level while ignoring the social information. How to elegantly formulate sentence-tweet relation and how to effectively generate high-quality summaries using social information are challenging questions.

Social context summarization can be solved by several approaches: topic modeling [3,9]; clustering [6,15]; graphical model [4,5]; or ranking [7,17]. Yang et al. proposed a dual wing factor graph model (DWFG) for incorporating tweets into the summarization [18]. The author used classification as a preliminary step in calculating weight of edges for building the graph. Wei et al. used ranking approach with 35 features trained by RankBoost for news highlight extraction [16]. However, lack of high-quality annotated data challenges supervised learning methods [16,18] to solve the summariation. In contrast, Wei et al. proposed a variation of LexRank, which used auxiliary tweet information in a heterogenous graph random walk (HGRW) to sumarize single documents [17].

The goal of this research is to automatically extract important sentences and representative tweets of a web document by incorporating its social information. This paper makes the following contributions:

- We propose to formally define sentence-tweet relation by Recognizing Textual Entailment (RTE). The relation is different in comparison to sentence-tweet representation in [16–18]. To the best of our knowledge, no existing methods solve social context summarization by RTE.
- We conduct a careful investigation to extract 14 features of RTE in the form of two groups: distance and statistical features. This provides a feature selection overview for the summarization using RTE.
- We propose a unified framework which utilizes RTE features for calculating sentence/tweet similarity. The framework is compared to several baselines and promising results indicate that our method can be successfully applied for summarizing web documents.

To solve the social context summarization, three hypothesis are considered: (1) *representation*: important sentences in a web document represent its content; (2) *reflection*: representative tweets written by readers reflect document content as well as important sentences and (3) *generation*: readers tend to use words/phrases appearing in a document to create their comments. Given a web document and tweets generated by readers after reading the document, the framework first calculates similarity score of each sentence by using RTE features with additional social information from tweets. Next, similarity score of each tweet is computed in the same mechanism using additional information from sentences. Finally, important sentences and representative tweets having the highest score are selected by a ranking algorithm as summaries.

The rest of this paper is organized as follows: Sect. 2 will show our approach to satisfy the goal along with idea, feature extraction and model; Sect. 3 will illustrate experimental results, and give discussion along with error analysis; the final section is conclusion.

2 Summarization by Ranking

This section shows our proposal of social context summarization by ranking in three steps: basic idea, feature selection and summarization.

2.1 Basic Idea

Cosine similarity can be used to incorporate document-social information, however, the noise of tweets e.g., hashtags, emoticons can badly affect the similarity calculation. We therefore propose to utilize RTE for representing sentence-tweet relation with rich features. Given a text T and hypothesis H, RTE is a task of deciding whether the meaning of H can be plausibly inferred from T in the same context and denoted by an unidirectional relation as $T \rightarrow H$ [1]. We extend RTE from unidirectional to bidirectional relation as $T \leftrightarrow H$, in which T is a sentence s_i and H is a tweet t_j. The "$\leftrightarrow$" means existing a content similarity[2] between s_i and t_j.

We define a variation of Dual Wing Entailment Graph (DWEG) from [18] for modeling sentence-tweet relation in a social context. In the DWEG, vertices are sentences and tweets; edges are the sentence-tweet relation; and weight is the RTE similarity value. Given a DWEG G_d, our goal is to select important sentences and the most representative tweets which mainly reflect the content of a document.

Our study has important differences in comparison to [18]. Firstly, our method is unsupervised (ranking) instead of classification. Secondly, we use RTE instead of three types of sentence-tweet relation. Our approach is similar to the method of Wei et al. [16,17] (using ranking) as well as the dataset. However, representing sentence-tweet by RTE and ranking to generate summaries are two key differences in comparison to [16], which used RankBoost, another supervised method. In addition, our method calculates inter-wing/dual-wing similarity with a set of RTE features instead of IDF-modified-cosine similarity in comparison to [17].

2.2 Feature Extraction

To detect the entailment, a straightforward method is to define a set of features for representing sentence-tweet relation. Term frequency - inverse document frequency (TF-IDF) or Cosine similarity can be used; however, they may be inefficient for the summarization due to the noise of data. As the main contribution, we proposed to use a set of features derived from [12] in the form of two groups: distance and statistical features shown in Table 1.

Distance Features: These features capture the distance aspect of a sentence-tweet pair, indicating that an important sentence should be close to a representative tweet rather than meaningless ones.

[2] The RTE term was kept instead of the similarity because all features were derived from RTE task.

Table 1. The features; italic in the second column is the distance features; S is a sentence, T is a tweet; LCS is the longest common sub string

Distance Features	Statistical Features
Manhattan	LCS between S and T
Euclidean	Inclusion-exclusion coefficient
Cosine similarity	% words of S in T
Word matching coefficient	% words of T in S
Dice coefficient	Word overlap coefficient
Jaccard coefficient	*Damerau-Levenshtein*
Jaro coefficient	*Levenshtein distance based on word*

Statistical Features: Statistical features capture word overlapping between a sentence and a tweet. An important sentence and a representative tweet usually contain common words (the *generation* hypothesis), indicating it has similar content.

2.3 Summarization

The goal of our approach is to select important sentences and representative tweets as summaries because they provide more information regarding the content of a document rather than only providing sentences. In our method, tweets are utilized to enrich the summary when calculating the score of a sentence, and sentences are also considered as mutual reinforcement information in computing the score of a tweet. More precisely, the weight of each instance is computed by the entailment relation using the features and the top K instances, which have the highest score will be selected as the summaries.

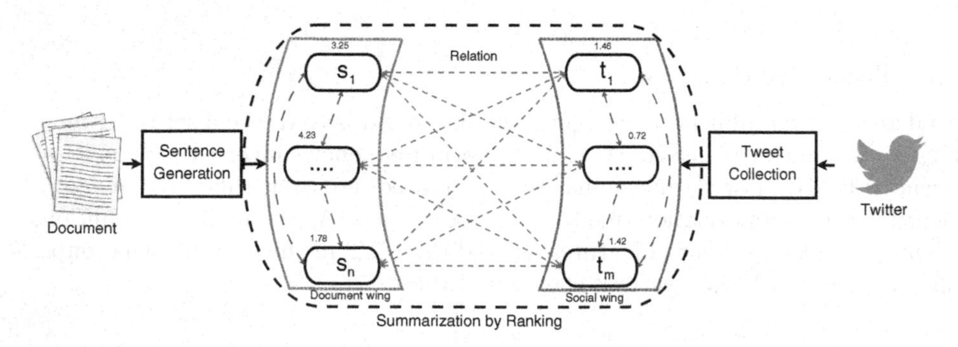

Fig. 1. The overview of summarization using DWEG; s_i and t_j denote a sentence and a tweet in document and social wing; red lines are inter-relation and blue lines are intra-relation; weight of each node (e.g., 3.25 at s_1) is the entailment value.

The framework in Fig. 1 calculates entailment weight by an iterative algorithm as an accumulative mechanism to decide whether a sentence is important or not. More precisely, we proposed two methods named *RTE-Sum inter wing* and *RTE-Sum dual wing*.

RTE-Sum Inter Wing: In this method, a sentence weight was computed by using additional information from tweets. For example, the weight of s_i was calculated by relevant tweet t_j on the tweet side. The score of t_j was also computed as the same mechanism. The calculation is shown in Eq. (1).

$$score(s_i) = \frac{1}{m}\sum_{j=1}^{m} rteScore(s_i, t_j) \tag{1}$$

where: $s_i \in S$, $t_j \in T$; $rteScore(s_i, t_j)$ returns the entailment value between sentence s_i and tweet t_j; m is the number of sentences/tweets corresponding to each document. The entailment score is calculated by Eq. (2).

$$rteScore(s_i, t_j) = \frac{1}{F}\sum_{k=1}^{F} f_k(s_i, t_j) \tag{2}$$

where: F is the number of features; f is the similarity function calculated by each feature.

RTE-Sum Dual Wing: In this method, a sentence RTE score was calculated by using remaining sentences as the main part (intra-relation) following tweets as auxiliary information (inter-relation) in an accumulative mechanism. For example, the score of s_i was calculated by s_1 to s_n; at the same time, the score was also computed by relevant tweets t_1 to t_m. Finally, RTE value of a sentence was average of all entailment values. The calculation is shown in Eq. (3).

$$score(s_i) = \delta * \sum_{k=1}^{n} rteScore(s_i, s_k) + (1-\delta) * \sum_{j=1}^{m} rteScore(s_i, t_j) \tag{3}$$

RTE value of a tweet was also computed as the same mechanism in Eq. (4).

$$score(t_j) = \delta * \sum_{k=1}^{m} rteScore(t_j, t_k) + (1-\delta) * \sum_{i=1}^{n} rteScore(t_j, s_i) \tag{4}$$

where δ is the damping factor; n and m are the number of sentences and tweets.

Ranking: Important sentences and representative tweets were found by selecting the highest score of vertices in the DWEG. The selection is denoted in Eq. (5).

$$S_r \leftarrow ranking(S_n);\ T_r \leftarrow ranking(T_m) \tag{5}$$

where *ranking()* returns a list of instances in a decreased weight order; *top-K* instances would be selected as the summaries from the S_r and T_r.

3 Results and Discussion

3.1 Experimental Setup

The dataset in [16][3] was used for evaluation. It contains 121 documents with 455 highlights and 78.419 tweets in 17 salient news events taken from CNN[4] and USAToday[5]. Each article includes three or four highlights which were manually selected by human. The detail can be seen in [16].

Comments less than five tokens were removed. Near-duplicate tweets (those containing similar content) were also removed by Simpson [13]; similar threshold 0.25 was obtained by running our experiments many times. The damping factor δ will be shown in Sect. 3.6. 5-fold validation with $K = 4$ is conducted in evaluation; stop words, hashtags, links were removed; and summary instances were also stemmed[6] [14].

3.2 Baselines

The following systems were used to compare to SoRTESum:

- **Random Method**: selects sentences and comments randomly.
- **SentenceLead:** chooses the first x sentences as the summarization [11].
- **LexRank:** summarizes a given news article using LexRank algorithm[7] [2].
- **L2R**: uses RankBoost with local and cross features [16], using RankLib[8].
- **Interwing-sent2vec:** uses Cosine similarity; by Eq. (1). A sentence to vector tool was utilized to generated vectors[9] ($size = 100$ and $window = 5$) with 10 million sentences from Wikipedia[10].
- **Dualwing-sent2vec:** uses Cosine similarity; by Eq. (3) and (4).
- **RTE One Wing:** uses one wing (document/tweet) to calculate RTE score.

3.3 Evaluation Method

Highlight sentences were used as standard summarization of evaluation by using ROUGE-N[11] (N=1, 2) [8] with stemming and removing stopwords.

3.4 Results and Discussion

Results in Tables 2 and 3 show that our method clearly outperforms the baselines by 0.51 %–8.8 % in the document side in ROUGE-1, except for CrossL2R.

---

[3] http://www1.se.cuhk.edu.hk/~zywei/data/hilightextraction.zip.
[4] http://edition.cnn.com.
[5] http://www.usatoday.com.
[6] http://snowball.tartarus.org/algorithms/porter/stemmer.html.
[7] https://pypi.python.org/pypi/sumy/0.3.0.
[8] https://people.cs.umass.edu/~vdang/ranklib.html.
[9] https://github.com/klb3713/sentence2vec/blob/master/demo.py.
[10] https://meta.wikimedia.org/wiki/Data_dump_torrents.
[11] http://kavita-ganesan.com/content/rouge-2.0-documentation.

Table 2. Document summarization; * is supervised methods; bold is the best value; italic is compared value to the best.

	ROUGE-1			ROUGE-2		
System	Avg-P	Avg-R	Avg-F	Avg-P	Avg-R	Avg-F
Random	0.140	0.205	0.167	0.031	0.044	0.037
Sentence Lead	0.196	0.341	0.249	0.075	0.136	0.096
LexRank	0.127	0.333	0.183	0.030	0.088	0.045
Interwing-sent2vec	0.208	0.315	0.250	0.069	0.116	0.086
Dualwing-sent2vec	0.148	0.194	0.168	0.044	0.058	0.050
RTE-Sum one wing	0.137	0.385	0.202	0.048	0.143	0.072
L2R* [16]	0.202	0.320	0.248	0.067	0.120	0.086
CrossL2R* [16]	0.215	0.366	**0.270**	0.086	0.158	**0.111**
SoRTESum inter wing	0.189	0.389	*0.255*	0.071	0.158	*0.098*
SoRTESum dual wing	0.186	0.400	*0.254*	0.068	0.162	0.096

Table 3. Tweet summarization, SentenceLead was not used

	ROUGE-1			ROUGE-2		
System	Avg-P	Avg-R	Avg-F	Avg-P	Avg-R	Avg-F
Random	0.138	0.179	0.156	0.049	0.072	0.059
LexRank	0.100	0.336	0.154	0.035	0.131	0.056
Interwing-sent2vec	0.177	0.222	0.197	0.055	0.071	0.062
Dualwing-sent2vec	0.153	0.195	0.171	0.039	0.055	0.046
RTE-Sum one wing	0.145	0.277	0.191	0.054	0.089	0.067
L2R* [16]	0.155	0.276	0.199	0.049	0.089	0.064
CrossL2R* [16]	0.165	0.287	**0.209**	0.053	0.099	*0.069*
SoRTESum inter wing	0.154	0.289	*0.201*	0.051	0.104	*0.068*
SoRTESum dual wing	0.161	0.296	**0.209**	0.056	0.111	**0.074**

In addition, the performance in the document side is better than that on the tweet side because comments are usually generated from document content (similarly [18]) supporting the *Reflection* hypothesis stated in Sect. 1.

SoRTESum outperforms L2R [16] in both ROUGE-1, 2, on document and tweet side even though it is a supervised method. This shows the efficiency of our approach as well as the features. In other words, our method performs comparably CrossL2R [16] in both ROUGE-1, 2. This is because (1) CrossL2R is also a supervised method; and (2) a salient score of an instance in [16] was computed by maximal ROUGE-1 between this instance and corresponding ground-truth highlight sentences. As the results, this model tends to select highly similar sentences and tweets with the highlights improving the overall performance of the model. However, even with this, our models still obtain comparable result of 0.255 vs. 0.270 in document side and the same result of 0.209 on tweet side of ROUGE-1. In ROUGE-2, although CrossL2R slightly dominates SoRTESum on document side (0.111 vs. 0.098), on tweet side, conversely, SoRTESum outperforms 0.05 % (0.074 vs. 0.069). This shows that out approach is also appropriate for tweet summarization and supports our hypothesis stated in Sect. 1.

We discuss some important different points with [17] (using same method - ranking and the dataset) due to experimental settings and re-running the experiments. Firstly, [17] uses *IDF-modified-cosine similarity*, thus the noise of tweets may badly affect to the summarization (same conclusion with [2]). The performance of CrossL2R-T and HGRW-T supports this conclusion (decreasing from 0.295 to 0.293, see [17]). On the other hand, our method combines a set of RTE features helping to avoid the tweet noise; hence, the performance increases from 0.201 to 0.209. In addition, the *IDF-modified-cosine similarity* needs a large corpus to calculate TF and IDF with a bag of words [2] whereas our approach only requires a single document and its social information to extract important sentences. This shows that our method is insensitive to the number of documents as well as tweets. In addition, new features e.g., word2vec similarity can be easy integrated into our model while adding new features into the *IDF-modified-cosine similarity* is still an open question. Finally, their work considers the impact of tweets volume and latency for sentence extraction. It is difficult to take these values for news comments as well as forum comments. In this sense, our method can be flexibly to adapt for other domains. Of course, tweets did not come from the news sources challenge both the two methods because there is no content consistency between sentences and tweets. However, we guess that even in this case, our method may be still effective because SoRTESum captures words/tokens overlapping based on a set of features while only using *IDF-modified-cosine similarity* may limit HGRW. In another words, both the two methods are inefficient in dealing informal tweets e.g., very short, abbreviated, or ungrammatical tweets. It is possibly solved by integrating a sophisticated preprocessing step.

The performance of *SoRTESum inter wing* is the same with *SoRTESum dual wing* on document side of ROUGE-1 (0.255 vs. 0.254); on tweet side, however, *SoRTESum dual wing* dominates *SoRTESum inter wing* (0.209 vs. 0.201). This is because a score of a tweet in *SoRTESum dual wing* was calculated by accumulating corresponding sentences and remaining tweets; therefore, long instances obtain higher scores. As a result, the model tends to select longer instances. However, the performance is not much different.

SoRTESum achieves a slight improvement of 0.51 % in comparison to Sentence Lead [11] because the highlights were generated by the same mechanism of Sentence Lead, taking some first sentences and changing some keywords. We guess that the results will change when SoRTESum is evaluated on other datasets where the highlights are selected from the original document instead of abstract generation.

SoRTESum one wing obtains acceptable (outperforms Random and LexRank) showing the efficiency of our features. *Interwing-sent2vec* yields comparable results of ROUGE-1 on both sides indicating vector representation can be used for summarization in an appropriate manner. In *Dualwing-sent2vec*, however, the performance is decreased because many negative values appearing in vectors leading the accumulative mechanism in Eqs. (3) and (4) is inefficient. This suggests deeper investigations of calculation should be considered.

3.5 The Role of the Features

We further examined the role of the features by removing each feature and keeping the remaining ones using 1-fold validation. The results are shown in Table 4; italic is the statistical features.

Table 4. Top five effective features; * is Inclusion-exclusion coefficient

Feature	Document		Feature	Tweet	
	ROUGE-1	ROUGE-2		ROUGE-1	ROUGE-2
Overlap-coefficent	0.23×10^{-2}	0.25×10^{-2}	Euclidean	0.4×10^{-3}	0.3×10^{-3}
Dice coefficient	0.21×10^{-2}	0.14×10^{-2}	Dice coefficient	0.38×10^{-4}	0.203×10^{-4}
*In-ex coeffi**	0.7×10^{-3}	0.7×10^{-3}	*In-ex coeffi**	0.33×10^{-4}	0.203×10^{-4}
Jaccard	0.4×10^{-3}	0.5×10^{-3}	Jaccard	0.33×10^{-4}	$0.203 \times 10^{-}4$
Matching coefficient	0.1×10^{-3}	0.3×10^{-3}	Manhattan	0.1×10^{-5}	0.4×10^{-5}

Both distance and statistical features affect the summarization. In a document, statistical features (in italic) play an important role. This shows that important sentences include important common words/phrases. On both sides, Dice coefficient, Inclusion-exclusion coefficient, and Jaccard have positive impact of the summarization. On the tweet side, however, distance features are more important than the remaining ones (only Inclusion-exclusion coefficient appearing).

The impact of *d-feature* and *s-feature* is illustrated in Fig. 2. Statistical features have a positive impact on both sides in generating summaries (big value of 0.05 and round 0.004) whereas d-feature has negative influence in tweet summarization in Fig. 2b. This concludes that *s-feature* plays an important role in document summarization. Although each distance feature in Table 4 has positive impact, combining them may lead to feature conflict. Note that negative values are very small (5.2×10^{-5}).

3.6 Tuning Damping Threshold

The impact of the damping factor in Eqs. (3) and (4) was considered by adjusting δ = [0.05..0.95], changing value = 0.1. Results from Fig. 3 show that when δ increases, auxiliary information benefits the performance of the model until some turning points. The performance is generally improved when δ closes to 0.85. After that, the performance slightly drops because with $\delta > 85$, the model is nearly same with *RTE one wing*. We therefore empirically selected $\delta = 0.85$. Note that the change is not much different among the tuning points because the RTE score is computed by averaging RTE features, hence the role of δ may be saturated.

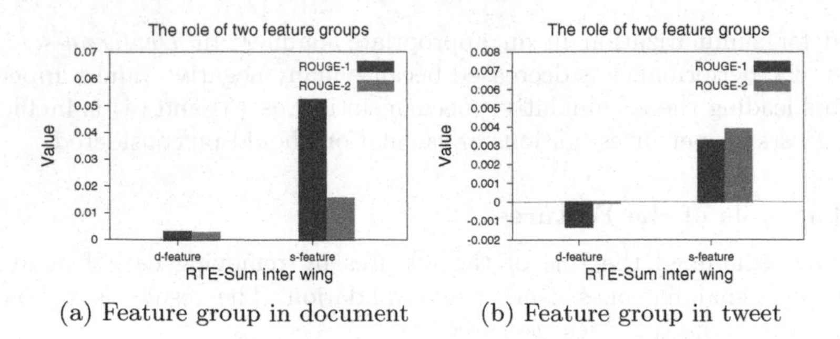

(a) Feature group in document (b) Feature group in tweet

Fig. 2. The impact of feature groups in our models

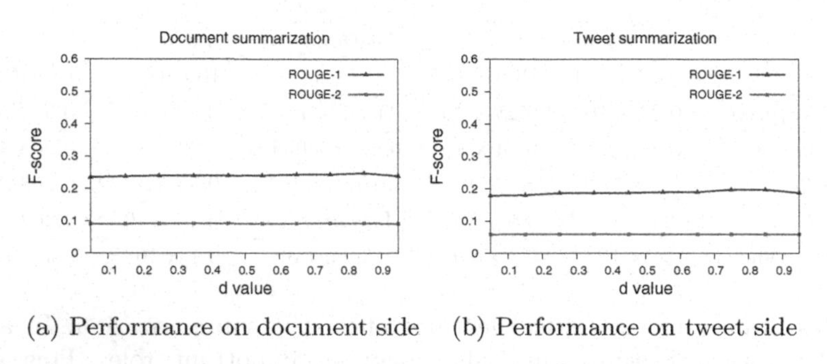

(a) Performance on document side (b) Performance on tweet side

Fig. 3. Parameter adjustment of δ of RTE-Sum dual wing

3.7 Error Analysis

In Table 5 (the web interface can be seen at SoRTESum system[12]), both models yield the same results in document summarization, in which *S1*, *S2*, and *S3* are important sentences. Clearly, the content of these sentences completely relate to the highlights, which mention about the death of Tamerlan Tsarnaev at the Boston bombing event or attending information in his college. In contrast, *S4* mentioning his father information has light relevance.

In tweet summarization, the two methods generate the same three tweets and the remaining one is different. The summarization contains one the same tweet (*T1* in *SoRTESum inter wing* and *T3* in *SoRTESum dual wing*); the other ones are different making the difference of summarization performance between the two models. They are quite relevant to this event, but do not directly mention the death of Tamerlan Tsarnaev e.g., *T2*. This leads to lower performance for both models.

Finally, although social information can help to improve summary performance, other irrelevant data can badly affect the generation. This is because a score of an instance was calculated by an accumulative mechanism; therefore, common information (sentences or tweets) can achieve high score. For example,

[12] http://150.65.242.101:9293.

Table 5. Summary example of our methods; bold style is important instances; [+] shows a strongly relevance and [-] is a light relevance.

Highlights	
+ HL1: Police identified Tamerlan Tsarnaev, 26, as the dead Boston bombing suspect	
+ HL2: Tamerlan studied engineering at Bunker Hill Community College in Boston	
+ HL3: He was a competitive boxer for a club named Team Lowell	
Summary Sentences	
+S1: Tamerlan Tsarnaev, the 26-year-old identified by police as the dead Boston bombing suspect, called his uncle Thursday night and asked for forgiveness, the uncle said	
+S2: Police have identified Tamerlan Tsarnaev as the dead Boston bombing suspect	
+S3: Tamerlan attended Bunker Hill Community College as a part-time student for three semesters, Fall 2006, Spring 2007, and Fall 2008	
-S4: He said Tamerlan has relatives in the United States and his father is in Russia	
Summary Tweets	
RTE-Sum inter wing	**RTE-Sum dual wing**
+T1: Before his death Tamerlan Tsarnaev called an uncle and asked for his forgiveness. Said he is married and has a baby	- T1: I proudly say I was the 1st 1 to write this on twitter. Uncle,Tamerlan Tsarnaev called, asked for forgiveness
-T2: I proudly say I was the 1st 1 to write this on twitter. Uncle,Tamerlan Tsarnaev called, asked for forgiveness	- T2: So apparently the dead suspect has a wife & baby? And beat his girlfriend enough to be arrested? (same woman?)
- T3: So apparently the dead suspect has a wife & baby? And beat his girlfriend enough to be arrested? (same woman?)	**+ T3: Before his death Tamerlan Tsarnaev called an uncle and asked for his forgiveness. Said he is married and has a baby**
+T4: Tamerlan Tsarnaev ID'd as dead Boston blast suspect - USA Today - USA TODAY, Tamerlan Tsarnaev ID'd as dead	**+T4: #BostonMarathon bomber Tamerlan called uncle couple of hours before he was shot dead said 'I love you and forgive me**

some tweets mention about the forgiveness of Tamerlan Tsarnaev's uncle e.g., *T2*. This, obviously, does not directly show the information of Tamerlan Tsarnaev's death, but this tweet received a lot of attention from readers when reading this event. More importantly, all sentences and tweets in Table 5 contain keywords that relate to Tamerlan Tsarnaev's event. This illustrates the efficiency of our method and suggests that the performance of the models can be improved based on informative phrases as *Generation* hypothesis in Sect. 1.

4 Conclusion

This paper presented *SoRTESum*, a summary framework using social information for summarization. Our framework utilizes a ranking approach to select important sentences and representative tweets in a novel similarity calculation. This paper also makes the contribution of formulating a sentence-tweet pair by RTE and proposes rich features to calculate the RTE score. Experimental results show that our approach achieves improvements of 0.51 % to 8.8 % over the unsupervised baselines and comparable results in comparison to supervised methods of ROUGE-1 in document summarization. In ROUGE-2, our models outperform all methods in tweet summarization.

For future direction, other important features e.g., NER or tree edit distance of the RTE task should be considered and integrated into the model. Another point is that our model should be compared to other supervised learning methods e.g., SVMs, CRF, or the model in [17,18]. An interesting point is that since deep learning has achieved promising results in NLP, we would like to adapt this technique to improve summary quality. Finally, abstract summarization should be considered in order to model the semantics in summarization.

Acknowledgment. We would like to thank to Preslav Nakov and Wei Gao for useful discussions and insightful comments on earlier drafts; Chien-Xuan Tran for building the web interface. We also thank to anonymous reviewers for their detailed comments for improving our paper. This work was partly supported by JSPS KAKENHI Grant number 3050941.

References

1. Dagan, I., Dolan, B., Magnini, B., Roth, D.: Recognizing textual entailment: rational, evaluation and approaches - erratum. Nat. Lang. Eng. **16**(1), 105–105 (2010)
2. Erkan, G., Radev, D.R.: Lexrank: graph-based lexical centrality as salience in text summarization. J. Artif. Intell. Res. **22**, 457–479 (2004)
3. Gao, W., Li, P., Darwish, K.: Joint topic modeling for event summarization across news, social media streams. In: CIKM, pp. 1173–1182 (2012)
4. Meishan, H., Sun, A., Lim, E.-P.: Comments-oriented blog summarization by sentence extraction. In: CIKM, pp. 901–904 (2007)
5. Meishan, H., Sun, A., Lim, E.-P.: Comments-oriented document summarization: understanding document with readers' feedback. In: SIGIR, pp. 291–298 (2008)
6. Po, H., Sun, C., Longfei, W., Ji, D.-H., Teng, C.: Social summarization via automatically discovered social context. In: IJCNLP pp. 483–490 (2011)
7. Huang, L., Li, H., Huang, L.: Comments-oriented document summarization based on multi-aspect co-feedback ranking. In: Wang, J., Xiong, H., Ishikawa, Y., Xu, J., Zhou, J. (eds.) WAIM 2013. LNCS, vol. 7923, pp. 363–374. Springer, Heidelberg (2013)
8. Lin, C.-Y., Hovy, E.H.: Automatic evaluation of summaries using n-gram cooccurrence statistics. In: HLT-NAACL, pp. 71–78 (2003)
9. Yue, L., Zhai, C.X., Sundaresan, N.: Rated aspect summarization of short comments. In: WWW, pp. 131–140 (2009)
10. Luhn, H.P.: The automatic creation of literature abstracts. IBM J. Res. Dev. **2**(2), 159–165 (1958)
11. Nenkova, A.: Automatic text summarization of newswire: lessons learned from the document understanding conference. In: AAAI pp. 1436–1441 (2005)
12. Nguyen, M.-T., Ha, Q.-T., Nguyen, T.-D., Nguyen, T.-T., Nguyen, L.-M.: Recognizing textual entailment in vietnamese text: an experimental study. In: KSE (2015). doi:10.1109/KSE.2015.23
13. Nguyen, M.-T., Kitamoto, A., Nguyen, T.-T.: TSum4act: a framework for retrieving and summarizing actionable tweets during a disaster for reaction. In: Cao, T., Lim, E.-P., Zhou, Z.-H., Ho, T.-B., Cheung, D., Motoda, H. (eds.) PAKDD 2015. LNCS, vol. 9078, pp. 64–75. Springer, Heidelberg (2015)
14. Porter, M.F.: Snowball: a language for stemming algorithms (2011)
15. Wan, X., Yang, J.: Multi-document summarization using cluster-based link analysis. In: SIGIR, pp. 299–306 (2008)
16. Wei, Z., Gao, W.: Utilizing microblogs for automatic news highlights extraction. In: COLING, pp. 872–883 (2014)
17. Wei, Z., Gao, W.: Gibberish, assistant, or master? Using tweets linking to news for extractive single-document summarization. In: SIGIR, pp. 1003–1006 (2015)
18. Yang, Z., Cai, K., Tang, J., Zhang, L., Zhong, S., Li, J.: Social context summarization. In: SIGIR, pp. 255–264 (2011)

A Graph-Based Approach to Topic Clustering for Online Comments to News

Ahmet Aker[1(✉)], Emina Kurtic[1], A.R. Balamurali[2], Monica Paramita[1], Emma Barker[1], Mark Hepple[1], and Rob Gaizauskas[1]

[1] University of Sheffield, Sheffield, UK
{ahmet.aker,e.kurtic,m.paramita,e.barker, m.r.hepple,r.gaizauskas} @sheffield.ac.uk

[2] LIF-CNRS Marseille, Marseille, France
balamurali.ar@lif.univ-mrs.fr

Abstract. This paper investigates graph-based approaches to labeled topic clustering of reader comments in online news. For graph-based clustering we propose a linear regression model of similarity between the graph nodes (comments) based on similarity features and weights trained using automatically derived training data. To label the clusters our graph-based approach makes use of DBPedia to abstract topics extracted from the clusters. We evaluate the clustering approach against gold standard data created by human annotators and compare its results against LDA – currently reported as the best method for the news comment clustering task. Evaluation of cluster labelling is set up as a retrieval task, where human annotators are asked to identify the best cluster given a cluster label. Our clustering approach significantly outperforms the LDA baseline and our evaluation of abstract cluster labels shows that graph-based approaches are a promising method of creating labeled clusters of news comments, although we still find cases where the automatically generated abstractive labels are insufficient to allow humans to correctly associate a label with its cluster.

1 Introduction

Online news outlets attract large volumes of comments every day. *The Huffington Post*, for example, received an estimated 140,000 comments in a 3 day period[1], while *The Guardian* has reported receiving 25,000 to 40,000 comments per day[2]. These figures suggest that online commenting forums are important for readers as a means to share their opinions on recent news. The resulting vast number of comments and information they contain makes them relevant to multiple stakeholders in the media business. All user groups involved in online commenting on news would profit from easier access to the multiple topics discussed within a large set of comments. For example, commenter posters would be able to gain

[1] http://goo.gl/3f8Hqu.

[2] http://www.theguardian.com/commentisfree/2014/aug/10/readers-editor-online-abuse-women-issues.

N. Ferro et al. (Eds.): ECIR 2016, LNCS 9626, pp. 15–29, 2016.
DOI: 10.1007/978-3-319-30671-1_2

a quick overview of topics already discussed and insert their contributions at a relevant place in the discussion. Journalists who wrote the news article would have access to multiple conversation topics that their article has triggered and would be able to engage with their readers in a more focused way. Editors would be able to monitor the topics that are most interesting to readers, comment forum moderators' work would be easier and marketers could use conversations grouped around topics for developing personalized marketing strategies.

In most current on-line commenting forums, comments are grouped into threads – micro-conversations within the larger set of comments on an article, initiated and expanded by users, occasionally with some intervention of moderators. However, threads do not correspond to topics. As in all freely developing conversations, threads tend to drift away from the topic first introduced and often end up addressing multiple topics. Furthermore, comments addressing a particular topic may occur in multiple threads or on their own. Thus, in the current thread-based setup there is no easy way for readers to get access to all comments pertaining to a particular topic. Such topic-based clusters would be highly useful, allowing users to get an overview of the conversation and to home in on parts of particular interest to themselves, particularly if good-quality and coherent labels were associated with the clusters, permitting them to quickly understand what the comments within a particular cluster were about.

In this paper we introduce a way to automatically generate end-user friendly topic clusters of reader comments to online news articles. We propose graph-based methods to address two tasks: (1) to group reader comments into topic clusters; and (2) to label the clusters for the topic(s) they represent.

These tasks present us with several challenges that our methods need to address. For instance: (1) the number of topics discussed in a conversation about a news article is always unknown; (2) a single reader comment can be multi-topical itself and therefore one comment can belong to different topic clusters; (3) comments that implement a conversational action like jokes, sarcastic remarks or short support items (e.g. 'Great') are typically difficult to assign to a topic cluster according to their contents; (4) assigning cluster labels is not an easy task as a cluster of comments can represent a topic along multiple dimensions of granularity and expression.

Related work has reported Latent Dirichlet Allocation (LDA) [4] as the best-performing approach for reader comment clustering. However, LDA has the limitation that it requires prior knowledge of the number of topics, which cannot be set in our domain as news articles vary in numbers of comments and topics they address. In this paper we investigate the Markov Clustering Algorithm (MCL) [20], a graph-based approach that is able to determine the number of clusters dynamically from the data. We adapt it to clustering of reader comments and show that it significantly outperforms LDA.

The cluster labels are generated from the MCL clusters. From each cluster we extract topic terms using LDA trained over reader comments, which are then used to create a concept-based graph using DBPedia.[3] Using graph centrality,

[3] http://wiki.dbpedia.org/.

abstract labels are created for topics, which in turn are projected onto comment clusters.

The paper is organised as follows. Section 2 reviews relevant previous work. In Sect. 3 we describe the dataset we work with, which we downloaded from *The Guardian* online news portal. Section 4 discusses our clustering and cluster labelling approaches. The experimental setup for the evaluation of the proposed methods on Guardian data is reported in Sect. 5. The results are reported in Sect. 6 and discussed in Sect. 7. In Sect. 8 we conclude the paper and outline directions for future work.

2 Related Work

2.1 Comment Clustering

Clustering of user comments has mostly been addressed in the context of automatic comment summarization, where LDA [4] has emerged as the best-performing clustering approach. For instance, Khabiri et al. [10] work on summarization of YouTube video comments. Prior to the summarization step, comments are grouped into clusters using K-Means clustering and LDA. Although the nature of LDA allows soft clustering[4], the authors convert LDA output to hard-clusters[5] by assigning a comment C to the most likely topic, i.e. the topic t_r that maximizes $P(C|t_r) * P(t_r)$, where r is the topic/cluster index. The authors claim that LDA is superior to the K-Means approach.

Another comment clustering approach is reported by Llewellyn et al. [14], who apply LDA and K-Means, as well as simple metrics such as the cosine measure to cluster the comments of a single news article. The authors report LDA to be the best performing system. Ma et al. [15] also report an LDA approach to clustering news comments, where cluster information is used to generate comment summaries. Their evaluation happens at the level of the final output, i.e. the summary built using the clusters, and clustering is not evaluated in its own right. Since the summaries are generated using only the three clusters with the most comments, it is not clear how the investigated methods perform on clustering only. A similar study is reported in Liu et al. [13], who treat clustering as a prior step to comment summarisation and do not directly assess the quality of clusters.

Graph-based clustering has not been considered so far for user comments. However, it has been applied to clustering snippets resulting from Web search [19]. Each result snippet is annotated using TAGME[6]. TAGME identifies short phrases or topics in the snippet and links them to a pertinent Wikipedia page. The nodes in the graph are the topics. Edges between nodes denote how the nodes are related to each other within Wikipedia. A detailed survey of similar approaches is provided by Carpineto et al. [5].

---

[4] Soft clustering methods allow one data item to be assigned to multiple clusters.
[5] I.e. one comment can be assigned to only one cluster.
[6] http://tagme.di.unipi.it/.

The good performance of graph-based methods on snippet clustering indicates that it may be well-suited for the comment clustering task. In addition, graph-based clustering allows us to build clusters without prior knowledge of the number of topics required – a feature needed for our task of comment clustering and one which LDA lacks.

2.2 Cluster Labelling

Labelling clusters can be seen as analogous to topic labelling. Most of the existing topic labelling procedures are extractive, i.e. they depend on extracting topic labels from within a text [11,12]. For example, given a set of topic words representing the cluster, one word can be chosen as a label based on topic coherence [17]. However, using extractive labeling it is not possible to obtain collective terms or concepts as cluster labels. For instance, it is not possible to obtain *color* as the label for topic terms *red, blue, green, yellow.* This can be achieved using abstractive labeling methods. For labeling clusters in this work, we modify the graph-based topic labeling algorithm described in Hulpus et al. [8]. A DBPedia concept graph is created and the center of the graph is selected as the label. We modify the way graph is created to increase the fine-grainedness of labels as compared to the original work.

3 Data

3.1 Training Data

For the graph-based clustering approach we present here, a regression model of similarity between graph nodes needs to be trained. Our training data was derived from a set of 3,362 online news articles and their associated comments downloaded from *The Guardian* online news portal over a period of two months (June–July 2014). *The Guardian* provides a specific RSS feed URL for each broad topic area, e.g. business, politics, etc. We manually collected RSS feeds for the topics: politics, health, sport, education, business, society, media, science, the-northerner, law, world-news, Scotland-news, money and environment. Using an in-house tool we visited the news published through the RSS feeds every 30 min, downloaded the article content and also recorded the news URL. Every recorded news URL was re-visited after a week (the time we found sufficient for an article to attract comments) to obtain its comments. Articles had between 1 and 6,223 associated comments, averaging 425.95 comments per article.

We build positive and negative training instances consisting of comment-comment pairs deemed to be topically similar/dissimilar from *The Guardian* data. To construct positive pairs we assume that if two or more comments associated with the same news article quote the same sentence in that article, then they are on the same topic and thus belong to the same topic cluster; i.e., positive pairs consist of comments that quote the same article sentence. When computing comment-comment similarity, if quotes are left in the comments, the similarity metric may be biased by the exact match as found in the quotes and may not be sensitive enough to capture similarity in comments that do not contain exactly matching quotes. For this reason, we expect that clustering results will be better if quotes are removed from the comments before computing similarity. To test this assumption we created two sets of training data. In the first set positive instances are comment pairs where the quote is left in the comments. In the second set positive instances are pairs of comments where we removed the shared quotes from the comments. For both training sets we set the topical similarity measure for each positive instance to be

$$quoteScore = len(quote_{C_1}) + len(quote_{C_2})/2 * len(sentence) \quad (1)$$

as the outcome. $len(X)$ returns the length of X in words and $quote_{C_i}$ is the segment of comment C_i quoted from *sentence* in the original article. When computing the *quoteScore* we make sure that the quoted sentence has at least 10 words. We add comment pairs to the positive training data whose *quoteScore* values are >= 0.5 – a value we obtained empirically.

The negative instances are created by pairing randomly selected comments from two different articles from *The Guardian*. They are used to present the linear regression algorithm with the instances of comment pairs that are not on the same topic or are only weakly topically related. The topical similarity measure for each such pair was set to 0. We have in total 14,700 positive pairs and the same number of negative instances.

3.2 Testing Data

For testing, clusters generated by human annotators are used as a gold standard data set. This data set was derived from 18 Guardian articles and associated comments distinct from those included in our training set. These articles and the first 100 comments associated with them served as the basis for a set of human-authored reference summaries of reader comments, created for the purpose of evaluating automatic comment summarization techniques. The procedure the summary authors were instructed to follow yielded comment clusters as a by-product: to facilitate the challenging task of writing summaries of 100 reader comments annotators were instructed to first group or cluster comments on the same topic and then to write the summary drawing on their clusters. We captured the clusters created in this process as well as the final summaries. At least two reference summaries plus related clusters were created for each article-comment set in the test set. The news article topics covered by this data are politics, sport, health, environment, business, scotland-news and science. On average, each annotator identified 8.97 clusters per news article.

4 Methods

4.1 Graph-based Clustering

Our graph-based clustering approach is based on the Markov Cluster Algorithm (MCL) [20] shown in Algorithm 1. The nodes (V) in the graph $G(V, E, W)$ are the comments. Edges (E) are created between the nodes and have associated weights (W). Each comment is potentially connected to every other comment using an undirected edge. An edge is present if the associated weight is greater than 0. Such a graph may be represented as a square matrix M of order $|V|$, whose rows and columns correspond to nodes in the graph and whose cell values $m_{i,j}$, where $m_{i,j} > 0$, indicate the presence of an edge of weight $m_{i,j}$ between nodes V_i and V_j. Following the recommendation in [20] we link all nodes to themselves with $m_{i,i} = 1$. Other edge weights are computed based on comment-comment similarity features described in the next section below.

Once such a graph is constructed, MCL repeats steps 11–13 in the Algorithm until the maximum number of iterations *iter* is reached[7]. First in step 11 the matrix is normalized and transformed to a column stohastic matrix, next expanded (step 12) and finally inflated (step 13). The expansion operator is responsible for allowing flow to connect different regions of the graph. The inflation operator is responsible for both strengthening and weakening this flow. These two operations are controlled by two parameters, the power $p -> 2$ results in too few clusters – and the inflation parameter $r - >= 2$ results in too many clusters. After some experimentation we set p to *2* and r to *1.5*, as these resulted in a good balance between too many and too few clusters.

Algorithm 1. MCL Algorithm

Require: a set of comments $\mathbf{C} = \{C_1 \dots C_n\}$, a square matrix M of order n, power parameter p, inflation parameter r, number of iterations *iter*, comment similarity measure *Sim_Score*, minimum similarity parameter

1: **for all** $m_{i,j}$ **do**
2:   **if** $i = j$ **then**
3:     $m_{i,j} = 1$
4:   **else if** $Sim\_Score(C_i, C_j) \geq Min\_Sim$ **then**
5:     $m_{i,j} = Sim\_Score$
6:   **else**
7:     $m_{i,j} = 0$
8:   **end if**
9: **end for**
10: **repeat**
11:   Normalize M $(m_{i,j} = m_{i,j} / \sum_{k=1}^{n} m_{k,j})$
12:   Expansion: Raise M to the pth power
13:   Inflation: $m_{ij} = (m_{ij})^r$
14: **until** current iteration > *iter*
15: Extract clusters from the final matrix

Once MCL terminates, the clusters are read off the rows of the final matrix (step 15 in Algorithm 1). For each row i in the matrix the comments in columns j are added to cluster i if the cell value $M_{i,j} > 0$ (the rows for items that belong

[7] MCL runs a predefined number of iterations. We ran MCL with 5000 iterations.

to the same cluster will each redundantly specify that cluster). In this setting the MCL algorithm performs hard clustering, i.e. assigns each comment to exactly one cluster.

4.2 Weighting Edges Between Comments

To weight an edge between two comments C_1 and C_2 we use the features below. When computing these features, except the $NE_{overlap}$ feature, we use terms to represent a comment instead of words, since we have found that terms are more suitable for computing similarity between short texts than single words [1]. Terms are noun phrase-like word sequences of up to 4 words. Terms are extracted using the TWSC tool [16], which uses POS-tag grammars to recognize terms.

- **Cosine Raw Count:** Cosine similarity [18] is the cosine of the angle between the vector representations of C_1 and C_2:
$$cosine(C_1, C_2) = \frac{V(C_1) \cdot V(C_2)}{|V(C_1)| * |V(C_2)|} \tag{2}$$
where $V(.)$ is the vector holding the raw frequency counts of terms from the comment.
- **Cosine TF*IDF:** Similar to the first cosine feature but this time we use the tf*idf measure for each term instead of the raw frequency counts. The idf values are obtained from the training data described in Sect. 3.1.
- **Cosine Modified:** Liu et al. [13] argue that short texts can be regarded as similar when they have already a predefined D terms in common. We have adopted their modified cosine feature:
$$cosine_{modified}(C_1, C_2) = \begin{cases} \frac{V(C_1) \cdot V(C_2)}{D}, & \text{if } V(C_1) \cdot V(C_2) \leq D \\ 1, & \text{otherwise} \end{cases} \tag{3}$$
We have set D experimentally to 5.
- **Dice:**
$$dice(C_1, C_2) = \frac{2 * |I(C_1, C_2)|}{|C_1| + |C_2|} \tag{4}$$
where $I(C_1, C_2)$ is the intersection between the set of terms in the comments C_1 and C_2.
- **Jaccard:**
$$jaccard(C_1, C_2) = \frac{|I(C_1, C_2)|}{|U(C_1, C_2)|} \tag{5}$$
where $U(C_1, C_2)$ is the union of sets of terms in the comments.
- **NE Overlap:**
$$NE_{overlap}(C_1, C_2) = \frac{|I(C_1, C_2)|}{|U(C_1, C_2)|} \tag{6}$$
where $I(C_1, C_2)$ is the intersection and $U(C_1, C_2)$ is the union set between the unique named entities (NEs) in the comments. We use the OpenNLP tools[8] to extract NEs.

[8] https://opennlp.apache.org/.

- **Same Thread**: Returns 1 if both C_1 and C_2 are within the same thread otherwise 0.
- **Reply Relationship**: If C_1 replies to C_2 (or vise versa) this feature returns 1 otherwise 0. Reply relationship is transitive, so that the reply is not necessarily direct, instead it holds: $reply(C_x, C_y) \wedge reply(C_y, C_z) \Rightarrow reply(C_x, C_z)$

We use a weighted linear combination of the above features to compute comment-comment similarity:

$$Sim\_Score(C_1, C_2) = \sum_{i=1}^{n} feature_i(C_1, C_2) * weight_i \tag{7}$$

To obtain the weights we train a linear regression[9] model using training data derived from news articles and comments as described in Sect. 3.1 above. The target value for positive instances is the value of *quoteScore* from Eq. 1 and for negative instances is 0.

We create an edge within the graph between comments C_i and C_j with weight $w_{i,j} = Sim\_Score(C_i, C_j)$ if *Sim_Score* is above 0.3, a minimum similarity threshold value set experimentally.

4.3 Graph-based Cluster Labelling

We aim to create abstractive cluster labels since abstractive labels can be more meaningful and can capture a more holistic view of a comment cluster than words or phrases extracted from it. We adopt the graph-based topic labelling algorithm of Hulpus et al. [8], which uses DBPedia [3], and modify it for comment cluster labelling.

Our use of the Hulpus et al. method proceeds as follows. An LDA model, trained on a large collection of Guardian news articles, plus their associated comments, was used to assign 5 (most-probable) topics to each cluster.[10] A separate label is created for each such topic, by using the top 10 words of the topic (according to the LDA model) to look up corresponding DBPedia concepts.[11] The individual concept graphs so-identified are then expanded using a restricted set of DBPedia relations,[12] and the resulting graphs merged, using the DBPedia merge operation. Finally, the central node of the merged graph is identified,

---

[9] We used Weka (http://www.cs.waikato.ac.nz/ml/weka/) implementation of linear regression.

[10] The number of topics (k) to assign was determined empirically, i.e. we varied $2<k<10$, and chose $k=5$ based on the clarity of the labels generated.

[11] We take the most-common sense. The 10 word limit is to reduce noise. Less than 10 DBPedia concepts may be identified, as not all topic words have an identically-titled DBPedia concept.

[12] To limit noise, we reduce the relation set c.f. Hulpus et al. to include only *skos:broader, skos:broaderOf, rdfs:subClassOf, rdfs.* Graph expansion is limited to two hops.

providing the label for the topic.[13] The intuition is that the label thus obtained should encompasses all the abstract concepts that the topic represents.[14] Thus, for example, a DBPedia concept set such as {*Atom, Energy, Electron, Quantum, Orbit, Particle*} might yield a label such as *Theoretical Physics.*

5 Experiments

We compare our graph-based clustering approach against LDA which has been established as a successful method for comment clustering when compared to alternative methods (see Sect. 2). We use two different LDA models: *LDA1* and *LDA2*[15]. The LDA1 model is trained on the entire data set described in Sect. 3.1. In this model we treat the news article and its associated comments as a single document. This training data set is large and contains a variety of topics. When we require the clustering method to identify a small number of topics, we expect these to be very general, so that the resulting comment clusters are less homogeneous than they would be if only comments of a single article are considered when training the LDA model, as do Llewellyn et al. [14].

Therefore we also train a second LDA model (LDA2), which replicates the setting reported in Llewellyn et al. [14]. For each test article we train a separate LDA2 model on its comments. In training we include the entire comment set for each article in the training data, i.e. both the first 100 comments that are clustered and summarised by human annotators, as well as the remaining comments not included in the gold standard. In building LDA2 we treated each comment in the set as separate document.

LDA requires a predetermined number of topics. We set the number of topics to 9 since the average number of clusters within the gold standard data is 8.97. We use 9 topics within both LDA1 and LDA2. Similar to Llewellyn et al. [14] we also set the α and β parameters to 5 and 0.01 respectively for both models.

Once the models are generated they are applied to the test comments for which we have gold standard clusters. LDA distributes the comments over the pre-determined number of topics using probability scores. Each topic score is the probability that the given comment was generated by that topic. Like [14] we select the most probable topic/cluster for each comment. Implemented in this way, the LDA model performs hard clustering.

For evaluation the automatic clusters are compared to the gold standard clusters described in Sect. 3.2. Amigo et al. [2] discuss several metrics to evaluate automatic clusters against the gold standard data. However, these metrics

[13] Several graph-centrality metrics were explored: *betweeness_centrality*, *load_centrality*, *degree_centrality*, *closeness_centrality*, of which the last was used for the results reported here.

[14] Hulpus et al. [8] merge together the graphs of multiple topics, so as to derive a single label to encompass them. We have found it preferable to provide a separate label for each topic, i.e. so the overall label for a cluster comprises 5 label terms for the individual topics.

[15] We use the LDA implementation from http://jgibblda.sourceforge.net/.

are tailored for hard clustering. Although our graph-based approach and baseline LDA models perform hard clustering, the gold standard data contains soft clusters. Therefore, the evaluation metric needs to be suitable for soft-clustering. In this setting hard clusters are regarded as a special case of possible soft clusters and will likely be punished by the soft-clustering evaluation method. We use fuzzy BCubed Precision, Recall and F-Measure metrics reported in [7,9]. According to the analysis of formal constraints that a cluster evaluation metric needs to fulfill [2], fuzzy BCubed metrics are superior to Purity, Inverse Purity, Mutual Information, Rand Index, etc. as they fulfill all the formal cluster constraints: *cluster homogeneity*, *completeness*, *rag bag* and *clusters size versus quantity*. The fuzzy metrics are also applicable to hard clustering.

To evaluate the association of comment clusters with labels created by the cluster labelling algorithm, we create an annotation task by randomly selecting 22 comment clusters along with their system generated labels. In the annotation bench for each comment cluster label, three random clusters are chosen along with the comment cluster for which the system generated the label. Three annotators (A, B, C) are chosen for this task. Annotators are provided with a cluster label and asked to choose the comment cluster that best describes the label from a list of four comment clusters. As the comment clusters are chosen at random, the label can correspond to more than one comment clusters. The annotators are free to choose more than one instance for the label, provided it abstracts the semantics of the cluster in some form.

In some instances, the comment label can be too generic or even very abstract. It can happen that a label does not correspond to any of the comment clusters. In such cases, the annotators are asked not to select any clusters. These instances are marked NA (not assigned) by the annotation bench. Inter-annotator agreement is measured using Fleiss Kappa metric [6]. We report overall agreement as well as agreement between all pairs of annotators. The output of the cluster labelling algorithm is then evaluated with the annotated set using standard classification metrics.

6 Results

Clustering results are shown in Table 1. A two-tailed paired t-test was performed for a pairwise comparison of the fuzzy Bcubed metrics across all four automatic systems and human-to-human setting.

Firstly, we observe that human-to-human clusters are significantly better than each of the automatic approaches in all evaluation metrics[16]. Furthermore, we cannot retain our hypothesis that the graph-based approach trained on the training data with quotes removed performs better than the one that is trained

[16] The difference in these results is significant at the Bonferroni corrected level of significance of $p < 0.0125$, adjusted for 4-way comparison between the human-to-human and all automatic conditions.

Table 1. Cluster evaluation results. The scores shown are macro averaged. For all systems the metrics are computed relative to the average scores over Human1 and Human2. *graphHuman* indicates the setting where similarity model for graph-based approach is trained with quotes included in the comments (see Sect. 4.1).

Metric	Human1-Human2	graph-Human	graph-Human-quotesRemoved	LDA1-Human	LDA2-Human
Fuzzy B^3Precision	0.41	0.29	0.30	0.25	0.23
Fuzzy B^3Recall	0.44	0.30	0.33	0.29	0.17
Fuzzy B^3FMeasure	0.40	0.29	0.31	0.24	0.18

Table 2. Annotator agreement (Fleiss Kappa) for comment labelling over 22 comment clusters

Annotators	A-B	B-C	C-A	Overall
Agreement	0.76	0.45	0.64	0.61

on data with quotes intact.[17] Although the results in the *quotes removed* condition are better for all metrics, none of the differences is statistically significant. We use the better performing model (graph without quotes) for comparisons with other automatic methods.

Secondly, the LDA1 baseline performs significantly better than the re-implementation of previous work, LDA2, in all metrics. This indicates that training LDA model on the larger data set is superior to training it on a small set of articles and their comments, despite the generality of topics that arises from compressing topics from all articles into 9 topic clusters for LDA1.

Finally, the *quotes removed* graph-based approach (column 4 in Table 1) significantly outperforms the better performing LDA1 baseline in all metrics. This indicates that graph-based method is superior to LDA, which has been identified as best performing method in several previous studies (cf. Sect. 2). In addition, clustering comments using graph-based methods removes the need for prior knowledge about the number of topics - a property of the news comment domain that cannot be considered by LDA topic modelling.

Tables 2 and 3 present results from the evaluation of the automatically generated comment cluster labels. Table 2 shows the agreement between pairs of annotators and overall, as measured by Fleiss' Kappa on the decision: given the label, which cluster does it describe best. Overall there is a *substantial agreement* of $\kappa = 0.61$ between the three annotators. The annotator pair B-C, however, achieves only *moderate agreement* of $\kappa = 0.45$, suggesting that some annotators make idiosyncratic choices when assigning more generic abstractive labels to clusters.

[17] We apply both models on comments regardless whether they contain quotes or not. However, in case of *graph-Human-quotesRemoved* before it is applied on the testing data we make sure that the comments containing quotes are also quotes free.

Table 3. Evaluation results of the cluster labeling system for each of the 3 annotators. NA corresponds to the number of labels not assigned.

Annotator	Precision	Recall	F-score	NA%
A	0.78	0.32	0.45	59.1 (13/22)
B	1.00	0.45	0.62	54.5 (12/22)
C	0.62	0.73	0.67	9.1 (2/22)
mean	0.80	0.50	0.58	40.9

Table 3 shows the evaluation scores for the automatically generated labels, given as precision, recall and F scores results, along with the percentage of labels not assigned (NA) to any cluster. Overall, annotators failed to assign labels to any cluster in 40.9 % of cases. In the remaining cases, where annotators *did* assign the labels to clusters, this was done with fairly high precision (0.8), and so as to achieve an overall average recall of 0.5, suggesting that meaningful labels had been created.

7 Discussion

The comment clustering results demonstrate that graph-based clustering is able to outperform the current state-of-the-art method LDA as implemented in previous work at the task of clustering reader's comments to online news into topics.

In addition to the quantitative study reported above we also performed a qualitative analysis of the results of the graph-based clustering approach. That analysis reveals that disagreements in human and automatic assignment of comments to clusters are frequently due to the current approach ignoring largely conversational structure and treating each comment as an independent document. Commenting forums, however, are conversations and as such they exhibit internal structuring where two comments are functionally related to each other, so that the first pair part (FPP) makes relevant the second pair part (SPP). In our automatic clusters we frequently found answers, questions, responses to compliments and other stand-alone FPPs or SPPs that were unrelated to the rest of an otherwise homogeneous cluster. For example, the comment *"No, just describing another right wing asshole"*. is found as the only odd comment in otherwise homogeneous cluster of comments about journalistic standards in political reporting. Its FPP *"Wait, are you describing Hillary Clinton?"* is assigned to a different cluster about US politician careers. We assume that our feature *reply relationship* was not sufficiently weighted to account for this, so that we need to consider alternative ways of training, which can help identify conversational functional pairs.

A further source of clustering disagreements is the fact that humans cluster both according to content and to the conversational action a comment performs, while the current system only clusters according to a comment's content. Therefore, humans have clusters labelled *jokes, personal attacks to commenters or empty sarcasm*, *support*, etc., in addition to the clusters with content labels. A few comments have been clustered by the annotators along both dimensions, content and action, and can be found in multiple clusters (soft clustering). Our graph-based method reported in this work produces hard clusters and is as such comparable with the relevant previous work. However, we have not addressed the soft-clustering requirement of the domain and gold standard data, which has most likely been partly reflected in the difference between human and automatic clustering results. When implementing soft-clustering in future one way to proceed would be to add automatic recognition of a comment's conversational action, which would make graph based clustering more human-like and therefore more directly comparable to the gold standard data we have.

Our evaluation of cluster labelling reveals that even though the labelling system has acceptable precision, recall is rather low, due, in large part, to the high number of NA labels. We qualitatively analysed those instances that were NA for more than one annotator. Barring three instances, where the system generated labels like *concepts in Metaphysics, Chemical elements, Water* with no obvious connection to the underlying cluster content, labels generated by the system describe the cluster in a meaningful way. However, in some cases annotators failed to observe the connection between the comment cluster and the label. This may be due to the fact that users expect a different level of granularity – either more general or more specific – for labeling. For instance, a comment talking about a *dry, arid room* can have a label like *laconium* but users may prefer having a label that corresponds to dryness. This is very subjective and poses a problem for abstractive labelling techniques in general.

The expansion of a graph using DBPedia relations encompasses related concepts. However, this expansion can also include abstract labels like *Construction, organs, monetary economics, Articles_containing_video_clips* etc. This happens due to merging of sub-graphs representing concepts too close to the abstract concepts. In these cases, the most common abstract node may get selected as the label. These nodes can be detrimental to the quality of the labels. This can be prevented by controlled expansion using more filtered DBPedia relations and a controlled merging.

8 Conclusion

We have presented graph-based approaches for the task of assigning reader comments on online news into labeled topic clusters. Our graph-based method is a novel approach to comment clustering, and we demonstrate that it is superior to LDA topic modeling – currently the best performing approach as reported in previous work. We model the similarity between graph nodes (comments) as a linear combination of different similarity features and train the linear regression model on an automatically generated training set consisting of comments containing article quotations.

For cluster labeling we implement a graph-based algorithm that uses DBPedia concepts to produce abstractive labels that generalise over the content of clusters in a meaningful way, thus enhancing readability and relevance of the labels.

User evaluation results indicate that there is a scope for improvement, although in general the automatic approach produces meaningful labels as judged by human annotators.

Our future work will address soft-clustering, improve feature weighting and investigate new features to better model conversational structure and dialogue pragmatics of the comments. Furthermore, we aim to create better training data. Currently, the quote-based approach for obtaining positive training instances yields few comment pairs that are in some reply structure – a comment replying to a previous comment is unlikely to quote the same sentence in the article and thus comment-pairs where one comment replies to the other are not taken to the training data. Due to this our regression model does not give much weight to the reply feature even though this feature is very likely to suggest comments are in the same topical structure. Finally, we also aim to improve the current DBPedia-based labeling approach, as well as explore alternative approaches to abstractive labeling to make cluster labels more appropriate.

Acknowledgements. The research leading to these results has received funding from the EU - Seventh Framework Program (FP7/2007–2013) under grant agreement n610916 SENSEI.

References

1. Aker, A., Kurtic, E., Hepple, M., Gaizauskas, R., Di Fabbrizio, G.: Comment-to-article linking in the online news domain. In: Proceedings of MultiLing, SigDial 2015 (2015)
2. Amigó, E., Gonzalo, J., Artiles, J., Verdejo, F.: A comparison of extrinsic clustering evaluation metrics based on formal constraints. Inf. Retrieval **12**(4), 461–486 (2009)
3. Auer, S., Bizer, C., Kobilarov, G., Lehmann, J., Cyganiak, R., Ives, Z.G.: DBpedia: a nucleus for a web of open data. In: Aberer, K., et al. (eds.) ASWC 2007 and ISWC 2007. LNCS, vol. 4825, pp. 722–735. Springer, Heidelberg (2007)
4. Blei, D.M., Ng, A.Y., Jordan, M.I.: Latent dirichlet allocation. J. Mach. Learn. Res. **3**, 993–1022 (2003)
5. Carpineto, C., Osiński, S., Romano, G., Weiss, D.: A survey of web clustering engines. ACM Comput. Surv. (CSUR) **41**(3), 17 (2009)
6. Fleiss, J.L.: Measuring nominal scale agreement among many raters. Psychol. bull. **76**(5), 378 (1971)
7. Hüllermeier, E., Rifqi, M., Henzgen, S., Senge, R.: Comparing fuzzy partitions: a generalization of the rand index and related measures. IEEE Trans. Fuzzy Syst. **20**(3), 546–556 (2012)
8. Hulpus, I., Hayes, C., Karnstedt, M., Greene, D.: Unsupervised graph-based topic labelling using dbpedia. In: Proceedings of the Sixth ACM International Conference on Web Search and Data Mining, WSDM 2013, pp. 465–474, NY, USA (2013). http://doi.acm.org/10.1145/2433396.2433454
9. Jurgens, D., Klapaftis, I.: Semeval-2013 task 13: Word sense induction for graded and non-graded senses. In: Second Joint Conference on Lexical and Computational Semantics (* SEM), vol. 2, pp. 290–299 (2013)
10. Khabiri, E., Caverlee, J., Hsu, C.F.: Summarizing user-contributed comments. In: ICWSM (2011)

11. Lau, J.H., Grieser, K., Newman, D., Baldwin, T.: Automatic labelling of topic models. In: Proceedings of the 49th Annual Meeting of the Association for Computational Linguistics: Human Language Technologies, vol. 1, pp. 1536–1545. Association for Computational Linguistics (2011)
12. Lau, J.H., Newman, D., Karimi, S., Baldwin, T.: Best topic word selection for topic labelling. In: Proceedings of the 23rd International Conference on Computational Linguistics: Posters, pp. 605–613. Association for Computational Linguistics (2010)
13. Liu, C., Tseng, C., Chen, M.: Incrests: Towards real-time incremental short text summarization on comment streams from social network services. IEEE Trans. Knowl. Data Eng. **27**, 2986–3000 (2015)
14. Llewellyn, C., Grover, C., Oberlander, J.: Summarizing newspaper comments. In: Eighth International AAAI Conference on Weblogs and Social Media (2014)
15. Ma, Z., Sun, A., Yuan, Q., Cong, G.: Topic-driven reader comments summarization. In: Proceedings of the 21st ACM International Conference on Information and Knowledge Management, pp. 265–274. ACM (2012)
16. Pinnis, M., Ljubešić, N., Ştefănescu, D., Skadina, I., Tadić, M., Gornostay, T.: Term extraction, tagging, and mapping tools for under-resourced languages. In: Proceedings of the 10th Conference on Terminology and Knowledge Engineering (TKE 2012), pp. 20–21 (2012)
17. Röder, M., Both, A., Hinneburg, A.: Exploring the space of topic coherence measures. In: Proceedings of the Eighth ACM International Conference on Web Search and Data Mining, pp. 399–408. ACM (2015)
18. Salton, G., Lesk, E.M.: Computer evaluation of indexing and text processing. J. ACM **15**, 8–36 (1968)
19. Scaiella, U., Ferragina, P., Marino, A., Ciaramita, M.: Topical clustering of search results. In: Proceedings of the Fifth ACM International Conference on Web Search and Data Mining, pp. 223–232. ACM (2012)
20. Van Dongen, S.M.: Graph clustering by flow simulation (2001)

Leveraging Semantic Annotations to Link Wikipedia and News Archives

Arunav Mishra(✉) and Klaus Berberich

Max Planck Institute for Informatics, Saarbrücken, Germany
{amishra,kberberi}@mpi-inf.mpg.de

Abstract. The incomprehensible amount of information available online has made it difficult to retrospect on past events. We propose a novel linking problem to connect excerpts from Wikipedia summarizing events to online news articles elaborating on them. To address this linking problem, we cast it into an information retrieval task by treating a given excerpt as a user query with the goal to retrieve a ranked list of relevant news articles. We find that Wikipedia excerpts often come with additional semantics, in their textual descriptions, representing the time, geolocations, and named entities involved in the event. Our retrieval model leverages text and semantic annotations as different dimensions of an event by estimating independent query models to rank documents. In our experiments on two datasets, we compare methods that consider different combinations of dimensions and find that the approach that leverages all dimensions suits our problem best.

1 Introduction

Today in this digital age, the global news industry is going through a drastic shift with a substantial increase in online news consumption. With new affordable devices available, general users can easily and instantly access online digital news archives using broadband networks. As a side effect, this ease of access to overwhelming amounts of information makes it difficult to obtain a holistic view on past events. There is thus an increasing need for more meaningful and effective representations of online news data (typically collections of digitally published news articles).

The free encyclopedia Wikipedia has emerged as a prominent source of information on past events. Wikipedia articles tend to summarize past events by abstracting from fine-grained details that mattered when the event happened. Entity profiles in Wikipedia contain excerpts that describe events that are seminal to the entity. As a whole, they give contextual information and can help to build a good understanding of the causes and consequences of the events.

Online news articles are published contemporarily to the events and report fine-grained details by covering all angles. These articles have been preserved for a long time as part of our cultural heritage through initiatives taken by media houses, national libraries, or efforts such as the Internet Archive. The archives of The New York Times, as a concrete example, go back until 1851.

N. Ferro et al. (Eds.): ECIR 2016, LNCS 9626, pp. 30–42, 2016.
DOI: 10.1007/978-3-319-30671-1_3

Individually, both Wikipedia and news articles are ineffective in providing complete clarity on multi-faceted events. On one hand, brief summaries in Wikipedia that abstract from the fine-grained details, make it difficult to understand all dimensions of an event. On the other hand, news articles that report a single story from a larger event do not make its background and implications apparent. What is badly missing are connections between excerpts from Wikipedia articles summarizing events and news articles. With these connections in place, a Wikipedia reader can jump to news articles to get the missing details.

Table 1. Examples of Wikiexcerpts

No.	Wikiexcerpt
1	**Jaber Al-Ahmad Al-Sabah:** After much discussion of a border dispute between **Kuwait** and **Iraq**, **Iraq** invaded its smaller neighbor on *August 2, 1990* with the stated intent of annexing it. Apparently, task of the invading **Iraqi** army was to capture or kill Sheikh Jaber.
2	**Guam:** The United States returned and fought the Battle of **Guam** on *July 21, 1944*, to recapture the island from Japanese military occupation. More than 18,000 Japanese were killed as only 485 surrendered. Sergeant Shoichi Yokoi, who surrendered in *January 1972*, appears to have been the last confirmed Japanese holdout in Guam.

We propose the following linking problem: *Given an excerpt from Wikipedia, coined Wikiexcerpt, summarizing an event, how can we identify past news articles providing contemporary accounts?* We cast this research question into a query-based retrieval task: given a source text, as a user query, retrieve a ranked list of documents that should be linked to it. In this task, the user poses the Wikiexcerpt as a query and the goal is to retrieve relevant articles from a news collection. Two concrete examples of Wikiexcerpts are given in Table 1.

Standard document retrieval models for keyword queries rely on syntactic matching and are ineffective for our task. Due to the verbosity of the Wikiexcerpts, they are prone to topic drift and result in lower retrieval quality. The Wikiexcerpts also contain additional semantics like temporal expressions, geolocations, and named entities which can be leveraged to identify relevant documents. Making the retrieval model aware of these semantic annotations so as to identify contemporary and relevant documents is not straightforward.

Our approach integrates text, time, geolocations, and named entities in a principled manner, treating them as independent dimensions of event query and ranks documents by comparing them to the query event along these dimensions.

Contributions made in this work are as follows: **(1)** a novel *linking task* to connect Wikipedia excerpts to news articles; **(2)** novel query modeling techniques to estimate independent models for *text*, *time*, *geolocations*, and *named entities* in a query; **(3)** a framework to combine independent query models to rank documents.

Organization. In Sect. 2, we first introduce our notations. Then Sect. 3, gives details on how we estimate the independent query models. Conducted experiments and their results are described in Sect. 4. Section 5 puts our work in context with existing prior research. Finally, we conclude in Sect. 6.

2 Model

Each document d in our document collection C consists of a textual part d_{text}, a temporal part d_{time}, a geospatial part d_{space}, and an entity part d_{entity}. As a bag of words, d_{text} is drawn from a fixed vocabulary V derived from C. Similarly, d_{time}, d_{space} and d_{entity} are bags of temporal expressions, geolocations, and named-entity mentions respectively. We sometimes treat the entire collection C as a single coalesced document and refer to its corresponding parts as C_{text}, C_{time}, C_{space}, and C_{entity}. In our approach, we use the Wikipedia Current Events Portal[1] to distinguish event-specific terms by coalescing into a single document d_{event}. Time unit or *chronon* τ indicates the time passed (to pass) since (until) a reference date such as the UNIX epoch. A temporal expression t is an interval $[tb, te] \in T \times T$, in time domain T, with begin time tb and end time te. Moreover, a temporal expression t is described as a quadruple $[tb_l, tb_u, te_l, te_u]$ [5] where tb_l and tb_u gives the plausible bounds for begin time tb, and te_l and te_u give the bounds for end time te. A geospatial unit l refers to a geographic point that is represented in the geodetic system in terms of latitude (lat) and longitude ($long$). A geolocation s is represented by its minimum bounding rectangle (MBR) and is described as a quadruple $[tp, lt, bt, rt]$. The first point (tp, lt) specifies the top-left corner, and the second point (bt, rt) specifies the bottom-right corner of the MBR. We fix the smallest MBR by setting the resolution $[resol_{lat} \times resol_{long}]$ of space. A named entity e refers to a location, person, or organization from the YAGO [15] knowledge base. We use YAGO URIs to uniquely identify each entity in our approach. A query q is derived from a given Wikiexcerpt in the following way: the text part q_{text} is the full text, the temporal part q_{time} contains explicit temporal expressions that are normalized to time intervals, the geospatial part q_{space} contains the geolocations, and the entity part q_{entity} contains the named entities mentioned. To distinguish contextual terms, we use the textual content of the source Wikipedia article of a given Wikiexcerpt and refer to it as d_{wiki}.

3 Approach

In our approach, we design a *two-stage* cascade retrieval model. In the first stage, our approach performs an initial round of retrieval with the text part of the query to retrieve top-K documents. It then treats these documents as pseudo-relevant and expands the *temporal*, *geospatial*, and *entity* parts of the query. Then, in the second stage, our approach builds independent query models using the expanded query parts, and re-ranks the initially retrieved K documents based on their divergence from the final integrated query model. As output it then returns top-k documents ($k < K$). Intuitively, by using pseudo-relevance feedback to expand query parts, we cope with overly specific (and sparse) annotations in the original query and instead consider those that are salient to the query event for estimating the query models.

[1] http://en.wikipedia.org/wiki/Portal:Currentevents.

For our linking task, we extend the KL-divergence framework [27] to the text, time, geolocation, and entity dimensions of the query and compute an overall divergence score. This is done in two steps: First, we independently estimate a query model for each of the dimensions. Let Q_{text} be the unigram *query-text model*, Q_{time} be the *query-time model*, Q_{space} be the *query-space model*, and Q_{entity} be the *query-entity model*. Second, we represent the overall query model Q as a joint distribution over the dimensions and exploit the additive property of the KL-divergence to combine divergence scores for query models as,

$$KL(Q \,||\, D) = KL(Q_{text} \,||\, D_{text}) + KL(Q_{time} \,||\, D_{time}) \\ + KL(Q_{space} \,||\, D_{space}) + KL(Q_{entity} \,||\, D_{entity}). \tag{1}$$

In the above equation, analogous to the query, the overall document model D is also represented as the joint distribution over D_{text}, D_{time}, D_{space}, and D_{entity} which are the independent document models for the dimensions.

The KL-divergence framework with the independence assumption gives us the flexibility of treating each dimension in isolation while estimating query models. This would include using different background models, expansion techniques with pseudo-relevance feedback, and smoothing. The problem thus reduces to estimating query models for each of the dimensions which we describe next.

Query-Text Model. Standard likelihood-based query modeling methods that rely on the empirical terms become ineffective for our task. As an illustration, consider the first example in Table 1. A likelihood-based model would put more stress on {*Iraq*} that has the highest frequency, and suffer from topical drift due to the terms like {*discussion*, *border*, *dispute*, *Iraq*}. It is hence necessary to make use of a background model that emphasizes event-specific terms.

We observe that a given q_{text} contains two factors, first, terms that give background information, and second, terms that describe the event. To stress on the latter, we combine a query-text model with a background model estimated from: **(1)** the textual content of the source Wikipedia article d_{wiki}; and **(2)** the textual descriptions of events listed in the Wikipedia Current Events portal, d_{event}. The d_{wiki} background model puts emphasis on the contextual terms that are discriminative for the event, like {*Kuwait*, *Iraq*, *Sheikh*, *Jaber*}. On the other hand, the background model d_{event} puts emphasis on event-specific terms like {*capture*, *kill*, *invading*}. Similar approaches that combine multiple contextual models have shown significant improvement in result quality [24,25].

We combine the query model with a background model by linear interpolation [28]. The probability of a word w from the Q_{text} is estimated as,

$$P(w|Q_{text}) = (1-\lambda)\cdot P(w|q_{text}) + \lambda\cdot \left[\beta\cdot P(w|d_{event}) + (1-\beta)\cdot P(w|d_{wiki})\right] . \tag{2}$$

A term w is generated from the background model with probability λ and from the original query with probability $1-\lambda$. Since we use a subset of the available terms, we finally re-normalize the query model as in [20]. The new generative probability $\hat{P}(w \mid Q_{text})$ is computed as,

$$\hat{P}(w \mid Q_{text}) = \frac{P(w \mid Q_{text})}{\sum_{w' \in V} P(w' \mid Q_{text})}. \tag{3}$$

Query-Time Model. We assume that a temporal expression $t \in q_{time}$ is sampled from the query-time model Q_{time} that captures the salient periods for an event in a given q. The generative probability of any time unit τ from the temporal query model Q_{time} is estimated by iterating over all the temporal expressions $t = [tb_l, tb_u, te_l, te_u]$ in q_{time} as,

$$P(\tau \mid Q_{time}) = \sum_{[tb,te] \in q_{time}} \frac{\mathbb{1}(\tau \in [tb_l, tb_u, te_l, te_u])}{|[tb_l, tb_u, te_l, te_u]|} \tag{4}$$

where the $\mathbb{1}(\cdot)$ function returns 1 if there is an overlap between a time unit τ and an interval $[tb_l, tb_u, te_l, te_u]$. The denominator computes the area of the temporal expression in $T \times T$. For any given temporal expression, we can compute its area and its intersection with other expressions as described in [5]. Intuitively, the above equation assigns higher probability to time units that overlap with a larger number of specific (smaller area) intervals in q_{time}.

The query-time model estimated so far has hard temporal boundaries and suffers from the issue of *near-misses*. For example, if the end boundary of the query-time model is "10 January 2014" then the expression "11 January 2014" in a document will be disregarded. To address this issue, we perform an additional *Gaussian smoothing*. The new probability is estimated as,

$$\hat{P}(\tau \mid Q_{time}) = \sum_{t \in T \times T} G_\sigma(t) \cdot P(\tau \mid Q_{time}) \tag{5}$$

where G_σ denotes a Gaussian kernel that is defined as,

$$G_\sigma(i) = \frac{1}{2\pi\sigma^2} \cdot exp\left(-\frac{(tb_l, tb_u)^2 + (te_l, te_u)^2}{2\sigma^2}\right). \tag{6}$$

Gaussian smoothing computes a weighted average of adjacent units with a weight decreasing with the spatial distance to center position τ in two dimensional space. The σ in the kernel defines the neighborhood size and can be empirically set. As a result of the Gaussian smoothing, the temporal boundaries are blurred, spilling some probability mass to adjacent time units. Finally, since we use only a subset of temporal expressions we re-normalize similar to Eq. 3.

Query-Space Model. We assume that a user samples a geolocation s from query-space model Q_{space} to generate q_{space}. The query-space model captures salient geolocations for the event in a given Wikiexcerpt. The generative probability of any spatial unit l from the query-space model Q_{space} by iterating over all geolocations $[tp, lt, bt, rt] \in q_{space}$ is estimated as

$$P(l \mid Q_{space}) = \sum_{(tp,lt,bt,rt) \in q_{space}} \frac{\mathbb{1}(l \in [tp, lt, bt, rt])}{|[tp, lt, bt, rt]|}. \tag{7}$$

Analogous to the Eq. 4, the $\mathbb{1}(\cdot)$ function returns 1 if there is an overlap between a space unit l and a MBR as $[tp, lt, bt, rt]$. Intuitively, query-space model assigns higher probability to l if it overlaps with a larger number of more specific (MBR with smaller area) geolocations in q_{space}. As the denominator, it is easy to compute $|\,[tp, lt, bt, rt]\,|$ as $|s| = (rt - lt + resol_{lat}) * (tp - bt + resol_{long})$. Addition of the small constant ensures that for all s, $|s| > 0$.

Similar to the query-time model, to address the issue of near misses we estimate $\hat{P}(l|Q_{Space})$ that additionally smooths $P(l|Q_{space})$ using a Gaussian kernel as described in Eq. 5 and also re-normalize as per Eq. 3.

Query-Entity Model. The query-entity model Q_{entity} captures the entities that are salient to an event and builds a probability distribution over an entity space. To estimate Q_{entity} we make use of the initially retrieved pseudo-relevant documents to construct a background model that assigns higher probability to entities that are often associated with an event. Let D_R be the set of pseudo-relevant documents. The generative probability of entity e is estimated as,

$$P(e \mid Q_{entity}) = (1 - \lambda) \cdot P(e \mid q_{entity}) + \lambda \cdot \sum_{d \in D_R} P(e \mid d_{entity}) \quad (8)$$

where $P(e \mid q_{entity})$ and $P(e \mid d_{entity})$ are the likelihoods of generating the entity from the original query and a document $d \in D_R$ respectively.

Document Model. To estimate the document models for each dimension, we follow the same methodology as for the query with an additional step of Dirichlet smoothing [28]. This has two effects: First, it prevents undefined KL-Divergence scores. Second, it achieves an IDF-like effect by smoothing the probabilities of expressions that occur frequently in the C. The generative probability of a term w from document-text model D_{text} is estimated as,

$$P(w \mid D_{text}) = \frac{\hat{P}(w \mid D_{text}) + \mu P(w \mid C_{text})}{|D_{text}| + \mu} \quad (9)$$

where $\hat{P}(w \mid D_{text})$ is computed according to Eq. 3 and μ is set as the average document length of our collection [28]. Similarly, we estimate D_{time}, D_{space}, and D_{entity} with C_{time}, C_{space}, and C_{entity} as background models to tackle the above mentioned issues. To estimate D_{time} and D_{space}, we follow methods similar to Eqs. 4 and 7. However, we do not apply the Gaussian smoothing (as described in Eq. 5) as it tends to artificially introduce temporal and spatial information into the document content.

4 Experiments

Next, we describe our experiments to study the impact of the different components of our approach. We make our experimental data publicly available[2].

[2] http://resources.mpi-inf.mpg.de/d5/linkingWiki2News/.

Document Collection. For the first set of experiments, we use The New York Times[3] Annotated Corpus (NYT) which contains about two million news articles published between 1987 and 2007. For the second set, we use the ClueWeb12-B13 (CW12) corpus[4] with 50 million web pages crawled in 2012.

Test Queries. We use the English Wikipedia dump released on February 3rd 2014 to generate two independent sets of test queries: **(1)** *NYT-Queries*, contains 150 randomly sampled Wikiexcerpts targeting documents from the NYT corpus; **(2)** *CW-Queries* contains 150 randomly sampled Wikiexcerpts targeting web pages from CW12 corpus. *NYT-Queries* have 104 queries, out of 150, that come with at least one temporal expression, geolocation, and named-entity mention. In the remaining 46 test queries, 17 do not have any temporal expressions, 28 do not have any geolocations, and 27 do not have any entity mentions. We have 4 test queries where our taggers fail to identify any additional semantics. *CW-Queries* have 119 queries, out of 150, that come with at least one temporal expression, geolocation and entity mention. 19 queries do not mention any geolocation, and 26 do not have entity mentions.

Relevance Assessments were collected using the CrowdFlower platform[5]. We pooled top-10 results for the methods under comparison, and asked assessors to judge a document as **(0)** irrelevant, **(1)** somewhat relevant, or **(2)** highly relevant to a query. Our instructions said that a document can only be considered highly relevant if its main topic was the event given as the query. Each query-document pairs was judged by three assessors. Both experiments resulted in 1778 and 1961 unique query-document pairs, respectively. We paid $0.03 per batch of five query-document pairs for a single assessor.

Effectiveness Measures. As a strict effectiveness measure, we compare our methods based on mean reciprocal rank (MRR). We also compare our methods using normalized discounted cumulative gain (NDCG) and precision (P) at cutoff levels 5 and 10. We also report the mean average precision (MAP) across all queries. For MAP and P we consider a document relevant to a query if the majority of assessors judged it with label **(1)** or **(2)**. For NDCG we plug in the mean label assigned by assessors.

Methods. We compare the following methods: **(1)** *txt* considers only the query-text model that uses the background models estimated from the current events portal and the source Wikipedia article (Eq. 2); **(2)** *txtT* uses the query-text and query-time model (Eq. 4); **(3)** *txtS* uses the query-text and query-space model (Eq. 7); **(4)** *txtE* uses the query-text and query-entity model (Eq. 8); **(5)** *txtST* uses the query-text, query-time and query-space model; **(6)** *txtSTE* uses all four query models to rank documents.

Parameters. We set the values for the different parameters in query and document models for all the methods by following [27]. For the NYT corpus, we treat

[3] http://corpus.nytimes.com.
[4] http://www.lemurproject.org/clueweb12.php/.
[5] http://www.crowdflower.com/.

top-100 documents retrieved in the first stage as pseudo-relevant. For CW12 corpus with general web pages, we set this to top-500. The larger number of top-K documents for the CW12 corpus is due to the fact that web pages come with fewer annotations than news articles. In Eq. 2 for estimating the Q_{text}, we set $\beta = 0.5$ thus giving equal weights to the background models. For the interpolation parameters, we set $\lambda = 0.85$ in Eqs. 2 and 8. For the Gaussian smoothing in Eq. 6 we set $\sigma = 1$. The smallest possible MBRs in Eq. 7 is empirically set as $resol_{lat} \times resol_{long} = 0.1 \times 0.1$.

Implementation. All methods have been implemented in java. To annotate named entities in the test queries and documents from the NYT corpus, we use the AIDA [16] system. For the CW12 corpus, we use the annotations released as Freebase Annotations of the ClueWeb Corpora[6]. To annotate geolocations in the query and NYT corpus, we use an open-source gazetteer-based tool[7] that extracts locations and maps them to GeoNames[8] knowledge base. To get geolocations for CW12 corpus we filter entities by mapping them from Freebase to GeoNames ids. Finally, we run Stanford Core NLP[9]on the test queries, NYT corpus and CW12 corpus to get the temporal annotations.

Results. Tables 2 and 3 compare the different methods on our two datasets. Both tables have two parts: **(a)** results on the entire query set; and **(b)** results on a subset of queries with at least one temporal expression, geolocation, and entity mention. To denote the significance of the observed improvements to the *txt* method, we perform one-sided paired student's T test at two alpha-levels: 0.05 (‡), and 0.10 (†), on the MAP, P@5, and P@10 scores [8]. We find that the *txtSTE* method is most effective for the linking task.

In Table 2 we report results for the NYT-Queries. We find that the *txtSTE* method that combines information in all the dimensions achieves the best result across all metrics except P@5. The *txt* method that uses only the text already gets a high MRR score. The *txtS* method that adds geolocations to text is able to add minor improvements in NDCG@10 over the *txt* method. The *txtT* method achieves a considerable improvement over *txt*. This is consistent for both NYT-Queries (a) and NYT-Queries (b). The *txtE* method that uses named-entities along with text shows significant improvement in P@5 and marginal improvements across other metrics. The *txtST* method that combines time and geolocations achieves significant improvements over *txt*. Finally, the *txtSTE* method proves to be the best and shows significant improvements over the *txt*.

In Table 3, we report results for the CW-Queries. We find that the *txtSTE* method outperforms other methods across all the metrics. Similar to previous results, we find that the *txt* method already achieves high MRR score. However, in contrast, the *txtT* approach shows improvements in terms of P@5 and NDCG@5, with a marginal drop in P@10 and MAP. The geolocations improve

[6] http://lemurproject.org/clueweb12/FACC1/.
[7] https://github.com/geoparser/geolocator.
[8] http://www.geonames.org/.
[9] http://nlp.stanford.edu/software/corenlp.shtml.

Table 2. Results for NYT-Queries

	NYT-Queries (a) - 150 queries						**NYT-Queries (b)** - 104 queries					
Measures	*txt*	*txtT*	*txtS*	*txtE*	*txtST*	*txtSTE*	*txt*	*txtT*	*txtS*	*txtE*	*txtST*	*txtSTE*
MRR	0.898	0.897	0.898	0.898	0.898	**0.902**	0.921	0.936	0.921	0.921	0.936	**0.942**
P@5	0.711	0.716	0.709	0.716 ‡	**0.719**	0.717	0.715	0.740 ‡	0.715	0.723 ‡	**0.742** ‡	0.740 ‡
P@10	0.670	0.679	0.669	0.671	0.679	**0.682** †	0.682	0.692	0.681	0.684	0.692	**0.696** †
MAP	0.687	0.700	0.687	0.688	0.701	**0.704** †	0.679	0.702 †	0.679	0.682	0.703 †	**0.708** ‡
NDCG@5	0.683	0.696	0.682	0.685	0.697	**0.697**	0.686	0.721	0.686	0.689	0.721	**0.723**
NDCG@10	0.797	0.813	0.798	0.796	0.814	**0.815**	0.794	0.823	0.795	0.795	0.823	**0.825**

Table 3. Results for CW12-Queries

	CW-Queries (a) - 150 queries						**CW-Queries (b)** - 119 queries					
Measures	*txt*	*txtT*	*txtS*	*txtE*	*txtST*	*txtSTE*	*txt*	*txtT*	*txtS*	*txtE*	*txtST*	*txtSTE*
MRR	0.824	0.834	0.831	0.827	0.833	**0.836**	0.837	0.855	0.846	0.842	0.854	**0.855**
P@5	0.448	0.460	0.451	0.456 †	0.467 ‡	**0.475** ‡	0.456	0.468	0.459 †	0.466 ‡	0.478 ‡	**0.488** ‡
P@10	0.366	0.349	0.366	0.375 ‡	0.367	**0.375** †	0.377	0.358	0.378	0.390 ‡	0.377	**0.386** ‡
MAP	0.622	0.616	0.628	0.640 ‡	0.640 †	**0.653** ‡	0.623	0.616	0.631	0.647 ‡	0.642 †	**0.661** ‡
NDCG@5	0.644	0.657	0.651	0.654	0.666	**0.673**	0.655	0.675	0.664	0.669	0.684	**0.695**
NDCG@10	0.729	0.723	0.734	0.746	0.744	**0.755**	0.736	0.736	0.744	0.759	0.755	**0.769**

the quality of the results in terms of MAP and significantly improve P@10. Though individually time and geolocations show only marginal improvements, their combination as the *txtST* method shows significant increase in MAP. We find that the *txtE* method performs better than other dimensions with a significant improvement over *txt* across all metrics. Finally, the best performing method is *txtSTE* as it shows the highest improvement in the result quality.

Discussion. As a general conclusion of our experiments we find that leveraging semantic annotations like time, geolocations, and named entities along with text improves the effectiveness of the linking task. Because all our methods that utilize semantic annotations (*txtS*, *txtT*, *txtE*, *txtST*, and *txtSTE*) perform better than the text-only (*txt*) method. However, the simple *txt* method already achieves a decent MRR score in both experiments. This highlights the effectiveness of the event-specific background model in tackling the verbosity of the Wikiexcerpts. Time becomes an important indicator to identify relevant news articles but it is not very helpful when it comes to general web pages. This is because the temporal expressions in the news articles often describe the event time period accurately thus giving a good match to the queries while this is not seen with web pages. We find that geolocations and time together can better identify relevant documents when combined with text. Named entities in the queries are not always salient to the event but may represent the context of the event. For complex queries, it is hard to distinguish salient entities which reduces the overall performance due to topical drifts on a news corpus. However, they prove to be effective to identify relevant web pages which can contain more general information thus also mentioning the contextual entities. The improvement of our method over a simple text-based method is more pronounced for the ClueWeb corpus than the news corpus because of mainly two reasons: firstly, the news corpus is too narrow with much smaller number of articles; and secondly, it is slightly easier to retrieve relatively short, focused, and high quality news articles. This is supported by the fact that all methods achieve much higher MRR scores for the NYT-Queries.

Gain/Loss Analysis. To get some insights into where the improvements for the *txtEST* method comes from, we perform a gain/loss analysis based on NDCG@5. The *txtSTE* method shows biggest gain (+0.13) in NDCG@5 for the following query in *NYT-Queries*:

West Windsor Township, New Jersey: The West Windsor post office was found to be infected with anthrax during the anthrax terrorism scare back in 2001-2002.

The single temporal expression *2001-2002* refers to a time period when there were multiple anthrax attacks in New Jersey through the postal facilities. Due to the ambiguity, the *txtT* and *txtS* methods become ineffective for this query. Their combination, however, as the *txtST* method becomes the second best method achieving NDCG@5 of 0.7227. The *txtEST* combines the entity *Anthrax* and becomes the best method by achieving NDCG@5 of 0.8539. This method suffers worst in terms of NDCG@5 (−0.464) for the following query in *CW-Queries*:

Human Rights Party Malaysia: The Human Rights Party Malaysia is a Malaysian human rights-based political party founded on 19 July 2009, led by human rights activist P. Uthayakumar.

The two entities, *Human Rights Party Malaysia* and *P. Uthayakumar* and one geolocation, *Malaysia*, do prove to be discriminative for the event. Time becomes an important indicator to identify relevant documents as *txtT* becomes most effective by achieving NDCG@5 of 0.9003. However, a combination of text, time, geolocations and named entities as leveraged by *txtEST* achieves a lower NDCG@5 of 0.4704.

Easy and Hard Query Events. Finally, we identify the easiest and the hardest query events across both our testbeds. We find that the following query, in the *CW-Queries*, gets the highest minimum P@10 across all methods:

Primal Therapy: In 1989, Arthur Janov established the Janov Primal Center in Venice (later relocated to Santa Monica) with his second wife, France.

For this query even the simple *txt* method gets a perfect P@10 score of 1.0. Terms *Janov*, *Primal*, and *Center* retrieve documents that are pages from the center's website, and are marked relevant by the assessors. Likewise, we identify the hardest query as the following one from the *NYT-Queries* set:

Police aviation in the United Kingdom: In 1921, the British airship R33 was able to help the police in traffic control around the Epsom and Ascot horse-racing events.
For this query none of the methods were able to identify any relevant documents thus all getting a P@10 score equal to 0. This is simply because this relatively old event is not covered in the NYT corpus.

5 Related Work

In this section, we put our work in context with existing prior research. We review five lines of prior research related to our work.

First, we look into efforts to link different document collections. As the earliest work, Henzinger et al. [14] automatically suggested news article links for an ongoing TV news broadcast. Later works have looked into linking related text across multiple archives to improve their exploration [6]. Linking efforts also go towards enriching social media posts by connecting them to news articles [26]. Recently, Arapakis et al. [1] propose automatic linking system between news articles describing similar events.

Next, we identify works that use time to improve document retrieval quality [23]. To leverage time, prior works have proposed methods that are motivated from cognitive psychology [21]. Time has also been considered as a feature for query profiling and classification [17]. In the realm of document retrieval, Berberich et al. [5] exploit explicit temporal expressions contained in queries to improve result quality. As some of the latest work, Peetz et al. [20] detect temporal burstiness of query terms, and Mishra et al. [19] leverage explicit temporal expressions to estimate temporal query models. Efron et al. [11] present a kernel density estimation method to temporally match relevant tweets.

There have been many prior initiatives [7,13] to investigate geographical information retrieval. The GeoCLEF search task examined geographic search in text corpus [18]. More recent initiatives like the NTCIR-GeoTime task [12] evaluated adhoc retrieval with geographic and temporal constraints.

We look into prior research works that use entities for information retrieval. Earlier initiatives like INEX entity ranking track [10] and TREC entity track [3] focus on retrieving relevant entities for a given topic. More recently, INEX Linked Data track [4] aimed at evaluating approaches that additionally use text for entity ranking. As the most recent work, Dalton et al. [9] show significant improvement for document retrieval.

Divergence-based retrieval models for text have been well-studied in the past. In their study, Zhai et al. [27,28] compare techniques of combining backgrounds models to query and documents. To further improve the query model estimation, Shen et al. [24] exploit contextual information like query history and click through history. Bai et al. [2] combine query models estimated from multiple contextual factors.

6 Conclusion

We have addressed a novel linking problem with the goal of establishing connections between excerpts from Wikipedia, coined Wikiexcerpts, and news articles. For this, we cast the linking problem into an information retrieval task and present approaches that leverage additional semantics that come with a Wikiexcerpt. Comprehensive experiments on two large datasets with independent test query sets show that our approach that leverages time, geolocations, named entities, and text is most effective for the linking problem.

References

1. Arapakis, I., et al.: Automatically embedding newsworthy links to articles: From implementation to evaluation. JASIST **65**(1), 129–145 (2014)
2. Bai, J., et al.: Using query contexts in information retrieval. In: SIGIR.(2007)
3. Balog, K., et al.: Overview of the TREC 2010 entity track. In: DTIC.(2010)
4. Bellot, P., et al.: Report on INEX 2013. ACM SIGIR Forum **47**(2), 21–32 (2013)
5. Berberich, Klaus, Bedathur, Srikanta, Alonso, Omar, Weikum, Gerhard: A Language Modeling Approach for Temporal Information Needs. In: Gurrin, Cathal, He, Yulan, Kazai, Gabriella, Kruschwitz, Udo, Little, Suzanne, Roelleke, Thomas, Rüger, Stefan, van Rijsbergen, Keith (eds.) ECIR 2010. LNCS, vol. 5993, pp. 13–25. Springer, Heidelberg (2010)
6. Bron, Marc, Huurnink, Bouke, de Rijke, Maarten: Linking Archives Using Document Enrichment and Term Selection. In: Gradmann, Stefan, Borri, Francesca, Meghini, Carlo, Schuldt, Heiko (eds.) TPDL 2011. LNCS, vol. 6966, pp. 360–371. Springer, Heidelberg (2011)
7. Cozza, Vittoria, Messina, Antonio, Montesi, Danilo, Arietta, Luca, Magnani, Matteo: Spatio-Temporal Keyword Queries in Social Networks. In: Catania, Barbara, Guerrini, Giovanna, Pokorný, Jaroslav (eds.) ADBIS 2013. LNCS, vol. 8133, pp. 70–83. Springer, Heidelberg (2013)
8. Croft, B., et al.: Search Engines: Information Retrieval in Practice. Addison-Wesley, Reading.(2010)
9. Dalton, J., et al.: Entity query feature expansion using knowledge base links. In: SIGIR.(2014)
10. Demartini, Gianluca, Iofciu, Tereza, de Vries, Arjen P.: Overview of the INEX 2009 Entity Ranking Track. In: Geva, Shlomo, Kamps, Jaap, Trotman, Andrew (eds.) INEX 2009. LNCS, vol. 6203, pp. 254–264. Springer, Heidelberg (2010)
11. Efron, M., et al.: Temporal feedback for tweet search with non-parametric density estimation. In: SIGIR.(2014)
12. Gey, F., et al.: NTCIR-GeoTime overview: Evaluating geographic and temporal search. In: NTCIR.(2010)
13. Hariharan, R., et al.: Processing spatial-keyword (SK) queries in geographic information retrieval (GIR) systems. In: SSDBM.(2007)
14. Henzinger, M.R., et al.: Query-free news search. World Wide Web **8**, 101–126 (2005)
15. Hoffart, J., et al.: YAGO2: A spatially and temporally enhanced knowledge base from Wikipedia. In: IJCAI.(2013)
16. Hoffart, J., et al.: Robust Disambiguation of Named Entities in Text. In: EMNLP.(2011)
17. Kulkarni, A., et al.: Understanding temporal query dynamics. In: WSDM.(2011)
18. Mandl, Thomas, Gey, Fredric C., Di Nunzio, Giorgio Maria, Ferro, Nicola, Larson, Ray R., Sanderson, Mark, Santos, Diana, Womser-Hacker, Christa, Xie, Xing: GeoCLEF 2007: The CLEF 2007 Cross-Language Geographic Information Retrieval Track Overview. In: Peters, Carol, Jijkoun, Valentin, Mandl, Thomas, Müller, Henning, Oard, Douglas W., Peñas, Anselmo, Petras, Vivien, Santos, Diana (eds.) CLEF 2007. LNCS, vol. 5152, pp. 745–772. Springer, Heidelberg (2008)
19. Mishra, A., et al.: Linking wikipedia events to past news. In: TAIA.(2014)
20. Peetz, M., et al.: Using temporal bursts for query modeling. Inf. retrieval **17**(1), 74–108 (2014)

21. Peetz, Maria-Hendrike, de Rijke, Maarten: Cognitive Temporal Document Priors. In: Serdyukov, Pavel, Braslavski, Pavel, Kuznetsov, Sergei O., Kamps, Jaap, Rüger, Stefan, Agichtein, Eugene, Segalovich, Ilya, Yilmaz, Emine (eds.) ECIR 2013. LNCS, vol. 7814, pp. 318–330. Springer, Heidelberg (2013)
22. Perea-Ortega, José M., Ureña-López, LAlfonso: Geographic Expansion of Queries to Improve the Geographic Information Retrieval Task. In: Bouma, Gosse, Ittoo, Ashwin, Métais, Elisabeth, Wortmann, Hans (eds.) NLDB 2012. LNCS, vol. 7337, pp. 94–103. Springer, Heidelberg (2012)
23. Ricardo, C., et al.: Survey of temporal information retrieval and related applications. ACM Comput. Surv. (CSUR) **47**(2), 1–41 (2014)
24. Shen, X., et al.: Context-sensitive information retrieval using implicit feedback. In: SIGIR.(2005)
25. Tan, B., et al.: Mining long-term search history to improve search accuracy. In: KDD.(2006)
26. Tsagkias, M., et al.: Linking online news and social media. In: WSDM.(2011)
27. Zhai, C., et al.: Model-based feedback in the language modeling approach to information retrieval. In: CIKM.(2001)
28. Zhai, C., et al.: Two-stage language models for information retrieval. In: SIGIR.(2002)

Machine Learning

Deep Learning over Multi-field Categorical Data – A Case Study on User Response Prediction

Weinan Zhang[1(✉)], Tianming Du[1,2], and Jun Wang[1]

[1] University College London, London, UK
{w.zhang,j.wang}@cs.ucl.ac.uk
[2] RayCloud Inc., Hangzhou, China
dutianming@quicloud.cn

Abstract. Predicting user responses, such as click-through rate and conversion rate, are critical in many web applications including web search, personalised recommendation, and online advertising. Different from continuous raw features that we usually found in the image and audio domains, the input features in web space are always of multi-field and are mostly discrete and categorical while their dependencies are little known. Major user response prediction models have to either limit themselves to linear models or require manually building up high-order combination features. The former loses the ability of exploring feature interactions, while the latter results in a heavy computation in the large feature space. To tackle the issue, we propose two novel models using deep neural networks (DNNs) to automatically learn effective patterns from categorical feature interactions and make predictions of users' ad clicks. To get our DNNs efficiently work, we propose to leverage three feature transformation methods, i.e., factorisation machines (FMs), restricted Boltzmann machines (RBMs) and denoising auto-encoders (DAEs). This paper presents the structure of our models and their efficient training algorithms. The large-scale experiments with real-world data demonstrate that our methods work better than major state-of-the-art models.

1 Introduction

User response (e.g., click-through or conversion) prediction plays a critical part in many web applications including web search, recommender systems, sponsored search, and display advertising. In online advertising, for instance, the ability of targeting individual users is the key advantage compared to traditional offline advertising. All these targeting techniques, essentially, rely on the system function of predicting whether a specific user will think the potential ad is "relevant", i.e., the probability that the user in a certain context will click a given ad [6]. Sponsored search, contextual advertising, and the recently emerged real-time bidding (RTB) display advertising all heavily rely on the ability of learned models to predict ad click-through rates (CTR) [32,41]. The applied CTR estimation models today are mostly linear, ranging from logistic regression [32] and naive Bayes [14] to FTRL logistic regression [28] and Bayesian probit regression [12],

N. Ferro et al. (Eds.): ECIR 2016, LNCS 9626, pp. 45–57, 2016.
DOI: 10.1007/978-3-319-30671-1_4

all of which are based on a huge number of sparse features with one-hot encoding [1]. Linear models have advantages of easy implementation, efficient learning but relative low performance because of the failure of learning the non-trivial patterns to catch the interactions between the assumed (conditionally) independent raw features [12]. Non-linear models, on the other hand, are able to utilise different feature combinations and thus could potentially improve estimation performance. For example, factorisation machines (FMs) [29] map the user and item binary features into a low dimensional continuous space. And the feature interaction is automatically explored via vector inner product. Gradient boosting trees [38] automatically learn feature combinations while growing each decision/regression tree. However, these models cannot make use of all possible combinations of different features [20]. In addition, many models require feature engineering that manually designs what the inputs should be. Another problem of the mainstream ad CTR estimation models is that most prediction models have shallow structures and have limited expression to model the underlying patterns from complex and massive data [15]. As a result, their data modelling and generalisation ability is still restricted.

Deep learning [25] has become successful in computer vision [22], speech recognition [13], and natural language processing (NLP) [19,33] during recent five years. As visual, aural, and textual signals are known to be spatially and/or temporally correlated, the newly introduced unsupervised training on deep structures [18] would be able to explore such *local* dependency and establish a *dense* representation of the feature space, making neural network models effective in learning high-order features directly from the raw feature input. With such learning ability, deep learning would be a good candidate to estimate online user response rate such as ad CTR. However, most input features in CTR estimation are of multi-field and are discrete categorical features, e.g., the user location city (`London`, `Paris`), device type (`PC`, `Mobile`), ad category (`Sports`, `Electronics`) etc., and their local dependencies (thus the sparsity in the feature space) are unknown. Therefore, it is of great interest to see how deep learning improves the CTR estimation via learning feature representation on such large-scale multi-field discrete categorical features. To our best knowledge, there is no previous literature of ad CTR estimation using deep learning methods thus far[1]. In addition, training deep neural networks (DNNs) on a large input feature space requires tuning a huge number of parameters, which is computationally expensive. For instance, unlike image and audio cases, we have about 1 million binary input features and 100 hidden units in the first layer; then it requires 100 million links to build the first layer neural network.

In this paper, we take ad CTR estimation as a working example to study deep learning over a large multi-field categorical feature space by using embedding methods in both supervised and unsupervised fashions. We introduce two types of deep learning models, called Factorisation Machine supported Neural Network (FNN) and Sampling-based Neural Network (SNN). Specifically, FNN with

[1] Although the leverage of deep learning models on ad CTR estimation has been claimed in industry (e.g., [42]), there is no detail of the models or implementation.

a supervised-learning embedding layer using factorisation machines [31] is proposed to efficiently reduce the dimension from sparse features to dense continuous features. The second model SNN is a deep neural network powered by a sampling-based restricted Boltzmann machine (SNN-RBM) or a sampling-based denoising auto-encoder (SNN-DAE) with a proposed negative sampling method. Based on the embedding layer, we build multiple layers neural nets with full connections to explore non-trivial data patterns. Our experiments on multiple real-world advertisers' ad click data have demonstrated the consistent improvement of CTR estimation from our proposed models over the state-of-the-art ones.

2 Related Work

Click-through rate, defined as the probability of the ad click from a specific user on a displayed ad, is essential in online advertising [39]. In order to maximise revenue and user satisfaction, online advertising platforms must predict the expected user behaviour for each displayed ad and maximise the expectation that users will click. The majority of current models use logistic regression based on a set of sparse binary features converted from the original categorical features via one-hot encoding [26,32]. Heavy engineering efforts are needed to design features such as locations, top unigrams, combination features, etc. [15].

Embedding very large feature vector into low-dimensional vector spaces is useful for prediction task as it reduces the data and model complexity and improves both the effectiveness and the efficiency of the training and prediction. Various methods of embedding architectures have been proposed [23,37]. Factorisation machine (FM) [31], originally proposed for collaborative filtering recommendation, is regarded as one of the most successful embedding models. FM naturally has the capability of estimating interactions between any two features via mapping them into vectors in a low-rank latent space.

Deep Learning [2] is a branch of artificial intelligence research that attempts to develop the techniques that will allow computers to handle complex tasks such as recognition and prediction at high performance. Deep neural networks (DNNs) are able to extract the hidden structures and intrinsic patterns at different levels of abstractions from training data. DNNs have been successfully applied in computer vision [40], speech recognition [8] and natural language processing (NLP) [7,19,33]. Furthermore, with the help of unsupervised pre-training, we can get good feature representation which guides the learning towards basins of attraction of minima that support better generalisation from the training data [10]. Usually, these deep models have two stages in learning [18]: the first stage performs model initialisation via unsupervised learning (i.e., the restricted Boltzmann machine or stacked denoising auto-encoders) to make the model catch the input data distribution; the second stage involves a fine tuning of the initialised model via supervised learning with back-propagation. The novelty of our deep learning models lies in the first layer initialisation, where the input raw features are high dimensional and sparse binary features converted from the raw categorical features, which makes it hard to train traditional DNNs in large scale.

Compared with the word-embedding techniques used in NLP [19,33], our models deal with more general multi-field categorical features without any assumed data structures such as word alignment and letter-n-gram etc.

3 DNNs for CTR Estimation Given Categorical Features

In this section, we discuss the two proposed DNN architectures in detail, namely Factorisation-machine supported Neural Networks (FNN) and Sampling-based Neural Networks (SNN). The input categorical features are field-wise one-hot encoded. For each field, e.g., `city`, there are multiple units, each of which represents a specific value of this field, e.g., `city=London`, and there is only one positive (1) unit, while all others are negative (0). The encoded features, denoted as $\boldsymbol{x}$, are the input of many CTR estimation models [26,32] as well as our DNN models, as depicted at the bottom layer of Fig. 1.

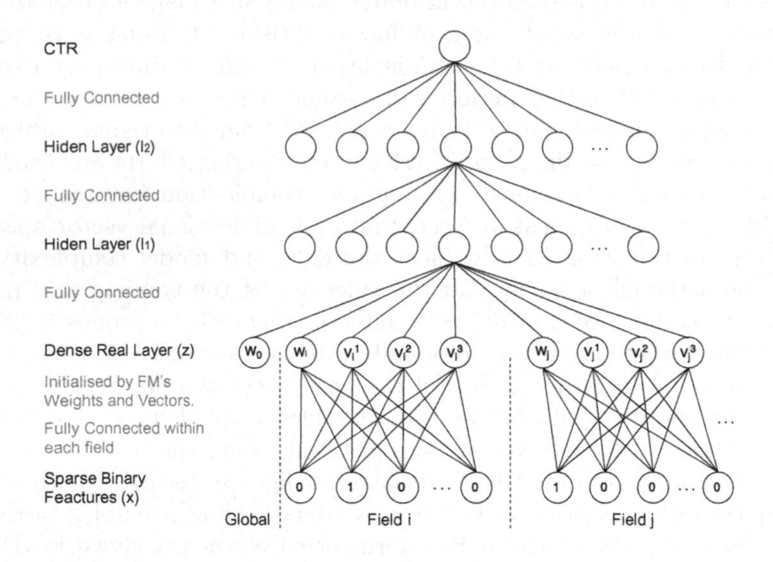

Fig. 1. A 4-layer FNN model structure.

3.1 Factorisation-Machine Supported Neural Networks (FNN)

Our first model FNN is based on the factorisation machine as the bottom layer. The network structure is shown in Fig. 1. With a top-down description, the output unit is a real number $\hat{y} \in (0, 1)$ as predicted CTR, i.e., the probability of a specific user clicking a given ad in a certain context:

$$\hat{y} = \text{sigmoid}(\boldsymbol{W}_3 \boldsymbol{l}_2 + b_3), \tag{1}$$

where $\text{sigmoid}(x) = 1/(1 + e^{-x})$ is the logistic activation function, $\boldsymbol{W}_3 \in \mathbb{R}^{1 \times L}$, $b_3 \in \mathbb{R}$ and $\boldsymbol{l}_2 \in \mathbb{R}^L$ as input for this layer. The calculation of $\boldsymbol{l}_2$ is

$$\boldsymbol{l}_2 = \tanh(\boldsymbol{W}_2 \boldsymbol{l}_1 + \boldsymbol{b}_2), \tag{2}$$

where $\tanh(x) = (1 - e^{-2x})/(1 + e^{-2x})$, $\boldsymbol{W}_2 \in \mathbb{R}^{L \times M}$, $\boldsymbol{b}_2 \in \mathbb{R}^L$ and $\boldsymbol{l}_1 \in \mathbb{R}^M$. We choose $\tanh(\cdot)$ as it has optimal empirical learning performance than other activation functions, as will be discussed in Sect. 4.3. Similarly,

$$\boldsymbol{l}_1 = \tanh(\boldsymbol{W}_1 \boldsymbol{z} + \boldsymbol{b}_1), \tag{3}$$

where $\boldsymbol{W}_1 \in \mathbb{R}^{M \times J}$, $\boldsymbol{b}_1 \in \mathbb{R}^M$ and $\boldsymbol{z} \in \mathbb{R}^J$.

$$\boldsymbol{z} = (w_0, \boldsymbol{z}_1, \boldsymbol{z}_2, ...\boldsymbol{z}_i, ..., \boldsymbol{z}_n), \tag{4}$$

where $w_0 \in \mathbb{R}$ is a global scalar parameter and n is the number of fields in total. $\boldsymbol{z}_i \in \mathbb{R}^{K+1}$ is a parameter vectors for the i-th field in factorisation machines:

$$\boldsymbol{z}_i = \boldsymbol{W}_0^i \cdot \boldsymbol{x}[\text{start}_i : \text{end}_i] = (w_i, v_i^1, v_i^2, \dots, v_i^K), \tag{5}$$

where start_i and end_i are starting and ending feature indexes of the i-th field, $\boldsymbol{W}_0^i \in \mathbb{R}^{(K+1) \times (\text{end}_i - \text{start}_i + 1)}$ and $\boldsymbol{x}$ is the input vector as described at beginning. All weights $\boldsymbol{W}_0^i$ are initialised with the bias term w_i and vector $\boldsymbol{v}_i$ respectively (e.g., $\boldsymbol{W}_0^i[0]$ is initialised by w_i, $\boldsymbol{W}_0^i[1]$ is initialised by v_i^1, $\boldsymbol{W}_0^i[2]$ is initialised by v_i^2, etc.). In this way, $\boldsymbol{z}$ vector of the first layer is initialised as shown in Fig. 1 via training a factorisation machine (FM) [31]:

$$y_{\text{FM}}(\boldsymbol{x}) := \text{sigmoid}\Big(w_0 + \sum_{i=1}^{N} w_i x_i + \sum_{i=1}^{N} \sum_{j=i+1}^{N} \langle \boldsymbol{v}_i, \boldsymbol{v}_j \rangle x_i x_j\Big), \tag{6}$$

where each feature i is assigned with a bias weight w_i and a K-dimensional vector $\boldsymbol{v}_i$ and the feature interaction is modelled as their vectors' inner product $\langle \boldsymbol{v}_i, \boldsymbol{v}_j \rangle$. In this way, the above neural nets can learn more efficiently from factorisation machine representation so that the computational complexity problem of the high-dimensional binary inputs has been naturally bypassed. Different hidden layers can be regarded as different internal functions capturing different forms of representations of the data instance. For this reason, this model has more abilities of catching intrinsic data patterns and leads to better performance.

The idea using FM in the bottom layer is ignited by Convolutional Neural Networks (CNNs) [11], which exploit spatially local correlation by enforcing a local connectivity pattern between neurons of adjacent layers. Similarly, the inputs of hidden layer 1 are connected to the input units of a specific field. Also, the bottom layer is not fully connected as FM performs a field-wise training for one-hot sparse encoded input, allowing local sparsity, illustrated as the dash lines in Fig. 1. FM learns good structural data representation in the latent space, helpful for any further model to build on. A subtle difference, though, appears between the product rule of FM and the sum rule of DNN for combination. However, according to [21], if the observational discriminatory information is highly ambiguous (which is true in our case for ad click behaviour), the posterior weights (from DNN) will not deviate dramatically from the prior (FM).

Furthermore, the weights in hidden layers (except the FM layer) are initialised by layer-wise RBM pre-training [3] using contrastive divergence [17], which effectively preserves the information in input dataset as detailed in [16,18]. The initial weights for FMs are trained by stochastic gradient descent (SGD), as detailed in [31]. Note that we only need to update weights which connect to the positive input units, which largely reduces the computational complexity. After pre-training of the FM and upper layers, supervised fine-tuning (back propagation) is applied to minimise loss function of cross entropy:

$$L(y, \hat{y}) = -y \log \hat{y} - (1 - y) \log(1 - \hat{y}), \tag{7}$$

where $\hat{y}$ is the predicted CTR in Eq. (1) and y is the binary click ground-truth label. Using the chain rule of back propagation, the FNN weights including FM weights can be efficiently updated. For example, we update FM layer weights via

$$\frac{\partial L(y, \hat{y})}{\partial \boldsymbol{W}_0^i} = \frac{\partial L(y, \hat{y})}{\partial \boldsymbol{z}_i} \frac{\partial \boldsymbol{z}_i}{\partial \boldsymbol{W}_0^i} = \frac{\partial L(y, \hat{y})}{\partial \boldsymbol{z}_i} \boldsymbol{x}[\text{start}_i : \text{end}_i] \tag{8}$$

$$\boldsymbol{W}_0^i \leftarrow \boldsymbol{W}_0^i - \eta \cdot \frac{\partial L(y, \hat{y})}{\partial \boldsymbol{z}_i} \boldsymbol{x}[\text{start}_i : \text{end}_i]. \tag{9}$$

Due to the fact that the majority entries of $\boldsymbol{x}[\text{start}_i : \text{end}_i]$ are 0, we can accelerate fine-tuning by updating weights linking to positive units only.

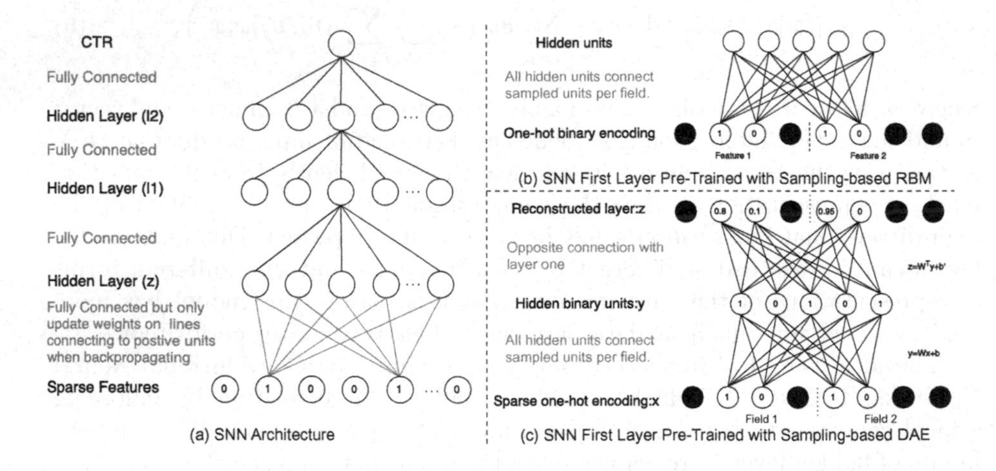

Fig. 2. A 4-layer SNN architecture and two first-layer pre-training methods.

3.2 Sampling-Based Neural Networks (SNN)

The structure of the second model SNN is shown in Fig. 2(a). The difference between SNN and FNN lies in the structure and training method in the bottom layer. SNN's bottom layer is fully connected with sigmoid activation function:

$$\boldsymbol{z} = \text{sigmoid}(\boldsymbol{W}_0 \boldsymbol{x} + \boldsymbol{b}_0). \tag{10}$$

To initialise the weights of the bottom layer, we tried both restricted Boltzmann machine (RBM) [16] and denoising auto-encoder (DAE) [4] in the pre-training stage. In order to deal with the computational problem of training large sparse one-hot encoding data, we propose a sampling-based RBM (Fig. 2(b), denoted as SNN-RBM) and a sampling-based DAE in (Fig. 2(c), denoted as SNN-DAE) to efficiently calculate the initial weights of the bottom layer.

Instead of modelling the whole feature set for each training instance set, for each feature field, e.g., `city`, there is only one positive value feature for each training instance, e.g., `city=London`, we sample m negative units, e.g., `city=Paris` when $m = 1$, randomly with value 0. Black units in Fig. 2(b) and (c) are unsampled and thus ignored when pre-training the data instance. With the sampled units, we can train an RBM via contrastive divergence [17] and a DAE via SGD with unsupervised approaches to largely reduce the data dimension with high recovery performance. The real-value dense vector is used as the input of the further layers in SNN.

In this way, computational complexity can be dramatically reduced and, in turn, initial weights can be calculated quickly and back-propagation is then performed to fine-tune SNN model.

3.3 Regularisation

To prevent overfitting, the widely used L2 regularisation term is added to the loss function. For example, the L2 regularisation for FNN in Fig. 1 is

$$\Omega(\boldsymbol{w}) = ||\boldsymbol{W}_0||_2^2 + \sum_{l=1}^{3} \left(||\boldsymbol{W}_l||_2^2 + ||\boldsymbol{b}_l||_2^2 \right). \tag{11}$$

On the other hand, *dropout* [35] is a technique which becomes a popular and effective regularisation technique for deep learning during the recent years. We also implement this regularisation and compare them in our experiment.

4 Experiment

4.1 Experiment Setup

Data. We evaluate our models based on iPinYou dataset [27], a public real-world display ad dataset with each ad display information and corresponding user click feedback. The data logs are organised by different advertisers and in a row-per-record format. There are 19.50 M data instances with 14.79 K positive label (click) in total. The features for each data instance are all categorical. Feature examples in the ad log data are `user agent`, partially masked `IP`, `region`, `city`, `ad exchange`, `domain`, `URL`, `ad slot ID`, `ad slot visibility`, `ad slot size`, `ad slot format`, `creative ID`, `user tags`, etc. After one-hot encoding, the number of binary features is 937.67 K in the whole dataset. We feed each compared model with these binary-feature data instances and the user click (1) and non-click (0) feedback as the ground-truth labels. In our experiments, we use training data from advertiser 1458, 2259, 2261, 2997, 3386 and the whole dataset, respectively.

Models. We compare the performance of the following CTR estimation models:

LR: Logistic Regression [32] is a linear model with simple implementation and fast training speed, which is widely used in online advertising estimation.

FM: Factorisation Machine [31] is a non-linear model able to estimate feature interactions even in problems with huge sparsity.

FNN: Factorisation-machine supported Neural Network is our proposed model as described in Sect. 3.1.

SNN: Sampling-based Neural Network is also our proposed model with sampling-based RBM and DAE pre-training methods for the first layer in Sect. 3.2, denoted as SNN-RBM and SNN-DAE respectively.

Our experiment code[2] of both FNN and SNN is implemented with `Theano`[3].

Metric. To measure the CTR estimation performance of each model, we employ the area under ROC curve (AUC)[4]. The AUC [12] metric is a widely used measure for evaluating the CTR performance.

4.2 Performance Comparison

Table 1 shows the results that compare LR, FM, FNN and SNN with RBM and DAE on 5 different advertisers and the whole dataset. We observe that FM is not significantly better than LR, which means 2-order combination features might not be good enough to catch the underlying data patterns. The AUC performance of the proposed FNN and SNN is better than the performance of

Table 1. Overall CTR estimation AUC performance.

	LR	FM	FNN	SNN-DAE	SNN-RBM
1458	70.42 %	70.21 %	**70.52 %**	70.46 %	70.49 %
2259	69.66 %	69.73 %	**69.74 %**	68.08 %	68.34 %
2261	62.03 %	60.97 %	62.99 %	**63.72 %**	**63.72 %**
2997	60.77 %	60.87 %	61.41 %	**61.58 %**	61.45 %
3386	80.30 %	79.05 %	**80.56 %**	79.62 %	80.07 %
all	68.81 %	68.18 %	**70.70 %**	69.15 %	69.15 %

[2] The source code with demo data: https://github.com/wnzhang/deep-ctr.

[3] `Theano`: http://deeplearning.net/software/theano/.

[4] Besides AUC, root mean square error (RMSE) is also tested. However, positive/negative examples are largly unbalanced in ad click scenario, and the empirically best regression model usually provides the predicted CTR close to 0, which results in very small RMSE values and thus the improvement is not well captured.

LR and FM on all tested datasets. Based on the latent structure learned by FM, FNN further learns effective patterns between these latent features and provides a consistent improvement over FM. The performance of SNN-DAE and SNN-RBM is generally consistent, i.e., the relative order of the results of the SNN are almost the same.

4.3 Hyperparameter Tuning

Due to the fact that deep neural networks involve many implementation details and need to tune a fairly large number of hyper-parameters, following details show how we implement our models and tune hyperparameters in the models.

We use stochastic gradient descent to learn most of our parameters for all proposed models. Regarding selecting the number of training epochs, we use early stopping [30], i.e., the training stops when the validation error increases. We try different learning rate from 1, 0.1, 0.01, 0.001 to 0.0001 and choose the one with optimal performance on the validation dataset.

For negative unit sampling of SNN-RBM and SNN-DAE, we try the negative sample number $m = 1, 2$ and 4 per field as described in Sect. 3.2, and find $m = 2$ produces the best results in most situations. For the activation functions in both models on the hidden layers (as Eqs. (3) and (2)), we try linear function, sigmoid function and tanh function, and find the result of tanh function is optimal. This might be because the hyperbolic tangent often converges faster than the sigmoid function.

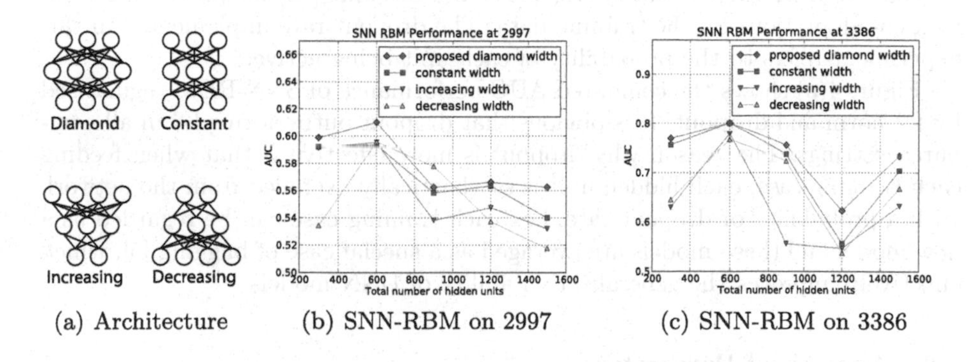

(a) Architecture (b) SNN-RBM on 2997 (c) SNN-RBM on 3386

Fig. 3. AUC Performance with different architectures.

4.4 Architecture Selection

In our models, we investigate architectures with 3, 4 and 5 hidden layers by fixing all layer sizes and find the architecture with 3 hidden layers (i.e., 5 layers in total) is the best in terms of AUC performance. However, the range of choosing their layer sizes is exponential in the number of hidden layers. Suppose there is a deep neural network with L hidden layers and each of the hidden layers is trained with a range of hidden units from 100 to 500 with increments of 100, thus there are 5^L models in total to compare.

Instead of trying all combinations of hidden units, in our experiment we use another strategy by starting tuning the different hidden layer sizes with the same number of hidden units in all three hidden layers[5] since the architecture with equal-size hidden layers is empirically better than the architecture with increasing width or decreasing width in [24]. For this reason, we start tuning layer sizes with equal hidden layer sizes. In fact, apart from increasing, constant, decreasing layer sizes, there is a more effective structure, which is the diamond shape of neural networks, as shown in Fig. 3(a). We compare our diamond shape network with other three shapes of networks and tune the total number of total hidden units on two different datasets shown in Fig. 3(b) and (c). The diamond shape architecture outperforms others in almost all layer size settings. The reason why this diamond shape works might be because this special shape of neural network has certain constraint to the capacity of the neural network, which provides better generalisation on test sets. On the other hand, the performance of diamond architecture picks at the total hidden unit size of 600, i.e., the combination of (200, 300, 100). This depends on the training data observation numbers. Too many hidden units against a limited dataset could cause overfitting.

4.5 Regularisation Comparison

Neural network training algorithms are very sensitive to the overfitting problem since deep networks have multiple non-linear layers, which makes them very expressive models that can learn very complicated functions. For DNN models, we compared L2 regularisation (Eq. (11)) and dropout [35] for preventing complex co-adaptations on the training data. The dropout rate implemented in this experiment refers to the probability of each unit being active.

Figure 4(a) shows the compared AUC performance of SNN-RBM regularised by L2 norm and dropout. It is obvious that dropout outperforms L2 in all compared settings. The reason why dropout is more effective is that when feeding each training case, each hidden unit is stochastically excluded from the network with a probability of dropout rate, i.e., each training case can be regarded as a new model and these models are averaged as a special case of bagging [5], which effectively improves the generalisation ability of DNN models.

4.6 Analysis of Parameters

As a summary of Sects. 4.4 and 4.5, for both FNN and SNN, there are two important parameters which should be tuned to make the model more effective: (i) the parameters of layer size decide the architecture of the neural network and (ii) the parameter of dropout rate changes generalisation ability on all datasets compared to neural networks just with L2 regularisation.

[5] Some advanced Bayesian methods for hyperparameter tuning [34] are not considered in this paper and may be investigated in the future work.

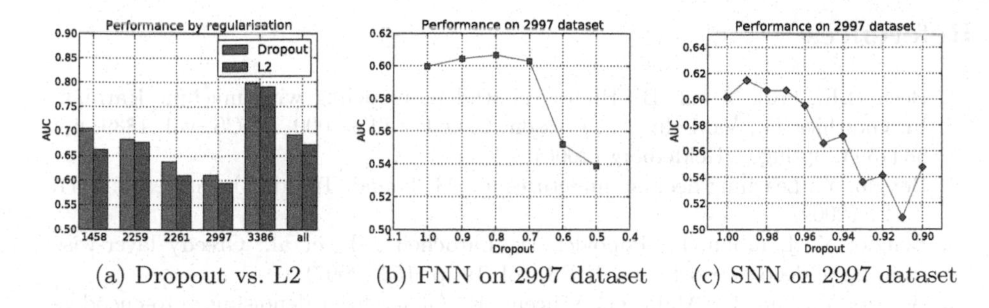

(a) Dropout vs. L2 (b) FNN on 2997 dataset (c) SNN on 2997 dataset

Fig. 4. AUC performance w.r.t difference regularisation settings.

Figure 4(b) and (c) show how the AUC performance changes with the increasing of dropout in both FNN and SNN. We can find that there is an upward trend of performance in both models at the beginning and then drop sharply with continuous decreasing of dropout rate. The distinction between two models is the different sensitivities of the dropout. From Fig. 4(c), we can see the model SNN is sensitive to the dropout rate. This might be caused by the connectivities in the bottom layer. The bottom layer of the SNN is fully connected with the input vector while the bottom layer for FNN is partially connected and thus the FNN is more robust when some hidden units are dropped out. Furthermore, the sigmoid activation function tend to more effective than the linear activation function in terms of dropout. Therefore, the dropout rates at the best performance of FNN and SNN are quite different. For FNN the optimal dropout rate is around 0.8 while for SNN is about 0.99.

5 Conclusion

In this paper, we investigated the potential of training deep neural networks (DNNs) to predict users' ad click response based on multi-field categorical features. To deal with the computational complexity problem of high-dimensional discrete categorical features, we proposed two DNN models: field-wise feature embedding with supervised factorisation machine pre-training, and fully connected DNN with field-wise sampling-based RBM and DAE unsupervised pre-training. These architectures and pre-training algorithms make our DNNs trained very efficiently. Comprehensive experiments on a public real-world dataset verifies that the proposed DNN models successfully learn the underlying data patterns and provide superior CTR estimation performance than other compared models. The proposed models are very general and could enable a wide range of future works. For example, the model performance can be improved by momentum methods in that it suffices for handling the curvature problems in DNN training objectives without using complex second-order methods [36]. In addition, the partial connection in the bottom layer could be extended to higher hidden layers as partial connectivities have many advantages such as lower complexity, higher generalisation ability and more similar to human brain [9].

References

1. Beck, J.E., Park Woolf, B.: High-level student modeling with machine learning. In: Gauthier, G., VanLehn, K., Frasson, C. (eds.) ITS 2000. LNCS, vol. 1839, pp. 584–593. Springer, Heidelberg (2000)
2. Bengio, Y.: Learning deep architectures for AI. Found. Trends Mach. Learn. **2**(1), 1–127 (2009)
3. Bengio, Y., Lamblin, P., Popovici, D., Larochelle, H., et al.: Greedy layer-wise training of deep networks. In: NIPS, vol. 19, p. 153 (2007)
4. Bengio, Y., Yao, L., Alain, G., Vincent, P.: Generalized denoising auto-encoders as generative models. In: NIPS, pp. 899–907 (2013)
5. Breiman, L.: Bagging predictors. Mach. Learn. **24**(2), 123–140 (1996)
6. Broder, A.Z.: Computational advertising. In: SODA, vol. 8, pp. 992–992 (2008)
7. Collobert, R., Weston, J., Bottou, L., Karlen, M., Kavukcuoglu, K., Kuksa, P.: Natural language processing (almost) from scratch. JMLR **12**, 2493–2537 (2011)
8. Deng, L., Abdel-Hamid, O., Yu, D.: A deep convolutional neural network using heterogeneous pooling for trading acoustic invariance with phonetic confusion. In: ICASSP, pp. 6669–6673. IEEE (2013)
9. Elizondo, D., Fiesler, E.: A survey of partially connected neural networks. Int. J. Neural Syst. **8**(05n06), 535–558 (1997)
10. Erhan, D., Bengio, Y., Courville, A., Manzagol, P.A., Vincent, P., Bengio, S.: Why does unsupervised pre-training help deep learning? JMLR **11**, 625–660 (2010)
11. Fukushima, K.: Neocognitron: a self-organizing neural network model for a mechanism of pattern recognition unaffected by shift in position. Biol. Cybern. **36**(4), 193–202 (1980)
12. Graepel, T., Candela, J.Q., Borchert, T., Herbrich, R.: Web-scale bayesian click-through rate prediction for sponsored search advertising in microsoft's bing search engine. In: ICML, pp. 13–20 (2010)
13. Graves, A., Mohamed, A., Hinton, G.: Speech recognition with deep recurrent neural networks. In: ICASSP, pp. 6645–6649. IEEE (2013)
14. Hand, D.J., Yu, K.: Idiot's bayes not so stupid after all? Int. Statist. Rev. **69**(3), 385–398 (2001)
15. He, X., Pan, J., Jin, O., Xu, T., Liu, B., Xu, T., Shi, Y., Atallah, A., Herbrich, R., Bowers, S., et al.: Practical lessons from predicting clicks on ads at facebook. In: ADKDD, pp. 1–9. ACM (2014)
16. Hinton, G.: A practical guide to training restricted boltzmann machines. Momentum **9**(1), 926 (2010)
17. Hinton, G.E.: Training products of experts by minimizing contrastive divergence. Neural comput. **14**(8), 1771–1800 (2002)
18. Hinton, G.E., Salakhutdinov, R.R.: Reducing the dimensionality of data with neural networks. Science **313**(5786), 504–507 (2006)
19. Huang, P.S., He, X., Gao, J., Deng, L., Acero, A., Heck, L.: Learning deep structured semantic models for web search using clickthrough data. In: CIKM, pp. 2333–2338 (2013)
20. Juan, Y.C., Zhuang, Y., Chin, W.S.: 3 idiots approach for display advertising challenge. In: Internet and Network Economics, pp. 254–265. Springer, Heidelberg (2011)
21. Kittler, J., Hatef, M., Duin, R.P., Matas, J.: On combining classifiers. PAMI **20**(3), 226–239 (1998)

22. Krizhevsky, A., Sutskever, I., Hinton, G.E.: Imagenet classification with deep convolutional neural networks. In: NIPS (2012)
23. Kurashima, T., Iwata, T., Takaya, N., Sawada, H.: Probabilistic latent network visualization: inferring and embedding diffusion networks. In: KDD, pp. 1236–1245. ACM (2014)
24. Larochelle, H., Bengio, Y., Louradour, J., Lamblin, P.: Exploring strategies for training deep neural networks. JMLR **10**, 1–40 (2009)
25. LeCun, Y., Bengio, Y., Hinton, G.: Deep learning. Nature **521**(7553) (2015)
26. Lee, K., Orten, B., Dasdan, A., Li, W.: Estimating conversion rate in display advertising from past performance data. In: KDD, pp. 768–776. ACM (2012)
27. Liao, H., Peng, L., Liu, Z., Shen, X.: ipinyou global rtb bidding algorithm competition dataset. In: ADKDD, pp. 1–6. ACM (2014)
28. McMahan, H.B., Holt, G., Sculley, D., Young, M., Ebner, D., Grady, J., Nie, L., Phillips, T., Davydov, E., Golovin, D., et al.: Ad click prediction: a view from the trenches. In: KDD, pp. 1222–1230. ACM (2013)
29. Oentaryo, R.J., Lim, E.P., Low, D.J.W., Lo, D., Finegold, M.: Predicting response in mobile advertising with hierarchical importance-aware factorization machine. In: WSDM (2014)
30. Prechelt, L.: Automatic early stopping using cross validation: quantifying the criteria. Neural Netw. **11**(4), 761–767 (1998)
31. Rendle, S.: Factorization machines with libfm. ACM TIST **3**(3), 57 (2012)
32. Richardson, M., Dominowska, E., Ragno, R.: Predicting clicks: estimating the click-through rate for new ads. In: WWW, pp. 521–530. ACM (2007)
33. Shen, Y., He, X., Gao, J., Deng, L., Mesnil, G.: A latent semantic model with convolutional-pooling structure for information retrieval. In: CIKM (2014)
34. Snoek, J., Larochelle, H., Adams, R.P.: Practical bayesian optimization of machine learning algorithms. In: NIPS, pp. 2951–2959 (2012)
35. Srivastava, N., Hinton, G., Krizhevsky, A., Sutskever, I., Salakhutdinov, R.: Dropout: A simple way to prevent neural networks from overfitting. JMLR **15**(1), 1929–1958 (2014)
36. Sutskever, I., Martens, J., Dahl, G., Hinton, G.: On the importance of initialization and momentum in deep learning. In: ICML, pp. 1139–1147 (2013)
37. Tang, J., Qu, M., Wang, M., Zhang, M., Yan, J., Mei, Q.: Line: Large-scale information network embedding. In: WWW, pp. 1067–1077 (2015)
38. Trofimov, I., Kornetova, A., Topinskiy, V.: Using boosted trees for click-through rate prediction for sponsored search. In: WINE, p. 2. ACM (2012)
39. Wang, X., Li, W., Cui, Y., Zhang, R., Mao, J.: Click-through rate estimation for rare events in online advertising. In: Online Multimedia Advertising: Techniques and Technologies, pp. 1–12 (2010)
40. Zeiler, M.D., Taylor, G.W., Fergus, R.: Adaptive deconvolutional networks for mid and high level feature learning. In: ICCV, pp. 2018–2025. IEEE (2011)
41. Zhang, W., Yuan, S., Wang, J.: Optimal real-time bidding for display advertising. In: KDD, pp. 1077–1086. ACM (2014)
42. Zou, Y., Jin, X., Li, Y., Guo, Z., Wang, E., Xiao, B.: Mariana: Tencent deep learning platform and its applications. VLDB **7**(13), 1772–1777 (2014)

Supervised Local Contexts Aggregation for Effective Session Search

Zhiwei Zhang[1(✉)], Jingang Wang[2], Tao Wu[1], Pengjie Ren[3], Zhumin Chen[3], and Luo Si[1]

[1] Department of Computer Science, Purdue University, West Lafayette, USA
{zhan1187,wu577,lsi}@purdue.edu

[2] School of Computer Science, Beijing Institute of Technology, Beijing, China
bitwjg@bit.edu.cn

[3] School of Computer Science and Technology, Shandong University, Jinan, China
Jay.Ren@outlook.com, chenzhumin@sdu.edu.cn

Abstract. Existing research on web search has mainly focused on the optimization and evaluation of single queries. However, in some complex search tasks, users usually need to interact with the search engine multiple times before their needs can be satisfied, the process of which is known as session search. The key to this problem relies on how to utilize the session context from preceding interactions to improve the search accuracy for the current query. Unfortunately, existing research on this topic only formulated limited modeling for session contexts, which in fact can exhibit considerable variations. In this paper, we propose Supervised Local Context Aggregation (SLCA) as a principled framework for complex session context modeling. In SLCA, the global session context is formulated as the combination of local contexts between consecutive interactions. These local contexts are further weighted by multiple weighting hypotheses. Finally, a supervised ranking aggregation is adopted for effective optimization. Extensive experiments on TREC11/12 session track show that our proposed SLCA algorithm outperforms many other session search methods, and achieves the state-of-the-art results.

Keywords: Session search · Context · Aggregation

1 Introduction

For the majority of existing studies on web search, queries are mainly optimized and evaluated independently [19,21]. However, this single query scenario is not the whole story in web search. When users have some complex search tasks (e.g. literature survey, product comparison), a single-query-search is probably insufficient [22]. Often, users iteratively interact with the search engine multiple times, until the task is accomplished. We call such a process a *session search* [9].

Formally, a session search $\mathcal{S}$ is defined as $\mathcal{S} = \{\mathcal{I}^t = \{q^t, \mathcal{D}^t, \mathcal{C}^t\} | t = 1 : T\}$, where $\mathcal{S}$ is composed of a sequence of interactions $\mathcal{I}^t$ between user and search engine. In each interaction $\mathcal{I}^t$, three steps are involved: (1) the user issues

N. Ferro et al. (Eds.): ECIR 2016, LNCS 9626, pp. 58–71, 2016.
DOI: 10.1007/978-3-319-30671-1_5

a query q^t; (2) the search engine retrieves the top-ranked documents $\mathcal{D}^t$; (3) and the user clicks a subset of documents $\mathcal{C}^t$ that he/she feels attractive. Then the user gradually refines the next query, iterates the above three steps, until the search is done. See Table 1 for an example.

Table 1. Example Session Search

Task: sunspot activity	
Queries	SAT Clicks
Q1. Sunspots Wikipedia	http://en.wikipedia.org/wiki/Sunspot
Q2. Sunspots Environmental effects	http://www.env-econ.net/2008/02/
Q3. Sunspot life cycle	http://astronomyonline.org/SolarSystem/
Q4. Are sunspots local burnouts	None
Q5 (Current Query): How do sunspots effect us	N/A

Example session search from TREC12 Session Track No. 64 [13]. "None" means no SAT clicks exist for Q4. "N/A" means not available yet. $\mathcal{D}^t$ is omitted to save space. Clicks with dwelling time $\geq$ 30 s are called satisfactory (SAT) clicks [4], which indicates user satisfaction.

Problem Analysis. The goal of session search is to provide search results for the *current query* (i.e. q^T, which has not been searched, namely $\mathcal{D}^T = \varnothing, \mathcal{C}^T = \varnothing$) so that the entire search task can be satisfied, based on preceding interactions within the same session, which we call *session context.* How to effectively utilize such session contexts is the key to session search, which can be challenging in real world applications. Consider the following two examples:

Example 1 ***(Recency Variation).*** *TREC11 session-60 has ten interactions before the current query. It is reasonable to believe the context from $\mathcal{I}^1$ is less important than the context from $\mathcal{I}^{10}$.*

Example 2 ***(Satisfaction Variation).*** *TREC12 session-85 has two interactions before the current query: $\mathcal{I}^1$ has one SAT click, while $\mathcal{I}^2$ has none. As SAT click is a strong indicator of user satisfaction, context from $\mathcal{I}^1$ should be more important than that from $\mathcal{I}^2$.*

Here we define *session context variations* as the variations of the importance of the context that each individual interaction contributes to the current query. The above two examples show that, *different session contexts can show great variations*, which requires delicated modeling. Besides variations on recency and satisfaction, we will explain more in Sect. 3.

The Literature. In general, there are two main categories of related algorithms on this topic. The first category mainly includes algorithms developed for TREC competitions (e.g. TREC11/12 session track [12,13]) in the traditional literature of IR. These algorithms usually treat session contexts as side information besides current query, which are combined in unsupervised ad hoc search algorithms such as language modeling [6,9,10,20,24]. The unsupervised nature, however, may not be optimal from the perspective of performance. The second category are mainly algorithms developed in industry (e.g. Bing, Yandex) that focused on personalized web search. Their works mainly utilized learning to rank algorithms with abundant session context features [2,16,26,27].

However, one major disadvantage shared by the above algorithms is, they either ignored session context variations by mixing all interactions indistinguishably [10,16,24,26,27], or only considered limited possibilities (primarily recency variation [2,6,20]) that are not powerful enough to maximize search accuracy. These observations motivate us to propose a better algorithm for session search.

Our Proposal. Seeing the limitations of existing algorithms, in this paper we propose a principled framework, named Supervised Local Context Aggregation (SLCA), which can model sophisticated session context variations. Global session context is decomposed into local contexts between consecutive interactions. We further propose multiple weighting hypotheses, each of which corresponds to a specific variation pattern in session context. Then the global session context can be modeled as the weighted combination of local contexts. A supervised ranking aggregation approach is adopted for effective and efficient optimization. Extensive experiments on TREC11/12 Session Tracks show that our proposed SLCA achieves the state-of-the-art results.

The rest of the paper is organized as follows: Sect. 2 reviews related literature; Sect. 3 elaborates all the details of SLCA; Sect. 4 gives all the experimental results; finally in Sect. 5, we conclude the paper.

2 Related Works

As mentioned above, previous related works on session search can be categorized as TREC-related algorithms and industry-related algorithms.

TREC-related algorithms are mainly designed for TREC Session Track competitions. For example, Jiang et al. [10] built several language models based on preceding interactions and current query, and utilized their linear combination for the ad hoc search. Similar works also include [9,24]. Yang's group [6,20] conducted several works by modeling session interactions as a markov decision process. Interactions are weighted by decaying factors w.r.t recency variation, and the final search is essentially language models combined with those weighted interactions. Their unsupervised nature, however, makes it difficult to do complex extensions to further improve search accuracy, as the number of parameters will grow rapidly.

Industry-related algorithms mainly focused on personalized web search by leveraging session contexts. For many of these works [16,25–27], session context variation is ignored, and all interactions are mixed indistinguishably. For example, many such works will count the frequency of documents being clicked in all preceding interactions, regardless of the difference among these interactions (e.g. whether they are distant or relevant to current query). We call such a strategy as *Global Context Modeling*. For the other works, limited session context variations is modeled, primarily focusing on recency variation. For example, [2,6,20] all belong to this case, where distant interactions are weighted in an exponentially decaying manner. We call such a strategy as *Single Variation Modeling*. In comparison, our method can be categorized as *Multiple Variation Modeling*, where multiple variation patterns are aggregated via supervised learning.

There also exist many works that are relevant in some aspects, but are not directly applicable to session search due to various reasons. Cao et al. [3] modeled the session contexts as a hidden markov model, but [27] pointed out this method cannot easily generalize to unseen queries. Works like [2,8,25,26] extensively utilized information beyond session contexts, such as search logs or users' long term search history etc., which are not accessible to us. [14,23] focused on context-aware non-search tasks such as recommendation or query suggestion, which are very different applications. Guo et al. [7] studied interaction features for query performance prediction; Liu et al. [17] analyzed what features between two consecutive queries indicate the finding of useful documents. However, their goals are again not session search, and did not involve session context modeling, which is crucial for session search.

For sake of easy comparison, Table 2 summarizes the difference between our work and the above related works. There, "-" means we haven't found previous studies belonging to that case, which is probably because unsupervised method cannot easily handle the potentially large number of parameters in Multiple Variations Modeling. Studies that do not focus on session search of current query are excluded.

Table 2. Session Search Methods Comparison

	Global Context Modeling	Single Variation Modeling	Multiple Variations Modeling
Unsupervised	[9,10,24]	[6,20]	-
Supervised	[16,25–27]	[2]	Our work

3 The Proposed Algorithm

This section presents the three main components in SLCA: local context modeling, local context weighting hypotheses and supervised ranking aggregation. At end, we also present the detailed search procedures using SLCA.

3.1 Local Context Modeling

The entire session context, which has multiple interactions, might be complicated to model as a whole. Instead, we break it down into smaller units called *local contexts*, which makes the problem easier. Specifically, a local context $\mathcal{X}^t$ is the context contained within two consecutive interactions $\mathcal{I}^{t-1}$ and $\mathcal{I}^t$. For sake of usage, such local context can be described as a feature vector jointly extracted from $\mathcal{I}^{t-1}$ and $\mathcal{I}^t$, which we call Local Context Feature (LCF). As $\mathcal{I}^{t-1}$ always happens ahead of $\mathcal{I}^t$, LCF actually is used as the representation for $\mathcal{I}^t$.

The following four kinds of LCF are explored in this paper.

(1) Added Information (Add). Query terms that appear in q^t but not in q^{t-1} represent information that the user pursues in q^t. We let those terms form a pseudo query Q_{Add}.
(2) Deleted Information (Del). Query terms that appear in q^{t-1} but not in q^t represent information that the user no longer desires in q^t. We let such terms form a pseudo query Q_{Del}.
(3) Implicit Information 1 (Imp1). If there is SAT clicks in $\mathcal{I}^{t-1}$, it means the user is satisfied by those clicked documents. We assume those SAT documents will produce some impact in the formation of q^t. Therefore, we utilize Lavrenko's Relevance Model (RM) [15] to extract m (empirically set as 10) terms from SAT documents, hoping that these extra terms will capture user's implicit information need. Those terms form a pseudo query Q_{Imp1}.
(4) Implicit Information 2 (Imp2). In case the SAT documents are too few to provide accurate Imp1 modeling, we further select top n Wikipedia documents from $\mathcal{D}^{t-1}$ in last interaction $\mathcal{I}^{t-1}$. Again we extract m terms using RM (m, n empirically set as 10). This strategy is inspired by the successful application of RM + Wikipedia in recent TREC competitions [1]. We let those terms form a pseudo query Q_{Imp2}. Notice we use the Wikipedia contained in the Clueweb-09 corpus for TREC11/12 session track, therefore we are not using external resources beyond TREC, thus assuring fair comparison with previous works.

For document d, we calculate its BM25 scores c w.r.t the above four pseudo queries. We then calculate the single query ranking score $h(q^t, d)$ between q^t and d (details will be given in Sect. 4). Now the overall local context feature for document d in $\mathcal{I}^t$ will be as follows:

$$f(d|\mathcal{I}^t) = [c_{Add}(d) \;\; c_{Del}(d) \;\; c_{Imp1}(d) \;\; c_{Imp2}(d) \;\; h(q^t, d)]' \tag{1}$$

where $'$ represents transposing a row vector into column vector.

As shown in later experiments, these features can already achieve excellent results. Since feature design is not our primary concern, we do not explore further possibilities, although this is quite straightforward.

3.2 SLCA Formulation

Recall in Sect. 1 we analyzed that the key challenge is how to model the session context variations, i.e. the importance of each interactions. Based on local context modeling, this problem can be naturally formulated as follows:

$$\mathcal{X} = \sum_{t=1}^{T} w^t \mathcal{X}^t \rightarrow F_d = \sum_{t=1}^{T} w^t f(d|\mathcal{I}^t) \quad (2)$$

The entire session context $\mathcal{X}$ is a weighted linear combination of local contexts $\mathcal{X}^t$. Accordingly, for document d, its session-level representation F_d can be expressed in the same way.

Based on that, now we can formulate SLCA algorithm. Suppose we have N training documents $D^{\mathcal{S}} = \{d_i^{\mathcal{S}} = \{F_i^{\mathcal{S}}, Y_i^{\mathcal{S}}\} | i = 1 : N\}$ for session $\mathcal{S}$, where $F_i^{\mathcal{S}}$ is the feature vector of document $d_i^{\mathcal{S}}$ based on Eq. 2, and $Y_i^{\mathcal{S}}$ is the groundtruth relevance label. We cast the session search problem into learning to rank framework. Let $\mathcal{L}(F, Y; \theta)$ denote the loss function for this learning problem, where θ is the ranking model to be learned. Then SLCA algorithm can be formulate as following optimization problem (OP):

$$\text{(OP 1)} \qquad \mathbf{w}, \theta = \arg\min_{\mathbf{w},\theta} \sum_{\mathcal{S}} \mathcal{L}(\{F_{d_i}^{\mathcal{S}}\}, \{Y_{d_i}^{\mathcal{S}}\}; \theta), \qquad F_{d_i}^{\mathcal{S}} = \sum_{t=1}^{T} w_{\mathcal{S}}^t f(d|\mathcal{I}_{\mathcal{S}}^t) \quad (3)$$

where $\mathbf{w}$ is the set of local context weights $w_{\mathcal{S}}^t$ for all interactions in all training sessions, and $\{\}$ indicates the feature/label set of documents for each $\mathcal{S}$.

If $\mathbf{w}$ is known, OP 1 reduces to a standard learning to rank problem, where for each document d we have a single feature vector $F_d^{\mathcal{S}}$ given by Eq. 2 and a relevance label $Y_d^{\mathcal{S}}$. Any learning to rank algorithms can be utilized, such as RankSVM [11] and GBDT [5]. Unfortunately, this is not the case. Even if we can derive some $\mathbf{w}$ (probably suboptimal due to large parameter number) for training sessions by optimizing Eq. 3, there is no way to predict $\mathbf{w}$ for unseen test sessions. To solve this problem, below we propose an effective alternative.

3.3 Local Context Weighting Hypotheses

We propose to construct a set of basic local context weighting hypotheses (WH), denoted as $\tilde{\mathbf{w}}$, so that any potential local context weighting sequence $\mathbf{w}$ can be expressed as the combination of K such basic WHs:

$$\mathbf{w} = \sum_{k=1}^{K} \alpha_k \tilde{\mathbf{w}}_k \quad (4)$$

Since $\tilde{\mathbf{w}}_k$ is given, the problem of optimizing $\mathbf{w}$ becomes the optimization of $\alpha = \{\alpha_1, ..., \alpha_K\}$, which has far less parameters to optimize. Moreover, it is easier to generalize to unseen testing sessions. These WHs are inborn session adaptive, which means even the same WH can derive different local contexts weights in different sessions. With α learned from training data, as well as those adaptive WHs, we can readily calculate local context weights for unseen testing sessions. In this paper, we design the following six WHs, each of which corresponds to one specific session context variation.

(1,2) Two Shrinking WHs for Recency Variation. For a session with T interactions, we denote $\tilde{\mathbf{w}}_{\mathcal{S}} = [\tilde{w}_s^t | t = 1 : T]$ as the local context weights. For shrinking hypothesis, we let $\tilde{w}_s^t = \gamma^{T-t}$. Here we empirically choose $\gamma = 0.6, 0.8$ (which

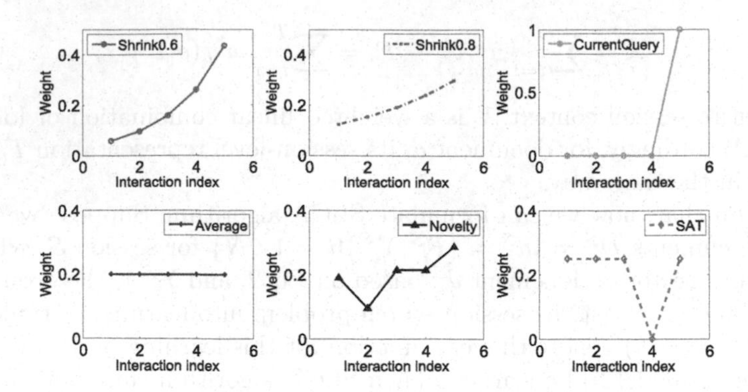

Fig. 1. Local context weights by applying the six WHs to the example session in Table 1.

obtain very good results in our preliminary experiments) to get two shrinking WHs: $\tilde{\mathbf{w}}_{S0.6}, \tilde{\mathbf{w}}_{S0.8}$. The closer one interaction $\mathcal{I}^t$ is to q^T, the more important its local context is, thus reflecting the recency variation of each local context.

(3) Current Query WH for Presence Variation. Similarly, we denote $\tilde{\mathbf{w}}_{Cur} = [\tilde{w}_{cur}^t | t = 1 : T]$ as the WH. For this WH, only the interaction at present (i.e. current query) is considered, i.e. $\tilde{w}_{cur}^T = 1$ and $\tilde{w}_{cur}^t = 0$ for $t < T$.

(4) Average WH for Equality Variation. We denote $\tilde{\mathbf{w}}_{Ave} = [\tilde{w}_{ave}^t | t = 1 : T]$, where $\tilde{w}_{ave}^t = \frac{1}{T}$. That is, all local contexts are treated equally, which is equivalent to previous works [16,27] that mixed all interactions indistinguishably.

(5) Novelty Variation. We denote $\tilde{\mathbf{w}}_{Nov} = [\tilde{w}_{nov}^t | t = 1 : T]$ as the weights. We believe, if a preceding query q^t is dissimilar with current query q^T, then q^t provides some novelty to the whole session, and should be emphasized. Otherwise, if q^t is very similar with q^T, or even identical, then its local context weight should be decreased. Specifically, we define $\tilde{w}_{nov}^t = 1 - Jaccard(q^T, q^t), t < T$ and $\tilde{w}_{nov}^T = 1$, where $Jaccard(q^T, q^t) = \frac{|terms \in q^T \cap terms \in q^t|}{|terms \in q^T \cup terms \in q^t|}$.

(6) SAT WH for Satisfaction Variation. If one interaction has many SAT clicks, it means this interaction satisfies the user well, and its local context should be more important. Denote $\tilde{\mathbf{w}}_{SAT} = [\tilde{w}_{sat}^t | t = 1 : T]$ as importance, then we have $\tilde{w}_{sat}^t = |\mathcal{C}^t|$, where $\mathcal{C}^t$ are SAT clicks in interaction $\mathcal{I}^t$. Hence w_{sat}^t equals the SAT number in $\mathcal{I}^t$. For current query, we let $\tilde{w}_{sat}^T = \max_{t=1:T-1} \tilde{w}_{sat}^t$.

For each of the above six WHs, we normalize $\tilde{\mathbf{w}}$ to have $\sum_{t=1}^T \tilde{w}^t = 1$.

Each WH represents one specific pattern of *session context variation*. With various α, the combination of six WHs produces the *Multiple Variations Modeling*, which can be much more expressive than Global Context Modeling and Single Variation Modeling (see Table 2). Extensions of designing other WHs is quite straightforward, while in this paper we will stick to the above six as they have already shown excellent performance. As an example, in Fig. 1 we show the six WHs for the example session in Table 1.

Algorithm 1. SLCA Training, i.e. Optimizing OP 3.

Input Training sessions $\{\mathcal{S}\}$, candidate documents $\{D\}$ and relevance labels $\{Y\}$. Learning to rank algorithm $\mathcal{L}$ and the off-the-shelf solver. Local context weighting hypotheses $WH_{1\sim K}$.

Output α, $\theta_1 \sim \theta_K$

1: For $d \in D$, set intermediate representation $A_d = [0, ..., 0]' \in R^{K\times 1}$.
2: **for** `k=1:K` **do**
3: Set $\alpha = [\underbrace{0, ..., 0,}_{k-1}\ 1,\ \underbrace{0, ..., 0}_{K-k}]'$.
4: Calculate $\tilde{F}_k^{\mathcal{S}} = \{\tilde{F}_{d,k}^{\mathcal{S}}\}$ w.r.t WH_k based on Eqs. 6, and 1.
5: Solve ranking problem $\theta_k = \arg\min_{\theta_k} \sum_{\mathcal{S}} \alpha_k \mathcal{L}(\tilde{F}_k^{\mathcal{S}}, Y^{\mathcal{S}}; \theta_k)$.
6: Set $A_d(k)$ as d's ranking score under θ_k.
7: **end for**
8: Treat A_d as the new feature vector for d. Along with label Y, again solve the following learning to rank problem: $\alpha = \arg\min_\alpha \sum_{\mathcal{S}} \mathcal{L}(A, Y; \alpha)$

3.4 Supervised Ranking Aggregation

With the above local context weighting hypotheses, now OP1 can be transformed into the following problem, with $\tilde{w}_{k,\mathcal{S}}^t$ being the kth WH for session $\mathcal{S}$:

$$\text{(OP 2)}: \alpha, \theta = \arg\min_{\alpha,\theta} \sum_{\mathcal{S}} \mathcal{L}(\{F_{d_i}^{\mathcal{S}}\}, \{Y_{d_i}^{\mathcal{S}}\}; \theta),\ \ F_{d_i}^{\mathcal{S}} = \sum_{t=1}^{T} (\sum_{k=1}^{K} \alpha_k \tilde{w}_{k,\mathcal{S}}^t) f(d|\mathcal{I}_{\mathcal{S}}^t) \tag{5}$$

As mentioned earlier, compared with OP 1, OP 2 has the two advantages: (1) besides θ, OP 2 has only K extra parameters (i.e. α) to be optimized, which is far less than OP 1; (2) OP 2 is much easier to generalize to unseen testing sessions than OP 1, as α is shared among training and testing sessions, and WH $\tilde{w}$ can adapt each individual sessions.

Nonetheless, OP 2 can still be difficult to solve, as the loss function $\mathcal{L}$ in Eq. 5 will contain α and θ tightly coupled together, which makes the existing solvers for learning to rank problems inapplicable. To solve this problem efficiently and effectively, we further relax OP 2 into the following formulation:

$$\text{(OP 3)}: \alpha, \theta_{1\sim K} = \arg\min \sum_{\mathcal{S}} \sum_{k=1}^{K} \alpha_k \mathcal{L}(\{\tilde{F}_{d_i,k}^{\mathcal{S}}\}, \{Y_{d_i}^{\mathcal{S}}\}; \theta_k), \tilde{F}_{d_i,k}^{\mathcal{S}} = \sum_{t=1}^{T} \tilde{w}_{k,\mathcal{S}}^t f(d|\mathcal{I}_{\mathcal{S}}^t) \tag{6}$$

In OP 3, the coupled α and θ are decoupled into a formulation of supervised ranking aggregation. Here again Eq. 1 is used as the document feature $f(d|\mathcal{I}_{\mathcal{S}}^t)$. The resulted advantage is, α no longer exists within $\mathcal{L}$, which makes existing solvers for learning to rank problems readily applicable. A small side effect is, we now have K ranking models $\theta_1 \sim \theta_K$ instead of one θ, but this can be easily handled in practice (e.g. We have small value $K = 6$).

Algorithm 2. Applying SLCA in Practical Session Search

1: $t \leftarrow 0$; session $\mathcal{S} = \varnothing$;
2: **while** user issues query q^t **do**
3: $t \leftarrow t + 1$.
4: Run q^t as single query retrieval; obtain top ranked documents d as candidates.
5: For document d, calculate $f(d|\mathcal{I}^t)$ based on Eq. 1.
6: Calculate all WHs $\tilde{w}_1 \sim \tilde{w}_K$ up to t, and $\tilde{F}^{\mathcal{S}}_{d,k}$ based on Eq. 6.
7: Calculate A_d based on $\tilde{F}^{\mathcal{S}}_{d,k}$ and θ_k, according to Algorithm 1.
8: Calculate d's ranking score as $\alpha' * A_d$; rerank all d to get search result $\mathcal{D}^t$.
9: Collect user clicks $\mathcal{C}^t$.
10: $\mathcal{S} \leftarrow \mathcal{S} \cup \mathcal{I}^t = \{q^t, \mathcal{D}^t, \mathcal{C}^t\}$
11: **end while**

To optimize OP 3, we propose a two-step strategy, as listed in Algorithm 1. First, in step $1 \sim 7$, we train the individual ranking model $\theta_1 \sim \theta_K$ w.r.t only one WH (via step 3). The intermediate ranking scores $A_d(1:K)$ w.r.t $\theta_1 \sim \theta_K$ are used as new representations for document d. Then, we treat α as the final aggregation model, which is solved again by $\mathcal{L}$. Both steps 5 and 8 are standard learning to rank problems, which can be readily solved by existing optimization algorithms. For example, in this paper we utilize the famous (linear) RankSVM [11] algorithm for $\mathcal{L}$, for which existing solvers are readily available[1]. The final ranking score for document d is $\sum_{k=1}^{K} \alpha_k A_d(k) = \alpha' * A_d$, where $A_d(k)$ is the ranking score of d w.r.t WH_k and θ_k. Notice both α and A_d are column vectors.

A merit of this two-step optimization is that each time, we only train a small ranking model. In the case of linear RankSVM, each time only $|\theta| = |\tilde{F}^{\mathcal{S}}_{d,k}|$ or $|\alpha| = K$ parameters are to be optimized. In comparison, another alternative is to concatenate all $\{\tilde{F}^{\mathcal{S}}_{d,1}, ..., \tilde{F}^{\mathcal{S}}_{d,K}\}$ as a single feature vector for d, and train a single ranking model with $|\tilde{F}^{\mathcal{S}}_{d,k}| \times K$ parameters being optimized simultaneously. This, unfortunately, incurs overfitting in our preliminary experiments, and performs worse than the two-step strategy. Therefore, in this paper we will stick on Algorithm 1.

3.5 Applying SLCA in Practical Session Search

In Algorithm 2 we show how SLCA is applied in practical session search, in which every newly arrived query is treated as the current query. Notice for academic datasets such as TREC11/12 Session Track, however, this problem is simplified due to the lack of live user interactions, in which the current query and preceding interactions are all fixed. Also for efficiency consideration, step 4 is introduced to reduce the potential calculation burden of SLCA, for which we usually apply simple single-query-retrieval methods such as KL-divergence or BM25 to retrieve the top-N (empirically set $N = 500$) documents for SLCA to apply.

[1] http://www.cs.cornell.edu/people/tj/svm_light/svm_rank.html.

4 Experiments

4.1 General Settings

Dataset. We utilize TREC11/12 Session Track data for experiments [12,13]. TREC11 includes 76 sessions, with average interaction number being 3.68 (including current query q^T as an interaction). TREC12 includes 98 sessions, with average interaction number being 3.03. Clueweb09 category B is utilized as document corpus (denoted as Cw09B), which includes 50 million documents. Due to the low quality of web pages, we apply Univ Waterloo's spam scores[2] to filter out spam documents with threshold being 70, leaving about 29 million documents. We utilize the well-known academic search engine Indri[3] to index Cw09B. Standard InQuery stopwords are removed, and Krovtzer stemmer is utilized.

Evaluation Scenarios. We adopt evaluation scenarios from official TREC reports [12,13]. For TREC11, two scenarios are utilized [12]: (1) measure if a document is whole-session relevant; (2) measure if a document is relevance to current query q^T. For TREC12, again two scenarios are utilized [13]: (1) the same whole-session relevance; (2) previously clicked documents are ignored when evaluating whole-session relevance. In all scenarios, NDCG@10 and ERR@10 are used as evaluation metrics. NDCG@10 of whole-session relevance is the major evaluation metric that ranks TREC competition teams (denoted as $\star$ below).

Training/Validation/Testing. As both TREC11/12 are available, we conduct the following strategy for dataset splitting. When we test on TREC12, we use TREC11 as training and validation set, where we do 3-fold cross validation to determine the optimal model and parameters (w.r.t main evaluation metric). Likewise when we test on TREC11, 3-fold cross validation is applied on TREC12.

Comparison. We conduct extensive comparisons with the following methods, based on our literature analysis in Sect. 2.

(1) BasicRet. KL divergence (KLD) is used as basic retrieval method to search current query q^T (i.e. step 4 in Algorithm 2). Dirichlet prior is empirically set as 5000.

(2) Single Query Ranking (SQRank). Traditional ranking scenario that only considers the current query q^T, for which we utilize RankSVM [11]. The corresponding single query ranking features are defined in Table 3. Notice SQRank is also used as $h(q, d)$ when we calculate local context feature in Eq. 1.

(3) TRECBest. Best results in TREC11/12 official reports [12,13] w.r.t main evaluation metric. Their submission IDs are wildcat2 [18] for TREC11 and PITTSHQM [10] for TREC12. As analyzed in Sect. 2, they both are unsupervised learning methods for Global Context Modeling.

(4) MDP. Markov decision process method proposed in [6], which is the state-of-the-art session search algorithm and achieves the best results on TREC11/12 so far. As our implementation is somehow worse than their results, for sake of fair

[2] https://plg.uwaterloo.ca/~gvcormac/clueweb09spam/.

[3] http://www.lemurproject.org/indri/.

comparison, we report the original performance from their paper. As reviewed in Sect. 2, MDP belongs to Single Variation Modeling with unsupervised learning.

(5,6) SinVar-Exp0.6/0.8. Now we consider Single Variation Modeling with supervised learning, such as [2]. In both [2,6], an exponentially decaying weighting strategy is applied to recency variation (see Example 1), which is equivalent to our shrinkage WH. Therefore, we take the two shrinkage WHs $\tilde{w}_{S0.6}, \tilde{w}_{S0.8}$ from SLCA to approximate this methodology (SinVar-Exp). It should be stronger than MDP as it utilizes supervised ranking while MDP is unsupervised. Compared with [2], although SinVar-Exp has some difference in feature design and detailed formulation, which is caused by the dataset difference (we use TREC while [2] used Bing data). the exponential decaying is kept the same, which is our primary concern. 0.6 and 0.8 are kept the same as in our SLCA, as their values perform best in our preliminary experiments.

(7,8) GlbCxtRank-V1/V2. For the category of Global Context Modeling with supervised learning, we apply the context-aware ranking method proposed in Xiang's [27]. There are two versions of implementation. In V1 version, we approximate their method by averaging local context features in Eq. 2, which equals to the Average WH. In V2 version, as in [27], we construct the global context features by transforming our design of local context features: "Add" feature is formed by terms t that $t \in q^T \setminus (q^1 \cup ... \cup q^{T-1})$; "Del" is formed by terms t that $t \in (q^1 \cup ... \cup q^{T-1}) \setminus q^T$; "Imp1" and "Imp2" utilize all SAT and top Wiki documents in preceding interactions.

Table 3. Single Query Ranking Features

(F1). Whether d comes from Wikipedia. (F2∼F3). The length of document title and body.
(F4). Ratio of stop terms in doc. (F5∼F6). Entropy of document title and body.
(F7∼F10) BM25/KLD scores between query/document-title, and query/document-body.
(F11∼F12) Ratio of covered query terms in document title and document body.
(F13) Frequency of query terms co-occurring as a consecutive pair in d.
(F14∼F17) Frequency of query terms co-occurrence in d within text window 5,10,15 and 20.

Except (3) (4) are taken directly from original papers, we implement the other methods, which rerank top-500 documents returned by BasicRet (see Algorithm 2).

4.2 Experiment Results

Major Results. The major experimental results are shown in Table 4. We can see that SLCA achieves the best performance (in bold) for almost all evaluation scenarios and metrics. Moreover, ‡, ♯, †, ⋄ indicate that our SLCA is significantly better than SinVarExp-0.6/0.8 and CxtRank-v1/v2 methods (t-test, $p < 0.05$). This verifies our motivation, that properly modeling session context variations (by weighting local contexts) is more effective than simple treatment that either mixes the entire session context indistinguishably or only models single variation.

The comparison shows that, supervised ranking algorithm can achieve better results than unsupervised non-ranking methods, and the aggregation of multiple WHs outperforms single variation modeling. This is because via supervised

Table 4. Results on TREC11 and TREC12

Dataset	TREC11				TREC12			
Scenario	WholeSession		CurrentQuery		WholeSession		IgnoreClick	
Metric	NDCG@10(⋆)	ERR@10	NDCG@10	ERR@10	NDCG@10(⋆)	ERR@10	NDCG@10	ERR@10
Basic Ret	0.368	0.2855	0.2492	0.1936	0.2537	0.1795	0.2461	0.173
TREC Best	0.454	-	0.2525	-	0.3221	-	0.3072	-
MDP	0.4821	-	-	-	0.3368	-	-	-
SQRank	0.4029	0.3075	0.2784	0.2197	0.28	0.1883	0.2706	0.1818
SinVar-Exp0.6	0.478	0.3482	0.2757	0.2126	0.332	0.2228	0.3105	0.2074
SinVar-Exp0.8	0.4875	0.3508	0.2816	0.2158	0.3257	0.221	0.3036	0.206
GlbCxt Rank-V1	0.487	0.3507	0.2843	0.2147	0.3206	0.2213	0.3004	0.2062
GlbCxt Rank-V2	0.4727	0.3535	0.2668	**0.2204**	0.3112	0.2121	0.296	0.202
SLCA	**0.4936**‡⋄	**0.359**‡♯†	**0.2862**‡⋄	0.2173	**0.3371**♯ † ⋄	**0.233**‡♯ † ⋄	**0.3189**♯ † ⋄	**0.2186**‡♯ † ⋄

(⋆) indicates main evaluation metric. Bold numbers are the highest in each column. ‡, ♯, †, ⋄ means SLCA is significantly better than SinVar-Exp0.6/0.8 and CxtRank-V1/V2, where t-test is used with $p < 0.05$. - represents unreported performance in the papers of TRECBest and MDP.

ranking, multiple session variations can be combined to generate some more sophisticated variation patterns that better fit data. - represents performance that is not reported in the original papers of TRECBest and MDP. Nor can we access their query level performance. Therefore, we cannot apply statistical testing between SLCA and them. However, as one can notice, on TREC11 SLCA is significantly better than MDP and TRECBest, while on TREC12, SLCA is similar with MDP and significantly outperforms TRECBest. We argue these observation has already shown the superiority of our SLCA algorithm.

Individual WH Investigation. In Fig. 2 we further show how each WH works individually, i.e. we only utilize one θ_k in Algorithm 1. We can see that, supervised ranking aggregation of multiple weighting hypotheses indeed improves the search accuracy. We can also observe that shrinking WHs are relatively better than other WHs, although the optimal shrinking coefficient (0.6 vs 0.8) varies on different datasets. CurrentQuery WH is always the worst. It is because this WH only considers the local context between $\mathcal{I}^{T-1}$ and $\mathcal{I}^{T}$, while ignoring all other contexts $\mathcal{I}^{1} \sim \mathcal{I}^{T-2}$ (if exist). Notice it is not the same as SQRank,

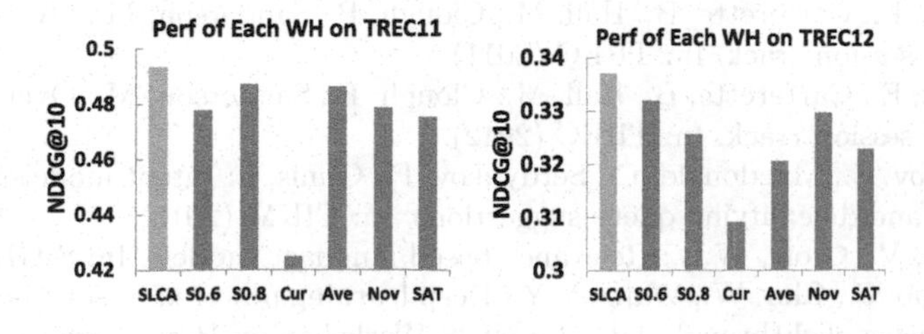

Fig. 2. Performance comparison of each WH.

as CurrentQuery WH indeed utilizes some context while SQRank does not. Ave, Nov and SAT IPs also perform reasonably, and are much better than CurrentQuery WH. Overall, these results verify the proposed SLCA algorithm.

5 Conclusion

The key to effective session search relies on how to effectively utilize preceding contexts of interactions to improve search for current query. Previous related research either ignored session context variations, or formulated single, simple modeling that is not powerful enough. In this paper, we proposed Supervised Local Context Aggregation (SLCA) algorithm, which learns a supervised ranking model based on aggregating multiple weighting hypotheses of local contexts. On TREC11/12 session track our algorithm has achieved the state-of-the-art results.

References

1. Bendersky, M., Fisher, D., Croft, W.B.: Umass at trec 2010 web track: term dependence, spam filtering and quality bias. In: TREC (2010)
2. Bennett, P.N., White, R.W., Chu, W., Dumais, S.T., Bailey, P., Borisyuk, F., Cui, X.: Modeling the impact of short- and long-term behavior on search personalization. In: SIGIR (2012)
3. Cao, H., Jiang, D., Pei, J., Chen, E., Li, H.: Towards context-aware search by learning a very large variable length hidden markov model from search logs. In: WWW (2009)
4. Collins-Thompson, K., Bennett, P.N., White, R.W., de la Chica, S., Sontag, D.: Personalizing web search results by reading level. In: CIKM (2011)
5. Friedman, J.H.: Greedy function approximation: a gradient boosting machine. In: Annals of Statistics (2001)
6. Guan, D., Zhang, S., Yang, H.: Utilizing query change for session search. In: SIGIR (2013)
7. Guo, Q., White, R.W., Dumais, S.T., Wang, J., Anderson, B.: Predicting query performance using query, result, and user interaction features. In: RIAO (2010)
8. Jiang, D., Leung, K.W.T., Ng, W.: Context-aware search personalization with concept preference. In: CIKM (2011)
9. Jiang, J., He, D., Allan, J.: Searching, browsing, and clicking in a search session: changes in user behavior by task and over time. In: SIGIR (2014)
10. Jiang, J., He, D., Han, S.: On duplicate results in a search session. In: TREC (2012)
11. Joachims, T.: Training linear svms in linear time. In: KDD (2006)
12. Kanoulas, E., Carterette, B., Hall, M., Clough, P., Sanderson, M.: Overview of the trec 2011 session track. In: TREC (2011)
13. Kanoulas, E., Carterette, B., Hall, M., Clough, P., Sanderson, M.: Overview of the trec 2012 session track. In: TREC (2012)
14. Kharitonov, E., Macdonald, C., Serdyukov, P., Ounis, I.: Intent models for contextualising and diversifying query suggestions. In: CIKM (2013)
15. Lavrenko, V., Croft, W.B.: Relevance-based language models. In: SIGIR (2001)
16. Li, X., Guo, C., Chu, W., Wang, Y.Y.: Deep learning powered in-session contextual ranking using clickthrough data. In: NIPS Workshop on Personalization: Methods and Applications (2014)

17. Liu, C., Gwizdka, J., Liu, J.: Helping identify when users find useful documents: examination of query reformulation intervals. In: IIiX (2010)
18. Liu, T., Zhang, C., Gao, Y., Xiao, W., Huang, H.: Bupt_wildcat at trec 2011 session track. In: TREC (2011)
19. Liu, T.Y.: Learning to Rank for Information Retrieval. Springer, Heidelberg (2011)
20. Luo, J., Zhang, S., Dong, X., Yang, H.: Designing states, actions, and rewards for using POMDP in session search. In: Hanbury, A., Kazai, G., Rauber, A., Fuhr, N. (eds.) ECIR 2015. LNCS, vol. 9022, pp. 526–537. Springer, Heidelberg (2015)
21. Manning, C.D., Raghavan, P., Schütze, H.: An Introduction to Information Retrieval. Springer, Heidelberg (2011)
22. Raman, K., Bennett, P.N., Collins-Thompson, K.: Toward whole-session relevance: exploring intrinsic diversity in web search. In: SIGIR (2013)
23. Rendle, S., Gantner, Z., Freudenthaler, C., Schmidt-Thieme, L.: Fast context-aware recommendations with factorization machines. In: SIGIR (2011)
24. Shen, X., Tan, B., Zhai, C.: Context-sensitive information retrieval using implicit feedback. In: SIGIR (2005)
25. Shokouhi, M., White, R.W., Bennett, P., Radlinski, F.: Fighting search engine amnesia: reranking repeated results. In: SIGIR (2013)
26. Ustinovskiy, Y., Serdyukov, P.: Personalization of web-search using short-term browsing context. In: CIKM (2013)
27. Xiang, B., Jiang, D., Pei, J., Sun, X., Chen, E., Li, H.: Context-aware ranking in web search. In: SIGIR (2010)

An Empirical Study of Skip-Gram Features and Regularization for Learning on Sentiment Analysis

Cheng Li(✉), Bingyu Wang, Virgil Pavlu, and Javed A. Aslam

College of Computer and Information Science, Northeastern University, Boston, MA, USA
{chengli,rainicy,vip,jaa}@ccs.neu.edu

Abstract. The problem of deciding the overall sentiment of a user review is usually treated as a text classification problem. The simplest machine learning setup for text classification uses a unigram bag-of-words feature representation of documents, and this has been shown to work well for a number of tasks such as spam detection and topic classification. However, the problem of sentiment analysis is more complex and not as easily captured with unigram (single-word) features. Bigram and trigram features capture certain local context and short distance negations—thus outperforming unigram bag-of-words features for sentiment analysis. But higher order n-gram features are often overly specific and sparse, so they increase model complexity and do not generalize well.

In this paper, we perform an empirical study of *skip-gram* features for large scale sentiment analysis. We demonstrate that skip-grams can be used to improve sentiment analysis performance in a model-efficient and scalable manner via regularized logistic regression. The feature sparsity problem associated with higher order n-grams can be alleviated by grouping similar n-grams into a single skip-gram: For example, "waste time" could match the n-gram variants "waste of time", "waste my time", "waste more time", "waste too much time", "waste a lot of time", and so on. To promote model-efficiency and prevent overfitting, we demonstrate the utility of logistic regression incorporating both L1 regularization (for feature selection) and L2 regularization (for weight distribution).

Keywords: Sentiment analysis · Skip-grams · Feature selection · Regularization

1 Introduction

The performance of sentiment analysis systems depends heavily on the underlying text representation quality. Unlike in traditional topical classification, simply applying standard machine learning algorithms such as Naive Bayes or SVM to *unigram* ("bag-of-words") features no longer provides satisfactory accuracy [19]. In sentiment analysis, unigrams cannot capture all relevant and informative features, resulting in information loss and suboptimal classification performance.

N. Ferro et al. (Eds.): ECIR 2016, LNCS 9626, pp. 72–87, 2016.
DOI: 10.1007/978-3-319-30671-1_6

For example, negation is a common linguistic construction that affects polarity but cannot be modeled by bag-of-words [24]. Finding a good feature representation for documents is central in sentiment analysis. Many rule and lexicon based methods are proposed to explicitly model negation relations [19,24]. However, rule and lexicon based approaches do not do well when the words' meanings change in specific domains. Researchers have found that applying machine learning algorithms to n-gram features captures some negations automatically and outperforms rule based systems [23]. For large datasets, frequent bigrams such as "not recommend" and "less entertaining" can model some negation-polarity word pairs. N-gram features are not only good at modeling short distance negations, but are also very useful in capturing subtle meanings, including implicit negations. For example, as mentioned in [19], the negative sentence "How could anyone sit through this movie?" contains no single negative unigram. However, the bigram "sit through" is a strong indicator for negative sentiment.

Although n-gram features are more powerful than unigrams, the diversity and variability of sentiment expressions sometimes makes strict n-gram matching hard to apply. Should the bigram "waste time" match the text "waste a lot of time"? A *skip-gram* [11] is an n-gram matched loosely in text, where looseness can be parameterized by a `slop` value, the number of additional words allowed in a matching span. For example with `slop=1` the skip-gram "waste time" would match "waste time", "waste of time", "waste more time", and "waste my time". With `slop=2` it would also match "waste of my time" and "waste too much time". With `slop=3` it can even match "waste a lot of time". The advantage of loose matching, informally, is that fewer features can match more phrases, which is good in several ways: First, it addresses the semantic matches that strict n-gram matching fails at, such as the n-gram "waste time" failing to match the text "waste my time". Second, higher order n-grams with $n > 3$ are often overly specific and sparse, and they only increase model complexity without generalizing well. Grouping similar n-grams to the same skip-gram alleviates this problem and makes learning more effective.

In this paper, the first research question we investigate is whether **skip-grams are good features** for large scale sentiment analysis when used by machine learning classifiers. We find that skip-gram features perform consistently better than unigram and n-gram features on all the data sets explored in our study. Skip-gram features also outperform word vector features on 2 of the 3 datasets we use, with inconclusive[1] results on the third (IMDB) dataset. We further investigate how varying the `slop` parameter affects sentiment analysis predictions. We generally find that `slop=1` helps as opposed to tight n-gram matches at `slop=0`, and increasing `slop` beyond 1 also helps, but to a lesser degree.

The second research question we investigate is an **appropriate learning mechanism that can handle such a large set of features**, addressing sparsity, speed, feature selection, and model-efficiency, all while retaining

[1] On the IMDB dataset, skip-grams perform worse than word vectors on the predefined test set, but better on randomly sampled test sets, as discussed in Sect. 3.

classification performance. An obvious concern when utilizing skip-grams with large size and `slop` is the large number of potential features generated. Even for the modest IMDB dataset with 50,000 reviews, we find that the number of potential skip-grams generated when using size up to 3 and `slop` up to 2 is nearly 2 million. For many learning algorithms, constructing a model from so many features is difficult. The model can overfit, and the prediction scores it produces are difficult to interpret. We investigate how to select a significantly smaller subset of features that yields good performance.

A natural way to deal with millions of features is to either employ feature selection *a priori* or to build selection into the training algorithm. The latter is often done via *regularization*. For example, L1-regularization [12] provides a convex surrogate to the L0 regularization, which linearly penalizes the number of features used in the model. Thus L1-regularization encourages a frugal use of features. In this paper, L1-regularization is our main mechanism for feature selection, demonstrated for Logistic Regression and SVM. In contrast, the other regularization often used—and one that we argue is necessary for sentiment analysis—is L2-regularization [12]. In the presence of correlated features, L2-regularization encourages the use of all correlates by explicitly penalizing large feature weights: small weights associated with multiple correlates will incur a lower penalty than a high weight associated with a single correlate. Having correlated features in the model can be very useful for generalizability and interpretability, at the cost of model efficiency as greater numbers of features are used. Hence, the L1- and L2-regularization trade-off: L1 encourages the frugal use of features, while L2 encourages judicious use of correlated features. For the problem of sentiment analysis, where sizable feature sets comprised of skip-grams generated with large n and `slop` are useful, we demonstrate the benefits of having a learning model that is both L1 and L2 regularized [12].

1.1 Related Work

Skip-Grams. For sentiment analysis on Twitter data, Fernández et al. [8] showed that using both n-grams and skip-grams give better performance than using n-grams alone. However, the dataset they used is small and tweets are usually short. It is unclear how many frequent skip-grams they extracted (the exact number is not mentioned). Also no feature selection is performed to pick informative skip-grams. As such, it is natural to ask whether skip-grams are still helpful on large datasets, where a huge number of frequent and possibly noisy skip-grams can be extracted. König and Brill [13] used skip-grams in deciding sentiments for movie review data and Microsoft customer feedback data via a multi-stage process. First, skip-gram candidates are generated based on a heuristic. Human assessors then review the skip-gram candidates and manually select the informative ones. At prediction time, a test document is checked to see if it matches any of the selected skip-grams; if it does, a label is assigned immediately based on the matched skip-gram; if not, a classifier trained on only n-grams is used to make a prediction. This hybrid approach is shown to work better than the standard method based purely on n-grams. However, it does not fully utilize the power of skip-grams: since manual assessment is time consuming, only a very small number (300 in their experiments) of skip-gram candidates are generated and presented to the human assessors, and an even smaller number of features is kept. For a small number of selected skip-grams to work well, it is essential for them to be orthogonal so that different aspects of the data can be covered and explained. But skip-grams are judged independently of each other in both the automatic generating procedure and the human assessment procedure; as a result, individually informative features, when put together, could be highly correlated and redundant for prediction purposes. Also, the skip-grams selected are not used in conjunction with n-grams to train classifiers. In our proposed method, the feature selection is done by regularized learning algorithms, which is a much cheaper solution compared with manual selection. This reduction in cost makes it possible to generate and evaluate a large number of skip-gram candidates. The feature selection algorithm considers all features simultaneously, making the selected feature set less redundant.

Word Vectors. Another related line of research performs sentiment analysis based on word vectors (or paragraph vectors) [14,15,18]. Typical word vectors have only hundreds of dimensions, and thus represent documents more concisely than skip-grams do. One common way of building word vectors is to train them on top of skip-grams. After this training step, skip-grams are discarded and only word vectors are used to train the final classifier. Classifiers trained this way are smaller compared with those trained on skip-grams. One should note, however, that training classifiers on a low-dimensional dense word vector representation is not necessarily faster than training classifiers on a high-dimensional sparse skip-gram representation, for two reasons: first, low-dimensional dense features often work best with non-linear classifiers while high-dimensional sparse features often work best with linear classifiers, and linear classifiers are much faster to train. Second, sparsity in the feature matrix can be explored in the latter case to further speed up training. Although the idea of building word vector representations on top of skip-grams is very promising, current methods have some limitations. Documents with word vector representations are compressed or decoded in a highly complicated way, and the learned models based on word vectors are much more difficult to interpret than those based directly on skip-grams. For example, to understand what Amazon customers care about in baby products, it is hard to infer any latent meaning from a word vector feature. On the other hand, it is very easy to interpret a high-weight skip-gram feature such as "no smell", which includes potential variants like "no bad smell", "no medicine-like smell" and "no annoying smell". Another limitation is that, while word vectors are trained on skip-grams, they do not necessarily capture all the information in skip-grams. In our method, the classifiers are trained directly on skip-grams, and thus can fully utilize the information provided by skip-grams. We exploit the sparsity in the feature matrix to speed up training, and feature selection is employed to shrink the size of the classifier. Experiments show that our method generally achieves both better performance and better interpretability.

2 Learning with Skip-Gram Features

Extracting skip-grams from documents and computing matching scores is an IR, or NLP preprocessing problem which should be solved before the learning. We divide it into four steps: First, we lemmatize documents into tokens with the Stanford NLP package [1]. All stop-words are kept as they are often useful for sentiment analysis tasks. Second, preprocessed documents are sent to the search engine ElasticSearch [2] and an inverted index is built. Third, skip-gram candidates which meet the size, `slop` and document frequency requirements are gathered from the training document collection. To save memory and computation, skip-grams with very low document frequencies are discarded.

2.1 Skip-Gram Matching using ElasticSearch

In the last preprocessing step, we determine the matched documents for each of the skip-gram candidates and their matching scores. There are several slightly different ways of computing the matching score, but the basic idea is the same: a phrase that matches the given n-gram tightly contributes more to the score than a phrase that matches the n-gram loosely, and if two documents have the same skip-gram frequency, the shorter document will receive a higher score.

An indexing service is needed for storage and matching, i.e., a service such as Lemur [3], Terrier [4], Lucene [5] or ElasticSearch [2]. Any such platform can be used for this purpose. For this study, we adopt the **"Span Near Query"** scoring function implemented in the open source search engine ElasticSearch, which matches the above criteria. For a given (n-gram g, `slop` s, document d) triple,

$$\mathbf{score}(g, s, d) = \sqrt{\frac{\mathbf{skipGramFreq}(g, s, d)}{\mathbf{length}(d)}}$$

where

$$\mathbf{skipGramFreq}(g, s, d) = \sum_{k=0}^{s} \frac{\mathbf{phraseFreq}(g, k, d)}{\mathbf{length}(g) + 1 + k}$$

and $\mathbf{phraseFreq}(g, k, d)$ is the number of phrases in d generated by inserting k extra words in the given n-gram. We further normalize $\mathbf{score}(g, s, d)$ to the range [0,1].

2.2 Learning Algorithms and Regularization

After matching each skip-gram against the document collection, we obtain a feature matrix which can be fed into most classification algorithms. We use regularized SVM with a linear kernel and regularized Logistic Regression (LR). Both are linear models, thus fast to train and less likely to overfit high dimensional data.

Regularized SVM minimizes thesum of hinge loss and a penalty term [7]. Specifically, for L2-regularized SVM, the objective is

$$\min_w \sum_{i=1}^{N} (\max(0, 1 - y_i w^T x_i))^2 + \lambda \frac{1}{2} ||w||_2^2,$$

and for L1-regularized SVM, the objective is

$$\min_w \sum_{i=1}^{N} (\max(0, 1 - y_i w^T x_i))^2 + \lambda ||w||_1,$$

where λ controls the strength of the regularization[2]. We use the LibLinear package [7] for regularized SVM. Using both L1 and L2 terms to regularize SVM has been proposed [22], but is not commonly seen in practice, possibly due to the difficult learning procedure; hence we do not consider it in this study.

Regularized LR [12] minimizes the sum of logistic loss and some penalty term. Specifically, for L2-regularized LR, the objective is

$$\min_w -\frac{1}{N} \sum_{i=1}^{N} y_i w^T x_i + \log(1 + e^{w^T x_i}) + \lambda \frac{1}{2} ||w||_2^2,$$

for L1-regularized LR, the objective is

$$\min_w -\frac{1}{N} \sum_{i=1}^{N} y_i w^T x_i + \log(1 + e^{w^T x_i}) + \lambda ||w||_1,$$

and for L1+L2-regularized LR, the objective is

$$\min_w -\frac{1}{N} \sum_{i=1}^{N} y_i w^T x_i + \log(1 + e^{w^T x_i}) + \lambda \alpha ||w||_1 + \lambda (1 - \alpha) \frac{1}{2} ||w||_2^2.$$

In L1+L2 LR, the L1 ratio α controls the balance between L1 regularization and L2 regularization. When $\alpha = 0$, the model has only the L2 penalty term; when $\alpha = 1$, the model has only the L1 penalty term.

Unlike the case for SVM, LR with both L1 and L2 penalties is widely adopted, possibly due to the efficient training and hyper-parameter tuning algorithms available [10]. However, we find that the most popular package glmnet [9] for L1+L2 LR does not scale well on our datasets which contain hundreds of thousands of documents with millions of skip-gram features. Thus we make use of our own Java implementation[3], which has a special optimization for sparse matrix representations and is more scalable than glmnet.

3 Experiments

Datasets and setup. To examine the effectiveness of skip-grams, we extract skip-gram features from three large sentiment analysis datasets and train several

[2] In the LibLinear package that we use, a different notation is used; there $C = 1/\lambda$.

[3] Our code is publicly available at https://github.com/cheng-li/pyramid.

machine learning algorithms on these extracted features. The datasets used are IMDB, Amazon Baby, and Amazon Phone. IMDB [15] contains 50,000 labeled movie reviews. Reviews with ratings from 1 to 4 are considered negative; and 7 to 10 are considered positive. Reviews with neutral ratings are ignored. The overall label distribution is well balanced (25,000 positive and 25,000 negative). IMDB comes with a predefined train/test split, which we adopt in our experiments. There are also another 50,000 unlabeled reviews available for unsupervised training or semi-supervised training, which we do not use. Amazon Baby (containing Amazon baby product reviews) and Amazon Phone (containing cell phone and accessory reviews) are both subsets of a larger Amazon review collection [17]. Here we use them for binary sentiment analysis the same manner as in IMDB dataset. By convention, reviews with rating 1–2 are considered negative and 4–5 are positive. The neutral ones are ignored. Amazon Baby contains 136,461 positive and 32,950 negative reviews. Amazon Phone contains 47,970 positive and 22,241 negative reviews. Amazon Baby and Amazon Phone do not have a predefined train/test partitioning. We perform stratified sampling to choose a random 20 % of the data as the test set. All results reported below on these two datasets are averaged across five runs.

For each dataset, we extract skip-gram features with max size n varying from 1 (unigram) to 5 (5-gram) and max `slop` varying from 0 (no extra words can be added) to 2 (maximal 2 words can be added). For example, when max n=2 and max `slop`=1, we will consider unigrams, bigrams, and skip-bigrams with `slop`=1. As a result, for each dataset, 13 different feature sets are created. The combinations (max n=1, max `slop`=1) and (max n=1, max `slop`=2) are essentially the same as (max n=1, max `slop`=0), and thus not considered. For each feature set, we run five learning algorithms on it and measure the accuracies on the test set. The algorithms considered are L1 SVM, L2 SVM, L1 LR, L2 LR and L1+L2 LR. In order to make the feature set the only varying factor, we use fixed hyper parameters for all algorithms across all feature sets. The hyper parameters are chosen by cross-validation on training sets with unigram features. For L2 SVM, $C = 1/\lambda = 0.0625$; for L1 SVM, $C = 1/\lambda = 0.25$; for LR, $\lambda = 0.00001$; and for L1+L2 LR, $\alpha = 0.1$. Performing all experiments took about five days using a cluster with six 2.80 GHz Xeon CPUs.

3.1 Main Results

Figure 1 shows how increasing max n and max `slop` of the skip-grams affects the logistic regression performance on Amazon Baby. In each sub-figure, the bottom line is the performance with standard n-gram (max `slop`=0) features. Along each bottom line, moving from unigrams (max n=1) to bigrams (max n=2) gives substantial improvement. Bigrams such as "not recommend" are effective at capturing short distance negations, which cannot be captured by unigrams.

Moving beyond bigrams (max n=2) to higher order n-grams, we can see some further improvement, but not as big as before. This observation is consistent with the common practice in sentiment analysis, where trigrams are not commonly

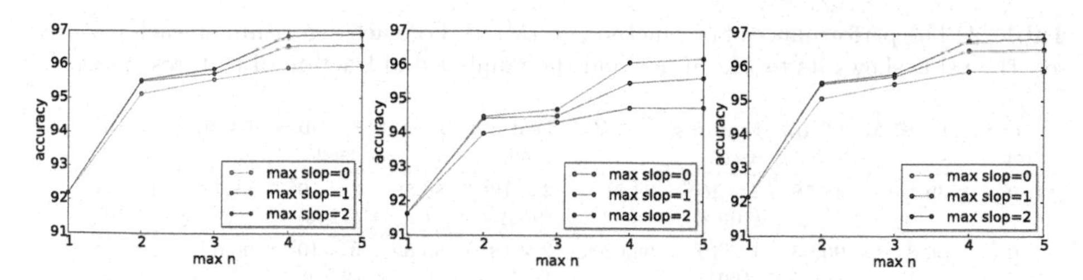

Fig. 1. Performance of LR on Amazon Baby with skip-gram features of varying sizes and slops. Left to right learning algorithms: L2-LR, L1-LR, L1+L2-LR.

Table 1. The performance of our method on Amazon Baby. For each algorithm on each feature set, the table shows its test accuracy and the number and fraction of features selected. The accuracies which are significantly better (at 0.05 level under t-test) than those by a corresponding `slop` 0 baseline are marked with *.

max n, slop	L2 SVM	L2 LR	Features used	L1 SVM	Features used	L1 LR	Features used	L1+L2 LR	Features used
1, 0	92.38	92.18	8×10^3 (100 %)	92.36	4×10^3 (49 %)	91.63	2×10^3 (23 %)	92.15	7×10^3 (88 %)
2, 0	95.12	95.13	6×10^4 (100 %)	94.89	8×10^3 (12 %)	93.98	3×10^3 (5 %)	95.07	3×10^4 (58 %)
3, 0	95.51	95.55	1×10^5 (100 %)	95.20	9×10^3 (6 %)	94.32	4×10^3 (2 %)	95.50	6×10^4 (45 %)
4, 0	95.88	95.89	1×10^5 (100 %)	95.59	9×10^3 (5 %)	94.74	4×10^3 (2 %)	95.88	7×10^4 (41 %)
5, 0	95.93	95.94	1×10^5 (100 %)	95.59	9×10^3 (5 %)	94.76	4×10^3 (2 %)	95.90	7×10^4 (40 %)
2, 1	95.51*	95.50*	1×10^5 (100 %)	95.22*	1×10^4 (6 %)	94.43*	4×10^3 (2 %)	95.50*	8×10^4 (45 %)
3, 1	95.70*	95.72*	5×10^5 (100 %)	95.36*	1×10^4 (2 %)	94.51*	5×10^3 (1 %)	95.72*	1×10^5 (25 %)
4, 1	96.51*	96.54*	6×10^5 (100 %)	96.19*	1×10^4 (1 %)	95.48*	5×10^3 (<1 %)	96.50*	1×10^5 (20 %)
5, 1	96.56*	96.56*	6×10^5 (100 %)	96.23*	1×10^4 (1 %)	95.63*	5×10^3 (<1 %)	96.54*	1×10^5 (19 %)
2, 2	95.51*	95.53*	3×10^5 (100 %)	95.19*	1×10^4 (3 %)	94.48*	6×10^3 (1 %)	95.54*	1×10^5 (32 %)
3, 2	95.89*	95.86*	1×10^6 (100 %)	95.41*	1×10^4 (1 %)	94.70*	6×10^3 (<1 %)	95.79*	1×10^5 (15 %)
4, 2	96.85*	96.84*	1×10^6 (100 %)	96.57*	1×10^4 (<1 %)	96.08*	6×10^3 (<1 %)	96.79*	1×10^5 (10 %)
5, 2	96.87*	96.89*	1×10^6 (100 %)	96.60*	1×10^4 (<1 %)	96.20*	7×10^3 (<1 %)	96.85*	1×10^5 (9 %)

used compared with bigrams, and n-grams beyond trigrams are rarely used. When we fix the max n and increase the max `slop`, we see the performance further improves. For $n \geq 4$, increasing max `slop` often brings more improvement than increasing max n. Similar observations can be made for SVM and for the other two datasets.

Table 2. The performance of our method on IMDB. For each algorithm on each feature set, the table shows its test accuracy and the number and fraction of features selected.

max n, slop	L2 SVM	L2 LR	Features used	L1 SVM	Features used	L1 LR	Features used	L1+L2 LR	Features used
1, 0	89.10	88.58	2×10^4 (100 %)	88.81	2×10^3 (9 %)	88.59	3×10^3 (12 %)	88.71	1×10^4 (75 %)
2, 0	90.81	90.63	1×10^5 (100 %)	90.02	2×10^3 (1 %)	89.80	3×10^3 (2 %)	90.62	6×10^4 (40 %)
3, 0	91.10	91.02	2×10^5 (100 %)	90.13	2×10^3 (<1 %)	89.84	3×10^3 (1 %)	90.89	8×10^4 (28 %)
4, 0	91.19	91.13	3×10^5 (100 %)	90.19	2×10^3 (<1 %)	89.85	3×10^3 (<1 %)	90.97	9×10^4 (25 %)
5, 0	91.21	91.16	4×10^5 (100 %)	90.18	2×10^3 (<1 %)	89.85	3×10^3 (<1 %)	90.96	9×10^4 (24 %)
2, 1	91.22	91.13	3×10^5 (100 %)	90.24	3×10^3 (<1 %)	90.01	3×10^3 (<1 %)	90.94	9×10^4 (25 %)
3, 1	91.46	91.44	9×10^5 (100 %)	90.26	3×10^3 (<1 %)	90.07	3×10^3 (<1 %)	91.20	1×10^5 (13 %)
4, 1	91.56	91.54	1×10^6 (100 %)	90.37	3×10^3 (<1 %)	90.11	3×10^3 (<1 %)	91.22	1×10^5 (11 %)
5, 1	91.65	91.64	1×10^6 (100 %)	90.36	3×10^3 (<1 %)	90.15	3×10^3 (<1 %)	91.24	1×10^5 (10 %)
2, 2	91.32	91.35	6×10^5 (100 %)	90.37	2×10^3 (<1 %)	90.07	4×10^3 (<1 %)	90.96	1×10^5 (17 %)
3, 2	91.65	91.60	2×10^6 (100 %)	90.40	3×10^3 (<1 %)	90.23	4×10^3 (<1 %)	91.25	1×10^5 (7 %)
4, 2	91.76	91.64	2×10^6 (100 %)	90.43	3×10^3 (<1 %)	90.26	4×10^3 (<1 %)	91.23	1×10^5 (5 %)
5, 2	91.71	91.63	3×10^6 (100 %)	90.43	3×10^3 (<1 %)	90.26	4×10^3 (<1 %)	91.26	1×10^5 (5 %)

Tables 1, 2, and 3 show more detailed results on these datasets. For each fixed n, we use a paired t-test (0.05 level) to check whether increasing max `slop` from 0 to 1 or 2 leads to significant improvement. On Amazon Baby, all improvements due to the increase of max `slop` are significant. On Amazon Phone, about half are significant. The significance test is not done on the IMDB dataset since only the predefined test set is used.

Tables 1, 2, and 3 also show how many features are selected by each learning algorithm. L2 regularized algorithms do best in terms of accuracy but at the cost of using all features. If that is acceptable in certain use cases, then L2 regularization is recommended. On the other hand, L1 regularization can greatly reduce the number of features used to below 1 %, sacrificing test accuracy by 1–2 %; if this drop in performance is acceptable, then L1 regularization is recommended for the extremely compact feature sets produced. Finally L1+L2 regularization is a good middle choice for reducing the number of features to about 5–20 % while at the same time maintaining test accuracy on par with L2 regularization.

Table 3. The performance of our method on Amazon Phone. For each algorithm on each feature set, the table shows its test accuracy and the number and fraction of features selected. The accuracies which are significantly better (at 0.05 level under t-test) than those by a corresponding `slop` 0 baseline are marked with *.

max n, `slop`	L2 SVM	L2 LR	Features used	L1 SVM	Features used	L1 LR	Features used	L1+L2 LR	Features used
1, 0	89.45	89.27	5×10^3 (100 %)	89.36	2×10^3 (45 %)	89.33	2×10^3 (41 %)	89.30	5×10^3 (96 %)
2, 0	92.03	91.89	3×10^4 (100 %)	91.82	5×10^3 (14 %)	91.85	4×10^3 (13 %)	91.94	2×10^4 (79 %)
3, 0	92.24	92.11	6×10^4 (100 %)	91.77	5×10^3 (8 %)	91.92	5×10^3 (8 %)	92.14	4×10^4 (70 %)
4, 0	92.44	92.26	8×10^4 (100 %)	91.94	5×10^3 (6 %)	91.99	5×10^3 (6 %)	92.32	5×10^4 (66 %)
5, 0	92.29	92.19	8×10^4 (100 %)	91.89	5×10^3 (6 %)	92.01	5×10^3 (6 %)	92.28	5×10^4 (65 %)
2, 1	92.39	92.25	9×10^4 (100 %)	92.19	6×10^3 (6 %)	92.30*	6×10^3 (7 %)	92.29	6×10^4 (65 %)
3, 1	92.31	92.33	2×10^5 (100 %)	92.09*	7×10^3 (3 %)	92.27	8×10^3 (3 %)	92.33*	9×10^4 (43 %)
4, 1	92.31	92.43	2×10^5 (100 %)	92.18	7×10^3 (2 %)	92.28	8×10^3 (2 %)	92.42	1×10^5 (38 %)
5, 1	92.33	92.37	3×10^5 (100 %)	92.15	7×10^3 (2 %)	92.31*	8×10^3 (2 %)	92.31	1×10^5 (36 %)
2, 2	92.57*	92.67*	4×10^5 (100 %)	92.13*	8×10^3 (1 %)	92.32*	1×10^4 (2 %)	92.53*	1×10^5 (28 %)
3, 2	92.53	92.64*	4×10^5 (100 %)	92.13	8×10^3 (1 %)	92.37*	1×10^4 (2 %)	92.55*	1×10^5 (28 %)
4, 2	92.59	92.74*	6×10^5 (100 %)	92.13	8×10^3 (1 %)	92.30	1×10^4 (1 %)	92.58*	1×10^5 (23 %)
5, 2	92.54*	92.67*	7×10^5 (100 %)	92.24*	8×10^3 (1 %)	92.40*	1×10^4 (1 %)	92.58*	1×10^5 (22 %)

3.2 Comparisons with Other Methods

For the IMDB dataset, public results on the predefined test set are listed in Table 4. Among the methods which only use labeled data, our method based on skip-grams achieved the highest accuracy. Paragraph vectors (based on word2vec) trained on both labeled data and unlabeled data achieve noticeably higher performance. In fact, using one public paragraph vector implementation[4], with only labeled data and a RBF SVM classifier[5], we are able to produce 93.56 % accuracy on the given test set. However, the performance of paragraph

[4] The paragraph vector implementation is from https://github.com/klb3713/sentence2vec/. The parameters we use are size=400, alpha=0.025, window=10, min_count=5, sample=0, seed=1, min_alpha=0.0001, sg=1, hs=1, negative=0, cbow_mean=0.

[5] After producing paragraph vectors, we run LIBSVM (https://www.csie.ntu.edu.tw/~cjlin/libsvm/) with c=32, g=0.0078. An RBF kernel performs better than a linear kernel.

Table 4. Our approach compared to other methods on the IMDB dataset.

Classifier	Features	Training documents	Accuracy
LR with dropout regularization [21]	bigrams	25,000 labeled	91.31
NBSVM [23]	bigrams	25,000 labeled	91.22
SVM with L2 regularization	structural parse tree features + unigrams [16]	25,000 labeled	82.8
LR L1+L2 regularization	5-grams selected by compressive feature learning [20]	25,000 labeled	90.4
SVM	word vectors trained by WRRBM [6]	25,000 labeled	89.23
SVM	word vectors [15]	25,000 labeled + 50,000 unlabeled	88.89
LR with dropout regularization [21]	bigrams	25,000 labeled + 50,000 unlabeled	91.98
LR	paragraph vectors [14]	25,000 labeled + 50,000 unlabeled	92.58
LR with L2 regularization	skip-grams	25,000 labeled	91.63
SVM with L2 regularization	skip-grams	25,000 labeled	91.71
LR with L1+L2 regularization	skip-grams	25,000 labeled	91.26

vectors seems quite sensitive to the specific training/testing partitioning. After re-partitioning the data randomly (50 %–50 % as before), the accuracy of paragraph vectors based on the same hyper-parameters dropped significantly to only around 85 %. By contrast, our method consistently produces high results on both the given test set and randomly sampled test set.

Amazon review datasets are often used differently by different researchers, which makes the published results not directly comparable. Here we train paragraph vectors on the same subset of documents and report the performance.[6] On Amazon Baby, paragraph vector gives 88.84 % while our method gives 96.85 %. On Amazon Phone, paragraph vector gives 85.38 % while our method gives 92.58 %.

4 Analysis of Skip-Grams

When designing a feature set, the primary concern is often generalizability, since good generalizability implies good prediction performance. In sentiment analysis

[6] The training parameters are the same as in IMDB.

data, people often express the same idea in many slightly different ways, which makes the prediction task harder as the algorithm has to learn many expressions with small variations. Skip-grams alleviate this problem by letting the algorithm focus on the important terms in the phrase and tolerate small changes in unimportant terms. Thus skip-grams perform feature grouping on top of n-grams without requiring any external domain knowledge. This not only improves generalizability but also interpretability. Several such skip-gram examples are shown in Table 5. They are selected by an L1+L2 regularized logistic regression model with high weights. For each skip-gram, we show its count in the entire collection and several n-gram instances that it matches. For each matched n-gram, the count in the collection is also listed in the table. We can see, for example, the skip-gram "only problem" (`slop`=1) could match bigram "only problem" and trigrams "only minor problem" and "only tiny problem". Although the bigram "only problem" is frequent enough in the collection, the trigram "only tiny problem" only occurs in four out of 169,411 reviews. It is hard for the algorithm to treat the trigram "only tiny problem" confidently as a positive sentiment indicator. After grouping all such n-gram variants into the same skip-gram, the algorithm can assign a large positive weight to the skip-gram as a whole, thus also handling the rare cases properly. This also provides more concise rules and facilitates user interpretation.

4.1 Feature Utility

We analyze to what extent skip-gram features contribute to overall performance. Take the Amazon Phone dataset as an example. The skip-gram features in it can be broken down into different types based on n and `slop` values. The left column of Fig. 2 shows, when max $n = 3$ and max `slop`=2, about 85 % of the extracted skip-gram features have non-zero `slop`s. In the middle column in Fig. 2, we only focus on features selected by L1+L2 logistic regression and recheck their count distribution. The fraction of unigrams increases, while the fraction of `slop`=2 trigrams decreases. One can imagine that many noisy/irrelevant `slop`=2 skip-trigrams are eliminated by the L1 regularization, and unigrams are less noisy. We further sum the logistic regression weights (absolute values, which are comparable since all features are normalized) for features within each type and display the results in the right column. The standard n-grams with `slop=0` only contribute to 20 % of the total weight, and the remaining 80 % is due to skip-grams with non-zero `slop`s.

4.2 Feature Selection for Skip-Grams

Grouping similar n-grams into skip-grams not only produces generalizable features but sometimes also noisy features. For example, in Table 5, "I have to return", "I have never had to return", "I finally have to return" and "I do not have to return" are all grouped into the skip-gram "I have to return" (`slop`=2).

Table 5. Examples of high weight skip-grams for LR.

Skipgram and count		Matched ngrams and count			
skip movie (slop 2)	42	skip this movie skip the movie skip watching this movie	28 8 1	skip this pointless movie skipping all the movies of this sort	1 1
it fail (slop 1)	358	it fails it even fails	279 5	it completely fails it simply fails	5 3
whole thing (slop 1)	729	whole thing whole damn thing	682 5	whole horrific thing	1
waste time (slop 1)	1562	waste time waste your time	109 4	waste of time waste more time	676 6
only problem (slop 1)	1481	only problem only minor problem	1378 11	only tiny problem	4
never leak (slop 2)	1053	never leak never have leak	545 86	never a urine leak problem never have any leak	1 77
no smell (slop 1)	445	no smell no bad smell	340 13	no medicine-like smell no annoying smell	1 5
it easy to clean and (slop 2)	314	it is easy to wipe clean and it is so easy to clean and	3 16	it is easy to keep clean and	3
I have to return (slop 2)	216	I have to return I have never had to return	151 1	I finally have to return I do not have to return	1 4
good service (slop 2)	209	good service good and fast service	131 2	good price and service	1

This is the worst kind of noise because the gap matches negation words and different instances of the skip-gram have opposite sentiments. Detecting and modeling the scope of negations is very challenging in general [24]. We do not deal with negations at skip-gram generation time; at learning time, we rely on feature selection to eliminate such noisy skip-grams. In this particular example, the noise is relatively low as the mismatched n-grams "I have never had to return" and "I do not have to return" are very rare in the document collection. Therefore logistic regression still assigns a large weight to this skip-gram. Some other skip-grams are more likely to include negations and are thus more noisy. For example, the skip-gram "I recommend" (`slop`=2) can match many occurrences of both "I highly recommend" and "I do not recommend". Our feature selection mechanism infers that this skip-gram does more harm than good and assigns a small weight to it. In practice, we find the denoising effect of feature selection to be satisfactory. Most of the classification mistakes are not caused by skip-gram mismatch but due to the inability to identify the subjects of the sentiment expressions: many reviews compare several movies/products and thus the algorithm gets confused as to which subject the sentiment expression should apply. Resolving this issue requires other NLP techniques and is beyond the scope of this study.

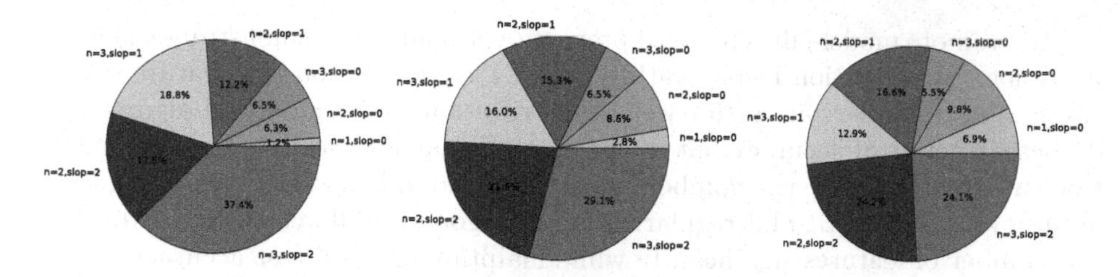

Fig. 2. L1+L2 LR selected features for Amazon Phone feature contribution analysis, max n =3. LEFT: feature count distribution in dataset; MIDDLE: feature count distribution of selected features; RIGHT: feature LR-weighted-distribution of selected features

From Tables 1, 2, and 3, it is very clear that L2 regularization achieves better overall accuracy than L1 regularization. This seems counter-intuitive because L1 regularization completely eliminates noisy features while L2 regularization only shrinks their weights. We believe there are two main reasons for this: First, the document collections are relatively big. The bigger the dataset is, the more parameters can be reliably estimated. L1 regularization is very successful at "large p, small n" problems where the sample size is often in the hundreds while the feature space could be in the millions. Our sentiment analysis datasets, however, are much larger, and this fact makes it possible for L2 logistic regression to estimate almost all parameters. In this case, assigning very low (not necessarily exactly 0) weights to noisy features will suffice. Second, in the presence of many highly correlated features, L1 regularization usually picks only one of them and discards the rest. But the same opinion/sentiment is often expressed in many different ways, which means L1 regularization's instability in handling correlated features can hurt the prediction performance.

But performance is not the only factor we care about. Having an L1 regularization can produce smaller models, which makes the prediction faster and the model more interpretable. L1+L2 regularization provides a good balance between model compactness and prediction accuracy, since a relatively small fraction of features is selected and the performance does not appreciably suffer. In all three datasets, if we limit the number of features used to be under 1×10^5, then the best performance is always achieved by L1+L2 LR, trained on skip-grams of maximum size 5 and slop 2.

5 Conclusion

We demonstrate that *skip-grams* can be used to improve large scale sentiment analysis performance in a model-efficient and scalable manner via regularized logistic regression. We show that although n-grams beyond trigrams are often very specific and sparse, many similar n-grams can be grouped into a single skip-gram which benefits both model-efficiency and classification performance.

To promote model-efficiency and prevent overfitting, we demonstrate the utility of logistic regression incorporating both L1 regularization (for feature selection) and L2 regularization (for weight distribution). L2 regularized algorithms do best in terms of accuracy but at the cost of using all features. L1 regularization can greatly reduce the number of features used to below 1 %, sacrificing test accuracy by 1–2 %. L1+L2 regularization is a good middle choice for reducing the number of features significantly while maintaining good test accuracy.

Acknowledgments. The research is supported by NSF grant IIS-1421399.

References

1. http://nlp.stanford.edu/software/
2. https://lucene.apache.org/
3. http://www.lemurproject.org/
4. http://terrier.org/
5. http://www.elasticsearch.org/
6. Dahl, G.E., Adams, R.P., Larochelle, H.: Training restricted Boltzmann machines on word observations. arXiv preprint (2012). arxiv:1202.5695
7. Fan, R.-E., Chang, K.-W., Hsieh, C.-J., Wang, X.-R., Lin, C.-J.: Liblinear: a library for large linear classification. J. Mach. Learn. Res. **9**, 1871–1874 (2008)
8. Fernández, J., Gutiérrez, Y., Gómez, J.M., Martınez-Barco, P.: Gplsi: supervised sentiment analysis in twitter using skipgrams. In: SemEval 2014, pp. 294–299 (2014)
9. Friedman, J., Hastie, T., Tibshirani, R.: glmnet: Lasso and elastic-net regularized generalized linear models. R package version, 1 (2009)
10. Friedman, J., Hastie, T., Tibshirani, R.: Regularization paths for generalized linear models via coordinate descent. J. Stat. Softw. **33**(1), 1 (2010)
11. Guthrie, D., Allison, B., Liu, W., Guthrie, L., Wilks, Y.: A closer look at skip-gram modelling. In: LREC-2006, pp. 1–4 (2006)
12. Hastie, T., Tibshirani, R., Friedman, J.: The Elements of Statistical Learning, vol. 2. Springer, New York (2009)
13. König, A.C., Brill, E.: Reducing the human overhead in text categorization. In: KDD, pp. 598–603. ACM (2006)
14. Le, Q.V., Mikolov, T.: Distributed representations of sentences and documents. arXiv preprint (2014). arxiv:1405.4053
15. Maas, A.L., Daly, R.E., Pham, P.T., Huang, D., Ng, A.Y., Potts, C.: Learning word vectors for sentiment analysis. In: ACL 2011, pp. 142–150. Association for Computational Linguistics (2011)
16. Massung, S., Zhai, C., Hockenmaier, J.: Structural parse tree features for text representation. In: ICSC, pp. 9–16. IEEE (2013)
17. McAuley, J., Leskovec, J.: Hidden factors and hidden topics: understanding rating dimensions with review text. In: Proceedings of the 7th ACM Conference on Recommender Systems, pp. 165–172. ACM (2013)
18. Mikolov, T., Chen, K., Corrado, G., Dean, J.: Efficient estimation of word representations in vector space. arXiv preprint (2013). arxiv:1301.3781

19. Pang, B., Lee, L., Vaithyanathan, S.: Thumbs up?: sentiment classification using machine learning techniques. In: Proceedings of the ACL-02 Conference on Empirical Methods in Natural Language Processing, vol. 10, pp. 79–86. Association for Computational Linguistics (2002)
20. Paskov, H.S., West, R., Mitchell, J.C., Hastie, T.: Compressive feature learning. In: NIPS, pp. 2931–2939 (2013)
21. Wager, S., Wang, S., Liang, P.S.: Dropout training as adaptive regularization. In: NIPS, pp. 351–359 (2013)
22. Wang, L., Zhu, J., Zou, H.: The doubly regularized support vector machine. Statistica Sinica **16**(2), 589 (2006)
23. Wang, S.I., Manning, C.D.: Baselines and bigrams: simple, good sentiment and topic classification. In: Proceedings of the ACL, pp. 90–94 (2012)
24. Wiegand, M., Balahur, A., Roth, B., Klakow, D., Montoyo, A.: A survey on the role of negation in sentiment analysis. In: Proceedings of the Workshop on Negation and Speculation in Natural Language Processing, pp. 60–68. Association for Computational Linguistics (2010)

Multi-task Representation Learning for Demographic Prediction

Pengfei Wang, Jiafeng Guo(✉), Yanyan Lan, Jun Xu, and Xueqi Cheng

CAS Key Laboratory of Network Data Science and Technology,
Institute of Computing Technology, Beijing, China
wangpengfei@software.ict.ac.cn,
{guojiafeng,lanyanyan,junxu,cxq}@ict.ac.cn

Abstract. Demographic attributes are important resources for market analysis, which are widely used to characterize different types of users. However, such signals are only available for a small fraction of users due to the difficulty in manual collection process by retailers. Most previous work on this problem explores different types of features and usually predicts different attributes independently. However, manually defined features require professional knowledge and often suffer from under specification. Meanwhile, modeling the tasks separately may lose the ability to leverage the correlations among different attributes. In this paper, we propose a novel Multi-task Representation Learning (MTRL) model to predict users' demographic attributes. Comparing with the previous methods, our model conveys the following merits: (1) By using a multi-task approach to learn the tasks, our model leverages the large amounts of cross-task data, which is helpful to the task with limited data; (2) MTRL uses a supervised way to learn the shared semantic representation across multiple tasks, thus it can obtain a more general and robust representation by considering the constraints among tasks. Experiments are conducted on a real-world retail dataset where three attributes (gender, marital status, and education background) are predicted. The empirical results show that our MTRL model can improve the performance significantly compared with the state-of-the-art baselines.

Keywords: Multi-task · Demographic prediction · Representation learning

1 Introduction

Acquiring users' demographic attributes is crucial for retailers to conduct market basket analysis [18], adjust marketing strategy [9], and provide personalized recommendations [20]. However, in practice, it is difficult to obtain users' demographic attributes, because most users are reluctant to offer their detailed information or even refuse to give their demographics due to privacy and other reasons.

N. Ferro et al. (Eds.): ECIR 2016, LNCS 9626, pp. 88–99, 2016.
DOI: 10.1007/978-3-319-30671-1_7

This is particularly true for traditional offline retailers[1], who collect users' demographic information mostly in a manual way (e.g. requiring costumers to provide demographic information when registering some shopping cards).

In this paper, we try to inference users' demographic attributes based on users' purchase history. Although some recent studies suggest that demographic attributes are predictable from different behavioral data, such as linguistics writing [5], web browsing [16], electronic communications [8,12] and social media [14,23], to our best knowledge, seldom practice has been conducted on purchase behaviors in the retail scenario.

The previous work about demographic prediction usually predicts demographic attributes separately based on manually defined features [3,17,19,22,23]. For example, Zhong et al. [23] predicted six demographic attributes (i.e., gender, age, education background, sexual orientation, marital status, blood type and zodiac sign) separately by merging spatial, temporal and location knowledge features into a continuous space. Obviously, manually defined features usually require professional knowledge and often suffer from under specification. Meanwhile, by taking each attribute as independent prediction task, some attributes may difficult to predict due to the insufficient data in training. Some recent studies proposed to take the relations between multiple attributes into account [3,22]. For example, Dong et al. [3] employed a Double Dependent-Variable Factor Graph model to predict gender and age simultaneously. Zhong et al. [22] attempted to capture pairwise relations between different tasks when predicting six demographic attributes from mobile data. However, these methods still rely on various human-defined features which are often costly to obtain.

To tackle the above problem, in this paper we propose a Multi-task Representation Learning(MTRL) model is used to predict users' gender, martial status, and education background based on users' purchase history. MTRL learns shared semantic representations across multiple tasks, which benefits from a more general representation for prediction. Specifically, we characterize each user by his/her purchase history using the bag-of-item representations. We then map all users' representations into semantic space learned by a multi-task approach. Thus we can obtain a more general shared representation to guide the prediction task separately. Compared with previous methods, the major contributions of our work are as follows:

- We make the first attempt to investigate the prediction power of users' purchase data for demographic prediction in the retail scenario.
- We apply a multi-task learning framework (MTRL) for our problem, which can learn a shared robust representation across tasks and alleviate the data sparse problem.
- We conduct extensive experiments on a real-world retail dataset to demonstrate the effectiveness of the proposed MTRL model as compared with different baseline methods.

[1] In our work, we mainly focus on traditional retailers in offline business rather than those in online e-commerce, where no additional behavioral data rather than transactions are available for analysis. Hereafter we will use retail/retailer for simplicity when there is no ambiguity.

The rest of the paper is organized as follows. After a summary of related work in Sect. 2, we describe the problem formalization of demographic prediction in the retail scenario in Sect. 3. In Sect. 4 we present our proposed model in detail. Section 5 concludes this paper and gives the future work.

2 Related Work

In this section we briefly review three research areas related to our work: demographic attribute prediction, multi-task approach, and representation learning.

2.1 Demographic Attribute Prediction

Demographic inference has been studied in different scenarios for more than fifty years. Early stage work on demographic prediction attempted to predict demographic attributes based on the linguistics writing and speaking. For example, Schler et al. [19] found that there are significant differences in both writing style and content between male and female bloggers as well as among authors of different ages. Otterbacher [17] used a logistic regression model to infer users' gender based on content of reviews.

Furthermore, researchers use internet information to predict demographic attributes [8,16]. For example, Torres [4] found a clear relation between the reading level of clicked pages and demographic attributes such as age and education background. Hu et al. [8] calculated demographic tendency of web pages, and modeled users' demographic attributes through a discriminative model. In [1], Bi et al. propose to infer the demographic attributes of search users based on the models trained on the independent social datasets. They demonstrated that by leveraging social and search data in a common representation, they can achieve better accuracy in demographic prediction.

Additionally, the fast development of online social networks and mobile computing technologies bring a new opportunity to identify users' demographic attributes. Mislove [14] found that users with common profiles were more likely to be friends and often formed a dense community. Zhong et al. [22] proposed a supervised learning framework to predict users' demographic attributes based on mobile data. Dong et al. [3] focused on micro-level analysis of the mobile networks to infer users' demographic attributes. Culotta et al. [2] fitted a regression model to predict users' demographic attributes using information on followers of each website on Twitter.

As we can see, most existing work on demographic prediction focused on designing different features for the prediction tasks. Besides, to the best of our knowledge, seldom practice has been conducted on demographic prediction based on purchase behaviors in the retail scenario.

2.2 Multi-task Approach

The advantage of multi-task approach is to improve the generalization performance by leveraging the information contained in the related tasks. A typical way of multi-task approach is to learn tasks in parallel with a shared representation [3,21,22]. Many algorithms have been proposed to solve multi-task learning tasks. For example, Micchelli et al. [13] discussed how various kernels can be used to model relations between tasks and presented linear multi-task learning algorithms. Evgeniou et al. [6] presented an approach to multi-task learning based on the minimization of regularization functionals.

2.3 Representation Learning

Learning representations of the data makes it easier to extract useful information when building classifiers or other predictors, without extracting features in a manual way. That is the reason why representation learning has attracted more and more attention and becomes a field in itself in the machine learning community.

Recently, plenty remarkable successes have been achieved based on representation learning in various applications in both academia and industry. For example, Alex Graves et al. [7] designed a deep recurrent neural network for speech recognition and obtained the best score on an evaluation benchmark. Krizhevsky et al. [10] proposed to use convolutional neural network to classify images, achieving record-breaking results. Mnih [15] proposed three graphical models to define the probability of observing next word in a sequence, leveraging distributed representations.

In this work, we propose to use the multi-task approach to learn a shared representation for demographic prediction in the retail scenario, a new application area where representation learning might be helpful, especially to the task with limited data.

3 Our Approach

In this section, we first give the motivation of our work, then we introduce the formalization of demographic prediction problem in the retail scenario. After that, we describe the proposed MTRL in detail. Finally, we present the learning procedure of MTRL.

3.1 Motivation

Obviously, a fundamental problem in demographic prediction based on users' behavioral data is how to represent users. Many existing work investigated different types of human defined features [3,17,22]. However, defining features manually costs time since expertise knowledge is required and one has to do the same process repeatedly. Moreover, human defined features may often suffer from under specification since it is difficult to identify those hidden complicated factors for prediction tasks. Recent work mainly employs unsupervised feature learning methods [8,12,23], like Singular Vector Decomposition (SVD), to automatically extract low-dimension features from the raw data. However, the features learned in an unsupervised manner may not be suitable for the prediction tasks. Therefore, concerning the weakness of extracting features humanly, in this paper we proposed to automatically learn representations of users for demographic prediction through a supervised method. Furthermore, some attributes are difficult to obtain(for example, only 8.96 % of users offer their education background in the BeiRen dataset we used). Thus the sparseness of data aggravates the difficulty of modeling the task separately [2,12,23]. In addition, modeling the tasks independently may ignore the correlations among these attributes.

Motivated by all these issues, inspired by [11], in this paper we propose a multi-task approach to learn a general representation to predict users' demographic attributes.

3.2 Problem Formalization

In this work, we try to predict multiple demographic attributes given users' behavioral data in the retail scenario. Specifically, each user can be characterized by his/her purchase history, i.e., a set of items. The demographic attributes we are interested include gender, marital status, and education ground, which are valuable signals for market analysis. The values of each attribute take are shown in Table 1. Given a user, based on his/her purchase history, we want to predict all the unknown attributes.

Table 1. List of demographic attributes

Demographic attributes	Values
gender	male, female
marital status	single, married
education background	doctor, master, bachelor, college, high school, middle school

Specifically, let $T = \{t_1, t_2, \ldots, t_n\}$ be a set of demographic prediction tasks (i.e., predicting demographic attributes). Let U be a set of users. Suppose the training set is composed of M instances, i.e.,

$$\{(x_{(1)}, y_{(1)}), (x_{(2)}, y_{(2)}), \ldots, (x_{(M)}, y_{(M)})\}$$

where $x_{(i)} \in X$ is a d-dimensional feature vector, representing the input of i-th user, and $y_{(i)}$ is the set of attribute labels of the i-th user. Note here $y^t_{(i)}$ denotes all the attribute labels under the t-th task $t \in T$ for the i-th user.

Based on the notations defined above, we try to learn a function to predict the unknown demographic attributes.

3.3 Multi-task Representation Learning Model

In this section, we now present the proposed MTRL model in detail. The feed-forward MTRL is shown in Fig. 1. In the retail scenario, each user is characterized

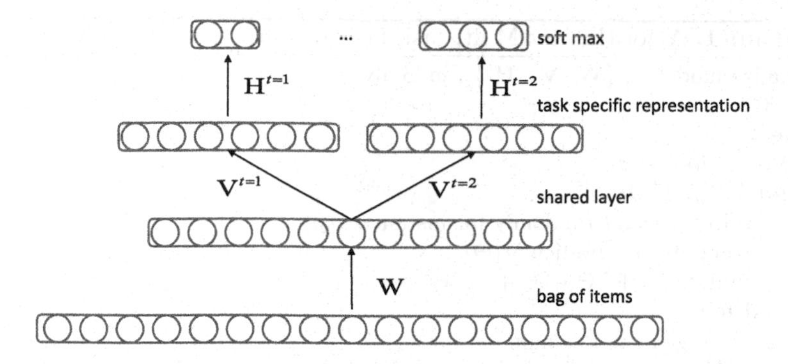

Fig. 1. The structure of Multi-task Representation Learning (MTRL) model. The lower two layers are shared across all the tasks, while top layers are task-specific. The input is represented as a bag of items. Then a non-linear projection **W** is used to generate a shared representation. Finally, for each task, additional non-linear projection **V** generates task-specific representations.

by his/her purchase history, i.e., a set of items. In MTRL, we take the bag-of-item representation as the user input $x_{(i)}$, then the shared layer is fully connected to the input layer with weights Matrix $\mathbf{W} = [w_{h,s}]$:

$$\mathcal{Y}_{(i),s} = f(\sum_h w_{h,s} \cdot x_{(i),h})$$

where matrix **W** is responsible for generating the cross-task representation, $\mathcal{Y}_{(i),s}$ is the value of s-th node on the shared layer, $x_{(i),h}$ is the h-th value of $x_{(i)}$, and $f(z)$ is the logistic nonlinear activation:

$$f(z) = \frac{1}{1+e^{-z}}$$

Based on the shared layer, for each task t, we use a transformation $V^t = [v_{s,j}]$ to map the shared representation into the task-specific representation by:

$$\mathcal{Y}^t_{(i),j} = f(\sum_s v^t_{s,j} \cdot \mathcal{Y}_{(i),s})$$

where t denotes the different tasks(gender, marital status, and education background), and $\mathcal{Y}^t_{(i),j}$ is the value of j-th node corresponding to the specific representation layer of task t.

After these, we use a softmax activation function to calculate the value of k-th node in the output layer:

$$\mathcal{Y}^t_{(i),k} = \frac{exp(\sum_j h^t_{j,k} \cdot \mathcal{Y}^t_{(i),j})}{\sum exp(\sum_j h^t_{j,k} \cdot \mathcal{Y}^t_{(i),j})}$$

where $H^t = [h^t_{j,k}]$ is the matrix that maps task specific representation to the output layer for task t, the k-th node in the output layer corresponds the value of the k-th label in task t.

Algorithm 1. Algorithm for Multi-task Learning Representation Model

1: Initialize model Θ: $\{\mathbf{W}, \mathbf{V}^t, \mathbf{H}^t\}$ randomly
2: iter=0
3: **repeat**
4:     $iter \leftarrow iter + 1$;
5:     **for** i=1,...,M **do**
6:         select a task t randomly for instance $x_{(i)}$
7:         compute the gradient$\nabla(\Theta)$
8:         update model $\Theta \leftarrow \Theta + \epsilon\nabla(\Theta)$
9:     **end for**
10: **until** converge or iter > num
11: **return** $\mathbf{W}, \mathbf{V}, \mathbf{H}$

The objective function of MTRL is then defined as the cross-entropy over the outputs of all the users and all tasks:

$$\ell_{MTRL} = \sum_t \sum_i \sum_k d^t_{(i),k} \ln \mathcal{Y}^t_{(i),k} + (1 - d^t_{(i),k}) \ln(1 - \mathcal{Y}^t_{(i),k}) - \lambda \|\Theta\|^2_F \quad (1)$$

where $d^t_{(i),k}$ is the real value of k-th node for user i under the task t, for example, for task t, if user i choose the k-th label, then $d^t_{(i),k} = 1$, else $d^t_{(i),k}$ equals 0. λ is the regularization constant and Θ are the model parameters (i.e. $\Theta = \{\mathbf{W}, \mathbf{V}^t, \mathbf{H}^t\}$).

3.4 Learning and Prediction

In order to learn parameters of MTRL model, we use the stochastic gradient decent algorithm, as shown in Algorithm 1. For each iteration, a task t is randomly selected, and parameters of the model is updated according to the task-specific objective.

With the learned parameters, the demographic prediction task is as follows. For each demographic task, we calculate the value of output layer through MTRL, and then the output node with the largest value is regarded as the label to the given user.

4 Experiments

In this section, we conduct empirical experiments to demonstrate the effectiveness of our proposed MTRL model on demographic attribute prediction in the retail scenario. We first introduce the experimental settings. Then we compare our MTRL model with the baseline methods to demonstrate the effectiveness of predicting users' demographic attributes in the retail scenario.

4.1 Experimental Settings

In this section, we introduce the experimental settings including the dataset, baseline methods, and evaluation metrics.

Dataset. We conduct our empirical experiments over a real world large scale retail dataset, named BeiRen dataset. This dataset comes from a large retailer[2] in China, which records its supermarket purchase histories during the period from 2012 to 2013. For research purpose, the dataset has been anonymized with all the users and items denoted by randomly assigned IDs for the privacy issue.

We first conduct some pre-process on the BeiRen dataset. We randomly collected 100000 users. We extract all the transactions related to these users to form their purchase histories, then we remove all the items bought by less than 10 times and the users with no labels. After pre-processing, the dataset contains 64097 distinct items and 80540 distinct users with at least on demographic attribute. In average, each user has bought about 225.5 distinct items.

Baseline Methods. We evaluate our model by comparing with several state-of-the-art methods on demographic attribute prediction:

- BoI-Single: Each user is represented by the items he/she has purchased with the Bag-of-Item representation and a logistic model[3] is learned to predict each demographic attribute separately.
- SVD-single: A singular value decomposition (SVD)[4] is first conducted over the user-item matrix to obtain low dimensional representations of users. Then a logistic model is learned over the low dimensional representation to predict each demographic attribute separately. This method has been widely used in the demographic attribute prediction task [8,16,23].
- SL: The Single Representation Learning model, which is a special case of MTRL when there is only one single task to learn. SL model has the same neural structure comparing with MTRL, just without considering the relationships among tasks.

For SVD-single method, we run several times with random initialization by setting the dimensionality as 200. For MTRL and SL, we set the dimensionality of shared representation layer and task-specific representation layer as 200 and 100 respectively. The parameters are initialized with uniform distribution in the range $(-\sqrt{6/(fan_{in}+fan_{out})}, \sqrt{6/(fan_{in}+fan_{out})})$.

For each experiment, we run several times, we then compare the best results of different methods and demonstrate the results in the following sections.

Evaluation Metrics. We follow the idea in [3] to use weighted F1 as an evaluation metric. For task t, the weighted F1 is computed as follows:

$$\text{wPrecision} = \sum_{y \in t} \frac{\sum_i I(y^t_{(i)} = \mathcal{Y}^t_{(i)} \& y^t_{(i)} = y)}{\sum_i I(y^t_{(i)} = y)} \cdot \frac{\sum_i I(\mathcal{Y}^t_{(i)} = y)}{|U|}$$

[2] http://www.brjt.cn/.
[3] http://www.csie.ntu.edu.tw/~cjlin/liblinear/.
[4] http://tedlab.mit.edu/~dr/SVDLIBC/.

$$\text{wRecall} = \frac{1}{|U|}\sum_{i} I(\mathcal{Y}^{t}_{(i)} = y^{t}_{(i)})$$

$$\text{wF1} = \frac{2 \times \text{wPrecision} \times \text{wRecall}}{\text{wPrecision} + \text{wRecall}}$$

where $\mathcal{Y}^{t}_{(i)}$ represents the calculated label of user i under task t, $I(\cdot)$ is an indicator function. Note here we use the weighted evaluation metrics because every class in task gender, marital status and education is as important as each other. As we can see, the weighted recall is the prediction accuracy in the user view.

The performance of different methods is shown in Fig. 2. We have the following observations:

- (1) Using SVD to obtain low-dimension representations of users can achieve a better performance than BoI on predicting each demographic attribute of users. This result is quite accordance with the previous finds [12,23].

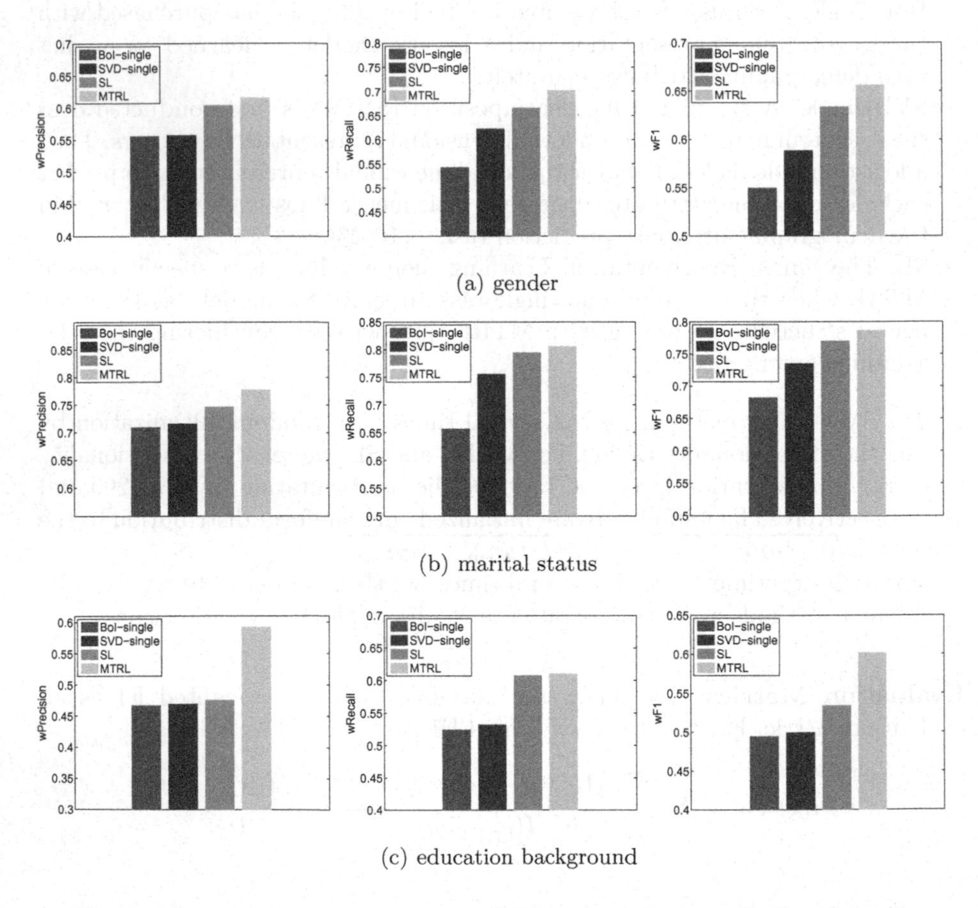

Fig. 2. The performance comparison of different methods on BeiRen dataset.

- (2) Both the deep models SL and MTRL perform better than SVD-single. The result demonstrates that the deep model can learn a better representation comparing with the shallow one(here we regard SVD-single as a shallow model).
- (3) MTRL can improve the performance of each demographic prediction task significantly, especially to the education prediction task with limited data.
- (4) By using a multi-task approach to learn a shared representation layer across tasks, we can obtain a better performance than SL, which proves that the correlations among demographic attributes are helpful. MTRL can achieve the best performance in terms of all the evaluation measures, for example, when compared with the second best method(SL), the improvement of weighted F1-Measure on gender, marital status, and education background is 2.6 %, 1.6 %, and 6.4 % respectively. By conducting the significant test, we find that the improvement of MTRL over the SL method is significant (*p-value* < 0.01) in terms of all the evaluation metrics.

To further investigate the performance of different methods, we split the users into three groups (i.e. inactive, medium and active) based on their activeness and conducted the comparisons on different user groups. We treat the user as inactive if there are less than 100 items in his/her purchase history, and active if there are more than 500 items in the purchase history. The remaining users are taken as medium. In this way, the proportions of inactive, medium and active are 45.1 %, 42.9 %, and 12.0 % respectively. The results are shown in Table 2.

From the results we can see that, not surprisingly, the BoI-single method is still the worst on all the groups. Furthermore, by reducing user representations into a low dimensional space, SVD-single, SL and MTRL can performance better than BoI-single in all groups. Finally, MTRL can achieve the best performances on most groups in terms of all the measures. The results demonstrate

Table 2. Performance comparison of different methods in terms of wF1 on BeiRen over different user groups.

user activeness	method	Gender	Martial Status	Education Background
unactive	BoI-single	0.522	0.660	0.403
	SVD-single	0.571	0.729	0.401
	SL	0.614	**0.747**	0.413
	MTRL	**0.678**	0.729	**0.415**
medium	BoI-single	0.589	0.686	0.506
	SVD-single	0.591	0.754	0.536
	SL	0.634	0.768	0.558
	MTRL	**0.645**	**0.802**	**0.647**
active	BoI-single	0.587	0.691	0.523
	SVD-single	0.568	**0.742**	0.526
	SL	0.646	0.716	0.533
	MTRL	**0.658**	0.732	**0.628**

that by learning a general representation using a multi-task approach, we can enjoy the relationships among demographic attributes, and complement each other to achieve a better performance.

5 Conclusion

In this paper, we try to predict users' demographic attributes given users' purchase behaviors. We propose a robust and practical representation learning algorithm MTRL based on multi-task objectives. Our MTRL can learn a shared representation across tasks, thus the sparseness problem can be avoided, especially for the task with limited data. Experiments on real-world purchase dataset demonstrate that our model can outperform the state-of-the-art baselines consistently under different evaluation metrics.

Although the MTRL model is proposed in this retail scenario, it is in fact a general model which can be applied on other multi-task multi-class problems. In the future, we would like to extend the usage of our MTRL model to model more demographic attributes to verify its effectiveness. Moreover, in this paper, we represent each user by simple bag of items as the raw input. It would be interesting to further explore the natural transaction structures in users' purchase data for a better demographic prediction.

Acknowledgment. This research work was funded by 863 Program of China award number under Grant 2014AA015204, 973 Program of China award number under Grant 2014CB340401, 2012CB316303, National Natural Science Foundation of China award numbers under Grant 61472401, 61433014, 61203298, 61425016, and Key Research Program of the Chinese Academy of Sciences under Grant NO.KGZD-EW-T03-2, and the Youth Innovation Promotion Association CAS under Grant no.20144310, and the Technology Innovation and Transformation Program of Shandong (Grant No.2014CGZH1103).

References

1. Bi, B., Shokouhi, M., Kosinski, M., Graepel, T.: Inferring the demographics of search users: Social data meets search queries. In: Proceedings of the 22nd International Conference on World Wide Web, pp. 131–140 (2013)
2. Cutler, J., Culotta, A., Ravi, N.K.: Predicting the demographics of twitter users from website traffic data. In: ICWSM, in press. AAAI Press, Menlo Park, California (2015)
3. Dong, Y., Yang, Y., Tang, J., Yang, Y., Chawla, N.V.: Inferring user demographics and social strategies in mobile social networks. In: Proceedings of the 20th ACM SIGKDD International Conference on Knowledge Discovery and Data Mining, pp. 15–24 (2014)
4. Torres, S.D., Weber, I.: What and how children search on the web. In: Proceedings of the 20th ACM International Conference on Information and Knowledge Management, CIKM 2011, pp. 393–402. ACM, New York (2011)
5. Eckert, P.: Gender and sociolinguistic variation. In: Readings in Language and Gender (1997)

6. Evgeniou, T., Pontil, M.: Regularized multi-task learning. In: Proceedings of the Tenth ACM SIGKDD International Conference on Knowledge Discovery and Data Mining, pp. 109–117. ACM (2004)
7. Hinton, G., Graves, A., Mohamed, A.: Speech recognition with deep recurrent neural networks. In: 2013 IEEE International Conference on Speech and Signal Processing (ICASSP) (2013)
8. Hu, J., Zeng, H.-J., Li, H., Niu, C., Chen, Z.: Demographic prediction based on user's browsing behavior. In: Proceedings of the 16th International Conference on World Wide Web, pp. 151–160. ACM (2007)
9. Putler, D.S., Kalyanam, K.: Incorporating demographic variables in brand choice models: An indivisible alternatives framework. Mark. Sci. **16**(2), 166–181 (1997)
10. Krizhevsky, A., Sutskever, I., Hinton, G.E.: Imagenet classification with deep convolutional neural networks. In: Advances in Neural Information Processing Systems (2012)
11. Liu, X., Gao, J., He, X., Deng, L., Duh, K., Wang, Y.-Y.: Representation learningusing multi-task deep neural networks for semantic classification and informationretrieval. In: Proceedings of the 2015 Conference of the North American Chapterof the Association for Computational Linguistics: Human Language Technologies (2015)
12. Stillwell, D., Kosinski, M., Graepel, T.: Private traits and attributes are predictable from digital records of human behavior. In: Proceedings of the National Academy of Sciences (2013)
13. Micchelli, C., Pontil, M.: Kernels for multi-task learning. In: NIPS (2005)
14. Mislove, A., Viswanath, B., Gummadi, K.P., Druschel, P.: You are who you know: Inferring user profiles in online social networks. In: WSDM, pp. 251–260 (2010)
15. Mnih, A., Hinton, G.: Three new graphical models for statistical language modelling. In: Proceedings of the 24th International Conference on Machine Learning, pp. 641–648 (2007)
16. Murray, D., Durrell, K.: Inferring demographic attributes of anonymous internet users. In: Masand, B., Spiliopoulou, M. (eds.) WebKDD 1999. LNCS (LNAI), vol. 1836, pp. 7–20. Springer, Heidelberg (2000)
17. Otterbacher, J.: Inferring gender of movie reviewers: exploiting writing style, content and metadata. In: Proceedings of the 19th ACM International Conference on Information and Knowledge Management, CIKM 2010, pp. 369–378. ACM, New York (2010)
18. Currim, I.S., Andrews, R.L.: Identifying segments with identical choice behaviors across product categories: an intercategory logit mixture model. Int. J. Res. Mark. **19**(1), 65–79 (2002)
19. Schler, J., Koppel, M., Argamon, S., Pennebaker, J.W.: Effects of age, gender on blogging. In: AAAI Spring Symposium: Computational Approaches to Analyzing Weblogs, pp. 199–205. AAAI (2006)
20. Sedhain, S., Sanner, S., Braziunas, D., Xie, L., Christensen, J.: Social collaborative filtering for cold-start recommendations. In: Proceedings of the 8th ACM Conference on Recommender Systems, pp. 345–348 (2014)
21. Sun, S., Ji, Y.: Multitask multiclass support vector machines. In: Data Mining Workshops (ICDMW) (2011)
22. Zhong, E., Tan, B., Mo, K., Yang, Q.: User demographics prediction based on mobile data. Pervasive Mob. Comput. **9**(6), 823–837 (2013)
23. Zhong, Y., Yuan, N.J., Zhong, W., Zhang, F., Xie, X.: You are where you go: Inferring demographic attributes from location check-ins. In: WSDM 2015, pp. 295–304. ACM, New York (2015)

Large-Scale Kernel-Based Language Learning Through the Ensemble Nyström Methods

Danilo Croce[(✉)] and Roberto Basili

Department of Enterprise Engineering, University of Roma, 00133 Roma, Tor Vergata, Italy
{croce,basili}@info.uniroma2.it

Abstract. Kernel methods have been used by many Machine Learning paradigms, achieving state-of-the-art performances in many Language Learning tasks. One drawback of expressive kernel functions, such as Sequence or Tree kernels, is the time and space complexity required both in learning and classification. In this paper, the Nyström methodology is studied as a viable solution to face these scalability issues. By mapping data in low-dimensional spaces as kernel space approximations, the proposed methodology positively impacts on scalability through compact linear representation of highly structured data. Computation can be also distributed on several machines by adopting the so-called Ensemble Nyström Method. Experimental results show that an accuracy comparable with state-of-the-art kernel-based methods can be obtained by reducing of orders of magnitude the required operations and enabling the adoption of datasets containing more than one million examples.

1 Introduction

Kernel methods [24] have been employed in many Machine Learning algorithms [5,25] achieving state-of-the-art performances in many tasks. One drawback of expressive but complex kernel functions, such as Sequence [2] or Tree kernels [4], is the time and space complexity of the learning and classification phases, that may prevent their adoption when large data volumes are involved. While important steps have been made forward in defining linear algorithms [10,16,22,23], the adoption of complex kernels is still limited. Some approaches have been defined to scale with kernel base methods, such as [11,13,26,28], but still specific to kernel formulations and learning algorithms.

A viable solution to scalability issues is the Nyström methodology [29] that maps original data in low-dimensional spaces and can be applied to the implicit space determined by the kernel function. These linear representations thus enable the application of scalable and performant learning methods, by capitalizing the existing large literature on linear methods. The idea is to use kernels to decouple representation of complex problems from the learning, and making use of the Nyström dimensionality reduction method to derive a linear mapping effectively. At the best of our knowledge, it is the first time this perspective is pursued in the area of language learning acting on discrete linguistic structures whose kernels

N. Ferro et al. (Eds.): ECIR 2016, LNCS 9626, pp. 100–112, 2016.
DOI: 10.1007/978-3-319-30671-1_8

have been largely discussed [6]. In [30] a different solution, namely Distributed Tree Kernel, has been proposed to approximate standard tree kernels [4] by defining an explicit function mapping trees to vectors. However, DTKs are designed to approximate specific tree kernel functions, while the proposed Nyström method can be applied to any kernel function.

In a nutshell, the Nyström method allows mapping a linguistic instance into a low-dimensional dense vector with up to l dimensions. Here, the representation of an instance o is obtained by selecting a set of l training instances, so-called *landmarks*, and the cost of projecting o is essentially $\mathcal{O}(lk)$, where k is the cost of a single kernel computation over linguistic objects such as o. This cost has been theoretically bounded [9] and the linguistic quality of the resulting space depends on the number of selected landmarks characterizing that space. The overall approach is highly applicable, without a bias toward input data, adopted kernels or learning algorithms. Moreover, the overall computational cost can be easily distributed across several machines, by adopting the Ensemble Nyström Method, presented in [17] as a possible learning scheme. In this variant, several representations of an example are created by selecting p subsets of landmarks. An approximation of the target kernel space is obtained by a linear combination of different spaces, acquired separately. A crucial factor influencing the scalability of our method is the cost of creating linear representations for complex (i.e. tree) structures. When no caching scheme is adopted, linear mappings should be invoked several times. Among the algorithms that bound the number of times a single instance is re-used during training, we investigated the Dual Coordinate Descent algorithm [15]: it is a batch learning algorithm whose achievable accuracy is made inversely dependent on the number of iterations over a training dataset. Online schemes are also very appealing, as they avoid to keep an entire dataset in memory; we also investigated the Soft Confidence-Weighted Learning [27], an online learning algorithm that shows state-of-the-art accuracy against most online algorithms. An experimental investigation on the impact of our learning methodology has been carried out: we adopted different robust and scalable algorithms over two kernel-based language learning tasks, i.e. Question Classification (QC) and Argument Boundary Detection (ABD) in Semantic Role Labeling. The compact linear approximations produced by our method achieve results comparable with their full kernel-based counterparts, by requiring a negligible fraction of the kernel computations w.r.t. standard methods. Moreover, we trained a kernel-based ABD classifier over about 1.4 millions of examples, a size that was hardly tractable before. In the rest of the paper, the adopted methodology is discussed in Sect. 2, the large-scale learning algorithms are presented in Sect. 3, while the empirical evaluation is discussed in Sect. 4.

2 Linearizing Kernel Functions

The Nyström method [29] allows reducing the computational cost of kernel-based learning algorithms by providing an approximation of the Gram Matrix underlying the used kernel function.

The Standard Nyström Method. Given an input training dataset $o_i \in \mathcal{D}$, a kernel function $K(o_i, o_j)$ allows surrogating the dot-product in an implicit space derived by applying a projection function $\boldsymbol{x}_i = \Phi(o_i) \in \mathbb{R}^n$. By applying the projection function over $\mathcal{D}$ we can derive a new representation $\boldsymbol{x}_i \in X$ and define the Gram Matrix as $G = XX^\top$, with the single element $G_{ij} = \Phi(o_i)\Phi(o_j) = K(o_i, o_j)$. Kernel-based methods [24] are appealing as they can be applied in these new spaces without requiring the $\Phi(\cdot)$ function, but only using the notion of dot-product between examples, i.e. $K(\cdot, \cdot)$. Moreover, the dimensionality n of the space underlying the kernel function can be very high, as for Tree Kernels [4] for which the computation of $\Phi(\cdot)$ may be prohibitive.

The aim of the Nyström method is to derive a new low-dimensional representation $\tilde{X} \in \mathbb{R}^l$, with $l \ll n$ so that $G \approx \tilde{G} = \tilde{X}\tilde{X}^\top$. This is obtained by generating an approximations of G using a subset of l columns of the matrix, i.e. a subset of examples $L \subseteq X$, called landmarks. Suppose we randomly sample l columns of G uniformly without replacement, and let C be the $n \times l$ matrix of these sampled columns. Then we can rearrange the columns and rows of G and define $X = [X_1 \;\; X_2]$ such that:

$$G = XX^\top = \begin{bmatrix} W & X_1^\top X_2 \\ X_2^\top X_1 & X_2^\top X_2 \end{bmatrix} \text{ and } C = \begin{bmatrix} W \\ X_2^\top X_1 \end{bmatrix} \tag{1}$$

where $W = X_1^\top X_1$. The Nyström approximation can be defined as:

$$G \approx \tilde{G} = CW^\dagger C^\top \tag{2}$$

where $W^\dagger$ denotes the Moore-Penrose inverse of W. The Singular Value Decomposition (SVD) is used in order to obtain $W^\dagger$ as it follows. First W is decomposed so that $W = USV^\top$ where U and V are both orthogonal matrices, and S is a diagonal matrix containing the (non-zero) singular values of W on its diagonal. Since W is symmetric and positive definite $W = USU^\top$. Then $W^\dagger = US^{-1}U^\top = US^{-\frac{1}{2}}S^{-\frac{1}{2}}U^\top$ and the Eq. 2 can be rewritten as

$$G \approx \tilde{G} = CUS^{-\frac{1}{2}}S^{-\frac{1}{2}}U^\top C^\top = (CUS^{-\frac{1}{2}})(CUS^{-\frac{1}{2}})^\top = \tilde{X}\tilde{X}^\top \tag{3}$$

Given an input example $\boldsymbol{o_i} \in \mathcal{D}$, a new low-dimensional representation $\tilde{\boldsymbol{x}}_i$ can be thus determined by considering the corresponding i-th item of C as

$$\tilde{\boldsymbol{x}}_i = \boldsymbol{c_i} US^{-\frac{1}{2}} \tag{4}$$

where $\boldsymbol{c_i}$ corresponds to a vector whose dimensions contain the evaluation of the kernel function between o_i and each landmark $o_j \in L$. While the method produce up to l-dimensional vector, no restriction is applied on the input dataset as long as a valid $K(\cdot, \cdot)$ is used. The Nyström method is often presented with an additional step whereby $W^\dagger$ in Eq. 2 is replaced by its rank-k approximation, $W_k^\dagger$, for some $k < l$, thus generating $\tilde{G}^k$, the rank-k Nyström approximation to G. Although the approximation of $\tilde{G}$ provided by $\tilde{G}^k$ is weaker but can be further improved, as discussed in [9], it is not applied in this work and is left for future work. Several policies have been defined to determine the best selection of landmarks to reduce the Gram Matrix approximation error. In this work the uniform sampling without replacement is adopted, as suggested by [18], where this policy has been theoretically and empirically shown to achieve results comparable with other (and more complex) selection policies.

The Ensemble Nyström Method. In order to minimize the bias introduced by the policy of landmark selection on the approximation quality, we apply a redundant approach, called Ensemble Nyström Method presented in [17]: the main idea is to treat each approximation generated by the Nyström method through a sample of l columns as an "*expert*" and to combine $p \geq 1$ such experts to derive an improved hypothesis, typically more accurate than any of the original experts. The Ensemble Nyström Method presented in [17] selects a collection of p samples, each sample containing l columns of W. The ensemble method combines the samples to construct an approximation in the form of

$$G_{m,p}^{ens} = \sum_{i=1}^{p} \lambda^{(i)} C^{(i)} W^{(i)\dagger} C^{(i)\top} \tag{5}$$

where $\lambda^{(i)}$ reflect the confidence of each expert, with $\sum_{i=1}^{p} \lambda^{(i)} = 1$. Typically, the ensemble Nyström method seeks to find out the weights by minimizing $||G - G^{ens}||_2$. A simple but effective strategy is to set the weights as $\lambda^{(1)} = \cdots = \lambda^{(t)} = \frac{1}{p}$ as shown in [17]. More details about the upper bound on the norm-2 error of the Nyström approximation $\|G - \tilde{G}\|_2 / \|G\|_2$ are reported in [9,17]. In practice, each expert is developed through a projection function that takes an instance o and its vector $\phi(o) = x$ and through Eq. 4 maps it into a l-dimensional vector $\tilde{x}^{(i)}$ according to the i-th independent sample, i.e. the choice of landmarks $L^{(i)}$: as a result we have p such vectors $\tilde{x}^{(i)}$ as $i = 1, ..., p$. A unified representation $\tilde{x}$ for the source instance o is thus derived through the simple concatenation of the different $\tilde{x}^{(i)}$: these are exactly p so that the overall dimensionality of $\tilde{x}$ is pl.

Complexity. The runtime of the Nyström method is $\mathcal{O}(l^3 + nl^2)$ as it depends on the SVD evaluation on W (i.e. $\mathcal{O}(l^3)$) and on the projection of the entire dataset through the multiplication by C (i.e. $\mathcal{O}(nl^2)$). The complexity of the Ensemble method is $O\big(p(l^3 + nl^2)\big)$. This analysis supposes that the kernel function has a cost comparable to the other operations. For several classes of kernels, such as Tree or Sequence Kernels [4], the above cost can be assumed negligible with respect to the cost of building vectors $\boldsymbol{c_i}$. Under this assumption, the computation cost is $\mathcal{O}(kl)$ with k the cost of a single kernel computation. Regarding the Ensemble method, it is worth noting that the construction of each projection can be distributed. The space complexity to derive $W^\dagger$ is $\mathcal{O}(l^2)$ while the projection of a dataset of size d is $\mathcal{O}(ld)$. In the Ensemble setting, the space complexity is $\mathcal{O}(pl^2)$ while the projection of a dataset of size d is $\mathcal{O}(pld)$.

3 Learning Algorithms for Linear Embeddings

In our language learning perspective, two training paradigms are investigated against the linear representations proposed in this work. In the *batch* learning paradigm, the complete training dataset is supposed to be entirely available during the learning phase. It means that the overall training set can be represented in the reduced space before training. The *online* learning paradigm differs from batch learning as individual examples are exploited as soon as they are available: it is appealing as the storage of the entire linearized dataset can be avoided. In the following, three linear learning algorithms are investigated over the representation obtained by the Ensemble Nyström approach thus resulting in large-scale kernel-based classification. Regardless of the algorithm, the classification of a single example corresponds to a dot-product in the l-dimensional space.

Dual Coordinate Descent Learning Algorithms. The Dual Coordinate Descent (DCD), defined in [15] is a batch and linear learning algorithm similar to the Support Vector Machine (SVM). Given d instances $o_i \in \mathcal{D}$, their labels $y_i \in \pm 1$ and their corresponding $\tilde{\boldsymbol{x}}_i \in \mathbb{R}^l$ low-dimensional counterparts obtained applying Eq. 4, the DCD acquires the function $f : X \rightarrow R$ which minimizes the misclassification error, e.g. by minimizing the probability that $y_i f(\tilde{\boldsymbol{x}}_i) = y_i \boldsymbol{w} \tilde{\boldsymbol{x}}_i \leq 0$, i.e. a binary classification function. The so-called primal formulation to determine $\boldsymbol{w}$ can be written as follows:

$$\underset{\boldsymbol{w} \in R^l}{\text{minimize}} \ \frac{1}{2}\|\boldsymbol{w}\|^2 + C \sum_{i=1}^{d} \max\{0, 1 - y_i \boldsymbol{w}^\top \tilde{\boldsymbol{x}}_i\} \tag{6}$$

The above problem can be rewritten in its dual form

$$\begin{aligned} &\underset{\alpha}{\text{minimize}} \ D(\alpha) := \frac{1}{2}\alpha^\top Q \alpha - \alpha^\top \mathbb{1} \\ &\text{subject to } 0 \leq \alpha \leq C\mathbb{1} \end{aligned} \tag{7}$$

Here, Q is an $d \times d$ matrix whose entries are given by $Q_{ij} = y_i y_j \tilde{\boldsymbol{x}}_i^\top \tilde{\boldsymbol{x}}_j$, and $\mathbb{1}$is the vector of all ones. The minimizer $\boldsymbol{w}^*$ of Eq. 6 and the minimizer α^* of Eq. 7 are related by the primal/dual connection: $\boldsymbol{w}^* = \sum_{i=1}^{d} \alpha_i^* y_i \tilde{\boldsymbol{x}}_i$. The dual problem in Eq. 7 is a Quadratic Program (QP) with box constraints, and the i-th coordinate α_i corresponds to the i-th instance $(\tilde{\boldsymbol{x}}_i, y_i)$.

According to [15], the following coordinate descent scheme can be used to minimize Eq. 7:

- Initialize $\alpha^1 = (0, \ldots, 0)$
- At iteration t select coordinate i_t
- Update α^t to α^{t+1} via

$$\begin{aligned} \alpha_{i_t}^{t+1} &= \underset{0 \leq \alpha_{i_t} \leq C}{\text{argmin}} \ D(\alpha^t + (\alpha_{i_t} - \alpha_{i_t}^t) e_{i_t}) \\ \alpha_i^{t+1} &= \alpha_i^t \ \forall i \neq i_t. \end{aligned} \tag{8}$$

Here, $\boldsymbol{e}_i$ denotes the i-th standard basis vector. Since $D(\alpha)$ is a QP, the above problem can be solved exactly:

$$\alpha_{i_t}^{t+1} = \min\{\max\{0, \alpha_{i_t}^t - \frac{\nabla_{i_t} D(\alpha^t)}{Q_{i_t} Q_{i_t}}\}, C\}. \quad (9)$$

Here, $\nabla_i D(\alpha)$ denotes the i_t-th coordinate of the gradient. The above updates are also closely related to implicit updates. If we maintain $\boldsymbol{w}^t := \sum_i^d \alpha_i^t y_i \boldsymbol{x}_i$, then the gradient $\nabla_{i_t} D(\alpha)$ can be computed efficiently using

$$\nabla_{i_t} D(\alpha) = \boldsymbol{e}_{i_t}^\top (Q\alpha - 1) = \boldsymbol{w}^t y_{i_t} \tilde{\boldsymbol{x}}_{i_t} - 1 \quad (10)$$

and kept related to α^{t+1} by computing $\boldsymbol{w}^{t+1} = \boldsymbol{w}^t + (\alpha_i^{t+1} - \alpha_i^t) y_i \tilde{\boldsymbol{x}}_i$. In each iteration, the entire dataset is used to optimize Eq. 6 and a practical choice is to randomly access examples. In [15], the proposed method is shown reaching an ϵ-accurate solution in $O\big(log(1/\epsilon)\big)$ iterations, so we can bound the number of iterations in order to fix a-priori also the computation cost of the training time.

Passive Aggressive. The Passive Aggressive (PA) learning algorithm [5] is one of the most popular online approaches. When an example is misclassified, the model is updated with the hypothesis most similar to the current one, among the set of classification hypotheses that correctly classify the example.

More formally, let $(\tilde{\boldsymbol{x}}_t, y_t)$ be the t-th example where $\tilde{\boldsymbol{x}}_t \in \mathbb{R}^l$ is a feature vector in a l-dimensional space and $y_t \in \pm 1$ is the corresponding label. Let $\boldsymbol{w}_t \in \mathbb{R}^l$ be the current classification hypothesis. As for the DCD, the PA classification function is linear, i.e. $f(\tilde{\boldsymbol{x}}) = \boldsymbol{w}^\top \tilde{\boldsymbol{x}}$. The learning procedure starts setting $\boldsymbol{w}_1 = (0, \ldots, 0)$, and after receiving $\tilde{\boldsymbol{x}}_t$, the new classification function $\boldsymbol{w}_{t+1}$ is the one that minimizes the following objective function[1]:

$$Q(\boldsymbol{w}) = \frac{1}{2}\|\boldsymbol{w} - \boldsymbol{w}_t\|^2 + C \cdot l(\boldsymbol{w}; (\tilde{\boldsymbol{x}}_t, y_t))^2 \quad (11)$$

where the first term $\|\boldsymbol{w} - \boldsymbol{w}_t\|$ is a measure of how much the new hypothesis differs from the old one, while the second term $l(\boldsymbol{w}, (\tilde{\boldsymbol{x}}_t, y_t))$ is a proper loss function[2] assigning a penalty cost to an incorrect classification. C is the aggressiveness parameter that balances the two competing terms in Eq. 11. Minimizing $Q(\boldsymbol{w})$ corresponds to solving a constrained optimization problem, whose closed form solution is the following:

$$\boldsymbol{w}_{t+1} = \boldsymbol{w}_t + \alpha_t \tilde{\boldsymbol{x}}_t, \quad \alpha_t = y_t \cdot \frac{H(\boldsymbol{w}_t; (\tilde{\boldsymbol{x}}_t, y_t))}{\|\tilde{\boldsymbol{x}}_t\|^2 + \frac{1}{2C}} \quad (12)$$

After a wrong prediction of an example $\tilde{\boldsymbol{x}}_t$, a new classification function $\boldsymbol{w}_{t+1}$ is computed. It is the result of a linear combination between the old $\boldsymbol{w}_t$ and the feature vector $\tilde{\boldsymbol{x}}_t$. The PA is extremely attractive as linearized example can be incrementally derived and used in the training process and, again,

[1] We are referring to the PA-II version in [5].

[2] In this work we will consider the hinge loss $H(\boldsymbol{w}; (\tilde{\boldsymbol{x}}_t, y_t)) = max(0, 1 - y_t \boldsymbol{w}^\top \tilde{\boldsymbol{x}}_t)$.

the classification and the updating steps have a computational complexity of $\mathcal{O}(l)$, i.e. a single dot product in the l-dimensional space.

Soft Confidence-Weighted Learning (SCW). The Soft Confidence-Weighted (SCW) online learning scheme [27] is a specific implementation of the family of the Confidence-Weighted (CW) learning methods [8]. This class of learning methods maintains a probability for each dimension of the representation space, i.e. the confidence on the contribution of each individual dimension. To better explore the underlying topological structure within the feature space, the CW algorithm assumes that the weight vector $\boldsymbol{w}$ follows a Gaussian distribution $\boldsymbol{w} \sim \mathcal{N}(\boldsymbol{\mu}, \Sigma)$ with mean vector $\boldsymbol{\mu} \in \mathbb{R}^l$ and covariance matrix $\Sigma \in \mathbb{R}^{l \times l}$. Less confident dimensions are updated more aggressively than more confident ones. Parameters of the Gaussian distribution are updated for each new training instance so that the probability of a correct classification for that instance under the updated distribution meets a specified confidence. The original Confidence-Weighted algorithm [8] is updated by minimizing the Kullback-Leibler divergence D_{KL} between the new weight distribution and the old one while ensuring that the probability of correct classification is greater than a threshold as follows:

$$\begin{aligned}(\boldsymbol{\mu}_{t+1}, \Sigma_{t+1}) &= \underset{\boldsymbol{\mu}, \Sigma}{\operatorname{argmin}}\; D_{KL}(\mathcal{N}(\boldsymbol{\mu}, \Sigma), \mathcal{N}(\boldsymbol{\mu}_t, \Sigma_t)) \\ &\text{subject to } Pr_{\boldsymbol{w} \sim (\boldsymbol{\mu}, \Sigma)}[y_t(\boldsymbol{w}^\top \tilde{\boldsymbol{x}}_t) \geq 0] \geq \eta \end{aligned} \tag{13}$$

where Σ_1 is set initially to the identity matrix and $\eta \in (0.5, 1]$ is the probability required for the updated distribution on the current instance.

Unfortunately, the CW method may adopt a too aggressive updating strategy where the distribution changes too much in order to satisfy constraints imposed by an individual input example. Although this speeds up the learning process, it could force wrong model updates (i.e. undesirable changes in the parameters of the distribution) caused by mislabeled instances. This makes the original CW algorithm to perform poorly in many noisy real-world applications. To overcome the above limitation, the Soft-Confidence extension of the standard CW learning has been proposed [27] with a more flexible handling of non-separable cases. After the introduction of the following loss function:

$$l^{\phi}\big(N(\boldsymbol{\mu}, \Sigma); (\tilde{\boldsymbol{x}}_t, y_t)\big) = \max\big(0, \phi\sqrt{\tilde{\boldsymbol{x}}_t^\top \Sigma \tilde{\boldsymbol{x}}_t} - y_t \boldsymbol{\mu} \cdot \tilde{\boldsymbol{x}}_t\big),$$

where $\phi = \Phi^{-1}(\eta)$ is the inverse cumulative function of the normal distribution, the optimization problem of the original CW can be re-written as follows [27]:

$$(\boldsymbol{\mu}_{t+1}, \Sigma_{t+1}) = \arg\min_{\boldsymbol{\mu}, \Sigma} D_{KL}\big(N(\boldsymbol{\mu}, \Sigma) \| N(\boldsymbol{\mu}_t, \Sigma_t)\big) + C l^{\phi}\big(N(\boldsymbol{\mu}, \Sigma); (\tilde{\boldsymbol{x}}_t, y_t)\big)^2$$

where C is a parameter to tradeoff the passiveness and aggressiveness. In [27], the above formulation of the Soft Confidence-Weighted algorithm is denoted by "SCW-II" for short and it employs a squared penalty. As for the PA learning algorithm, the above optimization problems can be solved in close form to determine the update weights of the hyperplane.

4 Experimental Evaluations

In the following experimental evaluations, we applied the proposed Ensemble Nyström methodology to two language learning tasks, i.e. Question Classification and Argument Boundary Detection (ABD) in Semantic Role Labeling. All the kernel functions and learning algorithms used in these experiments have been implemented and released in the KeLP framework[3] [12].

Question Classification. Question Classification (QC) is usually applied in Question Answering systems to map the question into one of k classes of answers, thus constraining the search. In these experiments, we used the UIUC dataset [19]. It is composed by a training set of 5,452 and a test set of 500 questions[4], organized in 6 classes (like `ENTITY` or `HUMAN`). It has been already shown the contribution of (structured) kernel-based learning within batch algorithms for this task, as in [31]. In these experiments the Smoothed Partial Tree Kernel (SPTK) is applied as it obtains state-of-the-art results over this task by directly acting over tree structures derived from the syntactic analysis of the questions [6]. The SPTK measures the similarity between two trees proportionally to the number of shared syntactic substructures, whose lexical nodes contribute according to a Distributional Lexical Semantic Similarity metrics between word vectors. In particular, lexical vectors are obtained through the distributional analysis of the UkWaC corpus, comprising 2 billions words, as discussed in [6]. While the learning algorithms discussed in Sect. 3 allow to acquire binary classifiers, the QC is a multi-classification task and a One-*vs*-All scheme is adopted to combine binary outcomes [21].

We acquired the linear approximation of the trees composing the dataset from 100 up to 500 dimensions. Landmarks have been selected by applying a random selection without replacement, as suggested in [18]; as the selection is random, the evaluations reported here are the mean results over ten different selection of landmarks. Moreover we applied several ensembles by using $p = 1, 2, 3$ experts. These linear approximations are used within the PA-II, SCW-II and DCD implementation of the SVM. A different numbers of iterations have been adopted for each algorithm. All the parameters of the kernel functions and the algorithms have been estimated over a development set.

In Table 1, results in terms of Accuracy, i.e. the percentage of correctly labeled examples over the test set, are reported: rows reflect the choice of p and l used in the kernel-based Nyström approximation of the tree structures. Columns reflect the learning algorithms and the number of iterations applied in the tests. The last row reports results obtained by standard kernel-based algorithms. This corresponds to a sort of upper bound of the quality achievable by the reference kernel function. In particular, the kernel-based C-SVM formulation [3] and the Kernel-based PA-II [5] are adopted. The SCW is not applied, as the kernel counterpart does not exist. The kernel-based C-SVM achieves the best result (93.8 %), comparable with the PA-II when five iterations are applied (93.4 %).

[3] http://sag.art.uniroma2.it/demo-software/kelp/.

[4] http://cogcomp.cs.illinois.edu/Data/QA/QC/.

Table 1. Results in terms of Accuracy in the QC task.

	l	PA-II			SCW-II		SVM		
		iter. 1	iter. 2	iter. 5	iter. 1	iter. 2	iter. 2	iter. 5	iter. 30
p=1	100	80.2 %	78.8 %	81.6 %	83.3 %	83.3 %	79.0 %	83.8 %	84.4 %
	200	84.6 %	84.6 %	86.4 %	87.8 %	87.8 %	85.4 %	87.7 %	88.8 %
	300	85.3 %	86.4 %	87.3 %	89.6 %	89.7 %	87.9 %	89.3 %	90.6 %
	400	86.1 %	87.4 %	88.2 %	90.3 %	90.6 %	88.9 %	90.0 %	90.9 %
	500	86.6 %	87.5 %	88.9 %	90.8 %	91.2 %	89.4 %	90.3 %	91.4 %
p=2	100	82.0 %	81.3 %	84.3 %	87.0 %	86.9 %	82.9 %	86.5 %	88.4 %
	200	84.9 %	85.0 %	87.3 %	89.4 %	90.1 %	86.8 %	89.1 %	90.4 %
	300	86.0 %	86.9 %	89.1 %	90.4 %	91.3 %	88.3 %	90.2 %	91.6 %
	400	87.0 %	87.6 %	89.5 %	90.9 %	91.6 %	89.2 %	90.8 %	91.9 %
	500	87.4 %	88.0 %	90.1 %	91.2 %	92.0 %	89.6 %	91.0 %	92.3 %
p=3	100	82.4 %	83.5 %	85.6 %	87.9 %	88.6 %	83.1 %	88.1 %	89.9 %
	200	85.8 %	85.9 %	88.5 %	89.7 %	90.5 %	87.4 %	89.7 %	91.1 %
	300	86.7 %	85.8 %	89.4 %	90.7 %	91.5 %	89.0 %	90.4 %	91.9 %
	400	87.6 %	86.4 %	89.9 %	90.6 %	91.6 %	89.6 %	90.8 %	92.3 %
	500	87.9 %	86.8 %	**90.5 %**	90.9 %	**92.0 %**	89.6 %	91.1 %	**92.4 %**
Kernel-based		86.6 %	93.2 %	**93.4 %**	-		**93.8 %**		

Linear counterparts are better when a higher number of dimensions and experts are used in the approximation. The PA-II learning algorithm seems weaker with respect to the SCW-II, while the SVM formulation achieves the best results. However, the number of iterations required from the DCD is higher with respect to the SCW-II. In fact the former requires 30 iteration (up to 92.4 %) that is approximated by the SCW-II with only 2 iterations (92.0 %). At a lower number of iterations the DCD perform worse than SCW-II. Results reported here are evaluated on the same test set used in [11,13], where best result is 91.1 % and 91.4 % respectively.

These results are remarkable as our method requires much less kernel computations, as shown in Table 2. We measured the total number of kernel operations required by all the above kernel-based settings[5], as reported in the last row of Table 2. The percentage of saved computation of the SCW-II is impressive: kernel-based methods, which achieve a comparable accuracy, require a considerable higher computational cost: the adoption of the Nyström linearization allows avoiding from 80 % to more than 95 % of the kernel computations.

Automatic Boundary Detection in Semantic Role Labeling. Semantic Role Labeling is a natural language processing task that can be defined over frame-based semantic interpretation of sentences. Frames are linguistic predicates providing a semantic description of real world situations.

[5] C-SVM [3] proposes a caching policy, here ignored for comparative purposes. Large-scale applications may impose prohibitive requirements on the required space.

Table 2. Saving of kernel operations obtained by the SCW-II compared with the kernel-based learning algorithms

		PA-II iter. 1	PA-II iter. 2	PA-II iter. 5	C-SVM
SCW-II iter. 1	100	95 %	98 %	99 %	99 %
	200	89 %	95 %	98 %	98 %
	300	84 %	93 %	98 %	97 %
	400	78 %	91 %	97 %	97 %
	500	73 %	89 %	96 %	96 %
SCW-II iter. 2	100	89 %	95 %	98 %	98 %
	200	78 %	91 %	97 %	97 %
	300	68 %	86 %	95 %	95 %
	400	57 %	82 %	94 %	93 %
	500	46 %	77 %	92 %	91 %
Kernel Comp		3.3E+07	7.9E+07	2.3E+08	2.1E+08

A frame is evoked in a sentence through the occurrence of specific *lexical units* (LU), i.e. words (such as nouns or verbs) that linguistically express the underlying situation. A frame characterizes the set of prototypical semantic roles, i.e. semantic arguments called *frame elements* (*fes*), describing all participants to the event for each lexical unit. For example, the following sentence evokes the DUPLICATION frame, through the LU "*copy*", while three *fes*, i.e. CREATOR, ORIGINAL and GOAL, are emphasized in the underlying frame: [*Bootleggers*]$_{\text{CREATOR}}$, *then* **copy** [*the film*]$_{\text{ORIGINAL}}$ [*onto hundreds of tapes*]$_{\text{GOAL}}$. SRL consists in the automatic recognizing of predicates and *fes* in sentences, recently pushed by the FrameNet project that made available a large set of about 130,000 annotated sentences from the British National Corpus (BNC) [1].

For our experiments, we targeted the *Automatic Boundary Detection* (ABD) task, i.e. the localization of segments in a sentence that correspond to a *fe*. In the previous example, the phrase "*the film*" expresses a role (i.e. the ORIGINAL), while "*film onto hundreds*" does not and refers to different *fes*. Obviously, given the tree reflecting the syntactic structure of a sentence, the ABD corresponds to a binary classification task over those subtrees that cover or not segments isomorphic to a role. The ABD task has been successfully tackled using tree kernels [20]: the syntactic information used to discriminate roles is in fact captured by the implicit feature space generated by the adopted tree kernel.

These experiments are run against collections of parse trees based on the dependency grammar formalism, as discussed in [6]. Each node in a parse tree is a candidate to cover a valid *fe* thus corresponding to a training instance. From the FrameNet 1.3 dataset, a set of about 60,000 sentences[6] is mapped into a set of about 1,400,000 trees, i.e. the overall number of labeled subtrees acting as positive and negative instances. The dataset is split in train and test

[6] Only sentences whose lexical unit corresponds to a verb are adopted in our tests.

Table 3. Automatic Boundary Detection results in terms of F1.

	l	STK - SCW-II		PTK - SCW-II	
		iter. 1	iter. 2	iter. 1	iter. 2
p=1	100	0.430	0.424	0.516	0.518
	200	0.459	0.458	0.581	0.581
	300	0.491	0.494	0.612	0.610
	400	0.515	0.512	0.631	0.627
	500	0.534	0.530	0.645	0.637
p=2	100	0.446	0.450	0.587	0.579
	200	0.510	0.513	0.641	0.629
	300	0.540	0.542	0.664	0.648
	400	0.558	0.559	0.682	0.665
	500	0.567	0.569	0.698	0.681
p=3	100	0.461	0.480	0.637	0.628
	200	0.523	0.533	0.678	0.664
	300	0.548	0.557	0.705	0.692
	400	0.567	0.576	0.715	0.702
	500	0.578	0.585	**0.724**	0.712

according to the 90/10 proportion. This size makes the straightforward application of a traditional kernel-based method unfeasible. We preserved the application of the Smoothed Partial Tree Kernel and investigated the same dimensions and sampling applied into the previous experimental settings. Given the size of the dataset, we adopted (only) an online learning scheme, by applying the SCW-II learning algorithm that achieved the best result in the previous QC tasks (see Table 1). For this binary task, we reported results in Table 3 through the standard F1 metrics. This is the first time a kernel-based method has been used to this dataset with the entire train set used for training. In order to have a comparison, we refer to [14] where the Budgeted Passive Aggressive learning algorithm and the Distributed Tree Kernel have been applied to a subset of up to 100,000 examples. In [14] authors approximate the Syntactic Tree Kernel (STK) proposed in [4], so we approximated also this kernel. Table 3 shows in the first two columns the results where the STK kernel is approximated, while the SPTK is used in the last columns. The SPTK is more robust than the STK and, compared with the size of the training material, we are able to outperform the solution proposed in [14], that achieved 0.645 with an approximation derived applying the Distributed Tree Kernel proposed in [30]. However, the Nyström ensemble approximation derived through the STK achieves 0.585 of F1, that is higher to all the results proposed in [14] at a similar dimensionality. In conclusion, the combination of the Nyström method with the SCW-II achieves 0.724 of F1 (with a relative improvement of 17 %). These outcomes suggest that applying structured learning to dataset of this size is effective as well as viable.

5 Conclusions

In this paper the Nyström methodology has been discusses as a viable solution to face scalability issues in kernel-based language learning. It allows deriving low-dimensional linear representations of training examples, regardless of their corresponding representations (e.g. vectors or discrete structures), by approximating the implicit space underlying a kernel function. These linear representations enable the application of scalable and performant linear learning methods. Large scale experimental results on two language learning tasks suggested that a classification quality comparable with original kernel-based methods can be obtained, even when a reduction of the computational cost up to 99 % is observed. We showed a successful application of these methods to a FrameNet datasets of about 1.4 million of training instances. At the best of our knowledge, this is the first application of this class of methods to language learning with several open lines of research. Other language learning problems where kernel-based learning has been previously applied can be investigated at a larger scale, such as Relation Extraction [7]. Moreover, other learning tasks can be investigated, such as linear regression, clustering or re-ranking. Further and more efficient learning algorithms, such as [22, 23], or more complex learning scheme, such as the stratified approach proposed in [13], can be also investigated.

References

1. Baker, C.F., Fillmore, C.J., Lowe, J.B.: The Berkeley FrameNet project. In: Proceedings of COLING-ACL. Montreal, Canada(1998)
2. Cancedda, N., Gaussier, É., Goutte, C., Renders, J.M.: Word-sequence kernels. J. Mach. Learn. Res. **3**, 1059–1082 (2003)
3. Chang, C.C., Lin, C.J.: Libsvm: a library for support vector machines. ACM Trans. Intell. Syst. Technol. **2**(3), 27:1–27:27 (2011)
4. Collins, M., Duffy, N.: Convolution kernels for natural language. In: Proceedings of Neural Information Processing Systems (NIPS 2001), pp. 625–632 (2001)
5. Crammer, K., Dekel, O., Keshet, J., Shalev-Shwartz, S., Singer, Y.: Online passive-aggressive algorithms. J. Mach. Learn. Res. **7**, 551–585 (2006)
6. Croce, D., Moschitti, A., Basili, R.: Structured lexical similarity via convolution kernels on dependency trees. In: Proceedings of EMNLP (2011)
7. Culotta, A., Sorensen, J.: Dependency tree kernels for relation extraction. In: Proceedings of ACL 2004. Stroudsburg, PA, USA (2004)
8. Dredze, M., Crammer, K., Pereira, F.: Confidence-weighted linear classification. In: Proceedings of ICML 2008. ACM, New York (2008)
9. Drineas, P., Mahoney, M.W.: On the nystrm method for approximating a gram matrix for improved kernel-based learning. J. ML Res. **6**, 2153–2175 (2005)
10. Fan, R.E., Chang, K.W., Hsieh, C.J., Wang, X.R., Lin, C.J.: Liblinear: a library for large linear classification. J. Mach. Learn. Res. **9**, 1871–1874 (2008)
11. Filice, S., Castellucci, G., Croce, D., Basili, R.: Effective kernelized online learning in language processing tasks. In: de Rijke, M., Kenter, T., de Vries, A.P., Zhai, C.X., de Jong, F., Radinsky, K., Hofmann, K. (eds.) ECIR 2014. LNCS, vol. 8416, pp. 347–358. Springer, Heidelberg (2014)

12. Filice, S., Castellucci, G., Croce, D., Basili, R.: Kelp: a kernel-based learning platform for natural language processing. In: Proceedings of ACL: System Demonstrations. Beijing, China, July 2015
13. Filice, S., Croce, D., Basili, R.: A stratified strategy for efficient kernel-based learning. In: AAAI Conference on Artificial Intelligence (2015)
14. Filice, S., Croce, D., Basili, R., Zanzotto, F.M.: Linear online learning over structured data with distributed tree kernels. In: Proceedings of ICMLA 2013 (2013)
15. Hsieh, C.J., Chang, K.W., Lin, C.J., Keerthi, S.S., Sundararajan, S.: A dual coordinate descent method for large-scale linear svm. In: Proceedings of the ICML 2008, pp. 408–415. ACM, New York (2008)
16. Joachims, T., Finley, T., Yu, C.N.: Cutting-plane training of structural SVMs. Mach. Learn. **77**(1), 27–59 (2009)
17. Kumar, S., Mohri, M., Talwalkar, A.: Ensemble nystrom method. In: NIPS, pp. 1060–1068. Curran Associates, Inc.(2009)
18. Kumar, S., Mohri, M., Talwalkar, A.: Sampling methods for the nyström method. J. Mach. Learn. Res. **13**, 981–1006 (2012)
19. Li, X., Roth, D.: Learning question classifiers: the role of semantic information. Nat. Lang. Eng. **12**(3), 229–249 (2006)
20. Moschitti, A., Pighin, D., Basili, R.: Tree kernels for semantic role labeling. Comput. Linguist. **34**, 193–224 (2008)
21. Rifkin, R., Klautau, A.: In defense of one-vs-all classification. J. Mach. Learn. Res. **5**, 101–141 (2004)
22. Shalev-Shwartz, S., Singer, Y., Srebro, N.: Pegasos: primal estimated sub-gradient solver for SVM. In: Proceedings of ICML. ACM, New York (2007)
23. Shalev-Shwartz, S., Zhang, T.: Stochastic dual coordinate ascent methods for regularized loss. J. Mach. Learn. Res. **14**(1), 567–599 (2013)
24. Shawe-Taylor, J., Cristianini, N.: Kernel Methods for Pattern Analysis. Cambridge University Press, New York (2004)
25. Vapnik, V.N.: Statistical Learning Theory. Wiley-Interscience, New York (1998)
26. Vedaldi, A., Zisserman, A.: Efficient additive kernels via explicit feature maps. IEEE Trans. Pattern Anal. Mach. Intell. **34**(3), 480–492 (2012)
27. Wang, J., Zhao, P., Hoi, S.C.: Exact soft confidence-weighted learning. In: Proceedings of the ICML 2012. ACM, New York (2012)
28. Wang, Z., Vucetic, S.: Online passive-aggressive algorithms on a budget. J. Mach. Learn. Res. Proc. Track **9**, 908–915 (2010)
29. Williams, C.K.I., Seeger, M.: Using the nyström method to speed up kernel machines. In: Proceedings of NIpPS 2000 (2001)
30. Zanzotto, F.M., Dell'Arciprete, L.: Distributed tree kernels. In: Proceedings of ICML 2012 (2012)
31. Zhang, D., Lee, W.S.: Question classification using support vector machines. In: Proceedings of SIGIR 2003, pp. 26–32. ACM, New York (2003)

Question Answering

Beyond Factoid QA: Effective Methods for Non-factoid Answer Sentence Retrieval

Liu Yang[1(✉)], Qingyao Ai[1], Damiano Spina[2], Ruey-Cheng Chen[2], Liang Pang[3], W. Bruce Croft[1], Jiafeng Guo[3], and Falk Scholer[2]

[1] Center for Intelligent Information Retrieval, University of Massachusetts Amherst, Amherst, MA, USA
{lyang,aiqy,croft}@cs.umass.edu
[2] RMIT University, Melbourne, Australia
{damiano.spina,ruey-cheng.chen,falk.scholer}@rmit.edu.au
[3] Institute of Computing Technology, Chinese Academy of Sciences, Beijing, China
pangliang@software.ict.ac.cn, guojiafeng@ict.ac.cn

Abstract. Retrieving finer grained text units such as passages or sentences as answers for non-factoid Web queries is becoming increasingly important for applications such as mobile Web search. In this work, we introduce the answer sentence retrieval task for non-factoid Web queries, and investigate how this task can be effectively solved under a learning to rank framework. We design two types of features, namely semantic and context features, beyond traditional text matching features. We compare learning to rank methods with multiple baseline methods including query likelihood and the state-of-the-art convolutional neural network based method, using an answer-annotated version of the TREC GOV2 collection. Results show that features used previously to retrieve topical sentences and factoid answer sentences are not sufficient for retrieving answer sentences for non-factoid queries, but with semantic and context features, we can significantly outperform the baseline methods.

1 Introduction

A central topic in developing intelligent search systems is to provide answers in finer-grained text units, rather than to simply rank lists of documents in response to Web queries. This can not only save the users' efforts in fulfilling their information needs, but also will improve the user experience in applications where the output bandwidth is limited, such as mobile Web search and spoken search. Significant progress has been made at answering factoid queries [18,22], such as "how many people live in Australia?", as defined in the TREC QA track. However, there are diverse Web queries which cannot be answered by a short fact, ranging from advice on fixing a mobile phone, to requests for opinions on some public issues. Retrieving answers for these "non-factoid" queries from Web documents remains a critical challenge in Web question answering (WebQA).

Longer answers are usually expected for non-factoid Web queries, such as sentences or passages. However, research on passage-level answer retrieval can

N. Ferro et al. (Eds.): ECIR 2016, LNCS 9626, pp. 115–128, 2016.
DOI: 10.1007/978-3-319-30671-1_9

be difficult due to both the vague definition of a passage and evaluation methods [10]. A more natural and direct approach is to focus on retrieving sentences that are part of answers. Sentences are basic expression units in most if not all natural languages, and are easier to define and evaluate compared with passages. Therefore, in this paper, we introduce answer sentence retrieval for non-factoid Web queries as a practical WebQA task. To facilitate research on this task, we have created a benchmark data set referred as *WebAP* using a Web collection (TREC GOV2). To investigate the problem, we propose the first research question:

RQ1. Could we directly apply existing methods like factoid QA methods and sentence selection methods to solve this task?

Methods on factoid QA and sentence retrieval/selection are closely related work for our task. Factoid QA has been studied for some time, facilitated by the TREC QA track, and a state-of-the-art method is to use a convolutional neural network (CNN) [22] based model. In our study, we first adopt the CNN based method for factoid QA [22] and a sentence selection method using machine learning technology [13] for our task. However, we obtain inferior results for these advanced models, as compared with traditional retrieval models (i.e., language model). The results indicate that automatically learned word features (as in CNN) and simple text matching features (as in the sentence selection method) may not be sufficient for the answer sentence retrieval task. This leads to our second research question:

RQ2. How could we design more effective methods for answer sentence retrieval for non-factoid Web queries?

Previous results show that retrieving sentences that are part of answers is a more challenging task, requiring more powerful features than traditional relevance retrieval. By analyzing the task, we make two key observations from the WebAP data:

1. Due to the shorter length of sentences compared with documents, the problem of vocabulary-mismatch may be even more severe in answer sentence retrieval for non-factoid Web queries. Thus in addition to text matching features, we need more features to capture the semantic relations between query and answer sentences.
2. Non-factoid questions usually require multiple sentences as answers, and these answer sentences do not scatter in documents but often form small clusters. Thus the context of a sentence may be an important clue for identifying answer sentences.

Based on these observations, we design and extract two new types of features for answer sentence retrieval, namely *semantic features* and *context features*. We adopt learning to rank (L2R) models for sentence ranking, which have been successfully applied to document retrieval, collaborative filtering and many other applications. The experimental results show that significant improvement over existing methods can be achieved by our ranking model with semantic and context features.

The contributions of this paper can be summarized as follows:

1. We formally introduce the answer sentence retrieval task for non-factoid Web queries, and build a benchmark data set (WebAP) using the TREC GOV2 collection. We show that a state-of-the-art method from research on TREC QA track data does not work for this task, indicating answer sentence retrieval for non-factoid Web queries is a challenging task that requires the development of new methods. We released this data set to the research community.
2. Based on the analysis of the WebAP data, we design effective new features including semantic and context features for non-factoid answer sentence retrieval. We analyze the performance of different feature combinations to show the relative feature importance.
3. We perform a thorough experimental study with sentence level answer annotations. The results show that MART with semantic and context features can significantly outperform existing methods including language models, a state-of-the-art CNN based factoid QA method and a sentence selection method using multiple features.

The rest of the paper is organized as follows. The next section, Sect. 2, is an introduction to related work. We describe task definition and data for non-factoid answer sentence retrieval in Sect. 3. Section 4 is a baseline experiment where we attempt to apply existing standard techniques to confirm whether a new set of techniques is needed for our task. Section 5 describes the proposed semantic features and context features. Section 6 is a systematic experimental analysis using the annotated TREC GOV2 collection. Section 7 gives the conclusions and discusses future work.

2 Related Work

Our work is related to several research areas, including answer passage retrieval, answer retrieval with translation models, answer ranking in community question answering (CQA) sites and answer retrieval for factoid questions.

Answer Passage Retrieval. Keikha et al. [9,10] developed an annotated data set for non-factoid answer finding using TREC GOV2 collections and topics. They annotated passage-level answers, revisited several passage retrieval models with this data, and came to the conclusion that the current methods are not effective for this task. Our research work departs from Keikha et al. [9,10] by developing methods for answer sentence retrieval.

Answer Retrieval with Translation Models. Some previous research on answer retrieval has been based on statistic translation models to find semantically similar answers [1,15,20]. Xue et al. [20] proposed a retrieval model that combines a translation-based language model for the question part with a query likelihood approach for the answer part. Riezler et al. [15] presented an approach to query expansion in answer retrieval that uses machine translation techniques to bridge the lexical gap between questions and answers. Berger et al. [1] studied multiple statistical methods such as query expansion, statistical translation, and latent variable models for answer finding.

Answer Ranking in CQA. Surdeanu et al. [16] investigated a wide range of feature types including similarity features, translation features, density / frequency features, Web correlation features for ranking answers to non-factoid questions in Yahoo! Answers. Jansen et al. [8] presented an answer re-ranking model for non-factoid questions that integrate lexical semantics with discourse information driven by two representations of discourse. Answer ranking in CQA sites is a somewhat different task than answer retrieval for non-factoid questions: answer sentences could come from multiple documents for general non-factoid question answering, and the candidate ranked answer set is much smaller for a typical question in CQA sites.

Answer Retrieval for Factoid Questions. There has also been substantial research on answer sentence selection with data from TREC QA track [17,18,21,22]. Yu et al. [22] proposed an approach to solve this task via means of distributed representations, and learn to match questions with answers by considering their semantic encoding. Although the target answers are also at the sentence level, this research is focused on factoid questions. Our task is different in that we investigate answer sentence retrieval for non-factoid questions. We also compare our proposed methods with a state-of-the-art factoid QA method to show the advantages of developing techniques specifically for non-factoid answer data.

3 Task Definition and Data

We now give a formal definition of our task. Given a set of non-factoid questions $\{Q_1, Q_2, \cdots, Q_n\}$ and Web documents $\{D_1, D_2, ...D_m\}$ that may contain answers, our task is to learn a ranking model $\mathbf{R}$ to rank the sentences in the Web documents to find sentences that are part of answers. The ranker is trained based on available features F_S and labels L_S to optimize a metric $\mathbf{E}$ over the sentence rank list.

Our task is different from previous research in the TREC QA track and answer retrieval in CQA sites like Yahoo! Answers. Unlike the TREC QA track, we are focusing on answer finding for non-factoid questions. Unlike the research on answer retrieval in CQA, we aim to find answers from general Web documents, not limited to CQA answer posts. As a consequence, this task is more challenging for two reasons: answers could be much longer than in factoid QA, and the search space is much larger than CQA answer posts.

A test collection of questions and multi-sentence (passage) answers has been created based on the TREC GOV2 queries and documents.[1] GOV2 is the test collection used for the TREC Terabyte Track and crawled from .gov

[1] The data set is publicly available at https://ciir.cs.umass.edu/downloads/WebAP/.

sites in early 2004. It contains 25 million documents. The annotated GOV2 data set was produced by the following process [9,10]. For each TREC topic that was likely to have passage-level answers (82 in total), the top 50 documents were retrieved using a state-of-the-art retrieval model [7,11]. From the retrieved documents, documents identified as relevant in the TREC judgments were annotated for answer passages. Passages were marked with labels "Perfect", "Excellent", "Good", "Fair". The annotators found 8027 answer passages to 82 TREC queries, which is 97 passages per query on average. Among the annotated passages, which exclude passages without annotations and are treated as negative instances, 43 % are perfect answers, 44 % are excellent, 10 % are good and the rest are fair answers.

To obtain the annotation for our answer sentence retrieval task, we let sentences in answer passages inherit the label of the passage. Then we map "Perfect", "Excellent", "Good", "Fair" to $4 \sim 1$ and assign 0 for all the other sentences. Note that there are some duplicate sentences in the previously annotated passage data set. Judgments over these duplicates are not entirely consistent. We fix this problem by a majority vote and break ties by favoring more relevant labels. The data after label inheritance and inconsistent judgment fixing is named *WebAP* data. There are 991233 sentences in the data set and the average length of sentences is 17.58. After label propagation from passage level to sentence level, 99.02 % (981510) sentences are labeled as 0 and less than 1 % sentences have positive labels (149 sentences are labeled as 1; 783 sentences are labeled as 2; 4283 sentences are labeled as 3; 4508 sentences are labeled as 4). Highly imbalanced labels make this task even more difficult.

To better understand our task and data, we show a comparison of sample questions and answers in TREC QA Track data and WebAP data in Table 1. We can clearly see the difference is that answers in TREC QA Track data are mostly short phrases like entities and numbers, whereas answers in WebAP data are longer sentences.

4 Baseline Experiments

Our first task on the WebAP data set is to seek solution of RQ1, in which we ask if a new set of techniques is needed. We address this question using a baseline experiment, in which we use some techniques that should be reasonable for retrieving non-factoid answers, and compare these techniques with a factoid question answering method.

We set up this experiment by including the following three classes of techniques:

1. **Retrieval Functions.** In this experiment, we considered query likelihood language model with Dirichlet smoothing (LM).
2. **Factoid Question Answering Method.** In this experiment, we use a more recent approach based on convolutional neural network (CNN) [22], whose performance is current on par with the best results on TREC QA Track data. This is a supervised method and needs to be trained with pre-defined

Table 1. Comparison of sample questions and answers in TREC QA Track data and WebAP data.

Sample Questions and Answers in TREC QA Track Data
Query 201:
Question: What was the name of the first Russian astronaut to do a spacewalk?
Answer: Aleksei A. Leonov
Answer Document ID: LA072490-0034
Query 202:
Question: Where is Belize located?
Answer: Central America
Answer Document ID: FT934-14974
Query 203:
Question: How much folic acid should an expectant mother get daily?
Answer: 400 micrograms
Answer Document ID: LA061490-0026

Sample Questions and Answer Sentences with Labels in WebAP Data
Query 704:
Question: What are the goals and political views of the Green Party?
Answer:
(*Perfect*) The Green Party promotes an ecological vision which understands that all life on our planet is interconnected; that cooperation is more essential to our well-being than competition; and that all people are connected to and dependent upon one another and upon the natural systems of our world.
(*Excellent*) The need for fairness for people and local communities in developing economic opportunities, instead of continually favoring big corporations and other concentrations of wealth and power; and a fair, equitable, progressive tax system.
(*None*) Statements were supplied by the political parties and have not been checked for accuracy by any official agency.
Query 847:
Question: What was the role of Portugal in World War II?
Answer:
(*Perfect*) The Allies began this economic war with some advantages, most notably a centuries-old Anglo-Portuguese alliance, coupled with close economic and trade relations (Britain was Portugal's leading trade partner), as well as Portugal's dependence on U.S. petroleum, coal, and chemical supplies.
(*Good*) Portugal, Spain, Sweden, and Turkey all sold Nazi Germany resources critical to sustaining its wartime industry.
(*Fair*) During WWII, Switzerland served as a curtain for other countries by creating multinational gold depository for neutral and non-aligned national states — Sweden, Portugal, Spain and Turkey — to use in trading money with the Axis.
(*None*) The records of claims reflect attempts at achieving diplomatic remedies to losses suffered in federal courts in suits handled by the Office of Alien Property.

word embeddings. Two variants are tested here, which are CNN with word count features and CNN without word count features.

3. **Summary Sentence Selection Method.** In this experiment, we test a L2R approach proposed by Metzler and Kanungo [13], which uses 6 simple features referred as *MK features* as described in Sect. 5 to address the lexical matching between the query and sentences. As suggested in the original paper, we use the MART ranking algorithm to combine these features.

The results are given in Table 2. LM and MK perform better than CNN based methods. LM achieves the best results under all the three metrics. CNN based methods perform poorly on this task. Using word count features in [22], CNN gets slightly better results. However, only using LM can achieve 170.73 % gain for MRR over CNN with word count features and the difference is statistically significant measured by the Student's paired t-test ($p < 0.01$). MK achieves better performance than CNN based methods, but it performs worse than LM. This result shows that automatically learned word features (as in CNN) and

Table 2. Baseline experiment results for non-factoid answer sentence retrieval of different methods. The best performance is highlighted in boldface.‡ means significant difference over CNN with word count features with $p < 0.01$ measured by the Student's paired t-test.

Feature Set	Model	NDCG@10	P@10	MRR
CNN(No word count)	CNN	0.0218	0.0341	0.0909
CNN(With word count)		0.0596	0.0646	0.1254
MK	MART	0.1163 ‡	0.1293 ‡	0.2677 ‡
LM	Base	**0.1340** ‡	**0.1451** ‡	**0.3395** ‡

simple combined text matching features (as in MK) may not be sufficient for our task, suggesting that a new set of techniques is needed for non-factoid answer sentence retrieval.

5 Beyond Factoid QA: Capturing Semantics and Context

In the previous experiment, we found that a state-of-the-art factoid question answering method is not very effective on our task. L2R with MK features and LM performs better than CNN based methods. We now construct a ranker using MK features which include LM. We further add semantic and context features for answer sentence retrieval of non-factoid questions. First, we give a detailed description of the MK features as follows.

ExactMatch. Exact match is a binary feature indicating whether the query is a substring of the sentence.

TermOverlap. Term overlap measures the number of terms that are both in the query and the sentence after stopping and stemming.

SynonymsOverlap. Synonyms overlap is the fraction of query terms that have a synonym (including the original term) in the sentence, computed by using Wordnet.[2]

LanguageModelScore. Language model score is computed as the log likelihood of the query being generated from the sentence. The sentence language model is smoothed using Dirichlet smoothing. This feature is essentially the query likelihood language model score [3]. The feature is computed as:

$$f_{LM}(Q,S) = \sum_{w \in Q} tf_{w,Q} log \frac{tf_{w,S} + \mu P(w|C)}{|S| + \mu} \quad (1)$$

where $tf_{w,Q}$ is the number of times that w occurs in the query, $tf_{w,S}$ is the number of times that w occurs in the sentence, $|S|$ is the length of sentences,

[2] http://wordnet.princeton.edu/.

$P(w|C)$ is the background language model and μ is a parameter for Dirichlet smoothing. Note that the parameter μ tends to be smaller than the case in document retrieval because the average length of sentences is much shorter than that of documents.

SentLength. Sentence length is the number of terms in the sentence after stopping.

SentLocation. Sentence location is the relative location of the sentence within the document, computed as the position divided by the total number of sentences.

5.1 Semantic Features

Short text units such as sentences are more likely to suffer from query mismatch. The same topics could get radically different wordings between the questions and answers, which leads to the "lexical chasm" problem [1]. Thus we consider the following semantic features to handle this problem.

ESA. Explicit Semantic Analysis (ESA) [6] is a method that represents text as a weighted mixture of a predetermined set of natural concepts defined by Wikipedia articles which can be easily explained. Semantic relatedness is computed as the cosine similarity between the query ESA vector and the sentence ESA vector. A recent dump of English Wikipedia (June 2015) is used to generate ESA representations for queries and sentences.

WordEmbedding. Word embeddings are continuous vector representations of words learned from large amount of text data using neural networks. Words with similar meanings are also in close distances in this vector space. Mikolov et al. [14] introduced the Skip-gram model as an efficient method for learning high quality vector representations of words from large amounts of unstructured text data. The implementation of the continuous bag-of-words and skip-gram architectures for computing vector representations of words is released as an open-source project Word2Vec.[3] We compute this feature as the average pairwise cosine similarity between any query-word vector and any sentence-word vector following previous work [2].

EntityLinking. Linking short texts to a knowledge base to obtain the most related concepts gives an informative semantic representation that can be used to represent queries and sentences. We generate such a representation through an entity linking system Tagme [4], which is able to efficiently augment a plain text with pertinent hyperlinks to Wikipedia pages. On top of this, we produce the following semantic feature: the Jaccard similarity between the set of Wikipedia pages linked by Tagme to the query q and the set of pages linked to a given sentence s (Eq. 2):

$$\text{TagmeOverlap}(q, s) = \frac{\text{Tagme}(q) \cap \text{Tagme}(s)}{\text{Tagme}(q) \cup \text{Tagme}(s)}. \tag{2}$$

[3] https://code.google.com/p/word2vec/.

ESA takes advantages of human labeled concepts from Wikipedia and explicitly presents topical differences between queries and sentences; **Word Embedding** captures word semantics from unstructured text data based on the hypothesis that words with similar meanings should appear in similar lingual contexts; and **EntityLinking** converts sentences to entity space where different words for the same entity, and different entities with similar Wikipedia pages are assigned with high similarity scores. Combined together, these semantic features compensate MK features and alleviate problems like query mismatching through the introduction of a semantic similarity score.

5.2 Context Features

Context features are features specific to the context of the candidate sentence. We define the context of a sentence as the adjacent sentence before and after it. The intuition is that the answer sentences are very likely to be surrounded by other answer sentences. Context features could be generated based on any sentence features. They include features in the following two types:

- **<Feature>SentenceBefore: MK** features and semantic features of the sentence before the candidate sentence.
- **<Feature>SentenceAfter: MK** features and semantic features of the sentence after the candidate sentence.

So applying this idea to a set of n features, we produce $2n$ new features. In our experiments, we define *context*() as a procedure which could be applied to MK features and semantic features. So there are 12 context features for MK features and 6 context features for semantic features.

5.3 Learning Models

With the computed features, we carry out sentence re-ranking with L2R methods using the following models:

- **MART** Multiple Additive Regression Trees [5], also known as gradient boosted regression trees, produces a prediction model in the form of an ensemble of weak prediction models.
- **LambdaMART** [19] combines the strengths of MART and LambdaRank which has been shown to be empirically optimal for a widely used information retrieval measure. It uses the LambdaRank gradients when training each tree, which could deal with highly non-smooth IR metrics such as DCG and NDCG.
- **Coordinate Ascent (CA)** [12] is a list-wise linear feature-based model for information retrieval which uses coordinate ascent to optimize the model's parameters. It optimizes the objective by iteratively updating each dimension while holding other dimensions fixed.

6 Experiments

6.1 Experimental Settings

As presented in Sect. 3, we use the aforementioned WebAP dataset as the benchmark. We follow the LETOR experiment protocol, partitioning data by queries and conducting 5-fold cross validation throughout. We use our own package *SummaryRank*[4] for feature extraction. We use RankLib[5] to implement MART and CA, and use *jforests*[6] to implement LambdaMART. For each experimental run, we optimize the hyperparameters using cross validation and report the best performance.[7] For evaluation, we compute a variety of metrics including NDCG@10, P@10 and MRR, focusing on the accuracy of top ranked answer sentences in the results.

6.2 Overall Analysis of Results

Table 3 shows the evaluation results for non-factoid answer sentence retrieval using different feature sets and learning models. We summarize our observations as follows: (1) For feature set comparison, the results are quite consistent across the three different learning models where the combination of MK, semantic features, and context features achieve the best results for all three learning model settings. For MRR, this combination achieves 5.4 % gain using CA, 68.55 % gain using MART, and 10.05 % gain using LambdaMART over the MK feature set. Similar gains are also observed under other metrics. In terms of relative feature importance, context features achieve larger gain compared to semantic features although adding both of them can improve the performance of the basic MK feature set. (2) For the learning model comparison, MART achieves the best performance with MK + Semantics + Context(All) features, with statistically significant differences with respect to both LM and MK(MART), which shows the effectiveness of MART for combining different features to learn a ranking model for non-factoid answer sentence retrieval.

---

[4] http://rmit-ir.github.io/SummaryRank.

[5] http://www.lemurproject.org/ranklib.php.

[6] https://code.google.com/p/jforests/.

[7] We just report the hyperparameters used for MK + Semantics + Context(All) feature sets given the space limit. The hyperparameters for other feature sets can be obtained by standard 5-fold cross validation. For parameter values in MART, we set the number of trees as 100, the number of leaves of each tree as 20, learning rate as 0.05, the number of threshold candidates for tree splitting as 256, min leaf support as 1, the early stop parameter as 100. For parameter values in CA, we set the number of random restarts as 3, the number of iterations to search in each dimension as 25, performance tolerance between two solutions as 0.001. For parameter values in LambdaMART, we set the number of trees as 1000, the number of leaves of each tree as 15, learning rate as 0.1, minimum instance percentage per leaf as 0.25, feature sampling ratio as 0.5. We empirically set the parameter $\mu = 10$ in the computation of the *LanguageModelScore* feature.

Table 3. Evaluation results for non-factoid answer sentence retrieval of different feature sets and learning models. "Context(All)" denotes context features for both MK and semantics features. The best performance is highlighted in boldface. † and ‡means significant difference over LM with $p < 0.05$ and $p < 0.01$ respectively measured by the Student's paired t-test. ** means significant difference over MK with the same learning model with $p < 0.01$ measured by the Student's paired t-test.

Feature Set	Model	NDCG@10	P@10	MRR
LM	Base	0.1340	0.1451	0.3395
MK	Coord. Ascent	0.1590	0.1524	0.3860
MK + Semantics		0.1486	0.1585	0.3781
MK + Semantics + Context(All)		**0.1795** ‡	**0.1817** †	**0.4070**
MK	MART	0.1163	0.1293	0.2677
MK + Semantics		0.1207	0.1341	0.2729
MK + Semantics + Context(All)		**0.1864** † **	**0.2024** † **	**0.4512** † **
MK	LambdaMART	0.1441	0.1537	0.3662
MK + Semantics		0.1591	0.1744	0.3439
MK + Semantics + Context(All)		**0.1798** †	**0.1939** †	**0.4030**

6.3 Effect of Semantic and Context Features

As mentioned previously, retrieval on short text such as sentences are vulnerable to problems such as "lexical chasm" [1]. By introducing semantic features, we attempt to address this problem through matching query and sentences in semantic space. The experiments indicate that semantic features indeed benefit retrieval effectiveness. As shown in Table 3, MK + Semantics with MART obtains 3.78 % gain in NDCG@10, 3.71 % in P@10, 1.94 % in MRR. Similar improvements can also be found in results with other L2R models including LambdaMART and CA except that we observe slightly loss under NDCG@10 and MRR for CA. Overall, the three simple but effective semantic features provide important information for non-factoid answer sentence retrieval.

Our intuition for the design of context features is that good answer sentences are likely to be surrounded by other answer sentences. Our experimental results also demonstrate the effectiveness of context features. In Table 3, we observe considerable improvement with context features: MK + Semantics + Context(All) outperforms MK + Semantics in NDCG@10, P@10 and MRR for 13.01 %, 11.18 %, 17.19 % with LambdaMART, 54.43 %, 50.93 %, 65.34 % with MART and 20.79 %, 14.63 %, 7.64 % with CA. Thus the performance of these three learning models could be improved by incorporating context features.

6.4 Examples of Top Ranked Sentences

We further show some examples and analysis of top-1 ranked sentences by different methods in Table 4. In general, the top-1 ranked sentences by MK + Semantics + Context(All) are better than the other two methods. Although there are lexical matches for top-1 sentences retrieved by all methods, sentences with lexical matches may be not answer sentences of high quality. For example, for query 808, MK features will be confused on "Korean", "North Korean" and "South Korean" appearing frequently in sentences since there is a common term among them. Semantic features play an important role here to alleviate this problem. Semantic features such as **EntityLinking** can differentiate "Korean" with "North Korean" by linking them to different Wikipedia pages.

Context features have the potential to guide the ranker in the right direction since correct non-factoid answer sentences are usually surrounded by other similar non-factoid answer or relevant sentences. Query 816 in Table 4 gives one example for their effects. Without context features, LambdaMART with MK+Semantics features retrieved a non-relevant sentence as the top-1 result. The retrieved sentence may seem relevant itself (as it indeed mentions USAID's efforts to support biodiversity), but if we pull out its context:

- SentenceBefore (None): ... *BIOFOR embarked upon a global study of lessons learned from USAID's community-based forest management projects since 1985.*
- SentenceRetrieved (None): *USAID forestry programs support biodiversity efforts by helping to protect habitats for forest inhabitants.*
- SentenceAfter (None): *The African continent contains approximately 650 million hectares of forests ...*

The sentences around the *SentenceRetrieved* are not relevant to the query and are labeled as "None". From its context, we can see that the *SentenceRetrieved* is actually not talking about Galapagos Islands. Comparing to that, the context of top-1 sentence retrieved by LambdaMART with MK + Semantics + Context(All) features are:

- SentenceBefore (None): *Relations between ... in the Galapagos have greatly improved ... and approval of the Galapagos Marine Reserve Management Plan.*
- SentenceRetrieved (Perfect): ... *USAID contributed to ... in the Galapagos Islands to guide the development of the sustainable, ... management of the Galapagos ecosystem.*
- SentenceAfter (Excellent): *USAID is also assisting ... for sustainable forestry development and biodiversity conservation in Ecuador.*

SentenceBefore and *SentenceAfter* are talking about Galapagos Islands and USAID's work for biodiversity conservation in Ecuador, and *SentenceAfter* is labeled as an "Excellent" answer sentence. With context features, *SentenceRetrieved* is promoted as the top-1 result due to its good context, and our empirical experiments show that these promotions are beneficial for model effectiveness (the promoted sentence in this case is indeed a perfect answer for query 816).

Table 4. Examples of top-1 ranked sentences by using different methods. Integers in the 2nd column are relevance levels (0 = not relevant). Lexically matched sentence terms are underlined.

770: *What is the state of Kyrgyzstan United States relations?*

MK	0	They seized hostages and several villages, allegedly seeking to create an Islamic state in south Kyrgyzstan as a springboard for a jihad in Uzbekistan.
MK+Sem	3	In 1994, Kyrgyzstan concluded a bilateral investment treaty with the United States, and in 1999 Kyrgyzstan became a member of the WTO.
MK+Sem+Context	4	The extension of unconditional normal trade relations treatment to the products of Kyrgyzstan will permit the United States to avail itself of all rights under the WTO with respect to Kyrgyzstan.

808: *What information is available on the involvement of the North Korean Government in counterfeiting of US currency?*

MK	0	U.S. firms report that, although the release of confidential business information is forbidden by Korean law, government officials have not sufficiently protected submitted information and in some cases, proprietary information has been made available to Korean competitors or to their trade associations.
MK+Sem	3	North Korean regime continues to export weapons and engage in state sponsored international crime to include narcotics trafficking, and counterfeiting U.S. currency.
MK+Sem+Context	3	While the State Department reports that North Korea's most profitable operation involves counterfeiting U.S. currency,(101) government-controlled operations also counterfeit a range of other countries' currencies and other trademarked products.

816: *Describe efforts made by USAID to protect the biodiversity in the Galapagos Islands in Ecuador.*

MK	0	efforts in the Galapagos Islands and other key areas of Ecuador, which has the greatest biodiversity per hectare of any country in South America.
MK+Sem	0	USAID forestry programs support biodiversity efforts by helping to protect habitats for forest inhabitants.
MK+Sem+Context	3	In addition, USAID contributed to the establishment of a capital fund for the benefit of the Charles Darwin Foundation in the Galapagos Islands to guide the development of the sustainable, scientifically based, participatory management of the Galapagos ecosystem.

7 Conclusions and Future Work

In this paper, we formally introduced the answer sentence retrieval task for non-factoid Web queries and investigated a framework based on learning to rank methods. We compared learning to rank methods with baseline methods including language models and a CNN based method. We found that both semantic and context features are useful for non-factoid answer sentences retrieval. In particular, the results show that MART with appropriate features outperforms all the baseline methods significantly under multiple metrics and provides a good basis for non-factoid answer sentence retrieval.

For future work, we would like to investigate more features such as syntactic features and readability features to further improve non-factoid answer sentence retrieval. Learning an effective representation of answer sentences for information retrieval is also an interesting direction to explore.

Acknowledgments. This work was partially supported by the Center for Intelligent Information Retrieval, by NSF grant #IIS-1160894, by NSF grant #IIS-1419693, by ARC Discovery grant DP140102655 and by ARC Project LP130100563. Any opinions, findings and conclusions expressed in this material are those of the authors and do not necessarily reflect those of the sponsor. We thank Mark Sanderson for the valuable comments on this work.

References

1. Berger, A., Caruana, R., Cohn, D., Freitag, D., Mittal, V.: Bridging the lexical chasm: statistical approaches to answer-finding. In: Proceedings of the SIGIR 2000 (2000)
2. Chen, R.-C., Spina, D., Croft, W.B., Sanderson, M., Scholer, F.: Harnessing semantics for answer sentence retrieval. In: Proceedings of ESAIR 2015 (2015)
3. Croft, W.B., Metzler, D., Strohman, T.: Search Engines: Information Retrieval in Practice, 1st edn. Addison-Wesley Publishing Company, Lebanon (2009)
4. Ferragina, P., Scaiella, U.: Fast and accurate annotation of short texts with wikipedia pages. IEEE Softw. **29**(1), 70–75 (2012)
5. Friedman, J.H.: Greedy function approximation: a gradient boosting machine. Ann. Stat. **29**, 1189–1232 (2000)
6. Gabrilovich, E., Markovitch, S.: Computing semantic relatedness using wikipedia-based explicit semantic analysis. In: Proceedings of IJCAI 2007 (2007)
7. Huston, S., Croft, W.B.: A comparison of retrieval models using term dependencies. In: Proceedings of CIKM 2014 (2014)
8. Jansen, P., Surdeanu, M., Clark, P.: Discourse complements lexical semantics for non-factoid answer reranking. In: Proceedings of ACL 2014, pp. 977–986 (2014)
9. Keikha, M., Park, J.H., Croft, W.B.: Evaluating answer passages using summarization measures. In: Proceedings of SIGIR 2014 (2014)
10. Keikha, M., Park, J.H., Croft, W.B., Sanderson, M.: Retrieving passages and finding answers. In: Proceedings of ADCS 2014, pp. 81–84 (2014)
11. Metzler, D., Croft, W.B.: A markov random field model for term dependencies. In: Proceedings of SIGIR 2005 (2005)
12. Metzler, D., Croft, W.B.: Linear feature-based models for information retrieval. Inf. Retr. **10**(3), 257–274 (2007)
13. Metzler, D., Kanungo, T.: Machine learned sentence selection strategies for query-biased summarization. In: Proceedings of SIGIR Learning to Rank Workshop (2008)
14. Mikolov, T., Chen, K., Corrado, G., Dean, J.: Efficient estimation of word representations in vector space. CoRR arxiv.org/abs/1301.3781 (2013)
15. Riezler, S., Vasserman, A., Tsochantaridis, I., Mittal, V., Liu, Y.: Statistical machine translation for query expansion in answer retrieval. In: Proceedings of ACL 2007 (2007)
16. Surdeanu, M., Ciaramita, M., Zaragoza, H.: Learning to rank answers on large online QA collections. In: Proceedings of ACL 2008, pp. 719–727 (2008)
17. Yih, W.-T., Chang, M.-W., Meek, C., Pastusiak, A.: Question answering using enhanced lexical semantic models. In: Proceedings of ACL 2013 (2013)
18. Wang, M., Smith, N.A., Mitamura, T.: What is the jeopardy model? A quasi-synchronous grammar for QA. In: EMNLP-CoNLL 2007, pp. 22–32 (2007)
19. Wu, Q., Burges, C.J., Svore, K.M., Gao, J.: Adapting boosting for information retrieval measures. Inf. Retr. **13**(3), 254–270 (2010)
20. Xue, X., Jeon, J., Croft, W.B.: Retrieval models for question and answer archives. In: Proceedings of SIGIR 2008, pp. 475–482 (2008)
21. Yao, X., Durme, B.V., Callison-burch, C., Clark, P.: Answer extraction as sequence tagging with tree edit distance. In: Proceedings of NAACL 2013 (2013)
22. Yu, L., Hermann, K.M., Blunsom, P., Pulman, S.: Deep Learning for Answer Sentence Selection. arXiv: 1412.1632, December 2014

Supporting Human Answers for Advice-Seeking Questions in CQA Sites

Liora Braunstain[1(✉)], Oren Kurland[1], David Carmel[2], Idan Szpektor[2], and Anna Shtok[1]

[1] Faculty of Industrial Engineering and Management, Technion, 32000 Haifa, Israel
liorab@campus.technion.ac.il, kurland@ie.technion.ac.il, annabel@tx.technion.ac.il

[2] Yahoo Labs, 31905 Haifa, Israel
{dcarmel,idan}@yahoo-inc.com

Abstract. In many questions in Community Question Answering sites users look for the advice or opinion of other users who might offer diverse perspectives on a topic at hand. The novel task we address is providing supportive evidence for human answers to such questions, which will potentially help the asker in choosing answers that fit her needs. We present a support retrieval model that ranks sentences from Wikipedia by their presumed support for a human answer. The model outperforms a state-of-the-art textual entailment system designed to infer factual claims from texts. An important aspect of the model is the integration of relevance oriented and support oriented features.

1 Introduction

Most questions posted on Community-based Question Answering (CQA) websites, such as Yahoo Answers, Answers.com and StackExchange, do not target simple facts such as "*what is Brad Pitt's height?*" or "*how far is the moon from earth?*". Instead, askers expect some human touch in the answers to their questions. Especially, many questions look for recommendations, suggestions and opinions, *e.g.* "*what are some good horror movies for Halloween?*", "*should you wear a jockstrap under swimsuit?*" or "*how can I start to learn web development?*". According to our analysis, based on editorial judgments of 12,000 Yahoo Answers questions, 70 % of all questions are advice or opinion seeking questions.

Examining answers for such advice-seeking questions, we found that quite often answerers do not provide supportive evidence for their recommendation, and that answers usually represent diverse perspectives of the different answerers for the question at hand. For example, answerers may recommend different horror movies. Still, the asker would like to choose only one or two movies to watch, and without additional supportive evidence her decision may be non-trivial.

In this paper we assume that askers would be happy to receive additional information that will help them in choosing the best fit for their need from the various suggestions or opinions provided in the CQA answers. More formally, we

N. Ferro et al. (Eds.): ECIR 2016, LNCS 9626, pp. 129–141, 2016.
DOI: 10.1007/978-3-319-30671-1_10

propose the novel task of retrieving sentences from the Web that provide support to a given recommendation or opinion that is part of an answer in a CQA site.

We refer to the part of the answer (*e.g.*, a sentence) that contains a recommendation as a *subjective claim* about the need expressed in the question (*e.g.*, a call for advice). For a sentence to be considered as *supporting the claim*, it should be relevant to the content of the claim and provide some supporting information; *e.g.*, examples, statistics, or testimony [1]. More specifically, a supporting sentence is one whose acceptance is likely to raise the confidence in the claim.

While supporting sentences may be part of the same answer containing the claim, or found in other answers given for the same question, in this paper we are interested in retrieving sentences from other sources which may provide different perspectives on the claim compared to content on CQA sites. For example, for the question "*what are some good horror movies?*", a typical CQA answer could be "*The Shining is a great movie; I love watching it every year*". On the other hand, a supporting sentence from external sites may contain information such as "*...in 2006, the Shining made it into Ebert's series of "Great Movie" reviews...*". Specifically, we focus on retrieving supporting sentences from Wikipedia, although our methods can be largely applied to other Web sites.

We present a general scheme of *Learning to Rank for Support*, in which the retrieval algorithm is directly optimized for ranking sentences by presumed support. Our feature set includes both relevance-oriented features, such as textual similarity, and support-oriented features, such as sentiment matching and similarity with language-model-based support priors.

We experimented with a new dataset containing 40 subjective claims from the Movies category of Yahoo Answers. For each claim, sentences retrieved from Wikipedia using relevance estimates were manually evaluated for relevance and support. The evaluated benchmark was then used to train and test our model. The results demonstrate the merits of integrating relevance-based and support-based features for the support ranking task. Furthermore, our model substantially outperforms a state-of-the-art Textual Entailment system used for support ranking. This result emphasizes the difference between prior work on supporting objective claims and our task of supporting subjective recommendations.

2 Ranking Sentences by Support

Our goal is to devise a sentence retrieval method that ranks sentences by the level of *support* they provide to a given *subjective* claim. For example, the sentence "*movie X received the Oscar academy award for the best film*" would be considered as providing strong support to the claim "*X is a good movie*".

We confine our treatment of the sentence retrieval task to claims about a single entity c_e — *e.g.* the movie X in the example above — since often advice-seeking CQA questions are about entities such as restaurants, movies, singers and products. For sentence s to provide support for a given claim c, s must be relevant to c and especially to the entity c_e that c is about. Hence, our approach for support ranking is based on an initial relevance ranking of sentences (Sect. 2.1).

Then, a set of features is used in a learning-to-rank method for re-ranking the top-retrieved sentences by their (presumed) support for c (Sect. 2.2).

2.1 Initial Relevance Ranking

Our first step is to rank sentences by their presumed relevance to claim c. Since these sentences are part of documents in a corpus D, we follow common practice in work on sentence retrieval [2] and first apply document retrieval with respect to c. Then, the sentences in the top ranked documents are ranked for relevance.

We assume that each document $d \in D$ is composed of a title, d_t, and a body, d_b. This is the case for Wikipedia, which is used in our experiment, as well as for most Web pages. The initial document retrieval, henceforth **InitDoc**, is based on the document score $S_{SDM}(c; d_b)$. This score is assigned to the body of document d with respect to the claim c by the state-of-the-art sequential dependence model (SDM) from the Markov Random Field framework [3]. For texts x and y,

$$S_{SDM}(x; y) \stackrel{def}{=} \lambda_T S_T(x; y) + \lambda_O S_O(x; y) + \lambda_U S_U(x; y); \qquad (1)$$

$S_T(x; y)$, $S_O(x; y)$ and $S_U(x; y)$ are the (smoothed) log likelihood values of the appearances of unigrams, ordered bigrams and unordered bigrams, respectively, of tokens from x in y; λ_T, λ_U, and λ_O are free parameters whose values sum to 1. We further bias the initial document ranking in favor of documents whose titles contain c_e — the entity the claim is about. Specifically, d is ranked by:

$$S_{InitDoc}(c; d) \stackrel{def}{=} \alpha S(c_e; d_t) + (1 - \alpha) S_{SDM}(c; d_b); \qquad (2)$$

$S(c_e; d_t)$ is the log of the Dirichlet smoothed maximum likelihood estimate, with respect to d's title, of the n-gram which constitutes the entity c_e [4]; smoothing is based on n-gram counts in the corpus[1]; α is a free parameter.

To estimate the relevance of sentence s to the claim c, we can measure their similarity using, again, the SDM model. We follow common practice in work on passage retrieval [2], and interpolate, using a parameter β, the claim-sentence similarity score with the retrieval score of document d which s is part of:

$$S_{InitSent}(c; s) \stackrel{def}{=} \beta S_{SDM}(c; s) + (1 - \beta) S_{InitDoc}(c; d). \qquad (3)$$

Equation 3 is used to rank the sentences in the top-N retrieved documents; N is a free parameter. The k most highly ranked sentences serve for $\mathcal{S}_{\text{init}}^{[k]}$ — the initial set of sentences to be ranked for support. Herein, **InitSent** denotes the sentence score assigned in Eq. 3 which is used to induce the initial sentence ranking.

[1] All SDM scoring function components in Eq. 1 also use the logs of Dirichlet smoothed estimates [3]. The smoothing parameter, μ, is set to the same value for all estimates.

2.2 Learning to Rank for Support

Next, we rank the sentences in $\mathcal{S}_{\text{init}}^{[k]}$ by the support they provide to the claim. To this end, we apply a learning-to-rank (LTR) approach [5] to construct a ranking function designed to optimize support. Specifically, we use a training set of claims, their respective sentences, and labels of the support level the sentences provide for the claims. Each pair of a claim and a sentence, (c, s), is represented as a feature vector. Below, we detail our feature set. In Sect. 3 we report the performance of three LTR methods applied with these features.

Language-Model Similarities. We use the initial retrieval scores, InitDoc (Eq. 2) and InitSent (Eq. 3), as relevance-estimate features. Additionally, we use several language-model-based similarity estimates. Let $p_{JM}^{[\psi]}(w|x)$ be the probability assigned to term w by a Jelinek-Mercer smoothed unigram language model induced from text x using the smoothing parameter ψ [4];[2] setting $\psi = 0$ amounts to the maximum likelihood estimate of w with respect to x. The similarity between texts x and y is estimated using the cross entropy, CE, between their induced language models: $sim_{LM}(x, y) \stackrel{def}{=} -CE\left(p_{JM}^{[0]}(\cdot|x) \;\middle|\middle|\; p_{JM}^{[\psi]}(\cdot|y)\right)$; higher values of CE correspond to reduced similarity.

We use the following similarity features: (i) **ClaimTitle**: between the claim and the document title ($sim_{LM}(c, d_t)$); (ii) **EntTitle**: between the entity and the document title ($sim_{LM}(c_e, d_t)$); (iii) **ClaimBody**: between the claim and the document body ($sim_{LM}(c, d_b)$); (iv) **EntBody**: between the entity and the document body ($sim_{LM}(s_e, d_b)$); (v) **ClaimSent**: between the claim and the sentence ($sim_{LM}(c, s)$); and, (vi) **EntSent**: between the entity and the sentence ($sim_{LM}(c_e, s)$). The entity is treated here as a bag of terms. These relevance-based similarity estimates, some of which are components of Eqs. 2 and 3, are weighed by the learning to rank method with respect to support ranking rather than relevance ranking, which helps to avoid metric divergence issues [3].

Semantic Similarities. Both the claim c and the candidate support sentence s can be short. Thus, to address potential vocabulary mismatch issues in textual similarity estimation, we also use semantic-based similarity measures that utilize word embedding [6]. Specifically, we use the word vectors, of dimension 300, trained over a Google news dataset with Word2Vec[3]. Let $\boldsymbol{w}$ denote the embedding vector representing term w. We measure the extent to which the terms in s "cover" the terms in c by **MaxSemSim**: $\sum_{w \in c} \max_{w' \in s} \cos(\boldsymbol{w}, \boldsymbol{w'})$. Additionally, we measure the similarity between the centroids of the claim and the sentence (cf. [7]), **CentSemSim**: $\cos(\frac{1}{|c|} \sum_{w \in c} \boldsymbol{w}, \frac{1}{|s|} \sum_{w' \in s} \boldsymbol{w'})$; $|c|$ and $|s|$ are the number of terms in the claim and sentence, respectively.

Sentiment Features. As the claim c is assumed to be subjective, we make the premise that a relevant sentence s is likely to also support c if the same sentiment

[2] Smoothing is performed using the term statistics in the document corpus D.

[3] https://code.google.com/p/word2vec/.

is expressed in c and s. We use the Stanford sentiment analyzer[4] [8], pre-trained with the Rotten Tomatoes movie reviews dataset [9]. This tool produces, for a given text, a probability distribution over a 1–5 sentiment scale; 1 stands for "very negative" and 5 stands for "very positive". As a sentiment similarity feature, **SentimentSim**, we use the Jensen Shannon (JS) divergence between the sentiment distributions for the claim and the sentence. Higher JS values correspond to lower similarity. Additionally, we compute **SentimentEnt**: the entropy of the sentiment distribution induced for s. This feature attests to the focus (or lack thereof) of the sentiment distribution induced from the sentence.

Quality-Oriented Language Models. In general, we expect to find differences between the language used to describe entities that are of "high quality" compared to those of "low quality". Still, to construct language models for such classes of entities, labeled examples are needed. This labeling is typically missing from most Web sources. Yet, for many domains there are sites that provide ratings for entities, *e.g.*, user feedback for local businesses in yelp.com. We propose to transfer such ratings to other sites as noisy quality labels. Specifically, our test claims are about movies, and the sentences ranked for support are extracted from Wikipedia which does not provide explicit ratings. Therefore, we automatically labeled each Wikipedia page about a movie with the 1–5 star grade review posted for this movie in IMDB[5] (if exists). Using this knowledge transfer, five unigram language models were induced, one per rating grade l. Specifically, all Wikipedia pages of movies with an IMDB review of a grade l were concatenated to yield the text: $Text_l$.[6] Then, for sentence s, the claim-independent features, denoted **Prior-l**, that correspond to quality levels $l \in \{1, \ldots, 5\}$ are: $\frac{sim_{LM}(s,Text_l)}{\sum_{l'=1}^{5} sim_{LM}(s,Text_{l'})}$.

Sentence Style. The **StopWords** feature is the fraction of terms in the sentence that are stop words. High occurrence of stop words potentially attests to rich use of language [10], and consequently, to sentence quality. Stop words are determined using the Stanford parser[7]. We also use the sentence length, **SentLength**, as a prior signal for sentence quality.

3 Empirical Evaluation

3.1 Dataset

There is no publicly available dataset for evaluating sentence ranking for support of subjective claims that originate from advice-seeking questions and corresponding answers. Hence, we created a novel dataset[8] as follows. Fifty subjective claims

[4] http://nlp.stanford.edu/software/corenlp.shtml.
[5] IMDB snapshot from 08/01/2014.
[6] The order of concatenation has no effect since unigram language models are used.
[7] http://nlp.stanford.edu/software/lex-parser.shtml.
[8] Available at http://iew3.technion.ac.il/~kurland/supportRanking.

Table 1. Examples of claims and supporting and non-supporting sentences.

Claim	The Pursuit of Happyness is one of the best inspirational movies
Support	"The Pursuit of Happyness" is an unexceptional film with exceptional performances
Non-support	The Pursuit of Happyness is a 2006 American biographical drama film based on Chris Gardner's nearly one-year struggle with homelessness
Claim	The Godfather is one of the top movies of all times
Support	Also in 2002, "The Godfather" was ranked the second best film of all time by Film4, after "Star Wars Episode V: The Empire Strikes Back"
Non-support	The opening shot of the film is a long, slow pullback, starting with a close-up of Bonasera, who is petitioning Don Corleone, and ending with the Godfather, seen from behind, framing the picture
Claim	Saving private Ryan is a favourite war movie
Support	In 2014, "Saving Private Ryan" was selected for preservation in the National Film Registry as per being deemed "culturally, historically, or aesthetically significant"
Non-support	Saving Private Ryan was released on home video in May 1999, earning $44 million from sales

about movies, which serve as the entities c_e, were collected from Yahoo Answers[9] by scanning its movies category. We looked for advice-seeking questions, which are common in the movies category, and selected answers that contain at least one movie title. Each pair of a question and a movie title appearing in an answer for the question was transformed to a claim by manually reformulating the question into an affirmative form and inserting the entity (movie title) as the subject. For example, the question "*any good science fiction movies?*" and the movie title "*Tron*" was transformed to the claim "*Tron is a good science fiction movie*".

The corpus used for sentence retrieval is a dump of the movies category of Wikipedia from March 2015, which contains 111, 164 Wikipedia pages. For each claim, 100 sentences were retrieved using the initial sentence retrieval approach, InitSent (Sect. 2.1). Each of these 100 sentences was categorized by five annotators from CrowdFlower[10] into: (1) not relevant to the claim, (2) strong non-support, (3) medium non-support, (4) neutral, (5) medium support, (6) strong support. The final label was determined by a majority vote.

We used the following induced scales: (a) ***binary relevance***: not relevant (category 1) vs. relevant (categories 2–6); (b) ***binary support***: non-support (categories 1–4) vs. support (categories 5–6); (c) ***graded support***: non-support (categories 1–4), weak support (category 5) and strong support (category 6). The Fleiss' Kappa inter-annotator agreement rates are: 0.68 (substantial) for

[9] answers.yahoo.com.
[10] www.crowdflower.com.

binary relevance, 0.592 (moderate) for *binary support* and 0.457 (moderate) for *graded support*. Table 1 provides examples of claims and relevant (on a binary scale) sentences that either support the claim or not (i.e., binary support scale is used).

Ten out of the fifty claims had no support sentences and were not used for evaluation. For the forty remaining claims, on average, half of the support sentences were weak support and the other half were strong support. On average, 23.5 % of the relevant sentences are supportive (binary scale). The median, average and standard deviation of relevant sentences, and of support sentences (binary scale), per claim are: 29, 40.5, 29.3 and 5.5, 7.4 and 6.7, respectively.

3.2 Methods

For the learning-to-rank methods (LTR) we used a linear SVMrank (LinearSVM) [11], a second-degree polynomial kernel SVMrank (PolySVM) [11], and LambdaMART [12], which is a state-of-the-art learning-to-rank method [5]. We used the LTR methods[11] with all the features described in Sect. 2.2 for ranking sentences by support and by relevance — i.e., we either optimized performance for support or for relevance. Leave-one-out cross validation, performed over queries, was used for training and testing.

The Indri toolkit[12] was used for experiments. Krovetz stemming was applied to claims and sentences only for inducing the initial document and sentence ranking and for computing the language-model-based similarity features described in Sect. 2.2. For these features, stopwords on the INQUERY list were removed only from claims. The number of documents (Wikipedia pages) initially retrieved using InitDoc (Eq. 2) for each claim was $N = 1000$; α was set to 0.66 to boost the ranking of the Wikipedia page about the target movie in the claim. Then, $k = 100$ sentences from these 1000 documents were retrieved using InitSent (Eq. 3) with $\beta = 0.5$. These 100 sentences constitute the set $\mathcal{S}_{\text{init}}^{[100]}$ which is reranked by the LTR methods. The SDM free parameters, λ_T, λ_O and λ_U were automatically set, in both InitDoc and InitSent, using the approach proposed in [13]. For language models, the Dirichlet smoothing parameter, μ, and the Jelinek-Mercer smoothing parameter, ψ, were set to the standard values of 1000 and 0.1, respectively [4]. We note that the free parameters of the initial document and sentence ranking could not be set using training data, as such data is only available for the initially retrieved sentence set, $\mathcal{S}_{\text{init}}^{[100]}$, as described above.

We view support ranking as a high-precision oriented task in which users are interested in seeing a few sentences that strongly support the claims at hand. Hence, for evaluation measures we use NDCG@1, NDCG@3, NDCG@10 and the precision of the top-5 sentences (p@5). The NDCG performance numbers for

[11] The implementations of LinearSVM and PolySVM are from http://www.cs.cornell.edu/people/tj/svm_light/svm_rank.html. The LambdaMART implementation is from http://sourceforge.net/p/lemur/wiki/RankLib/. All methods are used with default free-parameter values of the corresponding implementations.

[12] www.lemurproject.org.

Table 2. Main result table. Comparing the relevance-ranking and support-ranking performance of the three LTR methods with that of the initial sentence ranking (InitSent). Boldface: the best result in a column; 'i', 'l' and 'p' mark statistically significant differences with InitSent, LinearSVM and PolySVM, respectively.

	Relevance				Support			
	NDCG@1	NDCG@3	NDCG@10	p@5	NDCG@1	NDCG@3	NDCG@10	p@5
InitSent	.775	.766	.739	.730	.083	.165	.295	.215
LinearSVM	.800	.786	.782	.770	.441^{i}	.478^{i}	.519^{i}	.410^{i}
PolySVM	.800	.839	**.852**^{i,l}	**.835**^{i}	.525^{i}	.527^{i}	.564^{i,l}	.445^{i}
LambdaMART	**.825**	**.844**	.808	**.835**^{i}	**.608**^{i}	**.540**^{i}	**.593**^{i}	**.515**$^{i,l}_{p}$

support ranking are based on *graded support* scale, and those for p@5 are based on the *binary support* scale. All performance numbers for relevance ranking are based on the *binary relevance* scale. LambdaMART was trained for NDCG@10 as this yielded, in general, better support ranking performance across the evaluation measures than using NDCG@1 or NDCG@3. Statistically significant differences of performance are determined using the two tailed paired t-test with $p = 0.05$.

3.3 Results

Table 2 presents our main results. We see that all three LTR methods outperform the initial sentence ranking, InitSent, in terms of relevance ranking. Although few of these improvements are statistically significant, they attest to the potential merits of using the additional relevance-based features described in Sect. 2.2. More importantly, all LTR methods substantially, and statistically significantly, outperform the initial (relevance-based) sentence ranking in terms of support. This result emphasizes the difference between relevance and support and shows that our proposed features for support ranking are quite effective, especially when used in a non-linear ranker such as LambdaMART.

In Sects. 1 and 4 we discuss the difference between subjective and factoid claims. To further explore this difference, we compare our best performing LambdaMART method with the P1EDA[13] state-of-the-art textual entailment algorithm [14] when both are used for the support-ranking task we address here. P1EDA was designed for factual claims[14]. Specifically, given a claim and a candidate sentence, P1EDA produces a classification decision of whether the sentence entails the claim, accompanied with a confidence level. The confidence level was used for support (and relevance) ranking. We also tested the inclusion of P1EDA's output (confidence level) as an additional feature in LambdaMART, yielding the LMart+P1EDA method. Table 3 depicts the performance numbers.

We can see in Table 3 that P1EDA is (substantially) outperformed by both InitSent and LambdaMART, for both relevance and support ranking. Since the

[13] http://hltfbk.github.io/Excitement-Open-Platform/.

[14] We trained P1EDA using the SNLI data set [15], which contains 549,366 examples.

Table 3. Comparison and integration with a state-of-the-art textual entailment algorithm (P1EDA). LMart stands for "LambdaMart". Boldface: the best result in a column. Statistically significant differences with InitSent, P1EDA and LMart are marked with 'i', 'p', and 'm', respectively.

	Relevance				Support			
	NDCG@1	NDCG@3	NDCG@10	p@5	NDCG@1	NDCG@3	NDCG@10	p@5
InitSent	.775	.766	.739	.730	.083	.165	.295	.215
P1EDA	.525^{i}	.496^{i}	.462^{i}	.475^{i}	.066	.093	.129^{i}	.120^{i}
LMart	.825^{i,p}	**.844**^{i,p}	.808^{i,p}	**.835**^{i,p}	**.608**^{i,p}	.540^{i,p}	.593^{i,p}	**.515**^{i,p}
LMart+P1EDA	**.850**^{p}	.836^{p}	**.811**^{p}	.815^{p,m}	.600^{i,p}	**.571**^{i,p}	**.609**^{i,p}	.490^{i,p}

claims in our setting are simple, this finding implies that approaches for identifying texts that support (or "prove") a factoid claim may not be effective for the task of supporting subjective claims. The integration of P1EDA as a feature in LambdaMART improves performance (although not to a statistically significant degree) for some of the evaluation measures, including NDCG@10 for which the ranker was trained, and hurts performance for others — statistically significantly so in only a single case[15].

Integrating P1EDA with *only* our semantic-similarity features using LambdaMART, which is a conceptually similar approach to a classification method employed in some work on argument mining [16], resulted in support-ranking performance that is substantially worse than that of using all our proposed features in LambdaMART. Actual numbers are omitted due to space considerations and as they convey no additional insight.

Feature Analysis. To analyze the contribution of individual features to overall performance, Table 4 compares LambdaMART, used with all features, to using individual features alone for ranking. As LambdaMART was trained for NDCG@10 for support ranking, we explore the 10 features that yielded the highest NDCG@10 support-ranking performance.

Table 4 clearly shows that while a few features yield support-ranking performance that transcends that of the initial sentence ranking (InitSent), LambdaMART that integrates all features yields substantially, and statistically significantly, better support-ranking performance. This finding attests to the importance of integrating various features for support ranking. LambdaMART is also superior to almost all ten features for relevance ranking[16].

We see that quite a few of the top-10 features are (lexical) similarities between the claim and/or the entity it is about and the sentence and/or its ambient document. This shows that (direct) estimates of claim-sentence relevance can be quite important for support ranking, as is expected. Yet, integrating these estimates with support-oriented estimates is important for attaining highly effective support ranking performance as is evident in LambdaMART's performance.

[15] Integrating P1EDA in PolySVM did not yield support-ranking improvements.

[16] For relevance ranking, LambdaMART was trained for *binary relevance*.

Table 4. Using features alone (specifically, the 10 that yield the highest NDCG@10 support ranking) to rank the initial sentence list vs. integrating all features in LambdaMART. Boldface: the best result in a column; 'm': statistically significant difference with LambdaMART.

	NDCG@1		NDCG@3		NDCG@10	
	Relevance	Support	Relevance	Support	Relevance	Support
LambdaMART	**.825**	**.608**	**.844**	**.540**	.808	**.593**
ClaimTitle	.700	$.191^m$	.754	$.216^m$	.796	$.302^m$
EntTitle	.725	$.200^m$	.791	$.212^m$	**.811**	$.297^m$
InitSent	.775	$.083^m$	.766	$.165^m$	.739	$.295^m$
SentimentSim	$.475^m$	$.225^m$	$.460^m$	$.239^m$	$.433^m$	$.295^m$
InitDoc	$.600^m$	$.150^m$	$.669^m$	$.208^m$	$.684^m$	$.277^m$
ClaimSent	$.575^m$	$.291^m$	$.614^m$	$.272^m$	$.603^m$	$.276^m$
MaxSemSim	.700	$.183^m$	$.688^m$	$.195^m$	$.654^m$	$.228^m$
ClaimBody	$.450^m$	$.041^m$	$.612^m$	$.130^m$	$.696^m$	$.222^m$
Prior-5	$.500^m$	$.083^m$	$.479^m$	$.142^m$	$.506^m$	$.221^m$
EntSent	.650	$.066^m$	$.667^m$	$.113^m$	$.692^m$	$.188^m$

SentimentSim, the sentiment similarity between the claim and the sentence, is among the most effective features when used alone for support ranking. Additional ablation tests[17] reveal that removing SentimentSim from the set of all features results in the most severe performance degradation for all three learning-to-rank methods. Indeed, sentiment is an important aspect of subjective claims, and therefore, of inferring support for these claims.

We also found that ranking sentences by decreasing entropy of sentiment (SentimentEnt) is superior to ranking by increasing entropy for NDCG@1 and NDCG@3, while for NDCG@10 the reverse holds. The former finding is a conceptual reminiscent of those about using the entropy of a document term distribution for the document prior for Web search [10]: the higher the entropy, the "broader" the textual unit is – in our case, in terms of expressed sentiment — which presumably implies to a higher prior.

Finally, Table 4 also shows that Prior-5 is the most effective claim-independent feature[18]. It is the similarity between a language model of the sentence and that induced from Wikipedia movie pages which received high grade (5 stars) reviews in IMDB. This shows that although Wikipedia authors aim to be objective in their writing, the style and information for high rated movies is still quite different from that for lower rated ones, and it can potentially be modeled via the automatic knowledge transfer and labeling method proposed in Sect. 2.2.

[17] Actual numbers are omitted due to space considerations and as they convey no additional insight.

[18] Ablation tests reveal that removing this feature results in the second most substantial decrease of support-ranking performance among all features.

4 Related Work

A few lines of research are related to our work. The Textual Entailment task is inferring the truthfulness of a textual statement (*hypothesis*) from a given text [17]. A more specific incarnation of Textual Inference is automatic Question Answering (QA). Work on these tasks focused on factoid claims for which a clear correct/incorrect labeling should be inferred from supportive evidence. Thus, typical textual inference approaches are designed to find the claim (*e.g.* a candidate answer in QA) embedded in the supporting text, although it may be rephrased. In contrast, in this paper, claims originate from CQA users who provide subjective recommendations rather than state facts. Our model, designed for ranking sentences by support for a subjective claim, significantly outperforms for this task a state-of-the-art textual entailment method as shown in Sect. 3.3.

Blanco and Zaragoza [18] introduce methods for retrieving sentences that explain the relationship between a Web query and a related named entity, as part of the *entity ranking* task. In contrast, we rank sentences by support for a subjective claim. Kim et al. [19] present methods for retrieving sentences that explain reasons for sentiment expressed about an aspect of a topic. In contrast to these sentence ranking methods [18,19], ours utilizes a learning-to-rank method that integrates various relevance and support features not used in [18,19].

The task most related to ours is *argument mining* (e.g., [16,20–24]). Specifically, arguments supporting or contradicting a claim about a given debatable (often controversial) topic are sought. Some of the *types* of features we use for support ranking have also been used for argument mining; namely, semantic [16,24] and sentiment [24] similarities between the claim and a candidate argument. Yet, the actual estimates and techniques used here to induce these features are different than those in work on argument mining [16,24]. Furthermore, the knowledge-transfer-based features we utilize, and whose effectiveness was demonstrated in Sect. 3.3, are novel to this study.

Interestingly, while textual entailment features were found to be effective for argument mining [16,20], this is not the case for support ranking (see Sect. 3.3). This finding could be attributed to the fundamentally different nature of claims used in our work, and those used in argument mining. That is, our claims originate from answers to advice-seeking questions of subjective nature, rather than being about a given debatable/controversial topic. Also, additional information about the debatable topic which was utilized in work on argument mining [24] is not available in our setting.

Often, work on argument mining [24], similarly to that on question answering (e.g., [25]), focuses on finding supporting or contradicting evidence in the same document in which the claim appears. In contrast, we retrieve supporting sentences from the Web for claims originating from CQA sites. In fact, there has been very little work on using sentence retrieval for argument mining [22]. In contrast to our work, a Boolean retrieval method was used, different features were utilized, and relevance-based estimates were not integrated with support-based estimates using a learning-to-rank approach.

5 Conclusions and Future Work

We addressed a novel task: ranking sentences from the Web by the support they provide to a subjective claim. The claim originates from an answer provided in a community question answering (CQA) site to an advice-seeking question.

Our support-ranking model utilizes various features in a learning-to-rank method; some are relevance oriented while others are support oriented. Empirical evaluation performed using a new dataset of claims created from Yahoo Answers attested to the merits of our proposed approach.

For future work we intend to extend the set of features, explore additional data domains, and study the utilization of supportive sentences in answers posted for subjective questions in CQA sites.

Acknowledgments. We thank the reviewers for their helpful comments, and Omer Levy and Vered Shwartz for their help with the textual entailment tool used for experiments. This work was supported in part by a Yahoo! faculty research and engagement award.

References

1. Rieke, R., Sillars, M.: Argumentation and Critical Decision Making. Longman Series in Rhetoric and Society, Longman (1997)
2. Murdock, V.: Exploring Sentence Retrieval. VDM Verlag, Saarbrücken (2008)
3. Metzler, D., Croft, W.B.: A markov random field model for term dependencies. In: Proceedings of SIGIR, pp. 472–479 (2005)
4. Zhai, C., Lafferty, J.D.: A study of smoothing methods for language models applied to ad hoc information retrieval. In: Proceedings of SIGIR, pp. 334–342 (2001)
5. Liu, T.Y.: Learning to rank for information retrieval. Found. Trends Inf. Retrieval **3**(3), 225–331 (2009)
6. Mikolov, T., Sutskever, I., Chen, K., Corrado, G., Dean, J.: Distributed representations of words and phrases and their compositionality. In: Proceedings of NIPS, pp. 3111–3119 (2013)
7. Vulic, I., Moens, M.: Monolingual and cross-lingual information retrieval models based on (bilingual) word embeddings. In: Proceedings of SIGIR, pp. 363–372 (2015)
8. Socher, R., Perelygin, A., Wu, J., Chuang, J., Manning, C., Ng, A., Potts, C.: Recursive deep models for semantic compositionality over a sentiment treebank. In: Proceedings of EMNLP, pp. 363–372 (2013)
9. Pang, B., Lee, L.: Opinion mining and sentiment analysis. Found. Trends Inf. Retrieval **2**(1–2), 1–135 (2007)
10. Bendersky, M., Croft, W.B., Diao, Y.: Quality-biased ranking of web documents. In: Proceedings of WSDM, pp. 95–104 (2011)
11. Joachims, T.: Training linear SVMs in linear time. In: Proceedings of KDD, pp. 217–226 (2006)
12. Burges, C.J.: From RankNet to LambdaRank to LambdaMART: an overview. Technical report, Microsoft Research (2010)
13. Zhou, Y., Croft, B.: Query performance prediction in web search environments. In: Proceedings of SIGIR, pp. 543–550 (2007)

14. Noh, T.G., Pado, S., Shwartz, V., Dagan, I., Nastase, V., Eichler, K., Kotlerman, L., Adler, M.: Multi-level alignments as an extensible representation basis for textual entailment algorithms. In: Proceedings of SEM (2015)
15. Bowman, S.R., Angeli, G., Potts, C., Manning, C.D.: A large annotated corpus for learning natural language inference. In: Proceedings of EMNLP, pp. 632–642 (2015)
16. Boltužić, F., Šnajder, J.: Back up your stance: recognizing arguments in online discussions. In: Proceedings of the First Workshop on Argumentation Mining, Association for Computational Linguistics, pp. 49–58 (2014)
17. Dagan, I., Roth, D., Sammons, M., Zanzotto, F.M.: Recognizing textual entailment: models and applications. Synth. Lect. Hum. Lang. Technol. **6**(4), 1–220 (2013)
18. Blanco, R., Zaragoza, H.: Finding support sentences for entities. In: Proceedings of SIGIR, pp. 339–346 (2010)
19. Kim, H.D., Castellanos, M., Hsu, M., Zhai, C., Dayal, U., Ghosh, R.: Ranking explanatory sentences for opinion summarization. In: Proceedings of SIGIR, pp. 1069–1072 (2013)
20. Cabrio, E., Villata, S.: Combining textual entailment and argumentation theory for supporting online debates interactions. In: Proceedings of ACL, pp. 208–212 (2012)
21. Green, N., Ashley, K., Litman, D., Reed, C., Walker, V. (eds.) Proceedings of the First Workshop on Argumentation Mining, Baltimore, Maryland. Association for Computational Linguistics (2014). http://www.aclweb.org/anthology/W/W14/W14-21
22. Sato, M., Yanai, K., Yanase, T., Miyoshi, T., Iwayama, M., Sun, Q., Niwa, Y.: End-to-end argument generation system in debating. In: Proceedings of ACL-IJCNLP 2015 System Demonstrations (2015)
23. Cardie, C. (ed.): Proceedings of the 2nd Workshop on Argumentation Mining, Denver, CO. Association for Computational Linguistics (2015). http://www.aclweb.org/anthology/W15-05
24. Rinott, R., Dankin, L., Alzate Perez, C., Khapra, M.M., Aharoni, E., Slonim, N.: Show me your evidence - an automatic method for context dependent evidence detection. In: Proceedings of EMNLP, pp. 440–450 (2015)
25. Brill, E., Lin, J.J., Banko, M., Dumais, S.T., Ng, A.Y., et al.: Data-intensive question answering. In: Proceedings of TREC, vol. 56, p. 90 (2001)

Ranking

Does Selective Search Benefit from WAND Optimization?

Yubin Kim[1](✉), Jamie Callan[1], J. Shane Culpepper[2], and Alistair Moffat[3]

[1] Carnegie Mellon University, Pittsburgh, USA
yubink@cmu.edu
[2] RMIT University, Melbourne, Australia
[3] The University of Melbourne, Melbourne, Australia

Abstract. Selective search is a distributed retrieval technique that reduces the computational cost of large-scale information retrieval. By partitioning the collection into topical shards, and using a resource selection algorithm to identify a subset of shards to search, selective search allows retrieval effectiveness to be maintained while evaluating fewer postings, often resulting in 90+% reductions in querying cost. However, there has been only limited attention given to the interaction between dynamic pruning algorithms and topical index shards. We demonstrate that the WAND dynamic pruning algorithm is more effective on topical index shards than it is on randomly-organized index shards, and that the savings generated by selective search and WAND are additive. We also compare two methods for applying WAND to topical shards: searching each shard with a separate top-k heap and threshold; and sequentially passing a shared top-k heap and threshold from one shard to the next, in the order established by a resource selection mechanism. Separate top-k heaps provide low query latency, whereas a shared top-k heap provides higher throughput.

Keywords: Selective search · Distributed search · Dynamic pruning · Efficiency

1 Introduction

Selective search is a technique for large-scale distributed search in which the document corpus is partitioned into p topic-based shards during indexing. When a query is received, a resource selection algorithm such as Taily [1] or Rank-S [13] selects the most relevant k shards to search, where $k \ll p$. Results lists from those shards are merged to form a final answer listing to be returned to the user. Selective search has substantially lower computational costs than partitioning the corpus randomly and searching all index shards, which is the most common approach to distributed search [11,12].

Dynamic pruning algorithms such as *Weighted AND* (WAND) [3] and *term-bounded max score* (TBMS) [22] improve the computational efficiency of retrieval systems by eliminating or early-terminating score calculations for documents

N. Ferro et al. (Eds.): ECIR 2016, LNCS 9626, pp. 145–158, 2016.
DOI: 10.1007/978-3-319-30671-1_11

which cannot appear in the top-k of the final ranked list. But topic-based partitioning and resource selection change the environment in which dynamic pruning is performed, and query term posting lists are likely to be longer in shards selected by the resource selection algorithm than in shards that are not selected. As well, each topic-based shard should contain similar documents, meaning that it might be difficult for dynamic pruning to distinguish amongst them using only partial score calculations. Conversely, the documents in the shards that were *not* selected for search might be the ones that a dynamic pruning algorithm would have bypassed if it had encountered them. That is, while the behavior of dynamic pruning algorithms on randomly-organized shards is well-understood, the interaction between dynamic pruning and selective search is not. As an extreme position, it might be argued that selective search is simply achieving the same computational savings that dynamic pruning would have produced, but incurs the additional overhead of clustering the collection and creating the shards. To address these concerns, we investigate the behavior of the well-known *Weighted AND* (WAND) dynamic pruning algorithm in the context of selective search, considering two research questions:

RQ1: Does dynamic pruning improve selective search, and if so, why?
RQ2: Can the efficiency of selective search be improved further using a cascaded pruning threshold during shard search?

2 Related Work

Selective search is a cluster-based retrieval technique [6,19] that combines ideas from conventional distributed search and federated search [12]. Modern cluster-based systems use inverted indexes to store clusters that were defined using criteria such as broad topics [4] or geography [5]. The shards' vocabularies are assumed to be random and queries are sent to a single best shard, forwarding to additional shards as needed [5].

In selective search, the corpus is automatically clustered into query-independent *topic-based shards* with skewed vocabularies and distributed across resources. When a query arrives, a resource selection algorithm identifies a subset of shards that are likely to contain the relevant documents. The selected shards are searched in parallel, and their top-k lists merged to form a final answer. Because only a few shards are searched for each query, total cost per query is reduced, leading to higher throughput.

Previous studies showed that selective search accuracy is comparable to a typical distributed search architecture, but that efficiency is better [1,12], where computational cost is determined by counting the number of postings processed [1,12], or by measuring the execution time of a proof-of-concept implementation.

Resource Selection. Choosing which index shards to search for a query is critical to search accuracy. There are three broad categories of resource selection algorithm: term-based, sample-based, and classification-based. Term-based algorithms model the language distribution of a shard to estimate the relevance of the shard to a query, with the vocabulary of each shard typically treated as a bag of words. The estimation of relevance is accomplished by adapting an existing document scoring algorithm [8] or by developing a new algorithm specifically for resource selection [1,9,15,24]. Taily [1] is one of the more successful approaches, and fits a Gamma distribution over the relevance scores for each term. At query time, these distributions are used to estimate the number of highly scoring documents in the shard.

Sample-based algorithms extract a small (of the order of 1%) sample of the entire collection, and index it. When a query is received, the sample index is searched and each top-ranked document acts as a (possibly weighted) vote for the corresponding index shard [13,16,20,21,23]. One example is Rank-S [13], which uses an exponentially decaying voting function derived from the document's retrieval rank. The (usually small number of) resources with scores greater than 0.0001 are selected.

Classification-based algorithms use training data to learn models for resources using features such as text, the scores of term-based and sample-based algorithms, and query similarity to historical query logs [2,10]. While classification-based algorithms can be more effective than unsupervised methods, they require access to training data. Their main advantage lies in combining heterogeneous resources such as search verticals.

The Rank-S [13] and Taily [1] have both been used in prior work with similar effectiveness. However Taily is more efficient, because lookups for Gamma parameters are substantially faster than searching a sample index. We use both in our experiments.

Dynamic Pruning. *Weighted AND* (WAND) is a dynamic pruning algorithm that only scores documents that may become one of the current top k based on a preliminary estimate [3]. Dimopoulos et al. [7] developed a Block-Max version of WAND in which continuous segments of postings data are bypassed under some circumstances by using an index where each block of postings has a local maximum score. Petri et al. [17] explored the relationship between WAND-style pruning and document similarity formulations. They found that WAND is more sensitive than Block-Max WAND to the document ranking algorithm. If the distribution of scores is skewed, as is common with BM25, then WAND alone is sufficient. However, if the scoring regime is derived from a language model, then the distribution of scores is top-heavy, and BlockMax WAND should be used. Rojas et al. [18] presented a method to improve performance of systems combining WAND and a distributed architecture with random shards.

Term-Bounded Max Score (TBMS) [22] is an alternative document-at-a-time dynamic pruning algorithm that is currently used in the Indri Search Engine. The key idea of TBMS is to precompute a "topdoc" list for each term, ordered by the frequency of the term in the document, and divided by the document length. The algorithm uses the union of the topdoc lists for the terms to determine a candidate list of documents to be scored. The number of documents in the topdoc list for each term is experimentally determined, a choice that can have an impact on overall performance. Kulkarni and Callan [12] explored the effects of TBMS on selective search and traditional distributed search architectures. Based on a small set of queries they measured efficiency improvements of 23–40 % for a traditional distributed search architecture, and 19–32 % for selective search, indicating that pruning can improve the efficiency of both approaches.

3 Experiments

The observations of Kulkarni and Callan [12] provide evidence that dynamic pruning and selective search can be complementary. Our work extends that exploration in several important directions. First, we investigate whether there is a correlation between the rank of a shard and dynamic pruning effectiveness for that shard. A correlation could imply that dynamic pruning effectiveness depends on the number of shards searched. We focus on the widely-used WAND pruning algorithm, chosen because it is both efficient and versatile, particularly when combined with a scoring function such as BM25 that gives rise to skewed score distributions [7,17].

Experiments were conducted using the ClueWeb09 Category B dataset, containing 50 million web documents. The dataset was partitioned into 100 topical shards using k-means clustering and a KL-divergence similarity metric, as described by Kulkarni and Callan [11], and stopped using the default Indri stoplist and stemmed using the Krovetz stemmer. On average, the topical shards contain around 500k documents, with considerable variation, see Fig. 1. A second partition of 100 random shards was also created, a system in which exhaustive "all shards" search is the only way of obtaining effective retrieval. Each shard in the two systems was searched using BM25, with $k_1 = 0.9$, $b = 0.4$, and global corpus statistics for *idf* and average document length.[1]

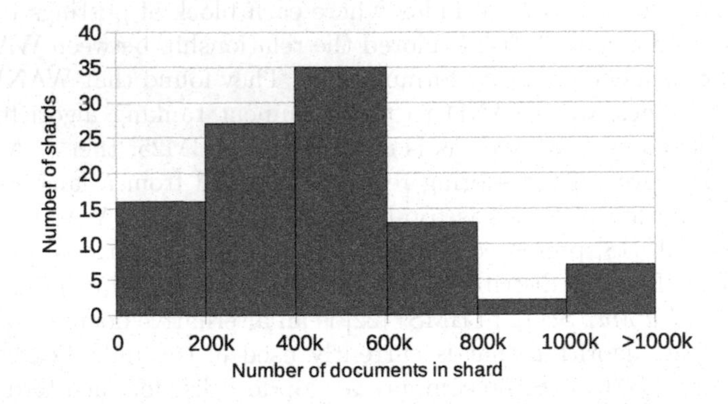

Fig. 1. Distribution of shard sizes, with a total of 100 shards.

[1] The values for b and k_1 are based on the parameter choices reported for Atire and Lucene in the 2015 IR-Reproducibility Challenge, see http://github.com/lintool/IR-Reproducibility.

Each selected shard returned its top 1,000 documents, which were merged by score to produce a final list of $k = 1{,}000$ documents. In selective search, deeper ranks are necessary because most of the good documents may be in one or two shards due to the term skew. Also, deeper k supports learning-to-rank algorithms. Postings lists were compressed and stored in blocks of 128 entries using the FastPFOR library [14], supporting fast block-based skipping during the WAND traversal.

Two resource selection algorithms were used: Taily [1] and Rank-S [13]. The Taily parameters were taken from Aly et al. [1]: $n = 400$ and $v = 50$, where v is the cut-off score and n represents the theoretical depth of the ranked list. The Rank-S parameters used are consistent with the values reported by Kulkarni et al. [13]. A decay base of $B = 5$ with a centralized sample index (CSI) containing 1% of the documents was used – approximately the same size as the average shard. We were unable to find parameters that consistently yielded better results than the original published values.

We conducted evaluations using the first 1,000 unique queries from each of the AOL query log[2] and the TREC 2009 Million Query Track. We removed single-term queries, which do not benefit from WAND, and queries where the resource selection process did not select any shards. Removing single-term queries is a common procedure for research with WAND [3] and allows our results to be compared with prior work. That left 713 queries from the AOL log, and 756 queries from MQT, a total of 1,469 queries.

Our focus is on the efficiency of shard search, rather than resource selection. To compare the efficiency of different shard search methods, we count the num-

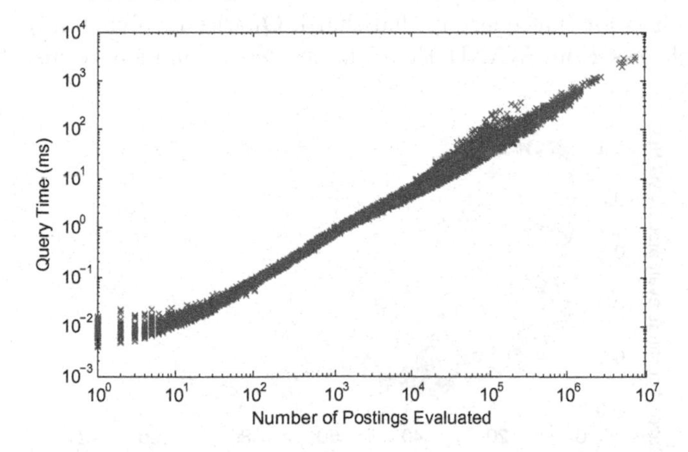

Fig. 2. Correlation between the number of postings processed for a query and the time taken for query evaluation. Data points are generated from MQT queries using both WAND and full evaluation, applied independently to all 100 topical shards and all 100 random shards. In total, $756 \times 200 \times 2 \approx 300{,}000$ points are plotted.

[2] We recognize that the AOL log has been withdrawn, but also note that it continues to be widely used for research purposes.

ber of postings scored, a metric that is strongly correlated with total processing time [3], and is less sensitive to system-specific tuning and precise hardware configuration than is measured execution time. As a verification of this relationship, Fig. 2 shows the correlation between processing time per query, per shard, and the number of postings evaluated. There is a strong linear relationship; note also that more than 99.9 % of queries completed in under 1 s with only a few extreme outliers requiring longer.

Pruning Effectiveness of WAND on Topical Shards. The first experiment investigated how WAND performs on the topical shards constructed by selective search. Each shard was searched independently, as is typical in distributed settings – parallelism is crucial to low response latency. w, the number of posting evaluations required in each shard by WAND-based query evaluation was recorded. The total length of the postings for the query terms in the selected shards was also recorded, and is denoted as b, representing the number of postings processed by an unpruned search in the same shard. The ratio w/b then measures the fraction of the work WAND carried out compared to an unpruned search. The lower the ratio, the greater the savings. Values of w/b can then be combined across queries in two different ways: micro- and macro-averaging. In micro-averaging, w and b are summed over the queries and a single value of w/b is calculated from the two sums. In macro-averaging, w/b is calculated for each query, and averaged across queries. The variance inherent in queries means that the two averaging methods can produce different values, although broad trends are typically consistent.

Figure 3 and Table 1 provide insights into the behavior of macro- and micro-averaging. Figure 3 uses the AOL queries and all 100 topical shards, plotting w/b values on a per query per shard basis as a function of the total length of the postings lists for that query in that shard. Queries involving only rare terms benefit much less from WAND than queries with common terms. Thus, the

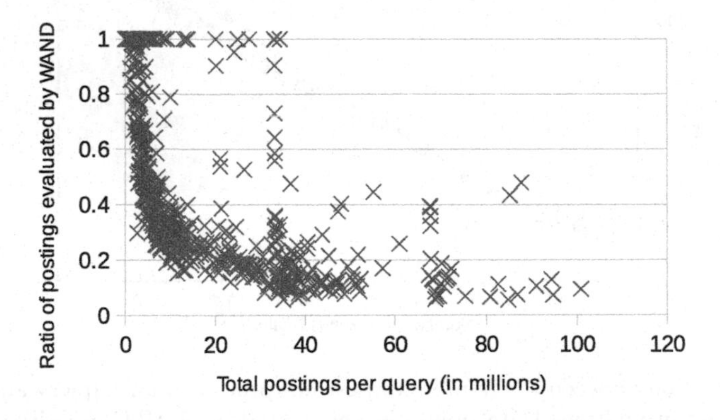

Fig. 3. Ratio of savings achieved by WAND as a function of the total postings length of each query in the AOL set, measured on a per shard basis. A total of $100 \times 713 \approx 71{,}000$ points are plotted. Queries containing only rare terms derive little benefit from WAND.

Table 1. Ratio of per shard per query postings evaluated and per shard per query execution time for WAND-based search, as ratios relative to unpruned search, averaged over 100 topical shards and over 100 randomized shards, and over two groups each of 700+ queries. The differences between the Topical and Random macro-averaged ratios are significant for both query sets and both measures (paired two-tailed t-test, $p < 0.01$).

		WAND postings cost ratio		WAND runtime cost ratio	
		Topical shards	Random shards	Topical shards	Random shards
AOL	micro-averaged	0.35	0.34	0.36	0.38
MQT	micro-averaged	0.36	0.36	0.39	0.43
AOL	macro-averaged	0.51	0.52	0.51	0.53
MQT	macro-averaged	0.60	0.63	0.58	0.63

macro-average of w/b is higher than the micro-average. Micro-averaging more accurately represents the total system savings, whereas macro-averaging allows paired significance testing. We report both metrics in Table 1. The second pair of columns gives millisecond equivalents of w/b, to further validate the postings-cost metric. These values are micro- and macro-averaged w_t/b_t ratios, where w_t is the time in milliseconds taken to process one of the queries on one of the 100 shards using WAND, and b_t is the time taken to process the same query with a full, unpruned search. A key result of Table 1 is that WAND is just as effective across the full set of topical shards as it is on the full set of randomly formed shards. Moreover, the broad trend of the postings cost ratios – that WAND avoids nearly half of the postings – is supported by the execution time measurements.

WAND and Resource Ranking Interactions. The second experiment compares the effectiveness of the WAND algorithm on the shards that the resource ranking algorithm would, and would not, select in connection with each query. The Taily and Rank-S resource selection algorithms were used to determine

Table 2. Average number of shards searched, and micro-averaged postings ratios for those selected shards and for the complement set of shards, together with the corresponding query time cost ratios, in each case comparing WAND-based search to unpruned search. Smaller numbers indicate greater savings.

	Shards searched	WAND postings cost ratio		WAND runtime cost ratio	
		Selected	Non-selected	Selected	Non-selected
Taily AOL	3.1	0.32	0.35	0.36	0.36
Taily MQT	2.7	0.23	0.37	0.30	0.40
Rank-S AOL	3.8	0.27	0.36	0.30	0.37
Rank-S MQT	3.9	0.24	0.37	0.30	0.40

Table 3. As for Table 2, but showing macro-averaged ratios. All differences between selected and non-selected shards are significant (paired two-tailed t-test, $p < 0.01$).

	WAND postings cost ratio		WAND runtime cost ratio	
	Selected	Non-selected	Selected	Non-selected
Taily AOL	0.42	0.52	0.45	0.52
Taily MQT	0.52	0.61	0.53	0.59
Rank-S AOL	0.42	0.53	0.44	0.52
Rank-S MQT	0.52	0.61	0.53	0.60

which shards to search. For each query the WAND savings were calculated for the small set of selected shards, and the much larger set of non-selected shards.

Table 2 lists micro-averaged w/b ratios, and Table 3 the corresponding macro-averaged ratios. While all shards see improvements with WAND, the selected shards see a greater efficiency gain than the non-selected shards, reinforcing our contention that resource selection is an important component in search efficiency. When compared to the ratios shown in Table 1, the selected shards see substantially higher benefit than average shards; the two orthogonal optimizations generate better-than-additive savings.

Figure 4a shows the distribution of the individual per query per shard times for the MQT query set, covering in the first four cases only the shards chosen by the two resource selection processes. The fifth exhaustive search configuration includes data for all of the 100 randomly-generated shards making up the second system, and is provided as a reference point. Figure 4b gives numeric values for

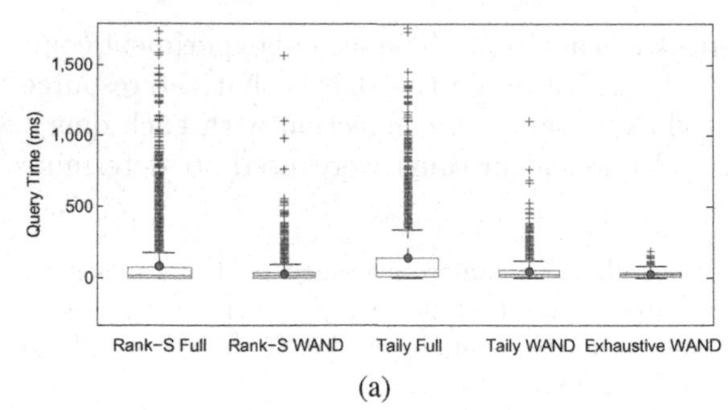

(a)

	Mean	Median
Rank-S Full	85.0	13.0
Rank-S WAND	28.5	11.3
Taily Full	134.0	34.2
Taily WAND	42.7	23.6
Exhaustive WAND	26.6	21.8

(b)

Fig. 4. Distribution of query response times for MQT queries on shards: (a) as a box plot distribution, with a data point plotted for each query-shard pair; (b) as a table of corresponding means and medians. In (a), the center line of the box indicates the median, the outer edges of the box the first and third quartiles, and the blue circle the mean. The whiskers extend to include all points within 1.5 times the inter-quartile range of the box. The graph was truncated to omit a small number of extreme points for both Rank-S Full and Taily-Full. The maximum time for both these two runs was 6,611 ms.

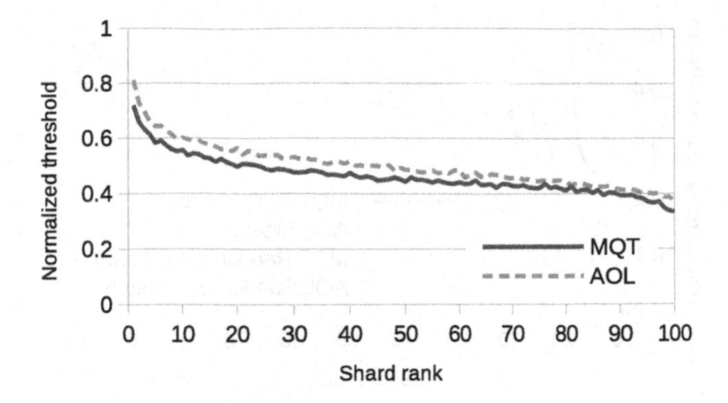

Fig. 5. Normalized 1,000 th document scores from shards, averaged over queries and then shard ranks, and expressed as a fraction of the collection-wide maximum document score for each corresponding query. The score falls with rank, as fewer high-scoring documents appear in lower-ranked shards.

the mean and median of each of the five distributions. When WAND is combined with selective search, it both reduces the average time required to search a shard and also reduces the variance of the query costs. Note the large differences between the mean and median query processing times for the unpruned search and the reduction in that gap when WAND is used; this gain arises because query and shard combinations that have high processing times due to long postings lists are the ones that benefit most from WAND. Therefore, in typical distributed environments where shards are evaulated in parallel, the slowest, bottleneck shard will benefit the most from WAND and may result in additional gains in latency reduction. Furthermore, while Fig. 4 shows similar per shard query costs for selective and exhaustive search, the total *work* associated with selective search is substantially less than exhaustive search because only 3–5 shards are searched per query, whereas exhaustive search involves all 100 shards. Taken in conjunction with the previous tables, Fig. 4 provides clear evidence that WAND amplifies the savings generated by selective search, answering the first part of RQ1 with a "yes". In addition, these experiments have confirmed that execution time is closely correlated with measured posting evaluations. The remaining experiments utilize postings counts as the cost metric.

We now consider the second part of RQ1 and seek to explain *why* dynamic pruning improves selective search. Part of the reason is that the postings lists of the query terms associated with the highly ranked shards are longer than they are in a typical randomized shard. With these long postings lists, there is more opportunity for WAND to achieve early termination. Figure 5 shows normalized final heap-entry thresholds, or equivalently, the similarity score of the 1,000 th ranked document in each shard. The scores are expressed as a fraction of the maximum document score for that query across all shards, then plotted as a function of the resource selector's shard ranking using Taily, averaged over queries. Shards that Taily did not score because they did not contain any query

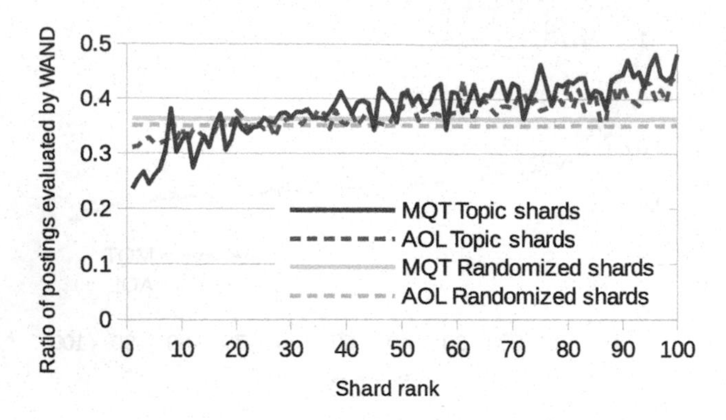

Fig. 6. The micro-average w/b ratio for WAND postings evaluations, as a function of the per query shard rankings assigned by Taily. Early shards generate greater savings.

terms were ordered randomly. For example, for the AOL log the 1,000 th document in the shard ranked highest by Taily attains, on average across queries, a score that is a little over 80 % of the maximum score attained by any single document for that same query. The downward trend in Fig. 5 indicates that the resource ranking process is effective, with the high heap-entry thresholds in the early shards suggesting – as we would hope – that they contain more of the high-scoring documents.

To further illustrate the positive relationship between shard ranking and WAND, w/b was calculated for each shard in the per query shard orderings, and then micro-averaged at each shard rank. Figure 6 plots the average as a function of shard rank, and confirms the bias towards greater savings on the early shards – exactly the ones selected for evaluation. As a reference point, the same statistic was calculated for a random ordering of the randomized shards (random since no shard ranking is applied in traditional distributed search), with the savings ratio being a near-horizontal line. If an unpruned full search were to be plotted, it would be a horizontal line at 1.0. The importance of resource selection to retrieval effectiveness has long been known; Fig. 6 indicates that effective resource selection can improve overall efficiency as well.

Improving Efficiency with Cascaded Pruning Thresholds. In the experiments reported so far, the rankings were computed on each shard independently, presuming that they would be executing in parallel and employing private top-k heaps and private heap-entry thresholds, with no ability to share information. This approach minimizes search latency when multiple machines are available, and is the typical configuration in a distributed search architecture. An alternative approach is suggested by our second research question: what happens if the shards are instead searched sequentially, passing the score threshold and top-k heap from each shard to the next? The heap-entry score threshold is then non-decreasing across the shards, and additional savings should result. While this approach would be unlikely to be used in an on-line system, it provides

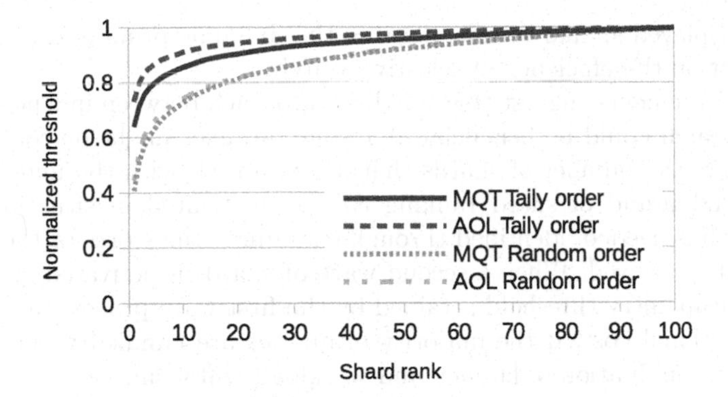

Fig. 7. Normalized 1,000 th document scores from shards relative to the highest score attained by any document for the corresponding query, micro-averaged over queries, assuming that shards are processed sequentially rather than in parallel, using the Taily-based ordering of topical shards and a random ordering of the same shards.

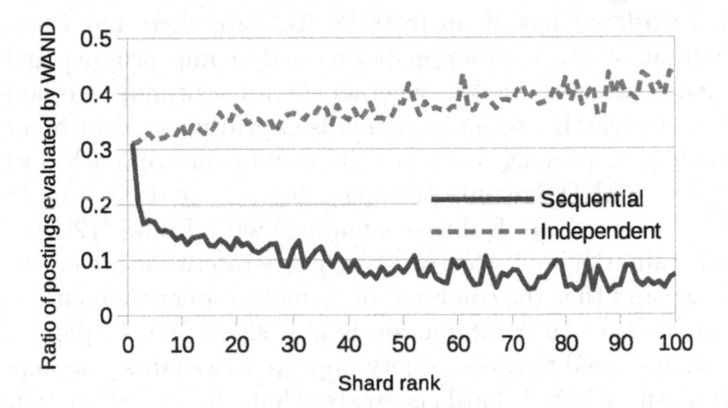

Fig. 8. Ratio of postings evaluated by WAND for independent shard search versus sequential shard search, AOL queries with micro-averaging. Shard ranking was determined by Taily.

an upper bound on the efficiency gains that are possible if a single heap was shared by all shards, and would increase throughput when limited resources are available and latency is not a concern: for example, in off-line search and text analytics applications.

Figure 7 demonstrates the threshold in the sequential WAND configuration, with shards ordered in two ways: by Taily score, and randomly. The normalized threshold rises quickly towards the maximum document score through the first few shards in the Taily ordering, which is where most of the documents related to the query are expected to reside. Figure 8 similarly plots the w/b WAND savings ratio at each shard rank, also micro-averaged over queries, and with shard ordering again determined by the Taily score. The independent and sequential configurations diverge markedly in their behavior, with a deep search in the latter processing far fewer postings than a deep search in the former. The MQT query set displayed similar trends. Sharing the dynamic pruning thresholds has a large effect on the efficiency of selective search.

Our measurements suggest that a hybrid approach between independent and sequential search could be beneficial. A resource-ranker might be configured to underestimate the number of shards that are required, with the understanding that a second round of shard ranking can be instigated in situations where deeper search is needed, identified through examining the scores or the quantity of documents retrieved. When a second wave of shards is activated, passing the maximum heap-entry threshold attained by the first-wave process would reduce the computational cost. If the majority of queries are handled within the first wave, a new combination of latency and workload will result.

4 Conclusion

Selective search reduces the computational costs of large-scale search by evaluating fewer postings than the standard distributed architecture, resulting in computational work savings of up to 90 %. To date there has been only limited consideration of the interaction between dynamic pruning and selective search [12], and it has been unclear whether dynamic pruning methods improve selective search, or whether selective search is capturing some or all of the same underlying savings as pruning does, just via a different approach. In this paper we have explored WAND dynamic pruning using a large dataset and two different query sets. In contrast to Kulkarni's findings with TBMS [12], we show that WAND-based evaluation and selective search generate what are effectively independent savings, and that the combination is more potent than either technique is alone – that is, that their interaction is a positive one. In particular, when resource selection is used to choose query-appropriate shards, the improvements from WAND on the selected shards is greater than the savings accruing on random shards, confirming that dynamic pruning further improves selective search – a rare situation where orthogonal optimizations are better-than-additive. We also demonstrated that there is a direct correlation between the efficiency gains generated by WAND and the shard's ranking. While it is well-known that resource selection improves effectiveness, our results suggest that it can also improve overall efficiency too.

Finally, two different methods of applying WAND to selective search were compared and we found that passing the top-k heap through a sequential shard evaluation greatly reduced the volume of postings evaluated by WAND. The significant difference in efficiency between this approach and the usual fully-parallel mechanism suggests avenues for future development in which hybrid models are used to balance latency and throughput in novel ways.

Acknowledgments. This research was supported by National Science Foundation (NSF) grant IIS-1302206; a Natural Sciences and Engineering Research Council of Canada (NSERC) Postgraduate Scholarship-Doctoral award; and the Australian Research Council (ARC) under the *Discovery Projects* scheme (DP140103256). Shane Culpepper is the recipient of an Australian Research Council (ARC) DECRA Research Fellowship (DE140100275).

References

1. Aly, R., Hiemstra, D., Demeester, T.: Taily: shard selection using the tail of score distributions. In: Proceedings of the 36th International ACM SIGIR Conference on Research and Development in Information Retrieval, pp. 673–682 (2013)
2. Arguello, J., Callan, J., Diaz, F.: Classification-based resource selection. In: Proceedings of the 18th ACM Conference on Information and Knowledge Management, pp. 1277–1286 (2009)
3. Broder, A.Z., Carmel, D., Herscovici, M., Soffer, A., Zien, J.: Efficient query evaluation using a two-level retrieval process. In: Proceedings of the 12th International Conference on Information and Knowledge Management, pp. 426–434 (2003)
4. Cacheda, F., Carneiro, V., Plachouras, V., Ounis, I.: Performance comparison of clustered and replicated information retrieval systems. In: Amati, G., Carpineto, C., Romano, G. (eds.) ECIR 2007. LNCS, vol. 4425, pp. 124–135. Springer, Heidelberg (2007)
5. Cambazoglu, B.B., Varol, E., Kayaaslan, E., Aykanat, C., Baeza-Yates, R.: Query forwarding in geographically distributed search engines. In: Proceedings of the 33rd International ACM SIGIR Conference on Research and Development in Information Retrieval, pp. 90–97 (2010)
6. Croft, W.B.: A model of cluster searching based on classification. Inf. Syst. **5**(3), 189–195 (1980)
7. Dimopoulos, C., Nepomnyachiy, S., Suel, T.: Optimizing top-k document retrieval strategies for block-max indexes. In: Proceedings of the of the Sixth ACM International Conference on Web Search and Data Mining, pp. 113–122 (2013)
8. Gravano, L., García-Molina, H., Tomasic, A.: GlOSS: Text-source discovery over the internet. ACM Trans. Database Syst. **24**, 229–264 (1999)
9. Ipeirotis, P.G., Gravano, L.: Distributed search over the hidden web: Hierarchical database sampling and selection. In: Proceedings of the 28th International Conference on Very Large Data Bases, pp. 394–405 (2002)
10. Kang, C., Wang, X., Chang, Y., Tseng, B.: Learning to rank with multi-aspect relevance for vertical search. In: Proceedings of the Fifth ACM International Conference on Web Search and Data Mining, pp. 453–462 (2012)
11. Kulkarni, A., Callan, J.: Document allocation policies for selective searching of distributed indexes. In: Proceedings of the 19th ACM International Conference on Information and Knowledge Management, pp. 449–458 (2010)
12. Kulkarni, A., Callan, J.: Selective search: Efficient and effective search of large textual collections. ACM Trans. Inf. Syst. **33**(4), 17:1–17:33 (2015)
13. Kulkarni, A., Tigelaar, A., Hiemstra, D., Callan, J.: Shard ranking and cutoff estimation for topically partitioned collections. In: Proceedings of the 21st ACM International Conference on Information and Knowledge Management, pp. 555–564 (2012)
14. Lemire, D., Boytsov, L.: Decoding billions of integers per second through vectorization. Soft. Prac. & Exp. **41**(1), 1–29 (2015)
15. Nottelmann, H., Fuhr, N.: Evaluating different methods of estimating retrieval quality for resource selection. In: Proceedings of the 26th Annual International ACM SIGIR Conference on Research and Development in Information Retrieval, pp. 290–297. ACM (2003)
16. Paltoglou, G., Salampasis, M., Satratzemi, M.: Integral based source selection for uncooperative distributed information retrieval environments. In: Proceedings of the 2008 ACM Workshop on Large-Scale Distributed Systems for Information Retrieval, pp. 67–74 (2008)

17. Petri, M., Culpepper, J.S., Moffat, A.: Exploring the magic of WAND. In: Proceedings of the Australian Document Computing Symposium, pp. 58–65 (2013)
18. Rojas, O., Gil-Costa, V., Marin, M.: Distributing effciently the block-max WAND algorithm. In: Proceedings of the 2013 International Conference on Computational Science, pp. 120–129 (2013)
19. Salton, G.: Automatic Information Organization and Retrieval. McGraw-Hill, New York (1968)
20. Shokouhi, M.: Central-Rank-Based Collection Selection in Uncooperative Distributed Information Retrieval. In: Amati, G., Carpineto, C., Romano, G. (eds.) ECIR 2007. LNCS, vol. 4425, pp. 160–172. Springer, Heidelberg (2007)
21. Si, L., Callan, J.: Relevant document distribution estimation method for resource selection. In: Proceedings of the 26th Annual International ACM SIGIR Conference on Research and Development in Informaion Retrieval, pp. 298–305 (2003)
22. Strohman, T., Turtle, H., Croft, W.B.: Optimization strategies for complex queries. In: Proceedings of the 28th Annual International ACM SIGIR Conference on Research and Development in Information Retrieval, pp. 219–225 (2005)
23. Thomas, P., Shokouhi, M.: Sushi: Scoring scaled samples for server selection. In: Proceedings of the 32nd ACM SIGIR Conference on Research and Development in Information Retrieval, pp. 419–426 (2009)
24. Yuwono, B., Lee, D.L.: Server ranking for distributed text retrieval systems on internet. In: Proceedings of the International Conference on Database Systems for Advanced Applications, pp. 41–49 (1997)

Efficient AUC Optimization for Information Ranking Applications

Sean J. Welleck(✉)

IBM, Austin, USA
swelleck@us.ibm.com

Abstract. Adequate evaluation of an information retrieval system to estimate future performance is a crucial task. Area under the ROC curve (AUC) is widely used to evaluate the generalization of a retrieval system. However, the objective function optimized in many retrieval systems is the error rate and not the AUC value. This paper provides an efficient and effective non-linear approach to optimize AUC using additive regression trees, with a special emphasis on the use of multi-class AUC (MAUC) because multiple relevance levels are widely used in many ranking applications. Compared to a conventional linear approach, the performance of the non-linear approach is comparable on binary-relevance benchmark datasets and is better on multi-relevance benchmark datasets.

Keywords: Machine learning · Learning to rank · Evaluation

1 Introduction

In various information retrieval applications, a system may need to provide a ranking of candidate items that satisfies a criteria. For instance, a search engine must produce a list of results, ranked by their relevance to a user query. The relationship between items (e.g. documents) represented as feature vectors and their rankings (e.g. based on relevance scores) is often complex, so machine learning is used to learn a function that generates a ranking given a list of items.

The ranking system is evaluated using metrics that reflect certain goals for the system. The choice of metric, as well as its relative importance, varies by application area. For instance, a search engine may evaluate its ranking system with Normalized Discounted Cumulative Gain (NDCG), while a question-answering system evaluates its ranking using precision at 3; a high NDCG score is meant to indicate results that are relevant to a user's query, while a high precision shows that a favorable amount of correct answers were ranked highly. Other common metrics include Recall @ k, Mean Average Precision (MAP), and Area Under the ROC Curve (AUC).

Ranking algorithms may optimize error rate as a proxy for improving metrics such as AUC, or may optimize the metrics directly. However, typical metrics such as NDCG and AUC are either flat everywhere or non-differentiable with respect to model parameters, making direct optimization with gradient descent difficult.

N. Ferro et al. (Eds.): ECIR 2016, LNCS 9626, pp. 159–170, 2016.
DOI: 10.1007/978-3-319-30671-1_12

LambdaMART [2] is a ranking algorithm that is able to avoid this issue and directly optimize non-smooth metrics. It uses a gradient-boosted tree model and forms an approximation to the gradient whose value is derived from the evaluation metric. LambdaMART has been empirically shown to find a local optimum of NDCG, Mean Reciprocal Rank, and Mean Average Precision [6]. An additional attractive property of LambdaMART is that the evaluation metric that LambdaMART optimizes is easily changed; the algorithm can therefore be adjusted for a given application area. This flexibility makes the algorithm a good candidate for a production system for general ranking, as using a single algorithm for multiple applications can reduce overall system complexity.

However, to our knowledge LambdaMART's ability to optimize AUC has not been explored and empirically verified in the literature. In this paper, we propose extensions to LambdaMART to optimize AUC and multi-class AUC, and show that the extensions can be computed efficiently. To evaluate the system, we conduct experiments on several binary-class and multi-class benchmark datasets. We find that LambdaMART with the AUC extension performs similarly to an SVM baseline on binary-class datasets, and LambdaMART with the multi-class AUC extension outperforms the SVM baseline on multi-class datasets.

2 Related Work

This work relates to two areas: LambdaMART and AUC optimization in ranking. LambdaMART was originally proposed in [15] and is overviewed in [2]. The LambdaRank algorithm, upon which LambdaMART is based, was shown to find a locally optimal model for the IR metrics NDCG@10, mean NDCG, MAP, and MRR [6]. Svore *et al.* [14] propose a modification to LambdaMART that allows for simultaneous optimization of NDCG and a measure based on click-through rate.

Various approaches have been developed for optimizing AUC in binary-class settings. Cortes and Mohri [5] show that minimum error rate training may be insufficient for optimizing AUC, and demonstrate that the RankBoost algorithm globally optimizes AUC. Calders and Jaroszewicz [3] propose a smooth polynomial approximation of AUC that can be optimized with a gradient descent method. Joachims [9] proposes an SVM method for various IR measures including AUC, and evaluates the system on text classification datasets. The SVM method is used as the comparison baseline in this paper.

3 Ranking Metrics

We will first provide a review of the metrics used in this paper. Using document retrieval as an example, consider n queries $Q_1 \dots Q_n$, and let $n(i)$ denote the number of documents in query Q_i. Let d_{ij} denote document j in query Q_i, where $i \in 1, \dots, n$, $j \in 1 \dots n(i)$.

3.1 Contingency Table Metrics

Several IR metrics are derived from a model's contingency table, which contains the four entries True Positive (TP), False Positive (FP), False Negative (FN), and True Negative (TN):

	$y = \ell_p$	$y = \ell_n$
$f(x) = \ell_p$	TP	FP
$f(x) = \ell_n$	FN	TN

where y denotes an example's label, $f(x)$ denotes the predicted label, ℓ_p denotes the class label considered positive, and ℓ_n denotes the class label considered negative.

Measuring the precision of the first k ranked documents is often important in ranking applications. For instance, *Precision*@1 is important for question answering systems to evaluate whether the system's top ranked item is a correct answer. Although precision is a metric for binary class labels, many ranking applications and standard datasets have multiple class labels. To evaluate precision in the multi-class context we use Micro-averaged Precision and Macro-averaged Precision, which summarize precision performance on multiple classes [10].

Micro-averaged Precision. Micro-averaged Precision pools the contingency tables across classes, then computes precision using the pooled values:

$$Precision_{micro} = \frac{\sum_{c=1}^{C} TP_c}{\sum_{c=1}^{C} TP_c + FP_c} \tag{1}$$

where C denotes the number of classes, TP_c is the number of true positives for class c, and FP_c is the number of false positives for class c.

$Precision_{micro}@k$ is measured by using only the first k ranked documents in each query:

$$Precision_{micro}@k = \frac{1}{n}\sum_{i=1}^{n} \frac{\sum_{c=1}^{C} |\{d_{ij}|y_j = c, j \in \{1, \ldots, k\}\}|}{(C)(k)} \tag{2}$$

Micro-averaged precision indicates performance on prevalent classes, since prevalent classes will contribute the most to the TP and FP sums.

Macro-averaged Precision. Macro-averaged Precision is a simple average of per-class precision values:

$$Precision_{macro} = \frac{1}{C}\sum_{c=1}^{C} \frac{TP_c}{TP_c + FP_c} \tag{3}$$

Restricting each query's ranked list to the first k documents gives:

$$Precision_{macro}@k = \frac{1}{C}\sum_{c=1}^{C}\sum_{i=1}^{n}\frac{|\{d_{ij}|y_j = c, j \in \{1,\ldots,k\}\}|}{k} \tag{4}$$

Macro-averaged precision indicates performance across all classes regardless of prevalence, since each class's precision value is given equal weight.

AUC. AUC refers to the area under the ROC curve. The ROC curve plots True Positive Rate ($TPR = \frac{TP}{TP+FN}$) versus False Positive Rate ($FPR = \frac{FP}{FP+TN}$), with TPR appearing on the y-axis, and FPR appearing on the x-axis.

Each point on the ROC curve corresponds to a contingency table for a given model. In the ranking context, the contingency table is for the ranking cutoff k; the curve shows the TPR and FPR as k changes. A model is considered to have better performance as its ROC curve shifts towards the upper left quadrant. The AUC measures the area under this curve, providing a single metric that summarizes a model's ROC curve and allowing for easy comparison.

We also note that the AUC is equivalent to the Wilcoxon-Mann-Whitney statistic [5] and can therefore be computed using the number of correctly ordered document pairs. Fawcett [7] provides an efficient algorithm for computing AUC.

Multi-class AUC. The standard AUC formulation is defined for binary classification. To evaluate a model using AUC on a dataset with multiple class labels, AUC can be extended to multi-class AUC (MAUC).

We define the *class reference* AUC value $AUC(c_i)$ as the AUC when class label c_i is viewed as positive and all other labels as negative. The multi-class AUC is then the weighted sum of class reference AUC values, where each class reference AUC is weighted by the proportion of the dataset examples with that class label, denoted $p(c_i)$ [7]:

$$MAUC = \sum_{i=1}^{C} AUC(c_i) * p(c_i). \tag{5}$$

Note that the class-reference AUC of a prevalent class will therefore impact the MAUC score more than the class-reference AUC of a rare class.

4 λ-Gradient Optimization of the MAUC Function

We briefly describe LambdaMART's optimization procedure here and refer the reader to [2] for a more extensive treatment. LambdaMART uses a gradient descent optimization procedure that only requires the gradient, rather than the objective function, to be defined. The objective function can in principal be left undefined, since only the gradient is required to perform gradient descent.

Each gradient approximation, known as a λ-gradient, focuses on document pairs (d_i, d_j) of conflicting relevance values (document d_i more or less relevant than document d_j):

$$\lambda_i = \sum_{j \in (d_i, d_j) | \ell_i \neq \ell_j} \lambda_{ij} \qquad \lambda_{ij} = S_{ij} \left| \Delta M_{IR_{ij}} \frac{\partial C_{ij}}{\partial o_{ij}} \right| \tag{6}$$

with $S_{ij} = 1$ when $l_i > l_j$ and -1 when $l_j < l_i$.

The λ-gradient includes the change in IR metric, $\Delta M_{IR_{ij}}$, from swapping the rank positions of the two documents, discounted by a function of the score difference between the documents.

For a given sorted order of the documents, the objective function is simply a weighted version of the RankNet [1] cost function. The RankNet cost is a pairwise cross-entropy cost applied to the logistic of the difference of the model scores. If document d_i, with score s_i, is to be ranked higher than document d_j, with score s_j, then the RankNet cost can be written as follows:

$$C(o_{ij}) = o_{ij} + \log(1 + e^{o_{ij}}) \tag{7}$$

where $o_{ij} = s_j - s_i$ is the score difference of a pair of documents in a query. The derivative of the RankNet cost according to the difference in score is

$$\frac{\partial C_{ij}}{\partial o_{ij}} = \frac{1}{(1 + e^{o_{ij}})}. \tag{8}$$

The optimization procedure using λ-gradients was originally defined using $\Delta NDCG$ as the ΔM_{IR} term in order to optimize NDCG. ΔMAP and ΔMRR were also used to define effective λ-gradients for MAP and MRR, respectively. In this work, we adopt the approach of replacing the ΔM_{IR} term to define λ-gradients for AUC and multi-class AUC.

4.1 λ-gradients for AUC and Multi-class AUC

λ-AUC. Defining the λ-gradient for AUC requires deriving a formula for ΔAUC_{ij} that can be efficiently computed. Efficiency is important since in every iteration, the term is computed for $O(n(i)^2)$ document pairs for each query Q_i.

To derive the ΔAUC_{ij} term, we begin with the fact that AUC is equivalent to the Wilcoxon-Mann-Whitney statistic [5]. For documents $d_{p_1}, \ldots, d_{p_m}$ with positive labels and documents $d_{n_1}, \ldots, d_{n_n}$ with negative labels, we have:

$$AUC = \frac{\sum_{i=1}^{m} \sum_{j=1}^{n} I(f(d_{p_i}) > f(d_{n_j}))}{mn}. \tag{9}$$

The indicator function I is 1 when the ranker assigns a score to a document with a positive label that is higher than the score assigned to a document with a negative label. Hence the numerator is the number of correctly ordered pairs, and we can write [9]:

$$AUC = \frac{CorrectPairs}{mn} \tag{10}$$

where

$$CorrectPairs = |\{(i,j) : (\ell_i > \ell_j) \ and \ (f(d_i) > f(d_j))\}|\,. \tag{11}$$

Note that a pair with equal labels is *not* considered a correct pair, since a document pair (d_i, d_j) contributes to *CorrectPairs* if and only if d_i is ranked higher than d_j in the ranked list induced by the current model scores, and $\ell_i > \ell_j$.

We now derive a formula for computing the ΔAUC_{ij} term in $O(1)$ time, given the ranked list and labels. This avoids the brute-force approach of counting the number of correct pairs before and after the swap, in turn providing an efficient way to compute a λ-gradient for *AUC*. Specifically, we have:

Theorem 1. *Let $d_1, \ldots, d_{m+n}$ be a list of documents with m positive labels and n negative labels, denoted $\ell_1, \ldots, \ell_{m+n}$, with $\ell_i \in \{0,1\}$. For each document pair (d_i, d_j), $i, j \in \{1, \ldots, m+n\}$,*

$$\Delta AUC_{ij} = \frac{(\ell_j - \ell_i)(j-i)}{mn}\,. \tag{12}$$

Proof. To derive this formula, we start with

$$\Delta AUC_{ij} = \frac{CP_{swap} - CP_{orig}}{mn} \tag{13}$$

where CP_{swap} is the value of *CorrectPairs* after swapping the scores assigned to documents i and j, and CP_{orig} is the value of *CorrectPairs* prior to the swap. Note that the swap corresponds to swapping positions of documents i and j in the ranked list. The numerator of ΔAUC_{ij} is the change in the number of correct pairs due to the swap. The following lemma shows that we only need to compute the change in the number of correct pairs *for the pairs of documents within the interval [i,j]* in the ranked list.

Lemma 1. *Let (d_a, d_b) be a document pair where at least one of $(a,b) \notin [i,j]$. Then after swapping documents (d_i, d_j), the pair correctness of (d_a, d_b) will be left unchanged or negated by another pair.*

Proof. Without loss of generality, assume $a < b$. There are five cases to consider.

Case $a \notin [i,j]$, $b \notin [i,j]$: Then the pair (d_a, d_b) does not change due to the swap, therefore its pair correctness does not change.

Note that unless one of a or b is an endpoint i or j, the pair (d_a, d_b) does not change. Hence we now assume that one of a or b is an endpoint i or j.

Case $a < i$, $b = i$: The pair correctness of (d_a, d_b) will change if and only if $\ell_a = 1$, $\ell_b = 1$, $\ell_j = 0$ prior to the swap. But then the pair correctness of (d_i, d_j) will change from correct to not correct, canceling out the change (see Fig. 1).

Case $a < i$, $b = j$: Then the pair correctness of (d_a, d_b) will change if and only if $\ell_a = 1$, $\ell_b = 1$, $\ell_i = 0$ prior to the swap. But then the pair correctness of (d_a, d_i) will change from correct to not correct, canceling out the change.

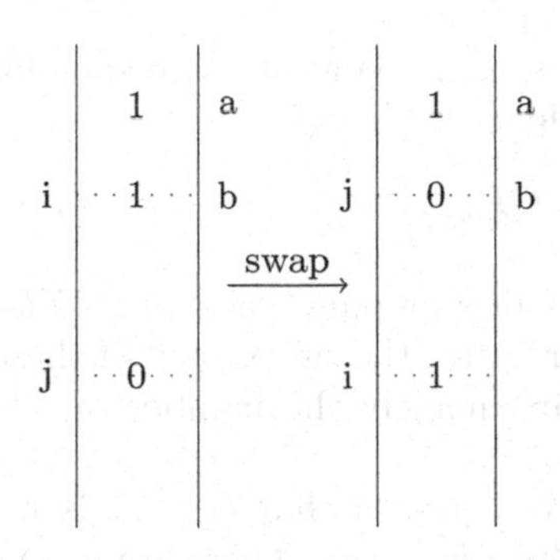

Fig. 1. Document swap for case $a < i$, $b = i$, with $\ell_a = 1$, $\ell_b = 1$, $\ell_j = 0$

Case $a = i$, $b > j$: Then pair correctness of (d_a, d_b) will change if and only if $\ell_a = 0$, $\ell_b = 0$, $\ell_j = 1$ prior to the swap. But then the pair correctness of (d_j, d_b) will change from correct to not correct, canceling out the change.

Case $a = j$, $b > j$: Then pair correctness of (d_a, d_b) will change if and only if $\ell_a = 0$, $\ell_b = 0$, $\ell_i = 1$ prior to the swap. But then the pair correctness of (d_i, d_b) will change from correct to not correct, canceling out the change.

Hence in all cases, either the pair correctness stays the same, or the pair (d_a, d_b) changes from not correct to correct and an additional pair changes from correct to not correct, thus canceling out the change with respect to the total number of correct pairs after the swap. □

Lemma 1 shows that the difference in correct pairs $CP_{swap} - CP_{orig}$ is equivalent to $CP_{swap_{[i,j]}} - CP_{orig_{[i,j]}}$, namely the change in the number of correct pairs within the interval [i,j]. Lemma 2 tells us that this value is simply the length of the interval [i,j].

Lemma 2. *Assume* $i < j$. *Then*

$$CP_{swap_{[i,j]}} - CP_{orig_{[i,j]}} = (\ell_j - \ell_i)(j - i)\,. \tag{14}$$

Proof. There are three cases to consider.

Case $\ell_i = \ell_j$: The number of correct pairs will not change since no document labels change due to the swap. Hence $CP_{swap_{[i,j]}} - CP_{orig_{[i,j]}} = 0 = (\ell_j - \ell_i)(j - i)$.

Case $\ell_i = 1$, $\ell_j = 0$: Before swapping, each pair (i, k), $i < k \leq j$ such that $\ell_k = 0$ is a correct pair. After the swap, each of these pairs is not a correct pair. There are $n_{l_0[i,j]}$ such pairs, namely the number of documents in the interval $[i, j]$ with label 0.

Each pair (k, j), $i \leq k < j$ such that $\ell_k = 1$ is a correct pair before swapping, and not correct after swapping. There are $n_{l_1[i,j]}$ such pairs, namely the number of documents in the interval $[i, j]$ with label 1.

Every other pair remains unchanged, therefore

$$n_{l_0[i,j]} + n_{l_1[i,j]} = j - i \tag{15}$$

pairs changed from correct to not correct, corresponding to a decrease in the number of correct pairs. Hence we have:

$$CP_{swap_{[i,j]}} - CP_{orig_{[i,j]}} = -(j-i) = (\ell_j - \ell_i)(j-i).$$

Case $\ell_i = 0$, $\ell_j = 1$: Before swapping, each pair (i,k), $i < k \leq j$ such that $\ell_k = 0$ is not a correct pair. After the swap, each of these pairs is a correct pair. There are $n_{l_0[i,j]}$ such pairs, namely the number of documents in the interval $[i,j]$ with label 0.

Each pair (k,j), $i \leq k < j$ such that $\ell_k = 1$ is not a correct pair before swapping, and is correct after swapping. There are $n_{l_1[i,j]}$ such pairs, namely the number of documents in the interval $[i,j]$ with label 1.

Each pair (i,k), $i < k \leq j$ such that $\ell_k = 1$ remains not correct. Each pair (k,j), $i \leq k < j$ such that $\ell_k = 0$ remains not correct. Every other pair remains unchanged. Therefore

$$n_{l_0[i,j]} + n_{l_1[i,j]} = j - i \tag{16}$$

pairs changed from not correct to correct, corresponding to an increase in the number of correct pairs. Hence we have:

$$CP_{swap_{[i,j]}} - CP_{orig_{[i,j]}} = (j-i) = (\ell_j - \ell_i)(j-i).$$

□

Therefore by Lemmas 1 and 2, we have:

$$\begin{aligned} \Delta AUC_{ij} &= \frac{CP_{swap} - CP_{orig}}{mn} \\ &= \frac{CP_{swap_{[i,j]}} - CP_{orig_{[i,j]}}}{mn} \\ &= \frac{(\ell_i - \ell_j)(j-i)}{mn} \end{aligned}$$

completing the proof of Theorem 1. □

Applying the formula from Theorem 1 to the list of documents sorted by the current model scores, we define the λ-gradient for AUC as:

$$\lambda_{AUC_{ij}} = S_{ij} \left| \Delta AUC_{ij} \frac{\partial C_{ij}}{\partial o_{ij}} \right| \tag{17}$$

where S_{ij} and $\frac{\partial C_{ij}}{\partial o_{ij}}$ are as defined previously, and $\Delta AUC_{ij} = \frac{(\ell_i - \ell_j)(j-i)}{mn}$.

λ-MAUC. To extend the λ-gradient for AUC to a multi-class setting, we consider the multi-class AUC definition found in Eq. (5). Since MAUC is a linear combination of class-reference AUC values, to compute $\Delta MAUC_{ij}$ we can compute the change in each class-reference AUC value $\Delta AUC(c_k)$ separately using Eq. (12) and weight each Δ value by the proportion $p(c_k)$, giving:

$$\Delta MAUC_{ij} = \sum_{k=1}^{C} \Delta AUC(c_k)_{ij} * p(c_k). \tag{18}$$

Using this term and the previously defined terms S_{ij} and $\frac{\partial C_{ij}}{\partial o_{ij}}$, we define the λ-gradient for $MAUC$ as:

$$\lambda_{MAUC_{ij}} = S_{ij} \left| \Delta MAUC_{ij} \frac{\partial C_{ij}}{\partial o_{ij}} \right|. \tag{19}$$

5 Experiments

Experiments were conducted on binary-class datasets to compare the AUC performance of LambdaMART trained with the AUC λ-gradient, referred to as LambdaMART-AUC, against a baseline model. Similar experiments were conducted on multi-class datasets to compare LambdaMART trained with the MAUC λ-gradient, referred to as LambdaMART-MAUC, against a baseline in terms of MAUC. Differences in precision on the predicted rankings were also investigated.

The LambdaMART implementation used in the experiments was a modified version of the JForests learning to rank library [8]. This library showed the best NDCG performance out of the available Java ranking libraries in preliminary experiments. We then implemented extensions required to compute the AUC and multi-class AUC λ-gradients. For parameter tuning, a learning rate was chosen for each dataset by searching over the values $\{0.1, 0.25, 0.5, 0.9\}$ and choosing the value that resulted in the best performance on a validation set.

As the comparison baseline, we used a Support Vector Machine (SVM) formulated for optimizing AUC. The SVM implementation was provided by the SVM-Perf [9] library. The ROCArea loss function was used, and the regularization parameter c was chosen by searching over the values $\{0.1, 1, 10, 100\}$ and choosing the value that resulted in the best performance on a validation set. For the multi-class setting, a binary classifier was trained for each individual relevance class. Prediction scores for a document d were then generated by computing the quantity $\sum_{c=1}^{C} cf_c(d)$, where C denotes the number of classes, and f_c denotes the binary classifier for relevance class c. These scores were used to induce a ranking of documents for each query.

5.1 Datasets

For evaluating LambdaMART-AUC, we used six binary-class web-search datasets from the LETOR 3.0 [13] Gov dataset collection, named td2003, td2004, np2003, np2004, hp2003, and hp2004. Each dataset is divided into five folds and contains feature vectors representing query-document pairs and binary relevance labels.

For evaluating LambdaMART-MAUC, we used four multi-class web-search datasets: versions 1.0 and 2.0 of the Yahoo! Learning to Rank Challenge [4] dataset, and the mq2007 and mq2008 datasets from the LETOR 4.0 [12] collection. The Yahoo! and LETOR datasets are divided into two and five folds, respectively. Each Yahoo! dataset has integer relevance scores ranging from 0 (not relevant) to 4 (very relevant), while the LETOR datasets have integer relevance scores ranging from 0 to 2. The LETOR datasets have 1700 and 800 queries, respectively, while the larger Yahoo! datasets have approximately 20,000 queries.

5.2 Results

AUC. On the binary-class datasets, LambdaMART-AUC and SVM-Perf performed similarly in terms of AUC and Mean-Average Precision. The results did not definitively show that either algorithm was superior on all datasets; LambdaMART-AUC had higher AUC scores on 2 datasets (td2003 and td2004), lower AUC scores on 3 datasets (hp2003, hp2004, np2004), and a similar score on np2003. In terms of MAP, LambdaMART-AUC was higher on 2 datasets (td2003 and td2004), lower on 2 datasets (np2004, hp2004), and similar on 2 datasets (np2003, hp2003). The results confirm that LambdaMART-AUC is an effective option for optimizing AUC on binary datasets, since the SVM model has previously been shown to perform effectively.

MAUC. Table 1 shows the MAUC scores on held out test sets for the four multi-class datasets. The reported value is the average MAUC across all dataset folds. The results indicate that in terms of optimizing Multi-class AUC, LambdaMART-MAUC is as effective as SVM-Perf on the LETOR datasets, and more effective on the larger Yahoo! datasets.

Additionally, the experiments found that LambdaMART-MAUC outperformed SVM-Perf in terms of precision in all cases. Table 2 shows the Mean Average Precision scores for the four datasets. LambdaMART-MAUC also had higher $Precision_{micro}@k$ and $Precision_{macro}@k$ on all datasets, for $k = 1, \ldots, 10$. For instance, Fig. 2 shows the values of $Precision_{micro}@k$ and $Precision_{macro}@k$ for the Yahoo! V1 dataset.

The class-reference AUC scores indicate that LambdaMART-MAUC and SVM-Perf arrive at their MAUC scores in different ways. LambdaMART-MAUC focuses on the most prevalent class; each $\Delta AUC(c_i)$ term for a prevalent class

Table 1. Summary of multi-class AUC on test folds

	Yahoo V1	Yahoo V2	mq2007	mq2008
LambdaMART-MAUC	**0.594**	**0.592**	**0.662**	0.734
SVM-Perf	0.576	0.576	0.659	**0.737**

Table 2. Summary of Mean Average Precision on test folds

	Yahoo V1	Yahoo V2	mq2007	mq2008
LambdaMART-MAUC	**0.862**	**0.858**	**0.466**	**0.474**
SVM-Perf	0.837	0.837	0.450	0.458

receives a higher weighting than for a rare class due to the $p(c_i)$ term in the λ_{MAUC} computation. As a result the λ-gradients in LambdaMART-MAUC place more emphasis on achieving a high $AUC(c_1)$ than a high $AUC(c_4)$. Table 3 shows the class-reference AUC scores for the Yahoo! V1 dataset. We observe that LambdaMART-MAUC produces better $AUC(c_1)$ than SVM-Perf, but worse $AUC(c_4)$, since class 1 is much more prevalent than class 4; 48 % of the documents in the training set with a positive label have a label of class 1, while only 2.5 % have a label of class 4.

Finally, we note that on the large-scale Microsoft Learning to Rank Dataset MSLR-WEB10k [11], the SVM-Perf training failed to converge on a single fold after 12 hours. Therefore training a model for each class for every fold was impractical using SVM-Perf, while LambdaMART-MAUC was able to train on all five folds in less than 5 hours. This further suggests that LambdaMART-MAUC is preferable to SVM-Perf for optimizing MAUC on large ranking datasets.

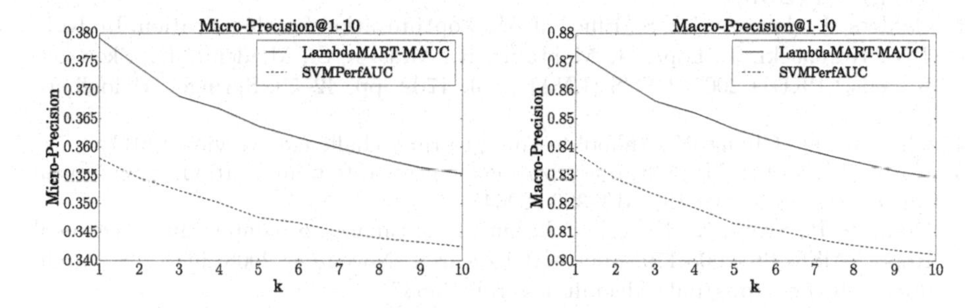

Fig. 2. Micro and Macro Precision@1-10 on the Yahoo! V1 test folds

Table 3. Summary of class-reference AUC scores on the Yahoo! V1 test folds

	AUC_1	AUC_2	AUC_3	AUC_4
LambdaMART-MAUC	**0.503**	**0.690**	0.757	0.831
SVM-Perf	0.474	0.682	**0.796**	**0.920**

6 Conclusions

We have introduced a method for optimizing AUC on ranking datasets using a gradient-boosting framework. Specifically, we have derived gradient approximations for optimizing AUC with LambdaMART in binary and multi-class settings, and shown that the gradients are efficient to compute. The experiments show that the method performs as well as, or better than, a baseline SVM method, and performs especially well on large, multi-class datasets. In addition to adding LambdaMART to the portfolio of algorithms that can be used to optimize AUC, our extensions expand the set of IR metrics for which LambdaMART can be used.

There are several possible future directions. One is investigating local optimality of the solution produced by LambdaMART-AUC using Monte Carlo methods. Other directions include exploring LambdaMART with multiple objective functions to optimize AUC, and creating an extension to optimize area under a Precision-Recall curve rather than an ROC curve.

Acknowledgements. Thank you to Dwi Sianto Mansjur for giving helpful guidance and providing valuable comments about this paper.

References

1. Burges, C., Shaked, T., Renshaw, E., Lazier, A., Deeds, M., Hamilton, N., Hullender, G.: Learning to rank using gradient descent. In: Proceedings of the 22Nd International Conference on Machine Learning ICML 2005, pp. 89–96. ACM, New York (2005)
2. Burges, C.J.: From ranknet to lambdarank to lambdamart: An overview. Learning **11**, 23–581 (2010)
3. Calders, T., Jaroszewicz, S.: Efficient AUC optimization for classification. In: Kok, J.N., Koronacki, J., Lopez de Mantaras, R., Matwin, S., Mladenič, D., Skowron, A. (eds.) PKDD 2007. LNCS (LNAI), vol. 4702, pp. 42–53. Springer, Heidelberg (2007)
4. Chapelle, O., Chang, Y.: Yahoo! learning to rank challenge overview (2011)
5. Cortes, C., Mohri, M.: AUC optimization vs. error rate minimization. Adv. Neural Inf. Process. Syst. **16**(16), 313–320 (2004)
6. Donmez, P., Svore, K., Burges, C.J.: On the optimality of lambdarank. Technical Report MSR-TR-2008-179, Microsoft Research, November 2008. http://research.microsoft.com/apps/pubs/default.aspx?id=76530
7. Fawcett, T.: An introduction to ROC analysis. Pattern Recogn. Lett. **27**(8), 861–874 (2006)
8. Ganjisaffar, Y., Caruana, R., Lopes, C.V.: Bagging gradient-boosted trees for high precision, low variance ranking models. In: Proceedings of the 34th international ACM SIGIR conference on Research and development in Information Retrieval, pp. 85–94. ACM (2011)
9. Joachims, T.: A support vector method for multivariate performance measures. In: Proceedings of the 22Nd International Conference on Machine Learning, pp. 377–384. ACM, New York (2005)
10. Manning, C., Raghavan, P., Schütze, H.: Introduction to Information Retrieval. Cambridge University Press, Cambridge (2008)
11. Microsoft learning to rank datasets. http://research.microsoft.com/en-us/projects/mslr/
12. Qin, T., Liu, T.: Introducing LETOR 4.0 datasets. CoRR abs/1306.2597 (2013). http://arxiv.org/abs/1306.2597
13. Qin, T., Liu, T.Y., Xu, J., Li, H.: Letor: A benchmark collection for researchon learning to rank for information retrieval. Inf. Retr. **13**(4), 346–374 (2010). http://dx.doi.org/10.1007/s10791-009-9123-y
14. Svore, K.M., Volkovs, M.N., Burges, C.J.: Learning to rank with multiple objective functions. In: Proceedings of the 20th International Conference on World Wide Web, pp. 367–376. ACM (2011)
15. Wu, Q., Burges, C.J., Svore, K.M., Gao, J.: Adapting boosting for information retrieval measures. Inf. Retrieval **13**(3), 254–270 (2010)

Modeling User Interests for Zero-Query Ranking

Liu Yang[1](✉), Qi Guo[2], Yang Song[3], Sha Meng[2], Milad Shokouhi[4], Kieran McDonald[2], and W. Bruce Croft[1]

[1] Center for Intelligent Information Retrieval, University of Massachusetts Amherst, Amherst, MA, USA
{lyang,croft}@cs.umass.edu
[2] Microsoft Bing, Bellevue, WA, USA
{qiguo,shmeng,kieran.mcdonald}@microsoft.com
[3] Microsoft Research Redmond, Redmond, WA, USA
yangsong@microsoft.com
[4] Microsoft Research Cambridge, Cambridge, UK
milads@microsoft.com

Abstract. Proactive search systems like Google Now and Microsoft Cortana have gained increasing popularity with the growth of mobile Internet. Unlike traditional reactive search systems where search engines return results in response to queries issued by the users, proactive systems actively push information cards to the users on mobile devices based on the context around time, location, environment (e.g., weather), and user interests. A proactive system is a zero-query information retrieval system, which makes user modeling critical for understanding user information needs. In this paper, we study user modeling in proactive search systems and propose a learning to rank method for proactive ranking. We explore a variety of ways of modeling user interests, ranging from direct modeling of historical interaction with content types to finer-grained entity-level modeling, and user demographical information. To reduce the feature sparsity problem in entity modeling, we propose semantic similarity features using word embedding and an entity taxonomy in knowledge base. Experiments performed with data from a large commercial proactive search system show that our method significantly outperforms a strong baseline method deployed in the production system.

1 Introduction

The recent boom of mobile internet has seen the emergence of proactive search systems like Google Now, Apple Siri and Microsoft Cortana. Unlike traditional reactive Web search systems where the search engines return results in response to queries issued by users, proactive search systems actively push information cards to users on mobile devices based on the context such as time, location, environment (e.g., weather), and user interests. Information cards are concise and informative snippets commonly shown in many intelligent personal assistant

L. Yang—Work primarily done when interning at Microsoft.

N. Ferro et al. (Eds.): ECIR 2016, LNCS 9626, pp. 171–184, 2016.
DOI: 10.1007/978-3-319-30671-1_13

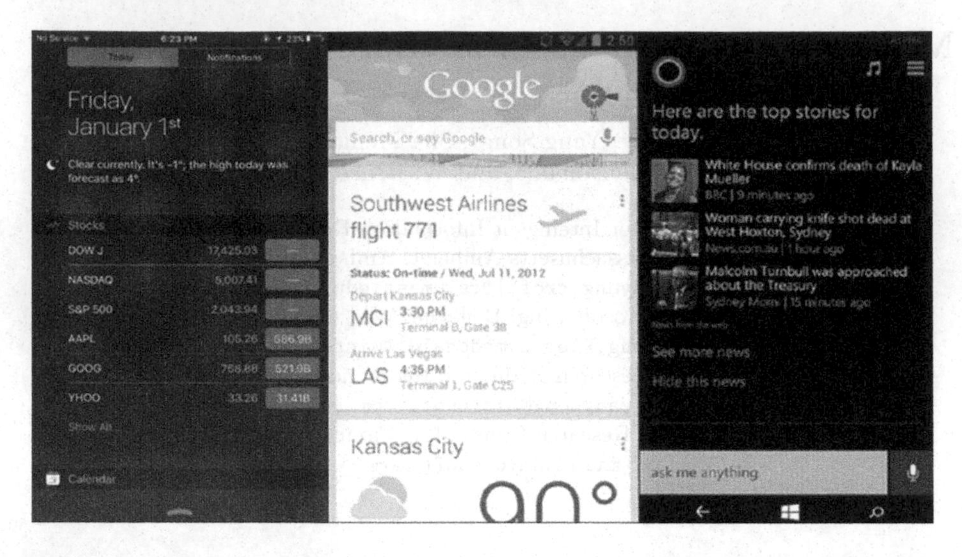

Fig. 1. Examples of proactive information cards presented in Apple Siri (stocks), Google Now (flight and weather) and Microsoft Cortana (news).

systems. Figure 1 shows examples of proactive information cards presented in Apple Siri (stocks), Google Now (flight and weather) and Microsoft Cortana (news). There are no explicit queries for these returned information cards which are triggered by some particular context.

The need for these proactive search systems increases in mobile environments, where the users' ability to interact with the system is hampered by the physical limitations of the devices [2]. Thus future information retrieval systems must infer user information needs and respond with information appropriate to the current context without the user having to enter a query. User modeling plays a critical role for understanding user information needs in the absence of query information.

Despite the abundance of research about user modeling and personalization for reactive Web search systems [1,4,19], little is known about how to model user interests to improve the effectiveness of the proactive systems. The closest research to our paper is the recent work by Shokouhi and Guo [15], where the authors looked at the usage patterns of proactive systems and aimed to understand the connections between reactive searching behavior and user interaction with the proactive systems.

In this paper, we explore a broader variety of ways and data sources in modeling user interests and compare the proposed method with a baseline that is similar and comparable to the Carre model presented in [15] for the application of improving ranking of information cards of the proactive systems. Our explorations include modeling at the coarser-level of card types, to finer-grained modeling of entities, and the capturing of variations lie in the variety of demographics. We develop entity based representations for user interests and card topics. Entities are extracted from user search/browse logs across multiple commercial platforms and devices to represent user interests. For card topics, entities are extracted from the associated URLs. To reduce the feature sparsity problem, we propose entity semantic similarity features based on word embedding and an entity taxonomy in knowledge base. The contributions of our work can be summarized as the follows:

- We present the first in-depth study on modeling user interests for improving the ranking of proactive search systems.
- We propose a variety of approaches of modeling user interests, ranging from coarser-grained modeling of card type preferences, to finer-grained modeling of entity preferences, and the variations in demographics.
- We perform a thorough experimental evaluation of the proposed methods with large-scale logs from a commercial proactive system, showing that our method significantly outperforms a strong baseline ranker deployed in the production system.
- We conduct in-depth feature analysis, which provides insights for guiding future feature design of the proactive ranking systems.

2 Related Work

There is a range of previous research related to our work that falls into different categories including proactive information retrieval, information cards, search personalization and recommender systems.

Proactive Information Retrieval. Rhodes and Maes [14] proposed the just-in-time information retrieval agent (JITIR agent) that proactively retrieves and presents information based on a person's local context. The motivation of modern proactive search system is very similar to JITIRs, but the monitored user context and presented content of modern proactive system are more extensive than traditional JITIRs.

Information Cards. Web search has seen rapid growth in mobile search traffic, where answer-like results on information cards are better choices than a ranked list to address simple information needs given the relative small size of screens on mobile devices. For some types of information cards like weather, users could directly find the target information from contents on cards without any clicks. This problem was introduced by Li et al. [9] as "good abandonment". Based on this problem, Guo et al. [7] proposed a study of modeling interactions on touch-enabled device for improving Web search ranking. Lagun et al. [8] studied the browser viewport which is defined as visible portion of a web page on mobile phones to provide good measurement of search satisfaction in the absence of clicks. For our experiments, we also consider viewport based dwell time to generate relevance labels for information cards to handle the good abandonment problem.

Search Personalization. Proactive systems recommend highly personalized contents to users based on their interest and context. Hence, our work is also related to previous research on search personalization [6,19,21]. Fox et al. [6] showed there was an association between implicit measures of user activity and the user's explicit satisfaction ratings. Agichtein et al. [1] showed incorporating implicit feedback obtained in a real web search setting can improve web search ranking. Bennett et al. [4] studied how short-term (session) behavior and long-term (historic) behavior interact, and how each may be used in isolation or in combination to optimally contribute to gains in relevance through search personalization. We also consider implicit feedback features based on user interactions with different card types and compare the relative importance of this feature group with other feature groups like entity based user interests features for proactive ranking.

Recommender Systems. Similar to recommender systems [13], we also push the most relevant content to the user based on user personal interests without a query issued by the user. However, the recommended items in the proactive system are for a smaller set of items that are highly heterogeneous and need to be personalized and contextualized in the ranking [15]. Unlike collaborative filtering methods [17] commonly used in recommender systems, we adopt a learning to rank framework that is suitable for combining multiple features derived from various user history information.

3 Method Overview

We adopt common IR terminology when we define the proactive search problem. A proactive impression consists of a ranked list of information cards presented to users together with the user interaction logs recording clicks and viewports. Given a set of information cards $\{C_1, C_2, ..., C_n\}$ and the corresponding relevance labels, our task is to learn a ranking model $\boldsymbol{R}$ to rank the cards based on available features θ and optimize a pre-defined metric E defined over the card ranked list.

We propose a framework for proactive ranking referred to as **UMPRanker** (User Modeling based Proactive Ranker). Firstly we mine user interests from multiple user logs. Each user is distinguished by a unique and anonymized identifier which is commonly used in these platforms. The information collected from these different platforms forms the basis of our user modeling. Then we derive multiple user interest features including entity based user interests, card type based implicit feedback and user demographics based on the collected information. Information cards are generated from multiple pre-defined information sources and templates including weather, finance, news, calendar, places, event, sports, flight, traffic, fitness, etc. We also extract card features from the associated URLs and card types. Given user features and card features, we can train a learning to rank model. Given a trigger context like particular time, location or event, information cards are ranked by the model and pushed to the user's device.

4 Mining User Interests from Logs

We can derive user interests from the short text that users specified and the textual content that users engaged with during their historical activity. Specifically, the information sources of user interests we consider include the following user behavior in logs:

1. Issued queries from the search behavior.
2. Satisfactory (SAT) clicked documents from the search behavior.
3. Browsed documents on an Internet browser and a Web portal.
4. Clicks and viewports on a personal assistant.

Note that users have the right to choose whether they would like the services to collect their behavior data. The logs we collected are from "opt-in" users only. To represent user interests, we extract entities from the text content specified by user behaviors. We can also represent information card topics by entities exacted from card URLs. Entities in user profiles and cards are linked with entities in a large scale knowledge base to get a richer representation of user interests.

5 Ranking Feature Extraction

5.1 Card Type Based Implicit Feedback Features (IF)

This feature group is based on statistics of user interactive history with different card types like average view time of each card type, accept ratio of each card type, etc. This group of features aims at capturing individual user preferences of particular card type, for example, news, based on the statistics of the historical interactions. Specifically, for each <user, card type> pair, features extracted include historical clickthrough rate (CTR), SAT CTR (i.e., clicks with more than 30 seconds dwell time on landing pages), SAT view rate (i.e., card views with more than 30 seconds), hybrid SAT rate (i.e., rate of either a SAT click or a SAT view), view rate, average view time, average view speed, accept ratio of the card suggestions, untrack ratio of the card type, ratio of the cards being a suggestion. The details of these features are explained in Table 1.

5.2 Entity Based User Interests Features (EF)

As described in Sect. 4, we can represent user interests by entities extracted from user behavior across multiple services and devices. For cards with URLs, we can also represent card topics with entities. So the next problem is how to measure the similarity between user entity sets and card entity sets. We consider features including exact match, term match, language models, word embedding and entity taxonomy in the knowledge base. In the following parts, we let U_i and C_j denote the entity set of the i-th user and the j-th card.

Exact Match. The first feature is *exact match*. It is computed based on the number of common entities matched by entity ID in U_i and C_j. We consider two variations: *RawMatchCount* and *EMJaccardIndex*. *RawMatchCount* uses the original match count as the feature value. *EMJaccardIndex* computes the Jaccard Index of U_i and C_j.

Table 1. Summary of card type based implicit feedback features (IF).

Feature	Description
CTR	Personal historical clickthrough rate of the card type
SATCTR	Personal historical SAT (landing page dwell time > 30 s) clickthrough rate of the card type
SATViewRate	Personal historical SAT (card view time > 30 s) view rate of the card type
ViewRate	Personal historical view rate of the card type
AverageViewTime	Personal historical average view time of the card type
AverageViewTimeSpeed	Personal historical average view time per pixel of the card type
AcceptRatio	Personal historical accept ratio when the card type was presented as a suggestion
UnTrackRatio	Personal historical ratio untrack the card type
SuggestionRate	Personal historical ratio of seeing the card type being presented as a suggestion

Term Match. *exact match* feature suffers from feature sparsity problem. A better method is to treat U_i and C_j as two term sets. Then we could get two entity term distributions over U_i and C_j. The cosine similarity between these two entity term distributions becomes *term match* feature.

Language Models. The feature *LMScore* is based on the language modeling approach to information retrieval [16]. We treat the card entity term set as the query and the user entity term set as the document. Then we compute the log likelihood of generating card entity terms using a language model constructed from user entity terms. We use Laplace smoothing in the computation of language model score.

Word Embedding. We extract semantic similarity features between entities based on word embeddings. Word embeddings [11,12] are continuous vector representations of words learned from very large data sets based on neural networks. The learned word vectors can capture the semantic similarity between words. In our experiment, we trained a Word2Vec model using the skip-gram algorithm with hierarchical softmax [12]. The training data was from the Wikipedia English dump obtained on June 6th, 2015. Our model outputs vectors of size 200. The total number of distinct words is 1,425,833. We then estimate entity vectors based on word vectors. For entities that are phrases, we compute the average vector of embedding of words within the entity phrase. After vector normalization, we use the dot product of entity vectors to measure entity similarity. To define features for the similarity of U_i and C_j, we consider feature variations inspired by hierarchical clustering algorithms as shown in Table 2.

Entity Taxonomy in Knowledge Base. Another way to extract semantic similarity features between entities is measuring the similarity of entity taxonomy [10].

Table 2. Summary of entity based user interests features (EF).

Feature	Description
RawMatchCount	The raw match count of entities by id in U_i and C_j
EMJaccardIndex	The Jaccard Index of entities matched by id in U_i and C_j
TMNoWeight	The cosine similarity between two entity term distributions in U_i and C_j
TMWeighted	Similar to TMNoWeight, but terms are weighted by impression count
LMScore	The log likelihood of generating terms in C_j using a language model constructed from terms in U_i
WordEBDMin	The similarity between U_i and C_j based on word embedding features(single-link)
WordEBDMax	The similarity between U_i and C_j based on word embedding features(complete-link)
WordEBDAvgNoWeight	The similarity between U_i and C_j based on word embedding features(average-link-noWeight)
WordEBDAvgWeighted	The similarity between U_i and C_j based on word embedding features(average-link-weighted)
KBTaxonomyLevel1	The similarity between U_i and C_j based on entity taxonomy similarity in level 1
KBTaxonomyLevel1Weighted	Similar to EntityKBTaxonomyLevel1 but each entity is weighted by its impression counts
KBTaxonomyLevel2	The similarity between U_i and C_j based on entity taxonomy similarity in level 2
KBTaxonomyLevel2Weighted	Similar to EntityKBTaxonomyLevel2 but each entity is weighted by its impression counts

As presented in Sect. 4, we link entities in the user interest profile with entities in a large scale knowledge base. From the knowledge base, we can extract the entity taxonomy which is the entity type information. Two entities without any common terms could have similarities if they share some common entity types.

Table 3 shows entity taxonomy examples for "Kobe Bryant" and "Byron Scott". We can see that these two entities share common taxonomies like "basketball. player", "award.winner". They also have their own special taxonomies. "Kobe Bryant" has "olympics.athlete" in the taxonomies whereas "Byron Scott" has the taxonomy named "basketball.coach". Based on this observation, we can measure the semantic similarity between two entities base on their taxonomies. Specifically, we measure the similarity of two entities based on the Jaccard index of the two corresponding taxonomy sets. Since all taxonomies only have two levels, we compute entity taxonomy similarity features in two different granularity. When we measure the similarity of the U_i and C_j, we can compute the average similar-

Table 3. Examples of entity taxonomy for "Kobe Bryant" and "Byron Scott".

Entity name	Kobe Bryant	Byron Scott
Taxonomy 1	award.nominee	award.winner
Taxonomy 2	award.winner	basketball.coach
Taxonomy 3	basketball.player	basketball.player
Taxonomy 4	celebrities.celebrity	event.agent
Taxonomy 5	event.agent	sports.sports_team_coach
Taxonomy 6	film.actor	film.actor
Taxonomy 7	olympics.athlete	tv.personality

ity of all entity pairs in this two entity sets. We compute a weighted version where each entity is weighted by its impression count and a non-weighted version for this features. In summary, in this feature group, we have 4 features that are listed in Table 2.

5.3 User Demographics Features (UD)

Part of user interests are influenced by their demographic information such as age and gender. The tastes of teenagers and adults are different. Men and women also have different preferences for information cards. Motivated by this intuition, we also extract features related to user demographic information. In addition to the raw user demographics features, we also add user demographics features in a *matched version*. We compute the matched features of the user demographics features between the user and users who clicked the card URLs. To achieve this, we need to compute the average age and average gender value for users that clicked on each card. The gender value is between 1 (male) and 2 (female), where the more the value is approaching 1, the more men clicked the URLs in the corresponding card. Based on this, we compute the differences between user demographic features. We distinguish zero distance with null cases by adding an offset to zero distance when we compute the matched version feature values. The details of these features are explained in the Table 4.

6 Experiments

6.1 Data Set and Experiment Settings

We use real data from a commercial intelligent personal assistant for the experiments. The training data is from one week between March 18th, 2015 and March 24th, 2015. The testing data is from one week between March 25th, 2015 and March 31st, 2015. The statistics of card impressions and users are shown in Table 5.

The user profiles represented by entities are built from multiple logs presented in Sect. 4. The time window from user profile is from March 18th, 2014

Table 4. Summary of user demographics features (UD).

Feature	Description
UserAge	Integer value of user's age
UserGender	Binary value of user's gender
UserLanguage	Integer value to denote user's language
UserRegisterYears	Integer value to denote the number of years since user's registration
CardAvgAge	Average age of all users who clicked the URLs on the card
CardAvgGender	Average gender value of all users who clicked the URLs on the card
AgeAbsDistance	The absolute distance for age between the user with all users who clicked card URLs
AgeRelDistance	The relative distance for age between the user with all users who clicked card URLs
GenderAbsDistance	The absolute distance for gender between the user with all users who clicked card URLs
GenderRelDistance	The relative distance for gender between the user with all users who clicked card URLs

to March 17 th, 2015. So there is no overlap time between the user profiles and training/testing data. Since most proactive impressions have only one card with a positive relevance label, we pick mean reciprocal rank(MRR) and NDCG@1 as the evaluation metric.

6.2 Relevance Labels Generation

Following previous research in reactive search personalization [3,6] and proactive information card ranking [15], we use the SAT-Hybrid method to generate the relevance labels in our experiments. This method considers all cards with a *SAT-Click* or a *SAT-View* as relevant cards. The definition of *SAT-Click* and *SAT-View* are as following.

SAT-Click: For each card in proactive impressions, we consider clicked cards with ≥ 30 s dwell time as relevant and other cards as non-relevant. This is a commonly used strategy for generating relevance labels in reactive search systems.

Table 5. Statistics of training data and testing data.

Item	Training data	Testing data
# Cards	8,499,640	9,400,779
# Cards with URLs	4,721,666(55.55 %)	4,920,380(52.34 %)
# Cards with entities	3,934,644(46.29 %)	3,960,484(42.13 %)
# Users	232,413	233,647
# Users with entities	210,139(90.42 %)	205,067(87.77 %)

SAT-View: Some types of cards do not require a click to satisfy users' information needs. For instance, users could scan the weather and temperature information on the cards without any clicking behavior. Stock cards could also tell users the real-time stock price of a company directly in the card content. Cards with viewport duration $\geq$ 10 s are labeled as relevant and the others are non-relevant.

6.3 Learning Models

We choose LambdaMART [20] as our learning model to rerank cards based on features extracted in Sect. 5. LambdaMART is an extension of LambdaRank [5]. This learning to rank method based on gradient boosted regression trees is one of the most effective models for the ranking task. It won Track 1 of the 2010 Yahoo! Learning to Rank Challenge and was commonly used in previous research on personalized ranking [3,4,18].

6.4 Comparison of Different Rankers

We compare the performance of different rankers. The baseline ranker is a production ranker which has been shipped to a commercial personal assistant system. This production ranker includes features that statically rank the different information cards based on their relative importance, and dynamic features that adjust their relevance scores based on the contextual information and the card content. This ranker is similar and comparable to the Carre model as described in [15]. We only report the relative gains and losses of other rankers against this production ranker to respect the proprietary nature of this ranker. The rankers which are compared with the baseline ranker include the following:

- UMPRanker-I (IF): The ranker from adding IF features on top of the features being used in the production ranker.
- UMPRanker-IE (IF + EF): The ranker from adding IF and EF features on top of the features being used in the production ranker.
- UMPRanker-IEU (IF + EF+ UD): The ranker from adding IF, EF and UD features on top of the features being used in the production ranker.

6.5 Experimental Results and Analysis

Table 6 summarizes the relative improvements of the different rankers against the baseline ranker. Starting from the base set of features used in the baseline model, we gradually add the three feature groups introduced in Sect. 5, namely, IF, EF and UD. IF is the feature group of directly modeling user historical interactions with the proactive cards, which is a coarser-level modeling, based on the card type, while EF is the finer-grained modeling at the level of entities. UD is the group of demographical features, which can be seen as a multiplier/conditioner on top of the first two feature groups for additional gains.

As we can see, with the IF features, we were able to capture the user interests nicely, resulting in significant improvements of 2.18 % in MRR and 2.25 % in

Table 6. Comparison of different rankers with the production ranker. The gains and losses are only reported in relative delta values to respect the proprietary nature of the baseline ranker. All differences are statistically significance ($p < 0.05$) according to the paired t-test.

Method	ΔMRR	ΔNDCG@1
UMPRanker-I (IF)	+2.18 %	+2.25 %
UMPRanker-IE (IF + EF)	+2.37 %	+2.38 %
UMPRanker-IEU (IF + EF+ UD)	+2.39 %	+2.39 %

NDCG@1 compared to the strong baseline ranker that was shipped to production, which is very substantial. On top of the baseline features and IF features, adding EF features, we were able to see significant larger gains of 2.37 % in MRR and 2.38 % in NDCG@1, demonstrating the substantial additional values in the entity-level modeling. Finally, with the UD features added, we were able to see additional statistically significant gains, even though to a lesser extent, making the total improvements of MRR to NDCG@1 both to 2.39 %.

6.6 Feature Importance Analysis

Next we perform feature importance analysis. By analyzing the relative importance, we can gain insights into the importance of each feature for the proactive ranking task. LambdaMART enables us to report the relative feature importance of each feature. Table 7 shows the top 10 features ordered by feature importance among IF, EF and UD feature groups. Half of the most important 10 features come from the IF feature group. 3 features come from the EF feature group and the rest are from the UD feature group. Features with the highest feature importance are *ViewRate*, *KBTaxonomyLevel1Weighted*, *CTR*, *AverageViewTime* and *LMScore*. Features like *ViewRate*, *CTR*, *AverageViewTime* can capture

Table 7. The most important features learnt by LambdaMART.

Feature name	Feature group	Feature importance
ViewRate	IF	1.0000
KBTaxonomyLevel1Weighted	EF	0.9053
CTR	IF	0.8593
AverageViewTime	IF	0.7482
LMScore	EF	0.6788
ViewRate	IF	0.4948
CardAvgAge	UD	0.1026
SATCTR	IF	0.0705
TMWeighted	EF	0.0628
GenderAbsDistance	UD	0.0486

users' preferences on different card types based on user historical interaction with the intelligent assistant system. Entity based features like *KBTaxonomyLevel*1 *Weighted* and *LMScore* are useful for improving proactive ranking through modeling user interests with user engaged textual content with term matching and semantic features. UD features, as shown in Table 7, are not as important as IF and EF features. However, they can still contribute to a better proactive ranking by capturing user preferences with user demographics information.

6.7 Case Studies of Re-Ranking

To better understand the improvements in ranking enabled through our UMP Ranker, we conduct case studies to look into the changes in the re-rankings of the proactive cards. From the examples, we find that the UMPRanker is able to identify the individual card types that each user prefers and rank them higher for the user (e.g., for users who like restaurant cards, the cards are promoted higher), thanks to the IF features; and provide customized ranking for different demographics, thanks to UD features (e.g., promoting sports cards for male users). And finally, we also observe the proposed EF features allow finer-grained improvements to adapt the ranking according to the user interests at the entity-level. Table 8 provides an example of this. As we can see, two News cards were promoted (i.e., News1 from 3rd to 1st, News2 from 4th to 2nd), while one News card (i.e., News3 from 2nd to 4th) was demoted. A closer look at the data reveals that the two promoted news cards are of higher weights learned in the EF representations of the user interests due to higher historical engagements with the entities (embedded in the news articles of News1 and News2) for the user, showing the benefits of finer-grained modeling such as EF on top of the coarser-grained user interests modeling at the card type level through IF.

Table 8. Examples of reranked cards in the testing data. "IsSuccess" denotes the inferred relevance labels based on SAT-Click or SAT-View with the timestamp denoted by "Time".

CardType	RankBefore	RankAfter	IsSuccess	Time
News1	3	1	TRUE	(3/28 8:11)
News2	4	2	TRUE	(3/28 8:13)
Calendar	1	3	FALSE	
News3	2	4	FALSE	
News4	5	5	FALSE	
Restaurant	7	6	FALSE	
Places1	8	7	FALSE	
Sports	10	8	FALSE	
Places2	9	9	FALSE	
Weather	6	10	FALSE	

7 Conclusion and Future Work

In this paper, we explore a variety of ways to model user interests, with the focus on improving the ranking of information cards for proactive systems such as Google Now and Microsoft Cortana. We propose a learning to rank framework and encode the various models as features, which include coarser-grained modeling of card type preferences directly mined from the historical interactions, finer-grained modeling of entity preferences, and features that capture the variations among demographics. Experiments performed with large-scale logs from a commercial proactive search system show that our method significantly outperforms a strong baseline method deployed in production, and show that the fine-grained modeling at the entity-level and demographics enable additional improvements on top of the coarser-grained card-type level modeling. In the future, we plan to experiment with different strategies, such as collaborative filtering, to further address the feature sparsity in entity-level modeling and contextualize the user interest modeling on factors such as time and location.

Acknowledgments. This work was done during Liu Yang's internship at Microsoft Research and Bing. It was supported in part by the Center for Intelligent Information Retrieval and in part by NSF grant #IIS-1419693. Any opinions, findings and conclusions or recommendations expressed in this material are those of the authors and do not necessarily reflect those of the sponsor. We thank Jing Jiang and Jiepu Jiang for their valuable and constructive comments on this work.

References

1. Agichtein, E., Brill, E., Dumais, S.: Improving web search ranking by incorporating user behavior information. In: SIGIR 2006, pp. 19–26. ACM, New York (2006)
2. Allan, J., Croft, B., Moffat, A., Sanderson, M.: Frontiers, challenges, and opportunities for information retrieval: Report from SWIRL 2012 the second strategic workshop on information retrieval in lorne. SIGIR Forum **46**(1), 2–32 (2012)
3. Bennett, P.N., Radlinski, F., White, R., Yilmaz, E.: Inferring and using location metadata to personalize web search. In: SIGIR 2011, July 2011
4. Bennett, P.N., White, R.W., Chu, W., Dumais, S.T., Bailey, P., Borisyuk, F., Cui, X.: Modeling the impact of short- and long-term behavior on search personalization. In: SIGIR 2012, pp. 185–194. ACM, New York (2012)
5. Burges, C., Ragno, R., Le, Q.: Learning to rank with non-smooth cost functions. In: NIPS 2007. MIT Press, Cambridge, January 2007
6. Fox, S., Karnawat, K., Mydland, M., Dumais, S., White, T.: Evaluating implicit measures to improve web search. ACM Trans. Inf. Syst. **23**(2), 147–168 (2005)
7. Guo, Q., Jin, H., Lagun, D., Yuan, S., Agichtein, E.: Mining touch interaction data on mobile devices to predict web search result relevance. In: SIGIR 2013, pp. 153–162 (2013)
8. Lagun, D., Hsieh, C.-H., Webster, D., Navalpakkam, V.: Towards better measurement of attention and satisfaction in mobile search. In: SIGIR 2014, pp. 113–122. ACM, New York (2014)
9. Li, J., Huffman, S., Tokuda, A.: Good abandonment in mobile and PC internet search. In: SIGIR 2009, pp. 43–50. ACM, New York (2009)

10. Lin, T., Mausam, Etzioni, O.: No noun phrase left behind: Detecting and typing unlinkable entities. In: EMNLP-CoNLL 2012, pp. 893–903. Association for Computational Linguistics, Stroudsburg (2012)
11. Mikolov, T., Chen, K., Corrado, G., Dean, J.: Efficient estimation of word representations in vector space. arXiv preprint (2013). arxiv:1301.3781
12. Mikolov, T., Sutskever, I., Chen, K., Corrado, G.S., Dean, J.: Distributed representations of words and phrases and their compositionality. In: NIPS 2013, pp. 3111–3119 (2013)
13. Resnick, P., Varian, H.R.: Recommender systems. Commun. ACM **40**(3), 56–58 (1997)
14. Rhodes, B.J., Maes, P.: Just-in-time information retrieval agents. IBM Syst. J. **39**(3–4), 685–704 (2000)
15. Shokouhi, M., Guo, Q.: From queries to cards: Re-ranking proactive card recommendations based on reactive search history. In: SIGIR 2015, May 2015
16. Song, F., Croft, W.B.: A general language model for information retrieval. In: CIKM 1999, pp. 316–321. ACM, New York (1999)
17. Su, X., Khoshgoftaar, T.M.: A survey of collaborative filtering techniques. Adv in Artif Intell **2009**, 4:2–4:2 (2009)
18. Wang, L., Bennett, P.N., Collins-Thompson, K.: Robust ranking models via risk-sensitive optimization. In: SIGIR 2012, pp. 761–770. ACM, New York (2012)
19. White, R.W., Chu, W., Hassan, A., He, X., Song, Y., Wang, H.: Enhancing personalized search by mining and modeling task behavior. In: WWW 2013, Republic and Canton of Geneva, Switzerland, pp. 1411–1420 (2013)
20. Wu, Q., Burges, C.J., Svore, K.M., Gao, J.: Adapting boosting for information retrieval measures. Inf. Retr. **13**(3), 254–270 (2010)
21. Xu, S., Jiang, H., Lau, F.C.-M.: Mining user dwell time for personalized web search re-ranking. In: IJCAI 2011, pp. 2367–2372 (2011)

Evaluation Methodology

Adaptive Effort for Search Evaluation Metrics

Jiepu Jiang and James Allan(✉)

Center for Intelligent Information Retrieval,
College of Information and Computer Sciences,
University of Massachusetts Amherst, Amherst, USA
{jpjiang,allan}@cs.umass.edu

Abstract. We explain a wide range of search evaluation metrics as the ratio of users' gain to effort for interacting with a ranked list of results. According to this explanation, many existing metrics measure users' effort as linear to the (expected) number of examined results. This implicitly assumes that users spend the same effort to examine different results. We adapt current metrics to account for different effort on relevant and non-relevant documents. Results show that such adaptive effort metrics better correlate with and predict user perceptions on search quality.

Keywords: Evaluation metric · Effort · Cost · User model · Adaptive model

1 Introduction

Searchers wish to find more but spend less. To accurately measure their search experience, we need to consider both the amount of relevant information they found (gain) and the effort they spent (cost). In this paper, we use *effort* and *cost* interchangeably because nowadays using search engines is mostly free of costs other than users' mental and physical effort (e.g., formulating queries, examining result snippets, and reading result web pages). Other costs may become relevant in certain scenarios – e.g., the price charged to search and access information in a paid database – but we only consider users' effort in this paper.

We show that a wide range of existing evaluation metrics can be summarized as some form of gain/effort ratio. These metrics focus on modeling users' gain [6,15] and interaction with a ranked list [2,6,13]—for example, nDCG [6], GAP [15], RBP [13], and ERR [2]. However, they use simple effort models, considering search effort as linear to the (expected) number of examined results. This implicitly assumes that users spend the same effort to examine every result. But evidence suggests that users usually invest greater effort on relevant results than on non-relevant ones, e.g., users are more likely to click on relevant entries [20], and they spend a longer time on relevant results [12].

To better model users' search experience, we adapt these metrics to account for different effort on results with different relevance grades. We examine two approaches: a parametric one that simply employs a parameter for the ratio

N. Ferro et al. (Eds.): ECIR 2016, LNCS 9626, pp. 187–199, 2016.
DOI: 10.1007/978-3-319-30671-1_14

of effort between relevant and non-relevant entries; and a time-based one that measures effort by the expected time to examine the results, which is similar to time-biased gain [17]. Both approaches model users' effort adaptively according to the results in the ranked list. We evaluate the adaptive effort metrics by correlating with users' ratings on search quality. Results show that the adaptive effort metrics can better predict users' ratings compared with existing ones.

2 Existing IR Evaluation Metrics

Much previous work [1,17] summarized search evaluation metrics in the form of $\sum_{i=1}^{k} d(i)g(i)$, where $g(i)$ is the ith result's gain, and $d(i)$ is the discount on the ith result. This framework does not explicitly consider users' effort. Instead, we summarize existing metrics as the ratio of users' gain to effort on the ranked list. We categorize these metrics into two groups:

M_1: E(gain)/E(effort). M_1 metrics separately measure the expected total gain (E(gain)) and effort (E(effort)) on the ranked list. They evaluate a ranked list by the ratio of E(gain) to E(effort). Existing M_1 metrics are usually implemented as Eq. 1: $P_{\text{examine}}(i)$ is the chance to examine the ith result; $g(i)$ and $e(i)$ are the gain and effort to examine the ith result. E(gain) and E(effort) simply sum up the expected gain and effort at each rank, until some cutoff k.

$$M_1 = \frac{E(\text{gain})}{E(\text{effort})} = \frac{\sum_{i=1}^{k} P_{\text{examine}}(i) \cdot g(i)}{\sum_{i=1}^{k} P_{\text{examine}}(i) \cdot e(i)} \tag{1}$$

M_2: E(gain/effort). M_2 metrics measure the expected ratio of gain to effort over different ways that users may interact with a ranked list. This is normally implemented by modeling the chances to stop at each rank when users examine results from top to bottom sequentially. M_2 metrics can be written as Eq. 2, where $P_{\text{stop}}(j)$ is the probability to stop after examining the jth result and $\sum_j P_{\text{stop}}(j) = 1$. Users' gain and effort for stopping at rank j, $g_{\text{stop}}(j)$ and $e_{\text{stop}}(j)$, simply sum up the gain and effort for all examined results.

$$M_2 = E(\frac{\text{gain}}{\text{effort}}) = \sum_{j=1}^{k} P_{\text{stop}}(j) \cdot \frac{g_{\text{stop}}(j)}{e_{\text{stop}}(j)} = \sum_{j=1}^{k} P_{\text{stop}}(j) \cdot \frac{\sum_{i=1}^{j} g(i)}{\sum_{i=1}^{j} e(i)} \tag{2}$$

Table 1 lists components for the M_1 and M_2 metrics discussed in this paper. Note that this is only one possible way of explaining these metrics, while other interpretations may also be reasonable. We use $r(i)$ for the relevance of the ith result and $b(i)$ for its binary version, i.e., $b(i) = 1$ if $r(i) > 0$, otherwise $b(i) = 0$.

2.1 Precision, AP, GAP, and RBP

P@k can be considered as an M_1 metric where users always examine the top k results. Each examined result provides $b(i)$ gain, and costs 1 unit effort.

Table 1. Components of existing M_1 and M_2 evaluation metrics.

Type	Metric	$P_{\text{examine}}(i)$ or $P_{\text{stop}}(j)$	$g(i)$	$e(i)$
M_1	P@k	1 if $i \leq k$, or 0	$b(i)$	1
	GP@k	1 if $i \leq k$, or 0	$\sum_{s=1}^{r(i)} g_s$	1
	DCG	$1/\log_2(i+1)$	$2^{r(i)} - 1$	$1/\sum_{i=1}^{k} \frac{1}{\log_2(i+1)}$
	RBP	$p^{(i-1)}$	$b(i)$	1
	GRBP	$p^{(i-1)}$	$\sum_{s=1}^{r(i)} g_s$	1
M_2	AP	$b(j)/N_r$	$b(i)$	1
	GAP *	$b(j)/N_r$	$\sum_{s=1}^{r(i)} g_s$	1
	RR	1 if $j = t$, or 0	1 if $i = j$, or 0	1
	ERR	$R(j)\prod_{m=1}^{j-1}(1 - R(m))$	1 if $i = j$, or 0	1

* GAP requires a normalization factor $N_r/E(N_r)$, as in Eq. 3

Following Robertson's work [14], we explain average precision (AP) as an M_2 metric in which users stop at each retrieved relevant result with an equal probability $1/N_r$. N_r is the total number of judged relevant results for the topic (or query). Therefore, the stopping probability at the jth result is $P_{\text{stop}}(j) = b(j)/N_r$. AP and P@$k$ share the same $g(i)$ and $e(i)$, as Table 1 shows.

Graded average precision (GAP) [15] generalizes AP to multi-level relevance judgments. It models that users may agree on a relevance threshold s with probability g_s—the probability that users only regard results with relevance grades $\geq s$ as relevant. Thus, the ith result has probability $\sum_{s=1}^{r(i)} g_s$ to be considered as relevant. The ith result's gain equates this probability, and $e(i) = 1$.

Similar to AP, we can explain GAP as an M_2 metric. Users stop at each retrieved relevant result with an equal probability $1/N_r$, regardless of the relevance grade. To obtain the original GAP, we need to further normalize the metric by $N_r/E(N_r)$, where $E(N_r)$ is the expected total number of results that users may consider as relevant, which takes into account the distribution of g_s. Equation 3 describes GAP, where: r_{max} is the highest possible relevance grade; N_m is the number of judged results with the relevance grade m.

$$\text{GAP} = \frac{N_r}{E(N_r)} \cdot \sum_{j=1}^{k} \frac{b(j)}{N_r} \cdot \frac{\sum_{i=1}^{j} \sum_{s=1}^{r(i)} g_s}{\sum_{i=1}^{j} 1}, \; E(N_r) = \sum_{m=1}^{r_{max}} N_m \sum_{s=1}^{m} g_s \quad (3)$$

Rank-biased precision (RBP) [13] models that after examining a result, users have probability p to examine the next result, and $1 - p$ to stop. Users always examine the first result. RBP is an M_1 metric. Users have p^{i-1} probability to examine the ith entry. RBP and P@k have the same gain and effort function. Note that the original RBP computes effort to an infinite rank ($E(\text{effort}) = \frac{1}{1-p}$). Here we measure both gain and effort to some cutoff k ($E(\text{effort}) = \frac{1-p^k}{1-p}$). This results in a slight numerical difference. But the two metrics are equivalent for evaluation purposes because they are proportional when p and k are predefined.

We also extend P@k and RBP to consider graded relevance judgments using the gain function in GAP ($g(i) = \sum_{s=1}^{r(i)} g_s$). We call the extensions graded P@k (GP@k) and graded RBP (GRBP).

2.2 Reciprocal Rank and Expected Reciprocal Rank

Reciprocal rank (RR) is an M_2 metric where users always and only stop at rank t (the rank of the first relevant result).

Expected reciprocal rank (ERR) [2] further models the chances that users stop at different ranks while sequentially examining a ranked list. ERR models that searchers, after examining the ith result, have probability $R(i)$ to stop, and $1 - R(i)$ to examine the next result. Chapelle et al. [2] define $R(i) = \frac{2^{r(i)}-1}{2^{r_{max}}}$. In order to stop at the jth result, users need to first have the chance to examine the jth result (they did not stop after examining results at higher ranks) and then stop after examining the jth result—$P_{\text{stop}}(j) = R(j) \prod_{m=1}^{j-1}(1 - R(m))$.

Both RR and ERR model that users always have 1 unit gain when they stop. They do not have an explicit gain function for individual results, but model stopping probability based on result relevance. To fit them into the M_2 framework, we define, when users stop at rank j, $g(i) = 1$ if $i = j$, otherwise $g(i) = 0$. For both metrics, $e(i) = 1$, such that stopping at rank j costs j unit effort.

$$\text{ERR@}k = \sum_{j=1}^{k} P_{\text{stop}}(j) \cdot \frac{1}{\sum_{i=1}^{j} 1}, \; P_{\text{stop}}(j) = R(j) \prod_{m=1}^{j-1} (1 - R(m)) \tag{4}$$

2.3 Discounted Cumulative Gain Metrics

Discounted cumulative gain (DCG) [6] sums up each result's gain in a ranked list, with a discount factor $1/\log_2(i+1)$ on the ith result. It seems that DCG has no effort factor. But we can also consider DCG as a metric where a ranked list of length k always costs the user a constant effort 1. We can rewrite DCG as an M_1 metric as Eq. 5. The log discount can be considered as the examination probability. Each examined result costs users $e(i)$ effort, such that $E(\text{effort})$ sums up to 1. $e(i)$ can be considered as a constant because it only depends on k.

$$\text{DCG@}k = \sum_{i=1}^{k} \frac{2^{r(i)} - 1}{\log_2(i+1)} = \frac{\sum_{i=1}^{k} \frac{2^{r(i)}-1}{\log_2(i+1)}}{\sum_{i=1}^{k} \frac{e(i)}{\log_2(i+1)}}, \; e(i) = \frac{1}{\sum_{i=1}^{k} \frac{1}{\log_2(i+1)}} \tag{5}$$

The normalized DCG (nDCG) metric [6] is computed as the ratio of DCG to IDCG (the DCG of an ideal ranked list). For both DCG and IDCG, $E(\text{effort})$ equates 1, which can be ignored when computing nDCG. However, $E(\text{effort})$ for DCG and IDCG can be different if we set $e(i)$ adaptively for different results.

3 Adaptive Effort Metrics

3.1 Adaptive Effort Vector

Section 2 explained many current metrics as users' gain/effort ratio with a constant effort on different results. This is oversimplified. Instead, we assign different effort to results with different relevance grades. Let $0, 1, 2, ..., r_{max}$ be the possible relevance grades. We define an effort vector $[e_0, e_1, e_2, ..., e_{r_{max}}]$, where e_r is the effort to examine a result with the relevance grade r.

We consider two ways to construct such effort vector in this paper. The first approach is to simply differentiate the effort on relevant and non-relevant results using a parameter $e_{r/nr}$. We set the effort to examine a relevant result to 1 unit. $e_{r/nr}$ is the ratio of effort on a relevant result to a non-relevant one—the effort to examine a non-relevant result is $\frac{1}{e_{r/nr}}$. For example, if we consider three relevance grades ($r = 0, 1, 2$), the effort vector is $[\frac{1}{e_{r/nr}}, 1, 1]$. Here we restrict $e_{r/nr} \geq 1$—relevant results cost more effort than non-relevant ones (because users are more likely to click on relevant results [20] and spend a longer time on them [12]).

The second approach estimates effort based on observed user interaction from search logs. Similar to time-biased gain [17], we measure effort by the amount of time required to examine a result. We assume that, when examining a result, users first examine its summary and make decisions on whether or not to click on its link. If users decide to click on the link, they further spend time reading its content. Equation 6 estimates $t(r)$, the expected time to examine a result with relevance r, where: t_{summary} is the time to examine a result summary; $t_{\text{click}}(r)$ is the time spent on a result with relevance r after opening its link; $P_{\text{click}}(r)$ is the chance to click on a result with relevance r after examining its summary.

$$t(r) = t_{\text{summary}} + P_{\text{click}}(r) \cdot t_{\text{click}}(r) \tag{6}$$

Table 2 shows the estimated time from a user study's search log [9]. We use this log to verify adaptive effort metrics. Details will be introduced in Sect. 4. The log does not provide t_{summary}. Thus, we use the reported value of t_{summary} in Smucker et al.'s article [17] (4.4 s). $P_{\text{click}}(r)$ and $t_{\text{click}}(r)$ are estimated from this log. The search log collected users' eye movement data such that we can estimate $P_{\text{click}}(r)$. The estimated time to examine *Highly Relevant*, *Relevant*, and *Non-relevant* results is 37.6, 23.0, and 9.8 s, respectively.

Table 2. Estimated time to examine results with each relevance grade.

Relevance (r)	t_{summary} [17]	$P_{\text{click}}(r)$	$t_{\text{click}}(r)$	$t(r)$
Non-relevant (0)	4.4 s	0.26	20.6 s	9.8 s
Relevant (1)	4.4 s	0.50	37.1 s	23.0 s
Highly Relevant (2)	4.4 s	0.55	60.3 s	37.6 s

We set the effort to examine a result with the highest relevance grade (2 for this search log) to 1 unit. We set the effort on a result with the relevance grade r to $\frac{t(r)}{t(r_{max})}$. The effort vector for this log is $[9.8/37.6, 23.0/37.6, 1] = [0.26, 0.61, 1]$.

3.2 Computation

The adaptive effort metrics are simply variants of the metrics in Table 1 using the effort vectors introduced in Sect. 3.1—we replace $e(i)$ by $e(r(i))$, i.e., the effort to examine the ith result only depends on its relevance $r(i)$. For example, let a ranked list of five results have relevance $[0, 0, 1, 2, 0]$. Equation 7 computes adaptive P@k and RR using an effort vector $[\frac{1}{e_{r/nr}}, 1, 1]$.

$$\mathrm{P}_{\text{adaptive}} = \frac{2}{2 + 3 \times \frac{1}{e_{r/nr}}}, \ \mathrm{RR}_{\text{adaptive}} = \frac{1}{\frac{1}{e_{r/nr}} + \frac{1}{e_{r/nr}} + 1} \tag{7}$$

When we set different effort to results with different relevance grades, users' effort is not linear to the (expected) number of examined results anymore, but further depends on results' relevance and positions in the ranked list. We look into the same example, and assume an ideal ranked list $[2, 2, 2, 1, 1]$. In such case, DCG has $E(\text{effort}) = \frac{1}{e_{r/nr}} + \frac{1}{e_{r/nr}\log_2 3} + \frac{1}{\log_2 4} + \frac{1}{\log_2 5} + \frac{1}{e_{r/nr}\log_2 6}$, but IDCG has $E(\text{effort}) = 1 + \frac{1}{\log_2 3} + \frac{1}{\log_2 4} + \frac{1}{\log_2 5} + \frac{1}{\log_2 6}$. The effort part is not trivial anymore when we normalize adaptive DCG using adaptive IDCG (adaptive nDCG).

Adaptive effort metrics have a prominent difference with static effort metrics. When we replace a non-relevant result with a relevant one, the gain of the ranked list does not increase for free in adaptive effort metrics. This is because the users' effort on the ranked list also increases (assuming relevant items cost more effort).

Equation 8 rewrites M_1 and M_2 metrics as $\sum_{i=1}^{k} g(i) \cdot d(i)$. It suggests that, when discounting a result's gain, adaptive effort metrics consider users' effort on each result in the ranked list. For M_1 metrics, $d(i) = \frac{P_{\text{examine}}(i)}{E(\text{effort})}$. Increasing the effort at any rank will increase $E(\text{effort})$, and penalize every result in the ranked list by a greater extent. For M_2 metrics, $d(i)$ depends on the effort to stop at rank i and each lower rank. Since $e_{\text{stop}}(j) = \sum_{m=1}^{j} e(m)$, $d(i)$ also depends on users' effort on each result in the ranked list. This makes the rank discounting mechanism in adaptive effort metrics more complex than conventional ones. We leave the analysis of such discounting mechanism for future work.

$$M_1 = \sum_{i=1}^{k} g(i) \cdot \frac{P_{\text{examine}}(i)}{E(\text{effort})}, \ M_2 = \sum_{i=1}^{k} g(i) \cdot \sum_{j=i}^{k} \frac{P_{\text{stop}}(j)}{e_{\text{stop}}(j)} \tag{8}$$

3.3 Relation to Time-Biased Gain and U-Measure

The time-based effort vector looks similar to the time estimation in time-biased gain (TBG) [17]. But we estimate t_{click} based on result relevance, while TBG uses a linear model that depends on document length. We made this choice because the former better correlates with t_{click} in the dataset used for evaluation.

Despite their similarity in time estimation, adaptive effort metrics and TBG are motivated differently. TBG models "the possibility that the user stops at some point by a decay function $D(t)$, which indicates the probability that the user continues until time t" [17]. The longer (the more effort) it takes to reach a result, the less likely that users are persistent enough to examine the result. Thus, we can consider TBG as a metric that models users' examination behavior (P_{examine}) adaptively according to the effort spent *prior to* examining a result.

U-measure [16] is similar to TBG. But the discount function $d(i)$ is dependent on the cumulative length of the texts users read *after* examining the ith result. The more users have read (the more effort users have spent) when they finish examining a result, the less likely the result will be useful. This seems a reasonable heuristic, but it remains unclear what the discounting function models.

In contrast to TBG, adaptive effort metrics retain the original examination models (P_{examine} and P_{stop}) in existing metrics, but further discount the results' gain by the effort spent. The motivation is that for each unit of gain users acquire, we need to account for the cost (effort) to obtain that gain (and assess whether or not it is worthwhile). Comparing to U-measure, our metrics are different in that a result's gain is discounted based on not only what users examined prior to the result and for that result, but also those examined afterwards (as Eq. 8 shows). The motivation is that user experience is derived from and measured for searchers' interaction with the ranked list as a whole—assuming a fixed contribution for an examined result regardless of what happened afterwards (such as in TBG and U-measure) seems oversimplified.

Therefore, we believe TBG, U-measure, and the proposed metrics all consider adaptive effort, but from different angles. This leaves room to combine them.

4 Evaluation

We evaluate a metric by how well it correlates with and predicts user perception on search quality. By modeling search effort adaptively, we expect the metrics can better indicate users' search experience. We use data from a user study [9] to examine adaptive effort metrics[1]. The user study asked participants to use search engines to work on some search tasks, and then rate their search experience and judge relevance of results. The dataset only collected user experience in a *search session*, so we must make some assumptions to verify metrics for a single query.

Relevance of results were judged at three levels: *Highly Relevant* (2), *Relevant* (1), or *Non-relevant* (0). Users rated search experience by answering: *how well do you think you performed in this task*? Options are: *very well* (5), *fairly well*

[1] The dataset and source code for replicating our experiments can be accessed at https://github.com/jiepujiang/ir_metrics/.

(4), *average* (3), *rather badly* (2), and *very badly* (1). Users rated 22 sessions as *very well*, 27 as *fairly well*, 22 as *average*, 7 as *rather badly*, and 2 as *very badly*.

When evaluating a metric, we first use it to score each query in a session. We use the average score of queries as an indicator for the session's performance. We assess the metric by how well the average score of queries in a session correlates with and predicts users' ratings on search quality for that session. This assumes that average quality of queries in a session indicates that session's quality.

We measure correlations using Pearson's r and Spearman's ρ. In addition, we evaluate a metric by how well it predicts user-rated performance. This approach was previously used to evaluate user behavior metrics [8]. For each metric, we fit a linear regression model (with intercept). The dependent variable is user-rated search performance in a session. The independent variable is the average metric score of queries for that session. We measure the prediction performance by normalized root mean square error (NRMSE). We produce 10 random partitions of the dataset, and perform 10-fold cross validation on each partition. This yields prediction results on 100 test folds. We report the mean NRMSE values of the 100 folds and test statistical significance using a two-tail paired t-test.

Table 3. Pearson's r and NRMSE for evaluation metrics.

		Pearson's r			**NRMSE (smaller is better)**		
	Metric	**static**	$e_{r/nr} = 4$	**time**	**static**	$e_{r/nr} = 4$	**time**
A	P@k	**0.326**	0.295	0.228	**0.246**	0.249 ↑↑	0.253 ↑↑↑ **
	AP	**0.065**	0.062	0.054	**0.257**	0.257	0.257
	RR	0.208	**0.236**	-0.052	0.253	**0.251**	0.256 *
B	GP@k	**0.371**	0.371	0.364	**0.241**	0.241	0.243 ↑
	GAP	**0.062**	0.061	0.055	0.257	**0.257**	0.257
C	RBP, $p = 0.8$	**0.331**	0.324	0.201	**0.245**	0.246	0.253 ↑↑↑ ***
	RBP, $p = 0.6$	0.305	**0.335**	0.154	0.247	**0.245**	0.255 ↑↑↑ ***
D	GRBP, $p = 0.8$	0.405	**0.440**	0.421	0.237	**0.233** ↓	0.236 *
	GRBP, $p = 0.6$	0.402	**0.463**	0.444	0.238	**0.230** ↓↓↓	0.233 ↓↓ *
	ERR	0.385	**0.427**	0.375	0.240	**0.236** ↓↓	0.242 ***
	DCG	0.398	**0.424**	0.418	0.238	**0.235**	0.237
	nDCG	0.352	0.398	**0.404**	0.243	0.238 ↓↓↓	**0.238** ↓↓
	TBG			**0.440**			**0.234** †††
	U-measure			**0.445**			**0.233** †
S	sDCG [7]	0.009			0.258 †††		
	nsDCG [10]	0.350			0.243 †††		
	esNDCG [10]	0.355			0.244 †		

Light , medium , and dark shading indicate Pearson's r is significant at 0.05, 0.01, and 0.001 levels, respectively. Arrow indicates NRMSE value is significantly different from **static**. * indicates significant difference between $e_{r/nr} = 4$ and **time**. † indicates significant difference comparing to GRBP ($p = 0.6$, $e_{r/nr} = 4$). One, two, and three symbols indicate $p < 0.05$, 0.01, and 0.001, respectively. **Bold font** and underline indicate the best value in its row and column, respectively.

5 Experiment

5.1 Parameters and Settings

For each metric in Table 1, we compare the metric using static effort with two adaptive versions using the parametric or time-based effort vector. We evaluate to a cutoff rank $k = 9$, because the dataset shows only 9 results per page.

For GP@k, GAP, and GRBP, we set the distribution of g_s to $P(s = 1) = 0.4$ and $P(s = 2) = 0.6$. This parameter yields close to optimal correlations for most metrics. For RBP and GRBP, we examine $p = 0.8$ (patient searcher) and $p = 0.6$ (less patient searcher). We set $e_{r/nr}$ to 4 (the effort vector is thus $[0.25, 1, 1]$).

We compare with TBG [17] and U-measure [16]. The original TBG predicted time spent using document length. However, in our dataset, we did not find any significant correlation between the two ($r = 0.02$). Instead, there is a weak but significant correlation between document relevance and time spent ($r = 0.274$, $p < 0.001$). We suspect this is because our dataset includes mostly web pages, while Smucker et al. [17] used a news corpus [18]. Web pages include many navigational texts, which makes it difficult to assess the size of the main content.

Thus, when computing TBG, we set document click probability and expected document examine time based on the estimation in Table 2. The dataset does not provide document save probability. Thus, we set this probability and parameter h by a brute force scan to maximize Pearson's r in the dataset. The save probability is set to $P_{\text{save}}(r = 1) = 0.2$ and $P_{\text{save}}(r = 2) = 0.8$. h is set to 31. Note that this corresponds to a graded-relevance version of TBG well tuned on our dataset.

To be consistent with TBG, we also compute U-measure [16] based on time spent. We set $d(i) = \max(0, 1 - \frac{t(i)}{T})$, where $t(i)$ is the expected total time spent after users examined the ith result, and T is a parameter similar to L in the original U-measure. We set T to maximize Pearson's r. T is set to 99 s.

5.2 Results

Table 3 reports Pearson's r and NRMSE for metrics. We group results as follows:

- Block A: metrics that do not consider graded relevance and rank discount.
- Block B: metrics that consider graded relevance, but not rank discount.
- Block C: metrics that consider rank discount, but not graded relevance.
- Block D: metrics that consider both graded relevance and rank discount.
- Block S: session-level metrics (for reference only).

Following Kanoulas et al.'s work [10], we set $b = 2$ and $bq = 4$ in sDCG and nsDCG. For esNDCG, we set parameters to maximize Pearson's r—$P_{\text{down}} = 0.7$ and $P_{\text{reform}} = 0.8$. As results show, for most examined metrics (Blocks A, B, C, and D), their average query scores significantly correlate with users' ratings on search quality in a session. The correlations are similarly strong compared with the session-level metrics (Block S). This verifies that our evaluation approach is reasonable—average query quality in a session does indicate the session's quality.

Adaptive Effort Vs. Static Effort. As we report in Table 3 (Block D), using a parametric effort vector ($e_{r/nr} = 4$) in GBRP, ERR, DCG, and nDCG can improve metrics' correlations with user-rated performance. The improvements in Pearson's r range from about 0.03 to 0.06. The adaptive metrics with $e_{r/nr} = 4$ also yield lower NRMSEs in predicting users' ratings compared with the static effort ones. The differences are significant except DCG ($p = 0.078$).

Such improvements seem minor, but are in fact a meaningful progress. Block A stands for metrics typically used before 2000. Since 2000, we witness metrics on modeling graded relevance (e.g., nDCG, GAP, and ERR) and rank discount (e.g., nDCG, RBP, and ERR). These work improve Pearson's r from 0.326 (P@k, the best "static" in Block A) to 0.405 (GRBP, $p = 0.8$, the best "static" in Blocks B, C, and D). The proposed adaptive effort metrics further improve Pearson's r from 0.405 to 0.463 (GRBP, $p = 0.6$, the best in the table). The magnitude of improvements in correlating with user-rated performance, as examined in our dataset, are comparable to those achieved by modeling graded relevance and rank discount. We can draw similar conclusions by looking at NRMSE. Although it requires further verification using larger datasets and query-level user ratings, our results at least suggest that the improvements are not negligible.

The best performing metric in our evaluation is adaptive GRBP ($p = 0.6$) with $e_{r/nr} = 4$. It also outperforms well-tuned TBG and U-measure. All these metrics consider adaptive search effort, but from different angles. The adaptive GRBP metric shows stronger Pearson's r with user-rated performance, and yields significantly lower NRMSE than TBG ($p < 0.001$) and U-measure ($p < 0.05$).

The preferable results of adaptive effort metrics confirms that users do not only care about how much relevant information they found, but are also concerned with the amount of effort they spent during search. By modeling search effort on relevant and non-relevant results, we can better measure users' search effort, which is the key to the improvements in correlating with user experience.

Parametric Effort Vector Vs. Time-Based One. Comparing the two ways of constructing effort vectors, results suggest that the time-based effort vector is not as good as the simple parametric one ($e_{r/nr} = 4$). Compared with the time-based effort vector, metrics using the parametric effort vector almost always yield stronger correlations and lower NRMSE in prediction. Compared with metrics using static effort, the time-based effort vector can still improve GRBP, DCG, and nDCG's correlations, but it fails to help ERR (Table 3, Block D). In addition, it can only significantly reduce NRMSE for GRBP ($p = 0.6$) and nDCG.

We suspect this is because time is only one aspect of measuring search effort. It has an advantage—we can easily measure time-based effort from search logs—but it does not take into account other effort such as making decisions and cognitive burden. Whereas other types of effort are also difficult to determine and costly to measure. Thus, it seems more feasible to tune parameters in the effort vector to maximize correlations with user-rated performance (or minimize prediction errors), if such data is available. In the presented results, we only differentiate the effort on relevant and non-relevant results. We also experimented assigning different effort to each relevance grade. But this yields not much better performance, probably due to the limited size of the dataset (80 sessions).

Adaptive Effort, Graded Relevance, and Rank Discount. Despite the success of adaptive effort in Block D, we did not observe consistent improvements in Blocks A, B, and C. The time-based effort vector even shows significantly worse prediction of user-rated performance for P@k, GP@k, and RBP. This suggests that adaptive effort needs to be applied together with graded relevance and rank discount. We suspect this is because: (1) when we apply adaptive effort vector to binary relevance metrics, they may prefer marginally relevant results over highly relevant ones in evaluation (for example, when using the time-based effort vector), which is problematic; (2) users are probably more concerned with search effort on top-ranked results, such that rank discount helps adaptive effort metrics. Although results in Blocks A, B, and C are negative, this is not a critical issue because all recent metrics (Block D) consider both factors. For these metrics, applying the adaptive effort vectors is consistently helpful.

Parameter Sensitivity. Figure 1 plots metrics' correlations when $e_{r/nr}$ varies from 1 to 10. $e_{r/nr} = 1$ (the leftmost points) stands for metrics using a static effort vector. The figure shows that when we increase the value of $e_{r/nr}$ (decrease the effort on non-relevant results, $1/e_{r/nr}$), all these metrics consistently achieve better correlations with user-rated performance. The adaptive effort metrics achieve near-to-optimal correlations with user-rated performance when we set $e_{r/nr}$ to about 3 to 5. In addition, some metrics seem quite stable when we set $e_{r/nr}$ to values greater than 5, such as GRBP ($p = 0.6$) and ERR.

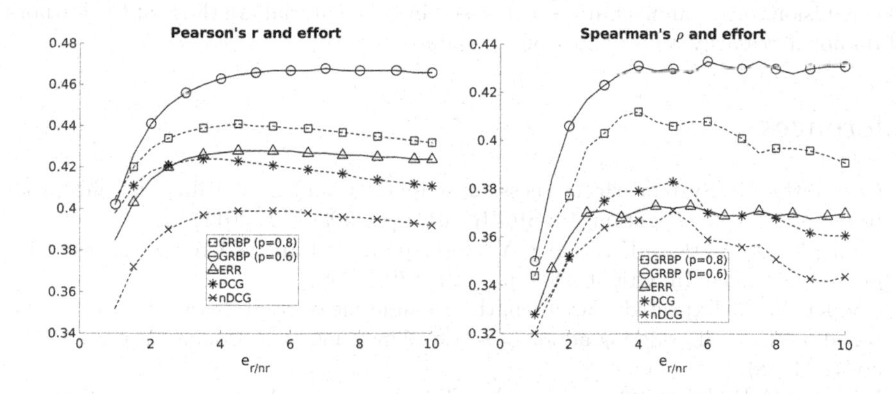

Fig. 1. Sensitivity of Pearson's r and Spearman's ρ to $e_{r/nr}$.

6 Discussion and Conclusion

Effort-oriented evaluation starts from expected search length [3], which measures the number of examined results to find a certain amount of relevant information.

Dunlop [5] extended the metric to measure the expected time required to find a certain amount of relevant results. Kazai et al. [11] proposed effort-precision, the ratio of effort (the number of examined results) to find the same amount of relevant information in the ranked list compared with in an ideal list. But these works all assume that examining different results involves the same effort.

This paper presents a study on search effectiveness metrics using adaptive effort components. Previous work on this topic is limited. TBG [17] considered adaptive effort, but applies it to the discount function. U-measure [16] is similar to TBG, and possesses the flexibility of handling SERP elements other than documents (e.g., snippets, direct answers). De Vries et al. [4] modeled searchers' tolerance to effort spent on non-relevant information before stopping viewing an item. Villa et al. [19] found that relevant results cost assessors more effort to judge than highly relevant and non-relevant ones. Yilmaz et al. [21] examined differences between searchers' effort (dwell time) and assessors' effort (judging time) on results, and features predicting such effort. Our study shows that the adaptive effort metrics can better indicate users' search experience compared with conventional ones (with static effort).

The dataset for these experiences was based on session-level user ratings and required that we make assumptions to verify query-level metrics. One important area of future research is to extend this study to a broader set of queries of different types to better understand the applicability of this research. Another direction for future research is to explore the effect of different effort levels, for example, assigning different effort for *Relevant* and *Highly Relevant* results rather than just to distinguish relevant from non-relevant.

Acknowledgment. This work was supported in part by the Center for Intelligent Information Retrieval and in part by NSF grant #IIS-0910884. Any opinions, findings and conclusions or recommendations expressed in this material are those of the authors and do not necessarily reflect those of the sponsor.

References

1. Carterette, B.: System effectiveness, user models, and user utility: a conceptual framework for investigation. In: SIGIR 2011, pp. 903–912 (2011)
2. Chapelle, O., Metlzer, D., Zhang, Y., Grinspan, P.: Expected reciprocal rank for graded relevance. In: CIKM 2009, pp. 621–630 (2009)
3. Cooper, W.S.: Expected search length: a single measure of retrieval effectiveness based on the weak ordering action of retrieval systems. Am. Documentation **19**(1), 30–41 (1968)
4. De Vries, A.P., Kazai, G., Lalmas, M.: Tolerance to irrelevance: a user-effort oriented evaluation of retrieval systems without predefined retrieval unit. RIAO **2004**, 463–473 (2004)
5. Dunlop, M.D.: Time, relevance and interaction modelling for information retrieval. In: SIGIR 1997, pp. 206–213 (1997)
6. Järvelin, K., Kekäläinen, J.: Cumulated gain-based evaluation of IR techniques. ACM Trans. Inf. Syst. **20**(4), 422–446 (2002)

7. Järvelin, K., Price, S.L., Delcambre, L.M.L., Nielsen, M.L.: Discounted cumulated gain based evaluation of multiple-query IR sessions. In: Macdonald, C., Ounis, I., Plachouras, V., Ruthven, I., White, R.W. (eds.) ECIR 2008. LNCS, vol. 4956, pp. 4–15. Springer, Heidelberg (2008)
8. Jiang, J., Hassan Awadallah, A., Shi, X., White, R.W.: Understanding and predicting graded search satisfaction. In: WSDM 2015. pp. 57–66 (2015)
9. Jiang, J., He, D., Allan, J.: Searching, browsing, and clicking in a search session: Changes in user behavior by task and over time. In: SIGIR 2014, pp. 607–616 (2014)
10. Kanoulas, E., Carterette, B., Clough, P.D., Sanderson, M.: Evaluating multi-query sessions. In: SIGIR 2011, pp. 1053–1062 (2011)
11. Kazai, G., Lalmas, M.: Extended cumulated gain measures for the evaluation of content-oriented xml retrieval. ACM Trans. Inf. Syst. **24**(4), 503–542 (2006)
12. Kelly, D., Belkin, N.J.: Display time as implicit feedback: Understanding task effects. In: SIGIR 2004, pp. 377–384 (2004)
13. Moffat, A., Zobel, J.: Rank-biased precision for measurement of retrieval effectiveness. ACM Trans. Inf. Syst. **27**(1), 2:1–2:27 (2008)
14. Robertson, S.E.: A new interpretation of average precision. In: SIGIR 2008, pp. 689–690 (2008)
15. Robertson, S.E., Kanoulas, E., Yilmaz, E.: Extending average precision to graded relevance judgments. In: SIGIR 2010, pp. 603–610 (2010)
16. Sakai, T., Dou, Z.: Summaries, ranked retrieval and sessions: a unified framework for information access evaluation. In: SIGIR 2013, pp. 473–482, (2013)
17. Smucker, M.D., Clarke, C.L.: Time-based calibration of effectiveness measures. In: SIGIR 2012, pp. 95–104(2012)
18. Smucker, M.D., Jethani, C.P.: Human performance and retrieval precision revisited. In: SIGIR 2010, pp. 595–602 (2010)
19. Villa, R., Halvey, M.: Is relevance hard work?: Evaluating the effort of making relevant assessments. In: SIGIR 2013, pp. 765–768 (2013)
20. Yilmaz, E., Shokouhi, M., Craswell, N., Robertson, S.E.: Expected browsing utility for web search evaluation. In: CIKM 2010, pp. 1561–1564 (2010)
21. Yilmaz, E., Verma, M., Craswell, N., Radlinski, F., Bailey, P.: Relevance and effort: an analysis of document utility. In: CIKM 2014, pp. 91–100 (2014)

Evaluating Memory Efficiency and Robustness of Word Embeddings

Johannes Jurgovsky(✉), Michael Granitzer, and Christin Seifert

Media Computer Science, Universität Passau, Passau, Germany
{Johannes.Jurgovsky,Michael.Granitzer,Christin.Seifert}@uni-passau.de
http://mics.fim.uni-passau.de

Abstract. Skip-Gram word embeddings, estimated from large text corpora, have been shown to improve many NLP tasks through their high-quality features. However, little is known about their robustness against parameter perturbations and about their efficiency in preserving word similarities under memory constraints. In this paper, we investigate three post-processing methods for word embeddings to study their robustness and memory efficiency. We employ a dimensionality-based, a parameter-based and a resolution-based method to obtain parameter-reduced embeddings and we provide a concept that connects the three approaches. We contrast these methods with the relative accuracy loss on six intrinsic evaluation tasks and compare them with regard to the memory efficiency of the reduced embeddings. The evaluation shows that low Bit-resolution embeddings offer great potential for memory savings by alleviating the risk of accuracy loss. The results indicate that post-processed word embeddings could also enhance applications on resource limited devices with valuable word features.

Keywords: Natural language processing · Word embedding · Memory efficiency · Robustness · Evaluation

1 Introduction

Word embeddings, also referred to as "word vectors" [7], capture syntactic and semantic properties of words solely from raw natural text corpora without human intervention or language dependent preprocessing. In natural language texts, the co-occurrence of words to appear together in the same context depends on the syntactic form and meaning of the individual words. In word embeddings the various nuances of word-to-word relations are distributed across several dimensions in vector space. These vector spaces are high-dimensional to provide enough degrees of freedom for hundreds of thousands of words to allow the relative arrangement of their embeddings reflect as many pairwise relations as possible out of the corpus statistics. However, embeddings carry a lot of information about words, which is hard to understand, interpret and quantify or may even be redundant and non-informative.

N. Ferro et al. (Eds.): ECIR 2016, LNCS 9626, pp. 200–211, 2016.
DOI: 10.1007/978-3-319-30671-1_15

The NLP community has been successfully exploiting these embeddings over the last years, e.g. [3,6]. However, the gain in task-accuracy brings the downside that high-dimensional continuous valued word vectors require a large amount of memory. Moreover, embeddings are trained by a fixed-size network architecture that sweeps through a huge text corpus. Consequently, the total number of parameters in the embedding matrix is implicitly defined a-priori. Further, there is no natural transition to more memory efficient embeddings, by which one could trade accuracy for memory. This is particularly limiting for NLP-applications on resource limited devices where memory is still a scarce resource. An embedding matrix with 150,000 vocabulary words can easily require 60–180 Megabytes of memory, which is rather inconvenient to be transferred to and stored in a browser or mobile application. This restriction gives rise to contemplate different types of post-processing methods in order to derive robust and memory-efficient word vectors from a trained embedding matrix.

In this paper, we investigate three post-processing methods for word vectors trained with the Skip-Gram algorithm that is implemented in the WORD2VEC software toolkit[1] [7]. The employed post-processing methods are (i) dimensionality reduction (PCA), (ii) parameter reduction (Pruning) and (iii) Bit-resolution reduction (Truncation). To isolate the effects on embeddings with different sizes, sparsity levels and resolutions, we employ intrinsic evaluation tasks based on word relatedness and abstain from extrinsic classification tasks. Our work makes the following contributions:

- We show through evaluation that Skip-Gram word embeddings are robust against linear dimensionality reduction, pruning and resolution-reduction on all tasks with only moderate loss of $< 10\,\%$ at a reduction of $40\,\%$.
- Our experiments reveal that higher-dimensional embeddings capture a larger fraction of redundant information, which can be exploited in favor of memory savings.
- We propose the resolution-based post-processing method as a means to gradually trade word vector quality for memory.

The remainder of this paper is structured as follows: First we present related work from the domain of language modeling and word representations. Next, we provide a conceptional overview of the post-processing methods used to reduce the amount of parameters in word embeddings. Then, we describe the experimental setup and the results in detail. A final discussion highlights the benefits of the different findings for practical applications.

2 Related Work

In computational linguistics, generating count-based language models has been an active research area since decades. The most common approach involves three parts: Collecting co-occurrence statistics of words from large text corpora,

[1] https://code.google.com/p/word2vec/.

transforming (e.g. tf-idf, Point-wise Mutual Information (PMI)) the counts to derive word association scores and finally applying a dimensionality reduction method (e.g. PCA, SVD). Dimensionality reduction is used for both smoothing sparsity and reducing the overall amount of parameters in order to obtain a low-dimensional and dense embedding matrix [1]. In this kind of approach, the quality of word vectors depends on the choice of methods used. In contrast, advances in recent years gave rise to new techniques [2–5], that implicitly model word co-occurrences by predicting context words from observed input words. Instead of first collecting co-occurrences of context words and then re-weighting these values with tf-idf or PMI, *predictive models* directly set the word vectors to optimally predict the contexts in which the corresponding words tend to appear. Since similar words appear in similar contexts, the classifier in a predictive model is trained to assign similar vectors to similar words. In an extensive evaluation, Baroni et al. [6] ascertain that embeddings of predictive models are superior to their count-based counterparts on word similarity tasks.

One particularly efficient representative of the family of predictive models is the Skip-Gram method, proposed by Mikolov et al. [7]. It offers the convenient property that the output of the model is a linear function of an input word vector and a context word vector, which not only results in meaningful nearest neighbours but also in informative relative positions of pairs of word embeddings. The intriguing finding is that arithmetic operations on word vectors in embedding space accurately reflect semantic and syntactic operations on words. We chose to use these embeddings in our experiments, since they encode a variety of language related information in both local and global neighborhoods. A thorough explanation of the rationale behind this technique was given by Levy and Goldberg [9,10].

Evaluations of word embeddings are published whenever new embedding methods are proposed. Besides manually inspecting 2D-projections of word vectors (e.g. t-SNE [11], PCA), it is difficult to associate meaning to individual dimensions. In language modeling, authors have traditionally employed perplexity to evaluate their models. In recent years, the common approach shifted towards testing the embeddings on various word similarity or word analogy test datasets [6,8]. In this domain, the work of Chen et al. [12] is the closest one to ours. Therein, they include a short section about information reduction capabilities of embeddings with limited experiments on other types of embeddings. We were particularly interested in preserving the linear structure in WORD2VEC-embeddings under limited memory conditions. So far, we are not aware of other experiments that explore ways to reduce word vectors in terms of memory.

3 Methodology

The word embeddings we use in our experiments, are obtained from Mikolov's Skip-Gram algorithm [7]. As a recent study [9] revealed, the algorithm factorizes an implicit word-context matrix, whose entries are the Point-wise Mutual Information of word-context pairs shifted by a constant offset. This PMI-matrix

$M \in \mathbb{R}^{|V| \times |V|}$ is factorized into a word embedding matrix $W \in \mathbb{R}^{|V| \times d}$ and a context matrix $C \in \mathbb{R}^{d \times |V|}$, where $|V|$ is the number of words in the vocabulary and d is the number of dimensions of each word vector. The context matrix is only required during training and usually discarded afterwards. The result of optimizing the Skip-Gram's objective is that word vectors (rows in W) have high similarity with respect to their cosine-similarity in case the words are syntactically or semantically similar. Besides that, the word vectors are dense and have significantly fewer dimensions than there are context words - columns in M. With sufficiently large d, the PMI-matrix could be perfectly reconstructed from its factors W and C, and thus provide the most accurate information about word co-occurrences in a corpus [6]. However, increasing the dimensionality d of word vectors also increases the amount of memory required to store the embedding matrix W. When using word embeddings in an application, we do not aim for a perfect reconstruction of the PMI-matrix but for reasonably accurate word vectors that reflect word similarities and word relations of language. Therefore, a more memory-efficient, yet accurate version of W would be desirable.

3.1 Memory Reduction with Post-Processing

More formally, we want to have a mapping τ from the full embedding matrix W to $\hat{W} = \tau(W)$, where $\hat{W}$ can be stored more efficiently while at the same time its word vectors are similarly accurate as the original vectors in W. For the vectors in $\hat{W}$ to have an accuracy loss as low as possible, word vectors in W must be robust against the mapping function τ. We consider W *robust* against the transformation τ, if the loss of $\tau(W)$ is small compared to W across very different evaluation tasks. A memory reduction through τ can be induced by reducing the number of dimensions, the amount of effective parameters or the parameters' Bit-resolution. Accordingly, we employed three orthogonal post-processing methods that can be categorized into *dimensionality-based*, *parameter-based* and *resolution-based* approaches:

- *Dimensionality-based:* Methods in this category can be described by the mapping $\tau_{dim} : \mathbb{R}^{|V| \times d} \to \mathbb{R}^{|V| \times \hat{d}}$, where $\hat{d} < d$. Fewer dimensions directly relate to less required memory. The dimensionality-based approaches provide a transformation that projects embeddings onto a lower-dimensional subspace while preserving the dominant properties of words. Both linear (e.g. PCA) and non-linear dimensionality (e.g. multilayer Autoencoder) reduction methods are applicable, as long as the inverse τ^{-1} of the transformation is available. In both variants, the computational overhead for computing τ and the memory overhead for storing the inverse τ^{-1} have to be considered. For linear transformations there is no memory overhead since the transformation can once be applied to the embedding matrix and subsequent methods can use the transformed embeddings alike. If there is reason to assume that the word vectors lie on a nonlinear manifold, nonlinear dimensionality reduction techniques can find a mapping to the components of the potentially low-dimensional

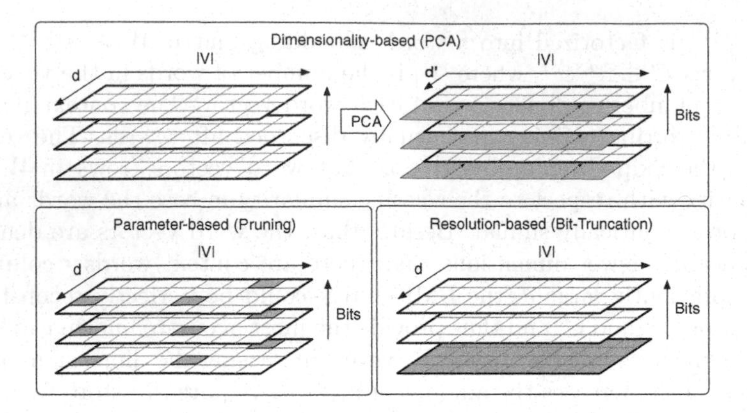

Fig. 1. Three methods for post-processing a word embedding matrix: PCA-Reduction (*top*), Pruning across all Bit-planes (*left*) and Truncation of the least significant Bits (*right*).

manifold. Both the computational overhead for estimating the nonlinear components and the memory overhead for storing the inverse of the mapping alongside with the transformed embeddings is high.

- *Parameter-based:* Whereas dimensionality-based methods change the bases of the embedding space, parameter-based methods leave the structure untouched but directly modify individual parameter values in the embedding matrix: $\tau_{par} : \mathbb{R}^{|V| \times d} \rightarrow \mathbb{R}^{|V| \times d}$, where τ_{par} is supposed to map most values to zero and leave only few non-zero elements in the output matrix. For instance, a simple pruning strategy can be used. Then, the output of τ_{par} is a sparse matrix that can be stored more efficiently.
- *Resolution-based:* $\tau_{res} : \mathbb{R}^{|V| \times d} \rightarrow \{0, \cdots, r-1\}^{|V| \times d}$, where $r \in \mathbb{N}^{+}$ is the resolution of the coordinate axes. With discrete coordinates, values can be stored at lower Bit-precision. Resolution-based methods discretize the coordinate axes into distinct intervals and thus reduce the resolution of the word vectors. For instance, the Bit-Truncation method subdivides the embedding space into regions of equal size.

In this work, we select one method of each category. In particular, we explore the robustness and memory efficiency of embeddings after applying *PCA*-reduction, *Pruning* and *Bit-Truncation*. In the following section we describe the selected methods along with the rationale for the selection (see Fig. 1).

3.2 Post-Processing Methods

We employ the following post-processing methods:

Linear Transformation. Dimensionality-based approaches assume that points are not uniformly scattered across the embedding space but exhibit certain directions of dominant variations. If there is some kind of structure in the data, it should be possible to exploit it by means of representing the same data with fewer dimensions. If the discarded dimensions only accounted for redundant information, we would obtain basis vectors that describe the word embeddings equally well but with fewer parameters. Since our evaluation tasks rely on vector arithmetic and cosine similarities, we do not use nonlinear dimensionality reduction methods as these operations would be meaningless on the transformed embeddings $\hat{L}$ produced by a nonlinear mapping. Therefore, we used the PCA-solution as a linear transformation to obtain lower dimensional embeddings.

Pruning. With *Pruning* we refer to a parameter-based method that discards a subset of the values in the embedding matrix by setting them to zero. With $\lambda \in [0, 1]$ we denote the *Pruning level.* Our naive pruning strategy is agnostic to word vectors since it determines a global threshold value p_λ from the whole matrix in such a way that $\lambda * 100\,\%$ of the matrix's values are greater than the threshold. All values w_{ij} below that threshold $|w_{ij}| < |p_\lambda|$ are set to zero. As a result of the pruning operation, we obtain a sparser embedding matrix with a degree of sparsity equivalent to $1 - \lambda$. Sparse matrices can be compressed more easily and thus require less memory than dense matrices. The rationale for using Pruning as reduction strategy arose from the observation that on normalized word vectors, pruning gradually projects points onto their closest coordinate axis. As we increase the pruning level, more points have coordinates that are aligned with the coordinate axes. Due to the normalization, this alignment gradually affects some but not all dimensions of individual word vectors. We hypothesize that up to a certain pruning level, the inaccuracy induced by Pruning has no qualitative effect on word vector arithmetic and word similarity computations.

Bit-Truncation. Besides a plain reduction of parameters by means of projection on fewer principle components, we explored a rather memory-focused approach that leaves the embedding dimensions untouched but migrates continuous word embeddings to discrete ones. The motivation is that in distributed embeddings the factors on all dimensions partially contribute to the meaning of a word. Thus, there should exist some degree of contribution which makes the meaning shift from one notion to another whereas for smaller contributions, the meaning is unaffected. We can exploit this relationship between the proximity of the embeddings' values and their similarities in meaning for purposes of memory efficiency by imposing resolution constraints on the value range along each coordinate. The Skip-Gram algorithm is defined on continuous valued word vectors which assumes each dimension to be real-valued. Figuratively, continuous embeddings allow for arbitrary positioning of a word's embedding in embedding space up to the precision of the datatype used. With *Bit-Truncation* we rasterize the embedding space uniformly by subdividing the range of values on each coordinate axis into distinct groups. Thus, all the factors of a distributed embedding still contribute to the meaning but only up to some pre-defined precision.

For the Bit-Truncation method, we adopt the approach described in [12] with slight adaptions. To reduce the resolution of the real numbers that make up the embedding matrix, first we shift the values to the positive range. Then we re-scale the values to the interval $[0, 1]$ and multiply them by 2^B, where B is the number of Bits we want to retain. Finally we cast the values to a 32-Bit Integer datatype. After casting to Integer, each coordinate axis has a resolution of $r = 2^B$ non-overlapping equally-spaced intervals. Consequently, the number of distinguishable regions in embedding space $R_\top = r^d$ is exponential in the number of dimensions d.

4 Experimental Setup

In all experiments we used word vectors estimated with the Skip-Gram method of the WORD2VEC-toolkit from a text corpus containing one billion words. The corpus was collected from the latest snapshot of English Wikipedia articles[2]. After removing words that appeared less than 100 times, the vocabulary contained 148,958 words, both uppercase and lowercase. We used a symmetric window covering $k = 9$ context words and chose the negative-sampling approximation to estimate the error from $neg = 20$ noise words. With this setup, we computed word vectors of several sizes in the range $d \in [50, 100, 150, 300, 500]$. After training, all vectors are normalized to unit length. To evaluate the robustness and efficiency of word vectors after applying post-processing, we compare PCA-reduction, Pruning and Bit-Truncation on three types of intrinsic evaluation tasks: word relatedness, word analogy and linguistic properties of words. In each of these tasks, we use two different datasets.

Word Relatedness: The *WordSim353*(WS353)[13] and *MEN*[14] datasets are used to evaluate pairwise word relatedness. Both consist of pairs of English words, each of which has been assigned a relatedness score by human evaluators. The *WordSim353* dataset contains 353 word pairs with scores averaged over judgments of at least 13 subjects. For the *MEN* dataset, a single annotator ranked each of the 3000 word-pairs relative to each of 50 randomly sampled word-pairs. The evaluation metric is the correlation (Spearman's ρ) between the human ratings and the cosine-similarities of word vectors.

Word Analogy: The word analogy task is more sensitive to changes of the global structure in embedding space. It is formulated as a list of questions of the form "a is to $\hat{a}$ as b is to $\hat{b}$", where $\hat{b}$ is hidden and has to be guessed from the vocabulary. The dataset we use here was proposed by Mikolov et al. [8] and consists of 19544 questions of this kind. About half of them are morphosyntactical (wa-syn) (*loud* is to *louder* as *tall* is to *taller*) and the other half semantic (wa-sem) questions (*Cairo* is to *Egypt* as *Stockholm* is to *Sweden*). It is assumed that the answer to a question can be retrieved by exploiting the relationship $a \rightarrow \hat{a}$ and applying it to b. Since WORD2VEC-embeddings exhibit a linear structure in embedding space, word relations are consistently reflected

[2] https://dumps.wikimedia.org/enwiki/20150112/.

in sums and differences of their vectors. Thus, the answer to an analogy question is given by the target word w_t whose embedding $\boldsymbol{w}_t$ is closest to $\boldsymbol{w}_q = \hat{\boldsymbol{a}} - \boldsymbol{a} + \boldsymbol{b}$ with respect to the cosine-similarity. The evaluation metric is the percentage of questions that have been answered with the expected word.

Linguistic Properties: Schnabel et al. [15] showed that results from intrinsic evaluations are not always consistent with results on extrinsic evaluations. Therefore, we include the recently proposed QVEC-evaluation[3] [16] as additional task. This evaluation uses two dictionaries of words, annotated with linguistic properties: a syntactic (QVEC-syn) dictionary (e.g., `ptb.nns`, `ptb.dt`) and a semantic (QVEC-sem) dictionary (e.g., `verb.motion`, `noun.position`). The proposed evaluation method assigns to each embedding dimension the linguistic property that has highest correlation across all mutual words. The authors showed that the sum over all correlation values can be used as an evaluation measure for word embeddings. Moreover, they showed that this score has high correlation with the accuracy the same embeddings achieve on real-world classification tasks.

5 Results

Since we evaluate the robustness of word embeddings against post-processing, we report the *relative loss* induced by applying a post-processing method. The loss is measured as the difference between the score of original embeddings and the score of post-processed embeddings. In case of the word relatedness task, the score is the Spearman correlation. On the word analogy task, the score is given as accuracy. And on the linguistic properties task, the score is the output of the QVEC evaluation method. We divide the loss by the score of the original embeddings to obtain a relative loss that is comparable across tasks.

5.1 Robustness of Word Vectors

In Fig. 2 we report the mean relative loss, averaged over the five word vector sizes on all datasets. The percentage of reduction refers to the fraction of principle components with lowest eigenvalues that were discarded after applying PCA and to the fraction of parameters that were set to zero after pruning, respectively.

The word embeddings show a similar trend for all three post-processing methods. A small relative reduction results in a small loss, whereas a large reduction leads to a large loss. For all methods, the loss increases exponentially with the percentage of reduction. On the word analogy datasets, the loss is consistently higher than on the word relatedness and QVEC datasets. In particular, the relative loss on word relatedness datasets is predominantly unaffected (relative loss $< 10\,\%$) by post-processing up to a reduction threshold of $40\,\%$. Compared to the naive Pruning approach, PCA-transformed embeddings suffer lower loss on all tasks. Actually, on WordSim353 and MEN, PCA-reduced embeddings exhibit slight negative loss ($< 3\,\%$). Bit-Truncation produces no loss on any dataset

[3] https://github.com/ytsvetko/qvec.

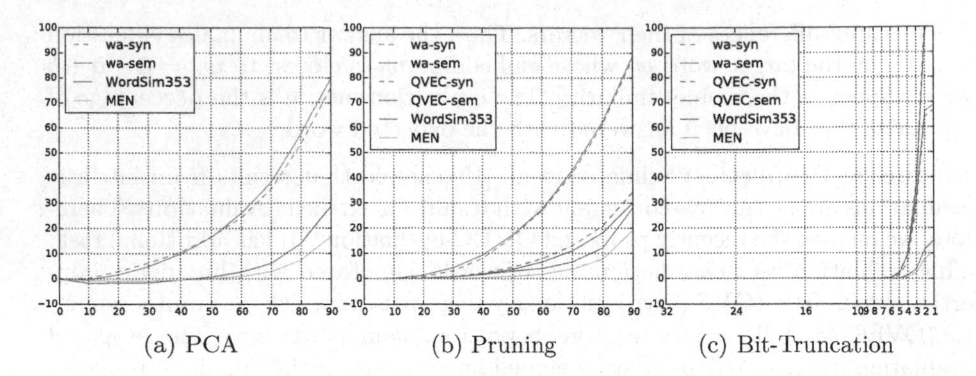

Fig. 2. Mean relative loss of embeddings after (a) PCA: percentage of removed dimensions, (b) Pruning: percentage of removed parameters and (c) Bit-Truncation: remaining Bits. Scores on the QVEC datasets are not shown for PCA since they are not comparable across different word vector sizes.

until the Bit-resolution of the parameters is lower than 8-Bit. To summarize, the Skip-Gram word embeddings are most robust against post-processing with resolution-based Bit-Truncation and the dimensionality-based PCA-reduction.

5.2 Memory Efficiency

The percentage of reduction achieved by PCA is directly proportional to memory savings induced by the smaller number of dimensions. There, the sweet spot is task-dependent and the relative reduction can be rather high before word vector quality suffers a loss. In contrast, the number of pruned values is not directly proportional to memory savings, since the coding of sparse matrices requires additional memory. For instance, the row compressed storage method [17] has, without further assumptions about the shape of the matrix, a memory complexity of $\mathcal{O}(3k)$, where k is the number of non-zero elements in the sparse matrix. Thus, the pruning method would only start to pay off in terms of memory consumption above a pruning level of $\frac{2}{3}$, which would result in serious quality-loss. Finally, post-processing the embedding matrix with Bit-Truncation does not cause any loss on any of the evaluated datasets up to 75 % reduction (24Bit). For resolutions below $r = 2^8$, all evaluated datasets respond to the low-precision embeddings with abruptly increasing loss.

Figure 3 shows that higher-dimensional embeddings ($d = 500$) can be reduced more aggressively than lower-dimensional ones before reaching the same level of relative loss. Since a similar behavior holds on all tasks (not shown due to space constraints), the observation is two-fold: First, it suggests that, the higher-dimensional the embedding space is, the more non-zero parameters there are and the higher their resolution is, the more redundant is the information that is captured in the embeddings. Secondly, the consistency across dimensionality-based and parameter-based methods indicates that neither the number of dimensions

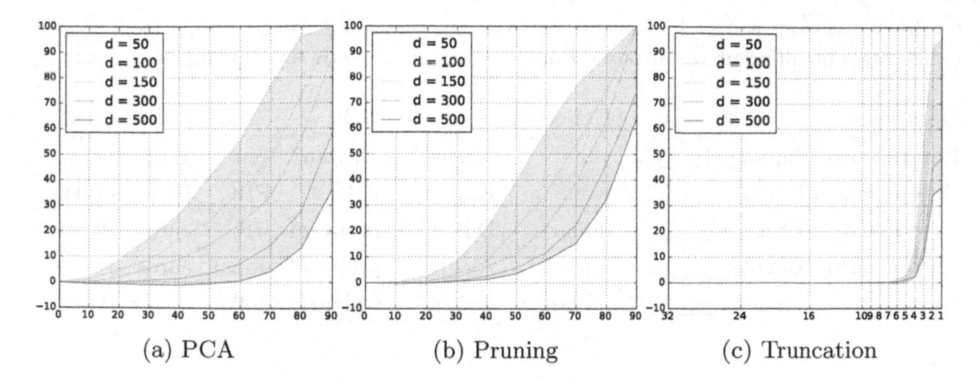

(a) PCA (b) Pruning (c) Truncation

Fig. 3. Relative loss of embeddings on the syntactic word analogy dataset (wa-syn) after PCA (a), Pruning (b) and Bit-Truncation (c).

or parameters nor the continuous values alone but the number of distinguishable regions in embedding space is crucial for accurate word embeddings.

With a sufficiently large Bit-resolution the accuracy of all embedding sizes approximates the same accuracy level as with continous values. Thus, we can confirm the finding in [12] also for Skip-Gram embeddings: The same accuracy can be achieved with discretized values at sufficiently large resolutions. Additionally, we state that this observation not only holds for cosine-similarity on word relatedness tasks but also for vector arithmetic on the word analogy task and for QVEC on the linguistic properties task.

To summarize, Skip-Gram word embeddings can be stored more efficiently using a post-processing method that reduces the number of distinguishable regions $R_{\top} = (2^B)^d$ in embedding space. Pruning does so by producing increasingly large zero-valued regions around each coordinate axis ($2^B - const.$). PCA does so by mapping the word vectors into an embedding space with fewer dimensions $\hat{d} < d$. And Bit-Truncation directly lowers the resolution of each coordinate by constraining the Bit-resolution $\hat{B} < B$.

6 Discussion

Memory Efficiency: If an application can take a loss in word vector accuracy in favor of memory or transmission times, Skip-Gram embeddings can be reduced with all three evaluated methods. Thereby, pruning is the least efficient method as the overhead introduced by sparse coding could only be compensated by pruning levels above $\frac{2}{3}$. Such an aggressive pruning strategy would result in an average accuracy loss of more than 30 %. In contrast, the linear dimensionality reduction technique worked well on our tasks and it allows for a consistent transition from higher to lower dimensional embeddings. The accuracy of PCA-reduced embeddings even improves over equivalently large original non-reduced embeddings on word relatedness and word analogy tasks

Table 1. Answer words for several country-currency analogy questions from word embeddings at different resolutions. Finally, all answer words are currencies.

Questions	Expected	3-Bit	4-Bit	5-Bit	6-Bit	7-Bit
Europe euro : Japan ______?	yen	Nagasaki	Taiwan	yen	yen	yen
Europe euro : Korea ______?	won	Kim	PRC	PRC	PRC	dollar
Europe euro : USA ______?	dollar	Dusty	proposal	US	euros	dollar
Europe euro : Brazil ______?	real	Alegre	proposal	dollar	euros	euros
Europe euro : Canada ______?	dollar	Calgary	Calgary	dollar	dollar	dollar

(see evaluation data online[4]). Thus, in terms of memory, it can be more efficient to first train high-dimensional embeddings and afterwards reduce them with PCA to the desired size. The resolution-based approach provides the greatest potential for memory savings. With only 8-Bit precision per value, there is no loss on any of the tasks. A straight-forward implementation can thus fit the whole embedding matrix in only 25 % of memory.

Resolution and Semantic Transition: Another observation is depicted in Table 1. On the word analogy dataset, the transition from lower to higher-precision values not only yields increasingly better average accuracy but also corresponds to a semantic transition from lower to higher relatedness. Even if the embeddings' values have only 3-Bit precision, the retrieved answer words are not totally wrong but still in some kind related to the expected answer word. It seems that some notions of meaning are encoded on a finer scale in embedding space and that these require more Bits to remain distinguishable.

For coarse resolutions (3-Bit) the regions in embedding space are too large to allow an identification of a country's currency. Because there are many words within the same distance to the target location, the most frequent one is retrieved as answer to the question. As the resolution increases, regions get smaller and thus more nuanced distances between word embeddings emerge, which yields not only increasingly accurate but also progressively more related answers.

7 Conclusion

In this paper, we explored three methods to post-process Skip-Gram word embeddings in order to identify means to reduce the amount of memory required to store the embedding matrix. Therefore, we evaluated the robustness of embeddings against a dimensionality-based (PCA), parameter-based (Pruning) and a resolution-based (Bit-Truncation) approach. The results indicate, that embeddings are most robust against Bit-Truncation and PCA-reduction and that preserving the number of distinguishable regions in embedding space is key

[4] http://tinyurl.com/jj-ecir2016-eval.

for obtaining memory efficient (75 % reduction) and accurate word vectors. Especially resource limited devices can benefit from these compact high-quality word features to improve NLP-tasks under memory constraints.

Acknowledgments. The presented work was developed within the EEXCESS project funded by the European Union Seventh Framework Programme FP7/2007-2013 under grant agreement number 600601.

References

1. Turney, P.D., Pantel, P.: From frequency to meaning: vector space models of semantics. J. Artif. Intell. Res. **37**(1), 141–188 (2010)
2. Bengio, Y., Ducharme, R., Vincent, P., Janvin, C.: A neural probabilistic language model. J. Mach. Learn. Res. **3**, 1137–1155 (2003)
3. Collobert, R., Weston, J., Bottou, L., Karlen, M., Kavukcuoglu, K., Kuksa, P.: Natural language processing (almost) from scratch. J. Mach. Learn. Res. **12**, 2493–2537 (2011)
4. Mnih, A., Hinton, G.: Three new graphical models for statistical language modelling. In: ICML, pp. 641–648. ACM, June 2007
5. Huang, E.H., Socher, R., Manning, C.D., Ng, A.Y.: Improving word representations via global context and multiple word prototypes. In: 50th Annual Meeting of the Association for Computational Linguistics, pp. 873–882. ACL, July 2012
6. Baroni, M., Dinu, G., Kruszewski, G.: Dont count, predict! a systematic comparison of context-counting vs. context-predicting semantic vectors. In: 52nd Annual Meeting of the Association for Computational Linguistics, pp. 238–247 (2014)
7. Mikolov, T., Chen, K., Corrado, G., Dean, J.: Efficient estimation of word representations in vector space (2013). arXiv.org
8. Mikolov, T., Yih, W.T., Zweig, G.: Linguistic regularities in continuous space word representations. In: HLT-NAACL, pp. 746–751, June 2013
9. Levy, O., Goldberg, Y.: Neural word embedding as implicit matrix factorization. In: Advances in Neural Information Processing Systems, pp. 2177–2185 (2014)
10. Goldberg, Y., Levy, O.: word2vec Explained: deriving Mikolov et al'.s negative-sampling word-embedding method. arXiv preprint (2014). arxiv:1402.3722
11. Van der Maaten, L., Hinton, G.: Visualizing data using t-SNE. J. Mach. Learn. Res. **9**(2579–2605), 85 (2008)
12. Chen, Y., Perozzi, B., Al-Rfou, R., Skiena, S.: The expressive power of word embeddings. In: Speech and Language Proceeding Workshop, ICML, Deep Learning for Audio (2013)
13. Finkelstein, L., Gabrilovich, E., Matias, Y., Rivlin, E., Solan, Z., Wolfman, G., Ruppin, E.: Placing search in context: the concept revisited. In: 10th International Conference on World Wide Web, pp. 406–414, ACM, April 2001
14. Bruni, E., Tran, N.K., Baroni, M.: Multimodal distributional semantics. J. Artif. Intell. Res. (JAIR) **49**, 1–47 (2014)
15. Schnabel, T., Labutov, I., Mimno, D., Joachims, T.: Evaluation methods for unsupervised word embeddings. In: Proc. Emp. Met. Nat. Lang. (2015)
16. Ling, Y., Dyer, G.L.C.: Evaluation of word vector representations by subspace alignment. In: Proc. Emp. Met. Nat. Lang., pp. 2049–2054, ACL, Lisbon (2015)
17. Saad, Y.: Iterative methods for sparse linear systems. Siam (2003)

Characterizing Relevance on Mobile and Desktop

Manisha Verma[✉] and Emine Yilmaz

University College London, London, UK

mverma@cs.ucl.ac.uk, emine.yilmaz@ucl.ac.uk

Abstract. Relevance judgments are central to Information retrieval evaluation. With increasing number of hand held devices at users disposal today, and continuous improvement in web standards and browsers, it has become essential to evaluate whether such devices and dynamic page layouts affect users notion of relevance. Given dynamic web pages and content rendering, we know little about what kind of pages are relevant on devices other than desktop. With this work, we take the first step in characterizing relevance on mobiles and desktop. We collect crowd sourced judgments on mobile and desktop to systematically determine whether screen size of a device and page layouts impact judgments. Our study shows that there are certain difference between mobile and desktop judgments. We also observe different judging times, despite similar inter-rater agreement on both devices. Finally, we also propose and evaluate display and viewport specific features to predict relevance. Our results indicate that viewport based features can be used to reliably predict mobile relevance.

Keywords: Information retrieval · Relevance judgment · Mobile · Evaluation

1 Introduction

The primary goal of Information Retrieval (IR) systems is to retrieve highly relevant documents for user's search query. Judges determine document relevance by assessing topical overlap between its content and user's information need. To facilitate repeated and reliable evaluation of IR systems, trained judges are asked to evaluate several documents (mainly shown on desktop computers) with respect to a query to produce large scale test collections (such as those produced at TREC[1] and NTCIR[2]). These collections are created with the following assumptions: (1) document rendering is device agnostic, i.e., a document appears the same whether it is viewed on a mobile or desktop. For example font size of text, headings and image resolutions remain unchanged with change in screen size. (2) document content is device agnostic, i.e. if some text if displayed on

[1] http://trec.nist.gov/.

[2] http://research.nii.ac.jp/ntcir/.

N. Ferro et al. (Eds.): ECIR 2016, LNCS 9626, pp. 212–223, 2016.

DOI: 10.1007/978-3-319-30671-1_16

desktop, will also be visible for instance on a mobile. Note that while the former assumes that document layout remains unchanged, the latter assumes that its content (for example, number of words, headings or paragraphs) remains the same across devices.

While this evaluation mechanism is robust, it is greatly challenged by explosion in devices with myriad resolutions. This requires optimization of pages for different resolutions. A popular website today has at least two views, one optimized for traditional desktops while the other tuned for mobiles or tablets, i.e. devices with smaller screen sizes. The small screen size limits both the layout and amount of content visible to the user at any given time. The continuous advancement in browsers and markup languages exacerbates this problem, as web developers today can collapse a traditionally multiple page website into a single web page with several sections. The same page can be optimized for both mobile and desktops with one style sheet and minimal change in HTML. Thus, with today's technology, a user may see separate versions of the same website on desktop and mobile, which in turn may greatly impact her judgment of the page with respect to a query.

To illustrate this further, Fig. 1 shows four web pages with their respective queries on mobile as well as desktop. Web pages in Fig. 1a and b are relevant to the query and have been optimized for mobiles. Judges in our study also marked these pages relevant on mobile and desktop. However, web pages in Fig. 1c and d are not suitable for mobile screens. In case of Fig. 1c, the whole page loads in the viewport[3] which in turn makes it extremely hard to read. Figure 1d has more ads on the viewport than the relevant content which prompts judges to assign lower relevance on mobile.

Thus, it needs to be determined whether device on which a document is rendered influences its evaluation with respect to a query. While some work [7] compares user search behavior on mobiles and desktop, we know little about how users judge pages on these two mediums and whether there is any significant difference between judging time or obtained relevance labels. We need to determine whether page rendering has any impact on judgments, i.e. different web page layouts (on mobile or desktop) translate into different relevance labels. We also need to verify whether viewport specific signals can be used to determine page relevance. If these signals are useful, mobile specific relevance can be determined using a classifier which in turn would reduce the overhead of obtaining manual judgments.

In this work we investigate above outlined problems. We collect and compare crowd sourced judgments of query-url pairs for mobile and desktop. We report the difference in agreements, judging time and relevance labels. We also propose novel viewport oriented features and use them to predict page relevance on both mobile and desktop. We analyze which features are strong signals to determine relevance. Our study shows that there are certain differences between mobile and desktop judgments. We also observe different judging times, despite similar

[3] Viewport is the framed area on a display screen of mobile or desktop for viewing information.

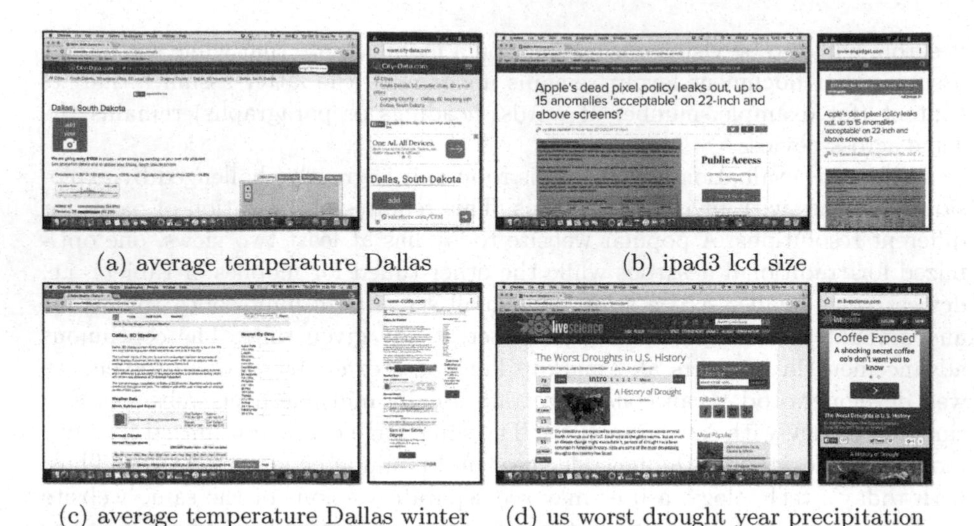

(a) average temperature Dallas (b) ipad3 lcd size

(c) average temperature Dallas winter (d) us worst drought year precipitation

Fig. 1. Sample queries and resulting web pages on desktop and mobile screens

inter-rater agreement on both devices. On mobiles, we also observe correlation between viewport features and relevance grades assigned by judges.

The remaining paper is organized as follows. We briefly cover the related work in Sect. 2. Section 3 describes the crowd sourced data collection and their comparison on mobiles and desktop. We describe the features and results of relevance prediction in Sect. 4. We summarize our findings and discuss future work in Sect. 5.

2 Related Work

While there exists large body of work that identifies factors affecting relevance on Desktop, only a fraction exists that characterizes search behavior or evaluates search engine result pages for user interaction on mobiles. We briefly discuss factors important for judging page relevance on desktop and contrast our work with existing mobile search studies.

Schamber *et al.* [13] concluded that relevance is a multidimensional concept that is affected by both internal and external factors. Since then, several studies have investigated and evaluated factors that constitute relevance. For instance, Xu *et al.* [18] and Zhang *et al.* [20] explored factors employed by users to determine page relevance. They studied impact of topicality, novelty, reliability, understandability, and scope on relevance. They found that topicality and novelty to be the most important relevance criteria. Borlund *et al.* [1] have shown that as search progresses, structure and understandability of document become important in determining relevance. Our work is quite different as we do not ask users explicit judgments on above mentioned factors. We compare relevance judgments obtained for same query-url pairs on mobile and desktop. Our primary focus is the difference in judging patterns on both mediums.

There is some work on mining large scale user search behavior in the wild. Several studies [3,5,6,8–10,17,19] report differences in search patterns across devices. For instance, Kamwar *et al.* [8] compare searches across computers and mobiles and conclude that smart phones are treated as extensions of users' computers. They suggested that mobiles would benefit from integration with computer based search interface. These studies found mobile queries to be short (2.3 – 2.5 terms) and high rate of query reformulation. One key result of Karen *et al.* [4] was that conventional desktop-based approach did not receive any click for almost 90 % of searches which they suggest maybe due to unsatisfactory search results. Song *et al.* [15] study mobile search patterns on three devices: mobile, desktop and tablets. Their study emphasizes that due to significant differences between user search patterns on these platforms use of similar web page ranking methodology is not an optimal solution. They propose a framework to transfer knowledge from desktop search such that search relevance on mobile and tablet can be improved. We train models for relevance prediction as opposed to search result ranking.

Other work includes abandonment prediction on mobile and desktop [12], query reformulation on mobile [14] and understanding mobile search intents [4]. Buchanan *et al.* [2] propose some ground rules to design web interfaces for mobile. All the above mobile related studies focus on search behavior, not on what constitutes page relevance on small screens. In this work our focus is not to study search behavior but to compare relevance judgments for same set of pages on different devices.

3 Mobile and Desktop Judgments

To understand whether user's device has any affect on relevance, we first collect judgments via crowd sourcing. We begin by describing the query-url pairs, the judging interface and the crowd sourcing experiment in detail. We use query-url pairs from Guo *et al.* [7], where the users were asked to perform seven search tasks similar to regular mobile search tasks from [4]. Their study collected 393 unique page views associated with explicit judgments. We filtered broken urls or search results pages. We also removed queries and corresponding pages that were temporal. First author found corresponding desktop urls manually for remaining urls. In total, we obtained crowdsourced judgments on desktop and mobile for 236 query-url pairs. We built two simple interfaces– one for desktop oriented judgments and the other for mobile specific judgments. We used Amazon Mechanical Turk (MTurk)[4] to host two sets of hits, one for each interface. Each interface had concise description of different relevance grades. We asked judges to rate each query-url pair on a scale of 1 to 4 with 4 being 'highly-relevant', 3 as 'relevant', 2 as 'somewhat-relevant' and 1 as 'non-relevant'. Each HIT was

[4] AMT (https://requester.mturk.com/) is a crowd sourcing marketplace to conduct experiments by recruiting multiple participants in exchange for compensation.

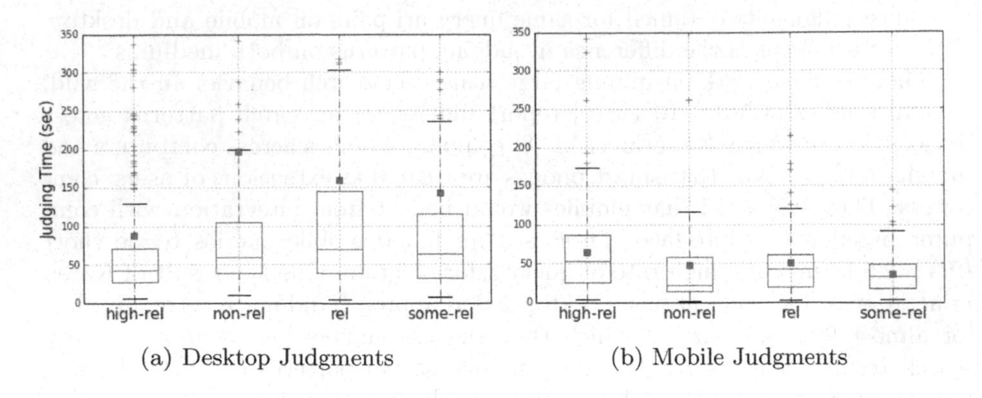

(a) Desktop Judgments (b) Mobile Judgments

Fig. 2. Relevance Grade vs Judging Time

payed $0.03. We collected three judgments per query-url pair. We ensured that query-url pairs were shown at random to avoid biasing the judge. We determined judge's browser type (and device) using javascript. The judgments performed on Android or iOS phones are used in our analysis. To help filter malicious workers, we restricted our HITs to workers with an acceptance rate of 95 % or greater and to ensure English language proficiency, to workers in the US. In total we collected 708 judgments from each interface. Desktop judgment hits were submitted by 41 workers and mobile judgment hits were completed by 28 workers on MTurk.

The final grade of each pair was obtained by taking majority of all three labels. We also group relevance labels[5] to form binary judgments from 4 scale judgments. The label distribution is as follows:

- **Desktop**: High-rel=108, rel=37, some-rel=47, non-rel=44
- **Mobile**: High-rel=86, rel=55, some-rel=64, non-rel=31

The inter-rater agreement (Fleiss kappa) for Desktop judgments was 0.28 (fair) on 4 scales and 0.42 (moderate) for binary grades. Similarly, inter-rater agreement for mobile judgments were 0.33 (fair) on scale of 4 grades and 0.53 (moderate) for binary grades. The agreement rate is comparable to that observed in previous relevance crowd sourcing studies [11]. However, Cohen's kappa between majority desktop and mobile relevance grades is only 0.127 (slight), indicating that judgments obtained on mobiles may differ greatly from those obtained from desktop. Kendall Tau is also low, only 0.114 (p-value=0.01), suggesting that judging device influences judges.

Boxplots for relevance grades assigned by crowd judges and their judging time on each interface is shown in Fig. 2. The average time (red squares) crowd judges spent labeling a highly relevant, relevant, somewhat relevant and non-relevant page on desktop was 88 s, 159 s, 142 s and 197 s respectively. Meanwhile,

[5] rel = high-rel+rel, non-rel=some-rel+non-rel.

the average mobile judging time for highly relevant, relevant, somewhat relevant and non-relevant page was 65 s, 51 s, 37 s and 48 s respectively. The plots show two interesting judging trends: Firstly, judges take longer to decide non-relevance on desktop as compared to mobile. This maybe due to several reasons. If a web page is not optimized for mobiles, it may be inherently difficult to find required information. Judges perhaps do not spend time zooming/pinching if information is not readily available on the viewport. It could also be a result of interaction fatigue. In the beginning, judges may thoroughly judge each page, but due the limited interaction on mobiles, they grow impatient as time passes and spend increasingly less time evaluating each page, thus giving up more quickly than a desktop judge. For optimized pages, the smaller viewport in mobile allows judges to quickly decide if the web page is relevant or not. For example, web pages with irrelevant ads (Fig. 1d) can be quickly marked as non-relevant. Secondly, judges spend more time analyzing highly relevant and relevant pages on mobile and desktop respectively. This is perhaps due to the time it takes to consume a page on mobile is longer, and with limited information on the viewport user has to tap, zoom or scroll several times to read the entire document.

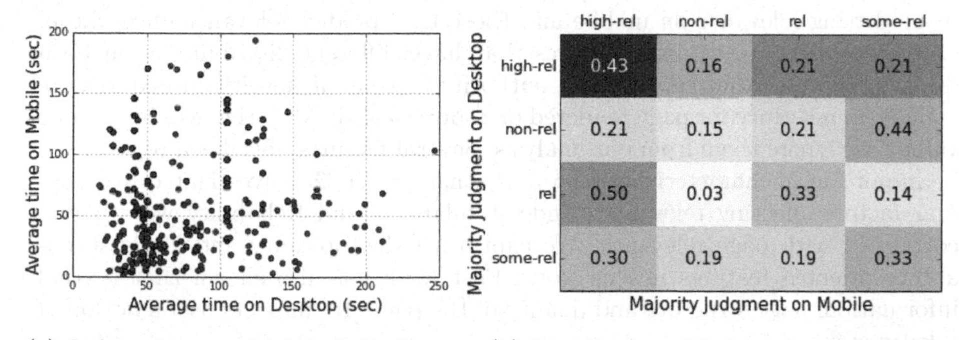

(a) Judging time on Mobile & Desktop (b) Distribution of Relevance judgments

Fig. 3. Time and Label comparison on Mobile and Desktop

Figure 3a shows the distribution of average judging time for documents on mobile and desktop. Since three assessors judged each document, we plot the mean judging time of each document. While a large fraction of pages is judged under 100 s on mobile and desktop, remaining documents take longer on both mediums. This is perhaps the result of the time it takes to judge non-relevance and relevance of a page on desktop and mobile. It may take longer to judge relevant documents on mobile and irrelevant documents on desktop. Figure 3b depicts the distribution of majority relevance grades on mobile and desktop. We see that several documents marked highly relevant in desktop are actually marked non-relevant on mobile. Again, this is due to the reasons (e.g. viewport full of ads, small fonts etc.) mentioned above.

4 Relevance Prediction

Relevance prediction is a standard but crucial problem in Information retrieval. There are several signals that are computed today to determine page relevance with respect to a user query. However, our goal is not to test or compare existing features. Our primary focus is to determine whether viewport and content specific features show different trends for relevance prediction on mobile and desktop. Given that non-relevant pages are small in number on 1–4 scale, we predict relevance on binary scale. We use several combinations of features to train Adaboost Classifier [21]. Given that our dataset is small, we perform 10-fold cross validation. We report average precision, recall and f1-score across 10 folds for mobile and desktop. We compute paired t-test to compute statistical significance. We begin by describing the features used to predict relevance in following subsection.

4.1 Features

We study whether there is a significant difference between features that are useful in predicting relevance on mobile and those that predict relevance on desktop. Our objective is to capture features that have different distributions on both devices. Features that capture link authority or content novelty will contribute equally to relevance for page rendered on mobile and desktop (i.e. will have same value), we ignore them from our analysis. Several features have been reported to be important in characterizing relevance. Zhang *et al.* [20] investigated five popular factors affecting relevance. Understandability and Reliability were highly correlated with page relevance. We capture both Topicality and Understandability oriented features in this work. Past work has also shown that textual information, page structure and quality of the page [16] impacts user's notion of relevance.

As mentioned before, screen resolutions of mobiles greatly affects the legibility of on-screen text. If websites are not optimized to render on small screens, users may get frustrated quickly due to repeated taps, pinching and zooming to understand its content. Our hypothesis is that relative position and size of text on page are important indicators as viewport size varies greatly between a mobile and desktop, thus affecting user's time and interaction with the page. We extract features that rely on visible content of the web page, the position of such elements and their rendered sizes. We evaluate some viewport i.e. interface oriented features to predict page relevance.

Both content and viewport specific features are summarized in Table 1. Content oriented features are calculated on two levels: the entire web page (*html*) and only the portion of page visible to user (*viewport*) when she first lands on the page. Geometric features capture mean, minimum and maximum of absolute coordinates of query terms and headings on screen. Display specific features capture the absolute size of query terms, headings and remaining words in the page in pixels. These features are calculated by simulating how these pages are

Table 1. Features calculated for webpages on mobile and desktop

Content Specific Features				
Number of words			html	viewport
Number of tables			html	viewport
Number of headings			html	viewport
Number of outlinks			html	viewport
Number of images			html	viewport
Number of sentences			html	viewport
Number of outlinks with query terms			html	viewport
Query term frequency			html	viewport
Unique tokens			html	viewport
Geometric Features				
Query term position	min	max	avg	x and y pos
Heading position	min	max	avg	x and y pos
Display Specific Features				
Query term size		min	max	avg
Heading size		min	max	avg
Word size		min	max	avg

rendered on mobile and desktop with the help of Selenium web browser automation.[6] This provides all information about rendered DOM elements in HTML at runtime.

Pearson's correlation (R) between top five statistically significant features (p-value < 0.05) and mobile/desktop judgments is shown in Table 2. As we can see content oriented features are highly correlated with desktop judgments but both view and size oriented features are correlated with mobile judgments.

Table 2. Pearson's R between features and judgments on mobile and desktop

Desktop			Mobile		
Feature	*R*	p-val	Feature	*R*	p-val
Number of sentences (html)	−0.13	0.04	Heading size (max)	0.16	0.01
Number of words (html)	−0.12	0.03	Number of headings (viewport)	0.16	0.02
Unique tokens (html)	−0.12	0.04	Number of words (viewport)	0.14	0.04
Query term size (mean)	0.26	0.00	Number of headings (html)	0.13	0.04
Query term position (min)	−0.17	0.01	Number of images (viewport)	0.10	0.04

4.2 Mobile Relevance Prediction

Several classifiers can be trained with large combination of features to determine relevance. We mainly focus on training Adaboost classifier with multiple feature combinations.

Table 3 shows the results for classification for several feature combinations. The row labeled *all* indicates the model trained with all the features. The rows

[6] http://www.seleniumhq.org/.

labeled *no.x* correspond to models using all the features but features of type x. The rows labeled *only.x* have metrics for models trained on features in group x. Finally, four pairs of features – geom.size, geom.view, geom.html and view.html correspond to models trained on features in either geometric, content (viewport (view) or html) or display group. Treating the model trained with *all* the features as baseline, the statistically significant (p-value < 0.05 for paired t-test) models have been marked with (*).

The classification accuracy for mobile relevance is significantly better than random. The best performing feature combination is **viewport features (only.view)** for mobile. The confusion matrix for best model (only.view) is given in Table 4. Surprisingly, the accuracy does not improve on using all features. There is also no improvement in performance when display specific features are taken into account. Binary classifier trained solely on *html* features does worse than the classifier trained using *viewport* features. In pairwise combinations, content specific features (view.html) perform the best, which is perhaps due to the presence of viewport based features in the model.

Table 3. Classification results for Mobile and Desktop

	Mobile				Desktop			
	Accuracy	Prec	Recall	F1-score	Accuracy	Prec	Recall	F1-score
all	0.76	0.83	0.84	0.839	0.60	0.70	0.73	0.71
no.html	0.75	0.83	0.83	0.82	0.65*	0.72	**0.79***	0.75*
no.view	0.73	0.81	0.82	0.81	0.61	0.68	**0.79***	0.73
no.geom	0.79*	0.85	0.86	0.86*	**0.67***	**0.75***	0.76	0.75*
no.display	0.78	0.857*	0.85	0.85	0.64*	0.71	**0.79***	0.74
only.html	0.74	0.78	**0.90***	0.83	0.61	0.70	0.75	0.72
only.view	**0.81***	**0.87***	0.88*	**0.87***	0.64*	0.70	0.78	0.74
only.geom	0.76	0.82	0.88*	0.84	0.54	0.63	0.72	0.67
only.display	0.71	0.79	0.81	0.80	0.64*	0.71	**0.79**	0.74
geom.display	0.75	0.84	0.82	0.82	0.61	0.68	0.76	0.72
geom.html	0.77	0.84	0.86	0.85	0.57	0.66	0.73	0.69
geom.view	0.78	0.84	0.87	0.85	0.62	0.70	0.77*	0.73
view.html	0.78	0.83	0.89*	0.86*	**0.67***	**0.75***	0.78*	**0.76***

Our hypothesis that geometric and display oriented features impact relevance are not supported by the results. Display features, for instance, when used alone for binary classification have the lowest accuracy amongst all feature combinations. It is worth noting that when viewport features are dropped (no.view) from training, the accuracy goes down by 3 % when compared to model trained on all the features.

Features that had highest scores while building the *viewport (only.view)* classifier are mentioned below. The most important set of features in decreasing order of their importance are: total tokens in view port (0.28), number of images in viewport (0.20), number of tables (0.16), number of sentences (0.10), number of outlinks with query terms (0.07), number of outlinks (0.06) and finally number of headings with query terms (0.06).

Table 4. Mobile Confusion matrix (only.view model)

	Rel	NonRel
Rel	0.88	0.12
Non-Rel	0.26	0.74

Table 5. Desktop Confusion matrix (view.html model)

	Rel	NonRel
Rel	0.68	0.32
Non-Rel	0.34	0.66

4.3 Desktop Relevance Prediction

The results for document relevance are shown in Table 3. The overall accuracy of relevance prediction on desktop pages is low. It is in fact lower than that observed on mobile. The best performing system is one trained with content based or **viewport and html features (view.html)**. The confusion matrix for best model (view.html) is given in Table 5. The difference between the accuracy of best performing model on mobile (only.view) and best performing system on desktop (view.html) is 17 %. It is interesting to note that viewport features are useful indicators of relevance regardless of judging device. The models with viewport features (only.view, geom.view and view.html) perform better than model built using all features. This suggests that users tend to deduce page relevance from immediate visible content once page finishes loading.

It is also surprising to note that classifier trained on features extracted from the entire document (only.html) performs worse the one trained using only viewport features (only.view). This could be due to the limited number of features used in our study. Perhaps, with more extensive set of page specific features, the classifier may perform better.

Our hypothesis that geometric features affect relevance is not supported by either experiment. Overall, geometric features are not useful in predicting relevance on desktop, the classifier trained only on geometric features achieves only 54 %, 10 % lower than model trained with all features. It is not surprising to observe the jump in accuracy once geometric features are removed from the model training. Thus, both the experiments suggest that position of query terms and headings, on both mobile and desktop, is not useful in predicting relevance.

Amongst models trained on a single set of features, the model with display specific features (only.display) performs the best with 64 % accuracy and 0.74 F1-score. It seems that font size of query terms, headings and other words is predictive of relevance. However, only.display model's accuracy (64 %) is still lower than that of view.html model. Amongst models trained on pairs of features, view.html performs best (67 %), closely followed by geom.view (62 %) model. This is perhaps due to the presence of viewport features in both models.

The most representative features in content based (view.html) classifier, in decreasing order of their importance are: content specific features such as number of headings (viewport) (0.23), number of images (viewport) (0.14) and query term frequency (html) (0.08), number of unique tokens (html) (0.06), number of tables (viewport) (0.06), number of words (html) (0.05) and finally number of outlinks (0.04).

Despite promising results, our study has several limitations. It is worth nothing that our study only contained 236 query-url pairs, with more data and an extensive set of features prediction accuracy would improve. We used query-url pairs from previous study, which had gathered queries for only seven topic descriptions or tasks. We shall follow up with a study containing more number and variety of topics to further analyze impact of device size on judgments.

Overall, our experiments indicate that viewport oriented features are useful in predicting relevance. However, model trained with viewport features on mobile judgments significantly outperforms the model trained on desktop judgments. Our experiments also show that features such as query term or heading positions are not useful in predicting relevance.

5 Conclusion

Traditional relevance judgments have always been collected via desktops. While existing work suggests that page layout and device characteristics have some impact on page relevance, we do not know whether page relevance changes with change in device. Thus, with this work we tried to determine whether device size and page rendering has any impact on judgments, i.e. different web page layouts (on mobile or desktop) translate into different relevance labels from judges. To that end, we systematically compared crowdsourced judgments of query-url pairs from mobile and desktop.

We analyzed different aspects of these judgments, mainly observing differences in how users evaluate highly relevant and non-relevant documents. We also observed strikingly different judging times, despite similar inter-rater agreement on both devices. We also used some viewport oriented features to predict relevance. Our experiments indicate that they are useful in predicting relevance on both mobiles and desktop. However, prediction accuracy on mobiles is significantly higher than that of desktop. Overall, our study shows that there are certain differences between mobile and desktop judgments.

There are several directions for future work. The first and foremost would be to scale this study and analyze the judging behavior more extensively to draw better conclusions. Secondly, it would be worthwhile to investigate further the role of viewport features on user interaction and engagement on mobiles and desktops.

References

1. Borlund, P.: The concept of relevance in IR. J. Am. Soc. Inf. Sci. Technol. **54**(10), 913–925 (2003)
2. Buchanan, G., Farrant, S., Jones, M., Thimbleby, H., Marsden, G., Pazzani, M.: Improving mobile internet usability. In: Proceedings of WWW. ACM (2001)
3. Church, K., Oliver, N.: Understanding mobile web and mobile search usein today's dynamic mobile landscape. In: Proceedings of MobileHCI. ACM (2011)
4. Church, K. Smyth, B.: Understanding the intent behind mobile information needs. In: Proceedings of IUI. ACM (2009)
5. Church, K., Smyth, B., Bradley, K., Cotter, P.: A large scale studyof european mobile search behaviour. In: Proceedings MobileHCI. ACM (2008)
6. Church, K., Smyth, B., Cotter, P., Bradley, K.: Mobile informationaccess: a study of emerging search behavior on the mobile internet. ACM Trans. Web **1**(1), 4 (2007)
7. Guo, Q., Jin, H., Lagun, D., Yuan, S., Agichtein, E.: Miningtouch interaction data on mobile devices to predict web search result relevance. In: Proceedings of SIGIR. ACM (2013)
8. Kamvar, M., Baluja, S.: A large scale study of wireless searchbehavior: Google mobile search. In: Proceedings SIGCHI. ACM (2006)
9. Kamvar, M., Baluja, S.: Deciphering trends in mobile search. Computer **40**(8), 58–62 (2007)
10. Kamvar, M., Kellar, M., Patel, R., Xu, Y.: Computers and iphonesand mobile phones, oh my!: a logs-based comparison of search users on differentdevices. In: Proceedings of WWW. ACM (2009)
11. Kazai, G., Kamps, J., Milic-Frayling, N.: An analysis of humanfactors and label accuracy in crowdsourcing relevance judgments. Inf. Retr. (2013)
12. Li, J., Huffman, S., Tokuda, A.: Good abandonment in mobile andpc internet search. In: Proceedings of SIGIR. ACM (2009)
13. Schamber, L., Eisenberg, M.: Relevance: The search for a definition (1988)
14. Shokouhi, M., Jones, R., Ozertem, U., Raghunathan, K., Diaz, F.: Mobile query reformulations. In: Proceedings SIGIR. ACM (2014)
15. Song, Y., Ma, H., Wang, H., Wang, K.: Exploring and exploiting user search behavior on mobile and tablet devices to improve search relevance. In: Proceedings of WWW (2013)
16. Tombros, A., Ruthven, I., Jose, J.M.: How users assess web pages for information seeking. J. Am. Soc. Inf. Sci. Technol. **56**(4), 327–344 (2005)
17. Tossell, C., Kortum, P., Rahmati, A., Shepard, C., Zhong, L.: Characterizing web use on smartphones. In: Proceedings of the SIGCHI. ACM (2012)
18. Xu, Y.C., Chen, Z.: Relevance judgment: What do informationusers consider beyond topicality? JASIST **57**(7), 961–973 (2006)
19. Yi, J., Maghoul, F., Pedersen, J.: Deciphering mobile search pat-terns: a study of yahoo! mobile search queries. In: Proceedings of the WWW. ACM (2008)
20. Zhang, Y., Zhang, J., Lease, M., Gwizdka, J.: Multidimensional relevance modeling via psychometrics and crowdsourcing. In: Proceedings of the SIGIR. ACM (2014)
21. Zhu, J., Zou, H., Rosset, S., Hastie, T.: Multi-class adaboost. Stat. Interface **2**(3), 349–360 (2009)

Probabilistic Modelling

Probabilistic Local Expert Retrieval

Wen Li[1](✉), Carsten Eickhoff[2], and Arjen P. de Vries[3]

[1] University College London, WC1E 6BT, London, UK
wen.li@ucl.ac.uk

[2] Department of Computer Science, ETH Zurich, Zurich, Switzerland
c.eickhoff@acm.org

[3] Faculty of Science, Radboud University, 6525 EC Nijmegen, The Netherlands
arjen@acm.org

Abstract. This paper proposes a range of probabilistic models of local expertise based on geo-tagged social network streams. We assume that frequent visits result in greater familiarity with the location in question. To capture this notion, we rely on spatio-temporal information from users' online check-in profiles. We evaluate the proposed models on a large-scale sample of geo-tagged and manually annotated Twitter streams. Our experiments show that the proposed methods outperform both intuitive baselines as well as established models such as the iterative inference scheme.

Keywords: Domain expertise · Geo-tagging · Twitter

1 Introduction

When visiting unfamiliar cities for the first time, visitors are confronted with a number of challenges related to finding the right spots to go, sights to see or even the most appropriate range of cuisine to sample. While residents quite naturally familiarize themselves with their surroundings, strangers often face difficulties in efficiently selecting the best location to suit their preferences. We refer to such location-specific, and frequently sought-after [11], knowledge as local expertise.

Local expertise can be acquired with the help of online resources such as review sites (e.g., yelp.com or tripadvisor.com) that rely on both paid professionals as well as user recommendations. General-purpose Web search engines, especially in the form of entertainment or food verticals, provide valuable information. However, these services merely return basic information and results are not specifically tailored to the individual. Ideally, a more effective way of solving this task would be to ask someone who is local and/or has the knowledge about the area in question. Seeking out this kind of people is an example of expert retrieval, and, more specifically, of local expert retrieval.

Location-based social networks allow users to post messages and document their whereabouts. When a user checks in at a given location, the action of *check-in* is not merely a user-place tuple. The physical attendance at the location also suggests that the user, at least to some extent, gets familiar with the

N. Ferro et al. (Eds.): ECIR 2016, LNCS 9626, pp. 227–239, 2016.
DOI: 10.1007/978-3-319-30671-1_17

location and its environment. The more frequently such evidence is observed, the more accurate our insights into the user's interests and expertise become. This paper introduces two novel contributions over the state of the art in local expert retrieval. **(1)** We propose a range of probabilistic models for estimating users' local expertise on the basis of geo-tagged social network streams. **(2)** In a large-scale evaluation effort, we demonstrate the merit of the methods on real-world data sampled from the popular microblogging service Twitter.

The remainder of this paper is structured as follows: Sect. 2 gives an overview of local expert retrieval methods as well as social question answering platforms. Section 3 derives a range of probabilistic models for local expert retrieval that are being evaluated on a concrete retrieval task in Sect. 4. Finally, Sect. 5 concludes with a brief discussion of our main findings as well as an outlook on future work.

2 Related Work

The task of expertise retrieval has first been addressed in the domain of enterprises managing and optimizing human resources (detailed in [1,17]). Early expertise retrieval systems required experts to manually fill out questionnaires about their areas of expertise to create the so-called expert profiles. Later, automated systems were employed for building and updating such profiles and probabilistic models were introduced for estimating a candidate's expertise based on the documents they authored, *e.g.*, [1,4,6,7,15]. These works inspire us to use probabilistic models for estimating local expertise. Since location information is not always presented in textual format, we approach the problem by building models based on candidates' check-in profiles. Li et al. [11] proposed the problem of local expert retrieval (using the term "geo-expertise") and investigated the main intuitions that could naturally support an automatic approach which considers user check-in profiles as evidence of having knowledge regarding locations they had visited. A preliminary empirical evaluation demonstrated its feasibility using three heuristic methods for automating local expert retrieval, however, without giving a formal derivation that would underline the soundness of these methods. In this paper, we follow up upon this line of research and investigate the probabilistic reasoning behind the methods proposed in the previous work. Cheng et al. [5] also proposed finding local experts as a retrieval problem, for which they combined models of local and topical authority to rank candidates based on data collected from Twitter. The authors rely on textual queries accompanied by a location to specify spatial constraints. In our setting, queries are phrased in terms of locations. This can for example be either a specific restaurant or a type of restaurants to which users are interested in paying a visit. The main difference between this study and the aforementioned one is that we focus on evidence of location knowledge in geo-spatial movement profiles while Cheng et al. introduce location constraints in text-based expertise retrieval.

Another domain closely related to expertise retrieval is found in the context of community question answering (CQA) platforms, such as Quora (http://www.quora.com), Stackoverflow (http://stackoverflow.com), or, Yahoo Answers (https://answers.yahoo.com). These services rely on routing questions to the most suitable potential answerers. These platforms also provide researchers with an opportunity to access large-scale topical expertise profiles. In particular, they provide data that can be used directly for evaluation, *i.e.*, whether the top-ranked candidates give satisfactory answers to the questions they have been retrieved for. Based on this kind of data, Liu et al. [13] proposed to use language models to profile candidates, Zhang et al. [18] used heuristic features from asker-answerer networks to rank candidates, and Horowitz et al. [9] relied on probabilistic models similar to Balog et al. [1] in their social search engine Aardvark. Aardvark is a CQA-platform-based instant messaging system including a location-sensitive classifier to decide whether a given question requires local expertise. Although the providers do not detail the algorithm of their classifier [9], their paper gives examples rendering the problem as a place entity recognition task. Studies based on CQA data focus on textual and social network features but do not currently explore candidates' movement history. We consider this type of information a crucial factor in local expert retrieval tasks, *i.e.*, modelling candidate's knowledge about locations.

There are several domain-specific studies of user expertise modelling. Bar-Haim et al. [3] aimed to identify stock experts on Twitter by testing candidates' predictions of stock prices (*e.g.*, buying or selling a stock at the right time) in their tweets. Whiting et al. [16] suggested using changes of Wikipedia pages as evidence for developing events. The authors retrieved tweets containing relevant terms and considered the authors of these tweets to be influencers for this topic. In a closely related effort, Bao et al. [2] proposed a method to finding local experts for location categories. They applied a hyperlink inference topic search algorithm on the user-location matrix. We have included their method as a performance baseline in this study.

3 Models of Local Expertise

In this paper, we define local expertise as knowledge regarding given (categories of) places of interest (POI). The POI information and POI categories are provided respectively by Twitter and Foursquare via their APIs. Consequently, a topic in local expert retrieval can be either a specific POI or a category of POIs within a geographical scope, *e.g.*, the [Blue Ribbon Fried Chicken] in New York or [Chinese Restaurants] in Los Angeles. The former is referred to as *POI topics*, which describe knowledge regarding a single location, such as opening times, or admittance fees. The latter is referred to as *category topics*, which describe knowledge regarding all locations in a specific category, such as different themes or decoration of locations in the category. High-ranking candidates should be able to answer questions about the locations or the category of locations in the topic. For simplicity, both types of topics are also referred to as locations in the rest of paper. In [11], check-ins are considered to be links between candidates' visits to a location and knowledge they may have about the location. They proposed three methods characterizing check-in profiles in three different aspects, *i.e.*, visiting *frequency*, *diversity* and *recency*. While the previous work captured these notions in a heuristic manner, the following sections pursue a formal derivation of these empirically proven notions. That is we develop a probabilistic model for each of these heuristic methods.

3.1 Within-Topic Activity (WTA)

The first approach we propose considers only the candidates' check-in frequency. This method focuses on knowledge about a single location or a single type of locations. We take a co-occurrence modelling approach, inspired by expert finding via text documents [6]. To be specific, we rank a candidate u by their probability of having local expertise in a given topic q, *i.e.*, $P(u|q)$. We estimate the conditional probability by aggregating over the user's check-ins at all locations (l), that is

$$P(u|q) = \sum_{l} P(u|l,q)P(l|q).$$

Assuming conditional independence of the candidate u and the query topic q given the location l, *i.e.*, $P(u|l,q) = P(u|l)$, we obtain

$$P(u|q) = \sum_{l} P(u|l)P(l|q).$$

As for $P(u|l)$, we apply Bayes' Rule which gives

$$P(u|l) \propto P(l|u)P(u)$$

where we assume a uniform prior for $P(l)$.
Putting these together, we obtain

$$P(u|q) \stackrel{\text{rank}}{=\!=} \sum_{l} P(l|q)P(l|u)P(u). \tag{1}$$

$P(u)$ is the query-independent confidence in the user model estimated by the number of check-ins observed for this user, *i.e.*,

$$P(u) = \frac{N_u}{N}$$

where N_u and N, respectively, are the number of check-ins posted by candidate u and the overall number of check-ins. Intuitively, the more data is available for a given user, the more we trust the model built from his/her check-in profiles to be accurate. The conditional probability $P(l|q)$ captures the user's query intent, *i.e.*, the possible range of locations users may be interested in. In our setting of local expertise, the query is actually a location or a type of locations. The final conditional probability can be estimated by

$$P(l|q) = \begin{cases} \frac{1}{|L_q|} & \text{if } l \in L_q, \\ 0 & \text{otherwise,} \end{cases}$$

where L_q is the set of locations matching the query. To estimate $P(l|u)$, we use

$$P(l|u) = \frac{N_{l,u}}{N_u},$$

where $N_{l,u}$ is the number of check-ins candidate u made at location l. The scoring function can be derived from simplifying Eq. 1, that is

$$\mathcal{S}_n(u,q) = \frac{1}{|L_q| \cdot N} \sum_{l \in L_q} N_{l,u} \stackrel{\text{rank}}{=\!=} \sum_{l \in L_q} N_{l,u}.$$

Intuitively, the more check-ins a candidate has at the queried location(s) in L_q, the more likely they are to be interested in those locations and knows about them.

3.2 Within-Topic Diversity (WTD)

Our second method uses the language model referred to as *Model 1* in [1], that is

$$P(\theta_u|q) \propto P(q|\theta_u)P(\theta_u).$$

where θ_u is a language model of a candidate based on his/her check-in profile. To estimate $P(q|\theta_u)$, we assume independence between individual locations. That is

$$P(q|\theta_u) = \prod_{l \in L_q} P(l|\theta_u) = \prod_{l \in L_q} \frac{N_{l,u}}{N_u}. \tag{2}$$

For the prior $P(\theta_u)$, we use

$$P(u) = \frac{N_u^{|L_q|}}{\sum_{u' \in U} N_{u'}^{|L_q|}},$$

so it will simplify the scoring function, that is

$$P(q|u) = \frac{N_u^{|L_q|}}{\sum_{u' \in U} N_{u'}^{|L_q|}} \prod_{l \in L_q} \frac{N_{l,u}}{N_u}.$$

By applying the logarithm (to avoid underflow in computation), we obtain

$$\mathcal{S}_d(u,q) = \log \frac{1}{\sum_{u' \in U} N_{u'}^{|L_q|}} + \sum_{l \in L_q} \log N_{l,u} \stackrel{\text{rank}}{=\!=} \sum_{l \in L_q} \log N_{l,u}.$$

The following smoothing function is used to differentiate the profiles containing visits to different numbers of locations but each location has been visited only once.

$$\mathcal{S}_d(u,q) = \sum_{l \in L_q} \log(N_{l,u} + 1).$$

The above scoring function indicates that check-ins at multiple distinct locations (within the queried location set) should increase the score more than repeatedly checking in at the same location. This means that a candidate will gain a high local expertise rating if he/she makes check-ins at a variety of relevant locations. This fits the intuition that candidates with experience at a variety of locations may know more about the essence of the topic rather than mere specifics of a single place within that category. For example, if we seek advice about Italian restaurants, individuals who have been to many Italian restaurants in town will be more suitable candidates than those who have been to the same restaurant a lot.

The prior $P(u)$ is selected for the scoring function so that the candidate-dependent denominator in the conditional probability $P(l|u)$ will be cancelled when combined with the prior. This accounts for the fact that language models represent users' topical focus rather than their knowledge, *i.e.*, they are biased towards shorter profiles, when two profiles have the same amount of relevant check-ins. Since check-ins are positive evidence of candidates knowing about a location, additionally knowing about other types of locations should not negatively affect the local expertise score. For example, if a candidate has visited two place categories A and B each for n times while another candidate only has been to A for n times, it is not reasonable to assume that the latter candidate should have more knowledge about A than the former candidate, even if the latter has focused on A more.

3.3 Within-Topic Recency (WTR)

Experts are humans and as such they rely on their memories to support their expertise. Therefore, we should take into account the fact that (1) people forget the knowledge they once gained and have not refreshed for a while, and (2) the world is changing as time goes by, *e.g.*, restaurants may have new chefs, and old buildings may have been replaced. The more time has passed since the creation of the memory, the more likely it will be forgotten or outdated. To incorporate such effects, we explicitly model the candidates' memory by $P(c|u)$, which indicates the probability that candidate u can recall his/her visit represented by the check-in c. As suggested in the domain of psychology, human memory can be assumed to decay exponentially [14]. Consequentially, we use an exponential decay function to represent the retention of individual check-ins, by which we obtain:

$$P_t(c|u) = \frac{e^{-\lambda(t-t_c)}}{\sum_{c \in C_u} e^{-\lambda(t-t_c)}},$$

where t is the time of query and t_c is the time when the user posted the check-in. Similarly, we define a prior for each candidate as follows

$$P_t(u) = \frac{\sum_{c \in C_u} e^{-\lambda(t-t_c)}}{\sum_{c \in C} e^{-\lambda(t-t_c)}}$$

The decay of the weight on check-ins models our belief on how up-to-date the information is, while the prior reflects the average recency of knowledge borne by the whole community on the social network. Then, for estimating the candidate's expertise, we weight each check-in according to its recency, *i.e.*, we marginalize the user's old check-ins.

$$P_t(l|u) = \sum_{c \in C_u} P(l|c)P_t(c|u) = \sum_{c \in C_u} \frac{\mathbf{1}(l_c = l)e^{-\lambda(t-t_c)}}{e^{-\lambda(t-t_c)}}, \quad (3)$$

where $\mathbf{1}(\cdot)$ is an indicator function, which equals 1 *if and only if* the condition in the parentheses evaluates to true. Given these two estimations, we have

$$\mathcal{S}_r(u,q) = P_t(u|q) = \sum_{l \in L} P_t(u|l)P(l|q) \stackrel{\text{rank}}{=\!=} \sum_{l \in L} P_t(l|u)P_t(u)P(l|q)$$
$$\stackrel{\text{rank}}{=\!=} \sum_{l \in L_q} \sum_{c \in C_u, l_c = l} e^{-\lambda(t-t_c)}.$$

As can be seen, $\mathcal{S}_r$ down-weights older check-ins' contribution to a candidate's expertise due to the fact that they may be vaguely memorized and become unreliable. The decay parameter λ is fixed to $\frac{1}{150}$ at a granularity of days and we leave the fine-tuning of this parameter for future work.

3.4 Combining Recency and Diversity (WTRD)

Diversity and recency of check-ins can both be important factors in estimating a candidate's local expertise. Thus, we propose a combination of the two features introduced in Sects. 3.2 and 3.3. In Eq. 2, the conditional probability can be transformed into

$$P(q|u) = \sum_{c \in C_u} P(l|c,u)P(c|u) = \sum_{c \in C_u} P(l|c)P(c|u) = \sum_{c \in C_u} \mathbf{1}(l_c = l)P(c|u)$$

in which we assume that candidate and location are conditionally independent given a check-in (*i.e.*, from the first equation to the second equation). Then, we estimate the conditional probability $P(c|u)$ with Eq. 3.

$$P(c|u) = \frac{e^{-\lambda(t-t_c)}}{\sum_{c \in C_u} e^{-\lambda(t-t_c)}}$$

Thus, we have

$$P(q|u) = \prod_{l \in L_q} \sum_{c \in C_u} \frac{\mathbf{1}(l_c = l)e^{-\lambda(t-t_c)}}{\sum_{c \in C_u} e^{-\lambda(t-t_c)}}$$

Similar to the prior probability used in the diversity method,

$$P(u) = \frac{(\sum_{c \in C_u} e^{(-\lambda(t-t_c))})^{|L_q|}}{(\sum_{c \in C} e^{-\lambda(t-t_u)})^{|L_q|}}$$

By replacing the counterparts with these into Eq. 2 and applying the logarithm on both sides of the equation, we obtain

$$\begin{aligned}\mathcal{S}_d(u,q) &= \log \frac{(\sum_{c\in C_u} e^{-\lambda(t-t_c)})^{|L_q|}}{(\sum_{c\in C} e^{-\lambda(t-t_u)})^{|L_q|}} \prod_{l\in L_q} \frac{\sum_{c\in C_u} \mathbf{1}(l_u = l) e^{-\lambda(t-t_c)}}{\sum_{c\in C_u} e^{-\lambda(t-t_c)}} \\ &= \log \frac{1}{(\sum_{c\in C} e^{-\lambda(t-t_c)})^{|L_q|}} + \sum_{l\in L_q} \log \sum_{c\in C_u} \mathbf{1}(l = l_c) e^{-\lambda(t-t_c)} \\ &\overset{\text{rank}}{=\!=} \sum_{l\in L_q} \log \sum_{c\in C_u} \mathbf{1}(l = l_c) e^{-\lambda(t-t_c)}\end{aligned}$$

The decay parameter λ is set to the same value as that in the WTR method.

3.5 Iterative Inference Model (HITS)

Bao et al. [2] propose a model for estimating local expertise based on the Hyperlink-Induced Topic Search (HITS) algorithm, an approach originally designed for link analysis of Web pages [10]. The model defines two properties for users and locations respectively, *i.e.*, hub scores for users and authority scores for locations. The hub score indicates how well a user can serve as an information source about a place and the authority score presents how popular a place is. We implement a normalized version of the algorithm and focus on hub scores for users (candidates) which are used as estimates of local expertise and are calculated as $\mathcal{S}_h(u,l) = \boldsymbol{h}_{u,l}^{(n+1)}$, where

$$\boldsymbol{h}^{(n+1)} = \frac{\mathbf{M}^T\mathbf{M} \cdot \boldsymbol{h}^{(n)}}{\| \mathbf{M}^T\mathbf{M} \cdot \boldsymbol{h}^{(n)} \|}.$$

3.6 Candidate Profiling

As a mainstream location-based social network, Foursquare attempts to increase user engagement by encouraging users, through gamification elements, to check in at a location far more times than they actually need to (*i.e.*, the user never left the location but checked in again in order to collect rewards). To mitigate the effect of this twisted relation between check-ins and actual visits, we define a different type of candidate profile, *i.e.*, the Active-Day Profile (referred to as +A while +C is used to refer to original check-in profiles). It is a subset of a user's check-ins which is defined as: $\{c | c \in C_u, \nexists c' \in C_u : l_c = l_{c'}, t_c < t_{c'} < \lceil t_c \rceil_D\}$ where $\lceil \cdot \rceil_D$ is a ceiling function towards midnights. Informally, the Active-Day profile contains only the last check-in at each place within each day, reducing the influence of multiple check-ins at the same place to at most one per day.

4 Evaluation

To evaluate the various discussed methods and profile types, we implement a configurable local expert retrieval system. The system accepts a topic which is composed of a scope of city and a (type of) location(s) and returns a list of related candidate experts. The dataset used in this study is an extended version of the collection of POI-tagged tweets proposed by Li et al. [12]. It comprises 1.3M check-ins from 8 K distinct users from New York, Chicago, Los Angeles, and San Francisco. The data collection process was set up such that each user's full check-in profile would be included. To filter out accounts that are solely used for branding and advertisement purposes (*e.g.*, by companies) we remove all users having a "*speed*" between consecutive check-ins higher than 700 kph (which corresponds approximately to the speed of a passenger aircraft). Similarly, users showing less than 5 geo-tagged tweets were excluded as well. As a consequence of this thresholding approach, Fig. 1, shows that the check-in distribution over users does not follow a complete power law which was observed in the previous dataset.

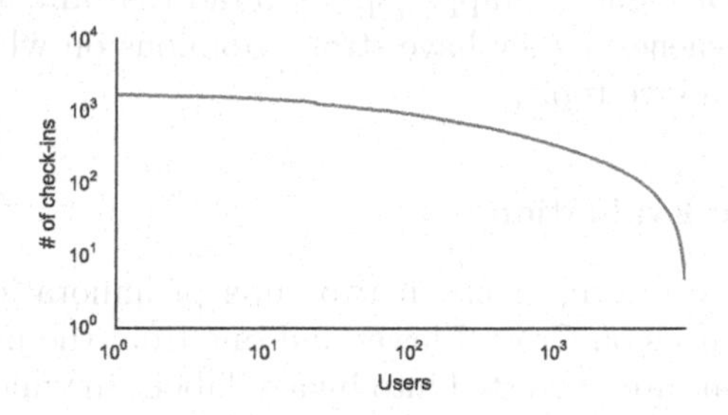

Fig. 1. Distribution of check-in frequencies across users.

To prepare a set of topics for evaluation, we rely on stratified sampling to identify a seed set of location categories and POIs, based on their popularity among the users. Two strata are composed respectively for popular POIs (top 10 %) and less popular POIs (remaining 90 %). Locations are selected randomly using a uniform distribution per stratum, and the number of samples is in accordance with the size of stratum. As for category topics, we include all 9 top categories from Foursquare's category hierarchy (*e.g.*, Food) and apply 10 %:90 % stratified sampling to the categories at the lower levels (*e.g.*, Chinese Food, Mexican Food). This results in a seed set totalling 275 topics for all 4 cities. To reduce the work load for human annotators, we eliminate entities which have less than 5 visitors and whose names are obscure (Building, Home – Private, Field, Professional & Other, Residence). As a result, we obtain 95 topics from 4 cities in total, among which there are 71 category topics and 24 POI topics. The top 5 returned candidates from each method are then pooled and annotated. This process resulted in a pool of 1588 distinct topic-candidate pairs across all methods. With the annotated topic-candidate pairs, we measure the performance of the proposed methods in terms of P@1, P@5 and MAP. A random selection approach and our implementation of [2]'s method are included as baselines.

4.1 Annotation

Topic-candidate pairs are annotated by human judges that assess each candidate's level of expertise about the given topics. To facilitate this process, an interactive annotation interface for displaying the candidate's historical check-ins has been designed and can be accessed at https://geo-expertise.appspot.com. For each topic-candidate pair, an annotator is asked to assign a value from 1 to 5 indicating their assessment of the candidate's knowledge about the topic, where "5" means the candidate knows the topic very well and "1" indicates the candidate knows barely anything about the topic. For greater reliability, we recruit assessors from different channels. (1) The first run has been carried out on the crowdsourcing platform CrowdFlower where each participant is paid 0.5 USD per task (each containing 10 topic-candidate pairs to annotate). Additionally, we invited students and staff from Delft University of Technology to contribute their assessments. Via Cohen's Kappa [8], we found that annotators are inclined to agree ($\kappa > 0.4$) whenever they have strong opinions on whether a candidate is a local expert on a give topic.

4.2 Quantitative Evaluation

We carry out separate evaluations on two runs of annotation, *i.e.*, one from the recruited annotators on CrowdFlower and one from the university staff and students. Annotations are converted into binary labels, in which topic-candidate pairs assigned with scores 4 or 5 are considered relevant (local experts) and those with scores 3 or below are considered as irrelevant (non-experts). Based on the binary relevance annotation, `trec_eval` (http://trec.nist.gov/trec_eval/) is used for measuring the performance of the proposed methods. We test the statistical significance of differences between the performance scores using a Wilcoxon Signed Rank Test ($\alpha < 0.05$).

According to the crowdsourced annotation shown in Table 1, the WTD method with Active-Day profiles performs the best under P@1 and P@5, while WTA with check-in profiles performs the best under MAP. All proposed methods with both types of profiles significantly outperform the random baseline. The HITS baseline performs significantly better than the random selection approach but is outperformed by the proposed movement profile methods even though this difference was not found to be significant.

The two types of profiles (+A and +C) were designed for comparing the potential influence of check-in gamification by Foursquare that might encourage users to check in as often as possible. In our evaluation, however, we do not observe a clear preference for either of the two profile types. This may suggest that the two types of profiles do not diverge much and check-in gamification does not have an observable influence on assessing candidates' local expertise.

Turning towards university annotations, we note that the annotators' relevance assessment is also in favour of configurations with WTD for P@1 and P@5 and WTA for MAP, although no significant differences between these two methods are observed. The proposed methods and the HITS method are all significantly better than the random baseline on this set of annotations. Different from the crowdsourced annotation, here, we observe a significant preference of all the proposed methods (WTD, WTA, or WTRD) over HITS. At the same time, we have not observed any significant differences between the three methods WTD, WTA and WTRD configured with either profile types.

Table 1. Local expert retrieval results.

Method	Profile	Crowdsourced annotation			University annotation		
		MAP	P@1	P@5	MAP	P@1	P@5
WTA	+A	0.2750	0.4211	0.3979	**0.3218**	0.5579	0.4463
	+C	**0.2771**	0.4316	0.3895	0.3147	0.5579	0.4337
WTD	+A	0.2340	**0.4789**	**0.4197**	0.2878	0.5775	0.4817
	+C	0.2280	0.4507	0.4169	0.2908	**0.5915**	**0.4845**
WTR	+A	0.2442	0.3789	0.3747	0.2814	0.5263	0.4042
	+C	0.2508	0.4211	0.3768	0.2824	0.5368	0.4063
WTRD	+A	0.2434	0.4316	0.3684	0.2862	0.5368	0.4042
	+C	0.2491	0.4421	0.3726	0.2919	0.5368	0.4168
HITS	+A	0.2327	0.4507	0.4113	0.2041	0.5211	0.3859
	+C	0.2363	0.4366	0.4028	0.2045	0.5493	0.3831
Rand	–	0.1343	0.2316	0.2063	0.1041	0.1579	0.1600

5 Conclusion

In this paper, we presented a range of probabilistic-model-based approaches to the task of local expert retrieval. Based on the existing theoretical work in expertise retrieval, we designed three models to capture the candidate's check-in profiles. We further designed a method for distilling users' check-in profile to test whether the gamification of online location-based social networks would affect the accuracy of geo-expertise estimation. To evaluate the proposed methods, we collected a large volume of check-ins via Twitter's and Foursquare's public APIs, for which we finally collected judgements from both online recruited annotators and university annotators. Our evaluation shows that the proposed methods do capture local expertise better than both random as well as refined baselines. During our experiments, we did not observe a significant difference between Active-Day profiles and the raw check-in profiles in the evaluation.

In the future, we propose to carry out this evaluation task *in-vivo* by building a dedicated local expert retrieval system. Such a system can access the Twitter/Foursquare APIs for users who authorize the application to analyse their check-in profiles as well as their friends' geo-tagged media streams to find the friend that is assumed to know most about the user's desired location. In consequence, it can be assumed to produce much more reliable expertise annotations since it would allow us to observe which recommendations are being followed-up on in practice, without the need of external assessment by judges. Additionally, it would be interesting to investigate the social ties between potential answerers and local expertise seekers, to ensure engagement of both parties and allow for greater personalization of answers.

References

1. Balog, K., Fang, Y., de Rijke, M., Serdyukov, P., Si, L.: Expertise retrieval. Found. Trends Inf. Retrieval **6**(2–3), 127–256 (2012)
2. Bao, J., Zheng, Y., Mokbel, M.F.: Location-based and preference-aware recommendation using sparse geo-social networking data. In: Proceedings of the 20th International Conference on Advances in Geographic Information Systems - SIGSPATIAL 2012, pp. 199–208 (2012)
3. Bar-Haim, R., Dinur, E., Feldman, R., Fresko, M., Goldstein, G.: Identifying and following expert investors in stock microblogs. In: Proceedings of the Conference on Empirical Methods in Natural Language Processing - EMNLP 2011, pp. 1310–1319 (2011)
4. Campbell, C.S., Maglio, P.P., Cozzi, A., Dom, B.: Expertise identification using email communications. In: Proceedings of the 12th International Conference on Information and Knowledge Management - CIKM 2003, pp. 528–531 (2003)
5. Cheng, Z., Caverlee, J., Barthwal, H., Bachani, V.: Who is the barbecue king of Texas? In: Proceedings of the 37th International ACM SIGIR Conference on Research and Development in Information Retrieval - SIGIR 2014, pp. 335–344 (2014)
6. Fang, H., Zhai, C.X.: Probabilistic models for expert finding. In: Amati, G., Carpineto, C., Romano, G. (eds.) ECiR 2007. LNCS, vol. 4425, pp. 418–430. Springer, Heidelberg (2007)
7. Fang, Y., Si, L., Mathur, A.P.: Discriminative models of integrating document evidence and document-candidate associations for expert search. In: Proceedings of the 33rd International ACM SIGIR Conference on Research and Development in Information Retrieval - SIGIR 2010, p. 683 (2010)
8. Fleiss, J.L.: Measuring nominal scale agreement among many raters. Psychol. Bull. **76**(5), 378–382 (1971)
9. Horowitz, D., Kamvar, S.D.: The anatomy of a large-scale social search engine. In: Proceedings of the 19th International Conference on World Wide Web - WWW 2010, pp. 431–440 (2010)
10. Kleinberg, J.M.: Authoritative sources in a hyperlinked environment. J. ACM **46**(5), 604–632 (1999)
11. Li, W., Eickhoff, C., de Vries, A.P.: Geo-spatial domain expertise in microblogs. In: de Rijke, M., Kenter, T., de Vries, A.P., Zhai, C.X., de Jong, F., Radinsky, K., Hofmann, K. (eds.) ECIR 2014. LNCS, vol. 8416, pp. 487–492. Springer, Heidelberg (2014)
12. Li, W., Serdyukov, P., de Vries, A.P., Eickhoff, C., Larson, M.: The where in the tweet. In: Proceedings of the 20th ACM International Conference on Information and Knowledge Management - CIKM 2011, pp. 2473–2476 (2011)
13. Liu, X., Croft, W.B., Koll, M.: Finding experts in community-based question-answering services. In: Proceedings of the 14th ACM International Conference on Information and Knowledge Management - CIKM 2005, pp. 315–316 (2005)

14. Loftus, G.R.: Evaluating forgetting curves. J. Exp. Psychol. Learn. Mem. Cogn. **11**(2), 397–406 (1985)
15. Wagner, C., Liao, V., Pirolli, P., Nelson, L., Strohmaier, M.: It's not in their tweets: modeling topical expertise of twitter users. In: SocialCom/PASSAT 2012, pp. 91–100 (2012)
16. Whiting, S., Zhou, K., Jose, J., Alonso, O., Leelanupab, T.: CrowdTiles: presenting crowd-based information for event-driven information needs. In: Proceedings of the 21st ACM International Conference on Information and Knowledge Management - CIKM 2012, pp. 2698–2700 (2012)
17. Yimam-Seid, D., Kobsa, A.: Expert-finding systems for organizations: problem and domain analysis and the DEMOIR approach. J. Organ. Comput. Electron. Commer. **13**(1), 1–24 (2003)
18. Zhang, J., Ackerman, M.S., Adamic, L.: Expertise networks in online communities: structure and algorithms. In: Proceedings of the 16th International Conference on World Wide Web - WWW 2007, pp. 221–230 (2007)

Probabilistic Topic Modelling with Semantic Graph

Long Chen(✉), Joemon M. Jose, Haitao Yu, Fajie Yuan, and Huaizhi Zhang

School of Computing Science, University of Glasgow,
Sir Alwyns Building, Glasgow, UK
long.chen@glasgow.ac.uk

Abstract. In this paper we propose a novel framework, topic model with semantic graph (TMSG), which couples topic model with the rich knowledge from DBpedia. To begin with, we extract the disambiguated entities from the document collection using a document entity linking system, i.e., DBpedia Spotlight, from which two types of entity graphs are created from DBpedia to capture local and global contextual knowledge, respectively. Given the semantic graph representation of the documents, we propagate the inherent topic-document distribution with the disambiguated entities of the semantic graphs. Experiments conducted on two real-world datasets show that TMSG can significantly outperform the state-of-the-art techniques, namely, author-topic Model (ATM) and topic model with biased propagation (TMBP).

Keywords: Topic model · Semantic graph · DBpedia

1 Introduction

Topic models, such as Probabilistic Latent Semantic Analysis (PLSA) [7] and Latent Dirichlet Analysis (LDA) [2], have been remarkably successful in analyzing textual content. Specifically, each document in a document collection is represented as random mixtures over latent topics, where each topic is characterized by a distribution over words. Such a paradigm is widely applied in various areas of text mining. In view of the fact that the information used by these models are limited to document collection itself, some recent progress have been made on incorporating external resources, such as time [8], geographic location [12], and authorship [15], into topic models.

Different from previous studies, we attempt to incorporate semantic knowledge into topic models. Exploring the semantic structure underlying the surface text can be expected to yield better models in terms of their discovered latent topics and performance on prediction tasks (e.g., document clustering). For instance, by applying knowledge-rich approaches (cf. Sect. 3.2) on two news articles, Fig. 1 presents a piece of global semantic graph. One can easily see that "United States" is the central entity (i.e., people, places, events, concepts, etc. in DBPedia) of these two documents with a large number of adjacent entities. It is also clear that a given entity only have a few semantic usage (connection to other

N. Ferro et al. (Eds.): ECIR 2016, LNCS 9626, pp. 240–251, 2016.
DOI: 10.1007/978-3-319-30671-1_18

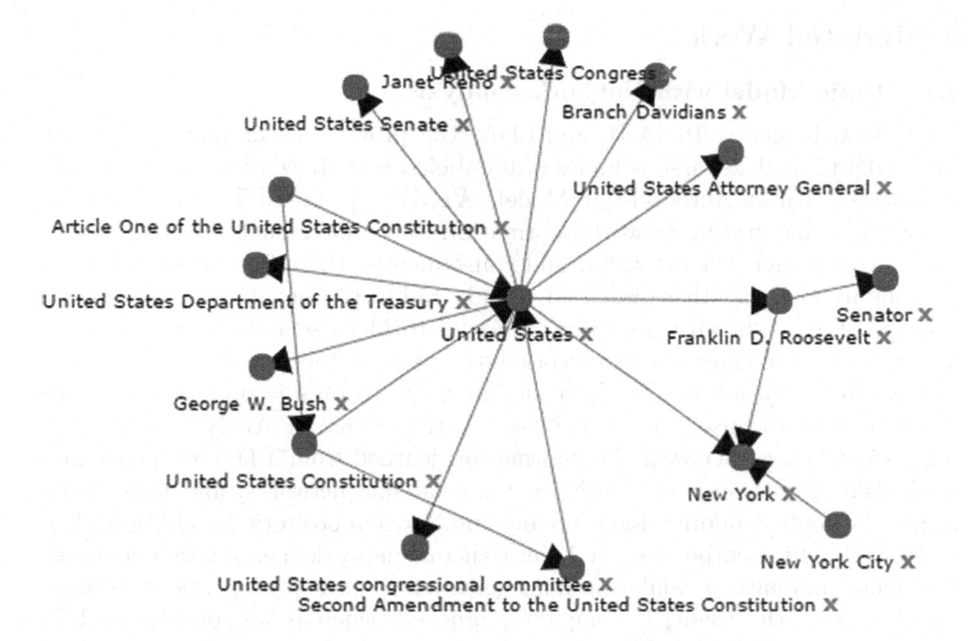

Fig. 1. A piece of global semantic graph automatically generated from two documents (178382.txt and 178908.txt of 20 Newsgroups dataset)

entities) and thus can only concentrate on a subset of topics, and utilization of this information can help infer the topics associated with each of the document in the collections. Hence, it is interesting to learn the interrelationships between entities in the global semantic graph, which allows an effective sharing of information from multiple documents. In addition to the global semantic graph, the inference of topics associated with a single document is also influenced by other documents that have the same or similar semantic graphs. For example, if two documents overlapped with their entities list, then it is highly possible that these two documents also have a common subset of topics. Following this intuition, we also construct local semantic graphs for each document in the collection with the hope to utilize their semantic similarity.

In a nutshell, the contribution of this paper are:

1. We investigate two types of graph-based representations of documents to capture local and global contextual knowledge, respectively, for enriching topic modelling with semantic knowledge.
2. We present a topic modelling framework, namely, Topic Models with Semantic Graph (TMSG), which can identify and exploit semantic relations from the knowledge repositories (DBpedia).
3. The experimental results on two real-world datasets show that our model is effective and can outperform state-of-the-art techniques, namely, author-topic Model (ATM) and topic model with biased propagation (TMBP).

2 Related Work

2.1 Topic Model with Network Analysis

Topic Model, such as PLSA [7] and LDA [16], provides an elegant mathematical model to analyze large volumes of unlabeled text. Recently, a large number of studies, such as Author-Topic Model (ATM) [15] and CFTM [4] have been reported for integrating network information into topic model, but they mostly focus on homogeneous networks, and consequently, the information of heterogeneous network is either discarded or only indirectly introduced. Besides, the concept of graph-based regularizer is related to Mei's seminal work [13] which incorporates a homogeneous network into statistic topic model to overcome the overfitting problem. The most similar work to ours is proposed by Deng et al. [5], which utilised the Probabilistic Latent Semantic Analysis (PLSA) [7] (cf. Sect. 3.1) together with the information learned from a heterogeneous network. But it was originally designed for academic networks, and thus didn't utilize the context information from any knowledge repository. In addition, their framework only incorporates the heterogeneous network (i.e., relations between document and entity), while the homogeneous network (i.e., relations between entity pairs with weight) is completely ignored, whereas we consider both of them in our framework.

2.2 Knowledge Rich Representations

The recent advances in knowledge-rich approaches (i.e., DBPedia[1] and Knowledge Graph[2]) provide new opportunities to gain insight into the semantic structure of a document collection. Although recent studies have already shown the effectiveness of knowledge-rich approaches in several NLP tasks such as document similarity [14], topic labelling [9], and question answering [3], its feasibility and effectiveness in topic modelling framework is mostly unknown. Hulpus et al. [9] reported a framework which extracts sub-graphs from DBpedia for labelling the topics obtained from a topic model. However, their graph construction process is relied on a small set of manually selected DBpedia relations, which does not scale and needs to be tuned each time given a different knowledge repository. Instead, we extend our semantic graphs by weighting the edges (see Sect. 3.2), which is similar to the spirit of [14]. However, there is a stark difference between their work and ours: the motivation of their work is to produce graph-representation of documents for the task of document ranking, while we aim to construct semantic graph for the task of topic modelling and documents clustering.

More generally, several semantic approaches [6,11] have been proposed to combine topic modelling with word's external knowledge. However, they either relied on a small-scale semantic lexicon, e.g., WordNet, or didn't consider the relationship of entities. In contrast, we used a larger widely-covered ontology with a general-purpose algorithm to propagate the inherent topic-entity distribution.

[1] http://wiki.dbpedia.org/.

[2] https://developers.google.com/freebase/.

3 Models

3.1 Probabilistic Topic Model

In PLSA, an unobserved topic variable $z_k \in \{z_1, ..., z_K\}$ is associated with the occurrence of a word $w_j \in \{w_1, ..., w_M\}$ in a particular document $d_i \in \{d_1, ..., d_N\}$. After the summation of variable z, the joint probability of an observed pair (d, w) can be expressed as

$$P(d_i, w_j) = P(d_i) \sum_{k=1}^{K} P(w_j|z_k)P(z_k|d_i) \tag{1}$$

where $P(w_j|z_k)$ is the probability of word w_j according to the topic model z_k, and $P(z_k|d_j)$ is the probability of topic z_k for document d_i. Following the likelihood principle, these parameters can be determined by maximizing the log likelihood of a collection C as follows:

$$L(C) = \sum_{i=1}^{N} \sum_{j=1}^{M} n(d_i, w_j) log \sum_{k=1}^{K} P(w_j|z_k)P(z_k|d_i) \tag{2}$$

The model parameters $\phi = P(w_j|z_k)$ and $\theta = P(z_k|d_i)$ can be estimated by using standard EM algorithm [7].

Thus PLSA provides a good solution to find topics of documents in a text-rich information network. However, this model ignores the associated heterogeneous information network as well as other interacted objects. Furthermore, in PLSA there is no constraint on the parameters $\theta = P(z_k|d_i)$, the number of which grows linearly with the data. Therefore, the model tends to overfitting the data. To overcome these problems, we propose to use a biased propagation algorithm by exploiting a semantic network.

3.2 Topic Modelling with Semantic Graph

In this section, we propose a biased propagation algorithm to incorporate the entity semantic network with the textual information for topic modelling, so as to estimate the probabilities of topics for documents as well as other associated entities, and consequently improve the performance of topic modelling. Given the topic probability of documents $P(z_k|d_i)$, the topic probability of an entity can be calculated by:

$$\begin{aligned} P(z_k|e) &= \frac{1}{2}\Big(\sum_{d_i \in D_e} P(z_k|d_i)P(d_i|e) + \sum_{e_j \in C_e} P(z_k|e_j)P(e_j|e)\Big) \\ &= \frac{1}{2}\Big(\sum_{d_i \in D_e} \frac{P(z_k|d_i)}{|D_e|} + \sum_{e_j \in C_e} P(z_k|e_j)P(e_j|e)\Big) \end{aligned} \tag{3}$$

where D_e is a set of documents that contain the entity e, C_e is a set of entities which are connected to entity e. $P(z_k|e_j)$ is the topic probability of entity e_j, which is estimated with a similar manner as $P(z_k|d_i)$ by using the EM algorithm (see Sect. 3.3). $P(e_j|e)$ is the highest weight between entity e_j and e (see Sect. 3.2). The underlying intuition behind the above equation is that the topic distribution of an entity is determined by the average topic distribution of connected documents as well as the connected entities of semantic graph. On the other hand, the topic distributions could be propagated from entities to documents, so as to reinforce the topic distribution of documents. Thus we propose the following topic-document propagation based on semantic graph:

$$P_E(z_k|d) = \xi P(z_k|d) + (1-\xi)\sum_{e \in E_d} \frac{P(z_k|e)}{|E_d|} \tag{4}$$

where E_d denotes the set of entities of document d, ξ is the biased parameter to strike the balance between inherent topic distribution $P(z_k|d)$ and entity topic distribution $P(z_k|e)$. If $\xi = 1$, the topics of documents retain the original ones. If $\xi = 0$, the topic of the documents are determined by the entity topic distribution. By replacing the $P(z_k|d)$ in $L(C)$ with $P_E(z_k|d)$ in Eq. 4, the log-likelihood of TMSG is given as:

$$L'(C) = \sum_{i=1}^{N}\sum_{j=1}^{M} n(d_i, w_j) log \sum_{k=1}^{K} P(w_j|z_k) P_E(z_k|d_i) \tag{5}$$

Semantic Graph Construction. When computing the $P(e_j|e)$ in the above, TMSG model, we adopt the method of [14] to construct the semantic graph. We start with a set of input entities C, which is found by using the off-the-shelf named entity recognition tool DBpedia Spotlight[3]. We then search a sub-graph of DBpedia which involes the entities we already identified in the document, together with all edges and intermediate entities found along all paths of maximal length L that connect them. In this work, we set $L = 2$, as we find when L is larger than 3 the model tends to produce very large graphs and introduce lots of noise.

Figure 2 illustrates an example of a semantic graph generated from the set of entities {**db:Channel, db:David Cameron, db:Ed Miliband**}, e.g. as found in the sentence "Channel 4 will host head-to-head debates between David Cameron and Ed Miliband." Starting from these seed entities, we conduct a depth-first search to add relevant intermediate entities and relations to G (e.g., **dbr:Conservative Party** or **foaf:person**). As a result, we obtain a semantic graph with additional entities and edges, which provide us with rich knowledge about the original entities. Notice that we create two versions of semantic graphs, namely, the local semantic graph and global semantic graph. The local entity graphs build a single semantic graph for each document, and it aims to capture

[3] https://github.com/dbpedia-spotlight/dbpedia-spotlight.

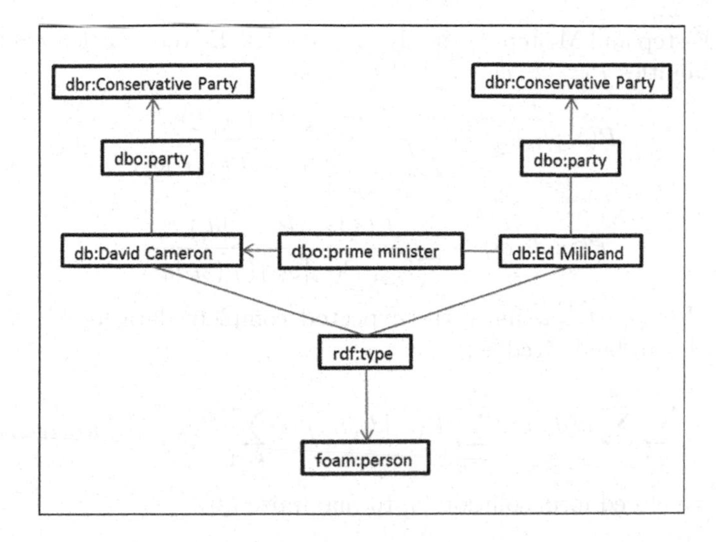

Fig. 2. A Sample Semantic Graph

the document context information. The global entity graph is constructed with the entities of the whole document collection, and we use it to detect the global context information.

Semantic Relation Weighting. So far, we simply traverse a set of input entities from DBpedia graph. However, DBpedia ontology contains many fine-grained semantic relations, which may not be equally informative. For example, in Fig. 2 seed entities **db:David Cameron** and **db:Ed Miliband** can be connected through both **rdf:type foaf:person** and **dbpprop:birthPlace**. However the former is less informative since it can apply to a large amount of entities (i.e., all persons in DBpedia). Weights can capture the degree of correlation between entities in the graph, and the core idea underlying our weighting scheme is to reward those entities and edges that are most specific to it. We define the weighting function as $W = -\log(P(W_{Pred}))$, where W is the weight of an edge, $P(W_{Pred})$ is the probability that the predicate W_{Pred} (such as rdf:type) describing the specific semantic relation. This measure is based on the hypothesis that specificity is a good proxy for relevance. We can compute the weights values for all types of predicates, as we have the whole DBpedia graph available and can query for all possible realizations of the variable X_{Pred}. We obtain the probability $P(W_{Pred})$ through maximum likelihood estimation, which is calculated by the frequency of W_{Pred} type divided by the overall counting of all the predicates.

3.3 Model Fitting with EM Algorithm

When a probabilistic model involves unobserved latent variables, the standard way is to employ the Expectation Maximization (EM) algorithm, which alternates two steps, E-step and M-step. Formally, we have the E-step to calculate the posterior probabilities $P(z_k|d_i, w_j)$ and $P(z_k|d_i, e_l)$:

$$P(z_k|d_i, w_j) = \frac{P(w_j|z_k)P_E(z_k|d_i)}{\sum_{k'=1}^{K} P(w_j|z_{k'})P_E(z_{k'}|d_i)} \tag{6}$$

$$P(z_k|d_i, e_l) = \frac{P(e_l|z_k)P_E(z_k|d_i)}{\sum_{k'=1}^{K} P(e_l|z_{k'})P_E(z_{k'}|d_i)} \tag{7}$$

In the M-step, we maximize the expected complete data log-likelihood for PLSA, which can be derived as:

$$Q_D = \sum_{i=1}^{N}\sum_{j=1}^{M} n(d_i, w_j) \sum_{k=1}^{K} P(z_k|d_i, w_j) log \sum_{k=1}^{K} P(w_j|z_k)P_E(z_k|d_i) \tag{8}$$

There is a closed-form solution [5] to maximize Q_D:

$$P(w_j|z_k) = \frac{\sum_{i=1}^{N} n(d_i, w_j)P(z_k|d_i, w_j)}{\sum_{j'=1}^{M}\sum_{i=1}^{N} n(d_i, w_{j'})P(z_k|d_i, w_{j'})} \tag{9}$$

$$P_E(z_k|d_i) = \xi \frac{\sum_{j=1}^{M} n(d_i, w_j)P(z_k|d_i, w_j)}{\sum_{j'=1}^{M} n(d_i, w_{j'})} + (1-\xi)\frac{\sum_{l=1}^{K} n(d_i, e_l)P(z_k|d_i, e_l)}{\sum_{l'=1}^{K} n(d_i, e_{l'})} \tag{10}$$

It is possible to employ more advanced parameter estimating methods, which is left for future work.

4 Experimental Evaluation

We conducted experiments on two real-world datasets, namely, DBLP and 20 Newsgroups. The first dataset, DBLP[4], is a collection of bibliographic information on major computer science journals and proceedings. The second dataset, 20 Newsgroups[5], is a collection of newsgroup documents, partitioned evenly across 20 different newsgroups. We experimented with topic modelling using a similar set-up as in [5]: For DBLP dataset, we select the records that belong to the following four areas: *database, data mining, information retrieval, and artificial intelligence.* For 20 Newsgroups dataset, we use the full dataset with 20 categories, such as *atheism, computer graphics*, and *computer windows X.*

For preprocessing, all the documents are lowercased and stopwords are removed using a standard list of 418 words. With the disambiguated entities (cf. 3.2), we create local and global entity collections, respectively, for constructing local and global semantic graphs. The creation process of entity collections is organized as a pipeline of filtering operations:

[4] http://www.informatik.uni-trier.de/~ley/db/.

[5] http://qwone.com/~jason/20Newsgroups/.

Table 1. Statistic of the DBLP and 20 Newsgroups datasets

	DBLP	20 Newsgroups
# of docs	40,000	20,000
# of entities (local)	89,263	48,541
# of entities (global)	9,324	8,750
# of links (local) docs	237,454	135,492
# of links (global) docs	40,719	37,713

1. The isolated entities, which have no paths with the other entities of the full entity collection in the DBpedia repository, are removed, since they have less power in the topic propagation process.
2. The infrequent entities, which appear in less than five documents when constructing the global entity collection, are discarded.
3. Similar to step 2, we discard entities that appear less than two times in the document when constructing the local entity collection.

The statistic of these two datasets along with their corresponding entities and links are shown in Table 1. We randomly split each of the dataset into a training set, a validation set, and a test set with a ratio 2:1:1. We learned the parameters in the semantic graph based topic model (TMSG) on the training set, tuned the parameters on the validation set and tested the performance of our model and other baseline models on the test set. The training set and the validation set are also used for tuning parameters in baseline models. To demonstrate the effectiveness of the TMSG method, we introduce the following methods for comparison:

- **PLSA:** The baseline approach which only employs the classic Probabilistic Latent Semantic Analysis [7].
- **ATM:** The state-of-the-art approach, Author Topic Model, which combines LDA with authorship network [15], in which authors are replaced with entities.
- **TMBP:** The state-of-the-art approach, Topic Model with Biased Propagation [5], which combines PLSA with an entity network (without the external knowledge, such as DBpedia).
- **TMSG:** The approach which described in Sect. 3, namely, Topic Model with Semantic Graph.

In order to evaluate our model and compare it to existing ones, we use accuracy (AC) and normalized mutual information (NMI) metrics, which are popular for evaluating effectiveness of clustering systems. The AC is defined as $AC = \frac{\sum_1^n \delta(a_i, map(l_i))}{n}$ [17], where n denotes the total number of documents, $\delta(x, y)$ is the delta function that equals one if $x = y$ and equals zero otherwise, and $map(l_i)$ is the mapping function that maps each cluster label l_i to the equivalent label from the data corpus. Given two set of documents, C and C', their mutual information metric $MI(C, C')$ is defined as: $MI(C, C') =$

$\sum_{c_i \in C, c'_j \in C'} p(c_i, c'_j) \cdot log_2 \frac{p(c_i, c'_j)}{p(c_i) \cdot p(c'_j)}$ [17], where $p(c_i)$ and pc'_j are the probabilities that a document arbitrarily selected from the corpus belongs to the clusters c_i and c'_j, respectively, and $p(c_i, c'_j)$ is the joint probability that arbitrarily selected document belongs to the cluster c_i and c'_j at the same time.

4.1 Experimental Results

Parameter Setting: For PLSA, we only use textual content for documents clustering with no additional entity information. For ATM, we use symmetric Dirichlet priors in the LDA estimation with $\alpha = 50/K$ and $\beta = 0.01$, which are common settings in the literature. For TMBP model, an entity-based heterogeneous network is constructed, and its parameter settings were set to be identical to [5]. Consistent to our previous setting of categories, we set the number of topics (K) to be four for DBLP and twenty for 20 Newsgroups as we need the data label for calculating the accuracy. The essential parameter in this work is ξ which controls the balance between the inherent textual information and semantic graph information (cf. Sect. 3.2). Figures 3 and 4 show how the performance varies with the bias parameter ξ. When $\xi = 1$, it is the baseline PLSA model. We see that the performance is improved over the baseline when incorporating

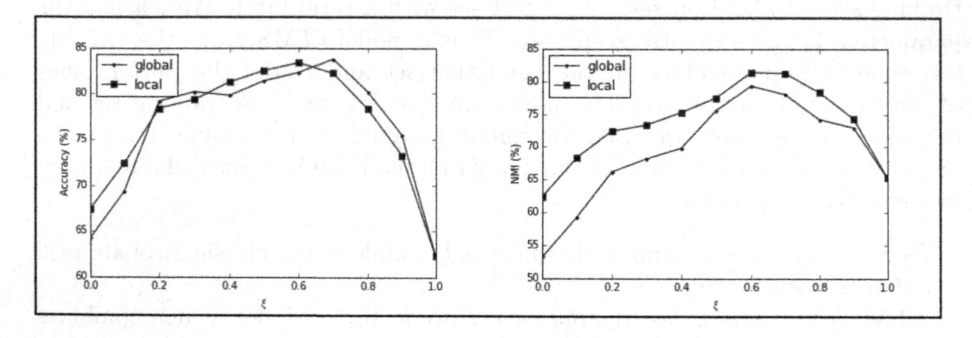

Fig. 3. The effect of varying parameter ξ in the TMSG framework on DBLP dataset.

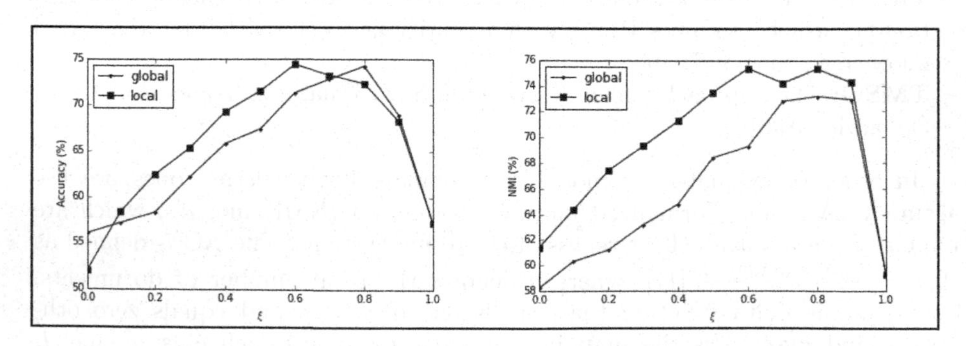

Fig. 4. The effect of varying parameter ξ in the TMSG framework on 20 Newsgroups dataset.

the semantic graph with $\xi < 0.6$. It is also notable that the TMSG with local semantic graphs (local TMSG) generally performs better then the TMSG with global semantic graph (global TMSG), which suggests that the local context is probably more important than the global one for document clustering task. We further tuned the parameters on the validation dataset. When comparing TMSG with other existing techniques, we empirically set the bias parameter $\xi = 0.6$ and the ratio between local and global TMSG is set as 0.6 : 0.4.

Table 2 depicts the clustering performance of different methods. For each method, 20 test runs are conducted, and the final performance scores were calculated by averaging the scores from the 20 tests. We can observe that ATM outperforms the baseline PLSA with additional entity network information. As expected, TMBP outperforms the ATM since it directly incorporates the heterogeneous network of the entities. More importantly, our proposed model TMSG can achieve better results than state-of-the-art ATM and TMBP algorithms. A comparison using the paired t-test is conducted for PLSA, ATM, and TMBP over TMSG, which clearly shows that our proposed TMSG outperforms all baseline methods significantly. This indicates that by considering the semantic graph information and integrating with topic modelling, TMSG can have better topic modelling power for clustering documents.

Table 2. The clustering performance of different methods on (a) DBLP and (b) 20 Newsgroups datasets (-*-* and -* indicate degraded performance compared to TMSG with p-value < 0.01 and p-value < 0.05, respectively).

(a) DBLP

	PLSA	**ATM**	**TMBP**	**TMSG**
AC	0.62-*-*	0.68-*	0.72-*	0.80
NMI	0.65-*-*	0.72-*	0.75-*	0.82

(b) 20 Newsgroups

	PLSA	**ATM**	**TMBP**	**TMSG**
AC	0.56-*-*	0.63-*-*	0.67-*	0.72
NMI	0.55-*-*	0.61-*-*	0.65-*	0.71

Table 3. The representative terms generated by PLSA, ATM, TMBP, and TMSG models. The terms are vertically ranked according to the probability $P(w|z)$.

	Topic 1 (DB)		Topic 2 (DM)		Topic 3 (IR)		Topic 4 (AI)	
PLSA	data	management	data	algorithm	information	learning	learning	knowledge
	database	processing	mining	performance	retrieval	search	algorithm	time
	memory	relational	learning	detection	web	**system**	application	logic
	system	processing	clustering	analysis	knowledge	language	human	**search**
	architecture	**feature**	classification	parameter	text	query	model	representation
ATM	data	management	mining	**multiple**	information	language	learning	algorithm
	database	software	data	algorithm	retrieval	text	knowledge	paper
	server	relational	classification	performance	search	web	logic	time
	system	function	learning	analysis	knowledge	**classification**	**image**	**method**
	query	processing	clustering	detection	**performance**	query	model	application
TMBP	data	software	data	parameter	information	learning	knowledge	paper
	database	relational	mining	algorithm	retrieval	query	application	**intelligence**
	management	architecture	classification	**result**	document	**estimation**	human	model
	algorithm	**text**	learning	analysis	query	**management**	algorithm	system
	server	processing	clustering	**time**	web	language	**compute**	performance
TMSG	data	**accelerator**	data	**analysis**	information	search	knowledge	logic
	database	function	mining	algorithm	retrieval	document	learning	system
	query	relational	classification	parameter	query	**semantic**	information	**data**
	system	software	clustering	**pattern**	knowledge	language	information	representation
	distributed	**performance**	learning	**information**	text	**user**	**reasoning**	uman

Since the DBLP dataset is a mixture of four areas, it is interesting to see whether the extracted topics could reflect this mixture. Shown in Table 3 are the most representative words of topics generated by PLSA, ATM, TMBP, and TMSG, respectively. For topic 2 and 3, although different models select slightly different terms, all these terms can describe the corresponding topic to some extent. For topic 1 (DB), however, the words "accelerator", "performance", and "distributed" of TMSG are more telling than "text" derived by TMBP, and "memory" and "feature" derived by PLSA. Similar subtle differences can be found for the topic 4 as well. Intuitively, TMSG selects more related terms for each topic than other methods, which shows the better performance of TMSG by considering the relationship of entities in the semantic graph.

5 Conclusion

The main contribution of this paper is to show the usefulness of semantic graph for topic modelling. Our proposed TMSG (Topic Model with Semantic Graph) supersedes the existing ones since it takes account both homogeneous networks (i.e., entity to entity relations) and heterogeneous networks (i.e., entity to document relations), and since it exploits both local and global representation of rich knowledge that go beyond networks and spaces.

There are some interesting future work to be continued. First, TMSG only relies on one of the simplest latent topic models (namely PLSA), which makes sense as a first step towards integrating semantic graphs into topic models. In the future, we will study how to integrate the semantic graph into other topic modeling algorithms, such as Latent Dirichlet Allocation. Secondly, it would be also interesting to investigate the performance of our algorithm by varying the weights of different types of entities.

Acknowledgements. We thank the anonymous reviewer for their helpful comments. We acknowledge support from the EPSRC funded project named **A Situation Aware Information Infrastructure Project** (EP/L026015) and the **Integrated Multimedia City Data** (IMCD), a project within the ESRC-funded **Urban Big Data Centre** (ES/L011921/1). This work was also partly supported by NSF grant #61572223. Any opinions, findings and conclusions or recommendations expressed in this material are those of the author(s) and do not necessarily reflect the view of the sponsor.

References

1. Bao, Y., Collier, N., Datta, A.: A partially supervised cross-collection topic model for cross-domain text classification. In: CIKM 2013, pp. 239–248 (2013)
2. Blei, D.M., Ng, A.Y., Jordan, M.I., Lafferty, J.: Latent dirichlet allocation. **3**, 459–565
3. Cai, L., Zhou, G., Liu, K., Zhao, J.: Large-scale question classification in cqa by leveraging wikipedia semantic knowledge. In: CIKM 2011, pp. 1321–1330 (2011)

4. Chen, X., Zhou, M., Carin, L.: The contextual focused topic model. In: KDD 2012, pp. 96–104 (2012)
5. Deng, H., Han, J., Zhao, B., Yintao, Y., Lin, C.X.: Probabilistic topic models with biased propagation on heterogeneous information networks. In: KDD 2011, pp. 1271–1279 (2011)
6. Guo, W., Diab, M.: Semantic topic models: Combining word distributional statistics and dictionary definitions. In: EMNLP 2011, pp. 552–561 (2011)
7. Hofmann, T.: Unsupervised learning by probabilistic latent semantic analysis. Mach. Learn. **45**, 256–269
8. Hong, L., Dom, B., Gurumurthy, S., Tsioutsiouliklis, K.: A time-dependent topic model for multiple text streams. In: KDD 2011, pp. 832–840 (2011)
9. Hulpus, I., Hayes, C., Karnstedt, M., Greene, D.: Unsupervised graph-based topic labelling using dbpedia. WSDM 2013, pp. 465–474 (2013)
10. Kim, H., Sun, Y., Hockenmaier, J., Han, J.: Etm: Entity topic models for mining documents associated with entities. In: ICDM 2012, pp. 349–358 (2012)
11. Li, F., He, T., Xinhui, T., Xiaohua, H.: Incorporating word correlation into tag-topic model for semantic knowledge acquisition. In: CIKM 2012, pp. 1622–1626 (2012)
12. Li, H., Li, Z., Lee, W.-C., Lee, D.L.: A probabilistic topic-based ranking framework for location-sensitive domain information retrieval. In: SIGIR 2009, pp. 331–338 (2009)
13. Mei, Q., Cai, D., Zhang, D., Zhai, C.: Topic modeling with network regularization. In: WWW 2008, pp. 342–351 (2008)
14. Schuhmacher, M., Ponzetto, S.P.: Knowledge-based graph document modeling. In: WSDM 2014, pp. 543–552 (2014)
15. Tang, J., Zhang, J., Yao, L., Li, J., Zhang, L., Zhong, S.: Arnetminer: extraction and mining of academic social networks. In: KDD 2008, pp. 428–437 (2008)
16. Xing Wei, W., Croft, B.: Lda-based document models for ad-hoc retrieval. In: SIGIR 2006, pp. 326–335 (2009)
17. Wei, X., Liu, X., Gong, Y.: Document clustering based on non-negative matrix factorization. In: SIGIR 2003, pp. 267–273 (2003)

Estimating Probability Density of Content Types for Promoting Medical Records Search

Yun He[2], Qinmin Hu[1,2](✉), Yang Song[2], and Liang He[1,2]

[1] Shanghai Key Laboratory of Multidimensional Information Processing, East China Normal University, Shanghai 200241, China
{qmhu,lhe}@cs.ecnu.edu.cn

[2] Department of Computer Science and Technology, East China Normal University, Shanghai 200241, China
heyunyun_2012@163.com, ysong@ica.stc.sh.cn

Abstract. Disease and symptom in medical records tend to appear in different content types: positive, negative, family history and the others. Traditional information retrieval systems depending on keyword match are often adversely affected by the content types. In this paper, we propose a novel learning approach utilizing the content types as features to improve the medical records search. Particularly, the different contents from the medical records are identified using a Bayesian-based classification method. Then, we introduce our type-based weighting function to take advantage of the content types, in which the weights of the content types are automatically calculated by estimating the probability density functions in the documents. Finally, we evaluate the approach on the TREC 2011 and 2012 Medical Records data sets, in which our experimental results show that our approach is promising and superior.

Keywords: Medical records search · Content types identification · Weighting function · Density estimation

1 Introduction

In general, the disease and symptom in electronic medical records (EMRs) can be divided into multiple content types: positive (e.g., "with lung cancer"), negative (e.g., "denies fever"), family history (e.g., "family history of lung cancer") and the others [1]. The traditional information retrieval (IR) systems usually treat all the EMRs equally such as keyword match, instead of considering different content types on different situations. For example, a query describes patients admitted with a diagnosis of dementia. Then, all EMRs including the keyword "dementia" will be retrieved by a classical IR model such as BM25 [2]. However, the real intention of the query is to find the patients with dementia, instead of the family member who had dementia or someone denies dementia.

In order to handle multiple content types, previous work mainly focuses on the following two ways: (1) transforming the format of the negative content, for example, changing "denies fever" to "nofever" as one word; (2) directly removing

N. Ferro et al. (Eds.): ECIR 2016, LNCS 9626, pp. 252–263, 2016.
DOI: 10.1007/978-3-319-30671-1_19

the negative and family history before indexing the EMRs. However, the above solutions still fail to promote the influence of the right content types in search process respectively based on the users' true desire. Therefore, we are motivated to utilize the content types as our relevance features for medical search.

In this paper, we propose a novel learning approach to promote the performance of the medical records search through probability density estimation over the content types. Specifically, we first present a Bayesian-based classification method to identify the different contents from the EMR data and queries. Then, a type-based weighting function is inferred, followed by estimating the kernel density function to compute the weights of the content types. After that, how to select the bandwidth of the density estimator is visualized.

We evaluate our approach on the TREC 2011 and 2012 Medical Records Tracks [3] (the track has been discontinued after 2012). The task of the TREC Medical Records Tracks requires an IR system to return a list of retrieved EMRs by the likelihood of satisfying the user's information desire. The retrieved EMRs are used to identify the patients who meet the criteria, described by the queries, for inclusion in possible clinical studies. The evaluation results show that the proposed approach outperforms the strong baselines.

In the rest of our paper, we briefly present the related work in Sect. 2. Then, we introduce the proposed approach in Sect. 3, including the identification of the content types, the definition of the weighting function and the density estimation. After that, we show the experiments in Sect. 4, followed by the discussion and analysis in Sect. 5. Finally, we draw the conclusions and describe the future work in Sect. 6.

2 Related Work

2.1 Content Types Processing

In medical records search, some content types, including the negative, history or experience of family members other than the patients, often have a bad effect on search performance, when users need the positive contents. Previous researchers endeavored to solve two problems: (1) how to detect the content types in clinical text? (2) how to prevent them from damaging the retrieval performance?

For the first problem, most algorithms have been designed based on regular expression. Chapman et al. [4] developed NegEx which utilized several phrases to indicate the negative content. NegEx has been widely applied to identify the negative content. Harkema et al. [5] proposed an algorithm called ConText which was an extension of NegEx. It was not only used to detect the negative content, but also the hypothetical, historical and experience of family members. In this paper, we apply the Bayesian-based classification method to discover the positive, negative and family history in EMRs and fit each query into one of them. The characteristic of our method is based on probability statistics rather than designing the regular expressions as detecting rules.

Averbuch et al. [6] proposed a framework to automatically identify the negative and positive content and assure the retrieved document in which at least one keyword appeared in its positive content. Limsopatham et al. [7] proposed a two-step method, where an algorithm called NegFlag transformed the format of the negative in medical records, and a term dependency approach obtained results candidates containing keywords in the negative. Furthermore, many participants in the TREC Medical Record Tracks removed all negative contents before they indexed the EMRs [8–12]. All of them assumed that the presence of the negative always harmed the retrieval effectiveness. The state-of-the-art study has been conducted by Koopman et al. [1], which aimed to understand the effect of negative and family history on medical search. They suggested that the content types were optimally handled by a per-query weighting. Specifically, they obtained the optimal weighting by performing an extensive parameter search in their empirical experiments. They also presented that assigning relatively low weight to negative content leads to a better performance.

Hence, we incorporate the features of the content types into retrieval to improve the performance.

2.2 Kernel Density Estimation

Kernel density estimation is a typical method of non-parametric statistics and widely used in supervised machine learning. Zhou et al. [13] proposed a length-based BM25 model, called BM25L, which incorporated the document length as the relevance feature. They investigated the distribution of the document length and its impact on relevance. In order to obtain the distribution, they visualized the kernel density estimator and analyzed the shapes of them. The main difference of our work lies in that we apply kernel density estimation to obtain the density functions of the content types. However, [13] obtained the length density function by estimating the parameters of the overall distribution, which is the traditional parametric statistics.

3 Methodology

First, we identify the content types in EMRs and queries by the Bayesian-based classification method. Then, we use kernel density estimation to obtain the density functions of the content types in the training set. After that, the weights of the content types, which utilize the kernel estimators, are incorporated into BM25 as the types-based weighting function. Finally, the EMRs are ranked by the weighting function.

3.1 Content Types Identification

Since each content type contains special terminology and format, which can be utilized as statistical features, we apply a Bayesian-based classifier to fit each sentence of EMRs into one of the defined four content types.

The notations are defined as follows. The vocabulary is composed of all words in the data set. n is the size of it. Each sentence is represented via a vector $w = (w_1, ...w_j...)$. $w_j = 1$ denotes that w contains the j-th word of the vocabulary, otherwise, we let the $w_j = 0$. $C = k$ denotes the content type of w, $k = 1$ denotes the family history, $k = 2$ denotes the positive, $k = 3$ denotes the negative and $k = 4$ denotes the others. The others represents the information which is irrelevant to the diagnosis of disease, for example, date and names of doctors.

Bayesian-Based Classification Method. We manually label the type of each sentence in 200 EMRs as the training set $\{w^{(i)}, C^{(i)}\}$, $i = 1, ..., m$. Then, we train the classifier by calculating probabilities by Eqs. 1 and 2.

$$P(w_j = 1|C = k) = \frac{\sum_{i=1}^{m} 1\{w_j^{(i)} = 1 \wedge C^{(i)} = k\} + 1}{\sum_{i=1}^{m} 1\{C^{(i)} = k\} + 4} \tag{1}$$

$$P(C = k) = \frac{\sum_{i=1}^{m} 1\{C^{(i)} = k\}}{m} \tag{2}$$

where m is the size of the training set and indicator function $1\{.\}$ takes on a value of 1 if its argument is true.$(1\{True\} = 1,\ 1\{False\} = 0)$

Using the classifier, w is classified into a content type which has the largest posterior probability obtained by Eq. 3.

$$P(C = k|w) = \frac{(\prod_{j=1}^{n} P(w_j = 1|C = k))P(C = k)}{\sum_{k=1}^{4}((\prod_{j=1}^{n} P(w_j = 1|C = k))P(C = k))} \tag{3}$$

Queries Analysis and Classification. We analyze the queries in the TREC Medical Records Tracks in 2011 and 2012. We find out that each query can be classified into one of the three content types according to its disease and symptom information. Hence, we apply the classification method, presented in the above section, to classify the total 85 queries in the 2011 and 2012 tracks. The examples and numbers of the queries are shown in Table 1.

Content Types Features. In the classification method, each EMR is segmented into four parts: positive, negative, family history and the others.

Table 1. Examples and numbers of queries in the TREC Medical Records Tracks in 2011 and 2012

Content types	Examples of the queries	Numbers
Family history	-	0
Positive diagnosis	Topic 104: Patients diagnosed with localized prostate cancer and treated with robotic surgery.	84
Negative diagnosis	Topic 179: Patients taking atypical antipsychotics without a diagnosis schizophrenia or bipolar depression	1
Others	-	0

Then, four indices are built separately based on the four parts. We implement the first round retrieval and extract the relevance scores of basic model (e.g., BM25) from these indices, referred as x_1, x_2, x_3, x_4. We use vector $X = (x_1, x_2, x_3, x_4)$ to represent the feature of the content types in EMRs, which is a continuous-valued random variable [14].

Correspondingly, each query is fitted into a type denoted by C_q as well. We use C_q to represent the feature of the content types in queries, which is a discrete-valued random variable.

3.2 Type-Based Weighting Function

Based on the "Probability Ranking Principle" described in [15], we obtain the optimum retrieval when EMRs are ranked by the decreasing values of the probability of relevance. Hence, we model the weights of the content types by $P(R = 1|X, C_q)$, where $R = 1$ indicates relevant and otherwise $R = 0$. The Bayes rule is adopted as:

$$P(R = 1|X, C_q) = \frac{P(X, C_q|R = 1)P(R = 1)}{P(X, C_q)} \tag{4}$$

Since X is the continuous-valued random variable, it is approximately to estimate the probability density functions $p(X, C_q|R = 1)$ and $p(X, C_q)$. That is:

$$P(R = 1|X, C_q) \approx \frac{p(X, C_q|R = 1)P(R = 1)}{p(X, C_q)} \tag{5}$$

Note that $P(R = 1)$ is a constant, given the EMRs collection and the query, having no impact on the ranking. We ignore $P(R = 1)$ in the Eq. 5 and define the weights of the content types as:

$$f(X, C_q) = log(\frac{p(X, C_q|R = 1)}{p(X, C_q)}) \tag{6}$$

where logarithmic function is used to normalize the weight.

In order to utilize the keyword match, by combining $f(X, C_q)$ with BM25 model, a type-based weighting function is proposed as follows:

$$F(X, C_q) = (1 - \theta) * f_{BM25} \oplus \theta * f(X, C_q) \tag{7}$$

where f_{BM25} is the relevance score of BM25, $\oplus$ denotes that the term $f(X, C_q)$ is added only once for each EMR. $F(X, C_q)$ is used as relevance ranking function.

3.3 Kernel Density Estimation

We use kernel density estimation to obtain $p(X, C_q|R = 1)$ and $p(X, C_q)$. The choice of bandwidth h strongly influence the shape of the estimator [16–19]. In order to select the suitable bandwidth, we visualize $p(X, C_q|R = 1)$ and $p(X, C_q)$.

For the visualization, we apply principal components analysis (PCA) to project the feature vector X into 1-dimensional subspace. In the rest of paper, x represents X in the 1-dimensional sub-space. Hence, we aim to obtain $p(x, C_q|R = 1)$ and $p(x, C_q)$. Collecting samples $x_1, x_2, ..., x_n$ from the training set, the kernel estimator of $p(x)$ is defined by:

$$\hat{p}(x) = \frac{1}{nh}\sum_{i=1}^{n}\omega_i = \frac{1}{nh}\sum_{i=1}^{n}K(\frac{x - x_i}{h}) \tag{8}$$

where $K(u)$ is the kernel function, and h is the bandwidth, ω_i is the weight of x_i to influence the $\hat{p}(x)$. Empirically, the choice of kernel function has almost no impact on the estimator. Hence, we choose the common Gaussian kernel, such that:

$$K(u) = \frac{1}{\sqrt{2\pi}}exp(-\frac{1}{2}u^2) \tag{9}$$

Bandwidth Selection. In our work, the *asymptotic mean integrated square error* (AMISE) between the estimators $\hat{p}(x; h)$ and actual $p(x)$ is utilized to obtain the theoretical optimal bandwidth h_{opt}. In general, since actual $p(x)$ is unknown, it is assumed as Gaussian distribution. AMISE is given as follows:

$$\begin{aligned} AMISE\{\hat{p}(x;h)\} &= E(\int\{\hat{p}(x;h) - p(x)\}^2 dx) \\ &= \int(\{E\hat{p}(x;h) - p(x)\}^2 + E\{\hat{p}(x;h) - E\hat{p}(x;h)\}^2)dx \\ &= \int[(Bias(\hat{p}(x;h)))^2 + Variance(\hat{p}(x;h))]dx \end{aligned} \tag{10}$$

Note that AMISE is divided into bias and variance. Based on AMISE, h_{opt} can be calculated by minimizing Eq. 10.

Except for h_{opt}, we investigate more variations of h_{opt}: $4 * h_{opt}$, $2 * h_{opt}$, $h_{opt}/2$, $h_{opt}/4$, $h_{opt}/8$. In Fig. 1, shapes of estimators based on different h are shown. It is presented that h has an obvious impact on the estimators, and the smoothness of estimators decreases with decreasing h. Empirically, a less smooth estimator indicates a low bias and high variance, and verse versa. Hence, the bias decreases with decreasing h. Our intuition is that a relatively small bandwidth, corresponding to an estimator which has low bias, will lead to a better performance on the TREC medical track data sets. The effect of the bandwidths is evaluated in our experiments.

4 Experiments

In this section, we present a set of experiments to evaluate the performance of our proposed type-based weighting function on the TREC Medical Records Tracks. Our evaluation baseline is the BM25 model. Moreover, we compare our approach with the popular method which removes the negation and family history before indexing the EMRs, referred as "negation removal" in the experiments.

4.1 Data Sets and Implementation

The data sets in the TREC 2011 and 2012 Medical Tracks are composed of de-identified medical records from the University of Pittsburgh NLP Repository. Each medical record is linked to a visit, which is the patients single stay at a hospital. The data sets contain 93,551 medical records which are linked to 17,264 visits. Queries were provided by physicians in the Oregon Health & Science University (OHSU) Biomedical Informatics Graduate Program. Each query is made up of symptom, diagnosis and treatment, matching a reasonable number of visits. All EMRs and queries are preprocessed by Potter's stemming and standard stopword removal. Terrier is applied for indexing and retrieval[1]. We first implement the initial search to obtain the relevance score of the content types as the learning features. The final result is achieved by ranking the EMRs according to Eq. 7.

We use all 35 queries in the 2011 track to obtain the kernel estimators, and all 50 queries in the 2012 track as the testing purpose. The relevance judgments for the test collection are evaluated on a 3-point scale: not relevant, normal relevant and highly relevant. In this work, we ignore the different degrees of relevance with regarding the highly relevant as the normal relevant in the training data.

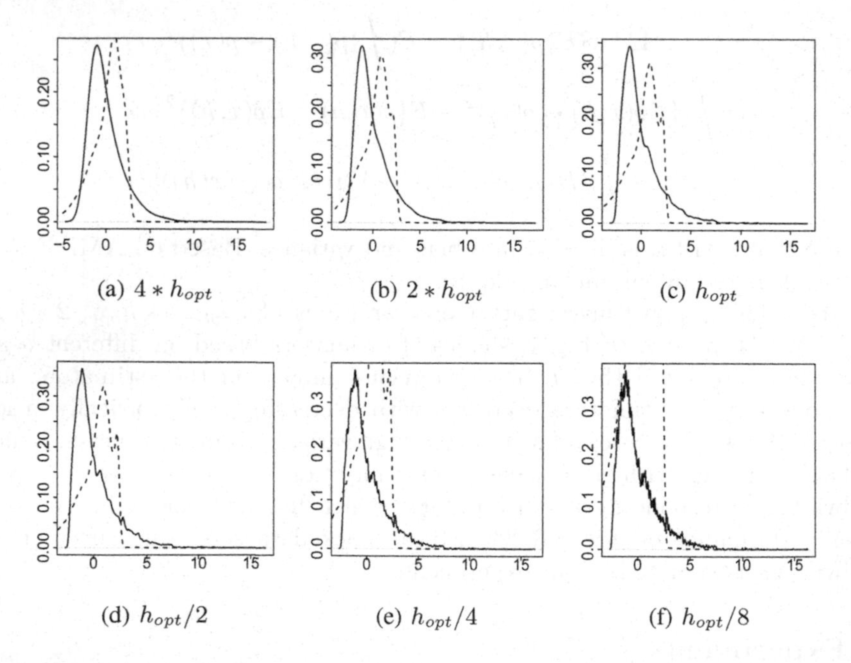

Fig. 1. Estimators based on the different bandwidths. The solid line denotes $\hat{p}(x, C_q = 2|R = 1)$ and the dashed line denotes $\hat{p}(x, C_q = 2)$.

[1] http://terrier.org.

4.2 Evaluation Metrics

Our evaluation measures include Precision at 5(P@5), 10(P@10), 15(P@15), 20(P@20), 30(P@30). We mainly focus on P@5 and P@10.

4.3 Results

Table 2 presents the evaluation results of the type-based weighting function. A-star "*" indicates that this result is the best one given a bandwidth. Two-star "**" denotes that this result is the best one over all bandwidths. We also show the improvements of the best result upon the baseline given the bandwidth. Table 3 compares our best result of the type-based weighting function with the popular the state-of-the-art "negation removal" method.

5 Discussion

Here we first investigate the influence of the proposed learning approach. Then, we analyze the effectiveness of removing negative and family history content. After that, the impact of the bandwidth on the kernel estimators is discussed. Finally, we show the empirically optimized parameter θ for the bandwidth.

5.1 Influence of the Proposed Learning Approach

As shown in Table 2, the results of our proposed learning approach outperform the BM25 baseline, in terms of all the evaluation metrics. The best result in terms of P@5 is 0.5149 which is 7.07 % higher than the baseline. The best result in terms of P@10 is 0.4979 which is 10.37 % higher than the baseline.

5.2 Impact of Negation

Table 3 displays the results of BM25, negation removal and our proposed approach. Here "negation removal" stands for the solution which removes the negative and family history content at all. Our best result is selected based on the tuned parameters and the optimized bandwidth.

The results of negation removal solution outperform BM25 in terms of P@5 and P@10. This explains that the keyword match in BM25 is not suitable for the medical task, since BM25 denies the differences among different content types in the EMRs. Hence, negation removal obtains better performance when the negation information is excluded for retrieval.

However, the negation information still has its own influence in the EMRs. Our approach which specifically identifies negation from the EMRs and queries, achieves the best results. Hence, we suggest to better make use of the negation information, instead of removing it arbitrarily.

Table 2. Evaluation results of the type-based weighting function in 2012 medical track.

Bandwidth	θ	p@5	p@10	p@15	p@20	p@30
BM25		0.4809	0.4511	0.427	0.4074	0.3773
$h_{opt} * 4$	0.2	0.5064*	0.483	0.4411	0.4309**	0.395
	0.4	0.5064*	0.483	0.4426	0.4245	0.3957
	0.6	0.5021	0.4894*	0.4454	0.4234	0.3965
	0.8	0.5021	0.4872	0.4525*	0.4277	0.3993**
		+5.30 %	+8.49 %	+5.97 %	+5.77 %	+5.83 %
$h_{opt} * 2$	0.2	0.5064*	0.4787	0.444	0.4266*	0.3957
	0.4	0.5021	0.4851	0.4411	0.4266*	0.3957
	0.6	0.4936	0.4915*	0.4468	0.4245	0.3972*
	0.8	0.5021	0.4872	0.4511*	0.4223	0.3908
		+5.30 %	+8.96 %	+5.64 %	+4.71 %	+5.27 %
h_{opt}	0.2	0.5064*	0.4809	0.444	0.4255	0.3965
	0.4	0.5021	0.4787	0.444	0.4266	0.3957
	0.6	0.4979	0.4872	0.4482	0.4287*	0.3979*
	0.8	0.4936	0.4979**	0.4553**	0.4191	0.3936
		+5.30 %	+10.37 %	+6.63 %	+5.23 %	+5.46 %
$h_{opt}/2$	0.2	0.5106*	0.483	0.444	0.4234*	0.3965*
	0.4	0.5064	0.4787	0.4496*	0.4213	0.3965*
	0.6	0.4979	0.4809	0.4482	0.4234*	0.3965*
	0.8	0.5106*	0.4894*	0.4468	0.4202	0.3936
		+6.18 %	+8.49 %	+5.29 %	+3.93 %	+5.09 %
$h_{opt}/4$	0.2	0.5106*	0.4766	0.4454	0.4245*	0.3972
	0.4	0.5064	0.4766	0.4482	0.4191	0.3979*
	0.6	0.5106*	0.4851*	0.4525*	0.4191	0.3972
	0.8	0.5064	0.4745	0.4454	0.4234	0.3922
		+6.18 %	+7.54 %	+5.97 %	+4.20 %	+5.46 %
$h_{opt}/8$	0.2	0.5106	0.4766	0.4468	0.4266*	0.3979*
	0.4	0.5106	0.4766	0.4482	0.4202	0.3965
	0.6	0.5149**	0.4787	0.4468	0.4213	0.395
	0.8	0.5106	0.4809*	0.4496*	0.4213	0.395
		+7.07 %	+6.61 %	+5.29 %	+4.71 %	+5.46 %

5.3 Impact of Bandwidth

Figure 2 shows the results based on different bandwidths. The optimal bandwidth achieves the best results in terms of P@10 and P@15 and the top results in terms of P@20 and P@30. These observations are theoretically proved by minimizing Eq. 10.

Table 3. Comparison between the type-based weighting function and the negation removal in 2012 medical track

	p@5	p@10	p@15	p@20	p@30
BM25	0.4809	0.4511	0.427	0.4074	0.3773
Negation removal	0.5277	0.4745	0.4128	0.3851	0.3589
	+9.73 %	+5.19 %	-3.33 %	-4.84 %	-4.87 %
Our best result	0.5149	0.4979	0.4553	0.4309	0.3993
	+7.07 %	+10.37 %	+6.63 %	+5.77 %	+5.83 %

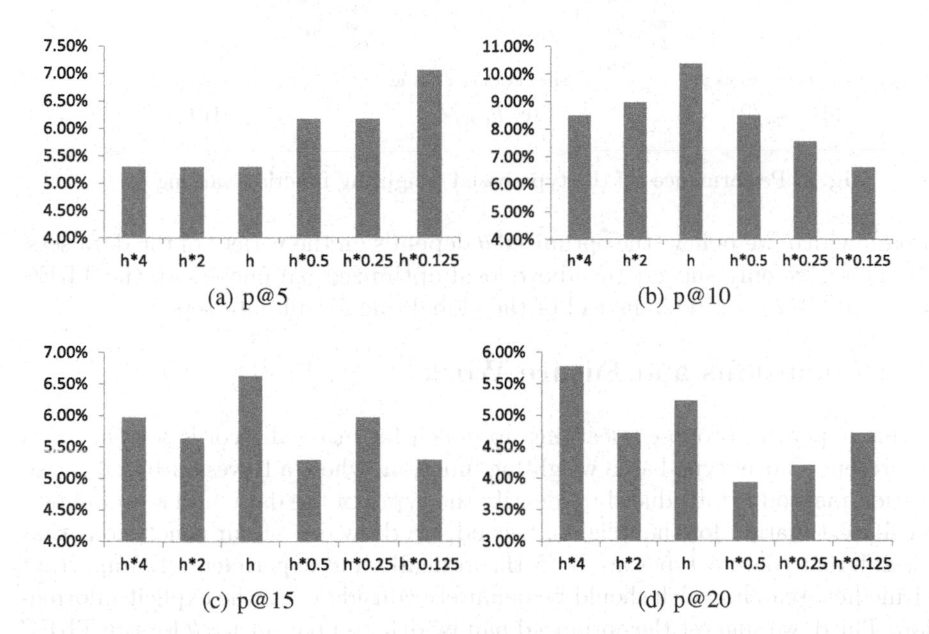

Fig. 2. Performances of the type-based weighting function varying on differen bandwidths

However, the performance of P@5 increases when the bandwidth decreased. We find that the estimator based on the relative h under-fits the training set, when h becomes bigger. In the other hand, the relative smaller h leads to $\hat{p}(x)$ with lower bias. Usually when the training and testing sets share the similar data distributions, we believe $\hat{p}(x)$ with the smaller h leads a high precision on the testing data, especially the top-ranked EMRs.

5.4 Impact of Parameter θ

In the method, θ denotes the influence of the content types in the type-based weighting function. Figure 3 shows the results which are obtained by the type-based weighting function varying on θ.

From Table 2, we can see that $\theta = 0.1$ obtains the best result in terms of P@5, $\theta = 0.8$ for P@10 and P@15, $\theta = 0.6$ for P@20 and P@30. As the discussion

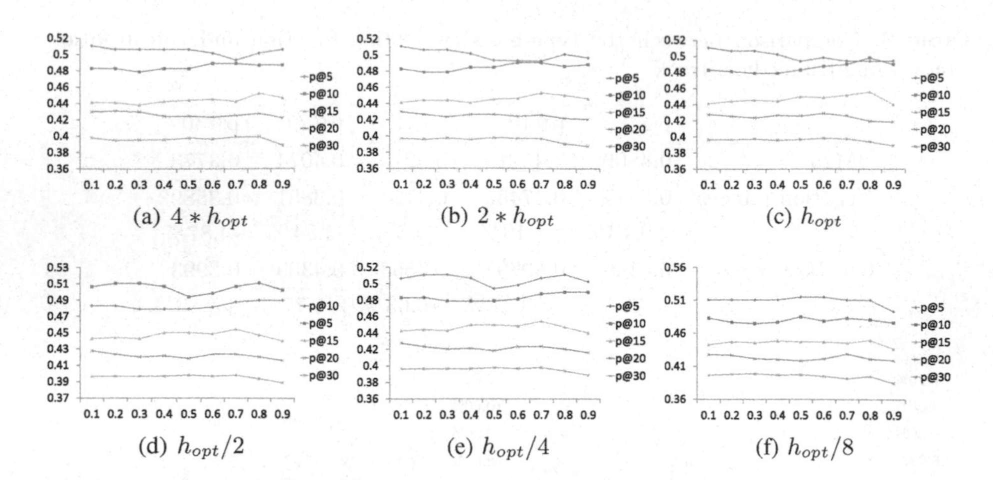

(a) $4 * h_{opt}$ (b) $2 * h_{opt}$ (c) h_{opt}

(d) $h_{opt}/2$ (e) $h_{opt}/4$ (f) $h_{opt}/8$

Fig. 3. Performances of the type-based weighting function varying on θ

of bandwidth, we believe the optimized θ depends on the variety of the data sets. Therefore, we only suggest the above local optimized parameters on the TREC 2011 and 2012 data sets, instead of the global one for all data sets.

6 Conclusions and Future Work

In this paper, we propose a learning approach for medical records search. First, we present a novel type-based weighting function, where a Bayesian-based classification method is introduced to identify the types of the data, and a probability density estimation for the weights. Second, we draw one of our conclusions that the different content types are both theoretically and experimentally important for medical search, which should be definitely considered as the explicit information. Third, we suggest the optimized bandwidth and parameter θ for the TREC 2011 and 2012 Medical data sets. Finally, we report our best results in terms of P@10 as 0.4979, which is 10.37 % better than the BM25 baseline.

In the future, we will continue on the investigation of multiple content types. In particular, we will focus on dealing with the queries, such as the feature representation of the queries in the learning approach, in order to identify the content types from every single query term instead of the whole query.

Acknowledgment. This research is funded by the Science and Technology Commission of Shanghai Municipality (No.15PJ1401700 and No.14511106803).

References

1. Koopman, B., Zuccon, G.: Understanding negation and family history to improve clinical information retrieval. In: Proceedings of the 37th International ACM SIGIR Conference on Research Development in Information Retrieval, pp. 971–974. ACM (2014)

2. Robertson, S., Zaragoza, H.: The Probabilistic Relevance Framework: BM25 and Beyond. Now Publishers Inc, Hanover (2009)
3. Voorhees, E., Tong, R.: Overview of the trec medical records track. In: Proceedings of TREC 2011 (2011)
4. Chapman, W.W., Bridewell, W., Hanbury, P., Cooper, G.F., Buchanan, B.G.: A simple algorithm for identifying negated findings and diseases in discharge summaries. J. Biomed. Inform. **34**(5), 301–310 (2001)
5. Harkema, H., Dowling, J.N., Thornblade, T., Chapman, W.W.: Context: an algorithm for determining negation, experiencer, and temporal status from clinical reports. J. Biomed. Inform. **42**(5), 839–851 (2009)
6. Averbuch, M., Karson, T., Ben-Ami, B., Maimon, O., Rokach, L.: Context-sensitive medical information retrieval. In: The 11th World Congress on Medical Informatics (MEDINFO 2004), San Francisco, CA, pp. 282–286. Citeseer (2004)
7. Limsopatham, N., Macdonald, C., McCreadie, R., Ounis, I.: Exploiting term dependence while handling negation in medical search. In: Proceedings of the 35th International ACM SIGIR Conference on Research and Development in Information Retrieval, pp. 1065–1066. ACM (2012)
8. Karimi, S., Martinez, D., Ghodke, S., Cavedon, L., Suominen, H., Zhang, L.: Search for medical records: Nicta at trec medical track. In: TREC 2011 (2011)
9. Amini, I., Sanderson, M., Martinez, D., Li, X.: Search for clinical records: rmit at trec medical track. In: Proceedings of the twentieth Text Retrieval Conference (TREC 2011). Citeseer (2011)
10. Córdoba, J.M., López, M.J.M., Díaz, N.P.C., Vázquez, J.M., Aparicio, F., de Buenaga Rodríguez, M., Glez-Peña, D., Fdez-Riverola, F.: Medical-miner at trec medical records track. In: TREC 2011 (2011)
11. King, B., Wang, L., Provalov, I., Zhou, J.: Cengage learning at trec medical track. In: TREC 2011 (2011)
12. Limsopatham, N., Macdonald, C., Ounis, I., McDonald, G., Bouamrane, M.: University of glasgow at medical records track: experiments with terrier. In: Proceedings of TREC 2011 (2011)
13. Zhou, X., Huang, J.X., He, B.: Enhancing ad-hoc relevance weighting using probability density estimation. In: Proceedings of the 34th international ACM SIGIR conference on Research and development in Information Retrieval, pp. 175–184 (2011)
14. Choi, S., Choi, J.: Exploring effective information retrieval technique for the medical web documents: Snumedinfo at clefehealth task 3. In: Proceedings of the ShARe/CLEF eHealth Evaluation Lab 2014 (2014)
15. Robertson, S.E.: The probability ranking principle in IR. J. Document. **33**, 294–304 (1977)
16. Gijbels, I., Delaigle, A.: Practical bandwidth selection in deconvolution kernel density estimation. Comput. Stat. Data Anal. **45**(2), 249–267 (2004)
17. Duraiswami, V.: Abstract fast optimal bandwidth selection for kernel density estimation. Fast optimal bandwidth selection for kernel density estimation. - ResearchGate (2006)
18. Jones, M.C.: A brief survey of bandwidth selection for density estimation. J. Am. Stat. Assoc. **91**(433), 401–407 (1996)
19. Comaniciu, D.: An algorithm for data-driven bandwidth selection. IEEE Trans. Pattern Anal. Mach. Intell. **25**, 281–288 (2003)

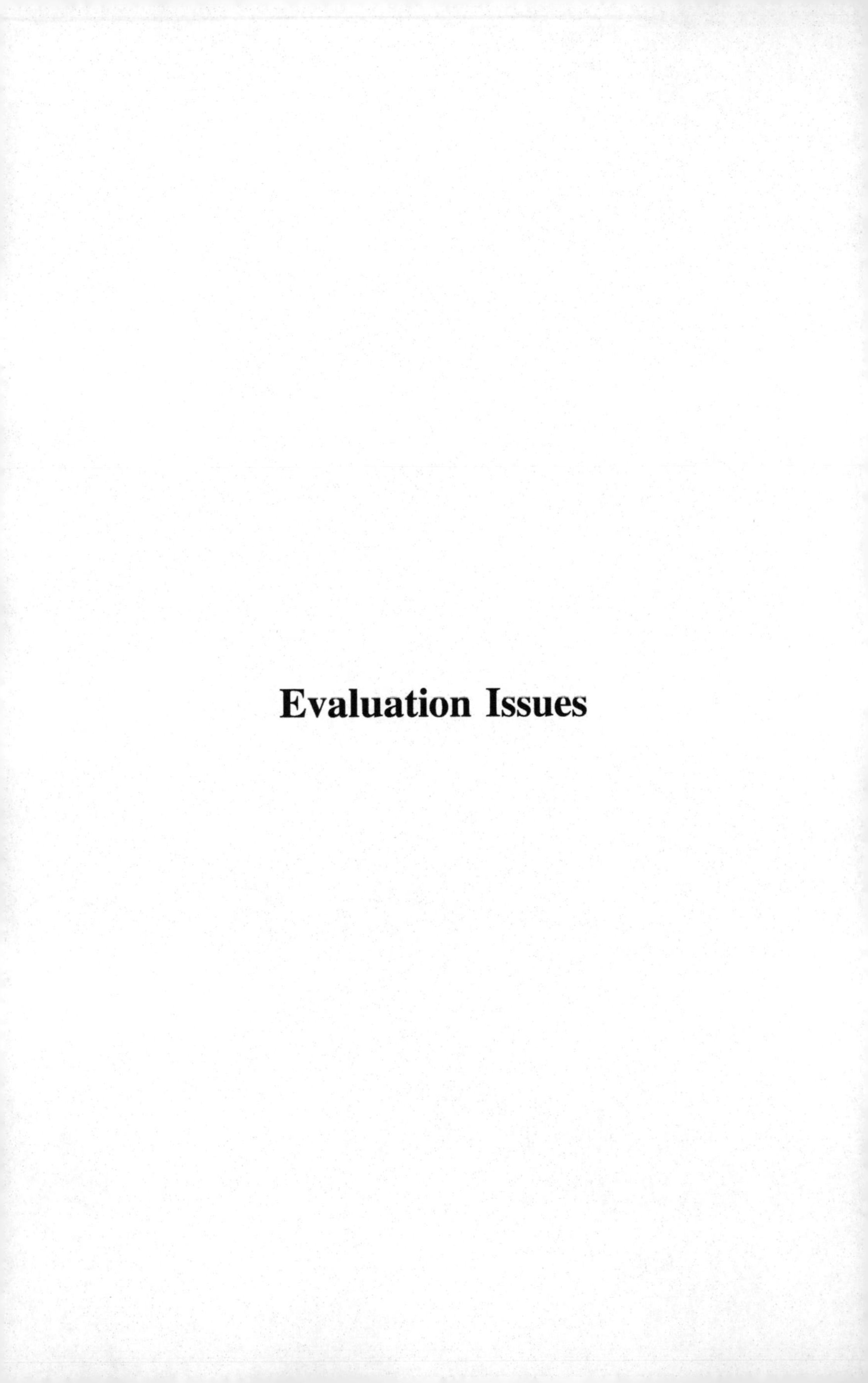

Evaluation Issues

The Curious Incidence of Bias Corrections in the Pool

Aldo Lipani(✉), Mihai Lupu, and Allan Hanbury

Institute of Software Technology and Interactive Systems (ISIS),
Vienna University of Technology, Vienna, Austria
{Lipani,Lupu,Hanbury}@ifs.tuwien.ac.at

Abstract. Recently, it has been discovered that it is possible to mitigate the Pool Bias of Precision at cut-off ($P@n$) when used with the fixed-depth pooling strategy, by measuring the effect of the tested run against the pooled runs. In this paper we extend this analysis and test the existing methods on different pooling strategies, simulated on a selection of 12 TREC test collections. We observe how the different methodologies to correct the pool bias behave, and provide guidelines about which pooling strategy should be chosen.

1 Introduction

An important issue in Information Retrieval (IR) is the offline evaluation of IR systems. Since the first Cranfield experiments in the 60s, the evaluation has been performed with the support of test collections. A test collection is composed of: a collection of documents, a set of topics, and a set of relevance assessments for each topic. Ideally, for each topic all the documents of the test collection should be judged, but due to the dimension of the collection of documents, and their exponential growth over the years, this praxis soon became impractical. Therefore, already early in the IR history, this problem has been addressed through the use of the pooling method [11]. The pooling method requires a set of runs provided by a set of IR systems having as input the collection of documents and the set of topics. Given these runs, the original pooling method consists, per topic, of: (1) collecting all the top n retrieved documents from each selected run in a so-called *pool*; (2) generating relevance judgments for each document in the pool. The benefit of this method is a drastic reduction of the number of documents to be judged, quantity regulated via the number d of documents selected. The aim of the pooling method, as pointed out by Spärck Jones, is to find an unbiased sample of relevant documents [6]. The bias can be minimized via increasing either the number of topics, or the number of pooled documents, or the number and variety of IR systems involved in the process. But albeit the first two are controllable parameters that largely depend on the budget invested in the creation of the test collection, the third, the number and

This research was partly funded by the Austrian Science Fund (FWF) project number P25905-N23 (ADmIRE).

N. Ferro et al. (Eds.): ECIR 2016, LNCS 9626, pp. 267–279, 2016.
DOI: 10.1007/978-3-319-30671-1_20

variety of the involved IR systems depends on the interest and participation of the IR community in the issued challenge.

In IR the need for more understandable metrics for practitioners has already been pointed out [4,7]. This led, on the one hand, to the development of new evaluation measures that 'make sense' and, on the other hand, to step back and focus on simple metrics such as Precision. Additionally Precision at cut-off ($P@n$) is a cornerstone for more complex and sophisticated evaluation measures in IR. This is why this study focuses exclusively on $P@n$.

Herein, we study how the reduced pool and the two pool bias correctors [7,16] behave when used on different pool strategies and configurations. We measure the bias using the Mean Absolute Error (MAE) on three pooling strategies, fixed-depth pool, uniformly sampled pool, and stratified pool for various parameter values. We provide insights about the two pool bias corrector approaches.

The remainder of the paper is structured as follows: in Sect. 2 we provide a summary of the related work on pooling strategies and on pooling correction. In Sect. 3 we generalize the pooling strategies, look at the existing pool bias correction approaches, and analyze their properties relating them to the studied pooling strategies. Section 4 confirms the theoretical observations experimentally. Results are discussed in Sect. 5. We conclude in Sect. 6.

2 Related Work

This section is divided into two parts. First we consider the work done in correcting the pool bias for the evaluation measure $P@n$. Second we consider the work conducted on the pooling strategies themselves. We will not cover the extensive effort in creating new metrics that are less sensitive to the pool bias (the work done for Bpref [2], followed by the work done by Sakai on the condensed lists [9] or by Yilmaz et al. on the inferred metrics [17,18]).

2.1 Pool Bias Estimators

Webber and Park [16] attempted to correct the bias by computing the Mean Absolute Error (MAE) of each run when pooled and not pooled, for a given evaluation measure and test collection, to be added as correction. Their method follows the assumption that the scores produced by the runs are normally distributed, a probably incorrect but common assumption. Although the method was presented only on Rank-Biased Precision, they pointed out that similar results were obtained also with $P@n$.

We [7] attempted to correct the pool bias with a more complex algorithm that estimates the correction by measuring the effect of a tested run against the pooled runs. Our method makes use of information that comes from both non-relevant and non-judged documents. The method works under the assumption that, if the correction is triggered, the adjustment needed is proportional to the average gain of non-judged documents on the affected pooled runs.

2.2 Pooling Strategies

Pooling was already used in the first TREC, in 1992, 17 years after it was introduced by Spärck Jones and van Rijsbergen [11], on the discussion of building an 'ideal' test collection that would allow reusability. The algorithm [5] is described as follows: (1) divide each set of results into results for a given topic; then, for each topic: (2) select the top 200 (subsequently generalized to d) ranked documents of each run, for input to the pool; (3) merge results from all runs; (4) sort results on document identifiers; (5) remove duplicate documents. This strategy is known as *fixed-depth pool*.

With the aim of further reducing the cost of building a test collection, Buckley and Voorhees [2] explored the uniformly sampled pool. At the time they observed that $P@n$ had the most rapid deterioration compared to a fully judged pool. The poor behavior of this strategy for top-heavy metrics was confirmed recently in Voorhees's [14] short comparison on pooling methods.

Another strategy is the stratified pool [18], a generalization of both the fixed-depth pool and the uniformly sampled pool. The stratified pool consists in layering the pool in different strata based on the highest rank obtained by a document in any of the given runs.

A comparison of the various pooling strategies has been recently reported by Voorhees [14]. We complement that report in several directions: First, and most importantly, we focus on bias correction methods and the effects of the pooling strategies on them rather than on the metrics themselves. Second, we generalize the stratified sampling method. Third, we expand the observations from 2 to 12 test collections. We also observe that the previous study does not distinguish between the effect of the number of documents evaluated with the effect of the different strategies (see Table 1 in [14]). In our generalization of the stratified pooling strategy we will ensure that the expected[1] number of judged documents is constant across different strategies.

3 Background Analysis

Here, the pooling method and its strategies are explained. Then the work conducted on the pool bias correction for the evaluation measure $P@n$ is analyzed. In this section, to simplify the notation, the average $P@n$ over the topics is denoted by g.

3.1 The Pooling Method

Three common strategies are used in the pooling method, listed in increasing order of generality: fixed-depth pool ($Depth@d$), uniformly sampled pool ($SampledDepth@d\&r$) and stratified ($Stratified$).

[1] Obviously, a guarantee on the actual number of judged documents cannot be provided without an a posteriori change in the sampling rates.

The simplest pooling strategy is *Depth@d*, which has been already described above. *SampledDepth@d&r* uses the *Depth@d* algorithm as an intermediary step. It produces a new pool by sampling without replacement from the resulting set at a given rate r. Obviously, if $r = 1$ the two strategies are equivalent. The *Stratified* further generalizes the pooling strategy, introducing the concepts of stratification and stratum. A stratification is a list of n strata, with sizes s_i and sample rates r_i: $z^n = [(s_0, r_0), ..., (s_n, r_n)]$. A stratum is a set of documents retrieved by a set of runs on a given range of rank positions. Which rank range ρ of the stratum j is: if $j = 1$ then $1 \leq \rho \leq s_1$ else if $j > 1$ then $\sum_{i=1}^{j-1} s_i < \rho \leq \sum_{i=1}^{j} s_i$. In this strategy, given a stratification z^n, we distinguish three phases: (1) pre-pooling: each document of each run is collected in a stratum based on its rank; (2) purification: for each stratum all the documents found on a higher rank stratum get removed; (3) sampling: each stratum is sampled without replacement based on its sample rate. Obviously, when the stratification is composed by only one stratum, it boils down to *SampledDepth@d&r*.

Which strategy to choose is not clear and sometimes it depends on the domain of study. Generally, the *Depth@d* is preferred because of its widespread use in the IR community, but for recall oriented domains the *Stratified* is preferred because of its ability to go deeper in the pool without explosively increasing the number of documents to be judged. The *SampledDepth@d&r* is generally neglected due to its lack in ability to confidently compare the performance of two systems, especially when used with top-heavy evaluation measures.

The main factor under the control of the test collection builder is the number of judged documents. This number depends both on the number of pooled runs and on the minimum number of judged documents per run. The following inequality shows the relation between these two components:

$$g(r, Q_{d+1}^{R_p}) - g(r, Q_d^{R_p}) \geq g(r, Q_{d+1}^{R_p \setminus \{r_p\}}) - g(r, Q_d^{R_p \setminus \{r_p\}}) \quad (1)$$

where r is a run, R_p is the set of runs used on the construction of the pool Q, $r_p \in R_p$, d is the minimum number of documents judged per run, and $g(r, Q)$ is the score of the run r evaluated on the pool Q. The proof is evident if we observe that: $Q_d^{R_p} \subseteq Q_{d+1}^{R_p}$, $Q_{d+1}^{R_p \setminus \{r_p\}} \subseteq Q_{d+1}^{R_p}$, $Q_d^{R_p \setminus \{r_p\}} \subseteq Q_{d+1}^{R_p \setminus \{r_p\}}$ and $Q_d^{R_p \setminus \{r_p\}} \subseteq Q_d^{R_p}$. When $r_p = r$, the inequality (Eq. 1) defines the *reduced pool* bias. In general however it shows that the bias is influenced by d, the minimum number of judged documents per run, and by $|R_p|$ the number of runs.

3.2 Pool Bias Correctors

Herein, we analyze the two pool bias correctors. Both attempt to calculate a coefficient of correction that is added to the biased score.

Webber and Park [16] present a method for the correction that computes the error introduced by the pooling method when one of the pooled runs is removed. This value is computed for each pooled run using a leave-one-out approach and then averaged and used as correction coefficient. Their correction coefficient for a run $r_s \notin R_p$ is the expectation:

$$\mathop{E}_{r_p \in R_p} \left[g(r_p, Q_d^{R_p}) - g(r_p, Q_d^{R_p \setminus \{r_p\}}) \right] \tag{2}$$

where R_p is the set of pooled runs, $r_p \in R_p$ and $Q_d^{R_p}$ is a pool constructed with d documents per each run in R_p. As done in a previous study [7] we evaluate the method using the mean absolute error (MAE). Equation 2 is simple enough that we can attempt to analytically observe how the method behaves with respect to the reduced pool, in the context of a *Depth@d* pool at varying d. We identify analytically a theoretical limitation of the Webber approach when used with a *Depth@d*. The maximum benefit, in expectation, is obtained when the cut-off value of the precision (n) is less or equal to d. After this threshold the benefit is lost.

We start analyzing the absolute error (AE) of the Webber approach for a run r_s:

$$\left| g(r_s, G) - \left[g(r_s, Q_d^{R_p}) + \mathop{E}_{r_p \in R_p} \left[g(r_p, Q_d^{R_p}) - g(r_p, Q_d^{R_p \setminus \{r_p\}}) \right] \right] \right|$$

where G is ground truth[2], $Q_d^{R_p}$ is the pool constructed using a *Depth@d* strategy where d is its depth and R_p is the set of pooled runs. We compare it to the absolute error of the reduced pool:

$$\left| g(r_s, G) - g(r_s, Q_d^{R_p}) \right| \tag{3}$$

We observe that when the depth of the pool d becomes greater or equal than n, $g(r_p, Q_d^{R_p})$ becomes constant. For the sake of clarity we substitute it with C_n. We substitute $g(r_s, G)$, which is also a constant, with C_G. Finally, we also rename the components $a(d) = g(r_s, Q_d^{R_p})$, $b(d) = E_{r_p \in R_p}[g(r_p, Q_d^{R_p \setminus \{r_p\}})]$, and call $f(d)$ the AE of the Webber method, and $h(d)$ the AE of the reduced pool:

$$f(d) = |C_G - [a(d) + C_n - b(d)]| \quad \text{and} \quad h(d) = |C_G - a(d)| \tag{4}$$

To study the behavior at varying of d, we define $\dot{g}$ as the finite difference of g with respect to d:

$$\dot{g}(r, Q_d^R) = g(r, Q_{d+1}^R) - g(r, Q_d^R) \tag{5}$$

We finitely differentiate the previous two equations, and since both are decreasing functions of d, to see where the margin between the two functions shrinks (the benefit decreases), it is sufficient to study when the inequality $\dot{f}(d) \geq \dot{g}(d)$ holds.

$$\dot{f}(d) = \begin{cases} -\dot{a}(d) + \dot{b}(d), & \text{if } C_G - [a(d) + C_n - b(d)] \geq 0 \\ \dot{a}(d) - \dot{b}(d), & \text{if } C_G - [a(d) + C_n - b(d)] < 0 \end{cases} \quad \text{and} \quad \dot{h}(d) = -\dot{a}(d)$$

[2] The ground truth is the pool using the maximum depth available in the test collection.

Therefore,

$$\dot{f}(d) \geq \dot{h}(d) \text{ iff } \begin{cases} \dot{b}(d) \geq 0, & \text{if } C_G - [a(d) + C_n - b(d)] \geq 0 \\ 2\dot{a}(d) \geq \dot{b}(d), & \text{if } C_G - [a(d) + C_n - b(d)] < 0 \end{cases}$$

While the first condition is always verified ($\dot{b}(d)$ is an average of positive quantities), the second tells us that if $\dot{b}(d)$ is less or equal to $2\dot{a}(d)$ the Webber method decreases more slowly than the reduced pool. This inequality, as a function of r_s does not say anything about its behavior as it can be different for each r_s. Therefore we study the MAE using its expectation. We define R_G as the set of runs of the ground truth G, in which $R_p \subset R_G$. Using the law of total expectation we can write:

$$\begin{aligned} \underset{r_s \in R_G}{E}[\dot{b}(d)] &= \underset{r_s \in R_G}{E}[\underset{r_p \in R_G \setminus \{r_s\}}{E}[g(r_p, Q_{d+1}^{R_G \setminus \{r_s, r_p\}}) - g(r_p, Q_d^{R_G \setminus \{r_s, r_p\}})]] = \\ &= \underset{r_{s_1}, r_{s_2} \in R_G : r_{s_1} \neq r_{s_2}}{E}[g(r_{s_1}, Q_{d+1}^{R_G \setminus \{r_{s_1}, r_{s_2}\}}) - g(r_{s_1}, Q_d^{R_G \setminus \{r_{s_1}, r_{s_2}\}})] \end{aligned} \quad (6)$$

Using the pool inequality in Eq. 1:

$$\begin{aligned} &\underset{r_{s_1}, r_{s_2} \in R_G : r_{s_1} \neq r_{s_2}}{E}[g(r_{s_1}, Q_{d+1}^{R_G \setminus \{r_{s_1}, r_{s_2}\}}) - g(r_{s_1}, Q_d^{R_G \setminus \{r_{s_1}, r_{s_2}\}})] \leq \\ &\leq \underset{r_{s_1} \in R_G}{E}[g(r_{s_1}, Q_{d+1}^{R_G \setminus \{r_{s_1}\}}) - g(r_{s_1}, Q_d^{R_G \setminus \{r_{s_1}\}})] = \underset{r_s \in R_G}{E}[\dot{a}(d)] \leq \underset{r_s \in R_G}{E}[2\dot{a}(d)] \end{aligned} \quad (7)$$

Therefore, in expectation, at increasing of depth of the pool d, for $P@n$ with $n \geq d$, the MAE of the Webber approach decreases more slowly than the MAE of the reduced pool.

We [7] introduced a method that attempts to correct the bias by measuring the effect of a run on the pooled runs, in terms of difference between the number of relevant, non-judged, and non-relevant documents. This information, averaged among all the pooled runs, in combination with the measurements made on the run itself, is used, first as a trigger to perform the correction and second as correction. Each pooled run r_p is effected using a merging function that averages the rank of all the shared documents between r_p itself and the selected run r_s, then uses the resulting average as a score to create a new reordered run r'_p. The trigger function is as follows, where ΔP_{r_s} and $\Delta \overline{P}_{r_s}$ is the average gain in precision and anti-precision (ratio of non-relevant documents) of the affected pooled runs.

$$\lambda = \hbar_{r_s}(\Delta P_{r_s} \overline{P}_{r_s} - \Delta \overline{P}_{r_s} P_{r_s})$$

For $\lambda > 0$ the correction is triggered, and the following correction added, where $\Delta \hbar_{r_s}$ is the gain on ratio of non-judged documents over n of the modified pooled runs, and $\hbar_{r_s}$ is the ratio of non-judged documents over n in the run to correct:

$$\hbar_{r_s} \cdot \max(\Delta \hbar_{r_s}, 0)$$

We observe that there exists a confounding factor that is the proportion of judged relevant to non-relevant documents. Assuming that all runs are ranked by some probability of relevance, i.e. that there is a higher probability to find relevant documents at the top than at the bottom of the runs, our approach (Lipani) is sensitive to the depth of the pool because at any one moment it compares one run, that is a set of d probably relevant documents and $|r_s| - d$ probably non-relevant documents with all the existing runs, i.e. a set of $d|R_p|$ probably relevant documents and $(E[|r_p|] - d)|R_p|$ probably non-relevant documents. The effects of this aggregation are difficult to formalize in terms of the proportion of relevant and non-relevant documents, and we explore them experimentally in the next section.

4 Experiments

To observe how the pool and the two pool bias correctors work in different contexts we used a set of 12 TREC test collections, sampled from different tracks: 8 from Ad Hoc, 2 from Web, 1 from Robust and 1 from Genomics. In order to make possible the simulation of the different pooling strategies, the test collections needed to have been built using a *Depth@d* strategy with depth $d \geq 50$.

For *Depth@d* and *SampledDepth@d&r*, all the possible combinations of parameters with a step size of 10 have been explored. Figure 1 shows the MAE of the different methods, for *Depth@d* at varying d. Figure 2 shows the MAE of the different methods for the *SampledDepth@d&r*, with fixed depth $d = 50$, at varying sample rate r from 10 % to 90 % in steps of 10.

For *Stratified*, due to its more flexible nature, we constrained the generation of the stratifications. We should note that there are practically no guidelines in the literature on how to define the strata. First, we defined the sizes of the strata for each possible stratification and then for each stratification we defined the sample rates of each stratum.

Given n, the number of strata to generate, and $s \in S$, a possible stratum's size, we find all the vectors of size n, $s^n = (s_0\, s_1 \ldots s_n)$ such that $\sum_{i=1}^{n} s_i = D$, where D is the maximum depth of the pool available, with sizes s_i chosen in increasing order, except for the last stratum, which may be a residual smaller than the second-last. For each $n \in \mathbb{N}^+$, and constraining the set of stratum sizes S to multiples of 10, when $D = 100$, we find only ten possible solutions.

To find the sample rates r_i to associate to each stratum s_i, we followed a more elaborated procedure. As pointed out by Voorhees [14], the best results are obtained fixing the sample rate of the first stratum to 100 %. From the second to the last stratum, when available, we sample keeping the expected minimal number of pooled documents for each run constant. This is done in order to allow a cross-comparison among the stratifications. However, for stratifications composed by 3 or more strata some other constraint is required. The TREC practice has shown that the sampling rate decreases fast, but so far decisions in this sense are very ad-hoc. Trying to understand how fast the rate should drop, we are led back to studies relating retrieval status values (RSV), i.e. scores, with probabilities of relevance. Intuitively, we would want our sampling rate to be related to the latter. Nottelmann and Fuhr [8] pointed out that mapping the RSV to the probability of relevance using a logistic function outperforms the mapping when a linear function is used. Therefore, to create the sampling rates, we define a logistic function with parameters $b_1 = 10/D$, $b_0 = D/2$ where D is the depth of the original pool (i.e. of the ground truth). b_1 defines the slope of the logistic function and is in this case arbitrary. b_0 is the minimal number of documents we want, on expectation, to assess per run. The sample rates are then the areas under the logistic curve for each stratum (Eq. 9). However, since, to keep in line with practice, we always force the first strata to sample at 100 %, we correct the remaining sampling rates proportionally (Eq. 10). To verify that the expected minimal number of sampled documents is b_0, it is enough to observe that the sum of the areas that define the sampling rates is b_0 (Eq. 8). The resulting stratifications are listed in Table 1 and the corresponding MAEs for the different methods are shown in Fig. 3.

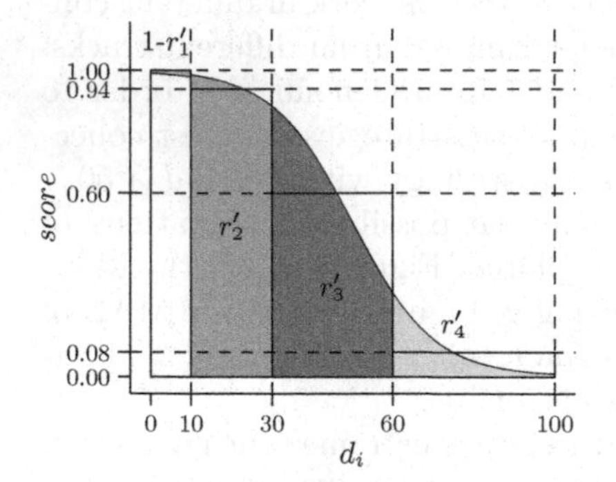

$$\int_0^D \frac{1}{1+e^{b_1(x-b_0)}}dx = b_0 \tag{8}$$

$$r_n' = \int_{\sum_{i=1}^{n-1} s_i}^{\sum_{i=1}^{n} s_i} \frac{1}{1+e^{b_1(x-b_0)}}dx \tag{9}$$

$$r_n = \frac{1}{s_n}\left[r_n' - (1-r_1')\frac{r_n'}{\sum_{i=2}^{N} r_i'}\right] \tag{10}$$

Table 1. List of the used stratifications z_i^n, where n is the size of the stratification, and i is the index of the solution found given the fixed constraints. $E[d]$ is the mean number of judged documents per run for all the test collections with respect to z_1^1. † indicates when the difference with respect to the previous stratification is statistically significant (t-test, $p < 0.05$).

n	i	z_i^n	$(s_1, r_1)...(s_n, r_n)$	E[d]
1	1	z_1^1	(100, 50)	50.00
2	1	z_1^2	(10, 100) (90, 44)	50.75†
	2	z_2^2	(20, 100) (80, 38)	52.11†
	3	z_3^2	(30, 100) (70, 29)	52.29†
	4	z_4^2	(40, 100) (60, 17)	52.45

n	i	z_i^n	$(s_1, r_1)...(s_n, r_n)$	E[d]
3	1	z_1^3	(10, 100) (20, 94) (70, 30)	51.71†
	2	z_2^3	(10, 100) (30, 90) (60, 22)	52.29†
	3	z_3^3	(10, 100) (40, 83) (50, 14)	52.32
	4	z_4^3	(20, 100) (30, 77) (50, 14)	52.37
4	1	z_1^4	(10, 100) (20, 94) (30, 60) (40, 8)	52.25

To measure the bias of the reduced pool and the two correcting approaches, we run a simulation[3] using only the pooled runs and a leave-one-organization-out approach, as done in previous studies [3,7]. The leave-one-organization-out approach consists in rebuilding the pool removing in sequence all the runs submitted

[3] The software is available on the website of the first author.

by an organization. Finally as performed in previous studies [1,7,10,12,13,15], to avoid buggy implementations of systems, the bottom 25 % of poorly performing runs are removed for each test collection.

5 Discussion

In case of *Depth@d* in Fig. 1 we observe that, as expected given the analytical observations of Sect. 3.2, the Webber approach slows down its correction with increasing depth d. The ratio between the error produced by the reduced pool and the method decreases systematically after d becomes greater than the cut-off value n of $P@n$. This trend sometimes leads to an inversion, as for Ad Hoc 2, 3, 6, 7 and 8, Web 9 and 10 and Robust 14. The Lipani approach, as expected, is less reliable when the depth d of the pool is less than the cut-off value of the precision. We see that very clearly in Web 9 and Web 11. It generally reaches a peak when d and n are equal, and then improves again. Comparing the two approaches, we see that in the majority of the cases the Lipani approach does better than the Webber approach.

The *SampledDepth@d&r*, shown in Fig. 2, does not display the same effects observed for the *Depth@d*. Both corrections do better than the reduced pool. The effect in Webber's method disappears because in this case (and also in the *Stratified* pool later) the constant C_n in Eq. 4 is no longer a constant here, even for $n > d$. The effect observed in Lipani's method is removed by the more non-relevant documents introduced on the top of the pooled runs, which reduce the effect of the selected run. The Lipani approach does generally better, sometimes with high margin as in: Ad Hoc 5 and 8, Web 9, 10 and 11, and Genomics 14.

Finally, in the *Stratified* case, Fig. 3, the effects observed on the *Depth@d* are also not visible. For $P@10$, the corrections perform much better if we sample more from the top, most notably for the stratifications of size 3, but the correction degrades when using $P@30$. This is particularly visible when comparing z_3^3 and z_4^3. Although they have essentially the same number of judged documents (with difference non statistically significant, Table 1), the stratification with a deeper first stratum makes a big difference in performance. Comparing the best stratification of size 2 (z_4^2) and the best stratification of size 3 (z_1^3) we observe that there is only a small difference in performance between them that could be justified by the smaller number of judged documents (with difference statistically significant, Table 1). The z_4^2 is the best overall stratification, confirming also the conclusion of Voorhees [14]. However a cheaper solution is the z_1^3, which, as shown in Table 1, evaluates fewer documents, but obtains a comparably low MAE.

Cross-comparing the three pooling strategies (observe the ranges on the y-scales), we see that the best performing strategy, fixing the number of judged documents, is *Depth@d*, then the *Stratified* and *SampledDepth@d&r*.

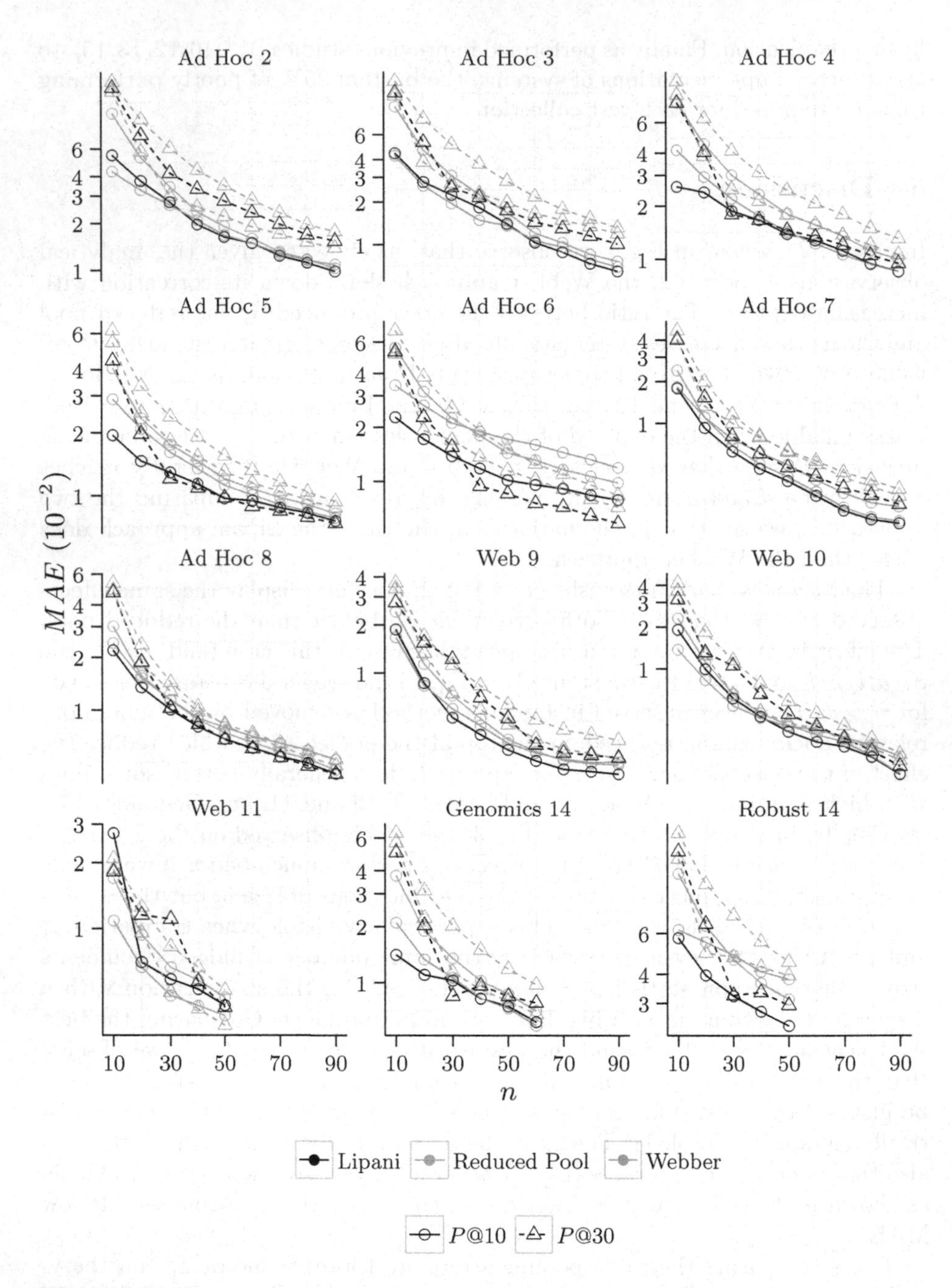

Fig. 1. MAE in logarithm scale of the ground truth (Depth@M, where M is the maximum depth of the test collection) against the *Depth@d* pool at varying of the depth n, for the evaluation measures, $P@10$ and $P@30$. MAE computed using the leave-one-organization out approach of pooled runs and removing the bottom 25 % poorly performing runs.

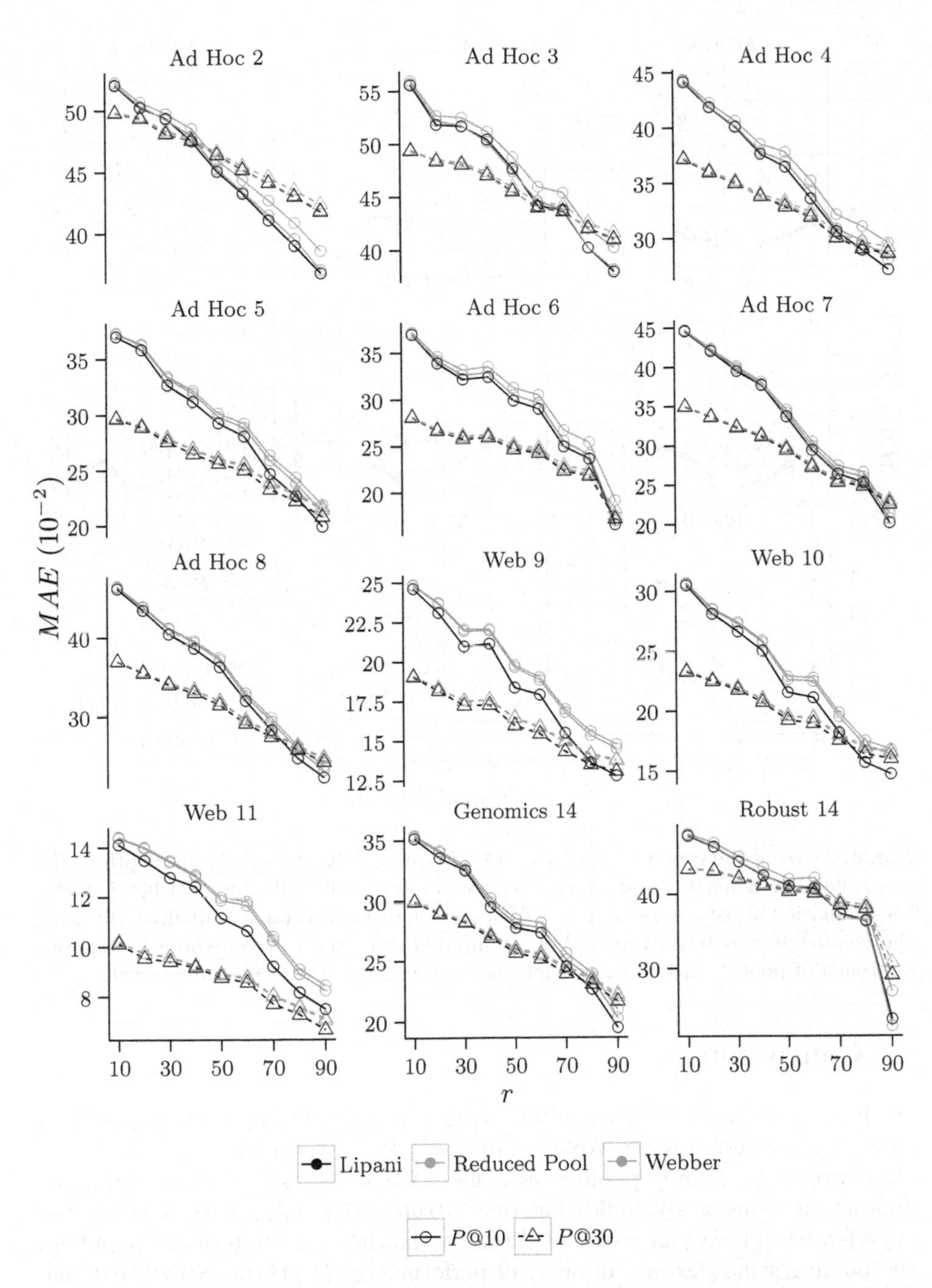

Fig. 2. MAE of the ground truth (Depth@M, where M is the maximum depth of the test collection) against the *SampledDepth@d&r* with fixed $d = 50$ at varying of the sample rate r, for the evaluation measures, $P@10$ and $P@30$. MAE computed using the leave-one-organization out approach of pooled runs and removing the bottom 25 % poorly performing runs.

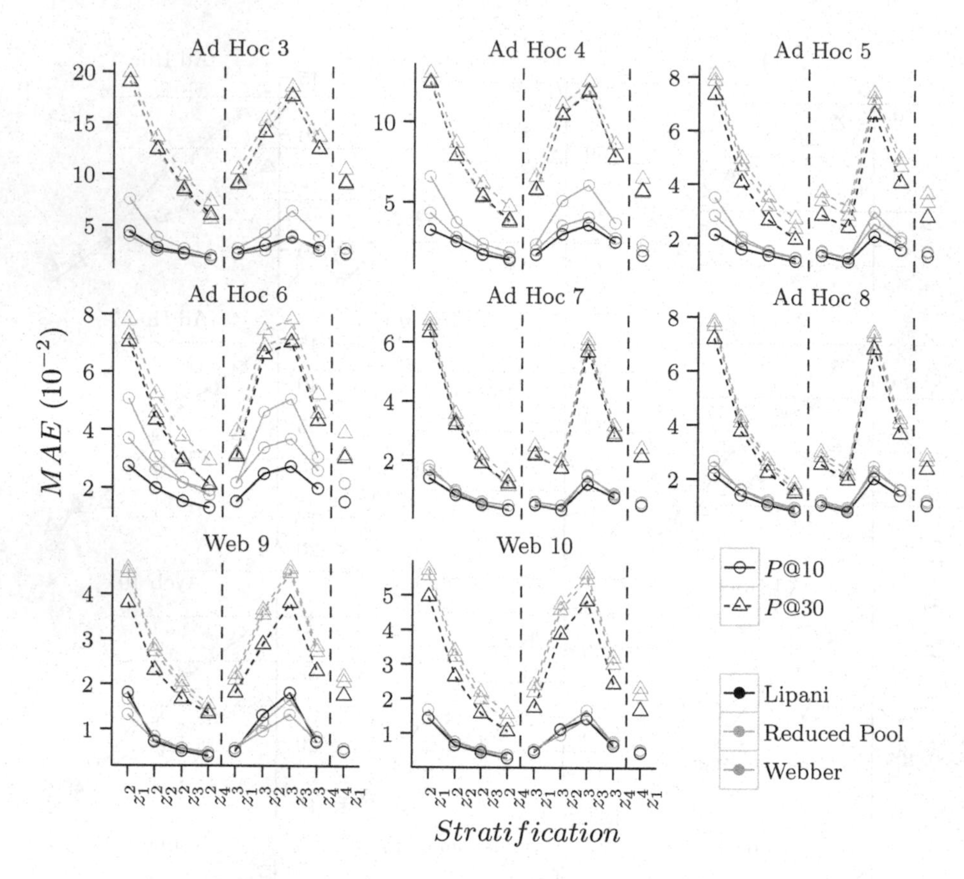

Fig. 3. MAE of the ground truth (*Depth@M*, where M is the maximum depth of the test collection) against the *Stratified* pool at varying of the different stratifications, for the evaluation measures, *P*@10 and *P*@30, only on test collections originally built using the *Depth*@100 pooling strategy. MAE computed using the leave-one-organization out approach of pooled runs and removing the bottom 25 % poorly performing runs.

6 Conclusion

We have confirmed a previous study [7] that the Lipani approach to pool bias correction outperforms the Webber approach. We have further expanded these observations to various pooling strategies. We have also partially confirmed another previous study indicating that *Stratified* pooling with a heavy top is preferable [14]. We have extended this by showing that, in terms of pool bias, the pooling strategies are, in order of performance, *Depth@d*, *Stratified*, and *SampledDepth@d&r*. Additionally, we made two significant observations on the two existing pool bias correction methods. We have shown, analytically and experimentally, that the Webber approach reduces its ability to correct the runs at increasing pool depth, when this is greater than the cut-off of the measured precision. Conversely, the Lipani approach sometimes manifests an instability when the depth of the pool is smaller than the cut-off of the measured precision. These opposite behaviors would make the Lipani estimator a better choice since it improves with increasing number of judged documents. Both of these side-effects are reduced when a sampled strategy is used.

References

1. Bodoff, D., Li, P.: Test theory for assessing ir test collections. In: Proceedings of SIGIR (2007)
2. Buckley, C., Voorhees, E.M.: Retrieval evaluation with incomplete information. In: Proceedings of SIGIR (2004)
3. Büttcher, S., Clarke, C.L.A., Yeung, P.C.K., Soboroff, I.: Reliable information retrieval evaluation with incomplete and biased judgements. In: Proceedings of SIGIR (2007)
4. Clarke, C.L.A., Smucker, M.D.: Time well spent. In: Proceedings of IIiX (2014)
5. Harman, D.: Overview of the first trec conference. In: Proceedings of SIGIR (1993)
6. Jones, K.S.: Letter to the editor. Inf. Process. Manage. **39**(1), 156–159 (2003)
7. Lipani, A., Lupu, M., Hanbury, A.: Splitting water: precision and anti-precision to reduce pool bias. In: Proceedings of SIGIR (2015)
8. Nottelmann, H., Fuhr, N.: From retrieval status values to probabilities of relevance for advanced ir applications. Inf. Retr. **6**(3–4), 363–388 (2003)
9. Sakai, T.: Alternatives to bpref. In: Proceedings of SIGIR (2007)
10. Sanderson, M., Zobel, J.: Information retrieval system evaluation: effort, sensitivity, andreliability. In: Proceedings of SIGIR (2005)
11. Jones, K.S., van Rijsbergen, C.J.: Report on the need for and provision of an "ideal" information retrieval test collection. British Library Research and Development Report No. 5266 (1975)
12. Urbano, J., Marrero, M., Martín, D.: On the measurement of test collection reliability. In: Proceedings of SIGIR (2013)
13. Voorhees, E.M.: Topic set size redux. In: Proceedings of SIGIR (2009)
14. Voorhees, E.M.: The effect of sampling strategy on inferred measures. In: Proceedings of SIGIR (2014)
15. Voorhees, E.M., Buckley, C.: The effect of topic set size on retrieval experiment error. In: Proceedings of SIGIR (2002)
16. Webber, W., Park, L.A.F.: Score adjustment for correction of pooling bias. In: Proceedings of SIGIR (2009)
17. Yilmaz, E., Aslam, J.A.: Estimating average precision with incomplete and imperfect judgments. In: Proceedings of CIKM (2006)
18. Yilmaz, E., Kanoulas, E., Aslam, J.A.: A simple and efficient sampling method for estimating AP and NDCG. In: Procedings of SIGIR (2008)

Understandability Biased Evaluation for Information Retrieval

Guido Zuccon(✉)

Queensland University of Technology (QUT), Brisbane, Australia
g.zuccon@qut.edu.au

Abstract. Although relevance is known to be a multidimensional concept, information retrieval measures mainly consider one dimension of relevance: topicality. In this paper we propose a method to integrate multiple dimensions of relevance in the evaluation of information retrieval systems. This is done within the gain-discount evaluation framework, which underlies measures like rank-biased precision (RBP), cumulative gain, and expected reciprocal rank. Albeit the proposal is general and applicable to any dimension of relevance, we study specific instantiations of the approach in the context of evaluating retrieval systems with respect to both the topicality and the understandability of retrieved documents. This leads to the formulation of understandability biased evaluation measures based on RBP. We study these measures using both simulated experiments and real human assessments. The findings show that considering both understandability and topicality in the evaluation of retrieval systems leads to claims about system effectiveness that differ from those obtained when considering topicality alone.

1 Introduction

Traditional information retrieval (IR) evaluation relies on the assessment of topical relevance: a document is topically relevant to a query if it is assessed to be on the topic expressed by the query. The Cranfield paradigm and its subsequent incarnations into many of the TREC, CLEF, NTCIR or FIRE evaluation campaigns have used this notion of relevance, as reflected by the collected relevance assessments and the retrieval systems evaluation measures, e.g., precision and average precision, recall, bpref, RBP, and graded measures such as discounted cumulative gain (DCG) and expected reciprocal rank (ERR).

Relevance is a complex concept and the nature of relevance has been widely studied [16]. A shared agreement has emerged that relevance is a multidimensional concept, with topicality being only one of the factors (or criteria) influencing the relevance of a document to a query [8,28]. Among others, core factors that influence relevance beyond topicality are: scope, novelty, reliability and understandability [28]. However, these factors are often not reflected in the evaluation framework used to measure the effectiveness of retrieval systems.

In this paper, we aim to develop a general evaluation framework for information retrieval that extends the existing one by considering the multidimensional

N. Ferro et al. (Eds.): ECIR 2016, LNCS 9626, pp. 280–292, 2016.
DOI: 10.1007/978-3-319-30671-1_21

nature of relevance. This is achieved by considering the gain-discount framework synthesised by Carterette [4]; this framework encompasses the widely-used DCG, RBP and ERR measures. Specifically, we focus on a particular dimension of relevance, understandability, and devise a family of measures that evaluate IR systems by taking into account both topicality and understandability. While the developed framework is general and could be used to model other factors of relevance, there are a number of compelling motivations for focusing on an extension to understandability only:

- even if a document is topically relevant, it is of no use to a user if it cannot be understood at all;
- understandability is a key factor when assessing relevance in many domain-specific scenarios, e.g., consumer health search [1,12,13,26,27];
- resources exist that allow us to assess the impact of evaluating multidimensional relevance when considering understandability, both through simulations and explicit human assessments of understandability.

Specifically, we aim to answer the following research questions: (RQ1) How can relevance dimensions (and specifically understandability) be integrated within IR evaluation? (RQ2) What is the impact of understandability biased measures on the evaluation of IR systems?

2 Related Work

Research on document relevance has shown that users' relevance assessments are affected by a number of factors beyond topicality, although topicality has been found to be the essential relevance criteria. Chamber and Eisenberg have synthesised four families of approaches for modelling relevance, highlighting its multidimensional nature [24]. Cosijn and Ingwersen investigated manifestations of relevance such as algorithmic, topical, cognitive, situational and socio-cognitive, and identified relation, intention, context, inference and interaction as the key attributes of relevance [8]. Note that relevance manifestations and attributes in that work are different from what we refer to as factors of relevance in this paper. Similarly, the dimensions described by Saracevic [23], which are related to those of Cosijn and Ingwersen mentioned above, differ in nature from the factors or dimensions of relevance we consider in this paper.

The actual factors that influence relevance vary across studies. Rees and Schulz [20] and Cuadra and Katter [9] identified 40 and 38 factors respectively. Xu and Chen proposed and validated a five-factor model of relevance which consists of novelty, reliability, understandability, scope, along with topicality [28]. Zhang et al. have further validated such model [33]. Their empirical findings highlight the importance of understandability, reliability and novelty along with topicality in the relevance judgements they collected. Barry also explored factors of relevance beyond topicality [2]; of relevance to this work is that these user experiments highlighted that criteria pertaining to user's experience and background, including the ability to understand the retrieved information, influence relevance assessments. Mizzaro offered a comprehensive account of previous work attempting to define and research relevance [16].

While dimensions of relevance are often ignored in the evaluation of IR systems, notable exceptions do exists. The evaluation of systems that promote the novelty and diversity of the retrieved information, for example, required the development of measures that account for both the topicality and novelty dimensions. This need has been satisfied by fragmenting the information need into subtopics, or nuggets, and evaluating the systems against relevance assessments performed explicitly for each of the subtopics of the query. This approach has lead to the formulation of measures such as subtopic recall and precision [31], α-nDCG [6], and D#-measures [22], among others [5]. Nevertheless, the formulation of novelty and diversity measures differ from that of the measures proposed in this paper because we combine the gains achieved from different dimensions of relevance, rather than summing gains contributed by the different subtopics.

In this paper, the integration of understandability within IR evaluation proposed is cast within the gain-discount framework [4]. This framework generalises the common structure of many evaluation measures, which often involve a sum over the product of a gain function, mapping relevance assessments to gain values, and a discount function, that serves to modulate the gain by a discount based on the rank position at which the gain is achieved.

Previous work on quantifying the importance of understandability when evaluating IR systems has also used the gain-discount framework and simulations akin to those of Sect. 5, although at a smaller scale [34]. That work motivated us to further develop multidimensional based evaluation of IR systems and specifically to further investigate evaluation measures that account for both relevance and understandability assessments.

3 Gain-Discount Framework

In the gain-discount framework [4] the effectiveness of a system, conveyed by a ranked list of documents, is measured by the evaluation measure M, defined as:

$$M = \frac{1}{\mathcal{N}} \sum_{k=1}^{K} d(k) g(d@k) \tag{1}$$

where $g(d@k)$ and $d(k)$ are respectively the gain function computed for the (relevance of the) document at rank k and the discount function computed for the rank k, K is the depth of assessment at which the measure is evaluated, and $1/\mathcal{N}$ is a (optional) normalisation factor, which serves to bound the value of the sum into the range [0,1] (see also [25]).

Without loss of generality, we can express the gain provided by a document at rank k as a function of its probability of relevance; for simplicity we shall write $g(d@k) = f(P(R|d@k))$, where $P(R|d@k)$ is the probability of relevance given the document at k. A similar form has been used for the definition of the gain function for time-biased evaluation measures [25]. Measures like RBP, nDCG and ERR can still be modelled in this context, where their differences with respect to $g(d@k)$ reflect on different $f(.)$ functions being applied to the estimations of $P(R|d@k)$.

Different measures within the gain-discount framework use different functions for computing gains and discounts. Often in RBP the gain function is binary-valued[1] (i.e., $g(d@k) = 1$ if the document at k is relevant, $g(d@k) = 0$ otherwise); while for nDCG $g(d@k) = 2^{P(R|d@k)} - 1$ and for ERR[2] $g(d@k) = (2^{P(R|d@k)} - 1)/2^{max(P(R|d))}$. The discount function in RBP is modelled by $d(k) = \rho^{k-1}$, where $\rho \in [0, 1]$ reflects user behaviour[3]; while in nDCG the discount function is given by $d(k) = 1/(\log_2(1 + k))$ and in ERR by $d(k) = 1/k$.

When only the topical dimension of relevance is modelled, as is in most retrieval systems evaluations, then $P(R|d@k) = P(T|d@k)$, i.e., the probability that the document at k is topically relevant (to a query). This probability is 1 for relevant and 0 for non-relevant documents, when considering binary relevance; it can be seen as the values of the corresponding relevance levels when applied to graded relevance.

4 Integrating Understandability

To integrate different dimensions of relevance in evaluation measures, we model the probability of relevance $P(R|d@k)$ as the joint distribution over all considered dimensions $P(D_1, \cdots, D_n|d@k)$, where each D_i represents a dimension of relevance, e.g., topicality, understandability, reliability, etc.

To compute the joint probability we assume that dimensions are compositional events and their probabilities independent, i.e., $P(D_1, \cdots, D_n|d@k) = \prod_{i=1}^{n} P(D_i|d@k)$. These are strong assumptions and are not always true. Eisenberg and Barry [10] highlighted that user judgements of document relevance are affected by order relationships, and proposals to model these dynamics have recently emerged, for example see Bruza et al. [3]. Nevertheless, Zhang et al. used crowdsourcing to prime a psychometric framework for multidimensional relevance modelling, where the relevance dimensions are assumed compositional and independent [33]. While the above assumptions are unrealistic and somewhat limitative, note that similar assumptions are common in information retrieval. For example, the Probability Ranking Principle assumes that relevance assessments are independent [21].

Following the assumptions above, the gain function with respect to different dimensions of relevance can be expressed in the gain-discount framework as:

$$g(d@k) = f(P(R|d@k)) = f\big(P(D_1, \cdots, D_n|d@k)\big) = f\Big(\prod_{i=1}^{n} P(D_i|d@k)\Big)$$

[1] Although there is no requirement for this to be the case and RBP can be used for graded relevance [17].

[2] Where $P(R|d@k)$ captures either binary ($P(R|d@k)$ either 0 or 1) or graded relevance and $max(P(R|d))$ is the highest relevance grade, e.g., 1 in case of binary relevance.

[3] High values representing persistent users, low values representing impatient users.

Evaluation measures developed within this framework would differ by means of the instantiations of $f\big(P(D_1, \cdots, D_n|d@k)\big)$, other than by which dimensions are modelled.

4.1 Understandability Biased Evaluation

In the remaining of this paper we investigate measures that limit the modelling of multidimensional relevance to only topicality, characterised by $P(T|d@k)$, and understandability, characterised by $P(U|d@k)$. In the following, $P(R|d@k)$ is thus modelled by the joint $P(T, U|d@k)$ that is in turn computed as the product $P(T|d@k)P(U|d@k)$ following the assumptions discussed above. This transforms the gain function into:

$$g(d@k) = f(P(R|d@k)) = f\big(P(T|d@k)P(U|d@k)\big) \tag{2}$$

For simplicity, we further assume that $f(.)$ satisfies the distributive property, such that $f\big(P(T|d@k)P(U|d@k)\big) = f\big(P(T|d@k)\big) \cdot f\big(P(U|d@k)\big)$; this is often the case for estimations of gain functions used in IR. For example, if the gain function used in RBP is applied as $f(.)$ to both topicality and understandability, then the equality above would be satisfied.

Next, we consider specific instantiations of a well-known IR measure, RBP, to the case of multidimensional relevance, and specifically when considering both topicality and understandability. Because topicality is a factor that is traditionally used to instantiate measures, we name the newly proposed measures as understandability biased, to highlight the fact that they model understandability, in addition to topicality, for evaluating the effectiveness of the systems. Nevertheless, the same approach can be applied to other dimensions of relevance.

Rank-biased precision (RBP) [17] is a well understood measure of retrieval effectiveness which fits within the gain-discount framework. In RBP, gain is measured by a function $r(d@k)$ which is 1 if $d@k$ is relevant and 0 otherwise; discount is measured by a geometric function of the rank, i.e., $d(k) = \rho^{k-1}$, and $1 - \rho$ acts as a normalisation component. Formally, RBP is defined as:

$$RBP = (1 - \rho) \sum_{k=1}^{K} \rho^{k-1} r(d@k) \tag{3}$$

Within the view presented in Sect. 3, $r(d@k)$ is an initialisation of $f(P(T|d@k))$, where $f(.)$ is the identity function and $r(d@k)$ estimates $P(T|d@k)$. To integrate understandability, we assume that $f(P(R|d@k)) = f(P(T|d@k) \cdot P(U|d@k))$ in line with Sect. 3, thus obtaining the understandability-biased RBP:

$$uRBP = (1-\rho) \sum_{k=1}^{K} \rho^{k-1} P(T|d@k) \cdot P(U|d@k) = (1-\rho) \sum_{k=1}^{K} \rho^{k-1} r(d@k) \cdot u(d@k) \tag{4}$$

where $r(d@k)$ is the function that transforms relevance values into the corresponding gains and $u(d@k)$ is the function that transforms understandability values into the corresponding gains.

This general expression for uRBP can be further instantiated by making explicit how the respective gain functions are computed. For example, $r(d@k)$ could be computed in the same way the corresponding function is computed in RBP. Similarly, the function responsible for translating understandability estimations into gains, i.e., $u(d@k)$, can be instantiated such that it returns a value of 1 if the document is assessed understandable ($P(U|d@k) = 1$) and a value of 0 if it is assessed as not understandable ($P(U|d@k) = 0$).

Alternatives may include collecting graded assessments of the understandability of documents, and associating different gains to different levels of understanding, akin to the use of graded relevance in measures like nDCG. This approach provides a graded variant of understandability-biased RBP, which we indicate as uRBPgr. Specifically, in Sect. 6 for uRBPgr we instantiate $u(d@k)$ as the function that provides a gain of 1 if $d@k$ is very easy to understand, 0.8 if it is somewhat easy to understand, 0.4 if it somewhat hard to understand and a gain of 0 (no gain) if it is very difficult to understand. Thus, if a document is very difficult (easy) to understand, its contribution to the value of uRBPgr would be 0 (1), regardless of the relevance of the document itself – this is in line with uRBP. However, when documents are somewhat easy or difficult to understand, the corresponding gains are used to modulate (or scale) the gains derived from the relevance assessments, in practice reducing these gains because of a partial lack in understandability.

5 Simulating Understandability Biased Evaluation

In the previous sections we have introduced a general framework for including understandability along with topicality in the evaluation of IR systems, and we have proposed instantiations of the framework based on the rank-biased precision measure (answering RQ1). Next, we aim to answer our second research question (RQ2): what is the impact of accounting for understandability in the evaluation of IR systems. To answer this, we instruct two empirical analyses.

The first analysis (Sect. 5) relies on simulations, where the understandability of documents is estimated using computational models of readability, which thus serves as a proxy to assess understandability. User understandability requirements are estimated using two (simple) user models. For this analysis we only consider the binary uRBP measure for brevity.

The second analysis (Sect. 6) relies on human provided assessments of understandability of documents, and considers both binary and graded uRBP.

Both analyses use the CLEF eHealth collection assembled to evaluate consumer health search tasks [12,13,18]. The collection consists of more than one million health related webpages. For the simulations we use the query topics distributed in 2013 and 2014 (for which there is no explicit understandability assessment) in addition to those distributed in 2015. Instead, for the experiments of Sect. 6 we use the query topics distributed in 2015 only as these come with explicit understandability assessments. Queries in this collection relate to the task of finding information about specific health conditions, treatments or diagnosis. We have chosen to study the impact of understandability biased evaluation using this collection because real-world tasks within consumer health search often require that the retrieved information can be understood by cohorts of users with different experience and understanding of health information [1,12,26,27,35]. Indeed, health literacy (the knowledge and understanding of health information) has been shown as a critical factor influencing the value of information consumers acquire through search engines [11].

Along with the queries, we also obtain the runs that were originally submitted to the relevant tasks at CLEF 2013–2015 [12,13,18][4]. Both the simulations and the analysis with real user assessments focus on the changes in system rankings obtained when evaluating using standard RBP and its understandability variants (uRBP and uRBPgr). System rankings are compared using Kendall rank correlation (τ) and AP correlation (τ_{AP}) [30], which assigns higher weights to changes that affect top performing systems.

In all our experiments the RBP parameter ρ which models user behaviour (RBP persistence parameter) was set to 0.8 for all variants of this measure, following the findings of Zhang et al. [32].

5.1 Readability as Estimation of Understandability

To computationally simulate the impact of understandability on the evaluation of search engines, we use readability as a proxy for understandability and we integrate this in the evaluation process, along with standard relevance assessments. Readability, although not providing a comprehensive account of understandability, is one of the aspects that influence the understanding of text [28].

To estimate readability (and thus understandability), we employ established general readability measures as those used in previous work that studied the readability of health information (including that returned by search engines [1,26,27]), e.g., SMOG, FOG and Flesch-Kincaid reading indexes. These measures consider the surface level of language in documents, i.e., the wording and syntax of sentences. Thus, long sentences, words containing many syllables and unpopular words, are each indicators of difficult language to read [15]. In this paper, we use the FOG measure to estimate the readability of a text; FOG is computed as

$$FOG(d) = 0.4 * (avgslen(d) + phw(d)) \tag{5}$$

where $avgslen(d)$ is the average length of sentences in a document d and $phw(d)$ is the percentage of hard words (i.e., words with more than two syllables) in d.

While often used in studies to evaluate the readability of health information, doubts have been casted on the suitability of these measures, especially to the specific health context. For example, Yan et al. [29] claimed that the highest readability difficulties are experienced at word level rather than at sentence level. Alternative approaches that measure language readability beyond the surface characteristics of language have been proposed, e.g., language models [7] and supervised support vector machine classifiers [14]. Nevertheless, these measures appear to be adequate for the purpose of the analysis reported here (study the impact of understandability on evaluation).

[4] Obtained from the CLEF eHealth repository, https://github.com/CLEFeHealth.

5.2 Modelling $P(U|d@k)$

Equation 5 provides document readability scores; we then transform readability scores into the probability of a document being understandable (i.e., $P(U|d@k)$) by means of user models. In this case, user models encode different ways in which users (and their capacity to understand retrieved information) are affected by different document readability levels.

Specifically, we consider two user models. In the first user model (characterised by the probability estimations $P_1(U|d@k)$), a user has a readability threshold th and every document that has a readability score below th is considered certainly understandable, i.e., $P_1(U|d@k) = 1$; while documents with readability above th are considered not understandable, i.e. $P_1(U|d@k) = 0$. This Heaviside step function is centred in th and its use to model $P(U|d@k)$ is akin to the gain function in RBP (also a step function). Thus, uRBP for this first user model can be rewritten as:

$$uRBP_1 = (1-\rho)\sum_{k=1}^{K}\rho^{k-1}r(k)u_1(k) \qquad (6)$$

where, for simplicity of notation, $u_1(k)$ indicates the value of $P_1(U|d@k)$ and $r(k)$ is the (topical) relevance assessment of document k (alternatively, the value of $P(T|d@k)$); thus $g(k) = f(P(T|d@k)P_1(U|d@k)) = P(T|d@k)P_1(U|d@k) = r(k)u_1(k)$.

In the second user model, the probability estimation $P_2(U|d@k)$ is similar to the previous step function, but it is smoothed in the surroundings of the threshed value. This provides a more realistic transition between understandable and non-understandable information. This behaviour is achieved by the following estimation:

$$P_2(U|d@k) \propto \frac{1}{2} - \frac{\arctan\Big(FOG(d@k) - th\Big)}{\pi} \qquad (7)$$

where arctan is the arctangent trigonometric function and $FOG(d@k)$ is the FOG readability score of the document at rank k. (Other readability scores could be used instead of FOG.) Equation 7 is not a probability distribution per se, but one such distribution can be obtained by normalising Eq. 7 by its integral between $[\min\big(FOG(d@k)\big), \max\big(FOG(d@k)\big)]$. However Eq. 7 is rank equivalent to such distribution, not changing the effect on uRBP. These settings lead to the formulation of a second simulated variant of uRBP, $uRBP_2$, which is based on this second user model and is obtained by substituting $u_2(k) = P_2(U|d@k)$ to $u_1(k)$ in Eq. 6.

5.3 Analysis of the Simulations

In the simulations we consider three thresholds to characterise the user models with respect to the FOG readability values: $th = 10, 15, 20$. In general, documents with a FOG score below 10 should be near-universally understandable, while documents with FOG scores above 15 and 20 increasingly restrict the audience able to understand the text. We performed simple cleansing of the HTML pages, although a more conscious pre-processing may be more appropriate [19].

In the following we report the results observed using the CLEF eHealth 2013 topics and assessments. Figure 1 reports RBP vs. uRBP for the 2013 systems. Table 1 reports the values of Kendall rank correlation (τ) and AP correlation (τ_{AP}) between system rankings obtained with RBP and uRBP.

Higher values of th produce higher correlations between systems rankings obtained with RBP and uRBP; this is regardless of the user model used in uRBP (Table 1). This is expected as the higher the threshold, the more documents will have $P(U|d@k) = 1$ (or ≈ 1 for $uRBP_2$): in this case uRBP degenerates to RBP. Overall, $uRBP_2$ is correlated with RBP more than $uRBP_1$ is to RBP. This is because of the smoothing effect provided by the arctan function. This function in fact increases the number of documents for which $P(U|d@k)$ is not zero, despite their readability score being above th. This in turn narrows the scope for ranking differences between systems effectiveness. These observations are confirmed in Fig. 1, where only few changes in the rank of systems are shown for $th = 20$ ($\times$ in Fig. 1), with more changes found for $th = 10$ ($\circ$) and $th = 15$ ($+$).

The simulations reported in Fig. 1 demonstrate the impact of understandability in the evaluation of systems for the considered task. The system ranked highest according to RBP (`MEDINFO.1.3.noadd`) is second to a number of systems according to uRBP if user understandability of up to FOG level 15 is wanted. Similarly, the highest $uRBP_1$ for $th = 10$ is achieved by

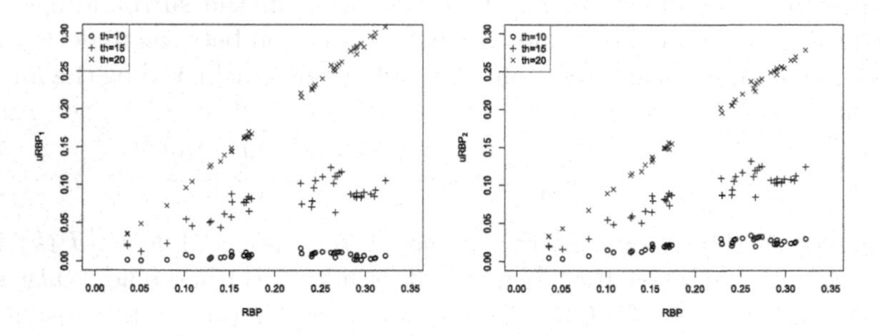

Fig. 1. RBP vs. **uRBP** for CLEF eHealth 2013 systems (left: $\mathbf{uRBP_1}$; right: $\mathbf{uRBP_2}$) at varying values of readability threshold ($\mathbf{th = 10, 15, 20}$).

Table 1. Correlation (τ and $\tau_{\mathbf{AP}}$) between system rankings obtained with **RBP** and $\mathbf{uRBP_1}$ or $\mathbf{uRBP_2}$ for different values of readability threshold on CLEF eHealth 2013.

	$th = 10$	$th = 15$	$th = 20$
RBP vs	$\tau = .1277$	$\tau = .5603$	$\tau = .9574$
$uRBP_1$	$\tau_{AP} = -.0255$	$\tau_{AP} = .2746$	$\tau_{AP} = .9261$
RBP vs	$\tau = .5887$	$\tau = .6791$	$\tau = .9574$
$uRBP_2$	$\tau_{AP} = .2877$	$\tau_{AP} = .4102$	$\tau_{AP} = .9407$

`UTHealth_CCB.1.3. noadd`, which is ranked 28th according to RBP, and for $th = 15$ by `teamAEHRC.6.3`, which is ranked 19th according to RBP and achieves the highest $uRBP_2$ for $th = 10, 15$.

We repeated the same simulations for the 2014 and 2015 tasks. While we omit to report all results here due to space constraints, we do report in Table 2 the results of the simulations for the first user model tested on the 2015 task, so that these values can be directly compared to those obtained using the real assessments (Sect. 6). The trends observed in these results are similar to those reported for the 2013 data (and for the 2014 data), i.e., the higher the threshold th, and the larger the correlation between RBP and uRBP becomes. However larger absolute correlations values between RBP and $uRBP_1$ are found when using the 2015 data, if compared to the correlations reported in Table 1 for the 2013 task. The full set of results, including high resolutions plots, is made available at http://github.com/ielab/ecir2016-UnderstandabilityBiasedEvaluation.

Table 2. Correlation (τ and $\tau_{\mathbf{AP}}$) between system rankings obtained with **RBP** and **uRBP₁** for different values of the readability threshold on CLEF eHealth 2015.

	$th = 10$	$th = 15$	$th = 20$
RBP vs	$\tau = .5931$	$\tau = .8898$	$\tau = .9986$
$uRBP_1$	$\tau_{AP} = .5744$	$\tau_{AP} = .8777$	$\tau_{AP} = .9990$

6 Evaluation with Real Judgements

Next, we study the impact of understandability in the evaluation of IR systems by considering judgements about document understandability and topicality provided by human assessors. To this aim, we consider topics and systems from CLEF eHealth 2015 Task 2 [18], in which both topicality and understandability assessments (binary and graded) were collected. We can thus compute uRBP according to its two instantiations in Sect. 4.1 and compare their system rankings with those of RBP.

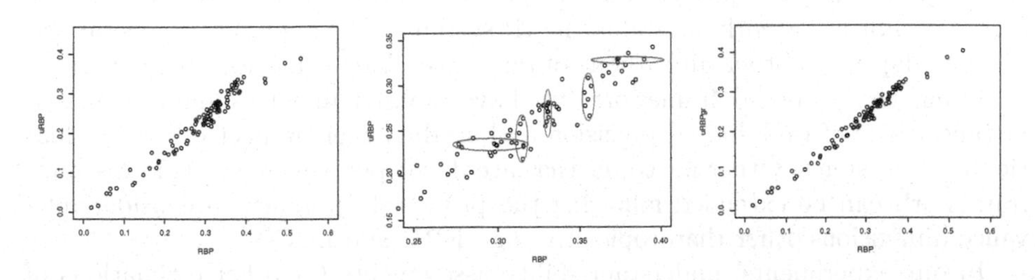

Fig. 2. RBP vs. **uRBP** for CLEF eHealth 2015 systems, with understandability judgements sourced from human assessors (binary uRBP left, uRBPgr (graded) right). Centre: a detail of the correlation between RBP vs. binary uRBP.

Figure 2 compares the evaluations of each CLEF system obtained with RBP and the two uRBP variants (binary, graded). In addition, the correlations between the measures are: RBP-uRBP, $\tau = 0.8666$, $\tau_{AP} = 0.8168$; RBP-uRBPgr, $\tau = 0.9077$, $\tau_{AP} = 0.8866$.

These results highlight that when human assessments of understandability are used, uRBP is generally strongly correlated with RBP. This is even more so for the graded uRBP variant because uRBPgr assigns a zero value of $P(U|d@k)$ to less documents than its binary counterpart, as documents that were assessed as somewhat hard to understand produce a small but not null gain (0.4) in uRBPgr, while they produce a zero gain in uRBP. When compared to the results of the simulations, the correlation trends between RBP and uRBP when real assessments are used is more in line with the findings obtained when the simulation used $th = 15$ than when other threshold values were used (Table 2).

Nevertheless, despite being highly correlated, system rankings obtained with RBP and uRBP do differ. In particular, in our experiments differences seem concentrated when RBP ranges between 0.25 and 0.40, and uRBP (or uRBPgr) ranges between 0.15 and 0.35: this is depicted in the central plot of Fig. 2 for the binary uRBP. Indeed, the analysis reveals that there is large variability in terms of uRBP for a number of systems, which instead appeared indistinguishable when evaluated using RBP: examples of such cases are highlighted in blue in the plot. These cases refer to situations in which there were a number of systems that returned a similar rank distribution of relevant documents (thus obtaining approximately the same RBP). However, these different systems retrieved different relevant documents and some of those documents are of no or limited understandability, and thus are discounted by uRBP. Similarly, in red we have highlighted examples where systems are evaluated as being equivalent in terms of uRBP, but are different in terms of RBP. This happens when the additional gains obtained by the systems that are superior in terms of RBP are due to documents that, despite being relevant, have been assessed as being of low or no understandability.

7 Conclusions

In this paper, we have proposed a method to integrate understandability in the gain-discount framework for evaluating IR systems. The approach is general and can be adapted to other dimensions of relevance. This is left for future work.

Using the proposed framework, we have devised understandability biased instantiations of rank-biased precision and studied their impact on the evaluation of IR systems. Other measures that are developed within the gain-discount framework can be extended following the proposed approach to consider relevance dimensions other than topicality, e.g., ERR and nDCG.

In our experiments, understandability assessments (or other estimations of the probability $P(U|d)$) were transformed into gains in a manner akin to how binary or graded relevance assessments are transformed into gains when computing gain-discount measures. Indeed, here topicality and understandability were given the same importance when determining the effectiveness of IR systems. However, different dimensions of relevance affect the perception of document relevance in different proportions. For example Xu and Chen [28] first, and Zhang et al. [33] later, have found that topicality is more influential than understandability. The weighting of different dimensions of relevance could be accomplished through a different $f(.)$ function for converting $P(T, U|d@k)$ into gain values. The exploration of this possibility and its implications for evaluation is left for future work.

Acknowledgements. The author is thankful to Bevan Koopman, Leif Azzopardi, Joao Palotti, Peter Bruza, Alistair Moffat and Lorraine Goeuriot for their comments on the ideas proposed in this paper.

References

1. Ahmed, O.H., Sullivan, S.J., Schneiders, A.G., McCrory, P.R.: Concussion information online: evaluation of information quality, content and readability of concussion-related websites. Br. J. Sports Med. **46**(9), 675–683 (2012)
2. Barry, C.L.: User-defined relevance criteria: an exploratory study. JASIS **45**(3), 149–159 (1994)
3. Bruza, P.D., Zuccon, G., Sitbon, L.: Modelling the information seeking user by the decision they make. In: Proceedings of MUBE, pp. 5–6 (2013)
4. Carterette, B.: System effectiveness, user models, and user utility: a conceptual-framework for investigation. In: Proceedings of SIGIR, pp. 903–912 (2011)
5. Clarke, C.L., Craswell, N., Soboroff, I., Ashkan, A.: A comparative analysis of cascade measures for novelty and diversity. In: Proceedings of WSDM, pp. 75–84 (2011)
6. Clarke, C.L., Kolla, M., Cormack, G.V., Vechtomova, O., Ashkan, A., Büttcher, S., MacKinnon, I.: Novelty and diversity in information retrieval evaluation. In: Proceedings of SIGIR, pp. 659–666 (2008)
7. Collins-Thompson, K., Callan, J.: Predicting reading difficulty with statistical language models. JASIST **56**(13), 1448–1462 (2005)
8. Cosijn, E., Ingwersen, P.: Dimensions of relevance. IP&M **36**(4), 533–550 (2000)
9. Cuadra, C.A., Katter, R.V.: Opening the black box of 'relevance'. J. Doc. **23**(4), 291–303 (1967)
10. Eisenberg, M., Barry, C.: Order effects: a study of the possible influence of presentation order on user judgments of document relevance. JASIS **39**(5), 293–300 (1988)
11. Friedman, D.B., Hoffman-Goetz, L., Arocha, J.F.: Health literacy and the world wide web: comparing the readability of leading incident cancers on the internet. Inf. Health Soc. Care **31**(1), 67–87 (2006)
12. Goeuriot, L., Jones, G., Kelly, L., Leveling, J., Hanbury, A., Müller, H., Salanterä, S., Suominen, H., Zuccon, G.: ShARe/CLEF eHealth Evaluation Lab 2013, Task 3: Informationretrieval to address patients' questions when reading clinical reports. In: Proceedings of CLEF (2013)
13. Goeuriot, L., Kelly, L., Lee, W., Palotti, J., Pecina, P., Zuccon, G., Hanbury, A., Gareth, H.M., Jones, J.F.: ShARe, CLEF eHealth Evaluation Lab 2014, Task 3: User-centred health information retrieval. In: Proceedings of CLEF Sheffield, UK (2014)

14. Larsson, P.: Classification into readability levels: implementation andevaluation. PhD thesis, Uppsala University (2006)
15. McCallum, D.R., Peterson, J.L.: Computer-based readability indexes. In: Proceedings of the ACM Conference, pp. 44–48 (1982)
16. Mizzaro, S.: Relevance: the whole history. JASIS **48**(9), 810–832 (1997)
17. Moffat, A., Zobel, J.: Rank-biased precision for measurement of retrieval effectiveness. TOIS **27**(1), 2 (2008)
18. Palotti, J., Zuccon, G., Goeuriot, L., Kelly, L., Hanbury, A., Jones, G.J., Lupu, M., Pecina, P.: Clef eHealth evaluation lab 2015, task 2: Retrieving informationabout medical symptoms. In: Proceedings of CLEF (2015)
19. Palotti, J., Zuccon, G., Hanbury, A.: The influence of pre-processing on the estimation of readability of web documents. In: Proceedings of CIKM (2015)
20. Rees, A.M., Schultz, D.G.: A field experimental approach to the study of relevance assessments in relation to document searching. Technical report, Case Western Reserve University (1967)
21. Robertson, S.E.: The probability ranking principle in IR. J. Doc. **33**(4), 294–304 (1977)
22. Sakai, T., Song, R.: Evaluating diversified search results using per-intent graded relevance. In: Proceedings of SIGIR, pp. 1043–1052 (2011)
23. Saracevic, T.: The stratified model of information retrieval interaction: extension and applications. Proceedings of ASIS, vol. 34, pp. 313–327 (1997)
24. Schamber, L., Eisenberg, M.: Relevance: the search for a definition. In: Proceedings of ASIS (1988)
25. Smucker, M.D., Clarke, C.L.: Time-based calibration of effectiveness measures. In: Proceedings of SIGIR, pp. 95–104 (2012)
26. Walsh, T.M., Volsko, T.A.: Readability assessment of internet-based consumer health information. Respir. Care **53**(10), 1310–1315 (2008)
27. Wiener, R.C., Wiener-Pla, R.: Literacy, pregnancy and potential oral health changes: the internetand readability levels. Matern. Child Health J. 1–6 (2013)
28. Xu, Y.C., Chen, Z.: Relevance judgment: what do information users consider beyond topicality? JASIST **57**(7), 961–973 (2006)
29. Yan, X., Song, D., Li, X.: Concept-based document readability in domain specific information retrieval. In: Proceedings of CIKM, pp. 540–549 (2006)
30. Yilmaz, E., Aslam, J.A., Robertson, S.: A new rank correlation coefficient for information retrieval. In: Proceedings of SIGIR, pp. 587–594 (2008)
31. Zhai, C.X., Cohen, W.W., Lafferty, J.: Beyond independent relevance: methods and evaluation metrics forsubtopic retrieval. In: Proceedings of SIGIR, pp. 10–17 (2003)
32. Zhang, Y., Park, L.A., Moffat, A.: Click-based evidence for decaying weight distributions in search effectiveness metrics. Inf. Retrieval **13**(1), 46–69 (2010)
33. Zhang, Y., Zhang, J., Lease, M., Gwizdka, J.: Multidimensional relevance modeling via psychometrics and crowdsourcing. In: Proceedings of SIGIR, pp. 435–444 (2014)
34. Zuccon, G., Koopman, B.: Integrating understandability in the evaluation of consumer health search engines. Proceedings of MedIR, pp. 32–35 (2014)
35. Zuccon, G., Koopman, B., Palotti, J.: Diagnose this if you can. In: Hanbury, A., Kazai, G., Rauber, A., Fuhr, N. (eds.) ECIR 2015. LNCS, vol. 9022, pp. 562–567. Springer, Heidelberg (2015)

The Relationship Between User Perception and User Behaviour in Interactive Information Retrieval Evaluation

Mengdie Zhuang(✉), Elaine G. Toms, and Gianluca Demartini

Information School, University of Sheffield, Sheffield S10 2TN, UK
{mzhuang1,e.toms,g.demartini}@sheffield.ac.uk

Abstract. Measures of user behaviour and user perception have been used to evaluate interactive information retrieval systems. However, there have been few efforts taken to understand the relationship between these two. In this paper, we investigated both using user actions from log files, and the results of the User Engagement Scale, both of which came from a study of people interacting with a novel interface to an image collection, but with a non-purposeful task. Our results suggest that selected behavioural actions are associated with selected user perceptions (i.e., focused attention, felt involvement, and novelty), while typical search and browse actions have no association with aesthetics and perceived usability. This is a novel finding that can lead toward a more systematic user-centered evaluation.

Keywords: User-centered evaluation · User perception evaluation · User behaviour evaluation

1 Introduction

Typically, interactive information retrieval (IIR) systems evaluations assess search processes and outcomes using a wide range of measures such as time-on-task, user satisfaction, and number-of-queries submitted. Some of these measures relate to *user perception* of the results, the search experience, or the interface; they use data from user responses to questions collected either during or after a search task is complete. Some measures relate to *user behaviour,* that is, the actions and selections made by the user while interacting with a system. These measures are calculated from data collected by system log files while the user is in the process of searching or browsing, and include, typically, time/date stamp, interface object used (e.g., mouse movements, search box), and keystrokes. Most evaluations will include a combination of these measures particularly in lab-based studies. In general we presume that both types of measures are indicative of performance, opinion and outcome.

This research uses an existing dataset that contains both perception and behavioural data to test the relationship between the two. This will be a first step toward testing the hypothesis that user behavioural actions predict user perceptions of IR systems. If this is indeed the case, the assessment of IIR evaluations can be significantly simplified for automatic data collection of essential measures. At the same time strong correlations

N. Ferro et al. (Eds.): ECIR 2016, LNCS 9626, pp. 293–305, 2016.
DOI: 10.1007/978-3-319-30671-1_22

(if they exist) among the various perception and behavioural measures will suggest that the measures are evaluating the same phenomena, which may lead to a more parsimonious set of measures. Surprisingly, we still do not know which measures are the more reliable and robust, and indicative of overall results.

This paper is structured as follows: Sect. 2 discusses how both user perception and behaviour are used in IIR evaluations. Section 3 describes the dataset used in this study, the measures extracted from the dataset, and our approach to the analysis. Sections 4-6 deal, respectively, with the results, discussion and conclusions.

2 Background

The evaluation of IR systems has puzzled the field for half a century. Initially relevance emerged as the measure of preference to assess primarily topical relevance using, e.g., mean average precision, mean reciprocal rank, and discounted cumulative gain [15]. But with *interactive* IR came a focus on users and their needs, which examined the effect of individual differences [6, 9] on search, and evaluated search outcomes [16], as well as user behaviour [24] and user perception [16] of the search process. More recently broader aspects of user context [5, 8] have been considered.

Due to the iterative nature of the search process, we do not know if and when an outcome meets a user's need. A user may assess an outcome immediately, but when the task that prompted the search is complex, that judgment may only come after a succession of search tasks (and other types of information tasks) and over a period of time. Individual differences such as age, gender, expertise, mental effort, and learning style may affect the process, but there is as yet definitive influential set [1, 6, 8].

The core measures used in evaluations to date have tended to combine elements of user behaviour (e.g., number of queries) and perception (e.g., satisfaction) as demonstrated by results of the various TREC, INEX and CLEF interactive tracks over the years. These have been characterized in multiple ways [14, 19, 25]. One of the few attempts to examine the interactions between these two dimensions is the work of Al-Maskari and Sanderson [1, 2], who examined the relationship between selected aspects of behaviour and perception, and found significant associations between user satisfaction and user effectiveness (e.g., completeness), and user satisfaction and system effectiveness (e.g., precision, recall). To our knowledge, there is only one measure that integrates user behaviour with user perception: Tague-Sutcliffe's informativeness measure [20] that assesses the performance of the system simultaneously with the perception of the user. But this is atypical and due to the effort (e.g., constant user feedback) required in implementation is rarely used [10].

2.1 User Perception

The multiple measures of user perception are often associated with measures of perceived usability. Satisfaction, for example, was borrowed from usability research and tends to be consistently deployed in IIR studies. Other measures include ease of use, perception of time, and usefulness of results. All are measured post the user's interaction with the system, and require user response to a set of questions or items.

One recent multi-dimensional measure is the User Engagement Scale (UES) [16] which calculates six dimensions (Table 1) of a user experience: Aesthetic Appeal, Novelty, Focused Attention, Felt Involvement, Perceived Usability, and Endurability (see definitions in Table 1). The scale contains 31 items; each item is presented as a statement using a 5 point scale from "strongly disagree" to "strongly agree". Unlike other measures, the model underpinning the UES shows how Endurability is explained either directly or indirectly by the other five dimensions. The UES has been used to evaluate multiple types of systems (e.g., e-shopping [16], wikiSearch [17], Facebook [4]). This scale follows standard psychometric scale development methods [7], and has been tested for reliability and validity. Although differences have emerged [17] in the various applications, it is the most tested measure of user perception of a system.

2.2 User Behaviour

How a user interacts with a search system is characterized typically by a set of low-level user actions and selections at the interface (see [2, 14, 18, 20]):

- frequency of interface object use, e.g., number of times search box has been used;
- counts of queries, categories viewed in a menu, mouse clicks, mouse rollovers;
- time spent using objects, viewing pages.

Multiple efforts have attempted to look for patterns in these actions, patterns that might have the capability to predict likelihood of a successful outcome [21, 24]. The challenge with user behaviour measures is that they are only *descriptive* of the outcome, and are not *interpretive* of the process. That is to say, they lack the rationale behind why those behaviours may lead to a successful outcome. The challenge with log files is the voluminous number of data points and the need to find a reliable approach to defining groups or sets based on behavioural patterns. Not all users are alike and nor do they all take the same approach to searching for the same things as evidenced by the TREC, INEX and CLEF interactive tracks.

2.3 Research Objectives

We hypothesise that behavioural patterns are indicative of a user perception of IIR system usage. That is, selected behavioural variables are associated with selected user perceptions of the user's interaction with that system. We test this hypothesis by isolating measures of user behaviour as represented by actions in a log file and examining the association with a user perception of their experience as measured by the UES.

3 Methods

3.1 Overview

We used the data collected by the CLEF 2013 Cultural Heritage Track (CHiC). This section briefly describes that dataset, the measures we extracted from the dataset, and how we approached the analysis, but see [12] for the details of that study.

3.2 Dataset

Application System. The system, an image *Explorer*, based on Apache Solr[1] contains about one million records from the Europeana Digital Library's English language collection. The Explorer was accessed using a custom-developed interface (see Fig. 1 [12]), adapted from wikiSearch [22], with three types of access: (1) hierarchical category browser, (2) search box, and (3) a metadata filter based on the Dublin core ontology although the labels were modified for better user understanding. The interface used a single display panel that brought items to the surface leaving the interface structure as a constant. Using one of the three access methods, participants searched or browsed the content, adding interesting items into a book-bag, and at the same time providing information about why the object was added using a popup box.

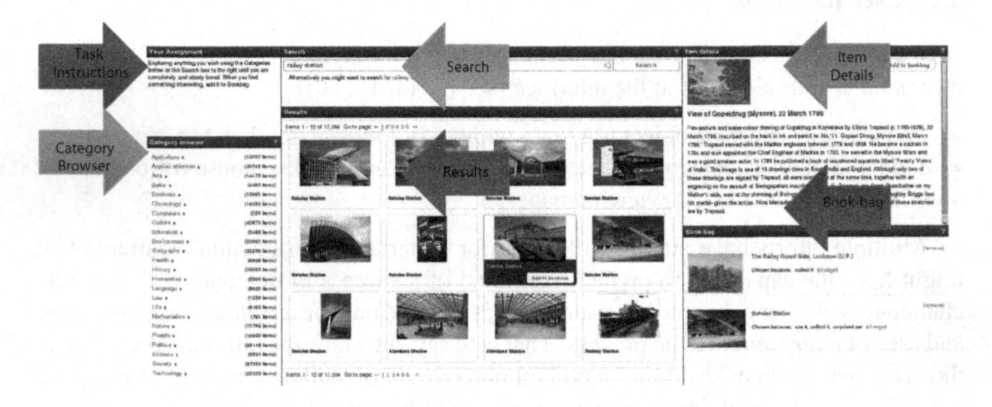

Fig. 1. CHiC Culture & Heritage Explorer user interface [12]

Task. Participants first read the scenario: "Imagine you are waiting to meet a friend in a coffee shop or pub or the airport or your office. While waiting, you come across this website and explore it looking at anything that you find interesting, or engaging, or relevant..." The next display, Fig. 1, presented the browse task with no explicit goals in the upper left corner: "Your Assignment: explore anything you wish using the Categories below or the Search box to the right until you are completely and utterly bored. When you find something interesting, add it to the Book-bag."

Participants. 180 participants volunteered with 160 on-line participants and 20 in-lab participants who were recruited via a volunteers' mailing list.

Procedure. Participants (both lab and online) used a web-based system, SPIRES [11] which guided them through the process. The only difference between the two is that lab participants were interviewed, which is outside the scope of this analysis. The SPIRES system started with an explanation of the study, acquired informed consent, and asked for a basic demographic profile and questions about culture before presenting the

[1] http://lucene.apache.org/solr/.

Explorer and the task to participants. Once participants had executed the task, and essentially were "bored," they moved on to the 31 item UES questionnaire [7, 16] about their perceptions of the search experience and the interface, and provided a brief explanation of objects in the book-bag, the metadata and the interface.

3.3 Measures

The following measures (see Table 1) were extracted from the CHiC study data:

Table 1. List of perception and behaviour measures

Variable	Definition
User Perception measures – the User Engagement Scale (UES)	
Aesthetic Appeal	Perception of the visual appearance of interface.
Felt Involvement	Feelings of being drawn in and entertained in interaction.
Focused Attention	The concentration of mental activity, flow an absorption.
Novelty	Curiosity evoked by content.
Perceived Usability	Affective and cognitive response to interface/content.
Endurability	Overall evaluation of the experience and future intentions.
User Behaviour measures	
Queries	Number of queries used
Query Time	Time spent issuing queries following the links
Items viewed (Queries)	Number of items viewed from queries
Bookbag (Queries)	Number of items added to Bookbag from queries
Topics	Number of categories used.
Topics Time	Time spent exploring categories and following links
Items viewed (Topics)	Number of items viewed from categories
Actions	Number of actions (e.g., keystrokes, mouseclicks)
Pages	Number of pages examined
Bookbag Time	Total time spent reviewing contents of Bookbag.
Bookbag (Total)	Number of items added to the Bookbag
Bookbag (Topics)	Number of items added to Bookbag from category.
Task Time	Total time user spent on the task.

1. User perception measures: the UES with six user perception dimensions [16];
2. User behaviour: 13 variables that represent typical user actions e.g., examining items, selecting categories, and deploying queries. Times were measured in seconds.

3.4 Data Analysis

Data Preparation. After extracting the data, each participant set was scanned for irregularities. Pilot participants and those who did not engage (e.g. left the interface for hours) were removed. 157 participants remained. The two datasets were saved into a spreadsheet or database for preliminary examination, and exported to SPSS.

User Perception. First, Reliability Analysis assessed the internal consistency [3] of the UES sub-scales using Cronbach's α. Second, the inter-item correlations among items were used to test the distinctiveness of the sub-scales. Third, Exploratory Factor Analysis using Maximum Likelihood with Oblique Rotation (as we assumed correlated factors [18]) to estimate factor loadings tested the underlying factors, to compare with previous UES analyses, and validate it for use in this research.

User Behaviour. First, the raw log file data were exported to a spreadsheet. A two-step data reduction process sorted 15396 user actions into 157 participant groups containing participant id, time stamp, action type and parameter. Next Exploratory Factor Analysis (using Maximum Likelihood with Oblique Rotation) was used to identify the main behavioural classes. These then were used to calculate the measure per participant for each variable listed in Table 1. Finally, Cluster Analysis extracted symbolic user archetypes across 157 participants.

Correlation Analysis. Correlation analysis using Pearson's r was then used to examine the relationship between user perception and user behaviour.

4 Results

The results first present the analysis of the user perception measures, then the user behaviour measures and finally the analysis of the relationship between the two.

4.1 User Perception

First, the Reliability Analysis resulted in Cronbach's $\alpha = 0.79$ to 0.90 indicating good internal consistency for each of the sub-scales; values between 0.7 and 0.9 are considered optimal [7]. Next, correlations among the UES subscales (see Table 2) were tested. Values <0.5 are indicate that the sub-scale should remain distinct while >0.5 indicates that the scale might be merged during Factor analysis.

Table 2. Correlations among UES sub-scales (**p<0.01)

Sub-scale	Aesthetics (AE)	EN	FA	FI	NO
Endurability (EN)	0.692*	1			
Focused Attention (FA)	0.370	0.621*	1		
Felt Involvement (FI)	0.558*	0.826*	0.793*	1	
Novelty (NO)	0.546*	0.715*	0.650*	0.824*	1
Perceived Usability (PU)	0.471*	0.596*	0.206	0.385	0.234

An initial examination of the *scree plot* (i.e., the eigenvalues of the principal components) that resulted from the Factor Analysis identified a four-factor solution that accounted for 59.8 % of the variance. A five-factor solution, albeit accounting for 63 % of the variance, was less appropriate as only two items were loaded on Factor 5 with lower absolute loading values than those on Factor 4. The four-factor model demonstrated a very high Kaiser-Meyer-Olkin Measure of Sampling Adequacy (KMO = 0.924), indicating that the factors are distinct. The statistically significant result from Bartlett's Test of Sphericity ($\chi^2 = 3609.9, df = 465, p < 0.001$) also suggested relationships existed amongst the dimensions. Table 3 summarises the four factors that were generated: Factor 1 contained 11 items from Novelty (3 of 3), Focused Attention (1 of 7), Felt Involvement (3 of 3), and Endurability (4 of 5). Factor 4 remained as in the original UES, Focused Attention (6 of 7) almost remained distinct (Factor 2), and Perceived Usability (8 of 8) plus 1 item from Endurability formed Factor 3. Factor 2-4 had good internal consistency as demonstrated by Cronbach's α. Correlation analysis resulted in significant, although moderate, correlations amongst the factors. Given the results, some of the overlapped items may be removed from Factor 1 (Cronbach's α > 0.95) (see Table 3). However, we used the original factors in our remaining analysis.

Table 3. Factors resulting from the Factor Analysis (**p<0.01)

					Factor		
Factor	Sub-scale	Cronbach's α	*M*	*n*	2	3	4
1	EN, FA, FI, NO	0.95	2.67	11	0.66**	0.45	0.59**
2	FA	0.90	2.19	6		-0.26**	0.36**
3	PU,EN	0.86	3.14	9			0.51**
4	AE	0.89	2.55	5			

4.2 User Behaviour

First, we performed Exploratory Factor Analysis on the behavioural measures listed in Table 1 to assess, first if they highly correlate and, second, to identify distinctive groups according to behavioual actions. The result demonstrated a mediocre Kaiser-Meyer-Olkin Measure of Sampling Adequacy (KMO = 0.634), and Bartlett's Test of Sphericity ($\chi^2 = 2736.4, df = 78, p < 0.001$) suggests that there were relationships amongst the items. This resulted in a three-factor solution, which accounted for approximately 76 % of the variance.

Table 4 displays the factor weights for three user behaviour factors. Factor 1 seems to represent search actions, and Factor 2, browsing actions. Factor 3 mainly contains general task-based actions. However, both Actions and Pages are present in both factors, and thus were excluded from further analysis. In order to test for other irrelevant variables in each factor, we performed a reliability analysis by factor ablation measuring *Cronbach's α if items Deleted*. Notably, the exclusion of Bookbag (Topics) from Factor 3 would yield an α value of 0.537, which makes it the most critical measure. Factor 1 (Cronbach's α = 0.846) and Factor 2 (Cronbach's α = 0.707) reflected good internal consistency. Correlation values between General behaviour and the other two behaviours are considered as moderate (i.e., 0.362 and 0.251 with 1 and 2 respectively). This indicates that searching and browsing behaviour had a moderate correlation with general behaviour. The correlation between Searching behaviour and Browsing behaviour was 0.621, which is considered significant. The resulting factor from this analysis suggests that participants' behaviours could be described from three main dimensions (Searching, Browsing, and General).

Table 4. Exploratory Factor Analysis of user behaviour data

	Factor 1 Searching	**Factor 2 Browsing**	**Factor 3 General**
Queries	0.970	-0.019	-0.103
Query Time	0.961	-0.079	0.051
Items viewed (Queries)	0.946	0.057	-0.135
Bookbag (Queries)	0.693	-0.221	0.421
Topics	0.060	0.998	-0.391
Topics Time	-0.162	0.887	0.196
Item viewed (Topics)	-0.077	0.793	0.114
Actions	0.519	0.616	0.102
Pages	0.307	0.394	0.300
Bookbag Time	-0.015	-0.230	1.037
Bookbag (Total)	0.225	0.003	0.824
Bookbag (Topics)	-0.398	0.380	0.749
Task Time	0.275	0.118	0.614

To assess how participants acted, one action item (i.e., the one with highest weight, shown in italics in Table 4), was selected from each factor and submitted to a Cluster Analysis using Ward's hierarchical clustering method [23]. The results were manually inspected including descriptive statistics for each action item, and the resulting dendrogram. The 157 participants best distributed into 3 clusters (see Table 5).

Table 5. Means of user behaviour variables in each cluster

Cluster	Label	*n*	Queries	Topics	Bookbag Time(s)
1	Explorers	10	18.2	9.8	821.6
2	Followers	98	2.7	10.10	29.8
3	Berrypickers	49	3.96	11.4	137.6

Each of the clusters represents a set of participants who exhibit certain types of behaviours illustrative of information seekers. The first represents *explorers*, who spent the longest time checking items in the book-bag, and used on average the most queries. They were clearly concerned about their results, and specific about what they were looking for. The second group contains directionless *followers*. They do not appear to have specific interests about the content and just trailed the inter-linked categories rather than using queries. They added fewer items to the bookbag, and appeared to stop early. The third group acted much like Bates' *berrypickers* [5]. Their search and browse activities interacted to sustain participants' interests in the collection. They seemed to obtain information by noticing and stopping to examine other contents, which are not strongly relating to the item that they currently viewing. Some used queries to refine their searches. The interpretation of three clusters suggests the three behavioural factors described the participants in this case. For the subsequent examination of the relationship between perception and behaviour, these three behaviour factors (Table 4) were used.

Table 6. Correlations between UES sub-scales and user behaviour factors

	Searching	Browsing	General
Aesthetics	0.057	0.09	0.097
Endurability	0.167	0.171	0.277
Focused Attention	0.149	0.233	0.354
Felt Involvement	0.232	0.221	0.383
Novelty	0.279	0.231	0.393
Perceived Usability	0.045	0.101	0.072

4.3 Relationship Between User Perception and User Behaviour

We tested the relationships among the three user behaviour factors and the six UES sub-scales (see Table 6). The user behaviour factors do not correlate with Aesthetics and Perceived Usability. Of the others, correlations between the searching and browsing behaviour factors and Endurability, Focused Attention, and Novelty were also insignificant. Only the general behaviour had a moderate correlation with Focused Attention, Felt Involvement, and Novelty.

5 Discussion

5.1 User Perception

The reliability analysis of all six original UES sub-scales demonstrated good internal consistency, which aligns with previous studies [16, 17]. In our correlation analysis, Perceived Usability had a positive and moderate relationship with Focused Attention, which is in contrast to the results of the wikiSearch study, which found a negative correlation between the two [17]. A key difference between the two studies is the interface and content, e.g., images versus Wikipedia, and multiple access tools versus only a search box.

The original six-dimensional UES structure was developed with e-shopping data [16]. However, our results identified four factors, which is consistent with the result obtained from the wikiSearch study [17] and Facebook [4]. This suggests that in a searching environment, the dimensions of UES structure may remain consistent regardless of data type (text data or image data), or perhaps it is due to the presence of rich information and interactivity. Novelty, Felt Involvement, and Endurability had been demonstrated to be reliable sub-scales in the e-shopping environment, and some of the items within these sub-scales were used successfully to measure website engagement and value as a consequence of website engagement in online travel planning [13]. This highlights the notion that different user perception dimensions may be more relevant to different interactive search systems. In our setting we observed that Endurability, Felt Involvement, and Novelty show the same information.

5.2 User Behaviour

Extracting types of user actions from the logfile resulted in three key behavioural classes that relate to users' search or browse behaviours and their general task-based actions. The searching behaviours were primarily associated with query actions. The browsing behaviours included actions related to using the categories as well as those related to keystroke and mouse activity and what could be construed as navigational activities. Actions and Pages, the items viewed, did not map well to any factor. While the third, which we call *general*, is more associated with actions related to the result and task. Notably actions associated with items selected as a result of using categories fit into this factor, whereas, those that resulted from using a query loaded with the other actions associated with a query.

In addition to examining and grouping the behavioural actions into usable sets, we found a novel set of user archetypes (explorers, followers, berrypickers) among our participant group. The explorers submitted sets of highly relevant queries. More specifically, subsequent queries were aimed at refining former ones. For instance, an explorer exhibited a closely related pathway: modern sculpture, modern british sculpture, hepworth, hepworth sculpture, henry moore, henry moore sculpture, family of man, family of man sculpture. In contrast, the query pathways input by followers and berrypickers are typically short (both pathway and query length), e.g., Scotland, Edinburgh. The user archetypes and pattern of query might be useful in evaluation simulations and in advancing log analysis techniques.

5.3 Relationship Between User Perception and User Behaviour

There are little indications of which measures are the reliable and robust. Therefore, as a first step to test the relationships between perception and behaviour measures, correlation values >0.35 should be considered. When we measured correlation of user behaviours with user perception, the results were not as anticipated. User behaviour appears to be not strongly related to a user's perception of Aesthetics and Perceived Usability. How people searched and browsed through the images seems unrelated to their subsequent perception of the system. This may be attributed to user expectations about aesthetics and usability that limit the degree of variation among individuals.

Similarly the searching and browsing behaviours have no strong correlation with Endurability, Focused Attention, Felt Involvement and Novelty. This suggests that single exploring behaviours could not comprehensively contribute to calculating user engagement. However, the general behaviours which had more to do with managing the results had a moderate correlation with Focused Attention, Felt Involvement and Novelty, which were combined into a single factor in our analysis of the UES. This indicates that system data that shows the general behaviour of users could contribute to these existing user engagement sub-scales; depending on the nature of the experiment, different user behaviour variables could be extracted from log files.

6 Conclusion

The key objective of our research was to assess whether a relationship exists between user behaviour and user perception of information retrieval systems. This was achieved by using actions from log files to represent *behaviour* and results from the UES to represent *perception.* The data came from a study in which people had no defined task while interacting with a novel interface to a set of images. In the past, studies have considered measures of behaviour and perception as two relatively independent aspects in evaluation. Our results showed that the aesthetics and usability perceptions of those searching and browsing appear un-influenced by their interactions with the system. However, general actions were associated with attention, involvement and novelty.

In addition, our research tested the UES scale, and like the wikiSearch results [17], we found four factors. This may be because both implementations were in information finding systems, and not the focused task of a shopper [16]. We also produced a novel set of information-seeking user archetypes (i.e., explorers, followers, and berrypickers), defined by their behavioural features which may be useful in testing evaluation simulations and build novel log analysis techniques that simulate user studies. Moreover, these user archetypes were reflective of search reality as behavioural measures were direct observables. On the other hand, user perception measures are based on a psychometric scale or descriptive data and thus are largely affected by context.

Our findings are preliminary and we need to replicate them using additional datasets. We have isolated selected behavioural variables that are significant to the analysis. The emerging relationship with the UES demonstrates that we may be able to isolate selected variables from log files that are indicative of user perception. Being able to do so would mean that IIR evaluations could be parsimoniously completed using only log file data.

This means that we need also to refine the UES so that the result consistently outputs distinctive reliable and valid factors that represent human perception. The additional part of the analysis lies with the task and with the user's background and personal experience, which may account for the remaining variance in the result.

References

1. Al-Maskari, A., Sanderson, M.: The effect of user characteristics on search effectiveness in information retrieval. Inf. Process. Manage. **47**, 719–729 (2011)
2. Al-Maskari, A., Sanderson, M.: A review of factors influencing user satisfaction in information retrieval. JASIST **61**, 859–868 (2010)
3. Aladwani, A.M., Palvia, P.C.: Developing and validating an instrument for measuring user-perceived web quality. Inf. Manage. **39**, 467–476 (2002)
4. Banhawi, F., Ali, N.M.: Measuring user engagement attributes in social networking application. In: Semantic Technology and Information Retrieval, pp. 297–301. IEEE (2011)
5. Bates, M.J.: The design of browsing and berrypicking techniques for the online search interface. Online Inf. Rev. **13**, 407–424 (1989)
6. Borgman, C.L.: All users of information retrieval systems are not created equal: an exploration into individual differences. Inf. Process. Manage. **25**, 237–251 (1989)
7. DeVellis, R.: Scale Development. Sage, Newbury Park, California (2003)
8. Dillon, A., Watson, C.: User analysis in HCI: the historical lessons from individual differences research. Int. J. Hum. Comput. Stud. **45**, 619–637 (1996)
9. Fenichel, C.H.: Online searching: Measures that discriminate among users with different types of experiences. JASIS **32**, 23–32 (1981)
10. Freund, L., Toms, E.G.: Revisiting informativeness as a process measure for information interaction. In: Proceedings of the WISI Workshop of SIGIR 2007, pp. 33–36 (2007)
11. Hall, M.M., Toms, E.: Building a common framework for iir evaluation. In: Forner, P., Müller, H., Paredes, R., Rosso, P., Stein, B. (eds.) CLEF 2013. LNCS, vol. 8138, pp. 17–28. Springer, Heidelberg (2013)
12. Hall, M.M., Villa, R., Rutter, S.A., Bell, D., Clough, P., Toms, E.G.: Sheffield submission to the chic interactive task: Exploring digital cultural heritage. In: Proceedings of the CLEF 2013 (2013)
13. Hyder, J.: Proposal of a Website Engagement Scale and Research Model: Analysis of the Influence of Intra-Website Comparative Behaviour. Ph.D. Thesis, University of Valencia (2010)
14. Kelly, D.: Methods for evaluating interactive information retrieval systems with users. Found. Trends Inf. Retrieval **3**, 1–224 (2009)
15. Manning, C.D., Raghavan, P., Schütze, H.: Introduction to information retrieval. Cambridge University Press, Cambridge (2008)
16. O'Brien, H.L., Toms, E.G.: The development and evaluation of a survey to measure user engagement. JASIST **61**, 50–69 (2010)
17. O'Brien, H.L., Toms, E.G.: Examining the generalizability of the User Engagement Scale (UES) in exploratory search. Info. Proc. Mgmt. **49**, 1092–1107 (2013)
18. Reise, S.P., Waller, N.G., Comrey, A.L.: Factor analysis and scale revision. Psychol. Assess. **12**, 287 (2000)
19. Su, L.T.: Special Issue: Evaluation Issues in Information Retrieval Evaluation measures for interactive information retrieval. Info. Proc. Mgmt. **28**, 503–516 (1992)

20. Tague-Sutcliffe, J.: Measuring information: an information services perspective. Academic Press, San Diego; London (1995)
21. Teevan, J., Liebling, D.J., Geetha, G.R.: Understanding and predicting personal navigation. In: Proceedings of the International Conference Web search and data mining, pp. 85–94. ACM, Hong Kong, China (2011)
22. Toms, E.G., Villa, R., McCay-Peet, L.: How is a search system used in work task completion? J. Inf. Sci. **39**, 15–25 (2013)
23. Ward Jr., J.H.: Hierarchical grouping to optimize an objective function. J. Am. Statist. Assoc. **58**, 236–244 (1963)
24. White, R.W., Drucker, S.M.: Investigating behavioral variability in web search. In: Proceedings of the 16th Int. Conf. WWW, pp. 21–30. ACM, Banff, Alberta, Canada (2007)
25. Yuan, W., Meadow, C.T.: A study of the use of variables in information retrieval user studies. JASIS **50**, 140–150 (1999)

Multimedia

Using Query Performance Predictors to Improve Spoken Queries

Jaime Arguello[1(✉)], Sandeep Avula[1], and Fernando Diaz[2]

[1] University of North Carolina at Chapel Hill, Chapel Hill, USA
jarguell@email.unc.edu
[2] Microsoft Research, New York, USA

Abstract. Query performance predictors estimate a query's retrieval effectiveness without user feedback. We evaluate the usefulness of pre- and post-retrieval performance predictors for two tasks associated with speech-enabled search: (1) predicting the most effective query transcription from the recognition system's n-best hypotheses and (2) predicting when to ask the user for a spoken query reformulation. We use machine learning to combine a wide range of query performance predictors as features and evaluate on 5,000 spoken queries collected using a crowd-sourced study. Our results suggest that pre- and post-retrieval features are useful for both tasks, and that post-retrieval features are slightly better.

1 Introduction

Speech-enabled search allows users to formulate queries using spoken language. The search engine transcribes the spoken query using an automatic speech recognition (ASR) system and then runs the textual query against the collection. Speech-enabled search is increasingly popular on mobile devices and is an important component in multimodal search interfaces and assistive search tools [12]. While speech is a natural means of communicating an information need, spoken queries pose a challenge for speech-enabled search engines. In a recent study, 55 % of all spoken queries had recognition errors that caused a significant drop in retrieval performance [7].

The goal of *query performance prediction* is to estimate a query's effectiveness without feedback. Current approaches are classified into *pre-* and *post-retrieval* measures. Pre-retrieval measures are computed without conducting a full retrieval from the collection and capture evidence such as the query terms' specificity [2,6,18] and topical relatedness [5]. Post-retrieval measures are computed from the query's retrieval from the collection and capture evidence such as the topical coherence of the top results [2] and the rank stability [1,17,19,20].

We investigate the usefulness of query performance predictors for two tasks associated with speech-enabled search: (1) re-ranking the ASR system's n-best list and (2) deciding when to ask for a spoken query reformulation. While ASR systems typically output the single most confident transcription of the input

N. Ferro et al. (Eds.): ECIR 2016, LNCS 9626, pp. 309–321, 2016.
DOI: 10.1007/978-3-319-30671-1_23

speech, internally they construct a ranked *n-best* list of the most confident hypotheses. In our n-best list re-ranking task, the input to the system is a spoken query's n-best list, and the goal of the system is to predict the candidate transcription from the n-best list that maximizes retrieval performance, which may not necessarily be the top candidate.

In certain situations, a spoken query may perform poorly due to ASR error or the user's failure to formulate an effective query. In our second predictive task, the input to the system is a spoken query (specifically, the top candidate from the ASR system's n-best list), and the goal of the system is to predict whether to run the input query or to ask for a reformulation.

For both tasks, we use machine learning to combine a wide range of performance predictors as features. We trained and tested models using a set of 5,000 spoken queries that were collected in a crowdsourced study. Our spoken queries were based on 250 TREC topics and were automatically transcribed using freely available APIs from AT&T and WIT.AI. We evaluate our models based on retrieval performance using the TREC 2004 Robust Track collection.

2 Related Work

The goal of *query performance prediction* is to estimate a query's performance without user feedback. Pre-retrieval measures capture evidence such as the query's specificity, topical coherence, and *estimated* rank stability [5]. In terms of query specificity, different measures consider the query terms' inverse document frequency (IDF) and inverse collection term frequency (ICTF) values [2,6,18]. Other specificity measures include the *query-scope*—proportional to the number of documents with at least one query term—and the *simplified clarity*—equal to the KL-divergence between the query and collection language models [6]. Topical coherence can be measured using the degree of co-occurrence between query terms [5]. Finally, the rank stability can be estimated using the query terms' variance of TF.IDF weights across documents in the collection [18].

Post-retrieval measures capture evidence such as the topical coherence of the top results, the *actual* rank stability, and the extent to which similar documents obtain a similar retrieval score. The *clarity* score measures the KL-divergence between the language model of the top documents and a background model of the collection [2]. Rank stability approaches perturb the query [17,20], the documents [19], or the retrieval system [1], and measure the degree of change in the output ranking. The assumption is that more effective queries produce more stable rankings. Finally, the auto-correlation score from Diaz [4] considers the extent to which documents with a high text similarity obtain similar scores. Pre- and post-retrieval performance predictors have been applied to IR tasks such as reducing natural language queries [8,16] and predicting the effectiveness of different query reformulations [3,13].

Prior work in the speech recognition domain also considered improving spoken query recognition using evidence similar to the query performance predictors mentioned above. Mamou *et al.* [10] focused on re-ranking the n-best list using term co-occurrence statistics in order to favor candidates with semantically related terms. Li *et al.* [9] combined language models generated from different query-click logs to bias the ASR output in favor of previous queries with clicks. Peng *et al.* [11] focused on re-ranking the n-best list using post-retrieval evidence such as the number of search results and the number of exact matches in the top results. We extend this prior work in three ways. First, in addition to re-ranking the n-best list, we consider the task of predicting when to ask for a spoken query reformulation. Second, we combine a wider range of pre- and post-retrieval performance predictors as features. Finally, we evaluate in terms of retrieval performance instead of recognition error.

3 Data Collection

In the next sections, we describe the user study that we ran to collect spoken queries, our search tasks, the ASR systems used, and our spoken queries.

User Study. Spoken queries were collected using Amazon's Mechanical Turk (MTurk). Each MTurk Human Intelligence Task (HIT) asked the participant to read a search task description and produce a recording of how they would request the information from a speech-enabled search engine.[1]

The study protocol proceeded as follows. Participants were first given a set of instructions and a link to a video explaining the steps required to complete the HIT. Participants were then asked to click a "start" button to open the main voice recording page in a new browser tab. While loading, the main page asked participants to grant access to their computer's microphone. Participants were required to grant access in order to continue. The main page provided participants with: (1) a button to display the search task description in a pop-up window, (2) Javascript components to record the spoken query and save the recording as a WAV file on the participant's computer, and (3) an HTML form to upload the WAV file to our server.

Within the main voice recording page, participants were first asked to click a "view task" button to display the search task description in a pop-up window. The task was displayed in a pop-up window to prevent participants from reading the task while recording their spoken query.[2] Participants were instructed to read the task carefully and to "imagine that you are looking for information on this specific topic and that you are going to ask a speech-enabled search engine for help in finding this information". Participants were asked to "not try to memorize the task description word-by-word". The instructions explained that our goal was to "learn how someone might formulate the information request as naturally as possible".

After reading the task, participants were asked to click a "record" button to record their spoken query and then a "save" button to save the recording

[1] Our source code and search task descriptions are available at: http://ils.unc.edu/~jarguell/ecir2016/.

[2] Participants had to close the pop-up window to continue interacting with the page.

as a WAV file on their computer. Next, participants were instructed to upload the saved WAV file to our server. The default WAV filename included the MTurk assignment ID, which is unique to each accepted MTurk HIT. The assignment ID was used by our server to check the validity of the uploaded WAV file. If the uploaded file was valid, the participant was then given a 10-character completion code. Finally, participants were asked to validate and submit the code to complete the HIT. As described in more detail below, we used a set of 250 search tasks and collected 20 spoken queries per search task, for a total of 5,000 spoken queries. Each HIT was priced at $0.15 USD.

Our HITs were restricted to workers with at least a 95 % acceptance rate and workers within the U.S. Also, in order to avoid having a few workers complete most of our HITs, workers were not allowed to do more than 100 of our HITs. We collected spoken queries from 167 participants.

Search Tasks. Our 250 search tasks were based on the 250 topics from the TREC 2004 Robust Track. We constructed our tasks using the TREC description and narrative as guidelines and situated each task within a simulated scenario that gave rise to the need for information:

> **TREC Topic 390:** You recently read a news article about the Orphan Drug Act, which promotes the development of drugs to treat "orphan" diseases that affect only a small number of people. Now you are curious to learn more. Find information about the Orphan Drug Act and how it is working on behalf of people who suffer from rare diseases.

ASR Systems. In this work, we treat the ASR system as a "black box" and used two freely available APIs provided by AT&T and WIT.AI.[3] Both APIs accept a WAV file as input and return one or more candidate transcriptions in JSON format. The AT&T API was configured to return an n-best list in cases where the API was less confident about the input speech. The AT&T API returned an n-best list with at most 10 candidates along with their ranks and confidence values. The WIT.AI API could not be configured to return an n-best list and simply returned the single most confident transcription without a confidence value.

Spoken Queries and ASR Output. In this section, we describe our spoken queries and ASR output. To conserve space, we focus on the ASR output from the AT&T API. The AT&T API was able to transcribe 4,905 of our 5,000 spoken queries due to the quality of the recording. Spoken queries had an average length of 5.86 ± 2.50 s and 10.04 ± 2.18 recognized tokens. The AT&T API returned an n-best list with more than one candidate for 70 % of the 4,905 transcribed spoken queries.

We were interested in measuring the variability between candidates from the same n-best list. To this end, we measured the similarity between candidate-pairs from the same n-best list in terms of their recognized tokens, top-10 documents retrieved, and retrieval performance. In terms of their recognized tokens,

[3] http://developer.att.com/apis/speech and https://wit.ai/.

after stemming and stopping, candidate-pairs had an average Jaccard correlation of 0.53 ± 0.23. In terms of their top-10 documents retrieved, candidate-pairs had an average Jaccard correlation of 0.28 ± 0.31. Finally, in terms of retrieval performance, candidate-pairs had an average P@10 difference of 0.11 ± 0.17. More importantly, the most confident candidate achieved the best P@10 performance only 82.71 % of the time. Together, these results suggest an opportunity to improve retrieval performance by re-ranking the n-best list.

We were also interested in measuring the variability between spoken queries from different study participants for the same TREC topic (using the most confident candidate from the ASR system's n-best list). In terms of their recognized tokens, spoken query-pairs had an average Jaccard correlation of 0.21 ± 0.22. In terms of their top-10 documents retrieved, the average Jaccard correlation was 0.12 ± 0.23. Finally, the average difference in P@10 performance was 0.19 ± 0.23. These measures suggest great variability in spoken query performance, either due to ASR error, background noise, or word choice. This helps motivate our second task of predicting when the input query is poor and the system should ask for a new spoken query.

4 Predictive Task Definitions

We investigate the effectiveness of existing query performance predictors on two tasks pertaining to speech-enabled search: (1) re-ranking the ASR system's n-best list and (2) predicting when to ask for a spoken query reformulation.

Re-ranking the N-Best List. While ASR systems often output the single most confident transcription, internally the system produces an n-best list of the most confident hypotheses. Off-the-shelf ASR systems such as the AT&T API can be configured to output the n-best list in cases where the system is less confident about the input speech.

We define the n-best list re-ranking task as follows. The input to the system is the spoken query's n-best list and the goal of the system is to predict the query transcription from the n-best that yields the greatest retrieval performance, which may not necessarily be the top candidate. The goal of the system is to maximize retrieval performance over a set of input n-best lists.

Predicting When to Ask for a Spoken Query Reformulation. In certain cases, a speech-enabled search engine may decide that the input spoken query is poor and may ask the user to reformulate the query. The input spoken query may be poor due to an ASR error or the user's word choice.

We define the spoken query reformulation task as follows. The input to the system is a spoken query (specifically, the top candidate from the ASR system's n-best list) and the goal of the system is to predict whether to show the results for the input query or to ask the user for a new spoken query. We assume that asking the user for a reformulation yields a more effective query, but incurs a cost. More formally, we assume that if the system decides to *not* ask for a reformulation, then the user experiences a gain equal to the retrieval performance of the original spoken query. Otherwise, if the system *does* decide to ask for a reformulation, then the user experiences a gain equal to the retrieval performance of the *new* query *discounted* by a factor denoted by α (in the range [0,1]). The system must decide whether to ask for a new spoken query without knowing the true performance of the original (e.g., using only pre- and post-retrieval performance predictors as evidence).

To illustrate, suppose that given an input spoken query, the system decides to ask for a reformulation. Furthermore, suppose that the original query achieves an average precision (AP) value of 0.15 and that the reformulated query achieves a AP value of 0.20. In this case, the user experiences a discounted gain of AP = $\alpha \times 0.20$. If we set $\alpha = 0.50$, then the discounted gain of the new query ($0.50 \times 0.20 = 0.10$) is less than the original (0.15), and so the system made the incorrect choice. Parameter α can be varied to simulate different costs of asking a user for a spoken query reformulation. The higher the α, the lower the cost. The goal of the system is to maximize the gain over a set of input spoken queries for a given value of α.

5 Features

For both tasks, we used machine learning to combine different types of evidence as features. We grouped our features into three categories. The numbers in parentheses indicate the number of features in each category.

N-best List Features (2). These features were generated from the ASR system's n-best list. We included two n-best list features: the rank of the transcription in the n-best list and its confidence value. These features were only available for the AT&T API and only used in the n-best list re-ranking task.

Pre-retrieval Features (27). Prior work shows that a query is more likely to perform well if it contains discriminative terms that appear in only a few documents. We included five types of features aimed to capture this type of evidence. Our inverse document frequency (IDF) and inverse collection term frequency (ICTF) features measure the IDF and ICTF values across query terms [2,6,18]. We included the min, max, sum, average, and standard deviation of IDF and ICTF values across query terms. The *query-collection similarity* (QCS) score measures the extent to which the query terms appear many times in only a few documents [18]. We included the min, max, sum, average, and standard deviation of QCS values across query terms. The *query scope* score is inversely proportional to the number of documents with at least one query term [6]. Finally, the *simplified clarity* score measures the KL-divergence between the query and collection language models [6].

Prior work also shows that a query is more likely to perform well if the query terms describe a coherent topic. We included one type of feature to capture this type of evidence. Our point-wise mutual information (PMI) features measure the degree of co-occurrence between query terms [5]. We included the min, max, sum, average, and standard deviation of PMI values across query-term pairs.

Finally, a query is more likely to perform well if it produces a stable ranking. We included one type of feature to capture this type of evidence. We estimate the pre-retrieval rank stability by considering the query terms' variance of TF.IDF weights across the documents in the collection [18]. We included the min, max, average, sum, and standard deviation of the variance across query terms.

Post-Retrieval Features (5). A query is more likely to perform well if the top-ranked documents describe a coherent topic. We included three types of features to model this type of evidence. The *clarity score* measures that KL-divergence between the language model of the top documents and a background model of the collection [2]. The *query feedback* score measures the degree of overlap between the top-ranked documents before and after query-expansion [20]. A greater overlap suggests that the original query is on-topic. Finally, we consider the *normalized query commitment* (NQC) score, which measures the standard deviation of the top document scores. Following Shtok *et al.* [14], we included three NQC scores: the standard deviation of the top document scores, the standard deviation of the scores *above* the mean top-document score, and the standard deviation of the scores *below* the mean top-document score.

6 Evaluation Methodology

Retrieval performance was measured by issuing spoken query transcriptions against the TREC 2004 Robust Track collection. In all experiments, we used Lucene's implementation of the query-likelihood model with Dirichlet smoothing. Queries and documents were stemmed using the Krovetz stemmer and stopped using the SMART stopword list. We evaluated in terms of average precision (AP), NDCG@30, and P@10.

Re-ranking the N-Best List. We cast this as a learning-to-rank (LTR) task, and trained models to re-rank an n-best list in descending order of retrieval performance. At test time, we re-rank the input n-best list and select the top query transcription as the one to run against the collection. We used the linear RankSVM implementation in the Sophia-ML toolkit and trained separate models for each retrieval performance metric.

Models were evaluated using 20-fold cross-validation. Recall that each TREC topic had 20 spoken queries from different study participants. To avoid training and testing on n-best lists for the same TREC topic (potentially inflating performance), we assigned all n-best lists for the same topic to the same fold. We report average performance across held-out folds and measure statistical significance using the approximation of Fisher's randomization test described in Smucker *et al.* [15]. We used the same cross-validation folds in all our experiments. Thus, when testing significance, the randomization was applied to the 20 pairs of performance values for the two models being compared. We normalized feature values to zero-min and unit-max for each spoken query (i.e., using the min/max values from the same n-best list).

We compare against two baseline approaches: (1) selecting the best-performing candidate from the n-best list (oracle) and (2) selecting the top candidate with the highest recognition confidence (top).

Predicting Spoken Query Reformulations. We cast this as a binary classification task. The input to the system is a spoken query's most confident transcription, and the goal of the system is to predict whether to run the input query or to ask for a spoken query reformulation. If the system decides to run the input query, then the user experiences a gain equal to the retrieval performance of the original query. Otherwise, if the system decides to ask for a reformulation, then the user experiences a gain equal to the retrieval performance of the new spoken query *discounted* by α.

We simulated the spoken query reformulation task as follows. Recall that each TREC topic had 20 spoken queries. For each topic, we used the top-performing spoken query to simulate the "reformulated" query and the remaining 19 spoken queries to simulate different inputs to the system. This produced $250 \times 19 = 4,750$ instances for training and testing.

While we cast this as a binary classification task, we decided to train a regression model to predict the difference between the performance of the input query and the discounted performance of the reformulated query. Our motivation was to place more emphasis on lower-performing training instances. At test time, we simply use the sign of the real-valued output to make a binary prediction. We used the linear SVM regression implementation in the LibLinear toolkit. We trained different models for different evaluation metrics (AP, NDCG@30, P@10) and different values of α. As in the n-best list re-ranking task, we evaluated using 20-fold cross-validation and assigned all spoken queries for the same TREC topic to the same fold. Similarly, we report average performance across held-out folds and measured statistical significance using Fisher's randomization test [15].

We compare against four different baselines: (1) always making the optimal choice between the input query and asking for a reformulation (oracle), (2) *always* asking for a reformulation (always), (3) *never* asking for a reformulation (never), and (4) asking for a reformulation randomly based on the training data probability that it is the optimal choice (random). The second and third baselines are expected to perform well for high values of α (low cost) and low values of α (high cost), respectively.

7 Results

Results for the n-best list re-ranking task are presented in Table 1. For this task, we used the n-best lists produced by the AT&T API. Furthermore, we focus on the subset of 3,414 (out of 5,000) spoken queries for which the AT&T API returned an n-best list with more than one transcription. The first and last rows in Table 1 correspond to our two baseline approaches: selecting the top-ranked candidate from the n-best list (top) and selecting the best-performing candidate for the corresponding metric (oracle). The middle rows correspond to the LTR model using all features (all), all features except for those in group x (no.x), and only those features in group x (only.x).

Table 1. Results for the n-best list re-ranking task. The percentages indicate percent improvement over top. A ▲ denotes a significant improvement compared to top, and for no.x and only.x, a ▽ denotes a significant performance drop compared to all. We used Bonferroni correction for multiple comparisons ($p < .05$).

	AP	NDCG@30	P@10
top	0.081	0.148	0.159
all	0.091 (13.75 %)▲	0.162 (12.50 %)▲	0.174 (12.26 %)▲
no.nbest	0.090 (12.50 %)▲▽	0.162 (12.50 %)▲	0.173 (11.61 %)▲▽
no.pre	0.089 (11.25 %)▲▽	0.155 (7.64 %)▲▽	0.166 (7.10 %)▲▽
no.post	0.085 (6.25 %)▲▽	0.154 (6.94 %)▲▽	0.168 (8.39 %)▲▽
only.nbest	0.080 (0.00 %)▽	0.144 (0.00 %)▽	0.155 (0.00 %)▽
only.pre	0.084 (5.00 %)▲▽	0.154 (6.94 %)▲▽	0.167 (7.74 %)▲▽
only.post	0.089 (11.25 %)▲▽	0.155 (7.64 %)▲▽	0.166 (7.10 %)▲▽
oracle	0.102 (27.50 %)▲	0.186 (29.17 %)▲	0.205 (32.26 %)▲

The results from Table 1 suggest five important trends. First, the LTR model using all features (all) significantly outperformed the baseline approach of always selecting the top-ranked transcription from the n-best list (top). The LTR model using all features had a greater than 10 % improvement across all metrics.

Second, our results suggest that both pre- and post-retrieval query performance predictors contribute useful evidence for this task. The LTR model using only pre-retrieval features (only.pre) and only post-retrieval features (only.post) significantly outperformed the top baseline across all metrics. Furthermore, in all cases, individually ignoring pre-retrieval features (no.pre) and post-retrieval features (no.post) resulted in a significant drop in performance compared to the LTR model using all features (all).

Third, there is some evidence that post-retrieval features were more predictive than pre-retrieval features. In terms of AP, ignoring post-retrieval features (no.post) and using *only* pre-retrieval features (only.pre) had the greatest performance drop compared to the model using all features (all). In terms of AP, post-retrieval features were more predictive in spite of having only 5 post-retrieval features versus 27 pre-retrieval features.

The fourth trend worth noting is that n-best list features contributed little useful evidence. In most cases, ignoring n-best list features (no.nbest) resulted in only a small drop in performance compared to the LTR model using all features (all). Furthermore, the LTR model using only n-best list features (only.nbest) was the worst-performing LTR model across all metrics and performed at the same level as the top baseline.

The final important trend is that there is still room for improvement. Across all metrics, the oracle performance was at least 25 % greater than the top baseline. While not shown in Table 1, the oracle outperformed all the LTR models and the top baseline across all metrics by a statistically significant margin ($p < .05$).

Results for the task of predicting when to ask for a spoken query reformulation are shown in Tables 2 and 3. To conserve space, we only show results in terms of AP. However, the results in terms of NDCG@30 and P@10 had the same trends. We show results using the most confident transcriptions from the AT&T API (Table 2) and the WIT.AI API (Table 3). Because the WIT.AI API only returned the most confident transcription without a confidence value, we ignore n-best lists features in this analysis. Results are presented for different values of α, with higher values indicating a higher cost of asking for a reformulation and therefore fewer cases where it was the correct choice. We show results for our four baselines (oracle, always, never, and random), as well as the regression model using all features (all), ignoring pre-retrieval features (no.pre), and ignoring post-retrieval features (no.post). The performance of never asking for a reformulation (never) is constant because it is independent of α. The performance of always asking for a reformulation (always) increases with α (lower cost).

The results in Tables 2 and 3 suggest three important trends. First, the model using all features (all) performed equal to or better than always, never, and random for both APIs and all values of α. The model performed at the same level as never for low values of α (high cost) and at the same level as always for high values of α (low cost). The model outpormed these three baselines for values of α in the mid-range ($0.4 \leq \alpha \leq 0.6$). For these values of α, the system had to be more selective about when to ask for a reformulation. These results show that pre- and post-retrieval performance predictors provide useful evidence for predicting when the input spoken query is relatively poor.

Second, post-retrieval features were more predictive than pre-retrieval features. This is consistent with the AP results from Table 1. For values of α in the mid-range, ignoring post-retrieval (no.post) features resulted in a greater drop in performance than ignoring pre-retrieval features (no.pre). The drop in performance was statistically significant for two values of α for the AT&T results

Table 2. Results for predicting when to ask for a spoken query reformulation: AT&T API, Average Precision. A $^{\blacktriangle}$ denotes a significant improvement compared to always, never, and random. A $^{\triangledown}$ denotes a significant performance drop in for no.pre and no.post compared to all. We report significance for $p < .05$ using Bonferroni correction.

discount (α)	oracle	always	never	random	all	no.pre	no.post
0.1	0.113	0.027	0.102	0.069	0.103	0.102 (-0.97 %)	0.102 (-0.97 %)
0.2	0.125	0.054	0.102	0.076	0.108$^{\blacktriangle}$	0.106 (-1.85 %)$^{\blacktriangle\triangledown}$	0.105 (-2.78 %)$^{\triangledown}$
0.3	0.138	0.082	0.102	0.092	0.122$^{\blacktriangle}$	0.119 (-2.46 %)$^{\blacktriangle}$	0.113 (-7.38 %)$^{\blacktriangle\triangledown}$
0.4	0.153	0.109	0.102	0.106	0.137$^{\blacktriangle}$	0.135 (-1.46 %)$^{\blacktriangle}$	0.130 (-5.11 %)$^{\blacktriangle}$
0.5	0.168	0.136	0.102	0.126	0.153$^{\blacktriangle}$	0.152 (-0.65 %)$^{\blacktriangle}$	0.149 (-2.61 %)$^{\blacktriangle}$
0.6	0.185	0.163	0.102	0.146	0.172$^{\blacktriangle}$	0.171 (-0.58 %)$^{\blacktriangle}$	0.167 (-2.91 %)$^{\blacktriangle}$
0.7	0.204	0.191	0.102	0.170	0.193	0.193 (0.00 %)	0.191 (-1.04 %)
0.8	0.224	0.218	0.102	0.198	0.218	0.218 (0.00 %)	0.218 (0.00 %)
0.9	0.247	0.245	0.102	0.231	0.245	0.245 (0.00 %)	0.245 (0.00 %)

Table 3. Results for predicting when to ask for a spoken query reformulation: WIT.AI API, Average Precision. Symbols ▲ and ▽ denote statistically significant differences as described in Table 2.

discount (α)	oracle	always	never	random	all	no.pre	no.post
0.1	0.180	0.031	0.176	0.149	0.176	0.176 (0.00 %)	0.176 (0.00 %)
0.2	0.186	0.062	0.176	0.142	0.176	0.176 (0.00 %)	0.176 (0.00 %)
0.3	0.194	0.093	0.176	0.149	0.179	0.178 (-0.56 %)▲	0.177 (-1.12 %)
0.4	0.203	0.124	0.176	0.154	0.185▲	0.182 (-1.62 %)	0.180 (-2.70 %)
0.5	0.214	0.156	0.176	0.165	0.196▲	0.192 (-2.04 %)▲	0.188 (-4.08 %)▲▽
0.6	0.227	0.187	0.176	0.182	0.208▲	0.207 (-0.48 %)▲	0.201 (-3.37 %)
0.7	0.243	0.218	0.176	0.203	0.225	0.224 (-0.44 %)	0.219 (-2.67 %)
0.8	0.261	0.249	0.176	0.229	0.250	0.250 (0.00 %)	0.247 (-1.20 %)
0.9	0.284	0.280	0.176	0.263	0.280	0.280 (0.00 %)	0.280 (0.00 %)

and one value of α for the WIT.AI results. Again, we observed this trend in spite of having fewer post-retrieval than pre-retrieval features.

Finally, we note that there is room for improvement. For both APIs, the oracle baseline (oracle) outperformed the model using all features (all) across all values of α. While not shown in Tables 2 and 3, all differences between oracle and all were statistically significant ($p < .05$).

8 Discussion

Our results from Sect. 7 show that the top candidate from an ASR system's n-best list is not necessarily the best-performing query and that we can use query performance predictors to find a lower-ranked candidate that performs better. A reasonable question is: Why is the most confident candidate not always the best query? We examined n-best lists where a lower-ranked candidate outperformed the most confident, and encountered cases belonging to three categories.

In the first category, the lower-ranked candidate was a more accurate transcription of the input speech. For example, the lower-ranked candidate 'protect children poison paint' (AP = 0.467) outperformed the top candidate 'protect children poison pain' (AP = 0.055). Similarly, the lower-ranked candidate 'prostate cancer detect treat' (AP = 0.301) outperformed the top candidate 'press cancer detect treat' (AP = 0.014). Finally, the lower-ranked candidate 'drug treat alzheimer successful' (AP = 0.379) outperformed the top candidate 'drug treat timer successful' (AP = 0.001). We do not know why the ASR system assigned the correct transcription a lower probability. It may be that the correct query terms ('paint', 'prostate', 'alzheimer') had a lower probability in the ASR system's language model than those in the top candidates ('pain', 'press', 'timer'). Such errors might be reduced by using a language model from the target collection. However, this may not be possible with an off-the-shelf ASR system.

In the second category, the user mispronounced an important word associated with the task. In these cases, the top candidate was a better match with the input speech, but a lower-ranked candidate had the correct spelling of the word. For example, the lower-ranked candidate 'articles lives nobel prize winner' (AP = 0.289) outperformed the top candidate 'articles lives noble prize winner' (AP = 0.007). Here, the participant mispronounced 'nobel' as 'noble'. Similarly, the lower-ranked candidate 'welsh devolution movement' (AP = 0.460) outperformed the top candidate 'welsh deevolution movement' (AP = 0.050). In this case, the participant mispronounced 'devolution' as 'de-evolution'.

In the third category, none of the candidates were a perfect transcription of the input speech, but a lower-ranked candidate had an ASR error that was less important for the search task. For example, the lower-ranked candidate 'resend relations britain argentina' (AP = 0.443) outperformed the top candidate 'recent relations brandon argentina' (AP = 0.065). The top candidate had 'brandon' versus 'britain', while the lower-ranked candidate had 'resend' versus 'recent'. While both candidates had exactly one ASR error and similar confidence values, the error in the lower-ranked candidate yielded a more effective query for this task. In fact, one might argue that the lower-ranked candidate describes a more coherent topic, as indicated by its higher clarity score (3.180 versus 2.612). Cases in this category possibly arise when the ASR system is tuned to minimize word error rate [12], without explicitly favoring candidates that describe a coherent topic with respect to the target collection.

9 Conclusion

We developed and evaluated models for two tasks associated with speech-enabled search: (1) re-ranking the ASR system's n-best hypotheses and (2) predicting when to ask for a spoken query reformulation. Our results show that pre- and post-retrieval performance predictors contribute useful evidence for both tasks. With respect to the first task, our analysis shows that lower-ranked candidates in the n-best list may perform better due to mispronunciation errors in the input speech or because the ASR system may not explicitly favor candidates that describe a coherent topic with respect to the target collection.

There are several directions for future work. In this work, we improved the input query by exploring candidates in the same n-best list. Future work might consider exploring a larger space, including reformations of the top candidate that are specifically designed for the speech domain (e.g., term substitutions with similar Soundex codes). Additionally, in this work, we predicted when to ask for a new spoken query. Future work might consider learning to ask more targeted clarification or disambiguation questions about the input spoken query.

Acknowledgments. This work was supported in part by NSF grant IIS-1451668. Any opinions, findings, conclusions, and recommendations expressed in this paper are the authors and do not necessarily reflect those of the sponsors.

References

1. Aslam, J.A., Pavlu, V.: Query Hardness Estimation Using Jensen-Shannon Divergence Among Multiple Scoring Functions. In: Amati, G., Carpineto, C., Romano, G. (eds.) ECiR 2007. LNCS, vol. 4425, pp. 198–209. Springer, Heidelberg (2007)
2. Cronen-Townsend, S., Zhou, Y., Croft, W.B.: Predicting query performance. In: SIGIR (2002)
3. Dang, V., Bendersky, M., Croft, W.B.: Learning to rank query reformulations. In: SIGIR (2010)
4. Diaz, F.: Performance prediction using spatial autocorrelation. In: SIGIR (2007)
5. Hauff, C.: Predicting the effectiveness of queries and retrieval systems. dissertation, Univeristy of Twente (2010)
6. He, B., Ounis, I.: Inferring Query Performance Using Pre-retrieval Predictors. In: Apostolico, A., Melucci, M. (eds.) SPIRE 2004. LNCS, vol. 3246, pp. 43–54. Springer, Heidelberg (2004)
7. Jiang, J., Jeng, W., He, D.: How do users respond to voice input errors?. SIGIR, Lexical and phonetic query reformulation in voice search. In (2013)
8. Kumaran, G., Carvalho, V.R.: Reducing long queries using query quality predictors. In: SIGIR (2009)
9. Li, X., Nguyen, P., Zweig, G., Bohus, D.: Leveraging multiple query logs to improve language models for spoken query recognition. In: ICASSP (2009)
10. Mamou, J., Sethy, A., Ramabhadran, B., Hoory, R., Vozila, P.: Improved spoken query transcription using co-occurrence information. In: INTERSPEECH (2011)
11. Peng, F., Roy, S., Shahshahani, B., Beaufays, F.: Search results based n-best hypothesis rescoring with maximum entropy classification. In: IEEE Workshop on Automatic Speech Recognition and Understanding (2013)
12. Schalkwyk, J., Beeferman, D., Beaufays, F., Byrne, B., Chelba, C., Cohen, M., Kamvar, M., Strope, B.: Your word is my command: Google search by voice: A case study. In: Neustein, A. (ed.) Advances in Speech Recognition. Springer, Heidelberg (2010)
13. Sheldon, D., Shokouhi, M., Szummer, M., Craswell, N.: Lambdamerge: Merging the results of query reformulations. In: WSDM (2011)
14. Shtok, A., Kurland, O., Carmel, D., Raiber, F., Markovits, G.: Predicting query performance by query-drift estimation. TOIS, 30(2) (2012)
15. Smucker, M.D., Allan, J., Carterette, B.: A comparison of statistical significance tests for information retrieval evaluation. In: CIKM (2007)
16. Xue, X., Huston, S., Croft, W.B.: Improving verbose queries using subset distribution. In: CIKM (2010)
17. Yom-Tov, E., Fine, S., Carmel, D., Darlow, A.: Learning to estimate query difficulty: Including applications to missing content detection and distributed information retrieval. In: SIGIR (2005)
18. Zhao, Y., Scholer, F., Tsegay, Y.: Effective Pre-retrieval Query Performance Prediction Using Similarity and Variability Evidence. In: Macdonald, C., Ounis, I., Plachouras, V., Ruthven, I., White, R.W. (eds.) ECIR 2008. LNCS, vol. 4956, pp. 52–64. Springer, Heidelberg (2008)
19. Zhou, Y., Croft, W.B.: A novel framework to predict query performance. In: CIKM (2006)
20. Zhou, Y., Croft, W.B.: Query performance prediction in web search environments. In: SIGIR (2007)

Fusing Web and Audio Predictors to Localize the Origin of Music Pieces for Geospatial Retrieval

Markus Schedl[1(✉)] and Fang Zhou[2]

[1] Department of Computational Perception,
Johannes Kepler University, Linz, Austria
markus.schedl@jku.at

[2] Center for Data Analytics and Biomedical Informatics,
Temple University, Philadelphia, USA
fang.zhou@temple.edu

Abstract. Localizing the origin of a music piece around the world enables some interesting possibilities for geospatial music retrieval, for instance, location-aware music retrieval or recommendation for travelers or exploring non-Western music – a task neglected for a long time in music information retrieval (MIR). While previous approaches for the task of determining the origin of music either focused solely on exploiting the audio content or web resources, we propose a method that fuses features from both sources in a way that outperforms standalone approaches. To this end, we propose the use of block-level features inferred from the audio signal to model music content. We show that these features outperform timbral and chromatic features previously used for the task. On the other hand, we investigate a variety of strategies to construct web-based predictors from web pages related to music pieces. We assess different parameters for this kind of predictors (e.g., number of web pages considered) and define a confidence threshold for prediction. Fusing the proposed audio- and web-based methods by a weighted Borda rank aggregation technique, we show on a previously used dataset of music from 33 countries around the world that the median placing error can be reduced from 1,815 to 0 kilometers using K-nearest neighbor regression.

1 Introduction

Predicting the location of a person or item is an appealing task given today's omnipresence and abundance of information about any topic on the web and in social media, which are easy to access through corresponding APIs. While a majority of research focuses on automatically placing images [5], videos [22], or social media users [1,7], we investigate the problem of placing music at its location of origin, focusing on the country of origin, which we define as the main country or area of residence of the artist(s). We approach the task by audio content-based and web-based strategies and eventually propose a hybrid

N. Ferro et al. (Eds.): ECIR 2016, LNCS 9626, pp. 322–334, 2016.
DOI: 10.1007/978-3-319-30671-1_24

method that fuses these two sources. We show that the fused method is capable of outperforming stand-alone approaches.

The availability of information about a music piece's or artist's origin opens interesting opportunities, not only for computational ethnomusicology [3], but also for location-aware music retrieval and recommendation systems. Examples include browsing and exploration of music from different regions in the world. This task seems particularly important as the strong focus on Western music in music information retrieval (MIR) research has frequently been criticized [13, 17]. Other tasks that benefit from information about the origin of music are trend analysis and prediction. If we understand better where a particular music trend emerges – which is strongly related to the music's origin – and how it spreads (e.g., locally, regionally, or globally), we could use this information for personalized and location-aware music recommendation or for predicting the future popularity of a song, album, artist, or music video [10, 25]. Another use case is automatically selecting music suited for a given place of interest, a topic e-tourism is interested in [6].

The remainder of this paper is structured as follows. Section 2 presents related work and highlights the main contributions of the paper at hand. Section 3 presents the proposed audio- and web-based methods as well the hybrid strategy. Section 4 outlines the evaluation experiments we conducted, presents and discusses their results. Eventually, Sect. 5 rounds off the paper with a summary and pointers to future research directions.

2 Related Work

The research task of automatically position a given multimedia item, such as an image [24], video [22], or text message [7] has received considerable attention in recent years. Also approaches to localize social media users, for instance via deep neural networks have been proposed [11]. Predicting the position or origin of a music entity, such as a music piece, composer, or performer, has been studied to a smaller extent so far.

However, identifying an artist's or piece's origin provides valuable clues about their background and musical context. For instance, a performer's geographic, political, and cultural context or a songwriter's lyrics might be strongly related to their origin. Our problem is to predict the origin of music, relating data values with their spatial location, which is one of the spatial statistics [12]. In the literature, two strategies have been followed to approach this goal: exploiting musical features extracted from the audio content and building predictors based on information harvested from the web.

Audio-based Approaches. One kind of approach is to automatically learn connections between audio features and geographic information of music. Even though this strategy may be considered the most straightforward one, to the best of our knowledge, there exists only one paper exploiting audio signal-based features for the task. Zhou et al. [26] first analyze geographical distribution of music by extracting and analyzing audio descriptors through the MARSYAS [23] software. They use spectral, timbral and chroma features. The authors then apply K-nearest neighbor (KNN) and random forest regression methods for prediction.

Web-based Approaches. Another category of methods approach the problem via web mining. Govaerts and Duval [4] search for occurrences of country names in biographies on Wikipedia[1] and Last.fm,[2] as well as in properties such as *origin*, *nationality*, *birthplace*, and *residence* on Freebase.[3] The authors then apply heuristics to predict the most probable country of origin for the target artist or band. For instance, one of their heuristics predict the country name that most frequently occurs in an artist's biography. Another one favors early occurrences of country names in the text. Govaerts and Duval show that combining the results of different data sources and heuristics yields superiors results. Schedl et al. propose three approaches that try to predict the country of origin from web pages identified by search engines [14,16]. One approach is a heuristic that compares the page count estimates returned by Google for queries of the form `''artist/band'' +country` and simply predicts the country with highest page count value for a given artist or band. Another approach takes into account the actual content of the web pages. For this purpose, up to 100 top-ranked web pages for each artist are downloaded and TF·IDF weights are computed. The country of origin for a given artist is then predicted as the country with highest TF·IDF score using the artist name as query. Their third approach computes text distances between country names and key terms such as "born" or "founded" in the set of web pages retrieved for the target artist. The country whose name appears closest to any of the key terms is eventually predicted. Schedl et al. show that the approach based on TF·IDF weighting outperforms the other two methods. A shortcoming of all of these web-based approaches is that they only operate on the level of artists, not on pieces. In this paper, by contrast, we build a web-based approach using as input the name of the music piece under consideration.

A related problem is to predict countries in which a music item or artist is particularly popular. This might correspond to their country of origin. Koenigstein et al. propose an approach based on localizing IP addresses of queries issued in peer-to-peer networks [10]. For the same task, they also look into the content of folders users share on peer-to-peer networks [9].

The main contributions of the paper at hand are (i) the investigation of the state-of-the-art audio similarity measure based on the block-level feature extraction framework [20,21] for the task of predicting the country of origin of individual music pieces, (ii) an extension and comprehensive evaluation of the state-of-the-art web-based method to predict the country of origin of artists, and (iii) a novel method that fuses audio- and web-based predictors. All of these methods are evaluated in classification and regression experiments.

[1] https://en.wikipedia.org.
[2] http://www.last.fm.
[3] http://www.freebase.com.

3 Localizing the Origin of Music Pieces

3.1 Music Features

For content-based description of the music pieces, we used a set of six features defined in the block-level framework (BLF) [21]. This choice is motivated by the fact that these features already proved to perform very well for music similarity and retrieval tasks [18], for music autotagging [19], and for content-based modeling in location-aware music recommender systems [6]. They have, however, never been exploited for the task at hand.

The BLF describes a music piece by defining overlapping blocks over the spectrogram of the audio signal, in which frequency is modeled on the Cent scale, as illustrated in Fig. 1 (top). Concretely, a window size of 2,048 samples per frame and a hop size of 512 samples are used to compute the short time Fourier transform on the Hanning-windowed frames of the audio signal. The resulting magnitude spectrum exhibits linear frequency resolution, it is mapped onto the logarithmic Cent scale to account for the human perception of music.

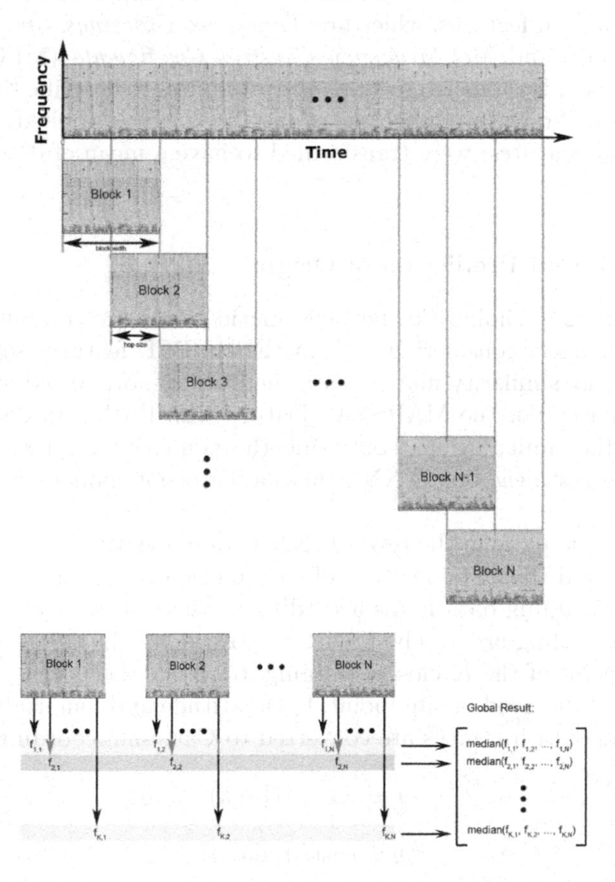

Fig. 1. Overview of the feature extraction (top) and summarization process (bottom) in the block-level framework.

From the resulting Cent spectrogram representation, several features computed on blocks of frames are inferred. Within the BLF, we use the following features: *Spectral Pattern* (SP) characterizes the frequency content, *Delta Spectral Pattern* (DSP) emphasizes note onsets, *Variance Delta Spectral Pattern* (VDSP) captures variations of onsets over time, *Logarithmic Fluctuation Pattern* (LFP) describes the periodicity of beats, *Correlation Pattern* (CP) models the correlation between different frequency bands, and *Spectral Contrast Pattern* (SCP) uses the difference between spectral peaks and valleys to identify tonal and percussive elements. Since the features for a given music piece are computed on the level of blocks, all features of the same kind are eventually aggregated using a statistical summarization function (typically, percentiles or variance), which is illustrated in Fig. 1 (bottom). After this aggregation, each music piece is described by six feature vectors of different dimensionality, totalling to 9,948 individual feature values.

For comparison to the previous audio-based state-of-the-art method [26], we also considered two other groups of audio features, NMdef and NMdefchrom, which were extracted by the program MARSYAS [23]. NMdef contains basic spectral and timbral features, which are *Time Zero Crossings*, *Spectral Centroid*, *Flux* and *Rolloff*, and *Mel-Frequency Cepstral Coefficients* (MFCC), whereas NMdefchrom includes additional *chromatic* attributes to describe the notes of the scale being used. No feature weighting or pre-filtering was applied. All numerical features (i.e. all features) were transformed to have a mean of 0 and a standard deviation of 1.

3.2 Audio-Based Prediction of Origin

As proposed in [21], similarities between music pieces are computed as inverse Manhattan distance, considering each of the six BLF features separately. The corresponding six similarity matrices are then Gauss-normalized and eventually linearly combined. For the MARSYAS feature sets, Euclidean distance is used to construct the similarity matrix. Using the similarity matrix, we apply the standard K-nearest neighbor (KNN) [26] as a regression model for the prediction of origin.

For each music piece in the test set, KNN computes the distance between its audio features and the audio features of each music item in the training set, and then sorts the training data items according to the feature distance to the test music in an ascending order. The predicted position of the target music piece is then the midpoint of the K closest training items' spatial position.

To calculate the geodesic midpoint, both latitude and longitude (ϕ, λ) in the top K training data instances are converted to Cartesian coordinates (x, y, z).

$$x = \cos(\phi)\cos(\lambda) \tag{1}$$

$$y = \cos(\phi)\sin(\lambda) \tag{2}$$

$$z = \sin(\phi) \tag{3}$$

The average coordinates $(\bar{x}, \bar{y}, \bar{z})$ are converted into the latitude and longitude (ϕ^p, λ^p) for the midpoint.

$$\phi^p = \arctan 2(\bar{z}, \sqrt{\bar{x}^2 + \bar{y}^2}) \tag{4}$$

$$\lambda^p = \arctan 2(\bar{y}, \bar{x}) \tag{5}$$

The quality of prediction is then measured by calculating the great circle distance from the true position $L^{Te} = (\phi^{Te}, \lambda^{Te})$ to the predicted position $L^p = (\phi^p, \lambda^p)$. The great circle distance $d(L^{Te}, L^p)$ between two points $(\phi^{Te}, \lambda^{Te})$ and (ϕ^p, λ^p) on the surface of the earth is defined as

$$\begin{aligned} d(L^{Te}, L^p) &= 2 \cdot R \cdot \arctan 2(\sqrt{a}, \sqrt{1-a}), \\ a &= \sin^2(\frac{\phi^p - \phi^{Te}}{2}) + \cos\phi^p \cos\phi^{Te} \sin^2(\frac{\lambda^p - \lambda^{Te}}{2}), \end{aligned} \tag{6}$$

where $R = 6{,}373$ kilometers.

3.3 Web-based Country of Origin Prediction

To make predictions for a given music piece p, we first fetch the top-ranked web pages returned by the Bing Search API[4] for several queries: `''piece'' music`, `''piece'' music biography`, and `''piece'' music origin`, in which `''piece''` refers to the exact search for the music piece's name.[5,6] In the following, we abbreviate these query settings by M, MB, and MO, respectively. We subsequently concatenate the content of the retrieved web pages for each p to yield a single document for p. Previous web-based approaches for the task at hand [14,16] only considered the problem at the artist level and only employed the query scheme `''artist'' music`.

Given a list of country names, we compute the term frequency (TF) of all countries in the document of p, and we predict the K countries with highest scores. We do not perform any kind of normalization, nor account for different overall frequencies of country names. This choice was made in accordance with previous research on the topic of country of origin detection, as [16] shows that TF outperforms TF·IDF weighting, and also outperforms more complex rule-based approaches.

In addition to different query settings (M, MB, and MO), we also consider fetching either 20 or 50 web pages per music piece. Knees et al. investigate the influence of different numbers of fetched web pages for the task of music similarity and genre classification [8]. According to the authors, considering more than 50 web pages per music item does not significantly improve results, in some cases

[4] https://datamarket.azure.com/dataset/bing/search.

[5] Please note that the obvious query scheme `''piece'' (music) country` does not perform well as it results in too many irrelevant pages about country music.

[6] Please further note that investigating queries in languages other than English is out of the scope of the work at hand, but will be addressed as part of future work.

even worsens them. Since overall best results were achieved when considering between 20 and 50 pages, we investigate these two numbers.

In order to control for uncertainly in made predictions, we further introduce a confidence parameter α. For each of the top K countries predicted for p, we relate its TF value to the sum of TF values of all predicted countries. We only keep a country c predicted for p if its resulting relative TF value is at least α, C_p being the set of top K countries predicted for p:

$$\frac{TF(p,c)}{\sum_{c \in C_p} TF(p,c)} \geq \alpha \tag{7}$$

3.4 Fusing Audio- and Web-based Predictions

In order to fuse the predictions of our audio-based and web-based algorithms, we propose a variant of the Borda rank aggregation technique [2] with a mixture parameter ξ for linear combination of scores. Variants of this aggregation technique have already been proven useful for music recommendation tasks [6]. Our approach first ranks separately the predictions made by the audio- and the web-based method for a given piece p and then converts ranks to scores, i.e., the top-ranked country among K receives a score of K, the second ranked a score of $K-1$, and so on. The individual scores for each country are then added up over the approaches to fuse. Since previous research on hybrid music similarity has shown that a simple linear weighting of individual similarities performs well [15], we use a parameter ξ that controls the weight of the audio-based scores. The whole scoring function is thus

$$s(p,c) = \xi \cdot s_{audio}(p,c) + (1-\xi) \cdot s_{web}(p,c). \tag{8}$$

4 Evaluation

We used the dataset presented by Zhou et al. [26], containing 1,059 pieces of music originating from 33 countries. Music was selected based on the following two criteria: First, no "Western" music is included, as its influence is global. We only consider the music that is strongly influenced by a specific location, namely only traditional, ethnic, or "World" music was included in this study, Second, any music that has ambiguous origin was removed from the dataset. The geographical origin was collected from compact disc covers. Since most location information is country names, we used the country's capital city (or the province of the area) to represent the absolute point of origin (represented as latitude and longitude), assuming that the political capital is also the cultural capital of the country or area. The country of origin is determined by the artist's or artists' main country or area of residence. If an artist has lived in several different places, we only consider the place that presumably had major influence. For example, if a Chinese artist is living in New York but composes a piece of traditional Chinese music, we take it as Chinese music.

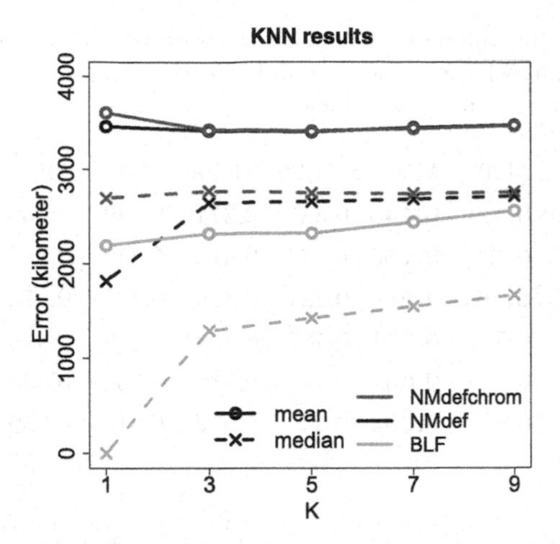

Fig. 2. Mean and median distance errors (kilometers) as a function of K values.

For evaluation, all music from the same country is equally distributed among 10 groups. We then apply 10-fold cross-validation and report the mean and median error distance (in kilometers) from the true positions to their corresponding predicted positions. We also measure prediction accuracy, i.e. the percentage of music pieces assigned to the correct class, treating countries as classes.

4.1 Prediction Performance of Audio-based Approach

We first compare the prediction performance of using different audio-based features, which is accessed by two criteria, mean distance (solid lines) between the true and the predicted geographical points and median distance (dashed lines). In Fig. 2, the black and blue lines are the results of using the baseline features from [26], extracted by MARSYAS, and the orange line represents the performance of the block-level features (cf. Sect. 3.1). The smallest mean distance error achieved using the NMdef and NMdefchrom features (cf. Sect. 3) is 3,410 km, and the smallest median distance error (1,815 km) is achieved using the NMdef features. In contrast, using BLF features the smallest mean distance error is 2,191 km, and median distance error is 0 km. All these results obtained considering only one nearest neighbor. Furthermore, Fig. 2 clearly shows that the BLF features yield the best results over the whole range of investigated K values when evaluated by both mean and median distance.

4.2 Prediction Performance of Web-based Approach

Table 1 shows the predictive accuracies for the different parameter settings (α, query scheme, and number of retrieved web pages). Obviously, the `''piece''` `music` (M) query setting yields the best results, compared to the other two query settings. This is in line with similar findings that strongly restricting the

Table 1. Accuracies for different variants of the web-based approach (query settings and number of web pages) and various confidence thresholds α. Settings yielding the highest performance are printed in boldface.

α	M20	M50	MB20	MB50	MO20	MO50
0.0	0.439	**0.445**	0.387	0.371	0.349	0.359
0.1	0.439	**0.450**	0.388	0.372	0.349	0.362
0.2	**0.461**	0.460	0.421	0.420	0.365	0.372
0.3	0.503	**0.551**	0.487	0.464	0.442	0.428
0.4	0.573	**0.636**	0.510	0.519	0.532	0.538
0.5	0.641	**0.684**	0.565	0.572	0.612	0.559

search may yield too specific results and in turn deteriorate performance [8]. As expected, for the same query setting, a higher confidence threshold improves the predictive accuracy.

Regarding the number of web pages, in general, using more web pages increases the amount of information considered. However, only the general query setting M seems to benefit from this. For a threshold of $\alpha = 0.5$, accuracy increases by 4.3 percentage points when comparing the M20 to the M50 setting. In contrast, for the other query settings MB and MO, no substantial increase (MB) or even a decrease (MO) can be observed, when using a large number of pages (and a high α). Taking a closer look at the fetched pages, we identified as a reason for this an increase of irrelevant or noisy pages using the more specific query settings. This might, however, also be influenced by the fact that we are dealing with "World" music. Therefore, many pieces in the collection are not very prominently represented on the web, meaning a rather small number of relevant pages is available.

4.3 Prediction Performance of Hybrid Approach

Figure 3 shows the predictive accuracy for 1-NN for the different parameters (confidence threshold α and mixture coefficient ξ) of the hybrid approach. When ξ equals 0, it means there is no audio-based prediction input. With increasing ξ values, the weight of the audio-based predictions is increasing; when ξ reaches 1, solely audio-based predictions are made. We can clearly observe from Fig. 3 a strong improvement of the web-based results when adding audio-based predictions, irrespective of the confidence threshold α. For large confidence thresholds, including only a small fraction of audio-based predictions actually increases accuracy the most. This means the more confident we are in the web-based predictions, the less audio-based predictions we need to include. Nevertheless, even when $\alpha = 0$, i.e., we consider all web-based predictions, including audio improves performance. We can also observe that different mixture coefficients ξ are required for different levels of α in order to reach peak performance.

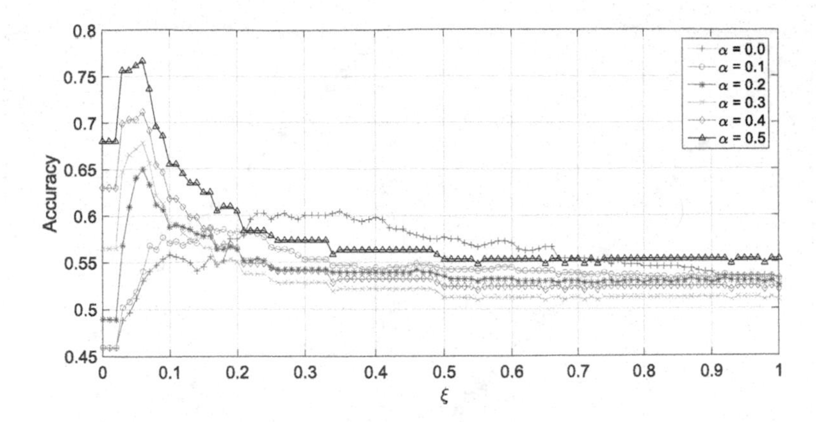

Fig. 3. Accuracies achieved by the proposed hybrid approach, for different values of parameters α (confidence threshold for web-based predictions) and ξ (mixture coefficient for Borda rank aggregation).

Table 2. Mean distance error (kilometers) for 1-NN predictions for the audio-based prediction and the hybrid approach with different confidence thresholds α and ξ.

α	ξ	BLF+M50	BLF
0.0	0.35	2,656	2,621
0.1	0.16	2,791	2,616
0.2	0.06	2,540	2,576
0.3	0.06	2,221	2,769
0.4	0.06	2,077	2,833
0.5	0.06	1,825	2,749

Based on the results in Fig. 3, we chose the corresponding ξ for different α levels, and then applied KNN regression on the dataset. The respective mean distance errors are shown in Table 2. Please note that the median distance error is 0 km for all settings since the accuracy is always $> 50\,\%$. The best result that the hybrid approach could reach is 1,824 km, whereas the minimum error of the web-based approach is 2,748 km.

4.4 Comparison of Approaches with Respect to Country Confusions

To further assess the types of mistakes made by the different approaches, we show in Fig. 4 the country-wise confusion matrices. Rows correspond to the true countries, columns correspond to the predicted countries. From the figure we can observe that the predicted locations of origin are spread across the whole matrix when using the audio-based approach (Fig. 4(a)), whereas the web-based approach tends to frequently make the same kinds of errors (Fig. 4(b)). The audio-based predictor obviously has particular problems correctly localizing Japanese

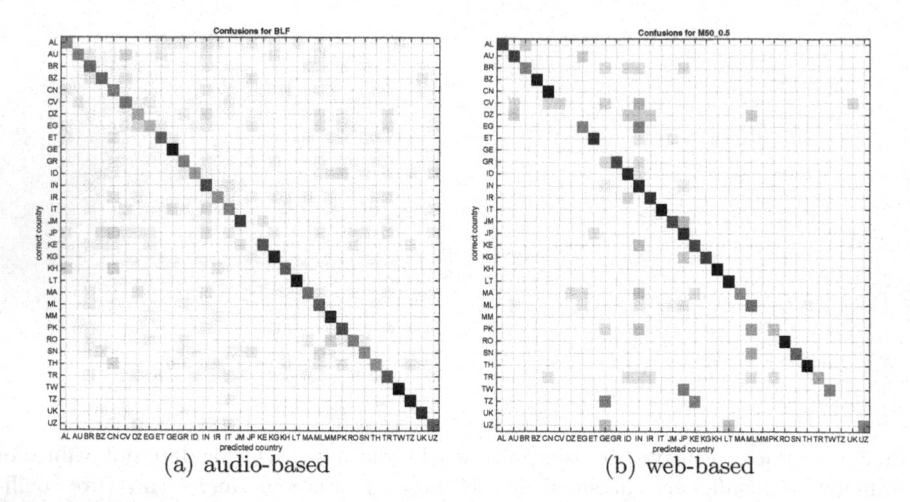

(a) audio-based (b) web-based

Fig. 4. Confusion matrix for the audio-based BLF approach and the web-based approach (M, p = 50, α = 0.5). Country names are encoded according to ISO 3166-1 alpha-2 codes.

(JP) music. The web-based method frequently misclassifies music as originating in Georgia (GE), India (IN), or Japan (JP), but on the other hand correctly classifies almost all music truly being from Japan. In contrast, Algerian (DZ) and Tanzanian (TZ) music is always misclassified by the web-based predictor.

Due to the different behavior of the audio- and web-based predictors in terms of errors made, fusing the results of both in the way we proposed in Sect. 3.4 yields better results than the stand-alone approaches.

5 Conclusion and Outlook

We have proposed a novel approach that fuses audio content-based and web-based strategies to predict the geographical origin of pieces of music. We further investigated for this task the use of block-level audio features (BLF), which are already known to perform well for music classification and autotagging. The proposed web-based predictor extends a previous approach, which we modified to (i) make predictions on the level of pieces, not only artists, (ii) consider different query schemes and (iii) numbers of fetched web pages, and (iv) include a confidence threshold for predictions based on relative TF weights.

We conducted KNN experiments on a standardized dataset consisting of 1,059 pieces of music originating from 33 countries. From the experimental results, we conclude that: (i) the audio-based approach that uses block-level features outperforms other standard audio descriptors, such as spectral, timbral, and chromatic features, for the task, (ii) for music from most countries, web-based results are superior to audio-based results, and (iii) the hybrid method produces substantially better results than the single audio-based and web-based approaches.

Given the large amount of non-Western music in the collection, we will look into multilingual extensions to our web-based approach. Furthermore, based on the finding that, for the used dataset, more specific query schemes deteriorate performance, rather then boost it, which is because of the small amount of relevant web pages, we will investigate whether this finding also holds for mainstream Western music. To this end, we will additionally investigate larger datasets, as ours is relatively small in comparison to the ones used in geolocalizing other kinds of multimedia material. We also plan to create more precise annotations for the origin of the pieces since the current granularity, i.e. the capital of the country or area, may introduce a distortion of results. Finally, we plan to look into data sources other than web pages, for instance social media, and to investigate aggregation techniques other than Borda rank aggregation.

Acknowledgments. This research is supported by the Austrian Science Fund (FWF): P25655. The authors would further like to thank Klaus Seyerlehner for his implementation of the block-level feature extraction framework and Ross D. King and the reviewers for their valuable comments on the manuscript.

References

1. Cheng, Z., Caverlee, J., Lee, K.: You are where you tweet: a content-based approach to geo-locating twitter users. In: Proceedings of CIKM, October 2010
2. de Borda, J.-C.: Mémoire sur les élections au scrutin. Histoire de l'Académie Royale des Sciences (1781)
3. Gómez, E., Herrera, P., Gómez-Martin, F.: Computational ethnomusicology: perspectives and challenges. J. New Music Res. **42**(2), 111–112 (2013)
4. Govaerts, S., Duval, E.: A web-based approach to determine the origin of an artist. In: Proceedings of ISMIR, October 2009
5. Hauff, C., Houben, G.-J.: Placing Images on the world map: a microblog-based enrichment approach. In: Proceedings of SIGIR, August 2012
6. Kaminskas, M., Ricci, F., Schedl, M.: Location-aware music recommendation using auto-tagging and hybrid matching. In: Proceedings of RecSys, October 2013
7. Kinsella, S., Murdock, V., O'Hare, N.: "I'm eating a sandwich in Glasgow": modeling locations with tweets. In: Proceedings of SMUC, October 2011
8. Knees, P., Schedl, M., Pohle, T.: A deeper look into web-based classification of music artists. In: Proceedings of LSAS, June 2008
9. Koenigstein, N., Shavitt, Y.: Song ranking based on piracy in peer-to-peer networks. In: Proceedings of ISMIR, October 2009
10. Koenigstein, N., Shavitt, Y., Tankel, T.: Spotting out emerging artists using geo-aware analysis of P2P query strings. In: Proceedings of KDD, August 2008
11. Liu, J., Inkpen, D.: Estimating user location in social media with stacked denoising auto-encoders. In: Proceedings of Vector Space Modeling for NLP, June 2015
12. Ripley, B.D.: Spatial Statistics. Wiley, New York (2004)
13. Schedl, M., Flexer, A., Urbano, J.: The neglected user in music information retrieval research. J. Intell. Inf. Syst. **41**, 523–539 (2013)
14. Schedl, M., Schiketanz, C., Seyerlehner, K.: Country of origin determination via web mining techniques. In: Proceedings of AdMIRe, July 2010

15. Schedl, M., Schnitzer, D.: Hybrid retrieval approaches to geospatial music recommendation. In: Proceedings of SIGIR, July–August 2013
16. Schedl, M., Seyerlehner, K., Schnitzer, D., Widmer, G., Schiketanz, C.: Three web-based heuristics to determine a person's or institution's country of origin. In: Proceedings of SIGIR, July 2010
17. Serra, X.: Data gathering for a culture specific approach in MIR. In: Proceedings of AdMIRe, April 2012
18. Seyerlehner, K., Schedl, M., Knees, P., Sonnleitner, R.: A refined block-level feature set for classification, similarity and tag prediction. In: Extended Abstract MIREX, October 2009
19. Seyerlehner, K., Schedl, M., Sonnleitner, R., Hauger, D., Ionescu, B.: From improved auto-taggers to improved music similarity measures. In: Nürnberger, A., Stober, S., Larsen, B., Detyniecki, M. (eds.) AMR 2012. LNCS, vol. 8382, pp. 193–202. Springer, Heidelberg (2014)
20. Seyerlehner, K., Widmer, G., Pohle, T.: Fusing block-level features for music similarity estimation. In: Proceedings of DAFx, September 2010
21. Seyerlehner, K., Widmer, G., Schedl, M., Knees, P.: Automatic music tag classification based on block-level features. In: Proceedings of SMC, July 2010
22. Trevisiol, M., Jégou, H., Delhumeau, J., Gravier, G.: Retrieving geo-location of videos with a divide & conquer hierarchical multimodal approach. In: Proceedings of ICMR, April 2013
23. Tzanetakis, G., Cook, P.: MARSYAS: a framework for audio analysis. Organ. Sound **4**, 169–175 (2000)
24. Workman, S., Souvenir, R., Jacobs, N.: Wide-area image geolocalization with aerial reference imagery. In: Proceedings of ICCV, December 2015
25. Yu, H., Xie, L., Sanner, S.: Views, Twitter-driven YouTube : beyond individual influencers. In: Proceedings of ACM Multimedia, November 2014
26. Zhou, F., Claire, Q., King, R.D.: Predicting the geographical origin of music. In: Proceedings of ICDM, December 2014

Key Estimation in Electronic Dance Music

Ángel Faraldo(✉), Emilia Gómez, Sergi Jordà, and Perfecto Herrera

Music Technology Group, Universitat Pompeu Fabra, Roc Boronat 138, 08018 Barcelona, Spain
{angel.faraldo,emilia.gomez,sergi.jorda,perfecto.herrera}@upf.edu

Abstract. In this paper we study key estimation in electronic dance music, an umbrella term referring to a variety of electronic music subgenres intended for dancing at nightclubs and raves. We start by defining notions of tonality and key before outlining the basic architecture of a template-based key estimation method. Then, we report on the tonal characteristics of electronic dance music, in order to infer possible modifications of the method described. We create new key profiles combining these observations with corpus analysis, and add two pre-processing stages to the basic algorithm. We conclude by comparing our profiles to existing ones, and testing our modifications on independent datasets of pop and electronic dance music, observing interesting improvements in the performance or our algorithms, and suggesting paths for future research.

Keywords: Music information retrieval · Computational key estimation · Key profiles · Electronic dance music · Tonality · Music theory

1 Introduction

The notion of tonality is one of the most prominent concepts in Western music. In its broadest sense, it defines the systematic arrangements of pitch phenomena and the relations between them, specially in reference to a main pitch class [9]. The idea of key conveys a similar meaning, but normally applied to a smaller temporal scope, being common to have several key changes along the same musical piece. Different periods and musical styles have developed different practices of tonality. For example, modulation (i.e. the process of digression from one local key to another according to tonality dynamics [21]) seems to be one of the main ingredients of musical language in euroclassical[1] music [26], whereas pop music tends to remain in a single key for a whole song or perform key changes by different means [3,16].

Throughout this paper, we use the term *electronic dance music* (EDM) to refer to a number of subgenres originating in the 1980's and extending into the present, intended for dancing at nightclubs and raves, with a strong presence

[1] We take this term from Tagg [26] to refer to European Classical Music of the so-called common practice repertoire, on which most treatises on harmony are based.

N. Ferro et al. (Eds.): ECIR 2016, LNCS 9626, pp. 335–347, 2016.
DOI: 10.1007/978-3-319-30671-1_25

of percussion and a steady beat [4]. Some of such styles even seem to break up with notions such as chord and harmonic progression (two basic building blocks of tonality in the previously mentioned repertoires) and result in an interplay between pitch classes of a given key, but without a sense of tonal direction.

These differences in the musical function of pitch and harmony suggest that computational key estimation, a popular area in the Music Information Retrieval (MIR) community, should take into account style-specific particularities and be tailored to specific genres rather than aiming at all-purpose solutions.

In the particular context of EDM, automatic key detection could be useful for a number of reasons, such as organising large music collections or facilitating *harmonic mixing*, a technique used by DJ's and music producers to mix and layer sound files according to their tonal content.

1.1 Template-Based Key Estimation Methods

One of the most common approaches to key estimation is based on pitch-class profile extraction and template matching. Figure 1 shows the basic architecture of such key estimation system. Regular methodologies usually convert the audio signal to the frequency domain. The spectral representation is then folded into a so-called pitch class profile (PCP) or chromagram, a vector representing perceptually equal divisions of the musical octave, providing a measure of the intensity of each semitone of the chromatic scale per time frame. For improved results, a variety of pre-processing techniques such as tuning-frequency finding, transient removal or beat tracking can be applied. It is also common to smooth the results by weighting neighbouring vectors. Lastly, similarity measures serve to compare the averaged chromagram to a set of templates of tonality, and pick the best candidate as the key estimate. We refer the reader to [7,18] for a detailed description of this method and its variations.

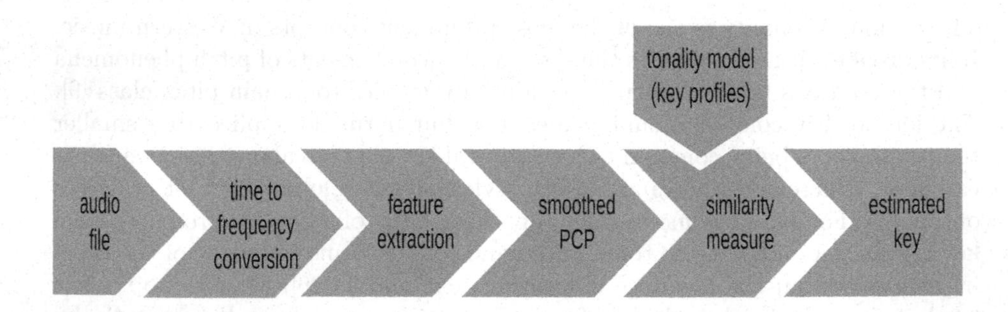

Fig. 1. Basic template-based key estimation system.

One of the most important aspects of such an approach is the model used in the similarity measure. Different key profiles have been proposed since the pioneering Krumhansl-Schmuckler algorithm, who proposed weighting coefficients derived from experiments involving human listeners ratings [13]. Most of them consist of deviations from the original coefficients to enhance performance on specific repertoires [6,23,27]. Among existing algorithms, *QM Key Detector* [19] and *KeyFinder* [23] deserve special attention, particularly since both are publicly available. The former has provided the best results in previous editions of MIREX[2], whereas the latter appears to be the only open-source algorithm specifically tailored to key detection in EDM.

2 Timbral and Tonal Characteristics of EDM

Most subgenres falling under the umbrella term EDM give a central role to percussion and bass, over which other pitched materials and sound effects are normally layered. This idiosyncrasy results in a number of generalisable spectral features that we list hereunder:

- The ubiquity of percussive sounds tends to flatten the spectrum, possibly masking regions with meaningful tonal content.
- Tonal motion often concentrates on the lower register, where algorithms normally offer less resolution.
- Some types of EDM are characterised by tonal effects such as *glissandi* and extreme timbral saturation, that could make difficult to identify pitch as quantised and stable units.

With regard to tonal practices in EDM, pitch relations are often freed from the tonal dynamics based on the building up of tension and its relaxation. Some idiomatic characteristics from empirical observation follow, that could be taken into account when designing methods of tonality induction for this repertoire:

- Beginnings and endings of tracks tend to be the preferred place for sonic experimentation, and it is frequent to find sound effects such as field recordings, musical quotations, un-pitched excerpts or atonal interludes without a sense of continuity with the rest of the music.
- There are likely occasional detuning of songs or excerpts from the standard tuning due to manual alterations in the pitch/speed control present in industry-standard vinyl players, such as the *Technics SL-1200*, for the purpose of adjusting the tempo between different songs.
- Euroclassical tonal techniques such as modulation are essentially absent.
- The dialectics between consonance and dissonance are often replaced by a structural paradigm based on rhythmic and timbral intensification [10,24].
- The music normally unfolds as a cumulative form made with loops and repetitions. This sometimes causes polytonality or conflicting modality due to the overlap of two or more riffs [24].

[2] The Music Information Retrieval Evaluation eXchange (MIREX) is an international committee born to evaluate advances in Music Information Retrieval among different research centres, by quantitatively comparing algorithm performance using test sets that are not available beforehand to participants.

- According to a general tendency observed in Western popular music since the 1960's, most EDM is in minor mode [22].
- Tritone and semitone relationships seem to be characteristics of certain subgenres [24], such as breakbeat or dubstep.
- In minor mode, there is hardly any appearance of the leading tone (♮VII) so characteristic of other tonal practices, favouring other minor scales, especially *aeolian* (♭VII) and *phrygian* (♭II) [25].
- It is also frequent to find pentatonic and hexatonic scales, instead of the major and minor heptatonic modes at the basis of most tonal theories [26].

3 Method

For this study, we gathered a collection of complete EDM tracks with a single key estimation per item. The main source was Sha'ath's list of 1,000 annotations, determined by three human experts[3]. However, we filtered out some non-EDM songs and completed the training dataset with other manually annotated resources from the internet[4], leaving us with a total of 925 tracks to extract new tonal profiles.

To avoid overfitting, evaluations were carried on an independent dataset of EDM, the so-called *GiantSteps key dataset* [12], consisting of 604 two-minute long excerpts from *Beatport*[5], a well-known internet music store for DJs and other EDM consumers. Additionally, we used Harte's dataset [15] of 179 songs by The Beatles reduced to a single estimation per song [20], to compare and test our method on other popular styles that do not follow the typical EDM conventions.

Despite the arguments presented in Sect. 2 about multi-modality in EDM, we decided to shape our system according to a major/minor binary model. In academic research, there has been little or no concern about tonality in electronic popular music, normally considered as a secondary domain compared to rhythm and timbre. In a way, the current paper stands as a first attempt at compensating this void. Therefore, we decided to use available methodologies and datasets (and all of these only deal with binary modality), to be able to compare our work with existing research, showing that current algorithms perform poorly on this repertoire. Furthermore, even in the field of EDM, commercial applications and specialised websites seem to ignore the modal characteristics referred and label their music within the classical paradigm.

Tonal Properties of the Datasets. The training dataset contains a representative sample of the main EDM subgenres, including but not limited to dubstep,

[3] http://www.ibrahimshaath.co.uk/keyfinder/KeyFinderV2Dataset.pdf.
[4] http://blog.dubspot.com/dubspot-lab-report-mixed-in-key-vs-beatport
http://www.djtechtools.com/2014/01/14/key-detection-software-comparison-2014-edition.
[5] https://pro.beatport.com/.

drum'n'bass, electro, hip-hop, house, techno and trance. The most prominent aspect is its bias toward the minor mode, which as stated above, seems representative of this kind of music. Compared to *beatles* dataset, of which only a 10.6 % is annotated in minor (considering one single key estimation per song), the training dataset shows exactly the inverse proportion, with 90.6 % of it in minor. The *GiantSteps* dataset shows similar statistics (84.8 % minor), confirming theoretical observation [22].

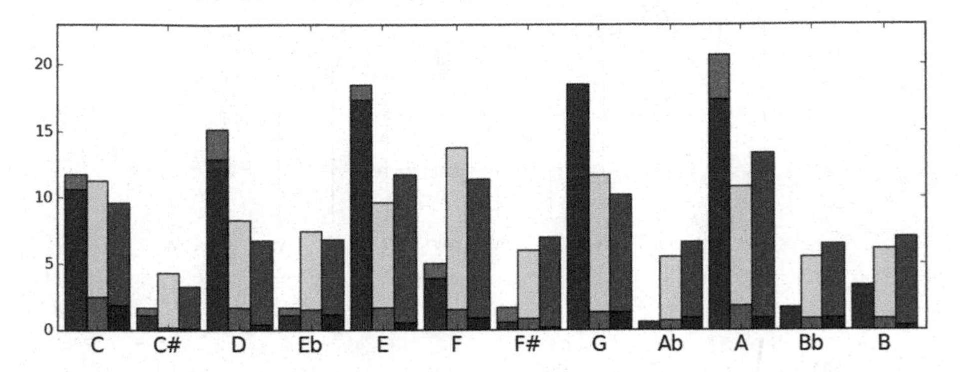

Fig. 2. Distribution of major (bottom) and minor (top) keys by tonal center in *beatles* (left), *GiantSteps* (center) and training (right) datasets.

Figure 2 illustrates the percentage distribution of keys in the three datasets according to the tonal centre of each song. We observe a tendency toward certain tonal centres in *beatles*, which correspond to guitar *open-string* keys (`C`, `D`, `E`, `G`, `A`), whereas the two EDM collections present a more even distribution among the 12 chromatic tones, probably as a result of music production with synthesisers and digital tools.

3.1 Algorithm

In the following, we propose modifications to a simple template-based key estimation method. We study the effect of different key profiles and create our own from a corpus of EDM. We modify the new templates manually, in the light of the considerations outlined in Sect. 2. Then, we incorporate a *spectral whitening* function as a pre-processing stage, in order to strengthen spectral peaks with presumable tonal content. Finally, taking into account the potential detuning of fragments of a given track due to hardware manipulations, we propose a simple *detuning correction* method.

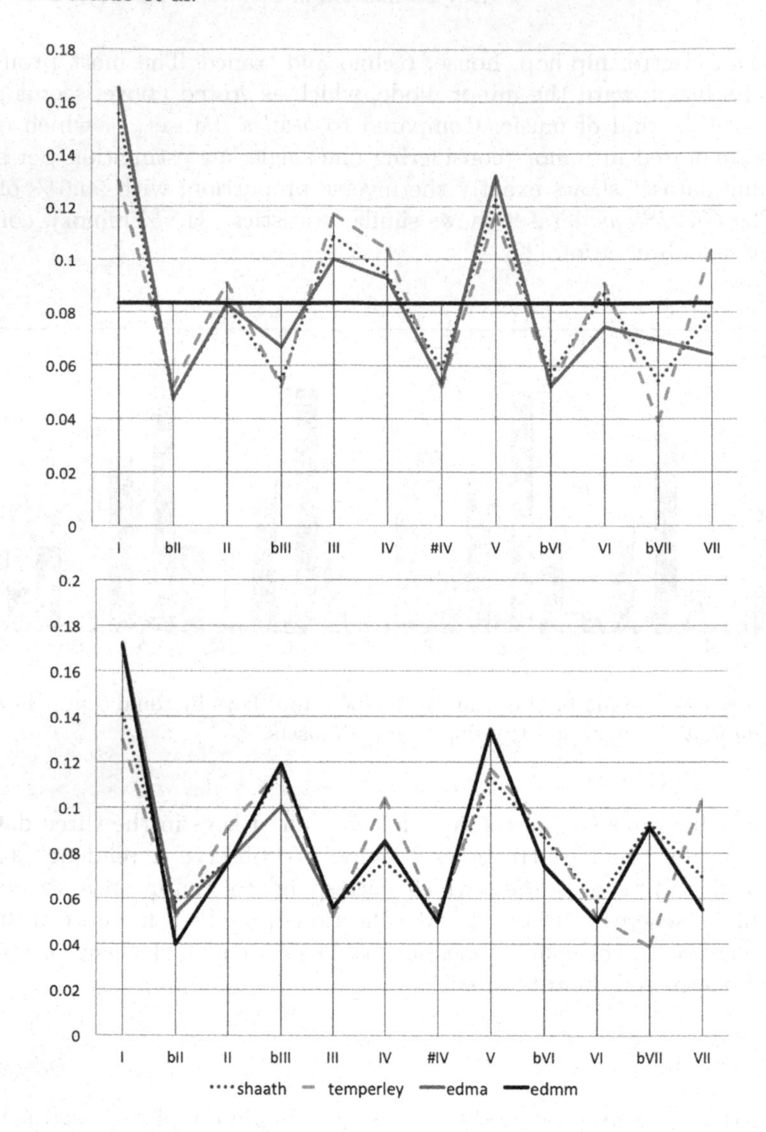

Fig. 3. The four major (above) and minor (below) key profiles. Note that the major profile of edmm is flat.

The simple method we chose is implemented with *Essentia*[6,7], a C++ library for audio information retrieval [1], and it is based on prior work by Gómez [6,7].

[6] http://essentia.upf.edu/.

[7] After informal testing, we decided to use the following settings in all the experiments reported: mix-down to mono; sampling rate: 44,100 Hz.; window size: 4,096 hanning; hop size: 16,384; frequency range: 25–3,500 Hz.; PCP size: 36 bins; weighting size: 1 semitone; similarity: cosine distance.

New Key Profiles. As explained above, one of the main ingredients in a template-based key estimator is the tonality model represented by the so-called key profile, a vector containing the relative weight of the different pitch classes for a given key. In this paper we compare four different profiles:

1. The ones proposed by Temperley [27], which are based on corpus analysis of euroclassical music repertoire.
2. Manual modifications of the original Krumhansl profiles [13] by Sha'ath [23], specifically oriented to EDM. The main differences in Sha'ath's profiles are (a) a slight boost of the weight for the ♮VII degree in major; and (b) a significant increment of the subtonic (♭VII) in minor. Other than these, the two profiles remain essentially identical.
3. Major and minor profiles extracted as the median of the averaged chromagrams of the training set. This provided best results compared to other generalisation methods (such as grand average, max average, etc.). Throughout this paper we refer to these profiles as *edma*.
4. Manual adjustments on the extracted profiles (referenced as *edmm*) accounting for some of the tonal characteristics described in Sect. 2, especially the prominence of the aeolian mode, and the much greater proportion of minor keys. In that regard, given the extremely low proportion of major tracks in the corpus, we decided to flatten the profile for major keys.

Figure 3 shows a comparison between these four profiles. They are all normalised so that the sum of each vector equals 1. It is visible how the profiles by Temperley favour the leading-tone in both modes, according to the euroclassical tonal tradition, whilst the other three profiles increase the weight for the subtonic. We can see that automatically generated profiles (edma) give less prominence to the diatonic third degree in both modes, reflecting the modal ambiguity present in much EDM. We compensated this manually, raising the ♭III in the minor profile, together with a decrement of the ♭II (edmm).

Spectral Whitening. We inserted a pre-processing stage that flattens the spectrum according to its spectral envelope, based on a method by Röbel and Rodet [28]. The aim was to increase the weight of the predominant peaks, so that notes across the selected pitch range contribute equally to the final PCP. This technique has been previously used by Gómez [7], and other authors have proposed similar solutions [11,14,17].

Detuning Correction. We noted that some of the estimations with the basic method produced tritone and semitone errors. Our hypothesis was that these could be due to possible de-tunings produced by record players with manual pitch/tempo corrections [23]. In order to tackle this, our algorithm uses a PCP resolution of 3 bins per semitone, as it is usual in key detection algorithms [8,19]. This allowed us to insert a post-processing stage that shifts the averaged PCP ±33 cents, depending on the position of the maximum peak in the vector.

Various tuning-frequency estimation methods have been proposed, mostly based on statistics [5,29]. Our approach is a simplification of that described in [8]. The algorithm finds the maximum value in the averaged chromagram and shifts the spectrum ±1 bin, depending on this unique position. This shift is done only once per track, after all the PCPs are averaged together.

Our approach is reference-frequency agnostic: it takes the reference pitch to be between 431.6 and 448.5 Hz (i.e. 1/3 of a semitone lower or higher than the pitch standard). We regard this margin as comfortable enough for the repertoire under consideration, assuming that most music would fit within the range mentioned above.

3.2 Evaluation Criteria

In order to facilitate reproducibility, we compared the performance of our method with two publicly available algorithms, already mentioned in Sect. 1.1: Sha'ath's *KeyFinder*[8] and the *QM Key Detector* vamp-plugin[9] by Noland and Landone [2], which we assume to be a close version of the best performing algorithm in MIREX 2014.

The MIREX evaluation has been so far carried on 30-second excerpts of MIDI renders of euroclassical music scores. This follows an extended practice of performing key estimation in fragments of short duration at the beginning or end of a piece of music. Contrary to this tendency, informal experiments suggest that computational key estimation in popular music provides better results when analysing full-length tracks. One of the motivations of observing the beginning of a piece of music is to skip modulational processes that can obstruct the global-key estimation task; however, modulation is not characteristic of EDM neither of pop music. Moreover, given the timbral complexity of most EDM, averaging the chromagrams over the full track likely provides a cleaner tonal profile, minimising the effect of transients and other unwanted spectral components. Based on these arguments, we performed all of our evaluations on complete tracks.

The ranking of the algorithms was carried out following the MIREX evaluation procedure, by which neighbouring keys are weighted by various factors and averaged into a final score.

4 Results

Table 1 presents the weighted scores of our basic algorithm with the variations we have described, now tested with 2 independent set collections, different than those used for the algorithm development. The top four rows show the effect of the key profiles discussed without further modifications. As expected, different profiles provide quite different responses, depending on the repertoire. Temperley's profiles perform well on the *beatles* set, whereas they offer poor performance

[8] http://www.ibrahimshaath.co.uk/keyfinder/.
[9] http://isophonics.net/QMVampPlugins.

for *GiantSteps*; Shaa'th's provide a moderate response in both datasets, whilst the two edm-variants raise the score on the EDM dataset, at the expense of a worse performance on *beatles.* This is especially true for the edmm profiles, provided the major profile is a flattened vector, and major keys are majority in the *beatles* collection (89.4 %).

Table 1. MIREX weighted scores for the two datasets with four different profiles and the proposed modifications: spectral whitening (sw) and detuning correction (dc). We additionally report the weighted scores on the training set on the third column.

	beatles	GiantSteps	training
temperley	62.0	47.6	*46.5*
sha'ath	55.8	55.8	*57.1*
edm-auto (edma)	50.1	60.1	*61.8*
edm-manual (edmm)	21.3	64.0	*66.7*
temperley + sw	66.4	48.9	*48.0*
sha'ath + sw	64.5	63.2	*64.5*
edma + sw	60.5	66.8	*67.3*
edmm + sw	22.7	71.5	*74.4*
temperley + dc	75.6	47.2	*46.7*
sha'ath + dc	71.3	56.1	*58.9*
edma + dc	68.7	60.8	*64.0*
edmm + dc	27.3	64.6	*69.0*
temperley + dc + sw	**81.2**	48.6	*48.4*
sha'ath + dc + sw	79.9	63.9	*66.1*
edma + dc + sw	76.0	67.3	*69.0*
edmm + dc + sw	28.8	**72.0**	*76.6*

We observe that spectral whitening offers improvement in all cases, from a slight increment of 1.5 points in the more extreme profiles (edmm and temperley) to a raise of 10 p.p. in the performance of edma in the *beatles* dataset. Profiles by Sha'ath get a fair boost in both collections.

Similarly, the detuning correction method alone pushes up all the scores except temperley's profiles on the EDM dataset. Significant improvement is only observed in *beatles*, with increments between 6 and 18.6 p.p. It is known that some of The Beatles' albums were recorded with deviations from the pitch standard (this is specially the case in *Please Please Me* and *Help!*) [8], and our method seems to detect and correct them. On the other hand, the neutrality of this parameter on *GiantSteps* suggests further experimentation with particular subgenres such as hip-hop, where tempo adjustments and vinyl-scratching techniques are commonplace, which is sparsely represented in the *GiantSteps key dataset* [12].

In any case, the combination of both processing stages gives the best results. It is noteworthy that these modifications address different problems in the key-estimation process, and consequently, the combined score results in the addition of the two previous improvements. With these settings, the edma profiles yield 25.9 p.p. over the default settings in *beatles*, on which all key profiles obtain significant improvement. On *GiantSteps*, however, we observe more modest improvement.

4.1 Evaluation

Table 2 shows the results of our evaluation following the MIREX convention. Results are organised separately for each dataset. Along with *QM Key Detector* (qm) and *KeyFinder* (kf), we present our algorithm (with spectral whitening and detuning correction) with three different profiles, namely Temperley's (edmt), automatically extracted profiles from our training dataset (edma), and the manually adjusted ones (edmm).

Both benchmarking algorithms were tested using their default settings. *KeyFinder* uses Sha'ath's own profiles presented above, and provides a single estimate per track. *QM Key Detector*, on the other hand, uses key profiles derived from analysis of J. S. Bach's *Well Tempered Klavier I* (1722), with window and hop sizes of 32,768 points, providing a key estimation per frame. We have reduced these by taking the most frequent estimation per track.

Edmt yields a weighted score of 81.2 in the *beatles* dataset, followed by edma (76), above both benchmarking algorithms. Most errors concentrate on the fifth, however, other common errors are minimised. Edmm produces 48 % parallel errors, identifying all major keys as minor due to its flat major profile. For *GiantSteps*, results are slightly lower. The highest rank is for edmm, with a weighted score of 72.0, followed by edma and *KeyFinder*.

Table 2. Performance of the algorithms on the two evaluation datasets. Our method is reported with spectral whitening and detuning correction, on three different profiles: temperley (edmt), edm-auto (edma) and edm-manual (edmm). Under the correct estimations, we show results for different types of common errors.

	beatles				
	qm	kf	edmt	edma	edmm
correct	61.4	46.9	**73.7**	67.0	10.1
fifth	17.3	17.3	**12.3**	12.3	6.7
relative	3.3	9.5	**2.2**	5.0	19.5
parallel	6.7	4.5	**3.4**	6.7	48.0
other	11.2	21.8	**8.4**	8.9	15.6
weighted	72.4	59.3	**81.2**	76.0	28.8

	GiantSteps				
	qm	kf	edmt	edma	edmm
correct	39.4	45.3	33.4	58.1	**64.2**
fifth	16.9	20.7	16.7	10.1	**10.8**
relative	13.4	6.8	15.7	06.6	**03.3**
parallel	5.1	7.8	10.6	10.6	**06.8**
other	25.2	19.4	23.5	14.6	**14.9**
weighted	52.9	59.3	48.6	.67.3	**72.0**

4.2 Discussion

Among all algorithms under comparison, edma provides the best compromise among different styles, scoring 76.0 points on *beatles* and 67.3 on *GiantSteps*. This suggests that the modifications described are style-agnostic, since they offer improvement over the compared methods in both styles. Spectral whitening and detuning correction address different aspects of the key estimation process, and their implementation works best in combination, independently of the key profiles used. However, results vary drastically depending on this last factor, evidencing that a method based on tonality profiles should be tailored to specific uses and hence not suitable for a general-purpose key identification algorithm. This is especially the case with our manually adjusted profiles, which are highly biased toward the minor modes.

5 Conclusion

In this paper, we adapted a template-based key estimation method to electronic dance music. We discussed some timbral and tonal properties of this metagenre, in order to inform the design of our method. However, although we obtained improved results over other publicly available algorithms, they leave room for improvement.

In future work, we plan to incorporate some of the tonal characteristics described more thoughtfully. In particular, we envision a model that expands the major/minor paradigm to incorporate a variety of modes (i.e. dorian, phrygian, etc.) that seem characteristic of EDM. Additionally, a model more robust to timbre variations could help us identifying major modes, in turn minimising the main flaw of our manually adjusted profiles. This would not only improve the performance on this specific repertoire, but also make it more generalisable to other musical genres.

Summarising, we hope to have provided first evidence that EDM calls for specific analysis of its particular tonal practices, and of computational methods informed by these. It could be the case that its subgenres also benefit from such kind of adaptations. In this regard, the bigger challenge would be to devise a method for adaptive key-template computation, able to work in agnostic genre-specific classification systems.

References

1. Bogdanov, D., Wack, N., Gómez, E., Gulati, S., Herrera, P., Mayor, O.: ESSENTIA: an open-source library for sound and music analysis. In: Proceedings 21st ACM-ICM, pp. 855–858 (2013)
2. Cannam, C., Mauch, M., Davies, M.: MIREX 2013 Entry: Vamp plugins from the Centre For Digital Music (2013). www.music-ir.org
3. Everett, W.: Making sense of rock's tonal systems. Music Theory Online, vol. 10(4) (2004)

4. Dayal, G., Ferrigno, E.: Electronic Dance Music. Grove Music Online. Oxford University Press, Oxford (2012)
5. Dressler, K., Streich, S.: Tuning frequency estimation using circular statistics. In: Proceedings of the 8th ISMIR, pp. 2–5 (2007)
6. Gómez, E.: Tonal description of polyphonic audio for music content processing. INFORMS J. Comput. **18**(3), 294–304 (2006)
7. Gómez, E.: Tonal description of music audio signals. Ph.D. thesis, Universitat Pompeu Fabra, Barcelona (2006)
8. Harte., C.: Towards automatic extraction of harmony information from music signals. Ph.D. thesis, Queen Mary University of London (2010)
9. Hyer, B.: Tonality. Grove Music Online. Oxford University Press, Oxford (2012)
10. James, R.: My life would suck without you / Where have you been all my life: Tension-and-release structures in tonal rock and non-tonal EDM pop. www.its-her-factory.com/2012/07/my-life-would-suck-without-youwhere-have-you-been-all-my-life-tension-and-release-structures-in-tonal-rock-and-non-tonal-edm-pop. Accessed 16th December 2014
11. Klapuri, A.: Multipitch analysis of polyphonic music and speech signals using an auditory model. IEEE Trans. Audio Speech Lang. Process. **16**(2), 255–266 (2008)
12. Knees, P., Faraldo, Á., Herrera, P., Vogl, R., Böck, S., Hörschläger, F., Le Goff, M.: Two data sets for tempo estimation and key detection in electronic dance music annotated from user corrections. In: Proceeings of the 16th ISMIR (2015)
13. Krumhansl, C.L.: Cognitive Foundations of Musical Pitch. Oxford Unversity Press, New York (1990)
14. Mauch, M., Dixon., S.: Approximate note transcription for the improvedidentification of difficult chords. In: Proceedings of the 11th ISMIR, pp. 135–140 (2010)
15. Mauch, M., Cannam, C., Davies, M., Dixon, S., Harte, C., Kolozali, S., Tidjar, D.: OMRAS2 metadata project 2009. In: Proceedings of the 10th ISMIR, Late-Breaking Session (2009)
16. Moore, A.: The so-called "flattened seventh" in rock. Pop. Music **14**(2), 185–201 (1995)
17. Müller, M., Ewert, S.: Towards timbre-invariant audio features for harmony-based music. IEEE Trans. Audio Speech Lang. Process. **18**(3), 649–662 (2010)
18. Noland, K.: Computational Tonality estimation: Signal Processing and Hidden Markov Models. Ph.D. thesis, Queen Mary University of London (2009)
19. Noland, K., Sandler, M.: Signal processing parameters for tonality estimation. In: Proceedings of the 122nd Convention Audio Engeneering Society (2007)
20. Pollack., A.W.: Notes on..series. (Accessed: 1 February 2015). www.icce.rug.nl/soundscapes/DATABASES/AWP/awp-notes_on.shtml
21. Saslaw, J.: Modulation (i). Grove Music Online. Oxford University Press, Oxford (2012)
22. Schellenberg, E.G., von Scheve, C.: Emotional cues in American popular music: five decades of the Top 40. Psychol. Aesthetics Creativity Arts **6**(3), 196–203 (2012)
23. Sha'ath., I.: Estimation of key in digital music recordings. In: Departments of Computer Science & Information Systems, Birkbeck College, University of London (2011)
24. Spicer, M.: (Ac)cumulative form in pop-rock music. Twentieth Century Music **1**(1), 29–64 (2004)
25. Tagg, P.: From refrain to rave: the decline of figure and raise of ground. Pop. Music **13**(2), 209–222 (1994)
26. Tagg., P.: Everyday tonality II (Towards a tonal theory of what most people hear). The Mass Media Music Scholars' Press. New York and Huddersfield (2014)

27. Temperley, D.: What's key for key? The Krumhansl-Schmuckler key-finding algorithm reconsidered. Music Percept. Interdiscip. J. **17**(1), 65–100 (1999)
28. Röbel, A., Rodet, X.: Efficient spectral envelope estimation and its application to pitch shifting and envelope preservation. In: Proceedings of the 8th DAFX (2005)
29. Zhu, Y., Kankanhalli, M.S., Gao., S.: Music key detection for musical audio. In: Proceedings of the 11th IMMC, pp. 30–37 (2005)

Summarization

Evaluating Text Summarization Systems with a Fair Baseline from Multiple Reference Summaries

Fahmida Hamid(✉), David Haraburda, and Paul Tarau

Department of Computer Science and Engineering,
University of North Texas, Denton, TX, USA
{fahmida.hamid,dharaburda,ptarau}@gmail.com

Abstract. Text summarization is a challenging task. Maintaining linguistic quality, optimizing both compression and retention, all while avoiding redundancy and preserving the substance of a text is a difficult process. Equally difficult is the task of evaluating such summaries. Interestingly, a summary generated from the same document can be different when written by different humans (or by the same human at different times). Hence, there is no convenient, complete set of rules to test a machine generated summary. In this paper, we propose a methodology for evaluating extractive summaries. We argue that the overlap between two summaries should be compared against the *average intersection size* of two random generated *baselines* and propose ranking machine generated summaries based on the concept of *closeness* with respect to reference summaries. The key idea of our methodology is the use of *weighted relatedness* towards the reference summaries, normalized by the relatedness of reference summaries among themselves. Our approach suggests a relative scale, and is tolerant towards the length of the summary.

Keywords: Evaluation technique · Baseline · Summarization · Random average · Reference summary · Machine-generated summary

1 Introduction

Human quality text summarization systems are difficult to design and even more difficult to evaluate [1]. The extractive summarization task has been most recently portrayed as *ranking sentences* based on their likelihood of being part of the summary and their salience. However different approaches are also being tried with the goal of making the ranking process more semantically meaningful, for example: using synonym-antonym relations between words, utilizing a semantic parser, relating words not only by their co-occurrence, but also by their semantic relatedness. Work is also on going to improve anaphora resolution, defining dependency relations, etc. with a goal of improving the *language understanding* of a system.

A series of workshops on text summarization (WAS 2000-2002), special sessions in ACL, CoLING, SIGIR, and government sponsored evaluation efforts in

N. Ferro et al. (Eds.): ECIR 2016, LNCS 9626, pp. 351–365, 2016.
DOI: 10.1007/978-3-319-30671-1_26

United States (DUC 2001-DUC2007) have advanced the technology and produced a couple of experimental online systems [15]. However there are no common, convenient, and repeatable evaluation methods that can be easily applied to support system development and comparison among different summarization techniques [8].

Several studies ([9,10,16,17]) suggest that *multiple human gold-standard summaries would provide a better ground for comparison.* Lin [5] states that multiple references tend to increase evaluation stability although human judgements only refer to single reference summary.

After considering the evaluation procedures of ROUGE [6], Pyramid [12], and their variants e.g., ParaEval [19], we present another approach to evaluating the performance of a summarization system which works with one or many reference summaries.

Our major contributions are:

- We propose *the average or expected size of the intersection of two random generated summaries* as a generic *baseline* (Sects. 3 and 4). Such a strategy was discussed briefly by Goldstein et al. [1]. However, to the best of our knowledge, we have found no direct use of the idea while scoring a summarization system. We use the baseline to find a *related (normalized)* score for each reference and machine-generated summaries.
- Using this baseline, we outline an approach (Sect. 5) to evaluating a summary. Additionally, we outline the rationale for a new measure of summary quality, detail some experimental results and also give an alternate derivation of the average intersection calculation.

2 Related Work

Most of the existing evaluation approaches use absolute scales (e.g., precision, recall, f-measure) to evaluate the performance of the participating systems. *Such measures can be used to compare summarization algorithms, but they do not indicate how significant the improvement of one summarizer over another is* [1].

ROUGE (Recall Oriented Understudy for Gisting Evaluation) [6] is one of the well known techniques to evaluate single/multi-document summaries. ROUGE is closely modelled after BLEU [14], a package for machine translation evaluation. ROUGE includes measures to automatically determine the quality of a summary by comparing it to other (ideal) summaries created by humans. The measures count the number of overlapping units such as n-gram, word sequences, and word pairs between the machine-generated summary and the reference summaries.

Among the major variants of ROUGE measures, e.g., ROUGE-N, ROUGE-L, ROUGE-W, and, ROUGE-S, three have been used in the Document Understanding Conference (DUC) 2004, a large-scale summarization evaluation sponsored by NIST. Though ROUGE shown to correlate well with human judgements, it considers fragments, of various lengths, to be equally important, a factor that rewards low informativeness fragments unfairly to relative high informativeness ones [3].

Nenkova [12] made two conclusions based on their observations:

- DUC scores cannot be used to distinguish a good human summarizer from a bad one
- The DUC method is not powerful enough to distinguish between systems

Another piece of work that we would like to mention is the Pyramid method [11]. A key assumption of the method is the need for multiple models, which taken together, yield a gold standard for system output. A pyramid represents the opinions of multiple human summary writers each of whom has written a model summary for the multiple set of documents. Each tier of the pyramid *quantitively* represents the agreements among human summaries based on *Summary Content Units (SCU)* which are content units, not bigger than a clause. SCUs that appear in more of the human summaries are weighted more highly, allowing differentiation between important content from less important one.

The original pyramid score is similar to a *precision metric*. It reflects *the number of content units that were included in a summary* under evaluation as highly weighted as possible and it penalizes the content unit when a more highly weighted one is available but not used. We would like to address following important aspects here -

- Pyramid method does not define a *baseline* to compare the degree of (dis)agreement between human summaries.
- High frequency units receive higher weights in the Pyramid method. Nenkova [13], in another work, stated that the frequency feature is not adequate for capturing all the contents. To include less frequent (but more informative) content into machine summaries is still an open problem.
- There is no clear direction about the summary length (or compression ratio).

Our method uses a unique baseline for all (system, and reference summaries) and it does not need the absolute scale (like f, p, r) to score the summaries.

3 A *Generic Baseline* for All

We need to ensure a single rating for each system unit [7]. Besides, we need a common ground for comparing available multiple references to reach a unique standard. Precision, Recall, and F-measure are not exactly good fit in such case.

Another important task for an evaluation technique is defining a *fair baseline*. Various ways (first sentence, last sentence, sentences overlapped mostly with the title, etc.) are being tried to generate the baseline. Nenkova [13] designed a baseline generation approach: *SumBasic*. It was applied over DUC 2004 dataset. But we need a generic way to produce the baseline for all types of documents. *The main task of a baseline is to define (quantify) the least possible result that can be compared with the competing systems to get a comparative scenario.*

Compression ratio plays an important role in summarization process. If the compression ratio is too high, it is difficult to cover the stated topic(s) in the summary. Usually the compression ratio is set to a fixed value (100 words, 75 bytes, etc.) So, the participants are not free to generate the summary as they might want. We believe the participants should be treated more leniently on selecting the size of summary. When it is allowed, we need to make sure *the evaluation is not affected due to the length.*

The following two sections discuss about our proposed *baseline*, its relationship with precision, recall, f-measure and how to use it for computing the integrity of a (both reference and system generated) summary.

3.1 Average (Expected) Size of Intersection of Two Sets

Given a set N of size n, and two randomly selected subsets $K_1 \subseteq N$ and $K_2 \subseteq N$ with k elements each, the average or expected size of the intersection ($|K_1 \cap K_2|$) is

$$avg(n,k)_{random} = \frac{\sum_{i=0}^{k} i\binom{k}{i}\binom{n-k}{k-i}}{\sum_{i=0}^{k}\binom{k}{i}\binom{n-k}{k-i}}. \tag{1}$$

For *two* randomly selected subsets $K \subseteq N$ and $L \subseteq N$ of sizes k and l (say, $k \leq l$) respectively this formula generalises to

$$avg(n,k,l)_{random} = \frac{\sum_{i=0}^{k} i\binom{k}{i}\binom{n-k}{l-i}}{\sum_{i=0}^{k}\binom{k}{i}\binom{n-k}{l-i}}. \tag{2}$$

For each possible size $i = \{0..k\}$ of an intersecting subset, the numerator sums the product of i and the number of different possible subsets of size i, giving the total number of elements in all possible intersecting subsets. For a particular size i there are $\binom{k}{i}$ ways to select the i intersecting elements from K, which leaves $n-k$ elements from which to choose the $k-i$ non-intersecting elements (or $l-i$ in the case of two randomly selected subsets). The denominator simply counts the number of possible subsets, so that the fraction itself gives the expected number of elements in a randomly selected subset.

Simplifying Equation 2: Equation 2 is expressed as a combinatorial construction, but the probabilistic one is perhaps simpler: the probability of any element x being present in both subset K and subset L is the probability that x is contained in the intersection of those two sets $I = L \cap K$.

$$\begin{aligned}\Pr(x \in K) \cdot \Pr(x \in L) & \\ &= \Pr(x \in (L \cap K)) \\ &= \Pr(x \in I)\end{aligned} \tag{3}$$

Putting another way, the probability that an element x is in K, L, or I is k/n, l/n and i/n respectively (where i is the number of elements in I). Then from Eq. 3 accordingly,

$$(k/n)(l/n) = i/n \tag{4}$$

$$i = \frac{kl}{n} \tag{5}$$

A combinatorial proof, relying on identities involving binomial coefficients shows that Eqs. 2 and 5 are equivalent and is contained in Appendix A.

3.2 Defining $f\text{-}measure_{expected}$

Recall and *Precision* are two re-known metrics to define the performance of a system. Recall (r) is the ratio of *number of relevant information received to the total number of relevant information in the system.* Precision (p), on the other hand, is the ratio of *number of relevant records retrieved to the total number (relevant and irrelevant) of records retrieved.* Assuming the subset with size k as the gold standard, we define recall, and precision for the randomly generated sets as:

$$r = \frac{i}{k} \qquad p = \frac{i}{l} \qquad f\text{-}measure = \frac{2pr}{p+r}$$

Therefore, f-measure (the balanced harmonic mean of p and r) for these two random sets is:

$$\begin{aligned} f\text{-}measure_{expected} &= 2pr/(p+r) \\ &= 2(l/n)(k/n)/(l/n + k/n) \\ &= 2(lk)/(n^2)/((l+k)/n) \\ &= 2(lk)/(n(l+k)) \\ &= 2i/(l+k) \\ &= i/((l+k)/2) \end{aligned} \tag{6}$$

3.3 Defining $f\text{-}measure_{observed}$, with Observed Size of Intersection 'ω'

Let, for a machine generated summary L and a reference summary K, the observed size of intersection, $|K \cap L|$ is ω.

$$r = \frac{|K \cap L|}{|K|} = \frac{\omega}{k} \qquad p = \frac{|K \cap L|}{|L|} = \frac{\omega}{l}$$

f-measure, in this case, can be defined as,

$$\begin{aligned} f\text{-}measure_{observed} &= 2pr/(p+r) \\ &= \frac{2 \cdot \omega^2}{k \cdot l} / \frac{(k+l) \cdot \omega}{k \cdot l} \\ &= 2\omega/(k+l) \\ &= \omega/((k+l)/2) \end{aligned} \tag{7}$$

4 The i-measure: A Relative Scale

A more direct comparison of an observed overlap, seen as the intersection size of two sets K and L, consisting of lexical units like unigrams or n-grams drawn from a single set N is provided by the *i-measure*:

$$i\text{-}measure(N, K, L) = \frac{observed\_size\_of\_intersection}{expected\_size\_of\_intersection} = \frac{|K \cap L|}{\frac{|K| \cdot |L|}{|N|}} = \frac{\omega}{\left(\frac{kl}{n}\right)} = \frac{\omega}{i} \quad (8)$$

By substituting ω and i using Eqs. 7 and 6, we get,

$$i\text{-}measure(N, K, L) = \frac{f\text{-}measure_{observed}}{f\text{-}measure_{expected}}$$

Interestingly, *i-measure* turned out as a ratio between the observed *f-measure* and the expected/ average *f-measure*. The *i-measure* is a form of *f-measure* with some tolerance towards the length of the summaries.

In the next section, we prepare an example to explain how *i-measure* adjusts the variation on lengths, yet produces comparable score.

4.1 Sample Scenario: *i-measure* Balances the Variation in Length

Suppose we have a document with $n = 200$ unique words, a reference summary composed of $k = 100$ unique words, and a set of machines $\{a, b, \ldots, h, i\}$. Each machine generates a summary with l unique words. Table 1 outlines some sample scenarios of *i-measure* scores that would allow one to determine a comparative performance of each of the systems.

For system b, e, and h, ω is the same, but the *i-measure* is highest for h as its summary length is smaller than the other two. On the other hand, systems e

Table 1. Sample cases: *i-measure*

case	n	k	l	i	ω	*i-measure*	sys. id
$k = l$	200	100	100	50	30	0.6	a
	200	100	100	50	45	0.9	b
	200	100	100	50	14	0.28	c
$k < l$	200	100	150	75	30	0.4	d
	200	100	150	75	45	0.6	e
	200	100	150	75	14	0.186	f
$k > l$	200	100	80	40	30	0.75	g
	200	100	80	40	45	1.125	h
	200	100	80	40	14	0.35	i

and a receive the same *i-measure*. Although ω is larger for e, it is penalized as its summary length is larger than a. We can observe the following properties of the *i-measure*:

- The system's summary size (l) does not have to be exactly same as the reference' summary size size (k); which is a unique feature. Giving this flexibility encourages systems to produce more informative summaries.
- If k and l are equal, *i-measure* follows the observed intersection, for example b wins over a and c. In this case i-measure shows a compatible behavior with *recall* based approaches.
- For two systems with different l values, but same intersection size, the one with smaller l wins (e.g., a,d, and g). It indicates that system g (in this case) was able to extract important information with greater compression ratio; this is compatible with the *precision* based approaches.

5 Evaluating a System's Performance with Multiple References

When multiple reference summaries are available, a fair approach is to compare the machine summary with each of them. *If there is a significant amount of disagreement among the reference (human) summaries, this should be reflected in the score of a machine generated summary. Averaging* the overlaps of machine summaries with human written ones does not *weigh* less informative summaries differently than more informative ones. Instead, the evaluation procedure should be modified so that it first compares the reference summaries among themselves in order to produce some weighting mechanism that provides a fair way to judge all the summaries and gives a unique measure to quantify the machine generated ones. In the following subsections we introduce the dataset, weighting mechanism for references, and finally, outline the scoring process.

5.1 Introduction to the Dataset and System

Our approach is generic and can be used for any summarization model that uses multiple reference summaries. We have used *DUC*-2004 structure as a model. We use *i-measure*(d, x_j, x_k) to denote the i-measure calculated for a particular document d using the given summaries x_j and x_k.

Let λ machines ($S = \{s_1, s_2, \ldots, s_\lambda\}$) participate in a *single document summarization task*. For each document, m reference summaries ($H = \{h_1, h_2, \ldots, h_m\}$) are provided. We compute the *i-measure* between $\binom{m}{2}$ pairs of reference summaries and normalize with respect to the best pair. We also compute the *i-measure* for each machine generated summary with respect to each reference summary and then normalize it. We call these *normalized i-measures* and denote them as

$$w_d(h_p, h_q) = \frac{i\text{-}measure(d,h_p,h_q)}{\mu_d} \\ w_d(s_j, h_p) = \frac{i\text{-}measure(d,s_j,h_p)}{\mu_{(d,h_p)}} \quad (9)$$

where,

$$\mu_d = max(i\text{-}measure(d, h_p, h_q)), \forall h_p \in H, h_q \in H, h_p \neq h_q \\ \mu_{(d,h_p)} = max(i\text{-}measure(d, s, h_p)), \forall s \in S$$

The next phase is to build a heterogeneous network of systems and references to represent the relationship.

5.2 Confidence Based Score

We assign each reference summary h_p a "confidence" $c_d(h_p)$ for document d by taking the average of its *normalized i-measure* with respect to every other reference summary:

$$c_d(h_p) = \frac{\sum_{q=1,p\neq q}^{m}(w_d(h_p, h_q))}{m-1}. \quad (10)$$

Taking the confidence factor associated with each reference summary allows us to generate a score for s_j:

$$score(s_j, d) = \sum_{p=1}^{m} c_d(h_p) \times w_d(s_j, h_p) \quad (11)$$

Given t different tasks (single documents) for which there are reference and machine generated summaries from the same sources, we can define the total performance of system s_j as

$$i\text{-}score(s_j) = \frac{\sum_{i=1}^{t} score(s_j, d_i)}{t}. \quad (12)$$

Table 2. Reference summaries (B,G,E,F) and three machine summaries on document $D30053.APW19981213.0224$

Reference	Summary
B	Clinton arrives in Israel, to go to Gaza, attempts to salvage Wye accord.
G	Mid-east Wye Accord off-track as Clintons visit; actions stalled, violence
E	President Clinton met Sunday with Prime Minister Netanyahu in Israel
F	Clinton meets Netanyahu, says peace only choice. Office of both shaky
90	ISRAELI FOREIGN MINISTER ARIEL SHARON TOLD REPORTERS DURING PICTURE-TAKIN=
6	VISIT PALESTINIAN U.S. President Clinton met to put Wye River peace accord
31	Clinton met Israeli Netanyahu put Wye accord

Table 3. Normalized i-measure of all possible reference pairs for document: $D30053.APW19981213.0224$

$Pair(p,q)$	n	k	l	ω	i	$i\text{-}measure$	$w_d(h_p,h_q)$
(G , F)	282	10	8	1	0.283687	3.525	0.375
(G, B)	282	10	9	3	0.319148	9.4	1.0
(G, E)	282	10	8	1	0.283687	3.525	0.375
(F, B)	282	8	9	1	0.255319	3.916	0.4166
(F, E)	282	8	8	2	0.226950	8.8125	0.9375
(E, B)	282	8	9	2	0.255319	7.8333	0.8333

Table 4. Confidence score

reference: h_p	confidence: $c_d(h_p)$
G	0.583
F	0.576
B	0.75
E	0.715

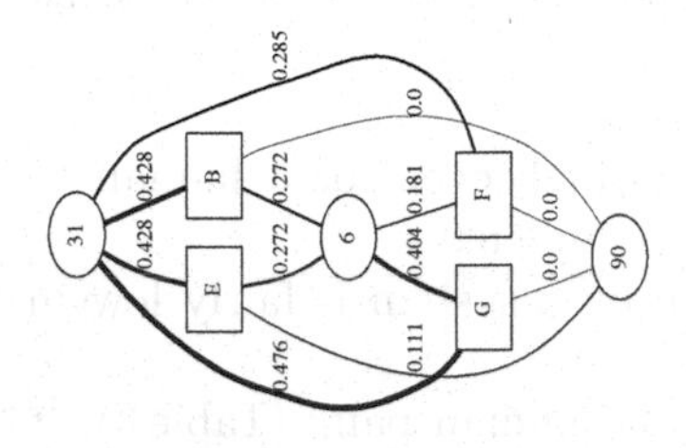

Fig. 1. System-reference graph: edge-weights represent the normalized *i-measure*

Table 5. Confidence based system score

System Id(s_j)	$score(s_j,d_i)$
31	0.2676
6	0.1850
90	0.0198

Table 2 shows four reference summaries (B, G, E, F) and three machine summaries $(31, 90, 6)$ for document *D30053.APW19981213.0224*. Table 3 shows the normalized *i-measure* for each reference pair. While comparing the summaries, we ignored the *stop-words* and *punctuations*. Tables 4 and Table 5, and Fig. 1 represents some intermediate calculation using Eqs. 10 and 11 for document D30053.APW19981213.0224.

6 Evaluating Multi-document Summary

Methodology defined in Sect. 5.2 can be adapted for evaluating *multi-document summaries* with minor modifications. Let, there are q clusters of documents, i.e. $D = \{D_1, D_2, \ldots, D_q\}$. Each cluster D_i contains t number of documents, $D_i = \{d_1, \ldots, d_t\}$. The system has a set of humans ($H = \{h_1, h_2, \ldots, h_z\}$) to generate gold summaries. For each D_i, a subset of humans ($H_{D_i} = \{h_1, h_2, \ldots, h_m\}, m \leq z$) write m different *multi-document summaries.*

We need to compute a score for system s_j among λ participating systems ($S = \{s_1, s_2, \ldots, s_\lambda\}$). We, first, compute *score*(s_j, D_i) for each D_i using formula 11. Then we use formula 12 to find the rank of s_j among all participants.

The only difference is at defining the *i-measure*. The value of n (total number of units like unigram, bi-gram etc.) comes from all the participating documents in D_i, other than a single document.

7 Experimental Results

We perform different experiments over the dataset. Section 7.1 describes how *i-measure* among the reference summaries can be used to find the confidence/

nearness/ similarity of judgements. In Sect. 7.2, we examine two types of rank-correlations (pair-based, distance based) generated by *i-score* and *ROUGE*-1. Section 7.3 states the correlation of *i-measure* based ranks with human assessors.

7.1 Correlation Between Reference Summaries

The *i-measure* works as a preliminary way to address some intuitive decisions. We discuss them in this section with two extreme cases.

- If the *i-measure* is too low (Table 6) for most of the pairs, some of the following issues might be true:-
  - The document discusses about diverse topics.
  - The compression ratio of the summary is too high even for a human to cover all the relevant topics discussed in the document.
  - The probability of showing high performance by a system is fairly low in this case.
- If the *i-measure* is fairly close among most of the human pairs (Table 3), it indicates:-
  - The compression ratio is adequate
  - The document is focused into some specific topic.
  - If a system shows good performance for this document, it is highly probable that the system is built on good techniques.

Therefore, the *i-measure* could be an easy technique to select ideal documents that are good candidates for summarization task. For example, Table 3 shows that all of the reference pairs have some words in common, hence their confidence score (Table 4) is fairly high. But Table 7 shows that most of the references do not share common words, hence *confidence* values of the references for document $D30015.APW19981005.1082$ is quite different from each other.

Table 6. Normalized i-measure of all possible reference pairs for $D30015.APW19981005.1082$

$Pair(p, q)$	n	k	l	ω	i	i-$measure$	$w_d(h_p, h_q)$
(A, H)	357	9	10	0	0.25210	0.0	0.0
(A, B)	357	9	10	3	0.25210	11.9	1.0
(A, E)	357	9	7	1	0.17647	5.66	0.4761
(H, B)	357	10	10	1	0.2801	3.57	0.3
(H, E)	357	10	7	0	0.19607	0.0	0.0
(B, E)	357	10	7	0	0.19607	0.0	0.0

Table 7. Confidence score

reference: h_p	confidence: $c_d(h_p)$
A	0.492
B	0.433
H	0.099
E	0.158

7.2 Correlation of Ranks: ROUGE-1 Vs. I-Score

To understand how the confidence based i-measures compare to the ROUGE-1 metric, we calculated *Spearman's* ρ [18] and *Kendall's* τ [4], (both of which are rank correlation coefficients) by ranking the machine and reference summary scores. Spearman's ρ considers the squared difference between two rankings while

Table 8. Rank correlations

i-score vs. ROUGE-1	Spearman's ρ	Kendall's τ
Task 1	0.786	0.638
Task 2	0.713	0.601
Task 5	0.720	0.579
i-score vs. f-measure		
Task 1	0.896	0.758
Task 2	0.955	0.838
Task 5	0.907	0.772

Table 9. Guess score for $D188$, assessor F

sys. id	*given_score*	*guess_score*
147	3	2
43	2	3
122	2	2
B	4	4
86	2	0
24	1	1
109	3	3
H	3	4

Kendall's τ is based on the number of concordant/discordant pairs (Table 8). Since the list of stopwords used by us can be different from the one used by ROUGE system, we also calculate pure *f-measure* based rank and report the correlation of with *i-score*. The results show, for both cases, *i*-measure is positively correlated, but not completely.

7.3 Correlation with Human Judgement: Guess the *RESPONSIVENESS* score

For multi-document summarization (DUC2004, task5), the special task (RESPONSIVENESS) was to assess the machine summaries per cluster (say, D_i) by a single human-assessor (h_a) and score between 0 to 4, to reflect the *responsiveness* on a given topic (question). We have used a *histogram* to divide the *i-score* based space into 5 categories ($\{0, 1, 2, 3, 4\}$). We found 341 decisions out of 947 responsiveness scores as an exact match (36.008 % accuracy) to the human assessor. Table 9 is a snapshot of the scenario.

The *Root Mean Square Error (RMSE)* based on *i-score* is 1.212 at the scale of 0 to 4. Once normalized over the scale, the error is 0.303

$$RMSE = \sqrt{1/n \sum_{i=1}^{n} (\hat{y}_i - y_i)^2}$$

7.4 Critical Discussion

After carefully analyzing the system generated summaries, rouge based scores, and i-score, we noticed that most of the systems are not producing well-formed sentences. Scoring based on weighted/un-weighted overlapping of *bag-of-important-phrases* is not the best way to evaluate a summarizer. Constraint on the length of the summary (byte/word) might be a trigger. As *i-measure* is lenient on lengths, we can modify Eq. 11 with the following to apply *extraction/generation of proper sentences* within a maximum word/sentence window as an impact factor.

$$score(s_j, d) = \left(\sum_{p=1}^{m} c_d(h_p) \times w_d(s_j, h_p) \right) \times \frac{c\_sen}{t\_sen} \tag{13}$$

where, t_sen is the total number of sentences produced/ extracted by s_j and c_sen is the number of grammatically *well-formed* sentences. For example, *"This is a meaningful sentence. It can be defined using english grammar."* is a delivered summary. Suppose, the allowed word-window-size is 8. So the output is chopped as *"This is a meaningful sentence. It can be"*. Now it contains 1 well-formed sentence out of 2. Then the *bag of words/phrases* model (e.g., *i-measure*) can be applied over it and a score can be produced using Eq. 13.

Standard sentence tokenizers, POS taggers, etc. can be used to analyze sentences. The word/ sentence window-size can be determined by some ratio of sentences (words) present in the original document. As we could not find any summary-evaluation conferences who follow similar rules (TREC, DUC, etc.), we were unable to generate results based on this hypothesis.

8 Conclusion

We present a mathematical model for defining a generic *baseline*. We also propose a new approach to evaluate machine-generated summaries with respect to multiple reference summaries, all normalized with the baseline. The experiments show comparable results with existing evaluation techniques (e.g., ROUGE). Our model correlates well with human decision as well.

The *i-measure* based approach shows some flexibility with summary length. Instead of using average overlapping of words/phrases, we define pair based *confidence* calculation between each reference. Finally, we propose an extension of the model to evaluate the quality of a summary by combining the bag-of-words like model to accredit *sentence structure* while scoring.

We will be extending the model, in future, so it works with semantic relations (e.g. synonym, hypernym etc.) We also need to investigate some more on the confidence defining approach for question-based/ topic-specific summary evaluation task.

A Appendix

The equivalence of Eqs. 2 and 5 can be shown using the following elementary identities on binomial coefficients: the *symmetry rule*, the *absorption rule* and *Vandermonde's convolution* [2].

Proof. Consider first the denominator of Eq. 2. The introduction of new variables makes it easier to see that identities are appropriately applied, and we do so here by letting $s = n - k$ and then swapping each binomial coefficient for its symmetrical equivalent (*symmetry rule*).

$$\sum_{i=0}^{k} \binom{k}{i} \binom{s}{l-i} = \sum_{i=0}^{k} \binom{k}{k-i} \binom{s}{s-l+i}$$

Substituting $j = s - l$ for clarity shows that *Vandermonde's convolution* can be applied to convert the sum of products to a single binomial coefficient, after which we back substitute the original variables, and finally apply the *symmetry rule.*

$$\sum_{i=0}^{k} \binom{s}{j+i}\binom{k}{k-i} = \binom{k+s}{j+k}$$
$$= \binom{k+n-k}{n-k-l+k}$$
$$= \binom{n}{n-l}$$
$$= \binom{n}{l}$$

The numerator of Eq. 2 can be handled in a similar fashion, after the i factor is removed using the *absorption rule.*

$$\sum_{i=0}^{k} i\binom{k}{i}\binom{n-k}{l-i} = k\sum_{i=0}^{k} \binom{k-1}{i-1}\binom{n-k}{l-i}$$

Applying *Vandermonde's convolution* yields:

$$k\sum_{i=0}^{k} \binom{k-1}{-1+i}\binom{n-k}{l-i} = k\binom{n-1}{l-1}$$

Eq. 2 has now been reduced to

$$k\binom{n-1}{l-1}\Big/\binom{n}{l}$$

A variation of the *absorption rule* allows the following transformation

$$k\binom{n-1}{l-1}\Big/\binom{n}{l} = k\binom{l}{1}\Big/\binom{n}{1}$$

which reduces to kl/n.

References

1. Goldstein, J., Kantrowitz, M., Mittal, V., Carbonell, J.:Summarizing text documents: sentence selection and evaluation metrics. In: Proceedings of the 22Nd Annual International ACM SIGIR Conference on Research and Development in Information Retrieval, SIGIR 1999, pp. 121–128. ACM, New York (1999). http://doi.acm.org/10.1145/312624.312665
2. Graham, R., Knuth, D., Patashnik, O.: Concrete Mathematics: A Foundation for Computer Science. Addison-Wesley, Boston (1994)
3. Hovy, E., Lin, C.-Y., Zhou, L., Fukumoto, J.: Automated summarization evaluation with basic elements. In: Proceedings of the Fifth Conference on Language Resources and Evaluation (LREC 2006) (2006)
4. Kendall, M.G.: A new measure of rank correlation. Biometrika **30**(1/2), 81–93 (1938). http://www.jstor.org/stable/2332226

5. Lin, C.Y.: Looking for a few good metrics: automatic summarization evaluation - how many samples are enough? In: Proceedings of the NTCIR Workshop 4 (2004)
6. Lin, C.Y.: Rouge: a package for automatic evaluation of summaries, pp. 25–26 (2004)
7. Lin, C.Y., Hovy, E.: Manual and automatic evaluation of summaries. In: Proceedings of the ACL-2002 Workshop on Automatic Summarization, AS 2002, vol. 4, pp. 45–51. Association for Computational Linguistics, Stroudsburg, PA, USA (2002). http://dx.doi.org/10.3115/1118162.1118168
8. Lin, C.Y., Hovy, E.: Automatic evaluation of summaries using n-gram co-occurrence statistics. In: Proceedings of the 2003 Conference of the North American Chapter of the Association for Computational Linguistics on Human Language Technology, NAACL 2003, vol. 1, pp. 71–78. Association for Computational Linguistics, Stroudsburg, PA, USA (2003). http://dx.doi.org/10.3115/1073445.1073465
9. Mani, I., Maybury, M.T.: Automatic summarization. In: Association for Computational Linguistic, 39th Annual Meeting and 10th Conference of the European Chapter, Companion Volume to the Proceedings of the Conference: Proceedings of the Student Research Workshop and Tutorial Abstracts, p. 5, Toulouse, France, 9-11 July 2001
10. Marcu, D.: From discourse structures to text summaries. In: Proceedings of the ACL Workshop on Intelligent Scalable Text Summarization, pp. 82–88 (1997)
11. Nenkova, A., Passonneau, R., McKeown, K.: The pyramid method: Incorporating human content selection variation in summarization evaluation. ACM Trans. Speech Lang. Process. 4(2) (2007). http://doi.acm.org/10.1145/1233912.1233913
12. Nenkova, A., Passonneau, R.J.: Evaluating content selection in summarization: the pyramid method. In: HLT-NAACL, pp. 145–152 (2004). http://acl.ldc.upenn.edu/hlt-naacl2004/main/pdf/91_Paper.pdf
13. Nenkova, A., Vanderwende, L.: The impact of frequency on summarization. Microsoft Research, Redmond, Washington, Technical report MSR-TR-2005-101 (2005)
14. Papineni, K., Roukos, S., Ward, T., Zhu, W.J.: Bleu: a method for automatic evaluation of machine translation. In: Proceedings of the 40th Annual Meeting on Association for Computational Linguistics, ACL 2002, pp. 311–318. Association for Computational Linguistics, Stroudsburg, PA, USA (2002). http://dx.doi.org/10.3115/1073083.1073135
15. Radev, D., Blair-Goldensohn, S., Zhang, Z., Raghavan, R.: Newsinessence: a system for domain-independent, real-time news clustering and multi-document summarization. In: Proceedings of the First International Conference on Human Language Technology Research (2001). http://www.aclweb.org/anthology/H01-1056
16. Rath, G.J., Resnick, A., Savage, T.R.: The formation of abstracts by the selection of sentences. Part I. Sentence selection by men and machines. Am. Documentation **12**, 139–141 (1961). http://dx.doi.org/10.1002/asi.5090120210
17. Salton, G., Singhal, A., Mitra, M., Buckley, C.: Automatic text structuring and summarization. Inf. Process. Manage. **33**(2), 193–207 (1997). http://dx.doi.org/10.1016/S0306-4573(96)00062-3

18. Spearman, C.: The proof and measurement of association between two things. Am. J. Psychol. **15**(1), 72–101 (1904). http://www.jstor.org/stable/1412159
19. Zhou, L., Lin, C.Y., Munteanu, D.S., Hovy, E.: Paraeval: using paraphrases to evaluate summaries automatically. Association for Computational Linguistics, April 2006. http://research.microsoft.com/apps/pubs/default.aspx?id=69253

Multi-document Summarization Based on Atomic Semantic Events and Their Temporal Relationships

Yllias Chali([✉]) and Mohsin Uddin

University of Lethbridge, Lethbridge, AB, Canada
chali@cs.uleth.ca, mdmohsin.uddin@uleth.ca

Abstract. Automatic multi-document summarization (MDS) is the process of extracting the most important information, such as events and entities, from multiple natural language texts focused on the same topic. In this paper, we experiment with the effects of different groups of information such as events and named entities in the domain of generic and update MDS. Our generic MDS system has outperformed the best recent generic MDS systems in DUC 2004 in terms of ROUGE-1 recall and f_1-measure. Update summarization is a new form of MDS, where novel yet salient sentences are chosen as summary sentences based on the assumption that the user has already read a given set of documents. We present an event based update summarization where the novelty is detected based on the temporal ordering of events, and the saliency is ensured by the event and entity distribution. To our knowledge, no other study has deeply experimented with the effects of the novelty information acquired from the temporal ordering of events (assuming that a sentence contains one or more events) in the domain of update multi-document summarization. Our update MDS system has outperformed the state-of-the-art update MDS system in terms of ROUGE-2 and ROUGE-SU4 recall measures. All our MDS systems also generate quality summaries which are manually evaluated based on popular evaluation criteria.

1 Introduction

Automatic multi-document summarization (MDS) extracts core information from the source text and presents the most important content to the user in a concise form [24]. The important information is contained in textual units or groups of textual units which should be taken into consideration in generating a coherent and salient summary. In this paper, we propose an event based model of the generic MDS where we represent the generic summarization problem as an atomic event extraction as well as a topic distribution problem. Another new type of summarization called update MDS, whose goal is to get a salient summary of the updated documents supposing that the user has read the earlier documents about the same topic. The best of the recent efforts to generate update summary use graph based algorithms with some additional features to explore the novelty of the document [9,20,36]. Maximal Marginal Relevance (MMR) based

N. Ferro et al. (Eds.): ECIR 2016, LNCS 9626, pp. 366–377, 2016.
DOI: 10.1007/978-3-319-30671-1_27

approach [3] is used to blindly filter out the new information. These approaches discard the sentences containing novel information if they contain some old information from the previous document sets [7].

Steinberger et al. [33] use the sentence time information in the Latent Semantic Analysis (LSA) framework to get the novel sentences. They only consider the first time expression as the anchored time of the sentence, but sentences may contain multiple time expressions from various chronologies. For instance, consider the sentence *"Two members of Basque separatist group ETA arrested while transporting half a tonne of explosives to Madrid just prior to the* ***March 2004*** *bombings received jail sentences of 22 years each on* ***Monday****"*. Here we get two[1] time expressions: *March 2004* and *Monday*. The first expression represents the very old information, and the second one represents the accurate anchoring of the sentence. If we consider the first time expression as the sentence's time, like Steinberger et al. [33] would, then it would give us false novel/update information. This is why we take into account all of the events of a sentence to calculate its anchored time. In this paper, we also design a novel approach by taking into account all of the events in a sentence and their temporal relations to ensure the novelty, as well as the saliency, in update summarization. We represent the novelty detection problem as a chronological ordering problem of the temporal events and time expressions. Our event based sentence ranking system uses a topic model to identify all of the salient sentences. The rest of the paper is organized as follows. Section 2 reviews previous related works in text summarization. Section 3 describes our proposed summarization models. Section 4 gives the evaluation of our systems. Section 5 presents some conclusions and future works.

2 Related Works

Every document covers a central theme or event. There are other sub-events which support the central event. There are also many words or terms across the whole document, which can act as an individual event, they contribute to the main theme. Named entities such as time, date, person, money, organizations, locations, etc., are also significant because they build up the document structure. Although events and named entities are terms or group of terms, they have a higher significance than the normal words or terms. Those events and named entities can help to generate high performing summaries. Filatova and Hatzivassiloglou [11] used atomic events in extractive summarization. They considered events as a triplet of two named entities and verb (or action noun), where the verb (or action noun) connects the two named entities. Several greedy algorithms based on the co-occurrence statistics of events are used to generate a summary. They showed that event-based summaries get a much better score than the summaries generated by *tf*idf* weighing of words. Li et al. [19] also defined the same complex structure as an event and the PageRank algorithm [29] is applied to

[1] Here **'22 years'** is a time period. Time periods do not carry important information for detecting novelty.

estimate the event relevance in a summary generation. Another recent summarization work based on the event semantics is done by Zhang et al. [37]. Their events may contain an unlimited number of entities. Due to the complex nature of all of the previous authors' defined events, it is hard to use their defined event concept in a topic model to get the semantic event distribution in text.

Our defined semantic event is an atomic term, which is similar to the TimeML [31]. Pustejovsky et al. [31] consider events as a cover term for situations that happen, occur, hold, or take place. Event spans can be a period of time. Aspect, intentional state, intentional action, perception, occurrence, and modal can be events. We consider some classes of event expressions such as verbs (e.g., launched, cultivated, resigned, won) and event nominals (e.g., Vietnam War, Military operation).

Events like deverbal nouns are used in G-FLOW [6] to identify discourse relations to ensure coherency in a summary. Our generic MDS system uses the event and entity distribution, obtained from a topic model in sentence ranking, to generate a quality summary.

Update summarization, the newest type of challenge for summarization communities, is introduced first in DUC'2007[2]. Several popular generic summarization approaches, such as LexRank [10], TextRank [26] were used in update summarization without paying attention on the novelty detection. Fisher and Roark [12] used a domain-independent supervised classification to rank sentences and then they extract all of the sentences containing old information by using some filtering rules. QCQPSum [21] involved the previous documents in an objective function formulation and a reinforcement propagation in the new documents. It did not try to extract the novel information at the semantic level. We can see a few semantic analysis based novelty detection approaches: the Iterative Residual Rescaling (IRR) based LSA framework [32] and the Bayesian multinomial probability distribution based approach [7]. The state-of-the-art update summarization system, h-uHDP model [16] used Hierarchical Dirichlet Process (HDP) [35] to get the history epoch and the update epoch distribution. They used Kullback-Leibler (KL) [15] divergence based greedy approach to select novel sentences. All of the above approaches neglected the semantic temporal information which is crucial in novelty detection.

3 Our Methodologies

3.1 Pre-processing of the Data Set

In this paper, we use Stanford CoreNLP[3] for tokenization, named entity recognition, and cross-document coreference resolution. We remove all of the candidate sentences containing quotations. We also remove the candidate sentences whose length are less than 11 words. Sentences containing quotations are not appropriate for summary and shorter sentences carry a small amount of relative

[2] http://duc.nist.gov/duc2007/tasks.html.

[3] http://nlp.stanford.edu/software/corenlp.shtml.

information [17]. After tokenization, we remove stop words. We use Porter Stemmer [30] for stemming. Stemmed words or terms are then fed to Latent Dirichlet Allocation (LDA) engine for further processing. We use ClearTK[4] system [2] for event and temporal relation extraction.

3.2 Generic Summarization

The generic summarization problem is formulated as follows. Any cluster c contains n documents and all of the documents are equiprobable. All of the documents in each cluster are sorted in the descending order of their creation time[5]. The topic probability of each topic T_j can be calculated by Eq. (1) where $j \in \{1, \ldots, K\}$ and K is the number of topics of Latent Dirichlet Allocation (LDA) Model.

$$P(T_j) = \sum_{d=1}^{n} P(T_j|D_d)P(D_d) \tag{1}$$

To increase the coherence of the summary, we calculate sentence position score, S_p. If SC_d is the number of sentences in Document D, S_p can be calculated by Eq. (2) where sentence position index, $i \in \{0, \ldots, SC_d - 1\}$.

$$S_p = 1 - \frac{i}{SC_d} \tag{2}$$

The score of a sentence can be computed by Eq. (3).

$$Score(S) = S_p \times \sum_{t \in S}(P(t) \times W_g) \tag{3}$$

In Eq. (3), W_g is the specific weight factor for each group of terms. TC_g is the number of terms in one group g where $g \in \{event(e), named\text{-}entity(n), other(o)\}$. We consider empirically W_g is 1 for the group called *other* (which is a set of normal terms other than events and named entities) and W_g for groups event and named entity can be calculated by Eq. (5).

$$M = max_g TC_g, g \in \{e, n, o\} \tag{4}$$

Here, M is the number of terms in the highest group.

$$W_g = \frac{M}{TC_g} \tag{5}$$

Our weight calculating scheme ensures larger weights for event and named entity groups and also prevents the high occurring group from scoring high. The steps of our generic MDS system are as follows:

[4] http://code.google.com/p/cleartk/.

[5] Document Creation Time (DCT) can be calculated from document name.

1. Apply the LDA topic model on the corpus of documents for a fixed number[6] of topics K.
2. Compute the probability of topic T_j by Eq. (1) and sort the topics in the descending order of their probabilities.
3. Pick the topic T_j from the sorted list in the order of the probabilities of T_j, i.e., $P(T_1), .., P(T_k)$.
4. For topic T_j, compute the score of all of the sentences by Eq. (3) where $P(t)$ is the unigram probability distribution obtained from the LDA topic model.
5. For topic T_j, pick up the sentence with the highest score and include it in the summary. If it is already included in the summary or it dissatisfies other requirements (cosine score between candidate sentence and already-included summary sentences crosses the certain range), then pick up the sentence with the next highest score for this topic T_j.
6. Each selected sentence is compressed according to the method described in Sect. 3.3.
7. If the summary reaches its desired length then terminate the operation, else continue from step 3.

3.3 Sentence Compression

The quality of a summary can be improved by sentence compression [13,18]. Consider the sentence *"The Amish school where a gunman shot 10 girls last week, killing five of them, is expected to be demolished on Thursday, a fire department official said"*. Here we can see the subclause "a fire department official said" does not have any significance in a summary. Removing this type of long unnecessary subclauses will improve summary quality and provide extra space to include new information in a fixed length summary. We mainly consider widely used reporting verbs such as *said, told, reported* etc., to find out subclauses like in the above example. In our experiments, we use the Stanford dependency parser [4] to parse each selected sentence. Sentences containing a reporting verb are always parsed following a fixed rule where the reporting verb is always the 'root' of the dependency tree. Then we traverse the parse tree to find the subclause related to that reporting verb.

3.4 Update Summarization

Time End Point Normalization. Time expression identification and normalization are integral parts for the temporal processing of raw text. We use Stanford SUTime [5], which is a rule-based temporal tagger, to extract all of the temporal expressions. SUTime is one of the best systems in capturing temporal expressions from a natural language text. It follows TimeML [31] formats (TIMEX3) for normalizing time expressions.

Consider the sentence: *"The Amish school where a gunman shot 10 girls* ***last week****, killing five of them, is expected to be demolished* ***Thursday****, a fire*

[6] Total 4 topics are taken into account, i.e. K is 4.

department official said". Here *last week* and *Thursday* are the time expressions of the sentence. SUTime output of the above text is mentioned below, where October 11th, 2006 is a reference date:

"The Amish school where a gunman shot 10 girls <TIMEX3 tid="t2" type="DATE" ***value="2006-W40"*** *> last week </TIMEX3>, killing five of them, is expected to be demolished <TIMEX3 tid="t3" type="DATE"* ***value= "2006-10-12"*** *>Thursday </TIMEX3 >, a fire department official said."*

SUTime extracts *2006-W40* and *2006-10-12* as the normalized date of *last week* and *Thursday*, respectively. We convert them into an absolute time end point on a universal timeline. We follow standard date and time format (*YYYY-MM-DD hh:mm:ss*) for the time end point. For example, after conversion of **2006-W40** and **2006-10-12**, we get 2006-09-23 23:59:59 and 2006-10-12 23:59:59, respectively.

Temporal Ordering of Events and Time Expressions. In update summarization, knowing the relative order of the events is very useful for merging and presenting information from various news sources [25]. Information, such as event occurrence time or what events occurred prior to a particular event, presuppose the ability to infer an event's temporal ordering in discourse [25]. Inferring relations of temporal entities and events is a crucial step towards update summarization task.

Unlike Denis and Muller [8], we anchor events to one time point only, which is the upper end point. We are concerned only about the relative ordering of the events. We use ClearTK-TimeML tool to extract events and temporal relations [2]. In ClearTK-TimeML, four types of temporal relations are predicted. They are **B**EFORE, **A**FTER, **I**NCLUDES, and **N**ORELATION. Our main goal is to solve the novelty problem by using relative events' anchored values. In order to saturate the event-event and event-time relations, we use Allen's [1] transitive closure rules. Some of them are given below:

A before B and B before C $\Longrightarrow$ *A before C*
A includes B and B includes C $\Longrightarrow$ *A includes C*
A after B and B after C $\Longrightarrow$ *A after C*

We anchor all of the events to absolute times based on the *'includes'* and *'is-included'* relations of the event-time links. The remaining events are anchored approximately, based on other relations which are *'before'* and *'after'*.

Temporal Score. To obtain temporal score, we use ClearTK system [2] for initial temporal relation extraction and some transitive rules as described earlier. First, we relax the original event time association problem by anchoring the event to an approximate time. Then, we calculate the temporal score of a sentence by taking an average time score of all of the events' anchored time. Then, all of the sentences are ordered in the descending order of their temporal scores except for the first sentence of each document. Then, we calculate the temporal position score (tp_s) of the temporally ordered sentences. The tp_s of the first sentence of a document is considered to be one. Temporal position scores of the remaining sentences can be calculated by the Eq. (6). D_s is the number of sentences in document D and the temporally ordered sentence position index, $i \in \{0, \ldots, D_s - 1\}$.

$$tp_s = 1 - \frac{\gamma \times i}{D_s} \tag{6}$$

The parameter (γ) is used to tune the weight of the relative temporal position of the sentences.

Sentence Ranking. From the Latent Dirichlet Allocation (LDA) topic model, we obtain a unigram (event or named entity) probability distribution, $P(t)$. For each topic, the sentence score can be computed using Eq. (7).

$$Score(s) = tp_s \times (\sum_{t \in S}(P(t) \times \alpha \times W_g) + \sum_{t \in S}(P(t) \times \beta \times W_g)) \tag{7}$$

In Eq. (7), W_g can be calculated using Eq. (5), tp_s is the temporal position score of a sentence obtained from Eq. (6), and α and β are the weight factors of the new terms and the topic title terms, respectively, which are learnt from TAC'2010 dataset. For each topic, one sentence is taken as a summary sentence from the ordered list of sentences (descending order of their score, $Score(s)$). We use cosine similarity score to remove the redundancy of the summary. Additionally, we use the same sentence compression technique as in the generic summarization.

4 Evaluation

4.1 ROUGE Evaluation: Generic Summarization

We use the DUC 2004 dataset to evaluate our generic MDS system. We perform our experiment on 35 clusters of 10 documents each. DUC 2004 Task-2 was to create short multi-document summaries no longer than 665 bytes. We evaluate the summaries generated by our system using the automatic evaluation toolkit ROUGE[7] [22]. We compare our system with some recent systems including, the best system in DUC 2004 (Peer 65), a conceptual units based model [34], and G-FLOW, a recent state-of-the-art coherent summarization system [6]. As shown in Table 1, our system outperforms those three systems. It also scores better than the recent submodular functions based state-of-the-art system[8] [23]

[7] ROUGE runtime arguments for DUC 2004:
ROUGE -a -c 95 *-b* 665 *-m -n* 4 *-w* 1.2.

[8] We do not compare our system with the recent topic model based system [14] because that system is significantly outperformed by Lin and Bilmes's [23] system in terms of both ROUGE-1 recall and f_1-measure.

Table 1. Evaluation on the DUC 2004 dataset (The best results are **bolded**)

Systems	*R*-1	F_1
Peer 65	0.3828	0.3794
Takamura and Okumura [34]	0.3850	-
G-FLOW	0.3733	0.3743
Lin	0.3935	0.3890
Our generic MDS System	**0.3953**	**0.3983**

Table 2. ROUGE-2 and ROUGE-SU4 scores with 95 % confidence on the DUC'2004 dataset

Systems	*ROUGE-2*	*ROUGE-SU4*
Our generic	**0.1017**	**0.142**
MDS System	**(0.0975-0.11)**	**(0.135-0.149)**

in terms of ROUGE-1 recall and f_1-measure. We also include ROUGE-2 and ROUGE-SU4 scores in Table 2 with 95 % confidence. For the DUC 2004 dataset, on average we find weight factors of the groups like events, named-entities, and others, which are 3, 1.14, and 1, respectively. That means that our summarization system assigns the highest priority to the events group and the lowest priority to the normal terms. Hence, it explains the importance of the semantic events in a summarization system. This also explains the importance of named entities over other tokens during summary generation.

4.2 ROUGE Evaluation: Update Summarization

To evaluate our update MDS system, we use the TAC'2011 dataset. TAC'2011 dataset contains two groups of data, A and B. Group A contains the old dataset. Group B contains the new dataset of the same topic as group A. We perform our experiment on 28 clusters of 10 documents each. TAC'2011 guided update summarization task was to create short multi-document summaries no longer than 100 words with the assumption that the user has already read the documents from group A. Table 3 tabulates ROUGE scores of our system and best performing systems in TAC'2011 update summarization task. Our model outperforms the current state-of-the-art system, which is h-uHDPSum, as well as the best update summarization system (peer 43) of TAC'2011 summarization track. 95 % confidence intervals in Table 4 show that our system obtains significant improvement over the two systems (h-uHDPSum and Peer 43) in terms of ROUGE-2 and ROUGE-SU4. The performance of our event and temporal relation based summarizer changes according to the type of documents we are considering to be summarized. Our system gets very high recall and f-measures for the documents that are well constituents of events. Our temporal relation based system reveals

Table 3. Evaluation on the TAC'2011 dataset

Systems	*ROUGE*-2	*ROUGE-SU*4
Our update MDS System	**0.1120**	**0.1460**
h-uHDPSum	0.1017	0.1364
Peer 43	0.0959	0.1309

Table 4. 95 % confidence for various systems on the TAC'2011 dataset

Systems	*ROUGE*-2	*ROUGE-SU*4
Our update MDS System	**0.1016-0.1244**	**0.1356-0.1587**
h-uHDPSum	0.0910-0.1034	0.1265-0.1473
Peer 43	0.0894-0.1029	0.1251-0.1366

all of the hidden novel information. At the same time, our event and named entity based scoring scheme ensures the saliency in update summarization.

4.3 Manual Evaluation

ROUGE evaluation is not enough to measure the quality of a summary properly. Human evaluation is necessary to get an accurate score of quality. In generic MDS, we use relevancy, non-redundancy, and overall responsivenes criteria to manually evaluate our generic summary. We randomly select 24 clusters from the DUC 2004 dataset and assign a total of 3 human assessors for the evaluation purpose. Each assessor examines the summaries from all 24 clusters that are generated by our system and gives a score of 1 (Very Poor) to 5 (Very Good). Finally average scores are calculated. Table 5 tabulates average scores of manual evaluation on DUC 2004 dataset.

Our event-based summarization system chooses the high relevance sentences as the summary sentences. We observe that the cosine similarity checking performs poorly in removing redundancy. It has been shown in the literature that highly responsive summary "would have redundancy to some extent" [34]. This may be the reason why our summarization system does not perform well in checking redundancy as we had hoped.

In update MSD, we use the following criteria to manually evaluate our update summaries: novelty (containing update information), readability/fluency,

Table 5. Manual evaluation on the DUC 2004 dataset

Relevancy	3.92
Non-redundancy	3.50
Overall responsiveness	3.70

Table 6. Manual evaluation on the TAC 2011 dataset

Novelty	4.13
Fluency	3.92
Overall responsiveness	4.07

and overall responsiveness (overall focus and content). We randomly select 21 clusters from TAC 2011 dataset. Table 6 tabulates average scores of manual evaluation on TAC 2011 dataset.

Our temporal summarization system chooses highly novel sentences as summary sentences without losing fluency and responsiveness.

5 Conclusion and Future Work

In this paper, we have shown a simple yet effective way of approaching the task of generating multi-document generic summaries. The importance of semantic events and named entities in generating summaries has been deeply analyzed using the LDA topic model. By dividing terms into different groups we achieve high ROUGE-1 recall and f_1 scores for generic MDS task. Our update summarization model can identify novel information based on temporal ordering of events. Our system outperforms the state-of-the-art update summarization system based on ROUGE-2 and ROUGE-SU4 recall measures. There is still much room to improve event-event and event-time ordering. Ordering temporal entity considering all possible 12 relations is an NP-complete problem. Denis and Muller [8] reduce the complexity of the problem by converting relations into end points, but they get only 41 % F1-score. By increasing the recall and precision of event-event and event-time relation extraction, it is possible to get better temporal ordering of sentences. This will eventually provide better update summarization. We believe that some recent works on temporal relation classification using dependency parses [27] and discourse analysis framework [28] can further improve our update summarization system performance.

References

1. James, F.: Allen.: maintaining knowledge about temporal intervals. Commun. ACM **26**(11), 832–843 (1983)
2. Bethard, S.: Cleartk-timeml: a minimalist approach to tempeval. In: Second Joint Conference on Lexical and Computational Semantics (* SEM), vol. 2, pp. 10–14 (2013)
3. Boudin, F., El-Bèze, M., Torres-Moreno, J. M.: A scalable MMR approach to sentence scoring for multi-document update summarization. COLING (2008)
4. Cer, D.M., De Marneffe, M.-C., Jurafsky, D., Manning, C.D.: Parsing to stanford dependencies: trade-offs between speed and accuracy. In: LREC (2010)
5. Chang, A.X., Manning, C.D.: Sutime: a library for recognizing and normalizing time expressions. In: Language Resources and Evaluation (2012)

6. Christensen, J., Mausam, S.S., Etzioni, O.: Towards coherent multi-document summarization. In: Proceedings of NAACL-HLT, pp. 1163–1173 (2013)
7. Delort, J.-Y., Alfonseca, E.: Dualsum: a topic-model based approach for update summarization. In: Proceedings of the 13th Conference of the European Chapter of the Association for Computational Linguistics, pp. 214–223 (2012)
8. Denis, P., Muller, P.: Predicting globally-coherent temporal structures from texts via endpoint inference and graph decomposition. In: Proceedings of the Twenty-Second International Joint Conference on Artificial Intelligence, vol. 3, pp. 1788–1793. AAAI Press (2011)
9. Pan, D., Guo, J., Zhang, J., Cheng, X.: Manifold ranking with sink points for update summarization. In: Proceedings of the 19th ACM International Conference on Information and Knowledge Management, pp. 1757–1760 (2010)
10. Erkan, G., Radev, D.R.: Lexrank: graph-based lexical centrality as salience in text summarization. J. Artif. Intell. Res. (JAIR) **22**(1), 457–479 (2004)
11. Filatova, E., Hatzivassiloglou, V.: Event-based extractive summarization. In: Proceedings of ACL Workshop on Summarization, vol. 111 (2004)
12. Fisher, S., Roark, B.: Query-focused supervised sentence ranking for update summaries. In: Proceeding of TAC 2008 (2008)
13. Gillick, D., Favre, B., Hakkani-Tur, D., Bohnet, B., Liu, Y., Xie, S.: The icsi/utd summarization system at tac. In: Proceedings of the Second Text Analysis Conference, Gaithersburg, Maryland, USA. NIST (2009)
14. Haghighi, A., Vanderwende, L.: Exploring content models for multi-document summarization. In: Proceedings of Human Language Technologies: The Annual Conference of the North American Chapter of the Association for Computational Linguistics, pp. 362–370. Association for Computational Linguistics (2009)
15. Kullback, S.: The kullback-leibler distance (1987)
16. Li, J., Li, S., Wang, X., Tian, Y., Chang, B.: Update summarization using a multi-level hierarchical dirichlet process model. In: COLING (2012)
17. Li, L., Heng, W., Jia, Y., Liu, Y., Wan, S.: Cist system report for acl multiling 2013-track 1: multilingual multi-document summarization. In: MultiLing 2013, p. 39 (2013)
18. Li, P., Wang, Y., Gao, W., Jiang, J.: Generating aspect-oriented multi-document summarization with event-aspect model. In: Proceedings of the Conference on Empirical Methods in Natural Language Processing, pp. 1137–1146 (2011)
19. Li, W., Mingli, W., Qin, L., Wei, X., Yuan, C.: Extractive summarization using inter-and intra-event relevance. In: Proceedings of the 21st International Conference on Computational Linguistics and the 44th Annual Meeting of the Association for Computational Linguistics, pp. 369–376. Association for Computational Linguistics (2006)
20. Li, X., Liang, D., Shen, Y.-D.: Graph-based marginal ranking for update summarization. In: SDM, pp. 486–497. SIAM (2011)
21. Li, Xuan, Liang, Du, Shen, Yi-Dong: Update summarization via graph-based sentence ranking. IEEE Trans. Knowl. Data Eng. **25**(5), 1162–1174 (2013)
22. Lin, C.-Y.: Rouge: a package for automatic evaluation of summaries. In: Text Summarization Branches Out: Proceedings of the ACL-2004 Workshop, pp. 74–81 (2004)
23. Lin, H., Bilmes, J.: A class of submodular functions for document summarization. In: ACL, pp. 510–520 (2011)
24. Mani, I.: Automatic Summarization, vol. 3. John Benjamins Publishing, Amsterdam (2001)

25. Mani, I., Schiffman, B., Zhang, J.: Inferring temporal ordering of events in news. In: Proceedings of the Conference of the North American Chapter of the Association for Computational Linguistics on Human Language Technology: Companion Volume of the Proceedings of HLT-NAACL 2003-Short Papers, vol. 2, pp. 55–57. Association for Computational Linguistics (2003)
26. Mihalcea, R., Tarau, P.: Textrank: Bringing order into texts. In: Proceedings of EMNLP, vol. 4, p. 275, Barcelona, Spain (2004)
27. Ng, J.-P., Kan, M.-Y.: Improved temporal relation classification using dependency parses and selective crowdsourced annotations. In: COLING, pp. 2109–2124 (2012)
28. Ng, J.-P., Kan, M.-Y., Lin, Z., Feng, W., Chen, B., Jian, S., Tan, C.L.: Exploiting discourse analysis for article-wide temporal classification. In: EMNLP, pp. 12–23 (2013)
29. Page, L., Brin, S., Motwani, R., Winograd, T.: The pagerank citation ranking: Bringing order to the web (1999)
30. Martin, F.: Porter.: an algorithm for suffix stripping. Program Electr. Libr. Inf. Syst. **14**(3), 130–137 (1980)
31. Pustejovsky, J., Castano, J.M., Ingria, R., Sauri, R., Gaizauskas, R.J., Setzer, A., Katz, G., Radev, D.R.: Timeml: robust specification of event and temporal expressions in text. In: New Directions in Question Answering, vol. 3, pp. 28–34 (2003)
32. Steinberger, J., Ježek, K.: Update summarization based on novel topic distribution. In: Proceedings of the 9th ACM symposium on Document Engineering, pp. 205–213 (2009)
33. Steinberger, J., Kabadjov, M., Steinberger, R., Tanev, H., Turchi, M., Zavarella, V.: Jrcs participation at tac: Guided and multilingual summarization tasks. In: Proceedings of the Text Analysis Conference (TAC) (2011)
34. Takamura, H., Okumura, M.: Text summarization model based on maximum coverage problem and its variant. In: Proceedings of the 12th Conference of the European Chapter of the Association for Computational Linguistics, pp. 781–789. Association for Computational Linguistics (2009)
35. Teh, Y.W., Jordan, M.I., Beal, M.J., Blei, D.M.: Hierarchical dirichlet processes. J. Am. Stat. Assoc. **101**, 1566–1581 (2004)
36. Wenjie, L., Wei Furu, L., Qin, H.Y.: Pnr 2: ranking sentences with positive and negative reinforcement for query-oriented update summarization. In: Proceedings of the 22nd International Conference on Computational Linguistics, pp. 489–496 (2008)
37. Zhang, R., Li, W., Qin, L.: Sentence ordering with event-enriched semantics and two-layered clustering for multi-document news summarization. In: Proceedings of the 23rd International Conference on Computational Linguistics, pp. 1489–1497 (2010)

Tweet Stream Summarization for Online Reputation Management

Jorge Carrillo-de-Albornoz, Enrique Amigó, Laura Plaza(✉), and Julio Gonzalo

NLP & IR Group, Universidad Nacional de Educación a Distancia (UNED), Madrid, Spain
{jcalbornoz,enrique,lplaza,julio}@lsi.uned.es

Abstract. Producing online reputation reports for an entity (company, brand, etc.) is a focused summarization task with a distinctive feature: issues that may affect the reputation of the entity take priority in the summary. In this paper we (i) propose a novel methodology to evaluate summaries in the context of online reputation which profits from an analogy between reputation reports and the problem of diversity in search; and (ii) provide empirical evidence that incorporating priority signals may benefit this summarization task.

Keywords: Summarization · Diversity · Tweets · Reputation management

1 Introduction

Since the advent of Social Media, an essential part of Public Relations (for organizations and individuals) is Online Reputation Management, which consists of actively listening online media, monitoring what is being said about an entity and deciding how to act upon it in order to preserve or improve the public reputation of the entity. Monitoring the massive stream of online content is the first task of online reputation experts. Given a client (e.g. a company), the expert must provide frequent (e.g. daily) reports summarizing which are the issues that people are discussing and involve the company.

In a typical workflow, the reputation experts start with a set of queries that try to cover all possible ways of referring to the client. Then they take the results set and filter out irrelevant content (e.g., texts about apple pies when looking for the Apple company). Next, they determine which are the different issues people are discussing, evaluate their priority, and produce a report for the client.

Crucially, the report must include any issue that may affect the reputation of the client (reputation alerts) so that actions can be taken upon it. The summary, therefore, is guided by the relative priority of issues. This notion of priority differs from the signals that are usually considered in summarization algorithms, and it depends on many factors, including popularity (How many people are commenting on the issue?), polarity for reputation (Does it have positive or

N. Ferro et al. (Eds.): ECIR 2016, LNCS 9626, pp. 378–389, 2016.
DOI: 10.1007/978-3-319-30671-1_28

negative implications for the client?), novelty (Is it a new issue?), authority (Are opinion makers engaged in the conversation?), centrality (Is the client central to the conversation?), etc. This complex notion of priority makes the task of producing reputation-oriented summaries a challenging and practical scenario.

In this context, we investigate two main research questions:

RQ1. Given the peculiarities of the task, what is the most appropriate evaluation methodology?

Our research is triggered by the availability of the RepLab dataset [1], which contains annotations made by reputation experts on tweet streams for 61 entities, including entity name disambiguation, topic detection and topic priority.

We will discuss two types of evaluation methodologies, and in both cases we will adapt the RepLab dataset accordingly. The first methodology sticks to the traditional summarization scenario, under the hypothesis that RepLab annotations can be used to infer automatically entity-oriented summaries of near-manual quality. The second evaluation methodology models the task as producing a ranking of tweets that maximizes both coverage of topics and priority. This provides an analogy with the problem of search with diversity, where the search system must produce a rank that maximizes both relevance and coverage.

RQ2. What is the relationship between centrality and priority?

The most distinctive feature of reputation reports is that issues related with the entity are classified according to their priority from the perspective of reputation handling (the highest priority being a *reputation alert*). We want to investigate how the notion of priority translates to the task of producing extractive summaries, and how important it is to consider reputational signals of priority when building and appropriate summary.

We will start by discussing how to turn the RepLab setting and datasets into a test collection for entity-oriented tweet stream summarization. Then we will introduce our experimental setting to compare priority signals with text quality signals and assess our evaluation methodology, discuss the results, link our study with related work, and finish with the main conclusions learned.

2 A Methodology to Evaluate Reputation-Oriented Tweet Stream Summarization

A reputation report is a summary – produced by an online reputation expert – of the issues being discussed online which involve a given client (a company, organization, brand, individual... in general, an entity). In reputation reports produced daily, microblogs (and Twitter in particular) are of special relevance, as they anticipate issues that may later hit other media. Typically, the reputation expert follows this procedure (with the assistance of more or less sophisticated software):

- Starts with a set of queries that cover all possible way of referring to the client.
- Takes the results set and filter out irrelevant content.
- Groups tweets according to the different issues (topics) people are discussing.

- Evaluates the priority of each issue, establishing at least three categories: reputation alerts (which demand immediate attention), important topics (that the company must be aware of), and unimportant content (refers to the entity, but do not have consequences from a reputational point of view).
- Produces a reputation report for the client summarizing the result of the analysis.

The reputation report must include any issue that may affect the reputation of the client (reputation alerts) so that action can be taken upon it. This (extractive) summary, therefore, is guided by the relative priority of issues. However, as we pointed out in the introduction, this notion of priority differs from the signals that are usually considered in summarization algorithms, and it depends on many factors, including: popularity, polarity for reputation, novelty, authority, and centrality. Thus, the task is novel and attractive from the perspective of summarization, because the notion of which are the relevant information nuggets is focused and more precisely defined than in other summarization tasks. Also, it explicitly connects the summarization problem with other Natural Language Processing tasks: there is a filtering component (because it is entity-oriented), a social media component (because, in principle, non-textual Twitter signals may help discovering priority issues), a semantic understanding component (to establish, for instance, polarity for reputation), etc.

2.1 The RepLab 2013 Dataset

The RepLab 2013 task is defined as (multilingual) topic detection combined with priority ranking of the topics. Manual annotations are provided for the following subtasks:

- *Filtering.* Systems are asked to determine which tweets are related to the entity and which are not. Manual annotations are provided with two possible values: related/unrelated. For our summarization task, we will use as input only those tweets that are manually annotated as related to the entity.
- *Polarity for Reputation Classification.* The goal is to decide if the tweet content has positive or negative implications for the company's reputation. Manual annotations are: positive/negative/neutral.
- *Topic Detection*: Systems are asked to cluster related tweets about the entity by topic with the objective of grouping together tweets referring to the same subject/event/conversation.
- *Priority Assignment.* It involves detecting the relative priority of topics. Manual annotations have three possible values: Alert, mildly_important, unimportant.

RepLab 2013 uses Twitter data in English and Spanish. The collection comprises tweets about 61 entities from four domains: automotive, banking, universities and music. We will restrict our study to the automotive and banking domains, because they consist of large companies which are the standard subject of reputation monitoring as it is done by experts: the annotation of *universities* and *music bands and artists* is more exploratory and does not follow widely adopted conventions as in the case of companies. Our subset of Replab 2013 comprises 71,303 tweets distributed as in the following table.

Table 1. Subset of RepLab 2013 dataset used in our experiments

	Automotive	Banking	Total
Entities	20	11	31
# Tweets (training)	15,123	7,774	22,897
# Tweets (test)	31,785	16,621	48,406
# Tweets (total)	46,908	24,395	71,303
# Tweets (EN)	38,614	16,305	54,919
# Tweets (ES)	8,294	8,090	16,384

2.2 Automatic Generation of Reference Summaries

We investigate two alternative ways of evaluating tweet stream summaries using RepLab data: the first one consists in automatically deriving "reference" or "model" summaries from the set of manual annotations provided by RepLab.

The goal of a reputation report is to cover all issues referring to the entity (in our dataset, a bank or a car manufacturer) which are relevant from a reputational perspective. RepLab manual annotations group relevant tweets according to fine-grained issues related to the company, and assign a three-valued priority to them. If we select only alerts and mildly important topics, and we pick randomly one tweet per topic, the result would be equivalent to a manual (extractive) summary under certain simplifying assumptions about the data:

- In a topic, all tweets are equally representative. This is a reasonable assumption in the RepLab dataset, because selected tweets are very focused, every tweet is independently assigned to a topic, and topics are fine-grained and therefore quite cohesive.
- A tweet is enough to summarize the content of an issue appropriately. This is certainly an oversimplification, and reputation experts will at least rewrite the content of a topic for a summary, and provide a logical structure to the different topics in a report. However, we may assume that, for evaluation purposes and as an average observation, most tweets are representative of the content of a topic.

Under this assumptions, variability between model summaries depends on which tweet we choose from each relevant topic. Therefore, we use a simplified user model where an expert may randomly pick any tweet, for every important topic (alerts and mildly relevant issues), to produce a reputation report. In our experiments, we generate 1,000 model summaries for every entity using this model. Note that the excess of simplification in our assumptions pays off, as we are able to generate a large number of model summaries with the manual annotations provided by the RepLab dataset.

Once we have created the models (1,000 per test case), automatic summaries can be evaluated using standard text similarity measures. In our experiments we use ROUGE [2], a set of evaluation metrics for summarization which measure the content overlap between a peer and one or more reference summaries. The most popular variant is ROUGE-2, due to its high correlation with human judges. ROUGE-2 counts the number of bigrams that are shared by the peer and reference summaries and computes a recall-related measure [2].

2.3 Tweet Summarization as Search with Diversity

Our second approach to evaluate summaries does not require model summaries. It reads the summary as a ranked list of tweets, and evaluates the ranking with respect to relevance and redundancy as measured with respect to the annotated topics in the RepLab dataset. The idea is making an analogy between the task of producing a summary and the task of document retrieval with diversity. In this task, the retrieval system must provide a ranked list of documents that maximizes both relevance (documents are relevant to the query) and diversity (documents reflect the different query intents, when the query is ambiguous, or the different facets in the results when the query is not ambiguous).

Producing an extractive summary is, in fact, a similar task: the set of selected sentences should maximize relevance (they convey essential information from the documents) and diversity (sentences should minimize redundancy and maximize coverage of the different information nuggets in the documents). The case of reputation reports using Twitter as a source is even more clear, as relevance is modeled by the priority of each of the topics. An optimal report should maximize the priority of the information conveyed and the coverage of priority entity-related topics (which, in turn, minimizes redundancy).

Let's think of the following user model for tweet summaries: the user starts reading the summary from the first tweet. At each step, the user goes on to the next tweet or stops reading the summary, either because she is satisfied with the knowledge acquired so far, or because she does not expect the summary to provide further useful information. User satisfaction can be modeled via two variables: (i) the probability of going ahead with the next tweet in the summary; (ii) the amount of information gained with every tweet. The amount of information provided by a tweet depends on the tweets that precede it in the summary: a tweet from a topic that has already appeared in the summary contributes less than a tweet from a topic that has not yet been covered by the preceding tweets. To compute the expected user satisfaction, the evaluation metric must also take into account that tweets deeper in the summary (i.e. in the rank) are less likely to be read, weighting the information gain of a tweet by the probability of reaching it. We propose to adapt Rank-Biased Precision (RBP) [3], an Information Retrieval evaluation measure which is defined as:

$$RBP = (1 - p) \sum_{i=1}^{d} r_i * p^{i-1}$$

where r_i is a known function of the relevance of document at position i, p is the probability of moving to the next document, and RBP is defined as utility/effort (expected utility rate), with utility being $\sum_{i=1}^{d} r_i * p^{i-1}$ and $1/(1-p)$ the expected number of documents seen, i.e. the effort.

We prefer RBP to other diversity-oriented evaluation metrics because it naturally fits our task, the penalty for redundancy can be incorporated without changing the formula (simply defining r_i), and because it has been shown to comply with more desired formal properties than all other IR measures in the literature [4], and can be naturally adapted to our task.

Indeed, the need to remove redundancy and the relevance of priority information can be incorporated via r_i. We will model r_i according to two possible scenarios. In the first scenario, incorporating more than one tweet from a single topic still contributes positively to the summary (but increasingly less than the first tweet from that topic). This is well captured by the reciprocal of the number of tweets already seen from a topic (although many other variants are possible):

$$r_i = \frac{1}{|\{k \in \{1 \ldots i-1\} | \text{topic}(i) = \text{topic}(k)\}|}$$

We will refer to RBP with this relevance formula as **RBP-SUM-R** (RBP applied to SUMmarization with a Reciprocal discount function for redundancy).

In the second scenario, each topic is exhaustively defined by one tweet, and therefore only the first tweet incorporated to the summary, for each topic, contributes to the informative value of the summary. Then the relevance formula is simply:

$$r_i = \begin{cases} 1 \text{ if } \forall k \in \{1..i-1\} \text{topic}(i) \neq \text{topic}(k) \\ 0 \text{ otherwise} \end{cases}$$

We will refer to RBP with this relevance formula as **RBP-SUM-B** (RBP applied to SUMmarization with a Binary discount function for redundancy). With respect to the parameter p (probability of going ahead reading the summary after reading a tweet), we must aim at large values, which better reflect the purpose of the summary. For instance, a value of $p = 0.95$ means that the user has only a 60 % chance of reading beyond the first ten tweets, and a value of $p = 0.5$ decreases that probability to only 0.1 %. Figure 1 shows how the probability of reading through the summary decays for different values of p. We will perform our experiments with the values $p = 0.9$ (which decays fast for a summarization task) and $p = 0.99$ (which has a slower but still representative decay).

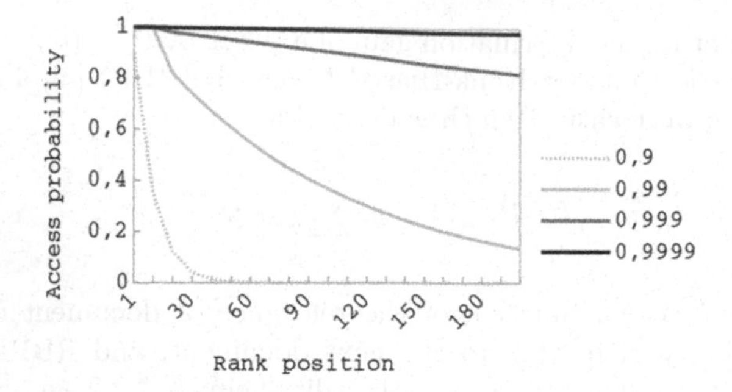

Fig. 1. Probability of reading through the summary for different p values

3 Experimental Design

Our first research question (how to evaluate the task) is partially answered in the previous section. We now want to compare how the two alternative evaluation metrics behave, and we want to investigate the second research question: what is the relationship between centrality and priority, and how priority signals can be used to enhance summaries. For this purpose, we will compare three approaches (two baselines and one contrastive system):

LexRank. As a standard summarization baseline, we use LexRank [5], one of the best-known graph-based methods for multi-document summarization based on lexical centrality. LexRank is executed through the MEAD summarizer [6] (http://www.summarization.com/mead/) using these parameters: `-extract -s -p 10 -fcp delete`. We build summaries at 5, 10, 20 and 30 % compression rate, for LexRank and also for the other approaches.

Followers. As a priority baseline, we simply rank the tweets by the number of followers of the tweet author, and then apply a technique to remove redundancy. The number of followers is a basic indication of priority: things being said by people with more followers are more likely to spread over the social networks. Redundancy is avoided using an iterative algorithm: a tweet from the ranking is included in the summary only if it has a vocabulary overlap less than 0.02, in terms of the Jaccard measure, with any of the tweets already included in the summary. Once the process is finished, if the resulting compression rate is higher than desired, discarded tweets are reconsidered and included by recursively increasing the threshold in 0.02 similarity points until the desired compression rate is reached.

Signal Voting. Our contrastive system considers a number of signals of priority and content quality. Each signal (computed using the training set) provides a ranking of all tweets for a given test case (an entity). We follow this procedure:

- Using the training part of the RepLab dataset, we compute two estimations of the quality of each signal: the ratio between average values within priority

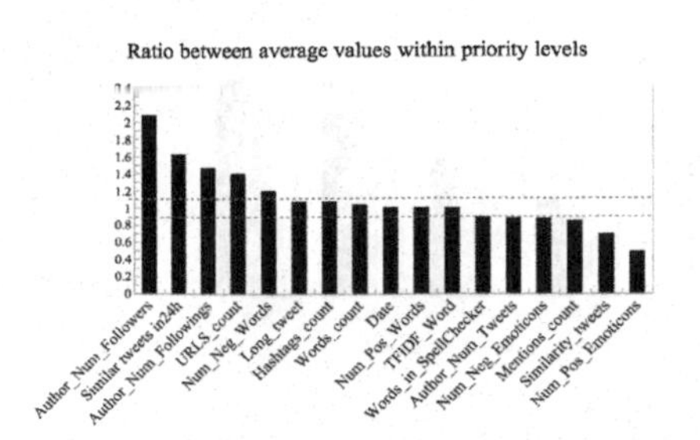

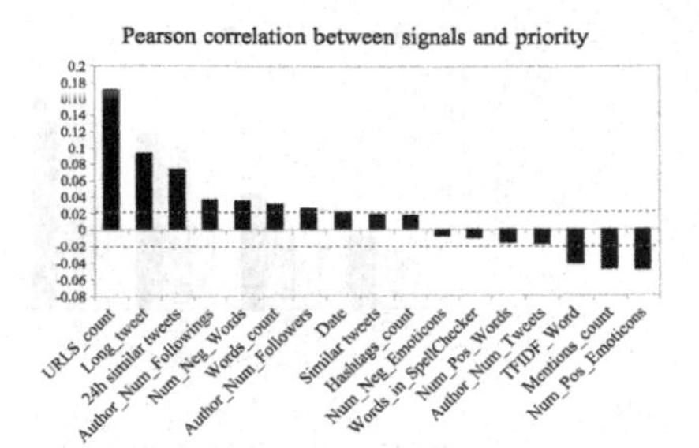

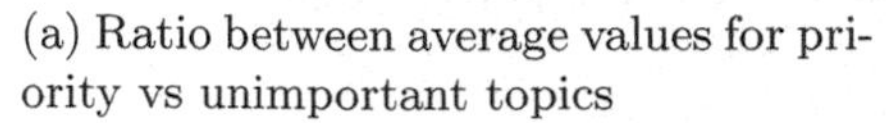
(a) Ratio between average values for priority vs unimportant topics

(b) Pearson correlation between signal values and manual priority

Fig. 2. Signal assessment

values (if priority tweets receive higher values than unimportant tweets, the signal is useful), and the Pearson correlation between the signal values and the manual priority values. The signals (which are self-descriptive) and the indicators are displayed in Fig. 2.

- We retain those signals with a Pearson correlation above 0.02 and with a ratio of averages above 10 %. The resulting set of signals is: `URLS count` (number of URLs in the tweet), `24h similar tweets` (number of similar tweets produced in a time span of 24 hours), `Author num followers` (number of followers of the author), `Author num followees` (number of people followed by the author), `neg words` (number of words with negative sentiment), `Num pos emoticons` (number of emoticons associated with a positive sentiment), and `Mentions count` (number of Twitter users mentioned).
- Each of the selected signals produces a ranking of tweets. We combine them to produce a final ranking using Borda count [7], a standard voting scheme to combine rankings.
- We remove redundancy with the same iterative procedure used in the *Followers* baseline.

4 Results and Discussion

We have evaluated all systems with respect to the test subset of RepLab 2013. Figure 3 (left) compares the results of LexRank, the followers baseline and the signal voting algorithm in terms of ROUGE-2. For each entity and for each compression rate, systems are compared with the set of 1,000 reference summaries automatically generated. Figure 3 (right) shows the recall of relevant topics at different compression ratios. Finally, Fig. 4 evaluates the summaries in terms of RBP-SUM-R and RBP-SUM-B directly with respect to the manual assessments in the RepLab 2013 dataset.

In terms of ROUGE, the combination of signals is consistently better than both LexRank and the Followers baseline at all compression levels. All differences are statistically significant according to the t-test, except at 5 % compression

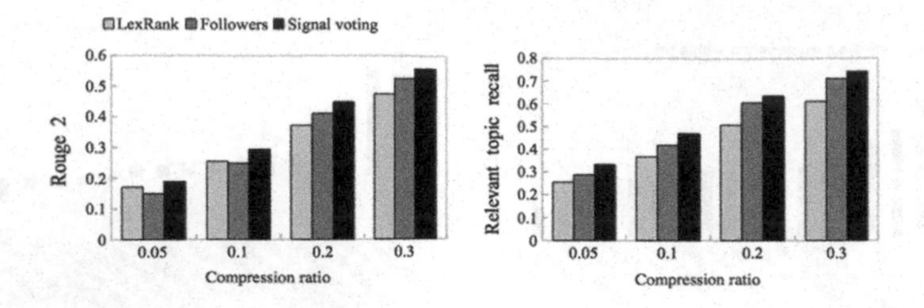

Fig. 3. Results in terms of Rouge and recall of important topics

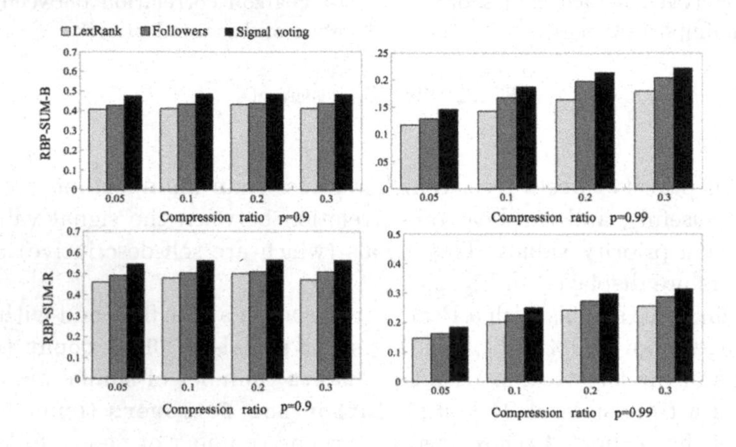

Fig. 4. Value of priority signals according to RBP-SUM

rate where the difference between signal voting and LexRank is not significant (p = 0.08). Remarkably, at 20 % and 30 % compression rates even the Followers baseline – which uses very little information and is completely unsupervised – outperforms the LexRank baseline. Altogether, these are clear indicators that priority signals play a major role for the task.

In terms of recall of relevant topics, the figure shows that Signal voting > Followers > LexRank at all compression ratios. In terms of RBP-SUM, results are similar. With both relevance scoring functions, signal voting outperforms the two baselines at all compression rates, and all differences are statistically significant. The only difference is that this evaluation methodology, which penalizes redundancy more heavily (tweets from the same topic receive an explicit penalty), gives the followers baseline a higher score than LexRank at all compression levels (with both relevance scoring functions).

Relative differences are rather stable between both p values and between both relevance scoring functions. Naturally, absolute values are lower for RBP-SUM-B, as the scoring function is stricter. Although experimentation with users would be needed to appropriately set the most adequate p value and relevance scoring schema, the measure differences seem to be rather stable with respect to both choices.

5 Related Work

5.1 Centrality Versus Priority-Based Summarization

Centrality has been one of the most widely used criteria for content selection [8]. Centrality refers to the idea of how much a fragment of text (usually a sentence) covers the main topic of the input text (a document or set of documents). However, the information need of users frequently goes far beyond centrality and should take into account other selection criteria such as diversity, novelty and priority. Although the importance of enhancing diversity and novelty in various NLP tasks has been widely studied [9,10], reputational priority is a domain-dependent concept that has not been considered before. Other priority criteria have been previously considered in some areas: In [11], concepts related to treatments and disorders are given higher importance than other clinical concepts when producing automatic summaries of MEDLINE citations. In opinion summarization, positive and negative statements are given priority over neutral ones. Moreover, different aspects of the product/service (e.g., technical performance, customer service, etc.) are ranked according to their importance to the user [12]. Priority is also tackled in query (or topic)-driven summarization where terms from the user query are given more weight under the assumption that they reflects the user relevance criteria [13].

5.2 Multi-tweet Summarization

There is much recent work focusing on the task of multi-tweet summarization. Most publications rely on general-purpose techniques from traditional text summarization along with redundancy detection methods to avoid the repetition of contents in the summary [14]. Social network specific signals (such as user connectivity and activity) have also been widely exploited [15].

Two different types of approaches may be distinguished: feature-based and graph-based. Feature-based approaches address the task as a classification problem, where the aim is to classify tweets into important/unimportant, so that only important tweets are used to generate the summary. Tweets are represented as sets of features, being the following the most frequently used: term frequency [16], time delay [16], user based features [17] and readability based features [15]. Graph-based approaches usually adapt traditional summarization systems (such as LexRank [5] and TextRank [18]) to take into consideration the particularities of Twitter posts [14,15,19]. These approaches usually include both content-based and network-based information into the text graph.

Concerning the subject of the input tweets, most works have focused on those related to sport and celebrity events [14,19]. These events are massively reported in social networks, so that the number of tweets to summarize is huge. In this context, simple frequency based summarizers perform well and even better than summarizers that incorporate more complex information [14]. The problem of summarizing tweets on a company's reputation has been, to the best of our knowledge, never tackled before and presents additional challenges derived from the less massive availability of data and the greater diversity of issues involved.

6 Conclusions

We have introduced the problem of generating reputation reports as a variant of summarization that is both practical and challenging from a research perspective, as the notion of reputational priority is different from the traditional notion of importance or centrality. We have presented two alternative evaluation methodologies that rely on the manual annotation of topics and their priority. While the first evaluation methodology maps such annotations into summaries (and then evaluates with standard summarization measures), the second methodology establishes an analogy with the problem of search with diversity, and adapts an IR evaluation metric to the task (RBP-SUM).

Given the high correlation between Rouge and RBP-SUM values, we advocate the use of the latter to evaluate reputation reports. There are two main reasons: first, it avoids the need of explicitly creating reference summaries, which is a costly process (or suboptimal if, as in our case, they are generated automatically from topic/priority annotations); the annotation of topics and priorities is sufficient. Second, it allows an explicit modeling of the patience of the user when reading the summary, and of the relative contribution of information nuggets depending on where in the summary they appear and their degree of redundancy with respect to already seen text.

As for our second research question, our experiments indicate that priority signals play a relevant role to create high-quality reputation reports. A straightforward voting combination of the rankings produced by useful signals consistently outperforms a standard summarization baseline (LexRank) at all compression rates and with all the evaluation metrics considered. In fact, the ranking produced by just one signal (number of followers) also may outperform LexRank, indicating that standard summarization methods are not competitive.

In future work we will consider including graded relevance with respect to priority levels in the data. In our setting, we have avoided such graded relevance to avoid bias in favor of priority-based methods, but RBP-SUM directly admits a more sophisticated weighting scheme via r_i.

Acknowledgments. This research was partially supported by the Spanish Ministry of Science and Innovation (VoxPopuli Project, TIN2013-47090-C3-1-P) and UNED (project 2014V/PUNED/0011).

References

1. Amigó, E., Carrillo-de-Albornoz, J., Chugur, I., Corujo, A., Gonzalo, J., Martín, T., Meij, E., de Rijke, M., Spina, D.: Overview of RepLab 2013: Evaluating online reputation monitoring systems. In: Forner, P., Müller, H., Paredes, R., Rosso, P., Stein, B. (eds.) CLEF 2013. LNCS, vol. 8138, pp. 333–352. Springer, Heidelberg (2013)
2. Lin, C.Y.: Rouge: A package for automatic evaluation of summaries. In: Proceedings of the ACL Workshop on Text Summarization Branches Out, pp. 74–81 (2004)

3. Moffat, A., Zobel, J.: Rank-biased precision for measurement of retrieval effectiveness. ACM Trans. Inf. Syst. (TOIS) **27**(1), 2 (2008)
4. Amigó, E., Gonzalo, J., Verdejo, F.: A general evaluation measure for document organization tasks. In: Proceedings of ACM SIGIR, pp. 643–652. ACM (2013)
5. Erkan, G., Radev, D.R.: Lexrank: Graph-based lexical centrality as salience in text summarization. J. Artif. Int. Res. **22**(1), 457–479 (2004)
6. Radev, D., Allison, T., Blair-Goldensohn, S., Blitzer, J., Çelebi, A., Dimitrov, S., Drabek, E., Hakim, A., Lam, W., Liu, D., Otterbacher, J., Qi, H., Saggion, H., Teufel, S., Topper, M., Winkel, A., Zhang, Z.: MEAD – A platform for multidocument multilingual text summarization. In: Proceedings of LREC (2004)
7. Van Erp, M., Schomaker, L.: Variants of the borda count method for combining ranked classifier hypotheses. In: Proceedings of Seventh International Workshop on Frontiers in Handwriting recognition. pp. 443–452 (2000)
8. Cheung, J.C.K., Penn, G.: Towards robust abstractive multi-document summarization: A caseframe analysis of centrality and domain. In: Proceedings of ACL, Sofia, Bulgaria. pp. 1233–1242 (2013)
9. Mei, Q., Guo, J., Radev, D.: Divrank: The interplay of prestige and diversity in information networks. In: Proceedings of ACM SIGKDD. pp. 1009–1018 (2010)
10. Clarke, C.L., Kolla, M., Cormack, G.V., Vechtomova, O., Ashkan, A., Büttcher, S., MacKinnon, I.: Novelty and diversity in information retrieval evaluation. In: Proceedings of ACM SIGIR 2008, pp. 659–666 (2008)
11. Fiszman, M., Demner-Fushman, D., Kilicoglu, H., Rindflesch, T.C.: Automatic summarization of medline citations for evidence-based medical treatment: A topic-oriented evaluation. J. Biomed. Inform. **42**(5), 801–813 (2009)
12. Pang, B., Lee, L.: Opinion mining and sentiment analysis. Found. Trends Inf. Retr. **2**(1–2), 1–135 (2008)
13. Nastase, V.: Topic-driven multi-document summarization with encyclopedic knowledge and spreading activation. In: Proceedings of EMNLP, pp. 763–772 (2008)
14. Inouye, D., Kalita, J.: Comparing twitter summarization algorithms for multiple post summaries. In: Proceedings of the IEEE Third International Conference on Social Computing, pp. 298–306 (2011)
15. Liu, X., Li, Y., Wei, F., Zhou, M.: Graph-based multi-tweet summarization using social signals. In: Proceedings of COLING 2012, pp. 1699–1714 (2012)
16. Takamura, H., Yokono, H., Okumura, M.: Summarizing a document stream. In: Clough, P., Foley, C., Gurrin, C., Jones, G.J.F., Kraaij, W., Lee, H., Mudoch, V. (eds.) ECIR 2011. LNCS, vol. 6611, pp. 177–188. Springer, Heidelberg (2011)
17. Duan, Y., Chen, Z., Wei, F., Zhou, M., Shum, H.Y.: Twitter topic summarization by ranking tweets using social influence and content quality. In: Proceedings of COLING 2012, Mumbai, India, pp. 763–780 (2012)
18. Mihalcea, R., Tarau, P.: Textrank: Bringing order into texts. In: Proceedings of EMNLP 2004, Barcelona, Spain pp. 404–411 (2004)
19. Sharifi, B., Hutton, M.A., Kalita, J.: Summarizing microblogs automatically. In: Proceedings of NAACL, pp. 685–688 (2010)

Reproducibility

Who Wrote the Web? Revisiting Influential Author Identification Research Applicable to Information Retrieval

Martin Potthast[1(✉)], Sarah Braun[2], Tolga Buz[3], Fabian Duffhauss[4], Florian Friedrich[5], Jörg Marvin Gülzow[6], Jakob Köhler[7], Winfried Lötzsch[8], Fabian Müller[9], Maike Elisa Müller[3], Robert Paßmann[10], Bernhard Reinke[10], Lucas Rettenmeier[5], Thomas Rometsch[11], Timo Sommer[12], Michael Träger[13], Sebastian Wilhelm[2], Benno Stein[1], Efstathios Stamatatos[14], and Matthias Hagen[1]

[1] Bauhaus-Universität Weimar, Weimar, Germany
martin.potthast@uni-weimar.de
[2] Technische Universität München, Munich, Germany
[3] Technical University of Berlin, Berlin, Germany
[4] RWTH Aachen University, Aachen, Germany
[5] Heidelberg University, Heidelberg, Germany
[6] University of Konstanz, Konstanz, Germany
[7] Free University of Berlin, Berlin, Germany
[8] Chemnitz University of Technology, Chemnitz, Germany
[9] Karlsruhe University of Applied Sciences, Karlsruhe, Germany
[10] University of Bonn, Bonn, Germany
[11] University of Michigan, Ann Arbor, USA
[12] Hamburg University of Technology, Hamburg, Germany
[13] University of Bamberg, Bamberg, Germany
[14] University of the Aegean, Mytilene, Greece

Abstract. In this paper, we revisit author identification research by conducting a new kind of large-scale reproducibility study: we select 15 of the most influential papers for author identification and recruit a group of students to reimplement them from scratch. Since no open source implementations have been released for the selected papers to date, our public release will have a significant impact on researchers entering the field. This way, we lay the groundwork for integrating author identification with information retrieval to eventually scale the former to the web. Furthermore, we assess the reproducibility of all reimplemented papers in detail, and conduct the first comparative evaluation of all approaches on three well-known corpora.

1 Introduction

Author identification is concerned with whether and how an author's identity can be inferred from their writing by modeling writing style. Author identification

N. Ferro et al. (Eds.): ECIR 2016, LNCS 9626, pp. 393–407, 2016.
DOI: 10.1007/978-3-319-30671-1_29

has a long history, the first known approach dating back to the 19th century [27]. Ever since, historians and linguists have tried to settle disputes over the authorship of important pieces of writing by manual authorship attribution, employing basic style markers, such as average sentence length, average word length, or hapax legomena (i.e., words that occur only once in a given context), to name only a few. It is estimated that more than 1,000 basic style markers have been proposed [31]. In the past two decades, author identification has become an active field of research for computer linguists as well, who employ machine learning on top of models that combine traditional style markers with new ones, the manual computation of which has been infeasible before. Author identification technology is evolving at a rapid pace. The field has diversified into many sub-disciplines where correlations of writing style with author traits are studied, such as age, gender, and other demographics. Moreover, in an attempt to scale their approaches, researchers apply them on increasingly large datasets with up to thousands of authors and tens of thousands of documents. Naturally, some of the document collections used for evaluation are sampled from the web, carefully ensuring that individual documents can be attributed with confidence to specific authors.

While applying this technology at web scale is still out of reach, we conjecture that it is only a matter of time until tailored information retrieval systems will index authorial style, retrieve answers to writing style-related queries as well as queries by example, and eventually, shed light on the question: Who wrote the web? Besides obvious applications in law enforcement and intelligence—a domain for which little is known about the state of the art of their author identification efforts—many other stakeholders will attempt to tap authorial style for purposes of targeted marketing, copyright enforcement, writing support, establishing trustworthiness, and of course as yet another search relevance signal. Many of these applications bring about ethical and privacy issues that need to be reconciled. Meanwhile, authorial style patterns already form a part of every text on the web that has been genuinely written by a human. At present, however, the two communities of information retrieval and author identification hardly intersect, whereas integration of technologies from both fields is necessary to scale author identification to the web.

The above observations led us to devise and carry out a novel kind of reproducibility study that has an added benefit for both research fields: we team up with a domain expert and a group of students, identify 15 influential author identification methods of the past two decades, and have each approach reimplemented by the students. By reproducing performance results from the papers' experiments, we aim at raising confidence that our implementations come close to those of the papers' authors. This paper surveys the approaches and reports on their reproducibility. The resulting source code is shared publicly. We further conduct comparative experiments among the reimplemented approaches, which has not been done before. The primary purpose of our reproducibility study is not to repeat *every* experiment reported in the selected papers, since it is unlikely that the most influential research is outright wrong. Rather, our goal is to release working implementations to both the information retrieval community as well as the author identification community, since only a few public implementations have surfaced to date. This lays the groundwork for future collaboration among both fields.

In what follows, Sect. 2 reviews related work and introduces the author identification papers selected, Sect. 3 overviews the setup of our study, Sect. 4 details the students' implementations and outlines reproducibility issues observed, and Sect. 5 reports on the first comparative evaluation of all approaches.

2 Background, Related Work, and Paper Selection

This section briefly reviews reproducibility-related research in computer science in general, and information retrieval in particular. Afterwards, we overview author identification paradigms and the papers selected for our reproducibility study.

2.1 Reproducibility in Computer Science and Information Retrieval

The reproducibility of research results that are obtained empirically determines whether the conclusions drawn from them may eventually be accepted as fact. While many of the empirical sciences have well-established best practices for reproducing research, this is not, yet, the case in the empirical branches of the comparably young field of computer science. Regardless, even sciences that have best practices currently face a reproducibility crisis: a number of studies made the news, revealing significant amounts of peer-reviewed research to be irreproducible. In the wake of these events, many computer scientists revisit their own reproducibility record and find it lacking in many respects. For brevity, we will not recite all causes for lack of reproducibility but focus on the one that relates to our contribution, namely computer science's primary research tool: software. Or rather, its absence: the vast majority of computer science research is about the development of software that solves problems of interest, but many researchers are reluctant to share their software.

Collberg et al. [9] recently assessed the availability of the pieces of software underlying 601 papers published at ACM conferences and journals; software could be collected for only 54 % of the papers.[1] No attempt was made to check whether the software actually works as advertised. To identify the reasons for not sharing software, Stodden [39] conducted a survey among 134 computer scientists and found, among others, the time to clean and polish the software (77.8 %), the time to deal with support questions (51.9 %), a fear of supporting competing colleagues without getting credit (44.8 %), and intellectual property constraints (40.0 %). After all, sharing software is voluntary, and scientometrics do not yet incorporate such community services. There are counterexamples, though, such as Weka [16] and LibSVM [8], which are used across disciplines,

[1] Interestingly, Collberg et al.'s study itself has been challenged for lack of rigor and has been reproduced more thoroughly: http://cs.brown.edu/~sk/Memos/Examining-Reproducibility/.

or Terrier [28] and Blei et al.'s LDA implementation [6], which have spread throughout information retrieval (IR). Various initiatives in IR have emerged simultaneously in 2015: the ECIR has introduced a dedicated track for reproducibility [17], a corresponding workshop has been organized at SIGIR [2], and the various groups that develop Evaluation-as-a-Service platforms for shared tasks have met for the first time [19].

One of the traditional forms of reproducibility research are meta studies, where existing research on a specific problem of interest is surveyed and summarized with special emphasis on performance. For example, in information retrieval, the meta study of Armstrong et al. [3] reveals that the improvements reported in various papers of the past decade on the ad hoc search task are void, since they employ too weak baselines. Recently, Tax et al. [40] have conducted a similar study for 87 learning-to-rank papers, where they summarize for the first time which of them perform best.

Still, meta studies usually do not include a reimplementation of existing methods. Reimplementation of existing research has been conducted by Ferro and Silvello [13] and Hagen et al. [15], the former aiming for exact replicability and the latter for reproducibility (i.e., obtaining similar results under comparable circumstances). Finally, Di Buccio et al. [11] and Lin [26] both propose the development of a central repository of baseline IR systems on standard tasks (e.g., ad hoc search). They observe that even the baselines referred to in most papers may vary greatly in performance when using different parameterizations, rendering results incomparable. A parameter model, repositories of runs, and executable baselines are proposed as a remedy. When open baseline implementations are available in a given research field such as IR, this is a sensible next step, whereas in the case of author identification, there are only a few publicly available baseline implementations to date. We are the first to provide them at scale.

2.2 Author Identification

Authorship analysis attempts to extract information from texts based on the personal writing style of their authors. The main focus of research in this area is on *author identification* and more specifically on authorship attribution, where given a set of candidate authors and some samples of their writing, a text of unknown or disputed authorship is attributed to one of them [20,36]. This can be viewed as either a closed-set classification task (i.e., realistic in most forensic cases where police investigations can define a small set of suspects) or an open-set classification task (i.e., realistic in web-based applications) [25]. An important variation of this task is *authorship verification* where the set of candidate authors is a singleton [38,42]. This can be viewed as a one-class classification problem where the negative class (i.e., texts written by other authors) is huge and heterogeneous. Another dimension gaining increasing attention is *author profiling* where the task is to extract information about the characteristics of the author (e.g., age, gender, educational level, personality, etc.) rather than their identity [30].

Following the practices of text categorization, all author identification approaches comprise two basic modules: feature extraction and classification. The former is much more challenging in comparison to topic-based text classification or sentiment analysis since writing style rather than topic or sentiment has to be quantified. Unfortunately, in general, there is a lack of style-specific words. The line of research dealing with the quantification of writing style is known as *stylometry*, it has a long history [27], and plenty of measures have been proposed so far [18]. These stylometric measures fall into the following categories [36]: *lexical* (e.g., word or sentence length distribution, vocabulary richness measures, function word frequencies), *character* (e.g., character type and character n-gram frequencies), *syntactic* (e.g., POS n-gram frequencies and rewrite rule frequencies), *semantic* (e.g., semantic relationship frequencies and semantic function frequencies), and *application-dependent* features (e.g., use of greetings in email messages or font size and color in HTML documents). Low-level features like function words and character n-grams have been reported to be the most effective while higher-level features related to syntactic parse trees or semantic information are useful complements [36]. The combination of measures from different categories can enhance the performance of authorship attribution approaches [10,43].

With respect to the classification methods, there are two main paradigms [36]: the *profile-based* approaches are author-centric and attempt to capture the cumulative style of the author by concatenating all available samples by that author and then extracting a single representation vector. Usually, generative models (e.g., naive Bayes) are used in profile-based approaches. On the other hand, *instance-based* methods are document-centric and attempt to capture the style of each text sample separately. In case only a single long document exists for one candidate author (e.g., a book), it is split into samples and each sample is represented separately. Usually, discriminative models (e.g., SVM) are exploited in instance-based approaches.

In order to reproduce a set of author identification approaches, we compiled an initial list of 30 influential papers published in the past two decades and meant to cover the main paradigms and approaches described above. Some well-known papers from the authorship attribution literature had to be excluded since their methods are based on NLP tools that are not publicly available making their reproduction infeasible within our study setup [14,37]. Finally, since the number of students participating in this study was limited, we assigned a paper to each student with the goal of maintaining the coverage of different paradigms, and, to match the complexity of a method with the student's background (computer science, mathematics, physics, engineering). The final list of selected papers alongside their basic characteristics is shown in Table 1.

Burrows' *Delta* [7] derives the deviation of function word frequencies from their norm. Keselj et al. [22] use character n-gram profiles, a method later modified for imbalanced datasets [35]. Benedetto et al. [5], Khmelev and Teahan [23], and Teahan and Harper [41] are exploiting compression models that are based on character sequences, while the approach of Peng et al. [29] can also use word

Table 1. Overview of papers selected for reimplementation. Tasks include closed-set attribution (cA), open-set attribution (oA), and verification (V). Features encode character (chr), lexical (lex), or syntactical (syn) information, or mixtures (mix) thereof. The paradigms implemented are profile-based (p) and instance-based (i). Complexity of implementation ranges from easy (*) via moderate (**) to hard (***). Citations as per Google Scholar (accessed September 29, 2015).

	Publication														
	[4]	[5]	[7]	[10]	[12]	[22]	[23]	[24]	[25]	[29]	[32]	[33]	[34]	[35]	[41]
Task	cA	cA	cA	cA	cA	cA	cA	V	oA	cA	cA	cA	cA	cA	cA
Features	lex	chr	lex	mix	chr	chr	chr	lex	chr	mix	lex	syn	lex	chr	chr
Paradigm	p	i	i	i	i	p	p	i	p	p	i	i	i	p	p
Complexity	**	*	*	*	***	*	**	**	*	**	***	**	*	*	**
Citations	14	377	213	366	41	267	60	75	89	201	17	44	26	43	80
Year	09	02	02	01	11	03	03	07	11	04	12	14	06	07	03

sequences. These compression-based methods have also been applied to tasks like topic detection, text genre recognition, or language identification. A combination of lexical, character, and application-dependent features suitable for the e-mail domain is described by de Vel et al. [10]. Also more complicated stylometric models are among our selection. Arun et al. [4] build a graph of function words using their proximity to estimate edge weights. Escalante et al. [12] propose local histograms representing the distribution of occurrences of character n-grams within a document. Seroussi et al. [32] describe an extension of LDA topic modeling using disjoint document and author topics. Sidorov et al. [33] make use of syntactic n-grams based on sequences of words or syntactic relations extracted from the parsing tree of sentences. Some of the selected methods focus on more complicated classification algorithms including feature subspace ensembles [25,34], and a meta-learning model [24].

3 Reproducibility Study

Our reproducibility study consists of seven steps: (1) paper selection, (2) student recruitment, (3) paper assignment and instruction, (4) implementation and experimentation, (5) auditing, (6) publication, and (7) post-publication rebuttal.

(1) *Paper Selection.* Every reproducibility study should supply justification for its selection of papers to be reproduced. For example, Ferro and Silvello [13] reproduce a method that has become important for performance measurement in IR in order to raise confidence in its reliability; Hagen et al. [15] reproduce the three best-performing approaches in a shared task, since shared task notebooks are often less well-written than other papers, rendering their reproduction difficult. Other justifications may include: comparison of a method with one's own approach, doubts whether a particular contribution works as advertised, completing a software library, using an approach as a sub-module to solve a different task, or identifying the best approach for an application.

The goal of our reproducibility study is a certain "coverage" of author identification. Given our limited human resources, we tried to cover different paradigms of author identification, whereas the papers selected were supposed to be influential for the field. In this regard, we considered it vitally important to consult with a domain expert to provide a selection of papers that satisfy these constraints, since hands-on experience is required to make such decisions. Particularly, the various paradigms to solve a problem typically emerge only with hindsight, whereas the terminology used in early papers may differ substantially from the present one. The number of citations that a paper received by itself turns out to be an insufficient yardstick, since this introduces a bias against recent papers. A total of 30 papers have been selected by our domain expert, whereas Table 1 overviews only those that were reproduced by the students recruited.

(2) *Student Recruitment.* To scale our reproducibility study, we employ students. Their recruitment for a task like this can be done in various ways within the context of a university, whereas proper incentives should be set for sufficient motivation. A dedicated course or project might be offered, or an extracurricular activity. The latter was what we offered to students from various universities. Altogether, we recruited 16 students with backgrounds in computer science (5), engineering (4), physics (3), and maths (4). Programming experience was in fact the only prerequisite for participation, which is why we did not restrict eligibility to computer science students only.

We were confident that a reproducibility study with students will work, since it resembles everyday work at universities, where advisors often pass tasks to students for implementation under guidance. Moreover, it tells a lot about any given paper whether or not it enables a student with basic training in programming to reproduce its results; ideally, the authors of technical papers ensure that even people outside their domain may follow up on their work. However, most papers omit the basics that are considered folklore in a given discipline, so that we tried to match students by their skill sets to papers, guiding them throughout the process.

(3) *Paper Assignment and Instruction.* The papers selected by our domain expert are of varying complexity, ranging from basic character-level string processing to dependency parsing to advanced statistical modeling (i.e., a customized LDA approach). Therefore, we did not assign papers at random but based on interviews about backgrounds and programming experiences of our students. More complex papers were assigned to students who have better chances of successfully implementing them. But matching students with papers is non-trivial, since interviews only paint an incomplete picture.

After paper assignment, we handed out papers to students alongside instructions what to do. After a brief explanation of the goals of the study (i.e., reimplementing influential approaches to author identification), the task was specified as follows:

1. Study the proposed main algorithmic contribution for author identification.
2. Implement the approach in a programming language of your choice.
3. Replicate at least one of the experiments described involving the approach.

Further, we asked students to take note of any imprecise, ambiguous, or missing details along the way. We did not ask students to repeat all experiments described in their papers, since we do not suspect the reported results to be false or entirely irreproducible. Rather, we use the papers' experiments as benchmarks to check the students' implementations.

(4) *Implementation and Experimentation.* In this step, students worked on their own, but were encouraged to ask questions. Our domain expert was accessible and we discussed technical questions with eleven of the students, most of which pertained to basic text processing, statistical computations, and performance optimization. Since the students lacked background in natural language processing, we pointed them to appropriate libraries that implement things like tokenization and dependency parsing. The students had ample time for implementation and experimentation, however, many started late before the deadline, and one failed to complete his task. To mitigate such issues, we recommend to engage students early on in (teleconference) meetings in this step.

(5) *Auditing.* After implementation, experts and students met for an auditing session. The purpose of this session was to ascertain that students had understood their paper at a fundamental, conceptual level so as to raise confidence in their implementations. Each approach was thoroughly discussed, highlighting the reproducibility issues observed. However, not everyone brought along flawless implementations; due to misunderstandings, some methods had to be amended. Therefore, a hackathon was organized to fix the issues, while encouraging group work and code sharing between compatible implementations. We were accompanying the students at all times during this step. Though we tried to finalize everything during auditing, some things were left for homework.

(6) *Publication.* Open sourcing the code is one of the main points of the exercise in order to provide baseline implementations to both the communities of author identification and information retrieval. We leave the choice of open source license at the discretion of the students. Since publishers are not yet ready to publish material alongside a scientific paper, we publish the code on our own.[2]

(7) *Post-Publication Rebuttal.* During steps (1)-(6), we specifically avoided to contact the authors of the selected papers. This was to prevent any bias entering our study or being influenced by the authors who might have been anxious about their approaches' performances. After our study has been accepted for publication, the authors were invited for a rebuttal, the outcome of which will be published as material alongside this paper.

[2] Materials and code of this study are available at www.uni-weimar.de/medien/webis/publications and the latest versions of the code in its GitHub repositories at www.github.com/pan-webis-de (for a convenient overview, see www.github.com/search?q=ECIR+2016+user:pan-webis-de).

4 Reproducibility Report

Each paper was assessed with regard to a number of reproducibility criteria pertaining to (1) approach clarity, (2) experiment clarity and soundness, (3) dataset availability or reconstructability, and (4) overall replicability, reproducibility, simplifiability (e.g., omitting preprocessing steps without harming performance), and improvability (e.g., with respect to runtime). The assessments result from presentations given by the students, a questionnaire, and subsequent individual discussions; Table 2 overviews the results.

(1) *Approach Clarity.* For none of the approaches source code (or executables) were available accompanying the papers (only ○ in row "Code available" of Table 2), so that all students had to start from scratch. The students

Table 2. Assessment of the individual approaches with respect to reproducibility criteria. A ○ indicates lacking reproducibility or information; a ◐ partial reproducibility or information; a ● sufficient reproducibility or information; a – indicates a criterion does not apply. Sizes are indicated as L(arge), M(edium), and S(mall), as judged by our domain expert. Programming languages Python and Java are abbreviated as Py and J.

Criterion	**Publication**														
	[4]	[5]	[7]	[10]	[12]	[22]	[23]	[24]	[25]	[29]	[32]	[33]	[34]	[35]	[41]
(1) *Approach clarity*															
Code available	○	○	○	○	○	○	○	○	○	○	○	○	○	○	○
Description sound	●	●	◐	◐	◐	●	●	●	●	●	●	●	●	●	●
Details sufficient	●	●	◐	◐	◐	●	●	●	●	◐	◐	◐	●	●	●
Paper self-contained	◐	●	●	◐	●	●	◐	●	●	●	○	◐	●	●	●
Preprocessing	○	●	●	●	–	–	–	◐	–	○	○	●	●	–	–
Parameter settings	–	◐	●	◐	●	●	–	●	●	●	●	○	●	●	○
Library versions	–	–	–	○	◐	–	–	◐	–	–	○	○	○	–	–
Reimplementation															
Language	Py	Py	Py	C++	J	Py	C++	Py	Py	C#	C++	J	Py	Py	Py
(2) *Experiment clarity / soundness*															
Setup clear	◐	●	◐	◐	●	●	●	●	●	●	●	◐	●	●	●
Exhaustiveness	◐	○	◐	○	◐	◐	○	◐	●	●	●	◐	●	●	○
Compared to others	○	○	○	●	●	◐	●	●	○	●	●	●	○	◐	●
Result reproduced	◐	◐	○	◐	◐	◐	●	○	◐	●	○	◐	●	●	●
(3) *Dataset reconstructability / availability*															
Text length	L	L	M	S	M	M	M	M	L	M	L	S	M	M	M
Candidate set	M	M	M	S	M	M	L	L	M	M	S	L	M	L	M
Origin given	●	●	◐	○	●	◐	●	●	●	●	●	●	●	●	○
Corpora available	○	○	○	○	●	◐	◐	○	○	●	●	○	●	◐	●
(4) *Overall assessment*															
Replicability	○	○	○	○	○	○	○	○	○	○	○	○	○	○	○
Reproducibility	●	◐	○	◐	◐	●	●	●	●	●	○	●	●	●	●
Simplifiability	●	●	●	○	○	○	○	○	○	○	○	○	○	○	●
Improvability	●	●	●	○	○	○	●	○	○	○	○	●	○	○	○

chose the programming language they are most familiar with, resulting in nine Python reimplementations, four reimplementations in a C dialect, and two Java reimplementations. Keeping in mind that most of the students had not worked in text processing before, it is a good sign that overall they had no significant problems with the approach descriptions. Some questions were answered by the domain expert, while some students also just looked up basic concepts like tokenizing or cosine similarity on their own. The students with backgrounds in math and theory mentioned a lack of formal rigor in the explanations of some papers (indicated by a ◐ in row "Description sound"); however, this was mostly a matter of taste and did not affect the understandability of the approaches. More problematic were two papers for which not even the references contained sufficient information, so that additional sources had to be retrieved by the students to enable them to reimplement the approach. The lack of details on how input should be preprocessed (○ in row "Preprocessing"), what parameter settings were used (○ in row "Parameter settings"), and missing version numbers of libraries employed (○ in row "Library versions") render the replication of seven out of the 15 selected papers' approaches difficult. This had an effect on the perceived approach clarity at an early stage of reimplementation.

(2) *Experiment Clarity / Soundness.* Since the students were asked to replicate or at least reproduce one of the experiments of their assigned papers, this gave us first-hand insights into the clarity of presentation of the experiments as well as their soundness. The most common problems we found were unclear splits between training and test data (◐ in row "Setup clear"). Another problem was that rather many approaches are evaluated only against simple baselines or only in small-scale experiments (○ in rows "Exhaustiveness" and "Comparison to others"). To rectify this issue, we conduct our own evaluation of all implemented approaches on three standard datasets in Sect. 5. Altogether, given the influential nature of the 15 selected approaches, it was not unexpected that in twelve cases the students succeeded in reproducing at least one result similar to those reported in the original papers (● or ◐ in row "Result reproduced").

(3) *Dataset Availability / Reconstructability.* We also asked students and our domain expert to assess the sizes of the originally used datasets. The approaches have been evaluated using different text lengths (S, M, and L indicate message, article, and book size in row "Text length") and different candidate set sizes (S, M, and L indicate below five, below 15, or more authors in row "Candidate set"). In eleven cases, the origin of the data was given, whereas in two cases each, the origin could only be indirectly inferred, or remained obscure. Corpora of which the datasets used for evaluation have been derived were available in four cases, whereas we tried to reconstruct the datasets in cases where sufficient information was given.

(4) *Overall Assessment and Discussion.* To complete the picture of our assessment, we have judged the overall replicability, reproducibility, simplifiability, and improvability of the original papers. Taking into account papers with only partially available information on preprocessing, parameter settings, and libraries

(ten papers) as well as the non-availability of the originally used corpora, none of the 15 publications' results are replicable. This renders the question of at least reproducing the results with a similar approach or using a similar dataset even more important. To this end, students were instructed to use the latest versions of the respective libraries with default parameter settings, and if nothing else helped, apply common sense. Regarding missing information on datasets, our domain expert suggested substitutions. With these remedies, all but one approach achieved results comparable to those originally reported (● or ◐ in row "Reproducibility"). The three partially reproducible papers are due to non-availability of the original data and the use of incomparable substitutions.

Only the reimplementation of the approach of Seroussi et al. [32] has been unsuccessful to date: it appears to suffer from an imbalanced text length distribution across candidate authors, resulting in all texts being attributed to authors with the fewest words among all candidates. This behavior is at odds with the paper, since Seroussi et al. do not mention any problems in this regard, nor that the evaluation corpora have been manually balanced. Since the paper is exceptionally well-written, leaving little to no room for ambiguity, we are unsure what the problem is and suspect a subtle error in our implementation. However, despite our best efforts, we have been unable to find this error to date. Perhaps the post-publication rebuttal phase or future attempts at reproducing Seroussi et al.'s work will shed light on this issue.[3]

In four cases, the respective students, while working on the reimplementations, identified possibilities of simplifying or even improving the original approaches (● in rows "Simplification" and "Improvability"). A few examples that concern runtime: when constructing the function word graph of Arun et al. [4], it suffices to take only the n last function words in a text window into account, where $n < 5$, instead of all previous ones. In Benedetto et al.'s approach [5], it suffices to only use the compression dictionary of the profile instead of recompressing profile and test text every time. In Burrows' approach [7], POS-tagging can be omitted, and in the approach of Teahan and Harper [41] one can refrain from actually compressing texts, but just compute entropy. For all of these improvements, the attribution performance was not harmed but often even improved while the runtime was substantially decreased.

On the upside, we can confirm that it is possible to reproduce almost all of the most influential work of a field when employing students to do so. On the downside, however, new ways of ensuring rigorous explanations of approaches and experimental setups should be considered.

5 Evaluation

To evaluate the reimplementations under comparable conditions we use the following corpora:

[3] Confer the repository of the reimplementation of Seroussi et al.'s approach to follow up on this.

Table 3. Evaluation results (classification accuracy) of the reimplemented approaches on three benchmark corpora. Best results (BR) are given as reported by the authors of [1,12,21]. Some approaches cannot be applied on all corpora (n/a) for reasons of runtime complexity or insufficient text lengths. One approach could not be successfully reproduced and was hence omitted (–).

Corpus	Publication															
	[4]	[5]	[7]	[10]	[12]	[22]	[23]	[24]	[25]	[29]	[32]	[33]	[34]	[35]	[41]	BR
C10	9.0	72.8	59.8	50.2	75.4	71.0	77.2	22.4	72.0	76.6	–	29.8	73.8	70.8	76.6	86.4
PAN11	0.1	29.6	5.4	13.5	43.1	1.8	32.8	n/a	20.2	46.2	–	n/a	7.6	34.5	65.0	65.8
PAN12	85.7	71.4	92.9	28.6	28.6	71.4	n/a	78.6	78.6	57.1	–	n/a	7.1	85.7	64.3	92.9

- *C10.* English news from the CCAT topic of the Reuters Corpus Volume 1 for 10 candidate authors (100 texts each). Best results reported by Escalante et al. [12].
- *PAN11.* English emails from the Enron corpus for 72 candidate authors with imbalanced distribution of texts. The corpus was used in the PAN 2011 shared task [1].
- *PAN12.* English novels for 14 candidate authors with three texts each. The corpus was used in the PAN 2012 shared task [21].

Parameters were set as specified in the original papers, unless they were not supplied, in which case parameters were optimized based on the training data. One exception is the approach of Escalante et al. [12] where a linear kernel was used instead of the diffusion kernel mentioned in that paper, since the latter could not be reimplemented in time.

Table 3 shows the evaluation results. As can be seen, some approaches are very effective on long texts (PAN12) but fail on short (C10) or very short texts (PAN11) [4,7]. Moreover, some approaches are considerably affected by imbalanced datasets (PAN11) [22]. It is interesting that in two out of the three corpora used (PAN12 and PAN11) at least one of the approaches competes with the best reported results to date. In general, the compression-based models seem to be more stable across corpora probably because they have few or none parameters to be fine-tuned [5,23,29,41]. The best macro-average accuracies on these corpora are obtained by Teahan and Harper [41] and Stamatatos [35]. Both follow the profile-based paradigm which seems to be more robust in case of limited text-length or limited number of texts per author. Moreover, they use character features which seem to be the most effective ones for this task.

6 Conclusion

To the best of our knowledge, a reproducibility study like ours, with the explicit goal of sharing working implementations of many important approaches, is unprecedented in information retrieval and in author identification, if not computer science as a whole. In this regard, we argue that employing students to systematically reimplement influential research and publish the resulting source code may prove to be a way of scaling the reproducibility efforts in many branches of computer science to a point at which a significant portion of research is covered. Conceivably, this would accelerate progress in the corresponding fields, since the entire community would have access to the state of the art. For students in their late education and early careers, reimplementing a given piece of influential research, and verifying its correctness by reproducing experimental results is definitely a worthwhile learning experience. Moreover, reproducing research from fields related to one's own may foster collaboration between both fields involved.

Acknowledgements. This study was supported by the German National Academic Foundation (German: Studienstiftung des deutschen Volkes). The foundation helped to recruit students among its scholars and organized our auditing workshop as part of its 2015 summer academy in La Colle-sur-Loup, France. We thank the foundation for their generous support. Our special thanks go to Dorothea Trebesius, Matthias Frenz, and Martina Rothmann-Stang who provided for our every need at the workshop.

References

1. Argamon, S., Juola, P.: Overview of the international authorship identification competition at PAN-. In: CLEF 2011 Notebooks (2011)
2. Arguello, J., Diaz, F., Lin, J., Trotman, A.: RIGOR @ SIGIR (2015)
3. Armstrong, T.G., Moffat, A., Webber, W., Zobel, J.: Improvements that don't add up: ad-hoc retrieval results since. In: CIKM 2009, pp. 601–610 (1998)
4. Arun, R., Suresh, V., Veni Madhavan, C.E.: Stopword graphs and authorship attribution in text corpora. In: ICSC, pp. 192–196 (2009)
5. Benedetto, D., Caglioti, E., Loreto, V.: Language trees and zipping. Phys. Rev. Lett. **88**, 048702 (2002)
6. Blei, D.M., Ng, A.Y., Jordan, M.I.: Latent Dirichlet allocation. J. Mach. Learn. Res. **3**, 993–1022 (2003)
7. Burrows, J.: Delta: a measure of stylistic difference and a guide to likely authorship. Lit. Ling. Comp. **17**(3), 267–287 (2002)
8. Chang, C.-C., Chih-Jen Lin, L.: A library for support vector machines. ACM TIST **2**, 27:1–27:27 (2011)
9. Collberg, C., Proebstring, T., Warren, A.M.: Repeatability, benefaction in computer systems research: a study and a modest proposal. TR 14-04, University of Arizona (2015)
10. de Vel, O., Anderson, A., Corney, M., Mohay, G.: Mining e-mail content for author identification forensics. SIGMOD Rec. **30**(4), 55–64 (2001)
11. Di Buccio, E., Di Nunzio, G.M., Ferro, N., Harman, D., Maistro, M., Silvello, G.: Unfolding off-the-shelf IR systems for reproducibility. In: RIGOR @ SIGIR (2015)
12. Escalante, H.J., Solorio, T., Montes-y Gómez, M.: Local histograms of character n-grams for authorship attribution. In: HLT 2011, pp. 288–298 (2011)
13. Ferro, N., Silvello, G.: Rank-biased precision reloaded: reproducibility and generalization. In: Hanbury, A., Kazai, G., Rauber, A., Fuhr, N. (eds.) ECIR 2015. LNCS, vol. 9022, pp. 768–780. Springer, Heidelberg (2015)
14. Gamon, M.: Linguistic correlates of style: authorship classification with deep linguistic analysis features. In: COLING (2004)

15. Hagen, M., Potthast, M., Büchner, M., Stein, B.: Twitter sentiment detection via ensemble classification using averaged confidence scores. In: Hanbury, A., Kazai, G., Rauber, A., Fuhr, N. (eds.) ECIR 2015. LNCS, vol. 9022, pp. 741–754. Springer, Heidelberg (2015)
16. Hall, M., Frank, E., Holmes, G., Pfahringer, B., Reutemann, P., Witten, I.H.: The WEKA data mining software: an update. SIGKDD Explor. **11**(1), 10–18 (2009)
17. Hanbury, A., Kazai, G., Rauber, A., Fuhr, N.: Proceedings of ECIR (2015)
18. Holmes, D.I.: The evolution of stylometry in humanities scholarship. Lit. Ling. Comp. **13**(3), 111–117 (1998)
19. Hopfgartner, F., Hanbury, A., Müller, H., Kando, N., Mercer, S., Kalpathy-Cramer, J., Potthast, M., Gollub, T., Krithara, A., Lin, J., Balog, K., Eggel, I.: Report on the Evaluation-as-a-Service (EaaS) expert workshop. SIGIR Forum **49**(1), 57–65 (2015)
20. Juola, P.: Authorship attribution. FnTIR **1**, 234–334 (2008)
21. Juola, P.: An overview of the traditional authorship attribution subtask. In: CLEF Notebooks (2012)
22. Keselj, V., Peng, F., Cercone, N., Thomas, C.: N-gram-based author profiles for authorship attribution. In: PACLING 2003, pp. 255–264 (2003)
23. Khmelev, D.V., Teahan, W.J.: A repetition based measure for verification of text collections and for text categorization. In: SIGIR 2003, pp. 104–110 (2003)
24. Koppel, M., Schler, J., Bonchek-Dokow, E.: Measuring differentiability: unmasking pseudonymous authors. J. Mach. Learn. Res. **8**, 1261–1276 (2007)
25. Koppel, M., Schler, J., Argamon, S.: Authorship attribution in the wild. LRE **45**(1), 83–94 (2011)
26. Lin, J.: The open-source information retrieval reproducibility challenge. In: RIGOR @ SIGIR (2015)
27. Mendenhall, T.C.: The characteristic curves of composition. Science **ns–9**(214S), 237–246 (1887)
28. Ounis, I., Amati, G., Plachouras, V., He, B., Macdonald, C., Lioma, C.: Terrier: a high performance and scalable information retrieval platform. In: OCIR @ SIGIR (2006)
29. Peng, F., Schuurmans, D., Wang, S.: Augmenting naive Bayes classifiers with statistical language models. Inf. Retr. **7**(3–4), 317–345 (2004)
30. Rangel, F., Rosso, P., Celli, F., Potthast, M., Stein, B., Daelemans, W.: Overview of the 3rd author profiling task at PAN. In: CLEF 2015 Notebooks (2015)
31. Rudman, J.: The state of authorship attribution studies: some problems and solutions. Comput. Humanit. **31**(4), 351–365 (1997)
32. Seroussi, Y., Bohnert, F., Zukerman, I.: Authorship attribution with author-aware topic models. In: ACL 2012, pp. 264–269 (2012)
33. Sidorov, G., Velasquez, F., Stamatatos, E., Gelbukh, A., Chanona-Hernández, L.: Syntactic n-grams as machine learning features for natural language processing. Expert Syst. Appl. **41**(3), 853–860 (2014)
34. Stamatatos, E.: Authorship attribution based on feature set subspacing ensembles. Int. J. Artif. Intell. Tools **15**(5), 823–838 (2006)
35. Stamatatos, E.: Author identification using imbalanced and limited training texts. In: DEXA 2007, pp. 237–241 (2007)
36. Stamatatos, E.: A survey of modern authorship attribution methods. JASIST **60**, 538–556 (2009)
37. Stamatatos, E., Fakotakis, N., Kokkinakis, G.: Automatic text categorization in terms of genre and author. Comput. Linguist. **26**(4), 471–495 (2000)

38. Stamatatos, E., Daelemans, W., Verhoeven, B., Stein, B., Potthast, M., Juola, P., Sánchez-Pérez, M.A., Barrón-Cedeño, A.: Overview of the author identification task at PAN. In: CLEF 2014 Notebooks (2014)
39. Stodden, V.: The scientific method in practice: reproducibility in the computational sciences. MIT Sloan Research Paper No. 4773–10 (2010)
40. Tax, N., Bockting, S., Hiemstra, D.: A cross-benchmark comparison of 87 learning to rank methods. IPM **51**(6), 757–772 (2015)
41. Teahan, W.J., Harper, D.J.: Using compression-based language models for text categorization, pp. 141–165. In: Language Modeling for Information Retrieval (2003)
42. van Halteren, H.: Linguistic profiling for author recognition and verification. In: ACL 2004, pp. 199–206 (2004)
43. Zheng, R., Li, J., Chen, H., Huang, Z.: A framework for authorship identification of online messages: writing-style features and classification techniques. JASIST **57**(3), 378–393 (2006)

Toward Reproducible Baselines: The Open-Source IR Reproducibility Challenge

Jimmy Lin[1(✉)], Matt Crane[1], Andrew Trotman[2], Jamie Callan[3], Ishan Chattopadhyaya[4], John Foley[5], Grant Ingersoll[4], Craig Macdonald[6], and Sebastiano Vigna[7]

[1] University of Waterloo, Waterloo, Canada
jimmylin@uwaterloo.ca
[2] eBay Inc., San Jose, USA
[3] Carnegie Mellon University, Pittsburgh, USA
[4] Lucidworks, Redwood City, USA
[5] University of Massachusetts Amherst, Amherst, USA
[6] University of Glasgow, Glasgow, UK
[7] Università degli Studi di Milano, Milan, Italy

Abstract. The Open-Source IR Reproducibility Challenge brought together developers of open-source search engines to provide reproducible baselines of their systems in a common environment on Amazon EC2. The product is a repository that contains all code necessary to generate competitive *ad hoc* retrieval baselines, such that with a single script, anyone with a copy of the collection can reproduce the submitted runs. Our vision is that these results would serve as widely accessible points of comparison in future IR research. This project represents an ongoing effort, but we describe the first phase of the challenge that was organized as part of a workshop at SIGIR 2015. We have succeeded modestly so far, achieving our main goals on the Gov2 collection with seven open-source search engines. In this paper, we describe our methodology, share experimental results, and discuss lessons learned as well as next steps.

Keywords: *ad hoc* retrieval · Open-source search engines

1 Introduction

As an empirical discipline, advances in information retrieval research are built on experimental validation of algorithms and techniques. Critical to this process is the notion of a competitive baseline against which proposed contributions are measured. Thus, it stands to reason that the community should have common, widely-available, reproducible baselines to facilitate progress in the field. The Open-Source IR Reproducibility Challenge was designed to address this need.

In typical experimental IR papers, scant attention is usually given to baselines. Authors might write something like "we used BM25 (or query likelihood) as the baseline" without further elaboration. This, of course, is woefully under-specified. For example, Mühleisen et al. [13] reported large differences in effectiveness across

N. Ferro et al. (Eds.): ECIR 2016, LNCS 9626, pp. 408–420, 2016.
DOI: 10.1007/978-3-319-30671-1_30

four systems that all purport to implement BM25. Trotman et al. [17] pointed out that BM25 and query likelihood with Dirichlet smoothing can actually refer to at least half a dozen different variants; in some cases, differences in effectiveness are statistically significant. Furthermore, what are the parameter settings (e.g., k_1 and b for BM25, and μ for Dirichlet smoothing)?

Open-source search engines represent a good step toward reproducibility, but they alone do not solve the problem. Even when the source code is available, there remain many missing details. What version of the software? What configuration parameters? Tokenization? Document cleaning and pre-processing? This list goes on. Glancing through the proceedings of conferences in the field, it is not difficult to find baselines that purport to implement the same scoring model from the same system on the same test collection (by the same research group, even), yet report different results.

Given this state of affairs, how can we trust comparisons to baselines when the baselines themselves are ill-defined? When evaluating the merits of a particular contribution, how can we be confident that the baseline is competitive? Perhaps the effectiveness differences are due to inadvertent configuration errors? This is a worrisome issue, as Armstrong et al. [1] pointed to weak baselines as one reason why *ad hoc* retrieval techniques have not really been improving.

As a standard "sanity check" when presented with a purported baseline, researchers might compare against previously verified results on the same test collection (for example, from TREC proceedings). However, this is time consuming and not much help for researchers who are trying to reproduce the result for their own experiments. The Open-Source IR Reproducibility Challenge aims to solve both problems by bringing together developers of open-source search engines to provide reproducible baselines of their systems in a common execution environment on Amazon's EC2 to support comparability both in terms of effectiveness *and* efficiency. The idea is to gather everything necessary in a repository, such that with a single script, anyone with a copy of the collection can reproduce the submitted runs. Two longer-term goals of this project are to better understand how various aspects of the retrieval pipeline (tokenization, document processing, stopwords, etc.) impact effectiveness and how different query evaluation strategies impact efficiency. Our hope is that by observing how different systems make design and implementation choices, we can arrive at generalizations about particular classes of techniques.

The Open-Source IR Reproducibility Challenge was organized as part of the SIGIR 2015 Workshop on Reproducibility, Inexplicability, and Generalizability of Results (RIGOR). We were able to solicit contributions from the developers of seven open-source search engines and build reproducible baselines for the Gov2 collection. In this respect, we have achieved modest success. Although this project is meant as an ongoing exercise and we continue to expand our efforts, in this paper we share results and lessons learned so far.

2 Methodology

The product of the Open-Source IR Reproducibility Challenge is a repository that contains everything needed to reproduce competitive baselines on standard

IR test collections[1]. As mentioned, the initial phase of our project was organized as part of a workshop at SIGIR 2015: most of the development took place between the acceptance of the workshop proposal and the actual workshop. To begin, we recruited developers of open-source search engines to participate. We emphasize the selection of developers—either individuals who wrote the systems or were otherwise involved in their implementation. This establishes credibility for the quality of the submitted runs. In total, developers from seven open-source systems participated (in alphabetical order): ATIRE [16], Galago [6], Indri [10,12], JASS [9], Lucene [2], MG4J [3], and Terrier [14]. In what follows, we refer to the developer(s) from each system as a separate team.

Once commitments of participation were secured, the group (on a mailing list) discussed the experimental methodology and converged on a set of design decisions. First, the test collection: we wished to work with a collection that was large enough to be interesting, but not too large as to be too unwieldy. The Gov2 collection, with around 25 million documents, seemed appropriate; for evaluation, we have TREC topics 701–850 from 2004 to 2006 [7].

The second major decision concerned the definition of "baseline". Naturally, we would expect different notions by each team, and indeed, in a research paper, the choice of the baseline would naturally depend on the techniques being studied. We sidestepped this potentially thorny issue by pushing the decisions onto the developers. That is, the developers of each system decided what the baselines should be, with this guiding question: "If you read a paper that used your system, what would you like to have seen as the baseline?" This decision allowed the developers to highlight features of their systems as appropriate. As expected, everyone produced bag-of-words baselines, but teams also produced baselines based on term dependence models as well as query expansion.

The third major design decision focused around parameter tuning: proper parameter settings, of course, are critical to effective retrieval. However, we could not converge on an approach that was both "fair" to all participants and feasible in terms of implementation given the workshop deadline. Thus, as a compromise, we settled on building baselines around the default "out of the box" experience—that is, what a naïve user would experience downloading the software and using all the default settings. We realize that in most cases this would yield sub-optimal effectiveness and efficiency, but at least such a decision treated all systems equitably. This is an issue we will revisit in future work.

The actual experiments proceeded as follows: the organizers of the challenge started an EC2 instance[2] and handed credentials to each team in turn. The EC2 instance was configured with a set of standard packages (the union of the needs of all the teams), with the Gov2 collection (stored on Amazon EBS) mounted at a specified location. Each team logged into the instance and implemented their baselines within a common code repository cloned from GitHub. Everyone agreed on a directory structure and naming conventions, and checked in their

[1] https://github.com/lintool/IR-Reproducibility/.

[2] We used the r3.4xlarge instance, with 16 vCPUs and 122 GiB memory, Ubuntu Server 14.04 LTS (HVM).

code when done. The code repository also contains standard evaluation tools (e.g., `trec_eval`) as well as the test collections (topics and qrels).

The final product for each system was an execution script that reproduced the baselines from end to end. Each script followed the same basic pattern: it downloaded the system from a remote location, compiled the code, built one or more indexes, performed one or more experimental runs, and printed evaluation results (both effectiveness and efficiency).

Each team got turns to work with the EC2 instance as described above. Although everyone used the same execution environment, they did not necessarily interact with the same instance, since we shut down and restarted instances to match teams' schedules. There were two main rounds of implementation—all teams committed initial results and then were given a second chance to improve their implementations. The discussion of methodology on the mailing list was interleaved with the implementation efforts, and some of the issues only became apparent after the teams began working.

Once everyone finished their implementations, we executed all scripts for each system from scratch on a "clean" virtual machine instance. This reduced, to the extent practical, the performance variations inherent in virtualized environments. Results from this set of experiments were reported at the SIGIR workshop. Following the workshop, we gave teams the opportunity to refine their implementations further and to address issues discovered during discussions at the workshop and beyond. The set of experiments reported in this paper incorporated all these fixes and was performed in December 2015.

3 System Descriptions

The following provides descriptions of each system, listed in alphabetical order. We adopt the terminology of calling a "count index" one that stores only term frequency information and a "positions index" one that stores term positions.

ATIRE. ATIRE built two indexes, both stemmed using an s-stripping stemmer; in both cases, SGML tags were pruned. The postings lists for both indexes were compressed using variable-byte compression after delta encoding. The first index is a frequency-ordered count index that stores the term frequency (capped at 255), while the second index is an impact-ordered index that stores pre-computed quantized BM25 scores at indexing time [8].

For retrieval, ATIRE used a modified version of BM25 [16] ($k_1 = 0.9$ and $b = 0.4$). Searching on the quantized index reduces ranking to a series of integer additions (rather than floating point calculations in the non-quantized index), which explains the substantial reduction in query latencies we observe.

Galago (Version 3.8). Galago built a count index and a positions index, both stemmed using the Krovetz stemmer and stored in document order. The postings consist of separate segments for documents, counts, and position arrays (if included), with a separate structure for skips every 500 documents or so. The indexes use variable-byte compression with delta encoding for ids and positions. Query evaluation uses the document-at-a-time MaxScore algorithm.

Galago submitted two sets of search results. The first used a query-likelihood model with Dirichlet smoothing ($\mu = 3000$). The second used a sequential dependence model (SDM) based on Markov Random Fields [11]. The SDM features included unigrams, bigrams, and unordered windows of size 8.

Indri (Version 5.9). The Indri index contains both a positions inverted index and `DocumentTerm` vectors (i.e., a forward index). Stopwords were removed and terms were stemmed with the Krovetz stemmer.

Indri submitted two sets of results. The first was a query-likelihood model with Dirichlet smoothing ($\mu = 3000$). The second used a sequential dependence model (SDM) based on Markov Random Fields [11]. The SDM features were unigrams, bigrams, and unordered windows of size 8.

JASS. JASS is a new, lightweight search engine built to explore score-at-a-time query evaluation on quantized indexes and the notion of "anytime" ranking functions [9]. It does not include an indexer but instead post-processes the quantized index built from ATIRE. The reported indexing times include both the ATIRE time to index and the JASS time to derive its index. For retrieval, JASS implements the same scoring model as ATIRE, but requires an additional parameter ρ, the number of postings to process. In the first submitted run, ρ was set to one billion, which equates to exhaustive processing. In the second submitted run, ρ was set to 2.5 million, corresponding to the "10 % of document collection" heuristic proposed by the authors [9].

Lucene (Version 5.2.1). Lucene provided both a count and a positions index. Postings were compressed using variable-byte compression and a variant of delta encoding; in the positions index, frequency and positions information are stored separately. Lucene submitted two runs, one over each index; both used BM25, with the same parameters as in ATIRE ($k_1 = 0.9$ and $b = 0.4$). The `English Analyzer` shipped with Lucene was used with the default settings.

MG4J. MG4J provided an index containing all tokens (defined as maximal subsequences of alphanumerical characters) in the collection stemmed using the Porter2 English stemmer. Instead of traditional gap compression, MG4J uses quasi-succinct indices [18], which provide constant-time skipping and uses the least amount of space among the systems examined.

MG4J submitted three runs. The first used BM25 to provide a baseline for comparison, with $k_1 = 1.2$ and $b = 0.3$. The second run utilized Model B, as described by Boldi et al. [4], which still uses BM25, but returns first the documents containing all query terms, then the documents containing all terms but one, and so on; quasi-succinct indices can evaluate these types of queries very quickly. The third run used Model B+, similar to Model B, but using positions information to generate conjunctive subqueries that are within a window two times the length of the query.

Terrier (Version 4.0). Terrier built three indexes, the count and positions indexes both use the single-pass indexer, while the "Count (inc direct)"—which includes a direct file (i.e., a forward index)—uses a slower classical indexer.

Table 1. Indexing results

System	Type	Size	Time	Threading	Terms	Postings	Tokens
ATIRE	Count	12 GB	41 m	Multi	39.9M	7.0B	26.5B
ATIRE	Count + Quantized	15 GB	59 m	Multi	39.9M	7.0B	26.5B
Galago	Count	15 GB	6 h 32 m	Multi	36.0M	5.7B	-
Galago	Positions	48 GB	26 h 23 m	Multi	36.0M	5.7B	22.3B
Indri	Positions	92 GB	6 h 42 m	Multi	39.2M		23.5B
JASS	ATIRE Quantized	21 GB	1 h 03 m	Multi	39.9M	7.0B	26.5B
Lucene	Count	11 GB	1 h 36 m	Multi	72.9M	5.5B	-
Lucene	Positions	40 GB	2 h 00 m	Multi	72.9M	5.5B	17.8B
MG4J	Count	8 GB	1 h 46 m	Multi	34.9M	5.5B	-
MG4J	Positions	37 GB	2 h 11 m	Multi	34.9M	5.5B	23.1B
Terrier	Count	10 GB	8 h 06 m	Single	15.3M	4.6B	-
Terrier	Count (inc direct)	18 GB	18 h 13 m	Single	15.3M	4.6B	-
Terrier	Positions	36 GB	9 h 44 m	Single	15.3M	4.6B	16.2B

The single-pass indexer builds partial posting lists in memory, which are flushed to disk when memory is exhausted, and merged to create the final inverted index. In contrast, the slower classical indexer builds a direct (forward) index based on the contents of the documents, which is then inverted through multiple passes to create the inverted index. While slower, the classical indexer has the advantage of creating a direct index which is useful for generating effective query expansions. All indexes were stemmed using the Porter stemmer and stopped using a standard stopword list. Both docids and term positions are compressed using gamma delta-gaps, while term frequencies are stored in unary. All of Terrier's indexers are single-threaded.

Terrier submitted four runs. The first was BM25 and used the parameters $k_1 = 1.2$, $k_3 = 8$, and $b = 0.75$ as recommended by Robertson [15]. The second run used the DPH ranking function, which is a hypergeometric parameter-free model from the Divergence from Randomness family of functions. The query expansion in the "DPH + Bo1 QE" was performed using the Bo1 divergence from randomness query expansion model, from which 10 terms were added from 3 pseudo-relevance feedback documents. The final submitted run used positions information in a divergence from randomness model called pBiL, which utilizes sequential dependencies.

4 Results

Indexing results are presented in Table 1, which shows both indexing time, the size of the generated index (1 GB = 10^9 bytes), as well as a few other statistics: the number of terms denotes the vocabulary size, the number of postings is equal to the sum of document frequencies of all terms, and the number of tokens

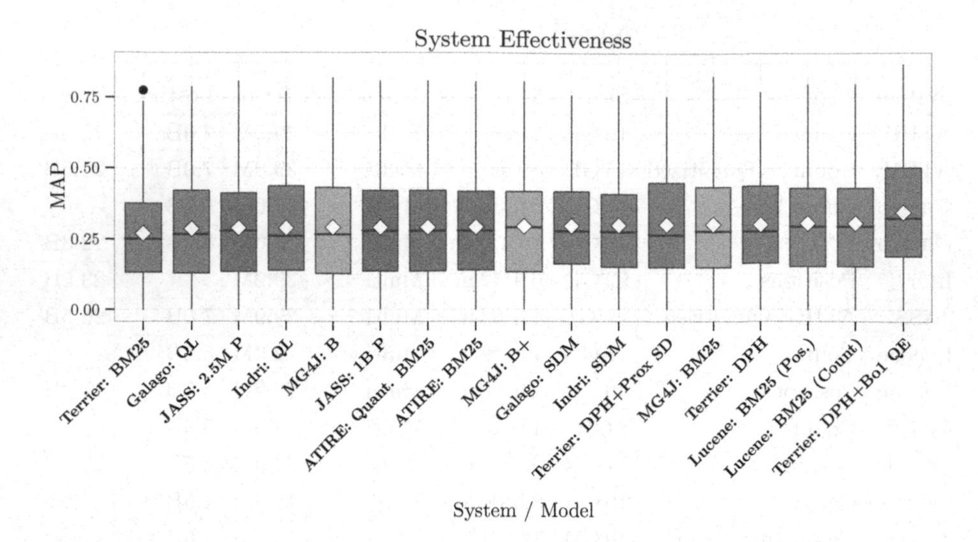

Fig. 1. Box-and-whiskers plot of MAP (all queries) ordered by mean (diamonds).

is the collection length (relevant only for positions indexes). Not surprisingly, for systems that built both positions and count indexes, the positions index took longer to construct. We observe a large variability in the time taken for index construction, some of which can be explained by the use of multiple threads. In terms of index size, it is unsurprising that the positions indexes are larger than the count indexes, but even similar types of indexes differed quite a bit in size, likely due to different tokenization, stemming, stopping, and compression.

Table 2 shows effectiveness results in terms of MAP (at rank 1000). Figure 1 shows the MAP scores for each system on all the topics organized as a box-and-whiskers plot: each box spans the lower and upper quartiles; the bar in the middle represents the median and the white diamond represents the mean. The whiskers extend to 1.5$\times$ the inter-quartile range, with values outside of those plotted as points. The colors indicate the system that produced the run.

We see that all the systems exhibit large variability in effectiveness on a topic-by-topic basis. To test for statistical significance of the differences, we used Tukey's HSD (honest significant difference) test with $p < 0.05$ across all 150 queries. We found that the "DPH + Bo1 QE" run of Terrier was statistically significantly better than all other runs and both Lucene runs significantly better than Terrier's BM25 run. All other differences were not significant. Despite the results of the significance tests, we nevertheless note that the systems exhibit a large range in scores, even though from the written descriptions, many of them purport to implement the same model (e.g., BM25). This is true even in the case of systems that share a common "lineage", for example, Indri and Galago. We believe that these differences can be attributed to relatively uninteresting differences in document pre-processing, tokenization, stemming, and stopwords. This further underscores the importance of having reproducible baselines to control for these effects.

Table 2. MAP at rank 1000.

System	Model	Index	Topics			
			701–750	751–800	801–850	All
ATIRE	BM25	Count	0.2616	0.3106	0.2978	0.2902
ATIRE	Quantized BM25	Count + Quantized	0.2603	0.3108	0.2974	0.2897
Galago	QL	Count	0.2776	0.2937	0.2845	0.2853
Galago	SDM	Positions	0.2726	0.2911	0.3161	0.2934
Indri	QL	Positions	0.2597	0.3179	0.2830	0.2870
Indri	SDM	Positions	0.2621	0.3086	0.3165	0.2960
JASS	1B Postings	Count	0.2603	0.3109	0.2972	0.2897
JASS	2.5M Postings	Count	0.2579	0.3053	0.2959	0.2866
Lucene	BM25	Count	0.2684	0.3347	0.3050	0.3029
Lucene	BM25	Positions	0.2684	0.3347	0.3050	0.3029
MG4J	BM25	Count	0.2640	0.3336	0.2999	0.2994
MG4J	Model B	Count	0.2469	0.3207	0.3003	0.2896
MG4J	Model B+	Positions	0.2322	0.3179	0.3257	0.2923
Terrier	BM25	Count	0.2432	0.3039	0.2614	0.2697
Terrier	DPH	Count	0.2768	0.3311	0.2899	0.2994
Terrier	DPH + Bo1 QE	Count (inc direct)	0.3037	0.3742	0.3480	0.3422
Terrier	DPH + Prox SD	Positions	0.2750	0.3297	0.2897	0.2983

Efficiency results are shown in Table 3: we report mean query latency (over three trials). These results represent query execution on a single thread, with timing code contributed by each team. Thus, these figures should be taken with the caveat that not all systems may be measuring exactly the same thing, especially with respect to overhead that is not strictly part of query evaluation (for example, the time to write results to disk). Nevertheless, to our knowledge this is the first large-scale efficiency evaluation of open-source search engines. Previously, studies typically consider only a couple of systems, and different experimental results are difficult to compare due to underlying hardware differences. In our case, a common platform moves us closer towards fair efficiency evaluations across many systems.

Figure 2 shows query evaluation latency in a box-and-whiskers plot, with the same organization as Fig. 1 (note the y axis is in log scale). We observe a large variation in latency: for instance, the fastest systems (JASS and MG4J) achieved a mean latency below 50 ms, while the slowest system (Indri's SDM model) takes substantially longer. It is interesting to note that we observe different amounts of per-topic variability in efficiency. For example, the fastest run (JASS 2.5M Postings) is faster than the second fastest (MG4J Model B) in terms of mean latency, but MG4J is actually faster if we consider the median—the latter is hampered by a number of outlier slow queries.

Table 3. Mean query latency (across three trials).

System	Model	Index	Topics 701–750	751–800	801–850	All
ATIRE	BM25	Count	132 ms	175 ms	131 ms	146 ms
ATIRE	Quantized BM25	Count + Quantized	91 ms	93 ms	85 ms	89 ms
Galago	QL	Count	773 ms	807 ms	651 ms	743 ms
Galago	SDM	Positions	4134 ms	5989 ms	4094 ms	4736 ms
Indri	QL	Positions	1252 ms	1516 ms	1163 ms	1310 ms
Indri	SDM	Positions	7631 ms	13077 ms	6712 ms	9140 ms
JASS	1B Postings	Count	53 ms	54 ms	48 ms	51 ms
JASS	2.5M Postings	Count	30 ms	28 ms	28 ms	28 ms
Lucene	BM25	Count	120 ms	107 ms	125 ms	118 ms
Lucene	BM25	Positions	121 ms	109 ms	127 ms	119 ms
MG4J	BM25	Count	348 ms	245 ms	266 ms	287 ms
MG4J	Model B	Count	39 ms	48 ms	36 ms	41 ms
MG4J	Model B+	Positions	91 ms	92 ms	75 ms	86 ms
Terrier	BM25	Count	363 ms	287 ms	306 ms	319 ms
Terrier	DPH	Count	627 ms	421 ms	416 ms	488 ms
Terrier	DPH + Bo1 QE	Count (inc. direct)	1845 ms	1422 ms	1474 ms	1580 ms
Terrier	DPH + Prox SD	Positions	1434 ms	1034 ms	1039 ms	1169 ms

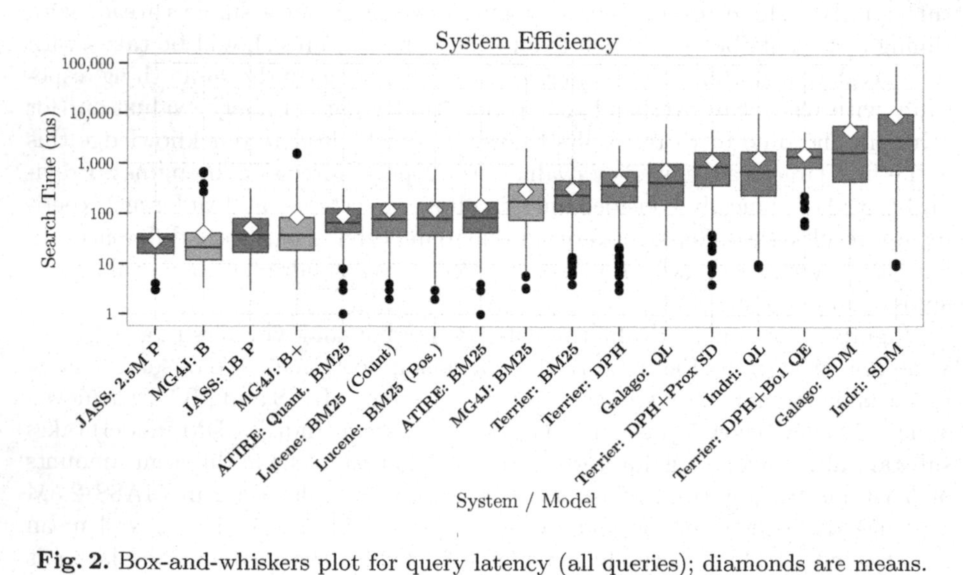

Fig. 2. Box-and-whiskers plot for query latency (all queries); diamonds are means.

Finally, Fig. 3 summarizes effectiveness/efficiency tradeoffs in a scatter plot. As expected, we observe a correlation between effectiveness and efficiency: $R^2 = 0.8888$ after a multi-variate regression of both MAP and system against log(time). Not surprisingly, faster systems tend to compromise quality.

5 Lessons Learned

Overall, we believe that the Open-Source IR Reproducibility Challenge achieved modest success, having accomplished our main goals for the Gov2 test collection. In this section, we share some of the lessons learned.

This exercise was a lot more involved than it would appear and the level of collective effort required was much more than originally expected. We were relying on the volunteer efforts of many teams around the world, which meant that coordinating schedules was difficult to begin with. Nevertheless, the implementations generally took longer than expected. To facilitate scheduling, the organizers asked the teams to estimate how long it would take to build their implementations at the beginning. Invariably, the efforts took more time than the original estimates. This was somewhat surprising because Gov2 is a standard test collection that researchers surely must have previously worked with before.

The reproducibility efforts proved more difficult than imagined for a number of reasons. In at least one case, the exercise revealed a hidden dependency—a pre-processing script that had never been publicly released. In at least two cases, the exercise exposed bugs in systems that were subsequently fixed. In multiple cases, the EC2 instance represented a computing environment that made different assumptions than the machines the teams originally developed on. It seemed that the reproducibility challenge helped the developers improve their systems, which was a nice side effect.

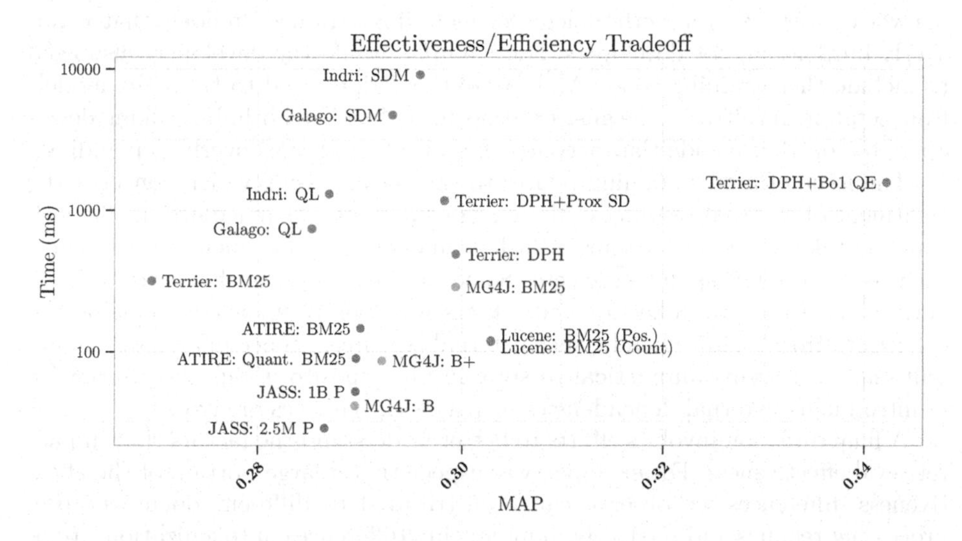

Fig. 3. Tradeoff between effectiveness and efficiency across all systems.

Another unintended consequence of the reproducibility challenge (that was not one of the original goals) is that the code repository serves as a useful teaching resource. In our experience, students new to information retrieval often struggle with basic tasks such as indexing and performing baseline runs. Our resource serves as an introductory tutorial that can teach students about the basics of working with IR test collections: indexing, retrieval, and evaluation.

6 Ongoing Work

The Open-Source IR Reproducibility Challenge is not intended to be a one-off exercise but a living code repository that is maintained and kept up to date. The cost of maintenance should be relatively modest, since we would not expect baselines to rapidly evolve. We hope that sufficient critical mass has been achieved with the current participants to sustain the project. There are a variety of motivations for the teams to remain engaged: developers want to see their systems "used properly" and are generally curious to see how their implementations stack up against their peers. Furthermore, as these baselines begin appearing in research papers, there will be further incentive to keep the code up to date. However, only time will tell if we succeed in the long term.

There are a number of ongoing efforts in the project, the most obvious of which is to build reproducible baselines for other test collections—work has already begun for the ClueWeb collections. We are, of course, always interested in including new systems into the evaluation mix.

Beyond expanding the scope of present efforts, there are two substantive (and related) issues we are currently grappling with. The first concerns the issue of training—from simple parameter tuning (e.g., for BM25) to a complete learning-to-rank setup. In particular, the latter would provide useful baselines for researchers pushing the state of the art in retrieval models. We have not yet converged on a methodology for including "trained" models that is not overly burdensome for developers. For example, would the developers also need to include their training code? And would the scripts need to train the models from scratch? Intuitively, the answer seems to be "yes" to both, but asking developers to contribute code that accomplishes all of this seems overly demanding.

The issue of model training relates to the second issue, which concerns the treatment of external resources. Many retrieval models (particularly in the web context) take advantage of sources such as anchor text, document-level features such as PageRank, spam score, etc. Some of these (e.g., anchor text) can be derived from the raw collection, but others incorporate knowledge outside the collection. How shall we handle such external resources? Since many of them are quite large, it seems impractical to store in our repository, but the alternative of introducing external dependencies increases the chances of errors.

A final direction involves efforts to better understand the factors that impact retrieval effectiveness. For example, we suspect that a large portion of the effectiveness differences we observe can be attributed to different document preprocessing regimes and relatively uninteresting differences in tokenization, stemming, and stopwords. We could explore this hypothesis by, for example, using a single document pre-processor. Such an experiment could be straightforwardly set up by creating a derived collection that every system then ingests, but it would be more efficient and architecturally cleaner to agree on a set of interfaces that allows different retrieval systems to inter-operate. This is similar to the proposal of Buccio et al. [5]: one difference, though, is that we would not prescribe these interfaces, but rather let them evolve based on community consensus. This might perhaps be a fanciful scenario, but the ability to mix-and-match different IR components would greatly accelerate research progress.

The Open-Source IR Reproducibility Challenge represents an ambitious effort to build reproducible baselines for use by the community. Although we have achieved modest success, there is much more to be done. We sincerely encourage participation from the community: both developers in contributing additional systems and everyone in terms of adopting our baselines in their work.

Acknowledgments. This work was supported in part by the U.S. National Science Foundation under IIS-1218043 and by Amazon Web Services. Any opinions, findings, conclusions, or recommendations expressed are those of the authors and do not necessarily reflect the views of the sponsors.

References

1. Armstrong, T.G., Moffat, A., Webber, W., Zobel, J.: Improvements that don't add up: Ad-hoc retrieval results since 1998. In: CIKM, pp. 601–610 (2009)
2. Białecki, A., Muir, R., Ingersoll, G.: Apache lucene 4. In: SIGIR 2012 Workshop on Open Source Information Retrieval (2012)
3. Boldi, P., Vigna, S.: MG4J at TREC 2005. In: TREC (2005)
4. Boldi, P., Vigna, S.: MG4J at TREC 2006. In: TREC (2006)
5. Buccio, E.D., Nunzio, G.M.D., Ferro, N., Harman, D., Maistro, M., Silvello, G.: Unfolding off-the-shelf IR systems for reproducibility. In: SIGIR 2015 Workshop on Reproducibility, Inexplicability, and Generalizability of Results (2015)
6. Cartright, M.A., Huston, S., Field, H.: Galago: A modular distributed processing and retrieval system. In: SIGIR 2012 Workshop on Open Source IR (2012)
7. Clarke, C., Craswell, N., Soboroff, I.: Overview of the TREC 2004 terabyte track. In: TREC (2004)
8. Crane, M., Trotman, A., O'Keefe, R.: Maintaining discriminatory power in quantized indexes. In: CIKM, pp. 1221–1224 (2013)
9. Lin, J., Trotman, A.: Anytime ranking for impact-ordered indexes. In: ICTIR, pp. 301–304 (2015)
10. Metzler, D., Croft, W.B.: Combining the language model and inference network approaches to retrieval. Inf. Process. Manage. **40**(5), 735–750 (2004)
11. Metzler, D., Croft, W.B.: A Markov random field model for term dependencies. In: SIGIR, pp. 472–479 (2005)
12. Metzler, D., Strohman, T., Turtle, H., Croft, W.B.: Indri at TREC 2004: Terabyte track. In: TREC (2004)
13. Mühleisen, H., Samar, T., Lin, J., de Vries, A.: Old dogs are great at new tricks: Column stores for IR prototyping. In: SIGIR, pp. 863–866 (2014)

14. Ounis, I., Amati, G., Plachouras, V., He, B., Macdonald, C., Lioma, C.: Terrier: A high performance and scalable information retrieval platform. In: SIGIR 2006 Workshop on Open Source IR (2006)
15. Robertson, S.E., Walker, S., Jones, S., Hancock-Beaulieu, M., Gatford, M.: Okapi at TREC-3. In: TREC (1994)
16. Trotman, A., Jia, X.F., Crane, M.: Towards an efficient and effective search engine. In: SIGIR 2012 Workshop on Open Source IR (2012)
17. Trotman, A., Puurula, A., Burgess, B.: Improvements to BM25 and language models examined. In: ADCS, pp. 58–65 (2014)
18. Vigna, S.: Quasi-succinct indices. In: WSDM, pp. 83–92 (2013)

Experiments in Newswire Summarisation

Stuart Mackie(✉), Richard McCreadie, Craig Macdonald, and Iadh Ounis

School of Computing Science, University of Glasgow, Glasgow G12 8QQ, UK
s.mackie.1@research.gla.ac.uk
{richard.mccreadie,craig.macdonald,iadh.ounis}@glasgow.ac.uk

Abstract. In this paper, we investigate extractive multi-document summarisation algorithms over newswire corpora. Examining recent findings, baseline algorithms, and state-of-the-art systems is pertinent given the current research interest in event tracking and summarisation. We first reproduce previous findings from the literature, validating that automatic summarisation evaluation is a useful proxy for manual evaluation, and validating that several state-of-the-art systems with similar automatic evaluation scores create different summaries from one another. Following this verification of previous findings, we then reimplement various baseline and state-of-the-art summarisation algorithms, and make several observations from our experiments. Our findings include: an optimised *Lead* baseline; indication that several standard baselines may be weak; evidence that the standard baselines can be improved; results showing that the most effective improved baselines are not statistically significantly less effective than the current state-of-the-art systems; and finally, observations that manually optimising the choice of anti-redundancy components, per topic, can lead to improvements in summarisation effectiveness.

1 Introduction

Text summarisation [15,19] is an information reduction process, where the aim is to identify the important information within a large document, or set of documents, and infer an essential subset of the textual content for user consumption. Examples of text summarisation being applied to assist with user's information needs include search engine results pages, where snippets of relevant pages are shown, and online news portals, where extracts of newswire documents are shown. Indeed, much of the research conducted into text summarisation has focused on multi-document newswire summarisation. For instance, the input to a summarisation algorithm being evaluated at the Document Understanding Conference[1] or Text Analysis Conference[2] summarisation evaluation campaigns is often a collection of newswire documents about a news-worthy event. Further, research activity related to the summarisation of news-worthy events has recently been conducted under the TREC Temporal Summarisation Track[3]. Given the

[1] duc.nist.gov.
[2] nist.gov/tac.
[3] trec-ts.org.

N. Ferro et al. (Eds.): ECIR 2016, LNCS 9626, pp. 421–435, 2016.
DOI: 10.1007/978-3-319-30671-1_31

current research interest in event summarisation [5,8,13], the reproduction, validation, and generalisation of findings from the newswire summarisation literature is important to the advancement of the field, and additionally, constitutes good scientific practice.

Hence, in this contribution, we begin by reproducing and validating two previous findings, over DUC 2004 Task 2. First, that the ROUGE-2 [9] metric is aligned with user judgements for summary quality, but generalising this finding in the context of crowd-sourcing. Second, that there exists measurable variability in the sentence selection behaviour of state-of-the-art summarisation algorithms exhibiting similar ROUGE-2 scores, but confirming such variability via a complementary form of analysis, adding to the weight of evidence of the original finding. Further, in this paper, we reproduce the *Random* and *Lead* baselines, over the DUC 2004 and TAC 2008 newswire summarisation datasets. Observations from such experiments include: a validation of the lower-bound on acceptable summarisation effectiveness; findings that the effectiveness of the simple *Lead* baseline used at DUC and TAC can be improved; and that the *Lead* baseline augmented with anti-redundancy components is competitive with several standard baselines, over DUC 2004. Finally, we reproduce a series of standard and state-of-the-art summarisation algorithms. Observations from these experiments include: optimisations to several standard baselines that improve effectiveness; results indicating that state-of-the-art techniques, using integer linear programming and machine learning, are not always more effective than simple unsupervised techniques; and additionally, that an oracle system optimising the selection of different anti-redundancy components, on a per-topic basis, can potentially lead to improvements in summarisation effectiveness.

The remainder of this paper is organised as follows: We report our experimental setup in Sect. 2, describing summarisation algorithms, datasets, and the evaluation process. In Sect. 3, we present the results from a user study reproducing and validating previous findings that the ROUGE-2 metric aligns with user judgements for summary quality. In Sect. 4, we reproduce and validate previous findings that, despite exhibiting similar ROUGE-2 scores, state-of-the-art summarisation algorithms vary in their sentence selection behaviour. In Sect. 5, we reproduce the *Random* and *Lead* baselines, making several observations over the DUC 2004 and TAC 2008 datasets. In Sect. 6, we reproduce standard baselines and state-of-the-art systems, making further observations over the DUC 2004 and TAC 2008 datasets. Finally, Sect. 7 summarises our conclusions.

2 Reproducible Experimental Setup

In this section, we briefly describe the summarisation algorithms that we investigate, with full details available in the relevant literature [7]. Then, we also describe the anti-redundancy components that aim to minimise repetition in the summary text. Following this, we provide details of the evaluation datasets and metrics used in our experiments.

Summarisation Algorithms – In general, each summarisation algorithm assigns scores to sentences, computing a ranked list of sentences where the highest-scoring sentences are most suitable for inclusion into the summary text. Some algorithms then pass the ranked list of scored sentences to an anti-redundancy component, described below, while other algorithms do not (i.e. handling redundancy internally). **FreqSum** [16] computes the probability of each word, over all the input sentences. Sentences are scored by summing the probabilities of each of its individual words, normalising by sentence length (i.e. average probability). The scored sentences (a ranked list) are passed to an anti-redundancy component for summary sentence selection. **TsSum** [2] relies on the computation of topic words [10], which are words that occur more often in the input text than in a large background corpus. The log-likelihood ratio test is applied, with a threshold parameter used to determine topic words from non-topic words. A further parameter of this algorithm is the background corpus to use; in our experiments we use the term frequencies of the 1,000,000 most common words in Wikipedia. Sentences are scored by taking the ratio of unique topic words to unique non-topic words. An anti-redundancy component is then applied to select novel sentences. **Centroid** [18] computes a centroid pseudo-document of all terms, and scores sentences by their cosine similarity to this centroid vector. This algorithm has a parameter, in that a vector weighting scheme must be chosen, e.g. *tf*idf*. Sentence selection is via an anti-redundancy component. **LexRank** [3] computes a highly-connected graph, where the vertices are sentences, and weighted edges represent the cosine similarity between vertices. Again, a vector weighting scheme, to represent sentences as vectors, must be chosen. Sentences are scored by using a graph algorithm (e.g. in-degree or PageRank) to compute a centrality score for each vertex. A threshold parameter is applied over the graph, disconnecting vertices that fall below a given cosine similarity, or, the edge weights may be used as transition probabilities in PageRank (i.e. *Cont. LexRank*). Further, an anti-redundancy component is then used to select novel sentences. **Greedy–KL** [6] computes the Kullback–Leibler divergence between each individual sentence and all other sentences. Then, summary sentences are chosen by greedily selecting the sentence that minimises the divergence between the summary text and all the original input sentences. This algorithm has a parameter, in the range $[0, 1]$, the Jelinek–Mercer smoothing λ value, used when computing the language models for the Kullback–Leibler divergence computation. **ICSISumm** [4] views the summarisation task as an optimisation problem, with a solution found via integer linear programming. An objective function is defined that maximises the presence of weighted concepts in the final summary text, where such concepts are computed over the set of input sentences (specifically, bi-grams valued by document frequency). In our experiments, we use an open source solver[4] to express and execute integer linear programs. Further, we also experiment with a **machine learned model**. The features used are the *FreqSum*, *TsSum*, *Centroid*, *LexRank*, and *Greedy–KL* baselines. The learned model is trained on the gold-standard of DUC 2002

[4] gnu.org/software/glpk/.

(manual sentence extracts), and tested on DUC 2004 and TAC 2008. For our experiments, we train a maximum entropy binary classifier[5], with feature values scaled in the range $[-1, 1]$. The probability estimates output from the classifier are used to score the sentences, producing a ranking of sentences that is passed through an anti-redundancy component for summary sentence selection.

Anti-redundancy Components – Each anti-redundancy component takes as input a list of sentences, previously ranked by a summarisation scoring function. The first, highest-scoring, sentence is always selected. Then, iterating down the list, the next highest-scoring sentence is selected on the condition that it satisfies a threshold. We experiment with the following anti-redundancy thresholding components, namely *NewWordCount*, *NewBigrams*, and *CosineSimilarity*. **NewWordCount** [1] only selects the next sentence in the list, for inclusion into the summary text, if that sentence contributes n new words to the summary text vocabulary. In our experiments, the value of n, the new word count parameter, ranges from $[1, 20]$, in steps of 1. **NewBigrams** only selects a sentence if that sentence contributes n new bi-grams to the summary text vocabulary. In our experiments, the value of n, the new bi-grams parameter, ranges from $[1, 20]$, in steps of 1. The **CosineSimilarity** thresholding component only selects the next sentence if that sentence is sufficiently dis-similar to all previously selected sentences. In our experiments, the value of the cosine similarity threshold ranges from $[0, 1]$ in steps of 0.05. As cosine similarity computations require a vector representation of the sentences, we experiment with different weighting schemes, denoted *Tf*, *Hy*, *Rt*, and *HyRt*. *Tf* is textbook *tf*idf*, specifically $\log(\text{tf}) * \log(\text{idf})$, where *tf* is the frequency of a term in a sentence, and *idf* is N/Nt, the number of sentences divided by the number of sentences containing the term *t*. *Hy* is a *tf*idf* variant, where the *tf* component is computed over all sentences combined into a pseudo-document, with *idf* computed as N/Nt. *Rt* and *HyRt* are *tf*idf* variants where we do not use log smoothing, i.e. raw *tf*. The 4 variants of weighting schemes are also used by *Centroid* and *LexRank*, to represent sentences as weighted vectors.

Summarisation Datasets – In our summarisation experiments, we use newswire documents from the Document Understanding Conference (DUC) and the Text Analysis Conference (TAC). Each dataset consists of a number of topics, where a topic is a cluster of related newswire documents. Further, each topic has a set of gold-standard reference summaries, authored by human assessors, to which system-produced summaries are compared in order to evaluate the effectiveness of various summarisation algorithms. The DUC 2004 Task 2 dataset has 50 topics of 10 documents per topic, and 4 reference summaries per topic. The TAC 2008 Update Summarization Task dataset has 48 topics, and also 4 reference summaries per topic. For each topic within the TAC dataset, we use the 10 newswire articles from document set A, and the 4 reference summaries

[5] mallet.cs.umass.edu/api/cc/mallet/classify/MaxEnt.html.

for document set A, ignoring the update summarisation part of the task (set B). Further, we use the TAC 2008 dataset for generic summarisation (ignoring the topic statements).

The Stanford CoreNLP toolkit is used to chunk the newswire text into sentences, and tokenise words. Individual tokens are then subjected to the following text processing steps: Unicode normalisation (NFD[6]), case folding, splitting of compound words, removal of punctuation, Porter stemming, and stopword removal (removing the 50 most common English words[7]). When summarising multiple documents for a topic, we combine all sentences from the input documents for a given topic into a single virtual document. The sentences from each document are interleaved one-by-one in docid order, and this virtual document is given as input to the summarisation algorithms.

Summarisation Evaluation – To evaluate summary texts, we use the ROUGE [9] evaluation toolkit[8], measuring n-gram overlap between a system-produced summary and a set of gold-standard reference summaries. Following best practice [7], the summaries under evaluation are subject to stemming, stopwords are retained, and we report ROUGE-1, ROUGE-2 and ROUGE-4 recall – measuring uni-gram, bi-gram, and 4-gram overlap respectively – with results ordered by ROUGE-2 (in bold), the preferred metric due to its reported agreement with manual evaluation [17]. Further, for all experiments, summary lengths are truncated to 100 words. The ROUGE parameter settings used are: "`ROUGE-1.5.5.pl -n 4 -x -m -l 100 -p 0.5 -c 95 -r 1000 -f A -t 0`". For summarisation algorithms with parameters, we learn the parameter settings via a five-fold cross validation procedure, optimising for the ROUGE-2 metric. Statistical significance in ROUGE results is reported using the paired Student's t-test, 95 % confidence level, as implemented in MATLAB. ROUGE results for various summarisation systems are obtained using SumRepo [7][9], which provides the plain-text produced by 5 standard baselines, and 7 state-of-the-art systems, over DUC 2004. Using this resource, we compute ROUGE results, over DUC 2004 only, for the algorithms available within SumRepo, obtaining reference results for use in our later experiments.

3 Crowd-Sourced User Study to Validate that the ROUGE-2 Metric Aligns with User Judgements of Summary Quality

Current best practice in summarisation evaluation [7] is to report ROUGE results using ROUGE-2 as the preferred metric, due to the reported agreement of ROUGE-2 with manual evaluation [17]. In this section, we now examine if the ROUGE-2 metric aligns with user judgements, reproducing and validating

[6] docs.oracle.com/javase/8/docs/api/java/text/Normalizer.html.
[7] en.wikipedia.org/wiki/Most_common_words_in_English.
[8] www.berouge.com.
[9] http://www.seas.upenn.edu/~nlp/corpora/sumrepo.html.

previous findings – but generalising to the context of crowd-sourcing. This provides a measure of confidence in using crowd-sourced evaluations of newswire summarisation, as has previously been demonstrated for microblog summarisation [11,12]. Our user study is conducted via CrowdFlower[10], evaluating 5 baseline systems and 7 state-of-the-art systems, over the DUC 2004 dataset using summary texts from SumRepo. A system ranking based on ROUGE-2 effectiveness is compared with a system ranking based on the crowd-sourced user judgements, in order to determine if the ROUGE-2 metric is aligned with user judgements.

Users are shown a summary text, and asked to provide a judgement on the quality of the summary, using a 10-point scale. The interface for soliciting summary quality assessments is shown in Fig. 1. Users are provided with minimal instructions, which they may opt to read, and although we provide criteria by which users could make judgements of summary quality[11], we make no attempt to simulate a complex work task. The total cost of the user study is $109.74, for 3,000 judgements (50 topics, 12 systems, each summary judged 5 times, approx. $0.036 per judgement). The per-system judgements provided by the users are aggregated first at the topic level (over 5 assessors) and then at the system level (over 50 topics). Table 1 provides results from the user study, where we compare a ranking of systems based on their ROUGE-2 effectiveness (denoted *Reference Results*) with a ranking of systems obtained from the mean of the 10-point scale user judgements (denoted *User Judgements*). Table 1 also includes the ROUGE-1 and ROUGE-4 scores of each system for the reference results, and the minimum, maximum, and median scores for the user judgements. The 12 systems under evaluation are separated into two broad categories [7], namely *Baselines* and *State-of-the-art*.

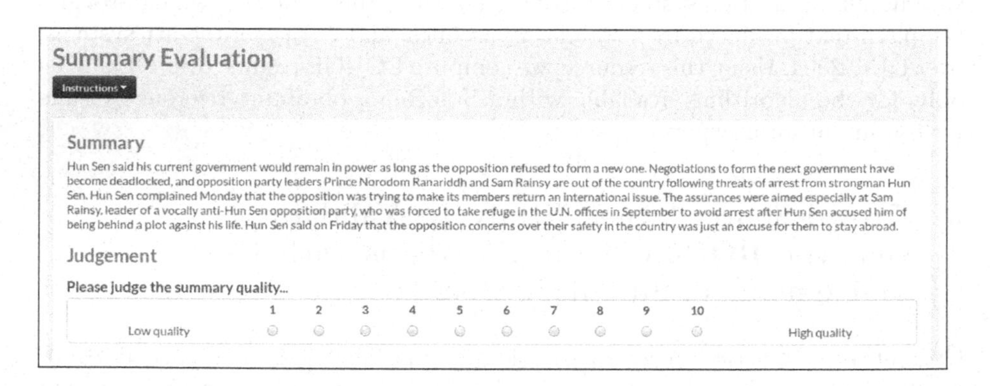

Fig. 1. The interface for our user study, soliciting summary judgements via CrowdFlower.

[10] crowdflower.com.

[11] www-nlpir.nist.gov/projects/duc/duc2007/quality-questions.txt.

Table 1. Reference ROUGE results, over DUC 2004, and results from our crowd-sourced user study validating ROUGE-2 is aligned with user judgements for summary quality.

Reference Results (*SumRepo*)					User Judgements (*CrowdFlower*)					
Baselines	**Rank**	R-1	**R-2**	R-4	**Baselines**	**Rank/**	**mean**	min	max	median
Cont. LexRank	1	36.00	7.51	0.83	*FreqSum*	3	7.16	3.40	9.00	7.40
Centroid	2	36.42	7.98	1.20	*CLASSY 04*	6	7.36	5.00	9.40	7.40
FreqSum	3	35.31	8.12	1.00	*TsSum*	4	7.60	5.00	9.20	7.60
TsSum	4	35.93	8.16	1.03	*Centroid*	2	7.64	5.00	9.40	7.60
Greedy–KL	5	38.03	8.56	1.27	*LexRank*	1	7.66	2.60	9.60	7.80
State-of-the-art	**Rank**	R-1	**R-2**	R-4	**State-of-the-art**	**Rank/**	**mean**	min	max	median
CLASSY 04	6	37.71	9.02	1.53	*OCCAMS_V*	10	7.70	3.80	9.60	7.90
CLASSY 11	7	37.21	9.21	1.48	*CLASSY 11*	7	7.71	5.20	9.20	7.80
Submodular	8	39.23	9.37	1.39	*Submodular*	8	7.75	5.60	9.40	7.80
DPP	9	39.84	9.62	1.57	*DPP*	9	7.80	5.20	9.80	8.00
OCCAMS_V	10	38.50	9.75	1.33	*RegSum*	11	7.85	6.00	9.60	8.00
RegSum	11	38.60	9.78	1.62	*Greedy–KL*	5	8.05	3.80	9.60	8.20
ICSISumm	12	38.44	9.81	1.74	*ICSISumm*	12	8.10	5.60	9.80	8.20

From Table 1, we observe that, generally, the crowd-sourced user judgements mirror the ROUGE-2 system ordering of baselines and state-of-the-art systems, i.e. that it is therefore possible for the crowd to distinguish between baseline algorithms and state-of-the-art systems. The two exceptions are *CLASSY 04*, which the crowd-sourced user judgements have rated less effective than the ROUGE-2 result, and *Greedy–KL*, which the crowd-sourced user judgements have rated more effective than the ROUGE-2 result. However, from Table 1, we can conclude that the ROUGE-2 metric is generally aligned with crowd-sourced user judgements, reproducing and validating previous findings [17], and generalising to the context of crowd-sourced summarisation evaluations.

4 Confirming Variability in Sentence Selection Behaviour of Summarisation Algorithms with Similar ROUGE-2 Scores

It has been previously reported [7], over DUC 2004 Task 2, that the top 6 state-of-the-art summarisation algorithms exhibit low overlap in the content selected for inclusion into the summary text, despite having no statistically significant difference in ROUGE-2 effectiveness (two-sided Wilcoxon signed-rank, 95 % confidence level). Content overlap was measured at the level of sentences, words, and summary content units, demonstrating that the state-of-the-art algorithms exhibit variability in summary sentence selection. In this section, we seek to reproduce and validate this finding, by investigating the variation in ROUGE-2 effectiveness of the state-of-the-art systems across topics. This analysis seeks to determine if, despite having very similar ROUGE-2 effectiveness, the sentence selection behaviour of the state-of-the-art systems varies over topics. This would confirm that the state-of-the-art systems are selecting different content for inclusion into the summary, reproducing and validating the previously published [7] results.

For our analysis, we examine the ROUGE-2 effectiveness of the state-of-the-art systems over the 50 topics of DUC 2004 Task 2, using the summary text from SumRepo. In Fig. 2, we visualise the distribution of ROUGE-2 scores over topics, for the top 6 state-of-the-art systems, with the topics on the x-axis ordered by the ROUGE-2 effectiveness of *ICSISumm*. In Table 2, we then quantify the ROUGE-2 effectiveness between the top 6 state-of-the-art systems, showing the Pearson's linear correlation coefficient of ROUGE-2 scores across the topics.

From Fig. 2, we observe that, for each of the top 6 state-of-the-art systems, there is variability in ROUGE-2 scores over different topics. Clearly, for some topics, certain systems are more effective, while for other topics, other systems are more effective. This variability is usually masked behind the ROUGE-2 score, which provides an aggregated view over all topics. Further, from Table 2 we observe that the per-topic ROUGE-2 scores of the top 6 state-of-the-art systems

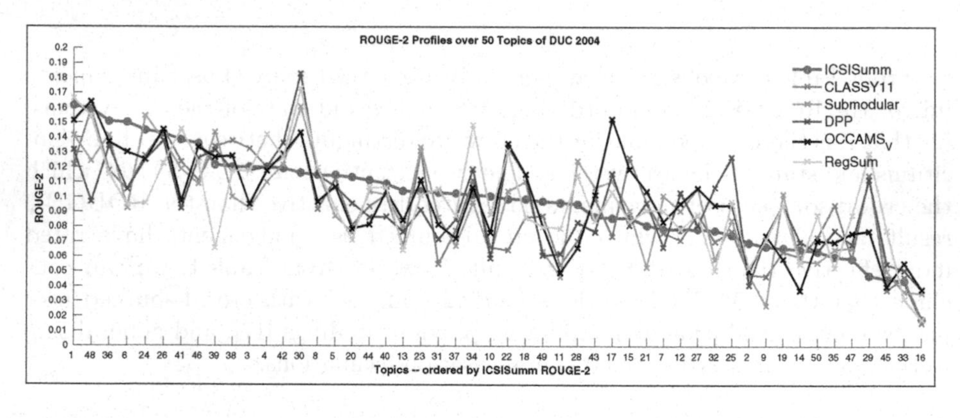

Fig. 2. ROUGE-2 effectiveness profiles, across the 50 topics of DUC 2004, for the top 6 state-of-the-art systems, with the x-axis ordered by the ROUGE-2 effectiveness of *ICSISumm*.

Table 2. Pearson's linear correlation coefficient of ROUGE-2 scores between the top six state-of-the-art systems, across the 50 topics of DUC 2004.

	CLASSY11	Submodular	DPP	OCCAMS_V	RegSum	ICSISumm
CLASSY11	1.0000	–	–	–	–	–
Submodular	0.7607	1.0000	–	–	–	–
DPP	0.6950	0.7605	1.0000	–	–	–
OCCAMS_V	0.7701	0.7456	0.7824	1.0000	–	–
RegSum	0.6721	0.7089	0.6849	0.6599	1.0000	–
ICSISumm	0.7385	0.6875	0.6089	0.7463	0.6516	1.0000

are not as highly correlated as indicated by these system's aggregated ROUGE-2 scores, which have no statistically significant difference. Indeed, we observe from Table 2 that the highest level of correlation is 0.7824, between *OCCAMS_V* and *DPP*, but falls to 0.6089, between *ICSISumm* and *DPP*. Given the visualisation of variability in Fig. 2, and the quantification of variability in Table 2, we conclude that, although these systems have very similar ROUGE-2 scores, they exhibit variability in sentence selection behaviour, validating the previous findings [7].

5 Reproducing the Random and Lead Baselines

In this section, we reproduce the *Random* and *Lead* baselines, making observations over DUC 2004 and TAC 2008. The *Random* baseline provides a lower-bound on acceptable effectiveness, i.e. an effective summarisation algorithm should out-perform a randomly generated summary. In our experiments, we generate 100 random summaries, per topic, and average the ROUGE-1, ROUGE-2 and ROUGE-4 scores to provide a final *Random* baseline result. The *Lead* baseline is reported to be very effective for newswire summarisation [14], due to journalistic convention of a news article's first sentence being very informative. We investigate the method used to derive the *Lead* baseline, and further, the results of augmenting the *Lead* baseline with different anti-redundancy components.

Table 3 gives the ROUGE results for the *Random* baseline, 2 variants of the *Lead* baseline (*recent doc* and *interleaved*), the *Lead (interleaved)* baseline passed through 6 anti-redundancy components, and also the results for the 5 standard baseline algorithms. In particular, Table 3 presents the ROUGE results for the *Lead (recent doc)* baseline used for the DUC and TAC evaluations, which consists of the lead sentences extracted from the most recent document in the collection of documents for a topic. We also show, in Table 3, the *Lead (interleaved)* baseline that results from the sentence interleaving of a virtual document, where the input sentences are arranged one-by-one from each document in turn. Further, Table 3 provides reference ROUGE results, over DUC 2004, for the 5 standard baselines computed using SumRepo (not available for TAC 2008).

From Table 3, we first observe the ROUGE effectiveness of the *Random* baseline, establishing a lower-bound on the acceptable performance over the two datasets. All of the standard baselines exceed the *Random* performance, however, *Lead (recent doc)* over TAC 2008 exhibits a ROUGE-1 score that is not significantly different from *Random*. This indicates that *Lead (recent doc)* may not be a strong baseline, over TAC 2008. Indeed, we observe a significant improvement in ROUGE results, shown in Table 3 using the "†" symbol, for the *Lead (interleaved)* baseline over the official *Lead (recent doc)* baselines used at DUC and TAC. From this, we conclude that using multiple lead sentences, from multiple documents, to construct a *Lead* baseline is more effective than simply using the first n sentences from the most recent document.

Further, from Table 3, we observe cases where the *Lead (interleaved)* baseline, when passed through an anti-redundancy component, achieves ROUGE

Table 3. ROUGE scores, over DUC 2004 and TAC 2008, for Random and Lead, the Lead baseline augmented with different anti-redundancy components, and 5 standard baselines.

DUC 2004				TAC 2008			
Lower-bound	R-1	**R-2**	R-4	**Lower-bound**	R-1	**R-2**	R-4
Random	30.27	4.33	0.35	*Random*	29.75	4.60	0.57
Lead (baselines)	R-1	**R-2**	R-4	**Lead (baselines)**	R-1	**R-2**	R-4
Lead (recent doc)	31.46	6.13	0.62	*Lead (recent doc)*	29.73	5.83	0.79
Lead (interleaved)	34.23†	**7.66†**	1.18†	*Lead (interleaved)*	33.18†	**7.69†**	1.44†
Lead (anti-redundancy)	R-1	**R-2**	R-4	**Lead (anti-redundancy)**	R-1	**R-2**	R-4
CosineSimilarityRt	35.67‡	7.91	1.20	*CosineSimilarityHy*	33.71	7.53	1.33
CosineSimilarityTf	36.02‡	7.97	1.20	*CosineSimilarityRt*	33.44	7.70	1.41
NewWordCount	35.54‡	8.02	1.22	*CosineSimilarityTf*	33.76	7.78	1.44
CosineSimilarityHyRt	35.91‡	8.08‡	1.24	*NewBigrams*	33.92	7.87	1.43
NewBigrams	36.05‡	8.11	1.18	*NewWordCount*	33.73	7.92	1.66
CosineSimilarityHy	36.38‡	**8.29‡**	1.29	*CosineSimilarityHyRt*	34.08‡	**8.10‡**	1.55
Baselines (*SumRepo*)	R-1	**R-2**	R-4	**SumRepo* baselines are not available over TAC 2008			
LexRank	36.00	7.51	0.83✗				
Centroid	36.42	7.98	1.20				
FreqSum	35.31	8.12	1.00				
TsSum	35.93	8.16	1.03				
Greedy–KL	38.03✔	**8.56**	1.27				

effectiveness scores that exhibit a significant improvement over the *Lead (interleaved)* baseline, as indicated by the "‡" symbol. In particular, over DUC 2004, *Lead (interleaved)* augmented with anti-redundancy filtering results in significant improvements in ROUGE-1 scores for all anti-redundancy components investigated, and significant improvements in ROUGE-2 scores using *CosineSimilarityHyRt* and *CosineSimilarityHy*. However, from Table 3, we observe that anti-redundancy filtering of *Lead (interleaved)* is not as effective over TAC 2008, where only *CosineSimilarityHyRt* exhibits significantly improved ROUGE-1 and ROUGE-2 scores. From these observations, we conclude that the optimal *Lead* baseline, for multi-document extractive newswire summarisation, can be derived by augmenting an interleaved *Lead* baseline with anti-redundancy filtering (such as cosine similarity).

Finally, from Table 3, we observe the 5 standard baselines, *LexRank*, *Centroid*, *FreqSum*, *TsSum*, and *Greedy–KL*, do not exhibit significant differences in ROUGE-2 scores, over DUC 2004, from *CosineSimilarityHy*, the most effective anti-redundancy processed interleaved *Lead* baseline. Indeed, only *Greedy–KL* exhibits a ROUGE-1 score ("✔") that is significantly more effective that *Lead interleaved* with *CosineSimilarityHy*, and further, *LexRank* shows a significant degradation in ROUGE-4 effectiveness ("✗"). From this, we conclude that the 5 standard baselines, over DUC 2004, may be weak baselines to use in future experiments, with any claimed improvements questionable.

6 Reproducing Standard and State-of-the-art Algorithms

In this section, we reproduce standard summarisation baselines, and state-of-the-art systems, making several observations over the DUC 2004 and TAC 2008 datasets. In particular, we reimplement the *LexRank*, *Centroid*, *FreqSum*, *TsSum*, and *Greedy–KL* standard baselines. Additionally, we investigate the state-of-the-art summarisation algorithms, that use integer linear programming (ILP) and machine learning techniques, reimplementing *ICSISumm*, and training a supervised machine learned model. Further, we investigate the optimisation of the selection of anti-redundancy components on a per topic basis, making observations regarding the best and worse cases, over DUC 2004 and TAC 2008, for our most effective reimplementations of the standard baselines.

Table 4 provides reference results for standard baselines and state-of-the-art systems, over DUC 2004 and TAC 2008, to which we compare our reimplementations of the various summarisation algorithms. In Table 4, the standard baselines and state-of-the-art reference results, over DUC 2004, are computed from SumRepo. The TAC 2008 reference results are computed from the participants submissions to TAC 2008, specifically *ICSISumm*, which were the most effective runs under ROUGE-2 for part A of the task (the non-update part). Table 5 presents results, over DUC 2004 and TAC 2008, that show the effectiveness of our reimplementations of the 5 standard baselines, our reimplementations of *ICSISumm*, and the machine learned model, *MaxEnt*.

From Table 5, we first observe the ROUGE results for our reimplementations of the standard baselines, where the standard baselines have been numbered (1) to (5). In Table 5, a "✔" symbol indicates a statistically signifi-

Table 4. Reference ROUGE results, for baselines and state-of-the-art systems.

DUC 2004			
Baselines	R-1	**R-2**	R-4
(1) *Cont. LexRank*	36.00	7.51	0.83
(2) *Centroid*	36.42	7.98	1.20
(3) *FreqSum*	35.31	8.12	1.00
(4) *TsSum*	35.93	8.16	1.03
(5) *Greedy–KL*	38.03	**8.56**	1.27

DUC 2004			
State-of-the-art	R-1	**R-2**	R-4
CLASSY 04	37.71	9.02	1.53
CLASSY 11	37.21	9.21	1.48
Submodular	39.23	9.37	1.39
DPP	39.84	9.62	1.57
OCCAMS_V	38.50	9.75	1.33
RegSum	38.60	9.78	1.62
ICSISumm	38.44	**9.81**	1.74

TAC 2008			
State-of-the-art	R-1	**R-2**	R-4
ICSISumm (13)	37.79	11.03	2.26
ICSISumm (43)	38.31	**11.13**	2.20

Table 5. Reimplementation ROUGE results, for baselines and state-of-the-art systems.

DUC 2004			
Baseline Reimplementation	R-1	**R-2**	R-4
(3) *Probability_NewWordCount*	37.52✔†	8.70	1.14
(4) *TopicWordsWp_CosineSimilarityTf*	37.54✔†	8.87	1.39✔†
(1) *GraphPRpriorsHy_CosineSimilarityHy*	38.05✔†	9.34✔†	1.44✔†
(2) *SimCentroidHy_NewWordCount*	37.79✔†	9.37✔†	1.59†
(5) *KLDivergence_CosineSimilarityHy*	38.44†	**9.59**✔†	1.56†
State-of-the-art Reimplementation	R-1	**R-2**	R-4
ILP_ICSISumm	37.77†	9.50†	1.56†
MaxEnt_NewBigrams	38.43†	**9.56**†	1.73†

TAC 2008			
Baseline Reimplementation	R-1	**R-2**	R-4
(3) *Probability_NewBigrams*	35.30	8.05	1.57
(4) *TopicWordsWp_NewWordCount*	36.92‡	9.27	1.93‡
(5) *KLDivergence_CosineSimilarityHyRt*	37.48‡	9.67	2.01‡
(2) *SimCentroidHy_NewWordCount*	36.92	9.77	2.16‡
(1) *GraphDegreeHyRt_NewWordCount*	37.42‡	**10.23**	2.22‡
State-of-the-art Reimplementation	R-1	**R-2**	R-4
MaxEnt_CosineSimilarityRt	36.38	9.51	2.04‡
ILP_ICSISumm	37.31‡	10.24‡	2.20‡

cant improvement of a baseline reimplementation over the standard baseline, while a "†" symbol indicates that there is no statistically significant difference to *ICSISumm* over DUC 2004, and a "‡" symbol indicates no statistically significant difference to *ICSISumm* over TAC 2008. Over DUC 2004, under the target metric ROUGE-2, *GraphPRpriorsHy_CosineSimilarityHy*, *SimCentroidHy_NewWordCount*, and *KLDivergence_CosineSimilarityHy* exhibit improvements over the standard baselines of *LexRank*, *Centroid*, and *Greedy–KL*, respectively, and these 3 baseline reimplementations exhibit similar effectiveness to a state-of-the-art algorithm, *ICSISumm*. We also note further improvements and state-of-the-art effectiveness for our baseline reimplementations under the ROUGE 1 and 4 metrics. For TAC 2008, we observe that reimplementations of *LexRank*, *TsSum*, and *Greedy–KL* exhibit ROUGE-1 effectiveness that is not statistically significantly different from *ICSISumm*.

The improvements for our reimplementations (optimising the standard baselines and closing the gap to the state-of-the-art) are attributed to variations in algorithm design. For example, most of the standard baselines use a cosine similarity anti-redundancy component [7], and altering the choice of anti-redundancy component can lead to improvements in effectiveness. Further, the most effective standard baseline reimplementation (over DUC 2004), *KLDivergence_CosineSimilarityHy*, is a variation of *Greedy–KL*. For this reimplementation, instead of greedily selecting the sentences that minimise divergence, our variation first scores sentences by their Kullback–Leibler divergence to all other sentences, then passes the ranked list to an anti-redundancy component. Other variations include altering the vector weighting scheme, such as the hybrid *tf*idf* vectors used by the *SimCentroidHy* baseline reimplementation. From the results presented in Table 5, we conclude that it is possible to optimise the standard baselines, even to the point where they exhibit similar effectiveness to a state-of-the-art system (*ICSISumm*).

Next, from Tables 4 and 5, we observe that our reimplementation of *ICSISumm*, and the machine learned model *MaxEnt*, exhibit state-of-the-art effectiveness over DUC 2004. In particular, the ROUGE-2 results from our reimplementations of *ICSISumm* and *MaxEnt* are not statistically significantly different from the reference results for the original *ICSIsumm*. Over TAC 2008, we observe similar results with our reimplementation of *ICSISumm*, in that it exhibits effectiveness that is not statistically significantly different to the original. However, we note that the learned model, trained on DUC 2002, is not as effective under ROUGE-2 over TAC 2008 as we observe over DUC 2004. From the results in Table 5, we conclude that our reimplementation of *ICSISumm* is correct, and, although our learned model performs effectively over DUC 2004, the learned model has not generalised effectively from DUC 2002 newswire to TAC 2008 newswire.

We now investigate the manual selection of the most effective anti-redundancy component, on a per topic basis. Taking effective standard baseline reimplementations, we compute ROUGE scores for an oracle system that selects the particular anti-redundancy component, per topic, which maximises

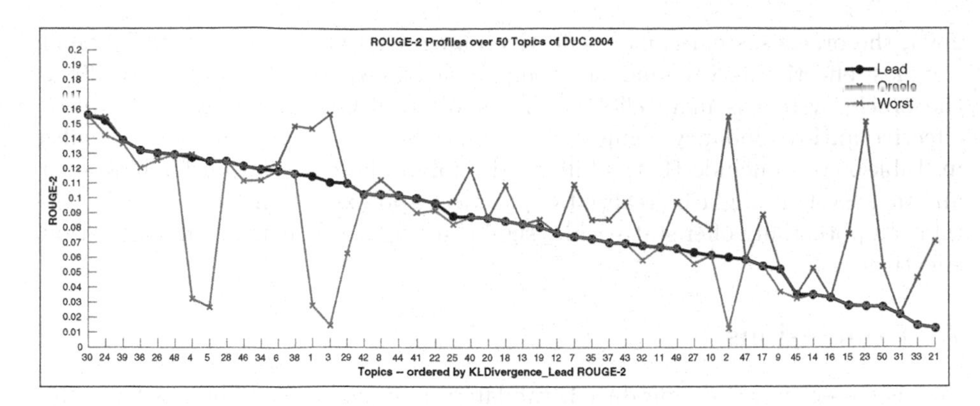

Fig. 3. ROUGE-2 effectiveness profiles, over DUC 2004, for KLDivergence_Lead, an oracle system optimising selection of anti-redundancy components over topics, and the worst case.

Table 6. Results over DUC 2004 and TAC 2008, showing the best/worst scores possible when manually selecting the most/least effective anti-redundancy components per-topic.

DUC 2004			
KLDivergence	R-1	**R-2**	R-4
CosineSimilarityHy	38.44	**9.59**	1.56
Oracle Score	39.06†	9.82	1.70†
Worst Case	35.28	8.04	1.27

TAC 2008			
GraphDegreeHyRt	R-1	**R-2**	R-4
NewWordCount	37.42	**10.23**	2.22
Oracle Score	41.32†	13.17†	4.07†
Worst Case	31.14	6.46	0.82

the ROUGE-2 effectiveness. Figure 3 visualises the distribution of ROUGE-2 scores, over the 50 topics of DUC 2004, for *KLDivergence_Lead* (no anti-redundancy filtering), and for the oracle system (best case), and additionally, the worst case (where the least effective anti-redundancy component is always chosen, per topic). Table 6 provides the ROUGE results for *KLDivergence* over DUC 2004, and *GraphDegreeHyRt* over TAC 2008, showing the most effective anti-redundancy component, the effectiveness of the oracle system, and the worst case.

From Fig. 3, we can observe that there exists best, and worst case, anti-redundancy component selection choices, per topic. This means, there are topics where we would wish to avoid a particular anti-redundancy component, and further, some topics where we would indeed wish to select a particular anti-redundancy component. If we create an oracle system that manually selects from the 6 different anti-redundancy components, optimising the ROUGE-2 metric over topics, we obtain the ROUGE scores we present in Table 6. From Table 6, we observe that the worst case is always significantly the least effective, over both DUC 2004 and TAC 2008. Further, from Table 6 we observe that the oracle system leads to statistically significant improvements over the most effective anti-redundancy component, indicated by the "† symbol. In particular, over DUC

2004, the oracle system is more effective under ROUGE-1 and ROUGE-4 than the most effective anti-redundancy component (shown in bold). Over TAC 2008, the oracle system is more effective under all ROUGE metrics than the most effective anti-redundancy component (again, shown in bold). From the results in Table 6, we conclude that, while we do not propose a solution for how such an oracle system might be realised in practice, approximations of the oracle system can potentially offer statistically significant improvements in summarisation effectiveness.

7 Conclusions

In this paper, we have reproduced, validated, and generalised findings from the literature. Additionally, we have reimplemented standard and state-of-the-art baselines, making further observations from our experiments. In conclusion, we have confirmed that the ROUGE-2 metric is aligned with crowd-sourced user judgements for summary quality, and confirmed that several state-of-the-art systems behave differently, despite similar ROUGE-2 scores. Further, an optimal *Lead* baseline can be derived from interleaving the first sentences from multiple documents, and applying anti-redundancy components. Indeed, an optimal *Lead* baseline exhibits ROUGE-2 effectiveness with no significant difference to standard baselines, over DUC 2004. Additionally, the effectiveness of the standard baselines, as reported in the literature, can be improved to the point where there is no significant difference to the state-of-the-art (as illustrated using *ICSISumm*). Finally, given that an optimal choice of anti-redundancy components, per-topic, exhibits significant improvements in summarisation effectiveness, we conclude that future work should investigate learning algorithm (or topic) specific anti-redundancy components.

Acknowledgements. Mackie acknowledges the support of EPSRC Doctoral Training grant 1509226. McCreadie, Macdonald and Ounis acknowledge the support of EC SUPER project (FP7-606853).

References

1. Allan, J., Wade, C., Bolivar, A.: Retrieval and novelty detection at the sentence level. In: Proceedings of SIGIR (2003)
2. Conroy, J.M., Schlesinger, J.D., O'Leary, D.P.: Topic-focused multi-document summarization using an approximate oracle score. In: Proceedings of COLING-ACL (2006)
3. Erkan, G., Radev, D.R.: LexRank: Graph-based lexical centrality as salience in text summarization. J. Artif. Intell. Res. **22**(1), 457–479 (2004)
4. Gillick, D., Favre, B.: A scalable global model for summarization. In: Proceedings of ACL ILP-NLP (2009)
5. Guo, Q., Diaz, F., Yom-Tov, E.: Updating users about time critical events. In: Serdyukov, P., Braslavski, P., Kuznetsov, S.O., Kamps, J., Rüger, S., Agichtein, E., Segalovich, I., Yilmaz, E. (eds.) ECIR 2013. LNCS, vol. 7814, pp. 483–494. Springer, Heidelberg (2013)

6. Haghighi, A., Vanderwende, L.: Exploring content models for multi-document summarization. In: Proceedings of NAACL-HLT (2009)
7. Hong, K., Conroy, J., Favre, B., Kulesza, A., Lin, H., Nenkova, A.: A repository of state of the art and competitive baseline summaries for generic news summarization. In: Proceedings of LREC (2014)
8. Kedzie, C., McKeown, K., Diaz, F.: Predicting salient updates for disaster summarization. In: Proceedings of ACL-IJCNLP (2015)
9. Lin, C.Y.: ROUGE: A package for automatic evaluation of summaries. In: Proceedings of ACL (2004)
10. Lin, C.Y., Hovy, E.: The automated acquisition of topic signatures for text summarization. In: Proceedings of COLING (2000)
11. Mackie, S., McCreadie, R., Macdonald, C., Ounis, I.: Comparing algorithms for microblog summarisation. In: Kanoulas, E., Lupu, M., Clough, P., Sanderson, M., Hall, M., Hanbury, A., Toms, E. (eds.) CLEF 2014. LNCS, vol. 8685, pp. 153–159. Springer, Heidelberg (2014)
12. Mackie, S., McCreadie, R., Macdonald, C., Ounis, I.: On choosing an effective automatic evaluation metric for microblog summarisation. In: Proceedings of IIiX (2014)
13. McCreadie, R., Macdonald, C., Ounis, I.: Incremental update summarization: Adaptive sentence selection based on prevalence and novelty. In: Proceedings of CIKM (2014)
14. Nenkova, A.: Automatic text summarization of newswire: Lessons learned from the document understanding conference. In: Proceedings of AAAI (2005)
15. Nenkova, A., McKeown, K.: Automatic summarization. Found. Trends Inf. Retrieval **5**(2–3), 103–233 (2011)
16. Nenkova, A., Vanderwende, L., McKeown, K.: A compositional context sensitive multi-document summarizer: Exploring the factors that influence summarization. In: Proceedings of SIGIR (2006)
17. Owczarzak, K., Conroy, J.M., Dang, H.T., Nenkova, A.: An assessment of the accuracy of automatic evaluation in summarization. In: Proceedings of NAACL-HLT WEAS (2012)
18. Radev, D.R., Jing, H., Styś, M., Tam, D.: Centroid-based summarization of multiple documents. Inf. Process. Manage. **40**(6), 919–938 (2004)
19. K, Spärck Jones: Automatic summarising: The state-of-the-art. Inf. Process. Manage. **43**(6), 1449–1481 (2007)

On the Reproducibility of the TAGME Entity Linking System

Faegheh Hasibi[1(✉)], Krisztian Balog[2], and Svein Erik Bratsberg[1]

[1] Norwegian University of Science and Technology, Trondheim, Norway
{faegheh.hasibi,sveinbra}@idi.ntnu.no
[2] University of Stavanger, Stavanger, Norway
krisztian.balog@uis.no

Abstract. Reproducibility is a fundamental requirement of scientific research. In this paper, we examine the repeatability, reproducibility, and generalizability of TAGME, one of the most popular entity linking systems. By comparing results obtained from its public API with (re)implementations from scratch, we obtain the following findings. The results reported in the TAGME paper cannot be repeated due to the unavailability of data sources. Part of the results are reproducible through the provided API, while the rest are not reproducible. We further show that the TAGME approach is generalizable to the task of entity linking in queries. Finally, we provide insights gained during this process and formulate lessons learned to inform future reducibility efforts.

1 Introduction

Recognizing and disambiguating entity occurrences in text is a key enabling component for semantic search [14]. Over the recent years, various approaches have been proposed to perform automatic annotation of documents with entities from a reference knowledge base, a process known as *entity linking* [7,8,10,12, 15,16]. Of these, TAGME [8] is one of the most popular and influential ones. TAGME is specifically designed for efficient ("on-the-fly") annotation of short texts, like tweets and search queries. The latter task, i.e., annotating search queries with entities, was evaluated at the recently held Entity Recognition and Disambiguation Challenge [1], where the first and second ranked systems both leveraged or extended TAGME [4,6]. Despite the explicit focus on short text, TAGME has been shown to deliver competitive results on long texts as well [8]. TAGME comes with a web-based interface and a RESTful API is also provided.[1] The good empirical performance coupled with the aforementioned convenience features make TAGME one of the obvious must-have baselines for entity linking research. The influence and popularity of TAGME is also reflected in citations; the original TAGME paper [8] (from now on, simply referred to as the TAGME paper) has been cited around 50 times according to the ACM digital library and nearly 200 times according to Google scholar, at the time of writing. The authors

[1] http://tagme.di.unipi.it/.

N. Ferro et al. (Eds.): ECIR 2016, LNCS 9626, pp. 436–449, 2016.
DOI: 10.1007/978-3-319-30671-1_32

have also published an extended report [9] (with more algorithmic details and experiments) that has received over 50 citations according to Google scholar.

Our focus in this paper is on the repeatability, reproducibility, and generalizability of the TAGME system; these are obvious desiderata for reliable and extensible research. The recent SIGIR 2015 workshop on Reproducibility, Inexplicability, and Generalizability of Results (RIGOR)[2] defined these properties as follows:

- *Repeatability*: "Repeating a previous result under the original conditions (e.g., same dataset and system configuration)."
- *Reproducibility*: "Reproducing a previous result under different, but comparable conditions (e.g., different, but comparable dataset)."
- *Generalizability*: "Applying an existing, empirically validated technique to a different IR task/domain than the original."

We address each of these aspects in our study, as explained below.

Repeatability. Although TAGME facilitates comparison by providing a publicly available API, it is not sufficient for the purpose of repeatability. The main reason is that the API works much like a black-box; it is impossible to check whether it corresponds to the system described in [8]. Actually, it is acknowledged that the API deviates from the original publication,[3] but the differences are not documented anywhere. Another limiting factor is that the API cannot be used for efficiency comparisons due to the network overhead. We report on the challenges around repeating the experiments in [8] and discuss why the results are not repeatable.

Reproducibility. TAGME has been re-implemented in several research papers, see, e.g., [2,3,11], these, however, do not report on the reproducibility of results. In addition, there are some technical challenges involved in the TAGME approach that have not always been dealt with properly in the original paper and accordingly in these re-implementations (as confirmed by some of the respective authors).[4] We examine the reproducibility of TAGME, as introduced in [8], and show that some of the results are not reproducible, while others are reproducible only through the TAGME API.

Generalizability. We test generalizability by applying TAGME to a different task: entity linking in queries (ELQ). This task has been devised by the Entity Recognition and Disambiguation (ERD) workshop [1], and has been further elaborated on in [11]. The main difference between conventional entity linking and ELQ is that the latter accepts that a query might have multiple interpretations, i.e., the output in not a single annotation, but (possibly multiple) sets of entities that are semantically related to each other. Even though TAGME has been developed for a different problem (where only a single interpretation is returned), we show that it is generalizable to the ELQ task.

[2] https://sites.google.com/site/sigirrigor/.

[3] http://tagme.di.unipi.it/tagme_help.html and is also mentioned in [5,18].

[4] Personal communication with authors of [3,8,11].

Before we proceed let us make a disclaimer. In the course of this study, we made a best effort to reproduce the results presented in [8] based on the information available to us: the TAGME papers [8,9] and the source code kindly provided by the authors. Our main goal with this work is to learn about reproducibility, and is in no way intended to be a criticism of TAGME. The communication with the TAGME authors is summarized in Sect. 6. The resources developed within this paper as well as detailed responses from the TAGME authors (and any possible future updates) are made publicly available at http://bit.ly/tagme-rep.

2 Overview of TAGME

In this section, we provide an overview of the TAGME approach, as well as the test collections and evaluation metrics used in the TAGME papers [8,9].

2.1 Approach

TAGME performs entity linking in a pipeline of three steps: (i) parsing, (ii) disambiguation, and (iii) pruning (see Fig. 1). We note that while Ferragina and Scaiella [8] describe multiple approaches for the last two steps, we limit ourselves to their final suggestions; these are also the choices implemented in the TAGME API.

Before describing the TAGME pipeline, let us define the notation used throughout this paper. Entity linking is the task of annotating an input text T with entities E from a reference knowledge base, which is Wikipedia here. T contains a set of entity mentions M, where each mention $m \in M$ can refer to a set of candidate entities $E(m)$. These need to be disambiguated such that each mention points to a single entity $e(m)$.

Parsing. In the first step, TAGME parses the input text and performs mention detection using a dictionary of entity surface forms. For each entry (surface form) the set of entities recognized by that name is recorded. This dictionary is built by extracting entity surface forms from four sources: anchor texts of Wikipedia articles, redirect pages, Wikipedia page titles, and variants of titles (removing parts after the comma or in parentheses). Surface forms consisting of numbers only or of a single character, or below a certain number of occurrences (2) are discarded. Further filtering is performed on the surface forms with low *link probability* (i.e., < 0.001). Link probability is defined as:

$$\mathrm{lp}(m) = P(\mathrm{link}|m) = \frac{link(m)}{freq(m)}, \quad (1)$$

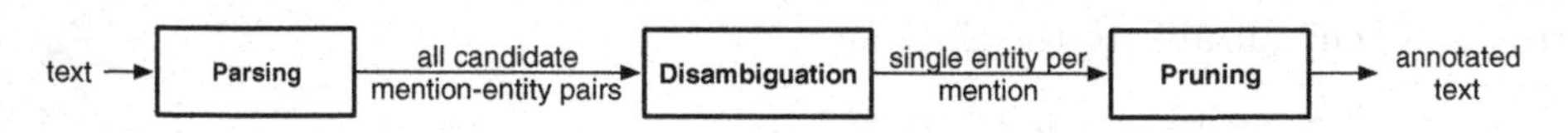

Fig. 1. Annotation pipeline in the TAGME system.

where $freq(m)$ denotes the total number of times mention m occurs in Wikipedia (as a link or not), and $link(m)$ is the number of times mention m appears as a link.

To detect entity mentions, TAGME matches all n-grams of the input text, up to $n = 6$, against the surface form dictionary. For an n-gram contained by another one, TAGME drops the shorter n-gram, if it has lower link probability than the longer one. The output of this step is a set of mentions with their corresponding candidate entities.

Disambiguation. Entity disambiguation in TAGME is performed using a voting schema, that is, the score of each mention-entity pair is computed as the sum of votes given by candidate entities of all other mentions in the text. Formally, given the set of mentions M, the relevance score of the entity e to the mention m is defined as:

$$\text{rel}(m, e) = \sum_{m' \in M - \{m\}} \text{vote}(m', e), \tag{2}$$

where vote(m', e) denotes the agreement between entities of mention m' and the entity e, computed as follows:

$$\text{vote}(m', e) = \frac{\sum_{e' \in E(m')} \text{relatedness}(e, e') \cdot \text{commonness}(e', m')}{|E(m')|}. \tag{3}$$

Commonness is the probability of an entity being the link target of a given mention [13]:

$$\text{commonness}(e', m') = P(e'|m') = \frac{link(e', m')}{link(m')}, \tag{4}$$

where $link(e', m')$ is the number of times entity e' is used as a link destination for m' and $link(m')$ is the total number of times m' appears as a link. *Relatedness* measures the semantic association between two entities [17]:

$$\text{relatedness}(e, e') = \frac{\log(\max(|in(e)|, |in(e')|)) - \log(|in(e) \cap in(e')|)}{\log(|E|) - \log(\min(|in(e)|, |in(e')|))}, \tag{5}$$

where $in(e)$ is the set of entities linking to entity e and $|E|$ is the total number of entities.

Once all candidate entities are scored using Eq. (2), TAGME selects the best entity for each mention. Two approaches are suggested for this purpose: (i) disambiguation by classifier (DC) and (ii) disambiguation by threshold (DT), of which the latter is selected as the final choice. Due to efficiency concerns, entities with commonness below a given threshold τ are discarded from the DT computations. The set of commonness-filtered candidate entities for mention m is $E_\tau(m) = \{e \in M(e) | \text{commonness}(m, e) \geq \tau\}$. Then, DT considers the top-ϵ

entities for each mention and then selects the one with the highest commonness score:

$$m(e) = \arg\max_{e}\{\text{commonness}(m, e) : e \in E_{\mathcal{T}}(m) \wedge e \in \text{top}_{\epsilon}[\text{rel}(m, e)]\}. \quad (6)$$

At the end of this stage, each mention in the input text is assigned a single entity, which is the most pertinent one to the input text.

Pruning. The aim of the pruning step is to filter out non-meaningful annotations, i.e., assign *NIL* to the mentions that should not be linked to any entity. TAGME hinges on two features to perform pruning: *link probability* (Eq. (1)) and *coherence*. The coherence of an entity is computed with respect to the candidate annotations of all the other mentions in the text:

$$\text{coherence}(e, T) = \frac{\sum_{e' \in E(T)-\{e\}} \text{relatedness}(e, e')}{|E(T)| - 1}, \quad (7)$$

where $E(T)$ is the set of distinct entities assigned to the mentions in the input text. TAGME takes the average of the link probability and the coherence score to generate a ρ score for each entity, which is then compared to the pruning threshold ρ_{NA}. Entities with $\rho < \rho_{NA}$ are discarded, while the rest of them are served as the final result.

2.2 Test Collections

Two test collections are used in [8]: Wiki-Disamb30 and Wiki-Annot30. Both comprise of snippets of around 30 words, extracted from a Wikipedia snapshot of November 2009, and are made publicly available.[5] In Wiki-Disamb30, each snippet is linked to a single entity; in Wiki-Annot30 all entity mentions are annotated. We note that the sizes of these test collections (number of snippets) deviate from what is reported in the TAGME paper: Wiki-Disamb30 and Wiki-Annot30 contain around 2M and 185K snippets, while the reported numbers are 1.4M and 180K, respectively. This suggests that the published test collections might be different from the ones used in [8].

2.3 Evaluation Metrics

TAGME is evaluated using three variations of precision and recall. The so-called *standard* precision and recall (P and R), are employed for evaluating the disambiguation phase, using the Wiki-Disamb30 test collection. The two other metrics, *annotation* and *topics* precision and recall are employed for measuring the end-to-end performance on the Wiki-annot30 test collection. The annotation metrics (P_{ann} and R_{ann}) compare both the mention and the entity against

[5] http://acube.di.unipi.it/tagme-dataset/.

the ground truth, while the topics metrics (P_{topics} and R_{topics}) only consider entity matches. The TAGME papers [8,9] provide little information about the evaluation metrics. In particular, the computation of the *standard* precision and recall is rather unclear; we discuss it later in Sect. 4.2. Details are missing regarding the two other metrics too: (i) How are overall precision, recall and F-measure computed for the annotation metrics? Are they micro- or macro-averaged? (ii) What are the matching criteria for the annotation metrics? Are partially matching mentions accepted or only exact matches? In what follows, we formally define the annotation and topics metrics, based on the most likely interpretation we established from the TAGME paper and from our experiments.

We write $\mathcal{G}(T) = \{(\hat{m}_1, \hat{e}_1), \ldots, (\hat{m}_m \hat{e}_m)\}$ for ground truth annotations of the input text T, and $\mathcal{S}(T) = \{(m_1, e_1), \ldots, (m_n, e_n)\}$ for the annotations identified by the system. Neither $\mathcal{G}(T)$ nor $\mathcal{S}(T)$ contains NULL annotations. The TAGME paper follows [12], which uses macro-averaging in computing annotation precision and recall:[6]

$$P_{ann} = \frac{|\mathcal{G}(T) \cap \mathcal{S}(T)|}{|\mathcal{S}(T)|}, \quad R_{ann} = \frac{|\mathcal{G}(T) \cap \mathcal{S}(T)|}{|\mathcal{G}(T)|}. \tag{8}$$

The annotation $(\hat{m}, \hat{e})$ matches (m, e) if two conditions are fulfilled: (i) entities match ($\hat{e} = e$), and (ii) mentions match or contain each other ($\hat{m} = m$ or $\hat{m} \in m$ or $m \in \hat{m}$). We note that the TAGME paper refers to "perfect match" of the mentions, while we use a more relaxed version of matching (by considering containment matches). This relaxation results in the highest possible P_{ann} and R_{ann}, but even those are below the numbers reported in [8] (cf. Sect. 4.2).

The topics precision and recall (P_{topics} and R_{topics}) [16] only consider entity matches ($\hat{e} = e$) and are micro-averaged over the set of all texts $\mathcal{F}$:

$$P_{topics} = \frac{\sum_{T \in \mathcal{F}} |\mathcal{G}(T) \cap \mathcal{S}(T)|}{\sum_{T \in \mathcal{F}} |\mathcal{S}(T)|}, \quad R_{topics} = \frac{\sum_{T \in \mathcal{F}} |\mathcal{G}(T) \cap \mathcal{S}(T)|}{\sum_{T \in \mathcal{F}} |\mathcal{G}(T)|}. \tag{9}$$

For all metrics the overall F-measure is computed from the overall precision and recall.

3 Repeatability

By definition (cf. Sect. 1), *repeatability* means that a system should be implemented under the same conditions as the reference system. In our case, the repeatability of the TAGME experiments in [8] is dependent on the availability of (i) the knowledge base and (ii) the test collections (text snippets and gold standard annotations).

[6] As explained later by the TAGME authors, they in fact used micro-averaging. This contradicts the referred paper [12], which explicitly defines P_{ann} and R_{ann} as being macro-averaged.

The reference knowledge base is Wikipedia, specifically, the TAGME paper uses a dump from November 2009, while the API employs a dump from July 2012. Unfortunately, neither of these dumps is available on the web nor could be provided by the TAGME authors upon request. We encountered problems with the test collections too. As we already explained in Sect. 2.2, there are discrepancies between the number of snippets the test collections (WIKI-DISAMB30 and WIKI-ANNOT30) actually contain and what is reported in the paper. The latter number is higher, suggesting that the results in [8] are based only on subsets of the collections.[7] Further, the WIKI-DISAMB30 is split into training and test sets in the TAGME paper, but those splits are not available.

Due to these reasons, which could all be classified under the general heading of *unavailability of data*, we conclude that the TAGME experiments in [8] are not repeatable. In the next section, we make a best effort at establishing the most similar conditions, that is, we attempt to reproduce their results.

4 Reproducibility

This section reports on our attempts to reproduce the results presented in the TAGME paper [8]. The closest publicly available Wikipedia dump is from April 2010,[8] which is about five months newer than the one used in [8]. On a side note we should mention that we were (negatively) surprised by how difficult it proved to find Wikipedia snapshots from the past, esp. from this period. We have (re)implemented TAGME based on the description in the TAGME papers [8,9] and, when in doubt, we checked the source code. For a reference comparison, we also include the results from (i) the TAGME API and (ii) the Dexter entity linking framework [3]. Even though the implementation in Dexter (specifically, the parser) slightly deviates from the original TAGME system, it is still useful for validation, as that implementation is done by a third (independent) group of researchers. We do not include results from running the source code provided to us because it requires the Wikipedia dump in a format that is no longer available for the 2010 dump we have access to; running it on a newer Wikipedia version would give results identical to the API. In what follows, we present the challenges we encountered during the implementation in Sect. 4.1 and then report on the results in Sect. 4.2.

4.1 Implementation

During the (re)implementation of TAGME, we encountered several technical challenges, which we describe here. These could be traced back to differences between the approach described in the paper and the source code provided by

[7] It was later explained by the TAGME authors that they actually used only 1.4M out of 2M snippets from WIKI-DISAMB30, as Weka could not load more than that into memory. From WIKI-ANNOT30 they used all snippets, the difference is merely a matter of approximation.

[8] https://archive.org/details/enwiki_20100408.

the authors. Without addressing these differences, the results generated by our implementation are far from what is expected and are significantly worse than those by the original system.

Link probability computation. Link probability is one of the main statistical features used in TAGME. We noticed that the computation of link probability in TAGME deviates from what is defined in Eq. (1): instead of computing the denominator $freq(m)$ as the number of occurrences of mention m in Wikipedia, TAGME computes the number of documents that mention m appears in. Essentially, document frequency is used instead of term (phrase) frequency. This is most likely due to efficiency considerations, as the former is much cheaper to compute. However, a lower denominator in Eq. (1) means that the resulting link probability is a higher value than it is supposed to be. In fact, this change in the implementation means that it is actually not link probability, but more like *keyphraseness* that is being computed. Keyphraseness [15] is defined as:

$$\text{kp}(m) = P(\text{keyword}|m) = \frac{key(m)}{df(m)}, \tag{10}$$

where $key(m)$ denotes number of Wikipedia articles where mention m is selected as a keyword, i.e., linked to an entity (any entity), and $df(m)$ is the number of articles containing the mention m. Since in Wikipedia a link is typically created only for the first occurrence of an entity ($link(m) \approx key(m)$), we can assume that the numerator of link probability and keyphraseness are identical. This would mean that TAGME as a matter of fact uses keyphraseness. Nevertheless, as our goal in this paper is to reproduce the TAGME results, we followed their implementation of this feature, i.e., $link(m)/df(m)$.[9]

Relatedness computation. We observed that the *relatedness* score, defined in Eq. (5), is computed as $1 - \text{relatedness}(e, e')$, furthermore, for the entities with zero inlinks or no common inlinks, the score is set to zero. These details are not explicitly mentioned in the paper, while they have significant impact on the overall effectiveness of TAGME.

Pruning based on commonness. In addition to the filtering methods mentioned in the parsing step (cf. Sect. 2.1), TAGME filters entities with commonness score below 0.001, but it is not documented in the TAGME papers. We followed this filtering approach, as it makes the system considerably faster.

4.2 Results

We report results for the intermediate disambiguation phase and for the end-to-end entity linking task. For all reproducibility experiments, we set the ρ_{NA} threshold to 0.2, as it delivers the best results and is also the recommended value in the TAGME paper.

[9] The proper implementation of link probability would result in lower values (as the denominator would be higher) and would likely require a different threshold value than what is suggested in [8]. This goes beyond the scope of our paper.

Table 1. Results of TAGME repeatability on the Wiki-Disamb30 test collection.

Method	P	R	F
Original paper [8]	0.915	0.909	0.912
TAGME API	0.775	0.775	0.775

Table 2. Results of TAGME repeatability on the Wiki-Annot30 test collection.

Method	P_{ann}	R_{ann}	F_{ann}	P_{topics}	R_{topics}	F_{topics}
Original paper [8]	0.7627	0.7608	0.7617	0.7841	0.7748	0.7794
TAGME API	0.6945	0.7136	0.7039	0.7017	0.7406	0.7206
TAGME-wp10 (our)	0.6143	0.4987	0.5505	0.6499	0.5248	0.5807
Dexter	0.5722	0.5959	0.5838	0.6141	0.6494	0.6313

Disambiguation phase. For evaluating the disambiguation phase, we submitted the snippets from the Wiki-Disamb30 test collection to the TAGME API, with the pruning threshold set to 0. This setting ensures that no pruning is performed and the output we get back is what is supposed to be the outcome of the disambiguation phase. We tried different methods for computing precision and recall, but we were unable to get the results that are reported in the original TAGME paper (see Table 1). We therefore relaxed the evaluation conditions in the following way: if any of the entities returned by the disambiguation phase matches the ground truth entity for the given snippet, then we set both precision and recall to 1; otherwise they are set to 0. This gives us an upper bound for the performance that can be achieved on the Wiki-disamb30 test collection; any other interpretation of precision or recall would result in a lower number. What we found is that even with these relaxed conditions the F-score is far below the reported value (0.775 vs. 0.912). One reason for the differences could be the discrepancy between the number of snippets in the test collection and the ones used in [8]. Given the magnitude of the differences, even against their own API, we decided not to go further to get the results for our implementation of TAGME. We conclude that this set of results is not reproducible, due to *insufficient experimental details* (test collection and metrics).

End-to-end performance. Table 2 shows end-to-end system performance according to the following implementations: the TAGME API, our implementation using a Wikipedia snapshot from April 2010, and the Dexter implementation using a Wikipedia snapshot from March 2013. For all experiments, we compute the evaluation metrics described in Sect. 2.3. We observe that the API results are lower than in the original paper, but the difference is below 10 %. We attribute this to the fact that the ground truth is generated from a 2009 version of Wikipedia, while the API is based on the version from 2012.

Concerning our implementation and Dexter (bottom two rows in Table 2) we find that they are relatively close to each other, but both of them are lower than the TAGME API results; the relative difference to the API results is –19 % for our implementation and –12 % for Dexter in F_{topics} score. Ceccarelli et al. [3] also report on deviations, but they attribute these to the processing of Wikipedia: "we observed that our implementation always improves over the WikiMiner online service, and that it behaves only slightly worse then TAGME after the top 5 results, probably due to a different processing of Wikipedia." The difference between Dexter and our implementation stems from the parsing step. Dexter relies on its own parsing method and removes overlapping mentions at the end of the annotation process. We, on the other hand, follow TAGME and delete overlapping mentions in the parsing step (cf. Sect. 2.1). By analyzing our results, we observed that this parsing policy resulted in early pruning of some correct entities and led accordingly to lower results.

Our experiments show that the end-to-end results reported in [8] are reproducible through the TAGME API, but not by (re)implementation of the approach by a third partner. This is due to *undocumented deviations from the published description.*

5 Generalizability

To test the generalizability of TAGME, we apply it to a (slightly) different entity linking task: entity linking in queries (ELQ). As discussed in [1,11], the aim of this task is to detect all possible entity linking *interpretations* of the query. This is different from conventional entity linking, where a single annotation is created. Let us consider the query "france world cup 98" to get a better understanding of the differences between the two tasks. In this example, both FRANCE and FRANCE NATIONAL FOOTBALL TEAM are valid entities for the mention "france." In conventional entity linking, we link each mention to a single entity, e.g., "france" ⇒ FRANCE, "world cup" ⇒ FIFA WORLD CUP. For the ELQ task, on the other hand, we detect all entity linking interpretations of the query, where each interpretation is a set of semantically related entities, e.g., {FRANCE, FIFA WORLD CUP} and {FRANCE NATIONAL FOOTBALL TEAM, FIFA WORLD CUP}. In other words, the output of conventional entity linking systems is a set of mention-entity pairs, while entity linking in queries returns set(s) of entity sets.

Applying a conventional entity linker to the ELQ task restricts the output to a single interpretation, but can deliver solid performance nonetheless [1]. TAGME has great potential to be generalized to the ELQ task, as it is designed to operate with short texts. We detail our experimental setup in Sect. 5.1 and report on the results in Sect. 5.2.

5.1 Experimental Setup

Implementations. We compare four different implementations to assess the generalizability of TAGME to the ELQ task: the TAGME API, our implementation of TAGME with two different Wikipedia versions, one from April 2010 and another from May 2012 (which is part of the ClueWeb12 collection), and Dexter's implementation of TAGME. Including results using the 2012 version of Wikipedia facilitates a better comparison between the TAGME API and our implementation, as they both use similar Wikipedia dumps. It also demonstrates how the version of Wikipedia might affect the results.

Datasets and evaluation metrics. We use two publicly available test collections developed for the ELQ task: ERD-dev [1] and Y-ERD [11]. ERD-dev includes 99 queries, while Y-ERD offers a larger selection, containing 2398 queries. The annotations in these test collections are confined to proper noun entities from a specific Freebase snapshot.[10] We therefore remove entities that are not present in this snapshot in a post-filtering step. In all the experiments, ρ_{NA} is set to 0.1, as it delivers the highest results both for the API and for our implementations, and is also the recommendation of the TAGME API. Evaluation is performed in terms of precision, recall, and F-measure (macro-averaged over all queries), as proposed in [1]; this variant is referred to as *strict* evaluation in [11].

5.2 Results

Table 3 presents the TAGME generalizability results. Similar to the reproducibility experiments, we find that the TAGME API provides substantially better results than any of the other implementations. The most fair comparison between Dexter and our implementations is the one against TAGME-wp12, as that has the Wikipedia dump closest in date. For ERD-dev they deliver similar results, while for Y-ERD Dexter has a higher F-score (but the relative difference is below 10 %). Concerning different Wikipedia versions, the more recent one performs better on the ERD-dev test collection, while the difference is negligible for Y-ERD. If we take the larger test collection, Y-ERD, to be the more representative one, then we find that TAGME API > Dexter > TAGME-wp10, which

Table 3. TAGME results for the entity linking in queries task.

Method	ERD-dev			Y-ERD		
	P_{strict}	R_{strict}	F_{strict}	P_{strict}	R_{strict}	F_{strict}
TAGME API	0.8352	0.8062	0.8204	0.7173	0.7163	0.7168
TAGME-wp10 (our)	0.7143	0.7088	0.7115	0.6518	0.6515	0.6517
TAGME-wp12 (our)	0.7363	0.7234	0.7298	0.6535	0.6532	0.6533
Dexter	0.7363	0.7073	0.7215	0.6989	0.6979	0.6984

is consistent with the reproducibility results in Table 2. However, the relative differences between the approaches are smaller here. We thus conclude that the TAGME approach can be generalized to the ELQ task.

[10] http://web-ngram.research.microsoft.com/erd2014/Datasets.aspx.

6 Discussion and Conclusions

TAGME is an outstanding entity linking system. The authors offer invaluable resources for the reproducibility of their approach: the test collections, source code, and a RESTful API. In this paper we have attempted to (re)implement the system described in [8], reproduce their results, and generalize the approach to the task of entity linking in queries. Our experiments have shown that some of the results are not reproducible, even with the API provided by the authors. For the rest of the results, we have found that (i) the results reported in the paper are higher than what can be reproduced using their API, and (ii) the TAGME API gives higher numbers than what is achievable by a third-party implementation (not only ours, but also that of Dexter [2]). Based on these findings, we recommend to use the TAGME API, much like a black-box, when entity linking is performed as part of a larger task. For a reliable and meaningful comparison between TAGME and a newly proposed entity linking method, the TAGME approach should be (re)implemented, like it has been done in some prior work, see, e.g., [2,11].

Post-acceptance responses from the TAGME authors. Upon the acceptance of this paper, the TAGME authors clarified some of the issues that surfaced in this study. This information came only after the paper was accepted, even though we have raised our questions during the writing of the paper (at that time, however, the reply we got only included the source code and the fact that they no longer have the Wikipedia dumps used in the paper). We integrated their responses throughout the paper as much as it was possible; we include the rest of them here. First, it turns out that the public API as well as the provided source code correspond to a newer, updated version ("version 2") of TAGME. The source code for the original version ("version 1") described in the TAGME papers [8,9] is no longer available. This means that even if we managed to find the Wikipedia dump used in the TAGME papers and ran their source code, we would have not been able to reproduce their results. Furthermore, TAGME performs additional non-documented optimizations when parsing the spots, filtering inappropriate spots, and computing relatedness, as explained by the authors. Another reason for the differences in performance might have to do with how links are extracted from Wikipedia. TAGME uses wiki page-to-page link records, while our implementation (as well as Dexter's) extracts links from the body of the pages. This affects the computation of relatedness, as the former source contains 20 % more links than the latter. (It should be noted that this file was not available for the 2010 and 2012 Wikipedia dumps.) The authors also clarified that all the evaluation metrics are micro-averaged and explained how the disambiguation phase was evaluated. We refer the interested reader to the online appendix of this paper for further details.

Lessons learned. Even though we have only partially succeeded in reproducing the TAGME results, we have gained invaluable insights about reproducibility requirements during the process. Lessons learned include the following: (i) all technical details that affect effectiveness or efficiency should be explained (or at least mentioned) in paper; sharing the source code helps, but finding answers in a large codebase can be highly non-trivial; (ii) if there are differences between the published approach and publicly made available source code or API (typically, the latter being an updated version), those should be made explicit; (iii) it is encouraged that authors keep all data sources used in a published paper (in particular, historical Wikipedia dumps, esp. in some specific format, are more difficult to find than one might think), so that these can be shared upon requests from other researchers; (iv) evaluation metrics should be explained in detail. Maintaining an "online appendix" to a publication is a practical way of providing some of these extra details that would not fit in the paper due to space limits, and would have the additional advantage of being easily editable and extensible.

Acknowledgement. We would like to thank Paolo Ferragina and Ugo Scaiella for sharing the TAGME source code with us and for the insightful discussions and clarifications later on. We also thank Diego Ceccarelli for the discussion on link probability computation and for providing help with the Dexter API.

References

1. Carmel, D., Chang, M.-W., Gabrilovich, E., Hsu, B.-J.P., Wang, K.: ERD'14: Entity recognition and disambiguation challenge. SIGIR Forum **48**(2), 63–77 (2014)
2. Ceccarelli, D., Lucchese, C., Orlando, S., Perego, R., Trani, S.: Dexter: An open source framework for entity linking. In: Proceedings of the Sixth International Workshop on Exploiting Semantic Annotations in Information Retrieval, pp. 17–20 (2013)
3. Ceccarelli, D., Lucchese, C., Orlando, S., Perego, R., Trani, S.: Learning relatedness measures for entity linking. In: Proceedings of CIKM 2013, pp. 139–148 (2013)
4. Chiu, Y.-P., Shih, Y.-S., Lee, Y.-Y., Shao, C.-C., Cai, M.-L., Wei, S.-L., Chen, H.-H.: NTUNLP approaches to recognizing and disambiguating entities in long and short text at the ERD challenge 2014. In: Proceedings of Entity Recognition & Disambiguation Workshop, pp. 3–12 (2014)
5. Cornolti, M., Ferragina, P., Ciaramita, M.: A framework for benchmarking entity-annotation systems. In: Proceedings of WWW 2013, pp. 249–260 (2013)
6. Cornolti, M., Ferragina, P., Ciaramita, M., Schütze, H., Rüd, S.: The SMAPH system for query entity recognition and disambiguation. In: Proceedings of Entity Recognition & Disambiguation Workshop, pp. 25–30 (2014)
7. Cucerzan, S.: Large-scale named entity disambiguation based on Wikipedia data. In: Proceedings of EMNLP-CoNLL 2007, pp. 708–716 (2007)
8. Ferragina, P., Scaiella, U.: TAGME: On-the-fly annotation of short text fragments (by Wikipedia entities). In: Proceedings of CIKM 2010, pp. 1625–1628 (2010)
9. Ferragina, P., Scaiella, U.: Fast and accurate annotation of short texts with Wikipedia pages. CoRR (2010). abs/1006.3498
10. Han, X., Sun, L., Zhao, J.: Collective entity linking in web text: A graph-based method. In: Proceedings of SIGIR 2011, pp. 765–774 (2011)
11. Hasibi, F., Balog, K., Bratsberg, S.E.: Entity linking in queries: tasks and evaluation. In: Proceedings of the ICTIR 2015, pp. 171–180 (2015)

12. Kulkarni, S., Singh, A., Ramakrishnan, G., Chakrabarti, S.: Collective annotation of Wikipedia entities in web text. In: Proceedings of KDD 2009, pp. 457–466 (2009)
13. Medelyan, O., Witten, I.H., Milne, D.: Topic indexing with Wikipedia. In: Proceedings of the AAAI WikiAI Workshop, pp. 19–24 (2008)
14. Meij, E., Balog, K., Odijk, D.: Entity linking and retrieval for semantic search. In: Proceedings of WSDM 2014, pp. 683–684 (2014)
15. Mihalcea, R., Csomai, A.: Wikify!: Linking documents to encyclopedic knowledge. In: Proceedings of CIKM 2007, pp. 233–242 (2007)
16. Milne, D., Witten, I.H.: Learning to link with Wikipedia. In: Proceedings of CIKM 2008, pp. 509–518 (2008)
17. Milne, D., Witten, I.H.: An effective, low-cost measure of semantic relatedness obtained from Wikipedia links. In: Proceedings of AAAI Workshop on Wikipedia and Artificial Intelligence: An Evolving Synergy, pp. 25–30 (2008)
18. Usbeck, R., Röder, M., Ngonga Ngomo, A.-C., Baron, C., Both, A., Brümmer, M., Ceccarelli, D., Cornolti, M., Cherix, D., Eickmann, B., Ferragina, P., Lemke, C., Moro, A., Navigli, R., Piccinno, F., Rizzo, G., Sack, H., Speck, R., Troncy, R., Waitelonis, J., Wesemann, L.: GERBIL: General entity annotator benchmarking framework. In: Proceedings of WWW 2015, pp. 1133–1143 (2015)

Twitter

Correlation Analysis of Reader's Demographics and Tweet Credibility Perception

Shafiza Mohd Shariff[1,2(✉)], Mark Sanderson[1], and Xiuzhen Zhang[1]

[1] School of Computer Science and IT, RMIT University, Melbourne, Australia
{shafiza.mohdshariff,mark.sanderson,xiuzhen.zhang}@rmit.edu.au
[2] Malaysian Institute of IT, Universiti Kuala Lumpur, Kuala Lumpur, Malaysia
shafiza@unikl.edu.my

Abstract. When searching on Twitter, readers have to determine the credibility level of tweets on their own. Previous work has mostly studied how the text content of tweets influences credibility perception. In this paper, we study reader demographics and information credibility perception on Twitter. We find reader's educational background and geolocation have significant correlation with credibility perception. Further investigation reveals that combinations of demographic attributes correlating with credibility perception are insignificant. Despite differences in demographics, readers find features regarding topic keyword and the writing style of a tweet to be independently helpful in perceiving tweets' credibility. While previous studies reported the use of features independently, our result shows that readers use combination of features to help in making credibility perception of tweets.

1 Introduction

Tweets from reliable news sources and trusted authors via known social links are generally trustworthy. However, when Twitter readers *search* for tweets regarding a particular topic, the returned messages require readers to determine the credibility of tweet content. How do readers perceive credibility, and what features (available on Twitter) do they use to help them determine credibility? Since Twitter readers come from all over the world, do demographic attributes influence their credibility perception?

There are several pieces of research regarding the automated detection of tweet credibility using various features, especially for news tweets and rumours [3,7,12,17]. However, these studies focus on building machine learned classifiers and not on the question of how readers perceive credibility. Other research that studies reader's credibility judgments were conducted on web blogs, Internet news media, and websites [5,6,23,24]. Quantitative studies were conducted on limited groups of participants to identify particular factors that influenced readers' credibility judgments. Since these user studies focused on certain factors, the subjects for readers' credibility assessment were controlled and limited.

We have found that there is a gap in understanding Twitter readers and their credibility judgments of news tweets. We aim to understand the features readers

N. Ferro et al. (Eds.): ECIR 2016, LNCS 9626, pp. 453–465, 2016.
DOI: 10.1007/978-3-319-30671-1_33

use when judging, especially when tweets are from authors unfamiliar to them. Therefore in this study, we address the following research questions:

1. Do Twitter readers' demographic profiles correlate with their credibility perception of news tweets?
2. Do the tweet features readers use for their credibility perception correlate with reader's demographic profiles?

To answer the research questions, we design a user study of 1,510 tweets returned by 15 search topics, which are judged by 754 participants. The study explores the correlation between readers' demographic attributes, credibility judgments, and features used to judge tweet credibility. We will focus only on tweet content features as presented by the Twitter platform and available directly to readers.

2 Related Work

A class of existing studies focus on tweet credibility prediction by supervised learning using tweet content and textual features, the tweet author's social network, and the source of retweets. The credibility of newsworthy tweets is determined by human annotators that are then used to predict the credibility of previously unseen tweets [3]. The tweet credibility model presented in [7] were used to rank the tweets by credibility. Both works used a current trending topics dataset. Other studies focused on the utility of individual features for automatically predicting credibility [17] and on the credibility verification of tweets for journalists based on the tweet authors' influence [19].

Another class of research has examined the features influencing readers' credibility perception of tweets. Examining only certain tweet features, Morris et al. [16] studied just under 300 readers from the US. The authors identified that a tweet written by authors with a topically related display name influenced reader credibility perception. Similar research was conducted [23], comparing readers from China and the US. People from different cultural background perceived the credibility of tweets differently in terms of what and how features were used. The differences in tweet credibility perception for different topics was also reported in [20]. The study found eight tweet-content features readers use when judging the credibility level of tweets.

Some research has considered credibility perception in media other than Twitter. In the work by [5], they discovered that different website credibility elements such as interface, expertise and security are influenced by users' demographic attributes. Another study found that the manipulation level of news photos influenced credibility perception of news media [6]. The study showed that people's demographics influenced the perception of media credibility.

A Taiwanese-based study of reader's credibility perception regarding news-related blogs found belief factors can predict user's perceived credibility [24]. They also found that reader's motivation in using news-related blogs as a news source influenced credibility perception. Demographic variables were also shown to affect credibility. In another study [11], demographic attributes are also found to correlate with visual features as information credibility factors for microblogs, especially by younger people.

3 Methodology

We describe the collection of credibility judgments and the techniques that we use to analyze the data.

3.1 Data Collection

Since we are aiming for broad participation in our study, a crowdsourcing platform was used to recruit participants. The use of crowdsourcing for annotating tweet credibility can be found in prior works [3,7,20]. We designed a questionnaire on the Crowdflower[1] platform. We divided the questionnaire into two parts. We first part of the questionnaire regards the basic demographic questions: gender, age, and education level. The country information is supplied to us by CrowdFlower platform as it is part of the workers' information upon their registration on the platform. The workers are regarded as tweet readers in this paper.

The second part of the questionnaire regards perceptions of the credibility of news-related tweets. We compiled tweets from three news categories: breaking news, political news, and natural disaster news, the same categories used in past studies [16,23]. Each news category consists five world news topics reported by news agencies including BBC, Reuters and CNN from 2011 until May 2014. We made sure the news topics were evenly divided between trending and not trending topics. Trends were determined from the trending list on Twitter and What the Trend[2]. The tweets were examined to ensure they were topic relevant tweets and unique (i.e. each tweet contained a different message about the particular topic).

Readers were shown tweets as they would be shown in a Twitter search result page, retrieved in response to a search topic. Workers were also shown the topic and a topic description. Without expanding the tweet to see any additional comments, the readers were asked to give their perception on the credibility level of the tweet. Four levels are listed: very credible, seem credible, not credible, and cannot decide [3,7]. Upon judging, readers were asked to describe what feature/s of the tweet they use to make the judgment. We prompted readers with a list of features reported in previous research [3,20] as well as encouraging them to describe other features in the free-text interface.

In the news tweet collection, two writing styles of tweets are included – a style expressing authors opinion or emotion towards the topic and another just reporting factual information. The writing styles were used after results from a pilot user study, which indicated that readers also find tweets expressing an author's feelings regarding a topic as credible.

[1] http://www.crowdflower.com/.

[2] http://whatthetrend.com/ a HootSuite Media company that lists Twitter's trending topic and explain why it is trending.

To ensure the quality of answers by readers, the readers were required to answer a set of gold questions at a minimum 80 % qualifying level before they were allowed to progress. The gold questions were standard awareness questions, e.g. determining whether a topic and a tweet message were about the same news topic. The gold questions were not counted as part of the user study. A number of pilot studies were run to determine the optimal number of tweet judgments readers were willing to make. Twelve judgments per reader was the figure chosen. The dataset ground truth is available at http://www.xiuzhenzhang.org/downloads/.

3.2 Statistical Analysis Method

The chi-square test of independence is used to establish if two categorical variables have significant correlation. The test calculates the difference between observed data counts and expected data counts. The cutoff acceptance for the relationship is based on the accepted probability value (p-value) of 0.05. The chi-square statistic test can be calculated as follows, where O_i and E_i are the observed value and expected value for cell i of the contingency table:

$$\chi^2 = \sum_i \frac{(O_i - E_i)^2}{E_i} \tag{1}$$

In this study, in addition to correlation analysis regarding a single demographic attribute and credibility judgments, we also aim to analyze how combinations of demographic attributes correlate with credibility judgments. Therefore, multi-way chi-square tests are also performed. Let V_1, ... , and V_k be k binary variables, the contingency table to calculate the χ^2 for these k binary variables is $(V_1, \bar{V_1}) \times (V_2, \bar{V_2}) \times ... \times (V_k, \bar{V_k})$. For example, when there are three binary variables A, B and C, to find out if variable A and B are correlated with variable C, the χ^2-statistic would be $\chi^2(\text{ABC}) + \chi^2(\text{AB}\bar{\text{C}})$ [1]. Note that the chi-square statistic is upward-closed, this means that the χ^2 value of ABC would always be greater than the χ^2 value of AB. Therefore, if AB is correlated, adding in variable C, ABC must also be correlated. Refer to [1] for proof of the theorem.

In our problem setting, we apply the theorem to prevent false discoveries for multi-way chi-square analysis. Assuming that A and B are independent variables for demographic attributes and C is the dependent variable for credibility levels. If A and B are correlated, even if A, B, and C are correlated, we would not be able to tell if the association between credibility level and the demographic attributes is due to an actual effect or to the non-independence of observations.

We first apply chi-square analysis between individual demographic attributes and the credibility judgments. If the result is insignificant, multi-way correlation analysis for combination of demographic attributes will be applied. To this end, the correlation for pairwise demographic attributes is first analyzed. If the attributes are significantly correlated, we will not continue the χ^2 test between the pair and credibility judgments. We similarly analyze the correlation between demographic attributes and features readers use for credibility judgments.

We also measure which cell in the contingency table influences the χ^2 value. The interest or dependence of a cell (c) is defined as I(c)= Oc/Ec. The further away the value is from 1, the higher influence it has on the χ^2 value. Positive dependence is when the interest value is greater than 1, and a negative dependence is those lower than 1 [1].

3.3 Slicing Reader Demographics

In this study the demographic data collected from the readers are used for chi-square analysis, refer to Table 1. The readers' demographic data, except for gender, are also categorized in binary and categorical setting based on other research [5,6] to examine any correlation of demographic attributes or combinations of demographic attributes with tweet credibility perception. The different ways of partitioning demographic data are as follows:

- Age: Binary {Young adult ($\leqslant$ 39 years old), Older adult ($\geqslant$ 40 years old)} and Categorical {Boomers (51–69 years old), Gen X (36–50 years old), Gen Y(21–35 years old), Gen Z (6–20 years old)} [14]
- Education: Binary {Below university level, University level} and Categorical {School level, Some college, Undergraduate, Postgraduate}
- Location: Binary {Eastern hemisphere, Western hemisphere} and Categorical {Asia-Pacific, Americas, Europe, Africa}

We conduct the correlation analysis for each single demographic attribute for all the different slicing with credibility judgments or features.

4 Results

A total of 10,571 credibility judgments for 1,510 news tweets were collected from the user study. Only 9,828 judgments from 819 crowdsource workers were accepted for this study because only those workers answered the demographic questions and completed all 12 judgments. For any credibility judgments that were found to not describe the features used to make the credibility judgment or gave nonsensical comments, all judgments of the reader were discarded. We also discarded judgments of two readers from Oceania continent and three readers that did not have any education background, due to their low values undermine the required minimal expected frequency to apply χ^2 analysis. We were left with a final dataset for analysis from 754 readers with 9,048 judgments.

4.1 Overall Demographics

Our final collection of data includes readers from 76 countries with the highest number of participants coming from India (15 %). We then group the countries into continents due to the countries' sparsity. Out of the 754 readers, the majority (69.0 %, n=521) of readers were male, similar to prior work that uses crowdsource workers for user study [11]. Most of the readers were in the age group of 20–29 years old (43.4 %, n=327). In regards to the readers' education background, the majority had a University degree (38.1 %, n=287). Table 1 shows the readers demographic profiles.

Table 1. Demographic profiles distribution

Demographic	Value	Frequency	%
Gender	Male	521	69.2
	Female	233	30.8
Age	16–19 years old	58	7.7
	20–29 years old	327	43.4
	30–39 years old	243	32.2
	40–49 years old	89	11.8
	50 years and older	37	4.9
Education	High school	127	16.8
	Technical training	58	7.7
	Diploma	81	10.7
	Professional certification	50	6.6
	Bachelor's degree	287	38.1
	Master's degree	137	18.2
	Doctorate degree	14	1.9
Location	Asia	275	36.5
	Europe	247	32.8
	South America	130	17.2
	North America	65	8.6
	Africa	37	4.9

4.2 Features

The features reported by readers are features of the tweet message itself, content-based and source-based. For features reported in free text, we applied a summative content analysis based on the list of features identified beforehand [9]. Table 2 (column 2) lists the features reported by readers when making their credibility judgments. Since the features are sparse, it is difficult to analyze their influence in the readers' credibility judgment. Therefore, we categorize the features into five categories and will be using the feature categories in all of our analysis related to the features:

- **Author**: features regarding the person who posted a tweet, including the Twitter ID, display name, and the avatar image;

- **Transmission**: features in a tweet message for broadcasting the messages on Twitter;
- **Auxiliary**: auxiliary information external to the textual message, including URL links, pictures, or videos;
- **Topic**: words and phrases indicating the search topic or news type, including search keywords and alert phrases such as "breaking news";
- **Style**: writing style of a tweet, including language style as well as message style as expressing opinion or stating facts.

Table 2. Features reported by readers to judge credibility for news tweets

Category	Feature	Description
Author	Tweet author	Twitter ID or display name e.g. `Sydneynewsnow`
Transmission	User mention	Other Twitter user's Twitter ID mentioned in the tweet starting with the @ symbol e.g. `@thestormreports`
	Hashtag	The # symbol used to categorise keywords in a tweet e.g. `#Pray4Boston`
	Retweet	Contain the letters RT (retweet) in the tweet and the retweet count
Auxiliary	Link	Link to outside source - URLs, URL shortener
	Media	Picture or video from other sources embedded within the tweet
Topic	Alert phrase	Phrase that indicate new or information update regarding a news topic - e.g. `Update`
	Topic keyword	The search keyword regarding a news topic e.g. `Hurricane Sandy`
Style	Language	The language construction of the tweet (formal or informal English)
	Author's opinion	Tweet that conveys the author's emotion or feeling towards the news topic
	Fact	Factual information on the tweet regarding the news topic

4.3 Findings

We report our findings based on the research questions.

RQ1: Do Twitter reader's demographic profiles contribute to the credibility perception of news tweets?

The correlation analysis for individual demographic attributes for each data setting (as described in Subsect. 3.2): Original (O), Binary (B), Categorical (C), and the credibility perceptions is shown in Table 3. At the original data setting, Education and Location are significantly correlated with credibility judgment, $\chi^2 = 49.43$, p<0.05 and $\chi^2 = 80.79$, p<0.05. Only Location is significantly correlated at all levels of partitioning. A post hoc analysis on the interest value of cells in the contingency table *Education* × *Credibility* for the original data found the cell that contributes most to the χ^2 value is readers with a 'Professional certification', who commonly gave 'not credible' judgments. In regards to the contingency table *Location* × *Credibility*, we found there was a correlation between the readers from the African continent and the 'cannot decide' credibility perception in the original and the categorical data setting with a positive dependence. Both cells interest values are far from 1, indicating strong dependence. In the contingency table for *Location* × *Credibility* in the binary data setting, the interest value in each cell is close to 1, therefore there are no strong dependence.

Table 3. Demographic profiles and credibility perception chi-square results

Demographic	Data setting	Credibility
Gender	Original	1.51
	Binary	1.51
	Categorical	1.51
Age	Original	4.87
	Binary	4.68
	Categorical	9.84
Education	Original	**49.43**
	Binary	4.78
	Categorical	12.29
Location	Original	**80.79**
	Binary	**39.62**
	Categorical	**80.33**

We then conduct multi-way correlation analysis between combinations of demographic attributes and credibility judgments. Since Location is significantly correlated at all data levels, due to the upward closeness of χ^2 statistics (Sect. 3.1), we will not analyze combinations including Location. The correlation result for the rest demographic attribute pairs is shown in Table 4. In analyzing the combination of demographic attributes, Bonferroni corrections of the *p-values* ($p < 0.003$) are applied. Table 4(b) shows that only for the binary setting the (Age, Education) pair is not significantly correlated. Therefore, we further analyze the correlation of the (Age, Education) pair with credibility judgments.

The correlation analysis outcome for *Age* × *Education* × *Credibility* is $\chi^2 = 3.70$, $p > 0.003$, accepting the null hypothesis. The result indicates that the joint independent demographic attributes of Age and Education in the binary setting do not correlate with the credibility judgments.

Table 4. Chi-square result for demographic attribute pairwise correlation

(a) (Age, Gender) & (Education, Gender)

		Gender
Age	O	**107.71**
	B	**77.40**
	C	**82.18**
Education	O	**105.89**
	B	**48.67**
	C	**61.80**

(b) Age, Education

	Education		
Age	O	B	C
O	**1791.23**	**763.96**	**1579.96**
B	**105.89**	2.18	**47.96**
C	**1732.96**	**749.53**	**1549.49**

RQ2: Do the tweet features readers use for their credibility perception of tweets correlate with reader's demographic profiles?

To answer this research question, we analyze the correlation between reader demographic attributes and the features readers reported for credibility judgments. From Table 5, all demographic attributes are significantly correlated with credibility perception features reported by readers. In the last column of Table 5, for the analysis of demographic attributes and the Transmission feature, as over 20 % of expected values of the contingency table have expected value of less than 5, Fisher's Exact Test is used [15]. Table 5 is based on demographic data at the original setting, and similar results are obtained for data at binary and categorical settings. As all demographic attributes are correlated with credibility perception features, due to the upward closeness of chi-square statistics, any combination of demographic attributes is also correlated with credibility perception features.

Topic and Style features have the most significant correlation with the demographic profiles while the Transmission feature has the least significant correlation with demographic attributes. *Age* and *Location* are significantly correlated with Author, and *Education* and *Location* are correlated with Auxiliary features. Meanwhile, only *Education* has significant correlation with Transmission. We are curious to know if there are combination of features readers reported to use when perceiving the credibility level of tweets. Using association mining to find the frequent combination of features [8], we found that Transmission, Author, and Auxiliary are frequently used with other features. Table 6 shows the frequent features that meet the support threshold of 1 % or 90 times. The support threshold refers to the feature/s frequency of occurrence in the dataset. A low support threshold would help to eliminate uninteresting patterns [22].

Table 5. The chi square correlation between demographics and features used in credibility perception

Demographic	Feature Categories				
	Author	Topic	Style	Auxiliary	Transmission
Gender	0.01	**18.15**	**23.27**	1.59	0.59[a]
Age	**16.63**	**26.65**	**41.99**	8.65	1.00[a]
Education	11.12	**31.87**	**50.12**	**16.53**	**0.03**[a]
Location	**46.87**	**83.81**	**67.35**	**13.60**	1.00[a]

[a] Calculated using Fisher's Exact Test

Table 6. Frequent pattern mining of feature category

Frequent patterns	Support (%)
Topic	14.1
Style	12.7
Topic, Style	6.1
Auxiliary, Style	5.2
Auxiliary, Topic	4.7
Auxiliary, Topic, Style	4.6
Auxiliary, Topic, Style, Transmission	3.7
Auxiliary, Topic, Transmission	2.7
Author	2.7
Author, Auxiliary, Topic, Style, Transmission	2.6
Topic, Style, Transmission	2.5
Style, Transmission	2.0
Auxiliary, Style, Transmission	1.9
Author, Topic, Style	1.8
Author, Style	1.8
Topic, Transmission	1.6

5 Discussion

In regards to our first research question, readers' education background and their geo-location have significant correlation with credibility judgments. This finding is different from [6,11,24], as these studies do not find any correlation between tweet credibility perception and the education background. From our analysis, readers with a 'Professional certificate' and who judge tweets as 'not credible' are the ones that contribute to the significant χ^2 result. It is likely that education background may be connected with experience and thus such readers are more careful in making credibility judgments. Another possible reason may be the absence or a low number of higher education level participants in past studies.

Although other researchers also found location correlated with credibility judgments, our dataset of international readership shows that readers from Africa have positive dependence with the 'cannot decide' credibility judgment. The political conflicts in countries on the Africa continent may have influenced the skeptical attitude towards media by the readers [4]. Therefore, tweets that readers find ambiguous resulted in their indecisive judgments on the tweet credibility judgements [18]. Other demographic attributes Age and Gender are not correlated with tweet credibility perception, which is a result similar to the work by [2]. Moreover the combination of Age and Gender does not have any significant correlation with tweet credibility perception either.

For the second research question, we find that all demographic attributes are significantly correlated with credibility perception features reported by readers. Especially the Topic features, including topic keyword and news alert phrase, and the tweet writing Style are important features used by readers for credibility perception. More than 26 % of credibility judgments rely on Topic and Style features.

Features that are used in broadcasting tweets, the Auxiliary feature and Author feature, seems to be not considered by readers when judging the tweets' credibility level. Our results show a perspective different from that in [3, 10, 13, 21]. We also find that Auxiliary and Author features are mostly combined with other feature categories when readers make credibility judgements of news tweets, a result that was missing in other works since they are previously studied separately.

6 Conclusion

Although research on Twitter information credibility has been reported, most work focuses on automatic predicting or detecting tweet credibility. Our focus is on understanding Twitter readers and what influences their credibility judgments. In this study, we provided new insights in the correlation of reader demographic attributes with credibility judgments of tweets and the features readers used to make those judgments. Furthermore, the richness of data collected for this study – derived from a wide range of demographic profiles and readers across countries – is the first to offer insights on Twitter reader's direct perception of credibility and the features readers use for credibility judgements. For future work, we plan to examine if the type of news tweets has any influence on a reader's credibility perception. We would also like to investigate deeper on the features readers use, and the type of credibility level relates to those features and news type.

Acknowledgment. This research is partially supported by Universiti Kuala Lumpur (UniKL), Majlis Amanah Rakyat (MARA), and by the ARC Discovery Project DP140102655.

References

1. Brin, S., Motwani, R., Silverstein, C.: Beyond market baskets: generalizing association rules to correlations. ACM SIGMOD Rec. **26**(2), 265–276 (1997)
2. Cassidy, W.P.: Online news credibility: an examination of the perceptions of newspaper journalists. J. Comput. Mediated Commun. **12**(2), 478–498 (2007)
3. Castillo, C., Mendoza, M., Poblete, B.: Information credibility on twitter. In: WWW 2011, pp. 675–684. ACM (2011)
4. Cozzens, M.D., Contractor, N.S.: The effect of conflicting information on media skepticism. Commun. Res. **14**(4), 437–451 (1987)
5. Fogg, B., Marshall, J., Laraki, O., Osipovich, A., Varma, C., Fang, N., Paul, J., Rangnekar, A., Shon, J., Swani, P., et al.: What makes web sites credible?: a report on a large quantitative study. In: SIGCHI 2001, pp. 61–68. ACM (2001)
6. Greer, J.D., Gosen, J.D.: How much is too much? assessing levels of digital alteration of factors in public perception of news media credibility. Vis. Commun. Q. **9**(3), 4–13 (2002)
7. Gupta, A., Kumaraguru, P.: Credibility ranking of tweets during high impact events. In: PSOSM 2012, pp. 2–8. ACM (2012)
8. Hahsler, M., Grün, B., Hornik, K.: Introduction to arules: mining association rules and frequent item sets. SIGKDD Explor. **2**, 4 (2007)
9. Hsieh, H.F., Shannon, S.E.: Three approaches to qualitative content analysis. Qual. Health Res. **15**(9), 1277–1288 (2005)
10. Hu, M., Liu, S., Wei, F., Wu, Y., Stasko, J., Ma, K.L.: Breaking news on twitter. In: SIGCHI 2012, pp. 2751–2754. ACM (2012)
11. Kang, B., Höllerer, T., O'Donovan, J.: Believe it or not? analyzing information credibility in microblogs. In: ASONAM 2015, pp. 611–616. ACM (2015)
12. Kang, B., O'Donovan, J., Höllerer, T.: Modeling topic specific credibility on twitter. In: IUI 2012, pp. 179–188. ACM (2012)
13. Liu, Z.: Perceptions of credibility of scholarly information on the web. Inf. Process. Manage. **40**(6), 1027–1038 (2004)
14. McCrindle, M., Wolfinger, E.: The ABC of XYZ: Understanding the Global Generations. University of New South Wales Press, Sydney (2009)
15. McDonald, J.H.: Handbook of Biological Statistics, vol. 3. Sparky House Publishing, Baltimore (2014)
16. Morris, M.R., Counts, S., Roseway, A., Hoff, A., Schwarz, J.: Tweeting is believing?: understanding microblog credibility perceptions. In: Proceedings of the ACM 2012 Conference on Computer Supported Cooperative Work, pp. 441–450. ACM (2012)
17. O'Donovan, J., Kang, B., Meyer, G., Höllerer, T., Adalii, S.: Credibility in context: an analysis of feature distributions in twitter. In: 2012 International Conference on Privacy, Security, Risk and Trust (PASSAT), and 2012 International Confernece on Social Computing (SocialCom), pp. 293–301. IEEE (2012)
18. Rassin, E., Muris, P.: Indecisiveness and the interpretation of ambiguous situations. Pers. Individ. Differ. **39**(7), 1285–1291 (2005)
19. Schifferes, S., Newman, N., Thurman, N., Corney, D., Göker, A., Martin, C.: Identifying and verifying news through social media: developing a user-centred tool for professional journalists. Digit. J. **2**(3), 406–418 (2014)
20. Mohd Shariff, S., Zhang, X., Sanderson, M.: User perception of information credibility of news on twitter. In: de Rijke, M., Kenter, T., de Vries, A.P., Zhai, C.X., de Jong, F., Radinsky, K., Hofmann, K. (eds.) ECIR 2014. LNCS, vol. 8416, pp. 513–518. Springer, Heidelberg (2014)

21. Sundar, S.S.: Effect of source attribution on perception of online news stories. J. Mass Commun. Q. **75**(1), 55–68 (1998)
22. Tan, P.N., Kumar, V.: Association analysis: basic concepts and algorithms. In: Introduction to Data Mining, Chap. 6. Addison-Wesley (2005)
23. Yang, J., Counts, S., Morris, M.R., Hoff, A.: Microblog credibility perceptions: comparing the usa and china. In: Proceedings of the 2013 Conference on Computer Supported Cooperative Work, pp. 575–586. ACM (2013)
24. Yang, K.C.C.: Factors influencing internet users perceived credibility of news-related blogs in taiwan. Telematics Inform. **24**(2), 69–85 (2007)

Topic-Specific Stylistic Variations for Opinion Retrieval on Twitter

Anastasia Giachanou[1(✉)], Morgan Harvey[2], and Fabio Crestani[1]

[1] Faculty of Informatics, Università della Svizzera italiana (USI), Lugano, Switzerland
{anastasia.giachanou,fabio.crestani}@usi.ch

[2] Department of Maths and Information Sciences, Northumbria University, Newcastle upon Tyne, UK
morgan.harvey@northumbria.ac.uk

Abstract. Twitter has emerged as a popular platform for sharing information and expressing opinions. Twitter opinion retrieval is now recognized as a powerful tool for finding people's attitudes on different topics. However, the vast amount of data and the informal language of tweets make opinion retrieval on Twitter very challenging. In this paper, we propose to leverage topic-specific stylistic variations to retrieve tweets that are both relevant and opinionated about a particular topic. Experimental results show that integrating topic specific textual meta-communications, such as emoticons and emphatic lengthening in a ranking function can significantly improve opinion retrieval performance on Twitter.

Keywords: Opinion retrieval · Microblogs · Stylistic variations

1 Introduction

Microblogs have emerged as a popular platform for sharing information and expressing opinion. Twitter attracts 284 million active users per month who post about 500 million messages every day[1]. Due to its increasing popularity, Twitter has emerged as a vast repository of information and opinion on various topics. However, all this opinionated information is hidden within a vast amount of data and it is therefore impossible for a person to look through all data and extract useful information.

Twitter opinion retrieval aims to identify tweets that are both relevant to a user's query and express opinion about it. Twitter opinion retrieval can be used as a tool to understand public opinion about a specific topic, which is helpful for a variety of applications. One typical example refers to enterprises that can capture the views of customers about their product or their competitors. This information can be then used to improve the quality of their services or products accordingly. In addition, it is possible for the government to understand the public view regarding different social issues and act promptly.

[1] See: https://about.twitter.com/company/.

N. Ferro et al. (Eds.): ECIR 2016, LNCS 9626, pp. 466–478, 2016.
DOI: 10.1007/978-3-319-30671-1_34

Retrieving tweets that are opinionated about a specific topic is a non-trivial task. One of the many reasons is the informal nature of the medium, which has effected the emergence of new stylistic conventions such as emoticons, emphatic lengthening and slang terms widely used on Twitter. These informal stylistic conventions can, however, be a valuable source of information when retrieving tweets that express opinion towards a topic. The use of emoticons usually implies an opinion [8] and emphatic lengthening has been shown to be strongly associated with opinionatedness [2]. For the rest of the paper, we use the phrases *stylistic conventions* and *stylistic variations* interchangeably to denote the emerged textual conventions in Twitter such as the emoticons and the emphatic lengthening. The stylistic conventions are only a subset of the writing style of users in Twitter. Writing style refers to a much wider manner that is used in writing [19].

The extent to which stylistic variations are used varies considerably among the different topics discussed on Twitter. That is, the number of the stylistic variations present in each tweet is dependent on its topic. For example, tweets about entertainment topics (i.e. movies, TV series) tend to use more stylistic variations than those that express opinion about social issues (i.e. immigration) or products (i.e. Google glass). This implies that stylistic variations do not have the same importance in revealing opinion across different topics.

Here we propose a Twitter opinion retrieval model which uses information about the topics of tweets to retrieve those that are relevant and contain opinion about a user's query. The proposed model calculates opinionatedness by combining information from the tweet's terms and the topic-specific stylistic variations that are extensively used in Twitter. We compare several combinations of stylistic variations, including emoticons, emphatic lengthening, exclamation marks and opinionated hashtags, and evaluate the proposed model on the opinion retrieval dataset proposed by Luo et al. [10]. Results show that stylistic variations are topic-specific and that incorporating them in the ranking function significantly improves the performance of opinion retrieval on Twitter.

2 Related Work

With the rapid growth of social media platforms, sentiment analysis and opinion retrieval has attracted much attention in the research community [14,15]. Early research focused on classifying documents as expressing either a positive or a negative opinion. The relevance of an opinionated document towards a topic was first considered by Yi et al. [23], while Eguchi and Lavrenko [3] were the first to consider ranking documents according to the opinion they contain about a topic. A comprehensive review of opinion retrieval and sentiment analysis can be found in a survey by Pang and Lee [15].

The increasing popularity of Twitter has recently stirred up research in the field of Twitter sentiment analysis. One of the first studies was carried out by Go et al. [4] treated the problem as one of binary classification, classifying tweets as either positive or negative. Due to the difficulty of manually tagging the sentiment of the tweets, they employed distant supervision to train a supervised machine learning classifier. The authors used a technique devised by Read [18] to collect the data, according to which emoticons can be used to differentiate the negative and positive tweets. They compared Naive Bayes (NB), Maximum Entropy (MaxEnt) and Support Vector Machines (SVM), among which SVM with unigrams achieved the best result. Following Go et al. [4], Pak and Paroubek [12] used emoticons to label training data from which they built a multinomial Naïve Bayes classifier which used N-gram and POS-tags as features.

Due to the informal language used on Twitter, which frequently contains unique stylistic features, a number of researchers explored features such as emoticons, abbreviations and emphatic lengthening, studying their impact on sentiment analysis. Brody and Diakopoulos [2] showed that the lengthening of words (e.g., cooool) in microblogs is strongly associated with subjectivity and sentiment. Kouloumpis et al. [8] showed that Twitter-specific features such as the presence or absence of abbreviations and emoticons improve sentiment analysis performance. None of these approaches considered, however, the possibility that stylistic features may depend on the topic of the tweet.

Topic-dependent approaches have been considered by researchers in relation to terms. Jiang et al. [7] used manually-defined rules to detect the syntactic patterns that showed if a term was related to a specific object. They employed a binary SVM to apply subjectivity and polarity classification and utilised microblog-specific features to create a graph which reflects the similarities of tweets. Canneyt et al. [20] introduced a topic-specific classifier to effectively detect the tweets that express negative sentiment whereas Wang et al. [21] leveraged the co-occurrence of hashtags to detect their sentiment polarity.

Twitter opinion retrieval was first considered by Luo et al. [10] who proposed a learning-to-rank algorithm for ranking tweets based on their relevance and opinionatedness towards a topic. They used SVM^{Rank} to compare different social and opinionatedness features and showed they can improve the performance of Twitter opinion retrieval. However, this improvement is over relevance baselines (BM25 and VSM retrieval models) and not over an opinion baseline. Our work is different as we propose to incorporate topic-specific stylistic variations into a ranking function to generate an opinion score for a tweet. To the best of our knowledge, there is no work exploring the importance of topic-specific stylistic variations for Twitter opinion retrieval. Another important difference is that we use both relevance and opinion baselines to compare the proposed topic-specific stylistic opinion retrieval method.

3 Topic Classification

Topic models aim to identify text patterns in document content. Standard topic models include Latent Dirichlet Allocation (LDA) [1] and Probabilistic Latent Semantic Indexing (pLSI) [5]. LDA, one of the most well known topic models, is a generative document model which uses a "bag of words" approach and treats each document as a vector of word counts. Each document is a mixture of topics and is represented by a multinomial distribution over those topics. More formally, each document d in the collection is associated with a multinomial distribution over K topics, denoted θ. Each topic z is associated with a multinomial distribution over words, denoted ϕ. Both θ and ϕ have Dirichlet prior with hyperparameters α and β respectively. For each word in a document d, a topic z is sampled from the multinomial distribution θ associated with the document and a word w from the multinomial distribution ϕ associated with topic z. This generative process is repeated N_d times, where N_d is the total number of words in the document d. LDA defines the following process for each document in the collection:

1. Choose $\theta_d \sim \text{Dir}(\alpha)$,
2. Choose $\phi_z \sim \text{Dir}(\beta)$,
3. For each of the N words w_n:

   (a) Pick a topic z_n from the multinomial distribution θ_d
   (b) Pick a word w_n from the multinomial distribution ϕ_z.

Topic models have been applied in a wide range of areas including Twitter. Hong and Davison [6] conducted an empirical study to investigate the best way to train models for topic modeling on Twitter. They showed that topic models learned from aggregated messages of the same user may lead to superior performance in classification problems. Zhao et al. [24] proposed the Twitter-LDA model which considered the shortness of tweets to compare topics discussed in Twitter with those in traditional media. Their results showed that Twitter-LDA works better than LDA in terms of semantic coherence. Ramage et al. [17] applied labeled-LDA in Twitter, a partially supervised learning model based on hashtags. Inspired by the popularity of LDA, Krestel et al. [9] proposed using LDA for tag recommendation. Based on the intuition that tags and words are generated from the same set of latent topics, they used the distributions of latent topics to represent tags and descriptions and to recommend tags.

In this work, we use LDA [1] to determine the topics of tweets, which are then used to learn the importance of the stylistic variations for each topic.

4 Twitter Opinion Retrieval

Twitter opinion retrieval aims to develop an effective retrieval function which retrieves and ranks tweets accordingly to the likelihood that express an opinion about a particular topic. The proposed approaches for opinion retrieval usually follow a three step framework. In the first step, traditional IR methods are applied to rank documents by their relevance to the query. In the second, opinion scores are generated for the documents that were retrieved during the first step and, in the last step, a final ranking of the documents is produced based both on their relevance and opinionatedness towards the query.

In this section, we propose a new opinion retrieval model which leverages topic-specific stylistic variations of short informal texts such as tweets to calculate their opinionatedness. The proposed model calculates the opinionatedness of a document by combining two different opinion scores. The *term-based* component is based on the opinionatedness of the document's terms, whereas the *stylistic-based* component instead considers the stylistic variations present in the document.

Let $S_d(o)$ be the opinion score of a document (tweet) d based on its terms and $S_{ls,d}(o)$ be the opinion score of a document d based on the stylistic variations that d contains. Then the opinionatedness of the document d is the weighted sum of the two opinion score components and is calculated as follows:

$$S_{q,d}(o) = \lambda * S_d(o) + (1 - \lambda) * S_{ls,d}(o)$$

where $\lambda \in [0, 1]$.

Term-Based Opinion Score. The presence of opinionated terms in a document, and their probability of expressing opinion, is a popular approach to calculate the document's opinionatedness. A simple method is to calculate this score as the average opinion score over all terms in the document, thus:

$$S_d(o) = \sum_{t \in d} opinion(t) p(t|d) \tag{1}$$

where $p(t|d) = c(t, d)/|d|$ is the relative frequency of term t in document d and $opinion(t)$ shows the opinionatedness of the term.

Since this is one of the most widely used methods to calculate the opinionatedness of a document, we also use this method as one of our baselines.

Stylistic-Based Opinion Score. Our method incorporates several stylistic variations of tweets into a ranking function to rank tweets according to their opinionatedness. The stylistic-based component of our model calculates an opinion score using the stylistic variations that a document contains. Let l be a stylistic variation taken from the list $L = (l_1, ..., l_i, ..., l_{|l|})$ which includes all the possible stylistic variations that reveal opinions. We then calculate the stylistic-based component as follows:

$$S_{ls,d}(o) = \sum_{l \in LS} SVF(l, d) * IDF(l)$$

where LS is a subset of stylistic variations ($LS \subset L$), $SVF(l, d)$ represents the frequency of the stylistic variation l in the document d and $IDF(l)$ represents the importance of the variation l, that is if the stylistic variation is common across the documents or not. The inverse frequency IDF of the stylistic variation l controls the amount of opinion information that the specific variation holds.

We explore various ways of calculating the frequency SVF of the stylistic variations. These are the following:

$$SVF_{Bool}(l,d) = \begin{cases} 0, & \text{if } f(l,d) = 0 \\ 1, & \text{if } f(l,d) > 0 \end{cases}$$
$$SVF_{Freq}(l,d) = f(l,d)$$
$$SVF_{Log}(l,d) = 1 + \log f(l,d)$$

where $f(l,d)$ is the number of occurrences of variation l in document d.

To model the relative importance of each stylistic variation l across the documents we consider the following methods:

$$IDF_{Inv}(l) = \log \frac{N}{1+n_l} \tag{2}$$

$$IDF_{Prob}(l) = \begin{cases} 0, & \text{if } N = n_l \\ \log \frac{N-n_l}{n_l}, & \text{if } N \neq n_l \end{cases} \tag{3}$$

where n_l can also be written as $|d \in D : l \in d|$ and denotes the number of documents that belong to collection D and contain the stylistic variation l.

Thus, the importance of a given stylistic variation l depends on how frequently it is used in the collection D.

Topic Specific Stylistic-Based Opinion Score. The assumption made in the existing literature, that the stylistic variations are used with the same frequency across the documents of different topics, is not accurate. The informal stylistic variations are used with differing frequencies depending on the topic discussed. For example, tweets that are relevant to a TV series probably contain more stylistic variations than those that are relevant to a social issue, such as immigration. That means that the probability that stylistic variations imply opinion depends on the topic of the tweet. In other words, if emoticons are extensively used in tweets about a specific topic, then their ability to imply opinion decreases.

Based on this assumption, we propose using topic-specific stylistic variations. To this end, we first apply topic modeling to determine the topic of a tweet and then we use this information to calculate the stylistic-based component of our approach, that is the opinionatedness of a tweet when it contains a specific stylistic variation. More formally, let $T = (T_1, ..., T_i, ..., T_{|T|})$ be the topics extracted after applying a topic model on the collection D, and $D_T = (d_1, ..., d_t)$ the documents that were assigned to the topic T_i. Then, the relative importance IDF of each stylistic variation l is calculated using the Eqs. 2 and 3 with the difference that n_l denotes the number of documents that belong to collection D_T and contain the stylistic variation l. In other words, n_l is calculated as $|d \in D_T : l \in d|$, where D_T is a collection of documents that were assigned the same topic T_i.

Combining Relevance and Opinion Scores. To generate the final ranking of documents according to their relevance and opinionatedness, we combine the relevance score with the opinionatedness of the tweet:

$$S_{o,q}(d) = S_d(q) * S_{q,d}(o)$$

where $S_d(q)$ is the relevance score of d given topic t and $S_{q,d}(o)$ is the opinionatedness of d. $S_d(q)$ can be estimated using any existing IR model.

5 Experimental Setup

Dataset. To evaluate our methods we used the dataset proposed by Luo et al. [10], which is, to the best of our knowledge, the only dataset that has been used for Twitter opinion retrieval. The original collection contains 50 topics and 5000 judged tweets crawled in November 2011. We note that there is another dataset which can be used for opinion retrieval in Twitter. This dataset was created by Paltoglou and Buckley [13] who annotated part of the Microblog dataset provided by TREC with subjectivity annotations. However, as this dataset has not yet been used in any study, we would not be able to make direct comparisons of our methods and therefore only consider the first.

Experimental Settings. Preprocessing was performed on the dataset. To create the index, we removed URLs, hashtag symbols (#) placed in front of some terms and character repetitions that appear more than twice in a row in a term. We indexed the collection with the Terrier IR system[2]. Our preprocessing also involves stop-word removal using the snowball stop word list[3] and stemming using the Porter stemmer [16].

To avoid overfitting the data we performed 5 fold cross-validation on the 50 queries. For each fold we used 40 queries for the training phase and 10 for testing. The training and test data was kept separate in all phases of our experiments. We perform our experiments under two different settings: *non topic-based* and *topic-based*. For the non topic-based settings, we apply the proposed method on the whole collection without considering the tweet's topic. For the topic-based settings we first apply LDA to detect the topics and then we apply the proposed method on tweets of the same topic. To estimate the LDA parameters we used a Gibbs sampler. Since the Gibbs sampler is a stochastic method, and therefore will produce different outputs by run, we report the mean performance of the methods based on ten runs.

Opinion Lexicon and Stylistic Variations. To identify the opinionated terms we use the AFINN Lexicon, as proposed by Nielsen [11]. AFINN contains more than 2000 words, each of which is assigned a valence from −5 to −1

[2] Available at: http://terrier.org/.

[3] Available at: http://snowball.tartarus.org/.

for terms with a negative sentiment or from 1 to 5 for terms with a positive sentiment. We chose this lexicon as it contains affective words that are used in Twitter. We took the absolute values of the scores since we do not consider sentiment polarity in our study. We use *MinMax* normalisation to convert the valence score of a term to opinion score. To avoid getting zero scores for terms with absolute score 1, we consider that the lexicon has also one term with no sentiment (assigned the score 0), so that 0 is the minimum score.

To calculate the stylistic-based component of our model, we identified, for each tweet, the number of emoticons, exclamation marks, terms under emphatic lengthening and opinionated hashtags as follows:

- *Emoticons*: Number of emoticons in a tweet. For the emoticons, we used the list provided on Wikipedia[4]. We consider all emoticons to be opinion-bearing. Therefore, we did not distinguish them by their subjectivity, sentiment or emotion they express.
- *Exclamation marks*: Number of exclamation marks in a tweet.
- *Emphatic lengthening*: Number of terms under emphatic lengthening in a tweet, that is terms that contain more than two repeated letters.
- *Opinionated hashtags*: Number of opinionated hashtags. As opinionated hashtags we considered any hashtag whose term is contained in the AFINN opinion lexicon. For example, the hashtag *#love* is considered an opinionated hashtag because the term *love* appears in the AFINN opinion lexicon.

Evaluation. We compare the proposed opinion retrieval method with two baselines. The first, *BM25*, is the method with the best performance in Twitter opinion retrieval according to the results presented in [10]. The *Relevance-Baseline* is based purely on topical relevance and does not consider opinion. As a second baseline, we use the term-based opinion score (Eq. 1). The *Opinion-Baseline* considers opinion and therefore it is a more appropriate baseline to compare our results with. To evaluate the methods, we report *Mean Average Precision* (MAP), which is the only metric reported in previous work [10] in Twitter opinion retrieval. To compare the different methods we used the Wilcoxon signed ranked matched pairs test with a confidence level of 0.05.

6 Results and Discussion

Topic Classification. In order to identify the topics discussed, we applied the LDA [1] topic model on the dataset proposed by Luo et al. [10]. For the analysis, we applied Gibbs sampling for the LDA model parameter estimation and inference as proposed in [22]. We considered each tweet to be a document. We tried a number of different values for the K parameter, which represents the number of topics, ranging from 1 to 200 with a step of 5. We set the number of iterations to 2000. The minimum log likelihood is obtained for 65 topics.

[4] See http://en.wikipedia.org/wiki/List_of_emoticons.

Table 1. Sample of topic descriptions when the number of topics is set to 65

Sample topics from Twitter
jennifer aniston lopez brad
steve jobs apple biography
disney world walt princess
music awards red carpet
biology chemistry science lab

Table 1 shows a list of five topics which were discovered in the collection of tweets when the number of topics was set to 65. We observe that LDA managed to group terms that are about the same topic together.

Twitter Opinion Retrieval. Table 2 presents the results of Twitter opinion retrieval when different stylistic variations are combined. Any of the approaches of calculating SVF and IDF presented in Sect. 4 can be used to evaluate the effectiveness of the different combinations. For the results displayed in Table 2 we applied SVF_{Log} and IDF_{Inv} under topic-based settings. We observe that all the three examined combinations (SVF$_{Log}$IDF$_{Inv}$-Emot-Excl, SVF$_{Log}$IDF$_{Inv}$-Emot-Excl-Emph, SVF$_{Log}$IDF$_{Inv}$-Emot-Excl-Emph-OpHash) perform significantly better than both the relevance and opinion baselines. Though there is no statistical difference between the different combinations of the stylistic variations, the best performance is achieved when we combined *emoticons*, *exclamation marks* and *emphatic lengthening*. This is a very interesting result that shows that integrating the most useful stylistic variations and the opinionatedness of the terms into a ranking function can be very effective for Twitter opinion retrieval.

Table 2. Performance results of the SVF$_{Log}$IDF$_{Inv}$ method under topic-based settings using different combinations of stylistic variations over the baselines. A star($*$) and dagger($\dagger$) indicate statistically significant improvement over the relevance and opinion baselines respectively.

	MAP
Relevance-Baseline	0.2835
Opinion-Baseline	0.3807$*$
SVF$_{Log}$IDF$_{Inv}$-Emot-Excl	0.4314$*$ $\dagger$
SVF$_{Log}$IDF$_{Inv}$-Emot-Excl-Emph	0.4413$*$ $\dagger$
SVF$_{Log}$IDF$_{Inv}$-Emot-Excl-Emph-OpHash	0.4344$*$ $\dagger$

Table 3 shows the performance of the proposed model on non topic-based and topic-based settings for Twitter opinion retrieval. We evaluate the effectiveness of different combinations of approaches in calculation of SVF and IDF. We observe that most of the approaches perform statistically better under the topic-based settings compared to the non topic-based settings. This is a very interesting result which shows that stylistic variations are indeed topic-specific and the amount of the opinion information they hold depends on the topic of the tweet. We also observe that there is no statistical difference between the different SVF and IDF approaches when they are compared under the same settings.

Table 3. Performance results of different SVF and IDF combinations, based on emoticons, exclamation marks and emphatic lengthening. A star(*) indicates statistically significant improvement over the non topic-based settings for the same approach.

SVF - IDF	Non topic-based	Topic-based
$\text{SVF}_{Bool}\text{IDF}_{Inv}$	0.4279	0.4419*
$\text{SVF}_{Freq}\text{IDF}_{Inv}$	0.4279	0.4398
$\text{SVF}_{Log}\text{IDF}_{Inv}$	0.4275	0.4413*
$\text{SVF}_{Bool}\text{IDF}_{Prob}$	0.4279	0.4427*
$\text{SVF}_{Freq}\text{IDF}_{Prob}$	0.4279	0.4421*
$\text{SVF}_{Log}\text{IDF}_{Prob}$	0.4275	0.4429*

In addition, we performed a per topic analysis to compare the model under topic-based against non topic-based settings. Figure 1 shows the increase and decrease in Average Precision (AP) when comparing the best run ($\text{SVF}_{Log}\text{IDF}_{Prob}$) of the proposed model under topic-based against non topic-based settings. The plot shows that the topic-based $\text{SVF}_{Log}\text{IDF}_{Prob}$ model has topics for which they can improve over the non topic-based $\text{SVF}_{Log}\text{IDF}_{Prob}$ model as well as topics for which the topic-based $\text{SVF}_{Log}\text{IDF}_{Prob}$ model is not helping. However, the topic-based $\text{SVF}_{Log}\text{IDF}_{Prob}$ model has more topics for which it improves performance compared to the number of topics that are hurt. This shows that in general considering topic-specific stylistic variations into ranking is helpful.

Table 4 shows the three topics that were helped or hurt the most using the $\text{SVF}_{Log}\text{IDF}_{Prob}$ model under the topic-based compared to the non topic-based settings. We observe that the topics that were helped are those that probably contain few informal stylistic variations as they are related to topics about products or politics. In future, we plan to do a thorough exploration to detect the possible reasons for the increase/decrease in the performance of the topics.

Finally, we compare the performance of our proposed approach with the performance of the best run presented by Luo et al. [10] and report the comparison result in Table 5. We observe that our best runs outperform their best reported result (denoted BM25_Best). Finally, we should mention that their method uses SVM^{Rank} and their best run (BM25_Best) is trained using a number of social

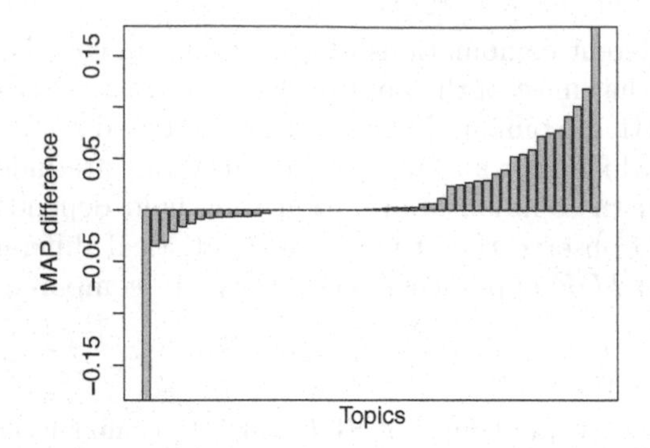

Fig. 1. Difference in performance between the topic-based SVF$_{Log}$IDF$_{Prob}$ and the non topic-based SVF$_{Log}$IDF$_{Prob}$ model. Positive/negative bars indicate improvement/decline over the non topic-based SVF$_{Log}$IDF$_{Prob}$ model in terms of MAP.

Table 4. Topics that are helped or hurt the most in the SVF$_{Log}$IDF$_{Prob}$ model under topic-based compared to non topic-based settings.

Helped		Hurt	
Title	Δ MAP	Title	Δ MAP
iran	0.1795	new start-ups	−0.1833
Lenovo	0.1185	iran nuclear	−0.0480
galaxy note	0.1017	big bang	−0.0319

Table 5. Results on Δ MAP for best runs over Opinion-Baseline

Run	Map	Δ MAP
Opinion-Baseline	0.3807	-
BM25_Best	0.4181	9.82 %
SVF$_{Log}$IDF$_{Prob}$-Emot-Excl-Emph	0.4429	16.33 %
SVF$_{Bool}$IDF$_{Prob}$-Emot-Excl-Emph	0.4427	16.28 %

features (URL, Mention, Statuses, Followers) together with BM25 score, and Query-Depedent opinionatedness (Q_D) features.

7 Conclusions and Future Work

In this paper, we considered the problem of Twitter opinion retrieval. We proposed a topic-based method that uses topic-specific stylistic variations to address the problem of opinion retrieval in Twitter. We studied the effect of different approaches and of the different stylistic variations in the performance of Twitter opinion retrieval. The experimental results showed that stylistic variations are good indicators for identifying opinionated tweets and that opinion retrieval performance is improved when emoticons, exclamation marks and emphatic lengthening are taken into account. Additionally, we demonstrated that the importance of stylistic variations in indicating opinionatedness is indeed topic dependent as our topic model-based approaches significantly outperformed those that assumed importance to be uniform over topics.

In future, we plan to extend the topic-based opinion retrieval method by investigating the effect of assigning different importance weights to each stylistic variation. We also plan to evaluate the performance of our method on other datasets that consider opinion retrieval on short texts that share similar stylistic variations to tweets such as MySpace and YouTube comments.

Acknowledgments. This research was partially funded by the Swiss National Science Foundation (SNSF) under the project OpiTrack.

References

1. Blei, D., Ng, A., Jordan, M.: Latent dirichlet allocation. J. Mach. Learn. Res. **3**, 993–1022 (2003)
2. Brody, S., Diakopoulos, N.: Cooooooollllllllll!!!!!!!!!!: using word lengthening to detect sentiment in microblogs. In: EMNLP 2011, pp. 562–570 (2011)
3. Eguchi, K., Lavrenko, V.: Sentiment retrieval using generative models. In: EMNLP 2006, pp. 345–354 (2006)
4. Go, A., Bhayani, R., Huang, L.: Technical report, Standford (2009)
5. Hofmann, T.: Probabilistic latent semantic indexing. In: SIGIR 1999, pp. 50–57 (1999)
6. Hong, L., Davison, B.D.: Empirical study of topic modeling in Twitter. In: SIGKDD Workshop on SMA, pp. 80–88 (2010)
7. Jiang, L., Yu, M., Zhou, M., Liu, X., Zhao, T.: Target-dependent twitter sentiment classification. In: ACL, HLT 2011, pp. 151–160 (2011)
8. Kouloumpis, E., Wilson, T., Moore, J.: Twitter sentiment analysis: The good the bad and the omg!. In: ICWSM 2011, pp. 538–541 (2011)
9. Krestel, R., Fankhauser, P., Nejdl, W.: Latent dirichlet allocation for tag recommendation. In: RecSys 2009, pp. 61–68 (2009)
10. Luo, Z., Osborne, M., Wang, T.: An effective approach to tweets opinion retrieval. In: WWW 2013, pp. 1–22 (2013)
11. Nielsen, F.: A new ANEW: Evaluation of a word list for sentiment analysis of microblogs. In: ESWC 2011 Workshop on 'Making Sense of Microposts': Big Things Come in Small Packages, pp. 93–98(2011)
12. Pak, A., Paroubek, P.: Twitter as a corpus for sentiment analysis and opinion mining. In: LREC 2010, pp. 1320–1326 (2010)
13. Paltoglou, G., Buckley, K.: Subjectivity annotation of the microblog 2011 realtime adhoc relevance judgments. In: Serdyukov, P., Braslavski, P., Kuznetsov, S.O., Kamps, J., Rüger, S., Agichtein, E., Segalovich, I., Yilmaz, E. (eds.) ECIR 2013. LNCS, vol. 7814, pp. 344–355. Springer, Heidelberg (2013)
14. Paltoglou, G., Giachanou, A.: Opinion retrieval: searching for opinions in social media. In: Paltoglou, G., Loizides, F., Hansen, P. (eds.) Professional Search in the Modern World. LNCS, vol. 8830, pp. 193–214. Springer, Heidelberg (2014)

15. Pang, B., Lee, L.: Opinion mining and sentiment analysis. Found. Trends Inf. Retr. **2**(1–2), 1–135 (2008)
16. Porter, M.: An algorithm for suffix stripping. Program **14**(3), 130–137 (1980)
17. Ramage, D., Dumais, S., Liebling, D.: Characterizing microblogs with topic models. In: ICWSM 2010, pp. 1–8 (2010)
18. Read, J.: Using emoticons to reduce dependency in machine learning techniques for sentiment classification. In: ACL Student Research Workshop, pp. 43–48 (2005)
19. Strunk, W.: The Elements of Style. Penguin, New York (2007)
20. Van Canneyt, S., Claeys, N., Dhoedt, B.: Topic-dependent sentiment classification on twitter. In: Hanbury, A., Kazai, G., Rauber, A., Fuhr, N. (eds.) ECIR 2015. LNCS, vol. 9022, pp. 441–446. Springer, Heidelberg (2015)
21. Wang, X., Wei, F., Liu, X., Zhou, M., Zhang, M.: Topic sentiment analysis in twitter: a graph-based hashtag sentiment classification approach. In: CIKM 2011, pp. 1031–1040 (2011)
22. Yao, L., Mimno, D., McCallum, A.: Efficient methods for topic model inference on streaming document collections. In: SIGKDD 2009, pp. 937–946 (2009)
23. Yi, J., Nasukawa, T., Bunescu, R., Niblack, W.: Sentiment analyzer: extracting sentiments about a given topic using natural language processing techniques. In: ICDM 2003, pp. 427–434 (2003)
24. Zhao, W.X., Jiang, J., Weng, J., He, J., Lim, E.-P., Yan, H., Li, X.: Comparing twitter and traditional media using topic models. In: Clough, P., Foley, C., Gurrin, C., Jones, G.J.F., Kraaij, W., Lee, H., Mudoch, V. (eds.) ECIR 2011. LNCS, vol. 6611, pp. 338–349. Springer, Heidelberg (2011)

Inferring Implicit Topical Interests on Twitter

Fattane Zarrinkalam[1,2(✉)], Hossein Fani[1,3], Ebrahim Bagheri[1], and Mohsen Kahani[2]

[1] Laboratory for Systems, Software and Semantics (LS3), Ryerson University, Toronto, Canada
fattane.zarrinkalam@gmail.com

[2] Department of Computer Engineering, Ferdowsi University of Mashhad, Mashhad, Iran

[3] Faculty of Computer Science, University of New Brunswick, New Brunswick, Canada

Abstract. Inferring user interests from their activities in the social network space has been an emerging research topic in the recent years. While much work is done towards detecting *explicit* interests of the users from their social posts, less work is dedicated to identifying *implicit* interests, which are also very important for building an accurate user model. In this paper, a graph based link prediction schema is proposed to infer implicit interests of the users towards emerging topics on Twitter. The underlying graph of our proposed work uses three types of information: user's followerships, user's explicit interests towards the topics, and the relatedness of the topics. To investigate the impact of each type of information on the accuracy of inferring user implicit interests, different variants of the underlying representation model are investigated along with several link prediction strategies in order to infer implicit interests. Our experimental results demonstrate that using topics relatedness information, especially when determined through semantic similarity measures, has considerable impact on improving the accuracy of user implicit interest prediction, compared to when followership information is only used.

Keywords: Implicit interest · Twitter · Topic relatedness · Collaborative filtering

1 Introduction

The growth of social networks such as Twitter has allowed users to share and publish posts on a variety of social events as they happen, in real time, even before they are released in traditional news outlets. This has recently attracted many researchers to analyze posts to understand the current emerging topics/events on Twitter in a given time interval by viewing each topic as a combination of temporally correlated words/terms or semantic concepts [2,4]. For instance, on 2 December 2010, Russia and Qatar were selected as the locations for the 2018 and 2022 FIFA World Cups. By looking at Twitter data on this

N. Ferro et al. (Eds.): ECIR 2016, LNCS 9626, pp. 479–491, 2016.
DOI: 10.1007/978-3-319-30671-1_35

day, a combination of keywords like *'FIFA World Cup'*, *'Qatar'*, *'England'* and *'Russia'* have logically formed a topic to represent this event.

The ability to model user interests towards these emerging topics provides the potential for improving the quality of the systems that work on the basis of user interests such as news recommender systems [21]. Most existing approaches build a user interest profile based on the explicit contribution of the user to the emerging topics [1,15]. However, such approaches struggle to identify a user's interests if the user has not explicitly talked about them. Consider the tweets posted by Mary:

- *"Qatar's bid to host the 2022 World Cup is gaining momentum, worrying the U.S., which had been the favorite* http://on.wsj.com/a8j3if"
- *"Russia rests 2018 World Cup bid on belief that big and bold is best | Owen Gibson (Guardian)* http://feedzil.la/g2Mpbs"

Based on the keywords explicitly mentioned by Mary in her tweets, one could easily infer that she is interested in the Russia and Qatar's selection as the hosts for the 2018 and 2022 FIFA World Cups. We refer to such interests that are directly derivable from a user's tweets as *explicit interests*. Expanding on this example, another topic emerged later in 2010, which was related to Prince William's engagement. Looking at Mary's tweets she never referred to this topic in her tweet stream. However, it is possible that Mary is British and is interested in both football and the British Royal family, although never explicitly tweeted about the latter. If that is in fact the case, then Mary's user profile would need to include such an interest. We refer to these concealed user topical interests as *implicit interests*, i.e., topics that the user never explicitly engaged with but might have interest in.

The main objective of our work in this paper is to determine implicit interests of a user over the emerging topics in a given time interval. To this end, we propose to turn the implicit interest detection problem into a graph-based link prediction problem that operates over a heterogeneous graph by taking into account *(i)* users' interest profile built based on their explicit contribution towards the extracted topics, *(ii)* theory of Homophily [12], which refers to the tendency of users to connect to users with common interests or preferences; and *(iii)* relationship between emerging topics, based on their similar constituent contents and user contributions towards them. More specifically, the key contributions of our work are as follows:

- Based on the earlier works [7,21], we model users' interests over the emerging topic on Twitter through a set of correlated semantic concepts. Therefore, we are able to infer finer-grained implicit interests that refer to real-world events.
- We propose a graph-based framework to infer the implicit interests of users toward the identified topics through a link prediction strategy. Our work considers a heterogeneous graph that allows for including three types of information: user followerships, user explicit interests and topic relatedness.
- We perform extensive experimentation to determine the impact of one or a combination of these information types on accurately predicting the implicit

interests of users on Twitter, which provides significant insight on how users are explicitly and implicitly inclined towards emerging topics.

The rest of this paper is organized as follows. In the next section, we review the related work. Our framework to infer users' implicit interests is introduced in Sect. 3. Section 4 is dedicated to the details of our empirical experimentation and our findings. Finally, in Sect. 5, we conclude the paper.

2 Related Work

In this paper, we assume that an existing state of the art technique such as those proposed in [2,4] can be employed for extracting and modeling the emerging topics on Twitter as sets of temporally correlated terms/concepts. Therefore, we will not be engaged with the process of identification of the topics and will only focus on determining the implicit interest of users towards the topics once they are identified. Given this focus, we review the works that are related to the problem of user interest detection from social networks.

There are different works for extracting users' interests from social networks through the analysis of the users' generated textual content. Yang et al. [19] have modeled the user interests by representing her tweets as a bag of words, and by applying a cosine similarity measure to determine the similarity between the users in order to infer common interests. Xu et al. [18] have proposed an author-topic model where the latent variables are used to indicate if the tweet is related to the author's interests.

Since Bag of Words and Topic Modeling approaches are designed for normal length texts, they may not perform so effectively on short, noisy and informal text such as tweets. There are insufficient co-occurrence frequencies between keywords in short posts to enable the generation of appropriate word vector representations [5]. Furthermore, bag of words approaches overlook the underlying semantics of the text. To address these issues, some recent works have tried to utilize external knowledge bases to enrich the representation of short texts [8,13]. Abel et al. [1] have enriched Twitter posts by linking them to related news articles and then modeled user's interests by extracting the entities mentioned in the enriched messages. DBpedia and Freebase are often used for enriching Tweets by linking their content with unambiguous concepts from these external knowledge bases. Such an association provides explicit semantics for the content of a tweet and can hence be considered to be providing additional contextual information about the tweet [8,10]. The work in [21] has inferred fine grained user topics of interest by extracting temporally related concepts in a given time interval.

While most of the works mentioned above have focused on extracting explicit interests through analysing only textual contents of users, less work has been dedicated to inferring *implicit* interests of the users. Some authors have shown interest in the Homophily theory [12] to extract implicit interests. Based on this theory, users tend to connect to users with common interests or preferences.

Mislove et al. [14] have used this theory to infer missing interests of a user based on the information provided by her neighbors. Wang et al. [16] have extended this theory by extracting user interests based on implicit links between users in addition to explicit relations. While these works incorporate the relationship between users, they do not consider the relationship between the emerging topics themselves. In our work, we are interested to explore if a holistic view that considers the semantics of the topics, the user followership information and the explicit interests of users towards the topics can provide an efficient platform for identifying users' implicit interests.

In another line of work, semantic concepts and their relationships defined in external knowledge bases are leveraged to extract implicit user interests. Kapanipathi et al. [10] have extracted implicit interests of the user by mapping her primitive interests to the Wikipedia category hierarchy using a spreading activation algorithm. Similarly, Michelson and Macskassy [13] have identified the high-level interests of the user by traversing and analyzing the Wikipedia categories of entities extracted from the user's tweets. The main difference between the problem we tackle here from the previously mentioned works is that we view each topic of interest as a combination of correlated concepts as opposed to just a single concept. So the relationship between two topics is not predefined in the external knowledge base and we need to provide a measure of topic similarity or relatedness.

3 Implicit User Interest Prediction

The objective of our work is to model and identify implicit interests of a user, within a specific time interval T, towards the emerging topics on Twitter. To address this challenge, we propose to turn the implicit interest prediction problem into a link prediction problem that operates over a heterogeneous graph. We believe that in addition to user explicit contributions toward the emerging topics, there are two other types of information that can be considered to infer implicit interests of users, namely user followership relations and the possible relation between the emerging topics themselves. By considering this information as our representation model, the main research question we are seeking to answer in this paper is: 'which or what combination of these three types of information are most effective in allowing us to accurately identify a user's implicit interests?' Therefore, we propose a comprehensive graph-based representation model that includes these three types of information and is used in order to model the implicit interest identification problem.

3.1 Representation Model

Our underlying representation model can be formalized as follows:

Definition 1 (Representation Model). Let T be a specified time interval. Given a set of emerging topics and individual users at time interval T denoted by $\mathbb{Z}$ and U, respectively, our representation model $G = (G_U \cup G_{U\mathbb{Z}} \cup G_{\mathbb{Z}})$, is a heterogeneous graph composed of three subgraphs, G_U, $G_{U\mathbb{Z}}$ and $G_{\mathbb{Z}}$. $G_U = (V_U, E_U)$ is unweighted and directed, which represents followership relations between users on Twitter, $G_{U\mathbb{Z}} = (V_{U\mathbb{Z}}, E_{U\mathbb{Z}})$ represents explicitly observable user-topic relations and $G_{\mathbb{Z}} = (V_{\mathbb{Z}}, E_{\mathbb{Z}})$ denotes potential relationships between emerging topics in $\mathbb{Z}$.

In line with earlier work in the literature [1,21], we view each emerging topic $z \in \mathbb{Z}$ as a set of temporally correlated semantic concepts derived from an external knowledge base, i.e., Wikipedia, and model each topic in the following form:

Definition 2 (Emerging Topic). An emerging topic z at time interval T, is defined as a set of weighted semantic concepts $z = \{(c, w(c, z)) | c \in C\}$, where $w(c, z)$ is a function that denotes the importance of concept c in topic z and C is the set of all semantic concepts observed at time interval T on Twitter.

In Definition 2, For instance, an emerging topic can be seen in our earlier example as a set $z_1 = \{$*'FIFA World Cup'*, *'Qatar'*, *'England'* and *'Russia'*$\}$, which is composed of four concepts from Wikipedia. Based on this topic representation model, the user-topic subgraph can be constructed based on the explicit mention of the topic by the user in her tweets.

Definition 3 (User-Topic Graph). A user-topic graph in time interval T, is a weighted directed graph $G_{U\mathbb{Z}} = (V_{U\mathbb{Z}}, E_{U\mathbb{Z}})$ where $V_{U\mathbb{Z}} = \mathbb{Z} \cup U$ and edges $E_{U\mathbb{Z}}$ are established by observing a user's explicit contributions towards any of the emerging topics. The weight of each edge $e_{uz} \in E_{U\mathbb{Z}}$ that ties user $u \in U$ to a topic $z \in \mathbb{Z}$ represents the degree of u's explicit interest in topic z in time interval T.

Our intuition for calculating the explicit interest of user $u \in U$ towards each topic z is that the more a user tweets about a certain topic, the more interested the user would be in that topic. We define the occurrence ratio of topic $z = \{(c, w(c, z))\}$ in tweet m, denoted $OR(z, m)$, as follows:

$$OR(z, m) = \frac{\sum_{c \in C} w(c, z) * \delta(c, m)}{\sum_{c \in C} w(c, z)} \tag{1}$$

where $\delta(c, m)$ is 1, if Tweet m is annotated with concept c, otherwise, $\delta(c, m) = 0$. The weight of each edge e_{uz} in $G_{U\mathbb{Z}}$ is calculated by averaging the value of $OR(z, m)$ over all tweets posted by the specific user u with regards to topic z.

Since we are interested in knowing whether potential relationships between topics can be used to infer implicit interests, the third type of information that we consider in our model is the relationship between the topics, i.e. topic-topic subgraph.

Definition 4 (Topic-Topic Graph). A topic-topic graph in time interval T, is a weighted undirected graph $G_{\mathbb{Z}} = (V_{\mathbb{Z}}, E_{\mathbb{Z}})$ where $V_{\mathbb{Z}}$ denotes the set of all emerging topics within time interval T, denoted by $\mathbb{Z}$, and $E_{\mathbb{Z}}$ denotes a set of edges representing the relationships between these topics. The weight of the edges between the topics in the topic-topic graph represents the degree of relatedness of the topics.

3.2 Topic Relatedness

There are three possible approaches through which the relation between the emerging topics can be identified in our model: *(i)* semantics relatedness, *(ii)* collaborative relatedness, and *(iii)* hybrid approach.

In the *semantic relatedness* approach, the relatedness of topics is determined based on the semantic similarity of their constituent concepts. In other words, two topics are considered to be similar if the concepts that make up the two topics are semantically similar. Given each topic in our model is composed of a set of Wikipedia concepts, the semantic relatedness of two emerging topics can be calculated by measuring the average pairwise semantic relatedness between the concepts of the two topics using a Wikipedia-based relatedness measure. In our experiments, we use WLM [17], which computes the concept relatedness through link structure analysis.

In the *collaborative relatedness* approach, the relatedness of two topics is determined based on a collaborative filtering strategy where relatedness is measured based on users' overlapping contributions toward these topics. Given a user-topic graph $G_{U\mathbb{Z}}$, we regard the problem of computing the collaborative relatedness of topics as an instance of a model-based collaborative filtering problem. To this end, we model the user-topic graph information as a user-item rating matrix R of size $|U| \times |\mathbb{Z}|$, in which an entry in R, denoted by r_{uz}, is used to represent the weight of the edge between user u and topic z in the user-topic graph $G_{U\mathbb{Z}}$, i.e., the degree of u's interest in topic z. By considering matrix R as the ground-truth item recommendation scores, our problem is to learn the relationship between topics in the form of an item similarity matrix. We adopt a factored item-item collaborative filtering method [9] that learns item-item similarities (topic relatedness) as a product of two rank matrices, P and Q. Two matrices P and Q denote latent factors of items. In our model, the rating for a given user u on topic z_i is estimated as:

$$\hat{r}_{ui} = b_u + b_i + (n_u^+)^{-\alpha} \sum_{j \in R_u^+} p_j q_i^T \tag{2}$$

where R_u^+ is the set of topics that user u is interested in, p_j and q_i are the learned topic latent factors, n_u^+ is the number of topics that user u is interested in and α is a user specified parameter between 0 and 1. According to [24], matrices P and Q can be learnt by minimizing a regularized optimization problem:

$$minimize \frac{1}{2} \sum_{u,i \in R} ||r_{ui} - \hat{r}_{ui}||_F^2 + \frac{\beta}{2}(||P||_F^2 + ||Q||_F^2) + \frac{\lambda}{2}||b_u||_2^2 + \frac{\gamma}{2}||b_i||_2^2 \tag{3}$$

where the vectors b_u and b_i correspond to the vector of user u and topic z_i biases.

The optimization problem can be solved using Stochastic Gradient Descent to learn two matrices P and Q. Given P and Q as latent factors of topics, the collaborative relatedness of two topics z_i and z_j is computed as the dot product between the corresponding factors from P and Q i.e., p_i and q_j.

While the collaborative relatedness measure can find the topic relatedness based on the user's contributions to the topics, it overlooks the semantic relatedness between the two topics. In the third approach, we develop a *hybrid relatedness measure* that considers both the semantic relatedness of the concepts within each topic as well as users' contributions towards the emerging topics. We follow the assumption of [20] for utilizing item attribute information to add the item relationship regularization term into Eq. (3). Based on this, two topic latent feature vectors would be considered similar if they are similar according to their attribute information. The topic relationship regularization term is defined as:

$$\frac{\delta}{2}\sum_{i=1}^{|\mathbb{Z}|}\sum_{i'=1}^{|\mathbb{Z}|} S_{ii'}(||q_i - q_{i'}||_F^2 + ||p_i - p_{i'}||_F^2) \quad (4)$$

where δ is a parameter to control the impact of topic information, S is a matrix in which $S_{ii'}$ denotes the similarity between topics z_i and $z_{i'}$ based on their attributes. In our approach, attributes of topics are their constituent concepts and $S_{ii'}$ is calculated by measuring the semantic relatedness of two topics as introduced earlier.

3.3 Implicit Interest Prediction

After building the representation model G, our problem is to infer whether a user $u \in U$ is implicitly interested in topic $z \in \mathbb{Z}$ for cases when no explicit interest between u and z is observed in G. In other words, we are going to find missing links of $G_{U\mathbb{Z}}$ by adopting an unsupervised link prediction strategy over observed links in G.

Most of the unsupervised link prediction strategies either generate scores based on vertex neighborhoods or path information [11]. Vertex neighborhood methods are based on the idea that two vertices x and y are more likely to have a link if they have many common neighbors. Path-based methods consider the ensemble of all paths between two vertices. All of these methods are based on a predictive score function for ranking links that are likely to occur. According to the experiments done in [11], there is no single superior method among existing work and their quality is dependent on the structure of the specific graph under study. Therefore, in our experiments, we exploit various well-known link prediction strategies for inferring implicit interests of a user. These strategies are introduced in Table 1.

4 Experiments

We perform our experimentation to answer the following research question: 'how and to what extent do the three types of information present in our representation model facilitate the identification of implicit user interests on Twitter?'.

Table 1. The five link prediction strategies chosen for user implicit interest prediction

Adamic/Adar	$score(x,y) = \sum_{z \in \Gamma(x) \cap \Gamma(y)} \frac{1}{log\lvert\Gamma(z)\rvert}$ $\Gamma(x)$: the set of neighbors of vertex x
Common neighbors	$score(x,y) = \Gamma(x) \cap \Gamma(y)$
Jaccard's coefficient	$score(x,y) = \lvert\Gamma(x) \cap \Gamma(y)\rvert / \lvert\Gamma(x) \cup \Gamma(y)\rvert$
Katz	$score(x,y) = \sum_{\ell=1}^{\infty} \beta^{\ell} \lvert path_{x,y}^{<\ell>}\rvert$ $\lvert path_{x,y}^{<\ell>}\rvert$: a set of all paths with length ℓ from x to y β: damping factor to give the shorter paths more weights
SimRank	$score(x,y) = sim(x,y)$ $sim(x,y) = \lambda(\sum_{a \in \Gamma(x)} \sum_{b \in \Gamma(y)} sim(a,b)) / \lvert\Gamma(x)\rvert\lvert\Gamma(y)\rvert$ $\lambda \in [0,1] and sim(x,x) = 1$

4.1 Experimental Setup

Dataset. Our experiments were conducted on the available Twitter dataset presented by Abel et al. [1]. It consists of approximately 3M tweets sampled between November 1 and December 31, 2010. Since we needed followership information to build the user-user graph, we used the Twitter RESTful API to crawl these relationships.

Evaluation Methodology and Metrics. Our evaluation strategy is based on the *leave-one-out method.* At each time, we divide our representation model into a training set and a test set by randomly picking one pair <user, topic> from user-topic graph $G_{U\mathbb{Z}}$ for test and the rest of the representation model for training. We repeat this procedure for all pairs. To evaluate the results, we use two metrics: the Area Under Receiver Operating Characteristic (AUROC) and the Area Under the Precision-Recall (AUPR) curves [6].

Parameter Setting. In Topic detection step, we follow the approach proposed in [7] to extract the emerging topics ($\mathbb{Z}$) within a given time interval T. After detecting $\mathbb{Z}$, based on Definition 2, we need to compute the weight of each concept c in each topic z, i.e., $w(c,z)$. To do so, we utilize the Degree Centrality of concept vertex c in topic z computed by summing the weights attached to the edges connected to c in topic z [21]. Further, in the learning step of computing the collaborative relatedness between topics, we use the default parameter settings of the Librec library and set $\beta = \lambda = \gamma = \delta = 0.001$. The learning rate is set to 0.01, the number of item latent factors is set to 10 and the number of iterations to 100.

4.2 Results and Discussion

To answer our research question, we conduct a set of experiments in which different link prediction strategies are applied on variants of our representation model. There are two main variation points which are incorporated in our representation model: *(i)* followership information (F) and *(ii)* the type of topics relatedness measure, i.e., semantic (S), collaborative (C) or hybrid (CS). By selecting and combining the different alternatives, we obtain 7 variants that we will systematically compare in this section. We include user's explicit interest information in all of the seven variants. As some brief example on how to interpret the models, Model F only uses user followership information in addition to users' explicit interests. The SF Model considers topic relationships computed using semantic relatedness in addition to user followership and user's explicit interests. The rest of the models can be interpreted similarly.

In order to make a fair comparison, we repeat the experimentation for all the selected link prediction strategies introduced in Table 1. The results in terms of AUROC and AUPR are reported in Table 2. Given AUROC and AUPR values can be misleading in some cases, we also visually inspect the ROC curves in addition to the area under the curve values. Due to space limitation and also the elaborate theorem proved in [6] that a curve dominates in ROC space if and only if it dominates in PR space, we only present the ROC curves in Fig. 1.

As illustrated in Table 2 and Fig. 1, we can clearly see that the SimRank link prediction method has not shown a good performance over none of the variants. Based on our results, SimRank acts as a random predictor because for most of the models its AUROC value is about 0.5 and its ROC curve is near y=x. Therefore, in the rest of this section, to investigate the influence of the different variants of our representation model on the performance of inferring implicit interests of users we ignore the results of the SimRank strategy.

Table 2. The AUROC/AUPR values showing the performance of different model variants

Model	Metric	Adamic/ Adar	Common neighbor	Jaccard coefficient	Katz $\beta = 0.0005$	Katz $\beta = 0.005$	Katz $\beta = 0.5$	SimRank $\lambda = 0.8$
F	AUROC	0.500	0.500	0.500	0.524	0.524	0.528	0.510
	AUPR	0.438	0.438	0.438	0.454	0.454	0.458	0.422
S	AUROC	**0.791**	0.790	0.774	0.790	0.790	0.788	0.500
	AUPR	**0.740**	0.739	0.723	0.740	0.739	0.734	0.438
SF	AUROC	**0.791**	0.790	0.762	0.757	0.753	0.720	0.520
	AUPR	**0.740**	0.739	0.707	0.660	0.652	0.602	0.430
C	AUROC	0.712	0.710	0.700	0.714	0.715	0.728	0.500
	AUPR	0.657	0.651	0.610	0.657	0.661	0.680	0.438
CF	AUROC	0.773	0.771	0.758	0.742	0.738	0.716	0.517
	AUPR	0.717	0.714	0.692	0.647	0.640	0.602	0.428
CS	AUROC	0.762	0.761	0.748	0.763	0.763	0.767	0.500
	AUPR	0.697	0.695	0.661	0.699	0.699	0.707	0.438
CSF	AUROC	0.762	0.761	0.738	0.736	0.732	0.707	0.520
	AUPR	0.697	0.695	0.652	0.640	0.632	0.595	0.428

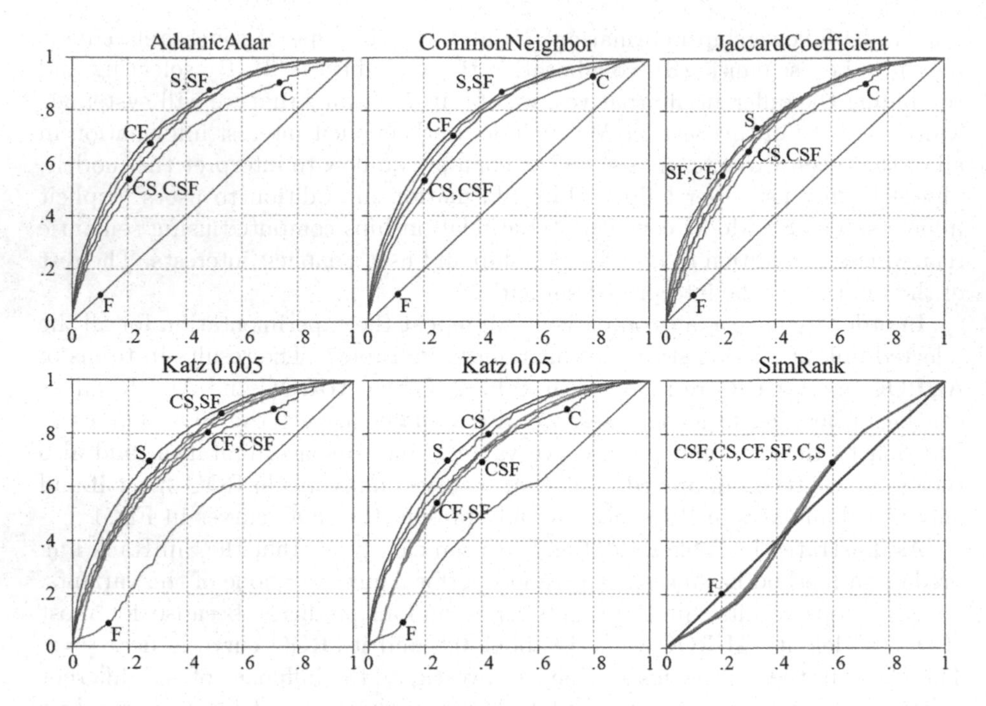

Fig. 1. The ROC curves for comparing the seven variants.

As mentioned earlier, Model F only considers followership information in addition to users' explicit interests to infer users' implicit interests. Instead, the models S, C and CS employ three different techniques for identifying topic relationships: model S uses semantic relatedness of the concepts included in the topics, model C uses collaborative relatedness and, model CS follows a hybrid approach. As depicted in Table 2, all these three models outperform Model F noticeably in terms of AUROC and AUPR. We can also see that the models S, C and CS dominate Model F in ROC space. This means that considering the relationships between the topics considerably improves the accuracy of inferring implicit interests in comparison with when only followership information is used.

By comparing S, C and CS themselves, it can be observed that using the semantic relatedness variant results in higher accuracy for the prediction of implicit interests compared to the collaborative and hybrid measures. This is an interesting observation that implies that users are predominantly interested in topics that are around similar topics. The three pairs of topics with the most relatedness obtained by the S model are shown in Fig. 2 (right). For an instance, the topics $z_1 = \{$*Chelsea F.C., Arsenal F.C.*$\}$ and $z_2 = \{$*FC Barcelona, Real Madrid C.F.*$\}$ refer to two derbies correspondingly in England and Spain. As confirmed by Wikipedia, these two competitions are among the most famous derbies in their countries and also in the world. As a result, it is reasonable to infer, with some lesser probability, that a user who is explicitly interested in one of these derbies, is probably interested also in the other one.

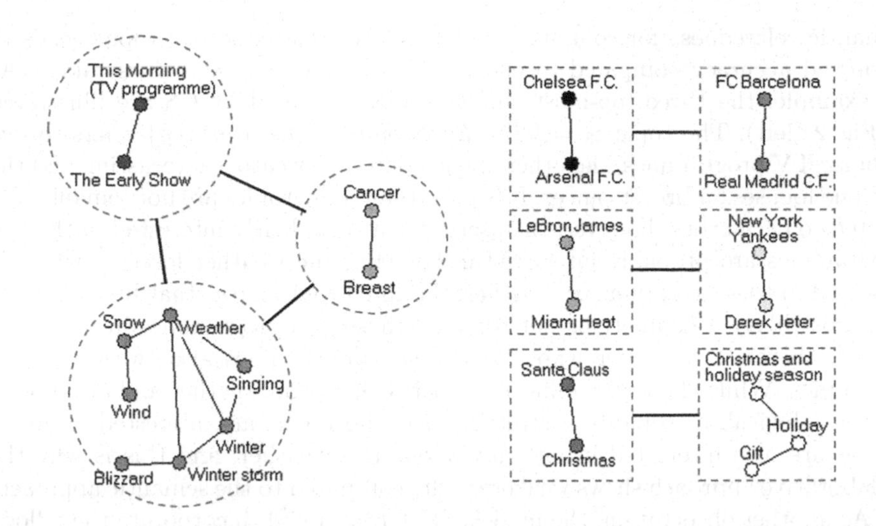

Fig. 2. Topmost related topics based on Hybrid (left) and semantic (right) measures

When looking at the results in Table 2, one can see that model C shows slightly weaker results compared to S, which can be the sign of two points: *(i)* semantic relatedness of topics is a more accurate indication of the tendency of users towards topics compared to collaborative relatedness of topics, and *(ii)* while C shows a weaker performance, its performance is in most cases only slightly weaker. This could mean that there is some degree of similarity between the results obtained by the two methods (C and S) pointing to the fact that even when using the collaborative relatedness measure, a comparable result to when the semantic relatedness measure is used can be obtained. Our explanation for this is that Twitter users seem to follow topics that are from similar domains or genres. This is an observation that is also reported in [3] and can be seen in the *Who Likes What* system. Therefore, when trying to predict a user's implicit interest, it would be logical to identify those that are on topics closely related to the user's explicit interests. Given this observation, the user's that are most similar within the context of collaborative filtering, are likely to also be following a coherent set of topics (not a variety of topics) and therefore, provide grounds for a reasonable estimation of the implicit interests.

The observation that S provides the best performance for predicting implicit interests is more appealing when the computational complexity involved in its computation is compared with the other methods. The computation of S only involves the calculation of the semantic similarity of the concepts in each pair of topics, which is quite an inexpensive operation, whereas the computation of C and CS require solving an optimization problem through Stochastic Gradient Descent. Additionally, by comparing C and CS, it can be concluded that adding semantic relatedness for computing collaborative relatedness of topics leads to improved accuracy compared to using only collaborative relatedness alone. As an example, the three top-most similar topics obtained by CS are illustrated in Fig. 2 (left). The topic z_3 = {*The Early Show, This Morning*} refers to two popular TV programmes, the other one is related to weather forecasting and the last one focuses on breast cancer. It is clear that these topics are not semantically related to each other, however, the users who are explicitly interested in the two programmes are probably interested in knowing the weather forecast which is reported in these programmes. Further, the third topic shows that breast cancer was most likely a contentious hot topic on these two programmes in that time period; therefore, the user who followed the programmes also tweeted about this topics. While the topic connections between z_3 and weather and also breast cancer is logical, it would be a stretch to say those who are interested in breast cancer are also interested in knowing about the weather, and this is why the collaborative approach shows weaker results compared to the semantic approach.

As another observation, the models SF, CF and CSF incorporate the followership information correspondingly in the S, C and CS models. As demonstrated in Table 2, no uniform observation can be made in any of the cases, i.e., the followership information does not seem to have a noticeable impact on the results. As a result, through our experiments we were not able to show the impact of homophily theory that suggests the user interests can be extracted from their relationship to other users. In summary, model S, which relies solely on the semantic relatedness of topics and user's explicit contributions to these topics shows the best performance across all seven variants. The SF model shows the same performance as S in which the additional followership information does not seem to have impacted the final results.

5 Conclusions and Future Work

In this paper, we studied the problem of inferring implicit interests of a user toward a set of emerging topics on Twitter. We model this problem as a link prediction task over a graph including three type of information: followerships, users explicit interests and topic relatedness. To investigate the influence of different types of information on the performance of the implicit interest detection problem, we proposed different variants of our representation model and applied some well-known link prediction strategies. The results showed that considering the relationships between the topics considerably improves the accuracy compared to using only followership information. Further, it was our observation that users on Twitter are predominantly interested in the coherent and semantically related topics and not on unrelated topics. As future work, we are investigating meta-path-based relationship prediction framework for heterogeneous graphs as our link prediction strategy. Further, based on the idea that user interests change over time, we intend to include temporal behavior of users toward topics in our implicit user interest identification problem.

References

1. Abel, F., Gao, Q., Houben, G.-J., Tao, K.: Analyzing user modeling on twitter for personalized news recommendations. In: Konstan, J.A., Conejo, R., Marzo, J.L., Oliver, N. (eds.) UMAP 2011. LNCS, vol. 6787, pp. 1–12. Springer, Heidelberg (2011)
2. Aiello, L.M., Petkos, G., Martin, C., Corney, D., Papadopoulos, S., Skraba, R., Goker, A., Kompatsiaris, I., Jaimes, A.: Sensing trending topics in twitter. IEEE Trans. Multimedia **15**(6), 1268–1282 (2013)
3. Bhattacharya, P., Muhammad, B.Z., Ganguly, N., Ghosh, S., Gummadi, K.P.: Inferring user interests in the twitter social network. In: RecSys 2014, pp. 357–360 (2014)
4. Cataldi, M., Di Caro, L., Schifanella, C.: Emerging topic detection on twitter based on temporal and social terms evaluation. In: IMDMKDD 2010, p. 4 (2010)
5. Cheng, X., Yan, X., Lan, Y., Guo, J.: Btm: Topic modeling over short texts. IEEE Trans. Knowl. Data Eng. **26**(12), 2928–2941 (2014)
6. Davis, J., Goadrich, M.: The relationship between precision-recall and ROC curves. In: ICML 2006, pp. 233–240 (2006)
7. Fani, H., Zarrinkalam, F., Zhao, X., Feng, Y., Bagheri, E., Du, W.: Temporal identification of latent communities on twitter (2015). arXiv preprint arxiv:1509.04227
8. Ferragina, P., Scaiella, U.: Fast and accurate annotation of short texts with wikipedia pages. IEEE Softw. **29**(1), 70–75 (2012)
9. Kabbur, S., Ning, X., Karypis, G.: FISM: factored item similarity models for top-N recommender systems. In: KDD 2013, pp. 659–667 (2013)
10. Kapanipathi, P., Jain, P., Venkataramani, C., Sheth, A.: User interests identification on twitter using a hierarchical knowledge base. In: Presutti, V., d'Amato, C., Gandon, F., d'Aquin, M., Staab, S., Tordai, A. (eds.) ESWC 2014. LNCS, vol. 8465, pp. 99–113. Springer, Heidelberg (2014)
11. Liben-Nowell, D., Kleinberg, J.: The link-prediction problem for social networks. J. Am. Soc. Inform. Sci. Technol. **58**(7), 1019–1031 (2007)
12. McPherson, M., Smith-Lovin, L., Cook, J.M.: Birds of a feather: Homophily in social networks. Ann. Rev. Sociol. 27, pp. 415–444 (2001)
13. Michelson, M., Macskassy, S.A.: Discovering users' topics of interest on twitter: A first look. In: AND 2010, pp. 73–80 (2010)
14. Mislove, A., Viswanath, B., Gummadi, K.P., Druschel, P.: You are who you know: inferring user profiles in online social networks. In: WSDM 2010, pp. 251–260 (2010)
15. Shin, Y., Ryo, C., Park, J.: Automatic extraction of persistent topics from social text streams. World Wide Web **17**(6), 1395–1420 (2014)
16. Wang, J., Zhao, W.X., He, Y., Li, X.: Infer user interests via link structure regularization. TIST **5**(2), 23 (2014)
17. Witten, I., Milne, D.: An effective, low-cost measure of semantic relatedness obtained from wikipedia links. In: WikiAI 2008, pp. 25–30 (2008)
18. Xu, Z., Lu, R., Xiang, L., Yang, Q.: Discovering user interest on twitter with a modified author-topic model. In: WI-IAT 2011, vol. 1, pp. 422–429 (2011)
19. Yang, L., Sun, T., Zhang, M., Mei, Q.: We know what@ you# tag: does the dual role affect hashtag adoption? In: WWW 2012, pp. 261–270 (2012)
20. Yu, Y., Wang, C., Gao, Y.: Attributes coupling based item enhanced matrix factorization technique for recommender systems. arXiv preprint. (2014). arxiv:1405.0770
21. Zarrinkalam, F., Fani, H., Bagheri, E., Kahani, M., Du, W.: Semantics-enabled user interest detection from twitter. In: WI-IAT 2015 (2015)

Topics in Tweets: A User Study of Topic Coherence Metrics for Twitter Data

Anjie Fang([envelope]), Craig Macdonald, Iadh Ounis, and Philip Habel

University of Glasgow, Glasgow, UK
a.fang.1@research.gla.ac.uk,
{craig.macdonald,iadh.ounis,philip.habel}@glasgow.ac.uk

Abstract. Twitter offers scholars new ways to understand the dynamics of public opinion and social discussions. However, in order to understand such discussions, it is necessary to identify coherent topics that have been discussed in the tweets. To assess the coherence of topics, several automatic topic coherence metrics have been designed for classical document corpora. However, it is unclear how suitable these metrics are for topic models generated from Twitter datasets. In this paper, we use crowdsourcing to obtain pairwise user preferences of topical coherences and to determine how closely each of the metrics align with human preferences. Moreover, we propose two new automatic coherence metrics that use Twitter as a separate background dataset to measure the coherence of topics. We show that our proposed Pointwise Mutual Information-based metric provides the highest levels of agreement with human preferences of topic coherence over two Twitter datasets.

1 Introduction

Twitter is an important platform for users to express their ideas and preferences. In order to examine the information environment on Twitter, it is critical for scholars to understand the topics expressed by users. To do this, researchers have turned to topic modelling approaches [1,2], such as Latent Dirichlet Allocation (LDA). In topic models, a document can belong to multiple topics, while a topic is considered a multinomial probability distribution over terms [3]. The examination of a topic's term distribution can help researchers to examine what the topic represents [4,5]. To present researchers with interpretable and meaningful topics, several topic coherence metrics have been previously proposed [6–8]. However, these metrics were developed based on corpora of news articles and books, which are dissimilar to corpora of tweets, in that the latter are brief (i.e. < 140 characters), contain colloquial statements or snippets of conversation, and use peculiarities such as hashtags. Indeed, while topic modelling approaches specific to Twitter have been developed (e.g. Twitter LDA [2]), the suitability of these coherence metrics for Twitter data has not been tested.

In this paper, we empirically investigate the appropriateness of ten automatic topic coherence metrics, by comparing how closely they align with human

N. Ferro et al. (Eds.): ECIR 2016, LNCS 9626, pp. 492–504, 2016.
DOI: 10.1007/978-3-319-30671-1_36

judgments of topic coherence. Of these ten metrics, three examine the statistical coherence of a topic at the term/document distributions levels, while the remaining seven consider if the terms within a topic exhibit semantic similarity, as measured by their alignment with external resources such as Wikipedia or WordNet. In this work, we propose two new coherence metrics based on semantic similarity, which use a separate background dataset of tweets.

To evaluate which coherence metrics most closely align with human judgments, we firstly use three different topic modelling approaches (namely LDA, Twitter LDA (TLDA) [2], and Pachinko Allocation Model (PAM) [9]) to generate topics on corpora of tweets. Then, for pairs of topics, we ask crowdsourcing workers to choose what they perceive to be the more coherent topic. By considering the pairwise preferences of the workers, we then identify the coherence metric that is best aligned with human judgments.

Our contributions are as follows: (1) we conduct a large-scale empirical crowdsourced user study to identify the coherence of topics generated by three different topic modelling approaches upon two Twitter datasets; (2) we use these pairwise coherence preferences to assess the suitability of 10 topic coherence metrics for Twitter data; (3) we propose two new topic coherence metrics, and show that our proposed coherence metric based on Pointwise Mutual Information using a Twitter background dataset is the most similar to human judgments.

The remainder of this paper is structured as follows: Sect. 2 provides an introduction to topic modelling; Sect. 3 reports the related work of evaluating topic models; Sect. 4 describes 10 topic coherence metrics; Sect. 5 shows how we compare automatic metrics to human judgments; Sect. 6 describes the Twitter datasets we use in the user study (Sect. 7), while the experimental setup and the results are discussed in Sects. 8 and 9. Finally, we provide concluding summaries in Sect. 10.

2 Background: Topic Modelling

Topic modelling approaches can be used to identify coherent topics of conversation in social media such as Twitter [1,2]. However, ensuring that the topic modelling approaches obtain coherent topics from tweets is challenging. Variants of LDA have been proposed to improve the coherence of the topics, while automatic metrics of topical coherence have also been proposed (see Sect. 3). However, as we argue in Sect. 3, the suitability of the automatic coherence metrics has not been demonstrated on Twitter data.

LDA [10] is one of the most popular topic modelling approaches. TLDA and PAM are two extensions of LDA. LDA is a Bayesian probabilistic topic modelling approach, where K latent topics (z) are identified, which are associated to both documents and terms, denoted as $P(z|d)$ and $P(w|z)$, respectively. PAM [9] is a 4-level hierarchical extension of LDA, where a document is represented by a multinomial distribution over super-topics θr, where a super-topic is a multinomial distribution θt over sub-topics. This structure helps to capture the relation between super-topics and sub-topics. PAM generates topics with higher coherence, improving the likelihood of held-out documents and improving the accuracy of classification [11]. On the other hand, Zhao et al. [2] recognised that due to their brevity, tweets can be challenging for obtaining coherent topic models. To counter this, they proposed TLDA, which employs a background Bernoulli term distribution, where a Bernoulli distribution π controls the selection between "real" topic terms and background terms. Moreover, Zhao et al. [2] assumed that a single tweet contained a single topic. Based on human judgments, they showed that TLDA outperformed the standard LDA for discovering topics in tweets. Indeed, both TLDA and PAM have been reported to produce more coherent topics than LDA. Hence, we apply the three aforementioned approaches to extract topics from Twitter corpora. In the following section, we review various topic coherence metrics.

3 Related Work: Evaluating Topic Models

The early work on evaluating topic models calculated the likelihood of held-out documents [12]. Chang et al. [13] deployed a user study for the interpretation of the generated topics, by comparing human judgments to the likelihood-based measures. However, it was shown that a model that had a good held-out likelihood performance can still generate uninterpretable topics.

Mei et al. [4] provided a method to interpret the topics from topic models. Their approach relied on the statistical analysis of a topic's term distribution. Similarly, AlSumait et al. [6] used another statistical analysis metric to evaluate the topics. In this paper, we compare their metrics to human judgments that assess the coherence of topics. Newman et al. [7,8] offered another way to evaluate the coherence of topics. They captured the semantically similar words among the top 10 terms in a topic and calculated the semantic similarity of the words using external resources, e.g. WordNet [14] and Wikipedia. They showed that the evaluation metric based on the Pointwise Mutual Information estimate of the word pairs generated from Wikipedia was the closest to human judgments.

The datasets used in [6–8] consisted of news articles and books; however Twitter data is different from the classical text corpora. Therefore, it is unclear how well these evaluation metrics perform when measuring the coherence of a topic in tweets. In the next section, we give more details about these metrics and our proposed new ones.

4 Automatic Topic Coherence Metrics

In this section, we describe the topic coherence metrics that we use to automatically evaluate the topics generated by topic modelling approaches. There are two types of coherence metrics: (1) metrics based on semantic similarity (introduced in [7,8]) and (2) metrics based on statistical analysis (introduced in [6]). We propose two new metrics based on semantic similarity, which use a Twitter background dataset.

4.1 Metrics Based on Semantic Similarity

In metrics based on semantic similarity, a topic is represented by the top 10 words ($\{w_1, w_2, ..., w_{10}\}$) ranked according to its term probabilities ($p(w|z)$) in the term distribution ϕ. A word pair of a topic is composed by any two words from the topic's top 10 words. The coherence of a topic is measured by averaging the semantic similarities of all word pairs [7,8] shown in Eq. (1) below. In this paper, the *Semantic Similarity SS* of a word pair is computed by using three external resources: WordNet, Wikipedia and a Twitter background dataset.

$$Coherence(topic) = \frac{1}{45} \sum_{i=1}^{10} \sum_{j=i+1}^{10} SS(w_i, w_j) \tag{1}$$

$$PMI(w_i, w_j) = log \frac{p(w_i, w_j)}{p(w_i), p(w_j)} \tag{2}$$

WordNet. WordNet groups words into synsets [14]. There are a number of semantic similarity and relatedness methods in the existing literature. Among them, the method designed by Leacock et al. [15] (denoted as LCH) and that designed by Jiang et al. [16] (denoted as JCN) are especially useful for discovering lexical similarity [8]. Apart from these two methods, Newman et al. [8] also showed that the method from Lesk et al. [17] (denoted as LESK) performs well in capturing the similarity of word pairs. Therefore, we select these 3 WordNet-based methods to calculate the semantic similarities of the topic's word pairs, and produce a topic coherence score.

Wikipedia. Wikipedia has been previously used as background data to calculate the semantic similarity of words [18,19]. In this paper, we select two popular approaches in the existing literature on calculating the semantic similarity of words: *Pointwise Mutual Information* (PMI) and *Latent Semantic Analysis* [20] (LSA). PMI is a popular method to capture semantic similarity [7,8,18]. Newman et al. [7,8] reported that the performance of PMI was close to human judgments when assessing the topic's coherence. Here the PMI data (denoted as W-PMI) is computed by using Eq. (2) consisting of the PMI score of word pairs from Wikipedia. On the other hand, since it has been reported that the performance of PMI is no better than LSA on capturing the semantic similarity of word pairs [19], in this paper we also use LSA to obtain the similarity of the word pairs. In the LSA model, a corpus is represented by a term-document matrix. The cells represent the frequency of a term occurring in a document. To reduce the dimension of this matrix, a *Singular Value Decomposition* is applied on the matrix using the k largest singular values. After the decomposition, each term is represented by a dense vector in the reduced LSA space. The semantic similarity of terms can be computed by the distance metrics (e.g. cosine similarity) between the terms' vectors. We use Wikipedia articles as background data and calculate the LSA space (denoted as W-LSA), which is a collection of term vectors in 300 dimensions described in [21].

Twitter Background Dataset. Since tweets contain abbreviations and hashtags[1], Wikipedia cannot capture their semantic similarity. Hence, we crawl an additional Twitter background corpus of 1 %–5 % random tweets from 1 Jan 2015 to 30 June 2015 on Twitter. The background collection is likely to better reflect the semantic similarity of words that occur on Twitter. We use the same method as for Wikipedia to obtain our proposed two new metrics: the reduced LSA space (300 dimensions, denoted as T-LSA) and the PMI score of word pairs (denoted as T-PMI) that appear in each tweet.

4.2 Metrics Based on Statistical Analysis

Properties of how the term or documents are assigned to the topics can be indicative of the coherence of a topic model. In this section, we describe the term/document distributions of 3 types of meaningless topics defined in [6]: a uniform distribution over terms; a semantically vacuous distribution over terms; and a background distribution over documents. We explain how these permit the measurement of the coherence of a topic.

Uniform Term Distribution. In a topic's term distribution, if all terms tend to have an equal and constant probability, this topic is unlikely to be meaningful nor easily interpreted. A typical uniform term distribution ϕ_{uni} is defined in Eq. (3), where i is the term index and N^k is the total number of terms in topic k.

$$\phi_{uni} = \{P(w_1), P(w_2), ..., P(w_{N^k})\},\ P(w_i) = \frac{1}{N^k} \tag{3}$$

Vacuous Term Distribution. A "real" topic should contain a unique collection of highly used words distinguishing this topic from the other topics. A topic is less coherent if a topic is mixed. A vacuous term distribution θ_{vac} represents a mixed term distribution, in which the term probability reflects the frequency of the term in the whole corpus. ϕ_{vac} is defined by Eq. (4), where d is the document index and D is the total number of documents.

$$\phi_{vac} = \{P(w_1), P(w_2), ..., P(w_{N^k})\},\ P(w_i) = \textstyle\sum_{k=1}^{K} \phi_{i,k} \times \frac{\sum_{d=1}^{D} \theta_{d,k}}{D} \tag{4}$$

Background Document Distribution. A "real" topic should represent documents within a semantically coherent theme. If a topic is close to most of the documents in the corpus, it is likely to be less meaningful and less coherent. Whereas the previous two distributions use terms to define the incoherent distribution of a topic, the topic distribution over documents can also reflect the quality of the topic [6]. A topic's document distribution ϑ^k is defined in Eq. (5) and a typical background document distribution ϑ_{gb} is defined in Eq. (6).

$$\vartheta^k = \{P(z = k|d_1), P(z = k|d_2), ..., P(z = k|d_D)\} \tag{5}$$

[1] Note that many hashtags are not recorded in Wikipedia.

$$\vartheta_{gb} = \{P(d_1), P(d_2), ..., P(d_D)\},\ P(d_i) = \frac{1}{D} \quad (6)$$

Given a topic k, the coherence of the topic is calculated by measuring the Kullback Leibler divergence between this topic and those three meaningless topics described above. A small divergence indicates that the topic is less coherent. Hereafter, we use U (uniform), V (vacuous) and B (background) to denoted three metrics corresponding to the coherence functions $Coherence^U(k)$, $Coherence^V(k)$ and $Coherence^B(k)$ in Eq. (7), respectively.

$$\begin{gathered} Coherence^U(k) = KL(\phi_{uni}||\phi^k), Coherence^V(k) = KL(\phi_{vac}||\phi^k) \\ Coherence^B(k) = KL(\vartheta_{gb}||\vartheta^k) \end{gathered} \quad (7)$$

In summary, in this paper we describe 7 metrics based on semantic similarity: LCH, JCN, LESK, W-LSA, W-PMI, T-LSA & T-PMI, and 3 metrics based on the statistical analysis of term/document distributions: U, V & B. Among them, T-LSA & T-PMI are our newly proposed metrics. In the following section, we present our approach to compare the discussed automatic coherence metrics to human judgments when assessing the coherence of topics.

5 Comparison of Coherence Metrics

In this section, we describe the methodology we use to identify whether the topic coherence metrics are aligned with human evaluations of topic coherence. It can be a challenging task for humans to produce graded coherence assessments of topics. Therefore, we apply a pairwise preference user study [22] to gather human judgments. A similar method has been previously used to compare summarisation algorithms [23]. In the rest of this section, we describe this comparison method.

Generating Topic Pairs. To compare the three topic modelling approaches, we divide the comparison task into three units: LDA vs. TLDA, LDA vs. PAM and TLDA vs. PAM. Each comparison unit consists of a certain number of topic pairs and each pair contains a topic from topic models T_1 and T_2, respectively (e.g. LDA vs. TLDA). To make the comparisons easier for humans, we present similar topics in a pair. Specifically, each topic model has a set of candidate topics, and each topic is represented as a topic vector using its term distribution. First, we randomly select a certain number of topics from topic model T_1. For each topic selected in T_1, we use Eq. (8) to select the closest topic in T_2 using cosine similarity. The selected topic pairs are denoted as Pairs($T_1 \rightarrow T_2$). Moreover, we also generate the same number of topic pairs Pairs($T_2 \rightarrow T_1$) for comparison unit(T_1,T_2). Therefore, every comparison unit has a set of topic pairs shown in Table 1.

$$closest(topic_j^{T_1}) = argmin_{i<K}\ (1 - cosine(Vector_{topic_j^{T_1}}, Vector_{topic_i^{T_2}})) \quad (8)$$

Automatic Topic Coherence Evaluation. We use the topic coherence metrics described in Sect. 4 to rank the three topic modelling approaches: LDA,

Table 1. Comparison task.

Comparison Unit	Topic Pairs in Unit
(1) Unit(LDA, TLDA)	Pairs(LDA→TLDA & TLDA→LDA)
(2) Unit(LDA, PAM)	Pairs(LDA→PAM & PAM→LDA)
(3) Unit(TLDA, PAM)	Pairs(TLDA→PAM & PAM→TLDA)

TLDA and PAM. For each topic in each topic pair, an automatic coherence metric gives a coherence score to each topic respectively. Thus, for each comparison unit, there is a group of data pairs. We apply the Wilcoxon signed-rank test to calculate the significance level of the difference between the two groups of data sample. For each comparison unit, an automatic coherence metric determines the better topic model between two approaches (e.g. LDA > TLDA), which results in a ranking order of the three topic modelling approaches. For instance, given the preferences LDA > TLDA, LDA > PAM & TLDA > PAM, we can obtain the ranking order LDA(1st) > TLDA(2nd) > PAM(3rd). However, while it is possible for the preference results of comparison units not to permit a ranking order to be obtained – i.e. a Condorcet paradox such as TLDA > LDA, LDA > PAM & PAM > TLDA – we did not observe any such paradoxes in our experiments.

Human Evaluation. Similarly as above, we also rank the three topic modelling approaches using the topic coherence assessments from humans described in Sect. 7. This obtained ranking order generated from humans is compared to that generated from the ten automatic coherence metrics to ascertain the most suitable coherence metric when assessing a topic's coherence.

6 Twitter Datasets

In our experiments, we use two Twitter datasets to compare the topic coherence metrics. The first dataset we use consists of tweets posted by 2,853 newspaper journalists in the state of New York from 20 May 2015 to 19 Aug 2015, denoted as NYJ. To construct this dataset, we tracked the journalists' Twitter handles using the Twitter Streaming API[2]. We choose this dataset due to the high volume of topics discussed by journalists on Twitter. The second dataset contains tweets related to the first TV debate during the UK General Election 2015. This dataset was collected by searching the TV debate-related hashtags and keywords (e.g. #TVDebate and #LeaderDebate) using the Twitter Streaming API. We choose this dataset because social scientists want to understand what topics people discuss. Table 2 reports the details of these two datasets. We describe our user study and experimental setups in Sects. 7 and 8, respectively.

[2] http://dev.twitter.com.

Table 2. The details of the two used Twitter datasets.

Name	Time Period	# of Users	# of Tweets
(1) NYJ	20/05/2015–19/08/2015	2,853	946,006
(2) TV debate	8pm–10pm 02/04/2015	121,594	343,511

7 User Preferences Study

In this section, we describe the method we use to obtain the human ground-truth ranking order of the three topic modelling approaches. As described in Sect. 5, the comparison task is divided into three comparison units. Each comparison unit has two sets of topic pairs from the NYJ and TV debate datasets respectively. We asked humans to conduct a pairwise preference evaluation, and we then used the obtained human' preferences of topics from the topic models to rank the three topic modelling approaches. For collecting human judgments, we used the CrowdFlower[3] crowdsourcing platform.

CrowdFlower Job Description. For each topic pair in our three comparison units, we present a worker (i.e. a human) with the top 10 highly frequent words from the two topics (a topic pair, generated from two topic modelling approaches) along with their associated 3 most retweeted tweets, which are likely to represent the topic. A CrowdFlower worker is asked to choose the more coherent topic from two topics using these 10 words. To help the workers understand and finish the task, we provide guidelines that define a more coherent topic as one that contains fewer discussions/events and that can be interpreted easily. We instruct workers to consider: (1) the number of semantically similar words among the 10 shown words, (2) whether the 10 shown words imply multiple topics and (3) whether the 10 shown words have more details about a discussion/event. If a decision cannot be made with these 10 words, a worker can then use the optional 3 associated tweets, shown in Fig. 1. We provide two guidelines for using these tweets for assistance: (1) consider the number of the 10 shown words from a topic that can be reflected by the tweets and (2) consider the number of tweets that are related with the topic. After the workers make their choices, they are asked to select the reasons, as shown in Fig. 1. The CrowdFlower workers were paid $0.05 for each judgment per topic pair. We gather 5 judgments for each topic pair from 5 different workers.

CrowdFlower Quality Control. To ensure the quality of the CrowdFlower judgments, we use several quality control strategies. First, we provide a set of test questions, where for each question workers are asked to choose a topic preference from a topic pair. The answers of the test questions are verified in advance. Only workers that pass the test are allowed to enter the task. Moreover, the worker must have maintained 70 % or more accuracy on the test questions in the task, otherwise their judgments are erased. Since the NYJ dataset is related

[3] http://crowdflower.com.

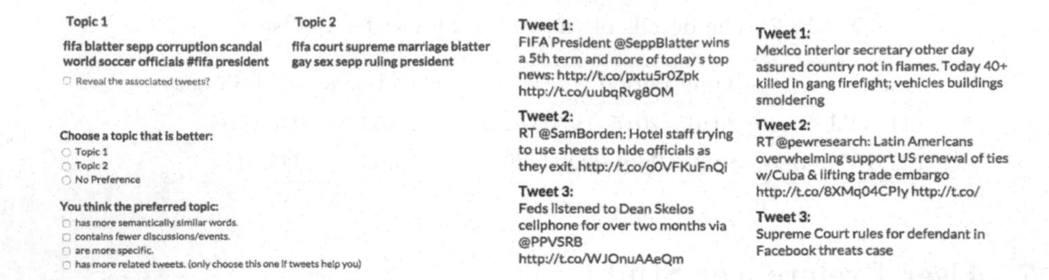

Fig. 1. The designed user interface and the associated tweets for two topics.

to the United States, we limit the workers country to the United States only. The TV debate dataset contains topics that can be easily understood, and thus we set the workers country to English speaking countries (e.g. United Kingdom, United States, etc.). Overall, 77 different trusted workers for the NYJ dataset and 91 for the TV debate dataset were selected, respectively.

Human Ground-Truth Ranking Order. As described above, we obtain 5 human judgments for each topic pair. A topic receives one vote if it is preferred by one worker. Thus, we assign each topic in each topic pair a fraction of the 5 votes received. A higher number of votes indicates that the topic is judged as being more coherent. Hence, for each comparison unit, we obtain a number of data pairs. Then, we apply the methodology described in Sect. 5 to obtain the human ground-truth ranking order of the three topic modelling approaches, i.e. 1st, 2nd, 3rd.

8 Experimental Setup

In this section, we describe the experimental setup for generating the topics and implementing the automatic metrics.

Generating Topics. We use Mallet[4] and Twitter LDA[5] to deploy the three topic modelling approaches on the two datasets (described in Sect. 6). The LDA hyper-parameters α and β are set to $50/K$ and 0.01 respectively, which work well for most corpora [3]. In TLDA, we follow [2] and set γ to 20. We set the number of topics K to a higher number, 100, for the NYJ dataset as it contains many topics. The TV debate dataset contains fewer topics, particularly as it took place only over a 2 hour period, and politicians were asked to respond to questions on specific themes and ideas[6]. Hence, we set K to 30 for the TV debate dataset. Each topic modelling approach is run 5 times for the two datasets. Therefore, for each topic modelling approach, we obtain 500 topics in the NYJ dataset and 150 topics in the TV debate dataset. We use the methodology described in Sect. 5

[4] http://mallet.cs.umass.edu.
[5] http://github.com/minghui/Twitter-LDA.
[6] http://goo.gl/JtzJDz.

to generate 100 topic pairs for each comparison unit. For example, for comparison Unit(LDA,TLDA), we generate 50 topic pairs of Pairs(LDA→TLDA) and 50 topic pairs of Pairs(TLDA→LDA).

Metrics Setup. Our metrics using WordNet (LCH, JCN & LESK) are implemented using the *WordNet::Similarity* package. We use the Wikipedia LSA space and the PMI data from the *SEMILAR* platform[7] to implement the W-LSA and W-PMI metrics. Since there are too many terms and tweets in our Twitter background dataset, we remove stopwords, terms occurring in less than 20 tweets, tweets with less than 10 terms and retweeting tweets. These steps help to reduce the computational complexity of LSA and PMI using this Twitter background dataset. After this pre-processing, the number of remaining tweets is 30,151,847. Table 3 shows the size of T-LSA space and the number of word pairs in T-PMI.

Table 3. The size of LSA space and the number of word pairs.

Model	Original size of matrix # of term × # of Doc	Model	# of word pairs
(1) W-LSA	1,096,192 × 3,873,895	(1)W-PMI	179,110,791
(2) T-LSA	609,878 × 30,151,847	(2) T-PMI	354,337,473

9 Results

We first compare the ranking order of the three topic modelling approaches using the automatic coherence metrics and human judgments. Then we show the differences between each of the automatic metric and human judgments.

Table 4 reports the average coherence score of the three topic models using the ten automatic metrics (displayed in white background). We also average the fraction of human votes of the three topic models, shown in Table 4 as column "human" (shown in grey background). We apply the methodology introduced in Sect. 5 to obtain the ranking orders shown in Table 4 as column "rank". By comparing the human ground-truth ranking orders of the three topic modelling approaches, we observe that the three topic modelling approaches perform differently over the two datasets.

Firstly, we observe that the ranking order from our proposed PMI-based metric using the Twitter background dataset (T-PMI) best matches the human ground-truth ranking order across our two Twitter datasets. This indicates that T-PMI can best capture the performance differences of the three topic modelling approaches. However, our other proposed metric T-LSA does not allow statistically distinguishable differences between topic modelling approaches to be identified (denoted by "×"). Second, for metrics based on semantic similarity, both W-PMI and W-LSA produce the same or a similar[8] ranking order

[7] http://semanticsimilarity.org.

[8] Part of the order matches the order from humans.

Table 4. The results of the automatic topic coherence metrics on the two datasets and the corresponding ranking orders. "×" means no statistically significant differences ($p \leq 0.05$) among the three topic modelling approaches. Two topic modelling approaches have the same rank if there are no significant differences between them.

NYJ												
	LCH	Rank	JCN	Rank	LESK	Rank	W-LSA	Rank	W-PMI	Rank	T-LSA	Rank
LDA	0.517		0.020		0.028		0.157	$1^{st}/2^{nd}$	0.205	1st	0.014	
TLDA	0.494	×	0.019	×	0.018	×	0.132	$1^{st}/2^{nd}$	0.190	2nd	0.004	×
PAM	0.544		0.021		0.009		0.073	3rd	0.150	3rd	0.011	
	T-PMI	Rank	U	Rank	V	Rank	B	Rank	Human	Rank		
LDA	1.63e-3	1st	0.092		0.548		1.365	1st	0.636	1st		
TLDA	1.52e-3	2nd	0.196	×	0.529	×	0.828	2nd	0.553	2nd		
PAM	4.53e-4	3rd	-0.074		0.542		-3.473	3rd	0.129	3rd		

TV debate												
	LCH	Rank	JCN	Rank	LESK	Rank	W-LSA	Rank	W-PMI	Rank	T-LSA	Rank
LDA	0.448		0.017		0.014		-0.019	$2^{nd}/3^{rd}$	0.134	$1^{st}/2^{nd}$	-0.033	
TLDA	0.434	×	0.016	×	0.014	×	0.064	1st	0.141	$1^{st}/2^{nd}$	-0.019	×
PAM	0.502		0.020		0.016		-0.041	$2^{nd}/3^{rd}$	0.127	3rd	-0.023	
	T-PMI	Rank	U	Rank	V	Rank	B	Rank	Human	Rank		
LDA	3.57e-4	$2^{nd}/3^{rd}$	0.293	$1^{st}/2^{nd}$	0.548		-1.31	$1^{st}/2^{nd}$	0.475	$2^{nd}/3^{rd}$		
TLDA	4.11e-4	1st	0.248	3rd	0.535	×	-0.606	$1^{st}/2^{nd}$	0.590	1st		
PAM	3.26e-4	$2^{nd}/3^{rd}$	0.304	$1^{st}/2^{nd}$	0.515		-2.092	3rd	0.431	$2^{nd}/3^{rd}$		

as humans on the two datasets. However, both W-PMI and W-LSA perform no better than T-PMI metric. On the other hand, for metrics based on statistical analysis, the B metric (statistical analysis on the document distribution) can also lead to a similar performance as W-LSA or W-PMI compared to human judgments. Moreover, our results show that the remaining metrics perform no better than T-PMI, W-PMI & W-LSA metrics according to the ranking orders, i.e. their ranking orders do not match the human ground-truth ranking order.

To further compare the automatic coherence metrics and human judgments, we use the sign test to determine whether the 10 automatic metrics perform differently than human judgments. Specifically, for an automatic metric or human judgments, we obtain 100 preference data points from 100 topic pairs for a comparison unit(e.g. Unit(T_1,T_2)), where "1"/"−1" represents that the topic from T_1/T_2 is preferred and "0" means no preference. Then, we hypothesise that there are no differences between the preference data points from an automatic metric and that from humans for a comparison unit (null hypothesis), and thus we calculate the p-values reported in Table 5. Each metric gets 6 tests (3 tests from the NYJ dataset and 3 tests from the TV debate dataset). If $p \leq 0.05$, the null hypothesis is rejected, which means that there are differences between the preferences of the same comparison unit between a given metric and humans.

We observe that the null hypotheses of 6 tests of T-PMI metric are not rejected across the two datasets. This suggests that T-PMI is the most aligned coherence metric with human judgments since there are no differences between T-PMI and human judgments for all the comparison units (shown in Table 5, $p \geq 0.05$). Moreover, only one test of W-PMI shows preference differences in a comparison unit (i.e. Unit(TLDA,PAM) in the NYJ dataset, where the null hypothesis is rejected) while W-LSA gets two tests rejected. Apart from these

Table 5. The obtained p-values from the sign tests.

NYJ										
	LCH	JCN	LESK	W-LSA	W-PMI	T-LSA	T-PMI	U	V	B
LDA vs. TLDA	0.104	0.133	**0.039**	0.783	0.779	0.097	0.410	**4.1e-11**	0.787	**2.2e-13**
TLDA vs. PAM	**2.7e-9**	**3.8e-10**	**0.0**	**1.8e-7**	**1.1e-4**	**1.7e-10**	1.0	**0.007**	**8.1e-13**	**0.007**
LDA vs. PAM	**2.2e-13**	**3.4e-11**	**7.2e-14**	**0.001**	0.210	**3.0e-11**	0.145	1.0	**2.4e-10**	**0.003**
TV debate										
	LCH	JCN	LESK	W-LSA	W-PMI	T-LSA	T-PMI	U	V	B
LDA vs. TLDA	**0.010**	0.104	0.075	0.999	0.401	0.651	0.999	**1.2e-6**	**2.0e-5**	**0.011**
TLDA vs. PAM	**0.003**	**0.007**	**0.005**	0.211	0.568	**0.010**	0.783	**4.7e-5**	**0.003**	**3.6e-12**
LDA vs. PAM	0.174	**0.007**	0.576	0.671	0.391	0.791	0.882	0.391	0.895	0.202

three metrics, the tests of the other metrics indicate that there are significant differences between these metrics and human judgments in most of comparison units. In summary, we find that the T-PMI metric demonstrates the best alignment with human preferences.

10 Conclusions

In this paper, we used three topic modelling approaches to evaluate the effectiveness of ten automatic topic coherence metrics for assessing the coherence of topic models generated from two Twitter datasets. Moreover, we proposed two new topic coherence metrics that use a separate Twitter dataset as background data when measuring the coherence of topics. By using crowdsourcing to obtain pairwise user preferences of topical coherences, we determined how closely each of the ten metrics align with the human judgments. We showed that our proposed PMI-based metric (T-PMI) provided the highest levels of agreement with the human assessments of topic coherence. Therefore, we recommend its use in assessing the coherence of topics generated from Twitter. If Twitter background data is not available, then we suggest one use PMI-based and LSA-based metrics using Wikipedia as a background (c.f. W-PMI & W-LSA). Among the metrics not requiring background data, the B metric (statistical analysis on the document distribution) is the most aligned with user preferences. For future work, we will investigate how to use the topic coherence metrics such that the topic modelling approaches can be automatically tuned to generate topics with high coherence.

References

1. Hong, L., Davison, B.D.: Empirical study of topic modeling in Twitter. In: Proceedings of SOMA (2010)
2. Zhao, W.X., Jiang, J., Weng, J., He, J., Lim, E.-P., Yan, H., Li, X.: Comparing Twitter and traditional media using topic models. In: Clough, P., Foley, C., Gurrin, C., Jones, G.J.F., Kraaij, W., Lee, H., Mudoch, V. (eds.) ECIR 2011. LNCS, vol. 6611, pp. 338–349. Springer, Heidelberg (2011)

3. Steyvers, M., Griffiths, T.: Probabilistic topic models. Handb. Latent Semant. Anal. **427**(7), 424–440 (2007)
4. Mei, Q., Shen, X., Zhai, C.: Automatic labeling of multinomial topic models. In: Proceedings of SIGKDD (2007)
5. Fang, A., Ounis, I., Habel, P., Macdonald, C., Limsopatham, N.: Topic-centric classification of Twitter user's political orientation. In: Proceedings of SIGIR (2015)
6. AlSumait, L., Barbará, D., Gentle, J., Domeniconi, C.: Topic significance ranking of LDA generative models. In: Proceedings of ECMLPKDD (2009)
7. Newman, D., Karimi, S., Cavedon, L.: External evaluation of topic models. In: Proceedings of ADCS (2009)
8. Newman, D., Lau, J.H., Grieser, K., Baldwin, T.: Automatic evaluation of topic coherence. In: Proceedings of NAACL (2010)
9. Li, W., McCallum, A.: Pachinko allocation: DAG-structured mixture models of topic correlations. In: Proceedings of ICML (2006)
10. Blei, D.M., Ng, A.Y., Jordan, M.I.: Latent dirichlet allocation. J. Mach. Learn. Res. **3**, 993–1022 (2003)
11. Li, W., Blei, D., McCallum, A.: Nonparametric bayes pachinko allocation. In: Proceedings of UAI (2007)
12. Wallach, H.M., Murray, I., Salakhutdinov, R., Mimno, D.: Evaluation methods for topic models. In: Proceedings of ICML (2009)
13. Chang, J., Gerrish, S., Wang, C., Boyd-Graber, J.L., Blei, D.M.: Reading tea leaves: how humans interpret topic models. In: Proceedings of NIPS (2009)
14. Fellbaum, C.: WordNet. Wiley Online Library, New York (1998)
15. Leacock, C., Chodorow, M.: Combining local context and WordNet similarity for word sense identification. WordNet Electr. Lexical Database **49**(2), 265–283 (1998)
16. Jiang, J.J., Conrath, D.W.: Semantic similarity based on corpus statistics and lexical taxonomy. In: Proceedings of ICRCL (1997)
17. Lesk, M.: Automatic sense disambiguation using machine readable dictionaries: how to tell a pine cone from an ice cream cone. In: Proceedings of SIGDOC (1986)
18. Rus, V., Lintean, M.C., Banjade, R., Niraula, N.B., Stefanescu, D.: SEMILAR: the semantic similarity toolkit. In: Proceedings of ACL (2013)
19. Recchia, G., Jones, M.N.: More data trumps smarter algorithms: comparing pointwise mutual information with latent semantic analysis. Behav. Res. Meth. **41**(3), 647–656 (2009)
20. Landauer, T.K., Foltz, P.W., Laham, D.: An introduction to latent semantic analysis. Discourse Processes **25**(2–3), 259–284 (1998)
21. Stefănescu, D., Banjade, R., Rus, V.: Latent semantic analysis models on wikipedia and TASA. In: Proceedings of LREC (2014)
22. Carterette, B., Bennett, P.N., Chickering, D.M., Dumais, S.T.: Here or there. In: Macdonald, C., Ounis, I., Plachouras, V., Ruthven, I., White, R.W. (eds.) ECIR 2008. LNCS, vol. 4956, pp. 16–27. Springer, Heidelberg (2008)
23. Mackie, S., McCreadie, R., Macdonald, C., Ounis, I.: On choosing an effective automatic evaluation metric for microblog summarisation. In: Proceedings of IIiX (2014)

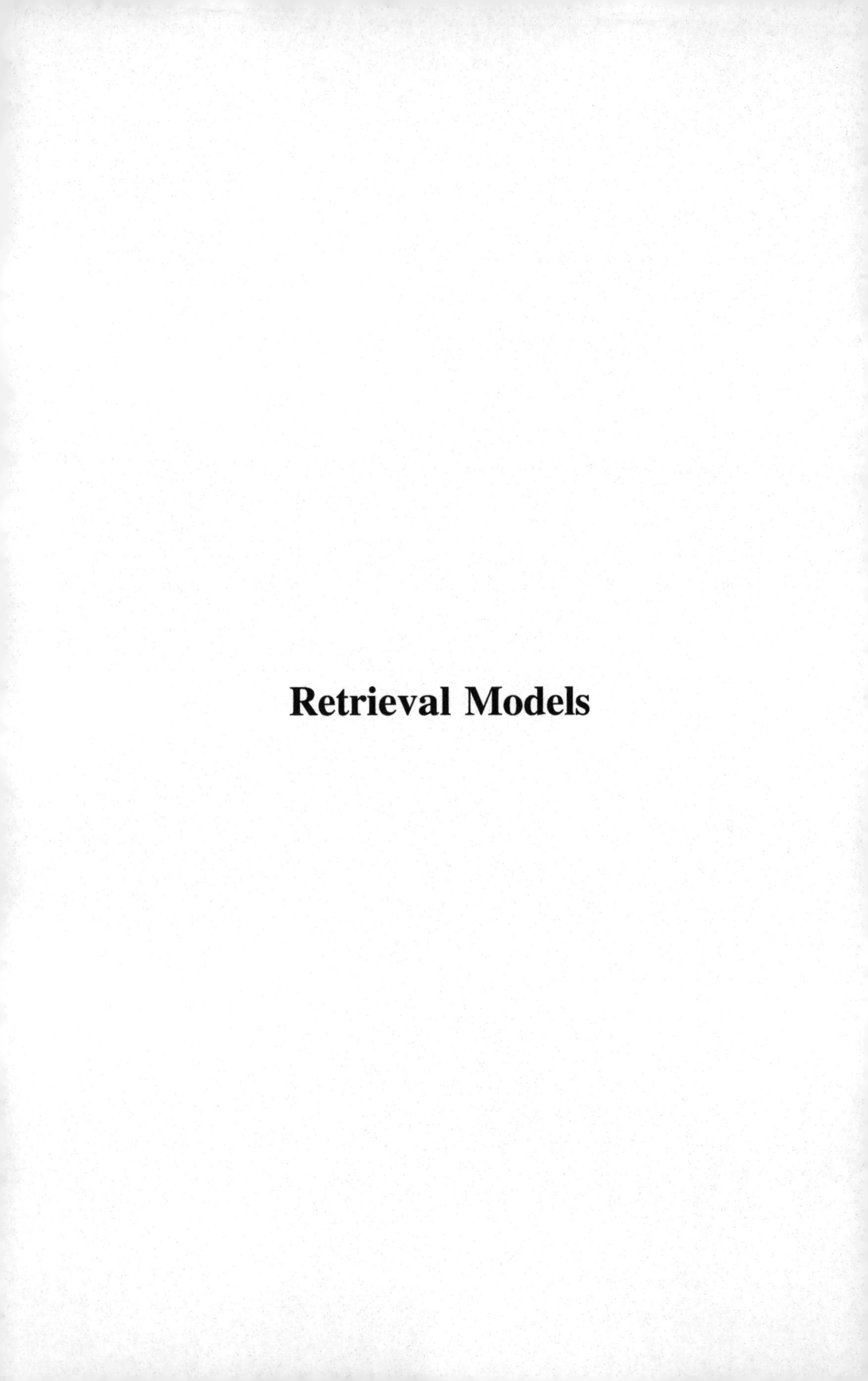

Retrieval Models

Supporting Scholarly Search with Keyqueries

Matthias Hagen(✉), Anna Beyer, Tim Gollub, Kristof Komlossy, and Benno Stein

Bauhaus-Universität Weimar, Weimar, Germany
{matthias.hagen,anna.beyer,tim.gollub,kristof.komlossy, benno.stein}@uni-weimar.de

Abstract. We deal with a problem faced by scholars every day: identifying relevant papers on a given topic. In particular, we focus on the scenario where a scholar can come up with a few papers (e.g., suggested by a colleague) and then wants to find "all" the other related publications. Our proposed approach to the problem is based on the concept of keyqueries: formulating keyqueries from the input papers and suggesting the top results as candidates of related work.

We compare our approach to three baselines that also represent the different ways of how humans search for related work: (1) a citation-graph-based approach focusing on cited and citing papers, (2) a method formulating queries from the paper abstracts, and (3) the "related articles"-functionality of Google Scholar. The effectiveness is measured in a Cranfield-style user study on a corpus of 200,000 papers. The results indicate that our novel keyquery-based approach is on a par with the strong citation and Google Scholar baselines but with substantially different results—a combination of the different approaches yields the best results.

1 Introduction

We tackle the problem of automatically supporting a scholar's search for related work. Given a research task, the term "related work" refers to papers on similar topics. Scholars collect and analyze related work in order to get a better understanding of their research problem and already existing approaches; a survey of the strengths and weaknesses of related work forms the basis for placing new ideas into context. In this paper, we show how the concept of keyqueries [9] can be employed to support search for related work.

Search engines like Google Scholar, Semantic Scholar, Microsoft Academic Search, or CiteSeerX provide a keyword-based access to their paper collections. However, since researchers usually have limited knowledge when they start to investigate a new topic, it is difficult to find all the related papers with one or two queries against such interfaces. Keyword queries help to identify a few promising initial papers, but to find further papers, researchers usually bootstrap their search from information in these initial papers.

Every paper provides two types of information useful for finding related work: content and metadata. Content (title, abstract, body text) is a good resource

N. Ferro et al. (Eds.): ECIR 2016, LNCS 9626, pp. 507–520, 2016.
DOI: 10.1007/978-3-319-30671-1_37

for search keywords. Metadata (bibliographic records, references) can be used to follow links to referenced papers. Recursively exploring the literature via queries or citations to and from some initial papers is common practice, although rather time-consuming. Support is provided by methods that automate the above procedure: graph-based methods exploit the citation network and content-based methods can generate queries.

Recently, Gollub et al. proposed the concept of keyqueries [9]: A *keyquery* for a document is a query that returns the document in the top result ranks against a (reference) search engine. Assuming that the top results returned by a document's keyqueries cover similar topics, we view the concept of keyqueries as promising for identifying related papers. Moreover, we believe that keyqueries can identify papers that graph-based methods might miss since the citation graph tends to be noisy and sparse [4]. Based on these assumptions, we address the research questions of whether keyqueries are useful for identifying related work and whether they complement other standard approaches.

Our contributions are threefold: (1) We develop a keyquery-based method identifying related work. (2) For the evaluation, we implement three strong baselines representing standard approaches: the graph-based Sofia Search by Golshan et al. [10], the query-based method by Nascimento et al. [24], and the "related articles"-feature of Google Scholar. (3) We conduct a Cranfield-style user study to compare the different approaches.

2 Related Work

Methods for identifying related work can be divided into citation-graph-based and content-based approaches. Only few of the content-based methods use queries such that we also investigate query formulation techniques for similar tasks.

2.1 Identifying Related Research Papers

Many variants of related work search are known: literature search, citation recommendation, research paper recommendation, etc. Some try to find references for a written text, others predict further "necessary" references given a subset of a paper's references. In our setting, the task is to find related papers according to a given set of papers. This represents the everyday use case of enlarging an initial related work research.

RELATED WORK SEARCH
Given: An input list $\mathcal{I} = \langle d_1, d_2, \ldots, d_n \rangle$ of papers
Task: Find an output list $\mathcal{O} = \langle d'_1, d'_2, \ldots, d'_m \rangle$ of related papers.

Given the initial knowledge of a scholar specified as the list $\mathcal{I}$ of input papers (could also be just one), most approaches to RELATED WORK SEARCH retrieve *candidate papers* in a first step. In a second step, the candidates are ranked to generate the final output list $\mathcal{O}$. More than 80 approaches are known for the problem of identifying related papers [1]. We concentrate on the recent and better performing approaches and classify them by the employed candidate retrieval method: *content* and/or *citations* can be used.

Citation-Graph-Based Methods. The network of citations forms the *citation graph.* If d cites d', we call d a *citing paper* of d' and d' a *cited paper* of d. Several approaches apply collaborative filtering (CF) using the adjacency matrix of the citation graph as the "rating" matrix [4,28]. A limitation of CF is that it has problems for poorly connected papers, known as the "cold-start-problem" [31]. Ekstrand et al. thus explore the additional application of link ranking algorithms like PageRank [6] that can also be applied stand-alone [19]. Since the citation graph tends to be noisy and sparse [4], by design, graph-based approaches favor frequently cited papers [31]. Methods that only use the citation graph easily miss papers that are rarely cited (e.g., very recent ones).

As a graph-based baseline, we select Sofia Search [10]. It very closely mimics the way how humans would identify candidates from the citation graph. Starting from an initial set of papers, the approach follows all links to cited and citing papers up to a given recursion depth or until a desired number of candidates is found. Note that this procedure conforms with CF methods and link ranking to some extent. Papers citing similar papers are linked via the cited papers, such that the CF candidates are included. Setting the recursion depth large enough, also all interesting results from PageRank random walk paths will be found—except the low-probability "clicks" on some random non-linked papers. Thus, Sofia Search forms a good representative of graph-based approaches.

Content-Based Techniques. Content-based approaches utilize the paper texts to find related work. Translation models are used to compute the probability of citing a paper based on citation contexts [15], the content of potential references [21], or via an embedding model [30]. Similar ideas are based on topic models: LDA was combined with PLSA to build a topic model from texts and citations [23]. Later improvements use only citation contexts instead of full texts [17,29]. Drawbacks of translation- and topic-model-based approaches are the long training phase and that re-training is necessary whenever papers dealing with new topics are added to the collection. Besides such efficiency aspects, topic and translation models do not resemble human behavior, and they cannot be used with the keyword query interfaces of existing scholarly search engines. We thus choose a query-based baseline to represent the content-based approaches.

While there are complete retrieval models for recommending papers for a given abstract using metadata and content-based features [3], we prefer a standard keyword query baseline since it can be used against any scholarly search interface. Nascimento et al. [24] propose such a method: given a paper, ten word bigrams are extracted from the title and the abstract and submitted as separate queries. Note that queries containing only two words are very general and return a large number of results. Still, Nascimento et al.'s idea is close to human behavior and forms a good content-based baseline.

Combined Approaches. Several methods combine citations and content. For instance, by querying with sentences from a given paper and then following references in the search results [13,14]. However, submitting complete sentences leads to very specific queries returning very few results only; a drawback we will avoid in our query formulation. Other combined approaches use topic models for citation weighting [7] or overcoming the cold-start-problem of papers without ratings in online reference management communities like CiteULike or Mendeley [31]. The CiteSight system [20] is supposed to recommend references while writing a manuscript using both graph- and content-based features to recommend papers the author cited in the past, or cited papers from references the author already added. Since our use case is different and considering only cited papers appears very restrictive, we do not employ this approach either.

As a representative of the combined approaches, we use Google Scholar's feature "related articles" to form an often used and very strong baseline. Even though the underlying algorithms are proprietary, it is very reasonable that content-based features (e.g., text similarity) and citation-based features (e.g., number of citations) are combined.

2.2 Query Formulation

Since the existing query techniques for related work search are rather simplistic, we briefly review querying strategies for other problems that inspired our approach.

There are several query-by-document approaches that derive "fingerprint" queries for a document in near-duplicate, text reuse, or similarity detection [2,5,32]. Hagen and Stein [12] further improve these query formulation strategies trying to satisfy a so-called *covering property* and the User-over-Ranking hypothesis [27]. Recently, Gollub et al. also introduced the concept of keyqueries for describing a document's content [9]. A query is a keyquery for a document if it returns the document in the top-ranked results when submitted to a reference search engine. Instead of just representing a paper by its keyqueries (as suggested by Gollub et al.), we further generalize the idea and assume that the other top-ranked results returned by a paper's keyqueries are highly related to the paper. We will adjust the keyquery formulation to our case of potentially more than one input paper and combine it with the covering property of Hagen and Stein [12] to derive a keyquery cover for the input papers.

3 Baselines and Our Approach

After describing three baselines, we introduce our novel keyquery-based approach and a straightforward interleaving scheme for combining the results of different methods.

3.1 Baselines

The baselines are chosen from the literature to mimic the strategies scholars employ: formulating queries, following citations, and Google Scholar's "related articles."

As a representative of the content-based strategies, we select a method by Nascimento et al. [24]. The approach submits as its ten queries the ten distinct consecutive bigrams of non-stopword terms from a paper's title and abstract that have the highest normalized *tf*-weights. For each query, the top-50 results are stored. The combined candidate set is then ranked according to the *tf*-weighted cosine similarity to the input paper. Note that the candidate retrieval and ranking are designed for only one input paper. We adapt the process to our use case by applying the retrieval phase for every input paper individually and then ranking the combined result sets of all input papers by the highest similarity to any of the input papers. The reason for taking the highest similarity and not the average is the better performance in pilot experiments. We also tried to combine the titles and abstracts to a single meta-paper in case of more than one input paper. Taking the highest similarity for individual input papers showed the best performance.

As a representative of the citation graph approaches, we select Sofia Search [10]. For each input paper d, both cited and citing papers are added to a candidate set $\mathcal{C}$. This routine is iterated using the candidates as new starting points until either enough (or no more) candidates are found or a specified recursion depth is reached (typically 2 or 3). Again, we rank the candidates by their highest similarity to any of the input papers.

Our third baseline is formed by Google Scholar's "related articles" feature. For a given research paper, a link in the Google Scholar interface yields a list of about 100 related articles. We collect all these related articles for each input paper individually, treating the underlying retrieval model as a black box. Since Google Scholar already presents ranked results, we do not re-rank them. In case of more than one input paper, we use a simple interleaving strategy: first the first rank for the first input paper, then the first rank for the second input paper, then the first rank for the third input paper, etc., then the second rank for the first input paper, etc. In case that a ranked result is already contained in the merged list it is not considered again.

3.2 New Keyquery-Based Approach

Our new approach combines the concepts of keyqueries [9] and query covers [12]. Generalizing the original single-document notion [9], a query q is a *keyquery* for a document set D with respect to a reference search engine S, if it fulfills the following conditions: (1) every $d \in D$ is in the top-k results returned by S on q, (2) q has at least l results, and (3) no subset $q' \subset q$ returns every $d \in D$ in its top-k results. The generality of a keyquery is defined by the parameters k and l (e.g., 10 or 50). We argue that the top-ranked results returned by a keyquery for D cover topics similar to the documents themselves—rendering keyqueries a promising concept for identifying related work.

The power set $\mathcal{Q} = 2^W$ forms the range of queries that can be formulated from the extracted keyphrases W of a document set D. A query q is said to *cover* the terms it contains. The more terms $w \in W$ are covered by q, and the more documents $d \in D$ are among q's top results, the better the query describes

Algorithm 1. Solving KEYQUERY COVER

Input: Sets D of documents and W of keywords, keyquery generality parameters k and l
Output: Set Q of keyqueries covering W
1: **for all** $w \in W$ **do**
2: **if** w returns less than l search results **then** $W \leftarrow W \setminus w$
3: $q \leftarrow \emptyset$
4: **for all** $w \in W$ **do**
5: $q \leftarrow q \cup \{w\}$
6: **if** q is keyquery for all $d \in D$ **then** $Q \leftarrow Q \cup \{q\}$, $q \leftarrow \emptyset$
7: **if** $q \neq \emptyset$ **then**
8: **for all** $w \in W$ **do**
9: **if** $\nexists\, q' \in Q : q' \subset q \cup \{w\}$ **then** $q \leftarrow q \cup \{w\}$
10: **if** q is keyquery for all $d \in D$ **then** $Q \leftarrow Q \cup \{q\}$, **break**
11: **return** Q

the topics represented by D's vocabulary W. The covering property [12] states that (1) in a proper set Q of queries for W, each term $w \in W$ should be contained in at least one query $q \in Q$ (i.e., the queries "cover" W in a set-theoretic sense) and (2) Q should be *simple* (i.e., $q_i \not\subseteq q_j$ for any $q_i, q_j \in Q$ with $i \neq j$), to avoid redundancy. The formal problem we tackle then is:

KEYQUERY COVER
Given: (1) A vocabulary W extracted from a set D of documents
(2) Levels k and l describing keyquery generality
Task: Find a simple set $Q \subseteq 2^W$ of queries that are keyquery for every $d \in D$ with respect to k and l and that together cover W.

The parameters k and l are typically set to 10, 50, or 100 but it will not always be possible to find a covering set of queries that are keyqueries for all documents in D. In such a case, we strive for queries that are keyqueries for a $|D| - 1$ subset of D.

Solving KEYQUERY COVER. Our approach has four steps (pseudocode in Algorithm 1). First, all keywords $w \in W$ are removed that return less than l results when submitted to the search engine S (lines 1–2); queries including such terms can not be keyqueries.

In a second step, the remaining terms $w \in W$ are iterated and added to an intermediate candidate query q (lines 3–5). If q is a keyquery for all papers $d \in D$, it is added to the set Q of keyqueries and q is emptied (line 6). Then the next query is formed from the remaining terms, etc., until the last term $w \in W$ has been processed.

After this first iteration, not all terms $w \in W$ are necessarily covered by the set Q of keyqueries. In this case, q is not empty (line 7) and we again go through the terms $w \in W$ (line 8) and consecutively add terms w to q—as long as no keyquery is found and simplicity of Q is not violated (line 9). According to the keyquery definition, keyqueries that are already in Q must not be contained in the candidate query q. We hence omit terms $w \in W$ that would cause q to contain a keyquery $q' \in Q$.

After this iteration, we do not further deepen the search but output Q although still not all terms $w \in W$ may be covered. This heuristic serves efficiency reasons and our experiments show that often all possible keywords are covered.

Solving RELATED WORK SEARCH. The pseudocode of our algorithm using keyquery covers for related work search is given as Algorithm 2. Using the KEYQUERY COVER algorithm as a subroutine, the basic idea is to first try to find keyqueries for all given input papers, and to add the corresponding results to the set $\mathcal{C}$ of candidates. If there are too few candidates after this step, KEYQUERY COVER is solved for combinations with $|\mathcal{I}| - 1$ input papers, then with $|\mathcal{I}| - 2$ input papers, etc.

The vocabulary combination (line 3) is based on the top-20 keyphrases per paper extracted by KP-Miner [8], the best unsupervised keyphrase extractor for research papers in SemEval 2010 [18]. The terms (keyphrases) in the combined vocabulary list W are ranked by the following strategy: First, all terms that appear in all papers ranked according to their mean rank in the different lists. Below these, all terms contained in $(|D| - 1)$-sized subsets ranked according to their mean ranks, etc.

In the next steps (lines 4–6), the KEYQUERY COVER-instance is solved for the subset D and its combined vocabulary W, the found candidate papers are added to the candidate set $\mathcal{C}$. In case that enough candidates are found, the algorithm stops (line 7). Otherwise, some other input subset is used in the next iteration. If not enough candidates can be found with keyqueries for more than one paper, the keyqueries for the single papers form the fallback option (also applies to single-paper inputs).

In our experiments, we will set $k, l = 10$ and $c = 100 \cdot |\mathcal{I}|$ (to be comparable, also the baselines are set to retrieve 100 candidate papers per input paper).

Algorithm 2. Solving RELATED WORK SEARCH

Input: List $\mathcal{I}$ of input papers, number c of desired related papers, keyquery generality parameters k and l
Output: Set $\mathcal{C}$ of candidate related papers

1: **for** $i \leftarrow |\mathcal{I}|$ down to 1 **do**
2:     **for all** $D : D \in 2^{\mathcal{I}}, |D| = i$ **do**
3:         $W \leftarrow$ combine vocabularies of documents in D
4:         $Q \leftarrow$ KEYQUERY COVER(D, W, k, l)
5:         $C \leftarrow$ combined at most top-l results of each $q \in Q$
6:         $\mathcal{C} \leftarrow \mathcal{C} \cup C$
7:         **if** $|\mathcal{C}| \geq c$ **then break**
8: **return** $\mathcal{C}$ ranked by highest similarity to any document in $\mathcal{I}$

3.3 Combining Approaches

To combine different RELATED WORK SEARCH algorithms (e.g., Google Scholar's related articles and our keyquery approach), we use a simple interleaving procedure. First, the results of the individual approaches are computed and ranked as described above. We then interleave the ranked lists by first taking the first rank from the first approach, then the first rank from the second approach and so on, then the second rank from the first approach, etc. Already contained papers are not added again.

4 Evaluation

To experimentally compare the algorithms, we conduct a Cranfield-style experiment on 200,000 computer science papers. Topics and judgments are acquired in a user study.

4.1 Experimental Design

In general, there are two different approaches to evaluate algorithms for RELATED WORK SEARCH. A widely used method is to take the reference lists of papers as ground truth for the purpose of evaluation. Some of the references of a single input paper are hidden and it is measured whether the RELATED WORK SEARCH algorithm is able to re-identify these. The second possible method uses relevance judgments assigned by scholars. These judgments then state whether a recommended paper is relevant to a specific topic or not. Since the first method is rather biased towards the citation graph and also not really representing the use case we have in mind, we choose the second approach and utilize human relevance judgments from a carefully designed user study.

Paper Corpus. We crawled a corpus of computer science papers starting from the 35,000 papers published at 20 top-tier conferences like SIGIR, CHI, CIKM, ACL, STOC, and iteratively including cited and citing papers until the desired size of 200,000 papers was reached. In the crawling process, not all papers had a full text available on the web (for 57 % of the corpus, we have an associated full text). In this case, only the abstracts and metadata were obtained when possible (contained for 43 % of the corpus) but often not even abstracts were available and the respective papers were not included. On average, only 7.0 of the 13.9 references in a paper and 6.8 of the 9.5 citations to a paper could be crawled. Thus, only about 60 % of the cited/citing papers are contained in our corpus. Not surprisingly, especially older papers often were not available. The papers included in our corpus have been published in the years 1962–2013 (more than 75 % published after 2000). Starting from the 20 seed conferences, we included papers from about 1,000 conferences/workshops and 500 journals.

Experimental Setup. Sofia Search is run on the citation information contained in the metadata of the papers. Note that due to the non-availability of 40 % of the cited/citing papers, our corpus might not be optimally suited for Sofia Search. Still, this was not intended in the corpus design and could not be avoided—it rather represents the realistic web scenario but should be kept in mind when analyzing Sofia Search's performance.

In order to run the query-based baseline and our keyquery cover algorithm, we indexed the corpus papers using Lucene 5.0 while treating title, abstract, and body text as separate fields. In case that no full text is available in the corpus, only title and abstract are indexed. The retrieval model is BM25F [26] with different boost factors: the title is the most important followed by the abstract and then the body text.

Whenever a corpus paper with only title and abstract is used as an input paper for our keyquery-based algorithm, the keyphrase extraction is done on these two fields only and text similarity of a candidate paper is also only measured against these two fields (similar to the Sofia Search and Nascimento et al. baselines).

User Study Design. Topics (information needs) and relevance judgments for a Cranfield-style analysis of the RELATED WORK SEARCH approaches are obtained from computer science students and scholars (other qualifications do not match the corpus characteristics). A study participant first specifies a topic by selecting a set of input papers (could just be one) and then judges the found papers of the different approaches with respect to their relevance. These judgments are the basis for our experimental comparison.

In order to ensure a smooth work flow, we have built a web interface that also allows a user to participate without being on site. The study itself consists of two steps with different interfaces. In the first step, a participant is asked to enter a research task they are familiar with and describe it with one or two sentences. Note that we request a familiar research task because expert knowledge is later required in order to judge the relevance of the suggested papers. After task description, the participants have to enter titles of input papers related to their task. While the user is entering a title, a background process automatically suggests title auto completions from our corpus. Whenever a title was not chosen from the suggestions but was manually entered, it is again checked whether the specified paper exists in our corpus or not. If the paper cannot be found in the collection, the user is notified to enter another title. After this two-phase topic formation (written description + input papers), the participants have to name at least one paper that they expect to be found (again with the help of auto completion). Last, the users should describe how they have chosen the input and expected papers to get some feedback of whether for instance Google Scholar was used which might bias the judgments.

After a participant has completed the topic formation, the different recommendation algorithms are run on the input papers and the pooled set of the top-10 results of each approach is displayed in random order for judgment (i.e., at most 40 papers to judge). For each paper, the fields title, authors, publication venue, and publication year are shown. Additionally, links to fade in the abstract and, if available in our corpus, to the respective PDF-file are listed. Thus, the participants can check the abstract or the full text if needed.

The participants assessed two criteria: relevance and familiarity. Relevance was rated on a 4-point scale (highly relevant, relevant, marginally, not relevant) while familiarity was a two-level judgment (familiar or not). Combining the two criteria, we can identify good papers not known to the participant before our study. This is especially interesting since we asked for research topics the participants are familiar with and in such a scenario algorithms identifying relevant papers not known before are very promising.

Characteristics of the User Study. In total, 13 experienced scholars and 7 graduate students have "generated" and judged 42 topics in our study—in a previous study we had another 25 topics with single-paper inputs only [11]. The scholars typically chose topics related to one of their paper projects while the graduate students chose topics from their Master's theses. On average, the topic creation took about 4 min while the judgment took about 27 min. For 23 topics, one or two input papers are given, while for 19 topics even three or up to five input papers were specified. For most topics (80 %), two or more expected papers were entered. The number of different papers returned by the four algorithms varies highly. For the topic with the most different results, the participant had to judge 37 papers, while the topic with the least different results required only 13 judgments. On average, a participant had to judge 28 papers. In total, 31 % of the results were judged as highly relevant, 22 % as relevant, 24 % as marginally relevant, and 23 % as not at all relevant to the topic. Interestingly, 79 % of the papers that are judged as relevant and 59 % of the highly relevant papers were unfamiliar to the user before the study. We will evaluate the algorithms with respect to their ability of retrieving unexpected good results when discussing the experimental results.

4.2 Experimental Results

We employ the standard retrieval effectiveness measures of nDCG [16] and precision for the top-10 results, and recall for evaluating the ability to retrieve the expected and also the relevant unexpected papers. The experimental results are reported in Table 1 (Left). The top four rows contain the results for the four individual algorithms while the bottom three rows contain the most promising combinations that can also be evaluated due to the pooling strategy. In the top-k results of our combination scheme described in Sect. 3.3 only results from the top-k of the combined algorithms are included.

General Retrieval Performance. We measure nDCG using the 4-point scale: high as 3, relevant as 2, marginal as 1, and not relevant as 0, and precision considering highly and relevant as the relevant class. The mean nDCG@10 and $prec$@10 over all topics are given in the second and third columns of Table 1 (Left). The top-10 of our new keyquery-cover-based approach KQC are the most relevant

Table 1. (Left) Performance values achieved by the different algorithms averaged over all topics. (Right) Percentage of top-10 rank overlap averaged over all topics.

Algorithm	nDCG@10	*prec*@10	rec_e@50	rec_{ur}@10
Sofia Search	0.60	0.59	0.33	0.20
Nascimento	0.58	0.56	0.34	0.16
Google Scholar	0.60	0.59	**0.43**	**0.21**
KQC	**0.62**	**0.60**	0.37	0.16
KQC+Google	**0.65**	0.63	**0.48**	0.21
KQC+Sofia	0.61	0.60	0.39	0.19
KQC+Sofia+Google	**0.65**	**0.64**	**0.48**	**0.24**

	Sofia	Nasc.	Google	KQC
Sofia Search	100	35	17	43
Nascimento	35	100	15	36
Google Scholar	17	15	100	18
KQC	43	36	18	100

among the individual methods on a par with Google Scholar and Sofia Search; both differences are not significant according to a two-sided paired t-test ($p = 0.05$). The overall best result is achieved by the KQC+Sofia+Google combination that significantly outperforms its components. Another observation is that most methods significantly outperform the Nascimento et al. baseline.

Recall of Expected Papers. To analyze how many papers were retrieved that the scholars entered as expected in the first step of our study, we use the recall of the expected papers denoted as rec_e. Since the computation of rec_e is not dependent on the obtained relevance judgments, we can compute recall for any k. We choose $k = 50$ since we assume that a human would often not consider many more than the top-50 papers of a single related work search approach. The fourth column of Table 1 (Left) shows the average rec_e@50-values for each algorithm over all topics. The best rec_e@50 among the individual methods is the 0.43 of Google Scholar. The combinations including KQC and Google Scholar achieve an even better result of 0.48. An explanation for the advantage of Google Scholar—beyond the probably good black-box model underlying the "related articles" functionality—can be found in the participants' free text fields of topic formation. For several topics, the study participants stated that they used Google Scholar to come up with the set of expected papers and this obviously biases the results.

Recall of Unfamiliar but Relevant Papers. We also measure how many "gems" the different algorithms recommend. That is, how many highly or relevant papers are found that the user was not familiar with before our study. Providing such papers is a very interesting feature that might eventually help a researcher to find "all" relevant literature on a given topic. Again we measure recall, but since we have familiarity judgments for the top-10 only, we measure the recall rec_{ur}@10 of unexpected but relevant papers listed in the rightmost column of Table 1 (Left). Again, the combination KQC+Sofia+Google finds the most of the unfamiliar but relevant papers.

Result Overlap. Since no participant had to judge 40 different papers, the top-10 results cannot be completely distinct; Table 1 (Right) shows the percentage of overlap for each combination. On average, the top-10 retrieved papers of two different algorithms share 2–4 papers meaning that the approaches retrieve a rather diverse set of related papers. This again suggests combinations of different approaches as the best possible system and combining the best query-based method (our new KQC), Google Scholar, and Sofia Search indeed achieves the best overall performance. Having in mind the "sparsity" of our crawled paper corpus' citation graph (only about 60 % of the cited/citing papers could be crawled), Sofia Search probably could diversify the results even more in a corpus containing all references. The combination of the three systems in our opinion also very well models the human way of looking for related work such that the KQC+Sofia+Google combination could be viewed as very close to automated human behavior.

A Word on Efficiency. The most costly part of our approach is the number of submitted queries. In our study, about 79 queries were submitted per topic; results for already submitted queries were cached such that no query was submitted twice (e.g., once when trying to find a keyquery for all input papers together, another time for a smaller subset again). On the one hand, 79 queries might be viewed more costly than the about 27 queries submitted by the Nascimento et al. baseline ($10 \cdot |\mathcal{I}|$ queries) or the about 30 requests submitted for the Google Scholar suggestions ($11 \cdot |\mathcal{I}|$ requests; one to find an individual input paper, ten to retrieve the 100 related articles). Also Sofia Search on a good index of the citation graph is much faster than our keyquery-based approach. On the other hand, keyqueries could be pre-computed at indexing time for every paper such that at retrieval time only a few postlists from a reverted index [25] have to be merged. This would substantially speed up the whole process, rendering a reverted-index-based variant of our keyquery approach an important step for deploying the first real prototype.

5 Conclusion

We have presented a novel keyquery-based approach to related work search. The addressed common scenario is a scholar who has already found a handful of papers in an initial research and wants to find "all" the other related papers—often a rather tedious task. Our problem formalization of Related Work Search is meant to provide automatic support in such situations. As for solving the problem, our new keyquery-based technique focuses on the content of the already found papers complementing most of the existing approaches that exploit the citation graph only. Our overall idea is to get the best of both worlds (i.e., queries and citations) from appropriate method combinations.

And in fact, in our effectiveness evaluations of a Cranfield-style experiment on a collection of about 200,000 computer science papers, the combination of keyqueries with the citation-based Sofia Search and Google Scholar's related article suggestions performed best on 42 topics (i.e., sets of initial papers). The top-10 results of each approach were judged by the expert who suggested the topic. Based on these relevance judgments, we have evaluated the different algorithms and identified promising combinations based on the rather different returned results of the individual approaches amongst which keyqueries slightly outperformed Sofia Search and Google Scholar suggestions.

Our experiments should be confirmed using other topic sets and paper corpora. The iSearch collection [22] may be suitable: it comprises a large set of research papers from the domain of physics and a set of topics with corresponding relevance judgments. Yet, the iSearch collection judgments may have been obtained using keyword queries, a fact that could give an advantage to our keyquery approach. Another promising future direction is to evaluate other background retrieval models, as well as keyphrase selection and weighting methods since our keyqueries heavily depend on these. Also the candidate ranking deserves further analyses. We have adopted simple text-based cosine similarity, which is also used in many other approaches but taking the number of citations or the publication venue into account may further improve the rankings.

References

1. Beel, J., Langer, S., Genzmehr, M., Gipp, B., Breitinger, C., Nürnberger, A.: Research paper recommender system evaluation: a quantitative literature survey. In: RepSys Workshop, pp. 15–22 (2013)
2. Bendersky, M., Croft, W.B.: Finding text reuse on the web. In: WSDM, pp. 262–271 (2009)
3. Bethard, S., Jurafsky, D.: Who should I cite: learning literature search models from citation behavior. In: CIKM, pp. 609–618 (2010)
4. Caragea, C., Silvescu, A., Mitra, P., Giles, C.L.: Can't see the forest for the trees? a citation recommendation system. In: JCDL, pp. 111–114 (2013)
5. Dasdan, A., D'Alberto, P., Kolay, S., Drome, C.: Automatic retrieval of similar content using search engine query interface. In: CIKM, pp. 701–710 (2009)
6. Ekstrand, M.D., Kannan, P., Stemper, J.A., Butler, J.T., Konstan, J.A., Riedl, J.: Automatically building research reading lists. In: RecSys, pp. 159–166 (2010)
7. El-Arini, K., Guestrin, C.: Beyond keyword search: discovering relevant scientific literature. In: KDD, pp. 439–447 (2011)
8. El-Beltagy, S.R., Rafea, A.A.: KP-Miner: a keyphrase extraction system for English and Arabic documents. Inf. Syst. **34**(1), 132–144 (2009)
9. Gollub, T., Hagen, M., Michel, M., Stein, B.: From keywords to keyqueries: content descriptors for the web. In: SIGIR, pp. 981–984 (2013)
10. Golshan, B., Lappas, T., Terzi, E.: Sofia search: a tool for automating related-work search. In: SIGMOD, pp. 621–624 (2012)
11. Hagen, M., Glimm, C.: Supporting more-like-this information needs: finding similar web content in different scenarios. In: Kanoulas, E., Lupu, M., Clough, P., Sanderson, M., Hall, M., Hanbury, A., Toms, E. (eds.) CLEF 2014. LNCS, vol. 8685, pp. 50–61. Springer, Heidelberg (2014)
12. Hagen, M., Stein, B.: Candidate document retrieval for web-scale text reuse detection. In: Grossi, R., Sebastiani, F., Silvestri, F. (eds.) SPIRE 2011. LNCS, vol. 7024, pp. 356–367. Springer, Heidelberg (2011)
13. He, Q., Kifer, D., Pei, J., Mitra, P., Giles, C.L.: Citation recommendation without author supervision. In: WSDM, pp. 755–764 (2011)
14. He, Q., Pei, J., Kifer, D., Mitra, P., Giles, C.L.: Context-aware citation recommendation. In: WWW, pp. 421–430 (2010)

15. Huang, W., Kataria, S., Caragea, C., Mitra, P., Giles, C.L., Rokach, L.: Recommending citations: translating papers into references. In: CIKM, pp. 1910–1914 (2012)
16. Järvelin, K., Kekäläinen, J.: Cumulated gain-based evaluation of IR techniques. ACM Trans. Inf. Syst. **20**(4), 422–446 (2002)
17. Kataria, S., Mitra, P., Bhatia, S.: Utilizing context in generative Bayesian models for linked corpus. In: AAAI, pp. 1340–1345 (2010)
18. Kim, S.N., Medelyan, O., Kan, M.-Y., Baldwin, T.: SemEval-2010 task 5: automatic keyphrase extraction from scientific articles. In: SemEval 2010, pp. 21–26 (2010)
19. Küçüktunç, O., Saule, E., Kaya, K., Catalyürek, Ü.V.: TheAdvisor: a webservice for academic recommendation. In: JCDL, pp. 433–434 (2013)
20. Livne, A., Gokuladas, V., Teevan, J., Dumais, S., Adar, E.: CiteSight: supporting contextual citation recommendation using differential search. In: SIGIR, pp. 807–816 (2014)
21. Lu, Y., He, J., Shan, D., Yan, H.: Recommending citations with translation model. In: CIKM, pp. 2017–2020 (2011)
22. Lykke, M., Larsen, B., Lund, H., Ingwersen, P.: Developing a test collection for the evaluation of integrated search. In: Gurrin, C., He, Y., Kazai, G., Kruschwitz, U., Little, S., Roelleke, T., Rüger, S., van Rijsbergen, K. (eds.) ECIR 2010. LNCS, vol. 5993, pp. 627–630. Springer, Heidelberg (2010)
23. Nallapati, R., Ahmed, A., Xing, E.P., Cohen, W.W.: Joint latent topic models for text and citations. In: KDD, pp. 542–550 (2008)
24. Nascimento, C., Laender, A.H.F., Soares da Silva, A., Gonçalves, M.A.: A source independent framework for research paper recommendation. In: JCDL, pp. 297–306 (2011)
25. Pickens, J., Cooper, M., Golovchinsky, G.: Reverted indexing for feedback and expansion. In: CIKM, pp. 1049–1058 (2010)
26. Robertson, S.E., Zaragoza, H., Taylor, M.J.: Simple BM25 extension to multiple weighted fields. In: CIKM, pp. 42–49 (2004)
27. Stein, B., Hagen, M.: Introducing the user-over-ranking hypothesis. In: Clough, P., Foley, C., Gurrin, C., Jones, G.J.F., Kraaij, W., Lee, H., Mudoch, V. (eds.) ECIR 2011. LNCS, vol. 6611, pp. 503–509. Springer, Heidelberg (2011)
28. Sugiyama, K., Kan, M.-Y.: Exploiting potential citation papers in scholarly paper recommendation. In: JCDL, pp. 153–162 (2013)
29. Tang, J., Zhang, J.: A discriminative approach to topic-based citation recommendation. In: Theeramunkong, T., Kijsirikul, B., Cercone, N., Ho, T.-B. (eds.) PAKDD 2009. LNCS, vol. 5476, pp. 572–579. Springer, Heidelberg (2009)
30. Tang, X., Wan, X., Zhang, X.: Cross-language context-aware citation recommendation in scientific articles. In: SIGIR, pp. 817–826 (2014)
31. Wang, C., Blei, D.M.: Collaborative topic modeling for recommending scientific articles. In: KDD, pp. 448–456 (2011)
32. Yang, Y., Bansal, N., Dakka, W., Ipeirotis, P.G., Koudas, N., Papadias, D.: Query by document. In: WSDM, pp. 34–43 (2009)

Pseudo-Query Reformulation

Fernando Diaz(✉)

Microsoft Research, New York, USA
fdiaz@microsoft.com

Abstract. Automatic query reformulation refers to rewriting a user's original query in order to improve the ranking of retrieval results compared to the original query. We present a general framework for automatic query reformulation based on discrete optimization. Our approach, referred to as *pseudo-query reformulation*, treats automatic query reformulation as a search problem over the graph of unweighted queries linked by minimal transformations (e.g. term additions, deletions). This framework allows us to test existing performance prediction methods as heuristics for the graph search process. We demonstrate the effectiveness of the approach on several publicly available datasets.

1 Introduction

Most information retrieval systems operate by performing a single retrieval in response to a query. Effective results sometimes require several manual reformulations by the user [11] or semi-automatic reformulations assisted by the system [10]. Although the reformulation process can be important to the user (e.g. in order to gain perspective about the domain of interest), the process can also lead to frustration and abandonment.

In many ways, the core information retrieval problem is to improve the initial ranking and user satisfaction and, as a result, reduce the need for reformulations, manual or semi-automatic. While there have been several advances in learning to rank given a fixed query representation [15], there has been somewhat less attention, from a formal modeling perspective, given to automatically reformulating the query before presenting the user with the retrieval results. One notable exception is pseudo-relevance feedback (PRF), the technique of using terms found in the top retrieved documents to conduct a second retrieval [4]. PRF is known to be a very strong baseline. However, it incurs a very high computational cost because it issues a second, much longer query for retrieval.

In this paper, we present an approach to automatic query reformulation which combines the iterated nature of human query reformulation with the automatic behavior of PRF. We refer to this process as *pseudo-query reformulation* (PQR). Figure 1 graphically illustrates the intuition behind PQR. In this figure, each query and its retrieved results are depicted as nodes in a graph. An edge exists between two nodes, q_i and q_j, if there is a simple reformulation from q_i to q_j; for example, a single term addition or deletion. This simulates the incremental query modifications a user might conduct during a session. The results

N. Ferro et al. (Eds.): ECIR 2016, LNCS 9626, pp. 521–532, 2016.
DOI: 10.1007/978-3-319-30671-1_38

in this figure are colored so that red documents reflect relevance. If we assume that a user is following a good reformulation policy, then, starting at q_0, she will select reformulations (nodes) which incrementally increase the number of relevant documents. This is depicted as the path of shaded nodes in our graph. We conjecture that a user navigates from q_i to q_j by using insights from the retrieval results of q_i (e.g. q_j includes a highly discriminative term in the results for q_i) or by incorporating some prior knowledge (e.g. q_j includes a highly discriminative term in general). PQR is an algorithm which behaves in the same way: issuing a query, observing the results, inspecting possible reformulations, selecting a reformulation likely to be effective, and then iterating.

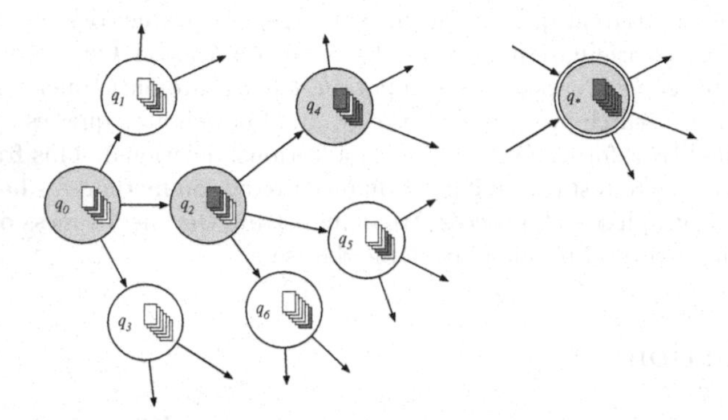

Fig. 1. Query reformulation as graph search. Nodes represent queries and associated retrieved results. Relevant documents are highlighted in red. Edges exist between nodes whose queries are simple reformulations of each other. The goal of pseudo-query reformulation is to, given a seed query q_0 by a user, automatically navigate to a better query (Color figure online).

PQR is a fundamentally new query reformulation model with several attractive properties. First, PQR directly optimizes performance for short, unweighted keyword interaction. This is important for scenarios where a searcher, human or artificial, is constrained by an API such as those found in many search services provided by general web search engines or social media sites. This constraint prevents the use of massive query expansion techniques such as PRF. Even if very long queries were supported, most modern systems are optimized (in terms of efficiency and effectiveness) for short queries, hurting the performance of massive query expansion. Second, our experiments demonstrate that PQR significantly outperforms several baselines, including PRF. Finally, PQR provides a framework in which to evaluate performance prediction methods in a grounded retrieval task.

2 Related Work

Kurland *et al.* present several heuristics for iteratively refining a language model query by navigating document clusters in a retrieval system [13]. The technique leverages specialized data structures storing document clusters derived from large scale corpus analysis. While related, the solution proposed by these authors violates assumptions in our problem definition. First, their solution assumes weighted language model style queries not supported by backends in our scenario. Second, their solution assumes access to the entire corpus as opposed to a search API.

Using performance predictors in order to improve ranking has also been studied previously, although in a different context. Sheldon *et al.* demonstrate how to use performance predictors in order to better merge result lists from pairs of reformulated queries [19]. This is, in spirit, quite close to our work and is a special case of PQR which considers only two candidate queries and a single iteration instead of hundreds of candidates over several iterations. In the context of learning to rank, performance predictors have been incorporated as ranking signals and been found to be useful [17]. From the perspective of query weighting, Lv and Zhai explored using performance predictors in order to set the optimal interpolation weight in pseudo-relevance feedback [16]. Similarly Xue and Croft have demonstrated how to use performance predictors in order to improve concept weighting in an inference network model [21]. Again, while similar to our work in the use of performance predictors for query reformulation, we focus on the discrete, iterated representation. The work of Xue and Croft focuses on a single iteration and a weighted representation. More generally, there has been some interest in detecting the importance of query terms in a long queries or in expanded queries [1,3].

The work of Kumaran and Carvahlo examines using performance predictors for *query reduction*, the task of selecting the best reformulation from the set of all $\mathcal{O}(2^n)$ possible subqueries of the original query [12]. Because of the size of this set, the authors use several heuristics to prune the search space. While related, our approach is substantially more scalable (in terms of number of reformulations considered) and, as a result, flexible (in terms of query transformations).

3 Motivation

As mentioned earlier, users often reformulate an initial query in response to the system's ranking [11]. Reformulation actions include adding, deleting, and substituting query words, amongst other transformations. There is evidence that manual reformulation can improve the quality of a ranked list for a given information need [11, Table 5]. However, previous research has demonstrated that humans are not as effective as automatic methods in this task [18].

In order to estimate an upper bound on the potential improvement from reformulation, we propose a simulation of an optimal user's reformulation behavior. Our simulator models an omniscient user based on query-document relevance judgments, known as qrels. Given a seed query, the user first generates a set of candidate reformulations based on one-word additions and deletions. The user then selects the query whose retrieval has the highest performance based on the qrels. The process then iterates up to some search depth. If we consider queries as nodes in a graph and edges between queries with single word additions or deletions, then the process can be considered a depth-limited graph search by an oracle.

We ran this simulation on a sample of queries described in Sect. 6. The results of these experiments (Table 1) demonstrate the range of performance for PQR. Our oracle simulator performs quite well, even given the limited depth of our search. Performance is substantially better than the baseline, with relative improvements greater than those in published literature. To some extent this should be expected since the oracle can leverage relevance information. Surprisingly, though, the algorithm is able to achieve this performance increase by adding and removing a small set of up to four terms.

Table 1. NDCG@30 for random and optimal (PQR*) pseudo-query reformulation compared to query likelihood (QL) and relevance model (RM3).

	trec12	Robust	Web
QL	0.4011	0.4260	0.1628
RM3	0.4578	0.4312	0.1732
Random	0.3162	0.2765	0.0756
PQR*	0.6482	0.6214	0.3053

4 Pseudo-Query Reformulation

We would like to develop algorithms that can approximate the behavior of our optimal policy *without having access to any qrels or an oracle*. As such, PQR follows the framework of the simulator from Sect. 3. The algorithm recursively performs candidate generation and candidate scoring within each recursion up to some depth and returns the highest scored query encountered.

4.1 Generating Candidates

Recall that our entire search space can be represented by a very large lattice of queries. Even if we were performing local graph search, the $O(|\mathcal{V}|)$ edges incident to any one node would make a single iteration computationally intractable. As a result, we need a method for pruning the full set of reformulation candidates to a smaller set that we can analyze in more detail. Fortunately, in many cases, we can establish heuristics so that we only consider those reformulations likely to improve performance. In our case, given q_t, we consider the following candidates, (a) all single term deletions from q_t, and (b) all single term additions from the n terms with the highest probability of occurring in relevant documents. Since we do not have access to the relevant documents at runtime, we approximate this distribution using the terms occurring in the retrieval for q_t. Specifically, we select the top n terms in the relevance model, $\theta_{\mathcal{R}_t}$, associated with q_t [14]. The relevance model is the retrieval score-weighted linear interpolation of retrieved document languages models. We adopt this approach for its computational ease and demonstrated effectiveness in pseudo-relevance feedback.

4.2 Scoring Candidates

The candidate generation process described in Sect. 4.1 provides a crude method for pruning the search space. Based on our observations with the random and oracle policies in Table 1, we know that inaccurately scoring reformulation candidates can significantly degrade the performance of an algorithm. In this section, we model the oracle by using established performance prediction signals.

Performance Prediction Signals. Performance prediction refers to the task of ordering a set of queries without relevance information so that the better performing queries are ordered above worse performing queries. With some exception, the majority of work in this area has focused on ranking queries coming from different information needs (i.e. one query per information need). We are interested in the slightly different task of ranking many queries for a single information need. Despite the difference in problem setting, we believe that, with some modifications discussed in Sect. 4.2, performance predictors can help model the oracle or, more accurately, the true performance of the reformulation. A complete treatment of related work is beyond the scope of this paper but details of approaches can be found in published surveys (e.g. [7]).

The set of performance predictors we consider can be broken into three sets: query signals, result set signals, and drift signals. Throughout this section, we will be describing signals associated with a candidate query q.

Query signals refer to properties of the terms in q alone. These signals are commonly referred to as 'pre-retrieval' signals since they can be computed without performing a costly retrieval. Previous research has demonstrated that queries including non-discriminative terms may retrieve non-relevant results. The inverse document frequency is one way to measure the discrimination ability of a term and has been used in previous performance prediction work [7]. Over all query terms in q, we consider the mean, maximum, and minimum IDF values. In addition to IDF, we use similarly-motivated signals such as Simplified Clarity (SC) and Query Scope (QS) [9].

Result set signals measure the quality of the documents retrieved by the query. These signals are commonly referred to as 'post-retrieval' signals. These features include the well-known Query Clarity (QC) measure, defined as the Kullback-Leibler divergence between the language model estimated from the retrieval results, $\theta_{\mathcal{R}_t}$, and the corpus language model, θ_C [5]. In our work, we use $\mathcal{B}(\mathcal{R}_t, \theta_C)$, the Bhattacharyya correlation between the corpus language model and the query language model [2]. This measure is in the unit interval and with low values for dissimilar pairs of language models and high values for similar pairs of language models. We use the Bhattacharyya correlation between these two distributions instead of the Kullback-Leibler divergence because the measure is bounded and, as a result, does not need to be rescaled across queries. We also use the score autocorrelation (SA), a measure of the consistency of scores of semantically related documents [6]. In our implementation, we again use the Bhattacharyya correlation to measure the similarity between all pairs of documents in $\mathcal{R}_t$, as represented by their maximum likelihood language models.

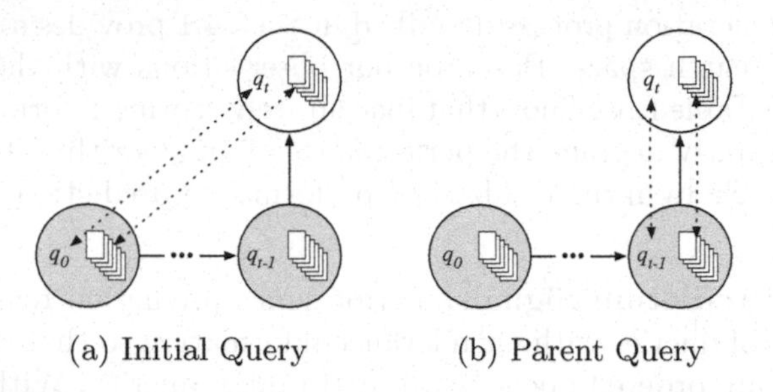

Fig. 2. Drift signal classes. Signals for q_t include comparisons with reference queries q_{t-1} and q_0 to prevent query drift.

Drift signals compare the current query q_t with its parent q_{t-1} and the initial query q_0 (Fig. 2). These signals can serve to anchor our prediction and avoid query drift, situations where a reformulation candidate appears to be high quality but is topically very different from the desired information need. One way to measure drift is to compute the difference in the query signals for these pairs. Specifically, we measure the aggregate IDF, SC, and QS values of the deleted, preserved, and introduced keywords. We also generate two signals comparing the results sets of these pairs of queries. The first measures the similarity of the ordering of retrieved documents. In order to do this, we compute the τ-AP between the rankings [22]. The τ-AP computes a position-sensitive version of Kendall's τ suitable for information retrieval tasks. The ranking of results for a reformulation candidate with a very high τ-AP will be indistinguishable from those of the reference query; the ranking of results for a reformulation candidate with a very low τ-AP will be quite different from the reference query. Our second result set signal measures drift by inspecting the result set language models. Specifically, it computes $\mathcal{B}(\theta_{\mathcal{R}_{t-1}}, \theta_{\mathcal{R}_t})$, the Bhattacharyya correlation between the result sets.

Performance Prediction Model. With some exception, the majority of performance prediction work has studied predictors independently, without looking at a combinations of signals. Several approaches to combine predictors focus on regressing against the absolute performance for a set of training queries [8]. Instead, we treat this as an ordinal regression problem. That is, we estimate a model which learns the correct ordering of reformulation candidates for a given information need. In practice, we *train* this model using true performance values of candidates encountered throughout a search process started at q_0; running this process over a number of training q_0's results in a large set of training candidates. We are agnostic about the precise functional form of our model and opt for a linear ranking support vector machine [15] due to its training and evaluation speed, something we found necessary when conducting experiments at scale. Precisely how this training set is collected will be described in the next section.

4.3 Searching Candidates

Considering reformulation as a graph, we still need to describe a process for searching for queries starting from q_0. We approach this process as a heuristic search problem, using the predicted performance as our heuristic. Unfortunately, algorithms such as A^* cannot be reliably used because our heuristic is not admissible. Similarly, the noise involved in our performance prediction causes greedy algorithms such as beam search or best first search to suffer from local maxima.

Motivated by our search simulator (Sect. 3), we propose an algorithm that recursively inspects n reformulation candidates at each q_i up to a certain depth, $d_{\max}$. We present this algorithm in Fig. 3. The algorithm differs from our simulation insofar as it executes several reformulation sessions simultaneously, keeping track of those reformulations with the highest predicted effectiveness. One attractive aspect of our algorithm is the broad coverage of the reformulation space unlikely to be visited in greedier algorithms.

At termination, the algorithm selects a small number (m) of candidate queries visited for final retrieval. These m retrievals are merged using a Borda count algorithm with constituent rankings weighted by predicted performance. This process allows the algorithm to be more robust to errors in scoring.

QUERYSEARCH($q, d, b, d_{\max}, m$)
1 **if** $d = d_{\max}$
2 **then return** q
3 $\mathcal{Q}^q \leftarrow$ GENERATECANDIDATES(q)
4 $\tilde{\mu} \leftarrow$ PREDICTPERFORMANCE($\mathcal{Q}^q$)
5 $\tilde{\mathcal{Q}}^q \leftarrow$ TOPQUERIES($\mathcal{Q}^q, \tilde{\mu}, b$)
6 $\hat{\mathcal{Q}}^q \leftarrow$ TOPQUERIES($\mathcal{Q}^q, \tilde{\mu}, m$)
7 **for** $q_i \in \tilde{\mathcal{Q}}^q$
8 **do** $\hat{\mathcal{Q}}^q \leftarrow \hat{\mathcal{Q}}^q \cup$ QUERYSEARCH($q_i, d+1, b, d_{\max}, m$)
9 $\hat{\mu} \leftarrow$ PREDICTPERFORMANCE($\hat{\mathcal{Q}}^q$)
10 **return** TOPQUERIES($\hat{\mathcal{Q}}^q, \hat{\mu}, m$)

Fig. 3. Query reformulation procedure. The search procedure recursively explores a reformulation graph and returns the top m highest scoring reformulations.

5 Training

The effectiveness of the search algorithm (Sect. 4.3) critically depends on the reliability of the performance predictor (Sect. 4.2). We can gather training data for this model by sampling batches of reformulations from the query graph. In order to ensure that our sample includes high quality reformulations, we run the oracle algorithm (Sect. 3) while recording the visited reformulations and their true performance values. Any bias in this data collection process can be addressed by gathering additional training reformulations with suboptimal performance predictor models (i.e. partially trained models).

6 Methods

We use three standard retrieval corpora for our experiments. Two news corpora, trec12 and robust, consist of large archives of news articles. The trec12 dataset consists of the Tipster disks 1 and 2 with TREC *ad hoc* topics 51–200. The robust dataset consists of Tipster disks 4 and 5 with TREC *ad hoc* topics 301–450 and 601–700. Our web corpus consists of the Category B section of the Clue Web 2009 dataset with TREC Web topics 1–200. We tokenized all corpora on whitespace and then applied Krovetz stemming and removed words in the SMART stopword list. We further pruned the web corpus of all documents with a Waterloo spam score less than 70. We use TREC title queries in all of our experiments. We randomly partitioned the queries into three sets: 60 % for training, 20 % for validation, and 20 % for testing. We repeated this random split procedure five times and present results averaged across the test set queries.

All indexing and retrieval was conducted using indri 5.7. Our SVM models were trained using liblinear 1.95. We evaluated final retrievals using NIST trec_eval 9.0. In order to support large parameter sweeps, each query reformulation in PQR performed a *re-ranking* of the documents retrieved by q_0 instead of a *re-retrieval* from the full index. Pilot experiments found that the effectiveness of re-retrieval was comparable with that of re-ranking though re-retrieval incurred *much* higher latency. Aside from the performance prediction model, our algorithm has the following free parameters: the number of term-addition candidates per query (n), the number of candidates to selection per query (b), and the maximum search depth ($d_{\max}$). Combined, the automatic reformulation and the multi-pass training resulted in computationally expensive processes whose runtime is sensitive to these parameters. Consequently, we fixed our parameter settings at relatively modest numbers ($n = 10, b = 3, d_{\max} = 4$) and leave a more thorough analysis of sensitivity for an extended manuscript. Although these numbers may seem small, we remind the reader that this results in roughly 800 reformulations considered within the graph search for a single q_0. The number of candidates to merge (m) is tuned throughout training on the validation set v_0 and ranges from five to twenty. All runs, including baselines, optimized NDCG@30. We used QL and RM3 as baselines. QL used Dirichlet smoothing with parameter tuned on the full training set using a range of values from 500 through 5000. We tuned the RM3 parameters on the full training set. The range of feedback terms considered was $\{5, 10, 25, 50, 75, 100\}$; the range of feedback documents was $\{5, 25, 50, 75, 100\}$; the range of λ was $[0, 1]$ with a step size of 0.1.

7 Results

We present the results for our experiments in Table 2. Our first baseline, QL, reflects the performance of q_0 alone and represents an algorithm which is representationally comparable with PQR insofar as it also retrieves using a short, unweighted query. Our second baseline, RM3, reflects the performance of a strong algorithm that also uses the retrieval results to improve performance, although with much richer representational power (the optimal number of terms often hover near 75–100). As expected, RM3 consistently outperforms QL in terms of MAP. And while the performance is superior across all metrics for trec12, RM3 is statistically indistinguishable from QL for higher precision metrics on our other two data sets. The random policy, which replaces our performance predictor with random scores, consistently underperforms both baselines for robust and web. Interestingly, this algorithm is statistically indistinguishable from QL for trec12, suggesting that this corpus may be easier.

Table 2. Comparison of PQR to QL and RM3 baselines for our datasets. Statistically significant difference with respect to QL ($\blacksquare$: better; $\square$: worse) and RM3 ($\blacklozenge$: better; $\lozenge$: worse) using a Student's paired t-test ($p < 0.05$ with a Bonferroni correction). The best performing run is presented in bold.

	NDCG@5	NDCG@10	NDCG@20	NDCG@30	NDCG	MAP
trec12						
QL	0.5442	0.5278	0.5066	0.4835	0.5024	0.2442
RM3	**0.6465**$^{\blacksquare}$	**0.6113**$^{\blacksquare}$	**0.5796**$^{\blacksquare}$	**0.5627**$^{\blacksquare}$	**0.5300**$^{\blacksquare}$	**0.2983**$^{\blacksquare}$
random	0.5690 $^{\lozenge}$	0.5563 $^{\lozenge}$	0.5257 $^{\lozenge}$	0.5089 $^{\lozenge}$	0.5120$^{\blacksquare\lozenge}$	0.2653$^{\blacksquare\lozenge}$
PQR	0.6112$^{\blacksquare\lozenge}$	0.5907$^{\blacksquare}$	0.5630$^{\blacksquare}$	0.5419$^{\blacksquare\lozenge}$	0.5216$^{\blacksquare\lozenge}$	0.2819$^{\blacksquare\lozenge}$
robust						
QL	0.4874	0.4559	0.4306	0.4172	0.5419	0.2535
RM3	0.4888	0.4553	0.4284	0.4176	0.5462	0.2726$^{\blacksquare}$
random	0.4240$^{\square\lozenge}$	0.3967$^{\square\lozenge}$	0.3675$^{\square\lozenge}$	0.3588$^{\square\lozenge}$	0.5143$^{\square\lozenge}$	0.2352$^{\square\lozenge}$
PQR	**0.5009**	**0.4713**$^{\blacksquare\blacklozenge}$	**0.4438**$^{\blacksquare\blacklozenge}$	**0.4315**$^{\blacksquare\blacklozenge}$	**0.5498**$^{\blacksquare}$	**0.2736**$^{\blacksquare}$
web						
QL	0.2206	0.2250	0.2293	0.2315	0.3261	0.1675
RM3	0.2263	0.2273	0.2274	0.2316	**0.3300**$^{\blacksquare}$	**0.1736**$^{\blacksquare}$
random	0.1559$^{\square\lozenge}$	0.1562$^{\square\lozenge}$	0.1549$^{\square\lozenge}$	0.1537$^{\square\lozenge}$	0.2790$^{\square\lozenge}$	0.1157$^{\square\lozenge}$
PQR	**0.2528**$^{\blacksquare\blacklozenge}$	**0.2501**$^{\blacksquare\blacklozenge}$	**0.2493**$^{\blacksquare\blacklozenge}$	**0.2435**$^{\blacksquare}$	**0.3300**	0.1690

Next, we turn to the performance of PQR. Across all corpora and across almost all metrics, PQR significantly outperforms QL. While this baseline might be considered low, it is a representationally fair comparison with PQR. So, this result demonstrates the ability of PQR to find more effective reformulations than q_0. The underperformance of the random algorithm signifies that the effectiveness of PQR is attributable to the performance prediction model as opposed to a merely walking on the reformulation graph. That said, PQR is statistically indistinguishable from QL for higher recall metrics on the web corpus (NDCG and MAP). In all likelihood, this results from the optimization of NDCG@30, as opposed to higher recall metrics. This outcome is amplified when we compare PQR to RM3. For the robust and web datasets, we notice *PQR significantly outperforming RM3 for high precision* metrics but showing weaker performance for high recall metrics. The weaker performance of PQR compared to RM3 for trec12 might be explained by the easier nature of the corpus combined with the richer representation of the RM3 model. Nevertheless, we stress that, for those metrics we optimized for and for the more precision-oriented collections, PQR dominates all baselines.

We can inspect the coefficient values in the linear model to determine the importance of individual signals in performance prediction. In Table 3, we present the most important signals for each of our experiments. Because our results are averaged over several runs, we selected the signals most often occurring amongst the highest weighted in these runs, using the final selected model (see Sect. 5). Interestingly, many of the top ranked signals are our drift features which compare the language models and rankings of the candidate result set with those of its parent and the first query. This suggests that the algorithm is successfully preventing query drift by promoting candidates that retrieve results similar to the original and parent queries. On the other hand, the high weight for Clarity suggests that PQR is simultaneously balancing ranked list refinement with ranked list anchoring.

Table 3. Top five highest weighted signals for each experiment.

trec12	Robust	Web
$\mathcal{B}(\theta_{\mathcal{R}_0}, \theta_{\mathcal{R}_t})$	$\mathcal{B}(\theta_{\mathcal{R}_0}, \theta_{\mathcal{R}_t})$	$\tau_{\text{AP}}(\mathcal{R}_0, \mathcal{R}_t)$
$\mathcal{B}(\theta_{\mathcal{R}_{t-1}}, \theta_{\mathcal{R}_t})$	$\mathcal{B}(\theta_{\mathcal{R}_{t-1}}, \theta_{\mathcal{R}_t})$	$\mathcal{B}(\theta_{\mathcal{R}_0}, \theta_{\mathcal{R}_t})$
$\tau_{\text{AP}}(\mathcal{R}_0, \mathcal{R}_t)$	Clarity	$\tau_{\text{AP}}(\mathcal{R}_{t-1}, \mathcal{R}_t)$
$\tau_{\text{AP}}(\mathcal{R}_{t-1}, \mathcal{R}_t)$	$\tau_{\text{AP}}(\mathcal{R}_{t-1}, \mathcal{R}_t)$	$\mathcal{B}(\theta_{\mathcal{R}_{t-1}}, \theta_{\mathcal{R}_t})$
Clarity	maxIDF	Clarity

8 Discussion

Although QL is the appropriate baseline for PQR, comparing PQR performance to that of RM3 helps us understand where improvements may be originating. The effectiveness of RM3 on trec12 is extremely strong, demonstrating statistically superior performance to PQR on many metrics. At the same time, the absolute metrics for QL on these runs is also higher than on the other two collections. This suggests that part of the effectiveness of RM3 results from the strong initial retrieval (i.e. QL). As mentioned earlier, the strength of the random run separately provides evidence of the initial retrieval's strength. Now, if the initial retrieval uncovered significantly more relevant documents, then RM3 will estimate a language model very close to the true relevance model, boosting performance. Since RM3 allows a long, rich, weighted query, it follows that it would outperform PQR's constrained representation. That said, it is remarkable that PQR achieves comparable performance to RM3 on many metrics with at most $|q_0| + d_{\max}$ words.

Despite the strong performance for high-precision metrics, the weaker performance for high-recall metrics was somewhat disappointing but should be expected given our optimization target (NDCG@30). Post-hoc experiments demonstrated that optimizing for MAP boosted the performance of PQR to 0.1728 on web, resulting in statistically indistinguishable performance with RM3. Nevertheless, we are not certain that human query reformulation of the type encountered in general web search would improve high recall metrics since users in that context rarely inspect deep into the ranked list.

One of the biggest concerns with PQR is efficiency. Whereas our QL baseline ran in a 100–200 ms, PQR ran in 10–20 s, even using the re-ranking approach. However, because of this approach, our post-retrieval costs scale modestly as corpus size grows, especially compared to massive query expansion techniques like RM3. To understand this observation, note that issuing a long RM3 query results in a huge slowdown in performance due to the number of postings lists that need to be evaluated and merged. We found that for the web collection, RM3 performed quite slow, often taking *minutes* to complete long queries. PQR, on the other hand, has the same overhead as RM3 in terms of an initial retrieval and fetching document vectors. After this step, though, PQR only needs to access the index for term statistic information, not a re-retrieval. Though even with our speedup, PQR is unlikely to be helpful for realtime, low-latency retrieval. However, there are several situations where such a technique may be permissible, including 'slow search' scenarios where users tolerate latency in order to receive better results [20], offline result caching, and document filtering.

9 Conclusion

We have presented a new formal model for information retrieval. The positive results on three separate corpora provide evidence that PQR is a framework worth investigating further. In terms of candidate generation, we considered only very simple word additions and deletions while previous research has demonstrated the effectiveness of applying multiword units (e.g. ordered and unordered windows). Beyond this, we can imagine applying more sophisticated operations such as filters, site restrictions, or time ranges. While it would increase our query space, it may also allow for more precise and higher precision reformulations. In terms of candidate scoring, we found that our novel drift signals allowed for effective query expansion. We believe that PQR provides a framework for developing other performance predictors in a grounded retrieval model. In terms of graph search, we believe that other search strategies might result in more effective coverage of the space.

References

1. Bendersky, M.: Information Retrieval with Query Hypergraphs. Ph.D. thesis, University of Massachusetts Amherst (2012)
2. Bhattacharyya, A.: On a measure of divergence between two statistical populations defined by probability distributions. Bull. Calcutta Math. Soc. **35**, 99–109 (1943)
3. Cao, G., Nie, J.Y., Gao, J., Robertson, S.: Selecting good expansion terms for pseudo-relevance feedback. In: SIGIR (2008)
4. Croft, W.B., Harper, D.J.: Using probabilistic models of document retrieval without relevance information. J. Documentation **35**(4), 285–295 (1979)
5. Cronen-Townsend, S., Zhou, Y., Croft, W.B.: Predicting query performance. In: SIGIR (2002)
6. Diaz, F.: Performance prediction using spatial autocorrelation. In: SIGIR (2007)
7. Hauff, C.: Predicting the Effectiveness of Queries and Retrieval Systems. Ph.D. thesis, University of Twente (2010)
8. Hauff, C., Azzopardi, L., Hiemstra, D.: The combination and evaluation of query performance prediction methods. In: Boughanem, M., Berrut, C., Mothe, J., Soule-Dupuy, C. (eds.) ECIR 2009. LNCS, vol. 5478, pp. 301–312. Springer, Heidelberg (2009)
9. He, B., Ounis, I.: Inferring query performance using pre-retrieval predictors. In: SPIRE (2004)
10. Huang, C.K., Chien, L.F., Oyang, Y.J.: Relevant term suggestion in interactive web search based on contextual information in query session logs. JASIST **54**, 638–649 (2003)
11. Huang, J., Efthimiadis, E.N.: Analyzing and evaluating query reformulation strategies in web search logs. In: CIKM (2009)
12. Kumaran, G., Carvalho, V.: Reducing long queries using query quality predictors. In: SIGIR (2009)
13. Kurland, O., Lee, L., Domshlak, C.: Better than the real thing?: iterative pseudo-query processing using cluster-based language models. In: SIGIR (2005)
14. Lavrenko, V., Croft, W.B.: Relevance based language models. In: SIGIR (2001)
15. Liu, T.Y.: Learning to Rank for Information Retrieval. Springer, Heidelberg (2009)
16. Lv, Y., Zhai, C.: Adaptive relevance feedback in information retrieval. In: CIKM (2009)
17. Macdonald, C., Santos, R.L., Ounis, I.: On the usefulness of query features for learning to rank. In: CIKM (2012)
18. Ruthven, I.: Re-examining the potential effectiveness of interactive query expansion. In: SIGIR (2003)
19. Sheldon, D., Shokouhi, M., Szummer, M., Craswell, N.: Lambdamerge: merging the results of query reformulations. In: WSDM (2011)
20. Teevan, J., Collins-Thompson, K., White, R.W., Dumais, S.: Slow search. Commun. ACM **57**(8), 36–38 (2014)
21. Xue, X., Croft, W.B., Smith, D.A.: Modeling reformulation using passage analysis. In: CIKM (2010)
22. Yilmaz, E., Aslam, J.A., Robertson, S.: A new rank correlation coefficient for information retrieval. In: SIGIR (2008)

VODUM: A Topic Model Unifying Viewpoint, Topic and Opinion Discovery

Thibaut Thonet(✉), Guillaume Cabanac, Mohand Boughanem, and Karen Pinel-Sauvagnat

IRIT, Université Paul Sabatier,
118 Route de Narbonne, 31062 Toulouse CEDEX 9, France
{thonet,cabanac,boughanem,sauvagnat}@irit.fr

Abstract. The surge of opinionated on-line texts provides a wealth of information that can be exploited to analyze users' viewpoints and opinions on various topics. This article presents VODUM, an unsupervised Topic Model designed to jointly discover viewpoints, topics, and opinions in text. We hypothesize that partitioning topical words and viewpoint-specific opinion words using part-of-speech helps to discriminate and identify viewpoints. Quantitative and qualitative experiments on the Bitterlemons collection show the performance of our model. It outperforms state-of-the-art baselines in generalizing data and identifying viewpoints. This result stresses how important topical and opinion words separation is, and how it impacts the accuracy of viewpoint identification.

1 Introduction

The surge of opinionated on-line texts raised the interest of researchers and the general public alike as an incredibly rich source of data to analyze contrastive views on a wide range of issues, such as policy or commercial products. This large volume of opinionated data can be explored through text mining techniques, known as Opinion Mining or Sentiment Analysis. In an opinionated document, a user expresses her **opinions** on one or several **topics**, according to her **viewpoint**. We define the key concepts of topic, viewpoint, and opinion as follows. A topic is one of the subjects discussed in a document collection. A viewpoint is the standpoint of one or several authors on a set of topics. An opinion is a wording that is specific to a topic and a viewpoint. For example, in the manually crafted sentence *Israel occupied the Palestinian territories of the Gaza strip*, the topic is *the presence of Israel on the Gaza strip*, the viewpoint is *pro-Palestine* and an opinion is *occupied*. Indeed, when mentioning the action of building Israeli communities on disputed lands, the pro-Palestine side is likely to use the verb *to occupy*, whereas the pro-Israel side is likely to use the verb *to settle*. Both sides discuss the same topic, but they use a different wording that conveys an opinion.

N. Ferro et al. (Eds.): ECIR 2016, LNCS 9626, pp. 533–545, 2016.
DOI: 10.1007/978-3-319-30671-1_39

The contribution of this article is threefold:

1. We first define the task of **Viewpoint and Opinion Discovery**, which consists in analyzing a collection of documents to identify the viewpoint of each document, the topics mentioned in each document, and the viewpoint-specific opinions for each topic.
2. To tackle this issue, we propose the *Viewpoint and Opinion Discovery Unification Model* (VODUM), an unsupervised approach to jointly model viewpoints, topics, and opinions.
3. Finally, we quantitatively and qualitatively evaluate our model VODUM on the Bitterlemons collection, benchmarking it against state-of-the-art baselines and degenerate versions of our model to analyze the usefulness of VODUM's specific properties.

The remainder of this paper is organized as follows. Section 2 presents related work and state-of-the-art Viewpoint and Opinion Discovery Topic Models. Our model's properties and inference process are described in Sect. 3. Section 4 details the experiments performed to evaluate VODUM. We conclude and give future directions for this work in Sect. 5.

2 Related Work: Viewpoint and Opinion Discovery Topic Models

Viewpoint and Opinion Discovery is a sub-task of Opinion Mining, which aims to analyze opinionated documents and infer properties such as subjectivity or polarity. We refer the reader to [6] for a general review of this broad research topic. While most Opinion Mining works first focused on product reviews, more recently, a surge of interest for sociopolitical and debate data led researchers to study tasks such as Viewpoint and Opinion Discovery. The works described in this section relate to LDA [2] and more generally to probabilistic Topic Models, as a way to model diverse latent variables such as viewpoints, topics, and opinions.

Several works modeled viewpoint-specific opinions [3,7] but they did not learn the viewpoint assignments of documents. Instead, they assumed these assignments to be known beforehand and leveraged them as prior information fed to their models. Some authors proposed a Topic Model to analyze culture-specific viewpoints and their associated wording on common topics [7]. In [3], the authors jointly modeled topics and viewpoint-specific opinions. They distinguished between topical words and opinion words based on part-of-speech: nouns were assumed to be topical words; adjectives, verbs, and adverbs were assumed to be opinion words.

Other works discovered document-level viewpoints in a supervised or semi-supervised fashion [4,8,12]. In [4], document-level and sentence-level viewpoints were detected using a supervised Naive Bayes approach. In [8], the authors defined the Topic-Aspect Model (TAM) that jointly models topics, and aspects, which play the role of viewpoints. Similarly, the Joint Topic Viewpoint (JTV)

Table 1. Comparison of our model VODUM against related work approaches.

Ref	Model is used without supervision	Topical words and opinion words are partitioned	Viewpoint assignments are learned	Model is independent of structure-specific properties
[7]	+	−	−	+
[3]	+	+	−	+
[4, 8, 9, 12]	−	−	+	+
[10, 11]	+	−	+	−
VODUM	+	+	+	+

model was proposed in [12] to jointly model topics and viewpoints. However, both TAM and JTV inferred parameters were only integrated as features into a SVM classifier to identify document-level viewpoints. TAM was extended to perform contrastive viewpoint summarization [9], but this extension was still weakly supervised as it leveraged a sentiment lexicon to identify viewpoints.

The task of viewpoint identification was also studied for user generated data such as forums, where users can debate on controversial issues [10, 11]. These works proposed Topic Models that however rely on structure-specific properties exclusive to forums (such as threads, posts, users, interactions between users), which cannot be applied to infer general documents' viewpoint.

The specific properties of VODUM compared to related work are summarized in Table 1. VODUM is totally unsupervised. It separately models topical words and opinion words. Document-level viewpoint assignments are learned. VODUM is also structure-independent and thus broadly applicable. These properties are further detailed in Sect. 3.

3 Viewpoint and Opinion Discovery Unification Model

3.1 Description

VODUM is a probabilistic Topic Model based on LDA [2]. VODUM simultaneously models viewpoints, topics, and opinions, i.e., it identifies topical words and viewpoint-specific topic-dependent opinion words. The graphical model of VODUM and the notation used in this article are provided in Fig. 1 and Table 2, respectively. The specific properties of VODUM are further detailed below.

Topical Words and Opinion Words Separation. In our model, topical words and opinion words are partitioned based on their part-of-speech, in line with several viewpoint modeling and Opinion Mining works [3, 5, 13]. Here, nouns are assumed to be topical words; adjectives, verbs and adverbs are assumed to be opinion words. While this assumption seems coarse, let us stress that a more accurate definition of topical and opinion words (e.g., by leveraging sentiment lexicons) could be used, without requiring any modification of our model. The part-of-speechtagging pre-processing step is further described in Sect. 4.2. The part-of-speech category is represented as an observed variable x which takes a value of 0 for topical words and 1 for opinion words. Topical words and opinion words are then drawn from distributions ϕ_0 and ϕ_1, respectively.

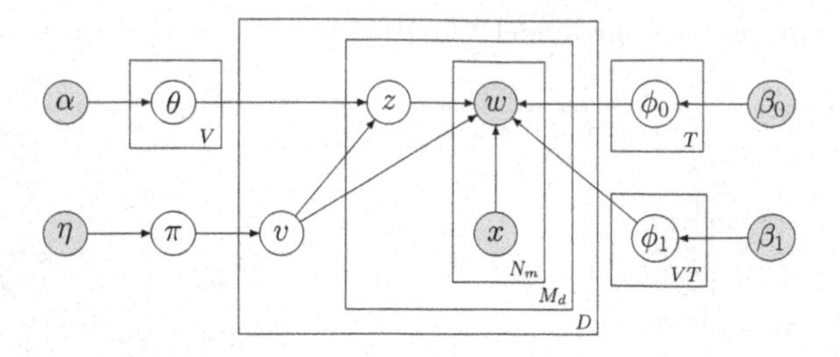

Fig. 1. Graphical model of VODUM.

Table 2. Notation for our model VODUM.

D, M_d, N_m	Number of documents in the collection, number of sentences in document d and number of words in sentence m, respectively
W	Number of words in the vocabulary
W_0, W_1	Number of topical words and opinion words in the vocabulary, respectively
T, V	Number of topics and viewpoints, respectively
$\mathcal{W}_0, \mathcal{W}_1$	Set of topical words and opinion words in the vocabulary, respectively
$w_{d,m,n}$	The n-th word of the m-th sentence from the d-th document
$x_{d,m,n}$	The part-of-speech category (0 or 1) of $w_{d,m,n}$
$z_{d,m}$	The topic assigned to the m-th sentence of the d-th document
v_d	The viewpoint assigned to the d-th document
$\boldsymbol{w}$, $\boldsymbol{x}$, $\boldsymbol{z}$, $\boldsymbol{v}$	Vector of all words, part-of-speech categories, topic assignments and view-point assignments, respectively
ϕ_0	$T \times W$ matrix of viewpoint-independent distributions over topical words
ϕ_1	$V \times T \times W$ matrix of viewpoint-dependent distributions over opinion words
θ	$V \times T$ matrix of viewpoint-dependent distributions over topics
π	V matrix of the distribution over viewpoints
β_0, β_1, α, η	Symmetric Dirichlet prior for ϕ_0, ϕ_1, θ and π, respectively
$n^{(i)}$	Number of documents in the collection assigned to viewpoint i
$n_i^{(j)}$	Number of sentences in the collection assigned to viewpoint i and topic j
$n_{0,j}^{(k)}$	Number of instances of topical word k assigned to topic j
$n_{1,i,j}^{(k)}$	Number of instances of opinion word k assigned to viewpoint i and topic j

Sentence-level Topic Variables. Most Topic Models define word-level topic variables (e.g., [2,3,8]). We hypothesize that using sentence-level topic variables, denoted by z, better captures the dependency between the opinions expressed in a sentence and the topic of the sentence. Indeed, coercing all words from a sentence to be related to the same topic reinforces the co-occurrence property leveraged by Topic Models. As a result, the topics induced by topical word distributions ϕ_0 and opinion word distributions ϕ_1 are more likely to be aligned.

Document-level Viewpoint Variables. Viewpoint variables v are defined at the document level and drawn from the distribution π. In previous works, viewpoint variables were allocated to words [8,9] or authors [10,11]. While it is reasonable to suppose that an author writes all her documents with the same viewpoint, the authorship information is not always available. On the other hand, allocating each word of a document to a potentially different viewpoint is meaningless. We thus modeled document-level viewpoint variables.

Viewpoint-specific Topic Distributions. In VODUM, topic distributions, denoted by θ in the graphical model, are viewpoint-specific instead of being document-specific as in other Topic Models [2,8,12]. This assumption comes from the observation in [10] that different viewpoints have different dominating topics. For example, opponents of same-sex marriage are more likely to mention religion than the supporting side.

Similarly to LDA and other probabilistic Topic Models, the probability distributions ϕ_0, ϕ_1, θ, and π are Multinomial distributions with symmetric Dirichlet priors β_0, β_1, α, and η, respectively.

The virtual generation of a document as modeled by VODUM is the following. The author writes the document according to her own viewpoint. Depending on her viewpoint, she selects for each sentence of the document a topic that she will discuss. Then, for each sentence, she chooses a set of topical words to describe the topic that she selected for the sentence, and a set of opinion words to express her viewpoint on this topic. Formally, the generative process of a document collection is performed as described in Fig. 2. In Sect. 3.2, we detail how we infer parameters ϕ_0, ϕ_1, θ, and π.

3.2 Model Inference

As for other probabilistic Topic Models, the exact inference of VODUM is not tractable. We thus rely on approximate inference to compute parameters ϕ_0, ϕ_1, θ, and π, as well as the document-level viewpoint assignments v. We chose collapsed Gibbs sampling as it was shown to be quicker to converge than approximate inference methods such as variational Bayes [1].

Collapsed Gibbs Sampling. Collapsed Gibbs sampling is a Markov chain Monte Carlo algorithm that generates a set of samples drawn from a posterior probability distribution, i.e., the probability distribution of latent variables (v and z in our model) given observed variables (w and x in our model). It does

1. Draw a viewpoint distribution π from Dirichlet(η).
2. Draw a viewpoint-independent topical word distribution $\phi_{0,j}$ from Dirichlet(β_0) for all topics $j \in [\![1, T]\!]$.
3. Draw a viewpoint-dependent opinion word distribution $\phi_{1,i,j}$ from Dirichlet(β_1) for all viewpoints $i \in [\![1, V]\!]$ and all topics $j \in [\![1, T]\!]$.
4. Draw a topic distribution θ_i from Dirichlet(α) for all viewpoints $i \in [\![1, V]\!]$.
5. For each document $d \in [\![1, D]\!]$
   (a) Draw a viewpoint v_d from Multinomial(π).
   (b) For each sentence $m \in [\![1, M_d]\!]$
      i. Draw a topic $z_{d,m}$ from Multinomial(θ_{v_d}).
      ii. For each word $n \in [\![1, N_m]\!]$
         A. Choose a part-of-speech category $x_{d,m,n}$ from $\{0, 1\}$, where category 0 denotes topical words (nouns) and category 1 denotes opinion words (adjectives, verbs and adverbs).
         B. If $x_{d,m,n} = 0$, draw a topical word $w_{d,m,n}$ from Multinomial($\phi_{0,z_{d,m}}$), else if $x_{d,m,n} = 1$, draw an opinion word $w_{d,m,n}$ from Multinomial($\phi_{1,v_d,z_{d,m}}$).

Fig. 2. Generative process for a collection as modeled by VODUM.

not require the actual computation of the posterior probability, which is usually intractable for Topic Models. Only the marginal probability distributions of latent variables (i.e., the probability distribution of one latent variable given all other latent variables and all observed variables) need to be computed in order to perform collapsed Gibbs sampling. For each sample, the collapsed Gibbs sampler iteratively draws assignments for all latent variables using their marginal probability distributions, conditioned on the previous sample's assignments. The marginal probability distributions used to sample the topic assignments and viewpoint assignments in our collapsed Gibbs sampler are described in (1) and (2), respectively. The derivation is omitted due to space limitation. The notation used in the equations is defined in Table 2. Additionally, indexes or superscripts $-d$ and $-(d, m)$ exclude the d-th document and the m-th sentence of the d-th document, respectively. Similarly, indexes or superscripts d and (d, m) include only the d-th document and the m-th sentence of the d-th document, respectively. A superscript $(\cdot)$ denotes a summation over the corresponding superscripted index.

$$p(z_{d,m} = j | v_d = i, \boldsymbol{v}_{-d}, \boldsymbol{z}_{-(d,m)}, \boldsymbol{w}, \boldsymbol{x}) \propto \frac{n_i^{(j),-(d,m)} + \alpha}{n_i^{(\cdot),-(d,m)} + T\alpha}$$
$$\times \frac{\prod_{k \in \mathcal{W}_0} \prod_{a=0}^{n_{0,j}^{(k),(d,m)}-1} n_{0,j}^{(k),-(d,m)} + \beta_0 + a}{\prod_{b=0}^{n_{0,j}^{(\cdot),(d,m)}-1} n_{0,j}^{(\cdot),-(d,m)} + W_0\beta_0 + b} \times \frac{\prod_{k \in \mathcal{W}_1} \prod_{a=0}^{n_{1,i,j}^{(k),(d,m)}-1} n_{1,i,j}^{(k),-(d,m)} + \beta_1 + a}{\prod_{b=0}^{n_{1,i,j}^{(\cdot),(d,m)}-1} n_{1,i,j}^{(\cdot),-(d,m)} + W_1\beta_1 + b} \quad (1)$$

$$p(v_d = i | \boldsymbol{v}_{-d}, \boldsymbol{z}, \boldsymbol{w}, \boldsymbol{x}) \propto \frac{n^{(i),-d} + \eta}{n^{(\cdot),-d} + V\eta}$$

$$\times \frac{\prod_{j=1}^{T} \prod_{a=0}^{n_i^{(j),d}-1} n_i^{(j),-d} + \alpha + a}{\prod_{b=0}^{n_i^{(\cdot),d}-1} n_i^{(\cdot),-d} + T\alpha + b} \times \prod_{j=1}^{T} \frac{\prod_{k \in \mathcal{W}_1} \prod_{a=0}^{n_{1,i,j}^{(k),d}-1} n_{1,i,j}^{(k),-d} + \beta_1 + a}{\prod_{b=0}^{n_{1,i,j}^{(\cdot),d}-1} n_{1,i,j}^{(\cdot),-d} + W_1\beta_1 + b} \quad (2)$$

Parameter Estimation. The alternate sampling of topics and viewpoints using (1) and (2) makes the collapsed Gibbs sampler converge towards the posterior probability distribution. The count variables $n^{(i)}$, $n_i^{(j)}$, $n_{0,j}^{(k)}$ and $n_{1,i,j}^{(k)}$ computed for each sample generated by the collapsed Gibbs sampler are used to estimate distributions π, θ, ϕ_0 and ϕ_1 as described in (3), (4), (5) and (6), respectively.

$$\pi^{(i)} = \frac{n^{(i)} + \eta}{n^{(\cdot)} + V\eta} \quad (3) \qquad \theta_i^{(j)} = \frac{n_i^{(j)} + \alpha}{n_i^{(\cdot)} + T\alpha} \quad (4)$$

$$\phi_{0,j}^{(k)} = \begin{cases} \frac{n_{0,j}^{(k)}+\beta_0}{n_{0,j}^{(\cdot)}+W_0\beta_0} & \text{if } k \in \mathcal{W}_0 \\ 0 & \text{otherwise} \end{cases} \quad (5) \qquad \phi_{1,i,j}^{(k)} = \begin{cases} \frac{n_{1,i,j}^{(k)}+\beta_1}{n_{1,i,j}^{(\cdot)}+W_1\beta_1} & \text{if } k \in \mathcal{W}_1 \\ 0 & \text{otherwise} \end{cases} \quad (6)$$

4 Experiments

We investigated the following hypotheses in our experiments:

- **(H1)** Using viewpoint-specific topic distributions (instead of document-level topic distributions, e.g., as in TAM, JTV, and LDA) has a positive impact on the ability of the model to identify viewpoints.
- **(H2)** The separation between opinion words and topical words has a positive impact on the ability of the model to identify viewpoints.
- **(H3)** Using sentence-level topic variables improves the ability of the model to identify viewpoints.
- **(H4)** Using document-level viewpoint variables helps the model to identify viewpoints.
- **(H5)** VODUM outperforms state-of-the-art models (e.g., TAM, JTV, and LDA) in the modeling and viewpoint identification tasks.

Note that an issue similar to (H1) was already addressed in [10,11]. The authors did not evaluate, however, the impact of this assumption on the viewpoint identification task. The rest of this section is organised as follows. In Sect. 4.1, we detail the baselines we compared VODUM against. Section 4.2 describes the dataset used for the evaluation and the experimental setup. In Sect. 4.3, we report and discuss the results of the evaluation.

4.1 Baselines

We compared VODUM against state-of-the-art models and degenerate versions of our model in order to answer the research questions underlying our five hypotheses. The state-of-the-art models we considered are TAM [8,9], JTV [12] and LDA [2]. These are used to investigate (H5). The four degenerate versions of VODUM are defined to evaluate the impact of each of our model's properties in isolation. The degenerate versions and their purpose are detailed below.

In VODUM-D, topic distributions are defined at the document level. In VODUM, topic distributions are instead viewpoint-specific and independent of documents. VODUM-D has been defined to study (H1).

VODUM-O assumes that all words are opinion words, i.e., all words are drawn from distributions that depend both on viewpoint and topic. On the contrary, VODUM distinguishes opinion words (drawn from viewpoint-specific topic-dependent distributions) from topical words (drawn from topic-dependent distributions). Comparing VODUM against VODUM-O answers (H2).

VODUM-W defines topic variables at the word level, instead of sentence level as in VODUM. This essentially allows a document to be potentially associated with more topics (one topic per word as opposed to one per sentence), and loosens the link between opinion words and topical words. VODUM-W has been defined to tackle (H3).

VODUM-S models sentence-level viewpoint variables, while, in VODUM, viewpoint variables are defined at the document level. Therefore, a document modeled by VODUM-S can contain sentences assigned to different viewpoints. Comparing VODUM against VODUM-S addresses (H4).

4.2 Dataset and Experimental Setup

We evaluated our model on a collection of articles published in the Bitterlemons e-zine.[1] It contains essays written by Israeli and Palestinian authors, discussing the Israeli-Palestinian conflict and related issues. It was first introduced in [4] and then used in numerous works that aim to identify and model viewpoints in text (e.g., [9,12]). This collection contains 297 essays written by Israeli authors and 297 written by Palestinian authors. Before using the collection, we performed the following pre-processing steps using the Lingpipe[2] Java library. We first filtered out tokens that contain numerical characters. We then applied stop

[1] http://www.bitterlemons.net/.
[2] http://alias-i.com/lingpipe/.

word removal and Porter stemming to the collection. We also performed part-of-speech tagging and annotated data with the binary part-of-speech categories that we defined in Sect. 3.1. Category 0 corresponds to topical words and contains common nouns and proper nouns. Category 1 corresponds to opinion words and contains adjectives, verbs, adverbs, qualifiers, modals, and prepositions. Tokens labeled with other part-of-speech were filtered out.

We implemented our model VODUM and the baselines based on the JGibbLDA[3] Java implementation of collapsed Gibbs sampling for LDA. The source code of our implementation and the formatted data (after all pre-processing steps) are available at https://github.com/tthonet/VODUM.

In the experiments, we set the hyperparameters of VODUM and baselines to the following values. The hyperparameters in VODUM were manually tuned: $\alpha = 0.01$, $\beta_0 = \beta_1 = 0.01$, and $\eta = 100$. The rationale behind the small α (θ's hyperparameter) and the large η (π's hyperparameter) is that we want a sparse θ distribution (i.e., each viewpoint has a distinct topic distribution) and a smoothed π distribution (i.e., a document has equal chance to be generated under each of the viewpoints). We chose the same hyperparameters for the degenerate versions of VODUM. The hyperparameters of TAM were set according to [9]: $\alpha = 0.1$, $\beta = 0.1$, $\delta_0 = 80.0$, $\delta_1 = 20.0$, $\gamma_0 = \gamma_1 = 5.0$, $\omega = 0.01$. For JTV, we used the hyperparameters' values described in [12]: $\alpha = 0.01$, $\beta = 0.01$, $\gamma = 25$. We manually adjusted the hyperparameters of LDA to $\alpha = 0.5$ and $\beta = 0.01$. For all experiments, we set the number of viewpoints (for VODUM, VODUM-D, VODUM-O, VODUM-W, VODUM-S, and JTV) and the number of aspects (for TAM) to 2, as documents from the Bitterlemons collection are assumed to reflect either the Israeli or Palestinian viewpoint.

4.3 Evaluation

We performed both quantitative and qualitative evaluation to assess the quality of our model. The quantitative evaluation relies on two metrics: held-out perplexity and viewpoint identification accuracy. It compares the performance of our model VODUM according to these metrics against the aforementioned baselines. In addition, the qualitative evaluation consists in checking the coherence of topical words and the related viewpoint-specific opinion words inferred by our model. These evaluations are further described below.

Held-Out Perplexity. Held-out perplexity is a metric that is often used to measure the generalization performance of a Topic Model [2]. Perplexity can be understood as the inverse of the geometric mean per-word likelihood. As computing the perplexity of a Topic Model is intractable, an estimate of the perplexity is usually computed using the parameters' point estimate provided by a Gibbs sampler, as shown in Sect. 3.2. The process is the following: the model is first learned on a training set (i.e., inference is performed to compute the parameters of the model), then the inferred parameters are used to compute the perplexity

[3] http://jgibblda.sourceforge.net/.

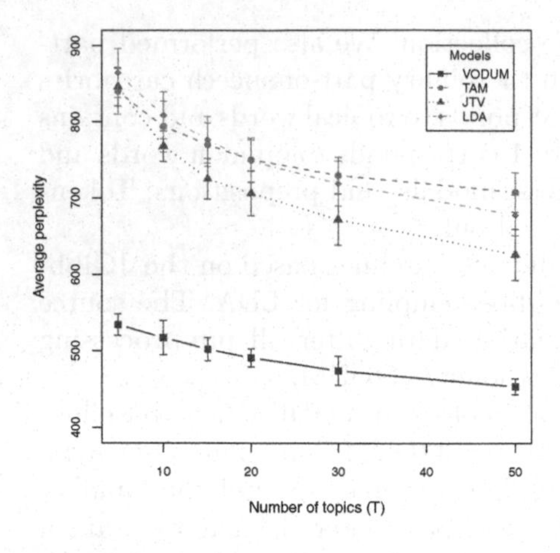

Fig. 3. Held-out perplexity of VODUM, TAM, JTV, and LDA computed for 5, 10, 15, 20, 30, and 50 topics (lower is better). Error bars denote standard deviation in the 10-fold cross-validation.

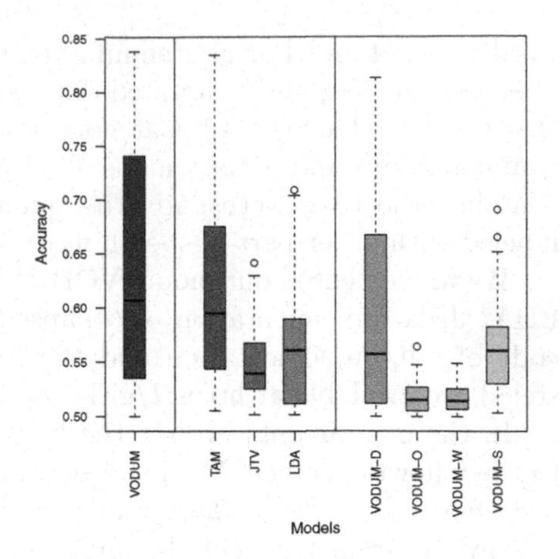

Fig. 4. Viewpoint identification accuracy of VODUM, TAM, JTV, LDA, VODUM-D, VODUM-O, VODUM-W, and VODUM-S (higher is better). Each boxplot is drawn from 50 samples.

of the test set (i.e., a set of held-out documents). A lower perplexity for the test set, which is equivalent to a higher likelihood for the test set, can be interpreted as a better generalization performance of the model: the model, learned on the training set, is less "perplexed" by the test set. In this experiment, we aimed to investigate (H5) and compared the generalization performance of our model VODUM against the state-of-the-art baselines. We performed a 10-fold cross-validation as follows. The model is trained on nine folds of the collection for 1,000 iterations and inference on the remaining, held-out test fold is performed for another 1,000 iterations. For both training and test, we only considered the final sample, i.e., the 1,000th sample. We finally report the held-out perplexity averaged on the final samples of the ten possible test folds. As the generalization performance depends on the number of topics, we computed the held-out perplexity of models for 5, 10, 15, 20, 30, and 50 topics.

The results of this experiment (Fig. 3) support (H5): for all number of topics VODUM has a significantly lower perplexity than TAM, JTV, and LDA. This implies that VODUM's ability to generalize data is better than baselines'. JTV presents slightly lower perplexity than TAM and LDA, especially for larger number of topics. TAM and LDA obtained comparable perplexity, TAM being slightly better for lower number of topics and LDA being slightly better for higher number of topics.

Viewpoint Identification. Another important aspect of our model is its ability to identify the viewpoint under which a document has been written. In this experiment, we aim to evaluate the viewpoint identification accuracy (VIA) of our model VODUM against our baselines, in order to investigate (H1), (H2), (H3), (H4), and (H5). As the Bitterlemons collection contains two different viewpoints (Israeli or Palestinian), viewpoint identification accuracy is here equivalent to binary clustering accuracy: each document is assigned to viewpoint 0 or to viewpoint 1. The VIA is then the ratio of well-clustered documents. As reported in [9], the viewpoint identification accuracy presents high variance for different executions of a Gibbs sampler, because of the stochastic nature of the process. For each model evaluated, we thus performed 50 executions of 1,000 iterations, and kept the final (1,000th) sample of each execution, resulting in a total of 50 samples. In this experiment, we set the number of topics for the different models as follows: 12 for VODUM, VODUM-D, VODUM-O, VODUM-W, and VODUM-S. The number of topics for state-of-the-art models was set according to their respective authors' recommendation: 8 for TAM (according to [9]), 6 for JTV (according to [12]). For LDA, the number of topics was set to 2: as LDA does not model viewpoints, we evaluated to what extent LDA is able to match viewpoints with topics.

VODUM, VODUM-D, VODUM-O, and VODUM-W provide documents' viewpoint assignment for each sample. We thus directly used these assignments to compute the VIA. However, VODUM-S only has sentence-level viewpoint assignments. We assigned each document the majority viewpoint assignment of its sentences. When the sentences of a document are evenly assigned to each viewpoint, the viewpoint of the document was chosen randomly. We proceeded similarly with TAM, JTV, and LDA, using their majority word-level aspect, viewpoint, and topic assignments, respectively, to compute the document-level viewpoint assignments. The results of the experiments are given in Fig. 4. The boxplots show that our model VODUM overall performed the best in the viewpoint identification task. More specifically, VODUM outperforms state-of-the-art models, thus supporting (H5). Among state-of-the-art models, TAM obtained the best results. We also observe that JTV did not outperform LDA in the viewpoint identification task. This may be due to the fact that the dependency between topic variables and viewpoint variables was not taken into account when we used JTV to identify document-level viewpoints – word-level viewpoint assignments in JTV are not necessarily aligned across topics. The observations of the degenerate versions of VODUM support (H1), (H2), (H3), and (H4). VODUM-O and VODUM-W performed very poorly compared to other models. The separation of topical words and opinion words, as well as the use of sentence-level topic variables – properties that were removed from VODUM in VODUM-O and VODUM-W, respectively – are then both absolutely necessary in our model to accurately identify documents' viewpoint, which confirms (H2) and (H3). The model VODUM-S obtained reasonable VIA, albeit clearly lower than VODUM. Document-level viewpoint variables thus lead to a better VIA than sentence-level viewpoint variables, verifying (H4). Among the degenerate versions of VODUM, VODUM-D overall yielded the highest VIA, but still slightly lower than VODUM. We conclude that the assumption made in [10,11], stating that the use of viewpoint-specific topic distributions (instead of document-specific topic distributions as in VODUM-D) improves viewpoint identification, was relevant, which in turn supports (H1).

Qualitative Evaluation. The qualitative evaluation of our model VODUM consists in studying the coherence of the topical words and the related viewpoint-specific opinion words. More specifically, we examine the most probable words in our model's viewpoint-independent distribution over words ϕ_0 and each viewpoint-specific distribution over words ϕ_1. This evaluation of VODUM is performed on the sample that obtained the best VIA. We report in Table 3 the most probable words for a chosen topic, which we manually labeled as *Middle East conflicts*. The most probable topical words are coherent and clearly relate to Middle East conflicts with words like *syria*, *jihad*, *war*, and *iraq*. The second and third rows of Table 3 show the opinion words used by the Israeli and Palestinian viewpoints, respectively. Not surprisingly, words like *islam*, *terrorist*, and *american* are used by the Israeli side to discuss Middle East conflicts. On the other hand, the Palestinian side remains nonspecific on the conflicts with words like *win*, *strong*, and *commit*, and does not mention Islam or terrorism. These observations confirm that the topical words and the opinion words are related and coherent.

Table 3. Most probable topical and opinion (stemmed) words inferred by VODUM for the topic manually labeled as *Middle East conflicts*. Opinion words are given for each viewpoint: Israeli (I) and Palestinian (P).

Middle East conflicts Topical words	israel	palestinian	syria	jihad	war	iraq	dai	suicid	destruct	iran
Middle East conflicts Opinion words (I)	islam	isra	terrorist	recent	militari	intern	like	heavi	close	american
Middle East conflicts Opinion words (P)	need	win	think	sai	don	strong	new	sure	believ	commit

5 Conclusion and Research Directions

This article introduced VODUM, an unsupervised Topic Model that enables viewpoint and opinion discovery in text. Throughout the experiments, we showed that our model outperforms state-of-the-art baselines, both in generalizing data and identifying viewpoints. We also analyzed the importance of the properties specific to our model. The results of the experiments suggest that the separation of opinion words and topical words, as well as the use of sentence-level topic variables, document-level viewpoint variables, and viewpoint-specific topic distributions improve the ability of our model to identify viewpoints. Moreover, the qualitative evaluation confirms the coherence of topical words and opinion words inferred by our model.

We expect to extend the work presented here in several ways. As the accuracy of viewpoint identification shows a high variance between different samples, one needs to design a method to automatically collect the most accurate sample or to deduce accurate viewpoint assignments from a set of samples. VODUM can also integrate sentiment labels to create a separation between positive and negative opinion words, using sentiment lexicons. This could increase the discrimination between different viewpoints and thus improve viewpoint identification. A viewpoint summarization framework can as well benefit from VODUM, selecting the most relevant sentences from each viewpoint and for each topic by leveraging VODUM's inferred parameters.

References

1. Asuncion, A., Welling, M., Smyth, P., Teh, Y.W.: On smoothing and inference for topic models. In: Proceedings of UAI 2009, pp. 27–34 (2009)
2. Blei, D.M., Ng, A.Y., Jordan, M.I.: Latent dirichlet allocation. In: Proceedings of NIPS 2001, pp. 601–608 (2001)
3. Fang, Y., Si, L., Somasundaram, N., Yu, Z.: Mining contrastive opinions on political texts using cross-perspective topic model. In: Proceedings of WSDM 2012, pp. 63–72 (2012)
4. Lin, W.H., Wilson, T., Wiebe, J., Hauptmann, A.: Which side are you on? Identifying perspectives at the document and sentence levels. In: Proceedings of CoNLL 2006, pp. 109–116 (2006)
5. Liu, B., Hu, M., Cheng, J.: Opinion observer: analyzing and comparing opinions on the web. In: Proceedings of WWW 2005, pp. 342–351 (2005)
6. Pang, B., Lee, L.: Opinion mining and sentiment analysis. Found. Trends Inf. Retrieval **2**(1–2), 1–135 (2008)
7. Paul, M.J., Girju, R.: Cross-cultural analysis of blogs and forums with mixed-collection topic models. In: Proceedings of EMNLP 2009, pp. 1408–1417 (2009)
8. Paul, M.J., Girju, R.: A two-dimensional topic-aspect model for discovering multi-faceted topics. In: Proceedings of AAAI 2010, pp. 545–550 (2010)
9. Paul, M.J., Zhai, C., Girju, R.: Summarizing contrastive viewpoints in opinionated text. In: Proceedings of EMNLP 2010, pp. 66–76 (2010)
10. Qiu, M., Jiang, J.: A latent variable model for viewpoint discovery from threaded forum posts. In: Proceedings of NAACL HLT 2013, pp. 1031–1040 (2013)
11. Qiu, M., Yang, L., Jiang, J.: Modeling interaction features for debate side clustering. In: Proceedings of CIKM 2013, pp. 873–878 (2013)
12. Trabelsi, A., Zaiane, O.R.: Mining contentious documents using an unsupervised topic model based approach. In: Proceedings of ICDM 2014, pp. 550–559 (2014)
13. Turney, P.D.: Thumbs up or thumbs down? Semantic orientation applied to unsupervised classification of reviews. In: Proceedings of ACL 2002, pp. 417–424 (2002)

Applications

Harvesting Training Images for Fine-Grained Object Categories Using Visual Descriptions

Josiah Wang[1(✉)], Katja Markert[2,3], and Mark Everingham[3]

[1] Department of Computer Science, University of Sheffield, Sheffield, UK
j.k.wang@sheffield.ac.uk

[2] L3S Research Center, Leibniz-University Hannover, Hannover, Germany
markert@l3s.de

[3] School of Computing, University of Leeds, Leeds, UK

Abstract. We harvest training images for visual object recognition by casting it as an IR task. In contrast to previous work, we concentrate on fine-grained object categories, such as the large number of particular animal subspecies, for which manual annotation is expensive. We use 'visual descriptions' from nature guides as a novel augmentation to the well-known use of category names. We use these descriptions in both the query process to find potential category images as well as in image reranking where an image is more highly ranked if web page text surrounding it is similar to the visual descriptions. We show the potential of this method when harvesting images for 10 butterfly categories: when compared to a method that relies on the category name only, using visual descriptions improves precision for many categories.

Keywords: Image retrieval · Text retrieval · Multi-modal retrieval

1 Introduction

Visual object recognition has advanced greatly in recent years, partly due to the availability of large-scale image datasets such as ImageNet [4]. However, the availability of image datasets for *fine-grained* object categories, such as particular types of flowers and birds [10,16], is still limited. Manual annotation of such training images is a notoriously onerous task and requires domain expertise.

Thus, previous work [2,3,6–9,12,14] has automatically harvested image datasets by retrieving images from online search engines. These images can then be used as training examples for a visual classifier. Typically the work starts with a keyword search of the desired category, often using the category name e.g. querying Google for "butterfly". As category names are often polysemous and, in addition, a page relevant to the keyword might also contain many pictures not of the required category, images are also filtered and reranked. While some work reranks or filters images using solely visual features [3,6,9,14], others

M. Everingham—who died in 2012—is included as a posthumous author of this paper for his intellectual contributions during the course of this work.

N. Ferro et al. (Eds.): ECIR 2016, LNCS 9626, pp. 549–560, 2016.
DOI: 10.1007/978-3-319-30671-1_40

Fig. 1. A visual description from eNature (http://www.enature.com/fieldguides) for the Monarch butterfly *Danaus plexippus*. We explore whether such descriptions can improve harvesting training images for fine-grained object categories.

have shown that features from the web pages containing the images, such as the neighbouring text and metadata information, are useful as well [2,7,8,12] (see Sect. 1.1 for an in-depth discussion). However, prior work has solely focused on basic level categories (such as "butterfly") and not been used for fine-grained categories (such as a butterfly species like "*Danaus plexippus*") where the need to avoid manual annotation is greatest for the reasons mentioned above.

Our work therefore focuses on the automatic harvesting of training images for fine-grained object categories. Although fine-grained categories pose particular challenges for this task (smaller number of overall pictures available, higher risk of wrong picture tags due to needed domain expertise, among others), at least for natural categories they have one advantage: their instances share strong visual characteristics and therefore there exist 'visual descriptions', i.e. textual descriptions of their appearances, in nature guides, providing a resource that goes far beyond the usual use of category names. See Fig. 1 for an example.

We use these visual descriptions for harvesting images for fine-grained object categories to (i) improve search engine querying compared to category name search and (ii) rerank images by comparing their accompanying web page text to the independent visual descriptions from nature guides as an expert source. We show that the use of these visual descriptions can improve precision over name-based search. To the best of our knowledge this is the first work using visual descriptions for harvesting training images for object categorization.[1]

1.1 Related Work

Harvesting Training Images. Fergus et al. [6] were one of the first to propose training a visual classifier by automatically harvesting (potentially noisy)

[1] Previous work [1,5,15] has used visual descriptions for object recognition without any training images but not for the discovery of training images itself.

training images from the Web, in their case obtained by querying Google Images with the object category name. Topic modelling is performed on the images, and test images are classified by how likely they are to belong to the best topic selected using a validation set. However, using a single best topic results in low data diversity. Li et al. [9] propose a framework where category models are learnt iteratively, and the image dataset simultaneously expanded at each iteration. They overcome the data diversity problem by retaining a small but highly diverse 'cache set' of positive images at each iteration, and using it to incrementally update the model. Other related work includes using multiple-instance learning to automatically de-emphasise false positives [14] and an active learning approach to iteratively label a subset of the images [3].

Harvesting Using Text and Images. The work described so far involves filtering only by images; the sole textual data involved are keyword queries to search engines. Berg and Forsyth [2] model both images *and* their surrounding text from Google *web* search to harvest images for ten animal categories. Topic modelling is applied to the *text*, and images are ranked based on how likely their corresponding text is to belong to each topic. Their work requires human supervision to identify relevant topics. Schroff et al. [12] propose generating training images *without* manual intervention. Class-independent text-based classifiers are trained to rerank images using binary features from web pages, e.g. whether the query term occurs in the website title. They demonstrated superior results to [2] on the same dataset without requiring any human supervision. George et al. [7] build on [12] by retrieving images iteratively, while Krapac et al. [8] add contextual features (words surrounding the image etc.) on top of the binary features of [12].

Like [2,7,8,12], our work ranks images by their surrounding text. However, we tackle fine-grained object categories which will allow the harvesting of training images to scale to a large number of categories. In addition, we do not only use the web text surrounding the image but use the visual descriptions in outside resources to rank accompanying web-text by their similarity to these visual descriptions. In contrast to the manual topic definition in [2], this method does then not require human intervention during harvesting.

1.2 Overview

We illustrate harvesting training images for ten butterfly categories of the Leeds Butterfly Dataset [15], using the provided eNature visual descriptions. Figure 2 shows the pipeline for our method, starting from the butterfly species' name and visual description. We obtain a list of candidate web pages via search engine queries (Sect. 2). These are parsed to produce a collection of images and text blocks for each web page, along with their position and size on the page (Sect. 3). Image-text correspondence aligns the images with text blocks on each web page (Sect. 4). The text blocks are then matched to the butterfly description (Sect. 5), and images ranked based on how similar their corresponding text blocks are to the visual description (Sect. 6). The ranked images are evaluated in Sect. 7, and conclusions offered in Sect. 8.

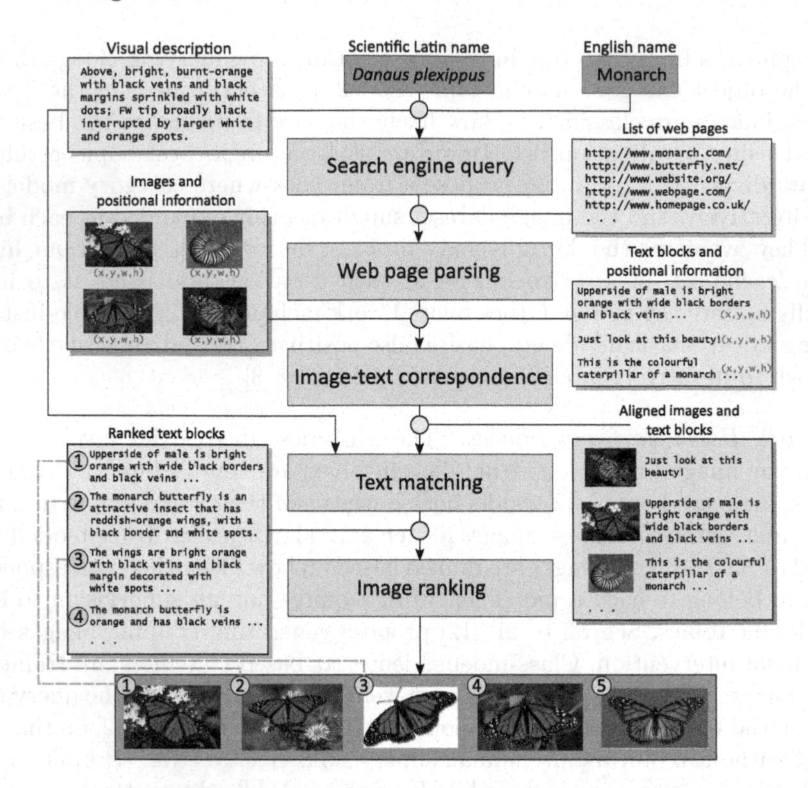

Fig. 2. General overview of the proposed framework, which starts from the butterfly species name (Latin and English) and description, and outputs a ranked list of images.

2 Search Engine Query

We use Google search to obtain as many candidate pages as possible containing images (along with textual descriptions) of the desired butterfly categories. To later compare our method using visual descriptions to one using category names only, we retrieve candidate pages by several different methods. First, we have four *base queries* mainly based on the category name. Here we use both the butterfly's (i) Scientific (Latin) name; (ii) common (English) name. As English names may be polysemous, the term "butterfly" is appended to these for better precision. To increase the recall of visual descriptions, additional queries are produced by appending "description OR identification" to the butterfly name. Our four *base queries* are: (i) "Latin name"; (ii) "English name" + butterfly; (iii) "Latin name" + (description OR identification); (iv) "English name" + butterfly + (description OR identification).

Besides the base queries, we aim to raise precision by also using phrases from the eNature textual descriptions themselves as *seed* terms for the queries; this returns web pages with similar phrases which could potentially include visual descriptions for the butterfly category. The seed phrases are restricted to *noun phrases* and *adjective phrases*, obtained via phrase chunking as in [15]. The number of seed phrases per category ranges from 5 to 17 depending on the length of the description; an example list is shown in Fig. 3. We query Google with the butterfly name augmented with each seed phrase individually, and with all possible combinations of seed phrase pairs and triplets (e.g. *'Vanessa atalanta' bright blue patch pink bar white spots*).

Two sets of seeded queries are used: one with the Latin and one with the English butterfly name. For each category, all candidate pages from the base and the seeded queries (54 to 1670 queries per category, mean 592) are pooled. For de-duplication, only one copy of pages with the same web address is retained.

Description: FW tip extended, clipped. Above, black with orange-red to vermilion bars across FW and on HW border. Below, mottled black, brown, and blue with pink bar on FW. White spots at FW tip above and below, bright blue patch on lower HW angle above and below.

Seed phrases: black brown and blue; bright blue patch; fw tip; hw border; lower hw angle; orange red to vermilion bars; pink bar; white spot

Fig. 3. Seed phrases for *Vanessa atalanta* extracted from its visual description.

3 Web Page Parsing

Previous work [2,7,8,12] performs image-text correspondence by parsing the HTML source code of a web page, and extracting any non-HTML text surrounding an image link, assuming that such text is positioned close to the image. However, this assumption is not always correct as the HTML source does not always dictate how a web page is *displayed*. The presentation of a web page is most often controlled by style sheets or scripts that dynamically change the web page's layout. As such, web page elements may be freely positioned independent of their sequence in the HTML source. Another example is the use of tables, where cells are defined from left-to-right and then top-to-bottom. Thus, text in a table cell might not be aligned to an image in the cell above since they may be positioned far apart from each other in the HTML source. These issues could be alleviated by using DOM trees, e.g. [17], but they still encode mainly structural and semantic information of web page elements and not positional information.

To address this issue, we match text and images by *where* they are located on the page as rendered to the user. Such positional information is not available from the HTML source or DOM tree, but is dependent on a browser layout engine which generates this information. We use QtWebKit[2], an implementation of the WebKit web browser engine in the Qt Framework. It provides details of all elements in a web page, including the tag name, content, horizontal and vertical positions, width, and height. The nature of the elements themselves also provide additional information, for example whether they are displayed at 'block level' (e.g. a paragraph) or 'inline level' (e.g. <span>, <a>, <i>). For our work, we

[2] http://trac.webkit.org/wiki/QtWebKit.

consider as *text blocks* all text within block-level elements (including tables and table cells) and those delimited by any images or the <br> element. All images and text blocks are extracted from web pages, along with their height, width, and (x, y) coordinates as would be rendered by a browser. The renderer viewport size is set as 1280×1024 across all experiments.

4 Image-text Correspondence

The list of images and text blocks with their positional information is then used to align text blocks to images (see Fig. 4 for an illustration). An image can correspond to multiple text blocks since we do not want to discard any good candidate visual descriptions by limiting ourselves to only one nearest neighbouring text. On the other hand, each text block may only be aligned to its closest image; multiple images are allowed only if they both share the same distance from the text block. This relies on the assumption that the closest image is more likely to correspond to the text blocks than those further away.

An image is a candidate for alignment with a text block only if all or part of the image is located directly above, below or either side of the text block. All candidate images must have a minimum size of 120×120. For each text block, we compute the perpendicular distance between the closest edges of the text block and each image, and select the image with the minimum distance subject to the constraint that the distance is smaller than a fixed threshold (100 pixels in our experiments). Text blocks without a corresponding image are discarded.

General description: Black; forewing with red median band a white subapical spots; hindwing with broad red marginal band; ventral hindwing mottled black, brown, blue and cream

Fig. 4. An illustration of the proposed image-text correspondence algorithm. The text block is matched to the two top images as they are both of the same distance from the text block. The image on the right is not matched as it is further away from the text block than the top two images. The caterpillar image on the bottom right is not considered as it is outside the 'candidate region' (shaded region in figure), i.e. it is not directly above or below, or directly to the left or right of the text block.

5 Text Matching

The text matching component computes how similar a text block is to the visual description from our outside resource, using IR methods. We treat the butterfly's visual description as a *query*, and the set of text blocks as a collection of *documents*. The goal is to search for documents which are similar to the query and assign each document a similarity *score*.

There are many different ways of computing text similarity, and we only explore one of the simplest in this paper, namely a bag of words, frequency-based vector model. It is a matter of future research to establish whether more sophisticated methods (such as compositional methods) will improve performance further. We represent each document as a vector of term frequencies (*tf*). Separate vocabularies are used per query, with the vocabulary size varying between 1649 to 9445. The vocabulary consists of all words from the *document* collection, except common stopwords and *Hapax legomena* (words occurring only once). Terms are case-normalised, tokenised by punctuation and Porter-stemmed [11]. We use the *lnc.ltc* weighting scheme of the SMART system [13], where the *query* vector uses the log-weighted term frequency with idf-weighting, while the *document* vector uses the log-weighted term frequency without idf-weighting. The relevance score between a query and a document vector is computed using the cosine similarity measure.

6 Image Ranking and Filtering

Each text block from Sect. 5 is treated as a candidate butterfly description, and assigned a similarity score with regard to the category visual description. Images are ranked by the *maximum* score among an image's neighbouring text blocks. Intuitively, for each image we choose the text block most likely to be a visual description and use this score to rerank the image collection. As many images from a web page may be irrelevant (e.g. page headers, icons, advertisements), we filter by retaining only images where their metadata (image file name, *alt* or *title* attribute) contains the butterfly name (Latin or English) and excludes a predefined list of 'negative' terms (e.g. caterpillar, pupa).

7 Experimental Results

We evaluate the image rankings via precision at selected recall levels. We compare our reranked images using visual descriptions to the Google ranking produced by name search only.

Annotation. For each category, we annotated the retrieved images as 'positive' (belonging to the category), 'negative' or 'borderline'. Borderline cases include non-photorealistic images, poor quality images, images with the butterfly being too small, images with major occlusions or extreme viewpoints, etc. Only positive and negative cases are considered during evaluation. For a fair evaluation we ignore borderline cases as they are not exactly 'incorrect' but are just poor examples; it would have been acceptable to have them classified either way.

Table 1. Statistics of annotated images, before and after pre-filtering.

Category	Number of retrieved images				Number of images after pre-filtering			
	Positive	Negative	Borderline	Total	Positive	Negative	Borderline	Total
Daxnaus plexippus	23.1 %	59.2 %	17.7 %	**12470**	42.7 %	34.4 %	23.0 %	**5240**
Heliconius charitonius	45.9 %	39.4 %	14.7 %	**2053**	70.8 %	8.2 %	21.0 %	**1025**
Heliconius erato	31.5 %	61.2 %	7.2 %	**1132**	37.9 %	55.2 %	6.8 %	**701**
Junonia coenia	45.5 %	39.5 %	15.0 %	**3055**	66.8 %	9.4 %	23.8 %	**1507**
Lycaena phlaeas	52.5 %	39.0 %	8.4 %	**1947**	73.5 %	15.8 %	10.7 %	**945**
Nymphalis antiopa	36.7 %	50.6 %	12.7 %	**3078**	60.7 %	18.2 %	21.1 %	**1297**
Papilio cresphontes	29.0 %	52.5 %	18.5 %	**3815**	48.9 %	18.0 %	33.0 %	**1571**
Pieris rapae	34.6 %	56.6 %	8.8 %	**2742**	59.7 %	27.9 %	12.4 %	**1112**
Vanessa atalanta	26.6 %	63.3 %	10.0 %	**6822**	63.8 %	16.2 %	20.0 %	**2150**
Vanessa cardui	19.4 %	72.6 %	8.0 %	**10301**	47.2 %	37.3 %	15.4 %	**3158**

Statistics and Filtering Evaluation. Table 1 provides the statistics for our annotations. The table shows the level of noise, where many images on the web pages are unrelated to the butterfly category. Filtering via metadata dramatically reduces the number of negative images without too strongly reducing the number of positive ones. The cases where the number of negative images is high after filtering are due to the categories being visually similar to other butterflies, which often have been confused by the web page authors.

Baselines. We use the four base queries (using predominantly category names) as independent baselines for evaluation. For each base query, we rank each image according to the rank of its web page returned by Google followed by its order of appearance on the web page. Images are filtered via category name appearance in metadata just as in our method. We also compare the results with two additional baselines, querying Google Images with (i) "Latin name"; (ii) "English name" + butterfly. These are ranked using the ranks returned by Google Images.

Results. We concentrate on the *precision* of images at early stages of recall, i.e. obtaining as many correct images as possible for top-ranked images. Figure 5 shows the precision-recall curves for our method against the baselines, up to a recall of 50 images. The precision for *Junonia coenia*, *Lycaena phlaeas*, *Pieris rapae* and *Vanessa atalanta* is consistently higher than all baselines across different recall levels. The precision of most remaining categories is relatively high, although not better than all baselines. There were some misclassifications at very early stages of recall for *Danaus plexippus* and *Papilio cresphontes*; however, the overall precision for these is high, especially at later stages of recall. The performance of *Heliconius charitonius* and *Nymphalis antiopa* is comparable to their best baselines. *Vanessa cardui* also gave higher precision than its baselines up to a recall of about 20 images. The only poor performance came from *Heliconius erato*: many subspecies of this butterfly exist which are visually different from the nature guide description, making ranking by similarity to description unsatisfactory. Our method needs categories with strong shared visual characteristics to work fully.

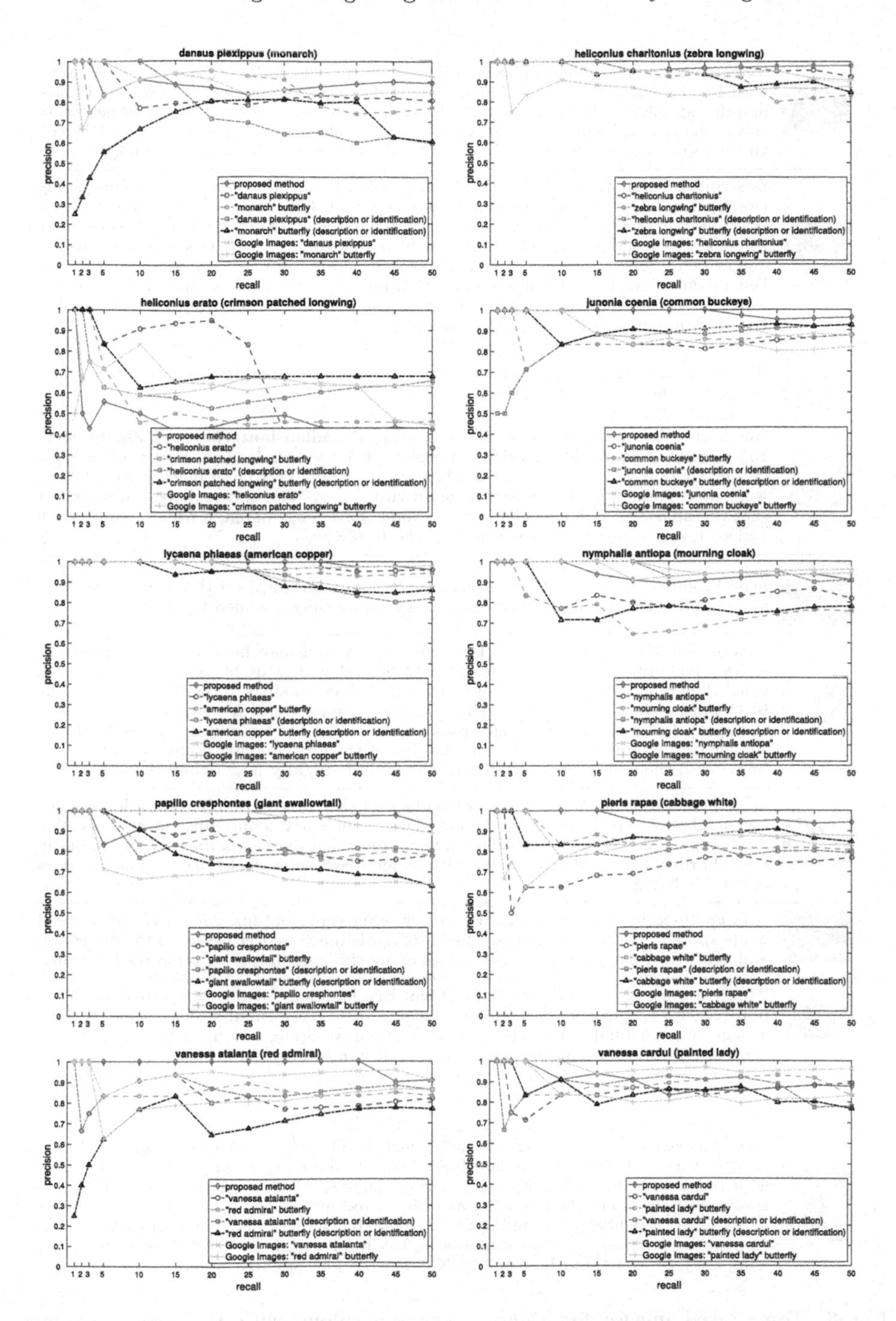

Fig. 5. Precision at selected levels of recall for the proposed method for ten butterfly categories, compared to baseline queries. The recall (x-axis) is in terms of *number* of images. For clarity we only show the precisions at selected recalls of up to 50 images.

1		Monarch Butterfly (Danaus plexippus) Description 3 1/2-4" (89-102 mm). Very large, with FW long and drawn out. Above, bright, burnt-orange with black veins and black margins sprinkled with white dots; FW tip broadly black interrupted by larger white and orange spots. Below, paler, duskier orange. 1 black spot appears between HW cell and margin on male above and below. Female darker with black veins smudged.
2		Description : Family: Nymphalidae, Brush-footed Butterflies view all from this family Description 3 1/2-4" (89-102 mm). Very large, with FW long and drawn out. Above, bright, burnt-orange with black veins and black margins sprinkled with white dots; FW tip broadly black interrupted by larger white and orange spots. Below, paler, duskier orange. 1 black spot appears between HW cell and margin on male above and below. Female darker with black veins smudged. Similar Species Viceroy smaller, has shorter wings and black line across HW. Queen and Tropic Queen are browner and smaller. Female Mimic has large white patch across black FW tips. ...
3		The wings are bright orange with black veins and black margin decorated with white spots. Female's veins are thicker.
4		Diagnosis: The Monarch is one of the largest Canadian butterflies (wingspan: 93 to 105 mm). The upperside is bright orange with heavy black veins, and a wide black border containing a double row of white spots. There is a large black area near the wing tip containing several pale orange or white spots. The underside is similar except that the hindwing is much paler orange. Males have a sex patch, a wider area of black scales on a vein just below the centre of the hindwing.
5		male bright orange w/oval black scent patch (for courtship) on HW vein above, and abdominal "hair-pencil;" female dull orange, more thickly scaled black veins
6		Description: This is a very large butterfly with a wingspan between 3 3/8 and 4 7/8 inches. The upperside of the male is bright orange with wide black borders and black veins. The hindwing has a patch of scent scales. The female is orange-brown with wide black borders and blurred black veins. Both sexes have white spots on the borders and the apex. There are a few orange spots on the tip of the forewings. The underside is similar to the upperside except that the tips of the forewing and hindwing are yellow-brown and the white spots are larger. The male is slightly larger than the female.
7		General description: Wings orange with black-bordered veins and black borders enclosing small white spots. Male with small black scent patch along inner margin. Ventral hindwing as above but paler yellow-orange and with more prominent white spots in black border. Female duller orange with wider black veins; lacks black scent patch on dorsal hindwing.
8		A large butterfly, mainly orange with black wing veins and margins, with two rows of white spots in the black margins. The Monarch is much lighter below on the hindwing, and males have a scent patch - a dark spot along the vein - in the center of the hindwing.
9		Wingspan: 3 1/2 to 4 inches Wings Open: Bright orange with black veins and black borders with white spots in the male. The male also has a small oval scent patch along a vein on each hind wing. The female is brownish-orange with darker veins Wings Closed: Forewings are bright orange, but hind wings are paler
		...
16	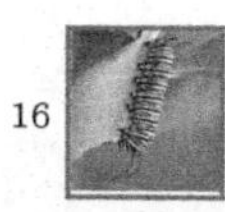	The Monarch's wingspan ranges from 3–4 inches. The upper side of the wings is tawny-orange, the veins and margins are black, and in the margins are two series of small white spots. The fore wings also have a few orange spots near the tip. The underside is similar but the tip of the fore wing and hind wing are yellow-brown instead of tawny-orange and the white spots are larger. The male has a black patch of androconial scales responsible for dispersing pheromones on the hind wings, and the black veins on its wing are narrower than the female's. The male is also slightly larger.

Fig. 6. Top ranked images for *Danaus plexippus*, along with their corresponding descriptions. A red border indicates that the image was misclassified. The first description is almost identical to the eNature description (Color figure online).

The main mistakes made by our method can be attributed to (i) the web pages themselves; (ii) our algorithm.

In the first case, the ambiguity of some web page layouts causes a misalignment between text blocks and images. In addition, errors arise from mistakes made by the page authors, for example confusing the Monarch (*Danaus plexippus*) with the Viceroy butterfly (*Limenitis archippus*).

For mistakes caused by our algorithm, the first involves the text similarity component. Apart from similar butterflies having similar visual descriptions, some keywords in the text can also be used to describe non-butterflies, e.g. "pale yellow" can be used to describe a caterpillar or butterfly wings. The second mistake arises from text-image misalignment as a side-effect of the filtering step: there were cases where a butterfly image does not contain the butterfly name in its metadata while a caterpillar image on the same page does. Since the butterfly image is discarded, the algorithm matches a text block with its next nearest image – the caterpillar. This could have been rectified by not matching text blocks associated with a previously discarded image, but it can be argued that such text blocks might still be useful in certain cases, e.g. when the discarded image is an advertisement and the next closest image is a valid image.

Figure 6 shows the top ranked images for *Danaus plexippus*, along with the retrieved textual descriptions. All descriptions at early stages of recall are indeed of *Danaus plexippus*. This shows that our proposed method performs exceptionally well given sufficient textual descriptions. The two image misclassifications that still are present are from image-text misalignment, as described above.

8 Conclusion

We have proposed methods for automatically harvesting training images for fine-grained object categories from the Web, using the category name and visual descriptions. Our main contribution is the use of visual descriptions for querying candidate web pages and reranking the collected images. We show that this method often outperforms the frequently used method of just using the category name on its own with regards to precision at early stages of recall. In addition, it retrieves further textual descriptions of the category.

Possible future work could explore different aspects: (i) exploring better language models and similarity measures for comparing visual descriptions and web page text; (ii) training generic butterfly/non-butterfly visual classifiers to further filter or rerank the images; (iii) investigating whether the reranked training set can actually induce better visual classifiers.

Acknowledgements. The authors thank Paul Clough and the anonymous reviewers for their feedback on an earlier draft of this paper. This work was supported by the EU CHIST-ERA D2K 2011 Visual Sense project (EPSRC grant EP/K019082/1) and the Overseas Research Students Awards Scheme (ORSAS) for Josiah Wang.

References

1. Ba, J.L., Swersky, K., Fidler, S., Salakhutdinov, R.: Predicting deep zero-shot convolutional neural networks using textual descriptions. In: Proceedings of the IEEE International Conference on Computer Vision (2015)
2. Berg, T.L., Forsyth, D.A.: Animals on the web. In: Proceedings of the IEEE Conference on Computer Vision & Pattern Recognition, vol. 2, pp. 1463–1470 (2006)
3. Collins, B., Deng, J., Li, K., Fei-Fei, L.: Towards scalable dataset construction: an active learning approach. In: Forsyth, D., Torr, P., Zisserman, A. (eds.) ECCV 2008, Part I. LNCS, vol. 5302, pp. 86–98. Springer, Heidelberg (2008)
4. Deng, J., Dong, W., Socher, R., Li, L.J., Li, K., Fei-Fei, L.: Imagenet: a large-scale hierarchical image database. In: Proceedings of the IEEE Conference on Computer Vision & Pattern Recognition, pp. 248–255 (2009)
5. Elhoseiny, M., Saleh, B., Elgammal, A.: Write a classifier: Zero-shot learning using purely textual descriptions. In: Proceedings of the IEEE Conference on Computer Vision & Pattern Recognition (2013)
6. Fergus, R., Fei-Fei, L., Perona, P., Zisserman, A.: Learning object categories from Google's image search. In: Proceedings of the IEEE International Conference on Computer Vision, vol. 2, pp. 1816–1823 (2005)
7. George, M., Ghanem, N., Ismail, M.A.: Learning-based incremental creation of web image databases. In: Proceedings of the 12th IEEE International Conference on Machine Learning and Applications (ICMLA 2013), pp. 424–429 (2013)
8. Krapac, J., Allan, M., Verbeek, J., Jurie, F.: Improving web-image search results using query-relative classifiers. In: Proceedings of the IEEE Conference on Computer Vision & Pattern Recognition, pp. 1094–1101 (2010)
9. Li, L.J., Wang, G., Fei-Fei, L.: OPTIMOL: Automatic Object Picture collecTion via Incremental MOdel Learning. In: Proceedings of the IEEE Conference on Computer Vision & Pattern Recognition, pp. 1–8 (2007)
10. Nilsback, M.E., Zisserman, A.: Automatedower classification over a large numberof classes. In: Proceedings of the Indian Conference on Computer Vision, Graphics and Image Processing, pp. 722–729 (2008)
11. Porter, M.F.: An algorithm for suffix stripping. Program **14**(3), 130–137 (1980)
12. Schroff, F., Criminisi, A., Zisserman, A.: Harvesting image databases from the Web. IEEE Trans. Pattern Anal. Mach. Intell. **33**(4), 754–766 (2011)
13. Singhal, A., Salton, G., Buckley, C.: Length normalization in degraded text collections. In: Proceedings of Fifth Annual Symposium on Document Analysis and Information Retrieval, pp. 149–162 (1996)
14. Vijayanarasimhan, S., Grauman, K.: Keywords to visual categories: Multiple-instance learning for weakly supervised object categorization. In: Proceedings of the IEEE Conference on Computer Vision & Pattern Recognition (2008)
15. Wang, J., Markert, K., Everingham, M.: Learning models for object recognition from natural language descriptions. In: Proceedings of the British Machine Vision Conference, pp. 2.1-2.11. BMVA Press (2009)
16. Welinder, P., Branson, S., Mita, T., Wah, C., Schroff, F., Belongie, S., Perona, P.: Caltech-UCSD Birds 200. Technical Report CNS-TR-2010-001. California Institute of Technology (2010)
17. Zhou, N., Fan, J.: Automatic image-text alignment for large-scale web image indexing and retrieval. Pattern Recogn. **48**(1), 205–219 (2015)

Do Your Social Profiles Reveal What Languages You Speak? Language Inference from Social Media Profiles

Yu Xu[1(✉)], M. Rami Ghorab[1], Zhongqing Wang[2], Dong Zhou[3], and Séamus Lawless[1]

[1] ADAPT Centre, Knowledge and Data Engineering Group, School of Computer Science and Statistics, Trinity College Dublin, Dublin, Ireland
{xuyu,rami.ghorab,seamus.lawless}@scss.tcd.ie

[2] Natural Language Processing Lab, Soochow University, Suzhou, China
wangzq.antony@gmail.com

[3] School of Computer Science and Engineering, Hunan University of Science and Technology, Xiangtan, China
dongzhou1979@hotmail.com

Abstract. In the multilingual World Wide Web, it is critical for Web applications, such as multilingual search engines and targeted international advertisements, to know what languages the user understands. However, online users are often unwilling to make the effort to explicitly provide this information. Additionally, language identification techniques struggle when a user does not use all the languages they know to directly interact with the applications. This work proposes a method of inferring the language(s) online users comprehend by analyzing their social profiles. It is mainly based on the intuition that a user's experiences could imply what languages they know. This is nontrivial, however, as social profiles are usually incomplete, and the languages that are regionally related or similar in vocabulary may share common features; this makes the signals that help to infer language scarce and noisy. This work proposes a language and social relation-based factor graph model to address this problem. To overcome these challenges, it explores external resources to bring in more evidential signals, and exploits the dependency relations between languages as well as social relations between profiles in modeling the problem. Experiments in this work are conducted on a large-scale dataset. The results demonstrate the success of our proposed approach in language inference and show that the proposed framework outperforms several alternative methods.

1 Introduction

As a result of globalization and cultural openness, it has become common for modern-day humans to speak multiple languages (polyglots) [1, 2]. Knowledge of what languages the online user *comprehends*[1] is becoming important for many Web applications to enable effective information services. For example, knowing the user's language

[1] In this paper, *comprehand* means the user is able to grasp information in that language to a good extent.

N. Ferro et al. (Eds.): ECIR 2016, LNCS 9626, pp. 561–574, 2016.
DOI: 10.1007/978-3-319-30671-1_41

information enables search engines to deliver the multilingual search services, machine translation tools to identify optional target translation languages, and advertisers to serve targeted international ads.

However, online users often choose not to explicitly provide their language information in real-world applications, even when they have a facilitated means to do so. For example, based on our analysis of 50,575 user profiles on LinkedIn, we found that only 11 % of users specified the languages that they speak, even though there are input fields available for this; Ghorab *et al.* [3] carried out an analysis where it was found that many users of *The European Library*[2] entered search queries in non-English languages and browsed documents in those languages, without bothering to use the drop down menu that allows them to change the interface language. Therefore, studies proposed to automatically acquire the users' language information through the Language Identification (LID) techniques [4]; they detect what languages a user comprehends by identifying what languages of texts the user read or wrote in interaction history. However, the challenge faced by this approach is the common cold-start problem, where there is only limited history of interactions available for a new user.

This work proposes the use of social profiles to infer the languages that a user comprehends. This idea is mainly based on three points of observation: (1) today most users maintain a profile on a number of Social Networking Sites (SNSs), such as Facebook, LinkedIn, which could include basic personal information like education and work experience; (2) the social profile provides first-hand information about the user to Third Party Applications (TPAs) of SNSs. For example, the popular social login techniques mostly authorize a TPA to access a user's social profile [5]; (3) there is a chance that the information in a user's social profile may implicitly suggest what languages the user comprehends. For example, if a user has conducted academic studies in Germany, this could imply that the user at least understands German to some extent.

Using automatic techniques to infer the user's language can serve to overcome the cold-start problem, and benefit numerous Web applications as mentioned previously. This work is also a first step towards automatically inferring the user's level of expertise in languages that they comprehend. Furthermore, this research can be integrated with other work in user profiling where numerous other characteristics of the user can be automatically inferred, such as the user's gender [6], location [7].

It is straightforward to cast the task of user language inference from social profiles as a standard text classification problem. In other words, predicting what languages a user comprehends relies on features defined from textual information of the social profiles, e.g. unigram features. However, the social profiles are usually incomplete, with critical information sometimes missing. For example, the location information of the work experience, which, in this research, is shown to be important evidence for language inference, is only provided by about half of the users in the collected dataset. Moreover, some languages can be *mutually intelligible* (i.e. speakers of different but related languages can readily understand each other without intentional study) or *regionally related* (i.e. multiple languages are spoken in one region). These languages may share many common features, which makes it hard to identify the discriminative features

[2] http://www.theeuropeanlibrary.org.

between them. Therefore, solely relying on the textual features to infer these languages may yield unsatisfactory results.

To address these challenges, this work investigates three factors to better model the problem: (1) Textural attributes in social profiles. They provide fundamental evidence about what languages a user may comprehend. This work also attempts to exploit external resources to enhance the textural attributes. It aims to import more information that is associated with the user and also may reflect user language information. (2) Dependency relations between languages. Languages may be related to each other in certain ways, e.g. mutually intelligible. This relation could reflect the possibility that a user comprehends other languages based upon a language we know the user understands. (3) Social relations between users through their social profiles. It is reasonable to assume that users with similar academic or professional backgrounds may comprehend one or many of the same languages. This relation information between users could be extracted from their social profiles. Finally, a language and social relation-based factor graph model (LSR-FGM) is proposed which predicts the user language information under the collective influence of the three factors.

Experiments are conducted on a large-scale LinkedIn profile dataset. Results show that LSR-FGM clearly outperforms several alternative models and is able to obtain an F1-score of over 84 %. Experiments also demonstrate that every factor contributes from a different perspective in the process of inference.

2 Related Work

Users' language information is an important input to multilingual applications. Previous studies primarily rely on Language Identification techniques to automatically acquire a user's language information [8, 9]. To different types of texts, distinct LID techniques may be used to harvest their language information, e.g., Web pages [10], search queries [11], social texts [12]. However, to the best of our knowledge, this work is the first attempt to acquire the users' language information via their social profiles. Specifically, the inference is based on the information stored in the profile itself and not on the posts or media shared by the user on their profile's space (e.g. the "wall" in Facebook). This approach is useful in cold-start scenarios where there are no prior user interactions available to base the LID decision upon.

From another angle, this work is also related to user profiling. The profiling of user interests has always been a hot research topic. Numerous studies explore user-centric data, like query logs [13], browsing/click behaviors [14, 15], social activities [16, 17], to model user interests from different perspectives, in order to serve different applications, such as personalized search [16], targeted advertisement [17]. In addition, a number of efforts have been made to investigate other personal information in user profiling. Mislove *et al.* exploited the friendship networks to infer Facebook users' attributes, such as colleges, majors [18]. Li *et al.* proposed a unified discriminative influence model to infer Twitter users' home locations by integrating social behaviors of both the user and her friends [7]. Most of these studies focus on the utilization of the historically interactive data of the user (and their connected users), while our work purely uses the textual content of static social profiles. Although a few studies were based on the content of user profiles, they mainly aimed to extract certain target information from the profile, like special skills [19], summary sentences [20], rather than to infer hidden knowledge about the user.

Due to the natural structure of social networks, the factor graph model attracts much attention from researchers in representing and mining social relationships between users in social networks [21, 22]. Tang *et al.* proposed to utilize the user interactions to infer the social relationship between users in social networks [23]. They defined multiple factors that related to the user relationship, and proposed a partially labeled pairwise factor graph model to infer a target social relationship between users. In social network analyses, Tang *et al.* defined several types of conformity from different levels, and modeled the effects of these conformities to users' online behavior using a factor graph model [24]. Our work is different from previous studies in two aspects. First, our work is based on self-managed profiles instead of the naturally connected social networks. Second, this paper proposes a factor graph model that collectively exploits local textual attributes in profiles and multiple types of relations between profiles to infer the user's language information.

3 User Language Information Inference Using Social Profiles

This section first gives some background information on the inference problem and then introduces two main challenges in modeling the problem as well as corresponding solutions. Finally, it gives the formal problem definition.

3.1 Background

In general, a social profile consists of multiple fields, each of which details a particular aspect of information about the user, such as education background, hobbies. Different platforms may use different fields to construct user profiles. Without loss of generality, this work considers three commonly used fields (SNSs, like Facebook, Google +, contain the three fields. But the mini-profiles in Twitter do not apply to our model.) in the social profile for the language information inference problem:

- ***Summary:*** Unstructured text where the users give a general introduction about themselves. Because there is no structure restriction, the focus of this field varies from user to user.
- ***Education Background:*** Structured text that details each study experience of the user by subsections. Each subsection could include attributes like school name, study major, etc.
- ***Work Experience:*** Structured text details the user's work experience by subsections. Each subsection could include attributes like company name, role, work period, etc.

In practice, the language information of **some users** is readily obtainable through certain means, e.g., it can be explicitly stated in the user's profile; or it can be easily predicted from the user's interactions with the system. Therefore the problem is how to infer the language information of the **remaining users** based on: (1) textual information of all the profiles; (2) known language information of the other part of users.

3.2 Challenges and Solutions

As discussed in Sect. 1, the attributes in a social profile may implicitly suggest what languages a user comprehends, especially for the location-related attributes. For example, if a user who stayed in multiple places that share a common language, it is reasonable to infer this user likely comprehends this language. Based on this assumption, we can model the language inference using social profiles as a supervised classification problem. However, there are two main challenges in modeling this problem:

(1) Users' online social profiles are usually incomplete and some profiles even miss critical information as analyzed in Sect. 1. Besides, some information is generally not asked for by the SNS platform in constructing the social profile but they may be important evidence for language information inference, e.g., the location of institutions in the field of Education Background.
    In order to alleviate this problem, this work proposes to correlate each experience of the user to the corresponding location (if missing) by exploiting external resources. For example, a university can be linked to its homepage, by which the location can be obtained. The specific strategy adopted in this work is detailed in Sect. 5.1.
(2) Some attributes in the profile may be misleading in the inference process. As languages could be regionally related or mutually intelligible, they may share similar discriminative features. When many of these languages are taken into consideration in the target language set, only considering textual attributes as the features may not be able to distinguish these languages.

In this work, Chinese, French, German, Hindi, Spanish are selected as target inference languages. In the five languages, Spanish, French and German are used in combination as official languages in a number of countries, e.g., both French and Spanish are official languages of Equatorial Guinea; French and German are official languages of Luxembourg[3]. Also, French has lexical similarities of 0.75 and 0.29 (1.0 is a total overlap between vocabularies) with Spanish and German respectively[4]. By contrast, Hindi and Chinese are spoken as an official language only in India and China respectively; they have no overlap with the above three languages in vocabulary. So lower prediction accuracy is expected on French, German and Spanish as it is more difficult to identify their discriminative features. This intuition is validated in our experiments.

This work takes two types of relation into consideration in modeling the language inference problem in an attempt to address this challenge: (1) **Dependency relation between languages**. The above example also can explain that if a user knows French, she has a much higher probability of also knowing Spanish than Chinese. Thus, this dependency relation between languages could be helpful in inferring users' language information.

---

[3] https://en.wikipedia.org/wiki/List_of_multilingual_countries_and_regions.

[4] https://en.wikipedia.org/wiki/Lexical_similarity.

(2) **Social relation between users**. Although new users have no direct friendship/followship with other users, they can be related through available information of their social profiles. This work focuses on the *same-experience* relation, i.e., two profiles share a study experience (studied the same major in the same institute) or work experience (worked as the same role in the same company), to help inference. For example, the fact that two users shared the same work experience may imply that they know a common language because communication is needed between employees in the department of the company. Thus, it is reasonable to assume that the users with the same-experience social relation are likely to know a common language.

Therefore, this paper proposes a model that collectively considers the three factors outlined above: enhanced textual attributes, language relation and social relation, to model the problem of language inference using social profiles.

3.3 Problem Definition

The input of N social profiles can be represented as $G = (U, L, \mathbf{W})$, where U is the set of $|U| = N$ users and L is the set of $|L| = K$ target inference languages; $\mathbf{W}$ is an attribute matrix associated with users in U in which each row corresponds to a user, each column an attribute of the profile, and an entry w_{ij} denotes the attribute value of the jth attribute in the profile of user u_i. The objective of this work is to learn a model that can effectively infer what languages a user comprehends.

This work defines the *correlation node* $v = (u_i, l_j)$ that is associated with a label y (binary value) to represent the output of the problem. It means user $u_i \in U$ comprehends language $l_j \in L$ if $y = 1$ or the opposite if $y = 0$. Each user in U is mapped to K correlation nodes with the K languages in L, so a set V with $K*N$ correlation nodes and a corresponding label set Y are obtained. As part of users' language information is known, label values of the corresponding correlation nodes are given, which is denoted as set Y^L. The remaining labels are denoted as set Y^U and $Y = Y^L \cup Y^U$. In addition, the correlation nodes are connected through the two types of relations defined above, which constitute an undirected edge set E. The definition on E is detailed in Sect. 4.

Therefore, given a partially labeled network $G = (V, U, L, Y^L, E, \mathbf{W})$, the objective of the language inference problem is to learn a predictive function: $f: G \rightarrow Y^U$.

4 Language and Social Relation-Based Factor Graph Model

This section details the construction of the Language and Social Relation-based Factor Graph Model (LSR-FGM) and proposes a method to learn the model.

4.1 Model Definition

The LSR-FGM collectively incorporates the three factors outlined above to better model the problem of user language inference. Its basic idea is to define these relations among users and languages using different factor functions in a graph. Thus, an objective function can be defined based on the joint probability of all factor functions. Learning the LSR-FGM is to estimate the model parameters, which can be achieved by maximizing the log-likelihood objective function based on the observation information. Below, we introduce the construction of the objective function in detail.

For simplicity, given a correlation node v_i, we use $v_i(u)$ and $v_i(l)$ to represent the user and language of this node respectively. Note each correlation node v_i in V is also associated with an attribute vector $\boldsymbol{x}_i$ which is from the attribute vector of user $v_i(u)$, and $\boldsymbol{X}$ is the attribute matrix corresponding to V. Then, we have a graph $G = (V, E, Y, X)$, in which the value of a label y_i depends on both the local attribute vector $\boldsymbol{x}_i$ and the connections related to v_i. Thus, we have the following conditional probability distribution over G:

$$P(Y|\boldsymbol{X}, E) = \frac{P(\boldsymbol{X}, E|Y)P(Y)}{P(\boldsymbol{X}, E)} \tag{1}$$

According to the Bayes' rule and assuming $X \perp E$ in LSR-FGM, we can further have:

$$P(Y|\boldsymbol{X}, E) \propto P(\boldsymbol{X}|Y)P(Y|E) \tag{2}$$

in which $P(\boldsymbol{X}|Y)$ represents the probability of generating attributes $\boldsymbol{X}$ associated to all correlation nodes given their labels Y, and $P(Y|E)$ denotes the probability of labels given all connections between correlation nodes. It is reasonable to assume that the generative probability of attributes given the label value of each correlation node is conditionally independent. Thus we can factorize Eq. (2) again:

$$P(Y|\boldsymbol{X}, E) \propto P(Y|E) \prod_i P(\boldsymbol{x}_i|y_i) \tag{3}$$

where $P(\boldsymbol{x}_i|y_i)$ is the probability of generating attribute vector $\boldsymbol{x}_i$ given label y_i. Now the problem is how to instantiate the probability $P(Y|E)$ and $P(\boldsymbol{x}_i|y_i)$. In principle, they can be instantiated in different ways. This work models them in a Markov random field, so the two probabilities can be instantiated based on the Hammersley-Clifford theorem [25]:

$$P(\boldsymbol{x}_i|y_i) = \frac{1}{Z_1} exp\left\{\sum_{j=1}^{d} \alpha_{(v_i(l),j)} \cdot f_{(v_i(l),j)}(x_{ij}, y_i)\right\} \tag{4}$$

$$\begin{aligned} P(Y|E) &= P(Y|E_{LANG}, E_{EXP}) \\ &= \frac{1}{Z_2} exp\left\{\sum_{v_i \in V}\left[\sum_{v_j \in LANG(v_i)} g(v_i, v_j) + \sum_{v_k \in EXP(v_i)} h(v_i, v_k)\right]\right\} \end{aligned} \tag{5}$$

in which, Z_1 and Z_2 are normalization factors. In Eq. (4), d is the length of the attribute vector; a feature function $f(x_{ij}, y_i)$ is defined for each attribute j (the jth attribute) of correlation node v_i for the language $v_i(l)$, and $\alpha_{(v_i(l),j)}$ is the weight of attribute j for language $v_i(l)$. In Eq. (5), E_{LANG} and E_{EXP} are edges between nodes in V through language dependency relations and same-experience relations respectively; two sets of relation factor functions $\boldsymbol{g}$ and $\boldsymbol{h}$ are defined which correspond to E_{LANG} and E_{EXP} respectively; $LANG(v_i)$ denotes the set of correlation nodes having the same user as v_i but with different languages (language dependency relation); $EXP(v_i)$ denotes the set of nodes in which the users have the same-experience relation with the user of v_i. Next, we will introduce the specific definitions of the feature functions $\boldsymbol{f}$, relation factor functions $\boldsymbol{g}$ and $\boldsymbol{h}$ adopted in LSR-FGM.

Local Textual Feature Functions: The unigram features of the textual information in social profiles are used to build the attribute vector space, and they are also used as binary features in the local feature function for each target language. For instance, if the profile of a user contains the jth word of the attribute vector space and specifies she knows language l, a feature $f_{(l,j)}(x\text{-}_{ij} = 1, y_i = 1)$ is defined and its value is 1; otherwise 0. This feature definition strategy is commonly used in graphical models like Conditional Random Field (CRF). Therefore, the conditional probability distribution $P(X|Y)$ over G can be obtained:

$$P(\boldsymbol{X}|Y) = \frac{1}{Z_1} exp\left\{ \sum_{v_i \in V} \sum_{j=1}^{d} \alpha_{(v_i(l),j)} f_{(v_i(l),j)}(x_{ij}, y_i) \right\} \tag{6}$$

Language Dependency Relation Factor: Any two nodes in V are connected through language dependency relation if they are from the same user. If nodes v_i and v_j have a language dependency connection, a language dependency relation factor is defined:

$$g\left(v_i, v_j\right) = \beta_{ij}(y_i - y_j)^2 \tag{7}$$

where β_{ij} represents the influence weight of node v_j on node v_i.

Same-Experience Social Relation Factor: Nodes in V are connected through same-experience relation if users of the nodes share a same work or study experience. Similarly, a same-experience relation factor is defined if two nodes have this connection:

$$h\left(v_i, v_j\right) = \gamma_{ij}(y_i - y_j)^2 \tag{8}$$

where γ_{ij} represents the influence weight of node v_j on v_i through social relation factor.

Finally, LSR-FGM can be constructed based on the above formulation. By combining Eqs. (3)-(8), we can define the objective likelihood function:

$$\begin{aligned} P(Y|\boldsymbol{X}, E) = \frac{1}{Z} exp \Bigg\{ & \sum_{v_i \in V} \sum_{j=1}^{d} \alpha_{(v_i(l),j)} \cdot f_{(v_i(l),j)}(x_{ij}, y_i) \\ & + \sum_{v_i \in V} \sum_{v_j \in LANG(v_i)} \beta_{ij}(y_i - y_j)^2 + \sum_{v_i \in V} \sum_{v_k \in EXP(v_i)} \gamma_{ik}(y_i - y_k)^2 \Bigg\} \end{aligned} \tag{9}$$

where $Z = Z_1 Z_2$ is a normalization factor and $\theta = (\{\alpha\},\{\beta\},\{\gamma\})$ represents a parameter configuration.

Thus, we build the LSR-FGM with the objective likelihood function Eq. (9). This model aims to best recover the label values Y, which can be represented by maximizing the objective likelihood function given the observation data.

4.2 Model Learning and Prediction

The last issue is to learn the LSR-FGM and to infer unknown label values Y^U in G. Learning the LSR-FGM is to estimate a parameter configuration of θ from a given partially labeled G, which maximizes the log-likelihood objective function $L(\theta) = \log P_\theta\,(Y^L|X, E)$, i.e.,

$$\theta^* = \arg\max \log P_\theta(Y^L|X,E) \tag{10}$$

This work uses a gradient descent method to solve the objective function. Taking γ as an example to explain how to learn the parameters, first the gradient of each γ_{ik} with regard to the objective function $L(\theta)$ can be obtained:

$$\frac{L(\theta)}{\gamma_{ik}} = \mathbb{E}\left[h\left(v_i, v_k\right)\right] + \mathbb{E}_{P_{\gamma_{ik}}(Y|X,E)}\left[h\left(v_i, v_k\right)\right] \tag{11}$$

in which, $\mathbb{E}\left[h\left(v_i, v_k\right)\right]$ is the expectation of factor function $h(v_i, v_k)$, i.e., the average value of $h(v_i, v_k)$ over the all same-experience connections in the training data; the second term is the expectation of $h(v_i, v_k)$ under the distribution $P_{\gamma_{ik}}(Y|\mathbf{X}, \mathbf{E})$ given by the estimated model. Similarly, the gradients of α and β can be derived.

As the graphical structure of G can be arbitrary and may contain cycles, it is intractable to directly calculate the second expectation. This work adopts the Loopy Belief Propagation (LBP) to approximate the gradients considering its ease of implementation and effectiveness [23]. In each iteration of the learning process, the LBP is employed twice, one for estimating the marginal distribution of unknown variables y_i and another for the marginal distribution over all connections. Then, the parameters θ are updated with the obtained gradients and a given learning rate in each iteration.

It is clear that a LBP is employed to infer the unknown Y^U in the learning process. Therefore, after convergence of the learning algorithm, all nodes in Y^U are labeled which maximizes the marginal probabilities. Correspondingly, the language information of unlabeled users is inferred.

5 Experiments and Results

This section first describes the dataset construction and the strategy of importing location information associated with the profile from external resources. It then introduces the comparative methods and, finally, presents and discusses the experimental results.

5.1 Dataset Construction and Location Information Enhancement

This work constructs a dataset using LinkedIn profiles, as most profiles in LinkedIn are publicly available and a field for language information is included. For privacy protection, the names of the profiles are not collected.

In total, 50,575 public profiles were collected from LinkedIn, among which 5906 profiles (11.7 %) specify the language field. In those profiles, over 70 different languages are found in the language field. But only 6 languages are encountered more than 300 times: English (3396 times/profiles), Chinese (1137), Spanish (1126), French (667), Hindi (657), German (473). As all profiles are written in English, English is not considered as a target language. Thus, the remaining five languages are used as target inference languages in our experiments. Our model is easily extended to other languages with the presence of corresponding training data. Finally, 3566 profiles are selected, each of which specifies the user knowing at least one of the five languages. Two thirds of the profiles are randomly sampled from this set as the training data, and the remaining one third are test data. Note that the positive and negative samples are imbalanced in the collected data where negative samples are much more frequent than positive samples. Therefore, balanced training and testing samples are selected in the experiments.

Like most other social platforms, location is not listed as an attribute of the Education field in LinkedIn profiles. However, this is important evidence for language information inference. This work considers a straightforward way to import the location of each study experience. We leave the deeper study of automatic location identification of study/work experiences using external resources as a future work.

LinkedIn has a homepage for many institutes in the world. These pages have information about the location of the institute. Thus, the location information associated with a person's study experience in the profiles can be harvested based on the institute names. In total 3212 different institutes are extracted from the profiles and the locations of 1658 of them are derived. The location information is attached as an additional attribute of profiles when needed in the experiments.

5.2 Baseline Methods

This work compares the LSR-FGM with the following methods of inferring what languages users comprehend based on their social profiles:

1. RM (Rule-based method): For each language, this method maintains a full list of countries/regions where this language is used as an official language. These lists are constructed based on the Wikipedia page[5] that lists the official language(s) of each country. This method makes an inference decision that a user comprehends a target language only if one of the country/region names in the corresponding list appears in her social profile.
2. RM-L: This method is almost the same as RM but the input attribute matrix is enhanced with the external location information, i.e., the additional location attribute of the institute.

[5] http://en.wikipedia.org/wiki/List_of_official_languages.

3. SVM (Support Vector Machine): This method uses the attribute vector of correlation nodes to train a classification model for each language, and predicts the language information by employing the model. The method is implemented with the SVM-light package[6] (linear kernel).
4. SVM-L: This method is almost the same as SVM but the input attribute matrix is enhanced with the external location information.

The LSR-FGM infers user language information by collectively considering the local textual attributes, user-user social relations and language-language dependency relations. The enhanced attribute matrix is applied on this method.

5.3 Performance Comparison and Analysis

The four metrics: accuracy, precision, recall and F1-score are used to measure the performance of these methods. Experimental results are listed in Table 1. It shows that the LSR-FGM method outperforms all other methods in overall on all target languages, and the import of location information remarkably improves the performance.

Table 1. Performance of language inference with different methods on different languages (%)

Language	Metrics	RM	RM-L	SVM	SVM-L	LSR-FGM
Chinese	Accuracy	67.79	86.39	83.69	87.74	**89.62**
	Precision	88.37	93.55	85.92	**94.30**	85.44
	Recall	40.97	78.17	80.59	80.32	**93.24**
	F1-score	55.98	85.17	83.17	86.75	**89.17**
French	Accuracy	59.77	65.81	63.95	69.30	**81.16**
	Precision	80.88	**85.42**	66.85	70.05	75.81
	Recall	25.58	38.14	55.35	67.44	**84.90**
	F1-score	38.87	52.73	60.56	68.72	**80.10**
German	Accuracy	62.16	66.89	66.55	68.58	**78.38**
	Precision	97.37	**98.08**	66.23	68.21	74.32
	Recall	25.00	34.46	67.57	69.59	**80.88**
	F1-score	39.79	51.00	66.89	68.89	**77.46**
Hindi	Accuracy	68.70	87.39	82.77	88.24	**92.86**
	Precision	98.90	**99.44**	87.86	95.05	91.60
	Recall	37.82	75.21	76.05	80.67	**93.97**
	F1-score	54.72	85.64	81.53	87.27	**92.77**
Spanish	Accuracy	56.22	58.07	74.47	77.12	**78.57**
	Precision	94.34	**95.52**	72.84	75.06	82.54
	Recall	13.23	16.93	78.04	**81.22**	76.47
	F1-score	23.21	28.76	75.35	78.02	**79.39**

Performance on Different Languages: First, the overall performance of all five methods is better on Chinese and Hindi than on French, German and Spanish. For example, LSR-FGM achieves a 92.77 % of F1-score on Hindi, while it gets a smaller

[6] http://svmlight.joachims.org/.

80.1 % of F1-score on the French. It demonstrates that the mix of related languages (refers to French, German and Spanish) in the target language set increases the inference difficulties to these languages. Then we compare the LSR-FGM with other methods on different languages. Table 1 shows that LSR-FGM outperforms all the four methods (in terms of F1-score) but with varying degree of improvement on different languages. For example, to the target language Chinese, LSR-FGM achieves a +2.42 % (F1-score) significant improvement, compared with SVM-L (p-value < 0.05). By comparison, LSR-FGM significantly outperforms SVM-L by 11.38 % (F1-score) on French (p-value < 0.05). This difference in performance between Chinese and French reflects two aspects of information as discussed in Sect. 3. First, again it shows that the discriminative features of Hindi and Chinese are easier to catch since they hardly share common characteristics with other languages. Second, the relations between languages and profiles significantly contribute to distinguishing the related languages.

Contribution of Additional Location Attributes: It is noted that the imported location information plays a crucial role in language inference. Both RM and SVM achieve a much better performance with the enhanced attribute matrix on all languages. For example, SVM-L outperforms SVM by 8.16 % (F1-score) on French.

It is observed that RM-L achieves the best precision among all the methods. This is because it tends to predict more negative cases (i.e. fail to find corresponding country names in the profile), thus would hurt the recall. For instance, RM-L achieves a high precision (95.52 %) on Spanish, while it gets an extremely low recall (16.93 %).

5.4 Factor Contribution Analysis

This subsection examines the contributions of the defined three factors in the LSR-FGM. Table 2 gives the overall performance (i.e. on all target languages) of the LSR-FGM considering different factors. Specifically, the two relation factors: language dependency relation and user social relation are removed and only the attribute factor is kept, and then each of the relation factors is added into the model separately. The experimental results show that both of the relation factors are useful for language inference task. It also indicates that they do contribute from different perspectives. This is demonstrated by the fact that the LSR-FGM with all three factors outperforms the instances which only consider one relation factor. For example, the same-experience factor could help for those profiles in which only a few study/work experiences are given and not enough

Table 2. Overall performance of LSR-RGM with different factors (%)

Factors	Accuracy	Precision	Recall	F1-score
Attributes	80.44	78.22	81.86	80.00
+Same-experience Rel.	80.78	79.48	81.68	80.54
+Language Dependency Rel.	84.04	82.59	85.05	83.80
All	**84.52**	**82.96**	**85.63**	**84.27**

discriminative features are available for inferring language information; the factor of language dependency relation would contribute for multilingual users whose profiles only contain enough evidence about certain languages.

6 Conclusion

This work studies the novel problem of inferring what languages a user comprehends based on their social profiles. This work precisely defines the problem and proposes a language and social relation-based factor graph model. This model collectively considers three factors in the inference process: textual attributes of the social profile; dependency relations between target languages; and social relations between users. Experiments on a real-world large-scale dataset show the success of the proposed model in inferring user language information using social profiles, and demonstrate that each one of the three factors makes a stand-alone contribution in the model from a different aspect. In addition, this work proposes to obtain information reflecting users' language information from external resources in order to help the inference, which is shown to be effective in the experiments. Future work involves exploiting more information related to the user, and exploring more features from the available information to infer the actual level of expertise that a user has in a language(s).

Acknowledgements. This research is supported by Science Foundation Ireland through the CNGL Programme (Grant 12/CE/I2267) in the ADAPT Centre (www.adaptcentre.ie) at Trinity College Dublin. The work is also supported by the National Natural Science Foundation of China under Project No. 61300129, and a project Sponsored by the Scientific Research Foundation for the Returned Overseas Chinese Scholars, State Education Ministry, China under grant number [2013] 1792.

References

1. Tucker, R.: A global perspective on bilingualism and bilingual education. In: Georgetown University Round Table on Languages and Linguistics, pp. 332–340 (1999)
2. Diamond, J.: The benefits of multilingualism. Sci. Wash. **330**(6002), 332–333 (2010)
3. Ghorab, M., Leveling, J., Zhou, D., Jones, G.J., Wade, V.: Identifying common user behaviour in multilingual search logs. In: Peters, C., Di Nunzio, G.M., Kurimo, M., Mandl, T., Mostefa, D., Peñas, A., Roda, G. (eds.) CLEF 2009. LNCS, vol. 6241, pp. 518–525. Springer, Heidelberg (2010)
4. Oakes, M., Xu, Y.: A search engine based on query logs, and search log analysis at the university of Sunderland. In: Proceedings of the 10th Cross Language Evaluation Forum (2009)
5. Kontaxis, G., Polychronakis, M., et al.: Minimizing information disclosure to third parties in social login platforms. Int. J. Inf. Secur. **11**(5), 321–332 (2012)
6. Burger, J.D., et al.: Discriminating gender on Twitter. In: EMNLP, pp. 1301–1309 (2011)
7. Li, R., Wang, S., Deng, H., et al.: Towards social user profiling: unified and discriminative influence model for inferring home locations. In: SIGKDD, pp. 1023–1031 (2012)
8. Dunning, T.: Statistical identification of language. Technical Report MCCS 940–273, Computing Research Laboratory, New Mexico State University (1994)

9. Xia, F., Lewis, W.D., Poon, H.: Language ID in the context of harvesting language data off the web. In: EACL, pp. 870–878 (2009)
10. Martins, B., et al.: Language identification in web pages. In: SAC, pp. 764–768 (2005)
11. Stiller, J., Gäde, M., Petras, V.: Ambiguity of queries and the challenges for query language detection. In: The proceedings of Cross Language Evaluation Forum (2010)
12. Carter, S., et al.: Microblog language identification: Overcoming the limitations of short, unedited and idiomatic text. Lang. Resour. Eval. **47**(1), 195–215 (2013)
13. Qiu, F., Cho, J.: Automatic identification of user interest for personalized search. In: WWW, pp. 727–736 (2006)
14. White, R.W., Bailey, P., Chen, L.: Predicting user interests from contextual information. In: SIGIR, pp. 363–370 (2009)
15. Liu, J., Dolan, P., Pedersen, E.R.: Personalized news recommendation based on click behavior. In: IUI, pp. 31–40 (2010)
16. Xu, S., et al.: Exploring folksonomy for personalized search. In: SIGIR, pp. 155–162 (2008)
17. Provost, F., Dalessandro, B., Hook, R., et al.: Audience selection for on-line brand advertising: privacy-friendly social network targeting. In: SIGKDD, pp. 707–716 (2009)
18. Mislove, A., Viswanath, B., Gummadi, K.P., Druschel, P.: You are who you know: inferring user profiles in online social networks. In: WSDM, pp. 251–260 (2010)
19. Maheshwari, S., Sainani, A., Reddy, P.: An approach to extract special skills to improve the performance of resume selection. In: Kikuchi, S., Sachdeva, S., Bhalla, S. (eds.) DNIS 2010. LNCS, vol. 5999, pp. 256–273. Springer, Heidelberg (2010)
20. Wang, Z., Li, S., Kong, F., Zhou, G.: Collective personal profile summarization with social networks. In: EMNLP, pp. 715–725 (2013)
21. Yang, Z., Cai, K., et al.: Social context summarization. In: SIGIR, pp. 255–264 (2011)
22. Dong, Y., Tang, J., Wu, S., et al.: Link prediction and recommendation across heterogeneous social networks. In: ICDM, pp. 181–190 (2012)
23. Tang, W., Zhuang, H., Tang, J.: Learning to infer social ties in large networks. In: Gunopulos, D., Hofmann, T., Malerba, D., Vazirgiannis, M. (eds.) ECML PKDD 2011, Part III. LNCS, vol. 6913, pp. 381–397. Springer, Heidelberg (2011)
24. Tang, J., Wu, S., Sun, J.: Confluence: Conformity influence in large social networks. In: SIGKDD, pp. 347–355 (2013)
25. Hammersley, J.M., Clifford, P.: Markov fields on finite graphs and lattices. Unpublished Manuscript (1971)

Retrieving Hierarchical Syllabus Items for Exam Question Analysis

John Foley(✉) and James Allan

Center for Intelligent Information Retrieval,
College of Information and Computer Sciences,
University of Massachusetts Amherst, Amherst, USA
{jfoley,allan}@cs.umass.edu

Abstract. Educators, institutions, and certification agencies often want to know if students are being evaluated appropriately and completely with regard to a standard. To help educators understand if examinations are well-balanced or topically correct, we explore the challenge of classifying exam questions into a concept hierarchy.

While the general problems of text-classification and retrieval are quite commonly studied, our domain is particularly unusual because the concept hierarchy is expert-built but without actually having the benefit of being a well-used knowledge-base.

We propose a variety of approaches to this "small-scale" Information Retrieval challenge. We use an external corpus of Q&A data for expansion of concepts, and propose a model of using the hierarchy information effectively in conjunction with existing retrieval models. This new approach is more effective than typical unsupervised approaches and more robust to limited training data than commonly used text-classification or machine learning methods.

In keeping with the goal of providing a service to educators for better understanding their exams, we also explore interactive methods, focusing on low-cost relevance feedback signals within the concept hierarchy to provide further gains in accuracy.

1 Introduction

Educators use exams to evaluate their students' understanding of material, to measure whether teaching methodologies help or hurt, or to be able to compare students across different programs. While there are many issues with exams and evaluations that could be and are being explored, we are interested in the question of coverage – whether an evaluation is complete, in the sense that it covers all the aspects or concepts that the designer of the evaluation hoped to cover.

We consider classifying multiple-choice questions into a known concept hierarchy. In our use case, an educator would upload or enter an exam into our system, and each question would be assigned to a category from the hierarchy. The results would allow the educator to understand and even visualize how the

N. Ferro et al. (Eds.): ECIR 2016, LNCS 9626, pp. 575–586, 2016.
DOI: 10.1007/978-3-319-30671-1_44

questions that make up the exam cover the overall hierarchy, making it possible to determine if this coverage achieves their goals for the examination: are all important topics covered?

This problem is traditionally treated as one of manual question creation and labeling, where an official, curated set of tests has been created and are to be used widely or repeatedly. Educators who use that exam are guaranteed "appropriate" coverage of the material. However, this centralized approach is only a partial solution to the problem of understanding coverage of exams since every institution and almost every teacher or professor is likely to have their own assignments, their own quizzes, their own exams. The global exam does not help those educators understand how their own material fits into the known set of topics.

For this study, our dataset is a medium-sized corpus of test questions classified into the American Chemical Society (ACS) hierarchy developed by their exams institute [12]. This dataset has been used for educational research [11,17], but as these are actual exams used by educators, it is not available publicly.

The problem is interesting because the hierarchy is crisply but very sparsely described and the questions are very short, on par with the size of microblog entries. In existing text classification datasets with a hierarchical components (e.g., Wikipedia categories, the Enron email folder dataset [14], and the Yahoo! Directory or Open Directory Project [26]) all of the labeled documents are quite dense, the categories were created with various levels of control, and the resulting categories are likely to be overlapping. In contrast, all of our information is sparse, the categories themselves were designed by experts in the field, and part of their goal was to have questions fall into a single category.

In this study, we explore methods for classifying exam questions into a concept hierarchy using information retrieval methods. We show that the best technique leverages both document expansion and concept-aware ranking methods, but that exploiting the structure of the questions is helpful but not shown to be an advantage in conjunction with our other approaches on this dataset.

Ideally this work would be repeated on additional sets of questions with their own hierarchy to show its broad applicability; unfortunately, such questions are carefully guarded[1] and difficult to come by so demonstrating the results on another dataset must be left for future work.

Although our evaluation dataset is not open, we believe the results will apply to any comparable collection of exam questions categorized into a known hierarchy and we hope that our success in this task will encourage other educators and institutions to open up their data and new problems to our community. Our key approach leverages structure present in this kind of dataset that is not available in standard retrieval collections, but we hope to explore its generality in future work.

[1] Even most standardized tests require test-takers to sign agreements not to distribute or mention the questions, even after the exam is taken.

2 Related Work

The problem we tackle in this study is classification of short text passages into a hierarchical concept hierarchy, sometimes with interaction. The classification of short texts is relevent even though we do not have sufficiently balanced training labels for our task. Additional prior work involves interactive techniques as well as hierarchical retrieval models.

Our domain is exam questions in chemistry. We have found very little existing work within this domain of education-motivated IR. Omar et al. [20] develop a rule-based system for classifying questions into a taxonomy of learning objectives (do students have knowledge, do they comprehend, etc.) rather than topics. They work with a small set of computer programming exam questions to develop the rules but do not actually evaluate their utility for any task.

The problem of question classification [18,30] seems related but refers to categorizing informational questions into major categories such as who, where, what, or when.

2.1 Short-Text Classification

There is a huge body of literature on the well known problem of text classification, with a substantial amount devoted to classifying short passages of text. We sketch the approaches of a sample of that work to give an idea of the major approaches. Rather than attempt to cover it here, we refer the interested reader to the survey by Aggarwal and Zhai [1].

Sun et al. [26] considers a problem similar to ours, classifying short web page descriptions into the Open Directory Project's hierarchy. In their work, classification is done in two steps: the 15 categories most similar to the text are selected from the larger set of over 100,000 categories, and then an SVM is used to build a classifier for just those 15 categories so that the text can be categorized. Their category descriptions are selected by tf·idf comparison as well as using "explicit semantic analysis" [8]. Following related earlier work by Xue et al. [29], they represent an inner node of the hierarchy by its own content as well as that of its descendants. We represent leaf nodes by the content of their ancestors as well as their descendants, and try this in conjunction with document expansion.

Ren et al. [23] consider the problem of classifying a stream of tweets into an overlapping concept hierarchy. They treat the problem as classification rather than ranking, and do not explore interactive possibilities. They expand the short texts using embedded links and references to named entities and address topic drive using time-aware topic modeling, approaches that have little utility when processing exam questions. Banerjee et al. [3] effectively expand text by retrieving articles from Wikipedia and using the titles of those articles as features. By contrast, we expand text using an unlabeled set of questions – that is, comparable instances of the items we are classifying, having found wikipedia to not be helpful in such a focused domain.

A similar result with information retrieval applications comes from Dumais and Chen [6], who consider the problem of classifying search engine snippets into a hierarchy with the goal of presenting an organization of the pages. They used SVM as a classifier, but worked only with the top levels of the category that had numerous training instances, unlike in our case where we have no training data.

While we have similarities to prior work in this space, we must reiterate that we used these works as inspiration and that bringing them to an *unsupervised setting* and validating the approaches in a new domain is a contribution.

2.2 Hierarchical Retrieval Models

The hierarchical retrieval models we propose and evaluate in this work draw inspiration from hierarchical classification. They also share some similarities with cluster-based retrieval [16], in the way that a document is represented by its terms and those of its cluster, we will represent nodes based on their features and the features belonging to their parents. Hierarchical language models show up in the task of expert finding as well, given the hierarchy of employees in the company [2,21]. Our task differs from expert retrieval in that the elements of our hierarchy are precisely defined by their own descriptions, but do not interact with documents in any way.

Lee et al. present an early work on leveraging a hierarchy in the form of a knowledge-base graph, constructed mostly of "is-a" relationships [15]. Ganesan et al. present a work on exploiting hierarchical relationships between terms or objects to compute similarity between objects that are expressed in terms of elements in the hierarchy [9], while relevant, this would be of more use if we were trying to match exams to other exams.

2.3 Interactive Learning

Active learning [24] is an approach to classification that allows the learning algorithm to select some instances of data for labeling, with the idea that some subset of labels is better for training than all of those available. Although this does reduce labeling effort, it is not typically directed at reducing user labels for a task.

Hoi et al. [10] explored batch active learning approaches for classification of web pages and news articles, all of which are much longer than the exam questions we consider. They explore the learning curves for 10 s or 100 s of labels rather than the single interaction we consider (we can't expect 10 s of labels per question a user wishes to classify, but one is more reasonable).

Bekkerman et al. [4] showed that a classifier could be improved by allowing a user to correct or augment the word features that were selected. If we consider the high-level concepts as added features, our approach is related, though they focus on document clustering rather than classification and use quite different collections.

3 Nodes, Questions, and Exams

The dataset we explore in this work is a collection of Chemistry exams created by the American Chemical Society (ACS) and a hierarchical taxonomy for those questions. We make the claim that these exams, in conjunction with the nodes in the hierarchy, are an interesting and challenging dataset. Although this dataset has been used in other studies [11,12,17], this paper must introduce it to our field. In this section, we discuss the format of the data, and some of our observations about its composition and distribution.

3.1 Concept Hierarchy

The concept hierarchy designed by the ACS has four levels, excluding the root "General Chemistry" node. Each level has a distinguishing numbering system. The top level of the hierarchy are identified as *Anchoring Concepts*, or Big Ideas. These are listed in Table 1.

Table 1. Top-level children in the ACS general chemistry hierarchy.

I	Atoms	**VI**	Energy and Thermodynamics
II	Bonding	**VII**	Kinetics
III	Structure and Function	**VIII**	Equilibrium
IV	Intermolecular Interactions	**IX**	Experiments
V	Chemical Reactions	**X**	Visualization

Each of the nodes described in the hierarchy has a succinct description, but only the nodes in Table 1 have titles, i.e.

X. Visualization. Chemistry constructs meaning interchangeably at the particulate and macroscopic levels.

X.A.2.a. Schematic drawings can depict key concepts at the particulate level such as mixtures vs. pure substance, compounds vs. elements, or dissociative processes.

The there are ten nodes at the first level, as already discussed, 61 at the level below that, 124 at the third level, and 258 leaf nodes. Of the middling nodes, there are between 1 and 10 children assigned to each, with most of the weight belonging to 1, 2 and 3 (72, 59, and 37 respectively). The average length of a node description is 18.3 terms, and there are 16.2 distinct terms per node.

3.2 Exam Questions

An exam question looks like the following, except it is slightly too broad and lacks multiple choice solutions:

> I know sulfuric acid is an important catalyzer and is used in various processes. My question is, how do I recover the remaining sulfuric acid? It will be impure, and I don't know how to do the "standard" procedure (is there one?)[2]

The exam question has three parts. The *context* "sulfuric acid is an important catalyzer" presents the background for the question, giving the background details that are needed to know what the question means and how to pick an answer. The *question statement* itself "how do I recover the remaining sulfuric acid?" is the actual statement. In many cases, a single context will occur with several different questions, a factor that complicates simple comparison of the entire exam question. Finally, the exam question has the *answers*, usually multiple choice and usually with only know of them a correct answer. We did not find question fields to be helpful in the presence of our other, less-domain-specific ideas.

The ACS dataset includes 1593 total questions, distributed across 23 exams, with an average of 69 questions per exam. One exam has only 58 questions, and the largest exam has 80.

The most frequently tested concepts are tested tens of times over all these exams, the most frequent occurring 47 times – on average twice per exam for 23 exams. This most common node belongs to the "experimental" sub-tree, and discusses the importance of schematic drawings in relation to key concepts. It is one of the more general nodes we have inspected. The other most frequent concepts include "quantitative relationships and conversions," "moles," and "molarity".

The labeled data itself is highly skewed overall. There are 65 nodes that have ten or more questions labeled to belong to them. There are 62 nodes that only have a single question and another 29 that only have two questions – the number of rarely-tested nodes are the reason we choose to eschew supervised approaches in this work.

4 Evaluation Measures

Our task is ultimately to classify an exam question into the correct leaf node of the concept hierarchy. In part to support reasonable interactive assistance, we treat this as a ranking problem. That is, rather than identify a single category for a question, we generate a ranked list of them and evaluate where the correct category appears in the list.

An individual question's ranking is measured by two metrics. We use reciprocal rank (RR), the inverse of the rank at which the correct category is found. If there are multiple correct categories (uncommon), the first one encountered in the list determines RR. We also use normalized discounted cumulative gain (NDCG) as implemented in the Galago search engine[3] and formulated by

---

[2] User Fiire; http://chemistry.stackexchange.com/questions/4250. This example displayed in lieu of the proprietary ACS data.

[3] http://lemurproject.org/galago.php.

Järvelin and Kekäläinen [13]. Additionally, we look at precision at rank 1, (P@1) because it represents the classification precision, if the rest of the ranking were to be ignored.

Since we are given exams as natural groupings of questions, and one of the key use-cases of our system will be the categorization of pre-existing exams for analysis, we evaluate our abilities on a per-exam level, rather than on a per-question level. That means that the accuracy for individual questions is averaged to create a per-exam average score. Formally, our reported scores are calculated as follows:

$$score = \frac{1}{|E|} \sum_{e \in E} \frac{1}{|Q_e|} \sum_{q \in Q_e} m(q)$$

where e is a single exam from E, the set of 23 exams, Q_E is the set of questions on exam e, and $m(q)$ is either RR, NDCG or P@1 for a query. This mean of averages is a macro-averaged score. We investigated whether micro-averaging (with each question treated equally rather than as part of an exam) made a difference, but there was no effect on the outcome of any experiment. As a result, we only report the score as described above.

5 Question-Framework Linking Methods

As mentioned previously, we consider our task to be one of retrieval, and not of classification, as we do not have training data for each of our labels. In this framework, each question is a query, and the corpus documents are the nodes or "labels" in the hierarchy (particularly the leaf nodes, but sometimes interior nodes). Therefore, we begin by using state-of-the-art retrieval models [19] and existing techniques like document expansion (Sect. 5.1). Our best improvement comes from an extension to our retrieval model which incorporates parent/child relations in the concept hierarchy (Sect. 5.3).

Our baseline is SDM, the sequential dependence model [19] which is known to be a highly effective ranking algorithm. Table 2 shows the results for the baseline in the top row. We also considered the query likelihood (QL) similarity [22], but SDM incorporates term dependencies in the context of bigrams and unordered window features. For all techniques, SDM was superior, so we do not report the unigram model (QL) numbers here. Our language model approach to retrieval is equivalent to a language-modeling approach to text-classification, but we present our ideas in the light of information retrieval for ease of implementation and evaluation.

5.1 Unsupervised Node Expansion

Both our corpus documents (concepts) and queries (questions) are short, so vocabulary mismatch – wherein a query and document are relevant but have little or no words in common – is quite likely. One way we address that is to expand the concept descriptions with synonyms and strongly related words or phrases.

We use document expansion to accomplish that. To apply document expansion, we look for highly similar "neighbor" documents in an additional, external data source to help to improve the representation of the original documents for retrieval. It has been used for numerous purposes and has been explored thoroughly in prior work [7,25,27].

We use a publicly-available Q&A dataset[4] where all questions and comments are likely to be on or near the topic of chemistry, and used it as our expansion corpus. We briefly explored leveraging Wikipedia as in related work [3,5,8,28], but initial experiments gave poor results: Wikipedia articles match too many nodes in our hierarchy; again, results are withheld for space.

For node expansion, we explored expansion with $k = \{1, 5, 10, 25, 50, 100\}$ Q&A comments or posts. We selected the neighbors using SDM, because it is known to perform well. Table 2 shows the substantial gain provided by node expansion (NX-50) before using SDM. We selected an expansion by 50 neighbors based on training data.

5.2 Question Context Model

Recall that the exam questions include three parts: the context, the statement, and the answers. We hypothesized that this structure could be leveraged to improve matching of exams. Indeed, the context can appear in multiple questions that are categorized differently, so although it is important, it also may be a distractor. We define the QCM similarity between two questions as:

$$QCM(q_i, q_{i+1}) = \lambda \text{SDM}_S(q_i, q_{i+1}) + (1 - \lambda)\text{SDM}_C(q_i, q_{i+1})$$

where q_i and q_{i+1} are two questions, SDM_S is the question statement similarity between them, and SDM_C is the similarity between the contexts.

5.3 A Hierarchy-Aware Retrieval Model

Drawing inspiration from hierarchical classification techniques, we propose a model of retrieval that takes into account the construction of the hierarchy, namely, that any node N in the hierarchy is described not just by its text, but also by the text of its ancestors and descendants. A low-level node about how to measure the density of a liquid is partially described by its highest level node, which encompasses all experimental techniques.

Hierarchical Node Scoring. The score of a leaf node given a query is given by a retrieval model. As mentioned above, we use the SDM approach for these experiments. However, if a query matches a leaf node well but does not match the parent of the leaf node, the match is suspect and should be down-weighted. To accommodate that, we use a hierarchical SDM scoring approach.

[4] The beta version of chemistry.stackexchange.com.

We first define an operator that returns the ancestors of a node, $A(N)$, excluding the root itself. This operator is defined inductively, using the operator $P(n)$ that returns the parent of node n.

$$A(N) = \begin{cases} \varnothing & N \text{ is a root node} \\ N \cup A(P(N)) & \text{otherwise} \end{cases}$$

We choose to exclude the root node because it has no description in our hierarchy.

Given the set of ancestors $A(N)$ of any node N, we can assign a joint score to the nodes based upon its score and that of its ancestors. If $\text{SDM}(q, N)$ is the SDM score for node N with query q, then:

$$\text{H-SDM}(q, N) = \prod_{n \in A(N)} \text{SDM}(q, n)$$

Descendant Node Expansion (NX). In addition to generating and combining scores for all nodes on a path to the root, we can accomplish a similar purpose by instead expanding nodes such that they are explicitly represented by the text of their descendants. We define an operator $D(N)$ which collects the set of descendants for a given node, given an operator $C(N)$ which returns a set of children the node N.

$$D(N) = \begin{cases} N & N \text{ is a leaf node} \\ N \cup \{D(c) \mid c \in C(N)\} & \text{otherwise} \end{cases}$$

We use this pattern to select nodes to expand the representation of the nodes in our model. This pattern leverages the intuition that experimental techniques (child IX of the root node) could be better represented by all of the experimental techniques available in the hierarchy in addition to its succinct description.

5.4 Experimental Results

This run is presented by "H-SDM (everything)" in Table 2 and clearly outperforms everything else; excepting the QCM part of it has no significant

Table 2. Evaluation of methods

Model	MRR	NDCG	P@1
SDM	0.179	0.311	0.090
QCM	0.188	0.319	0.090
SDM (NX-50)	0.263	0.398	0.144
H-SDM	0.244	0.369	0.133
H-SDM (QCM)	0.269	0.393	0.148
H-SDM (Desc)	0.253	0.377	0.138
H-SDM (NX-50, Desc)	0.318	0.440	0.188
H-SDM (everything)	0.322	0.445	0.180

benefit. In addition to the results reported above, we examined a few issues of pre-processing; we found no effect due to stemming or lemmatization, but found that removing stop-words actually harmed performance.

6 Interactive Methods

In this section we consider the possibility that the person using our algorithm would provide a small amount of information – perhaps indicating which top-level sub-tree is appropriate for the instance being considered, which we consider to be hierarchical relevance feedback, where we consider typical relevance feedback as considering our first 10 results. We expect that while users cannot remember hundreds of nodes in total, a working familiarity with the first level of the hierarchy (See Table 1) will be easier to learn and leverage in an interactive setting.

For each question to be classified, we simulate hierarchy feedback by removing concepts from our ranked list if they are not under the same top-level node in the concept hierarchy as the question. That is, we are simulating the case where a user selects the *correct* top-level category, so any candidates in other sub-trees can be automatically discarded. Table 3 shows that this simple approach ("Hierarchy") provides a substantial gain over no interaction, though it is not as helpful as having the correct question selected from the top 10.

Table 3. Performance with minimal feedback.

Feedback	MRR	NDCG	P@1
None	0.318	0.440	0.188
Hierarchy	0.458	0.564	0.287
RF / Success@10	0.598	0.650	0.568
Hierarchy + RF	0.812	0.830	0.781

While the results of this experiment may seem obvious, in lieu of having a user study to determine which of these techniques is easier, quantifying the gains that can be made with this kind of feedback is important. In the case of users familiar with the hierarchy, we expect that we can get a positive gain using both techniques, and for users who are less familiar with the hierarchy (we doubt anyone remembers all 258 leaf nodes), the ranking methods will hopefully provide a much smaller candidate set.

7 Conclusion

In this work we explored the challenge our users face of classifying exam questions into a concept hierarchy, but we explore it from an IR perspective due to the scarcity of labels available, and our desire to incorporate feedback. This problem was difficult because the exam questions are short and often quite similar and because the concepts in the hierarchy had quite short descriptions. We explored existing approaches, such as document expansion and typical retrieval models, as well as our own methods – especially a hierarchical transform for existing retrieval models that works well, and a model of question structure that provides gains over most baselines.

We hope that our promising results encourage more collaboration between education and information retrieval research, specifically in the identification and exploration of new tasks and datasets that may benefit both fields.

In future work, we hope to explore this problem with other subjects, more exams, and with expert humans in the loop to field-test the feasibility and helpfulness our overall retrieval methods, and our interactive methods.

Acknowledgments. The authors thank Prof. Thomas Holme of Iowa State University's Department of Chemistry for making the data used in this study available and Stephen Battisti of UMass' Center for Educational Software Development for help accessing and formatting the data.

This work was supported in part by the Center for Intelligent Information Retrieval and in part by NSF grant numbers IIS-0910884 and DUE-1323469. Any opinions, findings and conclusions or recommendations expressed in this material are those of the authors and do not necessarily reflect those of the sponsors.

References

1. Aggarwal, C.C., Zhai, C.: A survey of text classification algorithms. In: Aggarwal, C.C., Zhai, C. (eds.) Mining Text Data, pp. 163–222. Springer, New York (2012)
2. Balog, K., Azzopardi, L., de Rijke, M.: A language modeling framework for expert finding. Inf. Process. Manage. **45**(1), 1–19 (2009)
3. Banerjee, S., Ramanathan, K., Gupta, A.: Clustering short texts using wikipedia. In: SIGIR 2007, New York, NY, USA. ACM (2007)
4. Bekkerman, R., Raghavan, H., Allan, J., Eguchi, K.: Interactive clustering of text collections according to a user-specified criterion. In: Proceedings of IJCAI, pp. 684–689 (2007)
5. de Melo, G., Weikum, G.: Taxonomic data integration from multilingual wikipedia editions. Knowl. Inf. Syst. **39**(1), 1–39 (2014)
6. Dumais, S., Chen, H.: Hierarchical classification of web content. In: SIGIR 2000, pp. 256–263. ACM, New York, NY, USA (2000)
7. Efron, M., Organisciak, P., Fenlon, K.: Improving retrieval of short texts through document expansion. In: SIGIR 2012, pp. 911–920. ACM (2012)
8. Gabrilovich, E., Markovitch, S.: Computing semantic relatedness using wikipedia-based explicit semantic analysis. IJCAI **7**, 1606–1611 (2007)
9. Ganesan, P., Garcia-Molina, H., Widom, J.: Exploiting hierarchical domain structure to compute similarity. ACM Trans. Inf. Syst. **21**(1), 64–93 (2003)
10. Hoi, S.C., Jin, R., Lyu, M.R.: Large-scale text categorization by batch mode active learning. In: WWW 2006, pp. 633–642. ACM (2006)

11. Holme, T.: Comparing recent organizing templates for test content between ACS exams in general chemistry and AP chemistry. J. Chem. Edu. **91**(9), 1352–1356 (2014)
12. Holme, T., Murphy, K.: The ACS exams institute undergraduate chemistry anchoring concepts content Map I: general chemistry. J. Chem. Edu. **89**(6), 721–723 (2012)
13. Järvelin, K., Kekäläinen, J.: IR evaluation methods for retrieving highly relevant documents. In: SIGIR 2000, pp. 41–48. ACM (2000)
14. Klimt, B., Yang, Y.: Introducing the enron corpus. In: CEAS (2004)
15. Lee, J.H., Kim, M.H., Lee, Y.J.: Information retrieval based on conceptual distance in IS-A hierarchies. J. Doc. **49**(2), 188–207 (1993)
16. Liu, X., Croft, W.B.: Cluster-based retrieval using language models. In: SIGIR 2004, pp. 186–193. ACM, New York, NY, USA (2004)
17. Luxford, C.J., Linenberger, K.J., Raker, J.R., Baluyut, J.Y., Reed, J.J., De Silva, C., Holme, T.A.: Building a database for the historical analysis of the general chemistry curriculum using ACS general chemistry exams as artifacts. J. Chem. Edu. **92**, 230–236 (2014)
18. Metzler, D., Croft, W.: Analysis of statistical question classification for fact-based questions. Inf. Retr. **8**(3), 481–504 (2005)
19. Metzler, D., Croft, W.B.: A markov random field model for term dependencies. In: SIGIR 2005, pp. 472–479. ACM (2005)
20. Omar, N., Haris, S.S., Hassan, R., Arshad, H., Rahmat, M., Zainal, N.F.A., Zulkifli, R.: Automated analysis of exam questions according to bloom's taxonomy. Procedia - Soc. Behav. Sci. **59**, 297–303 (2012)
21. Petkova, D., Croft, W.B.: Hierarchical language models for expert finding in enterprise corpora. Int. J. Artif. Intell. Tools **17**(01), 5–18 (2008)
22. Ponte, J.M., Croft, W.B.: A language modeling approach to information retrieval. In: SIGIR 1998, pp. 275–281. ACM (1998)
23. Ren, Z., Peetz, M.-H., Liang, S., van Dolen, W., de Rijke, M.: Hierarchical multi-label classification of social text streams. In: SIGIR 2014, pp. 213–222. ACM, New York, NY, USA (2014)
24. Settles, B.: Active learning literature survey. Technical report, University of Wisconsin-Madison, Computer Sciences Technical report 1648, January 2010
25. Singhal, A., Pereira, F.: Document expansion for speech retrieval. In: SIGIR 1999, pp. 34–41. ACM (1999)
26. Sun, X., Wang, H., Yu, Y.: Towards effective short text deep classification. In: SIGIR 2011, pp. 1143–1144. ACM, New York, NY, USA (2011)
27. Tao, T., Wang, X., Mei, Q., Zhai, C.: Language model information retrieval with document expansion. In: NAACL 2006, pp. 407–414. ACL (2006)
28. Wang, P., Hu, J., Zeng, H.-J., Chen, Z.: Using wikipedia knowledge to improve text classification. Knowl. Inf. Syst. **19**(3), 265–281 (2009)
29. Xue, G.-R., Xing, D., Yang, Q., Yu, Y.: Deep classification in large-scale text hierarchies. In: SIGIR 2008, pp. 619–626. ACM, New York, NY (2008)
30. Zhang, D., Lee, W.S.: Question classification using support vector machines. In: SIGIR 2003, pp. 26–32. ACM, New York, NY, USA (2003)

Collaborative Filtering

Implicit Look-Alike Modelling in Display Ads

Transfer Collaborative Filtering to CTR Estimation

Weinan Zhang[1,2](✉), Lingxi Chen[1], and Jun Wang[1,2]

[1] University College London, London, UK
{w.zhang,lingxi.chen,j.wang}@cs.ucl.ac.uk
[2] MediaGamma Limited, London, UK

Abstract. User behaviour targeting is essential in online advertising. Compared with sponsored search keyword targeting and contextual advertising page content targeting, user behaviour targeting builds users' interest profiles via tracking their online behaviour and then delivers the relevant ads according to each user's interest, which leads to higher targeting accuracy and thus more improved advertising performance. The current user profiling methods include building keywords and topic tags or mapping users onto a hierarchical taxonomy. However, to our knowledge, there is no previous work that explicitly investigates the user online visits similarity and incorporates such similarity into their ad response prediction. In this work, we propose a general framework which learns the user profiles based on their online browsing behaviour, and transfers the learned knowledge onto prediction of their ad response. Technically, we propose a transfer learning model based on the probabilistic latent factor graphic models, where the users' ad response profiles are generated from their online browsing profiles. The large-scale experiments based on real-world data demonstrate significant improvement of our solution over some strong baselines.

1 Introduction

Targeting technologies have been widely adopted in various online advertising paradigms during the recent decade. According to the Internet advertising revenue report from IAB in 2014 [22], 51 % online advertising budget is spent on sponsored search (search keywords targeting) and contextual advertising (page content targeting), while 39 % is spent on display advertising (user demographics and behaviour targeting), and the left 10 % is spent on other ad formats like classifieds. With the rise of ad exchanges [19] and mobile advertising, user behaviour targeting has now become essential in online advertising.

Compared with sponsored search or contextual advertising, user behaviour targeting *explicitly* builds the user profiles and detects their interest segments via tracking their online behaviour, such as browsing history, search keywords and ad clicks etc. Based on user profiles, the advertisers can detect the users with similar interests to the known customers and then deliver the relevant ads to them. Such technology is referred as *look-alike modelling* [17], which efficiently provides

N. Ferro et al. (Eds.): ECIR 2016, LNCS 9626, pp. 589–601, 2016.
DOI: 10.1007/978-3-319-30671-1_43

higher targeting accuracy and thus brings more customers to the advertisers [29]. The current user profiling methods include building keyword and topic distributions [1] or clustering users onto a (hierarchical) taxonomy [29]. Normally, these inferred user interest segments are then used as target restriction rules or as features leveraged in predicting users' ad response [32].

However, the two-stage profiling-and-targeting mechanism is not optimal (despite its advantages of explainability). First, there is no flexible relationship between the inferred tags or categories. Two potentially correlated interest segments are regarded as separated and independent ones. For example, the users who like cars tend to love sports as well, but these two segments are totally separated in the user targeting system. Second, the first stage, i.e., the user interest segments building, is performed independently and with little attention of its latter use of ad response prediction [7,29], which is suboptimal. Third, the effective tag system or taxonomy structure could evolve over time, which makes it much difficult to update them.

In this paper, we propose a novel framework to *implicitly* and *jointly* learn the users' profiles on both the general web browsing behaviours and the ad response behaviours. Specifically, (i) Instead of building explicit and fixed tag system or taxonomy, we propose to directly map each user, webpage and ad into a latent space where the shape of the mapping is automatically learned. (ii) The users' profiles on general browsing and ad response behaviour are jointly learned based on the heterogeneous data from these two scenarios (or tasks). (iii) With a maximum a posteriori framework, the knowledge from the user browsing behaviour similarity can be naturally transferred to their ad response behaviour modelling, which in turn makes an improvement over the prediction of the users' ad response. For instance, our model could automatically discover that the users with the common behaviour on www.bbc.co.uk/sport will tend to click automobile ads. Due to its implicit nature, we call the proposed model *implicit look-alike modelling.*

Comprehensive experiments on a real-world large-scale dataset from a commercial display ad platform demonstrate the effectiveness of our proposed model and its superiority over other strong baselines. Additionally, with our model, it is straightforward to analyse the relationship between different features and which features are critical and cost-effective when performing transfer learning.

2 Related Work

Ad Response Prediction aims at predicting the probability that a specific user will respond (e.g., click) to an ad in a given context [4,18]. Such context can be either a search keyword [8], webpage content [2], or other kinds of real-time information related to the underlying user [31]. From the modelling perspective, many user response prediction solutions are based on linear models, such as logistic regression [14,24] and Bayesian probit regression [8]. Despite the advantage of high learning efficiency, these linear models suffer from the lack of feature interaction and combination [9]. Thus non-linear models such as tree models [9] and latent vector models [20,30] are proposed to catch the data non-linearity and interactions between features. Recently the authors in [12] proposed to first learn combination features from gradient boosting decision trees (GBDT) and, based on the tree leaves as features, learn a factorisation machine (FM) [23] to build feature interactions to improve ad click prediction performance.

Collaborative Filtering (CF) on the other hand is a technique for personalised recommendation [26]. Instead of exploring content features, it learns the user or/and item similarity based on their interactions. Besides the user(item)-based approaches [25,28], latent factor models, such as probabilistic latent semantic analysis [10], matrix factorisation [13] and factorisation machines [23], are widely used model-based approaches. The key idea of the latent factor models is to learn a low-dimensional vector representation of each user and item to catch the observed user-item interaction patterns. Such latent factors have good generalisation and can be leveraged to predict the users' preference on unobserved items [13]. In this paper, we explore latent models of collaborative filtering to model user browsing patterns and use them to infer users' ad click behaviour.

Transfer Learning deals with the learning problem where the learning data of the target task is expensive to get, or easily outdated, via transferring the knowledge learned from other tasks [21]. It has been proven to work on a variety of problems such as classification [6], regression [16] and collaborative filtering [15]. Different from multi-task learning, where the data from different tasks are assumed to drawn from the same distribution [27], transfer learning methods may allow for arbitrary source and target tasks. In online advertising field, the authors in a recent work [7] proposed a transfer learning scheme based on logistic regression prediction models, where the parameters of ad click prediction model were restricted with a regularisation term from the ones of user web browsing prediction model. In this paper, we consider it as one of the baselines.

3 Implicit Look-Alike Modelling

In performance-driven online advertising, we commonly have two types of observations about underlying user behaviours: one from their browsing behaviours (the interaction with webpages) and one from their ad responses, e.g., conversions or clicks, towards display ads (the interactions with the ads) [7]. There are two predictions tasks for understanding the users:

- **Web Browsing Prediction (CF Task).** Each user's online browsing behaviour is logged as a list containing previously visited publishers (domains or URLs). A common task of using the data is to leverage collaborative filtering (CF) [23,28] to infer the user's profile, which is then used to predict whether the user is interested in visiting any given new publisher. Formally, we denote the dataset for CF as D^{c} and an observation is denoted as $(\boldsymbol{x}^{c}, y^{c}) \in D^{c}$, where $\boldsymbol{x}^{c}$ is a feature vector containing the attributes from the user and the publisher and y^{c} is the binary label indicating whether the user visits the publisher or not.

- **Ad Response Prediction (CTR Task).** Each user's online ad feedback behaviour is logged as a list of pairs of ad impression events and their corresponding feedbacks (e.g., click or not). The task is to build a click-through rate (CTR) prediction model [5] to estimate how likely it is that the user will click a specific ad impression in the future. Each ad impression event consists of various information, such as user data (cookie ID, location, time, device, browser, OS etc.), publisher data (domain, URL, ad slot position etc.), and advertiser data (ad creative, creative size, campaign etc.). Mathematically, we denote the ad CTR dataset as D^{r} and its data instance as $(\boldsymbol{x}^{\mathrm{r}}, y^{\mathrm{r}})$, where $\boldsymbol{x}^{\mathrm{r}}$ is a feature vector and y^{r} is the binary label indicating whether the user clicks a given ad or not.

This paper focuses on the latter task: ad response prediction. We, however, observe that although they are different prediction tasks, the two tasks share a large proportion of users, publishers and their features. We can thus build a user-publisher interest model jointly from the two tasks. Typically we have a large number of observations about user browsing behaviours and we can use the knowledge learned from publisher CF recommendation to help infer display advertising CTR estimation.

3.1 The Joint Conditional Likelihood

In our solution, the prediction models on CF task and CTR task are learned jointly. Specifically, we build a joint data discrimination framework. We denote Θ as the parameter set of the joint model with prior $P(\Theta)$, and the *conditional* likelihood of an observed data instance is the probability of predicting the correct binary label given the features $P(y|\boldsymbol{x};\Theta)$. As such, the conditional likelihood of the two datasets are $\prod_{(\boldsymbol{x}^{\mathrm{c}},y^{\mathrm{c}})\in D^{\mathrm{c}}} P(y^{\mathrm{c}}|\boldsymbol{x}^{\mathrm{c}};\Theta)$ and $\prod_{(\boldsymbol{x}^{\mathrm{r}},y^{\mathrm{r}})\in D^{\mathrm{r}}} P(y^{\mathrm{r}}|\boldsymbol{x}^{\mathrm{r}};\Theta)$. Maximising a posteriori (MAP) estimation gives

$$\hat{\Theta} = \max_{\Theta} P(\Theta) \prod_{(\boldsymbol{x}^{\mathrm{c}},y^{\mathrm{c}})\in D^{\mathrm{c}}} P(y^{\mathrm{c}}|\boldsymbol{x}^{\mathrm{c}};\Theta) \prod_{(\boldsymbol{x}^{\mathrm{r}},y^{\mathrm{r}})\in D^{\mathrm{r}}} P(y^{\mathrm{r}}|\boldsymbol{x}^{\mathrm{r}};\Theta). \quad (1)$$

Just like most solutions on CF recommendation [10,13] and CTR estimation [14,24], in this discriminative framework, Θ is only concerned with the mapping from the features to the labels (the conditional probabilities) rather than modelling the prior distribution of features [11].

The details of the conditional likelihood $P(y^{\mathrm{c}}|\boldsymbol{x}^{\mathrm{c}};\Theta)$, $P(y^{\mathrm{r}}|\boldsymbol{x}^{\mathrm{r}};\Theta)$ and the parameter prior $P(\Theta)$ will be discussed in the latter subsections.

3.2 CF Prediction

For the CF task, we use a factorisation machine [23] as our prediction model. We further define the features $\boldsymbol{x}^{\mathrm{c}} \equiv (\boldsymbol{x}^{u}, \boldsymbol{x}^{p})$, where $\boldsymbol{x}^{u} \equiv \{x^{u}_{i}\}$ is the set of features for a user and $\boldsymbol{x}^{p} \equiv \{x^{p}_{j}\}$ is the set of features for a publisher[1]. The parameter

[1] All the features studied in our work are one-hot encoded binary features.

$\Theta \equiv (w_0^c, \boldsymbol{w}^c, \boldsymbol{V}^c)$, where $w_0^c \in \mathbb{R}$ is the global bias term and $\boldsymbol{w}^c \in \mathbb{R}^{I^c+J^c}$ is the weight vector of the I^c-dimensional user features and J^c-dimensional publisher features. Each user feature x_i^u or publisher feature x_j^p is associated with a K-dimensional latent vector $\boldsymbol{v}_i^c$ or $\boldsymbol{v}_j^c$. Thus $\boldsymbol{V}^c \in \mathbb{R}^{(I^c+J^c)\times K}$.

With such setting, the conditional probability for CF in Eq. (1) can be reformulated as:

$$\prod_{(\boldsymbol{x}^c, y^c)\in D^c} P(y^c|\boldsymbol{x}^c;\Theta) = \prod_{(\boldsymbol{x}^u, \boldsymbol{x}^p, y^c)\in D^c} P(y^c|\boldsymbol{x}^u, \boldsymbol{x}^p; w_0^c, \boldsymbol{w}^c, \boldsymbol{V}^c). \quad (2)$$

Let $\hat{y}_{u,p}^c$ be the predicted probability of whether user u will be interested in visiting publisher p. With the FM model, the likelihood of observing the label y^c given the features $(\boldsymbol{x}^u, \boldsymbol{x}^p)$ and parameters is

$$P(y^c|\boldsymbol{x}^u, \boldsymbol{x}^p; w_0^c, \boldsymbol{w}^c, \boldsymbol{V}^c) = (\hat{y}_{u,p}^c)^{y^c} \cdot (1 - \hat{y}_{u,p}^c)^{(1-y^c)}, \quad (3)$$

where the prediction $\hat{y}_{u,p}^c$ is given by an FM with a logistic function:

$$\hat{y}_{u,p}^c = \sigma\Big(w_0^c + \sum_i w_i^c x_i^u + \sum_j w_j^c x_j^p + \sum_i \sum_j \langle \boldsymbol{v}_i^c, \boldsymbol{v}_j^c \rangle x_i^u x_j^p\Big), \quad (4)$$

where $\sigma(x) = 1/(1+e^{-x})$ is the sigmoid function and $\langle \cdot, \cdot \rangle$ is the inner product of two vectors: $\langle \boldsymbol{v}_i, \boldsymbol{v}_j \rangle \equiv \sum_{f=1}^K v_{i,f} \cdot v_{j,f}$, which models the interaction between a user feature i and a publisher feature j.

3.3 CTR Task Prediction Model

For a data instance $(\boldsymbol{x}^r, y^r)$ in ad CTR task dataset D^r, its features $\boldsymbol{x}^r \equiv (\boldsymbol{x}^u, \boldsymbol{x}^p, \boldsymbol{x}^a)$ can be divided into three categories: the user features $\boldsymbol{x}^u$ (cookie, location, time, device, browser, OS, etc.), the publisher features $\boldsymbol{x}^p$ (domain, URL etc.), and the ad features $\boldsymbol{x}^a$ (ad slot position, ad creative, creative size, campaign, etc.). Each feature has potential influence to another one in a different category. For example, a mobile phone user might prefer square-sized ads instead of banner ads; users would like to click the ad on the sport websites during the afternoon etc.

By the same token as CF prediction, we leverage factorisation machine and the model parameter thus is $\Theta \equiv (w_0^r, \boldsymbol{w}^r, \boldsymbol{V}^r)$. Specifically, x_l^a is one of the L^r-dimensional ad features $\boldsymbol{x}^a$, w_l^r is the corresponding bias weight for the feature, and the feature is also associated with a K-dimensional latent vector $\boldsymbol{v}_l^r$. Thus $\boldsymbol{V}^r \in \mathbb{R}^{(I^r+J^r+L^r)\times K}$. Similar to CF task, the CTR data likelihood is:

$$\prod_{(\boldsymbol{x}^r, y^r)\in D^r} P(y^r|\boldsymbol{x}^r;\Theta) = \prod_{(\boldsymbol{x}^u, \boldsymbol{x}^p, \boldsymbol{x}^a, y^r)\in D^r} P(y^r|\boldsymbol{x}^u, \boldsymbol{x}^p, \boldsymbol{x}^a; w_0^r, \boldsymbol{w}^r, \boldsymbol{V}^r). \quad (5)$$

Then the factorisation machine with logistic activation function $\sigma(\cdot)$ is adopted to model the click probability over a specific ad impression:

$$P(y^r|\boldsymbol{x}^u, \boldsymbol{x}^p, \boldsymbol{x}^a; w_0^r, \boldsymbol{w}^r, \boldsymbol{V}^r) = (\hat{y}_{u,p,a}^r)^{y^r} \cdot (1 - \hat{y}_{u,p,a}^r)^{(1-y^r)}, \quad (6)$$

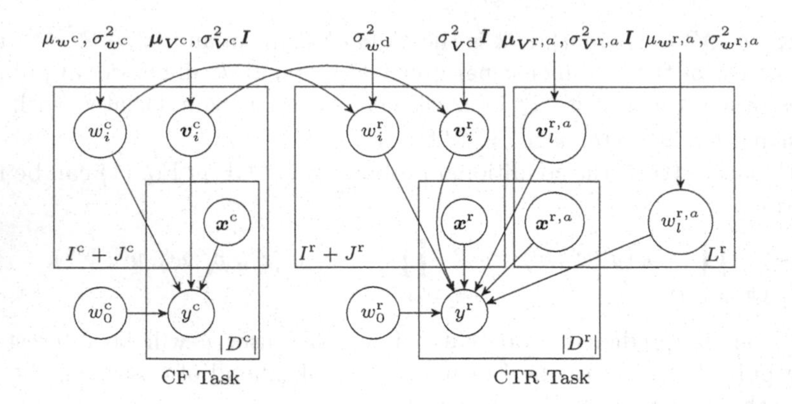

Fig. 1. Graphic model of transferred factorisation machines.

where $\hat{y}^{\mathrm{r}}_{u,p,a}$ is modelled by interactions among 3-side features

$$\hat{y}^{\mathrm{r}}_{u,p,a} = \sigma\Big(w^{\mathrm{r}}_0 + \sum_i w^{\mathrm{r}}_i x^u_i + \sum_j w^{\mathrm{r}}_j x^p_j + \sum_l w^{\mathrm{r}}_l x^a_l + \sum_i \sum_j \langle \boldsymbol{v}^{\mathrm{r}}_i, \boldsymbol{v}^{\mathrm{r}}_j \rangle x^u_i x^p_j + \sum_i \sum_l \langle \boldsymbol{v}^{\mathrm{r}}_i, \boldsymbol{v}^{\mathrm{r}}_l \rangle x^u_i x^a_l + \sum_j \sum_l \langle \boldsymbol{v}^{\mathrm{r}}_j, \boldsymbol{v}^{\mathrm{r}}_l \rangle x^p_j x^a_l \Big). \tag{7}$$

3.4 Dual-Task Bridge

To model the dependency between the two tasks, the weights of the user features and publisher features in CTR task are assumed to be generated from the counterparts in CF task (as a prior):

$$\boldsymbol{w}^{\mathrm{r}} \sim \mathcal{N}(\boldsymbol{w}^{\mathrm{c}}, \sigma^2_{\boldsymbol{w}^{\mathrm{d}}} \boldsymbol{I}), \tag{8}$$

where $\sigma^2_{\boldsymbol{w}^{\mathrm{d}}}$ is the assumed variance of the Gaussian generation process between each pair of feature weights of CF and CTR tasks and the weight generation is assumed to be independent across features. Similarly, the latent vectors of CTR task are assumed to be generated from the counterparts of CF task:

$$\boldsymbol{v}^{\mathrm{r}}_i \sim \mathcal{N}(\boldsymbol{v}^{\mathrm{c}}_i, \sigma^2_{\boldsymbol{V}^{\mathrm{d}}} \boldsymbol{I}) \tag{9}$$

where i is the index of a user or publisher feature; $\sigma^2_{\boldsymbol{V}^{\mathrm{d}}}$ is defined similarly.

The rational behind the above bridging model is that the users' interest towards webpage content is relatively general and the displayed ad can be regarded as a special kind of webpage content. One can infer user interests from their browsing behaviours, while their interests on commercial ads can be regarded as a modification or derivative from the learned general interests.

The graphic representation for the proposed *transferred factorisation machines* is depicted in Fig. 1. It illustrates the relationship among model parameters and observed data. The left part is for the CF task: $\boldsymbol{x}^{\mathrm{c}}$, w^{c}_0, $\boldsymbol{w}^{\mathrm{c}}$ and $\boldsymbol{V}^{\mathrm{c}}$

work together to infer our CF task target y^{c}, i.e., whether the user would visit a specific publisher or not. The right part illustrates the CTR task. Corresponding to CF task, $\boldsymbol{w}^{\mathrm{r}}$ and $\boldsymbol{V}^{\mathrm{r}}$ here represent user and publisher features' weights and latent vectors, while $\boldsymbol{w}^{\mathrm{r},a}$ and $\boldsymbol{V}^{\mathrm{r},a}$ are separately depicted to represent ad features' weights and latent vectors. All these factors work together to predict CTR task target y^{r}, i.e., whether the user would click the ad or not. On top of that, for each (user or publisher) feature i of the CF task, its weight w_i^{c} and latent vector $\boldsymbol{v}_i^{\mathrm{c}}$ act as a prior of the counterparts w_i^{r} and $\boldsymbol{v}_i^{\mathrm{r}}$ in CTR task while learning the model.

Considering the datasets of the two tasks might be seriously unbalanced, we choose to focus on the *averaged* log-likelihood of generating each data instance from the two tasks. In addition, we add a hyperparameter α for balancing the task relative importance. As such, the joint conditional likelihood in Eq. (1) is written as

$$\Big[\prod_{(\boldsymbol{x}^{\mathrm{c}},y^{\mathrm{c}})\in D^{\mathrm{c}}} P(y^{\mathrm{c}}|\boldsymbol{x}^{\mathrm{c}};\Theta)\Big]^{\frac{\alpha}{|D^{\mathrm{c}}|}}\cdot\Big[\prod_{(\boldsymbol{x}^{\mathrm{r}},y^{\mathrm{r}})\in D^{\mathrm{r}}} P(y^{\mathrm{r}}|\boldsymbol{x}^{\mathrm{r}};\Theta)\Big]^{\frac{1-\alpha}{|D^{\mathrm{r}}|}} \tag{10}$$

and its log form is

$$\begin{aligned}&\frac{\alpha}{|D^{\mathrm{c}}|}\sum_{(\boldsymbol{x}^{\mathrm{c}},y^{\mathrm{c}})\in D^{\mathrm{c}}}\Big[y^{\mathrm{c}}\log\hat{y}^{\mathrm{c}}_{u,p}+(1-y^{\mathrm{c}})\log(1-\hat{y}^{\mathrm{c}}_{u,p})\Big]\\&\quad+\frac{1-\alpha}{|D^{\mathrm{r}}|}\sum_{(\boldsymbol{x}^{\mathrm{r}},y^{\mathrm{r}})\in D^{\mathrm{r}}}\Big[y^{\mathrm{r}}\log\hat{y}^{\mathrm{r}}_{u,p,a}+(1-y^{\mathrm{r}})\log(1-\hat{y}^{\mathrm{r}}_{u,p,a})\Big].\end{aligned} \tag{11}$$

Moreover, from the graphic model, the prior of model parameters can be specified as

$$P(\Theta)=P(\boldsymbol{w}^{\mathrm{c}})P(\boldsymbol{V}^{\mathrm{c}})P(\boldsymbol{w}^{\mathrm{r}}|\boldsymbol{w}^{\mathrm{c}})P(\boldsymbol{V}^{\mathrm{r}}|\boldsymbol{V}^{\mathrm{c}})P(\boldsymbol{w}^{\mathrm{r},a})P(\boldsymbol{V}^{\mathrm{r},a}) \tag{12}$$

$$\begin{aligned}\log P(\Theta)=&\sum_i\log\mathcal{N}(w_i^{\mathrm{c}};\mu_{\boldsymbol{w}^{\mathrm{c}}},\sigma^2_{\boldsymbol{w}^{\mathrm{c}}})+\sum_i\log\mathcal{N}(\boldsymbol{v}_i^{\mathrm{c}};\boldsymbol{\mu}_{\boldsymbol{V}^{\mathrm{c}}},\sigma^2_{\boldsymbol{V}^{\mathrm{c}}}\boldsymbol{I})\\&+\sum_i\log\mathcal{N}(w_i^{\mathrm{r}};w_i^{\mathrm{c}},\sigma^2_{\boldsymbol{w}^{\mathrm{d}}})+\sum_i\log\mathcal{N}(\boldsymbol{v}_i^{\mathrm{r}};\boldsymbol{v}_i^{\mathrm{c}},\sigma^2_{\boldsymbol{V}^{\mathrm{d}}}\boldsymbol{I})\\&+\sum_l\log\mathcal{N}(w_l^{\mathrm{r},a};\mu_{\boldsymbol{w}^{\mathrm{r},a}},\sigma^2_{\boldsymbol{w}^{\mathrm{r},a}})+\sum_l\log\mathcal{N}(\boldsymbol{v}_l^{\mathrm{r},a};\boldsymbol{\mu}_{\boldsymbol{V}^{\mathrm{r},a}},\sigma^2_{\boldsymbol{V}^{\mathrm{r},a}}\boldsymbol{I}).\end{aligned} \tag{13}$$

3.5 Learning the Model

Given the detailed implementations of the MAP solution (Eq. (1)) components in Eqs. (11) and (13), for each data instance $(\boldsymbol{x},y)$, the gradient update of Θ is

$$\Theta\leftarrow\Theta+\eta\Big(\beta\frac{\partial}{\partial\Theta}\log P(y|\boldsymbol{x};\Theta)+\frac{\partial}{\partial\Theta}\log P(\Theta)\Big), \tag{14}$$

where $P(y|\boldsymbol{x};\Theta)$ is as Eqs. (3) and (6) for $(\boldsymbol{x}^{\mathrm{c}},y^{\mathrm{c}})\in D^{\mathrm{c}}$ and $(\boldsymbol{x}^{\mathrm{r}},y^{\mathrm{r}})\in D^{\mathrm{r}}$, respectively; η is the learning rate; β is the instance weight parameter depending on which task the instance belongs to, as given in Eq. (11). The detailed gradient for each specific parameter can be calculated routinely and thus are omitted here due to the page limit.

4 Experiments

4.1 Dataset

Our experiments are conducted based on a real-world dataset provided by Adform, a global digital media advertising technology company based in Copenhagen, Denmark. It consists of two weeks of online display ad logs across different campaigns during March 2015. Specifically, there are 42.1M user domain browsing events and 154.0 K ad display/click events. To fit the data into the joint model, we group useful data features into three categories: user features $\boldsymbol{x}^u$ (`user_cookie`, `hour`, `browser`, `os`, `user_agent` and `screen_size`), publisher features $\boldsymbol{x}^p$ (`domain`, `url`, `exchange`, `ad_slot` and `slot_size`), ad features $\boldsymbol{x}^a$ (`advertiser` and `campaign`). Detailed unique value numbers for each attribute are given as below.

ATTRIBUTE	user_cookie	hour	browser	os	user_agent	screen_size	domain
UNIQUE NUMBER	4,180,170	24	71	37	29,488	118	38,495
ATTRIBUTE	url	exchange	position	size	advertiser	campaign	
UNIQUE NUMBER	1,100,523	140	3	55	486	2,665	

In order to perform stable knowledge transfer, we have down-sampled the negative instances to make the ratio of positive over negative instances as 1:5.[2]

4.2 Experiment Protocol

We conduct a two-stage experiment to verify the effectiveness of our proposed models. First, in a very clean setting, we only focus on `user_cookie` and `domain` to check whether the knowledge of users' behaviour on webpage browsing can be transferred to model their behaviour on clicking the ads in these webpages. Second, we start to append various features in the first setting to observe the performance change and check which features lead to better transfer learning. Specifically, we try appending a single side feature into the baseline setting: 1. appending user feature $\boldsymbol{x}^u$, 2. appending publisher feature $\boldsymbol{x}^p$, 3. appending ad feature $\boldsymbol{x}^a$. Finally, all features are added into the model to perform the transfer learning.

For each experiment stage, there are three datasets: CF DATASET (D^c), CTR DATASET (D^r) and JOINT DATASET (D^c, D^r). Each dataset is split into two parts: the first week data as training data and the second one as test data.

[2] It is common to perform negative down sampling to balance the labels in ad CTR estimation [9]. Calibration methods [3] are then leveraged to eliminate the model bias.

4.3 Evaluation Metrics

To evaluate the performance of proposed model, area under the ROC curve (AUC) [8] and root mean square error (RMSE) [13] are adopted as performance metrics. As we focus on ad click prediction performance improvement, we only report the performance of the CTR estimation task.

4.4 Compared Models

We implement the following models for experimental comparison.

- Base: This baseline model only considers the ad CTR task, without any transfer learning. The parameters are learned by $\max_\Theta \prod_{(\boldsymbol{x}^{\mathrm{r}}, y^{\mathrm{r}}) \in D^{\mathrm{r}}} P(y^{\mathrm{r}}|\boldsymbol{x}^{\mathrm{r}};\Theta)P(\Theta)$.
- Disjoint: This method performs a knowledge transfer in a disjoint two-stage fashion. First, we train the CF task model to get the parameters $\boldsymbol{w}^{\mathrm{c}}$ and $\boldsymbol{V}^{\mathrm{c}}$ by $\max_\Theta \prod_{(\boldsymbol{x}^{\mathrm{c}}, y^{\mathrm{c}}) \in D^{\mathrm{c}}} P(y^{\mathrm{c}}|\boldsymbol{x}^{\mathrm{c}};\Theta)P(\Theta)$. Second, with the CF task parameters fixed, we train the CTR task using Eqs. (11) and (13). Note that α in Eq. (11) is still a hyperparameter for this method.
- DisjointLR: The transfer learning model proposed in [7] is considered as state-of-the-art transfer learning methods in display advertising. In this work, both source and target tasks adopt logistic regression as a behaviour prediction model, which uses the linear model to minimise the logistic loss from each observation sample:

$$\mathcal{L}_{\boldsymbol{w}}(\boldsymbol{x}, y) = -y \log \sigma(\langle \boldsymbol{w}, \boldsymbol{x} \rangle) - (1-y)\log(1-\sigma(\langle \boldsymbol{w}, \boldsymbol{x} \rangle)). \quad (15)$$

  In our context of regarding the CF task as source task and CTR task as target task, the learning objectives are listed below:

$$\text{CF TASK}: \overset{*}{\boldsymbol{w}}{}^{\mathrm{c}} = \arg\min_{\boldsymbol{w}^{\mathrm{c}}} \sum_{(\boldsymbol{x}^{\mathrm{c}}, y^{\mathrm{c}}) \in D^{\mathrm{c}}} \mathcal{L}_{\boldsymbol{w}^{\mathrm{c}}}(\boldsymbol{x}^{\mathrm{c}}, y^{\mathrm{c}}) + \lambda ||\boldsymbol{w}^{\mathrm{c}}||_2^2 \quad (16)$$

$$\text{CTR TASK}: \overset{*}{\boldsymbol{w}}{}^{\mathrm{r}} = \arg\min_{\boldsymbol{w}^{\mathrm{r}}} \sum_{(\boldsymbol{x}^{\mathrm{r}}, y^{\mathrm{r}}) \in D^{\mathrm{r}}} \mathcal{L}_{\boldsymbol{w}^{\mathrm{r}}}(\boldsymbol{x}^{\mathrm{r}}, y^{\mathrm{r}}) + \lambda ||\boldsymbol{w}^{\mathrm{r}} - \overset{*}{\boldsymbol{w}}{}^{\mathrm{c}}||_2^2. \quad (17)$$

  Besides the difference between the linear LR and non-linear FM, this method is a two-stage learning scheme, where the first stage Eq. (16) is disjoint with the second stage Eq. (17). Thus we denoted it as DisjointLR.
- Joint: Our proposed model, as summarised in Eq. (1), which performs the transfer learning when jointly learning the parameters on the two tasks.

Table 1. Overall AUC performance: DisjointLR vs Joint.

DisjointLR	Joint	Improvement
68.44 %	72.18 %	5.46 %

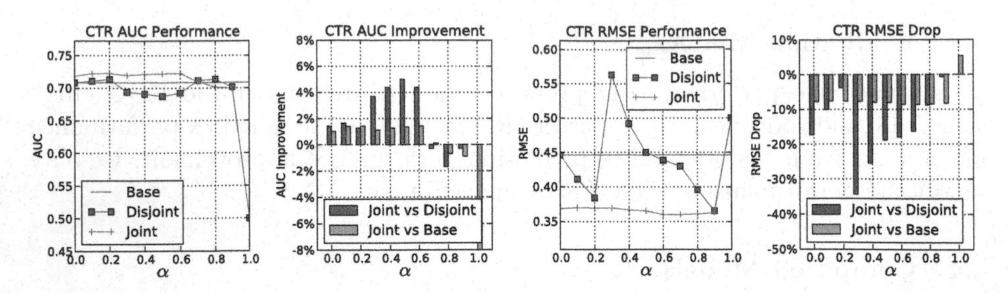

Fig. 2. Performance improvement with basic setting.

4.5 Result

Basic Setting Performance. Figure 2 presents the AUC and RMSE performance of Base, Disjoint and Joint and the improvement of Joint against the hyperparameter α in Eq. (11) based on the basic experiment setting. As can be observed clearly, for a large region of α, i.e., $[0.1, 0.7]$, Joint consistently outperforms the baselines Base and Disjoint on both AUC and RMSE, which demonstrates the effectiveness of our model to transfer knowledge from webpage browsing data to ad click data. Note that when $\alpha = 0$, the CF side model $\boldsymbol{w}^{\mathrm{c}}$ does not learn but Joint still outperforms Disjoint and Base. This is due to the different prior of $\boldsymbol{w}^{\mathrm{r}}$ and $\boldsymbol{V}^{\mathrm{r}}$ in Joint compared with those of Disjoint and Base. In addition, when $\alpha = 1$, i.e., no learning on CTR task, the performance of Joint reasonably gets back to initial guess, i.e., both AUC and RMSE are 0.5.

Table 1 shows the transfer learning performance comparison between Joint and the state-of-the-art DisjointLR with both models setting optimal hyperparameters. The improvement of Joint over DisjointLR indicates the success of (1) the joint optimisation on the two tasks to perform knowledge transfer and (2) the non-linear factorisation machine relevance model on catching feature interactions.

Appending Side Information Performance. From the Joint model as in Eq. (11) we see when α is large, e.g., 0.8, the larger weight is allocated on the CF task to optimise the joint likelihood. As such, if a large-value α leads to the optimal CTR estimation performance, it means the transfer learning takes effect. With such method, we try adding different features into the Joint model to obtain the optimal hyperparameter α leading to the highest AUC to check whether a certain feature helps transfer learning. On the contrary, if a low-value or 0 α leads to the optimal performance of Joint model when adding a certain feature, it means such feature has no effect of performing transfer learning.

Table 2 collects the AUC improvement of the Joint model for the conducted experiments. We observe that user browsing `hour`, ad slot `position` in the webpage are the most valuable features that help transfer learning, while the user `screen size` does not bring any transfer value. When adding all these features into Joint model, the optimal α is around 0.5 for AUC improvement and 0.6 for RMSE drop (see Fig. 3), which means these features along with the basic

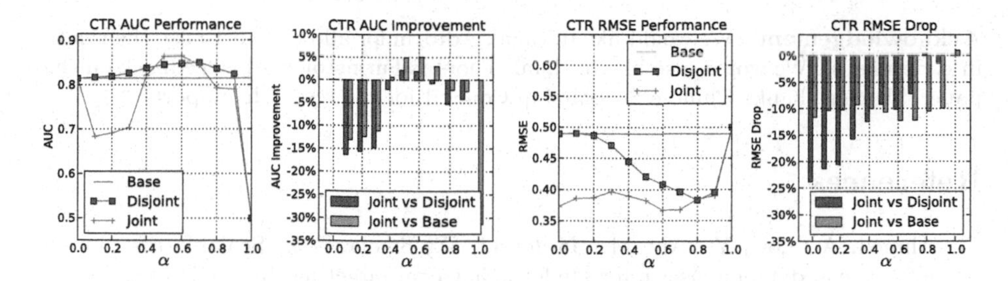

Fig. 3. Performance improvement with different side information.

Table 2. CTR task performance

	Joint vs Disjoint				Joint vs Base			
	$\overset{*}{\alpha}$	AUC Lift	Joint AUC	Disjoint AUC	$\overset{*}{\alpha}$	AUC Lift	Joint AUC	Base AUC(%)
BASIC SETTING	0.5	3.43 %	72.18 %	68.75 %	0.2	1.41 %	72.24 %	70.83 %
+ x^u: hour	0.8	2.44 %	89.35 %	86.91 %	0.6	1.99 %	89.35 %	87.36 %
+ x^u: browser	0.0	7.92 %	76.36 %	68.44 %	0.2	8.08 %	76.52 %	68.44 %
+ x^u: os	0.1	6.66 %	76.86 %	70.2 %	0.1	6.71 %	76.86 %	70.15 %
+ x^u: user_agent	0.0	2.57 %	67.12 %	64.55 %	0.8	4.31 %	68.86 %	64.55 %
+ x^u: screen_size	0.0	9.39 %	76.43 %	67.04 %	0.0	9.39 %	76.43 %	67.04 %
+ x^p: exchange	0.6	1.56 %	66.80 %	65.24 %	0.0	0.64 %	68.49 %	67.85 %
+ x^p: url	0.3	11.9 %	66.56 %	54.66 %	0.0	11.55 %	69.36 %	57.81 %
+ x^p: position	0.6	2.63 %	66.89 %	64.26 %	0.4	0.69 %	67.14 %	66.45 %
+ x^a: advertiser	0.4	2.39 %	84.98 %	82.59 %	0.5	0.87 %	85.07 %	84.20 %
+ x^a: campaign	0.2	1.29 %	85.81 %	84.52 %	0.1	0.48 %	85.91 %	85.43 %
+ x^a: size	0.0	0.59 %	69.16 %	68.57 %	0.0	0.59 %	69.16 %	68.57 %
+ ALL FEATURES	0.5	6.91 %	88.32 %	81.41 %	0.6	6.91 %	88.32 %	81.41 %

user, webpage IDs provide an overall positive value of knowledge transfer from webpage browsing behaviour to ad click behaviour.

5 Conclusion

In this paper, we proposed a transfer learning framework with factorisation machines to build implicit look-alike models on user ad click behaviour prediction task with the knowledge successfully transferred from the rich data of user webpage browsing behaviour. The major novelty of this work lies in the joint training on the two tasks and making knowledge transfer based on the non-linear factorisation machine model to build the user and other feature profiles. Comprehensive experiments on a large-scale real-world dataset demonstrated the effectiveness of our model as well as some insights of detecting which specific features help transfer learning. In the future work, we plan to explore on the user profiling utilisation based on the learned latent vector for each user. We also plan to extend our model to cross-domain recommendation problems.

Acknowledgement. We would like to thank Adform for allowing us to use their data in experiments. We would also like to thank Thomas Furmston for his feedback on the paper. Weinan thanks Chinese Scholarship Council for the research support.

References

1. Ahmed, A., Low, Y., Aly, M., Josifovski, V., Smola, A.J.: Scalable distributed inference of dynamic user interests for behavioral targeting. In: KDD (2011)
2. Broder, A., Fontoura, M., Josifovski, V., Riedel, L.: A semantic approach to contextual advertising. In: SIGIR, pp. 559–566. ACM (2007)
3. Caruana, R., Niculescu-Mizil, A.: An empirical comparison of supervised learning algorithms. In: ICML, pp. 161–168. ACM (2006)
4. Chapelle, O.: Modeling delayed feedback in display advertising. In: KDD, pp. 1097–1105. ACM (2014)
5. Chapelle, O., et al.: A simple and scalable response prediction for display advertising. ACM Trans. Intell. Syst. Technol. (TIST) **5**(4), 61 (2013)
6. Dai, W., Xue, G.R., Yang, Q., Yu, Y.: Transferring naive bayes classifiers for text classification. In: AAAI (2007)
7. Dalessandro, B., Chen, D., Raeder, T., Perlich, C., Han Williams, M., Provost, F.: Scalable hands-free transfer learning for online advertising. In: KDD (2014)
8. Graepel, T., Candela, J.Q., Borchert, T., Herbrich, R.: Web-scale bayesian click-through rate prediction for sponsored search advertising in microsoft's bing search engine. In: ICML, pp. 13–20 (2010)
9. He, X., Pan, J., Jin, O., Xu, T., Liu, B., Xu, T., Shi, Y., Atallah, A., Herbrich, R., Bowers, S., et al.: Practical lessons from predicting clicks on ads at facebook. In: ADKDD, pp. 1–9. ACM(2014)
10. Hofmann, T.: Collaborative filtering via gaussian probabilistic latent semantic analysis. In: SIGIR, pp. 259–266. ACM (2003)
11. Jebara, T.: Machine Learning: Discriminative and Generative, vol. 755. Springer Science & Business Media, New York (2012)
12. Juan, Y.C., Zhuang, Y., Chin, W.S.: 3 Idiots Approach for Display Advertising Challenge. Internet and Network Economics, pp. 254–265. Springer, Heidelberg (2011)
13. Koren, Y., Bell, R., Volinsky, C.: Matrix factorization techniques for recommender systems. Comput. **8**, 30–37 (2009)
14. Lee, K., Orten, B., Dasdan, A., Li, W.: Estimating conversion rate in display advertising from past performance data. In: KDD, pp. 768–776. ACM (2012)
15. Li, B., Yang, Q., Xue, X.: Transfer learning for collaborative filtering via a rating-matrix generative model. In: ICML, pp. 617–624. ACM (2009)
16. Liao, X., Xue, Y., Carin, L.: Logistic regression with an auxiliary data source. In: ICML, pp. 505–512. ACM (2005)
17. Mangalampalli, A., Ratnaparkhi, A., Hatch, A.O., Bagherjeiran, A., Parekh, R., Pudi, V.: A feature-pair-based associative classification approach to look-alike modeling for conversion-oriented user-targeting in tail campaigns. In: WWW, pp. 85–86. ACM (2011)
18. McAfee, R.P.: The design of advertising exchanges. Rev. Ind. Organ. **39**(3), 169–185 (2011)
19. Muthukrishnan, S.: Ad exchanges: research issues. In: Leonardi, S. (ed.) WINE 2009. LNCS, vol. 5929, pp. 1–12. Springer, Heidelberg (2009)

20. Oentaryo, R.J., Lim, E.P., Low, D.J.W., Lo, D., Finegold, M.: Predicting response in mobile advertising with hierarchical importance-aware factorization machine. In: WSDM (2014)
21. Pan, S.J., Yang, Q.: A survey on transfer learning. IEEE Trans. Knowl. Data Eng. **22**(10), 1345–1359 (2010)
22. PricewaterhouseCoopers: IAB internet advertising revenue report (2014). Accessed 29 July 2015. http://www.iab.net/media/file/PwC_IAB_Webinar_Presentation_HY2014.pdf
23. Rendle, S.: Factorization machines. In: ICDM, pp. 995–1000. IEEE (2010)
24. Richardson, M., Dominowska, E., Ragno, R.: Predicting clicks: estimating the click-through rate for new ads. In: WWW, pp. 521–530. ACM (2007)
25. Sarwar, B., Karypis, G., Konstan, J., Riedl, J.: Item-based collaborative filtering recommendation algorithms. In: WWW, pp. 285–295. ACM (2001)
26. Schafer, J.B., Frankowski, D., Herlocker, J., Sen, S.: Collaborative filtering recommender systems. In: Brusilovsky, P., Kobsa, A., Nejdl, W. (eds.) Adaptive Web 2007. LNCS, vol. 4321, pp. 291–324. Springer, Heidelberg (2007)
27. Taylor, M.E., Stone, P.: Transfer learning for reinforcement learning domains: a survey. J. Mach. Learn. Res. **10**, 1633–1685 (2009)
28. Wang, J., De Vries, A.P., Reinders, M.J.: Unifying user-based and item-based collaborative filtering approaches by similarity fusion. In: SIGIR (2006)
29. Yan, J., Liu, N., Wang, G., Zhang, W., Jiang, Y., Chen, Z.: How much can behavioral targeting help online advertising?. In: WWW, pp. 261–270. ACM (2009)
30. Yan, L., Li, W.J., Xue, G.R., Han, D.: Coupled group lasso for web-scale ctr prediction in display advertising. In: ICML, pp. 802–810 (2014)
31. Yuan, S., Wang, J., Zhao, X.: Real-time bidding for online advertising: measurement and analysis. In: ADKDD, pp. 3. ACM (2013)
32. Zhang, W., Yuan, S., Wang, J.: Real-time bidding benchmarking with ipinyou dataset. arXiv preprint.(2014). arxiv:1407.7073

Efficient Pseudo-Relevance Feedback Methods for Collaborative Filtering Recommendation

Daniel Valcarce(✉), Javier Parapar, and Álvaro Barreiro

Information Retrieval Lab, Computer Science Department,
University of A Coruña, A Coruña, Spain
{daniel.valcarce,javierparapar,barreiro}@udc.es

Abstract. Recently, Relevance-Based Language Models have been demonstrated as an effective Collaborative Filtering approach. Nevertheless, this family of Pseudo-Relevance Feedback techniques is computationally expensive for applying them to web-scale data. Also, they require the use of smoothing methods which need to be tuned. These facts lead us to study other similar techniques with better trade-offs between effectiveness and efficiency. Specifically, in this paper, we analyse the applicability to the recommendation task of four well-known query expansion techniques with multiple probability estimates. Moreover, we analyse the effect of neighbourhood length and devise a new probability estimate that takes into account this property yielding better recommendation rankings. Finally, we find that the proposed algorithms are dramatically faster than those based on Relevance-Based Language Models, they do not have any parameter to tune (apart from the ones of the neighbourhood) and they provide a better trade-off between accuracy and diversity/novelty.

Keywords: Recommender systems · Collaborative filtering · Query expansion · Pseudo-Relevance Feedback

1 Introduction

Recommender systems are recognised as a key instrument to deliver relevant information to the users. Although the problem that attracts most attention in the field of Recommender Systems is accuracy, the emphasis on efficiency is increasing. We present new Collaborative Filtering (CF) algorithms. CF methods exploit the past interactions between items and users. Common approaches to CF are based on nearest neighbours or matrix factorisation [17]. Here, we focus on probabilistic techniques inspired by Information Retrieval methods.

A growing body of literature has been published on applying techniques from Information Retrieval to the field of Recommender Systems [1,5,14,19–21]. These papers model the recommendation task as an item ranking task with an implicit query [1]. A very interesting approach is to formulate the recommendation problem as a profile expansion task. In this way, the users' profiles can be expanded with

N. Ferro et al. (Eds.): ECIR 2016, LNCS 9626, pp. 602–613, 2016.
DOI: 10.1007/978-3-319-30671-1_44

relevant items in the same way in which queries are expanded with new terms. An effective technique for performing automatic query expansion is Pseudo-Relevance Feedback (PRF). In [4,14,18], the authors proposed the use of PRF as a CF method. Specifically, they adapted a formal probabilistic model designed for PRF (Relevance-Based Language Models [12]) for the CF recommendation task. The reported experiments showed a superior performance of this approach, in terms of precision, compared to other recommendation methods such as the standard user-based neighbourhood algorithm, SVD and several probabilistic techniques [14]. These improvements can be understood if we look at the foundations of Relevance-Based Language Models since they are designed for generating a ranking of terms (or items in the CF task) in a principled way. Meanwhile, others methods aim to predict the users' ratings. However, it is worth mentioning that Relevance-Based Language Models also outperform other probabilistic methods that focus on top-N recommendation [14].

Nevertheless, the authors in [14] did not analyse the computational cost of generating recommendations within this probabilistic framework. For these reasons, in this paper we analyse the efficiency of the Relevance-Based Language Modelling approach and explore other PRF methods [6] that have a better trade-off between effectiveness and efficiency and, at the same time, do not require any type of smoothing as it is required in [14].

The contributions of this paper are: (1) the adaptation of four efficient Pseudo-Relevance Feedback techniques (Rocchio's weights, Robertson Selection Value, Chi-Squared and Kullback-Leibler Divergence) [6] to CF recommendation, (2) the conception of a new probability estimate that takes into account the length of the neighbourhood in order to improve the accuracy of the recommender system and (3) a critical study of the efficiency of these techniques compared to the Relevance-Based Language Models as well as (4) the analysis of the recommenders from the point of view of the ranking quality, the diversity and the novelty of the suggestions. We show that these new models improve the trade-off between accuracy and diversity/novelty and provide a fast way for computing recommendations.

2 Background

The first paper on applying PRF methods to CF recommendation established an analogy between the query expansion and the recommendation tasks [14]. The authors applied Relevance-Based Language Models [12] outperforming state-of-the-art methods. Next, we describe the PRF task and its adaptation to CF.

Pseudo-Relevance Feedback (PRF) is an automatic technique for improving the performance of a text retrieval system. Feedback information enables to improve the quality of the ranking. However, since explicit feedback is not usually available, PRF is generally a good alternative. This automatic query expansion method assumes that the top retrieval results are relevant. This assumption is reasonable because the goal of the system is to put the relevant results in the top positions of the ranking. Given this pseudo-relevant set of documents, the system extracts from them the best term candidates for query expansion and performs a second search with the expanded query.

The goal of a recommender is to choose for each user of the system ($u \in \mathcal{U}$) items that are relevant from a set of items ($\mathcal{I}$). Given the user u, the output of the recommender is a personalised ranked list L_u^k of k elements. We denote by $\mathcal{I}_u$ the set of items rated by the user u. Likewise, the set of users that rated the item i is denoted by $\mathcal{U}_i$.

The adaptation of the PRF procedure for the CF task [14] is as follows. Within the PRF framework, the users of the system are analogous to queries in IR. Thus, the ratings of the target user act as the query terms. The goal is to expand the original query (i.e., the profile of the user) with new terms that are relevant (i.e., new items that may be of interest to the user). For performing the query expansion process, it is necessary a pseudo-relevant set of documents, from which the expansion terms are extracted. In the context of recommender systems, the neighbours of the target user play the role of pseudo-relevant documents. Therefore, similar users are used to extract items that are candidates to expand the user profile. These candidate items conform the recommendation list.

Parapar et al. [14] experimented with both estimates of the Relevance-Based Language Models [12]: RM1 and RM2. However, as Eqs. 1 and 2 shows, they are considerably expensive. For each user u, they compute a relevance model R_u and they estimate the relevance of each item i under it, $p(i|R_u)$. V_u is defined as the neighbourhood of the user u. The prior probabilities, $p(v)$ and $p(i)$, are considered uniform. In addition, the conditional probability estimations, $p_\lambda(i|v)$ and $p_\lambda(j|v)$, are obtained interpolating the Maximum Likelihood Estimate (MLE) with the probability in the collection using Jelinek-Mercer smoothing controlled by the parameter λ (see Eq. 3). More details can be found in [14].

$$\text{RM1}: \quad p(i|R_u) \propto \sum_{v \in V_u} p(v) p_\lambda(i|v) \prod_{j \in \mathcal{I}_u} p_\lambda(j|v) \tag{1}$$

$$\text{RM2}: \quad p(i|R_u) \propto p(i) \prod_{j \in \mathcal{I}_u} \sum_{v \in V_u} \frac{p_\lambda(i|v) p(v)}{p(i)} p_\lambda(j|v) \tag{2}$$

$$p_\lambda(i|u) = (1 - \lambda) \frac{r_{u,i}}{\sum_{j \in \mathcal{I}_u} r_{u,j}} + \lambda \frac{\sum_{u \in \mathcal{U}} r_{u,i}}{\sum_{u \in \mathcal{U}, j \in \mathcal{I}} r_{u,j}} \tag{3}$$

3 New Profile Expansion Methods

Next, we describe our PRF proposals for item recommendation based on well-known methods in the retrieval community [6,23] that were never applied to CF. For each user, the following PRF methods assign scores to all the non-rated items of the collection. Neighbourhoods, V_u, are computed using k Nearest Neighbours (k-NN) and C denote the whole collection of users and items.

Rocchio's Weights. This method is based on the Rocchio's formula [16]. The assigned score is computed as the sum of the weights for each term of the pseudo-relevant set. This approach promotes highly rated items in the neighbourhood.

$$p_{Rocchio}(i|u) = \sum_{v \in V_u} \frac{r_{v,i}}{|V_u|} \tag{4}$$

Robertson Selection Value (RSV). The Robertson Selection Value (RSV) [15] technique computes a weighted sum of the item probabilities in the neighbourhood. The estimation of these probabilities is described below in this section.

$$p_{RSV}(i|u) = p(i|V_u) \sum_{v \in V_u} \frac{r_{v,i}}{|V_u|} \tag{5}$$

Chi-Squared (CHI-2). This method roots in the chi-squared statistic [6]. The probability in the neighbourhood plays the role of the observed frequency and the probability in the collection is the expected frequency.

$$p_{CHI-2}(i|u) = \frac{\left(p(i|V_u) - p(i|\mathcal{C})\right)^2}{p(i|\mathcal{C})} \tag{6}$$

Kullback-Leibler Divergence (KLD). KLD is a non-symmetric measure for assessing the relative entropy between two probability distributions. Carpineto et al. proposed its use for PRF [6] obtaining good results in the text retrieval task. The idea behind this method is to choose those terms of the pseudo-relevant set which diverge more from the collection in terms of entropy.

$$p_{KLD}(i|u) = p(i|V_u) \log \frac{p(i|V_u)}{p(i|\mathcal{C})} \tag{7}$$

From their equations, we can observe that the complexity of these methods is notably smaller than RM1 and RM2 and are parameter-free. These item ranking functions (except Rocchio's Weights) use probability estimations, $p(i|V_u)$ and $p(i|\mathcal{C})$. We compute these probabilities using the Maximum Likelihood Estimate (MLE) under a multinomial distribution of X. We represent by $\mathcal{U}_X$ the set of users that rated the items from the set X. Likewise, $\mathcal{I}_X$ denotes the set of items that were rated by the users of the set X.

$$p_{MLE}(i|X) = \frac{\sum_{u \in \mathcal{U}_X} r_{u,i}}{\sum_{u \in \mathcal{U}_X, j \in \mathcal{I}_X} r_{u,j}} \tag{8}$$

4 Neighbourhood Length Normalisation

When we use a hard clustering algorithm, the number of users in each cluster is variable. Even algorithms such as k-NN can lead to neighbourhoods with different sizes: a similarity measure based on the common occurrences among users may not be able to find k neighbours for all users when k is too high or when the collection is very sparse—we consider that a neighbour should have at least one common item. In these cases, the information of the neighbourhood is even more important since the user differs strongly from the collection. In IR, this situation would be associated with difficult queries that returned a very limited amount of documents. Therefore, the information of the relevant set should be promoted whilst the global collection information should be demoted.

We incorporated this intuition into the recommendation framework adding a bias to the probability estimate. Thus, we normalise the MLE by dividing the estimate by the number of users in the population as follows:

$$p_{NMLE}(i|X) \overset{\text{rank}}{=} \frac{1}{|\mathcal{U}_X|} \frac{\sum_{u \in \mathcal{U}_X} r_{u,i}}{\sum_{u \in \mathcal{U}_X, j \in \mathcal{I}_X} r_{u,j}} \tag{9}$$

This improvement does not make sense for the RSV item ranking function because the ranking would be the same (the scores will be rescaled by a constant); however, it can be useful for CHI-2 and KLD methods as it can be seen in Sect. 5.

5 Evaluation

We used three film datasets from GroupLens[1]: *MovieLens 100k*, *MovieLens 1M* and *MovieLens 10M*, for the efficiency experiment. Additionally, we used the *R3-Yahoo! Webscope Music*[2] dataset and the *LibraryThing*[3] book collection for the effectiveness tests. The details of the collections are gathered in Table 1. We used the splits provided by the collections. However, since Movielens 1M and LibraryThing do not offer predefined partitions, we selected 80 % of the ratings of each user for the training subset whilst the rest is included in the test subset.

5.1 Evaluation Methodology

In CF evaluation, a great variety of metrics have been applied. Traditionally, recommenders were designed as rating predictors and, thus, the evaluation was based on error metrics. However, there is a consensus among the scientific community that it is more useful to model recommendation as a ranking task (top-N recommendation) which leads to the use of precision-oriented metrics [2, 10, 13]. In addition, it was stated that not only accuracy but diversity and novelty are key properties of the recommendations [10]. For this reason, in this study we use metrics for these aspects.

We followed the *TestItems* approach described by Bellogín et al. [2] for estimating the precision of the recommendations. For each user, we compute a ranking for all the items having a test rating by some user and no training rating

[1] http://grouplens.org/datasets/movielens/.
[2] http://webscope.sandbox.yahoo.com.
[3] http://www.macle.nl/tud/LT/.

Table 1. Datasets statistics

Dataset	Users	Items	Ratings	Density
MovieLens 100 k	943	1682	100,000	6.305 %
MovieLens 1 M	6,040	3,952	1,000,209	4.190 %
MovieLens 10 M	71,567	10,681	10,000,054	1.308 %
R3-Yahoo!	15,400	1,000	365,703	2.375 %
LibraryThing	7,279	37,232	749,401	0.277 %

by the target user. It has been acknowledged that considering non-rated items as irrelevant may underestimate the true metric value (since non-rated items can be of interest to the user); however, it provides a better estimation of the recommender quality [2,13].

The employed metrics are evaluated at a specified cut-off rank, i.e., we consider only the top k recommendations of the ranking for each user because these are the ones presented to the user. For assessing the quality of the ranking we employed nDCG. This metric uses graded relevance of the ratings for judging the ranking quality. Values of nDGG increases when highly relevant documents are located in the top positions of the ranking. We used the standard formulation as described in [22]. We also employed the complement of the Gini index for quantifying the diversity of the recommendations [9]. The index is 0 when only a single item is recommended for every user. On the contrary, a value of 1 is achieved when all the items are equally recommended among the users. Finally, to measure the ability of a recommender system to generate unexpected recommendations, we computed the mean self-information (MSI) [25]. Intuitively, the value of this metric increases when unpopular items are recommended.

5.2 Baselines

To assess the performance of the proposed recommendation techniques, we chose a representative set of state-of-the-art recommenders. We used a standard user-based neighbourhood CF algorithm (labelled as UB): the neighbours are computed using k-NN with Pearson's correlation as the similarity measure [8]. We also tested Singular Value Decomposition (SVD), a matrix factorisation technique which is among the best methods for rating prediction [11]. Additionally, we included an algorithm which has its roots in the IR probabilistic modelling framework [20], labelled as UIR-Item. Finally, as the strongest baselines, we chose the RM1 and RM2 models [14]. Instead of employing Jelinek-Mercer smoothing as it was originally proposed [14], we used Absolute Discounting because recent studies showed that it is more effective stable than Jelinek-Mercer [18].

5.3 Efficiency Experiment

The principal motivation for this work was to propose more efficient PRF recommendation techniques than RM1 and RM2. To assess the efficiency of our proposals, we measured the user recommendation times on the MovieLens 100k, 1M and 10M datasets. The neighbourhoods are precomputed using k-NN with Pearson's correlation and $k = 100$. Since the time of computing the neighbours is common to each method, we can ignore it. We measured the algorithms in a desktop computer with an Intel i7-4790 @3.60GHz and 16 GB DDR3 1600 MHz.

Figure 1 illustrates the recommendation times on the three datasets. We report times (in logarithmic scale) for UIR-Item, RM1, RM2, RSV, Rocchio's Weights, CHI-2 and KLD. These results demonstrate that the proposed new methods are dramatically faster than RM1 and RM2 (our proposals obtain speed-ups up to 200x) meanwhile the variations in time among our proposed methods are small. Additionally, the differences in time between the probability estimates (MLE and NMLE) are insignificant. We do not report the recommendation time of UIR-Item on the MovieLens 10M collection because its performance was so poor that the experiment did not finish in a week.

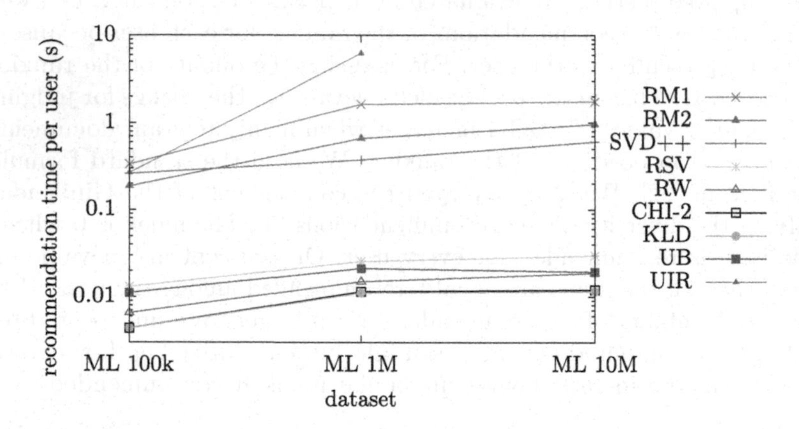

Fig. 1. Recommendation time per user (in logarithmic scale) using UIR-Item (UIR), RM1, RM2, RSV, Rocchio's Weights (RW), CHI-2 and KLD algorithms with NMLE as the probability estimate on the MovieLens 100k, 1M and 10M collections.

5.4 Effectiveness Experiment

We present now the results of our methods as well as the baselines on the MovieLens 100k, Movielens 1M, R3-Yahoo! and LibraryThing collections. We used k-NN with Pearson's similarity for computing the neighbourhoods and we tuned k from 50 to 950 neighbours (in steps of 50) for each method in the MovieLens 100k dataset. Those values were then used in the rest of the collections. We also tuned the number of latent factors of SVD and the λ parameter of UIR-Item. All parameters were tuned in order to optimise nDCG@10 using cross-validation with the five folders provided by the MovieLens 100k collection. In order to facilitate the reproducibility of these experiments we show, for each method, the optimal values for the tuned parameters in Table 2.

Table 2. Values of nDCG@10 for each recommender approach. Statistically significant improvements according to Wilcoxon Test ($p < 0.05$) with respect to the baselines UB, SVD, UIR-Item, RM1, RM2 are superscripted with a, b, c, d and e, respectively. The complementary statistically significant decreases are subscripted in the same way. The values in bold indicate the best recommender for the each dataset. The values underlined are not statistically different from the best value.

Algorithm		Tuned param.	ML 100 k	ML 1 M	R3-Yahoo!	LibraryThing
UB		$k = 50$	0.0468_{bcde}	0.0313_{bcde}	0.0108_{cde}	0.0055^{b}_{cde}
SVD		$factors = 400$	0.0936^{a}_{cde}	0.0608^{a}_{cde}	0.0101_{cde}	0.0015_{acde}
UIR-Item		$\lambda = 0.5$	0.2188^{ab}_{de}	0.1795^{abd}_{e}	0.0174^{abd}_{e}	0.0673^{abd}_{e}
RM1		$k = 400,\ \delta = 0.1$	0.2473^{abc}_{e}	0.1402^{ab}_{ce}	0.0146^{ab}_{ce}	0.0444^{ab}_{ce}
RM2		$k = 550,\ \delta = 0.1$	$\mathbf{0.3323}^{abcd}$	$\mathbf{0.1992}^{abd}$	$\mathbf{0.0207}^{abcd}$	0.0957^{abcd}
Rocchio's Weights		$k = 600$	0.2604^{abcd}_{e}	0.1557^{abd}_{ce}	0.0194^{abcd}_{e}	0.0892^{abcd}_{e}
RSV	MLE	$k = 600$	0.2604^{abcd}_{e}	0.1557^{abd}_{ce}	0.0194^{abcd}_{e}	0.0892^{abcd}_{e}
KLD	MLE	$k = 850$	0.2693^{abcd}_{e}	0.1264^{ab}_{cde}	$\underline{0.0197}^{abcd}$	$\mathbf{0.1576}^{abcde}$
	NMLE	$k = 700$	0.3120^{abcd}_{e}	0.1546^{ab}_{cde}	$\underline{0.0201}^{abcd}$	0.1101^{abcde}
CHI-2	MLE	$k = 500$	0.0777^{a}_{bcde}	0.0709^{ab}_{cde}	0.0149^{ab}_{ce}	0.0939^{abcd}
	NMLE	$k = 700$	0.3220^{abcd}_{e}	0.1419^{ab}_{cde}	$\underline{0.0204}^{abcd}$	0.1459^{abcde}

The obtained nDCG@10 values are reported in Table 2 with statistical significance tests (two-sided Wilcoxon test with $p < 0.05$). Generally, RM2 is the best recommender algorithm as it was expected—better probabilistic models should lead to better results. Nevertheless, it can be observed that in the R3-Yahoo! dataset, the best nDCG values of our efficient PRF methods are not statistically different from RM2. Moreover, in the LibraryThing collection, many of the proposed models significantly outperform RM2 with important improvements. This may be provoked by the sparsity of the collections which leads to think that RM2 is too complex to perform well under this more common scenario. Additionally, although we cannot improve the nDCG figures of RM2 on the MovieLens 100k, we significantly surpass the other baselines.

In most of the cases, the proposals that use collection statistics (i.e., KLD and the CHI-2 methods) tend to perform better than those that only use neighbourhood information (Rocchio's Weights and RSV). Regarding the proposed neighbourhood length normalisation, the experiments show that NMLE improves the ranking accuracy compared to the regular MLE in the majority of the cases. Thus, the evidence supports the idea that the size of the users' neighbourhoods is an important factor to model in a recommender system.

Now we take the best baselines (UIR-Item and RM2) and our best proposal (CHI-2 with NMLE) in order to study the diversity and novelty of the top ten recommendations. Note that we use the same rankings which were optimized for nDCG@10. The values of Gini@10 and MSI@10 are presented in Tables 3 and 4, respectively. In the case of Gini, we cannot perform paired significance analysis since it is a global metric.

Table 3. Gini@10 values of UIR-Item, RM2 and CHI-2 with NMLE (optimised for nDCG@10). Values in bold indicate the best recommender for the each dataset. Significant differences are indicated with the same criteria as in Table 2.

Algorithm	ML 100 k	ML 1 M	R3-Yahoo!	LibraryThing
UIR-Item	0.0124	0.0050	0.0137	0.0005
RM2	0.0256	0.0069	0.0207	0.0019
CHI-2 NMLE	**0.0450**	**0.0106**	**0.0506**	**0.0539**

Table 4. MSI@10 values of UIR-Item, RM1, RM2 and CHI-2 with NMLE (optimised for nDCG@10). Values in bold indicate the best recommender for the each dataset. Significant differences are indicated with the same criteria as in Table 2.

Algorithm	ML 100 k	ML 1 M	R3-Yahoo!	LibraryThing
UIR-Item	5.2337_e	8.3713_e	3.7186_e	17.1229_e
RM2	6.8273^c	8.9481^c	4.9618^c	$\mathbf{19.27343}^c$
CHI-2 NMLE	$\mathbf{8.1711}^{ec}$	$\mathbf{10.0043}^{ec}$	$\mathbf{7.5555}^{ec}$	8.8563

We observe that CHI-2 with NMLE generates more diverse recommendations than RM2, which is the strongest baseline in terms of nDCG. Also, CHI-2 with NMLE presents good novelty figures except for the LibraryThing collection. However, as we mentioned before, the performance of RM2 on the LibraryThing dataset is quite poor in terms of nDCG compared to the other models. It is easy to improve diversity and novelty decreasing the accuracy values [25]; however, we aim for an effective method in terms of all the metrics. In summary, the results showed that CHI-2 with NMLE is among the best performing studied methods with a good trade-off between accuracy and diversity/novelty.

The advantages in terms of the trade-offs among ranking precision and diversity and novelty are reported in Fig. 2 where we present the G-measure for both relations when varying the size of the neighbourhood. The G-measure is the geometric mean of the considered metrics which effectively normalizes the true positive class (in this case, relevant and diverse or relevant and novel). In this particular scenario, the use of other kind of means is not appropriate [7] due to the strong dependency and the difference in scale among the analysed variables. In the graphs, we observe that with values of $k > 400$, our proposal is even better than the strongest baseline, RM2, for both trade-offs. Therefore, we presented a competitive method in terms of effectiveness which is up to 200 times faster than previous PRF algorithms for CF.

6 Related Work

Exploring Information Retrieval (IR) techniques and applying them to Recommender Systems is an interesting line of research. In fact, in 1992, Belkin and Croft already stated that Information Retrieval and Information Filtering (IF)

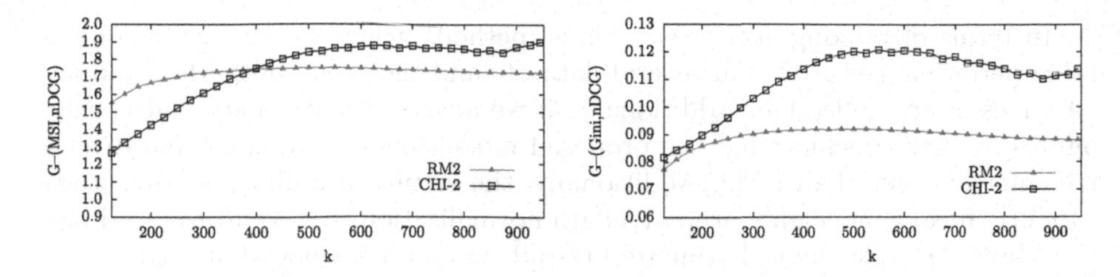

Fig. 2. Values of the G-measure in the MovieLens 100k collection plotted against the size of the neighbourhood (k), for the nDCG@10-MSI@10 (left) and the nDCG@10-Gini@10 (right) trade-offs.

are two sides of the same coin [1]. Recommenders are automatic IF systems: their responsibility lies in selecting relevant items for the users. Consequently, besides the work of Parapar et al. on applying Relevance-Based Language Models to CF recommendation [14], there is a growing amount of literature about different approaches that exploit IR techniques for recommendation [5,19–21].

Wang et al. derived user-based and item-based CF algorithms using the classic probability ranking principle [20]. They also presented a probabilistic relevance framework with three models [21]. Also, Wang adapted the language modelling scheme to CF using a risk-averse model that penalises less reliable scores [19].

Another approach is the one formulated by Bellogín et al. [5]. They devised a general model for unifying memory-based CF methods and text retrieval algorithms. They show that many IR methods can be used within this framework obtaining better results than classic CF techniques for the item ranking task.

Relevance-Based Language Models were also adapted to CF in a different manner. Bellogín et al. [4] formulate the formation of user neighbourhoods as a query expansion task. Then, by using the negative cross entropy ranking principle, they used the neighbours to compute item recommendations.

7 Conclusions and Future Work

Since Relevance Models [12] are an effective tool for item recommendation [14], the aim of this work was to assess if other faster PRF methods could be used for the same task. The results of this investigation revealed that, indeed, simpler and more efficient PRF techniques are suitable for this CF task. We have carried out experiments that showed that the proposed recommendation algorithms (Rocchio's Weigths, RSV, KLD and CHI-2) are orders of magnitude faster than the Relevance Models for recommendation. These alternatives offer important improvements in terms of computing time while incurring, in some cases, in a modest decrease of accuracy. Furthermore, these methods lack of parameters: they only rely on the neighbourhood information. In a large-scale scenario, a speed-up of 200x can lead to notable savings in computational resources.

In terms of ranking accuracy, various methods achieve statistically comparable performance to RM2 in several datasets and they even outperform all the baselines in one collection. Additionally, if we analyse the diversity and novelty figures, we can conclude that the proposed models offer more novel and diverse recommendations than RM2. Additionally, the empirical findings of this study support the idea of neighbourhood length normalisation that we introduced into the Maximum Likelihood Estimate. Overall, we can conclude that CHI-2 with NMLE provide highly precise and fast recommendations with a good trade-off between accuracy and diversity/novelty.

We think that exploring other state-of-the-art PRF techniques such as Divergence Minimization Models or Mixture Models [24] for recommendation may be a fruitful area for further research.

Moreover, a future study investigating different techniques for generating neighbourhoods would be very interesting. In this paper, we employed k-NN algorithm because of its efficiency. Nevertheless, exploring other clustering methods may produce important improvements. For example, the combination of Relevance-Based Language Models with Posterior Probability Clustering, a type of non-negative matrix factorisation, has been proved to generate highly precise recommendations [14]. Similarly, it may be of interest the use of Normalised Cut (a spectral clustering method) since it has been reported that it improves the effectiveness of the standard neighbourhood-based CF algorithms [3].

Acknowledgments. This work was supported by the *Ministerio de Economía y Competitividad* of the Government of Spain under grants TIN2012-33867 and TIN2015-64282-R. The first author also wants to acknowledge the support of *Ministerio de Educación, Cultura y Deporte* of the Government of Spain under the grant FPU014/01724.

References

1. Belkin, N.J., Croft, W.B.: Information filtering and information retrieval: two sides of the same coin? Commun. ACM **35**(12), 29–38 (1992)
2. Bellogín, A., Castells, P., Cantador, I.: Precision-oriented evaluation of recommender systems. In: RecSys 2011, p. 333. ACM (2011)
3. Bellogín, A., Parapar, J.: Using graph partitioning techniques for neighbour selection in user-based collaborative filtering. In: RecSys 2012, pp. 213–216. ACM (2012)
4. Bellogín, A., Parapar, J., Castells, P.: Probabilistic collaborative filtering with negative cross entropy. In: RecSys 2013, pp. 387–390. ACM (2013)
5. Bellogín, A., Wang, J., Castells, P.: Bridging memory-based collaborative filtering and text retrieval. Inf. Retr. **16**(6), 697–724 (2013)
6. Carpineto, C., de Mori, R., Romano, G., Bigi, B.: An information-theoretic approach to automatic query expansion. ACM Trans. Inf. Syst. **19**(1), 1–27 (2001)
7. Coggeshall, F.: The arithmetic, geometric, and harmonic means. Q. J. Econ. **1**(1), 83–86 (1886)
8. Desrosiers, C., Karypis, G.: A comprehensive survey of neighborhood-based recommendation methods. In: Ricci, F., Rokach, L., Shapira, B., Kantor, P.B. (eds.) Recommender Systems Handbook, pp. 107–144. Springer, Heidelberg (2011)

9. Fleder, D., Hosanagar, K.: Blockbuster culture's next rise or fall: the impact of recommender systems on sales diversity. Manage. Sci. **55**(5), 697–712 (2009)
10. Herlocker, J.L., Konstan, J.A., Terveen, L.G., Riedl, J.T.: Evaluating collaborative filtering recommender systems. ACM Trans. Inf. Syst. **22**(1), 5–53 (2004)
11. Koren, Y., Bell, R.: Advances in collaborative filtering. In: Ricci, F., Rokach, L., Shapira, B., Kantor, P.B. (eds.) Recommender Systems Handbook, pp. 145–186. Springer, Heidelberg (2011)
12. Lavrenko, V., Croft, W.B.: Relevance-based language models. In: SIGIR 2001, pp. 120–127. ACM (2001)
13. McLaughlin, M.R., Herlocker, J.L.: A collaborative filtering algorithm and evaluation metric that accurately model the user experience. In: SIGIR 2004, pp. 329–336. ACM (2004)
14. Parapar, J., Bellogín, A., Castells, P., Barreiro, A.: Relevance-based language modelling for recommender systems. Inf. Process. Manage. **49**(4), 966–980 (2013)
15. Robertson, S.E.: On term selection for query expansion. J. Doc. **46**(4), 359–364 (1990)
16. Rocchio, J.J.: Relevance feedback in information retrieval. In: Salton, G. (ed.) The SMART Retrieval System - Experiments in Automatic Document Processing, pp. 313–323. Prentice Hall (1971)
17. Schafer, J.B., Frankowski, D., Herlocker, J., Sen, S.: Collaborative filtering recommender systems. In: Brusilovsky, P., Kobsa, A., Nejdl, W. (eds.) Adaptive Web 2007. LNCS, vol. 4321, pp. 291–324. Springer, Heidelberg (2007)
18. Valcarce, D., Parapar, J., Barreiro, A.: A study of smoothing methods for relevance-based language modelling of recommender systems. In: Hanbury, A., Kazai, G., Rauber, A., Fuhr, N. (eds.) ECIR 2015. LNCS, vol. 9022, pp. 346–351. Springer, Heidelberg (2015)
19. Wang, J.: Language models of collaborative filtering. In: Lee, G.G., Song, D., Lin, C.-Y., Aizawa, A., Kuriyama, K., Yoshioka, M., Sakai, T. (eds.) AIRS 2009. LNCS, vol. 5839, pp. 218–229. Springer, Heidelberg (2009)
20. Wang, J., de Vries, A.P., Reinders, M.J.T.: A user-item relevance model for log-based collaborative filtering. In: Lalmas, M., MacFarlane, A., Rüger, S.M., Tombros, A., Tsikrika, T., Yavlinsky, A. (eds.) ECIR 2006. LNCS, vol. 3936, pp. 37–48. Springer, Heidelberg (2006)
21. Wang, J., de Vries, A.P., Reinders, M.J.T.: Unified relevance models for rating prediction in collaborative filtering. ACM Trans. Inf. Syst. **26**(3), 1–42 (2008)
22. Wang, Y., Wang, L., Li, Y., He, D., Chen, W., Liu, T.-Y.: A theoretical analysis of NDCG ranking measures. In: COLT 2013, pp. 1–30 (2013). JMLR.org
23. Wong, W.S., Luk, R.W.P., Leong, H.V., Ho, L.K., Lee, D.L.: Re-examining the effects of adding relevance information in a relevance feedback environment. Inf. Process. Manage. **44**(3), 1086–1116 (2008)
24. Zhai, C., Lafferty, J.: Model-based feedback in the language modeling approach to information retrieval. In: CIKM 2001, pp. 403–410. ACM (2001)
25. Zhou, T., Kuscsik, Z., Liu, J.-G., Medo, M., Wakeling, J.R., Zhang, Y.-C.: Solving the apparent diversity-accuracy dilemma of recommender systems. PNAS **107**(10), 4511–4515 (2010)

Language Models for Collaborative Filtering Neighbourhoods

Daniel Valcarce(✉), Javier Parapar, and Álvaro Barreiro

Information Retrieval Lab, Computer Science Department,
University of A Coruña, A Coruña, Spain
{daniel.valcarce,javierparapar,barreiro}@udc.es

Abstract. Language Models are state-of-the-art methods in Information Retrieval. Their sound statistical foundation and high effectiveness in several retrieval tasks are key to their current success. In this paper, we explore how to apply these models to deal with the task of computing user or item neighbourhoods in a collaborative filtering scenario. Our experiments showed that this approach is superior to other neighbourhood strategies and also very efficient. Our proposal, in conjunction with a simple neighbourhood-based recommender, showed a great performance compared to state-of-the-art methods (NNCosNgbr and PureSVD) while its computational complexity is low.

Keywords: Recommender systems · Language models · Collaborative filtering · Neighbourhood

1 Introduction

Recommender systems aim to provide useful items of information to the users. These suggestions are tailored according to the users's tastes. Considering the increasing amount of information available nowadays, it is hard to manually filter what is interesting and what is not. Additionally, users are becoming more demanding—they do not conform with traditional browsing or searching activities, they want relevant information immediately. Therefore, recommender systems play a key role in satisfying the users' needs.

We can classify recommendation algorithms in three main categories: content-based systems, which exploit the metadata of the items to recommend similar ones; collaborative filtering, which uses information of what other users have done to suggest items; and hybrid techniques, which combine both content-based and collaborative filtering approaches [15]. In this paper, we focus on the collaborative filtering scenario. Collaborative techniques ignore the content of the items since they merely rely on the feedback from other users. They tend to perform better than content-based approaches if sufficient historical data is available. We can distinguish two main types of collaborative methods. On the one hand, model-based techniques learn a latent factor representation from the data after a training process [10]. On the other hand, neighbourhood-based methods

N. Ferro et al. (Eds.): ECIR 2016, LNCS 9626, pp. 614–625, 2016.
DOI: 10.1007/978-3-319-30671-1_45

(also called memory-based algorithms) use the similarities among past user-item interactions [6]. Neighbourhood-based recommenders, in turn, are classified in two categories: user-based and item-based approaches depending on which type of similarities are computed. User-based recommenders rely on user neighbourhoods (i.e., they recommend items that similar users like). By contrast, item-based algorithms compute similarities between items (i.e., two items are related if users rate them in a similar way).

Neighbourhood-based approaches are simpler than their model-based counterparts because they do not require a previous training step—still, we need to compute the neighbourhoods. Multiple approaches to generate neighbourhoods exist in the literature [6] because this phase is crucial in the recommendation process. The effectiveness of these type of recommenders depends largely on how we calculate the neighbourhoods. A popular approach consists in computing the k Nearest Neighbours according to a pairwise similarity metric such as Pearson's correlation coefficient, adjusted cosine or cosine similarity.

Traditionally, recommender systems were designed as rating predictors; however, it has been acknowledged that it is more interesting to model the recommendation problem as an item ranking task [1,8]. Top-N recommendation is the term coined to name this new perspective [4]. For this task, the use of Information Retrieval techniques and models is attracting more and more attention [2,13,17,20]. The reason is that these methods were specifically conceived for ranking documents according to an explicit query. However, they can also rank items using the user's profile as an implicit query.

Previous work has found that the cosine similarity yields the best results in terms of accuracy metrics in the neighbourhood computation process [4]. In fact, it surpasses Pearson's correlation coefficient which is, by far, the most used similarity metric in the recommender system literature [6]. Thinking about cosine similarity in terms of retrieval models, we can note that it is the basic distance measure used in the Vector Space Model [16]. Following this analogy between Information Retrieval and Recommender Systems, if the cosine similarity is a great metric for computing neighbourhoods, it sounds reasonable to apply more sophisticated representations and measures to this task used in other more effective retrieval models. Thus, in this paper we focus on modelling the finding of user and item neighbourhoods as a text retrieval task. In particular, we propose an adaptation of the Language Modelling retrieval functions as a method for computing neighbourhoods. Our proposal leverages the advantages of this successful retrieval technique for calculating collaborative filtering neighbourhoods. Our proposal—which can be used in a user or item-based approach—in conjunction with a simple neighbourhood algorithm surpasses state-of-the-art methods (NNCosNgbr and PureSVD [4]) in terms of accuracy and is also very efficient.

2 Background

An extensive literature has studied several neighbourhood-based approaches because they are simple, interpretable and efficient [4–6,9]. After calculating the neighbourhoods, the recommendation process consists in computing correlations between item or user neighbours. Item-based approaches are usually preferred [4–6] because the number of items is usually smaller than the users. This enables efficient computation of the neighbourhoods. Also, they have been shown to report better results in terms of accuracy than user-based approaches [5,6]. Also, item-based recommendations are easy to justify with explanations such as "you would like item B because you liked item A". However, item-based methods may generate less serendipitous recommendations because they tend to recommender similar items to those rated by the user [6]. In contrast, user-based approaches recommend items that similar users enjoyed. Thus, it is possible to suggest items that strongly differ from the ones rated by the target user.

2.1 Non-normalised Cosine Neighbourhood (NNCosNgbr)

Non-Normalised Cosine Neighbourhood (NNCosNgbr) is an effective item-based neighbourhood algorithm presented in [4]. For computing the k Nearest Neighbours, this method uses cosine similarity instead of Pearson's correlation coefficient because the former is computed over all the ratings whilst the latter relies only on the shared ratings. Moreover, they introduced a shrinking factor based on common ratings into the similarity metric [9]. This modification penalises the similarity between very sparse vectors [4]. Additionally, NNCosNgbr removes user and item biases according to the definition in [9]. The predicted score $\hat{r}_{u,i}$ for the user u and the item i is given by the following expression:

$$\hat{r}_{u,i} = b_{u,i} + \sum_{j \in J_i} s_{i,j} \left(r_{u,j} - b_{u,j} \right) \tag{1}$$

where $b_{u,i}$ denotes the bias for the user u and the item i (computed as in [9]); $s_{i,j}$, the cosine similarity between items i and j; J_i, the neighbourhood of the item i, and $r_{u,j}$, the rating that the user u gave to the item j. The major difference between this method and the standard neighbourhood approach [6] is the absence of the normalising denominator. Since we are not interested in predicting ratings, we do not worry about getting scores in a fixed range. On the contrary, this method fosters those items with high ratings by many neighbours [4,5,9].

2.2 Language Models (LM)

Language Models (LM) represent a successful framework within the Information Retrieval (IR) field. Ponte and Croft presented the first approach of using Language Models for the text retrieval task in 1998 [14]. Nowadays, the use of Language Models has become so popular in the field that they have been improved to address several IR tasks achieving state-of-the-art performance [22]. Compared to previous techniques, the main contributions of these models are their solid statistical foundation and their interpretability [14,22].

The Language Modelling framework is a formal approach with a sound statistical foundation. It models the occurrences of words in the documents and queries as a random generative process—usually, using a multinomial distribution. Within this framework, we infer a language model for each document in the collection. To rank those documents according to a user's query, we estimate the posterior probability of each document d given the particular query q, $p(d|q)$:

$$p(d|q) = \frac{p(q|d)\, p(d)}{p(q)} \stackrel{\text{rank}}{=} p(q|d)\, p(d) \tag{2}$$

where $p(q|d)$ is the query likelihood and $p(d)$, the document prior. We can ignore the query prior $p(q)$ because it has no effect in the ranking for the same query. Usually, a uniform document prior is chosen and the query likelihood retrieval model is used. The most popular approach in IR to compute the query likelihood is to use a unigram model based on a multinomial distribution:

$$p(q|d) = \prod_{t \in q} p(t|d)^{c(t,d)} \tag{3}$$

where $c(t, d)$ denotes the count of term t in document d. The conditional probability $p(t|d)$ is computed via the maximum likelihood estimate (MLE) of a multinomial distribution smoothed with a background model [23].

3 WSR and LM for Neighbours

In this section, we explain how we designed our neighbourhood algorithm WSR within the Language Modelling framework. First, we propose our recommendation algorithm and, next, we explain how we compute neighbours.

3.1 Neighbourhood-Based Recommender for Ranking: WSR

Our recommendation algorithm stems from NNCosNgbr method and each modification described in this section was evaluated in Sect. 4.2. First, we kept the biases (see Eqs. 4 and 5), instead of removing them as in Eq. 1. Removing biases is very important in rating prediction recommenders because it allows to estimate ratings more accurately [6,9], however it is useless on the top-N recommendation because we are concerned about rankings. Moreover, this process adds an extra parameter to tune [9]. Next, we focused on the similarity metric. In [4], the authors introduced a shrinking factor into the cosine metric to promote those similarities that are based on many shared ratings. This shrinkage procedure has shown good results in previous studies based on error metrics [6,9] at the expense of putting an additional parameter into the model. However, we found that its inclusion is detrimental in our scenario. This is reasonable because the main advantage of cosine similarity over other metrics such as Pearson's correlation coefficient is that it considers non-rated values as zeroes. In this way, cosine already takes into account the amount of co-occurrence between vectors of ratings which makes unnecessary the use of a shrinkage technique.

In conclusion, the final formula of the recommendation algorithm is a weighted sum of the ratings of the neighbours, which we coined as WSR (Weighted Sum Recommender). Equations 4 and 5 are the user and item-based versions, respectively.

$$\hat{r}_{u,i} = \sum_{v \in V_u} s_{u,v} \, r_{v,i} \tag{4}$$

$$\hat{r}_{u,i} = \sum_{j \in J_i} s_{i,j} \, r_{u,j} \tag{5}$$

where s is the cosine similarity between the user or item vectors. V_u is the neighbourhood of user u, as J_i is the neighbourhood of item i.

3.2 Neighbourhoods Using Language Models

Preliminary tests showed that our algorithm (Eqs. 4 and 5) performs very well compared to more sophisticated ones using plain cosine similarity and evaluating ranking quality. In fact, techniques such as biases removal or similarity shrinkage worsened the performance and introduced additional parameters in the model. Major differences in terms of ranking accuracy metrics occur when varying the neighbourhood computation method. In particular, our experiments showed that cosine similarity is a great metric for computing k Nearest Neighbours (k-NN). This process is analogous to the document ranking procedure in the Vector Space Model [16] if the target user plays the role of the query and the rest of the users are the documents in the collection. The outcome of this model will be a list of neighbours ordered by decreasing cosine similarity with respect to the user. Thus, choosing the k nearest neighbours is the same as taking the top k results using the user as the query.

Language Models (LM) are a successful retrieval method [14,22] that deals with data sparsity [23], enables to introduce a priori information and performs document length normalisation [11]. Recommendation algorithms could benefit from LM because: user feedback is sparse, we may have a priori information and the profile sizes vary. We adapted LM framework to the task of finding neighbourhoods in a user or item-based manner. If we choose the former, we can model the generation of ratings by users as a random process given by a probability distribution (as Language Models do with the occurrences of terms). In this way, we can see documents and queries as users and terms as items. Thus, the retrieval procedure results in finding the nearest neighbours of the target user (i.e., the query). Analogously, we can flip to the item-based approach. In this case, the query plays the role of the target item while the rest of items play the role of the documents. In this way, a retrieval returns the most similar items.

The user-based analogy between the IR and recommendation tasks has already been stated. The consideration of a multinomial distribution of ratings has been used in [2,13] under the Relevance-Based Language Modelling framework for computing recommendations. In our Language Modelling adaptation for calculating neighbourhoods, we can estimate the probability of a candidate neighbour v given a user u as follows:

$$p(v|u) \stackrel{\text{rank}}{=} p(v)\, p(u|v) = p(v) \prod_{i \in \mathcal{I}_u} p(i|v)^{r_{u,i}} \tag{6}$$

where $\mathcal{I}_u$ are the items rated by user u. Here we only present the user-based approach for the sake of space: the item-based counterpart is derived analogously.

Language Modelling Smoothing. The probability of an item i given the user v, $p(i|v)$, is given by the MLE smoothed with the probability of an item in the collection. We explored three well-known smoothing methods (Absolute Discounting, Jelinek-Mercer and Dirichlet Priors [23]) that were recently applied to collaborative filtering [19] analysing their effects using Relevance-Based Language Models. For each method, the probability in the collection is computed as follows:

$$p(i|\mathcal{C}) = \frac{\sum_{v \in \mathcal{U}} \mathrm{r}_{v,i}}{\sum_{j \in \mathcal{I},\, v \in \mathcal{U}} \mathrm{r}_{v,j}} \tag{7}$$

Absolute Discounting (AD)

$$p_\delta(i|u) = \frac{\max(\mathrm{r}_{u,i} - \delta, 0) + \delta\, |\mathcal{I}_u|\, p(i|\mathcal{C})}{\sum_{j \in \mathcal{I}_u} \mathrm{r}_{u,j}} \tag{8}$$

Jelinek-Mercer (JM)

$$p_\lambda(i|u) = (1 - \lambda) \frac{\mathrm{r}_{u,i}}{\sum_{j \in \mathcal{I}_u} \mathrm{r}_{u,j}} + \lambda\, p(i|\mathcal{C}) \tag{9}$$

Dirichlet Priors (DP)

$$p_\mu(i|u) = \frac{\mathrm{r}_{u,i} + \mu\, p(i|\mathcal{C})}{\mu + \sum_{j \in \mathcal{I}_u} \mathrm{r}_{u,j}} \tag{10}$$

We employ this Language Modelling approach only for computing neighbourhoods. Cosine similarity is still used in our WSR algorithm as the similarity in Eqs. 4 and 5. In this way, we generate recommendations independently of the neighbourhood strategy. In fact, we can use our proposal for computing neighbours in any neighbourhood-based algorithm.

4 Experiments

We ran our experiments on four collections: MovieLens 100 k and 1 M[1] film dataset, the R3-Yahoo! Webscope Music[2] dataset and the LibraryThing[3] book dataset. We present the details of these collections in Table 1.

We used the splits that MovieLens 100 k and R3 Yahoo! provide for evaluation purposes. Since the MovieLens 1 M and LibraryThing collections do not include predefined splits, we put 80 % of the ratings of each user in the training subset and the rest in the test subset randomly.

[1] http://grouplens.org/datasets/movielens/.
[2] http://webscope.sandbox.yahoo.com.
[3] http://www.macle.nl/tud/LT/.

Table 1. Datasets statistics

Dataset	Users	Items	Ratings	Density
MovieLens 100 k	943	1,682	100,000	6.305 %
MovieLens 1 M	6040	3,706	1,000,209	4.468 %
R3-Yahoo!	15,400	1,000	365,703	2.375 %
LibraryThing	7,279	37,232	749,401	0.277 %

4.1 Evaluation Methodology

Instead of evaluating recommenders in the rating prediction task, we focused on the top-N recommendation task [4,8]. We used precision-oriented metrics [1] but also other diversity and novelty measures [8]. We followed the *TestItems* approach to create the rankings [1]: for each user, we rank all the items that have a rating in the test set and have not been rated by the target user.

We evaluated all the metrics at a specified cut-off rank because we wanted to focus on how the recommenders behave in top positions of the rankings—users seldom consider more than the top suggestions. We used Normalised Discounted Cumulative Gain (nDCG) for quantifying the quality of the ranking using the ratings in the test set as graded relevance judgements. In particular, we employed the *standard nDCG* as formulated in [21]. We measured diversity using the complement of the Gini index [7] (the index is 0 when a single item is recommended for every user, 1 when all the items are equally recommended among the users). Finally, we computed the mean self-information (MSI) to measure the ability of the system to generate unexpected—unpopular—recommendations [24].

4.2 Testing WSR Versus NNCosNgbr

In this section, we analyse the different options in WSR and NNCosNgr algorithms described in Sect. 3.1. Table 2 shows the best values of nDCG@10. We used cosine as the similarity metric in k-NN and tuned the number of nearest neighbours from $k = 50$ to 250 in steps of 50 neighbours. We chose a similarity shrinking factor of 100 as recommended in [4]. Biases were computed using L2 regularisation with a factor of 1 [9]. The first row corresponds to the NNCosNgbr algorithm [4]. The last two rows are the WSR method, our proposal for recommendation generation (Eqs. 4 and 5, respectively).

WSR variants performed the best and they significantly surpass NNCosNgbr (first row) in every dataset. The user-based approach reported the best figures on the dense film datasets while the item-based algorithm yielded the best results on the sparse songs and books collections. However, there are only statistically significant differences between these two methods in the LibraryThing dataset. This result agrees with the literature about neighbourhoods methods [4–6]: item-based approaches tend to work well on sparse datasets because they compute similarities among items which often contain more dense information than users.

Table 2. nDCG@10 best values on MovieLens 100 k, Movielens 1 M, R3-Yahoo! and LibraryThing datasets for different neighbourhood algorithms. The first column indicates whether a user-based (UB) or item-based (IB) approach is followed; the second, the pairwise similarity; the third, which treatment the user and item biases receive. Statistically significant differences (Wilcoxon two-sided test $p < 0.01$) with respect to the first, second, third and fourth method are superscripted with a, b, c and d, respectively. Values underlined are not statistically different from the best value.

Algorithm	Type	Similarity	Bias	ML 100 k	ML 1 M	R3	LT
NNCosNgr	IB	ShrunkCosine	Remove	0.1427	0.1042	0.0138	0.0550
NNCosNgr'	IB	ShrunkCosine	Keep	0.3704^{a}	0.3334^{a}	0.0257^{a}	0.2217^{ad}
WSR	IB	Cosine	Keep	0.3867^{ab}	0.3382^{ab}	$\mathbf{0.0274}^{ab}$	$\mathbf{0.2539}^{abd}$
WSR	UB	Cosine	Keep	$\mathbf{0.3899}^{ab}$	$\mathbf{0.3430}^{ab}$	0.0261^{a}	0.1906^{a}

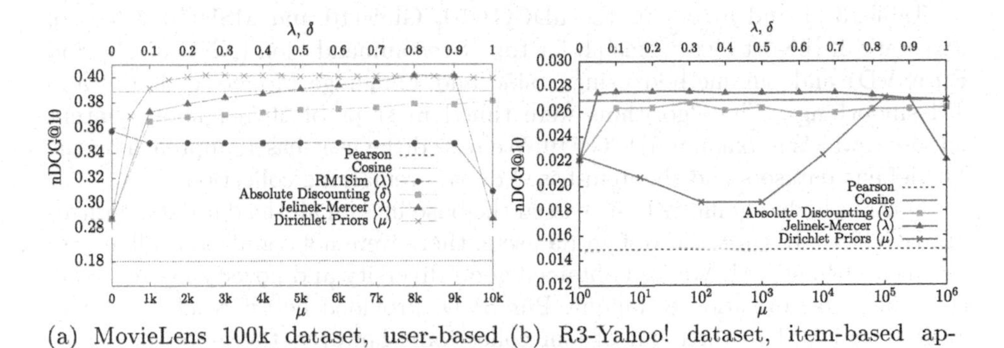

(a) MovieLens 100k dataset, user-based approach, $k = 50$ neighbours (b) R3-Yahoo! dataset, item-based approach, $k = 100$ neighbours

Fig. 1. Comparison in terms of nDCG@10 among different strategies (Pearson, cosine, RM1Sim and Language Models with AD, JM and DP) for computing neighbourhoods. Recommendations are computed using WSR algorithm.

4.3 Language Models for User Neighbourhoods

In this section, we compare our Language Modelling approach for computing neighbourhoods (see Eq. 6) against Pearson's correlation coefficient and cosine similarity. Recommendation are computed using WSR (Eqs. 4 and 5). Additionally, we implemented another baseline (RM1Sim [2]) for computing user neighbourhoods based on Relevance-Based Language Models. This method uses Jelinek-Mercer smoothing controlled by the interpolation parameter λ. Figure 1 presents the results of testing the different similarities (LM, cosine, Pearson and RM1Sim) experiments on the MovieLens 100 k and R3-Yahoo! collections using user and item-based approaches, respectively.

The experiments showed that DP and JM smoothings yielded the best results. They outperform our baselines (cosine, Pearson and RM1Sim) and also AD. The behaviour of DP varies between collections and the scale of the optimal μ is very different (4500 on MovieLens 100 k, 10000 on R3-Yahoo!). Accuracy figures of JM increased with a high amount of smoothing; however, we can observe a significant drop at $\lambda = 1$ which was expected because the estimate degenerates to the background model. On the other hand, AD is not competitive (cosine baseline is better). This also happens in Information Retrieval where DP and JM are the favourite methods [23].

In text retrieval, DP is preferred over JM, especially for short queries [23]. Nonetheless, since we are dealing with long profiles (the users' profiles contain multiple ratings), JM worked better than DP. However, in some cases, DP outperformed JM. The cause may be document length normalisation. Previous studies have found that DP applies a different amount of smoothing depending on the document length while JM smooths all documents in a length-independent manner [11,23]. This property is also very important for finding neighbourhoods because users and items have very diverse profile sizes. Thus, it would be interesting to consider this fact to leverage power users or popular items.

Tables 3, 4 and 5 present the nDCG@10, Gini@10 and MSI@10 values of two state-of-the-art algorithms [4] for top-N recommendation (NNCosNgbr and PureSVD) and our method using cosine and Language Models with DP and JM smoothings. The algorithms were tuned in steps of 50 neighbours/latent factors towards maximum nDCG@10. We tested the user-based approach on the MovieLens datasets and the item-based on the remaining collections.

Our methods significantly surpassed the baselines on all the datasets in terms of nDCG@10. In three out of four datasets, there were not significant differences between DP and JM. We also obtained good diversity and novelty figures. Still, the matrix factorisation technique, PureSVD, provided better results in terms of Gini and MSI in three out of four collections; however, the more sparse the dataset is, the less advantage PureSVD presented over our methods for these metrics. On the other hand, although NNCosNgbr also showed good results in these metrics, the accuracy figures were too low to be an effective alternative.

Finally, another benefit of our neighbourhood method is that inverted indexes enable the efficient computation of Language Models. These data structures are used in Information Retrieval to perform queries on web-scale scenarios. Thus, we can leverage search engines such as Indri or Terrier (which implement Language Models on inverted indexes) to compute neighbourhoods efficiently. In contrast, PureSVD needs to calculate a global matrix factorisation.

5 Related Work

The Language Modelling framework for collaborative filtering recommendation is recently attracting attention in the IR community. Bellogín et al. devised a model that uses any text retrieval method for generating recommendations in a user-based or item-based manner. They tested several techniques including Language Models using Dirichlet Priors and Jelinek-Mercer smoothings [3]. Although the framework is efficient and flexible, the accuracy figures are not outstanding.

Also, in the same line, literature about using Relevance-Based Language Models for collaborative filtering is growing. This model was designed for the

Table 3. nDCG@10 best values on the four datasets. Statistically significant improvements (Wilcoxon two-sided test $p < 0.01$) w.r.t. the first, second, third, fourth and fifth method are superscripted with a, b, c, d and e, respectively. Values underlined are not statistically different from the best value. The number of neighbours/latent factors used in each case is indicated in the right side.

Algorithm	ML 100 k		ML 1 M		R3-Yahoo!		LibraryThing	
NNCosNgbr	0.1427	300	0.1042	50	0.0138	50	0.0550	50
PureSVD	0.3595^{a}	50	0.3499^{ac}	50	0.0198^{a}	50	0.2245^{a}	450
Cosine-WSR	0.3899^{ab}	50	0.3430^{a}	50	$\underline{0.0274}^{ab}$	150	0.2476^{ab}	100
LM-DP-WSR	$\mathbf{0.4017}^{abc}$	50	0.3585^{abc}	100	$\underline{0.0271}^{ab}$	100	0.2464^{ab}	50
LM-JM-WSR	$\underline{0.4013}^{abc}$	50	$\mathbf{0.3622}^{abcd}$	100	$\mathbf{0.0276}^{ab}$	100	$\mathbf{0.2537}^{abcd}$	50

Table 4. Gini@10 values on the same settings as Table 3.

Algorithm	ML 100 k	ML 1 M	R3-Yahoo!	LibraryThing
NNCosNgbr	0.0910	0.0896	0.0256	0.0058
PureSVD	0.1364	0.0668	0.1335	0.0367
Cosine-WSR	0.0549	0.0400	0.0902	0.1025
LM-DP-WSR	0.0659	0.0435	0.1557	0.1356
LM-JM-WSR	0.0627	0.0435	0.1034	0.1245

Table 5. MSI@10 values on the same settings as Table 3.

Algorithm	ML 100 k	ML 1 M	R3-Yahoo!	LibraryThing
NNCosNgbr	18.4113	19.5975	43.4348	56.5973
PureSVD	14.2997	14.8416	30.9107	37.9681
Cosine-WSR	11.0579	12.4816	21.1968	41.1462
LM-DP-WSR	11.5219	12.8040	25.9647	46.4197
LM-JM-WSR	11.3921	12.8417	21.7935	43.5986

pseudo-relevance feedback task but it has been adapted for finding user neighbourhoods [2] and computing recommendations directly [13,18,19]. As a neighbourhood technique, our experiments showed that its accuracy is worse than our proposal in addition to being more computationally expensive. Regarding their use as a recommender algorithm, Relevance-Based Language Models have proved to be a very effective recommendation technique [13]. Since this method is based on user neighbourhoods, it would be interesting to combine it with our Language Modelling proposal. We leave this possibility as future work.

6 Conclusions and Future Work

In this paper, we presented a novel approach to find user or item neighbourhoods based on the LM framework. This method, combined with an adaptation of a neighbourhood-based recommender (WSR algorithm), yields highly accurate recommendations which surpass the ones from two state-of-the-art top-N recommenders (PureSVD and NNCosNgr) in term of nDCG with good values of diversity and novelty. This method is also very efficient and scalable. On the one hand, we can take advantage from inverted indexes for neighbours computation—these structures were designed for dealing with web-scale datasets in IR. On the other hand, WSR is simpler than NNCosNgbr and PureSVD without requiring a previous training step for computing biases or the latent factor representation.

Our experiments revealed that Jelinek-Mercer is the best choice for smoothing the estimate of Language Models for neighbourhoods although Dirichlet Priors is also a great choice. This result is analogous to the Information Retrieval task where JM works better than DP for long queries. To overcome the problem that JM does not vary the amount of smoothing applied depending on the document length (in contrast to DP), Losada et al. proposed the use of a length-based document prior [11]. This prior is equivalent to Linear Prior proposed for the Relevance-Based Language Modelling of collaborative filtering recommendations [18]. Testing the applicability of this prior combined with JM smoothing would be an interesting avenue for further work.

We also envision to expand this study to other language modelling approaches. The method proposed in this paper is based on multinomial distributions. A future study that explores the applicability of different probability distributions (such as the multivariate Bernoulli [12]) may be worthwhile.

Acknowledgments. This work was supported by the *Ministerio de Economía y Competitividad* of the Goverment of Spain under grants TIN2012-33867 and TIN2015-64282-R. The first author also wants to acknowledge the support of *Ministerio de Educación, Cultura y Deporte* of the Government of Spain under the grant FPU014/01724.

References

1. Bellogín, A., Castells, P., Cantador, I.: Precision-oriented evaluation of recommender systems. In: RecSys 2011, p. 333. ACM (2011)
2. Bellogín, A., Parapar, J., Castells, P.: Probabilistic collaborative filtering with negative cross entropy. In: RecSys 2013, pp. 387–390. ACM (2013)
3. Bellogín, A., Wang, J., Castells, P.: Bridging memory-based collaborative filtering and text retrieval. Inf. Retr. **16**(6), 697–724 (2013)
4. Cremonesi, P., Koren, Y., Turrin, R.: Performance of recommender algorithms on top-N recommendation tasks. In: RecSys 2010, pp. 39–46. ACM (2010)
5. Deshpande, M., Karypis, G.: Item-based top-N recommendation algorithms. ACM Trans. Inf. Syst. **22**(1), 143–177 (2004)

6. Desrosiers, C., Karypis, G.: A comprehensive survey of neighborhood-based recommendation methods. In: Ricci, F., Rokach, L., Shapira, B., Kantor, P.B. (eds.) Recommender Systems Handbook, pp. 107–144. Springer, Heidelberg (2011)
7. Fleder, D., Hosanagar, K.: Blockbuster culture's next rise or fall: the impact of recommender systems on sales diversity. Manage. Sci. **55**(5), 697–712 (2009)
8. Herlocker, J.L., Konstan, J.A., Terveen, L.G., John, T.: Riedl.: evaluating collaborative filtering recommender systems. ACM Trans. Inf. Syst. **22**(1), 5–53 (2004)
9. Koren, Y.: Factorization meets the neighborhood: a multifaceted collaborative filtering model. In: KDD 2008, pp. 426–434. ACM (2008)
10. Koren, Y., Bell, R.: Advances in collaborative filtering. In: Ricci, F., Rokach, L., Shapira, B., Kantor, P.B. (eds.) Recommender Systems Handbook, pp. 145–186. Springer, Heidelberg (2011)
11. Losada, D.E., Azzopardi, L.: An analysis on document length retrieval trends in language modeling smoothing. Inf. Retr. **11**(2), 109–138 (2008)
12. Losada, D.E., Azzopardi, L.: Assessing multivariate bernoulli models for information retrieval. ACM Trans. Inf. Syst. **26**(3), 17:1–17:46 (2008)
13. Parapar, J., Bellogín, A., Castells, P., Barreiro, Á.: Relevance-based language modelling for recommender systems. Inf. Process. Manage. **49**(4), 966–980 (2013)
14. Ponte, J.M., Bruce Croft, W.: A language modeling approach to information retrieval. In: SIGIR 1998, pp. 275–281. ACM (1998)
15. Ricci, F., Rokach, L., Shapira, B., Kantor, P.B.: Recommender Systems Handbook. Springer, Heidelberg (2011)
16. Salton, G., Wong, A., Yang, C.S.: A vector space model for automatic indexing. Commun. ACM **18**(11), 613–620 (1975)
17. Valcarce, D.: Exploring statistical language models for recommender systems. In: RecSys 2015, pp. 375–378. ACM (2015)
18. Valcarce, D., Parapar, J., Barreiro, Á.: A study of priors for relevance-based language modelling of recommender systems. In: RecSys 2015, pp. 237–240. ACM (2015)
19. Valcarce, D., Parapar, J., Barreiro, A.: A study of smoothing methods for relevance-based language modelling of recommender systems. In: Hanbury, A., Kazai, G., Rauber, A., Fuhr, N. (eds.) ECIR 2015. LNCS, vol. 9022, pp. 346–351. Springer, Heidelberg (2015)
20. Wang, J., de Vries, A.P., Reinders, M.J.T.: A user-item relevance model for log-based collaborative filtering. In: Lalmas, M., MacFarlane, A., Rüger, S.M., Tombros, A., Tsikrika, T., Yavlinsky, A. (eds.) ECIR 2006. LNCS, vol. 3936, pp. 37–48. Springer, Heidelberg (2006)
21. Wang, Y., Wang, L., Li, Y., He, D., Chen, W., Liu, T.-Y.: A theoretical analysis of NDCG ranking measures. In: COLT 2013, pp. 1–30 (2013). JMLR.org
22. Zhai, C.: Statistical Language Models for Information Retrieval. Synthesis Lectures on Human Language Technologies. Morgan & Claypool, San Rafael (2009)
23. Zhai, C., Lafferty, J.: A study of smoothing methods for language models applied to information retrieval. ACM Trans. Inf. Syst. **22**(2), 179–214 (2004)
24. Zhou, T., Kuscsik, Z., Liu, J.-G., Medo, M., Wakeling, J.R., Zhang, Y.-C.: Solving the apparent diversity-accuracy dilemma of recommender systems. PNAS **107**(10), 4511–4515 (2010)

Adaptive Collaborative Filtering with Extended Kalman Filters and Multi-armed Bandits

Jean-Michel Renders(✉)

Xerox Research Center Europe, Meylan, France
jean-michel.renders@xrce.xerox.com

Abstract. It is now widely recognized that, as real-world recommender systems are often facing drifts in users' preferences and shifts in items' perception, collaborative filtering methods have to cope with these time-varying effects. Furthermore, they have to constantly control the trade-off between exploration and exploitation, whether in a cold start situation or during a change - possibly abrupt - in the user needs and item popularity. In this paper, we propose a new adaptive collaborative filtering method, coupling Matrix Completion, extended non-linear Kalman filters and Multi-Armed Bandits. The main goal of this method is exactly to tackle simultaneously both issues – adaptivity and exploitation/exploration trade-off – in a single consistent framework, while keeping the underlying algorithms efficient and easily scalable. Several experiments on real-world datasets show that these adaptation mechanisms significantly improve the quality of recommendations compared to other standard on-line adaptive algorithms and offer "fast" learning curves in identifying the user/item profiles, even when they evolve over time.

1 Introduction

Real-world recommender systems have to capture the dynamic aspects of user and item characteristics: user preferences and needs obviously change over time, depending on her life cycle, on particular events and on social influences; similarly, item perception could evolve in time, due to a natural slow decrease in popularity or a sudden gain in interest after winning some award or getting positive appreciations of highly influential experts. Clearly, adopting a static approach, for instance through "static matrix completion" – as commonly designed and evaluated on a random split without considering the real temporal structure of the data – will fail to provide accurate results in the medium and long run. Intuitively, the system should give more weight to recent observations and should constantly update user and item "profiles" or latent factors in order to offer the adaptivity and flexibility that are required.

At the same time, these recommender systems typically have to face the "cold-start" problem: new users and new items constantly arrive in the system, without any historical information. In this paper, we assume no other external source of information than the time-stamped ratings, so that it is not possible

N. Ferro et al. (Eds.): ECIR 2016, LNCS 9626, pp. 626–638, 2016.
DOI: 10.1007/978-3-319-30671-1_46

to use external user or item features (including social and similarity relationships) to partly solve the cold-start issue. In the absence of such information, it is impossible to provide accurate recommendation at the very early stage and tackling the cold start problem consists then in solving the task (accurate recommendation), while trying to simultaneously uncover the user and item profiles. The problem amounts to optimally control the trade-off between exploration and exploitation, to which the Multi-Armed Bandit (MAB) setting constitutes an elegant approach.

This paper proposes to tackle both the adaptivity and the "cold start" challenges through the same framework, namely Extended Kalman Filters coupled with contextual Multi-Armed Bandits. One extra motivation of this work is focused on scalability and tractability, leading us to design rather simple and efficient methods, precluding us from implementing some complex inference methods derived from fully Bayesian approaches. The starting point of the framework is the standard Matrix Completion approach to Collaborative Filtering and the aim of the framework is to extend it to the adaptive, dynamic case, while controlling the exploitation/exploration trade-off (especially in the "cold start" situations). Extended Kalman Filters constitute an ideal framework for modelling smooth non-linear, dynamic systems with time-varying latent factors (called "states" in this case). They maintain, in particular, a covariance matrix over the state estimates or, equivalently, a posterior distribution over the user/item biases and latent factors, which will then be fully exploited by the MAB mechanism to guide its sampling strategy. Two different families of MAB are investigated: Thompson sampling, based on the probability matching principle, and UCB-like (Upper Confidence Bound) sampling, based on the principle of optimism in face of uncertainty.

In a nutshell, the method's principle is that, when a user u calls the system at time t to have some recommendations, an arm (i.e. an item i) is chosen that will simultaneously satisfy the user and improve the quality estimate of the parameters related to both the user u and the proposed item. The system then receives a new feedback ($< u, i, r, t >$ tuple) and updates the corresponding entries of the latent factor matrices as well as the posterior covariance matrices over the factor estimates. We will show that the problem could be solved by a simple algorithm, requiring only basic algebraic computations, without matrix inversion or singular value decomposition. The algorithm could easily update the parameters of the model and make recommendations, even with an arrival rate of several thousands ratings per second.

2 Related Work

One of the first works to stress the importance of temporal effects in Recommender Systems and to cope with it was the *timeSVD++* algorithm [8]. The approach consists in explicitly modelling the temporal patterns on historical rating data, in order to remove the "temporal drift" biases. This means that the time dependencies are modelled parametrically as time-series, typically in the form of linear trends, with a lot of parameters to be identified. As such, it could hardly extrapolate rating behaviour in the future, as it involves the discretisation of the timestamps into a finite set of "bins" and the identification of bin-specific parameters; in other words, it is impossible to predict ratings for future, unobserved bins.

Interestingly, tensor factorization approaches have also been adopted to model the temporal effects of the dynamic rating behaviour [17]: user, item and time constitute the 3 dimensions of the tensors. Variants of this general framework were also studied in [16,18], by introducing second-order interaction terms and a different definition of the time scale (user- or item-specific time scales, by considering the time interval since the user or item first entered into the system). Tensor factorization is useful for analysing the temporal evolution of user and item-related factors, but it suffers the same main drawback as *timeSVD++*: it is unable to extrapolate rating behaviour in the future.

Two earlier works [11,12] also proposed to incrementally update the item- or user-related factors corresponding to a new observation by performing a stochastic gradient step of a quadratic loss function, but allowing only one factor to be updated; the updating decision is taken based on the current number of observations associated to a user or to an item (for instance, a user with a high number of ratings will no longer be updated). The same approach was extended to a non-negative matrix completion setting in [7], assuming that the item-related factors are constant over time.

Kalman Filters for Collaborative Filtering were first introduced in [10], and then exploited further in [1,13]: this family of works rely on a Bayesian framework and on probabilistic matrix factorization, where a state-space model is introduced to model the temporal dynamics. The work in [14,15] enhances the approach by introducing an *EM*-like method based on Kalman smoothers – the non-causal extension of Kalman filters – to estimate the value of the hyperparameters. An important variant, proposed by [6], extends the approach by modelling the continuous-time evolution of the latent factors through Brownian motion. This constitutes a key concept that we use in the method we are proposing here. One of main advantages provided by this family of methods is that they could easily be extended to include additional user- or item-related features, addressing in this way the cold-start problem. But, in order to remain computationally tractable and to avoid to tackle non-linearities, they update only either the user factors, or the items factors, but never both factors simultaneously. Indeed, this amounts to consider only linear state-space models, for which standard (linear) Kalman Filters provide an efficient and adequate solution.

Recently, [5] proposed an incremental matrix completion method, that automatically allows the latent factors related to both users and items to adapt "on-line" based on a temporal regularization criterion, ensuring smoothness and consistency over time, while leading to very efficient computations. Even if this work addresses the issue of non-linearities induced by the coupled evolution of both item and user latent factors, it considers neither the cold-start problem, nor the different dynamics caused by "unfrequent" users or items (larger time intervals between successive observations should ideally allow for larger updates, while this approach does not capture this kind of implicit volatility).

On the side of Multi-Armed Bandits for item recommendation, the most representative works are [2,4,9]. They use linear contextual bandits, where a context is typically a user calling the system at time t and her associated feature vector; the reward (*ie.* the rating) is assumed to be a linear function of this feature vector and the coefficients of this linear function could be interpreted as the arm (or item) latent factors that are incrementally updated. Alternatively, they also consider binary ratings, with a logistic regression model for each arm (or item) and then use Thompson Sampling or UCB sampling to select the best item following an exploration/exploitation trade-off perspective. More recently, [19] combines Probabilistic Matrix Factorization and linear contextual bandits. Unfortunately, all these approaches do not offer any adaptive behaviour: the features associated to a user are assumed to be constant and known accurately; we could easily consider the dual problem, namely identifying the user as an arm and the item as the context (as in [19]), but then we have no adaptation to possible changes and drifts in the item latent factors and biases.

In a nutshell, to the best of our knowledge, there is no approach that simultaneously combines the dynamic tracking of both user and item latent factors with an adequate control of the exploration/exploitation trade-off in an on-line learning setting. The method that is proposed here is a first step to fill this gap.

3 Adaptive Matrix Completion

In this paper, we assume that the training data only consist in a sequence of tuples <user, item, rating, time-stamp> and we adopt the standard matrix factorization approach to Collaborative Filtering. In this setting, each observed rating is modelled as:

$$r_{u,i} = \mu + a_u + b_i + L_u.R_i^T + \epsilon$$

where a_u, b_i, L_u and R_i are respectively the user bias, the item popularity, the user latent factors and the item latent factors (μ is a constant which could be interpreted as the global average rating). Both L_u and R_i are row vectors with K components, K being the dimension of the latent space. The noise ϵ is assumed to be i.i.d. Gaussian noise, with mean equal to 0 and variance equal to σ^2. The matrix completion problem typically involves the minimization of the following loss function, combining the reconstruction error over the training set and regularization terms:

$$\mathcal{L}(a,b,L,R) = \sum_{(u,i)\in\Omega} \|r_{ui} - \mu - a_u - b_i - L_u.R_i^T\|^2 + \lambda_a\|a\|^2 + \lambda_b\|b\|^2 + \lambda_L\|L\|_F^2 + \lambda_R\|R\|_F^2$$

where Ω is the training set of observed tuples. Note that this loss could be interpreted in a Bayesian setting as the MAP estimate, provided that all parameters (a_u, b_i, L_u and R_i) have independent Gaussian priors, with diagonal covariance matrices. In this case, $\lambda_L = \frac{\sigma^2}{\sigma_L^2}$ where σ_L^2 is the variance of the diagonal Gaussian prior on L_u; the interpretation is similar for λ_a, λ_b and λ_R. The minimization of the loss function is typically solved by gradient descend (possibly accelerated by the L-BFGS quasi-Newton method) or by Alternating Least Squares (for L_u and R_i).

Departing from the static setting, we will now assume that the model parameters evolve over time, with their own dynamics. One standard approach to reconstruct the evolution of these parameters (considered as latent variables) from the sequence of observations relies on the use of Kalman Filters. Extended Kalman Filters are generalisation of recursive least squares for dynamic systems with "smooth" non-linearities and aim at estimating the current (hidden) state of the system:

$$\begin{aligned} x_t &= f(x_{t-1}) + w_t \\ y_t &= h(x_t) + z_t \\ x_0 &\sim N(x_0^*, \Lambda) \\ w_t &\sim N(0, Q_t) \\ z_t &\sim N(0, \sigma_t^2) \end{aligned}$$

where $N(m, C)$ denotes a multi-variate Gaussian distribution with mean m and covariance matrix C, x ($\in \mathbb{R}^K$) is the latent state of the system, y ($\in \mathbb{R}^N$) is the observable output, w and z are Gaussian white noises, while x_0 is the initial state of the system. Kalman Filters follow a general "predictor - corrector" scheme, which maintains a filter gain matrix K_t (applied to the prediction error at time t) and a covariance matrix $P_{t|t}$ of the posterior distribution of the state estimates:

Predictor Step:

$$\begin{aligned} \hat{x}_{t|t-1} &= f(\hat{x}_{t-1|t-1}) \\ P_{t|t-1} &= J_f(\hat{x}_{t|t-1}) P_{t-1|t-1} J_f(\hat{x}_{t|t-1})^T + Q_t \end{aligned}$$

Corrector Step:

$$\begin{aligned} K_t &= P_{t|t-1} J_h(\hat{x}_{t|t-1})^T (J_h(\hat{x}_{t|t-1}) P_{t|t-1} J_h(\hat{x}_{t|t-1})^T + \sigma_t^2)^{-1} \\ \hat{x}_{t|t} &= \hat{x}_{t|t-1} + K_t.[y_t - h(\hat{x}_{t|t-1})] \\ P_{t|t} &= [I - K_t J_h(\hat{x}_{t|t-1})] P_{t|t-1} \end{aligned}$$

where J_f and J_h are the Jacobian matrices of the functions f and h respectively ($J_t \doteq \frac{\partial f}{\partial x_t}$ and $J_h \doteq \frac{\partial h}{\partial x_t}$). Basically, $\hat{x}_{t|t-1}$ is the prediction of the state at time t, given all observations up to $t-1$, while $\hat{x}_{t|t}$ is the prediction that also includes the observation of the outputs at time t (y_t). In practice, we use the Iterated Extended Kalman Filter (IEKF), where the first two equations of the Corrector step are iterated till $\hat{x}_{t|t}^{(i)}$ is stabilised, gradually offering a better approximation of the non-linearity through the Jacobian matrices.

In order to apply these filters, let us now express our adaptive collaborative filtering as a continuous-time dynamic system with the following equations, assuming that we observe the tuple $< u, i, r_{u,i} >$ at time t:

$$\begin{aligned}
a_{u,t} &= a_{u,t-1} + w_{a,u}(t-1,t)\\
b_{i,t} &= b_{u,t-1} + w_{b,i}(t-1,t)\\
L_{u,t} &= L_{u,t-1} + W_{L,u}(t-1,t)\\
R_{i,t} &= R_{i,t-1} + W_{R,i}(t-1,t)\\
r_{u,i,t} &= \mu + a_{u,t} + b_{i,t} + L_{u,t}.R_{i,t}^T + \epsilon_t
\end{aligned}$$

with $a_{u,0} \sim N(0, \lambda_a)$, $b_{i,0} \sim N(0, \lambda_b)$, $L_{u,0} \sim N(0, \Lambda_L)$, $R_{i,0} \sim N(0, \Lambda_R)$ and $\epsilon_t \sim N(0, \sigma^2)$. Here $a_{u,t-1}$ denotes the value of the bias of user u when she appeared in the system for the last time before time t. Similarly, $b_{i,t-1}$ denotes the value of the popularity of item i when it appeared in the system for the last time before time t. So, the short-cut notation $(t-1)$ is contextual to an item or to a user. All the parameters a_u, b_i, L_u and R_i follow some kind of Brownian motion (the continuous counter-part of discrete random walk) with Gaussian transition process noises $w_{a,u}$, $w_{b,i}$, $W_{L,u}$ and $W_{R,i}$ respectively, whose variances are proportional to the time interval since a user / an item appeared in the system for the last time before the current time, denoted respectively as $\Delta_u(t-1,t)$ and $\Delta_i(t-1,t)$: $w_{a,u}(t-1,t) \sim N(0, \Delta_u(t-1,t).\gamma_a)$, $w_{b,i}(t-1,t) \sim N(0, \Delta_i(t-1,t).\gamma_b)$, $W_{L,u}(t-1,t) \sim N(0, \Delta_u(t-1,t).\Gamma_L)$ and $W_{R,i}(t-1,t) \sim N(0, \Delta_i(t-1,t).\Gamma_R)$. We call γ_a, γ_b, Γ_L and Γ_R the volatility hyper-parameters. It is assumed that the hyper-parameters λ_a, γ_a and the diagonal covariance matrices Λ_L, Γ_L are identical for all users, and independent from each other. The same is assumed for the hyper-parameters related to items.

With these assumptions, the application of the Iterated Extended Kalman filter equations gives:

Predictor Step:

$$\begin{aligned}
P^{a_u}_{t|t-1} &= P^{a_u}_{t-1|t-1} + \Delta_u(t,t-1)\gamma_a\\
P^{b_i}_{t|t-1} &= P^{b_i}_{t-1|t-1} + \Delta_i(t,t-1)\gamma_b\\
P^{L_u}_{t|t-1} &= P^{L_u}_{t-1|t-1} + \Delta_u(t,t-1)\Gamma_L\\
P^{R_i}_{t|t-1} &= P^{R_i}_{t-1|t-1} + \Delta_i(t,t-1)\Gamma_R
\end{aligned}$$

Corrector Step: Initialize $R_{i,t} \leftarrow R_{i,t-1}$ and $L_{u,t} \leftarrow L_{u,t-1}$
Iterate till convergence:

$$\omega = (\sigma^2 + P^{a_u}_{t|t-1} + P^{b_i}_{t|t-1} + R_{i,t}P^{L_u}_{t|t-1}R^T_{i,t} + L_{u,t}P^{R_i}_{t|t-1}L^T_{u,t})^{-1}$$

$$\begin{aligned}
K^{L_u}_t &= \omega P^{L_u}_{t|t-1}R^T_{i,t}\\
K^{R_i}_t &= \omega P^{R_i}_{t|t-1}L^T_{u,t}\\
L_{u,t} &= L_{u,t-1} + K^{L_u}_t(r_{u,i,t} - \mu - a_{u,t-1} - b_{i,t-1} - R_{i,t}L^T_{u,t-1})\\
R_{i,t} &= R_{i,t-1} + K^{R_i}_t(r_{u,i,t} - \mu - a_{u,t-1} - b_{i,t-1} - R_{i,t-1}L^T_{u,t})
\end{aligned}$$

Then:

$$\begin{aligned}
K_t^{a_u} &= \omega P_{t|t-1}^{a_u} \\
K_t^{b_i} &= \omega P_{t|t-1}^{b_i} \\
a_{u,t} &= a_{u,t-1} + K_t^{a_u}(r_{u,i,t} - \mu - a_{u,t-1} - b_{i,t-1} - R_{i,t-1}L_{u,t-1}^T) \\
b_{i,t} &= b_{i,t-1} + K_t^{b_j}(r_{u,i,t} - \mu - a_{u,t-1} - b_{i,t-1} - R_{i,t-1}L_{u,t-1}^T) \\
P_{t|t}^{a_u} &= P_{t|t-1}^{a_u}(1 - K_t^{a_u}) \\
P_{t|t}^{b_i} &= P_{t|t-1}^{b_i}(1 - K_t^{b_i}) \\
P_{t|t}^{L_u} &= (I - K_t^{L_u} R_{i,t}) P_{t|t-1}^{L_u} \\
P_{t|t}^{R_i} &= (I - K_t^{R_i} L_{u,t}) P_{t|t-1}^{R_i}
\end{aligned}$$

with $P_{0|0}^{a_u} = \lambda_a$, $P_{0,0}^{L_u} = \Lambda_L \quad \forall u$ and $P_{0|0}^{b_i} = \lambda_b$, $P_{0|0}^{R_i} = \Lambda_R \quad \forall i$. Note that ω, $K_t^{a_u}$, $K_t^{b_i}$, $P_{t|.}^{a_u}$ and $P_{t|.}^{b_i}$ are scalars.

In practice, at least with the datasets that were used in our experiments, the iterative part of the Corrector step converges in very few iterations (typically 2 or 3). It should be noted that, if a user is not well known (hight covariance P^{a_u} and P^{L_u} due to a low number of ratings or a long time since her last appearance), her weight – and so her influence – in adapting the item i is decreased, and vice-versa. Our independence and Gaussian assumptions make it simple to compute the posterior distribution of the rating of a new pair $< u, i >$ at time t: it is a Gaussian with mean $\mu + a_{u,t} + b_{i,t} + L_{u,t}R_{i,t}^T$ and variance $\sigma^2 + P_{t|t}^{a_u} + P_{t,t}^{b_i} + R_{i,t}P_{t|t}^{L_u}R_{i,t}^T + L_{u,t}P_{t|t}^{R_i}L_{u,t}^T$. Note also that one can easily extend the IEKF method to introduce any smooth non-linear link function (e.g. $r_{u,i,t} = g(\mu + a_{u,t} + b_{i,t} + L_{u,t}R_{i,t}^T)$ with $g(x)$ a sigmoid between the minimum and maximum rating values). The hyper-parameters could be learned from the training data through a procedure similar to the *EM* algorithm using Extended Kalman smoothers (a forward-backward version of the Extended Kalman Filters) as described in [14], or by tuning them on a development set, whose time interval is later than the training set.

4 Cold Start and the Exploration - Exploitation Trade-Off

Let us denote by θ the set of all parameters (biases a_u, b_i and latent factors L_u, R_i, $\forall u, i$). If the true parameters θ^* were known, for a given context (user u at time t), the system should recommend an item i^* such that $i^* = \text{argmax}_i \quad E(r|u, i, \theta^*)$ with $P(r|u, i, \theta^*) \sim N(\mu + L_u^* R_i^{T*} + a_u^* + b_i^*, \sigma^2)$. If θ^* is not known, we should marginalize over all possible θ through the use of the posterior $p(\theta|D)$ with $D=$ training data. This amounts to choose $i^* = \text{argmax}_i \quad \mu + L_u R_i^T + a_u + b_i$ if a Maximum Posterior solution (MAP) is adopted. But this is a "one-shot" approach, considered as pure exploitation. As our setting is multi-shot, we should balance exploitation and exploration, as materialized by the concept of "regret" (difference of expected rewards or ratings between a strategy that knows the true θ^* and the one based on a current estimate θ^t).

We will consider two different sampling strategies to control this trade-off: *Thompson sampling*, based on the "probability matching" principle, and *UCB* (Upper Confidence Bounds) sampling, based on the principle of optimism in the face of uncertainty (see for instance [2] for a good introduction to these strategies in the context of recommendation). Note that we are in a "contextual bandit" setting: at each time step t, we observe a context given by a single user u, characterised by an imperfect estimate of her bias a_u and latent factors L_u (some kind of noisy context) and the system should then recommend an arm (i.e. an item) such that, intuitively, the choice of this arm will simultaneously satisfy the user and improve the quality estimate of the parameters related to both the user u and the proposed item.

4.1 Thompson Sampling

The *Thompson Sampling* strategy could be expressed by the following algorithm:

D = some past data (possibly empty)
for $t = 1{:}\ T$ **do**
  Receive u_t;
  Draw $\tilde{\theta}_t = (\tilde{a}_{u,t}, \tilde{b}_{i,t}, \tilde{L}_{u,t}, \tilde{R}_{i,t})$ from $p(\theta|D)$ (multi-variate normal distribution with mean and covariance computed by IEKF)
  Select item i^*: $\text{argmax}_i \quad E(r|u, i, \theta_t) = \text{argmax}_i \quad \mu + \tilde{L}_{u,t}\tilde{R}_{i,t}^T + \tilde{a}_{u,t} + \tilde{b}_{i,t}$
  Observe rating r_t (for pair $< u_t, i^* >$), update D, update the parameters and the variances/covariances through IEKF.
end for

In practice, the "Optimistic Thompson sampling" variant is used [2] and the variance/covariance values/matrices are pre-multiplied by 0.5 to favour exploitation.

4.2 UCB-like Sampling

The *Upper-Confidence-Bounds* (UCB) algorithm is sketched in the following pseudo-code:

D = some past data (possibly empty)
for $t = 1{:}\ T$ **do**
  Receive u_t
  Select item i^*:

$$\text{argmax}_i \quad \mu + L_{u,t}R_{i,t}^T + a_{u,t} + b_{i,t} + \alpha\sqrt{\sigma^2 + P_{t|t}^{a_u} + P_{t,t}^{b_i} + R_{i,t}P_{t|t}^{L_u}R_{i,t}^T + L_{u,t}P_{t|t}^{R_i}L_{u,t}^T}$$

  Observe rating r_t (for pair $< u_t, i^* >$), update D, update the parameters and variances/covariances through IEKF
end for

The α parameter controls the trade-off between exploration and exploitation (in practice, $\alpha = 2$ is often used).

5 Experimental Results

Experiments have been performed on 2 datasets: MovieLens (10M ratings) and Vodkaster[1] (2.7M ratings), divided into 3 temporal, chronologically ordered, splits: Train (90 %), Development (5 %), Test (5 %). Note that we have beforehand removed the ratings corresponding to the early, non-representative (transient) time period from both datasets (2.5M ratings for MovieLens, 0.3M ratings for Vodkaster). The Development set is used to tune the different hyper-parameters of the algorithm. These two datasets show very different characteristics, as illustrated in Table 1, especially in the arrival rate of new users. We divided the experiments into two parts: one assessing separately the adaptive capacities of our method, the other evaluating the gain of coupling these adaptive capacities with Multi-Armed Bandits (*ie.* the full story).

Table 1. Dataset statistics

	MovieLens	Vodkaster
Number of ratings	7,501,601	2,428,163
Median number of ratings /user	92	11
Median number of ratings /item	121	7
% of users with at least 100 ratings	47	24
Total time span (years)	9.02	3.7
Duration Dev Set (months)	7	3
Duration Test Set (months)	7	3
% of new users in Dev Set	72.6	35.6
% of new items in Dev Set	4.2	4.3
% of new users in Test Set	79.4	35.4
% of new items in Test Set	3.2	2.2

5.1 Extended Kalman Filters for Adaptive Matrix Completion

The experimental protocol is the following: we run the Extended Kalman Filters from the beginning of the training dataset, initialising the item and user biases to 0 and the latent factors to small random values drawn from the Gaussian prior distributions with covariance matrices Λ_L and Λ_R for all users and all items respectively. The number of latent factors K is set to 20, without any tuning. Each user is associated to her own time origin (t=0 when the user enters in the system for the first time), and similarly for the items. To tune the values of the hyper-parameters (the 4 variances of the priors and the 4 volatility

[1] Vodkaster (http://www.vodkaster.com) is a French movie recommendation website, dedicated to rather movie-educated people.

values[2]), we choose the ones that optimize the Root-Mean-Squared-Error (RMSE) on the Development set. We then evaluate on the Test set the RMSE and the Mean Absolute Error (MAE) of the predictions, as well as the average Kendall's tau coefficient (for users with at least two ratings in the Test set); the latter focuses on the ranking quality of the predictions. The alternative methods that we consider are: (1) the static setting, where matrix factorization is derived from the ratings of the Training and Development sets (hyper-parameters tuned on the Development set) and the extracted models are then applied to the Test set; (2) Stochastic Gradient Descent applied to the biases and latent factors, with constant learning rates (4 different learning rates: one for a_u, one for b_i, one for L_u and one for R_i); (3) the on-line Passive-Aggressive algorithm to incrementally update the biases and latent factors as described in [3]; (4) Linear Kalman Filters applied to update only the user biases and latent factors (which is considered as the state-of-the-art). Finally, we check the statistical significance of the difference in performance of our proposed method with respect to the alternative approaches, through paired t-tests on the paired sequences of measures (squared residuals for RMSE, absolute residuals for MAE, and Kendall's tau for each user); numbers in bold indicate that the p-value of the corresponding test is smaller than 1 % (hypothesis H0: population with equal mean). The results show that the proposed method significantly improves the performances according to all RMSE, MAE and Kendall's tau metrics (Table 2). Trends are very similar for both MovieLens and Vodkaster datasets, despite their different characteristics. Remember that one extra advantage of our method is to maintain "for free" a posterior distribution over the parameters and the prediction itself, which is a key constituent for the sampling strategies of the MAB mechanism.

Table 2. Adaptive collaborative filtering performance

		RMSE	MAE	Kendall's Tau
MovieLens	(1) Static Setting	0.8903	0.6712	0.3261
	(2) SGD	0.8124	0.6115	0.3544
	(3) On-line Passive Aggressive	0.8035	0.6048	0.3589
	(4) Linear Kalman Filters (user only)	0.7884	0.5959	0.3662
	(5) Extended (non-linear) Kalman Filters	**0.7669**	**0.5724**	**0.3881**
Vodkaster	(1) Static Setting	0.8662	0.6598	0.4197
	(2) SGD	0.7874	0.5995	0.4481
	(3) On-line Passive Aggressive	0.7801	0.5929	0.4506
	(4) Linear Kalman Filters (user only)	0.7609	0.5827	0.4542
	(5) Extended (non-linear) Kalman Filters	**0.7465**	**0.5651**	**0.4636**

[2] It is easy to show that we can divide all values of the hyper-parameters by σ^2 without changing the predicted value; so we can fix σ^2 to 1.

5.2 Extended Kalman Filters Coupled with MAB

This second set of experiments is performed on the MovieLens(10M) dataset, because it spans a large time period and it is possible to follow the effects of both Multi-Armed Bandits and the adaptation mechanism in the medium and long runs. The experimental protocol is here a bit more subtle, especially because the evaluation of MAB strategies is always a tough problem. Basically, we assume that users enter in the system exactly as the initial datasets (so we keep the t and u values from the original sequence of tuples $< u, i, r, t >$), but we allow the system to propose another item than the one that was chosen in the original sequence. We also assume that all items are available from the beginning (this is of course a crude approximation of the reality). Each time the system proposes an item, it receives a "reward" or relevance feedback, which is 1 if the item was rated at least 4 and 0 otherwise. To be able to determine a reward value, during the item selection process we exclude the items that the user never rated. We compare different selection strategies: (1) the "pure exploitation" that greedily chooses the item not yet seen by the user that has the maximum predicted value, as given by the Extended Kalman Filters; (2) the UCB-sampling strategy (with α=2); (3) the Thompson sampling strategy (optimistic variant; pre-multiplying the variances/covariances by 0.5). Note that this protocol corresponds to an on-line learning setting, with no training data: the system learns from scratch; this means that methods that assume to know beforehand the user or the item latent factors from training data (such as in [19]) are not directly applicable. The metrics we use are the average precision – or equivalently the average reward – and the average recall after the system has presented n items to a user (n= 10, 50 and 100); the average is computed over all users who have rated at least 100 items (in order to be able to compute the precision and recall after the presentation of 100 items). We use the Extended Kalman Filters derived from the first set of experiments (Adaptive Matrix Completion), applied from the beginning of the dataset. Results are given in Table 3.

Paired t-tests indicate that both UCB and Thompson sampling strategies significantly outperform the greedy one at n=50 and n=100. Thompson sampling gives slightly better performance than UCB at n=100, but at the limit of the significance (p-value=0.051). For small values of n, the variances of the performances are high and there is no statistically significant differences between all the methods considered here, even if the greedy approach seems slightly less risky in the short run.

Table 3. Evolution of precision and recall (Learning Curves) on MovieLens(10M).

Strategies	Precision@n			Recall@n		
	P@10	P@50	P@100	R@10	R@50	R@100
Greedy	0.464	0.363	0.281	0.034	0.137	0.259
UCB sampling	0.457	0.367	**0.295**	0.030	**0.142**	**0.286**
Thompson sampling	0.459	**0.369**	**0.302**	0.031	**0.145**	**0.291**

6 Conclusion

We have proposed in this paper a single framework that combines the adaptive tracking of both user and item latent factors through Extended Non-linear Kalman filters and the exploration/exploitation trade-off required by the on-line learning setting (including cold-start) through Multi-Armed Bandits strategies. Experimental results showed that, at least for the datasets and settings that we considered, this framework constitutes an interesting alternative to more common approaches.

Of course, this is only a first step towards a more thorough analysis of the best models to capture the underlying dynamics of user and item evolution in real recommender systems. The use of more powerful non-linear state tracking techniques such as Particle Filters should be investigated, especially to overcome the limitations of the underlying Gaussian and independence assumptions. One promising avenue of research is to allow the volatility and prior variance hyperparameters to be user-specific (or item-specific) and to be themselves time-dependent. Moreover, it could be useful to take into account possible dependencies between the distributions of the latent factors, what was not at all considered here. All these topics will the subject of our future work.

Acknowledgement. This work was partially funded by the French Government under the grant <ANR-13-CORD-0020> (ALICIA Project).

References

1. Agarwal, D., Chen, B., Elango, P.: Fast online learning through offline initialization for time-sensitive recommendation. In: KDD 2010 (2010)
2. Chapelle, O., Li, L.: An empirical evaluation of thompson sampling. In: NIPS 2011 (2011)
3. Crammer, K., Dekel, O., Keshet, J., Shalev-Shwartz, S., Singer, Y.: Online passive-aggressive algorithms. J. Mach. Learn. Res. **7**, 551–585 (2006)
4. Mahajan, D., Rastogi, R., Tiwari, C., Mitra, A.: Logucb: an explore-exploit algorithm for comments recommendation. In: CIKM 2012 (2012)
5. Gaillard, J., Renders, J.-M.: Time-sensitive collaborative filtering through adaptive matrix completion. In: Hanbury, A., Kazai, G., Rauber, A., Fuhr, N. (eds.) ECIR 2015. LNCS, vol. 9022, pp. 327–332. Springer, Heidelberg (2015)
6. Gultekin, S., Paisley, J.: A collaborative kalman filter for time-evolving dyadic processes. In: ICDM 2014 (2014)
7. Han, S., Yang, Y., Liu, W.: Incremental learning for dynamic collaborative filtering. J. Softw. **6**(6), 969–976 (2011)
8. Koren, Y.: Collaborative filtering with temporal dynamics. Commun. ACM **53**(4), 89–97 (2010)
9. Li, L., Chu, W., Langford, J., Schapire, R.: A contextual-bandit approach to personalized news article recommendation. In: WWW 2010 (2010)
10. Lu, Z., Agarwal, D., Dhillon, I.: A spatio-temporal approach to collaborative filtering. In: RecSys 2009 (2009)

11. Ott, P.: Incremental matrix factorization for collaborative filtering. Science, Technology and Design 01/, Anhalt University of Applied Sciences, 2008 (2008)
12. Rendle, S., Schmidt-thieme, L.: Online-updating regularized kernel matrix factorization models for large-scale recommender systems. In: RecSys 2008 (2008)
13. Stern, D., Herbrich, R., Graepel, T.: Matchbox: large scale online bayesian recommendations. In: WWW 2009 (2009)
14. Sun, J., Parthasarathy, D., Varshney, K.: Collaborative kalman filtering for dynamic matrix factorization. IEEE Trans. Sig. Process. **62**(14), 3499–3509 (2014)
15. Sun, J., Varshney, K., Subbian, K.: Dynamic matrix factorization: A state space approach. In: IEEE International Conference on Acoustics, Speech and Signal Processing (ICASSP) (2012)
16. Xiang, L., Yang, Q.: Time-dependent models in collaborative filtering based recommender system. In: IEEE/WIC/ACM International Joint Conferences on Web Intelligence and Intelligent Agent Technologies, vol. 1 (2009)
17. Xiong, L., Chen, X., Huang, T.-K., Schneider, J., Carbonel, J.: Temporal collaborative filtering with bayesian probabilistic tensor factorization. In: Proceedings of the SIAM International Conference on Data Mining (SDM), vol. 10 (2010)
18. Yu, L., Liu, C., Zhang, Z.: Multi-linear interactive matrix factorization. Knowledge-Based Systems (2015)
19. Zhao, X., Zhang, W., Wang, J.: Interactive collaborative filtering. In: CIKM 2013 (2013)

Short Papers

A Business Zone Recommender System Based on Facebook and Urban Planning Data

Jovian Lin(✉), Richard J. Oentaryo, Ee-Peng Lim, Casey Vu, Adrian Vu, Agus T. Kwee, and Philips K. Prasetyo

Living Analytics Research Centre, Singapore Management University, Singapore, Singapore
jovian.lin@gmail.com,
{roentaryo,eplim,caseyanhthu,adrianvu,aguskwee,pprasetyo}@smu.edu.sg

Abstract. We present ZoneRec—a zone recommendation system for physical businesses in an urban city, which uses both public business data from Facebook and urban planning data. The system consists of machine learning algorithms that take in a business' metadata and outputs a list of recommended zones to establish the business in. We evaluate our system using data of food businesses in Singapore and assess the contribution of different feature groups to the recommendation quality.

Keywords: Facebook · Social media · Business · Location recommendation

1 Introduction

Location is a pivotal factor for retail success, owing to the fact that 94 % of retail sales are still transacted in physical stores [9]. To increase the chance of success for their stores, business owners need to know not only where their potential customers are, but also their surrounding competitors and potential allies. However, assessing a store location is a cumbersome task for business owners as numerous factors need to be considered that often require gathering and analyzing the relevant data. To this end, business owners typically conduct ground surveys, which are time-consuming, costly, and do not scale up well. Moreover, with the rapidly changing environment and emergence of new business locations, one has to continuously reevaluate the value of the store locations.

Fortunately, in the era of social media and mobile apps, we have an abundance of data that capture both online activities of users and offline activities at physical locations. For example, more than 1 billion people actively use Facebook everyday [8]. The availability of online user, location, and other behavioral data makes it possible now to estimate the value of a business location.

Accordingly, we develop ZoneRec, a business location recommender system that takes a user's description about his/her business and produces a ranked list of *zones* that would best suit the business. Such ranking constitutes a fundamental information retrieval (IR) problem [6, 7], where the user's description

N. Ferro et al. (Eds.): ECIR 2016, LNCS 9626, pp. 641–647, 2016.
DOI: 10.1007/978-3-319-30671-1_47

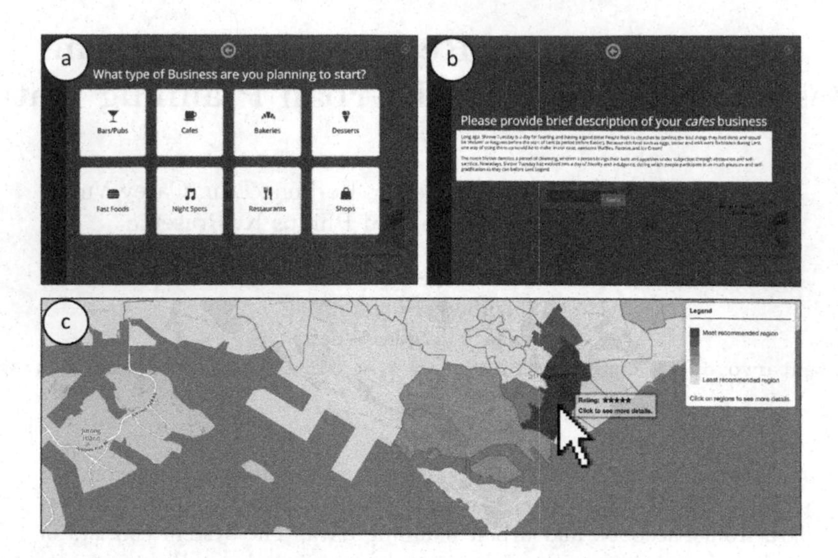

Fig. 1. Our zone recommendation system.

corresponds to the *query*, and the pairs of business profiles and zones are the *documents*. Our system is targeted at business owners who have little or no prior knowledge on which zone they should set up their business at. In our current work, the zones refer to the 55 urban planning areas, the boundaries of which are set by the Singapore government. While we currently focus on Singapore data, it is worth noting that our approach can be readily used in other urban cities.

Figure 1 illustrates how our `ZoneRec` system works. First, the system asks the user to define the *type* of his/her hypothetical (food) business (Fig. 1a), and to then provide some *description* of the business (Fig. 1b). In turn, our system analyzes the input data, based on which its recommendation algorithm produces a ranked list of zones. The ranking scores of the zones are represented by a heatmap overlaid on the Singapore map (Fig. 1c). Further details of each recommended zone can be obtained by hovering or clicking on the zone.

Related Works. Using social media data to understand the dynamics of a society is an increasingly popular research theme. For example, Chang and Sun [3] analyzed the "check-ins" data of Facebook users to develop models that can predict where users will check-in next, and in turn predict user friendships. Another close work by Karamshuk *et al.* [5] demonstrated the power of geographic and user mobility features in predicting the best placement of retail stores. Our work differs from [3] in that we use Facebook data to recommend locations instead of friendships. Meanwhile, Karamshuk *et al.* [5] discretized the city into a uniform grid of multiple circles. In contrast, we use more accurate, non-uniform area boundaries that are curated by government urban planning.

Contributions. In summary, our contributions are: (i) To our best knowledge, we are the first to develop a business zone recommendation method that fuses

Facebook business location and urban planning data to help business owners find the optimal zone placement of their businesses; (ii) We develop a user-friendly web application to realize our ZoneRec approach, which is now available online at http://research.larc.smu.edu.sg/bizanalytics/. (iii) We conduct empirical studies to compare different algorithms for zone recommendation, and assess the relevancy of different feature groups.

2 Datasets

In this work, we use two public data sources, which we elaborate below.

Singapore Urban Planning Data. To obtain the zone information, we retrieved the urban planning data from the Urban Development Authority (URA) of Singapore [10]. The data consist of 55 predetermined planning zones. To get the 55 zones, URA first divided Singapore into five regions: Central, West, North, North-East and East. Each region has a population of more than 500,000 people, and is a mix of residential, commercial, business and recreational areas. These regions are further divided into zones, each having a population of about 150,000 and being served by a town centre and several commercial/shopping centres.

Facebook Business Data. In this work, we focus on data from Facebook pages about food-related businesses that are located within the physical boundaries of Singapore. Our motivation is that food-related businesses constitute one of the largest groups in our Singapore Facebook data. From a total of 82,566 business profiles we extracted via Facebook's Graph API [4], we found 20,877 (25.2 %) profiles that are food-related. Each profile has the following attributes:

- **Business Name and Description.** This represents the name and textual description of the shop, respectively.
- **List of Categories.** From the 20,877 food-related businesses, we retrieve 357 unique categorical labels, as standardized by Facebook. These may contain not only food-related labels such as "bakery," "bar," and "coffee shop", but also non-food ones such as "movie theatre," "mall," and "train station." The existence of non-food labels for food businesses is Facebook's way of allowing the users to tag multiple labels for a business profile. For example, a Starbucks outlet near a train station in an airport may have both food and non-food labels, such as "airport," "cafe," "coffee shop," and "train station".
- **Location of Physical Store.** Each business profile has a `location` attribute containing the physical address and latitude-longitude coordinates (hereafter called "lat-long"). We map the lat-long information to the URA data to determine which of the 55 zones the target business is in. Note that we rule out business profiles that do not have explicit lat-long coordinates.
- **Customer Check-ins.** A check-in is the action of registering one's physical presence; and the total number of check-ins received by a business gives us a rough estimate of how popular and well-received it is.

3 Proposed Approach

We cast the zone recommendation as a classification task, where the input features are derived from the textual and categorical information of a business and the class labels are the zone IDs. This formulation corresponds to the *pointwise ranking* method for IR [6], whereby the ranking problem is transformed to a conventional classification task. Our approach consists of three phases:

Data Cleaning. For each business, we first extract its (i) *business name*, (ii) *business description*, and (iii) the *tagged categories* that it is associated with. As some business profiles may have few or no descriptive text, we set the minimum number of words in a description to be 20. This is to ensure that our study only includes quality business profiles, as the insertion of businesses with noisy "check-ins" will likely deterioriate the quality of the recommendations produced by our classification algorithms. We remove all stop words and words containing digits. Stemming is also performed to reduce inflectional forms and derivationally related forms of a word to a common base form (*e.g.*, car, cars, car's ⇒ car).

Feature Construction. Using the cleaned text from the previous stage, we construct a bag of words for each feature group, *i.e.*, the name, description, and categories of each business profile. As not all words in the corpus are important, however, we compute the term frequency-inverse document frequency (TF-IDF) [7] to measure how important a (set of) word or is to a business profile (*i.e.*, a document) in a corpus. We also include bigram features, since in some cases pairs of words make more sense than the individual words. With the inclusion of unigrams and bigrams, we have a total number of 51,397 unique terms. We set the minimum document frequency (DF) as 3, and retain the top 5000 terms based on their inverse document frequency (IDF) score.

Classification Algorithms. Based on the constructed TF-IDF features of a business profile as well as the zone (*i.e.*, class label) it belongs to, we can now craft the training data for our classification algorithms. Specifically, each classifier is trained to compute the *matching score* between a business profile and a zone ID. We can then apply the classifiers to the testing data and sort the matching scores in descending order, based upon which we pick the K highest scores that would constitute our top K recommended zones.

In this study, we investigate on three popular classification algorithms: (i) support vector machine (SVM) with linear kernel (SVM-Linear) [2], (ii) SVM with radial basis function kernel (SVM-RBF) [2], and (iii) random forest classifier (RF) [1]. The first two aim at maximizing the margin of separation between data points from different classes, which would imply a lower generalization error. Meanwhile, RF is an ensemble classifier that comprises an army of decision trees. It works based on bagging mechanism, *i.e.*, each tree is built from bootstrap samples drawn with replacement from the training data, and the final prediction is done via a majority voting of the decisions made by the constituent trees. Being an ensemble model, RF exhibits its high accuracy and robustness, and the bagging mechanism facilitates an efficient, parallelizable learning process.

4 Results and Analysis

Our experiment aims at evaluating the quality of zone recommendation for new businesses. To do so, we hide some of the known businesses and assess the accuracy of the recommended zones for those businesses. We measure the recommendation accuracy using three ranking metrics popularly used in IR, *i.e.*, Hit@10, MAP@10, and NDCG@10 [7]. The Hit@10 calculates whether the actual zone ID is in the top 10 recommended zones, irregardless of the position of the actual zone ID. The MAP@10 and NDCG@10 compute the mean average precision and normalized discounted cumulative gain at top 10 respectively, both of which give higher penalty when the actual zone ID has a lower position in the recommendation list. We evaluate our classifiers using 10-fold cross validation, whereby 90 % of the business profiles are used for training the models, and the remaining 10 % for testing the models' performances on unseen profiles. We record the Hit@10, MAP@10, and NDCG@10 for each fold, and then report the averaged results.

Performance Assessment. Table 1 shows the performances of different classifiers (with their corresponding best parameters). Here RF substantially outperforms the two SVMs for all metrics, at a significance level of 0.01 on the two-tailed *t*-test. The superiority of RF over SVM can be explained by comparing ensemble vs. single models. First, by taking a consensus from different decision trees, RF reduces the risk of using a wrong classifier. In effect, the combined decision of multiple trees is more robust than that of a single tree. Also, the bagging mechanism helps reduce the modeling *variance*—error from sensitivity to small fluctuations in the training data. Thus, RF is less prone to overfitting (*i.e.*, modeling random noise in the data) than single models such as SVM.

Comparing the two SVMs, we initially expected that SVM-RBF would outperform SVM-Linear, since the RBF kernel essentially maps the original data to an infinitely high-dimensional feature space, for which data from different classes should be more separable. It turns out, however, that SVM-Linear performs better than SVM-RBF. This may be attributed to our TF-IDF representation, which involves a sparse, fairly large number of features that is likely to be linearly separable already. In such case, using nonlinear kernel would not necessarily help improve the performance, and may instead increase the risk of overfitting.

Contribution of Features. As mentioned in Sect. 2, we divide our feature vectors into three groups: (i) business name, (ii) business description, and (iii) tagged categories. Here we evaluate the contribution of each feature group by performing an *ablation test* on the RF model. Table 2 consolidates the results

Table 1. Recommendation results of different algorithms

Algorithm	Hit@10	MAP@10	NDCG@10
SVM-Linear ($C = 1$)	0.502	0.300	0.348
SVM-RBF ($C = 1$, $\gamma = \frac{1}{\#\text{features}}$)	0.231	0.074	0.110
Random forest (#trees = 1000)	**0.721***	**0.430***	**0.499***

C: cost parameter, γ: kernel coefficient, *: significant at 0.01

Table 2. Feature ablation results for random forest classifier

Use name	Use description	Use categories	Hit@10	MAP@10	NDCG@10
-	-	Yes	0.537	0.212	0.287
-	Yes	-	0.685	0.398	0.465
Yes	-	-	0.554	0.240	0.313
Yes	-	Yes	0.557	0.239	0.313
Yes	Yes	-	**0.708**	**0.423**	**0.499**
-	Yes	Yes	0.694	0.404	0.472

of our ablation study. The first three rows show the results of ablating (removing) two feature groups, while the last three rows are for ablating one feature group.

From the first three rows, it is evident that the "description" is the most important feature group, consistently providing the highest Hit@10, MAP@10, and NDCG@10 scores compared to the other two. This is reasonable, as the "description" provides the richest set of features (in terms of word vocabulary and frequencies) representing a business, and some of these features provide highly discriminative inputs for our RF classifier. We can also see that the "name" group is more discriminative than the "categories" group for all the three metrics. Again, this can be attributed to the more fine-grained information provided by the business' name features as compared to the category features. Finally, we find that the results in the last three rows are consistent with those of the first three rows. That is, the "description" group constitutes the most informative features (for our RF model), followed by the "name" and "category" groups.

5 Conclusion

We put forward the `ZoneRec` recommender system that can help business owners decide which zones they should set their businesses at. Despite its promising potentials, there remains room for improvement. First, the zone-level recommendations may not provide sufficiently granular information for business owners, *e.g.*, where exactly a store should be set at and how the surrounding businesses may affect this choice. It is also fruitful to include more comprehensive residential and demographic information in our feature set, and conduct deeper analysis on the contribution of the individual features. To address these, we plan to develop a two-level location recommender system, whereby `ZoneRec` serves as the first level and the second level recommends the specific hotspots within each zone.

Acknowledgments. This research is supported by the Singapore National Research Foundation under its International Research Centre @ Singapore Funding Initiative and administered by the IDM Programme Office, Media Development Authority (MDA).

References

1. Breiman, L.: Random forests. Mach. Learn. **45**(1), 5–32 (2001)
2. Chang, C.-C., Lin, C.-J.: LIBSVM: a library for support vector machines. ACM-TIST **2**(27), 1–27 (2011)
3. Chang, J., Sun, E.: Location3: how users share and respond to location-baseddata on social networking sites. In: ICWSM, pp. 74–80 (2011)
4. Facebook. Graph API reference (2015). https://goo.gl/8ejSw0
5. Karamshuk, D., Noulas, A., Scellato, S., Nicosia, V., Mascolo, C.: Geo-spotting: mining online location-based services for optimal retail store placement. In: KDD, pp. 793–801 (2013)
6. Liu, T.-Y.: Learning to rank for information retrieval. Found. Trends Inf. Retrieval **3**(3), 225–331 (2009)
7. Manning, C.D., Raghavan, P., Schütze, H.: Introduction to Information Retrieval. Cambridge University Press, Cambridge (2008)
8. Smith, C.: 200+ amazing facebook user statistics (2016). http://goo.gl/RUoCxE
9. Thau, B.: How big data helps chains like starbucks pick store locations–an (unsung) key to retail success (2015). http://onforb.es/1k8VEQY
10. URA. Master plan: View planning boundaries (2015). http://goo.gl/GA3dR8

On the Evaluation of Tweet Timeline Generation Task

Walid Magdy[1(✉)], Tamer Elsayed[2], and Maram Hasanain[2]

[1] Qatar Computing Research Institute, HBKU, Doha, Qatar
wmagdy@qf.org.qa
[2] Computer Science and Engineering Department, Qatar University, Doha, Qatar
{telsayed,maram.hasanain}@qu.edu.qa

Abstract. Tweet Timeline Generation (TTG) task aims to generate a timeline of *relevant* but *novel* tweets that summarizes the development of a given topic. A typical TTG system first retrieves tweets then detects novel tweets among them to form a timeline. In this paper, we examine the dependency of TTG on retrieval quality, and its effect on having biased evaluation. Our study showed a considerable dependency, however, ranking systems is not highly affected if a common retrieval run is used.

1 Introduction

With the enormous volume of tweets posted daily and the associated redundancy and noise in such vibrant information sharing medium, a user can find it difficult to get updates about a topic or an event of interest. The Tweet Timeline Generation (TTG) task was recently introduced at TREC-2014 microblog track to tackle this problem. TTG aims at generating a timeline of *relevant* and *novel* tweets that summarizes the development of a topic over time [5].

In the TREC task, a TTG system is evaluated using variants of F_1 measure that combine precision and recall of the generated timeline against a gold standard of clusters of semantically-similar tweets. Different TTG approaches were presented in TREC-2014 [5] and afterwards [2,4]: *almost all* rely on an initial step of retrieval of a ranked list of potentially-relevant tweets, followed by applying novelty detection and duplicate removal techniques to generate the timeline [5]. In such design, the quality of generated timeline naturally relies on that of the initially-retrieved list. There is a major concern that the evaluation metrics do not fairly rank TTG systems since they start from *different* retrieved ranked lists. An effective TTG system that is fed low quality list may achieve lower performance compared to another low quality TTG system fed a high quality list; current TTG evaluation metrics lacks the ability to evaluate TTG independently from the retrieval effectiveness. This creates an evaluation challenge, especially for future approaches that use different retrieval models.

In this work, we examine the bias of TTG evaluation methodology introduced in the track [1]. We first empirically measure the dependency of TTG

N. Ferro et al. (Eds.): ECIR 2016, LNCS 9626, pp. 648–653, 2016.
DOI: 10.1007/978-3-319-30671-1_48

performance on retrieval quality, then examine the validity of using a single input retrieval list for ranking different TTG systems, and the consistency of ranking when using several retrieval lists with varying qualities. We ran experiments using 13 different ad-hoc retrieval runs and 8 TTG systems participated in TREC-2014. Our study shows considerable dependency of TTG systems performance on retrieval quality. Nonetheless, we found that using a single ad-hoc run for ranking different TTG systems could lead to a less-biased and stable ranking of TTG systems, regardless of which retrieval run is used. When a common retrieval run is not available, it is important to consider the final performance of the TTG system in the context of the quality of the used retrieval run.

2 Experimental Setup

A set of 55 queries and corresponding relevance judgments were provided by TREC [5]. For each query, a set of semantic clusters were identified; each consists of tweets relevant to an aspect of the topic but substantially similar to each other.

Precision, recall, and F_1 measures over the semantic clusters were used for evaluation. Precision (**P**) is defined as the proportion of tweets returned by a TTG system representing distinct semantic clusters. Recall (**R**) is defined as the proportion of the total semantic clusters that are covered by the returned tweets. Weighted Recall (**wR**) is measured similarly but weighs each covered semantic cluster by the sum of relevance grades[1] of its tweets. F_1 combines P and R, while wF_1 combines P and wR. Each of those measures is first computed over the returned timeline of *each* query and then averaged over all queries.

In our experiments, we used 12 officially-submitted ad-hoc runs by 3 of the top 4 participated groups in TREC-2014 TTG task [3,6,9]. Additionally, we used a baseline run directly provided by TREC search API [5]. This concludes a total of 13 ad-hoc runs for our study, denoted by the set A = $\{a_1, a_2, \dots, a_{13}\}$. The retrieval approaches used by those runs are mainly five: (1) direct search by TREC API (D), (2) using query expansion (QE), (3) using QE that utilizes the links in tweets (QE+Web), (4) using QE then learning to rank (QE+L2R), and (5) using relevance modeling (RM). Table 1 presents all ad-hoc runs and their retrieval performance.

We also used 8 different TTG systems (of two TREC participants) [3,6], denoted by T = $\{t_1, t_2, \dots, t_8\}$. Their approaches are summarized as follows:

- t_1 to t_4 applied 1NN-clustering (using modified versions of Jaccard similarity) on the retrieved tweets [6] and generated timelines using different retrieval depths, which made their performance results significantly different [5,6].
- t_5 is a simple TTG system that just returns the retrieved tweets after removing exact duplicates.
- t_6 to t_8 applied an incremental clustering approach that treats the retrieved tweets, sorted by their retrieval scores, as a stream and clusters each tweet based on cosine similarity to the centroids of existing clusters. They also used different number of top retrieved tweets and different similarity thresholds, and considered the top-scoring tweet in each cluster as its centroid [3].

[1] 1 for a relevant tweet and 2 for a highly-relevant tweet.

Table 1. Retrieval performance of ad-hoc runs.

Ad-hoc	MAP	P@30	P@100	R-Prec	Approach
a_1	0.477	0.669	0.500	0.501	QE+web
a_2	0.482	0.698	0.500	0.501	QE+L2R
a_3	0.464	0.668	0.491	0.498	QE
a_4	0.470	0.699	0.491	0.498	QE+L2R
a_5	0.490	0.670	0.505	0.508	QE
a_6	0.466	0.644	0.479	0.496	QE
a_7	0.406	0.647	0.461	0.445	QE
a_8	0.445	0.627	0.509	0.486	RM
a_9	0.385	0.624	0.473	0.436	RM
a_{10}	0.485	0.673	0.517	0.499	QE+web
a_{11}	0.497	0.681	0.518	0.512	QE+L2R
a_{12}	0.571	0.712	0.545	0.566	QE+L2R
a_{13}	0.398	0.646	0.468	0.439	D

Table 2. TTG systems performance with a_{13}.

TTG	R	wR	P	F_1	wF_1
$t_1(a_{13})$	0.342	0.535	0.320	0.245	0.330
$t_2(a_{13})$	0.260	0.463	0.411	0.241	0.371
$t_3(a_{13})$	0.159	0.261	0.364	0.153	0.226
$t_4(a_{13})$	0.137	0.261	0.444	0.150	0.255
$t_5(a_{13})$	0.353	0.552	0.263	0.231	0.297
$t_6(a_{13})$	0.315	0.511	0.354	0.252	0.350
$t_7(a_{13})$	0.191	0.365	0.484	0.215	0.355
$t_8(a_{13})$	0.334	0.537	0.311	0.246	0.328

Table 2 presents the performance of the 8 TTG systems when applied to a_{13}, which was selected as a sample to illustrate the quality of each TTG system. As shown, the quality of the 8 TTG systems varies significantly. In fact, by applying significance test on wF_1 using two-tailed t-test with α of 0.05, we found that all TTG system pairs but 6 were statistically significantly different.

Combinations of ad-hoc runs and TTG systems created a list of **104** TTG runs that we used to study the bias of the task evaluation. We aim to show whether the evaluation methodology used in the TREC microblog track is biased towards retrieval quality, and if there is a way to reduce possible bias.

To measure bias and dependency of TTG on the quality of the used ad-hoc runs, we use Kendall tau correlation (τ) and AP correlation (τ_{AP}) [10]. τ_{AP} is used besides τ since it is more sensitive to errors at higher ranks [10].

3 Dependency of TTG Performance on Retrieval Results

3.1 Correlation Between TTG Scores and Retrieval Scores

In this section, we try to answer the following research question: *"If we tried one TTG system with different ad-hoc runs, will the quality ranking of the resulting TTG timelines be correlated with the quality ranking of the ad-hoc runs?"*.

To answer this question, we compared the ranking of the ad-hoc runs (using retrieval scores) to the ranking of the resulting timelines from the same TTG system (using TTG scores). We repeated the process over each TTG system, and averaged the correlations as follows:

$$\sigma^* = \frac{1}{|T|} \sum_{t \in T} \sigma(\{S_r(a)|_{a \in A}\}, \{S_t(t(a))|_{a \in A}\}) \tag{1}$$

where σ^* is the average correlation, σ is τ or τ_{AP} correlation, $\{S_r(a)|_{a \in A}\}$ are the retrieval scores of the 13 ad-hoc runs, and $\{S_t(t(a))|_{a \in A}\}$ are the TTG scores of their corresponding timelines using the TTG system t.

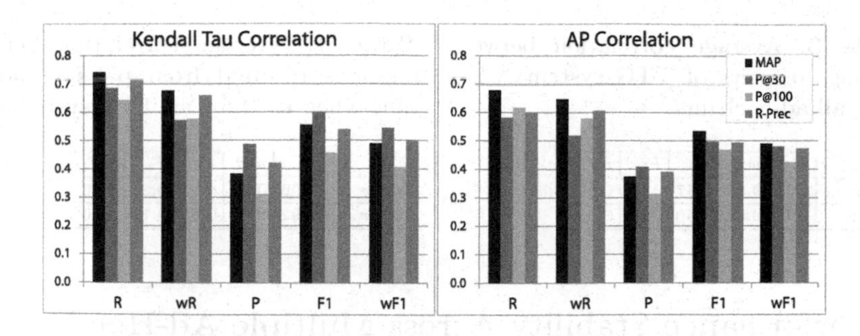

Fig. 1. τ and τ_{AP} between ad-hoc runs and their corresponding TTG timelines, averaged over TTG systems.

Figure 1 reports the *average* τ and τ_{AP} correlations using different retrieval and TTG performance metrics. As shown, there is always a positive correlation between the quality rankings of ad-hoc runs and the TTG timelines. Considering the main metrics for evaluating retrieval (MAP) and for evaluating TTG (wF_1), the correlation values are 0.49 for both τ and τ_{AP}. This indicates a considerable correlation, but it is not very strong as expected.

3.2 Correlation over TTG Scores

Since measuring correlation between systems ranking using measures of two different tasks may be sub-optimal, we continue to test the dependency of TTG output on the ad-hoc runs quality, but using TTG evaluation measures only.

Here we answer the following research question: *"If we tried a TTG system t_i with different ad-hoc runs, and we repeated that with another TTG system t_j, will the quality ranking of the resulting timelines of t_i be correlated with the quality ranking of the resulting timelines of t_j?"*.

Correlation is computed between the ranking of resulting timelines from TTG system t_i using different ad-hoc runs and the corresponding timelines from TTG system t_j. We apply this over the 8 TTG systems, creating a set of 28 pairwise comparisons. The average correlation is then computed as follows:

$$\sigma^* = \frac{2}{|T|(|T|-1)} \sum_{i=1}^{|T|} \sum_{j=i+1}^{|T|} \sigma(\{S_t(t_i(a))|_{a \in A}\}, \{S_t(t_j(a))|_{a \in A}\}) \qquad (2)$$

Table 3 reports the average τ and τ_{AP} correlations among all pairs of TTG systems. Achieved correlation scores align with the same ones in Fig. 1, but with slightly higher values. This also supports the finding that TTG system performance depends, to some extent, on the quality of input ad-hoc runs. This observation suggests that using different ad-hoc runs with different TTG systems makes it unlikely to have unbiased evaluation for the TTG systems, since the output of TTG systems, in general, depends on the quality of the retrieval run.

Table 3. Average correlation between rankings of pairs of TTG systems when using all ad-hoc runs.

σ^*	R	wR	P	F_1	wF_1
avg. τ	0.76	0.57	0.57	0.68	0.56
avg. τ_{AP}	0.71	0.52	0.50	0.63	0.51

Table 4. Average correlation between rankings resulted from pairs of ad-hoc runs when used for all TTG systems.

σ^*	R	wR	P	F_1	wF_1
avg. τ	0.96	0.97	0.93	0.86	0.85
avg. τ_{AP}	0.92	0.93	0.81	0.72	0.76

4 Performance Stability Across Multiple Ad-Hoc Runs

In this section, we study stability of performance of a TTG system using different ad-hoc runs. For example, we examine if the best-performing TTG system using one ad-hoc run would continue to be the best with other ad-hoc runs.

We specifically investigate the following research question: *"If we used an ad-hoc run a_i with different TTG systems, and we repeated that with another ad-hoc run a_j, will the quality ranking of the resulting timelines using a_i be correlated with the quality ranking of the resulting timelines using a_j?"*

We compute correlation between quality ranking of the resulting timelines of the 8 TTG systems when using ad-hoc run a_i and the corresponding ranking when using run a_j. We apply that over all pairs of the 13 ad-hoc runs, creating a set of 78 pairwise comparisons. Average correlation is computed as follows:

$$\sigma^* = \frac{2}{|A|(|A|-1)} \sum_{i=1}^{|A|} \sum_{j=i+1}^{|A|} \sigma(\{S_t(t(a_i))|_{t\in T}\}, \{S_t(t(a_j))|_{t\in T}\}) \quad (3)$$

Table 4 reports the average τ and τ_{AP} correlations of TTG rankings over all pairs of ad-hoc runs. It shows that there are strong correlation values for all of the evaluation metrics, especially recall and precision. There are some noticeable difference in the values of τ and τ_{AP}, where the latter is smaller. This is expected since τ_{AP} is more sensitive to changes on the ranks at the top of the list. According to Voorhees [8], a τ correlation over 0.9 "should be considered equivalent since it is not possible to be more precise than this. Correlations less than 0.8 generally reflect noticeable changes in ranking". A later study by Sanderson and Soborroff [7] showed that τ gets lower values when lists of smaller range of values are compared, which holds in our case. Thus, the correlation values achieved in Table 4 show that ranking of TTG systems is almost equivalent by all TTG evaluation scores regardless of the ad-hoc run used.

This finding is of high importance, since it suggests a possible solution to achieve less-biased evaluation of the TTG task, simply by using a common/standard ad-hoc run when evaluating new TTG systems.

One possible and straightforward ad-hoc retrieval run that can be used as a standard run for evaluating different TTG systems is the baseline run a_{13}. Such run is easy to construct through searching the tweets collection without any processing to the queries. Although the retrieval effectiveness of a_{13} is expected to be one of the poorest (see Table 1), when we calculated the average

τ and τ_{AP} correlation for ranking TTG systems with this run compared to the other 12 add-hoc runs using the wF_1 score, the values were 0.88 and 0.82 respectively. This is a high correlation according to Voorhees [8].

5 Discussion and Recommendation

In this study, we used a set of 13 ad-hoc retrieval runs and 8 TTG systems, resulting in a set of 104 different TTG outputs, which is a reasonable number for getting reliable results. Our main motivation behind the study was to investigate a potential bias of the currently-used TTG evaluation methodology, which is a critical and essential issue for future contributions to the task using the same dataset and evaluation methodology. The investigation confirmed the concern about the dependency of TTG output on the quality of the retrieval step. Nevertheless, we found that using one common ad-hoc retrieval run, fed to all TTG systems, might be sufficient for ranking these systems in a less-biased way using the current evaluation measures.

We recommend to use the baseline retrieval run (the one obtained using TREC search API) as the common run. It can be utilized in addition to other retrieval runs to allow for comparing TTG algorithms more fairly. Besides, using a high quality ad-hoc run continues to be highly recommended for understanding the performance of combining both retrieval and TTG methods for the best performing overall system pipeline.

Acknowledgments. This work was made possible by NPRP grant# NPRP 6-1377-1-257 from the Qatar National Research Fund (a member of Qatar Foundation). The statements made herein are solely the responsibility of the authors (The first author was not funded by the grant.).

References

1. Buckley, C., Voorhees, E.M.: Evaluating evaluation measure stability. In: SIGIR (2000)
2. Fan, F., Qiang, R., Lv, C., Xin Zhao, W., Yang, J.: Tweet timeline generation via graph-based dynamic greedy clustering. In: AIRS (2015)
3. Hasanain, M., Elsayed, T.: QU at TREC-2014: Online clustering with temporal and topical expansion for tweet timeline generation. In: TREC (2014)
4. Hasanain, M., Elsayed, T., Magdy, W.: Improving tweet timeline generation by predicting optimal retrieval depth. In: AIRS (2015)
5. Lin, J., Efron, M., Wang, Y., Sherman, G.: Overview of the TREC-2014 microblog track. In: TREC (2014)
6. Magdy, W., Gao, W., Elganainy, T., Wei, Z.: QCRI at TREC 2014: Applying the KISS principle for the TTG task in the microblog track. In: TREC (2014)
7. Sanderson, M., Soboroff, I.: Problems with Kendall's tau. In: SIGIR (2007)
8. Voorhees, E.M.: Evaluation by highly relevant documents. In: SIGIR (2001)
9. Xu, T., McNamee, P., Oard, D.W.: HLTCOE at TREC 2014: Microblog and clinical decision support. In: TREC (2014)
10. Yilmaz, E., Aslam, J.A., Robertson, S.: A new rank correlation coefficient for information retrieval. In: SIGIR (2008)

Finding Relevant Relations in Relevant Documents

Michael Schuhmacher[1(✉)], Benjamin Roth[2], Simone Paolo Ponzetto[1], and Laura Dietz[1]

[1] Data and Web Science Group, University of Mannheim, Mannheim, Germany
{Michael,Simone,Laura}@informatik.uni-mannheim.de

[2] College of Information and Computer Science, University of Massachusetts, Amherst, USA
beroth@cs.umass.edu

Abstract. This work studies the combination of a document retrieval and a relation extraction system for the purpose of identifying query-relevant relational facts. On the TREC Web collection, we assess extracted facts separately for correctness and relevance. Despite some TREC topics not being covered by the relation schema, we find that this approach reveals relevant facts, and in particular those not yet known in the knowledge base DBpedia. The study confirms that mention frequency, document relevance, and entity relevance are useful indicators for fact relevance. Still, the task remains an open research problem.

1 Introduction

Constructing knowledge bases from text documents is a well-studied task in the field of Natural Language Processing [3,5, *inter alia*]. In this work, we view task of *constructing query-specific knowledge bases* from an IR perspective, where a knowledge base of relational facts is to be extracted in response to a user information need. The goal is to extract, select, and present the relevant information directly in a structured and machine readable format for deeper analysis of the topic. We focus on the following task:

Task: Given a query Q, use the documents from a large collection of Web documents to extract binary facts, i.e., subject–predicate–object triples (S, P, O) between entities S and O with relation type P that are both correctly extracted from the documents' text and relevant for the query Q.

For example, a user who wants to know about the Raspberry Pi computer should be provided with a knowledge base that includes the fact that its inventor Eben Upton founded the Raspberry Pi Foundation, that he went to Cambridge University, which is located in the United Kingdom, and so on. This knowledge base should include all relational facts about entities that are of interest when understanding the topic according to a given relation schema, e.g., `Raspberry_Pi_Foundation`–*founded_by*–`Eben_Upton`. Figure 1 gives an example of such a query-specific resource, and shows how relations from text and those from a knowledge base (DBpedia, [1]) complement each other.

N. Ferro et al. (Eds.): ECIR 2016, LNCS 9626, pp. 654–660, 2016.
DOI: 10.1007/978-3-319-30671-1_49

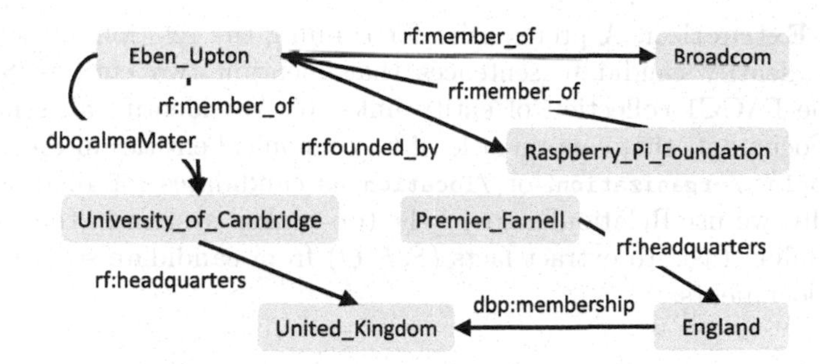

Fig. 1. Example of a knowledge base for the query "raspberry pi". `rf`: denotes relations extracted from documents, whereas `dbp`: and `dbo`: are predicates from DBpedia.

In addition to a benchmark dataset,[1] we present first experiments on building query-specific knowledge bases from a large-scale Web corpus by combining state-of-the-art retrieval models with a state-of-the-art relation extraction system [7]. This way we go beyond previous work on identifying relevant entities for Web queries [8] (where relations between entities were not considered), and query-agnostic knowledge base population (where determining fact relevance is not part of the task).

We aim at quantifying how well the direct application of a relation extraction system to a set of retrieved documents solves the task of extracting query-specific facts. This is different from the task of explaining relationships between entities in a knowledge base [9], since we include also yet unknown facts from documents. It is also different from explaining the relationship between entities and ad-hoc queries [2], since we look at relations between entities in documents. To isolate different kinds of errors, we evaluate the *correctness* of each fact extraction separately from the *relevance* of the fact for the query. We study the following research questions:

RQ1. Can the approach extract relevant facts for the queries?
RQ2. What are useful document- or KB-based indicators for fact relevance?
RQ3. Is relevance of entities and relevance of facts related?

2 Method

Document Retrieval. We use the Galago[2] search engine to retrieve documents D from the given corpus that are relevant for the query Q. We build upon the work of Dalton et al. [4] and rely on the same document pool and state-of-the-art content-based retrieval and expansion models, namely the sequential dependence model (SDM), the SDM model with query expansion through RM3 (SDM-RM3), and the SDM model with query expansion through the top-ranked Wikipedia article (WikiRM1).

[1] Dataset and additional information is available at http://relrels.dwslab.de.
[2] http://lemurproject.org/galago.php.

Relation Extraction. A prerequisite for running the relation extraction system is to identify candidate sentences that mention two entities S and O. We use the FACC1 collection of entity links [6]. We identify all sentences in retrieved documents that contain at least two canonical entities in Freebase with types `/people`, `/organization`, or `/location` as candidates for relation extraction. Finally, we use RelationFactory,[3] the top-ranked system in the TAC KBP 2013 Slot filling task, to extract facts (S, P, O) from candidate sentences of the retrieved documents.

3 Data Set and Assessments

To our knowledge, there exists no test collection for evaluating relational facts with respect to query-relevance. We augment existing test collections for document-relevance and entity-relevance with assessments on correctness and query-relevance of facts; and make the dataset publicly available. We base our analysis on the collection of test queries from the TREC Web track and documents from the ClueWeb12 corpus, which includes relevance assessments for documents, and the REWQ gold standard of query-relevant entities [8].[4]

The TREC Web track studies queries which fall into one of two categories: the query either constitutes an entity which is neither a person, organization, nor location i.e., "Raspberry Pi", or the query is about a concept or entity in a particular context, such as "Nicolas Cage movies". The closed relation extraction system only extracts relation types involving persons, organizations and locations. Due to this restriction, not all TREC Web queries can be addressed by relations in this schema, this is the case for TREC query 223 "Cannelini beans". We focus this study on the subset of 40 % of TREC Web queries such as "Raspberry Pi" for which anticipated relevant facts are covered by the relation schema.

For randomly selected 17 TREC queries, we assess the 40 most frequently mentioned facts and, in addition, all facts of which at least one of the entities was marked as relevant in the REWQ dataset. Due to the high number of annotations needed—914 facts and 2,658 provenance sentences were assessed in total—each item was inspected by only one annotator. We ask annotators to assess for each fact, (a) the correctness of the extraction from provenance sentences and (b) the relevance of the fact for the query. To assess relevance, assessors are asked to imagine writing an encyclopedic (i.e., Wikipedia-like) article about the query and mark the facts as relevant if they would mention them in the article, and non-relevant otherwise.

The number of provenance sentences per fact ranges from 1 to 82 with an average of 2.9. We define facts as correct when at least one extraction is correct. This leads to 453 out of 914 facts that are correctly extracted. The fact extraction correctness is thus at 49.6 %, which is higher than the precision obtained in the TAC KBP shared task, where about 42.5 % of extractions are correct. The assessment of relevance is performed on these 453 correctly extracted facts,

[3] https://github.com/beroth/relationfactory.

[4] http://rewq.dwslab.de.

leading to a dataset with 207 relevant facts and 246 non-relevant facts across all 17 queries, an average of 26.6 relevant facts per query. In this study we only consider queries with at least five correctly extracted facts (yielding 17 queries).

4 Evaluation

We evaluate here how well the pipeline of document retrieval and relation extraction performs for finding query-relevant facts. The *relevance* is separately evaluated from *extraction correctness*, as described in Sect. 3. In the following, we focus only on the 453 correctly extracted facts. For comparing different settings, we test statistical significant improvements on the accuracy measure through a two-sided exact binomial test on label agreements ($\alpha = 5\,\%$).

Applicability (RQ1). We report the results on fact relevance as micro-average across all facts (Table 1 bottom) and aggregated macro-averages per query (Table 1 top) to account for differences across queries. Among all correct facts, only every other fact is relevant for the query (0.45 micro-average precision, 0.47 macro-average precision). Factoring in the extraction precision of 0.51 we obtain one relevant out of four extracted facts on average. This strongly suggest that the problem of relevant relation finding (beyond correctness) is indeed an *open research problem.*

In about 60 % of TREC queries, such as "Cannelini beans", we found the relation schema of TAC KBP to not be applicable. Nevertheless, even with the schema limitations, the system found relevant facts for the (randomly) assessed 17 queries out of the remaining 40 queries.

Table 1. Experimental results for relation relevance (correctly extracted relations only) comparing different fact retrieval features: All facts (All), facts also included in DBpedia (DBp), fact mentioned three or more times ($\text{Frq}_{\geq 3}$), facts extracted from a relevant document (Doc). Significant accuracy improvements over "All" marked with †.

		All	$\text{Frq}_{\geq 3}$	DBp	Doc
Per query (macro-avg)	#Queries	17	10	17	10
	Precision	0.470	0.553	0.455	0.704
	Std Error	0.070	0.100	0.087	0.112
All facts (micro-avg)	#Retrieved Facts	453	106	145	46
	TP	207	58	64	30
	FP	246	48	81	16
	TN	-	198	165	230
	FN	-	149	143	177
	Precision	0.457	0.547	0.441	0.652
	Recall	1.000	0.280	0.309	0.145
	F_1	0.627	0.371	0.364	0.237
	Accuracy	0.457	0.565†	0.506	0.574†

Indicators for Fact Relevance (RQ2). We study several indicators that may improve the prediction of fact relevance. First, we confirm that the frequency of fact mentions indicates fact relevance. If we classify a correctly extracted fact as 'relevant' only when it is mentioned at least three times[5] then relevance accuracy is improved by 23.6 % from 0.457 to 0.565 (statistically significant). This also reduces the number of predicted facts to a fourth (see Table 1, column $Frq_{\geq 3}$).

Next, we compare the extracted facts with facts known to the large general-purpose knowledge base DBpedia. When classifying only extracted facts as relevant when they are confirmed—that is, both entities are related in DBpedia (independent of the relation type)—we do not obtain any significant improvements in accuracy or precision. Therefore, confirmation of a known fact in an external knowledge base does not indicate relevance. However, we notice that only 64 of the relevant facts are included in DBpedia, whereas another 143 new and relevant facts are extracted from the document-centric approach (cf. Table 1, column *DBp*). This indicates that extracting yet unknown relations (i.e., those not found in the knowledge base) from query-relevant text has the potential to provide the majority of relevant facts to the query-specific knowledge base.

Our study relies on a document retrieval system, leading to some non-relevant documents in the result list. We confirm that the accuracy of relation relevance improves significantly when we only consider documents assessed as relevant. However, it comes at the cost of retaining only a tenth of the facts (cf. Table 1, column *Doc*).

Fact Relevance vs. Entity Relevance (RQ3). We finally explore whether query-relevance of entities implies relevance of facts. In order to study this implication, we make use of the REWQ test collection on entity relevance [8] by studying the subset of the 108 correct facts where relevance assessments exist for both entities. Due to pooling strategies, this subset has a higher precision of 0.722. In Table 2 we consider the case where entity relevance is true for both entities (S $\wedge$ O) as well as at least one entity (S $\vee$ O).

For only 12 correct facts, both entities are assessed as non-relevant – these facts were also assessed as non-relevant by our (different) annotators. In contrast, for 45 facts both entities and the fact itself are assessed as relevant (we take this agreement also as a confirmation of the quality of our fact assessments). Using the entity assessments as an oracle for simulating a classifier, we obtain improvements in precision from 0.722 to 0.809 for either entity and 0.918 for both entities. While also accuracy improves for the case of either entity, it is actually much lower in the case of both entities. We conclude that the restriction to both entities being relevant misses 33 out of 78 relevant facts. In this set of 33 relevant facts with one relevant and one non-relevant entity, we find that the non-relevant entity is often too unspecific to be directly relevant for the query such as a country or city. For example, in Fig. 1 the `University_of_Cambridge` is relevant mostly because of the fact that `Eben_Upton` is a member.

[5] We chose ≥ 3 in order to be above the median of the number of sentences per fact, which is 2.

Table 2. Fact relevance when at least one entity ($S \vee O$) or both entities ($S \wedge O$) are relevant compared to all facts (All). Significant accuracy improvements over "All" marked with †.

	All	S ∨ O	S ∧ O
#Retrieved Facts	108	94	49
TP	78	76	45
FP	30	18	4
TN	-	12	26
FN	-	2	33
Precision	0.722	0.809	0.918
Recall	1.000	0.974	0.577
F_1	0.839	0.884	0.709
Accuracy	0.722	0.815†	0.657

5 Conclusion

We investigate the idea of extracting query relevant facts from text documents to create query-specific knowledge bases. Our study combines publicly available data sets and state-of-the-art systems for document retrieval and relation extraction to answer research questions on the interplay between relevant documents and relational facts for this task. We can summarize our key findings as follows:

(a) Query-specific documents contain relevant facts, but even with perfect extractions, only around half of the facts are actually relevant with respect to the query.
(b) Many relevant facts are not contained in a wide-coverage knowledge base like DBpedia, suggesting importance of extraction for query-specific knowledge bases.
(c) Improving retrieval precision of documents increases the ratio of relevant facts significantly, but sufficient recall is required for appropriate coverage.
(d) Facts that are relevant can contain entities (typically in object position) that are—by themselves—not directly relevant.

From a practical perspective, we conclude that the combination of document retrieval and relation extraction is a suitable approach to query-driven knowledge base construction, but it remains an open research problem. For further advances, we recommend to explore the potential of integrating document retrieval and relation extraction—as opposed to simply applying them sequentially in the pipeline architecture.

Acknowledgements. This work was in part funded by the Deutsche Forschungsgemeinschaft within the JOIN-T project (research grant PO 1900/1-1), in part by DARPA under agreement number FA8750-13-2-0020, through the Elitepostdoc program of the BW-Stiftung, an Amazon AWS grant in education, and by the Center for

Intelligent Information Retrieval. The U.S. Government is authorized to reproduce and distribute reprints for Governmental purposes notwithstanding any copyright notation thereon. Any opinions, findings and conclusions or recommendations expressed in this material are those of the authors and do not necessarily reflect those of the sponsor. We are also thankful for the support of Amina Kadry and the helpful comments of the anonymous reviewers.

References

1. Bizer, C., Lehmann, J., Kobilarov, G., Auer, S., Becker, C., Cyganiak, R., Hellmann, S.: DBpedia — A crystallization point for the web of data. J. Web Semant. **7**(3), 154–165 (2009)
2. Blanco, R., Zaragoza, H.: Finding support sentences for entities. In: Proceedings of SIGIR 2010, pp. 339–346 (2010)
3. Carlson, A., Betteridge, J., Kisiel, B., Settles, B., Hruschka, E.R., Mitchell, T.M.: Toward an architecture for never-ending language learning. In: Proceedings of AAAI 2010, pp. 1306–1313 (2010)
4. Dalton, J., Dietz, L., Allan, J.: Entity query feature expansion using knowledge base links. In: Proceedings of SIGIR-2014, pp. 365–374 (2014)
5. Fader, A., Soderland, S., Etzioni, O.: Identifying relations for open information extraction. In: Proceedings of EMNLP 2011, pp. 1535–1545 (2011)
6. Gabrilovich, E., Ringgaard, M., Subramanya, A.: FACC1: Freebase annotation of ClueWeb corpora, Version 1 (2013)
7. Roth, B., Barth, T., Chrupała, G., Gropp, M., Klakow, D.: Relationfactory: A fast, modular and effective system for knowledge base population. In: Proceedings of EACL 2014, p. 89 (2014)
8. Schuhmacher, M., Dietz, L., Ponzetto, S.P.: Ranking entities for web queries through text and knowledge. In: Proceedings of CIKM 2015 (2015)
9. Voskarides, N., Meij, E., Tsagkias, M., de Rijke, M., Weerkamp, W.: Learning to explain entity relationships in knowledge graphs. In: Proceedings of ACL 2015, pp. 564–574 (2015)

Probabilistic Multileave Gradient Descent

Harrie Oosterhuis, Anne Schuth(✉), and Maarten de Rijke

University of Amsterdam, Amsterdam, The Netherlands
harrie.oosterhuis@student.uva.nl, {a.g.schuth,derijke}@uva.nl

Abstract. Online learning to rank methods aim to optimize ranking models based on user interactions. The dueling bandit gradient descent (DBGD) algorithm is able to effectively optimize linear ranking models solely from user interactions. We propose an extension of DBGD, called probabilistic multileave gradient descent (P-MGD) that builds on probabilistic multileave, a recently proposed highly sensitive and unbiased online evaluation method. We demonstrate that P-MGD significantly outperforms state-of-the-art online learning to rank methods in terms of online performance, without sacrificing offline performance and at greater learning speed.

1 Introduction

Modern search engines are complex aggregates of multiple ranking signals. Such aggregates are learned using learning to rank methods. Online learning to rank methods learn from user interactions such as clicks [4,6,10,12]. Dueling Bandit Gradient Descent [16] uses interleaved comparison methods [1,3,6,7,10] to infer preferences and then learns by following a gradient that is meant to lead to an optimal ranker.

We introduce probabilistic multileave gradient descent (P-MGD), an online learning to rank method that builds on a recently proposed highly sensitive and unbiased online evaluation method, viz. probabilistic multileave. Multileave comparisons allow one to compare multiple but a still limited set of candidate rankers per user interaction [13]. The more recently introduced probabilistic multileave comparison method improves over this by allowing for comparisons of an unlimited number of rankers at a time [15]. We show experimentally that P-MGD significantly outperforms state-of-the-art online learning to rank methods in terms of online performance, without sacrificing offline performance and at greater learning speed than those methods. In particular, we include comparisons between P-MGD on the one hand and multiple types of DBGD and multileaved gradient descent methods [14, MGD] and candidate preselection [5, CPS] on the other. We answer the following research questions: (RQ1) Does P-MGD convergence on a ranker of the same quality as MGD and CPS? (RQ2) Does P-MGD require less queries to converge compared to MGD and CPS? (RQ3) Is the user experience during the learning process of P-MGD better than during that of MGD or CPS?

N. Ferro et al. (Eds.): ECIR 2016, LNCS 9626, pp. 661–668, 2016.
DOI: 10.1007/978-3-319-30671-1_50

2 Probabilistic Multileaving

Multileaving [13] is an online evaluation approach for inferring preferences between rankers from user clicks. Multileave methods take a set of rankers and when a query is submitted a ranking is computed for each of the rankers. These rankings are then combined into a single multileaved ranking. Team Draft Multileaving (TDM) assigns each document in this resulting ranking to a ranker. The user is then presented with this multileaved ranking and his interactions are recorded. TDM keeps track of the clicks and attributes every clicked document to the ranker to which it was assigned. Two important aspects of online evaluation methods are *sensitivity* and *bias*. TDM is more sensitive than existing interleaving methods [3,9,11], since it requires fewer user interactions to infer preferences. Secondly, empirical evaluation also showed that TDM has no significant bias [13]. Probabilistic multileave [15, PM] extends probabilistic interleave [3, PI]. Unlike TDM, PM selects documents from a distribution where the probability of being added correlates with the perceived relevance. It marginalizes over all possible team assignments, which makes it more sensitive and allows it to infer preferences within a virtually unlimited set of rankers from a single interaction. The increased sensitivity of PM and its lack of bias were confirmed empirically [15]. Our novel contribution is that we use PM instead of TDM for inferring preferences in our online learning to rank method, allowing the learner to explore a virtually unlimited set of rankers.

3 Online Learning to Rank Methods

Besides detailing learning to rank baselines, we introduce P-MGD, a variant of MGD.

Dueling Bandit Gradient Descent (DBGD) [16] uses an interleaving method (e.g. Team Draft Interleaving [10]) to infer a *relative* feedback signal: at each interaction with a user the algorithm uses interleaving to infer a preference between its current best ranker and a candidate ranker. If a preference for the candidate is inferred from the interaction DBGD updates the ranker accordingly. With $n = 1$, Algorithm 1 boils down to DBGD.

Multileave Gradient Descent (MGD) [14] is an extension to DBGD that infers preferences with a larger group of candidate rankers using multileaving, as described above. This allows the algorithm to learn and converge faster. MGD is outlined in Algorithm 1. The number of candidates compared at each iteration is set by the parameter n. MGD represents rankers by weight vectors. The ranker that MGD currently considers best is referred to as the current best ranker, initially $\mathbf{w}_0^0$, and is updated according to the user interactions, For each query issue, n candidate rankers are sampled from the unit sphere around the current best ranker. These candidate rankers and the current best ranker create rankings of documents that are subsequently multileaved and the resulting list is shown to the user. Clicks from the user are then interpreted by the multileaving method to infer a preference among the candidates. MGD allows multiple candidates to

Algorithm 1. Multileave Gradient Descent: MGD(n, α, δ, $\mathbf{w}_0^0$)

1: **for** $q_t, t \leftarrow 0..\infty$ **do**
2: $\quad \mathbf{l}^0 \leftarrow generate\_list(\mathbf{w}_t^0, q_t)$ // *ranking of current best*
3: $\quad$ **for** $i \leftarrow 1...n$ **do**
4: $\quad\quad \mathbf{u}^i \leftarrow sample\_unit\_vector()$
5: $\quad\quad \mathbf{w}_t^i \leftarrow \mathbf{w}_t^0 + \delta \mathbf{u}^i$ // *create a candidate ranker*
6: $\quad\quad \mathbf{l}^i \leftarrow generate\_list(\mathbf{w}_t^i, q_t)$ // *exploratory ranking*
7: $\quad \mathbf{m}_t, \mathbf{t}_t \leftarrow multileave(\mathbf{l})$ // *multileaving and teams*
8: $\quad \mathbf{b}_t \leftarrow infer\_winners(\mathbf{t}_t, receive\_clicks(\mathbf{m}_t))$ // *set of winning candidates*
9: $\quad \mathbf{w}_{t+1}^0 \leftarrow \mathbf{w}_t^0 + \alpha \frac{1}{|\mathbf{b}_t|} \sum_{j \in \mathbf{b}_t} \mathbf{u}^j$ // *update, note that* $\mathbf{b}_t$ *could be empty*

be preferred over the current best; we consider the Mean-Winner update approach [14] as it is the most robust; it updates the current best towards the mean of all preferred candidates. The algorithm repeats this for every incoming query, yielding an unending adaptive process.

Probabilistic Multileave Gradient Descent (P-MGD) is introduced in this paper. The novelty of this method comes from the usage of PM instead of TDM as its multileaving method. TDM needs to assign each document to a team in order to infer preferences. This limits the number of rankers that are compared at each interaction to the number of displayed documents. PM on the other hand allows for a virtually unlimited number of rankers to be compared. The advantage of P-MGD is that it can learn faster by having n, the number of candidates, in Algorithm 1 exceed the length of the result list.

Candidate Preselection (CPS) [5], unlike MGD, does not alter the number of candidates compared per impression. It speeds up learning by reusing historical data to select more promising candidates for DBGD. A set of candidates is generated by repeatedly sampling the unit sphere around the current best uniformly. Several rounds are simulated to eliminate all but one candidate. Each round starts by sampling two candidates between which a preference is inferred with Probabilistic Interleave [3, PI]. The least preferred of the two candidates is discarded; if no preference is found, one is discarded at random. The remaining candidate is then used by DBGD.

4 Experimental Setup

We describe our experiments, designed to answer the research questions posed in Sect. 1.[1] An experiment is based on a stream of independent queries submitted by users interacting with the

Table 1. Overview of instantiations of CCM [2].

	$P(click = 1\|R)$					$P(stop = 1\|R)$				
R	0	1	2	3	4	0	1	2	3	4
per	0.0	0.2	0.4	0.8	1.0	0.0	0.0	0.0	0.0	0.0
nav	0.05	0.3	0.5	0.7	0.95	0.2	0.3	0.5	0.7	0.9
inf	0.4	0.6	0.7	0.8	0.9	0.1	0.2	0.3	0.4	0.5

[1] Our experimental code is available at https://bitbucket.org/ilps/lerot.

system that is being trained. A result list of ten documents is displayed in response to each query. Users interact with the list by clicking on zero or more documents. The queries are sampled from several static datasets, clicks are simulated using click models. Four learning to rank datasets [8] were selected to cover a diverse set of tasks: named page finding (*NP2003*, 150 queries), topic distillation (*TD2003*, 50 queries), medical search (*OHSUMED*, 106 queries), and general web search (*MSLR-WEB10K*, 10K queries). Each dataset consists of a set of feature vectors of length 45–136, encoding query document relations, and manual relevance assessments for each document with respect to queries.

To simulate user interactions, we follow [14]. Clicks are produced by a *cascade click model* (CCM) [2]. Users are assumed to examine documents from top to bottom and click with probability $P(click = 1|R)$, conditioned on relevance grade R. The user then stops with probability $P(stop = 1|R)$. Table 1 lists the instantiations of CCM: a *perfect* (per) user with very reliable feedback, a *navigational* (nav) user looking for a singly highly relevant document, and an *informational* (inf) user whose interactions are noisier. Runs consist of 10,000 queries, exceeding the 1,000 queries used in previous work. Performance is evaluated offline and online. Offline NDCG is measured on held out data and represents the quality of the trained ranker. Online performance reflects the user experience during training and is measured by the discounted cumulative NDCG of the result lists shown to the user. A discount factor $\gamma = 0.9995$ was chosen so that queries beyond a horizon of 10,000 impressions have $< 1\,\%$ impact. Two tailed Student's t-tests are used for significance testing. All experiments ran 125 times (5 folds, 25 repetitions), results are averaged. Parameters come from previous work: $\mathbf{w}_0 = \mathbf{0}$, $\alpha = 0.01$, $\delta = 1$. We use two baselines: (PI-DBGD) – DBGD with PI, PI was chosen for a fair comparison with PM methods; and (TD-MGD-9c) – MGD with TDM, the number of candidate rankers is $n = 9$ so that each ranker is represented exactly once in the result list. Furthermore, two additional algorithms are compared to the baselines: (P-MGD-n) – MGD with PM, this algorithm is run with $n = 9$ and $n = 99$ candidates; the former matches the number of candidates with our TD-MGD baseline, the latter exploits the large number of candidates that PM enables; and CPS is run with the settings reported as best in [3]: $\eta = 6$ candidates, $\zeta = 10$ rounds, history length $\lambda = 10$; we use the unbiased version and discard historic interactions without clicks.

5 Results and Analysis

To investigate where P-MGD converges compared to TD-MGD and CPS (RQ1), we consider offline NDCG (Table 2). P-MGD significantly outperforms PI-DBGD on all runs. Compared to TD-MGD, P-MGD performs significantly worse on some datasets; no significant difference can be found for the majority of runs. The number of candidates in P-MGD does not seem to affect the offline performance strongly. Surprisingly, the offline performance of CPS is significantly worse than TD-MGD and DBGD on all runs except for four instances. Figure 1 shows that

Table 2. Offline score (NDCG) after 10,000 query impressions of each of the algorithms for the 3 instantiations of the CCM (see Table 1). Bold values indicate maximum performance per dataset and click model. Statistically significant improvements (losses) over the DBGD and TD-MGD baseline are indicated by $^{\triangle}(p < 0.05)$ and $^{\blacktriangle}(p < 0.01)$ ($^{\triangledown}$and $^{\blacktriangledown}$). Standard deviation in brackets.

		PI-DBGD	TD-MGD-9c	CPS	P-MGD-9c	P-MGD-99c
perfect	**TD2003**	0.330 (0.07)	0.327 (0.07)	0.288 (0.10) ▼▼	0.330 (0.07)	**0.334 (0.07)**
	OHSUMED	0.452 (0.06)	0.443 (0.06)	0.395 (0.06) ▼▼	0.459 (0.07)	**0.459 (0.07)**
	NP2003	0.718 (0.08)	0.727 (0.08)	0.702 (0.09) ▽	0.730 (0.08)	**0.732 (0.08)**
	MSLR-WEB10K	0.318 (0.03)	**0.351 (0.03)**	0.246 (0.03) ▼▼	0.325 (0.03) ▼	0.320 (0.03) ▼
navigational	**TD2003**	0.319 (0.08)	0.322 (0.07)	0.271 (0.10) ▼▼	0.324 (0.08)	**0.332 (0.07)**
	OHSUMED	0.436 (0.06)	**0.446 (0.06)**	0.389 (0.07) ▼▼	0.435 (0.06)	0.431 (0.06) ▽
	NP2003	0.706 (0.08)	0.721 (0.08)	0.702 (0.08)	0.719 (0.08)	**0.723 (0.08)**
	MSLR-WEB10K	0.305 (0.03)	**0.320 (0.03)**	0.238 (0.03) ▼▼	0.309 (0.03) ▼	0.308 (0.03) ▼
informational	**TD2003**	0.288 (0.09)	**0.323 (0.08)**	0.233 (0.10) ▼▼	0.310 (0.09)	0.313 (0.10) △
	OHSUMED	0.421 (0.06)	0.440 (0.06)	0.389 (0.07) ▼▼	**0.441 (0.07)** △	0.434 (0.07)
	NP2003	0.679 (0.08)	0.706 (0.08)	0.665 (0.09) ▼	0.704 (0.08) △	**0.707 (0.08)** ▲
	MSLR-WEB10K	0.283 (0.03)	**0.311 (0.03)**	0.213 (0.03) ▼▼	0.303 (0.03) ▲▽	0.306 (0.03) ▲

the offline performance of CPS drops after an initial peak. CPS seems to overfit because of the effect of historical data on candidate sampling. The other methods sample candidates uniformly, thus noisy false preferences are expected in all directions evenly; therefore, over time they will oscillate in the right direction. Conversely, CPS samples more candidates in the directions that historical data expects the best candidates to be, causing the method not to oscillate but drift due to noise. The increased sensitivity of CPS does not compensate for its bias in the long run.

To answer how the learning speed of P-MGD compares to our baselines (RQ2) we consider Fig. 1, which shows offline performance on the *NP-2003* dataset with the *informational* click model. P-MGD with 99 candidates and CPS perform substantially better than TD-MGD and DBGD during the first 1,000 queries and it takes around 2,000 queries before TD-MGD to reach a similar level of performance. From the substantial difference between P-MGD with 9 and 99 candidates, also present in the other runs, we conclude that P-MGD with a large number of candidates has a greater learning speed.

To answer (RQ3) we evaluate the user experience during learning. Table 3 displays the results of our online experiments. In all runs the online performance of P-MGD significantly improves over DBGD, again showing the positive effect of increasing the number of candidates. Compared to TD-MGD, P-MGD performs significantly better under the *informational* click model. We conclude that P-MGD is a definite improvement over TD-MGD when clicks contain a large amount of noise. We attribute this difference to the greater learning speed of P-MGD: fewer queries are required to find rankers of the same performance as TD-

Table 3. Online score (NDCG) after 10,000 query impressions of each of the algorithms for the 3 instantiations of the CCM (see Table 1). Notation is the same as that of Table 2.

		PI-DBGD	TD-MGD-9c	CPS	P-MGD-9c	P-MGD-99c
perfect	**TD2003**	499.1 (34.9)	557.0 (32.4)	503.7 (42.6) ▼	541.7 (32.3) ▲▼	**563.8** (30.8) ▲
	OHSUMED	780.3 (28.0)	781.4 (21.5)	764.7 (31.5) ▼▼	799.4 (25.6) ▲▲	**819.5** (23.8) ▲▲
	NP2003	1128.8 (27.9)	**1244.9** (31.6)	1222.8 (33.7) ▲▼	1169.3 (25.6) ▲▼	1191.9 (24.9) ▲▼
	MSLR-WEB10K	532.4 (18.5)	548.6 (9.4)	480.8 (41.5) ▼▼	548.3 (9.5) ▲	**560.5** (6.9) ▲▲
navigational	**TD2003**	453.5 (55.1)	**514.9** (33.9)	469.7 (61.0) △▼	480.6 (52.4) ▲▼	499.9 (54.7) ▲▽
	OHSUMED	747.4 (73.6)	**790.2** (27.3)	739.0 (72.4) ▼	776.1 (75.5) ▲	785.3 (75.6) ▲
	NP2003	1055.5 (103.4)	1173.0 (32.3)	**1206.7** (110.9) ▲▲	1085.0 (101.4) △▼	1114.0 (102.9) ▲▼
	MSLR-WEB10K	505.6 (22.9)	523.2 (12.3)	472.5 (54.9) ▼▼	523.2 (47.5) ▲	**538.8** (7.9) ▲▲
informational	**TD2003**	340.2 (76.2)	432.5 (53.2)	443.2 (42.8) ▲	430.1 (37.1) ▲	**490.4** (36.0) ▲▲
	OHSUMED	703.7 (49.8)	749.4 (73.4)	741.1 (31.6) ▲	764.5 (29.3) ▲△	**789.6** (24.7) ▲▲
	NP2003	849.0 (150.0)	1023.7 (101.6)	**1221.6** (27.5) ▲▲	996.0 (47.4) ▲▼	1119.5 (28.5) ▲▲
	MSLR-WEB10K	464.8 (56.0)	495.0 (46.8)	446.1 (46.6) ▼▼	514.3 (16.9) ▲▲	**536.7** (48.6) ▲▲

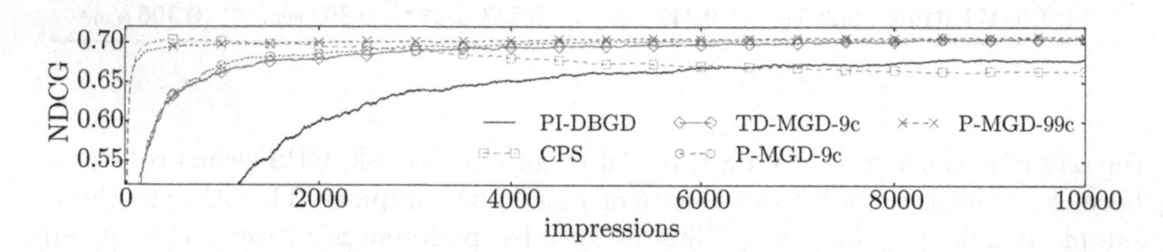

Fig. 1. Offline performance (NDCG) on the *NP-2003* dataset for the *informational* click model.

MGD. Consequently, the rankings shown to users are better during the learning process. When comparing CPS to TD-MGD we see no significant improvements except on the *informational* and *navigational* runs on the *NP-2003* dataset. This is surprising as CPS was introduced as an alternative to DBGD that improves the user experience. Thus, P-MGD is a better alternative of TD-MGD especially when clicks are noisy; CPS does not offer reliable benefits when compared to TD-MGD.

6 Conclusions

We have introduced an extension of multileave gradient descent (MGD) that uses a recently introduced multileaving method, probabilistic multileaving. Our extension, *probabilistic multileave gradient descent* (P-MGD) marginalizes over document assignments in multileaved rankings. P-MGD has an increased sensitivity as it can infer preferences over a large number of assignments. P-MGD can be run with a virtually unlimited number of candidates. We have compared P-MGD with dueling bandit gradient descent (DBGD), team-draft multileave gradient descent (TD-MGD), and candidate preselection (CPS), both offline and online. CPS overfits in terms of offline performance, due to bias introduced by the reuse of historical data. Online results for CPS did not show a convincing benefit over TD-MGD either. In contrast, P-MGD significantly improves over DBGD and TD-MGD in terms of online performance under noisy click models, without a significant decrease in offline performance. Moreover, P-MGD shows a greater learning speed than TD-MGD and DBGD, which becomes more evident as click model noise increases. Thus, P-MGD is a robust alternative for TD-MGD that is better able to deal with interaction noise.

Acknowledgements. This research was supported by Amsterdam Data Science, the Dutch national program COMMIT, Elsevier, the European Community's Seventh Framework Programme (FP7/2007-2013) under grant agreement nr 312827 (VOX-Pol), the ESF Research Network Program ELIAS, the Royal Dutch Academy of Sciences (KNAW) under the Elite Network Shifts project, the Microsoft Research Ph.D. program, the Netherlands eScience Center under project number 027.012.105, the Netherlands Institute for Sound and Vision, the Netherlands Organisation for Scientific Research (NWO) under project nrs 727.011.005, 612.001.116, HOR-11-10, 640.006.013, 612.066.930, CI-14-25, SH-322-15, 652.002.001, the Yahoo Faculty Research and Engagement Program, and Yandex. All content represents the opinion of the authors, which is not necessarily shared or endorsed by their respective employers and/or sponsors.

References

1. Chapelle, O., Joachims, T., Radlinski, F., Yue, Y.: Large-scale validation and analysis of interleaved search evaluation. ACM Trans. Inf. Syst. **30**(1), 41 (2012)
2. Guo, F., Liu, C., Wang, Y.M.: Efficient multiple-click models in web search. In: WSDM 2009. ACM (2009)
3. Hofmann, K., Whiteson, S., de Rijke, M.: A probabilistic method for inferring preferences from clicks. In: CIKM 2011. ACM (2011)
4. Hofmann, K., Whiteson, S., de Rijke, M.: Balancing exploration and exploitation in listwise and pairwise online learning to rank for information retrieval. Inf. Retr. **16**(1), 63–90 (2012)
5. Hofmann, K., Schuth, A., Whiteson, S., de Rijke, M.: Reusing historical interaction data for faster online learning to rank for IR. In: WSDM 2013. ACM (2013)
6. Joachims, T.: Optimizing search engines using clickthrough data. In: KDD 2002. ACM (2002)
7. Joachims, T.: Evaluating retrieval performance using clickthrough data. In: Franke, J., Nakhaeizadeh, G., Renz, I. (eds.) Text Mining. Physica/Springer, Heidelberg (2003)
8. Liu, T.-Y., Xu, J., Qin, T., Xiong, W., Li, H.: LETOR: benchmark dataset for research on learning to rank for information retrieval. In: LR4IR 2007 (2007)
9. Radlinski, F., Craswell, N.: Optimized interleaving for online retrieval evaluation. In: WSDM 2013. ACM (2013)
10. Radlinski, F., Kurup, M., Joachims, T.: How does clickthrough data reflect retrieval quality? In: CIKM 2008. ACM (2008)

11. Radlinski, F., Bennett, P.N., Yilmaz, E.: Detecting duplicate web documents using clickthrough data. In: WSDM 2011. ACM (2011)
12. Sanderson, M.: Test collection based evaluation of information retrieval systems. Found. Tr. Inform. Retr. **4**(4), 247–375 (2010)
13. Schuth, A., Sietsma, F., Whiteson, S., Lefortier, D., de Rijke, M.: Multileaved comparisons for fast online evaluation. In: CIKM 2014, pp. 71–80. ACM, November 2014
14. Schuth, A., Oosterhuis, H., Whiteson, S., de Rijke, M.: Multileave gradient descent for fast online learning to rank. In: WSDM 2016. ACM, February 2016
15. Schuth, A., et al.: Probabilistic multileave for online retrieval evaluation. In: SIGIR 2015 (2015)
16. Yue, Y., Joachims, T.: Interactively optimizing information retrieval systems as a dueling bandits problem. In: ICML 2009 (2009)

Real-World Expertise Retrieval: The Information Seeking Behaviour of Recruitment Professionals

Tony Russell-Rose(✉) and Jon Chamberlain

UXLabs, London, UK
tgr@uxlabs.co.uk, jchamb@essex.ac.uk

Abstract. Recruitment professionals perform complex search tasks in order to find candidates that match client job briefs. In completing these tasks, they have to contend with many core Information Retrieval (IR) challenges such as query formulation and refinement and results evaluation. However, despite these and other similarities with more established information professions such as patent lawyers and healthcare librarians, this community has been largely overlooked in IR research. This paper presents results of a survey of recruitment professionals, investigating their information seeking behaviour and needs regarding IR systems and applications.

1 Introduction

Research into how people find and share expertise can be traced back to the 1960s, with early studies focusing on knowledge workers such as engineers and scientists and the information sources they consult [1]. Since then, the process of finding human experts (or *expertise retrieval*) has been studied in a variety of contexts and become the subject of a number of evaluation campaigns (e.g. the TREC Enterprise track and Entity Track [2, 3]). This has facilitated the development of numerous research systems and prototypes, and led to significant advances in performance, particularly against a range of system-oriented metrics [4].

However, in recent years there has been a growing recognition that the effectiveness of expertise retrieval systems is highly dependent on a number of contextual factors [5]. This has led to a more human-centred approach, where the emphasis is on how people search for expertise in the context of a specific task. These studies have typically been performed in an enterprise context, where the aim is to utilize human knowledge within an organization as efficiently as possible (e.g. [5, 6]).

However, there is a more ubiquitous form of expertise retrieval that embodies expert finding in its purest, most elemental form: the work of the professional recruiter. The job of a recruiter is to find people that are the best match for a client brief and return a list of qualified candidates. This involves the creation and execution of complex Boolean expressions, including nested, composite structures such as the following:

```
Java AND (Design OR develop OR code OR Program) AND ("*
Engineer" OR MTS OR "* Develop*" OR Scientist OR technol-
ogist) AND (J2EE OR Struts OR Spring) AND (Algorithm OR
"Data Structure" OR PS OR Problem Solving)
```


N. Ferro et al. (Eds.): ECIR 2016, LNCS 9626, pp. 669–674, 2016.
DOI: 10.1007/978-3-319-30671-1_51

Over time, many recruiters create their own collection of queries and draw on these as a source of intellectual property and competitive advantage. Moreover, the creation of such expressions is the subject of many community forums (such as Boolean Strings and Undercover Recruiter) and the discussions that ensue involve topics that many IR researchers would recognise as wholly within their field of expertise (such as query expansion and optimisation, results evaluation, etc.).

However, despite these shared interests, the recruitment profession has been largely overlooked by the IR community. Even recent systematic reviews of professional search behaviour make no reference to this profession [7], and their information seeking behaviours remain relatively unstudied. This paper seeks to address that omission. We summarise the results of a survey of 64 recruitment professionals, examining their search tasks and behaviours, and the types of functionality that they value.

2 Background

We are aware of no prior work investigating the recruitment profession from an information seeking perspective. However, there are studies of other professions with related characteristics, such as Joho et al.'s [8] survey of patent searchers and Geschwandtner et al.'s [9] survey of medical professionals.

Unfilled vacancies have high impact on the economy, costing the UK £18bn annually [10]. Recruitment or *sourcing* is the process of finding capable applicants for those vacancies. It is a skill that is to some extent emulated by expert finding systems [4], although recruiters also must take into account contextual variables such as availability, previous experience, remuneration, etc.

Sourcing is also similar to people search on the web where the goal is to analyse large volumes of unstructured and noisy data to return a list of individuals who fit specific criteria [11]. The professional recruiter must normalise and disambiguate the returned results [2], and then apply additional factors to select a smaller group of qualified candidates. The gold standard for evaluation in this case is recommending one or more candidates that successfully fulfil a client brief.

3 Method

The survey instrument consisted of an online questionnaire of 40 questions divided into five sections[1]. It was designed to align with the survey instruments of Joho et al. [8] and wherever possible also with Geschwandtner et al. [9], to facilitate comparisons between the different professions. The five sections were as follows:

1. **Demographics**: The background and professional experience of the respondents.
2. **Search tasks**: The types of search task that respondents perform in their work.
3. **Query formulation**: The approaches and techniques used to construct queries.
4. **Evaluation**: How they assess and evaluate the results of their search tasks.
5. **Ideal search engine**: Any other features additional to those described above.

[1] Available from https://isquared.wordpress.com/.

The survey was designed to be completed in approximately 15 min. To obtain a large and representative sample we sent it out to interest groups via social media and also engaged with SurveyMonkey's panel of HR professionals based in North America. The survey ran from 09-Jun-2015 to 01-Aug-2015. We received 416 responses, of which 69 passed the qualifying question "*Is your primary job function to recruit and hire professionals for your organization or for clients*?" A further five were eliminated due to contradictory or nonsensical answers, which left 64 complete responses.

4 Results

4.1 Demographics

Of the 64 respondents, 69 % were female and 31 % male, with 54 % of respondents aged between 25 and 45 years - a profile that is more female-oriented and younger than the patent and medical search survey respondents. Most respondents worked full time (91 %), and the clients they worked for were predominantly external (48 %) rather than internal (34 %). This contrasts sharply with patent searchers, whose clients were predominantly internal (88 %). Most respondents had several years' experience as a recruiter, with a median of 10 years, which aligns with that of the patent searchers.

4.2 Search Tasks

We then examined the broader query lifecycle. In total, the majority of respondents (80 %) used examples or templates at least sometimes; suggesting that the value embodied in such expressions is recognised and re-used wherever possible. In addition, most respondents (57 %) were prepared to share queries with colleagues in their workgroup and a further 22 % would share more broadly within their organisation. However, very few (5 %) were prepared to share publicly, underlining the competitive nature of the industry. Job boards such as Monster, CareerBuilder and Indeed were the most commonly used databases (77 %), although a similar proportion (73 %) also targeted social networks such as LinkedIn, Twitter and Facebook.

Table 1 shows the amount of time that recruiters spend in completing their most frequently performed search task, the time spent formulating individual queries, and the number of queries they use. On average, it takes around 3 h to complete a search task which consists of roughly 5 queries, with each query taking around 5 min to formulate. This suggests that recruitment follows a largely iterative paradigm, consisting of successive phases of candidate search followed by other activities such as candidate selection and evaluation. Compared to patent search the task completion time is less (3 h vs. 12 h) but is longer than typical web search tasks [12]. Also, the number of queries is fewer (5 compared to 15) but the average query formulation time is the same (5 min).

Table 1. Search effort of recruitment professionals

	Min	Median	Max
Search task completion time (hrs)	0.06	3	30
Query formulation time (mins)	0.1	5	90
Number of queries submitted	1	5	50
Ideal number of results	1	33	1000
Number of results examined	1	30	100000
Time to assess relevance (mins)	1	5	50

4.3 Query Formulation

In this section we examine the mechanics of the query formulation process, by asking respondents to indicate a level of agreement to various statements using a 5-point Likert scale ranging from strong disagreement (1) to strong agreement (5). The results are shown in Fig. 1 as a weighted average across all responses.

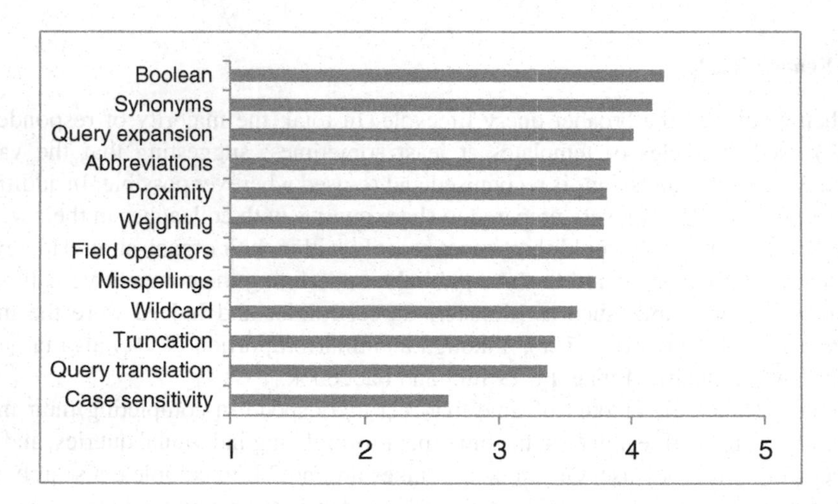

Fig. 1. Important query formulation features

The results suggest two observations in common with patent search. Firstly, the average of all but one of the features is above 3 (neutral), which suggests a willingness to adopt a wide range of search functionality to complete search tasks. Secondly, Boolean logic is shown to be the most important feature (4.25), closely followed by the use of synonyms (4.16) and query expansion (4.02). These scores indicate such functionality is desired by recruiters but the support offered by current search tools is highly variable. On the one hand, support for complex Boolean expressions is provided by many of the popular job boards. However, practical support for query formulation and synonym generation is much more limited, with most current systems still relying instead on the expertise and judgement of the recruiter.

4.4 Results Evaluation

In this section we examine the results evaluation process. Table 1 shows the ideal number of results returned, the number of results examined, and the time taken to assess relevance of a single result. Although the maximum probably represents outlier data, the median time to assess relevance of a single result is the same as that of the patent searchers (5 min). The number of results examined, however, is lower (30 vs. 100), suggesting that recruiters may adopt more of a satisficing strategy, evaluating only as many results as are required to create a shortlist of suitable candidates. This is supported by the median ideal number of returned results being similar (33).

We then asked respondents to indicate on a Likert scale how frequently they use various criteria to narrow down results. Job function was the most important (4.34), followed by location (4.29). These choices mirror some of the factors found to influence expert selection [6]. However, it contrasts with those of the medical searchers, who favoured content-based criteria such as type of source, date range, language, etc.

We also examined recruiters' strategies for interacting with results sets. The most popular approach was to start with the result that looked most relevant (56 %). The number of respondents who targeted the most trustworthy source was relatively low (9 %), which contrasts with the medical professionals and the claim [6] that "*source quality is the most dominant factor in the selection of human information sources*".

5 Discussion and Conclusions

This paper summarises the results of a survey of the information seeking behaviour of recruitment professionals, uncovering their search needs in a manner that allows comparison with other, better-studied professions. In this section we briefly discuss the findings with verbatim quotes from respondents shown in italics where applicable.

Sourcing is shown to be something of a hybrid search task. The goal is essentially a people search task, but, the objects being returned are invariably documents (e.g. CVs and resumes), so the practice also shares characteristics of document search. Recruiters' display a number of professional search characteristics that differentiate their behaviour from web search [13], such as lengthy search sessions, different notions of relevance, different sources searched separately, and the use of specific domain knowledge: "*The hardest part of creating a query is comprehending new information and developing a mental model of the ideal search result.*"

Recruitment professionals use complex search queries, and actively cultivate skills in the formulation of such expressions. The search tasks they perform are inherently interactive, requiring multiple iterations of query formulation and results evaluation "*it is the limitations of available technology that force them to downgrade their concept tree into a Boolean expression*". In contrast with patent searchers, recruiter search behaviour is characterised by satisficing strategies, in which the objective is to identify a sufficient number of qualified candidates in the shortest possible time *"Generally speaking, it's a trade-off between time and quality of results"*. The average time spent evaluating a typical result was 5 min, rather than the 7 s reported in previous eye tracking studies [14].

These findings also have important consequences for the IR community and the assumptions underlying many of its research priorities. For example, much academic research continues to assume that searches are formulated as natural language queries, but this study shows that many professions prefer to formulate their queries as Boolean expressions [15]. In closing, we would hope that these findings may inspire the creation of new test collections focused on recruitment tasks, and thus facilitate the translation of IR research into real-world impact on a growing information profession.

References

1. Menzel, H.: Information needs and uses in science and technology. Ann. Rev. Inf. Sci. Technol. **1**, 41–69 (1966)
2. Balog, K., Soboroff, I., Thomas, P., Craswell, N., de Vries, A.P., Bailey, P.: Overview of the TREC 2008 enterprise track. In: The Seventeenth Text Retrieval Conference Proceedings (TREC 2008), NIST (2009)
3. Balog, K., Serdyukov, P., de Vries, A.P.: Overview of the TREC 2011 entity track. In: Proceedings of the Twentieth Text REtrieval Conference (TREC 2011) (2012)
4. Balog, K., Fang, Y., de Rijke, M., Serdyukov, P., Si, L.: Expertise Retrieval. Found. Trends Inf. Retrieval **6**(2–3), 127–256 (2012)
5. Hofmann, K., Balog, K., Bogers, T., de Rijke, M.: Contextual factors for finding similar experts. Inf. Sci. Technol. **61**(5), 994–1014 (2010)
6. Woudstra, L.S.E., Van den Hooff, B.J.: Inside the source selection process: Selection criteria for human information sources. Inf. Process. Manage. **44**, 1267–1278 (2008)
7. Vassilakaki, E., Moniarou-Papaconstantinou, V.: A systematic literature review informing library and information professionals' emerging roles. New Librar. World **116**(1/2), 37–66 (2015)
8. Joho, H., Azzopardi, L., Vanderbauwhede, W.: A survey on patent users: An analysis of tasks, behavior, search functionality and system requirements. In: Proceedings of the 3rd Symposium on Information Interaction in Context (IIiX 2010) (2010)
9. Geschwandtner, M., Kritz, M., Boyer, C.: D8.1.2: Requirements of the health professional search. Technical report, Khresmoi Project (2011)
10. Cann, J.: IOR Recruitment Sector Report: Report No.1 (UK), Institute of Recruiters (2015)
11. Guan, Z., Miao, G., McLoughlin, R., Yan, X., Cai, D.: Co-occurrence-based diffusion for expert search on the web. IEEE Trans. Knowl. Data Eng. **25**(5), 1001–1014 (2013)
12. Broder, A.: A taxonomy of web search. SIGIR Forum **36**(2), 3–10 (2002)
13. Lupu, M., Salampasis, M., Hanbury, A.: Domain specific search. Professional search in the modern world. Springer International Publishing, pp. 96–117 (2014)
14. Evans, W.: Eye tracking online metacognition: Cognitive complexity and recruiter decision making, TheLadders (2012)
15. Tait, J.: An introduction to professional search, pp. 1–5. Professional search in the modern world. Springer International Publishing, Heidelberg (2014)

Compressing and Decoding Term Statistics Time Series

Jinfeng Rao[1], Xing Niu[1], and Jimmy Lin[2](✉)

[1] University of Maryland, College Park, USA
{jinfeng,xingniu}@cs.umd.edu
[2] University of Waterloo, Waterloo, Canada
jimmylin@uwaterloo.ca

Abstract. There is growing recognition that temporality plays an important role in information retrieval, particularly for timestamped document collections such as tweets. This paper examines the problem of compressing and decoding term statistics time series, or counts of terms within a particular time window across a large document collection. Such data are large—essentially the cross product of the vocabulary and the number of time intervals—but are also sparse, which makes them amenable to compression. We explore various integer compression techniques, starting with a number of coding schemes that are well-known in the information retrieval literature, and build toward a novel compression approach based on Huffman codes over blocks of term counts. We show that our Huffman-based methods are able to substantially reduce storage requirements compared to state-of-the-art compression techniques while still maintaining good decoding performance.

Keywords: Integer compression techniques · Huffman coding

1 Introduction

There is increasing awareness that time plays an important role in many retrieval tasks, for example, searching newswire articles [5], web pages [3], and tweets [2]. It is clear that effective retrieval systems need to model the temporal characteristics of the query, retrieved documents, and the collection as a whole. This paper focuses on the problem of efficiently storing and accessing term statistics time series—specifically, counts of unigrams and bigrams across a moving window over a potentially large text collection. These retrospective term statistics are useful for modeling the temporal dynamics of document collections. On Twitter, for example, term statistics can change rapidly in response to external events (disasters, celebrity deaths, etc.) [6]. Being able to store and access such data is useful for the development of temporal ranking models.

Term statistics time series are large—essentially the cross product of the vocabulary and the number of time intervals—but are also sparse, which makes them amenable to compression. Naturally, we would like to achieve as much

N. Ferro et al. (Eds.): ECIR 2016, LNCS 9626, pp. 675–681, 2016.
DOI: 10.1007/978-3-319-30671-1_53

compression as possible to minimize the storage requirements, but this needs to be balanced with decoding latencies, as the two desiderata are often in tension.[1] Our work explores this tradeoff.

The contribution of this paper is an exploration of compression techniques for term statistics time series. We begin with a number of well-known integer compression techniques and build toward a novel approach based on Huffman codes over blocks of term counts. We show that our Huffman-based techniques are able to substantially reduce storage requirements compared to state-of-the-art compression techniques while still maintaining good decoding performance. Our contribution enables retrieval systems to load large amounts of time series data into main memory and access term statistics with low latency.

2 Background and Related Work

We adopt the standard definition of a time series as a finite sequence of n real numbers, typically generated by some underlying process for a duration of n time units: $x = \{x_0, x_1, x_2, \ldots, x_n\}$, where each x_n corresponds to the value of some attribute at a point in time. In our case, these time series data correspond to counts on a stream of timestamped documents (tweets in our case) at fixed intervals (e.g., hourly). To be precise, these term statistics represent collection frequencies of unigrams and bigrams from a "temporal slice" of the document collection consisting of documents whose timestamps fall within the interval.

There has been much previous work, primarily in the database and data mining communities, on analyzing and searching time series data. We, however, focus on the much narrower problem of compressing and decoding time series data for information retrieval applications. There are a number of well-known integer coding techniques for compressing postings lists in inverted indexes: these include variable-byte encoding, γ codes, interpolative coding, the Simple-9 family [1], and PForDelta [9]. Various compression techniques represent different tradeoffs between degree of compression and decoding speed, which have been well studied in the indexing context. Note that our problem is different from that of postings compression: postings lists only keep track of documents that contain the term, and hence differ in length, whereas in our case we are also interested in intervals where a term *does not* appear.

3 Methods

In this work, we assume that counts are aggregated at five minute intervals, so each unigram or bigram is associated with $24 \times 60/5 = 288$ values per day. Previous work [7] suggests that smaller windows are not necessary for most applications, and coarser-grained statistics can always be derived via aggregation.

We compared five basic integer compression techniques: variable-byte encoding (VB) [8], Simple16 [1], PForDelta (P4D) [9], discrete wavelet transform

[1] We set aside compression speed since we are working with retrospective collections.

(DWT) with Haar wavelets, and variants of Huffman codes [4]. The first three are commonly used in IR applications, and therefore we simply refer readers to previous papers for more details. We discuss the last two in more detail.

Discrete Wavelet Transform (DWT). The discrete wavelet transform enables time-frequency localization to capture both frequency information and when (in time) those frequencies are observed. In this work, we use Haar wavelets. To illustrate how DWT with Haar wavelets work, we start with a simple example. Suppose we have a time series with four values: $X = \{7, 9, 5, 3\}$. We first perform pairwise averaging to obtain a lower resolution signal with the values: $\{8, 4\}$. The first value is obtained by averaging $\{7, 9\}$ and the second by averaging $\{5, 3\}$. To account for information lost in the averaging, we store detail coefficients equal to pairwise differences of $\{7, 9\}$ and $\{5, 3\}$, divided by two. This yields $\{-1, 1\}$, which allows us to reconstruct the original signal perfectly. Assuming a signal with 2^n values, we can recursively apply this transformation until we end up with an average of all values. The final representation of the signal is the final average and all the detail coefficients. This transformation potentially yields a more compact representation since the detail coefficients are often smaller than the original values. We further compress the coefficients using either variable-byte encoding or PForDelta. Since the coefficients may be negative, we need to store the signs (in a separate bit array).

Huffman Coding. A nice property of Huffman coding [4] is that it can find the optimal prefix code for each symbol when the frequency information of all symbols are given. In our case, given a list of counts, we first partition the list into several *blocks*, with each block consisting of eight consecutive integers. After we calculate the frequency counts of all blocks, we are able to construct a Huffman tree over the blocks and obtain a code for each block. We then concatenate the binary Huffman codes of all blocks and convert this long binary representation into a sequence of 32-bit integers. Finally, we can apply any compression method on top of these integer sequences. To decode, we first decompress the integer array into its binary representation. Then, this binary code is checked bit by bit to determine the boundaries of the original Huffman codes. Once the boundary positions are obtained, we can recover the original integer counts by looking up the Huffman code mapping. The decoding time is linear with respect to the length of Huffman codes after concatenation.

Beyond integer compression techniques, we can exploit the sparseness of unigram counts to reduce storage for bigram counts. There is no need to store the bigram count if any unigram of that bigram has a count of zero at that specific interval. For example, suppose we have count arrays for unigram A, B and bigram AB below: A: 00300523, B: 45200103, and AB: 00100002. In this case, we only need to store the 3rd, 6th, and 8th counts for bigram AB (that is, 102), while the other counts can be dropped since at least one of its unigrams has count zero in those intervals. To keep track of these positions we allocate a bit vector 288 bits long (per day) and store this bit vector alongside the compressed data. This truncation technique saves space but at the cost of an additional step during decoding. When recovering the bigram counts, we need to consult the bit vector, which is used to pad zeros in the truncated count array accordingly.

In terms of physical storage, we maintain a global array by concatenating the compressed representations for all terms across all days. To access the compressed array for a term on a specific day, we need its offset and length in the global array. Thus, we keep a separate table of the mapping from (term id, day) to this information. Although in our experiments we assume that all data are held in main memory, our approach can be easily extended to disk-based storage.

As an alternative, instead of placing data for all unigrams and bigrams for all days together, we could partition the global array into several shards with each shard containing term statistics for a particular day. The advantage of this design is apparent: we can select which data to load into memory when the global array is larger than the amount of memory available.

4 Experiments

We evaluated our compression techniques in terms of two metrics: size of the compressed representation and decoding latency. For the decoding latency experiments, we iterated over all unigrams or bigrams in the vocabulary, over all days, and report the average time it takes to decode counts for a single day (i.e., 288 integers). All our algorithms were implemented in Java and available open source.[2] Experiments were conducted on a server with dual Intel Xeon 4-core processors (E5620 2.4 GHz) and 128 GB RAM.

Our algorithms were evaluated over the Tweets2011 and Tweets2013 collections. The Tweets2011 collection consists of an approximately 1 % sample of tweets from January 23, 2011 to February 7, 2011 (inclusive), totaling approximately 16 m tweets. This collection was used in the TREC 2011 and TREC 2012 microblog evaluations. The Tweets2013 collection consists of approximately 243 m tweets crawled from Twitter's public sample stream between February 1 and March 31, 2013 (inclusive). This collection was used in the TREC 2013 and TREC 2014 microblog track evaluations. All non-ASCII characters were removed in the preprocessing phase. We set a threshold (by default, greater than one per day) to filter out all low frequency terms (including unigrams and bigrams). We extracted a total of 0.7 m unigrams and 7.3 m bigrams from the Tweets2011 collection; 2.3 m unigrams and 23.1 m bigrams from the Tweets2013 collection.

Results are shown in Table 1. Each row denotes a compression method. The first row "Raw" is the collection without any compression (i.e., each count is represented by a 32-bit integer). The row "VB" denotes variable-byte encoding; row "P4D" denotes PForDelta. Next comes the wavelet and Huffman-based techniques. The last row "Optimal" shows the optimal storage space with the lowest entropy to represent all Huffman blocks. Given the frequency information of all blocks, the optimal space can be computed by summing over the entropy bits

[2] https://github.com/Jeffyrao/time-series-compression.

consumed by each block (which is also the minimum bits to represent a block). The column "size" represents the compressed size of all data (in base two). To make comparisons fair, instead of comparing with the (uncompressed) raw data, we compared each approach against PForDelta, which is considered state of the art in information retrieval for coding sequences such as postings lists [9]. The column "percentage" shows relative size differences with respect to PForDelta. The column "time" denotes the decompression time for each count array (the integer list for one term in one day).

Table 1. Results on the Tweets2011 (top) and Tweets2013 (bottom) collections.

Tweets2011	Unigrams			Bigrams		
Method	size (MB)	percentage	time (μs)	size (MB)	percentage	time (μs)
Raw	4760			12800		
VB	1200	+442 %	1.9	3200	+318 %	1.1
Simple16	200	−9.50 %	1.1	653	−14.6 %	0.7
P4D	221	–	1.0	764	–	1.2
Wavelet+VB	1300	+488 %	2.3	3700	+384 %	2.3
Wavelet+P4D	352	+59.3 %	2.7	978	+28.0 %	2.3
Huffman	65	−70.6 %	7.8	396	−48.2 %	2.9
Huffman+VB	46	−79.2 %	8	180	−76.4 %	3.2
Optimal	32	−85.5 %	–	108	−85.9 %	-
Tweets2013	Unigrams			Bigrams		
Method	size (GB)	percentage	time (μs)	size (GB)	percentage	time (μs)
Raw	52.5			171.8		
VB	13.1	+446 %	3.8	43.0	+347 %	1.3
Simple16	2.2	−8.33 %	2.2	8.3	−13.5 %	0.8
P4D	2.4	–	1.9	9.6	–	1.2
Wavelet+VB	14.8	+517 %	6.4	49.0	+410 %	2.6
Wavelet+P4D	3.8	+58.3 %	4.7	12.9	+34.4 %	6.2
Huffman	0.71	−70.4 %	14.7	4.9	−49.0 %	6.2
Huffman+VB	0.48	−80.0 %	15.6	3.0	−68.7 %	6.3
Optimal	0.33	−86.2 %	-	0.95	−90.1 %	–

Results show that both Simple16 and PForDelta are effective in compressing the data. Simple16 achieves better compression, but for unigrams is slightly slower to decode. Variable-byte encoding, on the other hand, does not work particularly well: the reason is that our count arrays are aggregated over a relative small temporal window (five minutes) and therefore term counts are generally small. This enables Simple16 and PForDelta to represent the values using very few bits. In contrast, VB cannot represent an integer using fewer than eight bits.

We also noticed that the Wavelet+VB and Wavelet+P4D techniques require more space than just VB and PForDelta alone, which suggests that the wavelet transform is not effective. We believe this increase comes from: (1) DWT requires an additional array to store the sign bits of the coefficients, and (2) since the original counts were already sparse, DWT does not additionally help.

The decoding times for VB, Simple16, PForDelta, and the wavelet methods are all quite small, and it is interesting to note that decoding bigrams can be actually *faster* than decoding unigrams, which suggests that our masking mechanism is effective in reducing the length of the bigram count arrays.

Experiments show that we are able to achieve substantial compression with the Huffman-based techniques, up to 80 % reduction over PForDelta. Overall, our findings hold consistently over both the Tweets2011 and Tweets2013 collections. In fact, Huffman+VB is pretty close to the entropy lower bound. Entropy coding techniques like Huffman coding prefer highly non-uniform frequency distributions, and thus are perfectly suited to our time series data. Although our Huffman+VB technique also increases decoding time, we believe that this tradeoff is worthwhile, but of course, this is application dependent. We did not try to combine Huffman coding with Simple16 or PForDelta as we found that the integer lists transformed from Huffman codes were generally composed of large values, which are not suitable for word-aligned compression methods.

5 Conclusion

The main contribution of our work is an exploration of integer compression techniques for term statistics time series. We demonstrated the effectiveness of our novel techniques based on Huffman codes, which exploit the sparse and highly non-uniform distribution of blocks of counts. Our best technique can reduce storage requirements by a factor of four to five compared to PForDelta encoding. A small footprint means that it is practical to load large amounts of term statistics time series into memory for efficient access.

Acknowledgments. This work was supported in part by the U.S. National Science Foundation under IIS-1218043. Any opinions, findings, conclusions, or recommendations expressed are those of the authors and do not necessarily reflect the views of the sponsor.

References

1. Anh, V.N., Moffat, A.: Inverted index compression using word-aligned binary codes. Inf. Retrieval **8**(1), 151–166 (2005)
2. Busch, M., Gade, K., Larson, B., Lok, P., Luckenbill, S., Lin, J.: Earlybird: real-time search at Twitter. In: ICDE (2012)
3. Elsas, J.L., Dumais, S.T.: Leveraging temporal dynamics of document content in relevance ranking. In: WSDM (2010)
4. Huffman, D.A., et al.: A method for the construction of minimum redundancy codes. Proc. IRE **40**(9), 1098–1101 (1952)

5. Jones, R., Diaz, F.: Temporal profiles of queries. ACM TOIS **25**, Article no. 14 (2007)
6. Lin, J., Mishne, G.: A study of "churn" in tweets and real-time search queries. In: ICWSM (2012)
7. Mishne, G., Dalton, J., Li, Z., Sharma, A., Lin, J.: Fast data in the era of big data: Twitter's real-time related query suggestion architecture. In: SIGMOD (2013)
8. Williams, H.E., Zobel, J.: Compressing integers for fast file access. Comput. J. **42**(3), 193–201 (1999)
9. Zhang, J., Long, X., Suel, T.: Performance of compressed inverted list caching in search engines. In: WWW (2008)

Feedback or Research: Separating Pre-purchase from Post-purchase Consumer Reviews

Mehedi Hasan[1], Alexander Kotov[1(✉)], Aravind Mohan[1], Shiyong Lu[1], and Paul M. Stieg[2]

[1] Department of Computer Science, Wayne State University, Detroit, MI 48202, USA
{mehedi,kotov,aravind.mohan,shiyong}@wayne.edu
[2] Ford Motor Co., Dearborn, MI 48124, USA
pstieg@ford.com

Abstract. Consumer reviews provide a wealth of information about products and services that, if properly identified and extracted, could be of immense value to businesses. While classification of reviews according to sentiment polarity has been extensively studied in previous work, more focused types of review analysis are needed to assist companies in making business decisions. In this work, we introduce a novel text classification problem of separating post-purchase from pre-purchase review fragments that can facilitate identification of immediate actionable insights based on the feedback from the customers, who actually purchased and own a product. To address this problem, we propose the features, which are based on the dictionaries and part-of-speech (POS) tags. Experimental results on the publicly available gold standard indicate that the proposed features allow to achieve nearly 75 % accuracy for this problem and improve the performance of classifiers relative to using only lexical features.

Keywords: Text classification · Consumer reviews · E-commerce

1 Introduction

The content posted on online consumer review platforms contains a wealth of information, which besides positive and negative judgments about product features and services, often includes specific suggestions for their improvement and root causes for customer dissatisfaction. Such information, if accurately identified, could be of immense value to businesses. Although previous research on consumer review analysis has resulted in accurate and efficient methods for classifying reviews according to the overall sentiment polarity [8], segmenting reviews into aspects and estimating the sentiment score of each aspect [12], as well as summarizing both aspects and sentiments towards them [6,10,11], more focused types of review analysis, such as detecting the intent or the timing of reviews, are needed to better assist companies in making business decisions. One such problem is separating the reviews (or review fragments) written by the users

N. Ferro et al. (Eds.): ECIR 2016, LNCS 9626, pp. 682–688, 2016.
DOI: 10.1007/978-3-319-30671-1_53

after purchasing and using a product or a service (which we henceforth refer to as "post-purchase" reviews) from the reviews that are written by the users, who shared their expectations or results of research before purchasing and using a product (which we henceforth refer to as "pre-purchase" reviews).

We hypothesize that effective separation of these two types of reviews (or review fragments) can allow businesses to better understand the aspects of products and services, which the customers are focused on before and after the purchase and tailor their marketing strategies accordingly. It can also allow businesses to measure the extent to which the customer expectations are met by their existing products and services. Furthermore, "post-purchase" reviews, particularly the negative ones, can be considered as "high priority" reviews, since they provide customer feedback, which needs to be immediately acted upon by manufacturers. Such feedback typically contains reports of malfunctions, as well as poor performance of products that are already on the market. Pre-purchase reviews, on the other hand, are likely to be written for expensive products that are major purchasing decisions and require extensive research prior to purchase (e.g. cameras, motorcycles, boats, cars, etc.). Such products typically have communities of enthusiasts around them, who often post reviews of the product models they have only heard or read about.

In this work, we introduce a novel text classification problem of separating pre-purchase from post-purchase consumer review fragments. While, in some cases, the presence of past tense verb(s) or certain keywords in a given review fragment provides a clear clue about its timing with respect to purchase (e.g. "excellent vehicle, great price and the dealership provides very good service"), other cases require distinguishing subtle nuances of language use or making inferences. For example, although the past tense verbs in "The new Ford Explorer is a great looking car. I heard it has great fuel economy for an SUV" and "so far this is the best car I tested" indicate prior experience, these review fragments are written by the users, who didn't actually purchase these products. Despite an overall positive sentiment of these review fragments, they provide no specific information to the manufacturer about how these cars can be improved. On the other hand, while the fragment "If I could, I would have two" contains modal verbs, it is clearly post-purchase.

To address the proposed problem, we propose and evaluate the effectiveness of the features based on dictionaries and part-of-speech (POS) tags, in addition to the lexical ones.

2 Related Work

Although consumer reviews have been a subject of many studies over the past decade, a common trend of recent research is to move from detecting sentiments and opinions in online reviews towards the broader task of extracting actionable insights from customer feedback. One relevant recent line of work focused just on detecting wishes [5,9] in reviews or surveys. In particular, Goldberg et al. [5] studied how wishes are expressed in general and proposed a template-based method for detecting the wishes in product reviews and political discussion posts, while Ramanand et al. [9] proposed a method to identify suggestions in product reviews. Moghaddam [7] proposed a method based on distant supervision to detect the reports of defects and suggestions for product improvements.

Other non-trivial textual classification problems have also been recently studied the literature. For example, Bergsma et al. [2] used a combination of lexical and syntactic features to detect whether the author of a scientific article is a native English speaker, male or female, or whether an article was published in a conference or a journal, while de Vel et al. [3] used style markers, structural characteristics and gender-preferential language as features for the task of gender and language background detection.

3 Experiments

3.1 Gold Standard, Features and Classifiers

To create the gold standard for experiments in this work[1], we collected the reviews of all major car makes and models released to the market in the past 3 years from MSN Autos[2]. Then we segmented the reviews into individual sentences, removed punctuation except exclamation (!) and question (?) marks (since [1] suggest that retaining them can improve the results of some classification tasks), and annotated the review sentences using Amazon Mechanical Turk. In order to reduce the effect of annotator bias, we created 5 HITs per each label and used the majority voting scheme to determine the final label for each review sentence. In total, the gold standard consists of 3983 review sentences. Table 3 shows the distribution of these sentences over classes. We used unigram bag-of-words lexical feature representation for each review fragment as a baseline, to which we added four binary features based on the dictionaries and four binary features based on the POS tag patterns that we manually compiled as described in Sect. 3.2. We used Naive Bayes (NB), Support Vector Machine (SVM) with linear kernel implemented in Weka machine learning toolkit[3], as well as L2-regularized Logistic Regression (LR) implemented in LIBLINEAR[4] [4] as classification methods. All experimental results reported in this work were obtained using 10-fold cross validation and macro-averaged over the folds.

3.2 Dictionaries and POS Patterns

Each of the dictionaries contain the terms, which represent a particular concept related to product experience, such as negative emotion, ownership, satisfaction etc. To create the dictionaries, we first came up with a small set of seed terms,

1 Gold standard and dictionaries are available at http://github.com/teanalab/prepost.
2 http://www.msn.com/en-us/autos.
3 http://www.cs.waikato.ac.nz/ml/weka.
4 http://www.csie.ntu.edu.tw/~cjlin/liblinear.

Table 1. Dictionaries with associated words and phrases.

Dictionary	Words
OWNERSHIP	own, ownership, owned, mine, individual, personal, etc
PURCHASE	buy, bought, acquisition, purchase, purchased, etc
SATISFACTION	happy, cheerful, content, delighted, glad, etc
USAGE	warranty, guarantee, guaranty, cheap, cheaper, etc

Table 2. POS patterns with examples.

Pattern type	Patterns	Example
OWNERSHIP	**PRP$ CD**, PRP VBD, VBZ PRP$, VBD PRP$, etc.	this is **my third** azera from 2008 to 2010 until now a 2012
QUALITY	JJ, JJR, **JJS**	it is definitely the **best** choice for my family
MODALITY	**PRP MD**, IN PRP VBP	buy one **you will** love it
EXPERIENCE	VBD, **VBN**	i have **driven** this in the winter and the all wheel drive model

such as "buy", "own", "happy", "warranty", that capture the key lexical clues related to the timing of review creation regardless of any particular type of product. Then, we used on-line thesaurus[5] to expand the seed words with their synonyms and considered each resulting set of words as a dictionary.

Using similar procedure, we also created a small set of POS tag-based patterns that capture the key syntactic clues related to the timing of review creation with respect to the purchase of a product. For example, the presence of sequences of possessive pronouns and cardinal numbers (pattern "PRP$ CD", e.g. matching the phrases "my first", "his second", etc.), personal pronouns and past tense verbs (pattern "PRP VBD", e.g. matching "I owned") or modal (pattern "PRP MD", e.g. matching "I can", "you will", etc.) verbs, past participles (pattern "VBN", e.g. matching "owned or driven"), as well as adjectives, including comparative and superlative (patterns "JJ", "JJR" and "JJS") indicates that a review is likely to be post-purchase. More examples of dictionary words and POS patterns are provided in Tables 1 and 2.

4 Results and Discussion

4.1 Classification of Post-purchase vs. Pre-purchase Reviews Using only Lexical Features

Table 4 shows the performance of different classifiers for the task of separating post-purchase from pre-purchase reviews using only lexical features. From the

[5] http://www.thesaurus.com.

Table 3. Distribution of classes in experimental dataset.

Class	# Samp.	Fraction
pre-purchase	2122	53.28 %
post-purchase	1861	46.72 %
Total	**3983**	**100 %**

Table 4. Performance of different classifiers using only lexical features. The highest value of each performance metric among all classifiers is highlighted in boldface.

Method	Precision	Recall	F1	Accuracy
SVM	**0.734**	0.724	0.717	0.724
LR	0.729	**0.726**	**0.722**	**0.726**
NB	0.703	0.704	0.702	0.704

results in Table 4, it follows that LR outperforms SVM in terms of all performance metrics except precision and that both of them outperform Naive Bayes on average by 2.0 % across all performance metrics.

4.2 Classification of Post-purchase vs. Pre-purchase Reviews Using Combination of Lexical, Dictionary and POS Pattern Features

Results for the second set of experiments, aimed at determining the relative performance of SVM, NB and LR classifiers in conjunction with: (1) combination of lexical and POS pattern-based features; (2) combination of lexical and dictionary-based features; (3) combination of all three feature types (lexical, dictionary and POS pattern features) are presented in Table 5, from which several conclusions regarding the influence of non-lexical features on performance of different classifiers for this task can be made.

First, we can observe that SVM achieves the highest performance among all classifiers in terms of precision (0.752), recall (0.743) and accuracy (0.743), when a combination of lexical, POS pattern and dictionary-based features was used.

Table 5. Performance of classifiers using different combinations of lexical with dictionary and POS pattern based features. The percentage improvement is relative to using only lexical features by the same classifier. The highest value and largest improvement of each performance metric for a particular feature combination is highlighted in boldface and italic, respectively.

Method	Precision	Recall	F1 score	Accuracy
SVM + POS	**0.733**	0.727	0.722 (+0.70 %)	0.727 (+0.41 %)
LR + POS	**0.733**	**0.730**	**0.727** (+0.70 %)	**0.730** (+0.55 %)
NB + POS	0.709	0.710	0.709 (*+1.0 %*)	0.710 (*+0.85 %*)
SVM + Dictionary	**0.750**	**0.741**	**0.735** (*+2.51 %*)	**0.741** (*+2.35 %*)
LR + Dictionary	0.740	0.736	0.733 (+1.52 %)	0.736 (+1.38 %)
NB + Dictionary	0.713	0.714	0.713 (+1.57 %)	0.714 (+1.42 %)
SVM + POS + Dictionary	**0.752**	**0.743**	**0.738** (*+2.93 %*)	**0.743** (*+2.62 %*)
LR + POS + Dictionary	0.745	0.741	**0.738** (+2.22 %)	0.741 (+2.07 %)
NB + POS + Dictionary	0.717	0.718	0.717 (+2.14 %)	0.718 (+1.99 %)

Second, using POS pattern-based features in addition to lexical ones allowed LR to achieve the highest performance in terms of all metrics and resulted in the highest improvement for NB classifier, while using a combination of lexical, dictionary and POS pattern-based features is more effective for SVM than for both NB and LR. Overall, experimental results presented above indicate that dictionary and POS pattern features, as well as their combination, allow to improve the performance of all classifiers for the task of separating pre-purchase from post-purchase review fragments relative to using only lexical features.

5 Conclusion

In this paper, we introduced a novel problem of separating pre- from post-purchase consumer review fragments, which can facilitate identification of immediate actionable insights from customer feedback, and found out that combining lexical features with the ones based on dictionaries and POS patterns improves the performance of all classification models we experimented with to address this problem.

Acknowledgements. This work was supported in part by an unrestricted gift from Ford Motor Company.

References

1. Barbosa, L., Feng, J.: Robust Sentiment detection on Twitter from biased and noisy data. In: Proceedings of the 23rd COLING, pp. 36–44 2010)
2. Bergsma, S., Post, M., Yarowsky, D.: Stylometric analysis of scientific articles. In: Proceedings of the NAACL-HLT, pp. 327–337 (2012)
3. de Vel, O.Y., Corney, M.W., Anderson, A.M., Mohay, G.M.: Language and gender author cohort analysis of e-mail for computer forensics. In: Proceedings of the Digital Forensics Workshop (2002)
4. Fan, R.E., Chang, K.W., Hsieh, C.J., Wang, X.R., Lin, C.J.: LIBLINEAR: a library for large linear classification. J. Mach. Learn. Res. **9**, 1871–1874 (2008)
5. Goldberg, A.B., Fillmore, N., Andrzejewski, D., Xu, Z., Gibson, B., Zhu, X.: May all your wishes come true: a study of wishes and how to recognize them. In: Proceedings of the NAACL-HLT, pp. 263–271 (2009)
6. Hu, M., Liu, B.: Mining and summarizing customer reviews. In: Proceedings of the 10th ACM SIGKDD, pp. 168–177 (2004)
7. Moghaddam, S.: Beyond sentiment analysis: mining defects and improvements from customer feedback. In: Hanbury, A., Kazai, G., Rauber, A., Fuhr, N. (eds.) ECIR 2015. LNCS, vol. 9022, pp. 400–410. Springer, Heidelberg (2015)
8. Pang, B., Lee, L.: Opinion mining and sentiment analysis. Found. Trends Inf. Retrieval **2**(1–2), 1–135 (2008)
9. Ramanand, J., Bhavsar, K., Pedanekar, N.: Wishful thinking: finding suggestions and 'buy' wishes from product reviews. In: Proceedings of the NAACL-HLT Workshop on Computational Approaches to Analysis and Generation of Emotion in Text, pp. 54–61 (2010)

10. Titov, I., McDonald, R.T.: A joint model of text and aspect ratings for sentiment summarization. In: Proceedings of the 46th ACL, pp. 308–316 (2008)
11. Yang, Z., Kotov, A., Mohan, A., Lu, S.: Parametric and non-parametric user-aware sentiment topic models. In: Proceedings of the 38th ACM SIGIR, pp. 413–422 (2015)
12. Yu, J., Zha, Z.J., Wang, M., Chua, T.-S.: Aspect ranking: identifying important product aspects from online consumer reviews. In: Proceedings of the 49th ACL, pp. 1496–1505 (2011)

Inferring the Socioeconomic Status of Social Media Users Based on Behaviour and Language

Vasileios Lampos[1(✉)], Nikolaos Aletras[1], Jens K. Geyti[1], Bin Zou[1], and Ingemar J. Cox[1,2]

[1] Department of Computer Science, University College London, London, UK
v.lampos@ucl.ac.uk
[2] Department of Computer Science, University of Copenhagen, Copenhagen, Denmark

Abstract. This paper presents a method to classify social media users based on their socioeconomic status. Our experiments are conducted on a curated set of Twitter profiles, where each user is represented by the posted text, topics of discussion, interactive behaviour and estimated impact on the microblogging platform. Initially, we formulate a 3-way classification task, where users are classified as having an upper, middle or lower socioeconomic status. A nonlinear, generative learning approach using a composite Gaussian Process kernel provides significantly better classification accuracy (75 %) than a competitive linear alternative. By turning this task into a binary classification – upper vs. medium and lower class – the proposed classifier reaches an accuracy of 82 %.

Keywords: Social media · Twitter · User profiling · Socioeconomic status · Classification · Gaussian Process

1 Introduction

Online information has been used in recent research to derive new or enhance our existing knowledge about the *physical* world. Some examples include the use of social media or search query logs to model financial indices [1], understand voting intentions [10] or improve disease surveillance [4,8,9]. At the same time, complementary studies have focused on characterising individual users or specific groups of them. It has been shown that user attributes, such as age [15], gender [2], impact [7], occupation [14] or income [13], can be inferred from Twitter profiles. This automatic and often large-scale information extraction has commercial and research applications, from improving personalised advertisements to facilitating answers to various questions in the social sciences.

This paper presents a method for classifying social media users according to their socioeconomic status (SES). SES can be broadly defined as one's access to financial, social, cultural, and human capital resources; it also includes additional components such as parental and neighbourhood properties [3]. We focused our work on the microblogging platform of Twitter and formed a new data set of user

N. Ferro et al. (Eds.): ECIR 2016, LNCS 9626, pp. 689–695, 2016.
DOI: 10.1007/978-3-319-30671-1_54

profiles together with a SES label for each one of them. To map users to a SES, we utilised the Standard Occupation Classification (SOC) hierarchy, a broad taxonomy of occupations attached to socioeconomic categorisations in conjunction with the National Statistics Socio-Economic Classification (NS-SEC) [5,17].

Users are represented by a broad set of features reflecting their behaviour and impact on the social platform. The classification task uses a nonlinear, kernelised method that can more efficiently capture the divergent feature categories. Related work has looked into different aspects of this problem, such as inferring the job category [14] or the income (as a regression task [13]) of social media users. As with our work here, nonlinear methods showed better performance in these tasks as well. However, the previously proposed models did not jointly explore the various sets of features reported in this paper. The proposed classifier achieves a strong performance in both 3-way and binary classification scenarios.

2 Data Set and Task Description

Our analysis is conducted on a set of 1,342 Twitter user profiles located in the UK[1] and their corresponding tweets from February 1, 2014 to March 21, 2015 inclusive (2,082,651 tweets in total; denoted by $\mathcal{D}_1$). The user selection was performed by searching for occupation mentions in the profile description field of a pool of approximately 100,000 UK Twitter users. An extensive taxonomy of occupations was obtained from the SOC hierarchy. We have manually supervised this process, removing accounts where the assigned occupation was incorrect or uncertain. Accounts that were not related to individuals (e.g. representing an organisation) were not considered.

We have also created an additional data set by randomly sampling all UK tweets posted in the same exact period as $\mathcal{D}_1$ (159,101,560 tweets were sampled; denoted by $\mathcal{D}_2$). $\mathcal{D}_2$ was used to automatically compile a set of latent topics that Twitter users were communicating about.

From $\mathcal{D}_1$, we extracted the following five user **feature categories**:

$\mathbf{c_1}$: Platform-based **behaviour** as represented by the proportion (over the total number of tweets) of retweets, mentions of, unique mentions of and replies to other user accounts.

$\mathbf{c_2}$: Platform **impact** expressed by the number of accounts followed (followees), followed by (followers), times listed (bookmarked) as well as a user impact score (defined in [7]) that combines the previous metrics.

$\mathbf{c_3}$: Keywords (1-grams and 2-grams) present in a user's **profile description**.

$\mathbf{c_4}$: The frequency of the 1-grams present in a user's **tweets**. The frequency of a 1-gram x for a user i is defined as $z_i = |x_i|/N_i$, where N_i denotes the total number of tweets for i and $|x_i|$ is the number of appearances of x in them.

$\mathbf{c_5}$: A frequency distribution across a set of 200 latent **topics** represented by clusters of 1-grams. The frequency of a topic τ for a user i is defined as $\tau_i = \sum_{z_i \in \tau} z_i$, where $z_i \in \tau$ denotes the frequency (defined above) of a 1-gram that belongs to the cluster of 1-grams (topic) τ.

[1] Inferred from the location name provided in the user profile description.

Table 1. 1-gram samples from a subset of the 200 latent topics (word clusters) extracted automatically from Twitter data ($\mathcal{D}_2$).

Topic	Sample of 1-grams
Corporate	#business, clients, development, marketing, offices, product
Education	assignments, coursework, dissertation, essay, library, notes, studies
Family	#family, auntie, dad, family, mother, nephew, sister, uncle
Internet Slang	ahahaha, awwww, hahaa, hahahaha, hmmmm, loooool, oooo, yay
Politics	#labour, #politics, #tories, conservatives, democracy, voters
Shopping	#shopping, asda, bargain, customers, market, retail, shops, toys
Sports	#football, #winner, ball, bench, defending, footballer, goal, won
Summertime	#beach, #sea, #summer, #sunshine, bbq, hot, seaside, swimming
Terrorism	#jesuischarlie, cartoon, freedom, religion, shootings, terrorism

The dimensionality of user attributes $\mathbf{c}_3$ and $\mathbf{c}_4$, after filtering out stop words and n-grams occurring less than two times in the data, was equal to 523 (1-grams plus 2-grams) and 560 (1-grams) respectively. Thus, a Twitter user in our data set is represented by a 1, 291-dimensional feature vector.

We applied spectral clustering [12] on $\mathcal{D}_2$ to derive 200 (hard) clusters of 1-grams that capture a number of latent topics and linguistic expressions (e.g. 'Politics', 'Sports', 'Internet Slang'), a snapshot of which is presented in Table 1. Previous research has shown that this amount of clusters is adequate for achieving a strong performance in similar tasks [7,13,14]. We then computed the frequency of each topic in the tweets of $\mathcal{D}_1$ as described in feature category $\mathbf{c}_5$.

To obtain a SES label for each user account, we took advantage of the SOC hierarchy's characteristics [5]. In SOC, jobs are categorised based on the required skill level and specialisation. At the top level, there exist 9 general occupation groups, and the scheme breaks down to sub-categories forming a 4-level structure. The bottom of this hierarchy contains more specific job groupings (369 in total). SOC also provides a simplified mapping from these job groupings to a SES as defined by NS-SEC [17]. We used this mapping to assign an upper, middle or lower SES to each user account in our data set. This process resulted in 710, 318 and 314 users in the upper, middle and lower SES classes, respectively.[2]

3 Classification Methods

We use a composite Gaussian Process (GP), described below, as our main method for performing classification. GPs can be defined as sets of random variables, any finite number of which have a multivariate Gaussian distribution [16].

[2] The data set is available at http://dx.doi.org/10.6084/m9.figshare.1619703.

Formally, GP methods aim to learn a function $f : \mathbb{R}^d \rightarrow \mathbb{R}$ drawn from a GP prior given the inputs $\mathbf{x} \in \mathbb{R}^d$:

$$f(\mathbf{x}) \sim \mathcal{GP}(m(\mathbf{x}), k(\mathbf{x}, \mathbf{x}')) , \tag{1}$$

where $m(\cdot)$ is the mean function (here set equal to 0) and $k(\cdot, \cdot)$ is the covariance kernel. We apply the squared exponential (SE) kernel, also known as the radial basis function (RBF), defined as $k_{\mathrm{SE}}(\mathbf{x}, \mathbf{x}') = \theta^2 \exp\left(-\|\mathbf{x} - \mathbf{x}'\|_2^2/(2\ell^2)\right)$, where θ^2 is a constant that describes the overall level of variance and ℓ is referred to as the characteristic length-scale parameter. Note that ℓ is inversely proportional to the predictive relevancy of $\mathbf{x}$ (high values indicate a low degree of relevance). Binary classification using GPs 'squashes' the real valued latent function $f(\mathbf{x})$ output through a logistic function: $\pi(\mathbf{x}) \triangleq \mathrm{P}(y = 1|\mathbf{x}) = \sigma(f(\mathbf{x}))$ in a similar way to logistic regression classification. In binary classification, the distribution over the latent f_* is combined with the logistic function to produce the prediction $\bar{\pi}_* = \int \sigma(f_*)\mathrm{P}(f_*|\mathbf{x}, \mathbf{y}, x_*)df_*$. The posterior formulation has a non-Gaussian likelihood and thus, the model parameters can only be estimated. For this purpose we use the Laplace approximation [16,18].

Based on the property that the sum of covariance functions is also a valid covariance function [16], we model the different user feature categories with a different SE kernel. The final covariance function, therefore, becomes

$$k(\mathbf{x}, \mathbf{x}') = \left(\sum_{n=1}^{C} k_{\mathrm{SE}}(\mathbf{c}_n, \mathbf{c}_n')\right) + k_{\mathrm{N}}(\mathbf{x}, \mathbf{x}') , \tag{2}$$

where $\mathbf{c}_n$ is used to express the features of each category, i.e., $\mathbf{x} = \{\mathbf{c}_1, \ldots, \mathbf{c}_C,\}$, C is equal to the number of feature categories (in our experimental setup, $C = 5$) and $k_{\mathrm{N}}(\mathbf{x}, \mathbf{x}') = \theta_{\mathrm{N}}^2 \times \delta(\mathbf{x}, \mathbf{x}')$ models noise (δ being a Kronecker delta function). Similar GP kernel formulations have been applied for text regression tasks [7,9,11] as a way of capturing groupings of the feature space more effectively.

Although related work has indicated the superiority of nonlinear approaches in similar multimodal tasks [7,14], we also estimate a performance baseline using a linear method. Given the high dimensionality of our task, we apply logistic regression with elastic net regularisation [6] for this purpose. As both classification techniques can address binary tasks, we adopt the one–vs.–all strategy for conducting an inference.

4 Experimental Results

We assess the performance of the proposed classifiers via a stratified 10-fold cross validation. Each fold contains a random 10 % sample of the users from each of the three socioeconomic statuses. To train the classifier on a balanced data set, during training we over-sample the two less dominant classes (middle and lower), so that they match the size of the one with the greatest representation (upper). We have also tested the performance of a binary classifier, where the middle and lower classes are merged. The cumulative confusion matrices (all data

Table 2. SES classification mean performance as estimated via a 10-fold cross validation of the composite GP classifier for both problem specifications. Parentheses hold the SD of the mean estimate.

Num. of classes	Accuracy	Precision	Recall	F-score
3	75.09 % (3.28 %)	72.04 % (4.40 %)	70.76 % (5.65 %)	.714 (.049)
2	82.05 % (2.41 %)	82.20 % (2.39 %)	81.97 % (2.55 %)	.821 (.025)

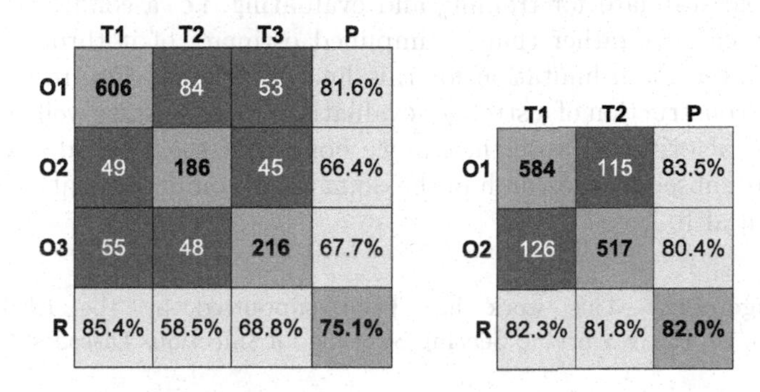

Fig. 1. The cumulative confusion matrices for the 3-way (left) and binary (right) classification tasks. Columns contain the **T**arget class labels and rows the **O**utput ones. The row and column extensions respectively specify the **P**recision and **R**ecall per class. The numeric identifiers (1–3) are in descending SES order (upper to lower).

from the 10 folds) for both classification scenarios and the GP-based classifier are presented in Fig. 1. Table 2 holds the respective mean performance metrics. The mean accuracy of the 3-way classification obtained by the GP model is equal to 75.09 % (SD = 3.28 %). The regularised logistic regression model yielded a mean accuracy of 72.01 % (SD = 2.45 %). A two sample t-test concluded that the 3.08 % difference between these mean performances is statistically significant ($p = 0.029$). The precision and recall per class are reported in the row and column extensions of the confusion matrices respectively. It is evident that it is more difficult to correctly classify users from the middle class (lowest precision and recall). The binary classifier is able to create a much better class separation, achieving a mean accuracy of 82.05 % (SD = 2.41 %) with fairly balanced precision and recall among the classes.

Looking at the occupation titles of the users, where false negatives occurred in the 3-way classification, we identified the following jobs as the most error-prone: 'sports players' for the upper class, 'photographers', 'broadcasting equipment operators', 'product/clothing designers' for the middle class, 'fitness instructors' and 'bar staff' for the lower class. Further investigation is needed to fully understand the nature of these errors. However, we note that SES is influenced by many factors, including income, education and occupation. In contrast, our classifier does not explicitly consider either income or education, and this may limit accuracy.

5 Conclusions and Future Work

We have presented the first approach for inferring the socioeconomic status of a social media user based on content (text, topics) and behaviour (interaction, impact). As in previous case studies [7,14], the multimodal feature space favoured a nonlinear classifier. Our method yielded an accuracy of 75 % and 82 % for the 3-way and binary classification scenarios respectively. The absence of a definitive gold standard for training and evaluating, i.e. a confirmed SES that represents each user rather than a simplified estimate of it through the SOC taxonomy, is the main limitation for this line of research. Future work should focus on the construction of a stronger evaluation framework, as well as improved classification algorithms. Nevertheless, we hope that the method outlined here will facilitate subsequent research in the domains of computational social science and digital health.

Acknowledgements. This work has been supported by the EPSRC grant EP/K031953/1 ("Early-Warning Sensing Systems for Infectious Diseases").

References

1. Bollen, J., Mao, H., Zeng, X.: Twitter mood predicts the stock market. J. Comput. Sci. **2**(1), 1–8 (2011)
2. Burger, D.J., Henderson, J., Kim, G., Zarrella, G.: Discriminating gender on twitter. In: EMNLP, pp. 1301–1309 (2011)
3. Cowan, C.D., et al.: Improving the measurement of socioeconomic status for the national assessment of educational progress: a theoretical foundation. Technical report, National Center for Education Statistics (2003)
4. Culotta, A.: Towards detecting influenza epidemics by analyzing Twitter messages. In: SMA, pp. 115–122 (2010)
5. Elias, P., Birch, M.: SOC2010: revision of the standard occupational classification. Econ. Labour Mark. Rev. **4**(7), 48–55 (2010)
6. Friedman, J., Hastie, T., Tibshirani, R.: Regularization paths for generalized linear models via coordinate descent. J. Stat. Softw. **33**(1), 1–22 (2010)
7. Lampos, V., Aletras, N., Preoţiuc-Pietro, D., Cohn, T.: Predicting and characterising user impact on Twitter. In: EACL, pp. 405–413 (2014)
8. Lampos, V., Cristianini, N.: Tracking the flu pandemic by monitoring the social web. In: CIP, pp. 411–416 (2010)
9. Lampos, V., Miller, A.C., Crossan, S., Stefansen, C.: Advances in nowcasting influenza-like illness rates using search query logs. Sci. Rep. **5**, 12760 (2015)
10. Lampos, V., Preoţiuc-Pietro, D., Cohn, T.: A user-centric model of voting intention from social media. In: ACL, pp. 993–1003 (2013)
11. Lampos, V., Yom-Tov, E., Pebody, R., Cox, I.: Assessing the impact of a health intervention via user-generated Internet content. Data Min. Knowl. Disc. **29**(5), 1434–1457 (2015)
12. von Luxburg, U.: A tutorial on spectral clustering. Stat. Comput. **17**(4), 395–416 (2007)

13. Preoţiuc-Pietro, D., Volkova, S., Lampos, V., Bachrach, Y., Aletras, N.: Studying user income through language, behaviour and affect in social media. PLoS ONE **10**(9), e0138717 (2015)
14. Preoţiuc-Pietro, D., Lampos, V., Aletras, N.: An analysis of the user occupational class through Twitter content. In: ACL, pp. 1754–1764 (2015)
15. Rao, D., Yarowsky, D., Shreevats, A., Gupta, M.: Classifying latent user attributes in Twitter. In: SMUC, pp. 37–44 (2010)
16. Rasmussen, C.E., Williams, C.K.I.: Gaussian Processes for Machine Learning. MIT Press, Cambridge (2006)
17. Rose, D., Pevalin, D.: Re-basing the NS-SEC on SOC2010: a report to ONS. Techincal report, University of Essex (2010)
18. Williams, C.K.I., Barber, D.: Bayesian classification with Gaussian processes. IEEE Trans. Pattern Anal. **20**(12), 1342–1351 (1998)

Two Scrolls or One Click: A Cost Model for Browsing Search Results

Leif Azzopardi[1](✉) and Guido Zuccon[2]

[1] School of Computing Science, University of Glasgow, Glasgow, UK
leif.azzopardi@glasgow.ac.uk
[2] Information Systems School, Queensland University of Technology (QUT), Brisbane, Australia
g.zuccon@qut.edu.au

Abstract. Modeling how people interact with search interfaces has been of particular interest and importance to the field of Interactive Information Retrieval. Recently, there has been a move to developing formal models of the interaction between the user and the system, whether it be to run a simulation, conduct an economic analysis, measure system performance, or simply to better understand the interactions. In this paper, we present a cost model that characterizes a user examining search results. The model shows under what conditions the interface should be more scroll based or more click based and provides ways to estimate the number of results per page based on the size of the screen and the various interaction costs. Further extensions to the model could be easily included to model different types of browsing and other costs.

1 Introduction

An emerging area of research within Interactive Information Retrieval (IIR) is the development of cost models that characterize the cost of interaction with a particular interface. Such cost models have been used for various purposes, such as: (i) controlling the number of interactions a simulated user can perform in a given period of time [4,5], (ii) approximating the time spent examining the results in a ranked list, e.g., within time biased measures [11,12], (iii) analyzing and empirically evaluating the costs and benefits of different interfaces [2,8,10] (iv) estimating the cost of different courses of interaction to determine the most efficient course of action [1,7,9]. Essentially, these lines of research have aimed at measuring the *cost* that the user incurs, as opposed to traditional performance measures, which look at measuring the *gain* or *benefit* that the user receives from such actions. Thus such models/measures of cost complement and extend the existing research on measuring the value of searching from a user perspective [3].

In this paper, we will focus on developing a cost model that captures the costs of a user interacting with a ranked list by browsing through the list until they find the document they would like to inspect. We use this model to estimate how many results per page should be shown under particular conditions.

N. Ferro et al. (Eds.): ECIR 2016, LNCS 9626, pp. 696–702, 2016.
DOI: 10.1007/978-3-319-30671-1_55

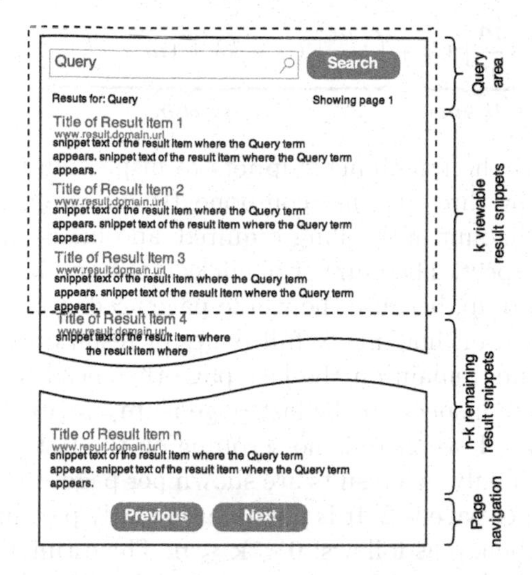

Fig. 1. The area marked by the dotted line shows how much of the page is initially visible, where **k** snippets can be seen. **k** will vary according to screen size. If the number of results per page **n** is larger than **k**, then **n** − **k** results are below the fold.

2 Cost Model

To develop a cost model for results browsing we assume that the user will be interacting with a standard search engine result page (SERP) with the following layout: a query box, a list of search results (snippets), and pagination buttons (see Fig. 1). Put more formally, the SERP displays **n** snippets, of which only **k** are visible above-the-fold. To view the remaining **n** − **k** snippets, i.e., those that are below-the-fold, the user needs to scroll down the page, while to see the next **n** snippets, the user needs to paginate (i.e., click next). And so we wonder whether is it better to scroll, click, or some combination of?

Here, we consider the case where the user wants the document at the **m**th result. However, **m** is not known a priori. To calculate the total browsing costs we assume that the user has just entered their query and has been presented with the result list. We further assume that there are three main actions the user can perform: inspecting a snippet, scrolling down the list, or clicking to go to the next page. Therefore, we are also assuming a linear traversal of the ranked list. Each action incurs a cost: $\mathbf{C_s}$ to inspect a snippet, $\mathbf{C_{scr}}$ to scroll to the next snippet[1], and when the user presses the 'next' button to see the subsequent **n** results, they incur a click cost $\mathbf{C_c}$. The click cost includes the time it takes the user to click and the time it takes the system to respond. Given these costs, we can now express a cost model for browsing to the **m**th result as follows:

[1] i.e., the scroll cost is the average cost to scroll the distance of one snippet, which includes the time to scroll and then focus on the next result.

$$C_b(n,k,m) = \underbrace{\lfloor \frac{m}{n} \rfloor . C_c}_{clicking} + \underbrace{\Big(\lfloor \frac{m}{n} \rfloor . (n-k) + (m_r - k) \Big) . C_{scr}}_{scrolling} + \underbrace{m . C_s}_{inspecting} \quad (1)$$

where $\mathbf{m_r}$ represents the remaining snippets to inspect on the last page. Equation 1 is composed by three distinct components: the number of clicks the user must perform, the amount of scrolling required, and finally the number of snippets they need to inspect. The number of clicks is the number of pages that need to be viewed rounded up, because the whole page needs to be viewed. The number of scrolls is based on how many full pages of results need to be examined, and how many results remain on the last page that need to be inspected. The remaining snippets to inspect on the last page is: $\mathbf{m_r} = (\mathbf{m} - \lfloor \frac{\mathbf{m}}{\mathbf{n}} \rfloor . \mathbf{n})$. However, if $\mathbf{m_r} < \mathbf{k}$, then $\mathbf{m_r} = \mathbf{k}$ as there is no scrolling on the last page. Note that $\mathbf{k}$ is bounded by $\mathbf{n}$, i.e., if only two results are shown per page, the maximum number of viewable results per page is 2. It is possible that only part of the result snippet is viewable, so $\mathbf{k}$ is bound as follows: $\mathbf{0} < \mathbf{k} \leq \mathbf{n}$. The estimated cost is based on the number of "clicks," "scrolls" and result snippets viewed (i.e., $\mathbf{m}$).

2.1 Application and Example

With this model it is possible to analyze the costs of various designs by setting the parameters accordingly. For example, a mobile search interface with a small screen size can be represented with a low $\mathbf{k}$, while a desktop search interface with a large screen can be characterized with a larger $\mathbf{k}$. The interaction costs for different devices can also be encoded accordingly.

Figure 2 shows an example of the cost of browsing $\mathbf{m}$ results when the number of results presented per page ($\mathbf{n}$) is 1, 3, 6 and 10 with up to $k = 6$ viewable results in the display window. Here we have approximated the costs of interaction as follows: 0.25 seconds to scroll, 2 seconds to click and 2.5 seconds to inspect a snippet[2].

Intuitively, displaying one result per page requires the most time as $(\mathbf{m} - \mathbf{1})$ clicks are needed to find the $\mathbf{m}$th document. Displaying three results per page is also costly as $\mathbf{m}$ increases requiring approximately $\mathbf{m/3}$ clicks. However, the difference in costs for 6 and 10 results per page vary depending on the specific number of results the user wants to inspect. For example, if $\mathbf{m} = \mathbf{12}$ then 6 results per page is lower in cost; whereas if $\mathbf{m} = \mathbf{13}$ then 10 results per page is lower in cost. In this example, since scrolling is relatively cheap, one might be tempted to conclude that the size of the result page should be as large as possible. However, using the model, we can determine the optimal size of the result page depending on the different parameters.

---

2 These values were based on the estimated time spent examining each snippet being between 1.7 and 3 seconds from [1], the GOMS timings for a mouse move (1.1 s), click (0.2 s) and system response (0.8 s) being approximately 2 seconds taken from [2] and based on [6]. For the scroll action on a wheeled mouse, we assume that it is similar to a key press (0.28 s) and was approximated by 0.25 s.

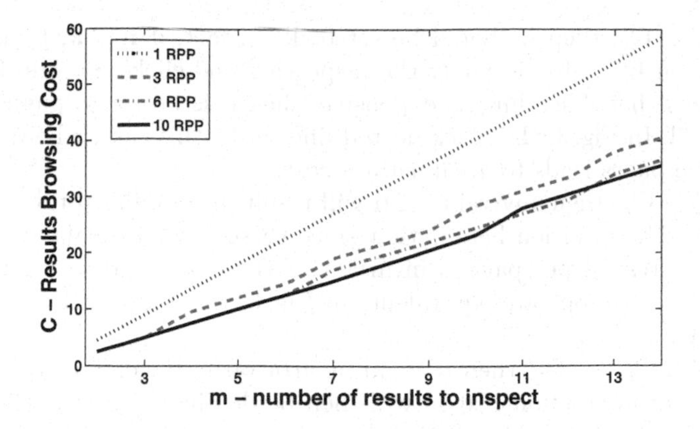

Fig. 2. The cost (total time in seconds) to examine m snippets for SERPs of different sizes ($c_{scr} = 0.25, c_c = 2, c_s = 2.5$ s).

2.2 Estimating SERP Size

To determine the optimal number of results to show per page, $\mathbf{n}^*$, we want to minimize $\mathbf{C_b}$. We can do this by differentiating the cost function with respect to $\mathbf{n}$. However, since $\mathbf{C_b}$ contains floor operators we need to use an approximation of Eq. 1:

$$\hat{C}_b(n,k,m) \approx \frac{m}{n}.C_c + \frac{m}{n}.(n-k).C_{scr} + C_s.m \tag{2}$$

which essentially provides a smoothed estimate of the clicking and scrolling costs (here we drop/ignore the inspect costs as they are constant). We can now differentiate this function, to obtain:

$$\frac{\partial \hat{C}_b}{\partial n} = -\frac{m}{n^2}.C_c + \frac{m}{n^2}.k.C_{scr} \tag{3}$$

and then solve the equation by setting $\frac{\partial \hat{C}_b}{\partial n} = 0$, in order to find what values minimize Eq. 2. The following is obtained:

$$\begin{aligned} -\frac{m}{n^2}.C_c + \frac{m}{n^2}.k.C_{scr} &= 0 \\ \frac{m}{n^2}.k.C_{scr} &= \frac{m}{n^2}.C_c \\ k.C_{scr} &= C_c \end{aligned} \tag{4}$$

Interestingly, $\mathbf{n}$ disappears from the equation. This at first seems counter intuitive, as it suggests that to minimize the cost of interaction $\mathbf{n}$ is not a factor. However, on closer inspection we see that the number of results to show per page depends on the balance between $\mathbf{k}$ and $\mathbf{C_{scr}}$, on one hand, and $\mathbf{C_c}$, on the other. If $\mathbf{k.C_{scr}}$ is greater than, equal to, or less than $\mathbf{C_c}$, then the influence on total cost in Eq. 1 results in three different cases (see below). To help illustrate these cases, we have plotted three examples in Fig. 3, where the user would like to inspect $\mathbf{m = 25}$ result snippets, and a maximum of $\mathbf{k = 6}$ result snippets are viewable per page.

1. if $\mathbf{k.C_{scr}} > \mathbf{C_c}$, then $\mathbf{n}$ should be set to $\mathbf{k}$. In fact, if $\mathbf{n} = \mathbf{k}$, then $\mathbf{C_{scr}}$ has no effect on Eq. 1, leaving only the inspecting and clicking costs. Intuitively, this means that if scrolling is expensive, then it is better to paginate rather than scroll. In Fig. 3, the black dotted line shows this case, where the cost of larger size pages leads to greater total cost.
2. if $\mathbf{k.C_{scr}} = \mathbf{C_c}$, then any value of $\mathbf{n}$ will result in a similar total cost, as long as $\mathbf{n} \geq \mathbf{k}$. The intuition here is that since these costs are balanced, then the number of results per page is invariant of the costs, and so scroll based or click based solutions are equivalent. In Fig. 3, the blue dot dash line shows when $\mathbf{k.C_{scr}} \approx \mathbf{C_c}$.
3. finally, if $\mathbf{k.C_{scr}} < \mathbf{C_c}$, then $\mathbf{n}$ could be arbitrarily large. This is because the total cost of interaction (Eq. 1) will then be dominated by the clicking cost. Intuitively, this means that if clicking is expensive, then it is better to scroll, rather than click; thus this advocates a solution such as infinite scrolling. In Fig. 3, the red solid line shows an example of this case, where increasing the size of pages leads to a lower total cost overall according to our model, where the total cost decreases at a diminishing rate. However, it is likely that other costs, which have not been modelled (i.e., cognitive costs, download costs, etc.) would lead to an increase in total cost at some point.

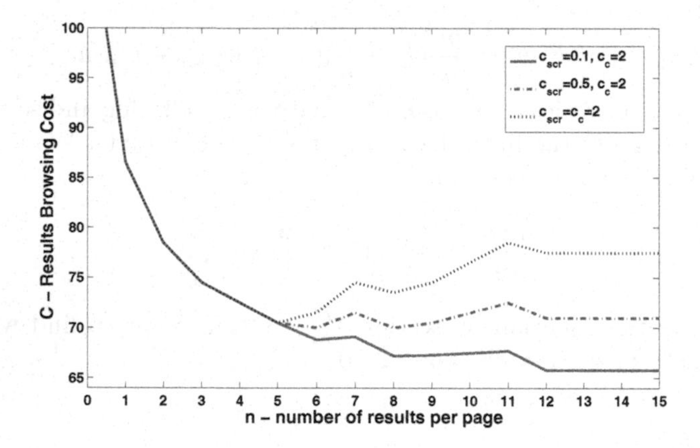

Fig. 3. An example of how the cost changes as the page size increases, when $m = 25$ and $k \leq 6$ for the three different cases of $k.C_{scr}$ versus C_c.

3 Discussion and Future Work

We have derived a general cost model of result browsing, which provides an indication of how the total cost of interaction changes with SERP size ($\mathbf{n}$), screen size ($\mathbf{k}$) and the associated interaction costs. In a desktop setting, as modeled above, where $\mathbf{k} \approx \mathbf{6}$ and the cost of click $\mathbf{C_c}$ is generally larger than the scroll cost $\mathbf{C_{scr}}$, it is likely that the total cost is approximated by the blue dot dashed line and the red solid line shown in Fig. 3. This is interesting because choosing a SERP size of $\mathbf{n} = \mathbf{10}$ (as done by most search engines when interacting from a PC), tends to be near or close to the minimum cost. While increasing the SERP size to beyond ten would lead to lower total costs, this is at a diminishing rate. In this model, we have assumed a fixed download cost, (within $\mathbf{C_c}$). However, a more realistic estimate of this cost would be proportional to $\mathbf{n}$, such that $\mathbf{C_c(n)}$, where a larger page takes longer to download. Another refinement of the model would be to condition scrolling on the number of results that need to be scrolled through; as users might find it increasingly difficult and cognitively taxing to scroll through long lists. Nonetheless, our model is still informative and a starting point for estimating the browsing costs. Future work, therefore, could: *(i)* extend the model to cater for these other costs in order to obtain a more accurate estimate of the overall cost, *(ii)* obtain empirical estimates for the different costs, on different devices (e.g., laptops, mobiles, desktops, tablets, etc.) as well as with different means of interaction (e.g., mouse with/without a scroll wheel, touchscreen, touchmouse, voice, etc.), and, *(iii)* incorporate such a browsing model into simulations, measures and analyses. A further extension would be to consider different types of layouts (e.g., grids, lists, columns, etc.) and different scenarios (e.g., finding an app on a tablet, mobile, etc.).

Acknowledgements. Thanks to all our tutorial participants who undertook this modelling exercise and helped us refine the model. Thanks to Diane Kelly & Kathy Brennan for the numerous conversations about clicks and scrolls which led to this model & title.

References

1. Azzopardi, L.: Modelling interaction with economic models of search. In: Proceedings of the 37th ACM SIGIR Conference, pp. 3–12 (2014)
2. Azzopardi, L., Kelly, D., Brennan, K.: How query cost affects search behavior. In: Proceedings of the 36th ACM SIGIR Conference, pp. 23–32 (2013)
3. Azzopardi, L., Zuccon, G.: An analysis of theories of search and search behavior. In: Proceedings of the 2015 International Conference on The Theory of Information Retrieval, pp. 81–90 (2015)
4. Baskaya, F., Keskustalo, H., Järvelin, K.: Time drives interaction: simulating sessions in diverse searching environments. In: Proceedings of the 35th ACM SIGIR Conference, pp. 105–114 (2012)
5. Baskaya, F., Keskustalo, H., Järvelin, K.: Modeling behavioral factors in interactive ir. In: Proceedings of the 22nd ACM SIGIR Conference, pp. 2297–2302 (2013)
6. Card, S.K., Moran, T.P., Newell, A.: The keystroke-level model for user performance time with interactive systems. Comm. ACM **23**(7), 396–410 (1980)
7. Kashyap, A., Hristidis, V., Petropoulos, M.: Facetor: Cost-driven exploration of faceted query results. In: Proceedings of the 19th ACM CIKM, pp. 719–728 (2010)
8. Kelly, D., Azzopardi, L.: How many results per page?: A study of serp size, search behavior and user experience. In: Proceedings of the 38th ACM SIGIR Conference, pp. 183–192. SIGIR 2015 (2015)

9. Pirolli, P., Card, S.: Information foraging. Psych. Rev. **106**, 643–675 (1999)
10. Russell, D.M., Stefik, M.J., Pirolli, P., Card, S.K.: The cost structure of sensemaking. In: Proceedings of the INTERACT/SIGCHI, pp. 269–276 (1993)
11. Smucker, M.D., Clarke, C.L.: Time-based calibration of effectiveness measures. In: Proceedings of the 35th ACM SIGIR Conference, pp. 95–104 (2012)
12. Smucker, M.D., Jethani, C.P.: Human performance and retrieval precision revisited. In: Proceedings of the 33rd ACM SIGIR Conference, pp. 595–602 (2010)

Determining the Optimal Session Interval for Transaction Log Analysis of an Online Library Catalogue

Simon Wakeling(✉) and Paul Clough

Information School, University of Sheffield, Sheffield, UK
{s.wakeling,p.d.clough}@sheffield.ac.uk

Abstract. Transaction log analysis at the level of a session is commonly used as a means of understanding user-system interactions. A key practical issue in the process of conducting session level analysis is the segmentation of the logs into appropriate user sessions (i.e., sessionisation). Methods based on time intervals are frequently used as a simple and convenient means of carrying out this segmentation task. However, little work has been carried out to determine whether the commonly applied 30-minute period is appropriate, particularly for the analysis of search logs from library catalogues. Comparison of a range session intervals with human judgements demonstrate that the overall accuracy of session segmentation is relatively constant for session intervals between 26 to 57 min. However, a session interval of between 25 and 30 min minimises the chances of one error type (incorrect collation or incorrect segmentation) predominating.

1 Introduction

Transaction log analysis (TLA) techniques provide a means of developing an in-depth account of a user's search behaviour when interacting with a given system. TLA typically focuses on one or more of three levels: term, query or session level [8]. At the term level, analysis is concerned with the frequency, diversity, or co-occurrence levels of particular text strings in user queries. Query level analysis broadens this approach to take entire queries as the base unit of analysis, and might seek to investigate patterns of query reformulation, query structure and complexity, or repeated queries. The session level widens the scope still further to encompass the entirety of interactions within a period of user-system interaction. This offers an opportunity for analysis that investigates issues of user intent and information seeking behaviour.

Recent years have seen increasing recognition of the importance and potential value of analysing search logs at the level of session [22]. However, in defining what exactly constitutes a 'session' from the perspective of TLA, we encounter contrasting views. For example, Jansen et al. define a session as *"a series of interactions by the user toward addressing a single information need"* [8] (p. 862). This definition is, however, problematic, since the usefulness of defining a session in relation to a "single information need" is called into question by studies showing the frequency with which users are found to address multiple work- and search-tasks in a single continuous period of interaction. An alternative definition, therefore, is a temporal one, whereby a session constitutes the sequence of searches and other actions undertaken by a user within a single 'episode' of engagement [20].

N. Ferro et al. (Eds.): ECIR 2016, LNCS 9626, pp. 703–708, 2016.
DOI: 10.1007/978-3-319-30671-1_56

For the practical purposes of TLA, a variety of methods for assigning sessions have been developed, with Gayo-Avello [4] providing a comprehensive summary of these approaches. Some researchers have advocated methods based on query reformulation [9], navigation patterns [12], and combinations of various metrics [10]. However, such methods are often complex and time-consuming to implement on the logs practically acquired from search systems. The simplest, and most widely used, method is the adoption of a session temporal cut-off interval, which segments sessions according to a set period of inactivity. Thus a new session is applied to logs originating from a single IP address if server transactions attributable to that IP address are separated by a pre-determined time interval, e.g. 30 min.

The work presented in this paper relates to a wider research project investigating the users and uses of WorldCat.org, an online union catalogue operated by OCLC, and containing the aggregated catalogues of over 70,000 libraries from around the world. In conducting TLA on search logs from WorldCat.org the challenge of determining an appropriate means of identifying and segmenting sessions within the logs arose. Whilst a 30 min inactive period is most commonly used for web search logs, there is little evidence to support the use of this period for the logs of an online library catalogue system. This paper describes an attempt to determine the optimum session interval for the WorldCat.org log by comparing the segmentation and collation of sessions for various session intervals with human judgements. Section 2 discusses related work; Sect. 3 describes the methodology used to study sessions; Sect. 4 presents and discusses results; and Sect. 5 concludes the paper.

2 Related Work

While there is a long history of applying TLA to library catalogues and other resource databases [19], little attention has been paid to the process of session segmentation. Despite the apparent advantages of session-level analysis, studies of library systems frequently chose to focus on query-level analysis (e.g., [7, 15]), thereby negating a need for session segmentation; whilst other studies that employ session level analysis do not specify how sessions have been segmented (e.g., [11, 18]). Other work in this area makes use of system-determined session delimitations, either through the use of client-side session cookies (e.g., [17]), or server-side system time-outs (e.g., [2, 14]); however, no details are provided regarding the precise details of the time-out periods. Only a small number of library system studies do define a session cut-off interval. At one extreme, Dogan et al.'s study of PubMed [3] specifies that all actions from a single user in a 24 h period are considered a single session. Lown [13] and Goodale and Clough [6] adopt a 30-minute session interval, with this period apparently based on the standard session interval applied to web search logs.

Given the paucity of discussion of this issue in studies relating to library systems, it is instructive to consider the greater body of literature relating to TLA for web search. Here a general consensus has emerged for a session interval of 30 min [10]. This figure is based on early search log work by Catledge and Pitkow [1] which showed that 25.5 min session interval meant that most events occurred within 1.5 standard deviations of the mean inactive period. Jones and Klinkner [10] used an analysis of manually annotated search sessions to argue that the 30-minute interval is not supported, and that any segmentation based solely on temporal factors achieves only 70-80 % accuracy. Other researchers have suggested both lower and higher session interval periods, ranging from 15 min [9] to 125 min [16]. In their study of Reuters Intranet logs, Goker and He [5] compared session boundaries created by a range of session intervals with human judgements. They also identified different error types: Type A errors being when adjacent log activity is incorrectly split into different sessions; and Type B errors when unrelated activities are incorrectly collated into the same session. Their findings indicate that whilst overall accuracy was relatively stable for intervals between 10 min and an hour, errors were split equally between the two error types between 10 and 15 min. Above 15 min, Type B errors were found to predominate. Their overall findings indicate an optimum session interval of between 11-15 min. The work presented in this paper applies this method to the logs of an online union catalogue, and represents the first attempt to establish an optimum session interval for segmentation of this type of log.

3 Methodology

The WorldCat.org log data contained the fields shown in Table 1. A random sample of 10,000 lines of the logs ordered by IP address was generated (representing 721 unique IP addresses), and all instances identified where lines of the log originating from the same IP address were separated by between 10 and 60 min. A total of 487 such instances were found. Each instance was then manually examined in the context of the full logs to determine whether the activity either side of the inactive period might reasonably be considered part of the same session, and coded accordingly ("Same session" or "Different Session"). Following [5], this judgement was primarily based on the subject

Table 1. Fields in log data supplied by OCLC

Field	Description
Anonymised IP Address	A random code assigned to each unique IP address present in the log
Country of origin	The country of origin of the IP address, as determined by an IP lookup service
Date	The date of the server interaction
Time	The time of the server interaction (hh:mm:ss)
URL	The URL executed by the server
OCLCID	The OCLC ID of the item being viewed (if applicable)
Referrer URL	The page from which the URL was executed
Browser	Technical information about the browser type and version

area of the queries executed and items viewed either side of the inactive period. Since this judgement was inherently subjective a third code was also used ("Unknown") to limit the likelihood of incorrect judgements. This was applied in circumstances where there was no reasonable way of judging whether the inactive period constituted a new session or not. A subset of 20 % of instances were coded by a second assessor, and results compared. Overall inter-coder reliability using Cohen's kappa coefficient was shown to be $\kappa = 0.86$, above the 0.80 required to indicate reliable coding [21].

The resulting dataset consisted of 487 inactive periods of between 10 and 60 min, and the code assigned to each period. 99 of these were coded "Unknown", and were not considered for further analysis. It was subsequently possible to simulate the effectiveness of a variety of potential session timeout durations based on the codes assigned to the 388 remaining inactive periods. Where i = the inactive period in the log, t = the proposed timeout duration, s = "Same session" and d = "Different session", we observe four potential outcomes:

1. $i > t$, s = Incorrect session segmentation (Type A error)
2. $i > t$, d = Correct session segmentation
3. $i < t$, s = Correct session collation
4. $i < t$, d = Incorrect session collation (Type B error)

Outcomes were calculated for each of the coded inactive periods in the log sample ($n = 388$) for session intervals at 30 s intervals between 10 and 60 min.

4 Results and Discussion

A session cut-off time of 39 min was found to provide the highest proportion of correctly segmented sessions (77.1 %), although there was little variation in the proportion of correctly segmented sessions between 26 and 57 min, with each session interval period producing correct outcomes for over 75 % of inactive periods. However, the results indicate that using a 10-minute cut-off time results in a high proportion (70 %) of sessions representing a Type A error (the sessions being incorrectly split). Naturally as the session cut-off period is extended, an increasing number of sessions are incorrectly collated. A session cut-off time of 28 min was found to produce an equal number of the two error types (Type A = 13 %; Type B = 13 %). Thus we can conclude that whilst session intervals of between 26 and 57 min have little effect on the overall accuracy of session segmentation, there is variation in the distribution of the error types. A session interval period of between 28 and 29 min is shown to reduce the likelihood of one error type predominating (see Fig. 1).

Whilst this goes some way to validating the commonly used 30-minute session interval, we suggest that attention needs to be paid to the overall aims of the TLA being undertaken. In particular, researchers may be conducting analysis for purposes where reducing one particular error type is preferable. In conducting TLA for the development of user-orientated learning techniques, for example, Goker and He [5] argue that minimising overall and Type B errors is a priority. Other situations, for example the investigation of rates of query reformulation, may demand a reduction in Type A errors.

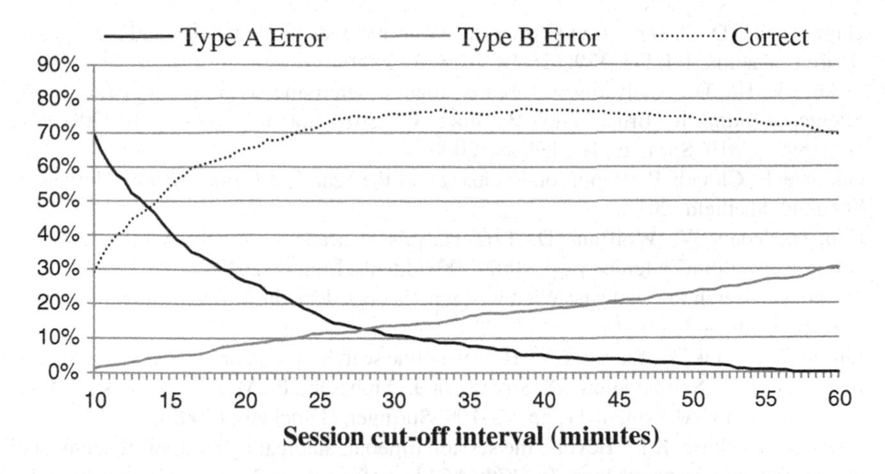

Fig. 1. Effects of different session intervals on session segmentation accuracy.

Thus whilst a 30-minute session interval provides the most effective means of mitigating the effects of any one error type, researchers investigating library catalogue logs should consider raising or lowering the session interval depending on their research goals.

5 Conclusions

This paper investigates the effects of using time intervals between 10 and 60 min for segmentating search logs from WorldCat.org into sessions for subsequent analysis of user searching behaviour. A period of 30 min is commonly used in the literature, particularly in the analysis of web search logs. However, this is often without sufficient justification. Analysis of library catalogue logs frequently also uses a 30 min cut-off, or does not employ sessionisation at all. Based on a manual analysis of sessions from WorldCat.org, our results indicate that the accuracy of segmenting sessions is relatively stable for time intervals between 26 and 57 min, with 28 and 29 min shown to reduce the likelihood of one error type (incorrect segmentation or incorrect collation) predominating. This work supports the use of a 30-minute timeout period in TLA studies, and is of particular value to researchers wishing to conduct session level analysis of library catalogue logs.

References

1. Catledge, L.D., Pitkow, J.E.: Characterizing browsing strategies in the World-Wide Web. Comput. Netw. Isdn. **27**(6), 1065–1073 (1995)
2. Cooper, M.D.: Usage patterns of a web-based library catalog. J. Am. Soc. Inf. Sci. Tec. **52**(2), 137–148 (2001)
3. Dogan, R.I., Murray, G.C., Névéol, A., Lu, Z.: Understanding PubMed® user search behavior through log analysis. Database 2009 (bap018) (2009)

4. Gayo-Avello, D.: A survey on session detection methods in query logs and a proposal for future evaluation. Inf. Sci. **179**(12), 1822–1843 (2009)
5. Göker, A., He, D.: Analysing web search logs to determine session boundaries for user-oriented learning. In: Brusilovsky, P., Stock, O., Strapparava, C. (eds.) AH 2000. LNCS, vol. 1892, p. 319. Springer, Heidelberg (2000)
6. Goodale, P., Clough, P.: Report on Evaluation of the Search25 Demo System. University of Sheffield, Sheffield (2012)
7. Han, H., Jeong, W., Wolfram, D.: Log analysis of academic digital library: User query patterns. In: 10th iConference, pp. 1002–1008. Ideals, Illinois (2014)
8. Jansen, B.: Search log analysis: What it is, what's been done, how to do it. Libr. Inform. Sci. Res. **28**(3), 407–432 (2006)
9. Jansen, B.J., Spink, A., Kathuria, V.: How to define searching sessions on web search engines. In: Nasraoui, O., Spiliopoulou, M., Srivastava, J., Mobasher, B., Masand, B. (eds.) WebKDD 2006. LNCS (LNAI), vol. 4811, pp. 92–109. Springer, Heidelberg (2007)
10. Jones, R., Klinkner, K.L.: Beyond the session timeout: automatic hierarchical segmentation of search topics in query logs. In: 17th ACM conference on Information and knowledge management, pp. 699–708. ACM, New York (2008)
11. Jones, S., Cunningham, S.J., McNab, R., Boddie, S.: A transaction log analysis of a digital library. Int. J. Digit. Libr. **3**(2), 152–169 (2000)
12. Kapusta, J., Munk, M., Drlík, M.: Cut-off time calculation for user session identification by reference length. In: 6th International Conference on Application of Information and Communication Technologies (AICT), pp. 1–6. IEEE, New York (2012)
13. Lown, C.: A transaction log analysis of NCSU's faceted navigation OPAC. School of Information and Library Science, University of North Carolina (2008)
14. Malliari, A., Kyriaki-Manessi, D.: Users' behaviour patterns in academic libraries' OPACs: a multivariate statistical analysis. New Libr. World **108**(3/4), 107–122 (2007)
15. Meadow, K., Meadow, J.: Search query quality and web-scale discovery: A qualitative and quantitative analysis. College Undergraduate Librar. **19**(2–4), 163–175 (2012)
16. Montgomery, A., Faloutsos, C.: Identifying web browsing trends and patterns. IEEE Comput. **34**(7), 94–95 (2007)
17. Nicholas, D., Huntington, P., Jamali, H.R.: User diversity: as demonstrated by deep log analysis. Electron. Libr. **26**(1), 21–38 (2008)
18. Niu, X., Zhang, T., Chen, H.L.: Study of user search activities with two discovery tools at an academic library. Int. J. Hum. Comput. Interact. **30**(5), 422–433 (2014)
19. Peters, T.A.: The history and development of transaction log analysis. Libr. Hi Tech. **11**(2), 41–66 (1993)
20. Spink, A., Park, M., Jansen, B.J., Pedersen, J.: Multitasking during web search sessions. Inform. Process. Manag. **42**(1), 264–275 (2006)
21. Yardley, L.: Demonstrating validity in qualitative psychology. In: Smith, J.A. (ed.) Qualitative psychology: A practical guide to research methods, pp. 235–251. SAGE, London, UK (2008)
22. Ye, C., Wilson, M.L.: A user defined taxonomy of factors that divide online information retrieval sessions. In: 5th Information Interaction in Context Symposium, pp. 48–57. ACM, New York (2014)

A Comparison of Deep Learning Based Query Expansion with Pseudo-Relevance Feedback and Mutual Information

Mohannad ALMasri(✉), Catherine Berrut, and Jean-Pierre Chevallet

LIG Laboratory, MRIM Group, Université Grenoble Alpes, Grenoble, France
{mohannad.almasri,catherine.berrut,jean-pierre.chevallet}@imag.fr

Abstract. Automatic query expansion techniques are widely applied for improving text retrieval performance, using a variety of approaches that exploit several data sources for finding expansion terms. Selecting expansion terms is challenging and requires a framework capable of extracting term relationships. Recently, several Natural Language Processing methods, based on Deep Learning, are proposed for learning high quality vector representations of terms from a large amount of unstructured text with billions of words. These high quality vector representations capture a large number of term relationships. In this paper, we experimentally compare several expansion methods with expansion using these term vector representations. We use language models for information retrieval to evaluate expansion methods. Experiments conducted on four CLEF collections show a statistically significant improvement over the language models and other expansion models.

1 Introduction

User queries are usually too short to describe the information need accurately. Important terms can be missing from the query, leading to a poor coverage of the relevant documents. To solve this problem, automatic query expansion techniques leveraging on several data sources and employ different methods for finding expansion terms [2]. Selecting such expansion terms is challenging and requires a framework capable of adding interesting terms to the query.

Different approaches have been proposed for selecting expansion terms. Pseudo-relevance feedback (PRF) assumes that the top-ranked documents returned for the initial query are relevant, and uses a sub set of the terms extracted from those documents for expansion. PRF has been proven to be effective in improving retrieval performance [4].

Corpus-specific approaches analyze the content of the whole document collection, and then generate correlation between each pair of terms by co-occurrence [6], mutual information [3], etc. Mutual information (MI) is a good measure to assess how much two terms are related, by analyzing the entire collection in order to extract the association between terms. For each query term, every term that has a high mutual information score with it is used to expand the user query.

N. Ferro et al. (Eds.): ECIR 2016, LNCS 9626, pp. 709–715, 2016.
DOI: 10.1007/978-3-319-30671-1_57

Many approaches exploit knowledge bases or thesauruses for query expansion, among them: WordNet [12], UMLS Meta thesaurus [13], Wikipedia [11], etc. The nature of these resources varies: linguistic like WordNet, domain specific like UMLS in the medical domain, or knowledge about named entities like Wikipedia.

Other approaches like semantic vectors and neural probabilistic language models, propose a rich term representation in order to capture the similarity between terms. In these approaches, a term is represented by a mathematical object in a high dimensional semantic space which is equipped with a metric. The metric can naturally encode similarities between the corresponding terms. A typical instantiation of these approaches is to represent each term by a vector and use a cosine or distance between term vectors in order to measure term similarity [1,7,10].

Recently, several efficient Natural Language Processing methods, based on Deep Learning, are proposed to learn high quality vector representations of terms from a large amount of unstructured text data with billions of words [5]. This high quality vector representation captures a large number of term relationships. In this paper, we propose to investigate these term vector representations in query expansion. We then experimentally compare this approach with two other expansion approaches: pseudo-relevance feedback and mutual information.

Our experiments are conducted on four CLEF medical collections. We use a language modeling framework to evaluate expanded queries. The experimental results show that the retrieval effectiveness can be significantly improved over the ordinary language models and pseudo-relevance feedback.

This paper is organized as follows. In Sect. 2, we present the query expansion method that we use. Our experimental set-up and results are presented in Sect. 3. Finally, Sect. 4 concludes the paper.

2 Query Expansion Method

We propose to investigate term vector representations in query expansion. In this section, we first present the source of these term vectors. Then, we describe how we use these term vectors for query expansion.

2.1 Expansion Terms

In this step, learning takes place from a large amount of unstructured text data, term vector representations are learned using Deep Learning. The resulting vectors carry relationships between terms, such as a city and the country it belongs to, e.g. France is to Paris what Germany is to Berlin [5]. Therefore, each term t is represented by a vector of a predefined dimension v_t[1]. In the rest of paper, we call this vector *Deep Learning Vector*. The similarity between two terms t_1 and t_2 is measured with the normalized cosine between their two vectors: v_{t_1} and v_{t_2}.

$$SIM(t_1, t_2) = \widetilde{cos}(v_{t_1}, v_{t_2}) \quad (1)$$

[1] A real-valued vector of a predefined dimension, 600 dimensions for exemple.

where $\widetilde{cos}(v_{t_1}, v_{t_2}) \in [0, 1]$ is the normalized cosine between the two term vectors v_{t_1} and v_{t_2}. Based on this normalized cosine similarity between terms, we now define the function that returns the k-most similar terms to a term t, $top_k(t)$:

$$top_k : V \rightarrow 2^V \tag{2}$$

where V is the set of all terms t.

2.2 Building Expanded Query

Let q be a user query represented by a bag of terms, $q = [t_1, t_2, ..., t_{|q|}]$. Each term in the query has a frequency $\#(t, q)$. In order to expand a query q, we follow these steps:

- For each $t \in q$, collect the k-most similar terms to t using the function $top_k(t)$, Eq. 2. The expanded query q' is defined as follows: $q' = q \bigcup_{t \in q} top_k(t)$.
- The frequency of each $t \in q$ still the same in the expanded query q'.
- The frequency of each expansion term $t' \in top_k(t)$ in the expanded query q' is given as follows:

$$\#(t', q') = \alpha \times \#(t, q') \times \widetilde{cos}(v_t, v_{t'}) \tag{3}$$

  where $\alpha \in [0, 1]$ is a tuning parameter that determines the importance of expansion terms.
  In the rest of paper, the expansion method based on deep learning vectors is denoted by VEXP.

3 Experiments

The first goal of our experiments is to analyze the effect of the number of expansion terms k on the retrieval performance using deep learning vectors. The second goal is to compare between the proposed expansion based on deep learning vectors (VEXP) with two existing expansion approaches: pseudo-relevance feedback (PRF) [4], and mutual information (MI) [3], which both have been proven to be effective in improving retrieval performance. In order to achieve the comparison between VEXP, PRF, and MI, we use a language model with no expansion as a baseline (NEXP).

Documents are retrieved using Indri search engine [9], and two smoothing methods of language models: Jelinek-Mercer and Dirichlet.

The optimization of the free parameter α (Eq. 3) for controlling expansion terms importance is done using 4-fold cross-validation with Mean Average Precision (MAP) as the target metric. We vary α values between [0.1, 1] with 0.1 as an interval. The best values of the tuning parameter α that indicate the importance of expansion terms are between [0.2, 0.4].

In our experiments, the statistical significance is determined using Fisher's randomization test with $p < 0.05$ [8].

3.1 Evaluation Data

Four medical corpora from CLEF[2] are used.

- Image2010, Image2011, Image2012: contain short documents and queries.
- Case2011: contains long documents and queries.

Table 1 shows some statistics about them, *avdl* and *avql* are average length of documents and queries, respectively. These medical collections provide a huge amount of medical text that we need in the training phase.

Table 1. Training and testing collections.

Corpus	#d	#q	avdl	avql
Image2009	74,901	25	62.16	3.36
Image2010	77,495	16	62.12	3.81
Image2011	230,088	30	44.83	4.0
Image2012	306,530	22	47.16	3.55
Case2011	55,634	10	2594.5	19.7
Case2012	74,654	26	2570.72	24.35

3.2 Learning Data and Tools

We use word2vec to generate deep learning vectors [5]. The word2vec tool takes a text corpus as input and produces the term vectors as output. It first constructs a vocabulary from the training text data and then learns the vector representation of terms. We build our training corpus using three different CLEF medical collection: Image2009, Case2011, Case2012. Our training corpus consists of about 400 millions words. The vocabulary size for this training corpus is about 350,000 different terms. We used the recommended setting for this training tool like the term vector dimension and the learning context window size.

3.3 Number of Expansion Terms Analysis

We first analyze the effect of number of expansion terms k on the retrieval performance of VEXP. Each query term is expanded with $k \in \{1, 2, 3, ..., 10\}$ terms. Stop words are not considered in the expansion. The optimal k value for the number of expansion terms vary depending on the test collections. All tested k values are given in Table 2. The best performance is presented in bold.

Similarly, we analyzed the best number of expansion terms for the two other approaches: PRF and MI:

- For PRF, we have tested several configurations with $k \in \{5, 10, ..., 50\}$ and the number of feedback documents $\#fbdocs \in \{5, 10, , ..., 50\}$.
- For MI, we have also tested several configurations with $k \in \{1, 2, ..., 25\}$.

Table 3 gives the best configurations for VEXP, PRF, and MI.

[2] www.clef-initiative.eu.

Table 2. VEXP performance using MAP on test collections. *k* is the number of expansion terms for each query term.

k	Jelinek-Mercer				Dirichlet			
	Image2010	Image2011	Image2012	Case2011	Image2010	Image2011	Image2012	Case2011
1	0.3286	0.2258	0.1997	0.1373	0.3397	0.2173	0.1947	0.1288
2	0.3298	0.2325	0.1988	0.1431	0.3361	0.2204	0.1890	0.1345
3	0.3395	0.2330	0.1996	0.1440	0.3411	0.2192	0.1902	0.1366
4	0.3399	0.2338	0.2002	0.1413	0.3561	0.2175	0.1909	0.1384
5	0.3323	0.2340	0.1909	**0.1634**	0.3519	0.2187	0.1787	0.1410
6	**0.3402**	0.2324	0.1909	0.1432	**0.3603**	0.2163	0.1798	**0.1451**
7	0.3397	0.2333	0.1881	0.1446	0.3599	0.2184	0.1778	0.1431
8	0.3397	**0.2353**	0.1895	0.1414	0.3584	0.2200	0.1813	0.1416
9	0.3365	0.2230	0.2004	0.1387	0.3544	**0.2221**	0.1953	0.1379
10	0.3362	0.2233	**0.2036**	0.1343	0.3510	0.2215	**0.1990**	.1357

Table 3. Best configurations for VEXP, PRF, and MI.

		Jelinek-Mercer				Dirichlet			
		Image 2010	Image 2011	Image 2012	Case 2011	Image 2010	Image 2011	Image 2012	Case 2011
PRF	k	15	10	20	10	15	10	10	10
	#fbdocs	10	10	20	10	10	10	10	10
MI	k	10	8	6	10	10	7	6	10
VEXP	k	6	4	10	4	5	9	10	5

3.4 Performance Comparison

In this section, we compare three expansion methods: VEXP, PRF, and MI, using a language model with no expansion as a baseline (NEXP). We use two tests for statistical significance: † indicates a statistical significant improvement over NEXP, and $*$ indicates a statistical significant improvement over PRF. Results are given in Table 4. We first observe that VEXP is always statistically better than NEXP for the four test collection, which is not the case for PRF and MI. VEXP shows a statistically significant improvement over PRF in five cases.

Deep learning vectors are a promising source for query expansion because they are learned from hundreds of millions of words, in contrast to PRF which is obtained from top retrieved document and MI which is calculated on the collection itself. Deep learning vectors are not only useful for collections that were used in the training phase, but also for other collections which contain similar documents. In our case, all collections deal with medical cases.

There are two architectures of neural networks for obtaining deep learning vectors: skip-gram and bag-of-words [5]. We only present the results obtained using the skip-gram architecture in our experiments. We have also evaluated the bag-of-words architecture, but there was no big difference in retrieval performance between the two architectures.

Table 4. Performance comparison using MAP on test collections. † indicates statistically significant improvement over NEXP. * indicates statistically significant improvement over PRF, $p < 0.05$.

	Jelinek-Mercer				Dirichlet			
	Image2010	Image2011	Image2012	Case2011	Image2010	Image2011	Image2012	Case2011
NEXP	0.3016	0.2113	0.1862	0.1128	0.3171	0.2033	0.1681	0.1134
PRF	0.3090	0.2136	0.1920	0.1256	0.3219	0.2126	0.1766	0.1267
MI	0.3239	0.2116	0.1974	0.1360	0.3338	0.2110	0.1775	0.1327
VEXP	0.3402†*	0.2340†	0.2036†	0.1634†*	0.3603†*	0.2221†	0.1990†*	0.1451†*

4 Conclusions

We explored the use of the relationships extracted from deep learning vectors for query expansion. We showed that deep learning vectors are a promising source for query expansion by comparing it with two effective methods for query expansion: pseudo-relevance feedback and mutual information. Our experiments on four CLEF collections showed that using this expansion source gives a statistically significant improvement over baseline language models with no expansion and pseudo-relevance feedback. In addition, it is better than the expansion method using mutual information.

Acknowledgements. This work was conducted as a part of the CHIST-ERA CAMOMILE project, which was funded by the ANR (Agence Nationale de la Recherche, France).

References

1. Bengio, Y., Schwenk, H., Sencal, J.-S., Morin, F., Gauvain, J.-L.: Neural probabilistic language models. In: Holmes, D.E., Jain, L.C. (eds.) Innovations in Machine Learning. Studies in Fuzziness and Soft Computing, vol. 194, pp. 137–186. Springer, Heidelberg (2006)
2. Carpineto, C., Romano, G.: A survey of automatic query expansion in information retrieval. ACM Comput. Surv. **44**(1), 1:1–1:50 (2012)
3. Jiani, H., Deng, W., Guo, J.: Improving retrieval performance by global analysis. In: ICPR 2006, pp. 703–706 (2006)
4. Lavrenko, V., Croft, W.B.: Relevance based language models. In: SIGIR 2001, pp. 120–127. ACM, New York (2001)
5. Mikolov, T., Sutskever, I., Chen, K., Corrado, G., Dean, J.: Distributed representations of words and phrases and their compositionality. CoRR (2013)
6. Peat, H.J., Willett, P.: The limitations of term co-occurrence data for query expansion in document retrieval systems. J. Am. Soc. Inf. Sci. **42**(5), 378–383 (1991)
7. Serizawa, M., Kobayashi, I.: A study on query expansion based on topic distributions of retrieved documents. In: Gelbukh, A. (ed.) CICLing 2013, Part II. LNCS, vol. 7817, pp. 369–379. Springer, Heidelberg (2013)
8. Smucker, M.D., Allan, J., Carterette, B.: A comparison of statistical significance tests for information retrieval evaluation. In: CIKM 2007. ACM (2007)

9. Strohman, T., Metzler, D., Turtle, H., Croft, W.B.: Indri: A language model-based search engine for complex queries. In: Proceedings of the International Conference on Intelligence Analysis (2004)
10. Widdows, D., Cohen, T.: The semantic vectors package: New algorithms and public tools for distributional semantics. In: ICSC, pp. 9–15 (2010)
11. Yang, X., Jones, G.J.F., Wang, B.: Query dependent pseudo-relevance feedback based on wikipedia. In: SIGIR 2009, Boston, MA, USA, pp. 59–66 (2009)
12. Zhang, J., Deng, B., Li, X.: Concept based query expansion using wordnet. In: AST 2009, pp. 52–55. IEEE Computer Society (2009)
13. Zhu, W., Xuheng, X., Xiaohua, H., Song, I.-Y., Allen, R.B.: Using UMLS-based re-weighting terms as a query expansion strategy. In: 2006 IEEE International Conference on Granular Computing, pp. 217–222, May 2006

A Full-Text Learning to Rank Dataset for Medical Information Retrieval

Vera Boteva[1], Demian Gholipour[1], Artem Sokolov[1(✉)], and Stefan Riezler[1,2]

[1] Computational Linguistics, Heidelberg University, Heidelberg, Germany
{boteva,gholipour,sokolov,riezler}@cl.uni-heidelberg.de
[2] IWR, Heidelberg University, Heidelberg, Germany

Abstract. We present a dataset for learning to rank in the medical domain, consisting of thousands of full-text queries that are linked to thousands of research articles. The queries are taken from health topics described in layman's English on the non-commercial www.NutritionFacts.org website; relevance links are extracted at 3 levels from direct and indirect links of queries to research articles on PubMed. We demonstrate that ranking models trained on this dataset by far outperform standard bag-of-words retrieval models. The dataset can be downloaded from: www.cl.uni-heidelberg.de/statnlpgroup/nfcorpus/.

1 Introduction

Health-related content is available in information archives as diverse as the general web, scientific publication archives, or patient records of hospitals. A similar diversity can be found among users of medical information, ranging from members of the general public searching the web for information about illnesses, researchers exploring the PubMed database[1], or patent professionals querying patent databases for prior art in the medical domain[2]. The diversity of information needs, the variety of medical knowledge, and the varying language skills of users [4] results in a lexical gap between user queries and medical information that complicates information retrieval in the medical domain.

In this paper, we present a dataset that bridges this lexical gap by exploiting links between queries written in layman's English to scientific articles as provided on the www.NutritionFacts.org (NF) website. NF is a non-commercial, public service provided by Dr. Michael Greger and collaborators who review state-of-the-art nutrition research papers and provide transcribed videos, blog articles and Q&A about nutrition and health for the general public. NF content is linked to scientific papers that are mainly hosted on the PubMed database. By extracting relevance links at three levels from direct and indirect links of queries to research articles, we obtain a database that can be used to directly learn ranking models for medical information retrieval. To our knowledge this is

[1] www.ncbi.nlm.nih.gov/pubmed.

[2] For example, the USPTO and EPO provide specialized patent search facilities at www.uspto.gov/patents/process/search and www.epo.org/searching.html.

N. Ferro et al. (Eds.): ECIR 2016, LNCS 9626, pp. 716–722, 2016.
DOI: 10.1007/978-3-319-30671-1_58

the first dataset that provides full texts for thousands of relevance-linked queries and documents in the medical domain. In order to showcase the potential use of our dataset, we present experiments on training ranking models, and find that they significantly outperform standard bag-of-words retrieval models.

2 Related Work

Learning-to-rank algorithms require a large amount of relevance-linked query-document pairs for supervised training of high capacity machine learning models. Such datasets have been made public[3] by search engine companies, comprising tens of thousands of queries and hundreds of thousands of documents at up to 5 relevance levels. The disadvantage of these datasets is the fact that they do not provide full texts but only pre-processed feature vectors. They are thus useful to compare ranking algorithms for given feature representations, but are of limited use for the development of complete learning approaches. Furthermore, Ohsumed, the only learning-to-rank dataset in the medical domain, contains only about a hundred of queries. A dataset for medical information retrieval comprising full texts has been made public[4] at the CLEF eHealth evaluations. This dataset contains approximately one million documents from medical and health domains, but only 55 queries, which makes this dataset too small for training learning-to-rank systems. Large full text learning-to-rank datasets for domains such as patents or Wikipedia have been used and partially made publicly available[5]. Similar to these datasets, the corpus presented in this paper contains full-text queries and abstracts of documents, annotated with automatically extracted relevance links at several levels (here: 3). The proposed dataset is considerably smaller than the above mentioned datasets from the patent and Wikipedia domain, however, it still comprises thousands of queries and documents.

3 Corpus Creation Methodology

The NF website contains three different content sources – videos, blogs, and Q&A posts, all written in layman's English, which we used to extract queries of different length and language style. Both the internal linking structure and the scientific papers citations establish graded relevance relations between pieces of NF content and scientific papers. Additionally, the internal NF topic taxonomy, used to categorize similar NF content that is not necessarily interlinked, is exploited to define the weakest relevance grade.

[3] www.research.microsoft.com/en-us/um/beijing/projects/letor, www.research.microsoft.com/en-us/projects/mslr, www.webscope.sandbox.yahoo.com.
[4] www.clefehealth2014.dcu.ie/task-3.
[5] www.cl.uni-heidelberg.de/statnlpgroup/boostclir/wikiclir.

Crawling Queries and Documents. The following text sections of NF content pages were extracted:

- **Videos**: *title*, *description* (short summary), *transcript* (complete transcript of the audio track), *"doctor's note"* (short remarks and links to related NF content), *topics* (content tags), *sources* (URLs to medical articles), *comments* (user comments).
- **Blog Articles** (usually summaries of a series of videos): *title*, *text* (includes links to other NF pages and medical articles), *topics*, *comments.*
- **Q&A**: *title*, *text* (the question and an answer with links to related NF pages and medical articles), *comments.*
- **Topic** pages listing NF material tagged with the topic: *title*, *text* (may include a topic definition, with links to NF content but not to medical articles).

Medical documents were crawled following direct links from the NF pages to:

- **PubMed**, where 86 % of all links led,
- **PMC** (PubMed Central) with 3 %,
- Neither PubMed nor PMC pages, i.e. links to pages of medical journals, 7 %,
- Direct links to PDF documents, 4 %.

Since PubMed pages could further link to full-texts on PMC and since extracting abstracts from these two types of pages was the least error-prone, we included titles and abstracts of only these two types into the documents side of the corpus.

Data. We focused on 5 types of queries that differ by length and well-formedness of the language. In particular we tested full queries, i.e., *all fields* of NF pages concatenated: titles, descriptions, topics, transcripts and comments), *all titles* of NF content pages, *titles of non-topic pages* (i.e., titles of all NF pages except topic pages), *video titles* (titles of video pages) and *video descriptions* (description from videos pages). The latter three types of queries often resemble queries an average user would type (e.g., "*How to Treat Kidney Stones with Diet*" or "*Meat Hormones and Female Infertility*"), unlike *all titles* that include headers of topics pages that often consist of just one word.

For each relevance link between a query and a document we randomly assigned 80 % of them to the training set and 10 % for dev and test subsets. Retrieval was performed over the full set of abstracts (3,633 in total, mean/median number of tokens was 147.1/76.0). Note that this makes the test PubMed abstracts (but not the queries) available during training. The same methodology was used in [1] who found that it only marginally affected evaluation results compared to the setting without overlaps. Basic statistics about the different query types are summarized in Table 1.

Extracting Relevance Links. We defined a special relation between queries and documents that did not exist in the explicit NF link structure. A directly linked document of query $\mathbf{q}$ is considered *marginally relevant* for query $\mathbf{q}'$ if the containment $|t(\mathbf{q}) \cap t(\mathbf{q}')|/|t(\mathbf{q})|$ between the sets of topics with which the queries

Table 1. Statistics of relevance-linked ranking data (without stop-word filtering).

type	# queries	mean/median # tokens per query	mean # docs per query		
			lev. 2	lev. 1	lev. 0
all fields	3244	1890.0/43.5	4.6	41.6	33.8
all titles	3244	3.6/1.5	4.6	41.6	33.8
titles of non-topic pages	1429	6.0/4.0	4.6	25.4	26.3
video titles	1016	5.5/6.0	4.9	23.6	27.1
video descriptions	1016	24.3/21.0	4.9	23.6	27.1

are tagged is at least 70 %. In general this relation may be considered as still weakly relevant and be preferred to, say, some completely out-of-domain (e.g. nutrition-unrelated) document from PubMed. However, we treat such documents as irrelevant but still in-domain in training and testing. The rationale is that we are mostly interested in learning a thin line between relevant and similar but yet irrelevant documents, as opposed to a simpler task of discerning them from completely out-of-domain documents.

We assign relevance levels to a query $\mathbf{q}$ with respect to a scientific document $\mathbf{d}$ from three possible values: The most relevant level (2) corresponds to a direct link from $\mathbf{q}$ to $\mathbf{d}$ from the cited sources section of a page, the next level (1) is used if there exists another query $\mathbf{q}'$ that directly links to $\mathbf{d}$ and also $\mathbf{q}$'s text contains an internal link to $\mathbf{q}'$. Finally, the lowest level of (0) is reserved for every marginally relevant $\mathbf{q}'$ and document $\mathbf{d}$.

Finally, once all links are known we excluded queries that wouldn't be of any use for learning, like queries without any text (e.g., many topic pages) and queries with no direct, indirect, or topic-based links to any documents.

4 Experiments

Systems. Our two baseline retrieval systems use the classical ranking scores: *tfidf* and Okapi *BM25*[6]. In addition, we evaluated two learning to rank approaches that are based on a matrix of query words times document words as feature representation, and optimize a pairwise ranking objective [1,7]: Let $\mathbf{q} \in \{0,1\}^Q$ be a query and $\mathbf{d} \in \{0,1\}^D$ be a document, where the nth vector dimension indicates the simple occurrence of the nth word for dictionaries of size Q and D. Both approaches learn a score function $f(\mathbf{q}, \mathbf{d}) = \mathbf{q}^\top W \mathbf{d} = \sum_{i=1}^{Q} \sum_{j=1}^{D} q_i W_{ij} d_j$, where $W \in \mathbb{R}^{Q \times D}$ encodes a matrix of word associations. Optimal values of W are found by pairwise ranking given supervision data in the form of a set $\mathcal{R}$ of tuples $(\mathbf{q}, \mathbf{d}^+, \mathbf{d}^-)$, where $\mathbf{d}^+$ is a relevant (or higher ranked) document and $\mathbf{d}^-$ an irrelevant (or lower ranked) document for query $\mathbf{q}$, the goal is to find W such that an inequality $f(\mathbf{q}, \mathbf{d}^+) > f(\mathbf{q}, \mathbf{d}^-)$ is violated for the fewest number of tuples

[6] BM25 parameters were set to $k_1 = 1.2$, $b = 0.75$.

from $\mathcal{R}$. Thus, the goal is to learn weights for all domain-specific associations of query terms and document terms that are useful to discern relevant from irrelevant documents by optimizing the ranking objectives defined below.

The first method [7] applies the *RankBoost* algorithm [2], where $f(\mathbf{q},\mathbf{d})$ is a weighted linear combination of T functions h_t such that $f(\mathbf{q},\mathbf{d}) = \sum_{t=1}^{T} w_t h_t(\mathbf{q},\mathbf{d})$. Here h_t is an indicator that selects a pair of query and document words. Given differences of query-document relevance ranks $m(\mathbf{q},\mathbf{d}^+,\mathbf{d}^-) = r_{\mathbf{q},\mathbf{d}^+} - r_{\mathbf{q},\mathbf{d}^-}$, RankBoost achieves correct ranking of $\mathcal{R}$ by optimizing the exponential loss

$$\mathcal{L}_{exp} = \sum_{(\mathbf{q},\mathbf{d}^+,\mathbf{d}^-)\in\mathcal{R}} m(\mathbf{q},\mathbf{d}^+,\mathbf{d}^-)e^{f(\mathbf{q},\mathbf{d}^-)-f(\mathbf{q},\mathbf{d}^+)}.$$

The algorithm combines batch boosting with bagging over independently drawn 10 bootstrap data samples from $\mathcal{R}$, each consisting of 100k instances. In every step, the single word pair feature h_t is selected that provides the largest decrease of $\mathcal{L}_{exp}$. The resulting models are averaged as a final scoring function. To reduce memory requirements we used random feature hashing with the size of the hash of 30 bits [5]. For regularization we rely on early stopping ($T = 5000$). An additional fixed-weight identity feature is introduced that indicates the identity of terms in query and document; its weight was tuned on the dev set.

The second method uses *stochastic gradient descent* (SGD) as implemented in the Vowpal Wabbit (VW) toolkit [3] to optimize the ℓ_1-regularized hinge loss:

$$\mathcal{L}_{hng} = \sum_{(\mathbf{q},\mathbf{d}^+,\mathbf{d}^-)\in\mathcal{R}} \left(f(\mathbf{q},\mathbf{d}^+) - f(\mathbf{q},\mathbf{d}^-)\right)_+ + \lambda||W||_1,$$

where $(x)_+ = \max(0, m(\mathbf{q},\mathbf{d}^+,\mathbf{d}^-) - x)$ and λ is the regularization parameter. VW was run on the same (concatenated) samples as the *RankBoost* using the same number of hashing bits. On each step, W is updated with a scaled gradient vector $\nabla_W \mathcal{L}_{hng}$ and clipped to account for ℓ_1-regularization; λ and the number of passes over the data were tuned on the dev set.

Table 2. MAP/NDCG results evaluated for different types of queries. Best NDCG results of learning-to-rank versus bag-of-words models are highlighted in **bold face**.

queries	RankBoost	SGD	tfidf	bm25
all fields	0.2632/0.5073	0.3831/**0.6064**	0.1360/0.3932	0.1627/**0.4169**
all titles	0.1549/**0.3475**	0.1360/0.3454	0.1233/0.2578	0.1251/**0.2582**
titles of non-topic pages	0.1615/**0.4039**	0.1775/0.3790	0.0972/0.2851	0.1124/**0.3032**
video descriptions	0.1312/**0.3826**	0.1060/0.3112	0.1110/0.3509	0.1262/**0.3765**
video titles	0.1350/**0.3804**	0.1079/0.3109	0.1010/0.2873	0.1127/**0.3042**

Experimental Results. Results according to the MAP and NDCG metrics on preprocessed data[7] are reported in the Table 2. Result differences between the best performing learning-to-rank versus bag-of-words models were found to be statistically significant [6]. As results show, learning-to-rank approaches outperform classical retrieval methods by a large margin, proving that the provided corpus is sufficient to optimize domain-specific word associations for a direct ranking objective. As shown in row 1 of Table 2, the *SGD* approach outperforms *RankBoost* in the evaluation on *all fields* queries, but performs worse with shorter (and fewer) queries as in the setups listed in rows 2–5. This is due to a special "pass-through" feature implemented in *RankBoost* that assigns a default feature to word identities, thus allowing to learn better from sparser data. The *SGD* implementation does not take advantage of such a feature, but it makes a better use of the full matrix of word associations which offsets the lacking pass-through if enough word combinations are observable in the data.

5 Conclusion

We presented a dataset for learning to rank in the medical domain that has the following key features: (1) full text queries of various length, thus enabling the development of complete learning models; (2) relevance links at 3 levels for thousands of queries in layman's English to documents consisting of abstracts of research article; (3) public availability of the dataset (with links to full documents for research articles). We showed in an experimental evaluation that the size of the dataset is sufficient to learn ranking models based on sparse word association matrices that outperform standard bag-of-words retrieval models.

Acknowledgments. We are grateful to Dr. Michael Greger for permitting crawling www.NutritionFacts.org. This research was supported in part by DFG grant RI-2221/1-2 "Weakly Supervised Learning of Cross-Lingual Systems".

References

1. Bai, B., Weston, J., Grangier, D., Collobert, R., Sadamasa, K., Qi, Y., Chapelle, O., Weinberger, K.: Learning to rank with (a lot of) word features. Inf. Retr. J. **13**(3), 291–314 (2010)
2. Collins, M., Koo, T.: Discriminative reranking for natural language parsing. Comput. Linguist. **31**(1), 25–69 (2005)
3. Goel, S., Langford, J., Strehl, A.L.: Predictive indexing for fast search. In: NIPS, Vancouver, Canada (2008)
4. Goeuriot, L., Kelly, L., Jones, G.J.F., Müller, H., Zobel, J.: Report on the SIGIR 2014 workshop on medical information retrieval (MedIR). SIGIR Forum **48**(2), 78–82 (2014)

[7] Preprocessing included lowercasing, tokenizing, filtering punctuation and stop-words, and replacing numbers with a special token.

5. Shi, Q., Petterson, J., Dror, G., Langford, J., Smola, A.J., Strehl, A.L., Vishwanathan, V.: Hash Kernels. In: AISTATS, Irvine, CA (2009)
6. Smucker, M.D., Allan, J., Carterette, B.: A comparison of statistical significance tests for information retrieval evaluation. In: CIKM, Lisbon, Portugal (2007)
7. Sokolov, A., Jehl, L., Hieber, F., Riezler, S.: Boosting cross-language retrieval by learning bilingual phrase associations from relevance rankings. In: EMNLP, Seattle (2013)

Multi-label, Multi-class Classification Using Polylingual Embeddings

Georgios Balikas(✉) and Massih-Reza Amini

University of Grenoble-Alpes, Grenoble, France
{georgios.balikas,massih-reza.amini}@imag.fr

Abstract. We propose a Polylingual text Embedding (PE) strategy, that learns a language independent representation of texts using Neural Networks. We study the effects of bilingual representation learning for text classification and we empirically show that the learned representations achieve better classification performance compared to traditional bag-of-words and other monolingual distributed representations. The performance gains are more significant in the interesting case where only few labeled examples are available for training the classifiers.

1 Introduction and Preliminaries

In this work we propose a mechanism for combining distributed representations of documents in different languages. Each document in a given language is first translated using an existing Machine Translation (MT) tool. The rationale behind is that translation offers the possibility to enrich and disambiguate the text, especially for short documents. Documents are then represented by aggregating the embeddings of their associated text spans in each language [7,9] using a non-linear auto-encoder (AE). The AE is trained on their concatenated representations and a classifier is finally trained in the polylingual space outputed by the auto-encoder. Our classification results in a subset of the publicly available Wikipedia show that our approach yields improved classification performance compared to the case where a classical bag-of-words space is used for document representation, especially in the case where the size of the training set is small.

Neural Networks have recently shown promising results in several machine learning and information extraction tasks [2,12,13]. For text classification, the use of embeddings as inputs or initializations to more complex architectures has been investigated and, for example, [4,5] study the benefits of embeddings of sentence-length spans (sentences and/or questions). In the multilingual setting, [3] proposed an approach to learn bilingual embeddings exploiting parallel and non-parallel text in the languages, [1] proposed to use correlated components analysis, together with small bi-lingual lexicons, to learn how to project embeddings in two separate languages into a common representation space and [6] proposed an approach similar to ours that uses an auto-encoder to learn bilingual representations.

In the next section we present our polylingual embedding strategy. In the experimental part (Sect. 3), we empirically show that the learned representations

N. Ferro et al. (Eds.): ECIR 2016, LNCS 9626, pp. 723–728, 2016.
DOI: 10.1007/978-3-319-30671-1_59

constitute better classification features compared to several baselines and their value can strongly benefit classification settings with few labeled examples. We discuss these results in Sect. 4 and conclude in Sect. 5.

2 The Proposed Approach

Monolingual distributed representations (DRs) project text spans into a language-dependent semantic space where spans with similar semantics are closer in that space. Here, we aim to combine two distributed representations of documents corresponding to the original document and its translation using an auto-encoder. We will refer to those combined representations as *Polylingual Embeddings* (PE). We suppose that the auto-encoder will disentangle the language-dependent factors and will learn robust representations on its hidden layer encoding as illustrated in Fig. 1. Given a document d_i in English, we first translate it into French using a commercial translator, we then generate the distributed representations of the document and its translation $\{\mathcal{G}^\ell(d_i)\}_{\ell=1}^2$, and then aggregate those DRs using an auto-encoder (Algorithm 1).

The auto-encoder is learned over all concatenated distributed representations of documents using a stochastic back-propagation algorithm. In this work we consider two strategies to create the DR of each document. The first one is based on average pooling, where word representations are first obtained using the word2vec tool [8]. DR of documents, i.e. functions $(\mathcal{G}^\ell)_{\ell\in\{1,2\}}$, are then obtained by averaging the vectors of words contained in them. In this study we consider the *continuous bag of words* (cbow) and the *skip-gram* models that generate word representations. The second strategy is based on the *distributed Memory Model of paragraph vectors* (DMMpv) and *distributed bag-of-words of paragraph vectors* (DBOWpv) models [7], that extend cbow and skip-gram respectively. In this case, $(\mathcal{G}^\ell)_{\ell\in\{1,2\}}$ are defined by the output of the models without further processing.

Require: $\{\mathcal{G}^\ell(d_i)\}_{\ell=1}^2$, a trained AE
1: **for** each document d_i **do**
2: Concatenate $\mathcal{G}^1(d_i)$ and $\mathcal{G}^2(d_i)$
3: Get PE representation of d_i as the hidden encoding of the AE fed with the concatenation
4: **end for**

Algorithm 1. The process of generating PE representations

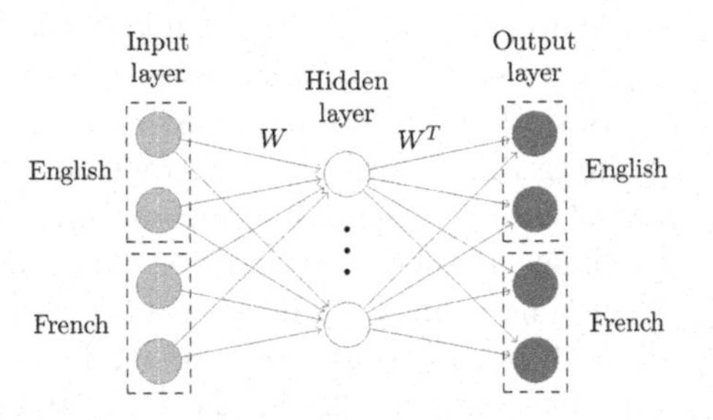

Fig. 1. An AE that generates the PE in its hidden layer. The dashed boxes denote the document DRs in the corresponding language.

Table 1. Statistics after pre-processing the datasets. The distributed representations dataset refers to the data used to train $\mathcal{G}$. The classification data refer to the supervised dataset used for classification purposes.

	Distributed Representations			Classification			
	Documents	Vocabulary	# Words	Documents	Vocabulary	Avg. Doc. Len	# Labels
English	6,358,467	490,122	198,213,780	12,670	56,886	115.32	1,17
French	6,358,467	713,171	177,766,544	12,670	58,678	132.29	1,17

3 The Experimental Framework

The Data. Training neural network models to generate distributed representations benefits by large amounts of free text. To train the models that generate DRs we used such free texts in English and French:[1] the left part of Table 1 (under "Distributed Representations") presents some basic statistics for those data. We used the same number of documents for the two languages to avoid any training bias. The raw text was pre-processed by applying lower-casing and space-padding punctuation. Similarly to previous studies [7,8], we kept the punctuation. Publicly available implementations of the models were used with their default parameters: the word2vec tool[2] for the cbow and skip-gram and the doc2vec for the DBOWpv and DMMpv from Gensim [11].

For the classification task we used the raw version of the Wikipedia dataset of the Large Scale Hierarchical Text Classification challenge [10]. The original dataset contains 60,252 categories; we restrict our study here in a fraction of the dataset with 12,670 documents belonging to the 100 most common categories. The right part of Table 1 presents basic statistics for this subset.

Baselines. We used as a first baseline Support Vectors Machines (SVM) fed with the tf-idf representation of the documents, which is commonly used in text classification problems (denoted by SVM_{BoW}). As a second baseline, we used k-Nearest Neighbours (k-NN) and SVMs learned on the monolingual space of the DRs of English documents (denoted respectively by SVM_{DR} and $k\text{-NN}_{\text{DR}}$). These baselines aim at evaluating the value of the fusion mechanism (PE) that we propose. k-NN and SVMs were adapted to the multi-label setting (denoted respectively by SVM_{PE} and $k\text{-NN}_{\text{PE}}$). For the former, given the labels of the k nearest training instances of a test document, the algorithm returns the labels that belong to at least $p\%$ of its nearest neighbours. For each run $k \in \{13, 14, 15\}$ and $p \in \{0.1, 0.2, 0.3\}$ are decided using 5-fold cross-validation on the training data. The SVMs were used in an one-vs-rest fashion; they return every label that has a positive distance from the separating hyperplane. The value of the hyper-parameter $C \in \{10^{-1}, \dots, 10^4\}$ that controls the importance of the regularization term in the optimization problem, is selected using 5-fold cross-validation over the training data.

[1] http://statmt.org/.

[2] https://code.google.com/p/word2vec/.

Table 2. F_1 measures of difference algorithms. The performance of 5-fold cross-validated SVM using the bag-of-words representation is 36.03

dim.	cbow				skip-gram			
	k-NN$_{DR}$	SVM$_{DR}$	k-NN$_{PE}$	SVM$_{PE}$	k-NN$_{DR}$	SVM$_{DR}$	k-NN$_{PE}$	SVM$_{PE}$
50	39.19	37.20	39.58	32.84	38.25	34.74	37.51	32.09
100	40.20	40.01	43.53	37.54	39.34	38.61	41.15	34.54
200	40.48	43.41	45.86	42.50	39.73	40.96	42.79	41.08
300	40.42	44.25	**46.33**	43.38	39.62	42.67	42.62	42.74
	DBOWpv				DMMpv			
50	24.45	25.06	30.26	24.08	24.47	25.56	29.55	24.94
100	31.20	28.53	34.63	26.88	24.74	29.31	31.21	27.22
200	27.73	29.80	36.02	30.80	18.22	30.04	29.01	32.10
300	27.79	29.92	38.71	30.82	15.98	30.49	25.20	32.01
SVM$_{BoW}$	36.03							

Our Approach. Using the above presented DR model, we first generated the document embeddings in English and French in a d-dimensional space with $d \in \{50, 100, 200, 300\}$. Then, for the AE we considered as activation functions the hyperbolic tangent and the sigmoid function. The sigmoid performed consistently better and thus we use it in the reported results. The AE was trained with tied weights using a stochastic back-propagation algorithm with mini-batchs of size 10 and the euclidean distance of the input/output as loss function. The number of neurons in the hidden layer was set to be 70 % of the size of the input.[3]

4 Experimental Results

Table 2 presents the scores of the F_1 measure when 10 % of the 12.670 documents were used for training purposes and the rest 90 % for testing. We report the classification performance with the four different DR models (cbow, skip-gram, DBOWpv and DMMpv) and 2 learning algorithms (k-NN and SVMs) for different input sizes. The columns labeled k-NN$_{DR}$ and SVM$_{DR}$ present the (baseline) performance of SVM and k-NN trained on the monolingual DRs. Also the last line of the table indicates the F_1 score of SVM with tf-idf representation (SVM$_{BoW}$). The best obtained result is shown in bold.

We first notice that the average pooling strategy (cbow and skip-gram) performs better compared to when the document vectors are directly learned (DBOWpv and DMMpv). In particular, cbow seems to be the best performing representation, both as a baseline model and when used as base model to generate the PE representations. On the other hand, DBOWpv and DMMpv perform significantly worse: in the baseline setting the best cbow performance achieved is 44.25 whereas the best DMMpv configuration achieves 30.49, 14 F_1 points less.

[3] The code is available at http://ama.liglab.fr/~balikas/ecir2015.zip.

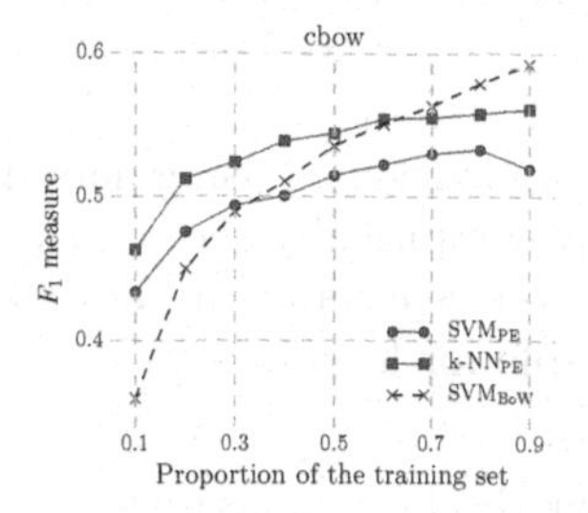

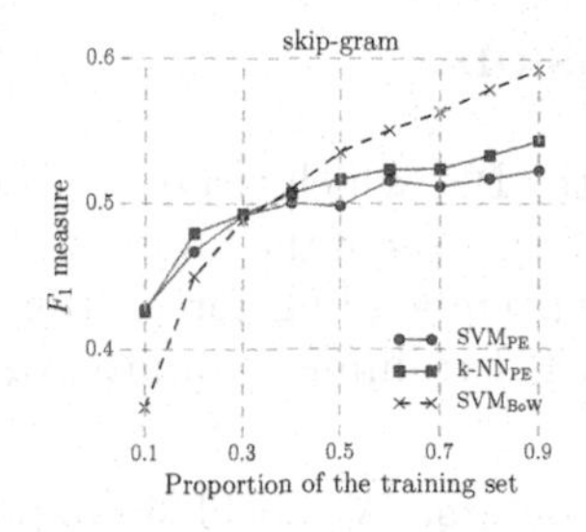

Fig. 2. Comparison of the performance of the learning algorithms learned on different representations with respect to the available labelled data. The dimension of the PE representations is 300.

The PE representations learned on top of the four base models improve significantly over the performance of the monolingual DRs, especially for k-NN. For instance, for cbow with base-model vector dimension 200, the baseline representation achieves 40.42 F_1 and its corresponding PE representation obtains 46.33, improving almost 6 points. In general, we notice such improvements between the base DR and its respective PE, especially when the dimension of the DR representation increases. Note that the PE improvements are independent of the methods used to generate the DRs: for instance k-NN$_{PE}$ over the 200-dimensional PE DMMpv representations gains more than 11 F_1 points compared to k-NN$_{DR}$. It is also to be noted that the baseline SVM$_{BoW}$ is outperformed by SVM$_{PE}$ especially when cbow and skip-gram DRs are used.

Comparing the two learning methods (k-NN$_{PE}$ and SVM$_{PE}$), we notice that k-NN$_{PE}$ performs best. This is motivated by the fact that distributed representations are supposed to capture the semantics in the low dimensional space. At the same time, the neighbours algorithm compares exactly this semantic distance between data instances, whereas SVMs tries to draw separating hyperplanes among them. Finally, it is known that SVMs benefit from high-dimensional vectors such as bag-of-words representations. Notably, in our experiments increasing the dimension of the representations consistently benefits SVMs.

We now examine the performance of the PE representations taking into account the amount of labeled training data. Figure 2 illustrates the performance of the SVM$_{BoW}$ and SVM$_{PE}$ and k-NN$_{PE}$ with PE representations when the fraction of the available training data varies from 10 % of the intial training set to 90 % and in the case where, cbow and skip-gram are used as DR representations with an input size of 300. Note that if only a few training documents are available, the learning approach is strongly benefited by the rich PE representations, that outperforms the traditional SVM$_{BoW}$ setting consistently. For instance, in the experiments with 300 dimensional PE representations with cbow DRs, when only 20 % of the data are labeled, the SVM$_{BoW}$ needs 20 % more data to achieve similar performance, a pattern that is observed in most of the runs in the figure. When, however, more training data are available the tf-idf copes with the complexity of the problem and levarages this wealth of information more efficiently than PE does.

5 Conclusion

We proposed the PE, which is a text embedding learned using neural networks by leveraging translations of the input text. We empirically showed the effectiveness of the bilingual embedding for classification especially in the interesting case where few labeled training data are available for learning.

Acknowledgements. We would like to thank the anonymous reviewers for their valuable comments. This work is partially supported by the CIFRE N 28/2015 and by the LabEx PERSYVAL Lab ANR-11-LABX-0025.

References

1. Faruqui, M., Dyer, C.: Improving vector space word representations using multilingual correlation. Association for Computational Linguistics (2014)
2. Gao, J., He, X., Yih, W.T., Deng, L.: Learning continuous phrase representations for translation modeling. In: Proceedings of ACL. Association for Computational Linguistics, June 2014
3. Gouws, S., Bengio, Y., Corrado, G.: Bilbowa: fast bilingual distributed representations without word alignments. arXiv preprint arXiv:1410.2455 (2014)
4. Kalchbrenner, N., Grefenstette, E., Blunsom, P.: A convolutional neural network for modelling sentences. arXiv preprint arXiv:1404.2188 (2014)
5. Kim, Y.: Convolutional neural networks for sentence classification. arXiv preprint arXiv:1408.5882 (2014)
6. Lauly, S., Larochelle, H., Khapra, M., Ravindran, B., Raykar, V.C., Saha, A.: An autoencoder approach to learning bilingual word representations. In: Advances in Neural Information Processing Systems, pp. 1853–1861 (2014)
7. Le, Q.V., Mikolov, T.: Distributed representations of sentences and documents. arXiv preprint arXiv:1405.4053 (2014)
8. Mikolov, T., Chen, K., Corrado, G., Dean, J.: Efficient estimation of word representations in vector space. arXiv preprint arXiv:1301.3781 (2013)
9. Mikolov, T., Sutskever, I., Chen, K., Corrado, G.S., Dean, J.: Distributed representations of words and phrases and their compositionality. In: Advances in Neural Information Processing Systems, pp. 3111–3119 (2013)
10. Partalas, I., Kosmopoulos, A., Baskiotis, N., Artieres, T., Paliouras, G., Gaussier, E., Androutsopoulos, I., Amini, M.R., Galinari, P.: Lshtc: a benchmark for large-scale text classification. arXiv preprint arXiv:1503.08581 (2015)
11. Řehůřek, R., Sojka, P.: Software Framework for Topic Modelling with Large Corpora. In: Proceedings of the LREC 2010 Workshop on New Challenges for NLP Frameworks, pp. 45–50. ELRA, Valletta, Malta, May 2010. http://is.muni.cz/publication/884893/en
12. Tang, D., Wei, F., Yang, N., Zhou, M., Liu, T., Qin, B.: Learning sentiment-specific word embedding for twitter sentiment classification. In: Proceedings of the 52nd Annual Meeting of the Association for Computational Linguistics. vol. 1, pp. 1555–1565 (2014)
13. Zhang, X., LeCun, Y.: Text understanding from scratch. arXiv preprint arXiv:1502.01710 (2015)

Learning Word Embeddings from Wikipedia for Content-Based Recommender Systems

Cataldo Musto(✉), Giovanni Semeraro, Marco de Gemmis, and Pasquale Lops

Department of Computer Science, University of Bari Aldo Moro, Bari, Italy
{cataldo.musto,giovanni.semeraro,marco.gemmis,pasquale.Lops}@uniba.it

Abstract. In this paper we present a preliminary investigation towards the adoption of *Word Embedding* techniques in a content-based recommendation scenario. Specifically, we compared the effectiveness of three widespread approaches as Latent Semantic Indexing, Random Indexing and Word2Vec in the task of learning a vector space representation of both items to be recommended as well as user profiles.

To this aim, we developed a content-based recommendation (CBRS) framework which uses textual features extracted from Wikipedia to learn user profiles based on such Word Embeddings, and we evaluated this framework against two state-of-the-art datasets. The experimental results provided interesting insights, since our CBRS based on Word Embeddings showed results comparable to those of well-performing algorithms based on Collaborative Filtering and Matrix Factorization, especially in high-sparsity recommendation scenarios.

1 Introduction

Word Embedding techniques recently gained more and more attention due to the good performance they showed in a broad range of natural language processing-related scenarios, ranging from *sentiment analysis* [10] and *machine translation* [2] to more challenging ones as learning a textual description of a given image[1].

However, even if some recent research gave new lymph to such approaches, Word Embedding techniques took their roots in the area of Distributional Semantics Models (DSMs), which date back in the late 60's [3]. Such models are mainly based on the so-called *distributional hypothesis*, which states that the meaning of a word depends on its *usage* and on the *contexts* in which it occurs. In other terms, according to DSMs, it is possible to infer the meaning of a term (e.g., leash) by analyzing the other terms it co-occurs with (dog, animal, etc.). In the same way, the correlation between different terms (e.g., leash and muzzle) can be inferred by analyzing the similarity between the contexts in which they are used. Word Embedding techniques have inherited the vision carried out by DSMs, since they aim to learn in a totally unsupervised way a *low-dimensional*

[1] http://googleresearch.blogspot.it/2014/11/a-picture-is-worth-thousand-coherent.html

N. Ferro et al. (Eds.): ECIR 2016, LNCS 9626, pp. 729–734, 2016.
DOI: 10.1007/978-3-319-30671-1_60

vector space representation of *words* by analyzing the usage of the terms in (very) large corpora of textual documents. Many popular techniques fall into this class of algorithms: Latent Semantic Indexing [1], Random Indexing [8] and the recently proposed Word2Vec [5], to name but a few.

In a nutshell, all these techniques carry out the learning process by encoding linguistic regularities (e.g., the co-occurrences between the terms or the occurrence of a term in a document) in a huge matrix, as a term-term or term-document matrix. Next, each Word Embedding technique adopts a different technique to reduce the overall dimension of the matrix by maintaining most of the *semantic nuances* encoded in the original representation. One of the major advantages that comes from the adoption of Word Embedding techniques is that the dimension of the representation (that is to say, the size of the vectors) is just a parameter of the model, so it can be set according to specific constraints or peculiarities of the data. Clearly, the smaller the vectors, the bigger the loss of information.

Although the effectiveness of such techniques (especially when combined with deep neural network architectures) is already taken for granted, just a few work investigated how well they do perform in recommender systems-related tasks. In [6], Musto et al. proposed a content-based recommendation model based on Random Indexing. Similarly, the effectiveness of LSI in a content-based recommendation scenario is evaluated in [4]. However, none of the current literature carried out a comparative analysis among such techniques: to this aim, in this work we defined a simple content-based recommendation framework based on *word embeddings* and we assessed the effectiveness of such techniques in a content-based recommendation scenario.

2 Methodology

2.1 Overview of the Techniques

Latent Semantic Indexing (LSI) [1] is a word embedding technique which applies Singular Value Decomposition (SVD) over a word-document matrix. The goal of the approach is to *compress* the original information space through SVD in order to obtain a smaller-scale word-*concepts* matrix, in which each column models a *latent concept* occurring in the original vector space. Specifically, SVD is employed to unveil the latent relationships between terms according to their usage in the corpus.

Next, Random Indexing (RI) [8] is an incremental technique to learn a low-dimensional word representation relying on the principles of the Random Projection. It works in two steps: first, a *context vector* is defined for each context (the definition of context is typically scenario-dependant. It may be a paragraph, a sentence or the whole document). Each context vector is ternary (it contains values in $\{-1, 0, 1\}$) very sparse, and its values are *randomly distributed*. Given such context vectors, the vector space representation of each word is obtained by just summing over all the representations of the contexts in which the word occurs. An important peculiarity of this approach is that it is incremental and scalable: if any new documents come into play, the vector space representation of the terms is updated by just adding the new occurrences of the terms in the new documents.

Finally, Word2Vec (W2V) is a recent technique proposed by Mikolov et al. [5]. The approach learns a vector-space representation of the terms by exploiting a two-layers neural network. In the first step, weights in the network are randomly distributed as in RI. Next, the network is trained by using the Skip-gram methodology in order to model fine-grained regularities in word usage. At each step, weights are updated through Stochastic Gradient Descent and a vector-space representation of each term is obtained by extracting the weights of the network at the end of the training.

2.2 Recommendation Pipeline

Our recommendation pipeline follows the classical workflow carried out by a content-based recommendation framework. It can be split into four steps:

1. Given a set of items I, each $i \in I$ is mapped to a Wikipedia page through a semi-automatic procedure. Next, textual features are gathered from each Wikipedia page and the extracted content is processed through a Natural Language Processing pipeline to remove noisy features. More details about this process are provided in Sect. 3.
2. Given a vocabulary V built upon the description of the items in I extracted from Wikipedia, for each word $w \in V$ a vector space representation w_T is learnt by exploiting a word embedding technique T.
3. For each item $i \in I$, a vector space representation of the item i_T is built. This is calculated as the centroid of the vector space representation of the words occurring in the document.
4. Given a set of users U, a user profile for each $u \in U$ is built. The vector space representation of the profile is learnt as the centroid of the vector space representation of the items the user previously liked
5. Given a vector space representation of both items to be recommended and user profile, recommendations are calculated by exploiting classic similarity measures: items are ranked according to their decreasing similarity and top-K recommendations are returned to the user.

Clearly, this is a very basic formulation, since more *fine-grained representations* can be learned for both items and users profiles. However, this work just intends to preliminarily evaluate the effectiveness of such representations in a simplified recommendation framework, in order to pave the way to several future research directions in the area.

3 Experimental Evaluation

Experiments were performed by exploiting two state-of-the-art datasets as MovieLens[2] and DBbook[3]. The first one is a dataset for movie recommendations,

[2] http://grouplens.org/datasets/movielens/.

[3] http://challenges.2014.eswc-conferences.org/index.php/RecSys.

while the latter comes from the ESWC 2014 Linked-Open Data-enabled Recommender Systems challenge and focuses on book recommendation. Some statistics about the datasets are provided in Table 1.

A quick analysis of the data immediately shows the very different nature of the datasets: even if both of them resulted as very sparse, MovieLens is more dense than DBbook (93.69 % vs. 99.83 % sparsity), indeed each Movielens user voted 84.83 items on average (against the 11.70 votes given by DBbook users). DBbook has in turn the peculiarity of being unbalanced towards negative ratings (only 45 % of positive preferences). Furthermore, MovieLens items were voted more than DBbook ones (48.48 vs. 10.74 votes for item, on average).

Experimental Protocol. Experiments were performed by adopting different protocols: as regards MovieLens, we carried out a 5-folds cross validation, while a single training/test split was used for DBbook. In both cases we used the splits which are commonly used in literature. Given that MovieLens preferences are expressed on a 5-point discrete scale, we decided to consider as *positive* ratings only those equal to 4 and 5. On the other side, the DBbook dataset is already available as *binarized*, thus no further processing was needed. Textual content was obtained by mapping items to Wikipedia pages. All the available items were successfully mapped by querying the title of the movie or the name of the book, respectively. The extracted content was further processed through a NLP pipeline consisting of a stop-words removal step, a POS-tagging step and a lemmatization step. The outcome of this process was used to learn the Word Embeddings. For each word embedding technique we compared two different sizes of learned vectors: 300 and 500. As regards the baselines, we exploited MyMediaLite library[4]. We evaluated User-to-User (U2U-KNN) and Item-to-Item Collaborative Filtering (I2I-KNN) as well as the Bayesian Personalized Ranking Matrix Factorization (BPRMF). U2U and I2I neighborhood size was set to 80, while BPRMF was run by setting the factor parameter equal to 100. In both cases we chose the optimal values for the parameters.

Table 1. Description of the datasets

	MovieLens	DBbook
Users	943	6,181
Items	1,682	6,733
Ratings	100,000	72,372
Sparsity	93.69 %	99.83 %
Positive Ratings	55.17 %	45.85 %
Avg. ratings/user ± stdev	84.83±83.80	11.70±5.85
Avg. ratings/item ± stdev	48.48±65.03	10.74±27.14

[4] http://www.mymedialite.net/.

Table 2. Results of the experiments. The best word embedding approach is highlighted in bold. The best overall configuration is highlighted in bold and underlined. The baselines which are overcame by at least a word embedding are reported in italics.

MovieLens	W2V		RI		LSI		U2U	I2I	BPRMF
Vector size	300	500	300	500	300	500			
F1@5	**0.5056**	0.5054	0.4921	0.4910	0.4645	0.4715	**0.5217**	*0.5022*	0.5141
F1@10	**0.5757**	0.5751	0.5622	0.5613	0.5393	0.5469	**0.5969**	0.5836	0.5928
F1@15	0.5672	**0.5674**	0.5349	0.5352	0.5187	0.5254	**0.5911**	0.5814	0.5876
DBbook	W2V		RI		LSI		U2U	I2I	BPRMF
Vector size	300	500	300	500	300	500			
F1@5	0.5183	**0.5186**	0.5064	0.5039	0.5056	0.5076	0.5193	*0.5111*	**0.5290**
F1@10	0.6207	0.6209	0.6239	0.6244	0.6256	**0.6260**	*0.6229*	*0.6194*	**0.6263**
F1@15	0.5829	0.5828	0.5892	0.5887	0.5908	**0.5909**	*0.5777*	*0.5776*	*0.5778*

Discussion of the Results. The first six columns of Table 2 provide the results of the comparison among the word embedding techniques. As regards MovieLens, W2V emerged as the best-performing configuration for all the metrics taken into account. The gap is significant when compared to both RI and LSI. Moreover, results show that the size of the vectors did not significantly affect the overall accuracy of the algorithms (with the exception of LSI). This is an interesting outcome since with an even smaller word representation, word embeddings can obtain good results. However, the outcomes emerging from this first experiments are controversial, since DBbook data provided opposite results: in this dataset W2V is the best-performing configuration only for F1@5. On the other side, LSI, which performed the worst on MovieLens data, overcomes both W2V and RI on F1@10 and F1@15. At a first glance, these results indicate non-generalizable outcomes. However, it is likely that such behavior depends on specific peculiarities of the datasets, which in turn influence the way the approaches learn their vector-space representations. A more thorough analysis is needed to obtain general guidelines which drive the behavior of such approaches.

Next, we compared our techniques to the above described baselines. Results clearly show that the effectiveness of word embedding approaches is directly dependent on the sparsity of the data. This is an expected behavior since content-based approaches can better deal with cold-start situations. In highly sparse dataset such as DBbook (99.13 % against 93.59 % of MovieLens), content-based approaches based on word embedding tend to overcome the baselines. Indeed, RI and LSI, overcome I2I and U2U on F1@10 and F1@15 and W2V overcomes I2I on F1@5 and I2I and U2U on F1@15. Furthermore, it is worth to note that on F1@10 and F@15 word embeddings can obtain results which are comparable (or even better on F1@15) to those obtained by BPRMF. This is a very important outcome, which definitely confirms the effectiveness of such techniques, even compared to matrix factorization techniques. Conversely, on less sparse datasets as MovieLens, collaborative filtering algorithms overcome their content-based counterpart.

4 Conclusions and Future Work

In this paper we presented a preliminary comparison among three widespread techniques in the task of learning Word Embeddings in a content-based recommendation scenario. Results showed that our model obtained performance comparable to those of state-of-the art approaches based on collaborative filtering. In the following, we will further validate our results by also further investigating both the effectiveness of novel and richer textual *data silos*, as those coming from the Linked Open Data cloud, and more expressive and complex Word Embedding techniques, as well as by extending the comparison to hybrid approaches such as those reported in [9] or in context-aware recommendation settings [7].

References

1. Deerwester, S., Dumais, S., Landauer, T., Furnas, G., Harshman, R.: Indexing by latent semantic analysis. JASIS **41**, 391–407 (1990)
2. Gouws, S., Bengio, Y., Corrado, G. Bilbowa: Fast bilingual distributed representations without word alignments (2014). arXiv:1410.2455
3. Harris, Z.S.: Mathematical Structures of Language. Interscience, New York (1968)
4. McCarey, F., Cinnéide, M.Ó., Kushmerick, N.: Recommending library methods: an evaluation of the vector space model (VSM) and latent semantic indexing (LSI). In: Morisio, M. (ed.) ICSR 2006. LNCS, vol. 4039, pp. 217–230. Springer, Heidelberg (2006)
5. Mikolov, T., Sutskever, I., Chen, K., Corrado, G., Dean, J.: Distributed representations of words and phrases and their compositionality. In: NIPS, pp. 3111–3119 (2013)
6. Musto, C., Semeraro, G., Lops, P., de Gemmis, M.: Random indexing and negative user preferences for enhancing content-based recommender systems. In: Huemer, C., Setzer, T. (eds.) EC-Web 2011. LNBIP, vol. 85, pp. 270–281. Springer, Heidelberg (2011)
7. Musto, C., Semeraro, G., Lops, P., de Gemmis, M.: Combining distributional semantics and entity linking for context-aware content-based recommendation. In: Dimitrova, V., Kuflik, T., Chin, D., Ricci, F., Dolog, P., Houben, G.-J. (eds.) UMAP 2014. LNCS, vol. 8538, pp. 381–392. Springer, Heidelberg (2014)
8. Sahlgren, M.: An introduction to random indexing. In: Methods and Applications of Semantic Indexing Workshop, TKE 2005 (2005)
9. Semeraro, G., Lops, P., Degemmis, M.: Wordnet-based user profiles for neighborhood formation in hybrid recommender systems. In: Fifth International Conference on Hybrid Intelligent Systems, HIS 2005, pp. 291–296. IEEE (2005)
10. Tang, D., Wei, F., Yang, N., Zhou, M., Liu, T., Qin, B.: Learning sentiment-specific word embedding for twitter sentiment classification. In: Proceedings of the 52nd Annual Meeting of the Association for Computational Linguistics, vol. 1, pp. 1555–1565 (2014)

Tracking Interactions Across Business News, Social Media, and Stock Fluctuations

Ossi Karkulahti, Lidia Pivovarova(✉), Mian Du, Jussi Kangasharju, and Roman Yangarber

Department of Computer Science, University of Helsinki, Helsinki, Finland
lidia.pivovarova@cs.helsinki.fi

Abstract. In this paper we study the interactions between how companies are mentioned in news, their presence on social media, and daily fluctuation in their stock prices. Our experiments demonstrate that for some entities these time series can be correlated in interesting ways, though for others the correspondences are more opaque. In this study, social media presence is measured by counting Wikipedia page hits. This work is done in a context of building a system for aggregating and analyzing news text, which aims to help the user track business trends; we show results obtainable by the system.

1 Introduction

The nature of the complex relationships among traditional news, social media, and stock price fluctuations is the subject of active research. Recent studies in the area demonstrate that it is possible to find some correlation between stock prices and news, when the news are properly classified [1,9]. A comprehensive overview of market data prediction from text can be found in [7]. In particular, [6] reported an increase in Wikipedia views for company pages and financial topics before stock market falls. Joint analysis of news and social media has been previously studied, *inter alia*, by [4,5,8]. The approach followed in these papers, as well as our approach [2], has two interrelated goals: to find information complementary to what is found in the news, and to control the amount of data that needs to be downloaded from social media.

We study the interplay among business news, social media, and stock prices. We believe that the combined analysis of information derived from news, social media and financial data can be of particular interest for specialists in various areas: business analysts, Web scientists, data journalists, etc. We use PULS [1] to collect on-line news articles from multiple sources and to identify the business entities mentioned in the news texts, e.g., companies and products, and the associated event types such as "product launch," "recall," "investment" [3]. Using these entities we then construct queries to get the corresponding social media content and its metadata, such as, Twitter posts, YouTube videos, or Wikipedia pages. We focus on analyzing the activity of users of social media in numerical terms, rather than on analyzing the content, polarity, sentiment, etc.

[1] The Pattern Understanding and learning System: http://puls.cs.helsinki.fi.

N. Ferro et al. (Eds.): ECIR 2016, LNCS 9626, pp. 735–740, 2016.
DOI: 10.1007/978-3-319-30671-1_61

US: 717,950 vehicles recalled by GM

A total of 717,950 vehicles have been recalled by US-based General Motors (GM) on 23 July 2014 in order to repair turn signals, seats and other parts. The six separate recalls include 124,008 vehicles that contain seat hook brackets that might have received an incomplete welding. The affected vehicles are the 2014 Chevrolet SS, 2014 Chevrolet Caprice, 2014-2015 GMC Sierra HD and LD and 2014-2015 Silverado HD and LD. Moreover, 120,426 2013 Chevrolet Malibus and 2011-2013 Buick Regals were recalled due to potentially faulty internal turn signal indicators. Besides that, 414,333 2011-2012 Buick LaCrosses, Camaros and Regals were recalled in order to replace a seat-height adjustment bolt.

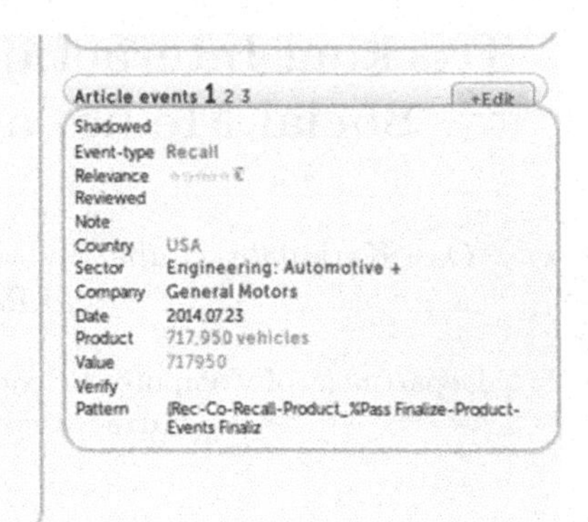

Fig. 1. A news text and a product recall event produced by the PULS IE engine.

The main contributions of this paper: we combine NLP with social media analysis, and discover interesting correlations between news and social media.

2 Process Overview

We now present the processing steps. First, the system collects unstructured text from multiple news sources on the Web. PULS uses over a thousand websites which provide news feeds related to business (Reuters Business News, New York Times Business Day, etc.). Next, the NLP engine is used to discover, aggregate, and verify information obtained from the Web. The engine performs Information Extraction (IE), which is a key component of the platform that transforms facts found in plain text into a structured form 2.

An example event is shown in Fig. 1. The text mentions a product recall event involving General Motors, in July 2014. For each event, the IE system extracts a set of *entities*: companies, industry sectors, products, location, date, and other attributes of the event. This structured information is stored in the database, for querying and broader analysis. Then PULS performs deeper semantic analysis and uses machine learning to infer some of the attributes of the events, providing richer information than general-purpose search engines.

Next, using the entities aggregated from the texts, the system builds queries for the social media sources, e.g. to search company and product names using Twitter API [2]. The role of the social media component is to enable investigation of how companies and products mentioned in the news are portrayed on social media. Our system supports content analysis from different social media services. In this paper, we focus on numerical measurement and analysis of the content. We count the number of Wikipedia views of the company and the number of its mentions in the news and then use time series correlation to demonstrate the correspondence between news and Wikipedia news. We also correlate these with upward vs. downward stock fluctuations.

We have complete Wikipedia page request history for all editions, starting from early 2008, updated daily. We can instantaneously access the daily hit-count

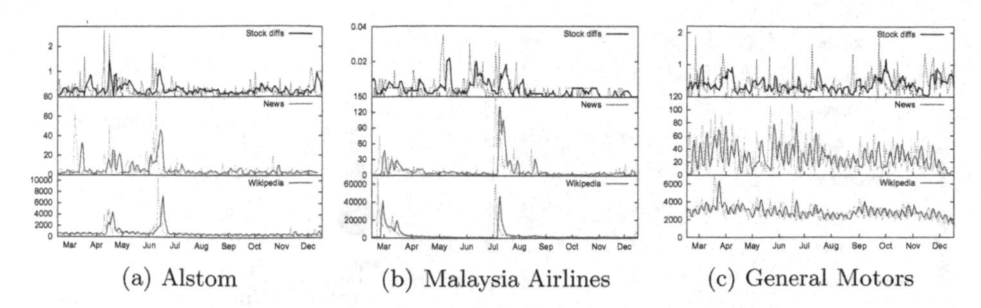

(a) Alstom (b) Malaysia Airlines (c) General Motors

Fig. 2. Daily differences in stock prices, number of mentions in PULS news and number of Wikipedia hits in 2014 for three companies.

history for any Wikipedia article. Mapping a name of an entity to a Wikipedia article is not always trivial to do automatically, but the mapping appears to be easy in the vast majority of cases. Thus, we have used the Wikipedia data to explore and demonstrate visibility in social media in the results presented in the following section.

3 Results

In this section we demonstrate results that can be obtained using this kind of processing. We present two types of results: A. visual analysis of correspondence between Wikipedia views, news hits and stock prices, and B. time-series correlations between news hits and Wikipedia views.

In the first experiment we chose three companies—Alstom, Malaysia Airlines, and General Motors. We present the number of mentions in the news collected by PULS, the number of views of the company's English-language Wikipedia page, and stock data, using data from March to December 2014.

In each figure, the top plot shows the daily *difference* in stock price—the absolute value of the opening price on a given day minus price on the previous day, obtained from Yahoo! Finance. The middle plot shows the number of mentions of the company in PULS news. The bottom plot shows the number of hits on the company's Wikipedia page. In each plot, the dashed line represents the daily values and the bold line is the value smoothed over three days.

Figure 2a plots the data for the French multinational Alstom. The company is primarily know for its train-, power-, and energy-related products and services. In the plot we can see a pattern where the stock price and news mentions seem to correlate rather closely. Wikipedia page hits show some correlation with the other plots. The news plot shows three major spikes, with two spikes in Wikipedia hits. The March peak corresponds to news about business events (investments), whereas the other peaks had a political aspect, which could trigger activity in social media; e.g., in June, the French government bought 20 % of Alstom shares, which caused an active public discussion.

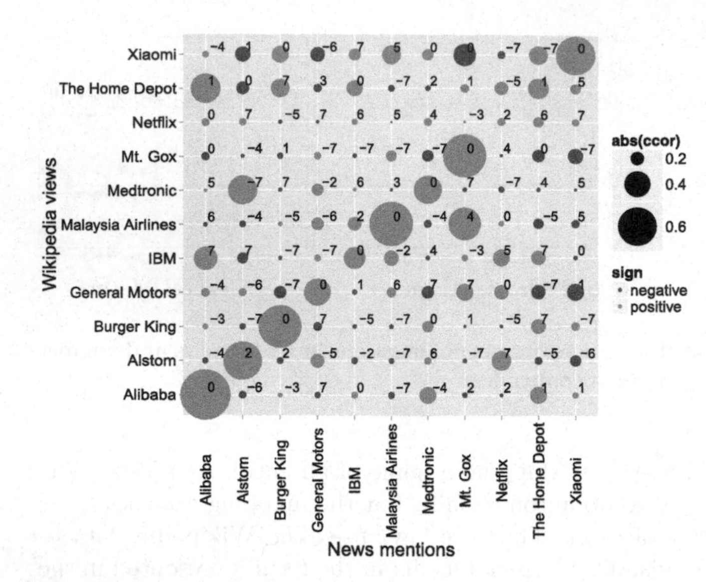

Circle width represents strength of correlation; colour represents sign of correlation: blue is positive, red is negative; the numbers indicate the time lag (in days) at which the correlation with the greatest magnitude is obtained for the given pair: positive lag means that Wikipedia views follow news mentions.

Fig. 3. Cross-correlation between Wikipedia views and mentions in PULS news for 11 companies.

Malaysia Airlines suffered two severe incidents in 2014. On March 8, they lost one aircraft over the Indian Ocean, and on July 17 another was shot down in Eastern Ukraine. Strong correlation in the patterns between news mentions and Wikipedia hits is clearly visible in Fig. 2b. The correlation with the stock price is less clear.

Figure 2c plots the data for General Motors, which was affected by numerous product recalls throughout the year. The company has been mentioned in the news and has been looked up on Wikipedia throughout the covered period. The stock price also oscillates over the entire year.

Although most of the local oscillations are due to normal fluctuations in the weekly flow of data on the Internet (with regular dips corresponding to the weekends), some broader-range correspondence is also discernible from the plots. Note, that the PULS IE system automatically assigns sentiment polarity to the news, classifying events as "positive" (e.g., investments, contracts, acquisitions) or "negative" (e.g., bankruptcies, layoffs, product recalls). This will form the basis for more detailed analysis of correlations with stock fluctuations in the future.

In the second experiment, we choose eleven big companies from different industry sectors, namely Alibaba, Alstom, Burger King, General Motors, IBM, Malaysia Airlines, Medtronic, Mt. Gox, Netflix, The Home Depot, and Xiaomi. For each of these companies we collect two time series: daily news mentions and Wikipedia views during time period from March to December 2014. Then we calculate the cross-correlation between all possible pairs in these dataset, for a total of 121 cross-correlations[2]. We limit the lag between time series by seven days, based on the assumption that if there exists a connection between news and Wikipedia views it should be visible within a week.

The results of this experiment are presented in Fig. 3, where the circle size represents correlation strength, the colours represents correlation size: blue means positive correlation, red negative; the numbers mean the time lag at which the highest correlation for a given company pair was obtained: positive lag means that Wikipedia views followed news mentions, negative lag means that news followed Wikipedia views.

It can be seen from the figure that the largest correlations and the lowest lags can be found on the diagonal, i.e., between news mention for a company and the number of views of the company Wikipedia page. Among the 11 companies there are two exceptions: The Home Depot and Netflix. For Netflix, news mentions and Wikipedia views do not seem to be strongly correlated with any time series. News about Alibaba show a surprising correlation with Wikipedia hits on Home Depot on the following day. At present we do not see a clear explanation for these phenomena; these can be accidental, or may indicate some hidden connections (they are both major on-line retailers).

The lag on the diagonal equals to zero in most cases, which means that in those cases the peaks occur on the same days. At a later time, we can investigate finer intervals (less than one day). We believe it would be interesting if a larger study confirmed that we can observe regular patterns in the correlations and the lags are stable—e.g., if a spike in the news regularly precedes a spike the Wikipedia views—since that would confirm that these models can have predictive power.

4 Conclusion and Future Work

We have presented a study of the interplay between company news, social media visibility, and stock prices. Information extracted from on-line news by means the of deep linguistic analysis is used to construct queries to various social media platforms. We expect that the presented framework would be useful for business professionals, Web scientists, and researchers from other fields.

The results presented in Sect. 3 demonstrate the utility of collecting and comparing data from a variety of sources. We were able to discover interesting correlations between the mentions of a company in the PULS news and the views of its page in Wikipedia. The correspondence with stock prices was less obvious. We continue work on refining the forms of data presentation. For example, we have found that plotting (absolute) differences in stock prices may in some cases provide better insights than using raw stock prices.

In future work, we plan to cover a wider range of data sources and social platforms, general-purpose (e.g., YouTube or Twitter) and business-specific ones (e.g., StockTwits). We plan to analyze the social media content as well, e.g.,

[2] We use standard R ccp function to calculate cross-correlation.

to determine the sentiment of the tweets that mention some particular company. Covering multiple sources is important due to the different nature of the social media. Tweets are short Twitter posts, where usually a user shares her/his impression about an entity (company or product), or posts a related link. Wikipedia, on the other hand, is used for obtaining more in-depth information about an entity. YouTube, in turn, is for both the consumption and creation of reviews, reports, and endorsements.

This phase faces some technical limitations. For example, while Twitter data can be collected through the Twitter API in near-real time, the API returns posts only from recent history (7–10 days). This means that keyword extraction and data collections should be done relatively soon after the company or product appears in the news; combined with Twitter API request limits, this poses challenges to having a comprehensive catalogue of the posts.

Our research plans include building accurate statistical models on top of the collected data, to explore the correlations, possible cause-effect relations, etc. We aim to find the particular event types (lay-offs, new products, lawsuits) that cause reaction on social media and/or in stock prices. We also aim to find predictive patterns of visibility on social media for companies and products, based on history or on typical behaviour for a given industry sector.

References

1. Boudoukh, J., Feldman, R., Kogan, S., Richardson, M.: Which news moves stock prices?. A textual analysis. Technical report, National Bureau of Economic Research (2013)
2. Du, M., Kangasharju, J., Karkulahti, O., Pivovarova, L., Yangarber, R.: Combined analysis of news and Twitter messages. In: Joint Workshop on NLP&LOD and SWAIE: Semantic Web, Linked Open Data and Information Extraction (2013)
3. Du, M., Pierce, M., Pivovarova, L., Yangarber, R.: Improving supervised classification using information extraction. In: Biemann, C., Handschuh, S., Freitas, A., Meziane, F., Métais, E. (eds.) NLDB 2015. LNCS, vol. 9103, pp. 3–18. Springer, Heidelberg (2015)
4. Guo, W., Li, H., Ji, H., Diab, M.T.: Linking tweets to news: A framework to enrich short text data in social media. In: Proceedings of ACL-2013 (2013)
5. Kwak, H., Lee, C., Park, H., Moon, S.: What is Twitter, a social network or a news media? In: Proceedings of the 19th International Conference on World Wide Web. ACM (2010)
6. Moat, H.S., Curme, C., Stanley, H., Preis, T.: Anticipating stock market movements with Google and Wikipedia. In: Matrasulov, D., Stanley, H.E., (eds.) Nonlinear Phenomena in Complex Systems: From Nano to Macro Scale, pp. 47–59 (2014)
7. Nassirtoussi, A.K., Aghabozorgi, S., Wah, T.Y., Ngo, D.C.L.: Text mining for market prediction: a systematic review. Expert Syst. Appl. **41**(16), 7653–7670 (2014)
8. Tanev, H., Ehrmann, M., Piskorski, J., Zavarella, V.: Enhancing event descriptions through Twitter mining. In: Sixth International AAAI Conference on Weblogs and Social Media (2012)
9. Tetlock, P.C.: Giving content to investor sentiment: the role of media in the stock market. J. Financ. **62**(3), 1139–1168 (2007)

Subtopic Mining Based on Three-Level Hierarchical Search Intentions

Se-Jong Kim(✉), Jaehun Shin, and Jong-Hyeok Lee

Department of Computer Science and Engineering,
Pohang University of Science and Technology (POSTECH), Pohang, South Korea
{sejong,rave0206,jhlee}@postech.ac.kr

Abstract. This paper proposes a subtopic mining method based on three-level hierarchical search intentions. Various subtopic candidates are extracted from web documents using a simple pattern, and higher-level and lower-level subtopics are selected from these candidates. The selected subtopics as second-level subtopics are ranked by a proposed measure, and are expanded and re-ranked considering the characteristics of resources. Using general terms in the higher-level subtopics, we make second-level subtopic groups and generate first-level subtopics. Our method achieved better performance than a state of the art method.

Keywords: Search intention · Subtopic mining · Popularity · Diversity · Hierarchical structure

1 Introduction

Many web queries are short and unclear. Some users do not choose appropriate words for a web search, and others omit specific terms needed to clarify search intentions, because it is not easy for users to express their search intentions explicitly through keywords. This intention gap between users and queries results in queries which are ambiguous and broad.

As one of the solutions for these problems, subtopic mining is proposed, which can find possible subtopics for a given query and return a ranked list of them in terms of their relevance, popularity, and diversity [1,2]. A subtopic is a query which disambiguates and specifies the search intention of the original query, and good subtopics must be relevant to the query and satisfy both high popularity and high diversity. The latest subtopic mining task [3] proposed new subtopic mining that the two-level hierarchy of subtopics consists of at most "five" first-level subtopics and at most "ten" second-level subtopics for each first-level subtopic. For example, if a query is "*apple*," its first-level subtopics are "*apple fruit*" and "*apple company*," and second-level subtopics for "*apple*

This work was partly supported by the ICT R&D program of MSIP/IITP (10041807), the SYSTRAN International corporation, the BK 21+ Project, and the National Korea Science and Engineering Foundation (KOSEF) (NRF-2010-0012662).

N. Ferro et al. (Eds.): ECIR 2016, LNCS 9626, pp. 741–747, 2016.
DOI: 10.1007/978-3-319-30671-1_62

company" are "*apple ipad*" and "*apple macbook.*" This hierarchy can present better the structure of diversified search intentions for the queries, and can also limit the number of subtopics to be shown to users by selecting only first-level subtopics because users do not want to see too many subtopics.

The state of the art methods [4,5] for the two-level hierarchy of subtopics used various external resources such as Wikipedia, suggested queries, and web documents from major web search engines instead of the provided resources from the subtopic mining task [3]. However, these methods did not consider the characteristics of resources in terms of popularity and diversity. Meanwhile, since the titles of web documents represent their overall subjects, these methods generated first-level subtopics using keywords which were extracted from only the titles. However, the title based first-level subtopics may not be enough to satisfy both high popularity and high diversity, because a title as a phrase or sentence is not more informative than a document.

To solve these issues, we propose a subtopic mining method based on three-level hierarchical search intentions of queries. Our method is a bottom-up approach which mines second-level subtopics first, and generates first-level subtopics. We extract various relevant subtopic candidates from web documents using a simple pattern, and select second-level subtopics from this candidate set. The selected subtopics are ranked by a proposed popularity measure, and are expanded and re-ranked considering the characteristics of provided resources. Using a topic modeling, we make five term-clusters and generate first-level subtopics consisting of the query and general terms. Our contributions are as follows:

- Our method uses only the limited resources (suggested queries, query dimensions[1], and web documents) provided from the subtopic mining task [3], and we reflect the characteristics of resources in terms of popularity and diversity to the second-level subtopic mining step.
- Our work divides "second-level" subtopics into "higher-level (level 2)" and "lower-level (level 3)" subtopics considering the hierarchical search intentions. Higher-level subtopics reflect wider search intentions than their lower-level ones. We generate high quality first-level (level 1) subtopics using words in higher-level subtopics as well as the titles of web documents.

2 Second-Level Subtopic Mining

2.1 Subtopic Extraction and Ranking

To extract various relevant subtopic candidates from web documents, a subtopic is assumed to consist of an original query and its one or more noun phrases that make the original query more specific. Since the ratio of nouns to real-query words is much higher than the ratio of other word classes, various subtopic

[1] Query dimensions are groups of items extracted from the style of lists such as tables in top retrieved documents [6]. Each dimension has a ranked list of its items.

candidates as new queries can be derived from nouns rather than others, and be useful in finding the hidden search intentions of the given query. From this assumption, we create a simple pattern:

$$((\text{adjective})^{?}(\text{noun})^{+}(\text{non-noun})^{*})^{?}(\text{query})((\text{non-noun})^{*}(\text{adjective})^{?}(\text{noun})^{+})^{?}$$

where the ? operator means "zero or one"; the + operator "one or more"; and the * operator "zero or more."

This pattern is applied to the top 1,000 relevant documents for the query, and the extracted subtopic candidates are truly relevant because of consisting of the whole query as well as noun phrases in the documents.

Next, we define that a higher-level subtopic reflects wide search intention, and its intention has clear distinction among search intentions. If highly relevant documents for a given query are assumed to represent user's all possible search intentions anyhow, and the appearance of subtopic candidates in documents is interpreted to mean that these subtopic candidates cover some search intentions, then a higher-level subtopic covers (appears in) many of the highly relevant documents (i.e., search intentions), and this document set has higher distinctness than the document sets for the other subtopics. Therefore, from the subtopic candidate set, to select higher-level subtopics satisfying both of the above two conditions, we propose a scoring measure, Selection Score (*SS*):

$$SS(st, US) = \frac{|D(st) \cap US^{c}|}{|\bigcup_{st' \in ST} D(st')|} \times CE(st), \tag{1}$$

where st is a subtopic candidate; ST is the set of all subtopic candidates; $D(st)$ is the set of documents containing st; US is the union of document sets containing the previously selected subtopics; and $CE(st)$ is the distinctness measure, Cluster Entropy, by regarding a document set containing st as a cluster [7].

Higher-level subtopics are continuously selected using *SS*. If $|US|$ is equal to $|\bigcup_{st' \in ST} D(st')|$, the selection process stops because the selected higher-level subtopics can cover all the highly relevant documents, which were originally covered by all subtopic candidates. For each of the higher-level subtopics, its lower-level subtopics can be selected in the same way recursively, except that we only use the relevant documents containing the higher-level subtopic.

The popularities of subtopics are estimated by their importance. Since the Term Frequency and Inverse Document Frequency (*TF-IDF*) is a general measure that indicates the importance of a target word (subtopic) to a document, we rank each type of second-level subtopics using the Sum of the values of *TF-IDF*.

2.2 Subtopic Addition and Re-ranking

We expand and re-rank the selected second-level subtopics considering the characteristics of relevant query dimensions and suggested queries. Note that the condition of relevant query dimensions is that two or more items include the query, or appear in the selected subtopics or suggested queries. Because each query dimension contains parallel items for a topic, if two or more items in a query dimension are relevant to the common query, we regard all items in this dimension as relevant items. For each resource, we assume that:

- Items (subtopics) in relevant query dimensions satisfy high diversity because they are distinct each other by the well-defined structures such as lists and tables.
- Suggested queries are good subtopics which satisfy high popularity because of real-query based resource.

From the first assumption, we add items of relevant query dimensions to the ranked list of second-level subtopics to improve the diversity of them (Fig. 1). (1) If a higher-level subtopic contains one of items in a relevant query dimension, the corresponding item is replaced with the higher-level subtopic, and the original place of the higher-level subtopic is replaced with the ranked list of items as higher-level subtopics. (2) If any higher-level subtopic does not contain items in a relevant query dimension, the top item in the dimension is added to the ranked list of second-level subtopics as the last-ranked higher-level subtopic. Meanwhile, from the second assumption, we reflect the high popularity of suggested queries to second-level subtopic re-ranking. If a higher-level subtopic contains the i-ranked suggested query, this subtopic is re-ranked as the i-ranked higher-level subtopic, and its lower-level subtopics are also re-ranked. The non-matched suggested query is deleted from the original suggested query list.

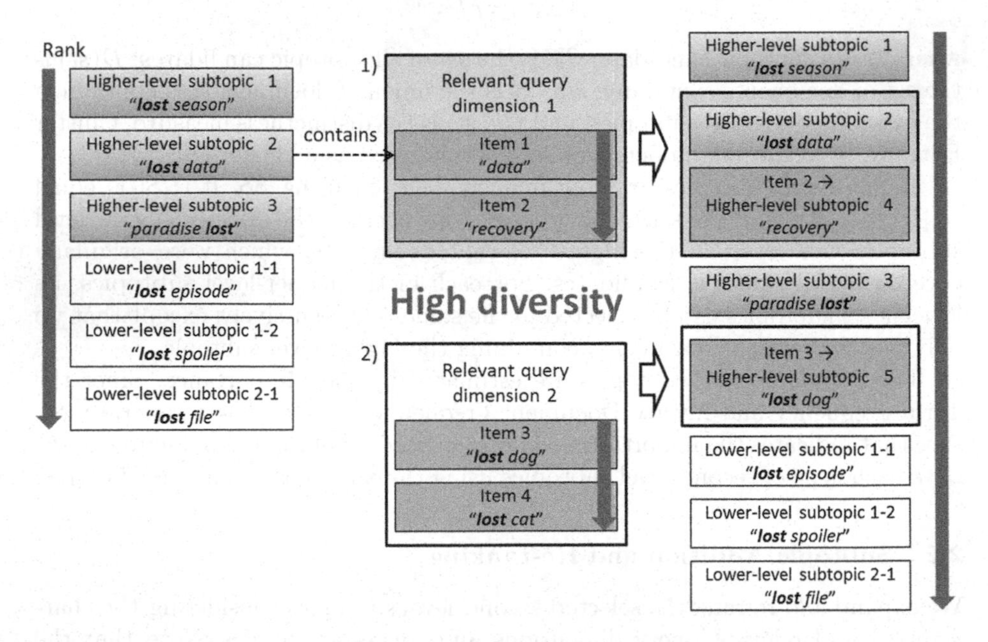

Fig. 1. A process of second-level subtopic addition using relevant query dimensions

3 First-Level Subtopic Mining

To achieve the high quality of first-level subtopics, we generate first-level subtopics using words in the higher-level subtopics. As mentioned earlier, since the higher-level subtopics reflect wide search intentions, these subtopics contain general terms which can cover other subtopics consisting of more specific terms, conceptually. Therefore, we make five term-clusters using the Latent Dirichlet Allocation (*LDA*) [8] based on words in the higher-level subtopics except the query words, and then, for each cluster, group at most ten second-level subtopics which are higher-level subtopics containing words of the term-cluster and their lower-level subtopics. We assume that the most general terms among words of the term-clusters appear in most of relevant documents for the second-level subtopics. From this assumption, for each term-cluster, the most general term is selected, and it is attached to the query. These expanded phrases are regarded as first-level subtopics and ranked by considering only the highest rank of second-level subtopics for each of them. In addition, to improve the quality of the hierarchical structure between first-level and second-level subtopics explicitly, if a second-level subtopic does not contain the most general term, the query part of the second-level subtopic is replaced with its first-level subtopic.

Table 1. Mean results of methods for relevance, popularity, and diversity of subtopics

Method name	First-level subtopic		Second-level subtopic	
	I-rec@5	D-nDCG@5	I-rec@50	D-nDCG@50
TD-E-BEST	0.6146	0.5194	0.6333	0.5596
BU-E-BEST	0.6052	0.4305	0.4440	0.5230
BU-E-PROP	**0.6427**	**0.6510**	**0.6623**	**0.6053**
TD-J-BEST	0.4235	0.1023	0.3449	0.2246
BU-J-PROP	**0.4853**	**0.5422**	**0.6592**	**0.6217**

Table 2. Mean results of methods for the overall quality of subtopics

Method name	*Hscore*	*Fscore*	*Sscore*	*H-measure*
TD-E-BEST	**0.9190**	0.5670	0.5964	0.5509
BU-E-BEST	0.8065	0.5179	0.4835	0.4257
BU-E-PROP	0.8719	**0.6469**	**0.6338**	**0.5763**
TD-J-BEST	0.2702	0.2629	0.2848	0.0845
BU-J-PROP	**0.3984**	**0.5137**	**0.6405**	**0.2600**

4 Results and Conclusion

We mined subtopics for 33 English and 34 Japanese queries of the NTCIR-11 subtopic mining task [3]. We used the search interface by Lemur project[2] and the BM25 model [9] to retrieve documents from the collection ClueWeb12-B13 (English) and ClueWeb09-JA (Japanese), respectively. To identify noun phrases, we used the English Stanford POS tagger[3] and the Japanese MeCab POS tagger[4]. As a Bottom-Up approach, our method names were "BU-E(English)/J(Japanese)-PROP." The state of the art Top-Down approach "KUIDL-S-E/J-1A [4]" and bottom-up approach "THUSAM-S-E-1A [5]" were newly named "TD-E/J-BEST" and "BU-E-BEST," respectively. The results were evaluated using *Hscore* (hierarchical relationship measure of first-level subtopic and its second-level subtopics), *Fscore*/*Sscore* (quality measure of first-level/second-level subtopics), I-rec (diversity measure of subtopics), D-nDCG (relevance and popularity measure of subtopics), and *H-measure* ("representative" measure) [3].

In Tables 1 and 2, the "bold" values represent the best performances among all the methods. Our proposed method achieved higher performances than the previous methods from almost every aspect. As compared to TD-E/J-BEST and BU-E-BEST, the mean *H-measure* of BU-E/J-PROP were improved by 4.61/207.69 % and 35.38 %, respectively. Furthermore, the performance differences of the means *Fscore* and *Sscore* between BU-E/J-PROP and the previous methods were statistically significant (2-tailed t-test, $p < 0.01$). However, our quality of the hierarchical structure between first-level and second-level subtopics was not the best, and the significance probability of the mean *H-measure* was 0.361 for TD-E-BEST, because a top-down approach using abundant English resources is more suitable to keep the high quality of the hierarchical structure between subtopics. Therefore, as for future work, we will find some criteria to improve *Hscore*, and design an appropriate method for other languages.

References

1. Song, R., Zhang, M., Sakai, T., Kato, M.P., Liu, Y., Sugimoto, M., Wang, Q., Orii, N.: Overview of the NTCIR-9 intent task. In: Proceedings of NTCIR-9 Workshop Meeting, pp. 82–105. National Institute of Informatics, Tokyo, Japan (2011)
2. Sakai, T., Dou, Z., Yamamoto, T., Liu, Y., Zhang, M., Song, R.: Overview of the NTCIR-10 INTENT-2 task. In: Proceedings of NTCIR-10 Workshop Meeting, pp. 94–123. National Institute of Informatics, Tokyo, Japan (2013)
3. Liu, Y., Song, R., Zhang, M., Dou, Z., Yamamoto, T., Kato, M., Ohshima, H., Zhou, K.: Overview of the NTCIR-11 imine task. In: Proceedings of NTCIR-11 Workshop Meeting, pp. 8–23. National Institute of Informatics, Tokyo, Japan (2014)

[2] http://lemurproject.org/clueweb12/.
[3] http://nlp.stanford.edu/software/tagger.shtml.
[4] http://mecab.sourceforge.net.

4. Yamamoto, T., Kato, M.P., Ohshima, H., Tanaka, K.: Kuidl at the NTCIR-11 imine task. In: Proceedings of NTCIR-11 Workshop Meeting, pp. 53–54. National Institute of Informatics, Tokyo, Japan (2014)
5. Luo, C., Li, X., Khodzhaev, A., Chen, F., Xu, K., Cao, Y., Liu, Y., Zhang, M., Ma, S.: Thusam at NTCIR-11 imine task. In: Proceedings of NTCIR-11 Workshop Meeting, pp. 55–62. National Institute of Informatics, Tokyo, Japan (2014)
6. Dou, Z., Hu, S., Luo, Y., Song, R., Wen, J.R.: Finding dimensions for queries. In: Proceedings of the 20th ACM International Conference on Information and Knowledge Management, pp. 1311–1320. Association for Computing Machinery, Glasgow, Scotland, UK (2011)
7. Zeng, H.J., He, Q.C., Chen, Z., Ma, W.Y., Ma, J.: Learning to cluster web search results. In: Proceedings of the 27th Annual International ACM SIGIR Conference on Research and Development in Information Retrieval, pp. 210–217. Association for Computing Machinery, Sheffield, South Yorkshire, UK (2004)
8. Blei, D.M., Ng, A.Y., Jordan, M.I.: Latent dirichlet allocation. J. Mach. Learn. Res. **3**, 993–1022 (2003)
9. Robertson, S., Zaragoza, H.: The probabilistic relevance framework: BM25 and beyond. Found. Trends Inf. Retr. **3**(4), 333–389 (2009)

Cold Start Cumulative Citation Recommendation for Knowledge Base Acceleration

Jingang Wang[1], Jingtian Jiang[2], Lejian Liao[1(✉)], Dandan Song[1], Zhiwei Zhang[3], and Chin-Yew Lin[2]

[1] School of Computer Science, Beijing Institute of Technology, Beijing 100081, China
{bitwjg,liaolj,sdd}@bit.edu.cn
[2] Knowledge Mining Group, Microsoft Research, Beijing 100080, China
{jiji,cyl}@microsoft.com
[3] Department of Computer Science, Purdue University, West Lafayette, IN 47906, USA
zhan1187@purdue.edu

Abstract. This paper studies cold start Cumulative Citation Recommen dation (CCR) for Knowledge Base Acceleration (KBA), whose objective is to detect potential citations for target entities without existing KB entries from a volume of stream documents. Unlike routine CCR, in which target entities are identified by a reference KB, cold start CCR is more common since lots of less popular entities do not have any KB entry in practice. We propose a two-step strategy to address this problem: (1) event-based sentence clustering and (2) document ranking. In addition, to build effective rankers, we develop three kinds of features based on the clustering results: time range, local profile and action pattern. Empirical studies on TREC-KBA-2014 dataset demonstrate the effectiveness of the proposed strategy and the novel features.

1 Introduction

Recent years have witnessed rapid growth of Knowledge Bases (KBs) such as Wikipedia. Currently, the maintenance of a KB mainly relies on human editors. To help the editors keep KBs up-to-date, Cumulative Citation Recommendation (CCR) is proposed by Text Retrieval Conference (TREC) in Knowledge Base Acceleration (KBA) track[1] since 2012. CCR aims to automatically filter vitally relevant documents from a chronological document collection and evaluate their citation-worthiness to target KB entities. The key objective of CCR is to identify "vital" documents, which would trigger updates to the entry page of target entities in the KB [6]. Therefore, CCR is also known as **vital filtering**.

Generally, the target entities are identified by a reference KB like Wikipedia, so their KB profiles can be employed to perform entity disambiguation and

This work was done when the first author was visiting Microsoft Research Asia.

[1] http://trec-kba.org/.

N. Ferro et al. (Eds.): ECIR 2016, LNCS 9626, pp. 748–753, 2016.
DOI: 10.1007/978-3-319-30671-1_63

further relevance estimation. Nevertheless, if a target entity does not have an entry in the reference KB yet, CCR becomes a cold start task, in which case we need to identity the entity solely by the entity's local context in the stream documents and a few labeled relevant documents.

The critical challenge of CCR is feature sparsity. In previous work, useful semantic and temporal features are extracted from both the reference KB and the stream documents [8,9]. For instance, daily views of each target entity's entry page in Wikipedia are utilized as effective temporal features. However, under cold start circumstances, we need to exploit more features to enrich the feature set.

There are two observations on vital documents in training data: (1) The vital information is usually organized as related events about the target entities. (2) Vital events are usually described in a single sentence or paragraph. Therefore, we propose a two-step strategy for cold start CCR. First, we extract sentences mentioning target entities from documents and cluster them by semantic similarity. The sentences in each cluster are supposed to represent a vital event related to the target entity. Then, we perform relevance estimation between documents and target entities at the sentence-level. In addition, we develop three kinds of features based on the clustering results: time range, local profile and action pattern. We evaluate these features in a random forest based ranking method on TREC-KBA-2014 dataset. The promising experimental results demonstrate the effectiveness of this strategy and the novel features.

As far as we know, this is the first work that considers cold start CCR. The main contributions can be summarized as follows. (1) We address CCR from the perspective of sentence-level instead of document-level to identify the vital information more accurately. (2) We propose novel features to resolve the feature sparsity problem in cold start CCR.

2 Two-Step Strategy for Cold Start CCR

In terms of cold start CCR, we perform relevance estimation for target entities without KB entries. We follow the four-point scale relevance settings of TREC KBA, i.e.,*vital*, *useful*, *unknown* and *non-referent*. The documents which contain timely information about target entities' current states, actions or situations are "vital" documents. Vital documents would motivate a change to an already up-to-date KB article.

2.1 Event-Based Sentence Clustering

Since vital signals are usually captured in the sentence or short passage surrounding entity mentions in a document [6], it would be better to consider the sentence mentioning a target entity instead of the whole document. Besides, if a document contains several sentences mentioning the target entity, we take the max rating of these sentences as the document's final rating.

Accordingly, we extract all sentences mentioning the target entities from the stream documents, and divide them into different clusters. Given a target entity, when a new sentence arrives, we compute the semantic (cosine) similarity between the new sentence and existing clusters. If the maximum similarity is larger than a predefined threshold, new sentence is assigned to the corresponding cluster; otherwise, it forms a new cluster by itself. Each cluster represents a relevant event of the given target entity.

2.2 Features Extraction

Various useful features have been exploited in previous work, including semantic and temporal features [1,3,8,9]. We adopt the same feature set used in [8] as **baseline features** in this paper, which were proven effective in previous TREC-KBA tracks. Besides, we develop three kinds of novel features especially for cold start CCR.

Time Range (TR). In vital filtering, we must assess the time lag between the relevant event and documents. It is intuitive to assume that the later a document occurs, the smaller relevance score it should get, even if the two documents report a same event. Hence, we penalize the later documents in event-based clusters by decreasing the feature value of a document over hours. The first document of a cluster get a feature value of 1.0, and later documents get smaller values, which can be expressed by a decay function $tr(d_i) = 1.0 - (h_i - h_0)/72.0$, where h_0 is the hour converted from the timestamp of the first document d_0 in the cluster, and h_i is the hour of the ith document d_i. The number 72.0 means 3 days we used in our experiments.

Local Profile (LP). Some entity mentions in stream documents are ambiguous. To solve this problem, we create local profile for target entities, which contains some profile information around the entity mentions in documents. Usually, when an entity is mentioned, its local profile (e.g., title and profession) is also mentioned to let the readers know who is this entity. For example, "*All the other stuff matters not,* ***Lions coach*** *Bill Templeton said*". From above sentence, we know the target entity (*Bill Templeton*) is a coach. Of course, if the mentioned entity is very popular, its title or profession is usually omitted. Nevertheless, most entities in cold start CCR are less popular entities.

In our approach, we calculate the cosine similarity between a target entity's local profile and the extracted local profile of its possible mention as a feature. Firstly, we need to construct the local profile for each target entity. We acquire the title/profession dictionaries from Freebase[2], containing 2,294 titles and 2,440 professions. Secondly, we extract the word-based n-grams (n=1, 2, 3, 4, 5) inside a sliding window around a target entity mention. These n-grams existing in the dictionaries form the local profile vector. Lastly, we construct the local profile vector for each target entity with the n-grams extracted from all vital and useful documents in the training data.

[2] https://www.freebase.com/.

Action Pattern (AP). Vital documents typically contain sentences that describe events in which the target entities carry out some actions, e.g. scored a goal, won an election. Therefore, an entity's action in a document is a key indicator that the document is vital to the target entity. We find that if a target entity involves in an action, it usually appears as the subject or object of the sentence. So we mine triples from sentences mentioning a target entity. If a triple is found in which the target entity occurs as subject or object, we consider this entity takes action in the sentence (event).

We adopt Reverb [4] to mine the triples, which is a state-of-the-art open domain extractor that targets verb-centric relations. Such relations are expressed in triples <subject, verb, object>. We run Reverb on each sentence mentioning a target entity and extract the triples. Then for each triple, we use the "entity + verb" and "verb + entity" as action patterns. Please note that the verb is stemmed in our experiments. For example, from the sentence "*Public Lands Commissioner Democrat Peter Goldmark won re-election*", the extracted triple is <*Peter Goldmark, won, re-election*>, and the action pattern is "*Peter Goldmark win*". In our system, each action pattern is used as a binary feature, if the sentence/document has an action pattern, the feature value is 1, otherwise 0.

2.3 Document Ranking

CCR task can be considered as a ranking problem. In our approach, we train a separate ranker for each target entity with the baseline features and the proposed novel features. As the random forest ranking method achieves the best results in previous work [2,8], we adopt random forest implemented in RankLib[3] as our document ranking method.

3 Experiments

3.1 Dataset

We use TREC-KBA-2014 dataset[4] to evaluate our approach. The dataset is composed of a target entity set and a stream of documents. The target entity set is composed of 71 entities, including facilities, persons and organizations. There are 38 entities that do not have corresponding Wikipedia entries. Our evaluation is performed on these 38 entities without KB entries. The stream corpus contains approximately 1.2 billion documents crawled from October 2011 to the end of April 2013. The cutoff between training and test is not consistent for different target entities, which promises each entity exists training instances.

[3] http://sourceforge.net/p/lemur/wiki/RankLib/.

[4] http://trec-kba.org/kba-stream-corpus-2014.shtml.

3.2 Evaluation Metrics

There are two metrics to evaluate CCR performance: $\max(F_1(avg(P), avg(R)))$ and $\max(SU)$. Scaled Utility (SU) is introduced in filtering track to evaluate the ability of a system to separate relevant and irrelevant documents in a stream [7]. The detailed calculations of these two measure can be found in [6]. The primary metric is $\max(F_1(avg(P), avg(R)))$.

3.3 Comparison Methods

- **Official Baseline** [6]. This baseline assigns a "vital" rating to each document that matches the canonical name of an entity.
- **RF**. A random forests based learning to rank method using the baseline features.
- **RF+TR**. A random forests based learning to rank method using baseline features and time range feature.
- **RF+TR+LP**. A random forests based learning to rank method using baseline features, time range, and local profile features.
- **RF+TR+LP+AP**. A random forests based learning to rank method using baseline features and all the proposed features.

Table 1. Results of all experimental methods. All measures are reported by the KBA official scorer with cutoff-step-size=10.

Method	$avg(P)$	$avg(R)$	$\max(F_1)$	$\max(SU)$
Official Baseline	.287	.948	.441	.267
RF	.342	.774	.474	.349
RF+TR	.367	.743	.491	.367
RF+TR+LP	.378	.744	.501	.377
RF+TR+LP+AP	.447	.702	.546	.464

3.4 Results and Discussion

Table 1 shows the results of the performance with adding features incrementally. We add time range feature first because it is easy to implement after sentence clustering. From the results, we can see that action pattern is the most effective which improved F_1 by 4.5 points. The time range feature is also powerful as it improved F_1 by 2 points. Local profile feature can improve the performance too, but is not so good as we expected. We analyze the results and find the reason is that there are not many ambiguous target entities in our evaluation dataset. Compared with the official baseline, our best method improves the performance by 10 points. This proves the effectiveness of the features in cold start CCR task. In addition, as pointed out in [5], improving precision is more difficult than improving recall in CCR. All the proposed features are helpful to enhance precision while keeping a satisfactory recall performance.

4 Conclusion

In this paper, we focus on cold start Cumulative Citation Recommendation for Knowledge Base Acceleration, in which the target entities do not exist in the reference KB. Since KB profile is unavailable in cold start CCR, we split sentences in the stream documents and cluster them chronologically to detect vital events related to the target entities. Based on the sentence clustering results, we then extract tree kinds of novel features: time range, local profile and action pattern. Moreover, we adopt random forest based ranking method to perform relevance estimation. Experimental results on TREC-KBA-2014 dataset have demonstrated that this two-step strategy can improve system performance under cold start circumstances.

Acknowledgement. The authors would like to thank Jing Liu for his valuable suggestions and the anonymous reviewers for their helpful comments. This work is funded by the National Program on Key Basic Research Project (973 Program, Grant No. 2013CB329600), National Natural Science Foundation of China (NSFC, Grant Nos. 61472040 and 60873237), and Beijing Higher Education Young Elite Teacher Project (Grant No. YETP1198).

References

1. Balog, K., Ramampiaro, H.: Cumulative citation recommendation: classification vs. ranking. In: SIGIR, pp. 941–944. ACM (2013)
2. K. Balog, H. Ramampiaro, N. Takhirov, and K. Nørvåg.: Multi-step classification approaches to cumulative citation recommendation. In: OAIR, pp. 121–128. ACM (2013)
3. Bonnefoy, L., Bouvier, V., Bellot, P.: A weakly-supervised detection of entity central documents in a stream. In: SIGIR, pp. 769–772. ACM (2013)
4. Fader, A., Soderland, S., Etzioni, O.: Identifying relations for open information extraction. In: EMNLP, pp. 1535–1545. ACL (2011)
5. Frank, J.R., Kleiman-Weiner, M., Roberts, D.A., Niu, F., Zhang, C., Re, C., Soboroff, I.: Building an entity-centric stream filtering test collection for TREC 2012. In: TREC, NIST (2012)
6. Frank, J.R., Kleiman-Weiner, M., Roberts, D.A., Voorhees, E., Soboroff, I.: Evaluating stream filtering for entity profile updates in TREC 2012, 2013, and 2014. In: TREC, NIST (2014)
7. Robertson, S.E., Soboroff, I.: The TREC 2002 filtering track report. In: TREC, NIST (2002)
8. Wang, J., Song, D., Lin, C.-Y., Liao, L.: BIT and MSRA at TREC KBA CCR track 2013. In: TREC, NIST (2013)
9. Wang, J., Song, D., Wang, Q., Zhang, Z., Si, L., Liao, L., Lin, C.-Y.: An entity class-dependent discriminative mixture model for cumulative citation recommendation. SIGIR **2015**, 635–644 (2015)

Cross Domain User Engagement Evaluation

Ali Montazeralghaem[1], Hamed Zamani[2], and Azadeh Shakery[1(✉)]

[1] School of Electrical and Computer Engineering, College of Engineering, University of Tehran, Tehran, Iran
{ali.montazer,shakery}@ut.ac.ir
[2] Center for Intelligent Information Retrieval, University of Massachusetts Amherst, Amherst, MA, USA
zamani@cs.umass.edu

Abstract. Due to the applications of user engagements in recommender systems, predicting user engagement has recently attracted considerable attention. In this task which is firstly proposed in ACM Recommender Systems Challenge 2014, the posts containing users' opinions about items (e.g., the tweets containing the users' ratings about movies in the IMDb website) are studied. In this paper, we focus on user engagement evaluation for cold-start web applications in the extreme case, when there is no training data available for the target web application. We propose an adaptive model based on transfer learning (TL) technique to train on the data from a web application and test on another one. We study the problem of detecting tweets with positive engagement, which is a highly imbalanced classification problem. Therefore, we modify the loss function of the employed transfer learning method to cope with imbalanced data. We evaluate our method using a dataset including the tweets of four popular and diverse data sources, i.e., IMDb, YouTube, Goodreads, and Pandora. The experimental results show that in some cases transfer learning can transfer knowledge among domains to improve the user engagement evaluation performance. We further analyze the results to figure out when transfer learning can help to improve the performance.

Keywords: User engagement · Transfer learning · Adaptive model · Cold-start

1 Introduction

Twitter is a popular micro-blogging platform, which allows users to share their opinions and thoughts as fast as possible in very short texts. This makes Twitter a rich source of information with high speed of information diffusion. Therefore, several web applications (e.g., IMDb) have been integrated with Twitter to let people express their opinions about items (e.g., movie) in a popular social network [2,10].

It is shown that the amount of users' interactions on tweets can be used to measure the users' satisfaction. In more detail, user engagements in Twitter has a strong positive correlation with the interest of users in the received tweets [2].

N. Ferro et al. (Eds.): ECIR 2016, LNCS 9626, pp. 754–760, 2016.
DOI: 10.1007/978-3-319-30671-1_64

In addition, the purpose of recommender systems is to increase the satisfaction of users and thus, measuring the user engagements of tweets which contain the opinions of users about items (or products) can be employed to improve recommender systems performance [9].

In addition to recommender systems, user engagement evaluation has several other usages. For instance, Uysal and Croft [8] designed a personalized content filter based on user engagements in Twitter. Petrovic et al. [4] predicted whether a tweet will be retweeted or not. These works have focused on tweets with arbitrary content, while we are interested in engagement evaluation of tweets with predefined content[1].

Regarding the importance of user engagement evaluation in recommender systems, ACM Recommender Systems Challenge 2014[2] [6] has focused on ranking tweets of each user based on their engagements. This challenge only considered the tweets that are tweeted using the IMDb website. Similar to this challenge, in this paper the "*engagement*" value is computed as the total number of *retweets* and *favorites* that a tweet has achieved.

Recently, Zamani et al. [9] proposed an adaptive user engagement evaluation model for different web applications. They considered four popular web applications (also called domains) with wide variety of items. They proposed to employ multi-task learning to train a generalized model using all domains to improve the user engagement evaluation performance for each individual domain. Although their method successfully transfers knowledge among domains, it cannot be employed for evaluating user engagement for cold-start domains.

In this paper, we propose a cross domain adaptive model to train on one domain (source domain) and test on another one (target domain). In fact, the proposed method would be useful when there is no training data available for the target domain, i.e., cold-start web applications. To do so, we consider adaptive regularization-based transfer learning (ARTL) [3], which considers both distribution adaptation and label propagation strategies for cross domain transfer learning. Since distribution of our data is highly imbalanced[3], we modify the loss function of the ARTL method by adding an instance weighting term to the loss function formulation. To the best of our knowledge, this is the first try to evaluate user engagement in the case of absence of training data from the target domain.

In our experiments, we consider a collection of tweets from four popular web applications with very different items: IMDb (movie), YouTube (video clip), Pandora (music), and Goodreads (book) [9]. In our experiments, we analyze when transfer learning can help to improve the user engagement evaluation performance.

---

[1] In each tweets, the user gives a rate to or likes/dislikes a product.
[2] "User Engagement as Evaluation" Challenge, http://2014.recsyschallenge.com/.
[3] There are lots of tweets with zero engagement and a few tweets with positive engagement.

2 Cross Domain Model for User Engagement Evaluation

In this section, we first briefly explain the employed transfer learning algorithm and describe how we deal with imbalanced data in transfer learning scenarios. We further introduce our features for user engagement evaluation.

2.1 Adaptive Regularization-based Transfer Learning

It is very difficult to induce a supervised classifier without any labeled data. Various transfer learning methods (also called domain adaptation methods) have been so far proposed to transfer knowledge from a source domain to a target domain, when there is no training data available for the target domain. In this paper, we employ adaptive regularization-based transfer learning (ARTL) [3], a cross domain transfer learning method whose goal is to improve classification performance for the unlabeled target domain using labeled data from the source domain.

Most existing transfer learning methods try to do one of the two following strategies: distribution adaptation and label propagation. ARTL framework considers both of these two strategies in its learning process. In fact, ARTL learns an adaptive classifier by optimizing the structural risk functional, the joint distribution matching between domains (J_s and J_t), and the manifold consistency underlying marginal distribution (P_s and P_t). Let $\{(x_1, y_1), \cdots, (x_n, y_n)\}$ be a set of n training instances from the source domain in which $y_i \in \mathbb{R}$ and $x_i \in \mathbb{R}^d$ respectively denote the label and the feature vector, where d is the number of features. The ARTL framework is formulated as:

$$f = \arg \min_{f \in H_K} \mathcal{L}(f(X), Y) + \sigma \|f\|_K^2 + \lambda D_{f,K}(J_s, J_t) + \gamma M_{f,K}(P_s, P_t)$$

where K, H_K, $M_{f,K}$, $D_{f,K}$, and $\mathcal{L}$ respectively denote the kernel function, Hilbert space, manifold regularization, joint distribution adaptation, and the loss function. σ, λ, and γ are positive regularization parameters. Squared loss function is used in ARTL formulation.

Since the distribution of data in our problem is highly skewed, we propose to assign higher weights to instances from the minority class and vice versa. To this end, we define an instance weighting matrix $W \in R^{n \times 1}$ where elements of the matrix correspond to the weight of individual training instances. The matrix W is computed as:

$$W_i = \frac{1/n^{(i)}}{\sum_{j=1}^{n} 1/n^{(j)}}$$

where $n^{(i)}$ denotes the number of training instances with label y_i. A similar idea for coping with imbalanced data has been previously proposed in [1] for single-task classification and in [7,9] for multi-task learning. We can now redefine the ARTL learning formulation as follows:

$$f = \arg \min_{f \in H_K} W\mathcal{L}(f(X), Y) + \sigma \|f\|_K^2 + \lambda D_{f,K}(J_s, J_t) + \gamma M_{f,k}(P_s, P_t)$$

2.2 Features

We extract 23 features from each tweet, that are partitioned into three categories: user-based, item-based, and tweet-based. Note that the contents of tweets in our task are predefined by the web applications and users usually do not edit tweets contents. These features are previously used in [9,10]. More details about the exact definition of features can be found in [10]. The list of our features are as follows:

User-Based Features. Number of followers, Number of followees, Number of tweets, Number of tweets about domain's items, Number of liked tweets, Number of lists, Tweeting frequency, Attracting followers frequency, Following frequency, Like frequency, Followers/Followees, Followers-Followees.

Item-Based features. Number of tweets about the item.

Tweet-Based Features. Mention count, Number of hash-tags, Tweet age, Membership age at the tweeting time, Hour of tweet, Day of tweet, Time of tweet, Holidays or not, Same language or not, English or not.

3 Experiments

3.1 Experimental Setup

In our evaluations, we use the dataset provided by [9], which is gathered from four diverse and popular web applications (domains): IMDb, YouTube, Goodreads, and Pandora which contain movies, video clips, books, and musics, respectively.[4] Statistics of the dataset are reported in Table 1.

To have a complete and fair evaluation, in our experiments all models are trained using the same number of training instances. For each domain, we randomly select 16, 361 and 32, 722 instances to create training and test sets, respectively. We repeat this process 30 times using random shuffling. We report the average of the results obtained on these 30 shuffles and classify tweets with positive engagement from the tweets with zero engagement.

Table 1. Dataset characteristics

	IMDb	YouTube	Goodreads	Pandora
# of tweets	100,206	239,751	65,445	98,212
# of users	6,852	6,480	3,813	3,312
# of items	13,502	154,041	31,558	32,321
Average engagement	0.1097	0.4737	0.1632	0.0778
% of tweets with positive engagement	4.139	14.193	6.931	6.285

[4] The dataset is freely available at http://ece.ut.ac.ir/node/100770.

According to Table 1, the data is highly imbalanced; percentage of data with positive engagement is by far lower than percentage of those with zero engagement. In our evaluations, we consider accuracy (as the most popular evaluation metric for classification) and balanced accuracy (BA) [5] (a widely used evaluation metric for imbalanced situations). BA is computed as the arithmetic mean of accuracy in each class.

For single-task learning (STL), we employ support vector machine (SVM) classifier, which has been shown to be highly effective in various tasks. The linear kernel is considered for both baseline and the proposed method. To set the parameters of each learning algorithm, we perform hyper-parameter optimization using grid search and stratified k-fold ($k = 5$) cross validation. In addition, we apply instance weighting for both baseline and the proposed method in all the experiments.[5] We use the t-test with 95 % confidence to capture the statistically significant differences between results.

3.2 Results and Discussion

The results obtained by STL and ARTL are reported in Table 2. In this table, the significant differences between results are shown by star. According to this table, in some cases STL performs better and in other cases ARTL outperforms STL. In the following, we analyze the obtained results for each target domain.

IMDb. In the case that IMDb is the target domain, ARTL significantly outperforms STL, in terms of BA; however, the accuracy values achieved by SVM are higher than those obtained by ARTL. This shows that ARTL can classify the minority class instances (tweets with positive engagement) significantly better than SVM, but it fails in classifying the instances belonging to the majority class. The reason is that IMDb is the most imbalanced domain in the dataset

Table 2. Accuracy and balanced accuracy achieved by single-task learning and transfer learning methods.

Train \ Test		IMDb		YouTube		Goodreads		Pandora	
		STL	ARTL	STL	ARTL	STL	ARTL	STL	ARTL
IMDb	BA	-	-	**0.6445***	0.6033	0.5802	**0.5911***	**0.5663***	0.5492
	Acc.			**0.7889***	0.6797	**0.8616***	0.6924	**0.8681***	0.6796
YouTube	BA	0.5378	**0.5542***	-	-	0.5534	**0.5582***	**0.5447***	0.5390
	Acc.	**0.9529***	0.9031			**0.9350***	0.9197	**0.9383***	0.9031
Goodreads	BA	0.5917	**0.5933**	**0.6767***	0.6506	-	-	**0.5752***	0.5572
	Acc.	**0.7830***	0.7008	0.5745	**0.6360***			**0.7557***	0.6720
Pandora	BA	0.5731	**0.5820***	**0.6602***	0.6368	0.5948	**0.5985**	-	-
	Acc.	**0.6835***	0.6525	0.5403	**0.6485***	**0.6769**	0.6682		

[5] The results without instance weighting is biased toward the majority class. For the sake of space, the results without instance weighting are not reported.

(see Table 1) and thus, STL cannot learn a proper model, when there is a large gap between the feature distribution of the source and the target domains. This is why the maximum difference between the performance of ARTL and STL is happened when YouTube is selected as the source domain.

YouTube. Unlike the previous case, when YouTube is chosen as the target domain, STL outperforms ARTL in terms of BA. In some cases (i.e., training on Goodreads and Pandora), ARTL achieves higher accuracy compared to STL. The reason is that other domains are much more imbalanced than YouTube and in that case, the trained STL model is more accurate in detecting instances from the minority class, which leads to the better BA, but worse accuracy.

Goodreads. The results achieved over the Goodreads domain are very similar to those obtained over the IMDb domain. In other words, ARTL is more successful than STL in detecting tweets with positive engagement, since it achieved higher balanced accuracy but lower accuracy. As shown in Table 2, the best performance over this target domain is achieved when the model is trained using the IMDb or the Pandora domains. The percentage of data with positive engagement in these two domains are much more similar to Goodreads, compared to YouTube. Thus, learning from these domains can achieve higher accuracy.

Pandora. According to Table 2, transferring knowledge do not help to improve the user engagement evaluation performance. The reason could be related to the different distributions of the data from Pandora and the other domains. As reported in Table 1, the average engagement in this domain is much lower than the other domains which leads to have a very different feature distribution.

4 Conclusions and Future Work

In this paper, we proposed an adaptive method based on adaptive regularization-based transfer learning for user engagement evaluation. To cope with imbalanced data, we modified the transfer learning objective function by adding an instance weighting matrix to its formulation. In our experiments, we considered four popular web applications: IMDb, YouTube, GoodReads, and Pandora. The experimental results show that in some cases, we can find some useful information to transfer knowledge between these very different domains. We analyzed the achieved results and discussed the situations that transfer learning can be applied to improve the user engagement evaluation performance. An interesting future direction is to also modify the manifold regularization and the joint distribution adaptation components in the transfer learning objective function to improve the classification performance, when the data is highly imbalanced.

Acknowledgements. This work was supported in part by the Center for Intelligent Information Retrieval. Any opinions, findings and conclusions or recommendations expressed in this material are those of the authors and do not necessarily reflect those of the sponsor.

References

1. Akbani, R., Kwek, S.S., Japkowicz, N.: Applying support vector machines to imbalanced datasets. In: Boulicaut, J.-F., Esposito, F., Giannotti, F., Pedreschi, D. (eds.) ECML 2004. LNCS (LNAI), vol. 3201, pp. 39–50. Springer, Heidelberg (2004)
2. Loiacono, D., Lommatzsch, A., Turrin, R.: An Analysis of the 2014 RecSys challenge. In: RecSysChallenge, pp. 1–6 (2014)
3. Long, M., Wang, J., Ding, G., Pan, S.J., Yu, P.S.: Adaptation regularization: a general framework for transfer learning. IEEE Trans. Knowl. Data Eng. **26**(5), 1076–1089 (2014)
4. Petrovic, S., Osborne, M., Lavrenko, V.: RT to win! predicting message propagation in twitter. In: ICWSM, pp. 586–589 (2011)
5. Powers, D.: Evaluation: from precision, recall and F-measure to ROC, informedness, markedness & correlation. J. Mach. Learn. Tech. **2**(1), 37–63 (2011)
6. Said, A., Dooms, S., Loni, B., Tikk, D.: Recommender systems challenge 2014. In: RecSys, pp. 387–388 (2014)
7. de Souza, J.G.C., Zamani, H., Negri, M., Turchi, M., Falavigna, D.: Multitask learning for adaptive quality estimation of automatically transcribed utterances. In: NAACL-HLT, pp. 714–724 (2015)
8. Uysal, I., Croft, W.B.: User oriented tweet ranking: a filtering approach to microblogs. In: CIKM, pp. 2261–2264 (2011)
9. Zamani, H., Moradi, P., Shakery, A.: Adaptive user engagement evaluation via multi-task learning. In: SIGIR, pp. 1011–1014 (2015)
10. Zamani, H., Shakery, A., Moradi, P.: Regression and learning to rank aggregation for user engagement evaluation. In: RecSysChallenge, pp. 29–34 (2014)

An Empirical Comparison of Term Association and Knowledge Graphs for Query Expansion

Saeid Balaneshinkordan and Alexander Kotov(✉)

Department of Computer Science, Wayne State University, Detroit, MI 48202, USA
{saeid.balaneshinkordan,kotov}@wayne.edu

Abstract. Term graphs constructed from document collections as well as external resources, such as encyclopedias (DBpedia) and knowledge bases (Freebase and ConceptNet), have been individually shown to be effective sources of semantically related terms for query expansion, particularly in case of difficult queries. However, it is not known how they compare with each other in terms of retrieval effectiveness. In this work, we use standard TREC collections to empirically compare the retrieval effectiveness of these types of term graphs for regular and difficult queries. Our results indicate that the term association graphs constructed from document collections using information theoretic measures are nearly as effective as knowledge graphs for Web collections, while the term graphs derived from DBpedia, Freebase and ConceptNet are more effective than term association graphs for newswire collections. We also found out that the term graphs derived from ConceptNet generally outperformed the term graphs derived from DBpedia and Freebase.

Keywords: Query expansion · Term graphs · Knowledge bases · Difficult queries

1 Introduction

Vocabulary gap, when searchers and the authors of relevant documents use different terms to refer to the same concepts, is one of the fundamental problems in information retrieval. In the context of language modeling approaches to IR, vocabulary gap is typically addressed by adding semantically related terms to query and document language models (LM), a process known as query or document expansion. Therefore, effective and robust query and document expansion requires information about term relations, which can be conceptualized as a term graph. The nodes in this graph are distinct terms, while the edges are weighed according to the strength of semantic relationship between pairs of terms.

Term association graph is constructed from a given document collection by calculating a co-occurrence based information theoretic measure, such as mutual information [7] or hyperspace analog to language [2], between each pair of terms in the collection vocabulary. Term graphs can also be derived from knowledge bases, such as DBpedia[1], a structured version of Wikipedia, Freebase[2],

[1] http://wiki.dbpedia.org/.
[2] http://freebase.com/.

N. Ferro et al. (Eds.): ECIR 2016, LNCS 9626, pp. 761–767, 2016.
DOI: 10.1007/978-3-319-30671-1_65

a popular graph-structured knowledge base created from different data sources, and ConceptNet[3], a large semantic network constructed via crowdsourcing.

Term association and knowledge graphs have their own advantages and disadvantages. The weights of edges between the terms in automatically constructed term graphs *are specific to each particular document collection.* On the other hand, methods that establish semantic term relatedness based only on co-occurrence require large amounts of data and often produce noisy term graphs. Semantic term associations in external resources (e.g. thesauri, encyclopedias, ontologies, semantic networks) are static and manually curated, but may result in a topic drift. It is also generally unknown which external resource would be the most effective for a particular collection type (e.g. shorter Web document versus longer news articles).

While the methods for retrieval from DBpedia [12] as well as query expansion utilizing ConceptNet [5], Freebase [9] and Wikipedia [10] in the context of pseudo-relevance feedback (PRF) have been examined in detail in previous studies, in this work, we focus on empirical comparison of retrieval effectiveness of term graphs derived from knowledge repositories with automatically constructed terms association graphs on the same standard IR collections of different type. Our work is also the first one to evaluate the effectiveness of DBpedia for query expansion at the level of individual terms without PRF.

2 Methods

2.1 Statistical Term Association Graphs

Statistical term association graphs are constructed by calculating a co-occurrence based information theoretic measure of similarity, such as Mutual information (MI) [7] or Hyperspace Analog to Language (HAL) [2], between each pair of terms in the vocabulary of a given document collection and considering the top-k terms with the highest value of that measure for each given term. The key difference between MI and HAL is in the size of contextual window to calculate co-occurrence. Term co-occurrences within entire documents are considered in MI calculation, whereas a sliding window of small size is used for HAL.

Mutual information measures the strength of association between a pair of terms based on the counts of their individual and joint occurrence. The higher the mutual information between the terms, the more often they tend to co-occur in the same documents, and hence the more semantically related they are.

Hyperspace Analog to Language is a representational model of high dimensional concept spaces, which was created based on the studies of human cognition. Previous work [8] has demonstrated that HAL can be effectively utilized in IR. Constructing the HAL space for an n-term vocabulary involves traversing a sliding window of width w over each term in the corpus. All terms within a sliding window are considered as part of the local context for the

[3] http://conceptnet5.media.mit.edu/.

term, over which the sliding window is centered. Each word in the local context is assigned a weight according to its distance from the center of the sliding window (words that are closer to the center receive higher weight). An $n \times n$ HAL space matrix $\mathbf{H}$, which aggregates the local contexts for all the terms in the vocabulary, is created after traversing an entire corpus. After that, the global co-occurrence matrix is produced by merging the row and column corresponding to each term in the HAL space matrix. Each distinct term w_i in the vocabulary of the collection corresponds to a row in the global co-occurrence matrix $\mathbf{H}_{w_i} = \{(w_{i1}, c_{i1}), \ldots, (w_{in}, c_{in})\}$, where $c_{i1}, \ldots, c_{in}$ are the number of co-occurrences of the term w_i with all other terms in the vocabulary. After the merge, each row $\mathbf{H}_{w_i}$ in the global co-occurrence matrix is normalized to obtain a HAL-based semantic term similarity matrix for the entire collection:

$$\mathbf{S}_{w_i} = \frac{c_{ij}}{\sum_{j=1}^{n} c_{ij}}$$

Due to the context window of smaller size, HAL-based term association graphs are typically less noisy than MI-based ones.

2.2 Knowledge Repositories

In addition to statistical term association graphs, we also experimented with the term graphs based on DBpedia, Freebase and ConceptNet. The key difference between DBpedia, Freebase and ConceptNet lies in the type of knowledge they provide.

DBpedia is a structured version of Wikipedia infoboxes, which provides descriptions of entities (people, locations, organizations, etc.) as RDF triplets. We used DBpedia 3.9[4] extended abstracts, which usually contain all words in the first section of the Wikipedia article corresponding to an entity, for term graph construction. Treating extended abstracts as documents, we generated two term graphs DB-MI and DB-HAL using MI and HAL as similarity measures, respectively. Those graphs were customized for each document collection by removing the words that are not in the index of a given collection.

Freebase, similar to DBpedia, provides descriptions of entities as RDF triplets, but features a more comprehensive list of concepts than DBpedia. We used the text property of documents (/common/document/text), which contains extended textual descriptions of entities, to generate the FB-MI and FB-HAL term graphs.

ConceptNet [6] codifies commonsense knowledge as subject-predicate-object triplets (e.g. "alarm clock", UsedFor, "wake up") and can be viewed as a semantic network, in which the nodes correspond to semi-structured natural language fragments (e.g., "food", "grocery store", "buy food", "at home") representing real or abstract concepts and the edges represent semantic relationships between the concepts. For experiments in this work, we used the weights

[4] http://wiki.dbpedia.org/Downloads39.

between the concepts provided by ConceptNet 5 (CNET)[5], as well as the ones calculated for each collection using MI (CNET-MI) and HAL (CNET-HAL). As in the case of DBpedia, we customized the term graph by removing the words that are not in the index of a given collection.

2.3 Retrieval Model and Query Expansion

We used the KL-divergence retrieval model with Dirichlet prior smoothing [11], according to which each document D in the collection is scored and ranked based on the Kullback-Leibler divergence between the query LM Θ_Q and document LM Θ_D. In language modeling approaches to IR, query expansion is typically performed via linear interpolation of the original query LM $p(w|Q)$ and query expansion LM $p(w|\hat{Q})$ with the parameter α:

$$p(w|\tilde{Q}) = \alpha p(w|Q) + (1 - \alpha)p(w|\hat{Q}) \quad (1)$$

Query expansion using a term graph involves finding a set of semantically related terms for each query term q_i (i.e. all direct neighbors of query terms in the term graph) and estimating $p(w|\hat{Q})$ according to the following formula:

$$p(w|\hat{Q}) = \frac{\sum_{i=1}^{k} p(w|q_i)}{\sum_{w \in V} \sum_{i=1}^{k} p(w|q_i)} \quad (2)$$

where $p(w|q_i)$ is the strength of semantic association between w and q_i according to a particular term graph.

3 Experiments

3.1 Datasets

For all experiments in this work we used AQUAINT, ROBUST and GOV datasets from TREC, which were pre-processed by removing stopwords and applying the Porter stemmer. To construct the term association graphs, all rare terms (that occur in less than 5 documents) and all frequent terms (that occur in more than 10 % of all documents in the collection) have been removed [3,4]. Term association graphs were constructed using either the top 100 most related terms or the terms with similarity metric greater than 0.001 for each distinct term in the vocabulary of a given collection. HAL term association graphs were constructed using the sliding window of size 20 [4]. The reported results are based on the optimal settings of the Dirichlet prior μ and interpolation parameter α empirically determined for all the methods and the baselines. Top 85 terms most similar to each query term were used for query expansion [1]. KL-divergence retrieval model with Dirichlet prior smoothing (**KL-DIR**) and document expansion based on translation model [3] (**TM**) were used as the baselines.

[5] http://conceptnet5.media.mit.edu/downloads/20130917/associations.txt.gz.

3.2 Results

Retrieval performance of query expansion using different types of term graphs and the baselines on different collections and query types is summarized in Tables 1, 2 and 3. The best and the second best values for each metric are highlighted in boldface and italic, while † and ‡ indicate statistical significance in terms of MAP ($p < 0.05$) using Wilcoxon signed rank test over the **KL-DIR** and **TM** baselines, respectively.

Table 1. Retrieval accuracy for (a) all queries and (b) difficult queries on AQUAINT dataset.

(a)

Method	MAP	P@20	GMAP
KL-DIR	0.1943	0.3940	0.1305
TM	0.2033	0.3980	0.1339
NEIGH-MI	0.2031†	0.3970	0.1326
NEIGH-HAL	0.1989†	0.3900	0.1319
DB-MI	**0.2073**†‡	**0.4160**	**0.1468**
DB-HAL	*0.2059*†‡	*0.4080*	*0.1411*
FB-MI	0.2055†‡	0.3990	0.1336
FB-HAL	0.2056†‡	0.3960	0.1384
CNET	0.2051†‡	0.3900	0.1388
CNET-MI	0.2042†	0.3920	0.1371
CNET-HAL	0.2058†‡	0.3920	0.1388

(b)

Method	MAP	P@20	GMAP
KL-DIR	0.0474	0.1250	0.0386
TM	0.0478	0.1250	0.0386
NEIGH-MI	0.0476	0.1375	0.0393
NEIGH-HAL	0.0474	0.1500	0.0378
DB-MI	0.0528†‡	**0.1906**	0.0452
DB-HAL	*0.0544*†‡	*0.1538*	*0.0455*
FB-MI	0.0534†‡	0.1333	0.0437
FB-HAL	**0.0564**†‡	0.1444	**0.0471**
CNET	0.0504†‡	0.1219	0.044
CNET-MI	0.0496†	0.1156	0.0422
CNET-HAL	0.0502†	0.1219	0.0436

Table 2. Retrieval accuracy for (a) all queries and (b) difficult queries on ROBUST dataset.

(a)

Method	MAP	P@20	GMAP
KL-DIR	0.2413	0.3460	0.1349
TM	0.2426	0.3488	0.1360
NEIGH-MI	0.2432	0.3460	0.1360
NEIGH-HAL	0.2431	0.3454	0.1333
DB-MI	0.2482†‡	0.3524	0.1397
DB-HAL	0.2426	0.3444	0.1349
FB-MI	0.2452†‡	0.3526	0.1232
FB-HAL	0.2476†‡	**0.3540**	0.1261
CNET	0.2452†	0.3472	0.1407
CNET-MI	*0.2495*†‡	*0.3530*	*0.1459*
CNET-HAL	**0.2503**†‡	0.3528	**0.1463**

(b)

Method	MAP	P@20	GMAP
KL-DIR	0.0410	0.1290	0.0261
TM	0.0458	0.1290	0.0267
NEIGH-MI	0.0429†	0.1323	0.0273
NEIGH-HAL	0.0419	0.1260	0.0265
DB-MI	0.0503†‡	0.1449	0.0301
DB-HAL	0.0474†	0.1437	0.0273
FB-MI	0.0381	0.1222	0.0200
FB-HAL	0.0393	0.1272	0.0211
CNET	**0.0559**†‡	**0.1487**	**0.0334**
CNET-MI	*0.0560*†‡	**0.1487**	*0.0326*
CNET-HAL	0.0558†‡	*0.1475*	0.0323

Examination of experimental results in Tables 1, 2 and 3 leads to the following major conclusions. First, relative retrieval performance of different types of term graphs varies by the collection. In particular, term graphs derived from external repositories are significantly more effective than term association graphs

Table 3. GOV dataset results on (a) all queries and (b) difficult queries.

(a)

Method	MAP	P@20	GMAP
KL-DIR	0.2333	0.0464	0.0539
TM	0.2399	0.0476	0.0551
NEIGH-MI	0.2415†‡	0.0489	0.0518
NEIGH-HAL	0.2419†‡	0.0456	0.0476
DB-MI	0.2346	0.0467	0.0529
DB-HAL	0.2404†	0.0467	0.053
FB-MI	*0.2420*†‡	0.0484	0.0573
FB-HAL	0.2404†	0.0476	0.0565
CNET	0.2407†	0.0489	0.0584
CNET-MI	0.2416†‡	*0.0504*	**0.0587**
CNET-HAL	**0.2428**†‡	**0.0516**	*0.0586*

(b)

Method	MAP	P@5	GMAP
KL-DIR	0.0311	0.0281	0.014
TM	0.0343	0.0304	0.0146
NEIGH-MI	0.0333†	0.0307	0.013
NEIGH-HAL	*0.0425*†‡	0.0293	0.0122
DB-MI	0.0312	0.0285	0.0136
DB-HAL	0.0306	0.0274	0.0134
FB-MI	0.0350†‡	*0.0319*	0.0154
FB-HAL	0.0339†	0.0293	*0.0152*
CNET	0.0407 †‡	0.0333	0.0172
CNET-MI	0.0427 †‡	0.0367	0.0176
CNET-HAL	**0.0453**†‡	**0.0385**	**0.0181**

for newswire datasets (AQUAINT and ROBUST) on both regular and difficult queries, with the HAL-based term association graph (NEIGH-HAL) outperforming the term graphs derived from DBpedia and Freebase (DB-HAL and FB-HAL) for all queries on the GOV collection. For difficult queries on the same dataset, NEIGH-HAL outperforms Freebase- and DBpedia-based terms graphs and has comparable performance with the term graphs derived from ConceptNet. We attribute this to the fact that the term graph for GOV is larger in size and less dense than the term graphs for AQUAINT and ROBUST, which results in less noisy term associations. Second, using MI and HAL-based weighs of edges in ConceptNet graph (CNET-MI and CNET-HAL) results in better retrieval accuracy that the original ConceptNet weights (CNET) in the majority of cases. This indicates the utility of tuning the weights in term graphs derived from external resources to particular collections. Finally, ConceptNet-based term graphs outperformed Freebase- and DBpedia-based ones on 2 out of 3 collections used in evaluation, which indicates the importance of commonsense knowledge in addition to information about entities.

References

1. Bai, J., Song, D., Bruza, P., Nie, J.-Y., Cao, G.: Query expansion using term relationships in language models for information retrieval. In: Proceedings of the 14th ACM CIKM, pp. 688–695 (2005)
2. Burgess, C., Livesay, K., Lund, K.: Explorations in context space: words, sentences and discourse. Discourse Process. **25**, 211–257 (1998)
3. Karimzadehgan, M., Zhai, C.: Estimation of statistical translation models based on mutual information for ad hoc information retrieval. In: Proceedings of the 33rd ACM SIGIR, pp. 323–330 (2010)
4. Kotov, A., Zhai, C.: Interactive sense feedback for difficult queries. In: Proceedings of the 20th ACM CIKM, pp. 163–172 (2011)
5. Kotov, A., Zhai, C.: Tapping into knowledge base for concept feedback: leveraging conceptnet to improve search results for difficult queries. In: Proceedings of the 5th ACM WSDM, pp. 403–412 (2012)

6. Liu, H., Singh, P.: Conceptnet - a practical commonsense reasoning tool-kit. BT Technol. J. **22**(4), 211–226 (2004)
7. Manning, C., Schütze, H.: Foundations of Statistical Natural Language Processing. MIT Press, Cambridge (1999)
8. Song, D., Bruza, P.: Towards context sensitive information inference. JASIST **54**(4), 321–334 (2003)
9. Xiong, C., Callan, J.: Query expansion with freebase. In: Proceedings of the 5th ACM ICTIR, pp. 111–120 (2015)
10. Xu, Y., Jones, G.J.F., Wang, B.: Query dependent pseudo-relevance feedback based on wikipedia. In: Proceedings of the 32nd ACM SIGIR, pp. 59–66 (2009)
11. Zhai, C., Lafferty, J.: Document language models, query models, and risk minimization for information retrieval. In: Proceedings of the 24th ACM SIGIR, pp. 111–119 (2001)
12. Zhiltsov, N., Kotov A., Nikolaev, F.: Fielded sequential dependence model for Ad-Hoc entity retrieval in the web of data. In: Proceedings of the 38th ACM SIGIR, pp. 253–262 (2015)

Deep Learning to Predict Patient Future Diseases from the Electronic Health Records

Riccardo Miotto(✉), Li Li, and Joel T. Dudley

Department of Genetics and Genomic Sciences,
Icahn School of Medicine at Mount Sinai, New York, USA
{riccardo.miotto,li.li,joel.dudley}@mssm.edu

Abstract. The increasing cost of health care has motivated the drive towards preventive medicine, where the primary concern is recognizing disease risk and taking action at the earliest stage. We present an application of deep learning to derive robust patient representations from the electronic health records and to predict future diseases. Experiments showed promising results in different clinical domains, with the best performances for liver cancer, diabetes, and heart failure.

Keywords: Disease prediction · Preventive medicine · Electronic health records · Medical information retrieval · Deep learning

1 Introduction

Developing predictive approaches to maintain health and to prevent diseases, disability, and death is one of the primary goals of preventive medicine. In this context, information retrieval applied to electronic health records (EHRs) has shown great promise in providing search engines that could support physicians in identifying patients at risk of diseases given their clinical status. Most of the works proposed in literature, though, focus on only one specific disease at a time (e.g., cardiovascular diseases [1], chronic kidney disease [2]) and patients are often represented using ad-hoc descriptors manually selected by clinicians. While appropriate for an individual task, this approach scales poorly, does not generalize well, and also misses the patterns that are not known.

EHRs are challenging to represent since they are high dimensional, sparse, noisy, heterogeneous, and subject to random errors and systematic biases [3]. In addition, the same clinical concept is usually reported in different ways. For example, a patient with "type 2 diabetes mellitus" can be identified by hemoglobin A1C lab values greater than 7.0, presence of 250.00 ICD-9 code, "diabetes mellitus" mentioned in the free-text clinical notes, and so on. Consequently, it is hard to automatically derive robust descriptors for effective patient indexing and retrieval. Representations based on raw vectors composed of all the descriptors available in the hospital data warehouse have also been used [4]. However, these representations are sparse, noisy, and repetitive, thus, not ideal to model the hierarchical information embedded in the EHRs.

N. Ferro et al. (Eds.): ECIR 2016, LNCS 9626, pp. 768–774, 2016.
DOI: 10.1007/978-3-319-30671-1_66

This paper applies deep learning to a large-scale EHR data warehouse to extract robust patient descriptors that can be effectively used to predict future patient diseases in different clinical domains. In particular, we first use a stack of denoising autoencoders to capture regularities and dependencies in the dataset, which, grouped together, lead to the deep patient representation. The latter aims to be domain free, lower-dimensional, dense, and easily applicable to various retrieval tasks. Second, we test this representation to predict the patient probability of developing new diseases within a year given their current clinical status using stand-alone classifiers as well as a fine-tuned supervised deep neural network.

Deep learning has been applied successfully to several fields, such as image retrieval, natural language processing, and speech recognition [5,6]. In medicine, large neural networks were recently used, e.g., to reconstruct brain circuits [7] and to predict the activity of potential drug molecules [8]. To the best of our knowledge, deep learning has not been used yet to derive patient representations from aggregated EHRs to benefit preventive medicine.

2 Deep Learning for Disease Prediction

EHRs are first extracted from the clinical data warehouse and grouped to be represented as one vector per patient[1]. The vectors obtained from all the patients are then processed by the unsupervised deep feature learning architecture, which derives a set of high level descriptors through a multi-layer neural network. This type of framework attempts to hierarchically combine the raw features into a more unified and compact representation through a sequence of non-linear transformations. Ideally, at every layer of the network, several overlapping descriptors are joined together to create a higher-level clinical concept (e.g., diseases, medications), leading to a representation that is non redundant and more effective to manipulate and process. We used a stack of denoising autoencoders (SDA), locally trained one layer at the time, to model EHRs. All the autoencoders in the deep architecture share the same structure, which is briefly reviewed in the following section (see [9] for more details).

The output of the last layer is the patient representation that can be used to predict future diseases[2]. On one hand, the representation can directly be the input of a stand-alone supervised algorithm, such as support vector machines (SVMs). On the other hand, a logistic regression layer can be added on top of the last autoencoder, yielding a deep neural network amenable to supervised learning. The parameters of all layers can then be simultaneously fine-tuned using a gradient-based procedure (e.g., stochastic gradient descent), leading to features specifically optimized for disease prediction.

[1] In this architecture, each patient can be described by just one single vector (as done in this study) or by a bag of vectors computed in, e.g., predefined temporal windows.

[2] While this study focuses on future disease prediction, it should be noted that the patient representation derived from the stack of denoising autoencoders can also be applied to unsupervised tasks (e.g., patient clustering and similarity) as well as to other supervised applications (e.g., personalized prescriptions).

2.1 Denoising Autoencoders

A denoising autoencoder takes an input $\boldsymbol{x} \in [0,1]^d$ and corrupts it to obtain a partially destroyed version $\tilde{\boldsymbol{x}}$, which is used during learning to prevent overfitting (i.e., denoising). We applied a masking noise corruption strategy, i.e., a fraction ν of the elements of $\boldsymbol{x}$ chosen at random were turned to zero [9]. This can be viewed as simulating the presence of missed components in the EHRs (e.g., medications or diagnoses not recorded), thus assuming that the input clinical data is a degraded or "noisy" version of the actual clinical situation.

The corrupted input $\tilde{\boldsymbol{x}}$ is then transformed (with an *encoder*) to a hidden representation $\boldsymbol{y} \in [0,1]^{d'}$ through a deterministic mapping:

$$\boldsymbol{y} = f_\theta(\tilde{\boldsymbol{x}}) = s(\boldsymbol{W}\tilde{\boldsymbol{x}} + \boldsymbol{b}), \tag{1}$$

parameterized by $\theta = \{\boldsymbol{W}, \boldsymbol{b}\}$, where $s(\cdot)$ is a non-linear activation function, $\boldsymbol{W}$ is a weight coefficient matrix, and $\boldsymbol{b}$ is a bias vector. Ideally, $\boldsymbol{y}$ is a distributed representation that captures the coordinates along the main factors of variation in the data.

The latent representation $\boldsymbol{y}$ is then mapped back (with a *decoder*) to a reconstructed vector $\boldsymbol{z} \in [0,1]^d$, such as:

$$\boldsymbol{z} = g_{\theta'}(\boldsymbol{y}) = s(\boldsymbol{W}'\boldsymbol{y} + \boldsymbol{b}'), \tag{2}$$

with $\theta' = \{\boldsymbol{W}', \boldsymbol{b}'\}$. We used tied weights (i.e., $\boldsymbol{W}' = \boldsymbol{W}^T$) and the sigmoid function as activation in both mappings.

The parameter of the model θ and θ' are optimized over the training set to minimize the difference between $\boldsymbol{x}$ and $\boldsymbol{z}$ (i.e., average reconstruction error $L(\boldsymbol{x}, \boldsymbol{z})$). We used reconstruction cross-entropy as the error function, i.e.,

$$L_H(\boldsymbol{x}, \boldsymbol{z}) = -\sum_{k=1}^{d} [x_k \log z_k + (1 - x_k)\log(1 - z_k)]. \tag{3}$$

Optimization is carried out by mini-batch stochastic gradient descent, which iterates through small subsets of the training patients and modifies the parameters in the opposite direction of the error gradient. Once trained, $f_\theta(\cdot)$ is applied to the input data (without corruption) to obtain the corresponding mapped representation.

3 Experimental Setup

This section describes the evaluation performed to validate the deep learning framework for future disease prediction using the Mount Sinai data warehouse.

3.1 Dataset

The Mount Sinai Health System generates a high volume of structured, semi-structured, and unstructured data as part of its healthcare and clinical operations. The entire EHR dataset is composed of approximately 4.2 million patients as of March 2015, with 1.2 million of them having at least one diagnosed disease expressed as a numerical ICD-9 code. In this context, we considered all the records till December 31, 2013 (i.e., "split-point") as training data and all the diseases diagnosed in 2014 as testing data.

We randomly selected 105,000 patients with at least one new disease diagnosed in 2014 and at least ten records before that (e.g., medications, lab tests, diagnoses). These patients composed validation (i.e., 5,000 patients) and test (i.e., 100,000 patients) sets. In particular, all the diagnoses in 2014 were used to validate the predictions computed using the patient data recorded before the split-point (i.e., clinical status). We then sampled another 350,000 different patients with at least ten records before the split-point to use as the training set.

The evaluation was performed on a vocabulary of 72 diseases, covering different clinical domains, such as oncology, endocrinology, and cardiology. This was obtained by initially using the ICD-9 codes to determine the diagnosis of a disease to a patient. However, since different codes can refer to the same disease, we mapped the codes to a categorization structure, which groups ICD-9s into a vocabulary of 231 general disease definitions [10]. This list was then filtered down to remove diseases not present in the data warehouse or not considered predictable using EHRs alone (e.g., physical injuries, poisoning), leading to the final vocabulary.

3.2 EHR Processing

The proposed framework allows flexible customization in terms of how to process and summarize patient EHRs[3]. For each patient in the dataset we retained some general demographic details (i.e., gender and race) as well as diagnoses (ICD-9 codes), medications, procedures, lab tests, and clinical notes recorded by the split-point. All the clinical features were pre-processed using the Open Biomedical Annotator [11] to obtain harmonized codes for procedures and lab tests, normalized medications based on brand name and dosages, and parsed representations of notes summarizing clinically relevant information extracted from the text.

For diagnoses, medications, procedures and lab tests, we then just counted the presence of each normalized code. Parsed clinical notes were further post-processed with latent Dirichlet allocation (i.e., topic modeling) to obtain a semantic abstraction of the embedded clinical information [12]. Each note was thus summarized as a multinomial of 200 topic probabilities; the number of topics was estimated through perplexity analysis of one million random notes. For each patient, we eventually retained one single topic-based representation averaged over all the notes available.

[3] While in this study we favored a basic pipeline to process EHRs, it should be noted that more sophisticated techniques might lead to better features as well as to better predictive results.

3.3 Evaluation

We first extracted all the descriptors available in the data warehouse related to the EHR categories mentioned in Sect. 3.2 and removed those that were either very frequent or rare in the training set. This led to vector-based patient representations of 41,072 entries (i.e., "raw").

We then applied a 3-layer SDA to the training set to derive the deep features. The autoencoders in the network shared the same configuration with 500 hidden units and a noise corruption factor $\nu = 0.1$. For comparison, we also derived features using principal component analysis (i.e., "PCA" with 100 principal components) and k-means clustering (i.e., "kMeans" with 500 centroids)[4].

Predictions were performed using random forests and SVMs with radial basis function kernel. Deep features were also fine-tuned adding a logistic regression layer on top of the last autoencoder as described in Sect. 2.1 (i.e., "sSDA"). Hence, for all the model combinations, we computed the probability of each test patient to develop every disease in the vocabulary and we evaluated how many of these predictions were correct in one year interval[5]. For each disease, we measured Area under the ROC curve (i.e., AUC-ROC) and F-score (with classification threshold equal to 0.6).

Table 1. Future disease prediction results averaged over 72 diseases and 100,000 patients. The symbols (†) and (*) after a numeric value mean that the difference with the corresponding second best measurement in the classification algorithm and overall, respectively, is statistically significant ($p \leq 0.05$, t-test).

	SVM				Random forest				
	raw	PCA	kMeans	SDA	raw	PCA	kMeans	SDA	sSDA
F-Score	0.076	0.116	0.104	**0.123**†	0.105	0.113	0.114	**0.149**†	**0.181***
AUC-ROC	0.690	0.729	0.716	**0.757**†	0.705	0.715	0.705	**0.766**†	**0.781**

4 Results

Table 1 shows the classification results averaged over all 72 diseases in the vocabulary. As it can be seen SDA features lead to significantly better predictions than "raw" as well as than PCA and kMeans with both classification models. In addition, fine-tuning the SDA features for the specific task further improved the final results, with 50 % and 10 % improvements over "raw" in F-score and AUC-ROC, respectively. Table 2 reports the top 5 performing diseases for sSDA based on AUC-ROC, showing promising results in different clinical domains. Some diseases in the vocabulary did not show high predictive power (e.g., HIV, ovarian

[4] All parameters in the feature learning models were identified through preliminary experiments, not reported here for brevity, on the validation set.

[5] This experiment only evaluates the prediction of new diseases for each patient, therefore not considering the re-diagnosis of a disease previously reported.

Table 2. Top 5 performing diseases for sSDA (with respect to AUC-ROC results).

Disease	F-score	AUC-ROC
Cancer of Liver	0.225	0.925
Regional Enteritis and Ulcerative Colitis	0.479	0.901
Diabetes Mellitus with Complications	0.464	0.889
Congestive Heart Failure	0.395	0.870
Chronic Kidney Disease	0.397	0.861

cancer), leading to almost random predictions. Additional EHR descriptors, such as social behavior and family history, should lead to patient representations more likely to obtain better results in these domains as well.

5 Conclusion

This article demonstrates the feasibility of using deep learning to predict patients' diseases from their EHRs. Future works will apply this framework to other clinical applications (e.g., therapy recommendation) and will incorporate additional EHR descriptors as well as more sophisticated pre-processing techniques.

References

1. Kennedy, E., Wiitala, W., Hayward, R., Sussman, J.: Improved cardiovascular risk prediction using non-parametric regression and electronic health record data. Med Care **51**(3), 251–258 (2013)
2. Perotte, A., Ranganath, R., Hirsch, J.S., Blei, D., Elhadad, N.: Risk prediction for chronic disease progression using heterogeneous electronic health record data and time series analysis. J Am Med Inform Assoc **22**(4), 872–880 (2015)
3. Jensen, P.B., Jensen, L.J., Brunak, S.: Mining electronic health records: towards better research applications and clinical care. Nat. Rev. Genet. **13**(6), 395–405 (2012)
4. Wu, J., Roy, J., Stewart, W.: Prediction modeling using EHR data: Challenges, strategies, and a comparison of machine learning approaches. Med. Care **48**(Suppl 6), 106–113 (2010)
5. Bengio, Y., Courville, A., Vincent, P.: Representation learning: A review and new perspectives. IEEE Trans. Pattern Anal. Mach. Intell. **35**(8), 1798–1828 (2013)
6. LeCun, Y., Bengio, Y., Hinton, G.: Deep learning. Nature **521**(7553), 436–444 (2015)
7. Helmstaedter, M., Briggman, K.L., Turaga, S.C., Jain, V., Seung, H.S., Denk, W.: Connectomic reconstruction of the inner plexiform layer in the mouse retina. Nature **500**(7461), 168–174 (2013)
8. Ma, J.S., Sheridan, R.P., Liaw, A., Dahl, G.E., Svetnik, V.: Deep neural nets as a method for quantitative structure-activity relationships. J. Chem. Inf. Model **55**(2), 263–274 (2015)
9. Vincent, P., Larochelle, H., Lajoie, I., Bengio, Y., Manzagol, P.A.: Stacked denoising autoencoders: learning useful representations in a deep network with a local denoising criterion. J. Mach. Learn Res. **11**, 3371–3408 (2010)

10. Cowen, M.E., Dusseau, D.J., Toth, B.G., Guisinger, C., Zodet, M.W., Shyr, Y.: Casemix adjustment of managed care claims data using the clinical classification for health policy research method. Med. Care **36**(7), 1108–1113 (1998)
11. LePendu, P., Iyer, S., Fairon, C., Shah, N.: Annotation analysis for testing drug safety signals using unstructured clinical notes. J. Biomed. Semant. **3**(S–1), S5 (2012)
12. Blei, D., Ng, A., Jordan, M.: Latent dirichlet allocation. J. Mach. Learn. Res. **3**, 993–1022 (2003)

Improving Document Ranking for Long Queries with Nested Query Segmentation

Rishiraj Saha Roy[1(✉)], Anusha Suresh[2], Niloy Ganguly[2], and Monojit Choudhury[3]

[1] Max Planck Institute for Informatics, Saarbrücken, Germany
rishiraj@mpi-inf.mpg.de
[2] Indian Institute of Technology (IIT), Kharagpur, India
{anusha.suresh,niloy}@cse.iitkgp.ernet.in
[3] Microsoft Research India, Bangalore, India
monojitc@microsoft.com

Abstract. In this research, we explore nested or hierarchical query segmentation (An extended version of this paper is available at http://research.microsoft.com/pubs/259980/2015-msri-tr-nest-seg.pdf), where segments are defined recursively as consisting of contiguous sequences of segments or query words, as a more effective representation of a query. We design a lightweight and unsupervised nested segmentation scheme, and propose how to use the tree arising out of the nested representation of a query to improve ranking performance. We show that nested segmentation can lead to significant gains over state-of-the-art flat segmentation strategies.

1 Introduction

Query segmentation [1–5] is one of the first steps towards query understanding where complex queries are partitioned into semantically coherent word sequences. Past research [1–5] has shown that segmentation can potentially lead to better IR performance. Till date, almost all the works on query segmentation have dealt with *flat* or *non-hierarchical segmentations*, such as: `windows xp home edition | hd video | playback`, where pipes (|) represent flat segment boundaries. For short queries of up to three or four words, such flat segmentation may suffice. However, slightly longer queries of about five to ten words are increasing over the years ($\simeq 27\,\%$ in our May 2010 Bing Australia log) and present a challenge to the search engine. One of the shortcomings of flat segmentation is that it fails to capture the relationships between segments which can provide important information towards the document ranking strategy, particularly in the case of a long query.

These relationships can be discovered if we allow nesting of segments inside bigger segments. For instance, instead of a flat segmentation, our running example query could be more meaningfully represented as Fig. 1. Here, the *atomic segments* – `windows xp` and `hd video`, are progressively joined with other words

This research was completed while the author was at IIT Kharagpur.

N. Ferro et al. (Eds.): ECIR 2016, LNCS 9626, pp. 775–781, 2016.
DOI: 10.1007/978-3-319-30671-1_67

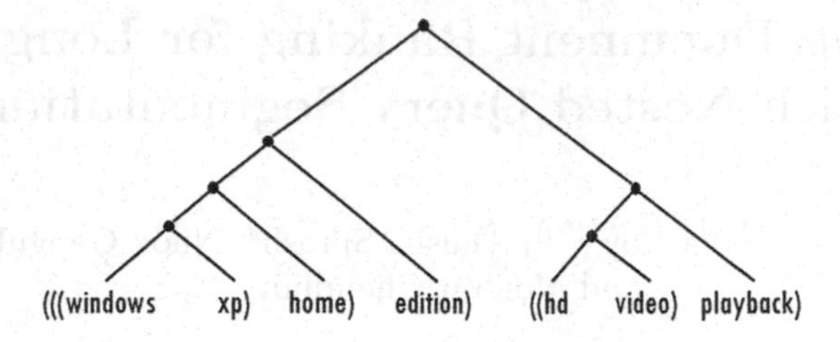

Fig. 1. Nested segmentation tree.

to produce larger segments – `windows xp home`, `windows xp home edition`, and `hd video playback`. It is intuitive from this representation that `windows xp` and `hd video` are non-negotiable (atomic units) when it comes to matching within documents, and the strength of ties between word pairs can be said to weaken as they move farther in terms of the (unique) path through the tree. We define some of the important concepts below.

Tree Distance. The tree distance[1] $td(t_1, t_2; n(q))$ between two terms t_1 and t_2 in $n(q)$, the nested segmentation of a query q, is defined as the shortest path (i.e., the number of hops) between t_1 and t_2 (or vice versa) through the nested segmentation tree for q (like Fig. 1). A tree ensures a unique shortest path between t_1 and t_2, which is through the common ancestor of t_1 and t_2. For example, td(`xp`, `video`; $n(q)$ in Fig. 1) = 7. The minimum possible tree distance between two terms is two. Note that td between t_1 and t_2 can vary for the same q, depending on $n(q)$.

Query Distance. The query distance $qd(t_1, t_2; q)$ between two terms t_1 and t_2 in a query q is defined as the difference between the positions of t_1 and t_2 in q, or equivalently, the number of intervening words plus one.

Document Distance. Let t_1 and t_2 be two terms in the query q, which are also present (matched) in a retrieved document $\mathcal{D}$. Let there be k instances of *ordered* pairwise occurrences of t_1 and t_2 (*ordered* pairs of positions of t_1 and t_2, (p_1, p_2) where $p_1 < p_2$) in $\mathcal{D}$ at minimum distances [6] $dist_i = dist_1, dist_2, \ldots, dist_k$, such that the $dist_i$-s are in ascending order. We combine the ideas of minimum distance and multiple occurrences of a term pair to formulate the following definition of accumulative inverse document distance ($AIDD$) for t_1 and t_2 in $\mathcal{D}$:

$$AIDD(t_1, t_2; \mathcal{D})_{t_1 \neq t_2} = \frac{1}{dist_1} + \frac{1}{dist_2} + \ldots + \frac{1}{dist_k} \quad (1)$$

By this method, a document with several (t_1, t_2) near to each other will have a high $AIDD$. Since our concept is based on minimum distance, we do not need a document length normalizer. A threshold on k is nevertheless necessary to avoid considering *all* pairwise distances of t_1 and t_2, as distant pairs could be

[1] For all distances, when the same word appears multiple times in a query, each word instance is treated as distinct during pairwise comparisons.

semantically unrelated. To avoid scoring unrelated occurrences of a term pair, we consider matches only if (t_1, t_2) occur within a given window size, win.

2 Algorithm and IR Application

2.1 Splitting and Joining Flat Segments

Since flat segmentation is a well-researched problem, we develop our algorithm for nested segmentation by starting with a flat segmentation of the query and trying to *split* within a flat segment and *join* adjacent flat segments recursively. Our main motivation for designing simple segment nesting strategies stems from the fact that most flat segmentation algorithms compute scores for n-grams as a key step of their respective methods (generally $n \leq 5$) [3–5]. In doing so, most often, the scores of the contiguous lower order n-grams $(n-1, n-2, \ldots)$ are also known. For **splitting a flat segment**, we exploit these scores to deduce the structure within a flat segment. In this work, we specifically use the state-of-the-art Co-occurrence Significance Ratio (CSR) measure [7] to score n-grams. We adopt a simple greedy approach in this research. The n-gram (here $n \leq 3$) that has the highest CSR score within a flat segment (where the number of words in the n-gram is less than the number of words in the corresponding flat segment) is immediately grouped together as a unit, i.e. a *sub-segment*. We define a *sub-segment* as a smaller segment created by the division of a larger segment. Recursively, this newly grouped sub-segment's left and right n-grams (possibly null) and the sub-segment itself are processed in the same greedy fashion till every string to be processed cannot be divided further.

Joining flat segments is essential to completing the nested segmentation tree, which in turn ensures a path between every pair of words in the query. The bigram at a flat segment boundary, i.e. the last word of a flat segment and the first word of the next flat segment, can be effectively used to take the segment joining decision. In our running example, if we wish to decide whether to join `windows xp home edition` and `hd video`, *or* `hd video` and `playback`, we check the relative order of the scores of the (ordered) bigrams formed by the underlined words only. The bigram with the higher score (in this case `video playback`) dictates which pair should be joined. This process is similarly repeated on the new parenthesized segments obtained until the whole query forms one unit. Here we use pointwise mutual information (PMI) [5] to score bigrams. It often happens that the last (or the first) word in a segment is a determiner, conjunction or preposition (DCP) (list used from http://goo.gl/RoleeA). In these cases, it is almost always meaningful to combine such a segment with the next segment (or the previous segment) to make a meaningful *super-segment* (a larger segment created by the joining of two smaller segments). Examples are `(bed and) (breakfast)` and `(sound) (of music)`. We prioritize such cases over the bigram scores during the joining process.

2.2 Using Nested Segmentation in IR

We define a score *Re-rank Status Value* ($RrSV$) of every document $\mathcal{D}$ that was retrieved in response to an unsegmented query q, determined by the following principle – *a pair of words that has a low tree distance in the nested representation of the query should not have a high document distance.* In other words, while re-ranking a document, the document distance Eq. (1) between a pair of words should be penalized by a factor *inversely* proportional to its tree distance. The $RrSV$ for a document $\mathcal{D}$ is thus defined as

$$RrSV_{\mathcal{D}} = \sum_{\substack{t_i, t_j \in q \cap \mathcal{D} \\ t_i \neq t_j \\ td(t_i, t_j; n(q)) < \delta}} \frac{AIDD(t_i, t_j; \mathcal{D})}{td(t_i, t_j; n(q))} \tag{2}$$

where t_i-s are query terms matched in the document and $n(q)$ is the nested segmentation for q. However, we do not wish to penalize $\mathcal{D}$ when the words are close by in the document and are relatively far in the tree. This analysis drives us to create a tree distance threshold (cut-off) parameter δ. In other words, if $td(a, b; n(q)) < \delta$, only then is the word pair a and b considered in the computation of $RrSV$. The original rank for a page (obtained using TF-IDF scoring, say) and the new rank obtained using $RrSV$ are fused using the method in Agichtein et al. [8], using w as a tuning parameter. We refer to this entire strategy as the *Tree* model. We use three re-ranking baselines: *Flat segmentation* (word pairs are limited to cases where both words come from a single flat segment), document distances only (no scaling using tree distance; *Doc* model), and query distances (scaling document distances using query distances (Sect. 1); *Query* model).

3 Datasets and Experimental Results

3.1 Datasets

Our nested segmentation algorithm requires a query log as the only resource, for computing the various n-gram scores. For our experiments, we use a query log sampled from a Bing Australia in May 2010. This raw data slice consists of 16.7 M queries (4.7 M unique). In order to ensure the replicability of our results, we report our IR evaluation on publicly available datasets only and use open source retrieval systems. We used the dataset released by Saha Roy et al. [2], and refer to it as SGCL12 (author last name initials and year), using Apache Lucene v.3.4.0 as the search engine. The first 250 queries were used as the development set for tuning model parameters (k, win, δ and w) and the last 250 queries were used as the test set. We also used a collection of 75 TREC queries sampled from the Web tracks of 2009 to 2012, with ≥ 3 words and at least one relevant document in the top-100 results. The Indri search engine was used along with the ClueWeb09 dataset. 35 queries were used as the development set for tuning model parameters and the remaining 40 queries were used as the test set, and the results are averaged over ten random 35-40 splits.

3.2 Experiments and Results

We used the outputs of three recent flat segmentation algorithms as input to the nested segmentation algorithm and final nested segmentations for these queries were obtained. Documents are retrieved using unsegmented queries, and re-ranked using the proposed technique and the baselines. Results are compared in terms of nDCG@k ($k = 5, 10, 20$; the IDCG is computed using the optimal ranking from all judgments for a query) and MAP (URLs with ratings > 0 were considered as relevant). For each dataset, the four parameters k, win, δ and w were optimized using the grid search technique for maximizing nDCG@10 on the development set. Partial data and complete code for this project is available at http://cse.iitkgp.ac.in/resgrp/cnerg/qa/nestedsegmentation.html.

(a) Improvements over Flat Segmentation: Some of the sample outputs from our nested segmentation algorithm are as follows: `((garden city) (shopping centre)) (brisbane qld)`, `(the ((chronicles of) riddick)) (dark athena)`, and `((sega superstars) tennis) ((nintendo ds) game)`. In Table 1, for each algorithm, *Flat* refers to the re-ranking strategy for flat segmentation by the corresponding algorithm, and *Nested* refers to the tree re-ranking strategy when applied to the nested segmentation generated when the corresponding flat segmentation was used as the start state. We observe that nested segmentation significantly outperforms the state-of-the-art flat segmentation algorithms in all cases. Importantly, improvements are observed for both datasets on all metrics. This indicates that one should not consider proximity measures for *only* pairs of terms that are within a flat segment. Thus, our experiments provide evidence against the hypothesis that a query is similar to a bag-of-segments [4]. We also note that both the flat and nested segmentations perform better than the unsegmented query, highlighting the general importance of query segmentation for IR.

Table 1. Comparison of flat and nested segmentations on SGCL12 and TREC-WT.

Dataset	Algo	Hagen et al. [5]		Mishra et al. [4]		Saha Roy et al. [2]		Huang et al. [9]
SGCL12	**Unseg**	**Flat**	**Nested**	**Flat**	**Nested**	**Flat**	**Nested**	**Nested**
nDCG@5	0.6839	0.6815	**0.6982**	**0.6977**	0.6976	0.6746	**0.7000**†	0.6996
nDCG@10	0.6997	0.7081	**0.7262**†	0.7189	**0.7274**	0.7044	**0.7268**†	0.7224
nDCG@20	0.7226	0.7327	**0.7433**†	0.7389	**0.7435**	0.7321	**0.7433**†	0.7438
MAP	0.8337	0.8406	**0.8468**†	0.8411	**0.8481**†	0.8423	**0.8477**	0.8456
TREC-WT	**Unseg**	**Flat**	**Nested**	**Flat**	**Nested**	**Flat**	**Nested**	**Nested**
nDCG@5	0.1426	0.1607	**0.1750**†	0.1604	**0.1752**†	0.1603	**0.1767**†	0.1746
nDCG@10	0.1376	0.1710	**0.1880**†	0.1726	**0.1882**†	0.1707	**0.1884**†	0.1845
nDCG@20	0.1534	0.1853	**0.1994**†	0.1865	**0.2000**†	0.1889	**0.2010**†	0.1961
MAP	0.2832	0.2877	**0.3298**†	0.3003	**0.3284**†	0.3007	**0.3296**†	0.3263

Statistical significance of nested segmentation (under the one-tailed paired t-test, $p < 0.05$) over flat segmentation *and* the unsegmented query is marked using †.

(b) Comparison with Past Work: We apply our re-ranking framework on the nesting output by Huang et al. [9] and show results in Table 1. We observed that their method is outperformed by the proposed nested segmentation (from all input flat segmentation strategies) on several metrics. We observed that while the average tree height is 2.96 for our nesting strategy, the same is about 2.23 for Huang et al. (SGCL12). Note that due to the strict binary partitioning at each step for Huang et al., one would normally expect a greater average tree height for this method. Thus, it is the inability of Huang et al. to produce a suitably deep tree for most queries (inability to discover fine-grained concepts) that is responsible for its somewhat lower performance. Most importantly, all nesting strategies faring favorably (none of the differences for Huang et al. with other nesting methods are statistically significant) with respect to flat segmentation bodes well for the usefulness of nested segmentation.

(c) Comparison of Re-ranking Strategies: We find the *Tree* model performs better than *Doc* and *Query* models. We observed that the number of queries on which *Doc*, *Query* and *Tree* perform the best (possibly multiple strategies) are 102, 94, 107 (SGCL12, 250 test queries) and 30, 29.7, 30.8 (TREC-WT, 40 test queries, mean over 10 splits) respectively.

4 Conclusions

This research is one of the first systematic explorations of nested query segmentation. We have shown that the tree structure inherent in the hierarchical segmentation can be used for effective re-ranking of result pages ($\simeq 7\,\%$ nDCG@10 improvement over unsegmented query for SGCL12 and $\simeq 40\,\%$ for TREC-WT). Importantly, since n-gram scores can be computed offline, our algorithms have minimal runtime overhead. We believe that this work will generate sufficient interest and several improvements over the present scheme would be proposed in the recent future. In fact, nested query segmentation can be viewed as the first step towards query parsing, and can lead to a generalized query grammar.

Acknowledgments. The first author was supported by Microsoft Corporation and Microsoft Research India under the Microsoft Research India PhD Fellowship Award.

References

1. Li, Y., Hsu, B.J.P., Zhai, C., Wang, K.: Unsupervised query segmentation using clickthrough for information retrieval. In: SIGIR 2011, pp. 285–294 (2011)
2. Saha Roy, R., Ganguly, N., Choudhury, M., Laxman, S.: An IR-based evaluation framework for web search query segmentation. In: SIGIR 2012, pp. 881–890 (2012)
3. Tan, B., Peng, F.: Unsupervised query segmentation using generative language models and Wikipedia. In: WWW 2008, pp. 347–356 (2008)
4. Mishra, N., Saha Roy, R., Ganguly, N., Laxman, S., Choudhury, M.: Unsupervised query segmentation using only query logs. In: WWW 2011, pp. 91–92 (2011)
5. Hagen, M., Potthast, M., Stein, B., Bräutigam, C.: Query segmentation revisited. In: WWW 2011, pp. 97–106 (2011)
6. Cummins, R., O'Riordan, C.: Learning in a pairwise term-term proximity framework for information retrieval. In: SIGIR 2009, pp. 251–258 (2009)

7. Chaudhari, D.L., Damani, O.P., Laxman, S.: Lexical co-occurrence, statistical significance, and word association. In: EMNLP 2011, pp. 1058–1068 (2011)
8. Agichtein, E., Brill, E., Dumais, S.: Improving web search ranking by incorporating user behavior information. In: SIGIR 2006, pp. 19–26 (2006)
9. Huang, J., Gao, J., Miao, J., Li, X., Wang, K., Behr, F., Giles, C.L.: Exploring web scale language models for search query processing. In: WWW 2010, pp. 451–460 (2010)

Sketching Techniques for Very Large Matrix Factorization

Raghavendran Balu[1(✉)], Teddy Furon[1], and Laurent Amsaleg[2]

[1] Inria Rennes Bretagne-Atlantique, Rennes, France
raghavendran.balu@inria.fr
[2] CNRS-IRISA, Rennes, France

Abstract. Matrix factorization is a prominent technique for approximate matrix reconstruction and noise reduction. Its common appeal is attributed to its space efficiency and its ability to generalize with missing information. For these reasons, matrix factorization is central to collaborative filtering systems. In the real world, such systems must deal with million of users and items, and they are highly dynamic as new users and new items are constantly added. Factorization techniques, however, have difficulties to cope with such a demanding environment. Whereas they are well understood with static data, their ability to efficiently cope with new and dynamic data is limited. Scaling to extremely large numbers of users and items is also problematic. In this work, we propose to use the count sketching technique for representing the latent factors with extreme compactness, facilitating scaling.

1 Introduction

Collaborative filtering is one of the successful techniques used in modern recommender systems. It uses the user-item rating relationship, materialized by a large and sparse matrix **R**, to provide recommendations. The common scenario in collaborative filtering systems is that maintaining the entire matrix **R** in memory is inefficient because it consumes way too much memory. Also, the observed elements in **R** are in many cases not even available as a static data beforehand, but instead as a stream of tuples Alternative representations for **R** have been invented, such as the popular *latent factor* model, which maps both users and items to a low dimensional representation, retaining pairwise similarity. Matrix factorization learns these representation vectors from some observed ratings, to predict (by inner product) the missing entries and thereby to fill the incomplete user-item matrix [4]. Compared to other methods, matrix factorization is simple, space efficient and can generalize well with missing information. It can be easily customized with different loss functions, regularization methods and optimization techniques. The storage requirements for the latent factors are significantly lower, however, the memory required for the factors grows linearly with the number of users and items. This becomes increasingly cumbersome when the numbers are in millions, a frequent real-world situation for domains like advertising and search personalization. The situation is complicated further

N. Ferro et al. (Eds.): ECIR 2016, LNCS 9626, pp. 782–788, 2016.
DOI: 10.1007/978-3-319-30671-1_68

when there are new incoming users and/or items. Overall, supporting real-world large-scale and dynamic recommendation applications asks for designing a much more compact representation to more efficiently manipulate updates while facilitating the insertion of new users and new items. This paper proposes to use sketching techniques to represent the latent factors in order to achieve the above goals. Sketching techniques enable to use extremely compact representations for the parameters, which help scaling. They also, by construction, facilitate updates and inserts. We find through experimental results that sketch based factorization improves storage efficiency without compromising much on prediction quality.

2 Background

2.1 Matrix Factorization

In a typical collaborative filtering system, the given data is a sparse matrix $\mathbf{R}$ with non zero entries $r_{u,i}$ representing the rating provided by the user $u \in \mathbb{U}$ for a given item $i \in \mathbb{I}$. Each row vector in $\mathbf{R}$ corresponds to a user u and column vector to an item i. In latent factor models, each user u (item i) is associated with a vector $\mathbf{p}_u(\mathbf{q}_i) \in \mathbb{R}^{\mathbf{d}}$. The latent vectors $\mathbf{p}_u$ and $\mathbf{q}_i$ of all users and items are represented together as $d \times |\mathbb{U}|$ matrix $\mathbf{P}$ and $d \times |\mathbb{I}|$ matrix $\mathbf{Q}$. Matrix factorization assumes a linear factor model and approximates $\mathbf{R}$ by the low rank matrix $\hat{\mathbf{R}} = \mathbf{P}^\top \mathbf{Q}$: rating $r_{u,i}$ is estimated by means of inner product estimation $\hat{r}_{u,i} = \mathbf{p}_u^\top \mathbf{q}_i$. A loss function $\mathcal{L}(r_{u,i}, \hat{r}_{u,i})$ (we use: $\frac{1}{2}(r_{u,i} - \hat{r}_{u,i})^2$) quantifies the deviation of the approximation to the observed value. Regularization terms controlled by a parameter λ are typically used to tune the model capacity and ensure generalization. We consider the one proposed by Koren et al. [4], which leads to minimization of the combined objective function: $\mathcal{R}_\lambda(\mathbf{P}, \mathbf{Q}) = \sum_{\text{observed: } <u,i,r_{u,i}>} \mathcal{L}(r_{u,i}, \mathbf{p}_u^\top \mathbf{q}_i) + \frac{\lambda}{2}(\|\mathbf{p}_u\|^2 + \|\mathbf{q}_i\|^2)$.

We use online/stochastic gradient descent to minimize the objective function $\mathcal{R}_\lambda(\mathbf{P}, \mathbf{Q})$. At each step, the stochastic gradient descent randomly picks an observed rating $r_{u,i}$ and optimizes the parameters with respect to that rating. This update relies on the gradient of the loss function with respect to parameters, controlled by the learning rate η:

$$\delta \mathbf{p}_u = -\eta \left(\mathcal{L}'_{u,i} \cdot \frac{\partial \hat{r}_{u,i}}{\partial \mathbf{p}_u} + \lambda \mathbf{p}_u \right), \ \delta \mathbf{q}_i = -\eta \left(\mathcal{L}'_{u,i} \cdot \frac{\partial \hat{r}_{u,i}}{\partial \mathbf{q}_i} + \lambda \mathbf{q}_i \right) \quad (1)$$

where $\mathcal{L}'_{u,i} = \partial \mathcal{L}(r_{u,i}, \hat{r}_{u,i}) / \partial \hat{r}_{u,i}$. As the algorithm is sequential, only the latent factors have to be stored in main memory. However, the number of latent factors is linear with the number of users and items, which is problematic for large-scale and dynamic environments. Allocating a d-dimensional vector to every sporadically recurring new user or item quickly becomes non-tractable. A better representation for the factors is therefore needed. We propose the use of count sketches that are typically used in other contexts.

2.2 Sketching Techniques

Sketching is an active area of development, particularly in a streaming setup. A data structure maintaining a particular synopsis of the data irrespective of the history of updates can be called a sketch [2]. Sketching technique have been applied for estimating the items frequency, matrix factorization [3], finding similar items and also limited numerical linear algebra operation. The popularity of sketching techniques is attributed to its runtime and space efficiencies. We are in particular interested in the count sketch [1].

Count sketch is a data structure originally designed to maintain approximate quantities $(q_{e_1}, \cdots, q_{e_N})$ related to items in $\mathcal{E} = \{e_1, \cdots, e_N\}$ appearing in a datastream, but with sub-linear space complexity. A count sketch is represented by a $k \times w$ matrix $\mathbf{C}$ and two sets of *pairwise independent* hash functions $\{h_j(\cdot), s_j(\cdot)\}_{j=1}^k$. The *address* hash function $h_j(\cdot)$ maps an element of $\mathcal{E}$ to the set $\{1, ..., w\}$ and the *sign* hash function s_j maps an element of $\mathcal{E}$ to $\{+1, -1\}$. Upon the reception at time t of the update $\delta q_{e,t}$ of the quantity related to item e, k entries of matrix $\mathbf{C}$ are updated: $c_{j,h_j(e)} \leftarrow c_{j,h_j(e)} + s_j(e)\delta q_{e,t}$, $1 \leq j \leq k$. Given a query item e, mean or median of $\{s_j(e) \cdot c_{j,h_j(e)}\}_{j=1}^k$ is returned as an approximation of q_e. Both update and query are of $O(k)$ complexity. The accuracy of the estimation is related to the size of the count sketch [2]. The estimate based on the mean operator is unbiased with variance σ^2/wk, where $\sigma^2 = \sum_{e \in \mathcal{E}} q_e^2$. The representational capacity N of count sketch can be controlled by (w, k).

3 Count Sketching for Large Matrix Factorization

3.1 Sketching Vectors

In regular matrix factorization, d-dimensional vectors $\{\mathbf{p}_u\}_{u \in \mathbb{U}}$ and $\{\mathbf{q}_i\}_{i \in \mathbb{I}}$ are stored as dense arrays $\mathbf{P}$ and $\mathbf{Q}$, contiguous in memory. This facilitates indexing on the two dimensional array by increments of d. We propose replacing this matrix representation with a single count sketch. Surprisingly, although user and item vectors carry different semantic, their underlying representations are the same, therefore we store both of them in the same structure, which should provide estimates for $N = d(|\mathbb{U}| + |\mathbb{I}|)$ elements. The storage improvement comes at the cost of increasing the retrieval complexity from $O(d)$ to $O(kd)$ for a d-dimensional vector. The trade-off is acceptable, supported by the observation that memory bound computations are more common than CPU bound computations. Lowering memory requirements also makes it possible to process huge sparse matrices using main memory alone and avoiding out-of-the-core computations, thereby improving run-time as well.

3.2 Sketch Based Factorization

The sketch based online factorization differs from regular online factorization (Sect. 2) in the latent factor retrieval and gradient updates merging. When a new tuple $< u, i, r_{u,i} >$ arrives, the count sketch is queried to approximately reconstruct user and item latent vectors. Both user ID u and component index l, $1 \leq l \leq d$, are used as inputs to the k pairs of address and sign hash functions $(h_j^{u,l} = h_j(u,l), s_j^{u,l} = s_j(u,l))$ to get a mean estimate of the vector component as follows (the same holds for item): $\forall l \in \{1...d\}$

$$\tilde{p}_{u,l} = \frac{1}{k}\sum_{j=1}^{k} s_j^{u,l} \cdot c_{j,h_j^{u,l}}, \quad \tilde{q}_{i,l} = \frac{1}{k}\sum_{j=1}^{k} s_j^{i,l} \cdot c_{j,h_j^{i,l}} \tag{2}$$

Yet for the same user or item, the kd accessed cells in the count sketch structure are not adjacent but at random locations addressed by the hash functions. The estimated rating $\tilde{r}_{u,i} = \tilde{\mathbf{p}}_u^\top \tilde{\mathbf{q}}_i$ is compared with $r_{u,i}$ to get the loss $\mathcal{L}(r_{u,i}, \tilde{r}_{u,i})$ and the derivative $\mathcal{L}'_{u,i}$ wrt $\tilde{r}_{u,i}$. The gradient updates for $\tilde{\mathbf{p}}_u$ and $\tilde{\mathbf{q}}_i$ are just computed as in (1). Then for each component of $\tilde{\mathbf{p}}_u$ (as well as $\tilde{\mathbf{q}}_i$), the k respective cells $\mathbf{C}$ are updated with their sign corrected gradients: $\forall l \in \{1...d\}$

$$\delta c_{j,h_j^{u,l}} = -\eta s_j^{u,l}\left(\mathcal{L}'_{u,i}.\tilde{q}_{i,l} + \lambda\tilde{p}_{u,l}\right), \; \delta c_{j,h_j^{i,l}} = -\eta s_j^{i,l}\left(\mathcal{L}'_{u,i}.\tilde{p}_{u,l} + \lambda\tilde{q}_{i,l}\right) \tag{3}$$

3.3 Approximation and Equivalence

Our approach leads to approximations compared to the original online algorithm. When we update a quantity, *i.e.* $p_{u,l}$ or $q_{i,l}$, the sketching technique inherently modifies this quantity. It means that a write directly followed by a read access of the count sketch sees a modification of the update. This hurts our algorithm twice: In (3), not only $\tilde{p}_{u,l}$ and $\tilde{q}_{i,l}$ are noisy versions of what was maintained along previous iterations, but also $\tilde{r}_{u,i}$ is different from $\hat{r}_{u,i}$. The true update depends on the derivative of the loss $\mathcal{L}(r_{u,i}, \hat{r}_{u,i})$ in (1), a quantity which is not computed in our scheme. Instead, the sketching technique yields a reconstructed loss $\mathcal{L}(r_{u,i}, \tilde{r}_{u,i})$. Our system is no longer linear and it is difficult to see how these double approximations cumulate along with the updates.

We show that optimizing $\mathcal{R}_\lambda(\mathbf{P},\mathbf{Q})$ is indeed equivalent to directly optimizing the count sketch structure: $\arg\min_{\mathbf{C}} \mathcal{R}_\lambda(\tilde{\mathbf{P}}(\mathbf{C}), \tilde{\mathbf{Q}}(\mathbf{C}))$ where $\tilde{\mathbf{P}}(\mathbf{C})$ and $\tilde{\mathbf{Q}}(\mathbf{C})$ are the reconstructed latent vectors (as given by (2)). Let us consider a particular cell $c_{j,m}$ of $\mathbf{C}$. Its update triggered by the observation $< u, i, r_{u,i} >$ is:

$$\delta c_{j,m} = -\eta\left(\frac{\partial R_\lambda(\tilde{\mathbf{P}},\tilde{\mathbf{Q}})}{\partial\tilde{\mathbf{p}}_u}^\top \cdot \frac{\partial\tilde{\mathbf{p}}_u}{\partial c_{j,m}} + \frac{\partial R_\lambda(\tilde{\mathbf{P}},\tilde{\mathbf{Q}})}{\partial\tilde{\mathbf{q}}_i}^\top \cdot \frac{\partial\tilde{\mathbf{q}}_i}{\partial c_{j,m}}\right)$$

The expression of vector $\partial\tilde{\mathbf{p}}_u/\partial c_{j,m}$ is derived from the read access to the count sketch (2): Its l-th component equals $s_j^{u,l} k^{-1} \mathbb{1}_{[h_j^{u,l}==m]}$ (same for $\partial\tilde{\mathbf{q}}_i/\partial c_{j,m}$). In the end, this stems into the following update rules: $\forall l \in \{1, \cdots, d\}$

$$\delta c_{j,m} = -\frac{\eta}{k}((s_j^{u,l}\mathcal{L}'_{u,i}\tilde{q}_{i,l} + \lambda\tilde{p}_{u,l})\mathbb{1}_{[h_j^{u,l}==m]} + (s_j^{i,l}\mathcal{L}'_{u,i}\tilde{p}_{u,l} + \lambda\tilde{q}_{i,l})\mathbb{1}_{[h_j^{i,l}==m]})$$

We find back the same update rules as in (3) up to a factor k^{-1} due to the read access to the count sketch based on the mean operator. However, this is not an issue as the gradient is at the end multiplied by the learning rate η.

4 Experiments

In this section, we benchmark our method against regular online matrix factorization and feature hashing based factorization [3] as it is a special case of our approach ($k = 1$). We use four publicly available datasets: Movielens1M and 10 M, EachMovie and Netflix. Data characteristics, along with results are in Table 1. The data is preprocessed and randomly partitioned into the training, validation and test sets with proportion [0.8, 0.1, 0.1]. User, item and global means are subtracted from rating to remove user and item bias. Ratings with user/item frequency < 10 are removed from the test and validation sets. The same procedure is repeated 10 times to obtain 10 fold dataset. We use root mean square error to measure the quality of recommendations: $RMSE(\mathbf{R}') = \sqrt{(\sum_{r_{u,i} \in \mathbf{R}'} (\tilde{\mathbf{p}}_u^\top \tilde{\mathbf{q}}_i - r_{u,i})^2 / \|\mathbf{R}'\|_0}$ where $\mathbf{R}'$ is the restriction of $\mathbf{R}$ to the testing set. We compare the performance for various configurations (w, k) of the sketch and different latent factor dimensions d. The sketch depth k is picked from $\{1, 4\}$ and the latent factor dimension d is chosen from $\{1, 2, 3, 4, 6, 8, 11, 16, 23, 32\}$. We measure the space gain γ by the ratio of space that the regular factorization would need for the same dimension d to the space actually utilized by sketch based factorization. We vary γ within $\{1, 2, 2.83, 4, 5.66, 8, 11.31, 16, 23, 32\}$. For a given setup (γ, d, k), we determine the sketch width as $w = \left\lceil \frac{(|\mathbb{U}|+|\mathbb{I}|)d}{\gamma k} \right\rceil$. We choose optimal parameters for learning rate η and regularization constant λ by a two stage line search in log-scale, based on validation set prediction score. We also initialize the parameters to uniformly sampled, small random values. We iterate for $T = 20$ epoch over the training set, before predicting on the testing set. Learning rate is scaled down at every iteration using the formula $\eta_t = \frac{\eta}{1+t/T}$.

Table 1. Dataset characteristics and RMSE on various real datasets

Dataset	$\lvert\mathbb{U}\rvert$	$\lvert\mathbb{I}\rvert$	$\lvert\mathbf{R}\rvert$	Regular factorization	Count sketch ($k = 4$)	Feature hashing
MovieLens 1 M	6,040	3,952	1,000,209	0.873	0.876	0.906
EachMovie	61,265	1,623	2,811,718	0.811	0.809	0.818
MovieLens 10 M	69,878	10,677	10,000,054	1.145	1.146	1.159
Netflix	480,136	17,167	96,649,938	0.854	0.855	0.862

4.1 Variation of RMSE with Factors Size and Space Gain

We now evaluate the effect of space gain and dimension d on RMSE measure. Results are displayed as heatmaps for different sketch depth values in Fig. 1a for MovieLens 1M and EachMovie datasets, where horizontal and vertical axes represent d and γ. As expected, the RMSE increases when d decreases, as it impacts the representation capacity of the model, and when space gain γ

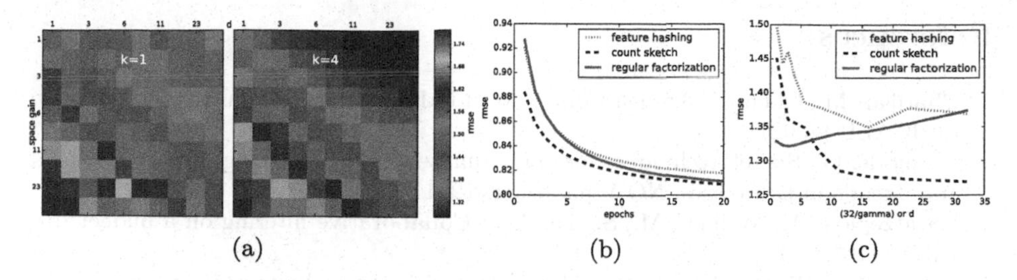

(a) (b) (c)

Fig. 1. (a) Heatmaps of RMSE for feature hashing (left) and count sketch (right) on EachMovie. (b) Convergence on MovieLens 10 M. (c) Dynamic setting on EachMovie.

increases because it implies smaller sketch width w and hence higher variance for $\tilde{\mathbf{p}}_u$ and $\tilde{\mathbf{q}}_i$. We can also observe that there is an improvement in RMSE score with higher k. This effect is more amplified for low γ values.

4.2 Variation of RMSE with Model Size on Dynamic Updates

We first evaluate the convergence of RMSE along the number of epochs on training data. Figure 1b shows the result for MovieLens 10M dataset for the same setting as in Table 1. The three algorithms take the same space. We observe that count sketch converges faster than the other two algorithms. Our explanation is the following: Every new observation $< u, i, r_{u,i} >$ stems into $2kd$ cell updates for the count sketch compared to only $2d$ cell updates for the other two methods, and this for the same space. To simulate a dynamic environment, we do one pass over the training data and report the results on test data. The space gain ranges from 32 to 1 while d_r goes from 1 to 32 in order to maintain the same space between the three methods. Figure. 1c shows that the performance of count sketch is better than other approaches for $d_r > 7$ (or $\gamma < 32/7$). We also observe that with increase in space, the performance of the other two techniques degrades, whereas count sketch saturates. We surmise that count sketch factorization can be more suitable to collaborative filtering systems with dynamic updates as it has better convergence properties.

5 Conclusion

The memory intensive nature of matrix factorization techniques calls for efficient representations of the learned factors. This work investigated the use of count sketch for storing the latent factors. Its compact and controllable representation makes it a good candidate for efficient storage of these parameters. We show that the optimization of the latent factors through the count sketch storage is indeed equivalent to finding the optimal count sketch structure for predicting the observed ratings. Experimental evaluations show the trade-off between performance and space and also reveal that count sketch factorization needs less data for training. This property is very useful in dynamic setting.

References

1. Charikar, M., Chen, K., Farach-Colton, M.: Finding frequent items in data streams. In: ICALP (2002)
2. Cormode, G.: Sketch techniques for approximate query processing. In: Foundations and Trends in Databases. NOW publishers (2011)
3. Karatzoglou, A., Weimer, M., Smola, A.J.: Collaborative filtering on a budget. In: AISTATS (2010)
4. Koren, Y., Bell, R., Volinsky, C.: Matrix factorization techniques for recommender systems. Computer **8**, 30–37 (2009)

Diversifying Search Results Using Time
An Information Retrieval Method for Historians

Dhruv Gupta[1,2] and Klaus Berberich[1(✉)]

[1] Max Planck Institute for Informatics, Saarbrücken, Germany
kberberi@mpi-inf.mpg.de
[2] Saarbrücken Graduate School of Computer Science, Saarbrücken, Germany
dhgupta@mpi-inf.mpg.de

Abstract. Getting an overview of a historic entity or event can be difficult in search results, especially if important dates concerning the entity or event are not known beforehand. For such information needs, users benefit if returned results covered diverse dates, thus giving an overview of what has happened throughout history. Such a method can be a building block for applications, for instance, in *digital humanities*. We describe an approach to diversify search results using temporal expressions (e.g., 1990s) from their contents. Our approach first identifies time intervals of interest to the given keyword query based on pseudo-relevant documents. It then re-ranks query results so as to maximize the coverage of identified time intervals. We present a novel and objective evaluation for our proposed approach. We test the effectiveness of our methods on The New York Times Annotated corpus and the Living Knowledge corpus, collectively consisting of around 6 million documents. Using history-oriented queries and encyclopedic resources we show that our method is able to present search results diversified along time.

1 Introduction

Large born-digital document collections are a treasure trove of historical knowledge. Searching these large longitudinal document collections is only possible if we take into account the temporal dimension to organize them. We present a method for diversifying search results using temporal expressions in document contents. Our objective is to specifically address the information need underlying *history-oriented* queries; we define them to be keyword queries describing a historical event or entity. An ideal list of search results for such queries should constitute a *timeline* of the event or portray the *biography* of the entity. This work shall yield a useful tool for scholars in history and humanities who would like to search large text collections for *history-oriented* queries without knowing relevant dates for them apriori.

No work, to the best of our knowledge, has addressed the problem of diversifying search results using temporal expressions in document contents. Prior approaches in the direction of diversifying documents along time have relied largely on publication dates of documents. However a document's publication

N. Ferro et al. (Eds.): ECIR 2016, LNCS 9626, pp. 789–795, 2016.
DOI: 10.1007/978-3-319-30671-1_65

date may not necessarily be the time that the text refers to. It is quite common to have articles that contain a historical perspective on a past event from the current time. Hence, the use of publication dates is clearly insufficient for history-oriented queries.

In this work, we propose a probabilistic framework to diversify search results using temporal expressions (e.g., 1990s) from their contents. First, we identify time intervals of interest to a given keyword query, using our earlier work [7], which extracts them from pseudo-relevant documents. Having identified time intervals of interest (e.g., [2000,2004] for the keyword query george w. bush), we use them as aspects for diversification. More precisely, we adapt a well-known diversification method [1] to determine a search result that consists of relevant documents which cover all of the identified time intervals of interest.

Evaluation of historical text can be highly subjective and biased in nature. To overcome this challenge; we view the evaluation of our approach from a statistical perspective and take into account an objective evaluation for automatic summarization to measure the effectiveness of our methods. We create a large history-oriented query collection consisting of long-lasting wars, important events, and eminent personalities from reliable encyclopedic resources and prior research. As a ground truth we use articles from *Wikipedia*[1] concerning the queries. We evaluate our methods on two large document collections, the New York Times Annotated corpus and the Living Knowledge corpus. Our approach is thus tested on two different types of textual data. One being highly authoritative in nature; in form of news articles. Another being authored by real-world users; in form of web documents. Our results show that using our method of diversifying search results using time; we can present documents that serve the information need in a history-oriented query very well.

2 Method

Notation. We consider a document collection $\mathcal{D}$. Each document $d \in \mathcal{D}$ consists of a multiset of keywords d_{text} drawn from vocabulary $\mathcal{V}$ and a multiset of temporal expressions d_{time}. Cardinalities of the multisets are denoted by $|d_{text}|$ and $|d_{time}|$. To model temporal expressions such as 1990s where the begin and end of the interval can not be identified, we utilize the work by Berberich et al. [3]. They allow for this uncertainty in the time interval by associating lower and upper bounds on begin and end. Thus, a temporal expression T is represented by a four-tuple $\langle b_l, b_u, e_l, e_u \rangle$ where time interval $[b, e]$ has its begin bounded as $b_l \leq b \leq b_u$ and its end bounded as $e_l \leq l \leq e_u$. The temporal expression 1990s is thus represented as $\langle 1990, 1999, 1990, 1999 \rangle$. More concretely, elements of temporal expression T are from time domain $\mathcal{T}$ and intervals from $\mathcal{T} \times \mathcal{T}$. The number of such time intervals that can be generated is given by $|T|$.

Time Intervals of Interest to the given keyword query q_{text} are identified using our earlier work [7]. A time interval $[b, e]$ is deemed *interesting* if its referred

[1] https://en.wikipedia.org/.

frequently by highly relevant documents of the given keyword query. This intuition is modeled as a two-step generative model. Given, a set of pseudo-relevant documents R, a time interval $[b, e]$ is deemed interesting with probability:

$$P([b,e] \mid q_{text}) = \sum_{d \in R} P([b,e] \mid d_{time}) P(d_{text} \mid q_{text}).$$

To diversify search results, we keep all the time intervals generated with their probabilities in a set q_{time}.

Temporal Diversification. To diversify search results we adapt the approach proposed by Agrawal et al. [1]. Formally, the objective is to maximize the probability that the user sees at least one result relevant to her time interval of interest. We thus aim to determine a query result $S \subseteq R$ that maximizes

$$\sum_{[b,e] \in q_{time}} \left(P([b,e] \mid q_{text}) \left(1 - \prod_{d \in S} (1 - P(q_{text} \mid d_{text}) P([b,e] \mid d_{time})) \right) \right).$$

The probability $P([b,e] \mid q_{text})$ is estimated as described above and reflects the salience of time interval $[b, e]$ for the given query. We make an independence assumption and estimate the probability that document d is relevant and covers the time interval $[b, e]$ as $P(q_{text} \mid d_{text})\, P([b,e] \mid d_{time})$. To determine the diversified result set S, we use the greedy algorithm described in [1].

3 Evaluation

Document Collections. We used two document collections one from a news archive and one from a web archive. The Living Knowledge[2] corpus is a collection of news and blogs on the Web amounting to approximately 3.8 million documents [8]. The documents are provided with annotations for temporal expressions as well as named-entities. The New York Times (NYT) Annotated[3] corpus is a collection of news articles published in *The New York Times*. It reports articles from 1987 to 2007 and consists of around 2 million news articles. The temporal annotations for it were done via SUTime [6]. Both explicit and implicit temporal expressions were annotated, resolved, and normalized.

Indexing. The document collections were preprocessed and subsequently indexed using the ElasticSearch software[4]. As an ad-hoc retrieval baseline and for retrieval of pseudo–relevant set of documents we utilized the state-of-the-art *Okapi-BM25* retrieval model implemented in ElasticSearch.

Collecting History-Oriented Queries. In order to evaluate the usefulness of our method for scholars in history, we need to find keyword queries that are highly

[2] http://livingknowledge.europarchive.org/.
[3] https://catalog.ldc.upenn.edu/LDC2008T19.
[4] https://www.elastic.co/.

ambiguous in the temporal domain. That is multiple interesting time intervals are associated with the queries. For this purpose we considered three categories of history-oriented queries: long-lasting wars, recurring events, and famous personalities. For constructing the queries we utilized reliable sources on the Web and data presented in prior research articles [7,9]. Queries for long-lasting wars were constructed from the *WikiWars* corpus [9]. The corpus was created for the purpose of temporal information extraction. For ambiguous important events we utilized the set of ambiguous queries used in our earlier work [7]. For famous personalities we use a list of most influential people available on the USA Today[5] website. The names of these famous personalities were used based on the intuition that there would important events associated with them at different points of time. The keyword queries are listed in our accompanying technical report [11]. The entire testbed is publicly available at the following url:

http://resources.mpi-inf.mpg.de/dhgupta/data/ecir2016/.

The objective of our method is to present documents that depict the historical timeline or biography associated with keyword query describing event or entity. We thus treat the set of diversified set of documents as a *historical summary* of the query. In order to evaluate this diversified summary we obtain the corresponding *Wikipedia* (see footnote 1) pages of the queries as ground truth summaries.

Baselines. We considered three baselines, with increasing sophistication. As a naïve baseline, we first consider the pseudo-relevant documents retrieved for the given keyword query. The next two baselines use a well-known implicit diversification algorithm *maximum marginal relevance* (MMR) [5]. Formally it is defined as: $\text{argmax}_{d \notin S} \left[\lambda \cdot \text{sim}_1 \left(q, d\right) - \left(1 - \lambda\right) \cdot \text{max}_{d' \in S} \, \text{sim}_2 \left(d', d\right) \right].$ MMR was simulated with sim_1 using query likelihoods and sim_2 using cosine similarity between the term-frequency vectors for the documents. The second baseline considered MMR with $\lambda = 0.5$ giving equal importance to query likelihood and diversity. While the final baseline considered MMR with $\lambda = 0.0$ indicating complete diversity. For all methods the summary is constructed by concatenating all the top-k documents into one large document.

Parameters. There are two parameters to our system: (i) The number of documents considered for generating time intervals of interest $|R|$ and (ii) The number of documents considered for *historical summary* $|S|$. We consider the following settings of these parameters: $|R| \in \{100, 150, 200\}$ and $|S| \in \{5, 10\}$.

Metrics. We use the ROUGE-N measure [12] (implementation[6]) to evaluate the *historical summary* constituted by diversified set of documents with respect to the ground truth. ROUGE-N is a recall-oriented metric which reports the number of n-grams matches between a candidate summary and a reference summary. The n in *ngram* is the length of the gram to be considered; we limit ourselves to $n \in \{1, 3\}$. We report the *recall*, *precision*, and $F_{\beta=1}$ for each ROUGE-N measure.

[5] http://usatoday30.usatoday.com/news/top25-influential.htm.

[6] http://www.berouge.com/Pages/default.aspx.

Table 1. Results for the New York Times Annotated corpus.

	Category	Historical Wars						Historical Events						Historical Entity					
	Metric	R		P		$F_{\beta=1.0}$		R		P		$F_{\beta=1.0}$		R		P		$F_{\beta=1.0}$	
	ROUGE-N	1	3	1	3	1	3	1	3	1	3	1	3	1	3	1	3	1	3
$\lvert R\rvert=100$ $\lvert S\rvert=5$	NAÏVE	30.5	12.0	62.7	23.5	33.9	13.2	43.3	18.0	42.4	15.7	21.0	8.4	19.9	7.9	74.6	29.8	24.4	9.8
	MMR (λ=0.5)	30.5	12.0	62.8	23.6	33.9	13.2	43.3	18.0	42.6	15.6	21.1	8.4	20.0	7.9	74.3	29.6	24.6	9.8
	MMR (λ=0.0)	30.5	12.0	62.8	23.6	33.9	13.2	43.3	18.0	42.6	15.6	21.1	8.4	20.0	7.9	74.3	29.6	24.6	9.8
	TIME-DIVERSE	46.4	17.5	55.7	21.1	41.0	15.5	56.7	22.0	35.9	13.0	26.3	9.9	35.3	13.4	67.0	25.3	34.5	13.1
$\lvert R\rvert=100$ $\lvert S\rvert=10$	NAÏVE	48.0	18.4	51.0	18.9	39.2	15.0	57.6	22.9	33.4	12.0	23.1	8.7	35.4	13.6	67.4	26.7	34.4	13.5
	MMR (λ=0.5)	48.4	18.5	50.6	18.8	39.2	15.0	57.5	22.9	33.4	11.9	23.1	8.7	35.8	13.7	67.2	26.8	34.7	13.6
	MMR (λ=0.0)	48.4	18.5	50.6	18.8	39.2	15.0	57.5	22.9	33.4	11.9	23.1	8.7	35.8	13.7	67.2	26.8	34.7	13.6
	TIME-DIVERSE	64.8	24.4	43.2	16.5	42.6	16.3	66.1	24.3	27.1	8.9	23.1	8.0	48.2	17.8	56.9	21.1	36.8	13.7
$\lvert R\rvert=150$ $\lvert S\rvert=5$	NAÏVE	30.5	12.0	62.7	23.5	33.9	13.2	43.3	18.0	42.4	15.7	21.0	8.4	19.9	7.9	74.6	29.8	24.4	9.8
	MMR (λ=0.5)	30.5	12.0	62.8	23.6	33.9	13.2	43.3	18.0	42.6	15.6	21.1	8.4	20.0	7.9	74.3	29.6	24.6	9.8
	MMR (λ=0.0)	30.5	12.0	62.8	23.6	33.9	13.2	43.3	18.0	42.6	15.6	21.1	8.4	20.0	7.9	74.3	29.6	24.6	9.8
	TIME-DIVERSE	48.2	18.6	55.1	21.1	42.0	16.2	58.1	22.6	33.4	12.2	25.7	9.6	38.0	14.1	65.3	23.9	36.7	13.7
$\lvert R\rvert=150$ $\lvert S\rvert=10$	NAÏVE	48.0	18.4	51.0	18.9	39.2	15.0	57.6	22.9	33.4	12.0	23.1	8.7	35.4	13.6	67.4	26.7	34.4	13.5
	MMR (λ=0.5)	48.5	18.6	50.7	18.8	39.3	15.1	57.5	22.9	33.4	11.9	23.1	8.7	35.7	13.7	67.3	26.8	34.7	13.7
	MMR (λ=0.0)	48.5	18.6	50.7	18.8	39.3	15.1	57.5	22.9	33.4	11.9	23.1	8.7	35.7	13.7	67.3	26.8	34.7	13.7
	TIME-DIVERSE	65.4	24.9	42.1	16.4	42.2	16.3	67.0	24.9	26.4	9.2	23.1	8.1	54.2	20.1	55.7	20.9	40.8	15.5
$\lvert R\rvert=200$ $\lvert S\rvert=5$	NAÏVE	30.5	12.0	62.7	23.5	33.9	13.2	43.3	18.0	42.4	15.7	21.0	8.4	19.9	7.9	74.6	29.8	24.4	9.8
	MMR (λ=0.5)	30.5	12.0	62.8	23.6	33.9	13.2	43.3	18.0	42.6	15.6	21.1	8.4	20.0	7.9	74.3	29.6	24.6	9.8
	MMR (λ=0.0)	30.5	12.0	62.8	23.6	33.9	13.2	43.3	18.0	42.6	15.6	21.1	8.4	20.0	7.9	74.3	29.6	24.6	9.8
	TIME-DIVERSE	51.7	20.0	53.2	20.3	43.7	16.8	59.4	23.0	34.8	12.7	27.7	10.4	39.6	15.2	64.6	23.8	37.6	14.5
$\lvert R\rvert=200$ $\lvert S\rvert=10$	NAÏVE	48.0	18.4	51.0	18.9	39.2	15.0	57.6	22.9	33.4	12.0	23.1	8.7	35.4	13.6	67.4	26.7	34.4	13.5
	MMR (λ=0.5)	48.5	18.6	50.7	18.8	39.3	15.1	57.5	22.9	33.4	11.9	23.1	8.7	35.7	13.7	67.3	26.8	34.7	13.7
	MMR (λ=0.0)	48.5	18.6	50.7	18.8	39.3	15.1	57.5	22.9	33.4	11.9	23.1	8.7	35.7	13.7	67.3	26.8	34.7	13.7
	TIME-DIVERSE	66.4	24.8	38.2	14.3	39.4	14.8	69.5	25.9	25.2	8.8	24.1	8.7	54.7	20.0	54.2	19.5	41.5	15.3

Results. Are shown for three different categories of history-oriented queries per document collection. All values reported are percentages of the metrics and averaged over all the queries in a group. The results for the New York Times Annotated corpus are presented in Table 1 and for the Living Knowledge corpus can be found in our accompanying technical report [11].

For The New York Times Annotated corpus we can clearly see that our method TIME-DIVERSE outperforms all three baselines by a large margin in recalling most important facts concerning the history-oriented queries. This shows that using retrieval method informed by temporal expressions presents documents that are *retrospectively relevant* for history-oriented queries. The slightly higher precision values for baseline system in all the findings above can be attributed to the fact that most of the baseline summaries tended to be of shorter length than the summaries produced by TIME-DIVERSE method. When increasing the size of $|R|$ we notice that recall also increases for TIME-DIVERSE as compared to the baselines. Since the increase in $|R|$ also implies a increase in the length of the summary; the precision also drops.

There is no clear correlation between a *good summary* and the number of top-k documents $|R|$ considered for generating time intervals of interest; in most cases though it seems increasing the size of pseudo-relevant set generation of time intervals hurts the performance of the diversification algorithm. Considering more documents that are presented to the user $|S|$ increases the performance; indicating that $|S| = 10$ for an optimal value.

Overall, the results show that using our diversification algorithm taking into account temporal expressions gives a better overview for history-oriented queries.

4 Related Work

Diversifying search results using time was first explored in [2]. In their preliminary study the authors limited themselves to using document publications dates, but posed the open problem of diversifying search results using temporal expressions in document contents and the challenging problem of evaluation. Both these aspects have been adequately addressed in our article. More recently, Nguyen and Kanhabua [10] diversify search results based on dynamic latent topics. The authors study how the subtopics for a multi-faceted query change with time. For this they utilize a time-stamped document collection and an external query log. However for the temporal analysis they limit themselves to document publication dates. The recent survey of temporal information retrieval by Campos et al. [4] also highlights the lack of any research that addresses the challenges of utilizing temporal expressions in document contents for search result diversification.

5 Conclusion

In this work, we considered the task of diversifying search results by using temporal expressions in document contents. Our proposed probabilistic framework utilized time intervals of interest derived from the temporal expressions present in pseudo-relevant documents and then subsequently using them as aspects for diversification along time. To evaluate our method we constructed a novel testbed of history-oriented queries derived from authoritative resources and their corresponding *Wikipedia* entries. We showed that our diversification method presents a more complete retrospective set of documents for the given history-oriented query set. This work is largely intended to help scholars in history and humanities to explore large born-digital document collections quickly and find relevant information without knowing time intervals of interest to their queries.

References

1. Agrawal, R., et al.: Diversifying search results. In: WSDM (2009)
2. Berberich, K., Bedathur, S.: Temporal diversification of search results. In: TAIA (2013)
3. Berberich, K., Bedathur, S., Alonso, O., Weikum, G.: A language modeling approach for temporal information needs. In: Gurrin, C., He, Y., Kazai, G., Kruschwitz, U., Little, S., Roelleke, T., Rüger, S., van Rijsbergen, K. (eds.) ECIR 2010. LNCS, vol. 5993, pp. 13–25. Springer, Heidelberg (2010)
4. Campos, R.: Survey of temporal information retrieval, related applications. ACM Comput. Surv. **47**(2), 15:1–15:41 (2014)
5. Carbonell, J.G., Goldstein, J.: The use of MMR, diversity-based reranking for reordering documents and producing summaries. In: SIGIR (1998)
6. Chang, A.X., Manning, C.D: A library for recognizing and normalizing time expressions. In: LREC, SUTIME (2012)
7. Gupta, D., Berberich, K.: Identifying time intervals of interest to queries. In: CIKM (2014)

8. Joho, H., et al.: NTCIR temporalia: A test collection for temporal information access research. In: WWW (2014)
9. Mazur, P.P., Dale, R.: A new corpus for research on temporal expressions. In: EMNLP, Wikiwars (2010)
10. Nguyen, T.N., Kanhabua, N.: Leveraging dynamic query subtopics for time-aware search result diversification. In: de Rijke, M., Kenter, T., de Vries, A.P., Zhai, C.X., de Jong, F., Radinsky, K., Hofmann, K. (eds.) ECIR 2014. LNCS, vol. 8416, pp. 222–234. Springer, Heidelberg (2014)
11. Gupta, D., Berberich, K.: Diversifying search results using time. Research Report MPI-I–5-001 (2016)
12. Lin, C.Y.: Rouge: A package for automatic evaluation of summaries. In: ACL (2004)

On Cross-Script Information Retrieval

Nada Naji[1(✉)] and James Allan[2]

[1] College of Computer and Information Science, Northeastern University, Boston, USA
najin@ccs.neu.edu
[2] Center for Intelligent Information Retrieval, College of Information and Computer Sciences, University of Massachusetts Amherst, Amherst, USA
allan@cs.umass.edu

Abstract. We address the problem of cross-script retrieval in the context of a microblog system such as Twitter. Specifically, we explore methods for using native Arabic script queries to retrieve Arabic tweets written in a Roman script known as Arabizi. For example, a query for "كتاب" would not match *"kitab"* even though an Arabic reader would see them as the same word. Moreover, because of the lack of Arabic script, automatic language identification methods fail to recognize the Arabizi text as Arabic and label it as English, Polish, or the like. We propose a cross-script retrieval system using automatic rule-based mapping and statistical selection of transliteration keywords. We show that our system can achieve effective cross-script retrieval with minimal knowledge of the target language and without the need to rely on external translation or transliteration tools or lexica. With minimal human annotation, our technique can be applied to other languages such as Hindi and Greek, which are commonly converted to a Roman character set similarly.

Keywords: Cross-script IR · CSIR · Social media retrieval · Arabic · Arabizi · Cross-language IR · CLIR · Mixed-script IR · MSIR · Transliteration

1 Introduction

The Web contains huge amounts of user-generated text in different writing systems and languages, but most popular platforms lack the mechanism of implicitly cross-matching Romanized versus native script texts. Twitter's language identifiers seem to only attempt to detect a language when written in its native/official character set. While it succeeds at identifying Arabic most of the time, Twitter does not detect nor identify Arabizi tweets as Arabic ones nor does it count Arabizi as a stand-alone language. Therefore, potentially novel and pertinent content is unreachable by simple search. Our proposed method for identifying Arabizi is intended to help with that challenge. The contributions of this paper are the following: (1) We describe an Arabic to Arabizi transliteration that works

N. Naji–This work was done while the author was at the University of Massachusetts Amherst, supported by the Swiss National Science Foundation Early Postdoc.Mobility fellowship project P2NEP2_151940.

N. Ferro et al. (Eds.): ECIR 2016, LNCS 9626, pp. 796–802, 2016.
DOI: 10.1007/978-3-319-30671-1_70

in the absence of lexica and parallel corpora. (2) We develop an approach to evaluate the quality of such a transliterator. (3) We demonstrate that our transliterator is superior to reasonable automatic baselines for identifying valid Arabizi transliterations. (4) We make the annotated data publicly available for future research[1].

2 Related Work

The problem of spelling variation in Romanized Arabic has been studied closely to perform Named Entity Recognition such as Machine Translation (MT) of Arabic names [8] and conversion of English to Arabic [9]. However, and to the best of our knowledge, no work has been done so far on cross-script Information Retrieval (CSIR) for the Arabic language. Some studies addressed dialect identification in Arabic or Arabizi [1, 3–5] and statistical MT from Arabizi to English via de-romanization to Arabic [10]. Arabic to Arabizi conversion has only been done as one-to-one mapping such as Qalam[2] and Buckwalter[3] resulting in Romanized vowel-less text. Darwish [2] uses a Conditional Random Field (CRF) to identify Arabizi from a corpus of mixed English and Arabic tweets with accuracy of 98.5 %. We are typically transcribing single words or short phrases, where the CRF rules do not work well. Gupta et al.'s work on mixed-script IR (MSIR) [6, 7] proposes a query expansion method to retrieve mixed text in English and Hindi using deep learning and achieving a 12 % increase in MRR over other baselines. In contrast to their work, we are using a transliteration-based technique that does not rely on lexica or datasets. Also, we are faced with very short documents lacking the redundancy that can be used to grasp language features. Bies et al. [11] released a parallel Arabic-Arabizi SMS and chat corpus of 45,246 case-sensitive tokens. Although it is a valuable resource, it only covers Egyptian Arabic and Arabizi.

3 Cross-Script Retrieval Task Description

Let q be a query in language l written in script s_1. A CSIR system retrieves documents from a corpus C in language l in response to q, where the documents are written in script s_1 or an alternative script s_2 or both s_1 and s_2, and where s_2 is an alternative writing system for l. The underlying corpus C may consist of documents in n languages and m scripts such that $n \geq 1$ and $m \geq 2$. Our definition of the CSIR problem is analogous to Gupta et al.'s definition of MSIR [6], but in their experimental setup, Gupta et al. focus on bilingual MSIR ($n = 2$ and $m = 2$). We address the problem of a both multi-lingual and multi-scripted corpus ($n \geq 2$, and $m \geq 2$) which is a complex task since vocabulary overlap between different languages is more likely to happen as more languages and more scripts co-exist in the searchable space. We describe our transliteration and statistical selection algorithms below:

[1] https://ciir.cs.umass.edu/downloads/.

[2] Webpage accessed January 3rd 2016, 19:17 http://langs.eserver.org/qalam.

[3] Webpage accessed January 3rd 2016, 19:18 http://languagelog.ldc.upenn.edu/myl/ldc/morph/buckwalter.html.

AR → ARZ Exhaustive Transliterator: We implement our word modeling algorithm to generate Arabizi forms for a given word in Arabic as described below:

1- Perform AR to ARZ mapping for stable consonants (Table 1). For example, ("كتاب") is mapped to "*ktb*". If the mapping is non-unique, enumerate all possible instances and apply the remaining steps to each candidate.

Table 1. Arabic to Arabizi mapping chart. Parenthesized letters are optional. '?' indicates an optional single character depending on the immediate subsequent character

AR	ARZ	AR	ARZ	AR	ARZ	AR	ARZ	AR	ARZ
ا	*a, 2*	خ	*5,(')7('), kh, x*	ش	*sh*	غ	*(')3('), g, gh*	ن	*n*
ب	*b*	د	*d*	ص	*9, s*	ف	*f*	ه	*h*
ت	*t*	ذ	*th, d, z*	ض	*(')9('), d*	ق	*q, g, k, a, 2, 8*	و	*w, o, u*
ث	*th, s, t*	ر	*r*	ط	*6, t*	ك	*k*	ي	*y, i, e*
ج	*j, g, ch*	ز	*z*	ظ	*(')6('), th, z*	ل	*l*	ء	*2 ([aeiou])?*
ح	*7, h*	س	*s*	ع	*3, (')?([aeiou])?*	م	*m*	ة	*t, h, a*

2- Map and handle long vowels, diphthongs and *hamza*: ('و') ,('ي') ,('ى' ,'ا') or ('ء' ,'ئ' ,'آ' ,'أ' ,'إ'), with an option to introduce '2' for hamza either alone or combined with a long vowel. Since ("كتاب") contains the long vowel ('ا') *'a'* is inserted accordingly "*ktab*".

3- Generate possible *tashdeed* (emphasis) instance(s) for the second and subsequent consonants or ('و') or ('ي'), then apply the remaining steps on all enumerated instances. "*kttab*", "*kttabb*", "*kttabb*".

4- Pad consecutive non-emphasis consonants or ('و') or ('ي') with an optional short vowel *(v)* (one of *'a'*, *'e'*, *'i'* *'o'*, or *'u'*). "*k(v)tab*", "*k(v)ttab*", "*k(v)tabb*", "*k(v)ttabb*" → *kitab, kuttab, ktabb, kattabb.*

Steps 3 and 4 allow accounting for the dropped diacritics in Arabic. For example, "مصر" can be found as "مِصْر" ("*misr*") (Egypt) and can also be written as "*masr*", "*m9r*", etc.

Arabizi Keyword Selection: To determine the potential Arabizi forms we need to quantify the adequacy of the elaborately produced transliterations. We propose *K score* which measures the "*Arabiziness*" of the resulting transliterations based on their occurrences and association with certain linguistic features across the corpus based on our hypothesis that if a word is Arabizi, it will frequently occur in the presence of other Arabizi words. In particular, it will occur in the presence of common function words such as stopwords. On the other hand, Arabizi candidates that rarely or never occur with other Arabizi words are likely to be words in other languages rather than Arabizi tokens. In operation, K score is systematically provided with the transliterations generated by our word modeling module then measures the Arabiziness of each input form according to the following algorithm:

1. Term Projection: Given the exhaustive set of Arabizi transliterations (***Word Transliteration***): $WT_{ARZ} = \{WT1_{ARZ}, .., WTn_{ARZ}\}$. For a given single-term Arabic W_{AR} intersect WT_{ARZ} with the set of actually occurring terms using the inverted index: $W_{ARZ} = Ix \cap WT_{ARZ} = \{W1_{ARZ}, W2_{ARZ}, ..., Wn_{ARZ}\}$ where $Ix = \{Ix_1, Ix_2, ..., Ix_N\}$

2. For each transliteration Wi_{ARZ} in W_{ARZ}, find the subset of tweets T_{Wi} that contain Wi_{ARZ} at least once: $T_{Wi} = \{t1_{Wi}, t2_{Wi}, .., tS_{Wi}\}$
3. For each tweet set T_{Wi}, find the union of all the tokens appearing in the tweets' set $T_{WiUnion}$
4. Given a predefined set of Arabizi stopwords *SW*, find the number of stopwords appearing in $T_{WiUnion}$: $K = | T_{WiUnion} \cap SW |$

A higher K value indicates the presence of more Arabizi stopwords in the tweet union when the transliteration form in question appears, hence reflecting more potential Arabiziness. A lower K means that there is less confidence that the word is in Arabizi. For example, let $WT_{ma9r} = \{WT1_{ma9r}, WT2_{ma9r}, ..., WTn_{ma9r}\}$ be the set of Arabizi transliterations of "مصر" generated by our AR → ARZ transliterator such that: WT_{ma9r} = *{"m9r", "ma9r", "masr", "masar", "miser", "misr", "mo9ur", "mu9irr"}*. First, WT_{ma9r} elements are projected against the inverted index's list of words *Ix*. Only *"mo9ur"* doesn't appear in *Ix* and is therefore excluded from the resulting W_{ma9r}. Each transliteration element in W_{ma9r} is then linked to the list of tweets in which it appears and a set of the words appearing in those tweets is formed. Assume that *"masr"* appeared in the following pseudo-tweets: $t1_{masr}$ = *"la fe masr.. ana fe masr delwaty fel beet"*, $t2_{masr}$ = *"salam keef el 2hal f masr"*, $t3_{masr}$ = *"creo que en brasil hay masr argentinos que brasileros"*. Whose term union yields the set: $T_{masrUnion}$ = *{"2hal", "ana", "argentinos", "beet", "brasil", "brasileros", "creo", "delwaty", "el", "en", "f", "fe", "fel", "hay", "keef", "la", "masr", "que", "salam"}*. The last step is to obtain the number of Arabizi stopwords that appear in $T_{masrUnion}$, in this case we have *"el", "f", "fe", "fel", and "la"*. Despite the fact that *"el"* and *"la"* overlap with other languages such as Spanish, the other stopwords do not which makes them distinctive features for Arabizi in this case. Finally, the K score is equal to the number of stopwords in $T_{masrUnion}$, hence $K_{masr} = 5$. The same process is repeated with the other transliterations to obtain their respective K values and the transliterations are then sorted accordingly to reflect their Arabiziness.

4 Evaluation and Discussion

Main corpus: Our dataset comprises around 72 M tweets that we automatically collected via an API over the period between mid-June and mid-July 2014 regardless of language. The content of "text:" was extracted to create an inverted index. **Queries:** We manually generated 50 single-term Arabic queries in neutral and dialectal forms. **Projected corpus:** The set of Arabic single-term queries is provided to our AR → ARZ transliterator, each keyword was then mapped to *n* transliterations (*n* > *1*) which were then *sifted* by term projection against the inverted index. **Relevance judgments and human assessment:** The transliterations are then manually judged by our annotators to determine whether each transliteration is a correct Arabizi transliteration (relevant) or not (non-relevant). Legitimate but non-matching Arabizi words were labeled as *edge* (neither relevant nor non-relevant). To ensure fair and abstract judgment, the annotators had to review the transliterations individually and without seeing the tweets. **Stopwords:**

Definite articles, prepositions, and conjunctions are attached to the word in Arabic script. Surprisingly, Arabizi writers tend to separate such articles from words [2]. We expanded the set of stopwords indicated by Darwish [2] to include more forms with dialectal variants (54 in total).

4.1 Evaluation Methodology and Baselines

Given an Arabic word, a system outputs a ranked list of Arabizi transliterations. For an Arabic word *A*, a system outputs *k* Arabizi words Z_1 to Z_k in ranked order. Our evaluation corpus has the complete list of correct Arabizi words, $Y = \{Y_1, \ldots, Y_m\}$. We calculate the well-known average precision (AP) measure. We average this value for all words in the test dataset to determine the system's MAP or mean AP score. We also provide standard interpolated recall/precision graphs and measure the reciprocal rank (RR) of the first valid Arabizi word in the ranked list. If Z_i is the best-ranked Arabizi word that is in *Y*, then the RR for that Arabic word is $1/i$. We average this score over all queries to determine MRR, the mean RR. We provide the following baselines to demonstrate that the K score-based approach is an improvement on obvious solutions to this task. **AllHuman** where only annotator-selected candidates are included. Since these are by definition correct, these results are perfect. (They are provided primarily for verification). **1stHuman** is a human-generated baseline, wherein we used the single best Arabizi transliteration for each Arabic word as provided by the pool of annotators. The remaining baselines are automatically generated: **allCommon** includes all Arabizi candidates generated as part of the algorithm described earlier. They are ordered by the number of tweets in which they appear. **1stCommon** is the first item from allCommon. We also evaluate a number of approaches: **K score** which is the set of all candidates ranked by the value of K (see Arabizi Keyword Selection) and **+K SW** which is the same as the K score, except that any Arabizi candidate that has fewer than *K* stopwords is discarded.

4.2 Results and Discussion

Our results are shown in Table 2 which reports the MRR and MAP values. As expected, allHuman performs perfectly. The allCommon run is our operational baseline. The K score results show that ranking by overlap of stopwords improves results: MAP increases from 56.28 % to 64.18 %, an almost 8 % absolute gain and a 14 % relative improvement over allCommon. The top-ranked choice improves with MRR increasing by just over 7 % absolute, or almost 11 % relative. We originally hypothesized that very low stopword overlap may indicate that a word is unlikely to be Arabizi. Dropping all terms with zero overlap (+1SW) causes a large drop in MAP and a modest drop in MRR. Each successful drop of candidates lowers both scores consistently. It seems that a weak (in terms of K score) match is better than no match at all. Both K score and +1SW returned matches for all 50 queries. However, K score clearly outperforms +1SW as it always returns relevant matches with 58 % percent of the time at ranks as early as the first one. The degradation in performance is proportional to the cutoff value *K*. A close examination of the results shows that unanswered queries are experienced starting at +2SW and gradually worsens as *K* increases (Fig. 1). The K score run is the second highest run at low recall and it maintains the highest precision across all levels of recall. As expected, the Buckwalter representation does not constitute a suitable real-life Arabizi transliteration system as can be seen from Table 2.

Table 2. K score and two baselines evaluation. * and † denote statistically significant difference with respect to allCommon and K score runs (two-tailed t-test, $\alpha = 5$ %)

	System	MAP	MRR
Baseline	1stComm	0.1051	0.5600
	allComm	0.5628	0.6757
K Score		0.6418*	0.7487
Human	1stHuman	0.2137	1.0000
	allHuman	1.0000	1.0000
Buckwalter		0.0424*†	0.3000*†

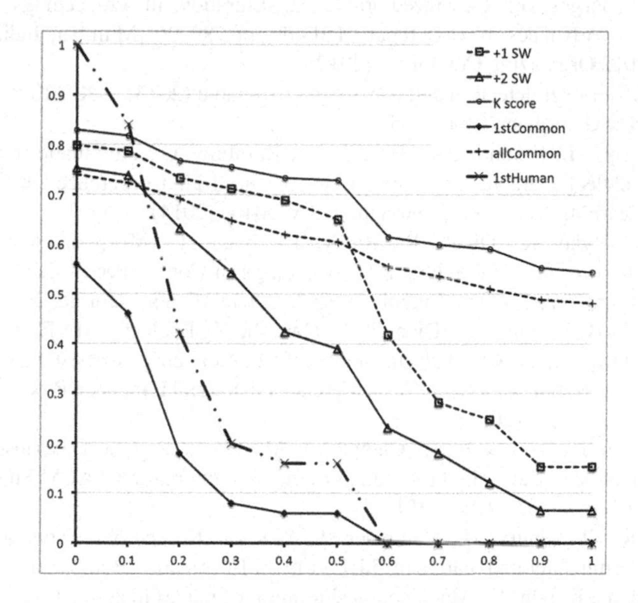

Fig. 1. Interpolated Precision-Recall curves for the K score and CSIR baselines.

5 Conclusion and Future Work

Our system can be seen as a module that existing search engines can integrate into their retrieval pipeline to cater for languages that are alternatively Romanized such as Arabic, Hindi, Russian, and the like. By doing so, relevant transliterated documents will be retrieved at an average rank as early as the second or first as opposed to not being retrieved at all. We plan to extend this work to handle multi-term queries, inflectional and morphological variants and attached articles and pronouns. We believe that it is fairly feasible to implement our work on other Romanizable languages given our preliminary work in other languages, in which non-linguist Arabizi users were able to cover about 80 % of the mapping and conversion rules within a reasonably short amount of time (less than 30 min) as opposed to the creation of parallel corpora – which is far more costly and time-consuming.

Acknowledgements. This work is supported by the Swiss National Science Foundation Early Postdoc.Mobility fellowship project P2NEP2_151940 and is supported in part by the Center for Intelligent Information Retrieval. Any opinions, findings and conclusions or recommendations expressed in this material are those of the authors and do not necessarily reflect those of the sponsor.

References

1. Chalabi, A., Gerges, H.: Romanized arabic transliteration. In: Proceedings of the Second Workshop on Advances in Text Input Methods, pp. 89–96 (Mumbai, India, 2012). The COLING 2012 Organizing Committee (2012)
2. Darwish, K.: Arabizi detection and conversion to Arabic (2013). arXiv:1306.6755 [cs.CL], arXiv. http://arxiv.org/abs/1306.6755
3. Al-Badrashiny, M., Eskander, R., Habash, N., Rambow, O.: Automatic transliteration of romanized dialectal arabic. In: Proceedings of the 18th Conference on Computational Language Learning (Baltimore, Maryland USA, 2014) (2014)
4. Habash, N., Ryan, R., Owen, R., Ramy, E., Nadt, T.: Morphological analysis and disambiguation for dialectal arabic. In: Proceedings of Conference of the North American Association for Computational Linguistics (NAACL) (Atlanta, Georgia, 2013) (2013)
5. Arfath, P., Al-Badrashiny, M., Diab, T.M., Habash, N., Pooleery, M., Rambow, O., Roth, M.R., Altantawy, M.: DIRA: Dialectal arabic information retrieval assistant. demo paper. In: Proceedings of the International Joint Conference on Natural Language Processing (IJCNLP) (Nagoya, Japan, 2013) (2013)
6. Gupta, P., Bali, P., Banchs, E., R., Choudhury, M., Rosso, P.: Query expansion for mixed-script information retrieval. In: Proceedings of the 37th International ACM SIGIR 2014. New York, NY, USA, pp. 677–686 (2014)
7. Saha Roy, R., Choudhury, M., Majumder, P., Agarwal, K.: Overview and datasets of FIRE 2013 track on transliterated search. In: 5th Forum for Information Retrieval Evaluation (2013)
8. Al-Onaizan, Y., Knight, K.: Machine transliteration of names in arabic text. In: Proceedings of the ACL Workshop on Computational Approaches to Semitic Languages (2002)
9. AbdulJaleel, N., Larkey, S.L.: Statistical transliteration for English-Arabic cross language information retrieval. In: Proceedings of the 12th International Conference on Information and Knowledge Management (CIKM 2003). ACM (New York, NY, USA, 2003), pp. 139–146 (2003)
10. May, J., Benjira, Y., Echihabi, A.: An arabizi-english social media statistical machine translation system. In: Proceedings of the Eleventh Biennial Conference of the Association for Machine Translation in the Americas, Vancouver, Canada (2014)
11. Bies, A., Song, Z., Maamouri, M., Grimes, S., Lee, H., Wright, J., Strassel, S., Habash, N., Eskander, R., Rambow, O.: Transliteration of arabizi into arabic orthography: developing a parallel annotated arabizi-arabic script SMS/Chat corpus. In: Proceedings of the EMNLP 2014 Workshop on Arabic Natural Langauge Processing (ANLP), pp. 93–103 (Doha, Qatar, 2014)

LExL: A Learning Approach for Local Expert Discovery on Twitter

Wei Niu(✉), Zhijiao Liu, and James Caverlee

Department of Computer Science and Engineering,
Texas A&M University, College Station, TX, USA
{wei,lzj,caverlee}@cse.tamu.edu

Abstract. In this paper, we explore a geo-spatial learning-to-rank framework for identifying local experts. Three of the key features of the proposed approach are: (i) a learning-based framework for integrating multiple factors impacting local expertise that leverages the fine-grained GPS coordinates of millions of social media users; (ii) a location-sensitive random walk that propagates crowd knowledge of a candidate's expertise; and (iii) a comprehensive controlled study over AMT-labeled local experts on eight topics and in four cities. We find significant improvements of local expert finding versus two state-of-the-art alternatives.

1 Introduction

Identifying *experts* is a critical component for many important tasks. For example, the quality of movie recommenders can be improved by biasing the underlying models toward the opinions of experts [1]. Making sense of information streams – like the Facebook newsfeed and the Twitter stream – can be improved by focusing on content contributed by experts. Along these lines, companies like Google and Yelp are actively soliciting *expert reviewers* to improve the coverage and reliability of their services [7].

Indeed, there has been considerable effort toward expert finding and recommendation, e.g., [2,3,6,10,11]. These efforts have typically sought to identify *general topic experts* – like the best Java programmer on github – often by mining information sharing platforms like blogs, email networks, or social media. However, there is a research gap in our understanding of *local experts*. Local experts, in contrast to general topic experts, have specialized knowledge focused around a particular location. Note that a local expert in one location may not be knowledgeable about a different location. To illustrate, consider the following two local experts:

- A "health and nutrition" local expert in Chicago is someone who may be knowledgeable about Chicago-based pharmacies, local health providers, local health insurance options, and markets offering specialized nutritional supplements or restricted diet options (e.g., for gluten allergies or strictly vegan diets).
- An "emergency response" local expert in Seattle is someone who could connect users to trustworthy information in the aftermath of a Seattle-based disaster, including evacuation routes and the locations of temporary shelters.

N. Ferro et al. (Eds.): ECIR 2016, LNCS 9626, pp. 803–809, 2016.
DOI: 10.1007/978-3-319-30671-1_71

Identifying local experts can improve location-based search and recommendation, and create the foundation for new crowd-powered systems that connect people to knowledgeable locals. Compared to general topic expert finding, however, there has been little research in uncovering these local experts or on the factors impacting local expertise.

Hence, we focus on developing robust models of *local expertise*. Concretely, we propose and evaluate a geo-spatial learning-to-rank framework called **LExL** for identifying local experts that leverages the fine-grained GPS coordinates of millions of Twitter users and their relationships in Twitter lists, a form of crowdsourced knowledge. The framework investigates multiple classes of features that impact local expertise including: (i) user-based features; (ii) list-based features; (iii) local authority features; and (iv) features based on a location-sensitive random walk that propagates crowd knowledge of a candidate's expertise.

Through a controlled study over Amazon Mechanical Turk, we find that the proposed local expert learning approach results in a large and significant improvement in Precision@10, NDCG@10, and in the average quality of local experts discovered versus two state-of-the-art alternatives. Our findings indicate that careful consideration of the relationships between the location of the query, the location of the crowd, and the locations of expert candidates can lead to powerful indicators of local expertise. We also find that high-quality local expert models can be built with fairly compact features.

2 Learning Approach to Local Expert Finding

In this section, we introduce the learning approach framework for finding local experts – **LExL**: **L**ocal **Ex**pert **L**earning. Given a query, composed of a topic and a location, the goal of LExL is to identify high-quality local experts. We assume there is a pool of local expert candidates $V = \{v_1, v_2, ..., v_n\}$, each candidate is described by a matrix of topic-location expertise scores (e.g., column i is Seattle, while row j is "web development"), and that each matrix element indicates to what extent the candidate is an expert on the corresponding topic in the corresponding location. Given a query q that includes both a topic t and a location l, our goal is to find the set of k candidates with the highest local expertise in query topic t and location l. For example, find the top experts on $t_q =$ "web development" in $l_q =$ Seattle, WA.

Learning Approach. We propose to address the local expert ranking problem with a supervised learning-to-rank framework that can combine any number of local expertise features, using a tool such as LambdaMART [12]. The basic idea of LambdaMART is to train an ensemble of weak models and to linearly combine the prediction of them into a final model which is more accurate. But what features should we investigate?

We propose four classes of features that potentially contribute to local topic expertise of a user. Compared to much existing work on expert finding that is content based (e.g., [9]), we focus on features that are independent of what a candidate has posted, instead relying on activity and network based features.

We anticipate integrating content-based features in our future work. In total, we focus on 25 features. Here we briefly introduce the features we used.

User-Based Features. This group of features capture user-oriented aspects that are independent of the query topic and query location. Three aspects of information are considered. User's network (for example, number of follower), user's activity (for example, number of status), and longevity (for example, how long the user has joined).

List-Based Features. We extract expertise evidence directly from the Twitter list, but ignoring the geo-spatial features of the lists (those aspects are part of the following two groups of features). Twitter lists have been recognized as a strong feature of expertise in previous work [6]. In particular, lists can shed light on a candidate from two perspectives: appearing on list and maintaining the list. We also defined a feature to characterise the quality of the list.

Local Autority Features. These features focus on the local authority of candidates revealed through the geo-located Twitter lists. The main idea is to capture the "localness" of these lists. Intuitively, a candidate who is well-recognized near a query location is considered more locally authoritative. We measure the local authority of a candidate in multiple ways, like Haversine distances among candidate, labeler and query location. We also adopt Candidate Proximity and Spread-Based Proximity as two features [4].

Distance-Biased Random Walk Features. We introduce a set of features that incorporate additional network context beyond these one-hop relationships in local authority features. Concretely, we explore features based on a random walk model that directly incorporates the query location, the location of a candidate expert, and the location of external evidence of a candidate's expertise (e.g., in the case of the Twitter lists, the location of the list labeler). The main intuition is to bias a random walker according to the distances between these different components (the query location, the labeler, the candidate) for propagating local expertise scores. In this way, each candidate can be enriched by the network formed around them via Twitter lists. We have a total of 7 features by considering distance among candidate, labeler and query location.

3 Evaluation

In this section, we present the experimental setup, including the collection of ground truth data via AMT, alternative local expert ranking methods, and metrics for comparing these methods, followed by experimental results and analysis.

3.1 Experimental Setup

Our experiments rely on the dataset described in [4], totaling 15 million list relationships in which the coordinates of labeler and candidate are known.

Queries. We adopt a collection of eight topics and four locations that reflect real information needs. The topics are divided into broader local expertise topics – "food", "sports", "business", and "health" – and into more specialized local expertise topics which correspond to each of the broader topics – "chefs", "football", "entrepreneur", and "healthcare". The locations are New York City, San Francisco, Houston and Chicago, which all have relatively dense coverage in the dataset for testing purposes. For each method tested below, we retrieve a set of candidates for ranking based on topics derived from list names.

Proposed Method: Local Expert Learning (LExL). There are a wide variety of learning-to-rank approaches possible; in this paper we evaluate four popular learning-to-rank strategies: Ranknet, MART, Random Forest and LambdaMART. We use an open source implementation of these methods in the RankLib toolkit. For each topic, we randomly partition the collected candidates together with their four categories of features into two equal-sized groups for training and testing. We use four-fold cross validation for reporting the results. We compare our proposed approach with two state of the art approaches for finding local experts: Cognos+ [6] and LocalRank [4].

Ground Truth. Since there is no publicly-available data that directly specifies a user's local expertise given a query (location + topic), we employ human raters (turkers) on Amazon Mechanical Turk to rate the level of local expertise for candidates via human intelligent tasks (HITs). In total, we collect 16 k judgments about user's local topic expertise in a scale of 5 (0–4) across the eight topics and four locations. To explore the validity of turker judgments, we measure the kappa statistic [5], where a value of 0.46 on average means "moderate agreement."

Evaluaton Metrics. To evaluate the quality of local expertise approaches, we adopt two well-known metrics Precision@k, NDCG@k. We also use Rating@k for a query pair to measure the average local expertise rating by the turkers for the top-k experts output by each approach, defined as: $R@k = \sum_{i=1}^{k} rating(c_i, q)/k$, where c is candidate and q is the query pair. In our scenario, we set k = 10.

3.2 Results

Comparison Versus Baselines. We begin by comparing the proposed learning method (LExL) versus the two baselines. Figure 1 shows the Precision@10, Recall@10, and NDCG@10 of each method averaged over all queries.[1] We consider the LambdaMART version of LExL, in addition to methods using Ranknet, MART and Random Forest. First, we observe that three versions of LExL clearly outperform all alternatives, resulting in a Precision@10 of around 0.78, an average Rating@10 of more than 3, and an NDCG of around 0.8.

[1] Note that the results reported here for LocalRank differ from the results in [4] as the experimental setups are different. First, our rating has 5 scales, which is intended to capture more detailed expertise level. Second, [4] only considers ideal ranking order for the top 10 results from LocalRank when calculating maximum possible (ideal) DCG@10, while we consider a much larger corpus.

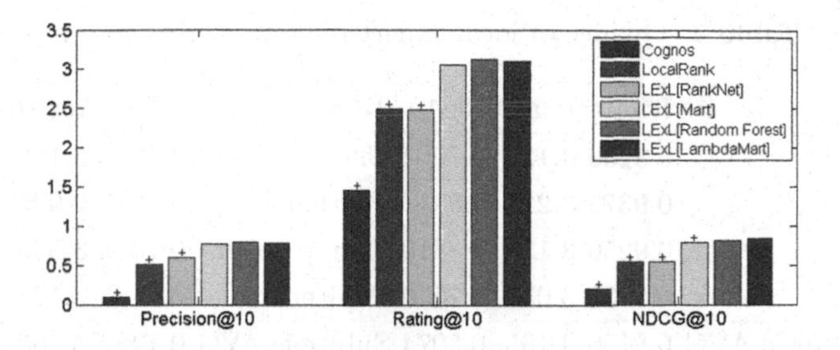

Fig. 1. Evaluating the proposed learning-based local expertise approach versus two alternatives. '+' marks statistical significant difference with LExL[LambdaMART] according to paired t-test at significance level 0.05 (Colour figure online).

Cognos has been shown to be effective at identifying topic experts. However, we see even a modified version to include distance factor is not compatible with local expert finding. For example, Cognos may identify a group of "healthcare" experts known nationwide, but it has difficulty uncovering local experts.

LocalRank has a much better Precision@10 of around 0.5 compared to Cognos+, which indicates that 50 percent of the candidates it identifies have at least "some local expertise" for the query. The average Rating@10 is 2.49, which means the candidates are generally rated between "a little expertise" and "some expertise". Since LocalRank explicitly builds on both topical and local signals (by exploiting the distance between a candidate's labelers and the query location), it performs much better than Cognos+. However, LocalRank is only a linear combination of these two factors, and so does not exploit either additional factors (like the random walk presented in this paper) nor take advantage of a learning approach for optimizing the weighting of these factors.

For the four LExL approaches, Ranknet performs comparably to LocalRank, but the remaining three all result in significantly better performance, with both Random Forest and LambaMART achieving comparably good results. These two methods have a Rating@10 of around 3.1, indicating that the local experts discovered have from "some local expertise" to "extensive local expertise". The Precision@10 and NDCG@10 also support the conclusion that these learning-based methods result in high-quality local experts. Since LambdaMART is significantly less computationally expensive (~1/6 of the computing time of Random Forest), we adopt it for the remainder of the paper.

Effectiveness Across Topics and Locations. We next turn to comparing the effectiveness of LExL with LamdaMART across topics.

We observe in Table 1 that NDCG@10 is consistently high for the four general topics, with an average value of 0.8074. Precision@10 and Rating@10 are also consistent for general topics except for the topic of "health" which has relatively low values. We attribute this poor showing due to data sparsity: through manual inspection we find that there are inherently only a limited number of candidates with high local expertise for the "health" topic in the training and

Table 1. Quality of local expert rankings across topics

Topics	P@10	R@10	N@10	Topics	P@10	R@10	N@10
Food	0.8250	3.125	0.7004	Chefs	0.8250	3.163	0.8554
Sports	0.9375	3.225	0.8913	Football	0.7220	2.925	0.8820
Business	0.9250	3.131	0.8810	Entrepreneurs	0.7333	3.040	0.7768
Health	0.5750	3.059	0.7570	Healthcare	0.7125	3.125	0.9423
General topic AVG	0.8156	3.135	0.8074	Subtopic AVG	0.7482	3.063	0.8641

testing datasets. However, since the learning framework is effective at identifying even those few local experts in "health", we see a high NDCG@10.

We observe comparable results for the four narrower topics. The Precision@10 is lower than for the general topics (0.74 versus 0.82), but the NDCG@10 is higher (0.86 versus 0.81). Part of the higher NDCG results may be attributed to the decrease in the denominator of NDCG for these narrower topics (the Ideal DCG), so the ranking method need only identify some of a pool of moderate local experts rather than identify a few superstar local experts.

4 Conclusion

In this paper, we have proposed and evaluated a geo-spatial learning-to-rank framework for identifying local experts that leverages the fine-grained GPS coordinates of millions of Twitter user and carefully curated Twitter list data. We introduced four categories of features for learning model, including a group of location-sensitive graph random walk features that captures both the dynamics of expertise propagation and physical distances. Through extensive experimental investigation, we find the proposed learning framework produces significant improvement compared to previous methods.

References

1. Amatriain, X., Lathia, N., Pujol, J.M., et al.: The wisdom of the few: a collaborative filtering approach based on expert opinions from the web. In: SIGIR (2009)
2. Balog, K., Azzopardi, L., De Rijke, M.: Formal models for expert finding in enterprise corpora. In: SIGIR (2006)
3. Campbell, C.S., Maglio, P.P., et al.: Expertise identification using email communications. In: CIKM (2003)
4. Cheng, Z., Caverlee, J., Barthwal, H., Bachani, V.: Who is the barbecue king of texas?: a geo-spatial approach to finding local experts on twitter. In: SIGIR (2014)
5. Fleiss, J.L., et al.: Large sample standard errors of kappa and weighted kappa. Psychol. Bull. **72**(5), 323 (1969)
6. Ghosh, S., Sharma, N., et al.: Cognos: crowdsourcing search for topic experts in microblogs. In: SIGIR (2012)
7. Google: overview of local guides (2015). https://goo.gl/NFS0Yz

8. Kleinberg, J.M.: Authoritative sources in a hyperlinked environment. J. ACM **46**(5), 604–632 (1999)
9. Lappas, T., Liu, K., Terzi, E.: A survey of algorithms and systems for expert location in social networks. In: Aggarwal, C.C. (ed.) Social Network Data Analytics. Springer, Heidelberg (2011)
10. Pal, A., Counts, S.: Identifying topical authorities in microblogs. In: WSDM (2011)
11. Weng, J., Lim, E.P., Jiang, J., He, Q.: Twitterrank: finding topic-sensitive influential twitterers. In: WSDM (2010)
12. Wu, Q., Burges, C.J., Svore, K.M., Gao, J.: Ranking, boosting, and model adaptation. Technical report, MSR-TR-2008-109 (2008)

Clickbait Detection

Martin Potthast[(✉)], Sebastian Köpsel, Benno Stein, and Matthias Hagen

Bauhaus-Universität Weimar, Weimar, Germany
{martin.potthast,sebastian.koepsel,benno.stein,
matthias.hagen}@uni-weimar.de

Abstract. This paper proposes a new model for the detection of clickbait, i.e., short messages that lure readers to click a link. Clickbait is primarily used by online content publishers to increase their readership, whereas its automatic detection will give readers a way of filtering their news stream. We contribute by compiling the first clickbait corpus of 2992 Twitter tweets, 767 of which are clickbait, and, by developing a clickbait model based on 215 features that enables a random forest classifier to achieve 0.79 ROC-AUC at 0.76 precision and 0.76 recall.

1 Introduction

Clickbait refers to a certain kind of web content advertisement that is designed to entice its readers into clicking an accompanying link. Typically, it is spread on social media in the form of short teaser messages that may read like the following examples:

- A Man Falls Down And Cries For Help Twice. The Second Time, My Jaw Drops
- 9 Out Of 10 Americans Are Completely Wrong About This Mind-Blowing Fact
- Here's What Actually Reduces Gun Violence

When reading such and similar messages, many get the distinct impression that something is odd about them; something unnamed is referred to, some emotional reaction is promised, some lack of knowledge is ascribed, some authority is claimed. Content publishers of all kinds discovered clickbait as an effective tool to draw attention to their websites. The level of attention captured by a website determines the prize of displaying ads there, whereas attention is measured in terms of unique page impressions, usually caused by clicking on a link that points to a given page (often abbreviated as "clicks"). Therefore, a clickbait's *target link* alongside its *teaser message* usually redirects to the sender's website if the reader is afar, or else to another page on the same site. The content found at the linked page often encourages the reader to share it, suggesting clickbait for a default message and thus spreading it virally. Clickbait on social media has been on the rise in recent years, and even some news publishers have adopted this technique. These developments have caused general concern among many outspoken bloggers, since clickbait threatens to clog up social media channels, and since it violates journalistic codes of ethics.

N. Ferro et al. (Eds.): ECIR 2016, LNCS 9626, pp. 810–817, 2016.
DOI: 10.1007/978-3-319-30671-1_72

In this paper, we present the first approach to automatic clickbait detection. Our contributions are twofold: (1) we collect and annotate the first publicly available clickbait corpus of 3000 Twitter tweets, sampled from the top Twitter publishers, and (2) we develop and evaluate the first clickbait detection model. After discussing related work in Sects. 2 and 3 reports on corpus construction, Sect. 4 on our clickbait model, and Sect. 5 on its evaluation.

2 Related Work

The rationale why clickbait works is widely attributed to teaser messages opening a so-called "curiosity gap," increasing the likelihood of readers to click the target link to satisfy their curiosity. Loewenstein's information-gap theory of curiosity [19] is frequently cited to provide a psychological underpinning (p. 87): "the information-gap theory views curiosity as arising when attention becomes focused on a gap in one's knowledge. Such information gaps produce the feeling of deprivation labeled curiosity. The curious individual is motivated to obtain the missing information to reduce or eliminate the feeling of deprivation." Loewenstein identifies stimuli that may spark involuntary curiosity, such as riddles or puzzles, event sequences with unknown outcomes, expectation violations, information possessed by others, or forgotten information. The effectiveness by which clickbait exploits this cognitive bias results from data-driven optimization. Unlike with printed front page headlines, for example, where feedback about their potential contribution to newspaper sales is indirect, incomplete, and delayed, clickbait is optimized in real-time, recasting the teaser message to maximize click-through [16]. Some companies allegedly rely mostly on clickbait for their traffic. Their success on social networks recently caused Facebook to take action against clickbait as announced by El-Arini and Tang [8]. Yet, little is known about Facebook's clickbait filtering approach; no corresponding publications have surfaced. El-Arini and Tang's announcement mentions only that context features such as dwell time on the linked page and the ratio of clicks to likes are taken into account.

To the best of our knowledge, clickbait has been subject to research only twice to date, both times by linguists: first, Vijgen [26] studies articles that compile lists of things, so-called "listicles." Listicles are often under suspicion to be clickbait. The authors study 720 listicles published at BuzzFeed in two weeks of January 2014, which made up about 30 % of the total articles published in this period. The titles of listicles, which are typically shared as teaser messages, exert a very homogeneous structure: all titles contain a cardinal number—the number of items listed—and 85 % of the titles start with it. Moreover, these titles contain strong nouns and adjectives to convey authority and sensationalism. Moreover, the main articles consistently achieve easy readability according to the Gunning fog index [10]. Second, Blom and Hansen [3] study phoricity in headlines as a means to arouse curiosity. They analyze 2000 random headlines from a Danish news website and identify two common forms of forward-references: discourse deixis and cataphora. The former are references at discourse level ("*This* news will blow your mind".), and the latter at phrase level ("*This* name is hilarious".). Based on a dictionary of basic deictic and cataphoric expressions, the share of such phoric expressions at 10 major Danish news websites reveals that they occur mostly in commercial, ad-funded, and tabloid news websites. However, no detection approach is proposed.

Besides, some dedicated individuals have taken the initiative: Gianotto [9] implements a browser plugin that transcribes clickbait teaser messages based on a rule set so that they convey a more "truthful," or rather ironic meaning. We employ the rule set premises as features and as a baseline for evaluation. Beckman [2], Mizrahi [20], Stempeck [24], and Kempe [15] manually re-share clickbait teaser messages, adding spoilers. Eidnes [7] employs recurrent neural networks to generate nonsense clickbait for fun.

3 A Twitter Clickbait Corpus

To sample our corpus, we focus on Twitter as a social media platform used by many content publishers. To obtain an unbiased choice of publishers, we sample from the top 20 most prolific publishers on Twitter as determined by their influence in terms of re-tweets. Table 1 (left) overviews these publishers. Well-known English-speaking newspapers are among them, but also publishers which have been pointed out for making excessive use of clickbait, including Business Insider [11], the Huffington Post [20], and BuzzFeed [1]; BuzzFeed has publicly opposed the allegations [23]. BBC News has been the most prolific publisher throughout 2014, increasing their number of re-tweets steadily from 2.7 million in January to more than 3.7 million in December for a total of 39.6 million. The New York Times comes in second with a total of 23.8 million. On third rank, the online-only news publisher Mashable is listed, showing that these companies compete with traditional media.

For our corpus, we collected tweets sent by the publishers in week 24 of 2015 that included links, as shown in Table 1 (right). We randomly sampled 150 tweets per publisher for a total of 2992 tweets (one publisher sent only 142 tweets in that time). Each tweet was annotated independently by three assessors who rated them being clickbait or not. Judgments were made only based on the tweet's plain text and image (i.e., the teaser message), and not by clicking on links. We obtain a "fair" inter-annotator agreement with a Fleiss' κ of 0.35. Taking the majority vote as ground truth, a total of 767 tweets (26 %) are considered clickbait. Table 1 (right, column "Clickbait") shows the distribution of clickbait across publishers. According to our annotation, Business Insider sends 51 % clickbait, followed by Huffington Post, The Independent, BuzzFeed, and the Washington Post with more than 40 % each. Most online-only news publishers (Business Insider, Huffington Post, BuzzFeed, Mashable) send at least 33 % clickbait, Bleacher Report being the only exception with a little less than 10 %. TV networks (CNN, NBC, ABC, Fox) are generally at the low end of the distribution. Altogether, these figures suggest that all of the top 20 news publishers employ clickbait on a regular basis, supporting the allegations raised by bloggers.

Table 1. Left: Top 20 publishers on Twitter according to NewsWhip [21] in 2014. The darker a cell, the more prolific the publisher; white cells indicate missing data. Right: Our clickbait corpus in terms of tweets with links posted in week 24, 2015, tweets sampled for manual annotation, and tweets labeled as clickbait (absolute and relative) by majority vote of three assessors.

Publisher	Twitter re-tweets ($\times 10^6$)												Σ	Clickbait corpus			
	Jan	Feb	Mar	Apr	May	Jun	Jul	Aug	Sep	Oct	Nov	Dec	2014	Tweets	Sample	Clickbait	
BBC News	2.70	2.48	2.71	2.87	3.12	3.25	3.56	3.39	3.79	4.02	3.96	3.75	39.6	694	150	25	17%
New York Times		1.28	1.84	2.00	2.11	2.28	2.48	2.35	2.42	2.60	2.22	2.18	23.8	875	150	32	21%
Mashable	1.42	1.46	1.66	1.83	1.77	1.83	1.95	1.66	1.86	1.82	1.78	1.60	20.6	803	150	49	33%
ABC News	0.79	1.15	1.76	1.56	1.62	1.80	1.91	1.68	1.67	1.36	1.28	1.06	17.6	279	150	13	9%
CNN	1.18	1.16	1.21	1.25	1.25	1.17	1.39	1.31	1.35	1.53	1.27	0.97	15.0	345	150	25	17%
The Guardian		1.07	1.16	1.13	1.23	1.32	1.42	1.27	1.37	1.51	1.35	1.19	14.0	744	150	22	15%
Huffington Post	0.96	0.72	0.85	0.77	0.83	0.93	1.14	1.14	1.12	1.17	1.02	0.90	11.6	770	150	69	46%
Forbes	0.80	0.81	0.96	0.89	0.96	1.03	1.11	1.13	1.18	1.10	0.78	0.75	11.5	721	150	57	38%
Bleacher Report	0.58	0.57	0.68	0.72	0.74	0.84	0.84	0.83	1.09	1.12	1.18	1.04	10.2	196	150	13	9%
Fox News	0.59	0.54	0.68	0.82	0.83	0.92	0.95	0.96	0.97	1.04	0.99	0.87	10.2	378	150	12	8%
BuzzFeed	0.76	0.80	0.81	0.84	0.74	0.74	0.99	0.85	0.86	0.90	0.86	0.82	10.0	695	150	63	42%
NBC News	0.60	0.64	0.78	0.75	0.72	0.75	0.86	0.89	0.95	0.95	0.93	0.87	9.7	408	150	21	14%
Yahoo!		0.57	0.66	0.59	0.67	0.77	0.90	1.00	0.84	0.82	0.77	0.58	8.2	195	150	34	23%
Daily Mail	0.51	0.48	0.51	0.54		0.65	0.73	0.67	0.75	0.69	0.79	0.63	6.9	516	150	33	22%
ESPN	0.44	0.40	0.50	0.52	0.53	0.49	0.58	0.53	0.76	0.78	0.69	0.67	6.9	142	142	34	24%
Wall Street Journal		0.56	0.64	0.63	0.66	0.68	0.81	0.77	0.88	0.90			6.5	747	150	28	19%
Business Insider	0.46	0.50	0.54	0.49	0.47	0.50	0.55	0.52	0.59	0.71	0.56	0.63	6.5	779	150	76	51%
The Telegraph		0.50	0.59				0.89	0.89	0.90	0.94	0.91	0.82	6.4	699	150	32	21%
Washington Post		0.39	0.40	0.41	0.42	0.51	0.62	0.73	0.75	0.80	0.70	0.64	6.4	691	150	62	41%
The Independent	0.34			0.39	0.46	0.45	0.69	0.67	0.60	0.75	0.82	0.67	5.8	530	150	67	45%
													Σ	11207	2992	767	26%

4 Clickbait Detection Model

Our clickbait detection model is based on 215 features; Table 2, column "Feature (type)," gives an overview. The features divide into three categories pertaining to (1) the teaser message, (2) the linked web page, and (3) meta information.

(1) Teaser message. Our primary feature engineering focus is on capturing the characteristics of a clickbait's teaser message, which is why most features are in this category. We subdivide the teaser message features into three subcategories: the first subcategory (1a) comprises basic text statistics. Features 1–9 are bag-of-words features, where Features 7 and 8 are Twitter-specific and Feature 9 consists of automatically generated image tags for images sent as part of a tweet, obtained from the Imagga tagging service [13]. Feature 10 computes the sentiment polarity of a tweet using the Stanford NLP library, and Features 11–13 measure a tweet's readability, where Features 12 and 13 are based on the Terrier stop word list [22] and the Dale-Chall list of easy words [5]. Features 14–16 quantify contractions and punctuation use, and Features 17–19 length statistics. The second and third subcategory (1b) and (1c) of teaser message features comprise dictionary features, where each feature encodes whether or not a tweet contains a word from a given dictionary of specific words or phrases. Features 20 and 21 are two dictionaries obtained from Gianotto [9], where the first contains common clickbait phrases and the second clickbait patterns in the form of regular expressions. Finally, Features 22–203 are all 182 General Inquirer dictionaries [25].

(2) Linked web page. Analyzing the web pages linked from a tweet, Features 204–209 are again bag-of-words features, whereas Features 210 and 211 measure readability and length of the main content when extracted with Boilerpipe [17].

(3) Meta information. Feature 212 encodes a tweet's sender, Feature 213 whether media (e.g., an image or a video) has been attached to a tweet, Feature 214 whether a tweet has been retweeted, and Feature 215 the part of day in which the tweet was sent (i.e., morning, afternoon, evening, night).

5 Evaluation

We randomly split our corpus into datasets for training and testing at a 2:1 training-test ratio. To avoid overfitting, we discard all features that have non-trivial weights in less than 1 % of the training dataset only. The features listed in Table 2 remained, whereas many individual features from the bag-of-words feature types were discarded (see the feature IDs marked with a *). Before training our clickbait detection model, we balance the training data by oversampling clickbait. We compare the three well-known learning algorithms logistic regression [18], naive Bayes [14], and random forest [4] as implemented in Weka 3.7 [12] using default parameters. To assess detection performance, we measure precision and recall for the clickbait class, and the area under the curve (AUC) of the receiver operating characteristic (ROC). We evaluate the performance of all features combined, each feature category on its own, and each individual feature (type) in isolation. Table 2 shows the results.

All features combined achieve a ROC-AUC of 0.74 with random forest, 0.72 with logistic regression, and 0.69 with naive Bayes. The precision scores on all features do not differ much across classifiers, the recall ranges from 0.66 with naive Bayes to 0.73 with random forest. Interestingly, the teaser message features (1a) alone compete or even outperform all features combined in terms of precision, recall, and ROC-AUC, using naive Bayes and random forest. The character n-gram features and the word 1-gram feature (IDs 1–4) appear to contribute most to this performance. Character n-grams are known to capture writing style, which may partly explain their predictive power for clickbait. The other features from category (1a) barely improve over chance as measured by ROC-AUC, yet, some at least achieve high precision, recall, or both. We further employ feature selection based on the χ^2 test to study the dependency of performance on the number of high-performing features. Selecting the top 10, 100, and 1000 features, overall performance with random forest outperforms that of feature category (1a) with 0.79 ROC-AUC. Features from all categories are selected, but mostly n-gram features from the teaser message and the linked web page.

Finally, as a baseline for comparison, the Downworthy rule sets [9] achieve about 0.69 recall at about 0.64 precision, whereas their ROC-AUC is only 0.54. This baseline is not only outperformed by combinations of other features, but also individual features, such as the General Inquirer dictionary "You" (9 pronouns indicating another person is being addressed directly) as well as

Table 2. Evaluation of our clickbait detection model. Some features are feature types that expand to many individual frequency-weighted features (i.e., IDs 1–9 and IDs 204–209). As classifiers, we evaluate logistic regression (LR), naive Bayes (NB), and random forest (RF).

Feature (type)		Precision			Recall			ROC-AUC		
ID	Description	LR	NB	RF	LR	NB	RF	LR	NB	RF
	all features	0.70	0.71	0.70	0.70	0.66	0.73	0.72	0.69	0.74
	top 10 as per χ^2 ranking	0.70	0.70	0.68	0.67	0.72	0.65	0.71	0.70	0.66
	top 100 as per χ^2 ranking	0.71	0.72	0.72	0.65	0.65	0.71	0.73	0.72	0.76
	top 1000 as per χ^2 ranking	0.64	0.70	0.76	0.58	0.65	0.76	0.60	0.69	0.79
(1a) *Teaser message*		0.60	0.74	0.71	0.55	0.72	0.73	0.54	0.74	0.73
1*	character 1-grams	0.71	0.68	0.71	0.65	0.56	0.71	0.72	0.68	0.71
2*	character 2-grams	0.64	0.73	0.71	0.60	0.70	0.72	0.60	0.75	0.74
3*	character 3-grams	0.63	0.74	0.74	0.58	0.74	0.75	0.61	0.76	0.77
4*	word 1-grams	0.70	0.74	0.72	0.66	0.66	0.71	0.70	0.76	0.72
5*	word 2-grams	0.64	0.63	0.61	0.68	0.45	0.46	0.58	0.58	0.55
6*	word 3-grams	0.55	0.55	0.55	0.69	0.69	0.69	0.50	0.50	0.50
7	hashtags	0.64	0.65	0.65	0.32	0.32	0.32	0.50	0.49	0.50
8	@ mentions	0.71	0.72	0.71	0.37	0.37	0.37	0.53	0.53	0.53
9	image tags as per Imagga [13]	0.55	0.59	0.57	0.41	0.50	0.51	0.48	0.52	0.51
10	sentiment polarity (Stanford NLP)	0.64	0.64	0.64	0.57	0.57	0.57	0.58	0.58	0.58
11	readability (Flesch-Kincaid)	0.63	0.63	0.61	0.54	0.55	0.54	0.59	0.59	0.56
12	stop words-to-words ratio	0.67	0.67	0.60	0.59	0.62	0.48	0.65	0.65	0.57
13	easy words-to-words ratio	0.09	0.09	0.49	0.30	0.30	0.70	0.50	0.50	0.50
14	has abbreviations	0.57	0.57	0.55	0.48	0.59	0.47	0.50	0.48	0.47
15	number of dots	0.63	0.64	0.63	0.42	0.37	0.42	0.54	0.54	0.54
16	starts with number	0.72	0.72	0.72	0.72	0.72	0.72	0.55	0.55	0.55
17	length of longest word	0.62	0.61	0.60	0.49	0.55	0.44	0.57	0.57	0.55
18	mean word length	0.58	0.57	0.61	0.51	0.56	0.56	0.50	0.48	0.54
19	length in characters	0.67	0.64	0.64	0.59	0.61	0.56	0.62	0.62	0.58
(1b) *Teaser message: Downworthy*		0.64	0.64	0.64	0.69	0.69	0.69	0.54	0.54	0.54
20	common clickbait phrases	0.65	0.65	0.65	0.70	0.70	0.70	0.54	0.54	0.54
21	common clickbait patterns	0.58	0.58	0.58	0.69	0.69	0.69	0.50	0.50	0.50
(1c) *Teaser message: General Inquirer (GI)*		0.66	0.70	0.67	0.60	0.64	0.67	0.65	0.68	0.70
22	GI dict. You	0.71	0.71	0.71	0.73	0.73	0.73	0.58	0.58	0.58
23	GI dict. POLIT	0.63	0.70	0.63	0.52	0.44	0.52	0.58	0.58	0.58
24	GI dict. Intrj	0.67	0.67	0.67	0.71	0.71	0.71	0.57	0.57	0.57
25	GI dict. HU	0.63	0.66	0.63	0.52	0.40	0.52	0.57	0.57	0.57
26	GI dict. Space	0.63	0.62	0.62	0.56	0.55	0.52	0.57	0.56	0.56
27	GI dict. Understated	0.64	0.58	0.65	0.67	0.32	0.69	0.56	0.55	0.56
28	GI dict. PowTot	0.65	0.65	0.65	0.49	0.49	0.49	0.59	0.59	0.56
...	+ 175 further GI dictionaries									
(2) *Linked web page*		0.64	0.64	0.64	0.67	0.67	0.67	0.56	0.56	0.56
204*	character 1-grams	0.63	0.57	0.62	0.55	0.62	0.61	0.60	0.54	0.61
205*	character 2-grams	0.58	0.62	0.61	0.49	0.62	0.62	0.50	0.59	0.61
206*	character 3-grams	0.58	0.67	0.62	0.49	0.59	0.64	0.52	0.63	0.61
207*	word 1-grams	0.60	0.72	0.65	0.50	0.64	0.65	0.54	0.71	0.64
208*	word 2-grams	0.58	0.70	0.67	0.48	0.60	0.65	0.51	0.70	0.66
209*	word 3-grams	0.56	0.65	0.63	0.44	0.55	0.58	0.46	0.61	0.63
210	main content readability (Flesch-Kincaid)	0.57	0.59	0.60	0.59	0.61	0.54	0.45	0.54	0.55
211	main content word length	0.61	0.63	0.58	0.56	0.68	0.50	0.54	0.56	0.51
(3) *Meta information*		0.62	0.74	0.74	0.55	0.72	0.75	0.54	0.74	0.77
212	sender name	0.65	0.65	0.65	0.60	0.60	0.60	0.67	0.67	0.67
213	has media attachment	0.55	0.55	0.55	0.53	0.53	0.53	0.47	0.47	0.47
214	is retweet	0.60	0.60	0.60	0.39	0.39	0.39	0.51	0.51	0.51
215	part of day as per server time	0.60	0.60	0.60	0.53	0.53	0.53	0.51	0.51	0.51

several others. Furthermore, sentiment analysis alone appears to be insufficient to detect clickbait (Feature 10), whereas in feature combinations it does possess some predictive power.

6 Conclusion

This paper presents the first machine learning approach to clickbait detection: the goal is to identify messages in a social stream that are designed to exploit cognitive biases to increase the likelihood of readers clicking an accompanying link. Clickbait's practical success, and the resulting flood of clickbait in social media, may cause it to become another form of spam, clogging up social networks and being a nuisance to its users. The adoption of clickbait by news publishers is particularly worrisome. Automatic clickbait detection would provide for a solution by helping individuals and social networks to filter respective messages, and by discouraging content publishers from making use of clickbait. To this end, we contribute the first evaluation corpus as well as a strong baseline detection model. However, the task is far from being solved, and our future work will be on contrasting clickbait between different social media, and improving detection performance.

References

1. Ajani, S.: A full 63% of buzzfeed's posts are clickbait (2015). http://keyhole.co/blog/buzzfeed-clickbait/
2. Beckman, J.: Saved you a click—don't click on that. I already did (2015). https://twitter.com/savedyouaclick
3. Blom, J.N., Hansen, K.R.: Click bait: forward-reference as lure in online news headlines. J. Pragmat. **76**, 87–100 (2015)
4. Breiman, L.: Random forests. Mach. Learn. **45**(1), 5–32 (2001)
5. Rocca, J.: Dale-Chall easy word list (2013). http://countwordsworth.com/download/DaleChallEasyWordList.txt
6. Davis, J., Goadrich, M.: The relationship between precision-recall and ROC curves. In: Proceedings of ICML 2006, pp. 233–240 (2006)
7. Eidnes, L.: Auto-generating clickbait with recurrent neural networks (2015). http://larseidnes.com/2015/10/13/auto-generating-clickbait-with-recurrent-neural-networks/
8. El-Arini, K., Tang, J.: News feed FYI: click-baiting (2014). http://newsroom.fb.com/news/2014/08/news-feed-fyi-click-baiting/
9. Gianotto, A.: Downworthy—a browser plugin to turn hyperbolic viral headlines into what they really mean (2014). http://downworthy.snipe.net
10. Gunning, R.: The fog index after twenty years. J. Bus. Commun. **6**(2), 3–13 (1969)
11. Hagey, K.: Henry Blodget's Second Act (2011). http://www.wsj.com/articles/SB10000872396390444840104577555180608254796
12. Hall, M., Frank, E., Holmes, G., Pfahringer, B., Reutemann, P., Witten, I.H.: The WEKA data mining software: an update. SIGKDD Explor. **11**(1), 10–18 (2009)
13. Imagga Image Tagging Technology (2015). http://imagga.com
14. John, G.H., langley, P.: Estimating continuous distributions in bayesian classifiers. In: Proceedings of UAI 1995, pp. 338–345 (1995)
15. Kempe, R.: Clickbait spoilers—channeling traffic from clickbaiting sites back to reputable providers of original content (2015). http://www.clickbaitspoilers.org
16. Koechley, P.: Why the title matters more than the talk (2012). http://blog.upworthy.com/post/26345634089/why-the-title-matters-more-than-the-talk

17. Kohlschütter, C., Fankhauser, P., Nejdl, W.: Boilerplate detection using shallow text features. In: Proceedings of WSDM 2010, pp. 441–450 (2010)
18. le Cessie, S., van Houwelingen, J.C.: Ridge estimators in logistic regression. Appl. Stat. **41**(1), 191–201 (1992)
19. Loewenstein, G.: The psychology of curiosity: a review and reinterpretation. Psychol. Bull. **116**(1), 75 (1994)
20. Mizrahi, A.: HuffPo spoilers—I give in to click-bait so you don't have to (2015). https://twitter.com/huffpospoilers
21. NewsWhip Media Tracker (2015). http://www.newswhip.com
22. Ounis, I., Amati, G., Plachouras, V., He, B., Macdonald, C., Lioma, C.: Terrier: a high performance and scalable information retrieval platform. In: OSIR @ SIGIR (2006)
23. Smith, B.: Why buzzfeed doesn't do clickbait (2015). http://www.buzzfeed.com/bensmith/why-buzzfeed-doesnt-do-clickbait
24. Stempeck, M.: Upworthy spoiler—words that describe the links that follow (2015). https://twitter.com/upworthyspoiler
25. Stone, P.J., Dunphy, D.C., Smith, M.S., Inquirer, T.G.: A Computer Approach to Content Analysis. MIT Press, Cambridge (1966)
26. Vijgen, B.: The listicle: an exploring research on an interesting shareable new media phenomenon. Stud. Univ. Babes-Bolyai-Ephemerides **1**, 103–122 (2014)

Informativeness for Adhoc IR Evaluation: A Measure that Prevents Assessing Individual Documents

Romain Deveaud[1,3], Véronique Moriceau[2], Josiane Mothe[1], and Eric SanJuan[1,3](✉)

[1] IRIT-CNRS UMR5505, Université de Toulouse, Toulouse, France
{romain.deveaud,josiane.mothe}@irit.fr
[2] LIMSI-CNRS, Université de Paris-Sud, Université de Paris-Saclay, Paris, France
moriceau@limsi.fr
[3] LIA, Agorantic, Université d'Avignon, Avignon, France
romain.deveaud@gmail.com, eric.sanjuan@univ-avignon.fr

Abstract. Informativeness measures have been used in interactive information retrieval and automatic summarization evaluation. Indeed, as opposed to *adhoc* retrieval, these two tasks cannot rely on the Cranfield evaluation paradigm in which retrieved documents are compared to static query relevance document lists. In this paper, we explore the use of informativeness measures to evaluate *adhoc* task. The advantage of the proposed evaluation framework is that it does not rely on an exhaustive reference and can be used in a changing environment in which new documents occur, and for which relevance has not been assessed. We show that the correlation between the official system ranking and the informativeness measure is specifically high for most of the TREC *adhoc* tracks.

Keywords: Information retrieval · Evaluation · Informativeness · Adhoc retrieval

1 Introduction

Information Retrieval (IR) aims at retrieving the relevant information from a large volume of available documents. Evaluating IR implies to define evaluation frameworks. In *adhoc* retrieval, Cranfield framework is the prevailing framework [1]; it is composed of documents, queries, relevance assessments and measures. Moreover, document relevance is considered as independent from the document rank and generally as a Boolean function (a document is relevant or not to a given query) even though levels of relevance can be used [7]. Effectiveness measurement is based on comparing the retrieved documents with the reference list of relevant documents. Moreover, it is based on the assessment assumption, that is the relevance of documents is known in advance for each query. It implies that the collection is static since it is assessed by humans. Cranfield paradigm facilitates reproductibility of experiments: at any time it is possible to evaluate a new IR method and to compare it against previous results; this is one of its main

N. Ferro et al. (Eds.): ECIR 2016, LNCS 9626, pp. 818–823, 2016.
DOI: 10.1007/978-3-319-30671-1_73

strengths. However, such a framework is not usable in changing environments when new documents are continuously added.

As opposed to Cranfield document relevance independency assumption, informativeness expresses the dependency of document relevance and takes into account the interactive nature of IR [8]. Indeed, one limitation of Cranfield-based evaluation is that relevance is encoded by documents [4]. Moreover, document relevance assessment is a clear limitation in dynamic context, when new documents are continuously added.

Nugget-based evaluation has been introduced to tackle this problem: rather than considering document relevance, it considers information relevance [4]. This method makes it possible to consider documents that have not been evaluated to be labeled as relevant or not, simply because they contain relevant information or not. Similar assumption is considered in automatic translation and automatic summarization evaluation. However this type of measure has not been intensively used in *adhoc* retrieval evaluation.

Our goal is to develop a method to evaluate *adhoc* IR using an informativeness measure to ensure reproductibility in dynamic document collections. To evaluate our method, we compare the system rankings we obtained using the informativeness measure proposed in [6] with the official system rankings based on document relevance, considering various TREC collections on *adhoc* tasks.

We show that the correlation between the official system rankings and the informativeness measure is specifically high for most of the TREC *adhoc* tracks.

The rest of the paper is organized as follows. Section 2 presents our evaluation framework which makes use of n-grams for informativeness-based evaluation applied to *adhoc* retrieval. We also present the *adhoc* retrieval collections we will be using. Section 3 presents and discusses the comparison of system rankings when using informativeness-based measures with the official ranking for the various *adhoc* retrieval collections/sub-tasks. Finally, Sect. 4 concludes this paper.

2 N-Gram Based Measure for Adhoc Retrieval

The evaluation method we developed makes an informativeness measure being usable in the case of *adhoc* retrieval. We use it on various TREC adhoc tracks.

We use the generic Log Similarity (LogSim) informativeness measure initially introduced to evaluate tweet contextualization in 2011 CLEF-INEX QA task [5]. LogSim is based on pools of relevant passages extracted from the document collection (Wikipedia in the CLEF/INEX lab case) called t-rels. t-rels are chunks of texts that are marked-up as relevant by human assessors. By considering each n-gram word in these t-rels as a relevant item, the LogSim normalized measure is based on n-gram precision and graded using log frequencies.

Given a reference R and a summary S, the Log Similarity on n-grams ($LogSim$) measure stands as:

$$LogSim(S|R) = \sum_{w \in \mathcal{F}^n_{St}(R) \cap \mathcal{F}^n_{St}(S)} \frac{log(\min(P(w|S), P(w|R)).|R| + 1)}{log(\max(P(w|S), P(w|R)).|R| + 1)}.P(w|R) \quad (1)$$

where $P(w|X) = \frac{f_X(w)}{|X|}$ corresponds to the frequency $f_X(w)$ of n-gram w in X over the length $|X|$ of text X, $\cap\mathcal{F}^n_{St}(S(X)$ is the set of n-grams of stem words from X, and X is either R or S.

To build such textual references over document ad-hoc q-rels in order to easily apply informativeness to *adhoc* IR tasks, one approach consists in extracting from documents information nugget candidates; it has been shown that this is possible over non-spammed document collections like TREC robust track or Gov collections [4]. This paper aims at showing that similar results can be obtained without requiring a prior extraction of relevant nuggets. Indeed we propose a direct conversion of relevant documents into a textual reference and experiment plain informativeness measures over it.

For that, we introduce the concept of content interpolated precision at length λ (cP_λ). Assuming that a user reads the retrieved documents following the ranking given by an IR system, cP_λ evaluates the informativeness of the reading after λ words.

2.1 Evaluating Precision Based on Document Content

Consider an *adhoc* task and its document q-rels. Assume that runs are ranked according to Mean Average Precision or Interpolated Precision at several recall levels. Runs can be converted into textual outcomes by concatenating ranked documents and q-rels can be converted into a textual reference by merging together all relevant documents per topic. Runs can be then evaluated by applying informativeness metrics to measure the overlap between submission and reference at various recall levels.

Let $D = (D_i)_{1 \leq i \leq d}$ be a ranked list of d documents. We consider as text $T = (t_1, ..., t_n)$ the concatenation of these documents (where $n = \sum_{i=1}^{i=d} |D_i|$). For each integer λ, we denote by T_λ the truncated text $T_\lambda = (t_1, ..., t_\lambda)$ and by $D_\lambda^{n,k}$ the set of n-grams with gap k. $D_\lambda^{n,0}$ or D_λ^n being the set of n-grams.

Similarly, given a set $R = \{R_i : 1 \leq i \leq r\}$ of r relevant documents we shall consider: $R^{n,k}$ the set of n-grams with gap k occurring in at least one reference.

In the case of TREC tracks, D is a run, each D_λ^n is the set of n-grams occurring in one of the m top ranked documents such that $\sum_{i=1}^{i=m} |D_i| \leq \lambda$ meanwhile R^n is the set of n-grams appearing at least once in the relevant documents from the corresponding q-rels.

We apply the English Porter stemming algorithm[1] to all documents after removing all stop words and all document identifiers like TREC doc-ID, etc. This is not only to reduce data, but to convert q-rels into reusable textual relevance judgments (t-rels) than can be applied to non official runs including documents not in the initial collection.

Given a run D and a reference R, we define for $n \in \{1, 2\}$, $k \in \{0, 2\}$ the content interpolated precision $cP_\lambda^{n,k}$ as:

$$cP_\lambda^{n,k}(D, R) = LS(D_\lambda^{n,k} | R^{n,k}) \qquad (2)$$

[1] http://snowball.tartarus.org/algorithms/english/stemmer.html.

Observe that, if $D_j \in R$ then:

$$cP^{1,0}_{\sum_{i=1}^{i=d}|D_i|}(D,R) \geq \frac{|D_j|}{\sum_{i=1}^{i=d}|D_i|}$$

but conversely, $D \cap R = \emptyset$ does not imply $cP_\lambda(D) = 0$ since there can be some overlap between n-grams in documents and in the reference.

So our approach based on document contents instead of document IDs does not require exhaustive references and therefore, can be applied to incomplete references based on pools of relevant documents. However, meanwhile *adhoc* IR returns a ranked list of documents independently of their respective lengths, relevance judgments can be used to automatically generate text references (*t-rels*) by concatenating the textual content of relevant documents.

2.2 Data Sets and Ground Truth

Among international evaluation collections, we chose TREC collections composed of news articles (Robust2004) and Web (Web and Terabyte). Doing so, we also focused on the quality of the collections with large amounts of runs and a comprehensive set of relevance judgments. The number of retrieval systems to rank ranges from 56 to 129, while the number of topics is typically 50 and increases to 150 for Terabyte2006 and 250 for Robust2004 (Table 1). All runs can be downloaded from the TREC web site, and document collections can be obtained on the web site for active participants or through track organizers.

We take the same experimental approach as in [2,3], and reproduced the official rankings of all these retrieval systems for these various collections using

Table 1. Summary of TREC test collections and size in tokens of generated t-rels used for evaluation.

Name	# Runs	# Topics	Corpus	$\lvert D_n^{1,1} \cup R^{1,1}\rvert$
TREC-5	106	50	TREC Vol. 4+5	15×10^6
TREC-6	107	50	TREC Vol. 4+5	12×10^6
TREC-7	103	50	TREC Vol. 4+5	28×10^6
TREC-8	129	50	TREC Vol. 4+5	35×10^6
Web2000	104	50	WT10g	137×10^6
Web2001	97	50	WT10g	195×10^6
Robust2004	110	250	TREC Vol. 4+5	150×10^6
Terabyte2004	70	50	GOV2	46×10^6
Terabyte2005	58	50	GOV2	46×10^6
Terabyte2006	80	150	GOV2	46×10^6
Web2010	56	50	ClueWeb09-B	66×10^6
Web2011	62	50	ClueWeb09-B	64×10^6

the official measure. For all collections, the official measure is the Mean Average Precision (MAP), except for Web2010 and Web2011 where Expected Reciprocal Rank (ERR@20) was preferred. These official rankings constitute the ground truth ranking, against which we will compare the rankings produced by:

- cP_{10^3} based on the 1000 tokens of each run and on their log frequencies.
- cP_n based on all tokens of each run.

By comparing averaged measures, we evaluate if the average informativeness of all documents retrieved by a system is correlated to the official ranking. Let us emphasize that this does not necessarily imply that document informativeness is correlated to individual document relevance for a given query. We use the Kendall's τ rank correlation coefficient to identify correlations between the ground truth ranking and the informativeness ranking.

3 Results and Discussion

In this section we report the correlation results of the ground truth ranking (TREC official measure depending on the track) and the content-based ranking produced by the cP_λ informativeness measure. All correlations reported are significantly different from zero with a p-value < 0.001. While we chose Kendall's τ as the correlation measure, we also report the Pearson's linear correlation coefficient for convenience. A $\tau > 0.5$ typically indicates a strong correlation since it implies an agreement between the two measures over more than half of all ordered pairs.

Table 2. Retrieval systems ranking correlations between the official ground truth and the cP_λ informativeness measure. $cP_\lambda^{1,1}$ stands for uniterms while $cP_\lambda^{2,2}$ corresponds to bigrams with skip. We use either 10^3 terms or all the terms from the ordered list of retrieved documents.

Track	$cP_{10^3}^{1,1}$		$cP_{10^3}^{2,0}$		$cP_{10^3}^{2,2}$		$cP_n^{1,1}$	
	τ	ρ	τ	ρ	τ	ρ	τ	ρ
TREC-5	57.91 %	71.04 %	56.22 %	56.30 %	56.15 %	56.25 %	**69.91 %**	**88.62 %**
TREC-6	72.49 %	84.18 %	76.50 %	92.90 %	**76.50 %**	**93.01 %**	58.78 %	68.18 %
TREC-7	61.27 %	82.17 %	70.18 %	90.59 %	**70.27 %**	**90.60 %**	63.45 %	53.49 %
TREC-8	54.80 %	84.04 %	65.79 %	92.25 %	65.85 %	**92.30 %**	**67.46 %**	72.94 %
Web2000	46.48 %	68.00 %	61.42 %	83.40 %	62.05 %	77.82 %	**70.83 %**	**86.68 %**
Web2001	31.65 %	56.51 %	36.94 %	57.89 %	36.66 %	56.66 %	**77.45 %**	**88.90 %**
Robust2004	40.97 %	64.33 %	58.67 %	85.43 %	59.50 %	86.22 %	**74.71 %**	**90.88 %**
Terabyte2004	48.56 %	60.12 %	61.25 %	73.65 %	61.45 %	74.75 %	**76.37 %**	**86.12 %**
Terabyte2005	59.45 %	85.34 %	69.65 %	88.98 %	69.80 %	**89.18 %**	**76.01 %**	84.28 %
Terabyte2006	41.14 %	50.24 %	54.75 %	70.70 %	55.28 %	70.90 %	**65.04 %**	**89.37 %**
Web	28.56 %	44.00 %	**44.38 %**	**69.11 %**	44.17 %	69.09 %	-	-
Web2011	56.03 %	**80.98%**	55.68 %	79.14 %	**56.35 %**	79.42 %	**34.50 %**	-

When looking at Table 2, we see that cP_λ accurately reproduces official ranking based on MAP for early TREC tracks (TREC6-7-8, Web2000) as well as for Robust2004, Terabyte2004-5 and Web2011. *LogSim*-score applied to all tokens in runs is often the most effective whenever systems are ranked based on MAP. However on early TREC tracks, $cP^{2,2}_{10^3}$ can perform better even though only the first 1000 tokens of each run are considered after concatenating ranked retrieved documents. Indeed, the traditional TREC *adhoc* and Robust tracks used newspaper articles as document collection. Since a single article often deals with a single subject, relevant concepts are likely to occur together, which might be less the case in web pages for example. A relevant news article is very likely to contain only relevant information, whereas a long web document that deals with several subjects might not be relevant as a whole.

4 Conclusion

In this paper, we proposed a framework to evaluate adhoc IR using the LogSim informativeness measure based on token n-grams. To evaluate this measure, we compared the ranks of the systems we obtained with the official rankings based on document relevance, considering various TREC collections on *adhoc* tasks. We showed that (1) rankings obtained based on n-gram informativeness and with Mean Average Precision are strongly correlated; and (2) *LogSim* informativeness can be estimated on top ranked documents in a robust way. The advantage of this evaluation framework is that it does not rely on an exhaustive reference and can be used in a changing environment in which new documents occur, and for which relevance has not been assessed. In future work, we will evaluate various *LogSim* parameters influence.

References

1. Cleverdon, C.: The cranfield tests on index language devices. In: Aslib Proceedings, vol. 19. No. 6. MCB UP Ltd (1967)
2. Hauff, C.: Predicting the effectiveness of queries and retrieval systems. Ph.D thesis, Enschede, SIKS Dissertation Series No. 2010-05, January 2010
3. Nuray, R., Can, F.: Automatic ranking of information retrieval systems using data fusion. Inf. Process. Manage. **42**(3), 595–614 (2006)
4. Pavlu, V., Rajput, S., Golbus, P.B., Aslam, J.A.: IR system evaluation using nugget-based test collections. In: Proceedings of WSDM (2012)
5. SanJuan, E., Moriceau, V., Tannier, X., Bellot, P., Mothe, J.: Overview of the INEX 2011 question answering track (QA@INEX). In: Geva, S., Kamps, J., Schenkel, R. (eds.) INEX 2011. LNCS, vol. 7424, pp. 188–206. Springer, Heidelberg (2012)
6. SanJuan, E., Moriceau, E., Tannier, X., Bellot, P., Mothe, J.: Overview of the INEX 2012 tweet contextualization track. In: CLEF (2012)
7. Spink, A., Greisdorf, H., Bateman, J.: From highly relevant to not relevant: examining different regions of relevance. Inf. Process. Manage. **34**(5), 599–621 (1998)
8. Tague-Sutcliffe, J.: Measuring the informativeness of a retrieval process. In: Proceedings of the International ACM SIGIR Conference on Research and Development in Information Retrieval, pp. 23–36 (1992)

What Multimedia Sentiment Analysis Says About City Liveability

Joost Boonzajer Flaes[1(✉)], Stevan Rudinac[2(✉)], and Marcel Worring[2(✉)]

[1] Twitter Inc., London, UK
jflaes@twitter.com
[2] Informatics Institute, University of Amsterdam, Amsterdam, The Netherlands
{s.rudinac,m.worring}@uva.nl

Abstract. Recent developments allow for sentiment analysis on multimodal social media content. In this paper we analyse content posted on microblogging and content-sharing platforms to estimate sentiment of the city's neighbourhoods. The results of sentiment analysis are evaluated through investigation into the existence of relationships with the indicators of city liveability, collected by the local government. Additionally, we create a set of sentiment maps that may help discover existence of possible sentiment patterns within the city. This study shows several important findings. First, utilizing multimedia data, i.e., both visual and text content leads to more reliable sentiment scores. The microblogging platform Twitter further appears more suitable for sentiment analysis than the content-sharing website Flickr. However, in case of both platforms, the computed multimodal sentiment scores show significant relationships with the indicators of city liveability.

Keywords: Multimodal sentiment analysis · Semantic concept detection · Social multimedia · City liveability

1 Introduction

Posting messages on social networks is a popular means for people to communicate and to share thoughts and feelings about their daily lives. Previous studies investigated the correlation between sentiment extracted from user-generated text and various demographics [6]. However, as technology improves, the bandwidth available for users also increases. As a result, users can share images and videos with greater ease. This led to a change in types of media being shared on these online networks. More particularly, user-generated content often consist of a combination of modalities, e.g., text, images, video and audio. As a result, more recent studies have tried to predict sentiment from visual content too [2].

A recent study on urban computing conducted by Zheng et al. underlines the potential of utilizing user-generated content for solving various challenges a modern metropolis is facing with, ranging from urban planning and transportation

J.B. Flaes—This research was performed while the first author was a student at the UvA.

N. Ferro et al. (Eds.): ECIR 2016, LNCS 9626, pp. 824–829, 2016.
DOI: 10.1007/978-3-319-30671-1_74

to public safety and security [7]. In this paper we investigate whether the sentiment analysis of spontaneously generated social multimedia can be utilized for detecting areas of the city facing such problems. More specifically, we aim at creating a sentiment map of Amsterdam that may help paint a clearer picture of the city liveability. Since visual content may contain complementary information to text, in our approach we choose to utilize them jointly. Additionally, we make use of automatically captured metadata (i.e., geotags) to analyse the messages in context of the locations where they were posted.

A direct evaluation of our results would require a user study in which the participants would be asked about their sentiment on different neighbourhoods of a city. As conducting such study would be both extremely time consuming and labour intensive, here we propose the use of open data as an indirect ground-truth. We consider a large number of demographic, economic and safety parameters comprising the liveability index of a neighbourhood and investigate their association with the automatically produced sentiment scores in different scenarios.

2 Research Methodology

Our approach consists of three steps outlined in Fig. 1 and described below.

2.1 Data Collection

The data used in this research comes from two different online social networks, which both include textual and visual content. The emphasis of the analysis will be on modalities dominant on a particular platform, i.e., textual data in case of Twitter and visual data for Flickr. However, visual data shared on Twitter and text shared on Flickr will also be used for a better understanding of the content hosted on these platforms and an increased quality of the sentiment analysis.

For about two months, 64 thousand tweets were collected within a 10-mile radius of the city center of Amsterdam. The dataset only includes tweets that have a geo-location available. A total of 64 thousand images were downloaded from Flickr that are taken in and around the city of Amsterdam.

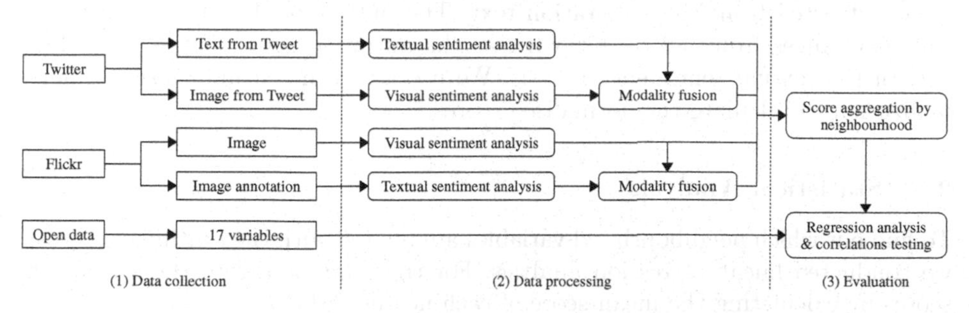

Fig. 1. The approach overview.

From open data as provided by the city of Amsterdam we utilize the following neighbourhood variables: *percentages of non-western immigrants*, *western immigrants*, *autochthonous inhabitants*, *income*, *children in households with minimum incomes*, *people working*, *people living on welfare*, *people with low, average and high level of education*, *recreation area size*, *housing prices*, *physical index*, *nuisance index*, *social index*, and *liveability index* [5].

2.2 Calculating Sentiment

Textual Data. Sentiment for the textual data is calculated by using two different Python packages. The first is NLTK which makes use of a naive Bayes classifier to predict sentiment [1]. For each tweet, the sentiment is calculated ranging from -1 to $+1$. The second is Pattern [3], which uses stemming and part-of-speech tagging to predict sentiment and includes both a Dutch and English based lexicon.

In our approach, first the language is detected using the Google Translate API. Then, if the language is not English the text is translated into English. However, for the Pattern package we also included the Dutch lexicon if a tweet was predicted to be Dutch. To compare the two packages for sentiment analysis, we manually annotated a random sample of 150 tweets as either positive or negative. On these tweets the Pattern package significantly outperformed NLTK, which is why we decided to use it for further evaluation.

Visual Content. We analyse visual content of the images using SentiBank [2] and detect 1200 adjective noun pairs (ANPs). For example, using SentiBank a 'happy person' can be detected, which combines the adjective 'happy' with the noun 'person'. Each of the ANPs detected has a sentiment score associated with it and for each image, we compute the average of the top 10 detected ANPs.

Combined Score. To obtain more reliable scores, we combine the results of visual and textual sentiment analysis. For Twitter this means that a combined (i.e., average) score will be calculated if the tweet contains a direct link to an image. For Flickr, the sentiment scores of the images are combined with sentiment extracted from the annotation text. The output of the visual sentiment classifier ranges from -2 to $+2$, whereas the output ranges from -1 to $+1$ in case of the textual sentiment analysis. We use a zero mean unit variance normalization to calculate the normalised scores.

2.3 Statistical Analysis

To find out which neighbourhood variables are related with the sentiment scores, we conducted linear regression analysis. For that, we aggregate the sentiment scores by calculating the mean score of each neighbourhood.

To identify which neighbourhood variables were significantly associated with the sentiment scores, single regression analyses were conducted for each of the neighbourhood variables and the sentiment scores. Regression coefficients were assessed for significance with statistical significance set at $p \leq 0.05$.

3 Experimental Results

We conduct several experiments in order to evaluate the proposed research methodology described in Sect. 2.3.

3.1 Flickr Sentiment Analysis

Our analysis shows that no significant relationship can be found between the visual sentiment scores from Flickr and the selected liveability indicators. The most significant relationships are found with the *percentage of people living on governmental welfare checks* and the *level of education*.

The combined sentiment scores showed more promising results. The *safety index* and the *people living on governmental welfare* showed significant associations with the sentiment scores ($p = 0.037$ and $p = 0.028$, cf. Fig. 2).

3.2 Twitter Sentiment Analysis

To compute reliable scores, only neighbourhoods with more than 40 tweets were taken into account. However, no significant relationship is found between the open data and the scores based on the analysis of textual content.

However, the combined score shows multiple significant relationships with the liveability indicators. The first interesting relationship is found between ethnic composition of the neighbourhood and sentiment scores. More particularly, there is a positive association between sentiment scores and the percentage of native Dutch inhabitants (cf. Fig. 3). Similarly, a positive association can be found between level of education and the sentiment scores. This is not surprising as these two variables are strongly correlated.

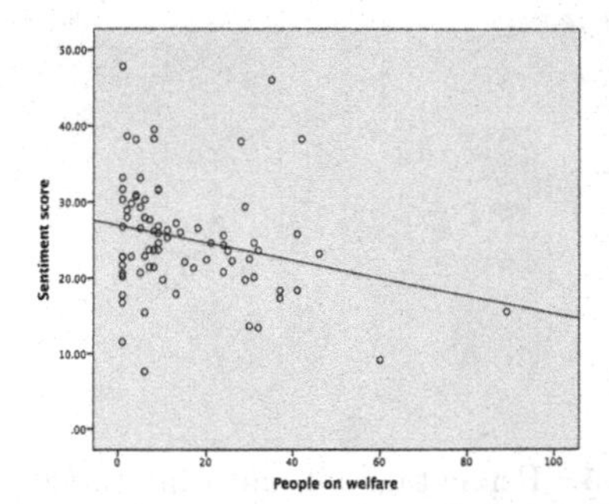

Fig. 2. Relationship between Flickr sentiment scores and percentage of people living on welfare checks.

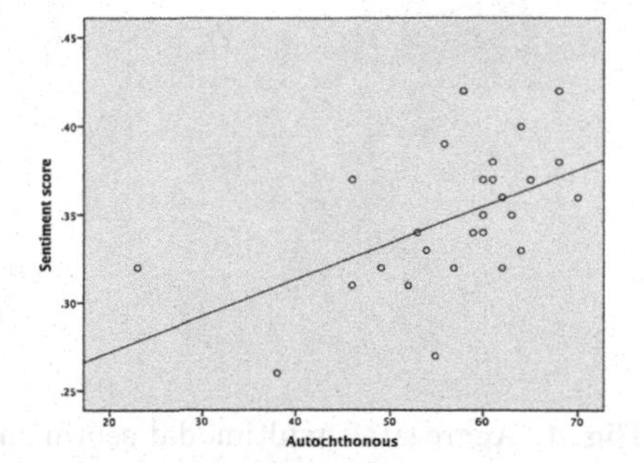

Fig. 3. Relationship between Twitter sentiment scores and percentage of autochthonous population.

3.3 Sentiment Maps

To facilitate easier gaining of new insights about the developments in the neighbourhoods and in order to visualize findings of our study, we created a simple data exploration interface[1]. The interface features interactive maps showing the city of Amsterdam and visualizing the sentiment scores of each neighbourhood generated according to the different methods evaluated in Sects. 3.1 and 3.2 (cf. Fig. 4). Additionally, the maps are visualising various indicators of city liveability for easier assessment of possible relationships (cf. Fig. 5). Finally, the interface also includes the possibility to view a sub-sample of the tweets uploaded in Amsterdam.

4 Discussion and Conclusion

In this paper we investigate if the sentiment scores derived from the analysis of social multimedia data relate to the geographic location in which they are posted. We use state-of-the-art sentiment analysis methods in order to process multimodal content from two very different social media platforms, a content-sharing website Flickr and a microblogging platform Twitter.

Our research reveals significant relationships between automatically extracted sentiment and the indicators of city liveability. Namely, both sentiment scores from Flickr and Twitter showed significant relationships with the open data when multimodal content is analysed. However, in case of both analysed platforms, we found no significant relationships when using a single modality for sentiment analysis. This confirms our assumption that the multimodal sentiment analysis provides for higher accuracy.

The detected sentiment mostly correlates with the demographics of the inhabitants. The percentage of people living on welfare checks shows a negative linear association with sentiment scores from the Flickr data. Ethnic background, income and education level show significant relationships with the sentiment

Fig. 4. Aggregated multimodal sentiment scores based on Flickr data.

Fig. 5. Percentage of autochthonous inhabitants in Amsterdam neighbourhoods.

[1] http://goo.gl/DAj9y2.

scores based on Twitter data. Further research is needed to investigate the nature of these relationships. However, it is interesting to observe that for both platforms the economic indicators (i.e., people living on welfare checks and income) show significant relationships with our computed sentiment scores. On the other hand, the liveability or social index of a neighbourhood showed no significant relationship. Since these indices are designed to measure the subjective feelings of the inhabitants, we would have expected these to be more significant in our research.

Using the Twitter data shows more significant relationships than using the data from Flickr. A possible explanation for this might be that people do not tend to share opinions or feelings on this platform but mainly use it as a method to share their photographs.

To further improve this research, it would also be interesting to see if the sentiment prediction could be adjusted to factors that are important for residents of a city. The examples are the detectors created specifically for urban phenomena like noise nuisance or graffiti, known to influence the liveability of the city [4]. Finally, combining sentiment scores from user-generated data and open data allows for new research opportunities.

Our research shows that sentiment scores may give additional insights in a geographic area. The big advantage of training on social multimedia data is that it provides for real-time insights. Additionally, sentiment in these areas can prove to be an indication of important factors like crime rate or infrastructure quality. This may be useful for government services to know what area to improve or for new businesses to find a convenient location.

References

1. Bird, S.: Nltk: the natural language toolkit. In: Proceedings of the COLING/ACL on Interactive Presentation Sessions, pp. 69–72. Association for Computational Linguistics (2006)
2. Borth, D., Ji, R., Chen, T., Breuel, T., Chang, S.-F.: Large-scale visual sentiment ontology and detectors using adjective noun pairs. In: Proceedings of the 21st ACM International Conference on Multimedia, pp. 223–232. ACM (2013)
3. De Smedt, T., Daelemans, W.: Pattern for python. J. Mach. Learn. Res. **13**, 2063–2067 (2012)
4. Dignum, K., Jansen, J., Sloot, J.: Growth and decline - demography as a driving force. PLAN Amsterdam **17**(5), 25–27 (2011)
5. C.G.A.O. for Research and Statistics. Fact sheet leefbaarheidsindex periode 2010–2013, Febraury 2014. https://www.amsterdam.nl/publish/pages/502037/fact_sheet_6_leefbaarheidsindex_2010_-_2013_opgemaakt_def.pdf
6. Mitchell, L., Frank, M.R., Harris, K.D., Dodds, P.S., Danforth, C.M.: The geography of happiness: connecting twitter sentiment and expression, demographics, and objective characteristics of place. PLoS ONE **8**(5), e64417 (2013)
7. Zheng, Y., Capra, L., Wolfson, O., Yang, H.: Urban computing: concepts, methodologies, and applications. ACM Trans. Intell. Syst. Technol. (TIST) **5**(3), 38 (2014)

[illegible] within the [illegible]. And [illegible] about City [illegible]

[illegible] based on Twitter data. Further research is needed to investigate the nature [illegible] however [illegible] observed that for both [illegible] with [illegible] and [illegible] show [illegible] sentiment scores. On the other hand, the [illegible] social [illegible] showed no significant relationship. [illegible] measure the [illegible] feelings [illegible] we would [illegible] to be more significant in our research.

Using the [illegible] data shows [illegible] relationships [illegible] using the data from [illegible]. A possible explanation for this might be that people do not tend to share [illegible] on this platform but mainly use it as a method [illegible].

[illegible] it would also be interesting [illegible] further [illegible] that the [illegible] specifically [illegible] the [illegible] sentiment scores [illegible] generated data and [illegible].

Our research shows that sentiment [illegible] give additional insights [illegible] that would [illegible] social [illegible]. [illegible] important [illegible] quality [illegible] useful for [illegible] to keep [illegible] and [illegible] location.

References

[illegible] Proceedings of the [illegible] Association for Computational Linguistics [illegible] (2008)

[illegible] sentiment [illegible] In: Proceedings of the 21st ACM [illegible] Conference [illegible], pp. [illegible] ACM (2009) [illegible]

[illegible]

[illegible], pp. 417 (2013) [illegible]

[illegible] ACM Trans. [illegible] (2010)

Demos

Scenemash: Multimodal Route Summarization for City Exploration

Jorrit van den Berg[1](✉), Stevan Rudinac[2](✉), and Marcel Worring[2](✉)

[1] TNO, Den Haag, The Netherlands
jorrit.vandenberg@tno.nl
[2] Informatics Institute, University of Amsterdam, Amsterdam, The Netherlands
{s.rudinac,m.worring}@uva.nl

Abstract. The potential of mining tourist information from social multimedia data gives rise to new applications offering much richer impressions of the city. In this paper we propose Scenemash, a system that generates multimodal summaries of multiple alternative routes between locations in a city. To get insight into the geographic areas on the route, we collect a dataset of community-contributed images and their associated annotations from Foursquare and Flickr. We identify images and terms representative of a geographic area by jointly analysing distributions of a large number of semantic concepts detected in the visual content and latent topics extracted from associated text. Scenemash prototype is implemented as an Android app for smartphones and smartwatches.

1 Introduction

When visiting a city, tourists often have to rely on travel guides to get information about interesting places in their vicinity or between two locations. Existing crowdsourced tourist websites, such as TripAdvisor primarily focus on providing point of interest (POI) reviews. The available data on social media platforms allows for new use-cases, stemming from a much richer impression about places. Efforts to utilize richness of social media for tourism applications have been made by e.g., extracting user demographics from visual content of the images [3], modelling POIs and user mobility patterns by analysing Wikipedia pages and image metadata [2] or by representing users and venues by topic modelling in both text and visual domains [7].

We propose Scenemash[1], a system that supports way-finding for tourists by automatically generating multimodal summaries of several alternative routes between locations in a city and describing geographic area around a given location. To represent geographic areas along the route, we make use of user-contributed images and their associated annotations. For this purpose, we systematically collect information about venues and the images depicting them from location-based social networking platform Foursquare and we turn to content sharing website Flickr for a richer set of images and metadata capturing

J. van den Berg—This research was performed while the first author was a student at the UvA.

[1] Scenemash demo: https://www.youtube.com/watch?v=oAnj6A1oq2M.

N. Ferro et al. (Eds.): ECIR 2016, LNCS 9626, pp. 833–836, 2016.
DOI: 10.1007/978-3-319-30671-1_75

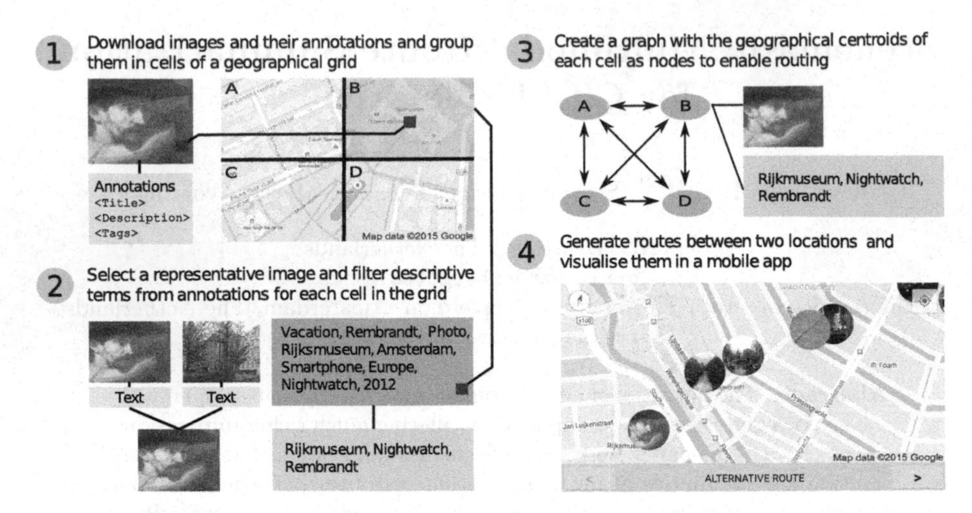

Fig. 1. Overview of our approach to multimodal summarization of tourist routes.

a wide range of aspects users find interesting. We create summaries by jointly analysing distributions of semantic concepts detected in the visual channel of the images and the latent topics extracted from their associated annotations. To demonstrate the effectiveness of our approach, using Amsterdam as a showcase, we implement Scenemash prototype as an Android app for smartphones and smartwatches.

2 Approach Overview

The pipeline of our approach is illustrated in Fig. 1. In this section we describe data collection and analysis steps as well as the procedure for generating multimodal representations of the geographic areas (i.e., steps 1 and 2).

2.1 Data Collection and Analysis

We first queried the Facebook Graph and Foursquare APIs and compiled a list of all POIs within the radius of 9 km from the centre of Amsterdam. Then, we crawled georeferenced Flickr images along with their annotations (i.e., title, description and tags) within 500 m from each POI. To further enlarge the collection, we downloaded more images taken in Amsterdam by already known Flickr users. Finally, we crawled images of all verified Foursquare venues. The resulting dataset consisted of 157,000 images and their associated annotations.

We represent visual content of each image in the collection with a distribution of 15,293 semantic concept scores output by a customised implementation of Google "Inception" net [6]. We tokenize the text associated with the images and remove stopwords, unique words and HTML markup. After preprocessing step, we represent each image in the text domain using 100 LDA topics extracted using Gensim framework [8].

Fig. 2. Smartphone and smartwatch user interfaces.

2.2 Summarization of a Geographic Area

We use a rectangular geographical grid with 125 × 125 meter cells to define geographic areas and group the images. For each grid cell, we compute pairwise cosine similarity matrices for both distributions of visual concepts and LDA topics, extracted as described in previous section. We then combine such created unimodal similarity matrices using the weighting fusion scheme proposed by Ah-Pine et al. [1]. The resulting multimodal similarities serve as an input into the affinity propagation clustering, which aims at automatically selecting an optimal number of clusters [4]. Finally, we sort clusters in decreasing order of their size and select the first centroid available under a Creative Commons license as a representative image for a given geographic area.

As a starting point for generating description of the area we make use of pre-processed text associated with the images (cf. Sect. 2.1). To identify the terms representative of a particular geographic area, we apply tf-idf weighting, considering each grid cell as a single document. In general, tf-idf discriminates well between the terms that are used on a certain location and those used in the entire city. However, it also has the tendency to give a high weight to rare (often unwanted) terms. To mitigate this effect and improve alignment between visual and text representation of the area, we utilize tag ranking approach similar to the one introduced by Li et al. [5] and weight the terms by their frequency of occurrence in the k-nearest visual neighbours of the selected representative image. The ranked lists of terms produced by the two above-mentioned weighting schemes are combined and the top-10 ranked terms are selected.

3 Scenemash Prototype Design

We implement the Scenemash prototype as a native Android app for smartphones and smartwatches, which the attendees will be able to test at the conference. The app interface allows the user to query locations in the city or use the current location provided by GPS sensor. Scenemash features "explore" and "get route" functions. On the server side, a graph illustrated in step 3 of Fig. 1 is used to get the neighbouring nodes/grid cells of a node containing user coordinates when explore function is selected. In the get route mode, we apply the breadth-first search algorithm on the same graph for computing a route between two locations. Alternative routes are computed by selecting different neighbour nodes of the origin node. To give the users an opportunity to avoid crowded places, we create a weighted version of the same graph which uses the number of images captured in a geographic cell as a proxy for crowdedness. If the crowd avoidance feature is selected, we deploy Dijkstra's shortest path algorithm for computing the route between two locations.

The data collection and analysis steps described in Sect. 2 are precomputed offline, in order to reduce online computation load. Figure 2 illustrates the user interfaces. Each relevant geographic area (i.e., on the route or in user's vicinity) is represented by a circular thumbnail displayed in Google Maps. If the smartphone is paired with a smartwatch, the images are shown as a slideshow on the smartwatch. When a user interacts with the map by tapping on one of the images, the image is enlarged and an info-box with the most relevant terms for the area is shown. The effectiveness of the prototype gives us confidence that the Scenemash could be implemented in other cities as well.

References

1. Ah-Pine, J., Clinchant, S., Csurka, G., Liu, Y.: Xrce's participation in imageclef. In: Working Notes of CLEF 2009 Workshop Co-located with the 13th European Conference on Digital Libraries (ECDL 2009) (2009)
2. Brilhante, I., Macedo, J.A., Nardini, F.M., Perego, R., Renso, C.: TripBuilder: a tool for recommending sightseeing tours. In: de Rijke, M., Kenter, T., de Vries, A.P., Zhai, C.X., de Jong, F., Radinsky, K., Hofmann, K. (eds.) ECIR 2014. LNCS, vol. 8416, pp. 771–774. Springer, Heidelberg (2014)
3. Cheng, A.-J., Chen, Y.-Y., Huang, Y.-T., Hsu, W.H., Liao, H.-Y.M.: Personalized travel recommendation by mining people attributes from community-contributed photos. In: Proceedings of the 19th ACM International Conference on Multimedia, MM 2011, pp. 83–92. ACM, New York (2011)
4. Frey, B.J., Dueck, D.: Clustering by passing messages between data points. Science **315**(5814), 972–976 (2007)
5. Li, X., Snoek, C., Worring, M.: Learning social tag relevance by neighbor voting. IEEE Trans. Mult. **11**(7), 1310–1322 (2009)
6. Szegedy, C., Liu, W., Jia, Y., Sermanet, P., Reed, S., Anguelov, D., Erhan, D., Vanhoucke, V., Rabinovich, A.: Going deeper with convolutions. In: 2015 IEEE Conference on Computer Vision and Pattern Recognition (CVPR), pp. 1–9, June 2015
7. Zahálka, J., Rudinac, S., Worring, M.: New yorker melange: interactive brew of personalized venue recommendations. In: Proceedings of the ACM International Conference on Multimedia, MM 2014, pp. 205–208. ACM, New York (2014)
8. Řehůřek, R., Sojka, P.: Software framework for topic modelling with large corpora. In: Proceedings of LREC Workshop New Challenges for NLP Frameworks, pp. 46–50. University of Malta, Valletta (2010)

Exactus Like: Plagiarism Detection in Scientific Texts

Ilya Sochenkov[1,2], Denis Zubarev[2](✉), Ilya Tikhomirov[2], Ivan Smirnov[2], Artem Shelmanov[2], Roman Suvorov[2], and Gennady Osipov[2]

[1] Peoples' Friendship University of Russia, Moscow, Russia
[2] Federal Research Center "Computer Science and Control" of Russian Academy of Sciences, Moscow, Russia
{sochenkov,zubarev,tih,ivs,shelmanov,suvorov,gos}@isa.ru

Abstract. The paper presents an overview of Exactus Like – a plagiarism detection system. Deep parsing for text alignment helps the system to find moderate forms of disguised plagiarism. The features of the system and its advantages are discussed. We describe the architecture of the system and present its performance.

1 Introduction

Plagiarism is a serious problem in education and science. Improper citations, textual borrowings, and plagiarism often occur in student and research papers. Academics, peer reviewers, and editors of scientific journals should detect plagiarism in all forms and prevent substandard works from being published [1].

Numerous computer-assisted plagiarism detection systems (CaPD) were recently developed: Turnitin, Antiplagiat.ru, The Plagiarism Checker, PlagScan, Chimpsky, Copyscape, PlagTracker, Plagiarisma.ru. The difference between these systems lies in search engines used to find similar textual fragments, ranking schemas, and result presentations. Most of the aforementioned systems implement simple techniques to detect "copy-and-paste" borrowings based on exact textual matching or w-shingling algorithms [2,3]. Such an approach shows good computational performance, but it cannot find heavily disguised plagiarism [4].

In this demonstration we present Exactus Like[1] – an applied plagiarism detection system, which finds besides simple copy-and-paste plagiarism also moderately disguised borrowings (word/phrase reordering, substitution of some words with synonyms). To do this, the system leverages deep parsing techniques.

2 System User Interface and Features

Exactus Like is a web application. The start page contains fields to input a suspicious text or upload a file. Most of the popular file formats are supported: Adobe PDF, Microsoft Word, RTF, ODT, HTML, etc. One can specify the year

[1] The demo is available online at http://like.exactus.ru/index.php/en/.

N. Ferro et al. (Eds.): ECIR 2016, LNCS 9626, pp. 837–840, 2016.
DOI: 10.1007/978-3-319-30671-1_76

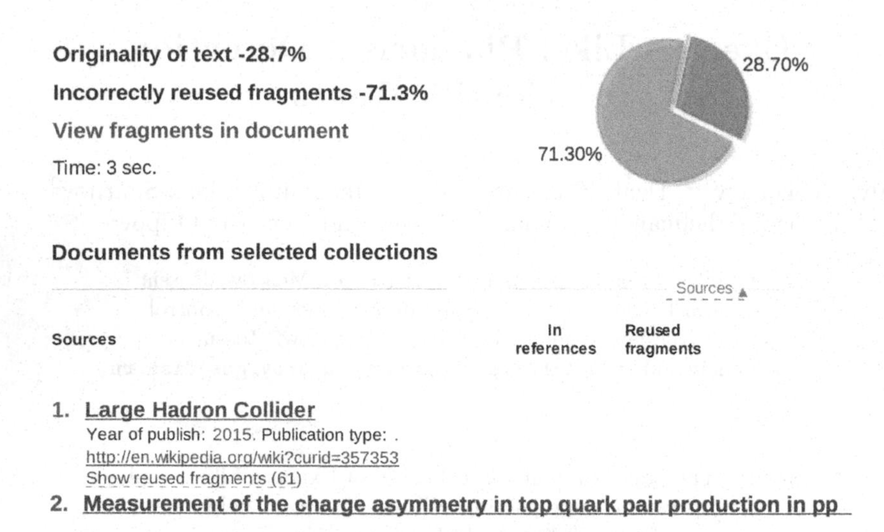

Fig. 1. Visualization of plagiarism detection results

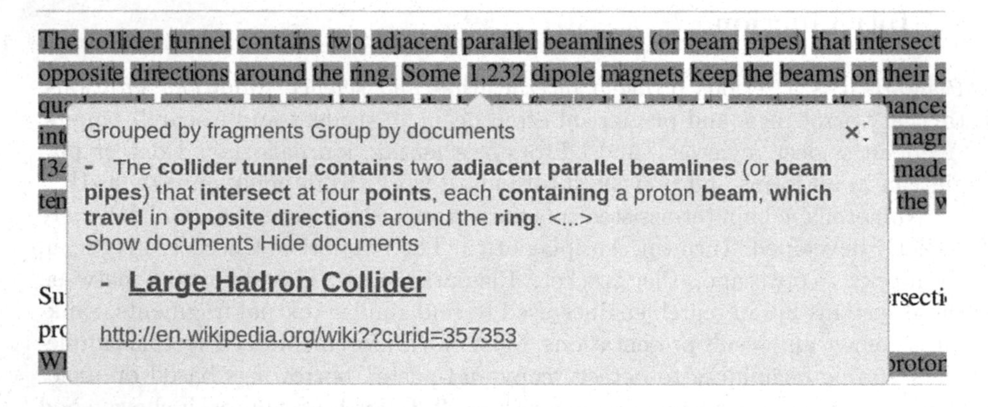

Fig. 2. Found reused fragments in the checked document and their sources

of the publishing to distinguish the sources of borrowings from the documents that reuse fragments from the checked document. The fragment of the user interface with plagiarism detection results is presented in Fig. 1.

The diagram shows the percentage of original fragments, the percentage of potentially incorrectly borrowed fragments, and the percentage of the fragments that are found in documents from the bibliography of the checked text. One can compare fragments from the checked and source documents one by one or use a convenient tool shown in Fig. 2 for visualization of the non-original content in the uploaded document. This tool presents the checked document divided into pages with highlighted sentences that might be reused from the found documents.

Exactus Like extracts bibliographic references from the uploaded document and matches them with titles and authors of found source documents. Successfully matched documents have a mark in the "In references" field. The fragments from these documents are considered not to be incorrectly reused fragments. The system detects well-known fragments (that are shared by at least 10 documents). They are also presented on the results page.

Currently Exactus Like indexes about 3 million documents in Russian (PhD theses, student essays, etc.) and 5.5 million documents in English (ArXiv, ACL, the dump of Wikipedia from June 2015). The size of the index database is about 300 GB. Users are not restricted to the collections provided by the system, one can search in the whole web. This functionality becomes available only after searching the collections. Only limited amount of sentences (200), for which nothing was found in the collections, are sent to the Yandex search engine.

3 Architecture and Implemented Approach

The architecture of Exactus Like comprises the following main subsystems: (a) crawling subsystem; (b) linguistic analysis subsystem; (c) index database subsystem; (d) search subsystem (e) web user interface.

The crawling subsystem downloads documents and extracts texts and the corresponding metadata from documents and side web-pages (i.e. sitemaps) using XPath rules and regular expressions. The linguistic analysis subsystem performs deep parsing of texts, which includes postagging, syntactic parsing, semantic role labeling, and semantic relation extraction [5]. The index database subsystem contains a set of incrementally updatable indexes, which provide an effective data access for the search subsystem.

The search subsystem implements the following approach. First, we use the inverted spectral index for searching for documents on the topic of the suspicious document. This index stores a mapping from single words and two-word noun phrases to their TF-IDF weights [6] (as the modification of the inverted index described in [7]). IDF weights are calculated based on word and phrase frequencies in the all collections. The 600 most similar documents are retrieved on this stage. We will refer to them as candidates. The following operations are performed only on the candidates. Second, we choose sentences from the suspicious document. For the text alignment, we select top 2000 weighted sentences using various kinds of filters: a TF-IDF weight threshold, a length of a sentence, a complexity of a syntactic structure, etc. Third, we intersect each selected sentence from the suspicious document with all other sentences from the candidates. We use fast set intersection algorithm [8] to exclude irrelevant sentences with unmatched lexis. Pairs of sentences that share at least 50 % of words are passed to the next stage. Fourth, the calculation of a sentence similarity is performed on the basis of the similarity evaluation of the two graphs that present the syntax and semantic structures of the sentences [9].

For search on the whole web, we use the approach that was evaluated at PAN CLEF 2014 and scored at the level of the top-rated systems [9].

Internally Exactus Like is a distributed system currently running on 4 servers (quad-core CPU, 16 GB RAM, HDD RAID). The mean processing time for a document (250 selected sentences on average) under the stress testing with 20 active parallel checks is about 20 s (47 % – linguistic analysis, 48 % – search, 5 % – other operations).

4 Conclusion

The demo of Exactus Like is available online at http://like.exactus.ru/index.php/en/. We are working on computational performance of our linguistic tools to provide a faster detection. Our current research is focused on the detection of heavily disguised plagiarism.

Acknowledgments. The reported study was partially funded by RFBR, according to the research projects No. 14-07-31149 mol_a and 16-37-60048 mol_a_dk.

References

1. Osipov, G., Smirnov, I., Tikhomirov, I., Sochenkov, I., Shelmanov, A., Shvets, A.: Information retrieval for R&D support. In: Paltoglou, G., Loizides, F., Hansen, P. (eds.) Professional Search in the Modern World. LNCS, vol. 8830, pp. 45–69. Springer, Heidelberg (2014)
2. Stein, B.: Fuzzy-fingerprints for text-based information retrieval. In: Proceedings of the 5th International Conference on Knowledge Management, pp. 572–579 (2005)
3. Brin, S., Davis, J., Garcia-Molina, H.: Copy detection mechanisms for digital documents. In: Proceedings of the 1995 ACM SIGMOD International Conference on Management of Data, vol. 24, pp. 398–409 (1995)
4. Hagen, M., Potthast, M., Stein, B.: Source retrieval for plagiarism detection from large web corpora: recent approaches. In: Working Notes of CLEF 2015 - Conference and Labs of the Evaluation Forum (2015)
5. Osipov, G., Smirnov, I., Tikhomirov, I., Shelmanov, A.: Relational-situational method for intelligent search and analysis of scientific publications. In: Proceedings of the Workshop on Integrating IR technologies for Professional Search, in conjunction with the 35th European Conference on Information Retrieval, vol. 968, pp. 57–64 (2013)
6. Shvets, A., Devyatkin, D., Sochenkov, I., Tikhomirov, I., Popov, K., Yarygin, K.: Detection of current research directions based on full-text clustering. In: Proceedings of Science and Information Conference, pp. 483–488. IEEE (2015)
7. Elsayed, T., Lin, J., Oard, D.W.: Pairwise document similarity in large collections with mapreduce. In: Proceedings of the 46th Annual Meeting of the Association for Computational Linguistics on Human Language Technologies: Short Papers, pp. 265–268 (2008)
8. Takuma, D., Yanagisawa, H.: Faster upper bounding of intersection sizes. In: Proceedings of the 36th International ACM SIGIR Conference on Research and Development in Information Retrieval, pp. 703–712 (2013)
9. Zubarev, D., Sochenkov, I.: Using sentence similarity measure for plagiarism source retrieval. In: Working Notes for CLEF 2014 Conference, pp. 1027–1034 (2014)

Jitter Search: A News-Based Real-Time Twitter Search Interface

Flávio Martins[1](✉), João Magalhães[1], and Jamie Callan[2]

[1] NOVA LINCS, DI, Faculty of Science and Technology, Universidade NOVA de Lisboa, Caparica, Portugal
flaviomartins@acm.org, jm.magalhaes@fct.unl.pt

[2] School of Computer Science, LTI, Carnegie Mellon University, Pittsburgh, USA
callan@cs.cmu.edu

Abstract. In this demo we show how we can enhance real-time microblog search by monitoring news sources on Twitter. We improve retrieval through query expansion using pseudo-relevance feedback. However, instead of doing feedback on the original corpus we use a separate Twitter news index. This allows the system to find additional terms associated with the original query to find more "interesting" posts.

1 Introduction

This demo presents a real-time search system that monitors news sources on the social network Twitter and provides an online search interface that performs query expansion on the users' queries using terms extracted from news headlines. We tackle a specific problem related to news aggregated search, which deals with the integration of fresh content extracted from news article collections into the microblog retrieval process. In this work, news published by reputable news sources on Twitter are indexed into different shards to be used in query expansion. The demo is available online at https://jitter.io.

2 Jitter Search

Our system listens to the Twitter 1 % public sample stream and indexes this information in real-time into Lucene[1] indexes and provides facilities for this data to be searched in near real-time. To improve results and provide an improved experience the system also indexes all the tweets published by a curated list of news sources and other media producers, which are later used for query expansion. A manually curated source-topic mapping was created to build topical shards i.e., each shard indexes a set of sources and corresponding documents by topic. Different topical shards are created inspired by the topics suggested by Twitter in their account creation process and the subjects of the TREC Microblog queries. We arrived at the following manually curated topic-based

[1] https://lucene.apache.org.

N. Ferro et al. (Eds.): ECIR 2016, LNCS 9626, pp. 841–844, 2016.
DOI: 10.1007/978-3-319-30671-1_77

categories: sports, entertainment, movies, music, politics, news, breaking, technology, and science.

We repurposed distributed information retrieval methods to select the topical shards used for the expansion of a given query. Specifically, we use the class of resource selection algorithms that represent the shard's content with a central sample index (CSI). These approaches have been demonstrated to be highly effective and provide an efficient shard ranking process by several previous studies. We used CRCS [2], which uses a strategy similar to the ReDDE [3] algorithm, where the query is submitted to a CSI and the top n retrieved documents are assumed to be relevant. However, CRCS focuses on selecting the collections with high-quality documents (high-precision) while ReDDE selects the collections with the highest number of relevant documents (high recall).

2.1 Frontpage

The front page (see Fig. 1) presents a large traditional search box where the users can enter keyword queries. In addition, a live news stream is shown as a timeline below the search box with the most recent news appearing at the top. It shows the news posted on Twitter by known media producers hinting that the system is capable of monitoring Twitter in real-time. The user can observe what news are being disseminated at this moment by different media outlets on a variety of topics: places and events, brands and products, and other entities. This feature enhances discoverability by letting the user peek into the stream of news.

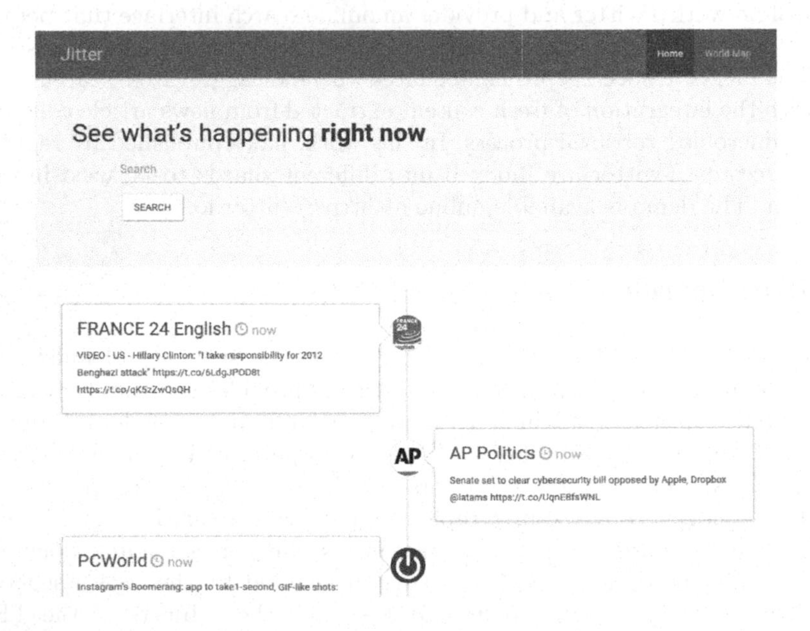

Fig. 1. Jitter search front page.

2.2 Searching

Once the user submits a query, the system shows the results page (see Fig. 2). The system retrieves a ranked list of the most interesting tweets and presents them in the main area of the results page in the middle. Each tweet is accompanied by its age and authorship information including the username and a profile image.

This ranked list of tweets is obtained by augmenting the original query with additional high quality terms extracted from a Twitter news corpus. First, the system uses the original query to retrieve a set of candidate pseudo-relevant documents from the indexed news stream. Second, the system uses standard resource selection algorithms over this candidate set to obtain a rank of the most probable topics for the query. Then the most probable topics are selected and the candidate set is filtered to remove news not belonging to these topics. Finally, the filtered candidate set of news is given as input to a pseudo-relevance feedback method and the terms obtained are added to the original query to obtain the final expanded query, which is used to retrieve the results from the tweets index.

A sidebar on the left shows the top 5 topics and the top news sources detected for that query. The user is able override the topics selected automatically and fine tune their search to one of the topics presented. He can select the desired topic and obtain a new ranked list of tweets. The sidebar on the right presents tweets from the news stream, which are filtered by the topics selected at the time i.e., the topics selected automatically by the system or the user's selected topic.

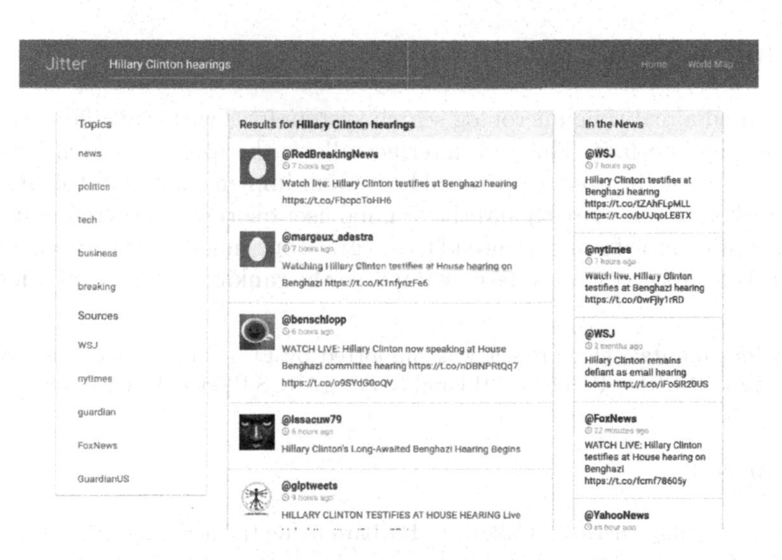

Fig. 2. Presentation of search results on Jitter search.

2.3 World Map

In Fig. 3 we can see a world map showing live Twitter activity. A heatmap is updated continuously with new locations extracted from geotagged tweets.

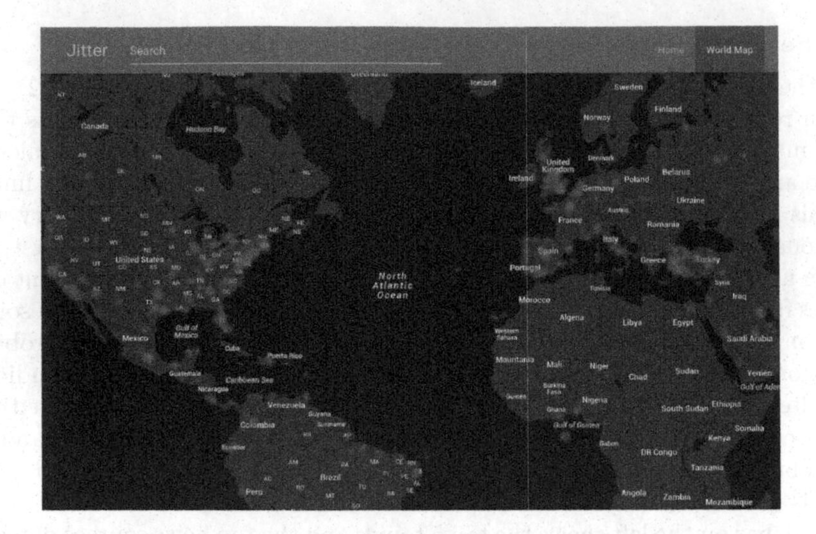

Fig. 3. Jitter world map: real-time heatmap of geotagged tweets (Color figure online).

The regions of the heatmap with a higher volume of tweets show up in shades of red and other locations show up in shades of green.

3 Summary and Future Work

We presented a real-time microblog search system for Twitter available through an online web interface. This web interface allows the users to search Twitter's 1 % public sample in near real-time. The system implements an automatic relevance feedback method to expand the original user queries and therefore improve search results. In the future we intend to integrate our time-aware learning to rank method [1] into the system to largely improve the ranking of temporal queries.

Acknowledgments. This project was supported by FCT/MEC under the projects: GoLocal CMUP-ERI/TIC/0046/2014 and NOVA LINCS PEst UID/CEC/04516/2013.

References

1. Martins, F., Magalhães, J., Callan, J.: Barbara made the news: mining the behavior of crowds for time-aware learning to rank. In: Proceedings of the WSDM 2016. ACM, San Francisco (2016)
2. Shokouhi, M.: Central-rank-based collection selection in uncooperative distributed information retrieval. In: Amati, G., Carpineto, C., Romano, G. (eds.) ECiR 2007. LNCS, vol. 4425, pp. 160–172. Springer, Heidelberg (2007)
3. Si, L., Callan, J.: Relevant document distribution estimation method for resource selection. In: Proceedings of SIGIR 2003, pp. 298–305. ACM, New York (2003)

TimeMachine: Entity-Centric Search and Visualization of News Archives

Pedro Saleiro[1,2(✉)], Jorge Teixeira[1,2,3], Carlos Soares[1,2,4], and Eugénio Oliveira[1,2,3]

[1] Labs Sapo UP, University of Porto, Rua Dr. Roberto Frias s/n, Porto, Portugal
pssc@fe.up.pt
[2] DEI-FEUP, University of Porto, Rua Dr. Roberto Frias s/n, Porto, Portugal
[3] LIACC, University of Porto, Rua Dr. Roberto Frias s/n, Porto, Portugal
[4] INESC-TEC, University of Porto, Rua Dr. Roberto Frias s/n, Porto, Portugal

Abstract. We present a dynamic web tool that allows interactive search and visualization of large news archives using an entity-centric approach. Users are able to search entities using keyword phrases expressing news stories or events and the system retrieves the most relevant entities to the user query based on automatically extracted and indexed entity profiles. From the computational journalism perspective, TimeMachine allows users to explore media content through time using automatic identification of entity names, jobs, quotations and relations between entities from co-occurrences networks extracted from the news articles. TimeMachine demo is available at http://maquinadotempo.sapo.pt/.

1 Introduction

Online publication of news articles has become a standard behavior of news outlets, while the public joined the movement either using desktop or mobile terminals. The resulting setup consists of a cooperative dialog between news outlets and the public at large. Latest events are covered and commented by both parties in a continuous basis through the social networks, such as Twitter. At the same time, it is necessary to convey how story elements are developed over time and to integrate the story in the larger context. This is extremely challenging when journalists have to deal with news archives that are growing everyday in a thousands scale. Never before has computation been so tightly connected with the practice of journalism. In recent years, computer science community have researched [1–8] and developed[1] new ways of processing and exploring news archives to help journalists perceiving news content with an enhanced perspective.

TimeMachine, as a computational journalism tool, brings together a set of Natural Language Processing, Text Mining and Information Retrieval technologies to automatically extract and index entity related knowledge from the news articles [5–11]. It allows users to issue queries containing keywords and phrases about news stories or events, and retrieves the most relevant entities mentioned

[1] NewsExplorer (IBM Watson): http://ibm.co/1OsBO1a.

N. Ferro et al. (Eds.): ECIR 2016, LNCS 9626, pp. 845–848, 2016.
DOI: 10.1007/978-3-319-30671-1_78

in the news articles through time. TimeMachine provides readable and user friendly insights and visual perspective of news stories and entities evolution along time, by presenting co-occurrences networks of public personalities mentioned on news, following a force atlas algorithm [12] for the interactive and real-time clustering of entities.

2 News Processing Pipeline

The news processing pipeline, depicted in Fig. 1, starts with a news cleaning module which performes the boilerplate removal from the news raw files (HTML/XML). Once the news content is processed we apply the NERD module which recognizes entity mentions and disambiguates each mention to an entity using a set of heuristics tailored for news, such as job descriptors (e.g. "Barack Obama, president of USA") and linguistic patterns well defined for the journalistic text style. We use a bootstrap approach to train the NER system [11]. Our method starts by annotating persons names on a dataset of 50,000 news items. This is performed using a simple dictionary-based approach. Using such training set we build a classification model based on Conditional Random Fields (CRF). We then use the inferred classification model to perform additional annotations of the initial seed corpus, which is then used for training a new classification model. This cycle is repeated until the NER model stabilizes. The entity snippet extraction consists of collecting sentences containing mentions to a given entity. All snippets are concatenated generating an entity document, which will be indexed in the entity index. The entity index represents the frequency of co-occurrence of each entity with each term that it occurs with in the news. Therefore, by relying on the redundancy of news terms and phrases associated with an entity we are able to retrieve the most relevant entity to a given input keyword or phrase query. As we also index the snippet datetime it is possible to filter query results based on a time span. For instance, the keyword "corruption" might retrieve a different entity list results in different time periods. Quotations are typically short and very informative sentences, which may directly or indirectly quote a given entity. Quotations are automatically extracted (refer to "Quotations Extraction" module) using linguistic patterns, thus enriching the information extracted for each entity. Finally, once we have all mentioned entities in a given news articles we extract entity tuples representing co-occurrences of entities in a given news article and update the entity graph by incrementing the number of occurrences of a node (entity) and creating/incrementing the number of occurrences of the edge (relation) between any two mentions.

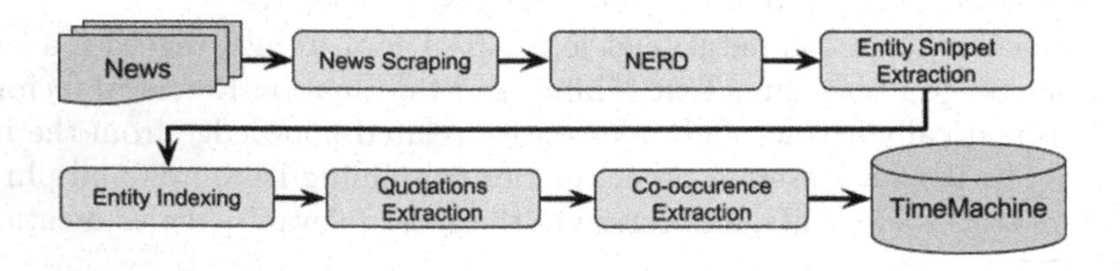

Fig. 1. News processing pipeline.

3 Demonstration

The setup for demonstration uses a news archive of Portuguese news. It comprises two different datasets: a repository from the main Portuguese news agency (1990–2010), and a stream of online articles provided by the main web portal in Portugal (SAPO) which aggregates news articles from 50 online newspapers. By the time of writing this paper, the total number of news articles used in this demonstration comprises over 12 million news articles. The system is working on a daily basis, processing articles as they are collected from the news stream. TimeMachine allows users to explore its news archive through an entity search box or by selecting a specific date. Both options are available on the website homepage and in the top bar on every page. There are a set of "stories" recommendations on the homepage suited for first time visitors. The entity search box is designed to be the main entry point to the website as it is connected to the entity retrieval module of TimeMachine.

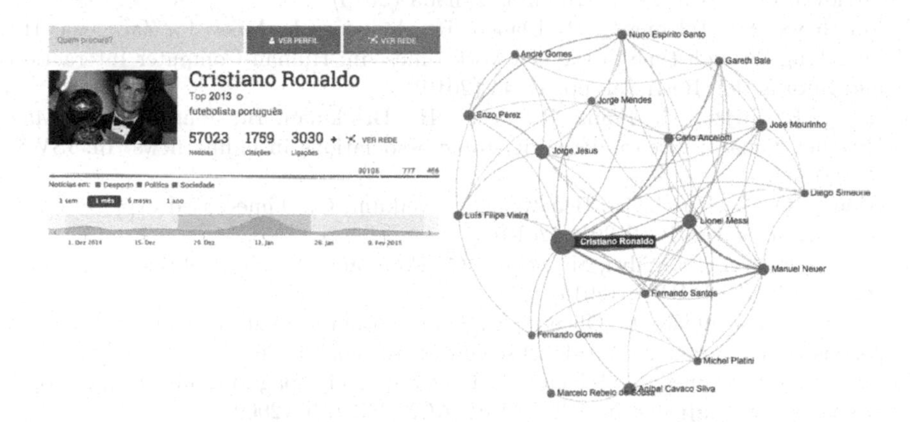

Fig. 2. Cristiano Ronaldo page headline (left) and egocentric network (right).

Users may search for surface names of entities (e.g. "Cristiano Ronaldo") if they know which entities they are interested to explore in the news, although the most powerful queries are the ones containing keywords or phrases describing topics or news stories, such as "eurozone crisis" or "ballon d'or nominees". When selecting an entity from the ranked list of results, users access the entity profile page which containing a set of automatically extracted entity specific data: name, profession, a set of news articles, quotations from the entity and related entities. Figure 2, left side, represents an example of the entity profile headline. The entity timeline allows users to navigate entity specific data through time. By selecting a specific period, different news articles, quotations and related entities are retrieved. Furthermore, users have the option of "view network" which consists in a interactive network depicting connections among entities mentioned in news articles for the selected time span. This visualization is depicted in Fig. 2, right side, and it is implemented using the graph drawing library Sigma JS, together with "Force Atlas" algorithm for the clustering of entities. Nodes consist of entities and edges represent a co-occurrence of mentioned entities in the same news article. The size of the nodes and the width of edges is proportional to the number of mentions and co-occurrences, respectively. Different node colors represent specific news topics where entities were mentioned. By selecting a date interval on the homepage, instead of issuing a query, users get a global interactive network of mentions and co-occurrences of the most frequent entities mentioned in the news articles for the selected period of time.

As future work we plan to enhance TimeMachine with semantic extraction and retrieval of relations between mentioned entities.

References

1. Demartini, G., Missen, M.M.S., Blanco, R., Zaragoza, H.: Taer: time aware entity retrieval. In: CIKM. ACM, Toronto, Canada (2010)
2. Matthews, M., Tolchinsky, P., Blanco, R., Atserias, J., Mika, P., Zaragoza, H.: Searching through time in the new york times. In: Human-Computer Interaction and Information Retrieval, pp. 41–44 (2010)
3. Balog, K., Rijke, M., Franz, R., Peetz, H., Brinkman, B., Johgi, I.: and Max Hirschel. Sahara: discovering entity-topic associations in online news. In: ISWC (2009)
4. Alonso, O., Berberich, K., Bedathur, S., Weikum, G.: Time-based exploration of news archives. In: HCIR 2010 (2010)
5. Saleiro, P., Amir, S., Silva, M., Soares, C.: Popmine: Tracking political opinion on the web. In IEEE IUCC (2015)
6. Teixeira, J., Sarmento, L., Oliveira, E.: Semi-automatic creation of a reference news corpus for fine-grained multi-label scenarios. In: CISTI (2011)
7. Sarmento, L., Nunes, S., Teixeira, J., Oliveira, E.: Propagating fine-grained topic labels in news snippets. In: IEEE/WIC/ACM WI-IAT (2009)
8. Abreu, C., Teixeira, J., Oliveira, E.: Encadear encadeamento automático de notícias. Linguistica, Informatica e Traducao: Mundos que se Cruzam, Oslo Studies in Language **7**(1), 2015 (2015)
9. Saleiro, P., Rei, L., Pasquali, A., Soares, C.: Popstar at replab 2013: name ambiguity resolution on twitter. In: CLEF (2013)
10. Saleiro, P., Sarmento, L.: Piaf vs Adele: classifying encyclopedic queries using automatically labeled training data. In OAIR (2013)
11. Teixeira, J., Sarmento, L., Oliveira, E.: A bootstrapping approach for training a NER with conditional random fields. In: Antunes, L., Pinto, H.S. (eds.) EPIA 2011. LNCS, vol. 7026, pp. 664–678. Springer, Heidelberg (2011)
12. Jacomy, M., Venturini, T., Heymann, S., Bastian, M.: Forceatlas2, a continuous graph layout algorithm for handy network visualization designed for the gephi software. PLoS ONE (2014)

OPMES: A Similarity Search Engine for Mathematical Content

Wei Zhong and Hui Fang(✉)

University of Delaware, Newark, DE, USA
{zhongwei,hfang}@udel.edu

Abstract. This paper presents details about a new mathematical search engine, i.e., *OPMES*. This search engine leverages operator trees in both representation and relevance modeling of the mathematical content. More specifically, *OPMES* represents mathematical expressions using operator trees, and then indexes each expression based on all the leaf-root paths of the generated operator tree. Such data structures enable *OPMES* to implement an efficient two-stage query processing technique. The system first identifies structurally relevant expressions based on the matching of the leaf-root paths, and then further ranks them based on their symbolic similarity to the query.

1 Introduction

Mathematical content is widely used in technical documents such as the publications and course materials from STEM fields. To better utilize such a valuable digitalized mathematical asset, it is important to offer search users the ability to find similar mathematical expressions. For example, some students may want to collect additional information about the formula that they have learned in the class, and others may want to find an existing proof for an equation. Unfortunately, major search engines do not support the similarity-based search for mathematical content.

The goal of this paper is to present our efforts on developing *OPMES* (Operator-tree Pruning based Math Expression Search), a similarity-based search engine for mathematical content. Given a query written as a mathematical expression, the system will return a ranked list of relevant math expressions from the underlying math collections.

Compared with existing mathematical search systems, such as MIaS[1], Tangent[2], and Zentralblatt math from Math Web Search[3] (MWS), the developed *OPMES* is unique in that operator trees [1] are leveraged in all the system components to enable efficient and effective search.

More specifically, *OPMES* parses an math expression into an operator tree, and then extracts leaf-root paths from the operator tree to represent structural

[1] https://mir.fi.muni.cz/mias/.
[2] http://saskatoon.cs.rit.edu/tangent/random.
[3] http://search.mathweb.org.

N. Ferro et al. (Eds.): ECIR 2016, LNCS 9626, pp. 849–852, 2016.
DOI: 10.1007/978-3-319-30671-1_79

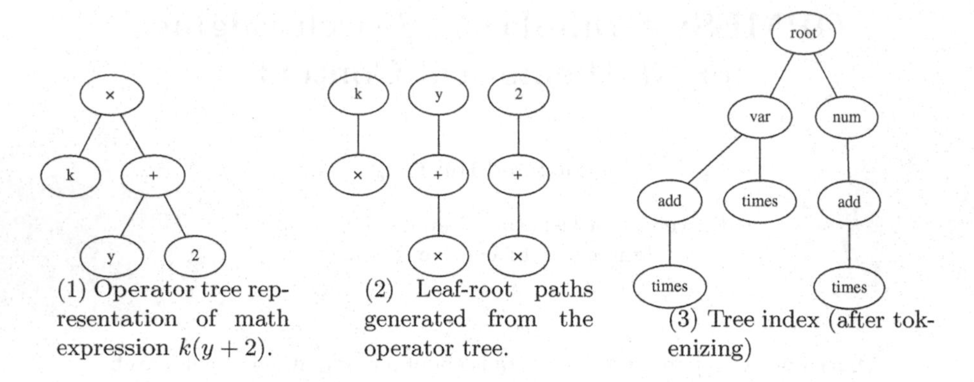

(1) Operator tree representation of math expression $k(y + 2)$.

(2) Leaf-root paths generated from the operator tree.

(3) Tree index (after tokenizing)

Fig. 1. Operator tree index example

information. An example of a math expression operator tree representation and its extracted leaf-root paths is illustrated in (1) and (2) of Fig. 1. The intuition is that no matter how operands are ordered, an operator tree uniquely determines the leaf-root paths decomposed from the tree. This property implies the advantage to use leaf-root paths from operator tree as indices or keys to retrieve math expressions (as previously proposed by [2–4]) in which a large portion of commutative operators is present. Built on top of this idea, *OPMES* further constructs a global tree index from all indexed math expressions, by continuously inserting tokenized leaf-root paths into this persistent tree index, as shown by Fig. 1 (3). Using this index, *OPMES* is able to search for structurally relevant expressions efficiently through a pruning method. Then, *OPMES* evaluates symbolic similarity between query and document expressions to finally rank search results based on symbol set similarity. For example, $E = mc^2$ is considered more meaningful when exact symbols are used rather than just being structurally identical with $y = ax^2$. We also need to rank documents higher if they are *α-equivalent* to query, since changes of symbols in an expression preserve more syntactic similarity when these changes are made by substitution, e.g. for query $x(1 + x)$, expression $a(1 + a)$ are considered more relevant than $a(1 + b)$.

The demo page of the *OPMS* is avaialble at http://tkhost.github.io/opmes, and the source code can be downloaded at https://github.com/t-k-/OPMES. Students or scholars who have the need to search mathematical Q&A website or math-content articles are potential users of our search engine.

2 System Description

We now provide the details for three major components in our system: parser, indexer and searcher.

OPMES parser is responsible to extract math mode LaTeX markups from HTML files. A LALR (look-ahead LR) parser implemented by Bison/Flex is used to transform math mode LaTeX markup into in-memory operator tree from

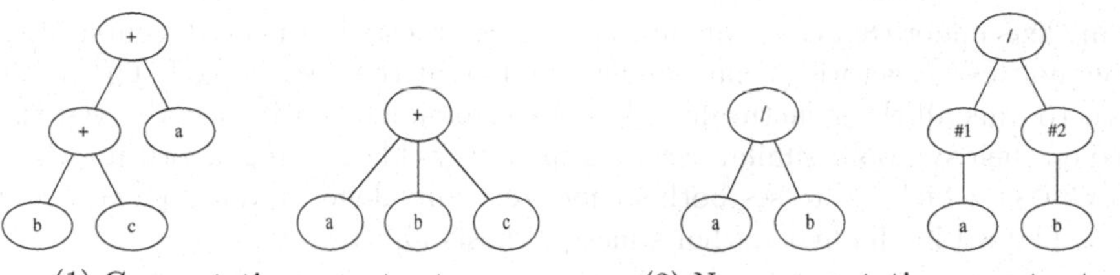

(1) Commutative operator tree transformation for $a + b + c$.

(2) Non-commutative operator tree transformation for $\frac{a}{b}$.

Fig. 2. Operator tree transformation example

bottom up. In the case of constructing an operator tree with a commutative nodes, if a commutative node has a father operator who is also commutative, the node will pass its children to its father and delete itself (see (1) of Fig. 2) so that we make sure any sub-expression also represents a subtree in operator tree. On the other hand, when an operator tree of non-commutative node is constructed, it will insert different pseudo nodes (in our implementation we use the rank of the corresponding child) on top of its children (see (2) of Fig. 2) to distinguish their operands order in generated leaf-root paths.

The indexer then uses this operator tree to extract leaf-root paths. Note that a subtree of an operator tree (corresponding to a subexpression) would have its leaf-root paths being prefix of some leaf-root paths from this operator tree. To show this, use the same example in Fig. 1, in which $k(y+2)$ has a leaf-root path set $A = \{k\times, y+\times, 2+\times\}$, while the subexpression $y+2$ has a leaf-root path set $B = \{y+, 2+\}$ where each element is prefix of A's subset $\{y+\times, 2+\times\}$ respectively. OPMES indexer therefore tokenizes and inserts leaf-root paths into a "prefix tree" index as shown in Fig. 1 (3), to speed the lookup for another similar leaf-root path. The indexer also attaches (appends) an expression ID (or exprID) to the bottom node (of the newly inserted leaf-root path) every time a tokenized leaf-root path is inserted. In the case of Fig. 1 (3), all the three times nodes are linked to a separate "posting list", the new exprID is appended after each. For implementation simplicity, we use file system directories to realize the index tree, the path name of each directory corresponds to a tokenized leaf-root path, and the posting list file of each node is stored at corresponding directory which represents the node it belongs to (every node has only one posting file).

OPMES searcher takes a query, parses it into operator tree, and decomposes the operator tree into leaf-root paths. Query processing step is divided into two stages. The first stage is to search for structurally relevant expressions. Instead of searching each leaf-root path one by one, the searcher searches simultaneously along the way of all leaf-root paths in the index tree, and merges the exprIDs from posting lists in all the corresponding directories of the query paths. Moreover, if all query leaf-root paths can go one deeper level in the index tree, and the deeper level nodes have a common node (with the same tokenized name), the searcher will simultaneously go into the common node from all query paths and merge the posting lists under that common directory. This process is repeated recursively to prune indexes (directories) that are not common at the deeper level. The second step is to rank all the structurally relevant expressions identified in the first step based on their symbolic similarities with the query. The scoring algorithm MARK AND CROSS, which addresses both symbol set equivalence and α-equivalence, is fully explained in the first author's master thesis [5].

3 Demonstration Plan

In our demo, we first illustrate some key ideas mentioned above. We will choose a simple query, show its operator tree representation in ASCII graph as well as its leaf-root paths (through the output of parser). Then we demonstrate the structure of our index tree, and walk through the steps and directories where the searcher goes and finds relevant expressions for input query. Users are invited to enter queries and experience our search engine based on a collection (with over 8 million math expressions scrawled from Math Stack Exchange website) that contains most frequently used and elementary math expressions.

4 Conclusion and Future Work

The paper describes a new similarity-based mathematical search engine, i.e., *OPMES*. Operator trees are used in almost all the system components to facilitate the representation, query processing and relevance modeling of the mathematical content.

As for the future work, we plan to enhance the system with MathML parser and wildcard support. Moreover, we also plan to integrate text search ability into our math-only search method. Finally, equivalent math-expression transformations (such as $a + \frac{1}{b} = \frac{ab+1}{b}$) can also be introduced into pre-query process to further improve the usefulness of math similarity search engine.

References

1. Zanibbi, R., Blostein, D.: Recognition and retrieval of mathematical expressions. Int. J. Doc. Anal. Recogn. (IJDAR) **15**(4), 331–357 (2012)
2. Ichikawa, H., Hashimoto, T., Tokunaga, T., Tanaka, H.: New methods of retrieve sentences based on syntactic similarity. IPSJ SIG Technical Reports, pp. 39–46 (2005)
3. Hijikata, Y., Hashimoto, H., Nishida, S.: An investigation of index formats for the search of mathml objects. In: Web Intelligence/IAT Workshops, pp. 244–248. IEEE (2007)
4. Yokoi, K., Aizawa, A: An approach to similarity search for mathematical expressions using MathML. In: Towards a Digital Mathematics Library, Grand Bend, Ontario, Canada (2009)
5. Zhong, W.: A Novel Similarity-Search Method for Mathematical Content in LaTeX-Markup and Its Implementation (2015). http://tkhost.github.io/opmes/thesis-ref.pdf

SHAMUS: UFAL Search and Hyperlinking Multimedia System

Petra Galuščáková(✉), Shadi Saleh, and Pavel Pecina

Faculty of Mathematics and Physics, Institute of Formal and Applied Linguistics, Charles University in Prague, Prague, Czech Republic
{galuscakova,saleh,pecina}@ufal.mff.cuni.cz

Abstract. In this paper, we describe SHAMUS, our system for an easy search and navigation in multimedia archives. The system consists of three components. The *Search* component provides a text-based search in a multimedia collection, the *Anchoring* component determines the most important segments of videos, and segments topically related to the anchoring ones are retrieved by the *Hyperlinking* component. In the paper, we describe each component of the system as well as the online demo interface http://ufal.mff.cuni.cz/shamus which currently works with a collection of TED talks.

1 Introduction

The problem of navigation in large multimedia collections is especially emerging recently, as the sizes of multimedia archives are rapidly growing [9,10]. Efficient navigation methods are especially needed by professional archive users (e.g. historians, librarians, content producers), who are searching for particular information and need to explore topics in detail.

The text-based *Search* is probably the most common way how can users find desired information. In the SHAMUS system, we apply the *Search* on available subtitles. Instead of keyword spotting, frequently used in speech retrieval applications [5], we run an information retrieval and retrieve relevant segments of videos. In addition to the text-based *Search*, we also provide the *Hyperlinking* which recommends links to other related segments on the fly. Unlike the most multimedia recommendation systems, which are typically focused on entertaining users [1], our *Hyperlinking* system is more focused on retrieving links between segments of videos based on their topical similarity. While recommendation systems often make use of collaborative filtering methods which retrieve videos based on users' opinions [8], we strictly rely on the content of recordings. This approach can not only be useful in the case of the cold start when no information about the user or video are available but it can also be more appropriate for professional searchers and in the case of exploratory search.

The strategy diagram of the SHAMUS system is displayed in Fig. 1.

If a user wants to find particular information in the multimedia collection, the retrieval of full videos is often insufficient. The videos could be very long,

N. Ferro et al. (Eds.): ECIR 2016, LNCS 9626, pp. 853–856, 2016.
DOI: 10.1007/978-3-319-30671-1_80

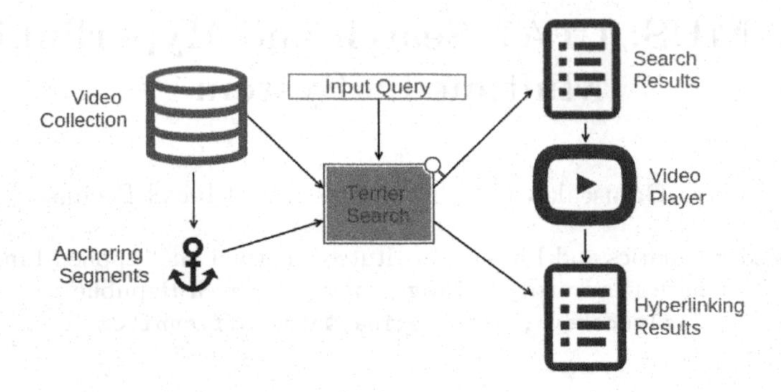

Fig. 1. SHAMUS - strategy diagram of the system.

and the user needs to skim through the retrieved recordings, what can be very tedious. Therefore, our system retrieves relevant video segments instead of full videos. Relevant segments are retrieved in both *Search* and *Hyperlinking* components, and the most informative segments are suggested by the *Anchoring* component.

2 Search, Hyperlinking, and Anchoring

The information retrieval framework Terrier [6] is used in all system components. We segment all recordings into 60-second long passages first and use them as documents in our further setup. In the *Search* component, the retrieval is run on the list of created segments and the segments relevant to the user-typed query are retrieved. In the *Hyperlinking* component, the currently played segment is converted into a text-based query using the subtitles and the same setup as the one in the *Search* component is used. To formulate the *Hyperlinking* query, we use the 20 most frequent words lying inside the query segment and filter out the most common words using the stopwords list.

Both retrieval components, including the segmentation method, were tuned on almost 2700 hours of BBC TV programmes provided in the Search and Hyperlinking Task at the MediaEval 2014 Benchmark. The setup used in the *Search* and *Hyperlinking* components is in more detail described in the Task report papers [2,3]. Even though the proposed methods are simple, they outperformed the rest of the methods submitted to the Benchmark. For the *Anchoring*, we use our system proposed for the MediaEval 2015 Search and Anchoring Task [4]. We assume that the most informative segments of videos are often similar to the video description as the description usually provides the summary of the document. Therefore, we convert the available metadata description of each file into a textual query. Information retrieval is then applied on the video segments and the highest retrieved segments are considered as the *Anchoring* ones. The list of the *Anchoring* segments is pre-generated in advance for each video.

Hyperlinking is then run on detected *Anchoring* segments. The list of the related segments is automatically regenerated on the fly when a new *Anchoring* segment begins.

3 User Interface

The user interface consists of three main sites. The first one serves for the *Search* query input. The second site displays the results of the *Search* including the metadata, transcript of the beginning of the retrieved segment and time of this segment. The video with its title, description, source, marked *Anchoring* segments, and list of related segments are displayed in the third site (see Fig. 2).

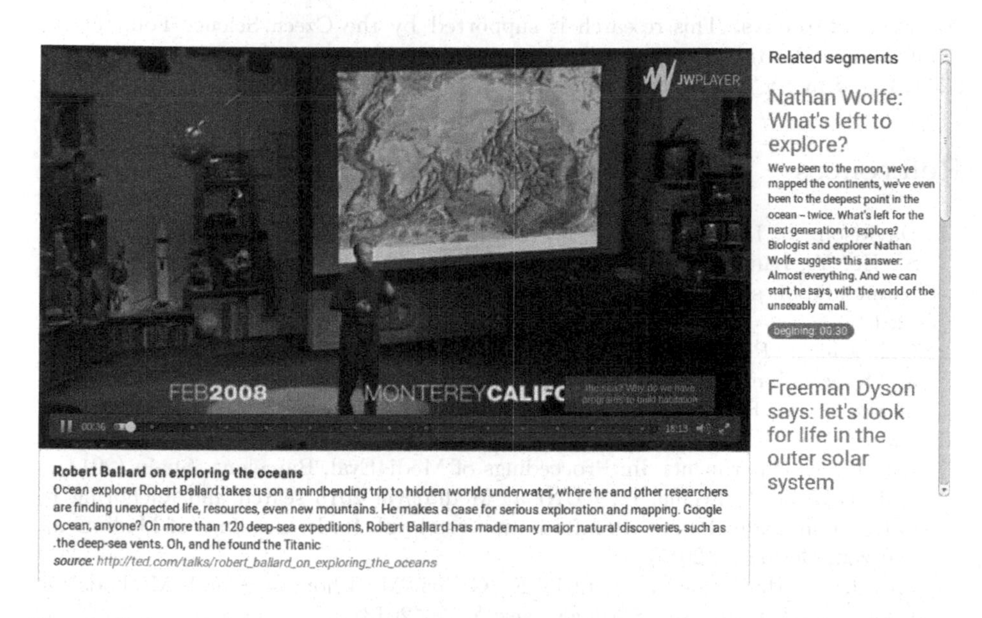

Fig. 2. Video player with detected most informative *Anchoring* segments and recommended links to related video segments.

The JWPlayer is used for the video playback. The *Anchoring* segments are marked as individual chapters of the video. The transcript of the beginning of each segment is retrieved and used instead of the chapter name – users can thus overview the most important segments of the video without the need to navigate there. The list of three most related segments is displayed on the right side, next to the video player.

SHAMUS demo interface currently works with a collection of 1219 TED talks. We used a list of talks available in the TED dataset [7] and downloaded available subtitles and videos for each talk. However, the system should be easily adaptable to work with any collection of videos for which subtitles or automatic transcripts and metadata descriptions are available.

4 Conclusion and Future Work

In the paper, we presented the user interface to the content-based video retrieval system SHAMUS. Unlike the most retrieval systems, SHAMUS works with segments of videos instead of full videos. It thus enables text-based retrieval of relevant segments, recommendation of topically related video segments and an automatic detection of the most informative segments of videos.

In the future, we plan to extend SHAMUS with the visual similarity, which we have already successfully used in our *Hyperlinking* experiments. We would also like to examine diversification of results which could be beneficial for different user groups and for users with more varied search intentions.

Acknowledgments. This research is supported by the Czech Science Foundation, grant number P103/12/G084, Charles University Grant Agency GA UK, grant number 920913, and by SVV project number 260 224.

References

1. Davidson, J., Liebald, B., Liu, J., Nandy, P., Van Vleet, T., Gargi, U., Gupta, S., He, Y., Lambert, M., Livingston, B., Sampath, D.: The YouTube video recommendation system. In: Proceedings of RecSys, pp. 293–296, Barcelona, Spain (2010)
2. Galuščáková, P., Kruliš, M., Lokoč, J., Pecina, P.: CUNI at MediaEval 2014 search and hyperlinking task: visual and prosodic features in hyperlinking. In: Proceedings of MediaEval, Barcelona, Spain (2014)
3. Galuščáková, P., Pecina, P.: CUNI at MediaEval 2014 search and hyperlinking task: search task experiments. In: Proceedings of MediaEval, Barcelona, Spain (2014)
4. Galuščáková, P., Pecina, P.: CUNI at MediaEval 2015 search and anchoring in video archives: anchoring via information retrieval. In: Proceedings of MediaEval, Wurzen, Germany (2015)
5. Moyal, A., Aharonson, V., Tetariy, E., Gishri, M.: Phonetic Search Methods for Large Speech Databases. Springer, New York (2013)
6. Ounis, I., Amati, G., Plachouras, V., He, B., Macdonald, C., Lioma, C.: Terrier: a high performance and scalable information retrieval platform. In: Proceedings of ACM SIGIR 2006 Workshop on Open Source Information Retrieval, pp. 18–25, Seattle, Washington, USA (2006)
7. Pappas, N., Popescu-Belis, A.: Combining content with user preferences for TED lecture recommendation. In: Proceedings of CBMI, pp. 47–52. IEEE, Veszprém, Hungary (2013)
8. Schafer, J.B., Frankowski, D., Herlocker, J., Sen, S.: Collaborative filtering recommender systems. In: Brusilovsky, P., Kobsa, A., Nejdl, W. (eds.) Adaptive Web 2007. LNCS, vol. 4321, pp. 291–324. Springer, Heidelberg (2007)
9. Schoeffmann, K.: A user-centric media retrieval competition: the video browser showdown 2012–2014. IEEE Multimedia **21**(4), 8–13 (2014)
10. Schoeffmann, K., Hopfgartner, F., Marques, O., Boeszoermenyi, L., Jose, J.M.: Video browsing interfaces and applications: a review. Spie rev. **1**(1), 018004 (2010)

Industry Day

Industry Day Overview

Omar Alonso[1](✉) and Pavel Serdyukov[2]

[1] Microsoft, Mountain View, CA, USA
omalonso@microsoft.com
[2] Yandex, Moscow, Russia
pavser@yandex-team.ru

Abstract. The Industry track aims to bring together information retrieval researchers, practitioners and analysts from industry and academia. Since ECIR 2006, these events have been very successful and provided many interesting talks.

1 Introduction

The goal of the Industry Day track is to bring an exciting programme that contains a mix of invited talks by industry leaders with presentations of novel and innovative ideas from the search industry. The final program consists of four invited talks by Domonkos Tikk (Gravity), Etienne Sanson (Criteo), Debora Donato (StumbleUpon) and Nicola Montecchio (Spotify), and four accepted proposal talks.

1.1 Invited Talks Abstracts

We invited technical leaders who are working on exciting and diverse topics with an emphasis on industrial system design and implementation. Domonkos Tikk's keynote talk abstract can be found on a separate section on this proceedings.

An Overview of the Challenges R&D Had to Face over the Last 10 Years (Criteo). Criteo is the leading company in the performance retargeting field. It makes sense of digital user behavior, across any device, to deliver relevant, personalized ads that drive incremental sales to our clients, using a transparent cost-per-click model and measuring value purely on post-click sales. This demanding model is supported by an engine that has to choose the most relevant products to display to a user from over billions of candidate products and to evaluate the exact value of showing this ad to the user. The engine has to do that in a few milliseconds, running millions prediction per second. In order to that, we developed a prediction engine that analyzes huge volume of data using scalable state-of-the-art machine learning algorithms. The R&D team had to tackle lots of technical challenges to build this engine. Several factors influenced the design of our engine: the business pivots in the early years of the company, the advertising industry standards evolution, the hyper-growth of the team and of the business, and the need for fast and continuous improvement to stay ahead

N. Ferro et al. (Eds.): ECIR 2016, LNCS 9626, pp. 859–862, 2016.
DOI: 10.1007/978-3-319-30671-1_81

of competition. We'll see how we addressed these challenges, both in innovative and pragmatic ways, and the lessons we learned from this ongoing journey.

Improving User Engagement Through Genre Diversification (StumbleUpon). Diversity is recognized by the Recommendation Systems community as one of the key engagement factors beyond accuracy and indeed user studies have shown that topic diversity influences content consumption across different typologies of social media. However content diversification only focus on content topics and no attention has been so far devoted to understand how diversification in document genre (blogs, news, recipes, listicles, media) affects engagement or retention. The talk will present a novel semi-supervised approach for Genre Detection, the challenges to deploy the algorithm in an industrial environment and the results of A/B tests.

Music Search, Personalization and Discovery (Spotify). Spotify is a key player in the music streaming industry, currently serving a collection of circa 30M recordings to over 75M users around the world. The simplest user interaction mode resembles the traditional IR framework of a query-answering system, in which a user chooses a particular song from a jukebox. However, in order to make the product compelling, it is vital to take into account the very nature of the collection, which prompts the development of tools that move past traditional IR techniques and exploit the information contained in usage patterns or intrinsic to the media content itself. In this talk we'll walk through various facets of the Spotify product and explore how ML and IR techniques enable these features. An overview of the ranking strategies that power the search feature, with a focus on explicit personalization, provides a starting point for a broader discussion of personalization challenges at Spotify, leading to two popular user-facing features focused on new content discovery: Fresh Finds and Discover Weekly.

1.2 Proposed Talks Abstracts

Talk proposals were selected based on problem space, real-world system readiness, and technical depth. The list of accepted proposed talks are:

1. Real-time multilingual document categorization using open source software by Alan Said (Recorded Future).
2. Thinking outside the search box: a visual approach to complex query formulation by Tony Russell-Rose (UXLabs).
3. Get on with it! Recommender system industry challenges move towards real-world, online evaluation by Roberto Turrin (Moviri/ContentWise), Martha Larson (Delft University of Technology), and Daniel Kohlsdorf (XING).
4. Multilingual query categorization by Michal Laclavik (Magnetic), Marek Ciglan (Magnetic), Sam Steingold (Magnetic), and Alex Dorman (Magnetic).

Real-Time Multilingual Document Categorization Using Open Source Software. Document categorization allows for more specific analysis of documents harvested from the Web. For instance, when attempting to identify the events that occur in a newspaper article we can assume that aggressive events (attack, hit, tackle) in sports related articles are not assigned the semantic meaning of similarly named events in warfare, etc. In this talk, we will present how we use state of the art multilingual generative topic modeling using LDA in combination with multiclass SVM classifiers to categorize documents into seven base categories. This talk will focus on the engineering aspects of building a classification system. In the talk we will present the software architecture, the combination of open source software used, and the problems encountered while implementing, testing and deploying the system. The focus will be on the usage open source software in order to build and deploy state of the art information retrieval systems

Thinking Outside the Search Box: A Visual Approach to Complex Query Formulation. UXLabs is developing 2dSearch: a radical alternative to traditional keyword search. Instead of a one dimensional search box, users express and manipulate concepts as objects on a two-dimensional canvas, using a visual syntax that is simpler and more transparent than traditional query formulation methods. This guides the user toward the formulation of syntactically correct expressions, exposing their semantics in a comprehensible manner. Concepts may be combined to form aggregate structures, such as lists (unordered sets sharing a common operator) or composites (nested structures containing a combination of sub-elements). In this way 2dSearch supports the modular creation of queries of arbitrary complexity. 2dSearch is deployed as a framework and interactive development environment for managing complex queries and search strategies. It provides support for all stages in the query lifecycle, from creation and editing, through to sharing and execution. In so doing, it offers the potential to improve query quality and efficiency and promote the adoption and sharing of templates and best practices.

Get on with it! Recommender System Industry Challenges Move Towards Real-World, Online Evaluation. Recommender systems have enormous commercial importance in connecting people with content, information, and services. Historically, the recommender systems community has benefited from industry-academic collaborations and the ability of industry challenges to drive forward the state of the art. However, today's challenges look very different from the NetFlix Prize in 2009. This talk features speakers representing two ongoing recommendation challenges that typify the direction in which such challenges are rapidly evolving. The first direction is the move to leverage information beyond the user-item matrix. Success requires algorithms capable of integrating multiple sources of information available in real-world scenarios. The second direction is the move from evaluation on offline data sets to evaluation with online systems. Success requires algorithms that are able to produce recommendations that satisfy users, but that are also able to satisfy technical constraints (i.e., response time, system availability) as well as business metrics (i.e., item coverage). The challenges covered in this talk are supported by the EC-funded project CrowdRec, which studies stream recommendation and user engagement for next-generation recommender systems. We describe each in turn.

Multilingual Query Categorization. Classification of short text into a predefined hierarchy of categories is a challenge. The need to categorize short texts arises in multiple domains: keywords and queries in online advertising, improvement of search engine results, analysis of tweets or messages in social networks, etc. We leverage community moderated, freely available data sets (Wikipedia, DBPedia, Wikidata) and open source tools (Hadoop, Spark, Solr, Lucene) to build a flexible and extensible classification model for English and other languages (Spanish, French, German). We will share our experiences in building a real world data science system working with four languages that scales to production data volumes of more than 8 billion keyword classifications per month. We will touch on some aspect of unlocking knowledge hidden in Wikipedia with help of IR and simple NLP techniques. We will also discuss multilingual customization effort and challenges.

Workshops

Bibliometric-Enhanced Information Retrieval: 3rd International BIR Workshop

Philipp Mayr[1(✉)], Ingo Frommholz[2], and Guillaume Cabanac[3]

[1] GESIS – Leibniz Institute for the Social Sciences, Unter Sachsenhausen 6-8, 50667 Cologne, Germany
philipp.mayr@gesis.org

[2] Department of Computer Science and Technology, University of Bedfordshire, Luton, UK
ingo.frommholz@beds.ac.uk

[3] Department of Computer Science, IRIT UMR 5505 CNRS, University of Toulouse, 118 Route de Narbonne, 31062 Toulouse Cedex 9, France
guillaume.cabanac@univ-tlse3.fr

Abstract. The BIR workshop brings together experts in Bibliometrics and Information Retrieval. While sometimes perceived as rather loosely related, these research areas share various interests and face similar challenges. Our motivation as organizers of the BIR workshop stemmed from a twofold observation. First, both communities only partly overlap, albeit sharing various interests. Second, it will be profitable for both sides to tackle some of the emerging problems that scholars face today when they have to identify relevant and high quality literature in the fast growing number of electronic publications available worldwide. Bibliometric techniques are not yet used widely to enhance retrieval processes in digital libraries, although they offer value-added effects for users. Information professionals working in libraries and archives, however, are increasingly confronted with applying bibliometric techniques in their services. The first BIR workshop in 2014 set the research agenda by introducing each group to the other, illustrating state-of-the-art methods, reporting on current research problems, and brainstorming about common interests. The second workshop in 2015 further elaborated these themes. This third BIR workshop aims to foster a common ground for the incorporation of bibliometric-enhanced services into scholarly search engine interfaces. In particular we will address specific communities, as well as studies on large, cross-domain collections like Mendeley and ResearchGate. This third BIR workshop addresses explicitly both scholarly and industrial researchers.

Keywords: Bibliometrics · Scientometrics · Informetrics · Information retrieval · Digital libraries

1 Introduction

IR and Bibliometrics are two fields in Information Science that have grown apart in recent decades. But today 'big data' scientific document collections (e.g., Mendeley, ResearchGate) bring together aspects of crowdsourcing, recommendations, interactive

N. Ferro et al. (Eds.): ECIR 2016, LNCS 9626, pp. 865–868, 2016.
DOI: 10.1007/978-3-319-30671-1_82

retrieval, and social networks. There is a growing interest in revisiting IR and bibliometrics to provide cutting-edge solutions that help to satisfy the complex, diverse, and long-term information needs that scientific information seekers have, in particular the challenge of the fast growing number of publications available worldwide in workshops, conferences and journals that have to be made accessible to researchers. This interest was shown in the well-attended recent workshops, such as "Computational Scientometrics" (held at iConference and CIKM 2013), "Combining Bibliometrics and Information Retrieval" (at the ISSI conference 2013) and the previous BIR workshops at ECIR. Exploring and nurturing links between bibliometric techniques and IR is beneficial for both communities (e.g., Abbasi and Frommholz, 2015; Cabanac, 2015, Wolfram, 2015). The workshops also revealed that substantial future work in this direction depends on a rise in ongoing awareness in both communities, manifesting itself in tangible experiments/exploration supported by existing retrieval engines.

It is also of growing importance to combine bibliometrics and information retrieval in real-life applications (see Jack et al., 2014; Hienert et al., 2015). These include monitoring the research front of a given domain and operationalizing services to support researchers in keeping up-to-date in their field by means of recommendation and interactive search, for instance in 'researcher workbenches' like Mendeley /ResearchGate or search engines like Google Scholar that utilize large bibliometric collections. The resulting complex information needs require the exploitation of the full range of bibliometric information available in scientific repositories. To this end, this third edition of the BIR workshop will contribute to identifying and explorating further applications and solutions that will bring together both communities to tackle this emerging challenging task.

The first two bibliometric-enhanced Information Retrieval (BIR) workshops at ECIR 2014[1] and ECIR 2015[2] attracted more than 40 participants (mainly from academia) who engaged in lively discussions and future actions. For the third BIR workshop[3] we build on this experience.

2 Goals, Objectives and Outcomes

Our workshop aims to engage the IR community with possible links to bibliometrics. Bibliometric techniques are not yet widely used to enhance retrieval processes in digital libraries, yet they offer value-added effects for users (Mutschke, et al., 2011). To give an example, recent approaches have shown the possibilities of alternative ranking methods based on citation analysis can lead to an enhanced IR.

Our interests include information retrieval, information seeking, science modelling, network analysis, and digital libraries. Our goal is to apply insights from bibliometrics, scientometrics, and informetrics to concrete, practical problems of information retrieval and browsing. More specifically we ask questions such as:

---

1 http://www.gesis.org/en/events/events-archive/conferences/ecirworkshop2014/.
2 http://www.gesis.org/en/events/events-archive/conferences/ecirworkshop2015/.
3 http://www.gesis.org/en/events/events-archive/conferences/ecirworkshop2016/.

- The tectonics of IR and bibliometrics: convergent, divergent, or transform boundaries?
- How feasible and effective is bibliometric-enhanced IR in accomplishing specific complex search tasks, such as literature reviews and literature-based discovery (Bruza and Weeber, 2008)?
- How can we build scholarly information systems that explicitly use bibliometric measures at the user interface?
- How can models of science be interrelated with scholarly, task-oriented searching?
- How can we combine classical IR (with emphasis on recall and weak associations) with more rigid bibliometric recommendations?
- How can we develop evaluation schemes without being caught in too costly setting up large scale experimentation?
- Identifying suitable testbeds (like iSearch corpus[4])

3 Format and Structure of the Workshop

The workshop will start with an inspirational keynote by Marijn Koolen "Bibliometrics in Online Book Discussions: Lessons for Complex Search Tasks" to kick-start thinking and discussion on the workshop topic (for the keynote from 2015 see Cabanac, 2015). This will be followed by paper presentations in a format that we found to be successful at BIR 2014 and 2015: each paper is presented as a 10 min lightning talk and discussed for 20 min in groups among the workshop participants followed by 1-minute pitches from each group on the main issues discussed and lessons learned. The workshop will conclude with a round-robin discussion of how to progress in enhancing IR with bibliometric methods.

4 Audience

The audiences (or clients) of IR and bibliometrics partially overlap. Traditional IR serves individual information needs, and is– consequently – embedded in libraries, archives and collections alike. Scientometrics, and with it bibliometric techniques, has a matured serving science policy.

We propose a half-day workshop that should bring together IR and DL researchers with an interest in bibliometric-enhanced approaches. Our interests include information retrieval, information seeking, science modelling, network analysis, and digital libraries. The goal is to apply insights from bibliometrics, scientometrics, and informetrics to concrete, practical problems of information retrieval and browsing.

The workshop is closely related to the BIR workshops at ECIR 2014 and 2015 and strives to feature contributions from core bibliometricians and core IR specialists who already operate at the interface between scientometrics and IR. In this workshop, however, we focus more on real experimentations (including demos) and industrial participation.

[4] http://www.gesis.org/fileadmin/upload/issi2013/BMIR-workshop-ISSI2013-Larsen.pdf.

5 Output

The papers presented at the BIR workshop 2014 and 2015 have been published in the online proceedings http://ceur-ws.org/Vol-1143, http://ceur-ws.org/Vol-1344. We plan to set up online proceedings for BIR 2016 again. Another output of our BIR initiative was prepared after the ISSI 2013 workshop on "Combining Bibliometrics and Information Retrieval" as a special issue in Scientometrics. This special issue has attracted eight high quality papers and will appear in early 2015 (see Mayr and Scharnhorst, 2015). We aim to have a similar dissemination strategy for the proposed workshop, but now oriented towards core-IR. In this way we shall build up a sequence of explorations, visions, results documented in scholarly discourse, and create a sustainable bridge between bibliometrics and IR.

References

Abbasi, M.K., Frommholz, I.: Cluster-based polyrepresentation as science modelling approach for information retrieval. Scientometrics (2015). doi:10.1007/s11192-014-1478-1

Bruza, P., Weeber, M.: Literature-based discovery. In: Information Science and Knowledge Management series, vol. 15. Springer, Berlin (2008)

Cabanac, G.: In praise of interdisciplinary research through scientometrics. In: Proceedings of the 2nd Workshop on Bibliometric-enhanced Information Retrieval (BIR2015), pp. 5–13. Vienna, Austria: CEUR-WS.org (2015)

Hienert, D., Sawitzki, F., Mayr, P.: Digital library research in action – supporting information retrieval in sowiport. D-Lib. Mag. **21**(3/4), 150–161 (2015). doi:10.1045/march2015-hienert

Mayr, P., Scharnhorst, A., Larsen, B., Schaer, P., Mutschke, P.: Bibliometric-Enhanced information retrieval. In: de Rijke, M., Kenter, T., de Vries, A.P., Zhai, C., de Jong, F., Radinsky, K., Hofmann, K. (eds.) ECIR 2014. LNCS, vol. 8416, pp. 798–801. Springer, Heidelberg (2014)

Mayr, P., Scharnhorst, A.: Scientometrics and Information Retrieval - weak-links revitalized. Scientometrics (2015). doi:10.1007/s11192-014-1484-3

Mutschke, P., Mayr, P., Schaer, P., Sure, Y.: Science models as value-added services for scholarly information systems. Scientometrics **89**(1), 349–364 (2011). doi:10.1007/s11192-011-0430-x

Wolfram, D.: The symbiotic relationship between information retrieval and informetrics. Scientometrics (2015). doi:10.1007/s11192-014-1479-0

MultiLingMine 2016: Modeling, Learning and Mining for Cross/Multilinguality

Dino Ienco[1], Mathieu Roche[2], Salvatore Romeo[3], Paolo Rosso[4], and Andrea Tagarelli[5](✉)

[1] IRSTEA, LIRMM, Montpellier, France
dino.ienco@irstea.fr
[2] CIRAD, LIRMM, Montpellier, France
mathieu.roche@cirad.fr
[3] Qatar Computing Research Institute, Doha, Qatar
sromeo@qf.org.qa
[4] Universitat Politecnica de Valencia, Valencia, Spain
prosso@dsic.upv.es
[5] DIMES, University of Calabria, Rende, Italy
tagarelli@dimes.unical.it

Abstract. The increasing availability of text information coded in many different languages poses new challenges to modern information retrieval and mining systems in order to discover and exchange knowledge at a larger world-wide scale. The 1st International Workshop on Modeling, Learning and Mining for Cross/Multilinguality (dubbed MultiLingMine 2016) provides a venue to discuss research advances in cross-/multilingual related topics, focusing on new multidisciplinary research questions that have not been deeply investigated so far (e.g., in CLEF and related events relevant to CLIR). This includes theoretical and experimental on-going works about novel representation models, learning algorithms, and knowledge-based methodologies for emerging trends and applications, such as, e.g., cross-view cross-/multilingual information retrieval and document mining, (knowledge-based) translation-independent cross-/multilingual corpora, applications in social network contexts, and more.

1 Motivations

In the last few years the phenomenon of multilingual information overload has received significant attention due to the huge availability of information coded in many different languages. We have in fact witnessed a growing popularity of tools that are designed for collaboratively editing through contributors across the world, which has led to an increased demand for methods capable of effectively and efficiently searching, retrieving, managing and mining different language-written document collections. The multilingual information overload phenomenon introduces new challenges to modern information retrieval systems. By better searching, indexing, and organizing such rich and heterogeneous information, we can discover and exchange knowledge at a larger world-wide scale.

N. Ferro et al. (Eds.): ECIR 2016, LNCS 9626, pp. 869–873, 2016.
DOI: 10.1007/978-3-319-30671-1_83

However, since research on multilingual information is relatively young, important issues still remain uncovered:

- how to define a translation-independent representation of the documents across many languages;
- whether existing solutions for comparable corpora can be enhanced to generalize to multiple languages without depending on bilingual dictionaries or incurring bias in merging language-specific results;
- how to profitably exploit knowledge bases to enable translation-independent preserving and unveiling of content semantics;
- how to define proper indexing and multidimensional data structures to better capture the multi-topic and/or multi-aspect nature of multi-lingual documents;
- how to detect duplicate or redundant information among different languages or, conversely, novelty in the produced information;
- how to enrich and update multi-lingual knowledge bases from documents;
- how to exploit multi-lingual knowledge bases for question answering;
- how to efficiently extend topic modeling to deal with multi/cross-lingual documents in many languages;
- how to evaluate and visualize retrieval and mining results.

2 Objectives, Topics, and Outcomes

The aim of the *1st International Workshop on Modeling, Learning and Mining for Cross/Multilinguality* (dubbed *MultiLingMine 2016*)[1], held in conjunction with the 2016 ECIR Conference, is to establish a venue to discuss research advances in cross-/multilingual related topics. MultiLingMine 2016 has been structured as a *full-day* workshop. Its program schedule includes invited talks as well as a panel discussion among the participants. It is mainly geared towards students, researchers and practitioners actively working on topics related to information retrieval, classification, clustering, indexing and modeling of multilingual corpora collections. A major objective of this workshop is to focus on research questions that have not been deeply investigated so far. Special interest is devoted to contributions that aim to consider the following aspects:

- Modeling: methods to develop suitable representations for multilingual corpora, possibly embedding information from different views/aspects, such as, e.g., tensor models and decompositions, word-to-vector models, statistical topic models, representational learning, etc.
- Learning: any unsupervised, supervised, and semi-supervised approach in cross/multilingual contexts.
- The use of knowledge bases to support the modeling, learning, or both stages of multilingual corpora analysis.
- Emerging trends and applications, such as, e.g., cross-view cross-/multilingual IR, multilingual text mining in social networks, etc.

[1] http://events.dimes.unical.it/multilingmine/.

Main research topics of interest in MultiLingMine 2016 include the following:

- Multilingual/cross-lingual information access, web search, and ranking
- Multilingual/cross-lingual relevance feedback
- Multilingual/cross-lingual text summarization
- Multilingual/cross-lingual question answering
- Multilingual/cross-lingual information extraction
- Multilingual/cross-lingual document indexing
- Multilingual/cross-lingual topic modeling
- Multi-view/Multimodal representation models for multilingual corpora and cross-lingual applications
- Cross-view multi/cross-lingual information retrieval and document mining
- Multilingual/cross-lingual classification and clustering
- Knowledge-based approaches to model and mine multilingual corpora
- Social network analysis and mining for multilinguality/cross-linguality
- Plagiarism detection for multilinguality/cross-linguality
- Sentiment analysis for multilinguality/cross-linguality
- Deep learning for multilinguality/cross-linguality
- Novel validity criteria for cross-/multilingual retrieval and learning tasks
- Novel paradigms for visualization of patterns mined in multilingual corpora
- Emerging applications for multilingual/cross-lingual domains

The ultimate goal of the MultiLingMine workshop is to increase the visibility of the above research themes, and also to bridge closely related research fields such as information access, searching and ranking, information extraction, feature engineering, text mining and machine learning.

3 Advisory Board

The scientific significance of the workshop is assured by a Program Committee which includes 20 research scholars, coming from different countries and widely recognized as experts in cross/multi-lingual information retrieval:
Ahmet Aker, Univ. Sheffield, United Kingdom
Rafael Banchs, I2R Singapore
Martin Braschler, Zurich Univ. of Applied Sciences, Switzerland
Philipp Cimiano, Bielefeld University, Germany
Paul Clough, Univ. Sheffield, United Kingdom
Andrea Esuli, ISTI-CNR, Italy
Wei Gao, QCRI, Qatar
Cyril Goutte, National Research Council, Canada
Parth Gupta, Universitat Politcnica de Valncia, Spain
Dunja Mladenic, Jozef Stefan International Postgraduate school, Slovenia
Alejandro Moreo, ISTI-CNR, Italy
Alessandro Moschitti, Univ. Trento, Italy; QCRI, Qatar
Matteo Negri, FBK - Fondazione Bruno Kessler, Italy
Simone Paolo Ponzetto, Univ. Mannheim, Germany
Achim Rettinger, Institute AIFB, Germany

Philipp Sorg, Institute AIFB, Germany
Ralf Steinberger, JRC in Ispra, Italy
Marco Turchi, FBK - Fondazione Bruno Kessler, Italy
Vasudeva Varma, IIIT Hyderabad, India
Ivan Vulic, KU Leuven, Belgium

4 Related Events

A COLING'08 workshop [1] was one of the earliest events that emphasized the importance of analyzing multilingual document collections for information extraction and summarization purposes. The topic also attracted attention from the semantic web community: in 2014, [2] solicited works to discuss principles on how to publish, link and access mono and multilingual knowledge data collections; in 2015, another workshop [3] took place on similar topics in order to allow researchers continue to address multilingual knowledge management problems. A tutorial on Multilingual Topic Models was presented at WSDM 2014 [4] focusing on how statistically model document collections written in different languages. In 2015, a WWW workshop aimed at advancing the state-of-the-art in Multilingual Web Access [5]: the contributing papers covered different aspects of multilingual information analysis, leveraging attention on the lack of current information retrieval techniques and the necessity of new techniques especially tailored to manage, search, analysis and mine multilingual textual information.

The main event related to our workshop is the CLEF initiative [6] which has long provided a premier forum for the development of new information access and evaluation strategies in multilingual contexts. However, differently from MultiLingMine, it does not have emphasized research contributions on tasks such as searching, indexing, mining and modeling of multilingual corpora.

Our intention is to continue the lead of previous events about multilingual related topics, however from a broader perspective which is relevant to various information retrieval and document mining fields. We aim at soliciting contributions from scholars and practitioners in information retrieval that are interested in Multi/Cross-lingual document management, search, mining, and evaluation tasks. Moreover, differently from previous workshops, we would emphasize some specific trends, such as cross-view cross/multilingual IR, as well as the growing tightly interaction between knowledge-based and statistical/algorithmic approaches in order to deal with multilingual information overload.

References

1. Bandyopadhyay, S., Poibeau, T., Saggion, H., Yangarber, R.: Proceedings of the Workshop on Multi-source Multilingual Information Extraction and Summarization (MMIES). ACL (2008)
2. Chiarcos, C., McCrae J.P., Montiel, E., Simov, K., Branco, A., Calzolari, N., Osenova, P., Slavcheva, M., Vertan, C.: Proceedings of the 3rd Workshop on Linked Data in Linguistics: Multilingual Knowledge Resources and NLP (LDL) (2014)

3. McCrae, J.P., Vulcu, G.: CEUR Proceedings of the 4th Workshop on the Multilingual Semantic Web (MSW4), vol. 1532 (2015)
4. Moens, M.-F., Vulié, I.: Multilingual probabilistic topic modeling and its applications in web mining and search. In: Proceedings of the 7th ACM WSDM Conference (2014)
5. Steichen, B., Ferro, N., Lewis, D., Chi, E.E.: Proceedings of the International Workshop on Multilingual Web Access (MWA) (2015)
6. The CLEF Initiative. http://www.clef-initiative.eu/

Proactive Information Retrieval: Anticipating Users' Information Need

Sumit Bhatia[1(✉)], Debapriyo Majumdar[2], and Nitish Aggarwal[3]

[1] IBM Watson Research, New York, USA
sumit.bhatia@us.ibm.com
[2] Indian Statistical Institute, Kolkata, India
debapriyo@isical.ac.in
[3] Insight-Centre, National University of Ireland, Galway, Ireland
nitish.aggarwal@insight-center.org

Abstract. The ultimate goal of an IR system is to fulfill the user's information need. Traditional search systems have been *reactive* in nature wherein the search systems react to an input query and return a set of ranked documents most probable to contain the desired information. Due to the inability of, and efforts required by users to create efficient queries expressing their information needs, techniques such as query expansion, query suggestions, using relevance feedback and click-through information, personalization, etc. have been used to better understand and satisfy users' information needs. Given the increasing popularity of smartphones and Internet enabled wearable devices, how can the information retrieval systems use the additional data, and better interact with the user so as to better understand, and even *anticipate* her precise information needs? Building such *zero query* or *minimum user effort* systems require research efforts from multiple disciplines covering algorithmic aspects of retrieval models, user modeling and profiling, evaluation, context modeling, novel user interfaces design, etc. The proposed workshop intends to gather together the researchers from academia and industry practitioners with these diverse backgrounds to share their experiences and opinions on challenges and possibilities of developing such proactive information retrieval systems.

1 Motivation

The ultimate goal of an IR system is to fulfill the user's information need. Traditionally, IR systems have been reactive in nature, wherein the system would react only after the information need has been expressed by the user as a query. Oftentimes, users are unable to express their needs clearly using a few keywords leading to considerable research efforts to close the gap between the information need in the user's mind and the data residing in the system's index by approaches such as query expansion, query suggestions, relevance feedback and click-through information, and using personalization techniques to fine tune the search results to users' liking. These techniques, though helpful, increase the complexity of the system and many times also require additional efforts from the

N. Ferro et al. (Eds.): ECIR 2016, LNCS 9626, pp. 874–877, 2016.
DOI: 10.1007/978-3-319-30671-1_84

user, especially when they are using a mobile device (typing queries, providing feedback, selecting and trying multiple suggestions, etc.). Given that the search traffic from mobile devices is set to overtake desktop/PCs in near future[1], and ever increasing amount of user data provided by the growing popularity of Internet enabled wearable devices (smartwatches, smart wrist bands, Google Glass, etc.), it behooves the IR community to move beyond the keyword query-10 blue links paradigm and to develop systems that are more proactive in nature, and can anticipate and fulfill user's information need with minimal efforts from the user. Increasingly, *push models have replaced pull models in various platforms*, the result pages of commercial search engines have changed to display results in ways that are more easily consumable by the user.

Systems like Google Now, Apple Siri and IBM Watson proactively provide useful and personalized information such as weather updates, flight and traffic alerts, trip-planning[2] etc.; and even try to provide answers to often noisy and underspecified user questions. For example, given a user's location and preferences (vegetarian, specific cuisines, etc.) suggesting nearby restaurants, suggesting fitness articles/diet plans based on exercise logs, suggesting possible recipients based on an email content, etc.

The underlying theme of the workshop will be *systems that can proactively anticipate and fulfill the information needs of the user.* To cover the various aspects of proactive IR systems, the topic areas of the workshop include (but are not limited to):

- Zero query or minimum effort IR systems
- Proactive query understanding and anticipating information need
- Utilizing multi-device multi-modal data for building user profiles
- User studies, user and task models, user effort estimation
- Interaction analysis
- Novel user interfaces for proactive IR systems
- Session analysis
- Search interface design and result presentation
- Personalization, social and context aware search
- Search applied to Internet of things

2 Workshop Objectives, Goals, and Expected Outcome

The prime objective of the workshop will be to attract attention of the IR community to, and spark discussions about, *systems that can proactively anticipate and fulfill the information needs of the user.* Building such systems require research efforts from various areas covering algorithmic aspects of retrieval models, user modeling and profiling, evaluation, context modeling, etc. It also

[1] http://searchengineland.com/matt-cutts-mobile-queries-may-surpass-pc-year-186816.

[2] http://www.tnooz.com/article/startup-pitch-wayblazer-aims-travel-insights-service/.

requires research efforts in UI design, result presentation and visualization. In order to achieve this goal, the workshop will strive to:

- provide a platform to researchers from academia and industry to share latest research, discuss current shortcomings, explore different use cases;
- brainstorm about future research directions towards developing systems that can intelligently anticipate users' information needs;
- build upon the current research in information retrieval, natural language processing, semantic analysis, personalization, etc.;
- identify killer applications and key industry drivers (bringing theory into practice);
- explore means of developing benchmark test collections and evaluation metrics for evaluation.

3 Workshop Program Format

The tentative schedule of the half-day workshop is as follows:

- **Introduction (10 min):** Short introduction by the organizers outlining the technical program and reiterating the goals of the workshop.
- **Keynote Talk (50 min):** A keynote from a distinguished researcher on a topic relevant to the workshop theme.
- **Technical Presentations (100–120 min):** Technical presentations of 3–4 selected papers from among the submissions received. We plan to allocate 20 min for the presentations and 10 min for questions and discussions.
- **Brainstorming Session (30–40 min):** We plan to have a brainstorming session in the end focused solely on group discussions among the participants to discuss about the technical issues, challenges and lay out future research directions. Organizers will take meeting notes to be shared among the participants and later, through a workshop report in SIGIR forum.
- **Concluding Session (10 min)**

4 Intended Audience

We expect to bring together researchers from both industry and academia with diverse backgrounds spanning information retrieval, natural language processing, speech processing, user modeling and profiling, data mining, machine learning, human computer interface design, to propose new ideas, identify promising research directions and potential challenges. We also wish to attract practitioners who seek novel ideals for applications. **Participation and Selection Process:** We plan to encourage attendance and attract quality submissions by:

- Instituting a best paper cash award (depending on sponsor approval)
- Inviting papers from established researchers
- Inviting position papers and application papers to stimulate discussions

We plan to attract attendance and submissions from both academia and industry, and will accept regular research papers (8 pages) as well as short papers (4 pages). Each paper will be reviewed by at least three program committee members and selected based on their relevance to the workshop theme, novelty and boldness of the ideas, and ability to spark discussions.

First International Workshop on Recent Trends in News Information Retrieval (NewsIR'16)

Miguel Martinez-Alvarez[1(✉)], Udo Kruschwitz[2], Gabriella Kazai[3], Frank Hopfgartner[4], David Corney[1], Ricardo Campos[5], and Dyaa Albakour[1]

[1] Signal Media, London, UK
{miguel.martinez,david.corney,dyaa.albakour}@signal.uk.com
[2] University of Essex, Colchester, UK
udo@essex.ac.uk
[3] Lumi, London, UK
gabs@lumi.do
[4] University of Glasgow, Glasgow, Scotland
frank.hopfgartner@glasgow.ac.uk
[5] LIAAD-INESC TEC, Instituto Politécnico de Tomar, Tomar, Portugal
ricardo.campos@ipt.pt
http://www.signal.uk.com, http://www.lumi.do

Abstract. The news industry has gone through seismic shifts in the past decade with digital content and social media completely redefining how people consume news. Readers check for accurate fresh news from multiple sources throughout the day using dedicated apps or social media on their smartphones and tablets. At the same time, news publishers rely more and more on social networks and citizen journalism as a frontline to breaking news. In this new era of fast-flowing instant news delivery and consumption, publishers and aggregators have to overcome a great number of challenges. These include the verification or assessment of a source's reliability; the integration of news with other sources of information; real-time processing of both news content and social streams in multiple languages, in different formats and in high volumes; deduplication; entity detection and disambiguation; automatic summarization; and news recommendation. Although Information Retrieval (IR) applied to news has been a popular research area for decades, fresh approaches are needed due to the changing type and volume of media content available and the way people consume this content. The goal of this workshop is to stimulate discussion around new and powerful uses of IR applied to news sources and the intersection of multiple IR tasks to solve real user problems. To promote research efforts in this area, we released a new dataset consisting of one million news articles to the research community and introduced a data challenge track as part of the workshop.

1 Background and Motivation

News from mainstream media outlets is often one of the most relevant, influential and powerful sources of information. This ranges from the influence that

N. Ferro et al. (Eds.): ECIR 2016, LNCS 9626, pp. 878–882, 2016.
DOI: 10.1007/978-3-319-30671-1_85

newspapers may have on elections to the reputational damage that a negative article in a well-known magazine can cause to a brand. The process of consuming news itself is constantly changing. We receive a continuous influx of news information from different sources (e.g., traditional newspapers, blogs and social media) and this has had a massive impact on the nature of information systems.

Some of the current challenges we are facing are the integration of news data with other sources of information such as social media [1]; real-time analytics [2]; processing text in multiple languages; automatic temporal summarization [3]; and scalable processing of millions of articles on a daily basis.

Following discussions at ECIR 2015 we created a forum[1] to discover if there was enough interest within the IR community for a workshop focusing on traditional media, and news data in particular. We were very happy to see that around 40 members joined the forum straightaway and that several fruitful discussions started. This was a clear indication for the strong interest in the community for organizing such a workshop. Furthermore, the discussion in the forum illustrated the diversity of topics that this workshop could explore, including:

- Traditional and social media integration
- Temporal aspects of news
- Credibility, readability and controversy
- Bias and plurality in news
- Event and anomaly detection
- Diversification
- Summarization of multiple documents
- User-generated content (e.g., using comments to enhance news retrieval)
- News recommendation
- De-duplication and clustering of news articles
- Author identification and disambiguation
- Evaluation
- Data Visualization

2 Workshop Goals

The main goal of the workshop is to bring together scientists conducting relevant research in the field of news and information retrieval. In particular, scientists can present their latest breakthroughs with an emphasis on the application of their findings to research from a wide range of areas including: information retrieval; natural language processing; journalism (including data journalism); network analysis; and machine learning. This will facilitate discussion and debate about the problems we face and the solutions we are exploring, hopefully finding common grounds and potential synergies between different approaches. We aim to have a substantial representation from industry, from small start-ups to large enterprises, to strengthen their relationships with the academic community. This also represents a unique opportunity to understand the different problems

[1] https://groups.google.com/forum/#!forum/news-ir.

and priorities of each community and to recognize areas that are not currently receiving much academic attention but are nonetheless of considerable commercial interest. Finally, to accompany the workshop, we have released a new dataset suitable for conducting research on news IR. We describe the dataset in the next section. Detailed information about the workshop can be found on the workshop website[2].

3 The Signal Media One-Million News Articles Dataset

To stimulate workshop participation (and more generally to provide a useful resource for researchers in the area), we have prepared a new dataset of one million recent news articles from a wide range of sources (The Signal Media One-Million News Articles Dataset)[3]. In contrast to many existing collections (such as Reuters-21578 and Reuters RCV1), our new dataset include news articles from a wide range of sources including global, national and local newspapers, along with magazines and blogs. This dataset is released under the standard Creative Commons license[4] to encourage re-use in diverse non-commercial research projects. Furthermore, in the call for papers, we introduced a 'data challenge track' to encourage submissions of experimental results on our new dataset. We believe that one or more shared tasks or challenges will emerge and that, with suitable refinement, these may form the basis of future workshops. Possible challenges include but are not limited to:

- detecting and summarizing events over time;
- identifying bias in news sources to different topics and/or different entities;
- identifying influencers in media coverage and visualizing information flow;
- sentiment analysis on media coverage.

4 Keynotes and Panel

We have invited two keynote speakers who can provide insights into the topic from both an industry and an academic point of view. The industry keynote speaker is **Dr. Jochen Leidner**. Jochen is currently Director of Research at Thomson Reuters, where he heads the London (UK) R&D site, which he established. He has worked in many areas including information extraction from legal, news and financial documents, search engine technology and its application to legal information retrieval, automated proofing support for contracts, sentiment analysis, rule based systems, citation analysis and social media. The academic keynote speaker is **Dr. Julio Gonzalo**. Julio is an assistant professor at UNED (Universidad Nacional de Educacin a Distancia). Julio has been recently involved in organizing the CLEF RepLab, which is an evaluation campaign for online reputation management.

[2] http://research.signalmedia.co/newsir16.
[3] http://research.signalmedia.co/newsir16/signal-dataset.html.
[4] https://creativecommons.org/.

The workshop also includes a panel discussion with members drawn from academia, from large companies and from SMEs. This includes Dr. Jochen Leidner (Thomson Reuters), Dr. Gabriella Kazai (Lumi) and Dr. Julio Gonzalo (UNED). This panel focuses on the commonalities and differences between the communities as they face related challenges in news-based information retrieval.

5 Programme Committee

The Programme Committee (PC) is formed by key researchers from industry and academia. We thank all the PC members, whose names and affiliations are listed below.

- Ramkumar Aiyengar, Bloomberg, UK
- Marco Bonzanini, Bonzanini Consulting Ltd
- Omar Alonso, Microsoft, USA
- Alejandro Bellogin Kouki, UAM, Spain
- Horatiu-Sorin Bota, University of Glasgow, UK
- Igor Brigadir, Insight Centre for Data Analytics, Ireland
- Toine Bogers, Aalborg University Copenhagen (AAU-CPH), Denmark
- Ivan Cantador, UAM, Spain
- Arjen De Vries, Centrum Wiskunde & Informatica (CWI), The Netherlands
- Ernesto Diaz Aviles, IBM Research, Ireland
- Angel Castellanos Gonzalez , UNED, Spain
- Julio Gonzalo, UNED, Spain
- David Graus, University of Amsterdam, The Netherlands
- Jon Atle Gulla, NTNU, Norway
- Charlie Hull, Flax, UK
- Alípio Jorge, University of Porto, Portugal
- Jussi Karlgren, Gavagai, Sweden
- Marijn Koolen, University of Amsterdam, The Netherlands
- David D. Lewis, David D. Lewis Consulting, USA
- Stefano Mizzaro, University of Udine, Italy
- Elaheh Momeni, University of Vienna, Austria
- Miles Osborne, Bloomberg, UK
- Filipa Peleja, Yahoo! Research, Spain
- Vassilis Plachouras, Thomson Reuters, UK
- Barbara Poblete, University of Chile, Chile
- Muhammad Atif Qureshi, National University of Ireland, Ireland
- Paolo Rosso, Universidad Politecnica de Valencia, Spain
- Alan Said, Recorded Future, Sweden
- Damiano Spina, RMIT, Australia
- Jeroen Vuurens, TU Delft, The Netherlands
- Colin Wilkie, University of Glasgow, UK
- Arjumand Younus, National University of Ireland, Ireland
- Arkaitz Zubiaga, University of Warwick, UK

References

1. De Francisci, G., Morales, A.G., Lucchese, C.: From chatter to headlines: harnessing the real-time web for personalized news recommendation. In: Proceedings of WSDM (2012)
2. Mathioudakis, M., Koudas, N.: Twittermonitor: trend detection over the Twitter stream. In: Proceedings of SIGMOD (2010)
3. Aslam, J., Ekstrand-Abueg, M., Pavlu, V., Diaz, F., Sakai, T.: TrREC temporal summarization. In: Proceedings of TREC (2013)

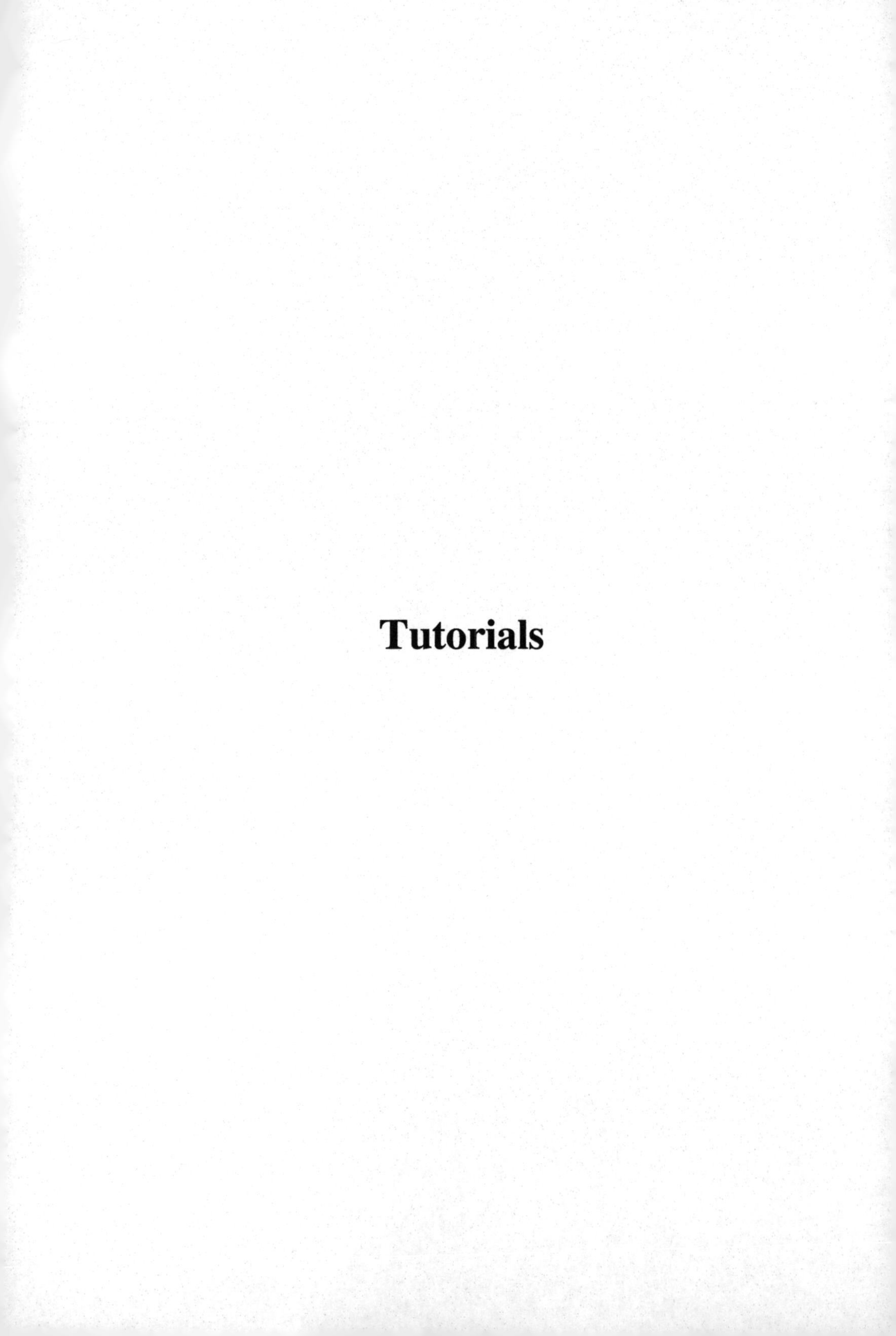

Tutorials

Collaborative Information Retrieval: Concepts, Models and Evaluation

Lynda Tamine[1] and Laure Soulier[2,3](✉)

[1] University of Toulouse UPS, IRIT, 118 Route de Narbonne, 31062 Toulouse Cedex 9, France
tamine@irit.fr

[2] Sorbonne Universités, UPMC University of Paris 06, UMR 7606, LIP6, 75005 Paris, France

[3] CNRS, UMR 7606, LIP6, 75005 Paris, France
Laure.Soulier@lip6.fr

Abstract. Recent work have shown the potential of collaboration for solving complex or exploratory search tasks allowing to achieve synergic effects with respect to individual search, which is the prevalent information retrieval (IR) setting this last decade. This interactive multi-user context gives rise to several challenges in IR. One main challenge relies on the adaptation of IR techniques or models [8] in order to build algorithmic supports of collaboration distributing documents among users. The second challenge is related to the design of Collaborative Information Retrieval (CIR) models and their effectiveness evaluation since individual IR frameworks and measures do not totally fit with the collaboration paradigms. In this tutorial, we address the second challenge and present first a general overview of collaborative search introducing the main underlying notions. Then, we focus on related work dealing with collaborative ranking models and their effectiveness evaluation. Our primary objective is to introduce these notions by highlighting how and why they should be different from individual IR in order to give participants the main clues for investigating new research directions in this domain with a deep understanding of current CIR work.

Keywords: Collaborative information retrieval · Collaboration · Search process · Ranking model · Evaluation

1 Introduction and Tutorial Objectives

Traditional conceptualizations of an IR task generally rely on an individual user's perspective. Accordingly, a great amount of research in the IR domain mostly dealt with both the design of enhanced document ranking models and a deep user's behavior understanding with the aim of improving an individual search effectiveness. However, in practice, collaboration among a community of users is increasingly acknowledged as an effective mean for gathering the complementary skills and/or knowledge of individual users in order to solve complex shared

N. Ferro et al. (Eds.): ECIR 2016, LNCS 9626, pp. 885–888, 2016.
DOI: 10.1007/978-3-319-30671-1_86

search tasks, such as fact-finding tasks (e.g., travel planning) or exploratory search tasks [13,18]. Collaboration allows the group achieving a result that is more effective than the simple aggregation of the individual results [14]. This class of complex search settings frequently occurs within a wide-range of domain-applications, such as the medical domain, the legal domain or the librarian domain to cite but a few. CIR results in collaborative information behavior processes, such as information sharing, evaluation, synthesis and sense-making. Two fundamental research challenges are faced by the design of CIR systems [8]: (1) allowing effective communication and coordination among the collaborators and (2) achieving high synergic effectiveness of the search results.

This tutorial focuses on the second challenge and pay a great deal of attention to how collaboration could be integrated in IR models and effectiveness evaluation processes. Our goal is to provide concepts and motivation to researchers so that participants could investigate this emerging IR domain as well as giving them some clues on how to experiment their models. More specifically, the tutorial aims to:

1. Give an overview of the concepts underlying collaborative information behavior and retrieval;
2. Present state-of-the art retrieval techniques and models that tackle the search effectiveness challenge;
3. Synthesize the metrics used for the evaluation of the effectiveness of CIR systems.

2 Outline

Part 1: Collaborative Information Retrieval Fundamental Notions. In this part, our primary objective is specifically to propose a broad review of collaborative search by presenting a detailed notion of collaboration in a search context including its definition [6,20], dimensions [2,5], paradigms [4,9], and underlying behavioral search process [3,7].

1. **Notion of Collaboration in Information Seeking and Retrieval**
2. **Dimensions of Collaboration**
3. **Collaboration Paradigms**
4. **Behavior Processes**

Part 2: Models and Techniques for Collaborative Document Seeking and Retrieval. CIR models provide an algorithmic mediation that enables to leverage from collaborators' actions in order to enhance the effectiveness of the search process [19]. In this context, previous work have been proposed, characterized by two common axes based on the relevance judgment integration and the division of labor paradigm. While the integration of relevance judgments is issued from interactive and contextual search, division of labor is an intrinsic feature of collaboration and represents the most common paradigm integrated in CIR models since it avoids redundant actions between collaborators. Among the multiple possible types of division of labor, only the algorithmic and the role-based ones are appropriate for the CIR domain [9]. One of the main approaches relies on the search strategy differences between collaborators using roles [12,15–17,19], where the algorithmic-based division of labor considers that users have similar objectives (in this case, users could be seen as peers) [4,11]. These approaches are contrasted to the ones surrounded by a division of labor guided by collaborators' roles in which users are characterized by asymmetric roles with distinct search strategies or intrinsic peculiarities.

1. **Algorithmic-Driven Division of Labor-Based Document Ranking Models**
2. **Role-Based Document Ranking Models**

Part 3: Effectiveness Evaluation of Collaborative Document Seeking and Retrieval. Due to the complexity of the collaborative search setting, its evaluation is challenging and constitutes an opened perspective in CIS and CIR [13]. Indeed, the high-level on heterogeneous interactions engaged for the coordination and the collective sense-making necessary to solve a shared information need raise new issues not yet tackled in interactive and contextual search. Concerning the IR aspects, the goal of this evaluation is no longer limited to the assessment of the document relevance with respect to a query, but rather the collective relevance in response to the information need expressed by all users. While the evaluation of individual IR depends only on the query and, in the case of personalized IR, the user, evaluating collaborative ranking models and techniques should consider the aspects connected to the collaboration. We present in this section, for both CIS and CIR, the evaluation framework listing the existing protocols and the evaluation metrics.

1. **Taxonomy of Evaluation Methodologies**
2. **Evaluation Metrics**

Part 4: Perspectives. Collaborative search rises several perspectives. Some of them are outlined in this tutorial. More particularly, we focus on the use of roles in collaborative search [17], the leveraging of social media [10] as well as the standardization of evaluation framework [1].

1. **User-Driven CIR Models**
2. **Community-Based and Social-Media-Based Collaborative Information Information Retrieval Systems**
3. **Standardization and CIR Evaluation Campaigns**

Part 5: Questions and Discussion with the Instructors. We end with an open discussion with participants.

References

1. Azzopardi, L., Pickens, J., Sakai, T., Soulier, L., Tamine, L.: Ecol: first international workshop on the evaluation on collaborative information seeking and retrieval. In: CIKM 2015, pp. 1943–1944 (2015)
2. Capra, R., Velasco-Martin, J., Sams, B.: Levels of "working together" in collaborative information seeking and sharing. In: CSCW 2010, ACM (2010)
3. Evans, B.M., Chi, E.H.: An elaborated model of social search. Inf. Process. Manag. (IP&M) **46**(6), 656–678 (2010)
4. Foley, C., Smeaton, A.F.: Synchronous collaborative information retrieval: techniques and evaluation. In: Boughanem, M., Berrut, C., Mothe, J., Soule-Dupuy, C. (eds.) ECIR 2009. LNCS, vol. 5478, pp. 42–53. Springer, Heidelberg (2009)
5. Golovchinsky, G., Qvarfordt, P., Pickens, J.: Collaborative information seeking. IEEE Comput. **42**(3), 47–51 (2009)
6. Gray, B.: Collaborating: Finding Common Ground for Multiparty Problems. Jossey Bass Business and Management Series. Jossey-Bass, San Francisco (1989)
7. Hyldegärd, J.: Beyond the search process - exploring group members' information behavior in context. IP&M **45**(1), 142–158 (2009)
8. Joho, H., Hannah, D., Jose, J.M.: Revisiting IR techniques for collaborative search strategies. In: Boughanem, M., Berrut, C., Mothe, J., Soule-Dupuy, C. (eds.) ECIR 2009. LNCS, vol. 5478, pp. 66–77. Springer, Heidelberg (2009)
9. Kelly, R., Payne, S.J.: Division of labour in collaborative information seeking: current approaches and future directions. In: CIS Workshop at CSCW 2013, ACM (2013)
10. Morris, M.R.: Collaborative search revisited. In: CSCW 2013, pp. 1181–1192. ACM (2013)
11. Morris, M.R., Teevan, J., Bush, S.: Collaborative web search with personalization: groupization, smart splitting, and group hit-highlighting. In: CSCW 2008, pp. 481–484. ACM (2008)
12. Pickens, J., Golovchinsky, G., Shah, C., Qvarfordt, P., Back, M.: Algorithmic mediation for collaborative exploratory search. In: SIGIR 2008, pp. 315–322. ACM (2008)
13. Shah, C.: Collaborative Information Seeking - The Art and Science of Making the Whole Greater than the Sum of All. pp. I–XXI, 1–185. Springer, Heidelberg (2012)
14. Shah, C., González-Ibáñez, R.: Evaluating the synergic effect of collaboration in information seeking. In: SIGIR 2011, pp. 913–922. ACM (2011)
15. Shah, C., Pickens, J., Golovchinsky, G.: Role-based results redistribution for collaborative information retrieval. Inf. Process. Manag. (IP&M) **46**(6), 773–781 (2010)
16. Soulier, L., Shah, C., Tamine, L.: User-driven system-mediated collaborative information retrieval. In: SIGIR 2014, pp. 485–494. ACM (2014)
17. Soulier, L., Tamine, L., Bahsoun, W.: On domain expertise-based roles in collaborative information retrieval. Inf. Process. Manag. (IP&M) **50**(5), 752–774 (2014)
18. Spence, P.R., Reddy, M.C., Hall, R.: A survey of collaborative information seeking practices of academic researchers. In: SIGGROUP Conference on Supporting Group Work, GROUP 2005, pp. 85–88. ACM (2005)
19. Tamine, L., Soulier, L.: Understanding the impact of the role factor in collaborative information retrieval. In: CIKM 2015, ACM, October 2015
20. Twidale, M.B., Nichols, D.M., Paice, C.D.: Browsing is a collaborative process. Inf. Process. Manag. (IP&M) **33**(6), 761–783 (1997)

Group Recommender Systems: State of the Art, Emerging Aspects and Techniques, and Research Challenges

Ludovico Boratto(✉)

Dipartimento di Matematica e Informatica,
Università di Cagliari, Via Ospedale, 72 - 09124 Cagliari, Italy
ludovico.boratto@unica.it

Abstract. A *recommender system* aims at suggesting to users items that might interest them and that they have not considered yet. A class of systems, known as *group recommendation*, provides suggestions in contexts in which more than one person is involved in the recommendation process. The goal of this tutorial is to provide the ECIR audience with an overview on group recommendation. We will first illustrate the recommender systems principles, then formally introduce the problem of producing recommendations to groups, and present a survey based on the tasks performed by these systems. We will also analyze challenging topics like their evaluation, and present emerging aspects and techniques in this area. The tutorial will end with a summary that highlights open issues and research challenges.

Keywords: Group recommendation · Algorithms · Evaluation · Research challenges

1 Tutorial Outline

Recommender systems are designed to provide information items that are expected to interest a user [11]. Given their capability to increase the revenue in commercial environments, nowadays they are employed by the most relevant websites, such as Amazon and Netflix.

Group recommender systems are a class of systems designed for contexts in which more than one person is involved in the recommendation process [7]. Group recommendation has been highlighted as a challenging research area, with the first survey on the topic [7] being placed in the Challenges section of the widely-known book "The Adaptive Web", and recent research indicating it as a future direction in recommender systems, since it presents numerous open issues and challenges [10].

With respect to classic recommendation, a system that works with groups has to complete a set of specific and additional tasks. This tutorial will present how the state-of-the-art approaches in the literature handle these tasks in order to produce recommendations to groups.

N. Ferro et al. (Eds.): ECIR 2016, LNCS 9626, pp. 889–892, 2016.
DOI: 10.1007/978-3-319-30671-1_87

The evaluation of the accuracy of a system for a group is not a trivial aspect [10], so we will also analyze the evaluation techniques for group recommender systems.

Recent studies are characterized by advanced recommendation techniques and novel aspects, such as social data and temporal features, so the tutorial will also cover these emerging aspects and techniques.

In conclusion, the open issues and research challenges in this area will be presented.

This tutorial will cover these topics in six sections. In detail, the outline of the tutorial is the following.

1. **Recommender Systems Principles**
   - *Definition and application domains*
   - *Main classes of systems*
2. **Group Recommendation Introduction**
   - *Definition and application domains*
   - *Problem statement*
3. **Tasks and State of the Art Survey**
   - *Types of groups.* In [2], we highlighted that the type of group handled by a system is one of the characterizing aspects of a group recommender system and we provided a classification of the different types of groups, which was also adopted by Carvalho and Macedo in their WWW'13 paper [4].
   - *Preference acquisition.* A group recommender system might acquire the preferences by considering only those expressed by the individual users, or by allowing the groups to express them.
   - *Group modeling* [9] is the process adopted to combine the individual preferences in a unique model that represents the group.
   - *Rating prediction* is the most characterizing aspect in all the types of recommender systems, and also plays an important role when working with groups, since the ratings might be predicted for the individual users or specifically for the groups [7].
   - *Help the members to achieve consensus.* This task is adopted in order to find an agreement on what should be proposed to the group.
   - *Explanation of the recommendations*, i.e., the task performed by some of the systems to justify why an item has been suggested to the group.
4. **Evaluation Methods**
   - *Offline methods*, which evaluate a system on existing datasets.
   - *User surveys* that test the effectiveness of a system by asking users to answer questionnaires.
   - *Live systems* that work in real-world domains, like the social networks.
5. **Emerging Aspects and Techniques**
   - *Advanced recommendation techniques applied to group recommendation.* The last advances in recommendation techniques, such as generative and stochastic models, have recently been employed in group recommender systems too [8,12].
   - *Temporal aspects in group recommendation.* Recently, the temporal factor has been considered in the recommendation process [1,5].

- *Social group recommender systems.* The widespread relevance of social media has recently had an impact also on this research area [6,8].
- *Group recommendation with automatic detection of groups.* There are scenarios in which groups do not exist, but the recommendations have to be proposed to groups because of limitations on the number of recommendation lists that can be produced (i.e., it is not possible to suggest a different list of items to each user), so a clustering of the users specifically designed for recommendation purposes has to be performed [3].

6. **Summary**
   - *Open issues*
   - *Research challenges*

2 Target Audience

This tutorial is aimed at anyone interested in the topic of producing recommendation to groups of users, from data mining and machine learning researchers to practitioners from industry. For those not familiar with group recommendation or recommender systems in general, the tutorial will cover the necessary background material to understand these systems and will provide a state-of-the-art survey on the topic. Additionally, the tutorial aims to offer a new perspective that will be valuable and interesting even for more experienced researchers that work in this domain, by providing the recent advances in this area and by illustrating the current research challenges.

3 Instructor

Ludovico Boratto is a research assistant at the University of Cagliari (Italy). His main research area are Recommender Systems, with special focus on those that work with groups of users and in social environments. In 2010 and 2014, he spent 10 months at the Yahoo! Lab in Barcelona as a visiting researcher.

References

1. Amer-Yahia, S., Omidvar-Tehrani, B., Basu Roy, S., Shabib, N.: Group recommendation with temporal affinities. In: Proceedings of 18th International Conference on Extending Database Technology (EDBT), pp. 421–432. OpenProceedings.org (2015)
2. Boratto, L., Carta, S.: State-of-the-art in group recommendation and new approaches for automatic identification of groups. In: Soro, A., Vargiu, E., Armano, G., Paddeu, G. (eds.) IRMDE 2010. SCI, vol. 324, pp. 1–20. Springer, Heidelberg (2010)
3. Boratto, L., Carta, S.: Using collaborative filtering to overcome the curse of dimensionality when clustering users in a group recommender system. In: ICEIS 2014 - Proceedings of 16th International Conference on Enterprise Information Systems, vol. 2. pp. 564–572. SciTePress (2014)

4. Carvalho, L.A.M., Macedo, H.T.: Generation of coalition structures to provide proper groups' formation in group recommender systems. In: Proceedings of 22nd International Conference on World Wide Web Companion, pp. 945–950. International World Wide Web Conferences Steering Committee (2013)
5. Chen, J., Liu, Y., Li, D.: Dynamic group recommendation with modified collaborative filtering and temporal factor. Int. Arab J. Inf. Technol. (IAJIT) (2014)
6. Christensen, I.A., Schiaffino, S.: Social influence in group recommender systems. Online Inf. Rev. **38**(4), 524–542 (2014)
7. Jameson, A., Smyth, B.: Recommendation to groups. In: Brusilovsky, P., Kobsa, A., Nejdl, W. (eds.) Adaptive Web 2007. LNCS, vol. 4321, pp. 596–627. Springer, Heidelberg (2007)
8. Kim, H.N., Saddik, A.E.: A stochastic approach to group recommendations in social media systems. Inf. Syst. **50**, 76–93 (2015)
9. Masthoff, J.: Group recommender systems: combining individual models. In: Ricci, F., Rokach, L., Shapira, B., Kantor, P.B. (eds.) Recommender Systems Handbook, pp. 677–702. Springer, Heidelberg (2011)
10. Ricci, F.: Recommender systems: models and techniques. In: Alhajj, P., Rokne, J. (eds.) Encyclopedia of Social Network Analysis and Mining, pp. 1511–1522. Springer, New York (2014)
11. Ricci, F., Rokach, L., Shapira, B.: Recommender systems: introduction and challenges. In: Ricci, F., Rokach, L., Shapira, B. (eds.) Recommender Systems Handbook, pp. 1–34. Springer, Heidelberg (2015). doi:10.1007/978-1-4899-7637-6_1
12. Yuan, Q., Cong, G., Lin, C.Y.: COM: a generative model for group recommendation. In: Proceedings of 20th ACM SIGKDD International Conference on Knowledge Discovery and Data Mining, KDD 2014, pp. 163–172. ACM (2014)

Living Labs for Online Evaluation: From Theory to Practice

Anne Schuth[1(✉)] and Krisztian Balog[2]

[1] University of Amsterdam, Amsterdam, The Netherlands
anne.schuth@uva.nl
[2] University of Stavanger, Stavanger, Norway
krisztian.balog@uis.no

Abstract. Experimental evaluation has always been central to Information Retrieval research. The field is increasingly moving towards *online* evaluation, which involves experimenting with real, unsuspecting users in their natural task environments, a so-called *living lab*. Specifically, with the recent introduction of the Living Labs for IR Evaluation initiative at CLEF and the OpenSearch track at TREC, researchers can now have direct access to such labs. With these benchmarking platforms in place, we believe that online evaluation will be an exciting area to work on in the future. This half-day tutorial aims to provide a comprehensive overview of the underlying theory and complement it with practical guidance.

1 Motivation and Overview

Experimental evaluation has always been a key component in Information Retrieval research. Most commonly, systems are evaluated following the Cranfield methodology [4,22]. Using this approach, systems are evaluated in terms of document relevance for given queries, which is assessed by trained experts. While the Cranfield methodology ensures high internal validity and repeatability of experiments, it has been shown that the users' search success and satisfaction with an IR system are not always accurately reflected by standard IR metrics [29,31]. One reason is that the relevance judges typically do not assess queries and documents that reflect their own information needs, and have to make assumptions about relevance from an assumed user's point of view. Because the true information need can be difficult to assess, this can cause substantial biases [11,30,34]. To address these shortcomings, the field is increasingly moving towards *online* evaluation, which involves experimenting with real, unsuspecting users in their natural task environments. Essentially, the production search engine operates as a "living lab." For a long time, this type of evaluation was only available to those working within organizations that operate a search engine. But this is about to change. For one thing, the need to involve real users is know openly and widely acknowledged in our community (as witnessed, e.g., by the panel discussion at ECIR'15 and the Salton Award keynote lecture of Belkin at SIGIR'15 [2]). For another thing, pioneering efforts to realize the idea of living

N. Ferro et al. (Eds.): ECIR 2016, LNCS 9626, pp. 893–896, 2016.
DOI: 10.1007/978-3-319-30671-1_88

labs in practice are now in place and are available to the community. Specifically the Living Labs for IR Evaluation (LL4IR)[1] initiative runs as a benchmarking campaign at CLEF, but also operates monthly challenges so that people do not have to wait for a yearly evaluation cycle. The most recent initiative is the OpenSearch track at TREC[2], which focuses on *academic literature search.*

Understanding the differences between online and offline evaluation is still a largely unexplored area of research. There is a lot of fundamental research to happen in this space that has not happened yet because of the lack availability of experimental resources to the academic community. With recent developments, we believe that online evaluation will be an exciting area to work on in the future. The motivation for this tutorial is twofold: (1) to raise awareness and promote this form of evaluation (i.e., online evaluation with living labs) in the community, and (2) to help people get started by working through all the steps of the development and deployment process, using the LL4IR evaluation platform.

This half-day tutorial aims to provide a comprehensive overview of the underlying theory and complement it with practical guidance. The tutorial is organized in two 1,5 hours sessions with a break in between. Each session interleaves theoretical, practical, and interactive elements to keep the audience engaged. For the practical parts, we break with the traditional format by using hands-on instructional techniques. We will make use of an online tool, called DataJoy,[3] that proved invaluable in our previous classroom experience. This allows participants to (1) run Python code in a browser window without having to install anything locally, (2) follow the presenter's screen on their own laptop and, (3) at the same time, have their own private copy of the project on a different browser tab.

2 Target Audience and Learning Objectives

The primary target audience are graduate students and lecturers/professors teaching IR classes. Engineers from companies operating search engines might also find the tutorial useful. Our learning objectives include the following topics.

We will start our tutorial with an extensive overview of online evaluation methods. We begin with *A/B Testing* [16], which compares two systems by showing system A to one group of users and system B to another group. A/B testing then tries to infer a difference between the systems from differences in observed behavior. We describe many ways of measuring observed behavior: (1) click through rate (CTR) [14]; (2) dwell time [34]; (3) satisfied clicks [15]; (4) tabbed browsing [13]; (5) abandonment [18,28]; (6) query reformulation [8]; (7) skips [32]; (8) mouse movement [5–7,10,33]; and (9) in-view time [17].

While providing flexibility and control, A/B comparisons typically require a large number of observations. *Interleaved comparison methods* reduce the variance of measurement by presenting users with a result list that combines the

---

[1] http://living-labs.net.

[2] http://trec-open-search.org/.

[3] http://getdatajoy.com.

rankings of systems A and B. We provide a comprehensive overview of the following interleaving methods: (1) balanced interleave (BI) [14]; (2) team draft interleave (TDI) [21]; (3) document constraints (DC) [9]; (4) probabilistic interleave (PI) [12]; (5) optimized interleave (OI) [20]; (6) team draft multileave (TDM) [27]; and (7) probabilistic multileave (PM) [24].

Next, we discuss a comparison of interleaving and A/B metrics [25]. We then turn to simulating user interactions [26] using *click models* [3]. Finally, we touch on *learning to rank* in two variants: *offline* learning to rank [19] and *online* learning to rank [35], of which the latter requires the aforementioned evaluation methods.

Having provided the necessary theoretical background, we introduce the *living labs for IR* (LL4IR) [1] evaluation platform in depth. We will focus on two specific use-cases [23] from the CLEF lab: product search and web search. the practical sessions, participants will gain hands-on experience with the LL4IR platform [1], which includes: (1) registering and obtaining an API key; (2) getting queries and candidate items; (3) generating and uploading a ranking; and (4) obtaining feedback and outcomes. API documentation and course material are available at http://living-labs.net.

References

1. Balog, K., Kelly, L., Schuth, A.: Head first: living labs for ad-hoc search evaluation. In: CIKM 2014, pp. 1815–1818. ACM Press, New York, USA, November 2014
2. Belkin, N.J.: Salton award lecture: people, interacting with information. In: Proceedings of 38th International ACM SIGIR Conference on Research and Development in Information Retrieval, SIGIR 2015, pp. 1–2. ACM (2015)
3. Chuklin, A., Markov, I., de Rijke, M.: Click Models for Web Search. Synthesis Lectures on Information Concepts, Retrieval, and Services. Morgan & Claypool Publishers, San Rafael (2015)
4. Cleverdon, C.W., Keen, M.: Aslib Cranfield research project-factors determining the performance of indexing systems; Volume 2, Test results, National Science Foundation (1966)
5. Diaz, F., White, R., Buscher, G., Liebling, D.: Robust models of mouse movement on dynamic web search results pages. In: CIKM, pp. 1451–1460. ACM Press, October 2013
6. Guo, Q., Agichtein, E.: Understanding "abandoned" ads: towards personalized commercial intent inference via mouse movement analysis. In: SIGIR-IRA (2008)
7. Guo, Q., Agichtein, E.: Towards predicting web searcher gaze position from mouse movements. In: CHI EA, 3601p, April 2010
8. Hassan, A., Shi, X., Craswell, N., Ramsey, B.: Beyond clicks: query reformulation as a predictor of search satisfaction. In: CIKM (2013)
9. He, J., Zhai, C., Li, X.: Evaluation of methods for relative comparison of retrieval systems based on clickthroughs. In: CIKM 2009, ACM (2009)
10. He, Y., Wang, K.: Inferring search behaviors using partially observable markov model with duration (POMD). In: WSDM (2011)
11. Hersh, W., Turpin, A.H., Price, S., Chan, B., Kramer, D., Sacherek, L., Olson, D.: Do batch and user evaluations give the same results? In: SIGIR, pp. 17–24 (2000)
12. Hofmann, K., Whiteson, S., de Rijke, M.: A probabilistic method for inferring preferences from clicks. In: CIKM 2011, ACM (2011)

13. Jeff, H., Thomas, L., Ryen, W.: No search result left behind. In: WSDM, 203p (2012)
14. Joachims, T., Granka, L.A., Pan, B., Hembrooke, H., Radlinski, F., Gay, G.: Evaluating the accuracy of implicit feedback from clicks and query reformulations in web search. ACM Trans. Inf. Syst. **25**(2), 7 (2007)
15. Kim, Y., Hassan, A., White, R., Zitouni, I.: Modeling dwell time to predict click-level satisfaction. In: WSDM (2014)
16. Kohavi, R.: Online controlled experiments: introduction, insights, scaling, and humbling statistics. In: Proceedings of UEO 2013 (2013)
17. Lagun, D., Hsieh, C.H., Webster D., Navalpakkam, V.: Towards better measurement of attention and satisfaction in mobile search. In: SIGIR (2014)
18. Li, J., Huffman, S., Tokuda, A.: Good abandonment in mobile and pc internet search. In: SIGIR 2009, pp. 43–50 (2009)
19. Liu, T.-Y.: Learning to Rank for Information Retrieval. Springer, Heidelberg (2011)
20. Radlinski, F., Craswell, N.: Optimized interleaving for online retrieval evaluation. In: WSDM 2013, ACM (2013)
21. Radlinski, F., Kurup, M., Joachims, T.: How does clickthrough data reflect retrieval quality? In: CIKM 2008, ACM (2008)
22. Sanderson, M.: Test collection based evaluation of information retrieval systems. Found. Trends Inf. Retrieval **4**(4), 247–375 (2010)
23. Schuth, A., Balog, K., Kelly, L.: Overview of the living labs for information retrieval evaluation (ll4ir) clef lab. In: Mothe, J., Savoy, J., Kamps, J., Pinel-Sauvagnat, K., Jones, G.J.F., SanJuan, E., Cappellato, L., Ferro, N. (eds.) CLEF 2015. LNCS, pp. 484–496. Springer, Heidelberg (2015)
24. Schuth, A., Bruintjes, R.-J., Büttner, F., van Doorn, J., Groenland, C., Oosterhuis, H., Tran, C.-N., Veeling, B., van der Velde, J., Wechsler, R., Woudenberg, D., de Rijke, M.: Probabilistic multileave for online retrieval evaluation. In: Proceedings of SIGIR (2015)
25. Schuth, A., Hofmann, K., Radlinski, F.: Predicting search satisfaction metrics with interleaved comparisons. In: SIGIR 2015 (2015)
26. Schuth, A., Hofmann, K., Whiteson, S., de Rijke, M.: Lerot: an online learning to rank framework. In: LivingLab 2013, pp. 23–26. ACM Press, November 2013
27. Schuth, A., Sietsma, F., Whiteson, S., Lefortier, D., de Rijke, M.: Multileaved comparisons for fast online evaluation. In: CIKM 2014 (2014)
28. Song, Y., Shi, X., White, R., Hassan, A.: Context-aware web search abandonment prediction. In: SIGIR (2014)
29. Teevan, J., Dumais, S., Horvitz, E.: The potential value of personalizing search. In: SIGIR, pp. 756–757 (2007)
30. Turpin, A., Hersh, W.: Why batch and user evaluations do not give the same results. In: SIGIR, pp. 225–231 (2001)
31. Turpin, A., Scholar, F.: User performance versus precision measures for simple search tasks. In: SIGIR, pp. 11–18 (2006)
32. Wang, K., Walker, T., Zheng, Z.: PSkip: estimating relevance ranking quality from web search clickthrough data. In: KDD, pp. 1355–1364 (2009)
33. Wang, K., Gloy, N., Li, X.: Inferring search behaviors using partially observable Markov (POM) model. In: WSDM (2010)
34. Yilmaz, E., Verma, M., Craswell, N., Radlinski, F., Bailey, P.: Relevance and effort: an analysis of document utility. In: CIKM (2014)
35. Yue, Y., Joachims, T.: Interactively optimizing information retrieval systems as a dueling bandits problem. In: ICML 2009, pp. 1201–1208 (2009)

Real-Time Bidding Based Display Advertising: Mechanisms and Algorithms

Jun Wang[1], Shuai Yuan[2](✉), and Weinan Zhang[1]

[1] University College London, London, UK
{j.wang,weinan.zhang}@cs.ucl.ac.uk
[2] MediaGamma, London, UK
shuai.yuan@mediagamma.com

Abstract. In display and mobile advertising, the most significant development in recent years is the Real-Time Bidding (RTB), which allows selling and buying in real-time one ad impression at a time. The ability of making impression level bid decision and targeting to an individual user in real-time has fundamentally changed the landscape of the digital media. The further demand for automation, integration and optimisation in RTB brings new research opportunities in the IR fields, including information matching with economic constraints, CTR prediction, user behaviour targeting and profiling, personalised advertising, and attribution and evaluation methodologies. In this tutorial, teamed up with presenters from both the industry and academia, we aim to bring the insightful knowledge from the real-world systems, and to provide an overview of the fundamental mechanism and algorithms with the focus on the IR context. We will also introduce to IR researchers a few datasets recently made available so that they can get hands-on quickly and enable the said research.

1 Introduction

Interested readers could go to http://tutorial.computational-advertising.org for more information and supplement materials.

This tutorial aims to provide not only a comprehensive and systemic introduction to RTB and computational advertising in general, but also the emerging challenges, tools, and datasets in order to facilitate the research. Compared to previous Computational Advertising tutorials in relevant top-tier conferences, this tutorial takes a fresh, neutral, and the latest look of the field and focuses on the fundamental changes brought by RTB in the context of information retrieval research. We expect the audience, after attending the tutorial, to understand the real-time online advertising mechanisms and the state of the art techniques, as well as to grasp the research challenges in this field related to information retrieval. Our motivation is to help the audience acquire domain knowledge and obtain relevant datasets, and to promote research activities in RTB and computational advertising in general.

N. Ferro et al. (Eds.): ECIR 2016, LNCS 9626, pp. 897–901, 2016.
DOI: 10.1007/978-3-319-30671-1_89

2 Outlines

The tutorial will be structured as follows:

1. Background
   (a) The history and evolution of computational advertising;
   (b) The emergence of RTB;
2. The Framework and platform
   (a) Auction mechanisms
   (b) The current eco-system of RTB;
   (c) Mining RTB auctions;
3. Research problems and techniques
   (a) Dynamic pricing and information matching with economics constraints
   (b) Click-through rate and conversion prediction
   (c) Bidding strategies
   (d) Attribution models
   (e) Fraud detection
4. Datasets and evaluations
   (a) Datasets and evaluation methodologies
   (b) Live test and APIs
5. Panel discussion: research challenges and opportunities for the IR community

3 Relevance to ECIR and IR Community

The tutorial is strongly related to IR/DM/ML areas such as information matching and retrieval, CTR estimation, behaviour targeting, knowledge extraction, user log analysis and modelling, information retrieval, text mining, recommender systems and personalisation. These areas have been well studied by researchers attending ECIR conferences. Besides, computational advertising has received great development in recent years. Although relevant tutorials have been presented before (5 years ago or more), there have been lots of update in the research field. ECIR is a forum to inform that. In addition, ECIR has the tradition to encourage young researchers. This tutorial builds up a strong link between traditional IR topics (like document ranking, text retrieval, collaborative filtering etc.) to the emerging topics in advertising (lookalike modelling, behaviour targeting, CTR/conversion prediction, dynamics, bid optimisation, economic constraints etc.) and enables their research in those emerging topics by providing the required knowledge, datasets and needed evaluation methodologies.

4 Description of Topics

Online advertising is now one of the fastest advancing areas in IT industry. In display and mobile advertising, the most significant development in recent years is the growth of Real-Time Bidding (RTB), which allows selling and buying online display advertising per ad impression in real time [16]. Since then, RTB has fundamentally changed the landscape of the digital media market by scaling the buying process across a large number of available inventories. It also encourages behaviour targeting, audience extension, look-alike modelling etc. and makes a significant shift toward buying focused on user data, rather than contextual data.

Scientifically, the further demand for automation, integration and optimisation in RTB brings new research opportunities in the information retrieval, data mining and machine learning fields. For instance, the much enhanced flexibility of allowing advertisers and agencies to maximize impact of budgets by more optimised buys based on their own or third party (user) data [6] makes the online advertising market a step closer to the financial markets [1,12], where unification and interconnection are strongly promoted. This trend across webpages, advertisers, and users require significant research on statistical machine learning, data mining, information retrieval, behaviour targeting and their links to game theory [14], economics and optimisation. Despite its rapid growth and huge potential, many aspects of RTB remain unknown to the research community, particularly the Information Retrieval community, for a variety of reasons. In this tutorial, we aim to bring the insightful knowledge from the real-world systems, to bridge the gaps between industry and academia, and to provide an overview of the fundamental infrastructure, algorithms, and technical and research challenges of the new frontier of computational advertising.

This tutorial covers the following IR related topics:

1. Response prediction (CTR and conversion) in the context of real-time bidding, e.g., [8,11].
2. Bid optimisation and information match with economic constraints, e.g., [3,4].
3. Collaborative filtering approaches to user behaviour targeting and look-alike modelling, e.g., [13,15].
4. User profiling and segmentation, e.g., [2,5].
5. Fraud detection, e.g., [10,17].
6. Attribution analysis and evaluation methodologies, e.g., [7,9].

For the purpose of evaluation and promote research in the field, we will also cover a few datasets which are publicly available.

5 Target Audience and Prerequisites

The content of the tutorial is intermediate and is targeted to Ph.D. students, general researchers and practitioners in information retrieval and its related areas on data mining and knowledge management. The audience is expected to have basic knowledge of Information Retrieval, Data Mining, Machine Learning, and good understanding of Probability and Statistics.

References

1. Chen, B., Yuan, S., Wang, J.: A dynamic pricing model for unifying programmatic guarantee and real-time bidding in display advertising. In: Proceedings of the 8th International Workshop on Data Mining for Online Advertising (ADKDD 2014), p. 9 (2014). http://arxiv.org/abs/1405.5189

2. Elmeleegy, H., Li, Y., Qi, Y., Wilmot, P., Wu, M., Kolay, S., Dasdan, A., Chen, S.: Overview of turn data management platform for digital advertising. Proc. VLDB Endowment **6**(11), 1138–1149 (2013)
3. Ghosh, A., Rubinstein, B.: Adaptive bidding for display advertising. In: Proceedings of the 18th International Conference on World Wide Web, pp. 251–260 (2009). http://dl.acm.org/citation.cfm?id=1526744
4. Gummadi, R., Key, P.B., Proutiere, A.: Optimal bidding strategies in dynamic auctions with budget constraints. In: 2011 49th Annual Allerton Conference on Communication, Control, and Computing, Allerton 2011, p. 588 (2011)
5. Harvey, M., Crestani, F., Carman, M.J.: Building user profiles from topic models for personalised search. In: Proceedings of the 22nd ACM International Conference on Information & Knowledge Management - CIKM 2013, pp. 2309–2314 (2013). http://dl.acm.org/citation.cfm?d=2505515.2505642
6. Jaworska, J., Sydow, M.: Behavioural targeting in on-line advertising: an empirical study. In: Bailey, J., Maier, D., Schewe, K.-D., Thalheim, B., Wang, X.S. (eds.) WISE 2008. LNCS, vol. 5175, pp. 62–76. Springer, Heidelberg (2008)
7. Kitts, B., Wei, L., Au, D., Powter, A., Burdick, B.: Attribution of conversion events to multi-channel media. In: Proceedings - IEEE International Conference on Data Mining, ICDM, pp. 881–886 (2010)
8. McMahan, H.B., Holt, G., Sculley, D., Young, M., Ebner, D., Grady, J., Nie, L., Phillips, T., Davydov, E., Golovin, D., Chikkerur, S., Liu, D., Wattenberg, M., Hrafnkelsson, A.M., Boulos, T., Kubica, J.: Ad click prediction: a view from the trenches. In: Proceedings of the 19th ACM SIGKDD International Conference on Knowledge Discovery and Data Mining, pp. 1222–1230 (2013). http://dl.acm.org/citation.cfm?id=2488200
9. Shao, X., Li, L.: Data-driven multi-touch attribution models. In: Proceedings of the 17th ACM SIGKDD, pp. 258–264 (2011)
10. Stitelman, O., Perlich, C., Dalessandro, B., Hook, R., Raeder, T., Provost, F.: Using co-visitation networks for detecting large scale online display advertising exchange fraud. In: Proceedings of the 19th ..., pp. 1240–1248 (2013). http://dl.acm.org/citation.cfm?id=2487575.2488207
11. Tagami, Y., Ono, S., Yamamoto, K., Tsukamoto, K., Tajima, A.: CTR prediction for contextual advertising. In: Proceedings of the Seventh International Workshop on Data Mining for Online Advertising - ADKDD 2013 (2013)
12. Wang, J., Chen, B.: Selling futures online advertising slots via option contracts. In: WWW, vol. 1000, pp. 627–628 (2012). http://dl.acm.org/citation.cfm?id=2188160
13. Yan, J., Liu, N., Wang, G., Zhang, W.: How much can behavioral targeting help online advertising? In: Proceeding WWW 2009 Proceedings of the 18th International Conference on World Wide Web, pp. 261–270 (2009). http://portal.acm.org/citation.cfm?id=1526745
14. Yuan, S., Chen, B., Wang, J., Mason, P., Seljan, S.: An empirical study of reserve price optimisation in real-time bidding. In: Proceeding of the 20th ACM SIGKDD Conference (2014)
15. Yuan, S., Wang, J.: Sequential selection of correlated ads by POMDPs. In: Proceedings of the 21st ACM International Conference on Information and Knowledge Management - CIKM 2012, p. 515 (2012). http://dl.acm.org/citation.cfm?id=2396761.2396828

16. Yuan, S., Wang, J., Zhao, X.: Real-time bidding for online advertising: measurement and analysis. In: Proceedings of the Seventh International Workshop on Data Mining for Online Advertising (2013). http://dl.acm.org/citation.cfm?id=2501980
17. Zhang, L., Guan, Y.: Detecting click fraud in pay-per-click streams of online advertising networks. In: Proceedings - The 28th International Conference on Distributed Computing Systems, ICDCS 2008, pp. 77–84 (2008)

Author Index

Aggarwal, Nitish 874
Ai, Qingyao 115
Aker, Ahmet 15
Albakour, Dyaa 878
Aletras, Nikolaos 689
Allan, James 187, 575, 796
ALMasri, Mohannad 709
Alonso, Omar 859
Amigó, Enrique 378
Amini, Massih-Reza 723
Amsaleg, Laurent 782
Arguello, Jaime 309
Aslam, Javed A. 72
Avula, Sandeep 309
Azzopardi, Leif 696

Bagheri, Ebrahim 479
Balamurali, A.R. 15
Balaneshinkordan, Saeid 761
Balikas, Georgios 723
Balog, Krisztian 436, 893
Balu, Raghavendran 782
Barker, Emma 15
Barreiro, Álvaro 602, 614
Basili, Roberto 100
Berberich, Klaus 30, 789
Berrut, Catherine 709
Beyer, Anna 507
Bhatia, Sumit 874
Boonzajer Flaes, Joost 824
Boratto, Ludovico 889
Boteva, Vera 716
Boughanem, Mohand 533
Bratsberg, Svein Erik 436
Braun, Sarah 393
Braunstain, Liora 129
Buz, Tolga 393

Cabanac, Guillaume 533, 865
Callan, Jamie 145, 408, 841
Campos, Ricardo 878
Carmel, David 129
Carrillo-de-Albornoz, Jorge 378
Caverlee, James 803
Chali, Yllias 366
Chamberlain, Jon 669
Chattopadhyaya, Ishan 408
Chen, Lingxi 589
Chen, Long 240
Chen, Ruey-Cheng 115
Chen, Zhumin 58
Cheng, Xueqi 88
Chevallet, Jean-Pierre 709
Choudhury, Monojit 775
Clough, Paul 703
Corney, David 878
Cox, Ingemar J. 689
Crane, Matt 408
Crestani, Fabio 466
Croce, Danilo 100
Croft, W. Bruce 115, 171
Culpepper, J. Shane 145

de Gemmis, Marco 729
de Rijke, Maarten 661
de Vries, Arjen P. 227
Demartini, Gianluca 293
Deveaud, Romain 818
Diaz, Fernando 309, 521
Dietz, Laura 654
Du, Mian 735
Du, Tianming 45
Dudley, Joel T. 768
Duffhauss, Fabian 393

Eickhoff, Carsten 227
Elsayed, Tamer 648
Everingham, Mark 549

Fang, Anjie 492
Fang, Hui 849
Fani, Hossein 479
Faraldo, Ángel 335
Foley, John 408, 575
Friedrich, Florian 393
Frommholz, Ingo 865
Furon, Teddy 782

Gaizauskas, Rob 15
Galuščáková, Petra 853
Ganguly, Niloy 775
Geyti, Jens K. 689
Gholipour, Demian 716
Giachanou, Anastasia 466
Gollub, Tim 507
Gómez, Emilia 335
Gonzalo, Julio 378
Granitzer, Michael 200
Gülzow, Jörg Marvin 393
Guo, Jiafeng 88, 115
Guo, Qi 171
Gupta, Dhruv 789

Habel, Philip 492
Hagen, Matthias 393, 507, 810
Hamid, Fahmida 351
Hanbury, Allan 267
Haraburda, David 351
Harvey, Morgan 466
Hasan, Mehedi 682
Hasanain, Maram 648
Hasibi, Faegheh 436
He, Liang 252
He, Yun 252
Hepple, Mark 15
Herrera, Perfecto 335
Hopfgartner, Frank 878
Hu, Qinmin 252

Ienco, Dino 869
Ingersoll, Grant 408

Jiang, Jiepu 187
Jiang, Jingtian 748
Jordà, Sergi 335
Jose, Joemon M. 240
Jurgovsky, Johannes 200

Kahani, Mohsen 479
Kangasharju, Jussi 735
Karkulahti, Ossi 735
Kazai, Gabriella 878
Kim, Se-Jong 741
Kim, Yubin 145
Köhler, Jakob 393
Komlossy, Kristof 507
Köpsel, Sebastian 810
Kotov, Alexander 682, 761
Kruschwitz, Udo 878
Kurland, Oren 129
Kurtic, Emina 15
Kwee, Agus T. 641

Lampos, Vasileios 689
Lan, Yanyan 88
Lawless, Séamus 561
Lee, Jong-Hyeok 741
Li, Cheng 72
Li, Li 768
Li, Wen 227
Liao, Lejian 748
Lim, Ee-Peng 641
Lin, Chin-Yew 748
Lin, Jimmy 408, 675
Lin, Jovian 641
Lipani, Aldo 267
Liu, Zhijiao 803
Lops, Pasquale 729
Lötzsch, Winfried 393
Lu, Shiyong 682
Lupu, Mihai 267

Macdonald, Craig 408, 421, 492
Mackie, Stuart 421
Magalhães, João 841
Magdy, Walid 648
Majumdar, Debapriyo 874
Markert, Katja 549
Martinez-Alvarez, Miguel 878
Martins, Flávio 841
Mayr, Philipp 865
McCreadie, Richard 421
McDonald, Kieran 171
Meng, Sha 171
Miotto, Riccardo 768
Mishra, Arunav 30
Moffat, Alistair 145
Mohan, Aravind 682
Mohd Shariff, Shafiza 453
Montazeralghaem, Ali 754
Moriceau, Véronique 818
Mothe, Josiane 818
Müller, Fabian 393
Müller, Maike Elisa 393
Musto, Cataldo 729

Naji, Nada 796
Nguyen, Minh-Le 3
Nguyen, Minh-Tien 3
Niu, Wei 803
Niu, Xing 675

Oentaryo, Richard J. 641
Oliveira, Eugénio 845
Oosterhuis, Harrie 661
Osipov, Gennady 837
Ounis, Iadh 421, 492

Pang, Liang 115
Paramita, Monica 15
Parapar, Javier 602, 614
Paßmann, Robert 393
Pavlu, Virgil 72
Pecina, Pavel 853
Pinel-Sauvagnat, Karen 533
Pivovarova, Lidia 735
Plaza, Laura 378
Ponzetto, Simone Paolo 654
Potthast, Martin 393, 810
Prasetyo, Philips K. 641

Rami Ghorab, M. 561
Rao, Jinfeng 675
Reinke, Bernhard 393
Ren, Pengjie 58
Renders, Jean-Michel 626
Rettenmeier, Lucas 393
Riezler, Stefan 716
Roche, Mathieu 869
Romeo, Salvatore 869
Rometsch, Thomas 393
Rosso, Paolo 869
Roth, Benjamin 654
Rudinac, Stevan 824, 833
Russell-Rose, Tony 669

Saha Roy, Rishiraj 775
Saleh, Shadi 853
Saleiro, Pedro 845
Sanderson, Mark 453
SanJuan, Eric 818
Schedl, Markus 322
Scholer, Falk 115
Schuhmacher, Michael 654
Schuth, Anne 661, 893
Seifert, Christin 200
Semeraro, Giovanni 729
Serdyukov, Pavel 859
Shakery, Azadeh 754
Shelmanov, Artem 837
Shin, Jaehun 741
Shokouhi, Milad 171
Shtok, Anna 129
Si, Luo 58
Smirnov, Ivan 837
Soares, Carlos 845
Sochenkov, Ilya 837
Sokolov, Artem 716
Sommer, Timo 393
Song, Dandan 748
Song, Yang 171, 252
Soulier, Laure 885
Spina, Damiano 115
Stamatatos, Efstathios 393
Stein, Benno 393, 507, 810
Stieg, Paul M. 682
Suresh, Anusha 775
Suvorov, Roman 837
Szpektor, Idan 129

Tagarelli, Andrea 869
Tamine, Lynda 885
Tarau, Paul 351
Teixeira, Jorge 845
Thonet, Thibaut 533
Tikhomirov, Ilya 837
Toms, Elaine G. 293
Träger, Michael 393
Trotman, Andrew 408

Uddin, Mohsin 366

Valcarce, Daniel 602, 614
van den Berg, Jorrit 833
Verma, Manisha 212
Vigna, Sebastiano 408
Vu, Adrian 641
Vu, Casey 641

Wakeling, Simon 703
Wang, Bingyu 72
Wang, Jingang 58, 748
Wang, Josiah 549
Wang, Jun 45, 589, 897
Wang, Pengfei 88
Wang, Zhongqing 561
Welleck, Sean J. 159

Wilhelm, Sebastian 393
Worring, Marcel 824, 833
Wu, Tao 58

Xu, Jun 88
Xu, Yu 561

Yang, Liu 115, 171
Yangarber, Roman 735
Yilmaz, Emine 212
Yu, Haitao 240
Yuan, Fajie 240
Yuan, Shuai 897

Zamani, Hamed 754
Zarrinkalam, Fattane 479
Zhang, Huaizhi 240
Zhang, Weinan 45, 589, 897
Zhang, Xiuzhen 453
Zhang, Zhiwei 58, 748
Zhong, Wei 849
Zhou, Dong 561
Zhou, Fang 322
Zhuang, Mengdie 293
Zou, Bin 689
Zubarev, Denis 837
Zuccon, Guido 280, 696

Printed in the United States
by Bookmasters